FLUID MECHANICS

FIFTH EDITION

Founders of Modern Fluid Dynamics

Ludwig Prandtl
(1875-1953)

G. I. Taylor
(1886-1975)

(Biographical sketches of Prandtl and Taylor are given in Appendix C.)

Photograph of Ludwig Prandtl is reprinted with permission from the *Annual Review of Fluid Mechanics*, Vol. 19. Copyright 1987 by Annual Reviews: www.AnnualReviews.org.

Photograph of Geoffrey Ingram Taylor at age 69 in his laboratory reprinted with permission from the AIP Emilio Segrè Visual Archieves. Copyright, American Institute of Physics, 2000.

FLUID MECHANICS

FIFTH EDITION

Pijush K. Kundu
Ira M. Cohen
David R. Dowling

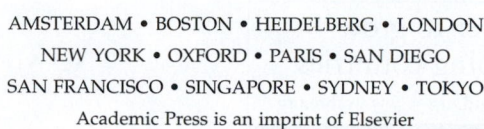
AMSTERDAM • BOSTON • HEIDELBERG • LONDON
NEW YORK • OXFORD • PARIS • SAN DIEGO
SAN FRANCISCO • SINGAPORE • SYDNEY • TOKYO
Academic Press is an imprint of Elsevier

Academic Press is an imprint of Elsevier
225 Wyman Street, Waltham, MA 02451, USA
The Boulevard, Langford Lane, Kidlington, Oxford, OX5 1GB, UK

Library of Congress Cataloging-in-Publication Data
Kundu, Pijush K.
    Fluid mechanics / Pijush K. Kundu, Ira M. Cohen, David R. Dowling. – 5th ed.
        p. cm.
    Includes bibliographical references and index.
    ISBN 978-0-12-382100-3 (alk. paper)
    1. Fluid mechanics. I. Cohen, Ira M. II. Dowling, David R. III. Title.
    QA901.K86 2012
    620.1'06–dc22

                                                           2011014138

British Library Cataloguing-in-Publication Data
A catalogue record for this book is available from the British Library.

For information on all Academic Press publications
visit our website at www.elsevierdirect.com

Printed in the United States of America

11 12 13 14   10 9 8 7 6 5 4 3 2 1

Working together to grow
libraries in developing countries

www.elsevier.com | www.bookaid.org | www.sabre.org

ELSEVIER   BOOK AID
             International   Sabre Foundation

Dedication

This revision to this textbook is dedicated to my wife and family who have patiently helped chip many sharp corners off my personality, and to the many fine instructors and students with whom I have interacted who have all in some way highlighted the allure of this subject for me.

D.R.D.

In Memory of Pijush Kundu

Pijush Kanti Kundu was born in Calcutta, India, on October 31, 1941. He received a BS degree in Mechanical Engineering in 1963 from Shibpur Engineering College of Calcutta University, earned an MS degree in Engineering from Roorkee University in 1965, and was a lecturer in Mechanical Engineering at the Indian Institute of Technology in Delhi from 1965 to 1968. Pijush came to the United States in 1968, as a doctoral student at Penn State University. With Dr. John L. Lumley as his advisor, he studied instabilities of viscoelastic fluids, receiving his doctorate in 1972. He began his lifelong interest in oceanography soon after his graduation, working as Research Associate in Oceanography at Oregon State University from 1968 until 1972. After spending a year at the University de Oriente in Venezuela, he joined the faculty of the Oceanographic Center of Nova Southeastern University, where he remained until his death in 1994.

During his career, Pijush contributed to a number of sub-disciplines in physical oceanography, most notably in the fields of coastal dynamics, mixed-layer physics, internal waves, and Indian-Ocean dynamics. He was a skilled data analyst, and, in this regard, one of his accomplishments was to introduce the "empirical orthogonal eigenfunction" statistical technique to the oceanographic community.

I arrived at Nova Southeastern University shortly after Pijush, and he and I worked closely together thereafter. I was immediately impressed with the clarity of his scientific thinking and his thoroughness. His most impressive and obvious quality, though, was his love of science, which pervaded all his activities. Some time after we met, Pijush opened a drawer in a desk in his home office, showing me drafts of several chapters to a book he had always wanted to write. A decade later, this manuscript became the first edition of *Fluid Mechanics*, the culmination of his lifelong dream, which he dedicated to the memory of his mother, and to his wife Shikha, daughter Tonushree, and son Joydip.

Julian P. McCreary, Jr.,
University of Hawaii

In Memory of Ira Cohen

Ira M. Cohen earned his BS from Polytechnic University in 1958 and his PhD from Princeton University in 1963, both in aeronautical engineering. He taught at Brown University for three years prior to joining the University of Pennsylvania faculty as an assistant professor in 1966. He served as chair of the Department of Mechanical Engineering and Applied Mechanics from 1992 to 1997.

Professor Cohen was a world-renowned scholar in the areas of continuum plasmas, electrostatic probe theories and plasma diagnostics, dynamics and heat transfer of lightly ionized gases, low current arc plasmas, laminar shear layer theory, and matched asymptotics in fluid mechanics. Most of his contributions appear in the *Physics of Fluids* journal of the American Institute of Physics. His seminal paper, "Asymptotic theory of spherical electrostatic probes in a slightly ionized, collision dominated gas" (1963; *Physics of Fluids, 6,* 1492–1499), is to date the most highly cited paper in the theory of electrostatic probes and plasma diagnostics.

During his doctoral work and for a few years beyond that, Ira collaborated with a world-renowned mathematician/physicist, the late Dr. Martin Kruskal (recipient of National Medal of Science, 1993) on the development of a monograph called "Asymptotology." Professor Kruskal also collaborated with Professor Cohen on plasma physics. This was the basis for Ira's strong foundation in fluid dynamics that has been transmitted into the prior editions of this textbook.

In his forty-one years of service to the University of Pennsylvania before his death in December 2007, Professor Cohen distinguished himself with his integrity, his fierce defense of high scholarly standards, and his passionate commitment to teaching. He will always be remembered for his candor and his sense of humor.

Professor Cohen's dedication to academics was unrivalled. In addition, his passion for physical fitness was legendary. Neither rain nor sleet nor snow would deter him from his daily bicycle commute, which began at 5:00 AM, from his home in Narberth to the University of Pennsylvania. His colleagues grew accustomed to seeing him drag his forty-year-old bicycle, with its original three-speed gearshift, up to his office. His other great passion was the game of squash, which he played with extraordinary skill five days a week at the Ringe Squash Courts at Penn, where he was a fierce but fair competitor. During the final year of his life, Professor Cohen remained true to his bicycling and squash-playing schedule, refusing to allow his illness get in the way of the things he loved.

Professor Cohen was a member of Beth Am Israel Synagogue, and would on occasion lead Friday night services there. He and his wife, Linda, were first married near Princeton, New Jersey, on February 13, 1960, when they eloped. They were married a second time four months later in a formal ceremony. He is survived by his wife, his two children, Susan Cohen Bolstad and Nancy Cohen Cavanaugh, and three grandchildren, Melissa, Daniel, and Andrew.

Senior Faculty
Department of Mechanical Engineering
and Applied Mechanics
University of Pennsylvania

About the Third Author

David R. Dowling was born in Mesa, Arizona, in 1960 but grew up in southern California where early practical exposure to fluid mechanics—swimming, surfing, sailing, flying model aircraft, and trying to throw a curve ball—dominated his free time. He attended the California Institute of Technology continuously for a decade starting in 1978, earning a BS degree in Applied Physics in 1982, and MS and PhD degrees in Aeronautics in 1983 and 1988, respectively. After graduate school, he worked at Boeing Aerospace and Electronics and then took a post-doctoral scientist position at the Applied Physics Laboratory of the University of Washington. In 1992, he started a faculty career in the Department of Mechanical Engineering at the University of Michigan where he has since taught and conducted research in fluid mechanics and acoustics. He has authored and co-authored more than 60 archival journal articles and more than 100 conference presentations. His published research in fluid mechanics includes papers on turbulent mixing, forced-convection heat transfer, cirrus clouds, molten plastic flow, interactions of surfactants with water waves, and hydrofoil performance and turbulent boundary layer characteristics at high Reynolds numbers. From January 2007 through June 2009, he served as an Associate Chair and as the Undergraduate Program Director for the Department of Mechanical Engineering at the University of Michigan. He is a fellow of the American Society of Mechanical Engineers and of the Acoustical Society of America. He received the Student Council Mentoring Award of the Acoustical Society of America in 2007, the University of Michigan College of Engineering John R. Ullrich Education Excellence Award in 2009, and the Outstanding Professor Award from the University of Michigan Chapter of the American Society for Engineering Education in 2009. Prof. Dowling is an avid swimmer, is married, and has seven children.

Contents

About the DVD xvii

Preface xix

  Companion Website xx

Acknowledgments xxi

Nomenclature xxii

1. Introduction 1

1.1. Fluid Mechanics 2

1.2. Units of Measurement 3

1.3. Solids, Liquids, and Gases 3

1.4. Continuum Hypothesis 5

1.5. Molecular Transport Phenomena 5

1.6. Surface Tension 8

1.7. Fluid Statics 9

1.8. Classical Thermodynamics 12

    First Law of Thermodynamics 13

    Equations of State 14

    Specific Heats 14

    Second Law of Thermodynamics 15

    Property Relations 16

    Speed of Sound 16

    Thermal Expansion Coefficient 16

1.9. Perfect Gas 16

1.10. Stability of Stratified Fluid Media 18

    Potential Temperature and Density 19

    Scale Height of the Atmosphere 21

1.11. Dimensional Analysis 21

    Step 1. Select Variables and Parameters 22

    Step 2. Create the Dimensional Matrix 23

    Step 3. Determine the Rank of the Dimensional Matrix 23

    Step 4. Determine the Number of Dimensionless Groups 24

    Step 5. Construct the Dimensionless Groups 24

    Step 6. State the Dimensionless Relationship 26

    Step 7. Use Physical Reasoning or Additional Knowledge to Simplify the Dimensionless Relationship 26

Exercises 30

Literature Cited 36

Supplemental Reading 37

2. Cartesian Tensors 39

2.1. Scalars, Vectors, Tensors, Notation 39

2.2. Rotation of Axes: Formal Definition of a Vector 42

2.3. Multiplication of Matrices 44

2.4. Second-Order Tensors 45

2.5. Contraction and Multiplication 47

2.6. Force on a Surface 48

2.7. Kronecker Delta and Alternating Tensor 50

2.8. Vector, Dot, and Cross Products 51

2.9. Gradient, Divergence, and Curl 52

2.10. Symmetric and Antisymmetric Tensors 55

2.11. Eigenvalues and Eigenvectors of a Symmetric Tensor 56

2.12. Gauss' Theorem 58

2.13. Stokes' Theorem 60

2.14. Comma Notation 62

Exercises 62

Literature Cited 64

Supplemental Reading 64

3. Kinematics 65

3.1. Introduction and Coordinate Systems 65

3.2. Particle and Field Descriptions of Fluid Motion 67

3.3. Flow Lines, Fluid Acceleration, and Galilean Transformation 71

3.4. Strain and Rotation Rates 76

    Summary 81

3.5.   Kinematics of Simple Plane Flows   82
3.6.   Reynolds Transport Theorem   85
Exercises   89
Literature Cited   93
Supplemental Reading   93

4. Conservation Laws   95

4.1.   Introduction   96
4.2.   Conservation of Mass   96
4.3.   Stream Functions   99
4.4.   Conservation of Momentum   101
4.5.   Constitutive Equation for a Newtonian
        Fluid   111
4.6.   Navier-Stokes Momentum Equation   114
4.7.   Noninertial Frame of Reference   116
4.8.   Conservation of Energy   121
4.9.   Special Forms of the Equations   125
        Angular Momentum Principle for a
          Stationary Control Volume   125
        Bernoulli Equations   128
        Neglect of Gravity in Constant Density
          Flows   134
        The Boussinesq Approximation   135
        Summary   137
4.10. Boundary Conditions   137
        Moving and Deforming Boundaries   139
        Surface Tension Revisited   139
4.11. Dimensionless Forms of the Equations and
        Dynamic Similarity   143
Exercises   151
Literature Cited   168
Supplemental Reading   168

5. Vorticity Dynamics   171

5.1.   Introduction   171
5.2.   Kelvin's Circulation Theorem   176
5.3.   Helmholtz's Vortex Theorems   179
5.4.   Vorticity Equation in a Nonrotating
        Frame   180
5.5.   Velocity Induced by a Vortex Filament: Law
        of Biot and Savart   181

5.6.   Vorticity Equation in a Rotating Frame   183
5.7.   Interaction of Vortices   187
5.8.   Vortex Sheet   191
Exercises   192
Literature Cited   195
Supplemental Reading   196

6. Ideal Flow   197

6.1.   Relevance of Irrotational Constant-Density
        Flow Theory   198
6.2.   Two-Dimensional Stream Function and
        Velocity Potential   200
6.3.   Construction of Elementary Flows in Two
        Dimensions   203
6.4.   Complex Potential   216
6.5.   Forces on a Two-Dimensional Body   219
        Blasius Theorem   219
        Kutta-Zhukhovsky Lift Theorem   221
6.6.   Conformal Mapping   222
6.7.   Numerical Solution Techniques in Two
        Dimensions   225
6.8.   Axisymmetric Ideal Flow   231
6.9.   Three-Dimensional Potential Flow and
        Apparent Mass   236
6.10. Concluding Remarks   240
Exercises   241
Literature Cited   251
Supplemental Reading   251

7. Gravity Waves   253

7.1.   Introduction   254
7.2.   Linear Liquid-Surface Gravity Waves   256
        Approximations for Deep and Shallow
          Water   265
7.3.   Influence of Surface Tension   269
7.4.   Standing Waves   271
7.5.   Group Velocity, Energy Flux, and
        Dispersion   273
7.6.   Nonlinear Waves in Shallow and Deep
        Water   279
7.7.   Waves on a Density Interface   286

7.8. Internal Waves in a Continuously Stratified Fluid   293
    Internal Waves in a Stratified Fluid   296
    Dispersion of Internal Waves in a Stratified Fluid   299
    Energy Considerations for Internal Waves in a Stratified Fluid   302
Exercises   304
Literature Cited   307

8. Laminar Flow   309

8.1. Introduction   309
8.2. Exact Solutions for Steady Incompressible Viscous Flow   312
    Steady Flow between Parallel Plates   312
    Steady Flow in a Round Tube   315
    Steady Flow between Concentric Rotating Cylinders   316
8.3. Elementary Lubrication Theory   318
8.4. Similarity Solutions for Unsteady Incompressible Viscous Flow   326
8.5. Flow Due to an Oscillating Plate   337
8.6. Low Reynolds Number Viscous Flow Past a Sphere   338
8.7. Final Remarks   347
Exercises   347
Literature Cited   359
Supplemental Reading   359

9. Boundary Layers and Related Topics   361

9.1. Introduction   362
9.2. Boundary-Layer Thickness Definitions   367
9.3. Boundary Layer on a Flat Plate: Blasius Solution   369
9.4. Falkner-Skan Similarity Solutions of the Laminar Boundary-Layer Equations   373
9.5. Von Karman Momentum Integral Equation   375
9.6. Thwaites' Method   377
9.7. Transition, Pressure Gradients, and Boundary-Layer Separation   382

9.8. Flow Past a Circular Cylinder   388
    Low Reynolds Numbers   389
    Moderate Reynolds Numbers   389
    High Reynolds Numbers   392
9.9. Flow Past a Sphere and the Dynamics of Sports Balls   395
    Cricket Ball Dynamics   396
    Tennis Ball Dynamics   398
    Baseball Dynamics   399
9.10. Two-Dimensional Jets   399
9.11. Secondary Flows   407
Exercises   408
Literature Cited   418
Supplemental Reading   419

10. Computational Fluid Dynamics   421
    HOWARD H. HU

10.1. Introduction   421
10.2. Finite-Difference Method   423
    Approximation to Derivatives   423
    Discretization and Its Accuracy   425
    Convergence, Consistency, and Stability   426
10.3. Finite-Element Method   429
    Weak or Variational Form of Partial Differential Equations   429
    Galerkin's Approximation and Finite-Element Interpolations   430
    Matrix Equations, Comparison with Finite-Difference Method   431
    Element Point of View of the Finite-Element Method   434
10.4. Incompressible Viscous Fluid Flow   436
    Convection-Dominated Problems   437
    Incompressibility Condition   439
    Explicit MacCormack Scheme   440
    MAC Scheme   442
    Θ-Scheme   446
    Mixed Finite-Element Formulation   447
10.5. Three Examples   449
    Explicit MacCormack Scheme for Driven-Cavity Flow Problem   449
    Explicit MacCormack Scheme for Flow Over a Square Block   453

Finite-Element Formulation for
Flow Over a Cylinder Confined in
a Channel   459
10.6.   Concluding Remarks   470
Exercises   470
Literature Cited   471
Supplemental Reading   472

11. Instability   473

11.1.   Introduction   474
11.2.   Method of Normal Modes   475
11.3.   Kelvin-Helmholtz Instability   477
11.4.   Thermal Instability: The Bénard
Problem   484
11.5.   Double-Diffusive Instability   492
11.6.   Centrifugal Instability: Taylor Problem   496
11.7.   Instability of Continuously Stratified Parallel
Flows   502
11.8.   Squire's Theorem and the Orr-Sommerfeld
Equation   508
11.9.   Inviscid Stability of Parallel Flows   511
11.10. Results for Parallel and Nearly Parallel
Viscous Flows   515
Two-Stream Shear Layer   515
Plane Poiseuille Flow   516
Plane Couette Flow   517
Pipe Flow   517
Boundary Layers with Pressure
Gradients   517
11.11. Experimental Verification of Boundary-Layer
Instability   520
11.12. Comments on Nonlinear Effects   522
11.13. Transition   523
11.14. Deterministic Chaos   524
Closure   531
Exercises   532
Literature Cited   539

12. Turbulence   541

12.1.   Introduction   542
12.2.   Historical Notes   544

12.3.   Nomenclature and Statistics for Turbulent
Flow   545
12.4.   Correlations and Spectra   549
12.5.   Averaged Equations of Motion   554
12.6.   Homogeneous Isotropic Turbulence   560
12.7.   Turbulent Energy Cascade and
Spectrum   564
12.8.   Free Turbulent Shear Flows   571
12.9.   Wall-Bounded Turbulent Shear Flows   581
Inner Layer: Law of the Wall   584
Outer Layer: Velocity Defect Law   585
Overlap Layer: Logarithmic Law   585
Rough Surfaces   590
12.10. Turbulence Modeling   591
A Mixing Length Model   593
One-Equation Models   595
Two-Equation Models   595
12.11. Turbulence in a Stratified Medium   596
The Richardson Numbers   597
Monin-Obukhov Length   598
Spectrum of Temperature Fluctuations   600
12.12. Taylor's Theory of Turbulent Dispersion   601
Rate of Dispersion of a Single Particle   602
Random Walk   605
Behavior of a Smoke Plume in the Wind   606
Turbulent Diffusivity   607
12.13. Concluding Remarks   607
Exercises   608
Literature Cited   618
Supplemental Reading   620

13. Geophysical Fluid Dynamics   621

13.1.   Introduction   622
13.2.   Vertical Variation of Density in the
Atmosphere and Ocean   623
13.3.   Equations of Motion   625
13.4.   Approximate Equations for a Thin Layer on
a Rotating Sphere   628
f-Plane Model   630
β-Plane Model   630
13.5.   Geostrophic Flow   630
Thermal Wind   632
Taylor-Proudman Theorem   632

13.6.   Ekman Layer at a Free Surface   633
        Explanation in Terms of Vortex Tilting   637
13.7.   Ekman Layer on a Rigid Surface   639
13.8.   Shallow-Water Equations   642
13.9.   Normal Modes in a Continuously Stratified
        Layer   644
        Boundary Conditions on ψ_n 646
        Vertical Mode Solution for Uniform N 646
        Summary   649
13.10. High- and Low-Frequency Regimes
        in Shallow-Water Equations   649
13.11. Gravity Waves with Rotation   651
        Particle Orbit   652
        Inertial Motion   653
13.12. Kelvin Wave   654
13.13. Potential Vorticity Conservation in
        Shallow-Water Theory   658
13.14. Internal Waves   662
        WKB Solution   664
        Particle Orbit   666
        Discussion of the Dispersion Relation   668
        Lee Wave   670
13.15. Rossby Wave   671
        Quasi-Geostrophic Vorticity Equation   671
        Dispersion Relation   673
13.16. Barotropic Instability   676
13.17. Baroclinic Instability   678
        Perturbation Vorticity Equation   679
        Wave Solution   681
        Instability Criterion   682
        Energetics   684
13.18. Geostrophic Turbulence   685
Exercises   688
Literature Cited   690
Supplemental Reading   690

14. Aerodynamics   691

14.1.   Introduction   692
14.2.   Aircraft Terminology   692
        Control Surfaces   693
14.3.   Characteristics of Airfoil Sections   696
        Historical Notes   701
14.4.   Conformal Transformation for
        Generating Airfoil Shapes   702

14.5.   Lift of a Zhukhovsky Airfoil   706
14.6.   Elementary Lifting Line Theory for
        Wings of Finite Span   708
        Lanchester Versus Prandtl   716
14.7.   Lift and Drag Characteristics of
        Airfoils   717
14.8.   Propulsive Mechanisms of Fish
        and Birds   719
14.9.   Sailing against the Wind   721
Exercises   722
Literature Cited   728
Supplemental Reading   728

15. Compressible Flow   729

15.1.   Introduction   730
        Perfect Gas Thermodynamic Relations   731
15.2.   Acoustics   732
15.3.   Basic Equations for One-Dimensional
        Flow   736
15.4.   Reference Properties in Compressible
        Flow   738
15.5.   Area-Velocity Relationship in
        One-Dimensional Isentropic Flow   740
15.6.   Normal Shock Waves   748
        Stationary Normal Shock Wave in a
           Moving Medium   748
        Moving Normal Shock Wave in a
           Stationary Medium   752
        Normal Shock Structure   753
15.7.   Operation of Nozzles at Different
        Back Pressures   755
        Convergent Nozzle   755
        Convergent–Divergent Nozzle   757
15.8.   Effects of Friction and Heating in
        Constant-Area Ducts   761
        Effect of Friction   763
        Effect of Heat Transfer   764
15.9.   Pressure Waves in Planar Compressible
        Flow   765
15.10. Thin Airfoil Theory in Supersonic Flow   773
Exercises   775
Literature Cited   778
Supplemental Reading   778

16. Introduction to Biofluid
        Mechanics   779
    PORTONOVO S. AYYASWAMY

16.1. Introduction   779
16.2. The Circulatory System in the Human
       Body   780
       The Heart as a Pump   785
       Nature of Blood   788
       Nature of Blood Vessels   793
16.3. Modeling of Flow in Blood Vessels   796
       Steady Blood Flow Theory   797
       Pulsatile Blood Flow Theory   805
       Blood Vessel Bifurcation: An Application of
           Poiseuille's Formula and Murray's Law   820
       Flow in a Rigid-Walled Curved Tube   825
       Flow in Collapsible Tubes   831
       Laminar Flow of a Casson Fluid in a Rigid-
           Walled Tube   839

Pulmonary Circulation   841
The Pressure Pulse Curve in the Right
    Ventricle   842
Effect of Pulmonary Arterial Pressure on
    Pulmonary Resistance   843
16.4. Introduction to the Fluid Mechanics
       of Plants   844
Exercises   849
Acknowledgment   850
Literature Cited   851
Supplemental Reading   852

Appendix A   853
Appendix B   857
Appendix C   869
Appendix D   873
Index   875

About the DVD

We are pleased to include a free copy of the DVD *Multimedia Fluid Mechanics*, 2/e, with this copy of *Fluid Mechanics, Fifth Edition*. You will find it in a plastic sleeve on the inside back cover of the book. If you are purchasing a used copy, be aware that the DVD might have been removed by a previous owner.

Inspired by the reception of the first edition, the objectives in *Multimedia Fluid Mechanics*, 2/e, remain to exploit the moving image and interactivity of multimedia to improve the teaching and learning of fluid mechanics in all disciplines by illustrating fundamental phenomena and conveying fascinating fluid flows for generations to come.

The completely new edition on the DVD includes the following:

- Twice the coverage with new modules on turbulence, control volumes, interfacial phenomena, and similarity and scaling

- Four times the number of fluid videos, now more than 800
- Now more than 20 virtual labs and simulations
- Dozens of new interactive demonstrations and animations

Additional *new* features:

- Improved navigation via sidebars that provide rapid overviews of modules and guided browsing
- Media libraries for each chapter that give a snapshot of videos, each with descriptive labels
- Ability to create movie playlists, which are invaluable in teaching
- Higher-resolution graphics, with full or part screen viewing options
- Operates on either a PC or a Mac OSX

Preface

In the fall of 2009, Elsevier approached me about possibly taking over as the lead author of this textbook. After some consideration and receipt of encouragement from faculty colleagues here at the University of Michigan and beyond, I agreed. The ensuing revision effort then tenaciously pulled all the slack out of my life for the next 18 months. Unfortunately, I did not have the honor or pleasure of meeting or knowing either prior author, and have therefore missed the opportunity to receive their advice and guidance. Thus, the revisions made for this *5th Edition* of *Fluid Mechanics* have been driven primarily by my experience teaching and interacting with undergraduate and graduate students during the last two decades.

Overall, the structure, topics, and technical level of the *4th Edition* have been largely retained, so instructors who have made prior use of this text should recognize much in the *5th Edition*. This textbook should still be suitable for advanced-undergraduate or beginning-graduate courses in fluid mechanics. However, I have tried to make the subject of fluid mechanics more accessible to students who may have only studied the subject during one prior semester, or who may need fluid mechanics knowledge to pursue research in a related field.

Given the long history of this important subject, this textbook (at best) reflects one evolving instructional approach. In my experience as a student, teacher, and faculty member, a textbook is most effective when used as a supporting pedagogical tool for an effective lecturer. Thus my primary revision objective has been to improve the text's overall utility to students and instructors by adding introductory material and references to the first few chapters, by increasing the prominence of engineering applications of fluid mechanics, and by providing a variety of new exercises (more than 200) and figures (more than 100). For the chapters receiving the most attention (1–9, 11–12, and 14) this has meant approximately doubling, tripling, or quadrupling the number of exercises. Some of the new exercises have been built from derivations that previously had appeared in the body of the text, and some involve simple kitchen or bathroom experiments. My hope for a future edition is that there will be time to further expand the exercise offerings, especially in Chapters 10, 13, 15, and 16.

In preparing this *5th Edition*, some reorganization, addition, and deletion of material has also taken place. Dimensional analysis has been moved to Chapter 1. The stream function's introduction and the dynamic-similarity topic have been moved to Chapter 4. Reynolds transport theorem now occupies the final section of Chapter 3. The discussion of the wave equation has been placed in the acoustics section of Chapter 15. Major topical additions are: apparent mass (Chapter 6), elementary lubrication theory (Chapter 8), and Thwaites method (Chapter 9). The sections covering the laminar shear layer, and boundary-layer theory from a purely mathematical perspective, and coherent structures in wall-bounded turbulent flow have been removed. The specialty chapters (10, 13, and 16) have been left largely untouched except for a few language changes and appropriate renumbering of equations. In addition, some sections have been combined to save space, but this has been offset by an expansion of nearly every figure caption and the introduction of a nomenclature section with more than 200 entries.

Only a few notation changes have been made. Index and vector notation predominate throughout the text. The comma notation for derivatives now only appears in Section 5.6. The notation for unit vectors has been changed from bold **i** to bold **e** to conform to other texts in physics and engineering. In addition, a serious effort was made to denote two- and three-dimensional coordinate systems in a consistent manner from chapter to chapter. However, the completion of this task, which involves retyping literally hundreds of equations, was not possible in the time available. Thus, cylindrical coordinates (R, φ, z) predominate, but (r, θ, x) still appear in Table 12.1, Chapter 16, and a few other places.

And, as a final note, the origins of many of the new exercises are referenced to individuals and other sources via footnotes. However, I am sure that such referencing is incomplete because of my imperfect memory and record keeping. Therefore, I stand ready to correctly attribute the origins of any problem contained herein. Furthermore, I welcome the opportunity to correct any errors you find, to hear your opinion of how this book might be improved, and to include exercises you might suggest; just contact me at *drd@umich.edu*.

David R. Dowling
Ann Arbor, Michigan
April 2011

COMPANION WEBSITE

An updated errata sheet is available on the book's companion website. To access the errata, visit www.elsevierdirect.com/9780123821003 and click on the companion site link. Instructors teaching with this book may access the solutions manual and image bank by visiting www.textbooks.elsevier.com and following the online instructions to log on and register.

Acknowledgments

The current version of this textbook has benefited from the commentary and suggestions provided by the reviewers of the initial revision proposal and the reviewers of draft versions of several of the chapters. Chief among these reviewers is Professor John Cimbala of the Pennsylvania State University. I would also like to recognize and thank my technical mentors, Professor Hans W. Liepmann (undergraduate advisor), Professor Paul E. Dimotakis (graduate advisor), and Professor Darrell R. Jackson (post-doctoral advisor); and my friends and colleagues who have contributed to the development of this text by discussing ideas and sharing their expertise, humor, and devotion to science and engineering.

Nomenclature

$\bar{f}$ = principle-axis version of f, background or quiescent-fluid value of f, or average or ensemble average of f

$\hat{f}$ = complex amplitude of f

$\tilde{f}$ = full field value of f

f' = derivative of f with respect to its argument, or perturbation of f from its reference state

f^* = complex conjugate of f, dimensionless version of f, or the value of f at the sonic condition

f^+ = the dimensionless, law-of-the-wall value of f

f_{cr} = critical value of f

f_{CL} = centerline value of f

f_0 = reference, surface, or stagnation value of f

f_∞ = reference value of f or value of f far away from the point of interest

Δf = change in f

SYMBOLS*

α = contact angle, thermal expansion coefficient (1.20), angle of rotation, angle of attack, Womersley number (16.12), angle in a toroidal coordinate system, area ratio

a = triangular area, cylinder radius, sphere radius, amplitude

*Relevant equation numbers appear in parentheses

a_0 = initial tube radius

$\mathbf{a}$ = generic vector, Lagrangian acceleration (3.1)

$\mathbf{A}$ = generic second-order (or higher) tensor

A, A = a constant, an amplitude, area, surface, surface of a material volume, planform area of a wing

A^* = control surface, sonic throat area

A_o = Avogadro's number

A_0 = reference area

A_{ij} = representative second-order tensor

β = angle of rotation, coefficient of density change due to salinity or other constituent, variation of the Coriolis frequency with latitude, camber parameter

$\mathbf{b}$ = generic vector, control surface velocity (3.35)

B, B = a constant, Bernoulli function (4.70), log-law intercept parameter (12.88)

$\mathbf{B}, B_{ij}$ = generic second-order (or higher) tensor

Bo = Bond number (4.118)

c = speed of sound (1.19, 15.6), phase speed (7.4), chord length (14.2), pressure pulse wave speed, concentration of solutes

c_j = pressure pulse wave speed in tube j

$\mathbf{c}$ = phase velocity vector (7.8)

$c_g, \mathbf{c_g}$ = group velocity magnitude (7.68) and vector (7.144)

χ = scalar stream function

$^\circ C$ = degrees centigrade

C = a generic constant, hypotenuse length, closed contour

Ca = Capillary number (4.119)

C_f = skin friction coefficient (9.32)

C_p = coefficient of pressure (4.106, 6.32)

C_p = specific heat capacity at constant pressure (1.14)

C_D = coefficient of drag (4.107, 9.33)

C_L = coefficient of lift (4.108)

C_v = specific heat capacity at constant volume (1.15)

C_{ij} = matrix of direction cosines between original and rotated coordinate system axes (2.5)

d = diameter, distance, fluid layer depth

$\mathbf{d}$ = dipole strength vector (6.29), displacement vector

δ = Dirac delta function (B.4.1), similarity-variable length scale (8.32), boundary-layer thickness, generic length scale, small increment, flow deflection angle (15.53), tube radius divided by tube radius of curvature

$\bar{\delta}$ = average boundary-layer thickness

δ^* = boundary-layer displacement thickness (9.16)

δ_{ij} = Kronecker delta function (2.16)

δ_{99} = 99% layer thickness

D = distance, drag force, diffusion coefficient, Dean number (16.179)

D_i = lift-induced drag (14.15)

D/Dt = material derivative (3.4) or (3.5)

D_T = turbulent diffusivity of particles (12.127)

$\mathcal{D}$ = generalized field derivative (2.31)

ε = roughness height, kinetic energy dissipation rate (4.58), a small distance, fineness ratio h/L (8.14), downwash angle (14.14)

$\bar{\varepsilon}$ = average dissipation rate of the turbulent kinetic energy (12.47)

$\bar{\varepsilon}_T$ = average dissipation rate of the variance of temperature fluctuations (12.112)

ε_{ijk} = alternating tensor (2.18)

e = internal energy per unit mass (1.10)

$\mathbf{e}_i$ = unit vector in the i-direction (2.1)

$\bar{e}$ = average kinetic energy of turbulent fluctuations (12.47, 12.49)

Ec = Eckert number (4.115)

E_k = kinetic energy per unit horizontal area (7.39)

E_p = potential energy per unit horizontal area (7.41)

E = average energy per unit horizontal area (7.43), Ekman number (13.18), Young's modulus

$\bar{E}$ = kinetic energy of the average flow (12.46)

$\widehat{E}_1$ = total energy dissipation in a blood vessel

f = generic function, Helmholtz free energy per unit mass, longitudinal correlation coefficient (12.38), Coriolis frequency (13.8), dimensionless friction parameter (15.45)

ϕ = velocity potential (6.10), an angle

$\mathbf{f}$ = surface force vector per unit area (2.15, 4.13)

F = force magnitude, generic flow field property, average energy flux per unit length of wave crest (7.44), generic or profile function

$\mathbf{F}$ = force vector, average wave energy flux vector

Φ = body force potential (4.18), undetermined spectrum function (12.53)

F_D = drag force

F_L = lift force

Fr = Froude number (4.104)

γ = ratio of specific heats (1.24), velocity gradient, vortex sheet strength, generic dependent-field variable

$\dot{\gamma}$ = shear rate

$\mathbf{g}$ = body force per unit mass (4.13)

g = acceleration of gravity, undetermined function, transverse correlation coefficient (12.38)

g' = reduced gravity (7.188)

Γ = vertical temperature gradient or lapse rate, circulation (3.18)

Γ_a = adiabatic vertical temperature gradient (1.30)

$\Gamma_\mathbf{a}$ = circulation due to the absolute vorticity (5.33)

G = gravitational constant, pressure-gradient pulse amplitude, profile function

G_n = Fourier series coefficient

G = center of mass, center of vorticity

h = enthalpy per unit mass (1.13), height, gap height, viscous layer thickness, grid size, tube wall thickness

η = free surface shape, waveform, similarity variable (8.25, 8.32), Kolmogorov microscale (12.50), radial tube-wall displacement

η_T = Batchelor microscale (12.114)

H = atmospheric scale height, water depth, shape factor (9.46), profile function, Hematocrit

i = an index, imaginary root

I = incident light intensity, bending moment of inertia

j = an index

J, J_s = jet momentum flux per unit span (9.61)

J_i = Bessel function of order i

$\mathbf{J}_m$ = diffusive mass flux vector (1.1)

φ = a function, azimuthal angle in cylindrical and spherical coordinates

k = thermal conductivity (1.2), an index, wave number (7.2), wave number component

κ = thermal diffusivity, von Karman constant (12.88), Dean number (16.171)

κ_s = diffusivity of salt

κ_T = turbulent thermal diffusivity (12.95)

κ_m = mass diffusivity of a passive scalar in Fick's law (1.1)

κ_{mT} = turbulent mass diffusivity (12.96)

k_B = Boltzmann's constant (1.21)

Kn = Knudsen number

K = a generic constant, magnitude of the wave number vector (7.6), lift curve slope, Dean Number (16.178)

K_p = constant proportional to tube wall bending stiffness

K = compliance of a blood vessel, degrees Kelvin (16.48)

$\mathbf{K}$ = wave number vector, stiffness matrix

l = molecular mean free path, spanwise dimension, generic length scale, wave number component (7.5, 7.6), shear correlation in Thwaites method (9.45), length scale in turbulent flow

l_T = mixing length (12.98)

L, L = generic length dimension, generic length scale, lift force

L_M = Monin-Obukhov length scale (12.110)

λ = wavelength (7.1, 7.7), laminar boundary-layer correlation parameter (9.44), flow resistance ratio

λ_m = wavelength of the minimum phase speed

λ_t = temporal Taylor microscale (12.19)

λ_f, λ_g = longitudinal and lateral spatial Taylor microscale (12.39)

Λ = lubrication-flow bearing number (8.16), Rossby radius of deformation, wing aspect ratio

Λ_f, Λ_g = longitudinal and lateral integral spatial scales (12.39)

Λ_t = integral time scale (12.18)

μ = dynamic or shear viscosity (1.3), Mach angle (15.49)

μ_v = bulk viscosity (4.37)

m = molecular mass (1.22), generic mass, an index, two-dimensional source strength, moment order (12.1), wave number component (7.5, 7.6)

M, M = generic mass dimension, mass, Mach number (4.111), apparent or added mass (6.108)

M_w = molecular weight

n = number of molecules (1.21), an index, generic integer number

$\mathbf{n}$ = normal unit vector

n_s = index of refraction

N = Brunt-Väisälä or buoyancy frequency (1.29, 7.128), number, number of pores in a sieve plate

N_A = basis or interpolation functions

ν = kinematic viscosity (1.4), cyclic frequency, Prandtl-Meyer function (15.56)

ν_T = turbulent kinematic viscosity (12.94)

$\hat{\nu}$ = Poisson's ratio

O = origin

p = pressure

p_{atm} = atmospheric pressure

p_i = inside pressure

p_o = outside pressure

p_0 = reference pressure at $z = 0$

p_∞ = reference pressure far upstream or far away

$\bar{p}$ = average or quiescent pressure in a stratified fluid

P = average pressure

P = normalized pressure in a collapsible tube

Π = wake strength parameter

Pr = Prandtl number (4.116)

$\mathbf{q}$, q_i = heat flux (1.2)

q_n = generic parameter in dimensional analysis

q = heat added to a system (1.10), volume flux per unit span, dimensionless heat addition parameter (15.45)

Q = thermodynamic heat per unit mass, volume flux in two or three dimensions

θ = potential temperature (1.31), unit of temperature, angle in polar coordinates, momentum thickness (9.17), local phase, an angle, angle in a toroidal coordinate system

ρ = mass density (1.1)

ρ_m = mass density of a mixture

$\bar{\rho}$ = average or quiescent density in a stratified fluid

ρ_θ = potential density (1.33)

r = matrix rank, distance from the origin, distance from the axis

$\mathbf{r}$ = particle trajectory (3.1, 3.8)

R = distance from the cylindrical axis, radius of curvature, gas constant (1.23), generic nonlinearity parameter, total peripheral resistance (16.9), tube radius of curvature

R = viscous resistance per unit length, reflection coefficient (16.204), (16.153)

R_u = universal gas constant (1.22)

R_i = radius of curvature in direction i (1.5)

$\mathbf{R}$, R_{ij} = rotation tensor (3.13), correlation tensor (12.13, 12.23)

Ra = Rayleigh number (11.21)

Re = Reynolds number (4.103)

Ri = Richardson number, gradient Richardson number (11.66, 12.108)

Rf = flux Richardson number (12.107)

Ro = Rossby number (13.13)

σ = surface tension (1.5), interfacial tension, vortex core size (3.28, 3.29), temporal growth rate (11.1), shock angle

s = entropy (1.16), arc length, salinity, wingspan (14.1), dimensionless arc length

σ_{ij} = viscous stress tensor (4.27)

S = salinity, scattered light intensity, an area, dimensionless speed index, entropy

S_e = one-dimensional temporal longitudinal energy spectrum (12.20)

S_{11} = one-dimensional spatial longitudinal energy spectrum (12.45)

S_T = one-dimensional temperature fluctuation spectrum (12.113, 12.114)

$\mathbf{S}$, S_{ij} = strain rate tensor (3.12), symmetric tensor

St = Strouhal number (4.102)

t = time

$\mathbf{t}$ = tangent vector

T, T = temperature (1.2), generic time dimension, period, transmission coefficient (16.153)

Ta = Taylor number (11.52)

T_o = free stream temperature

T_w = wall temperature

T_i = tension in the i-direction

τ = shear stress (1.3), time lag

τ, τ_{ij} = stress tensor (2.15)

τ_0 = wall or surface shear stress

v = specific volume = $1/\rho$

u = horizontal component of fluid velocity (1.3)

$\mathbf{u}$ = generic vector, fluid velocity vector (3.1)

u_i = fluid velocity components, fluctuating velocity components

u_* = friction velocity (12.81)

$\mathbf{U}$ = generic uniform velocity vector

U_i = ensemble average velocity components

U = generic velocity, average stream-wise velocity

ΔU = characteristic velocity difference

U_e = local free-stream flow speed above a boundary layer (9.11), flow speed at the effective angle of attack

U_{CL} = centerline velocity (12.56)

U_∞ = flow speed far upstream or far away

v = component of fluid velocity along the y axis

$\mathbf{v}$ = generic vector

V = volume, material volume, average stream-normal velocity, average velocity, variational space, complex velocity

V^* = control volume

w = complex potential (6.42), vertical component of fluid velocity, function in the variational space, downwash velocity (14.13)

W = thermodynamic work per unit mass, wake function

$\dot{W}$ = rate of energy input from the average flow (12.49)

We = Weber number (4.117)

ω = temporal frequency (7.2)

$\boldsymbol{\omega}, \omega_i$ = vorticity vector (3.16)

Ω = oscillation frequency, computational domain, rotation rate, rotation rate of the earth

$\boldsymbol{\Omega}$ = angular velocity of a rotating frame of reference

x = first Cartesian coordinate

$\mathbf{x}$ = position vector (2.1)

x_i = components of the position vector (2.1)

ξ = generic spatial coordinate, integration variable, similarity variable (12.84), axial tube wall displacement

y = second Cartesian coordinate

Y = mass fraction (1.1)

Y_{CL} = centerline mass fraction (12.69)

Y_i = Bessel function of order i, admittance

ψ = stream function (6.3, 6.75), water potential

Ψ = Reynolds stress scaling function (12.57), generic functional solution

$\boldsymbol{\Psi}$ = vector potential, three-dimensional stream function (4.12)

z = third Cartesian coordinate, complex variable (6.43)

ζ = interface displacement, angular tube-wall displacement, relative vorticity

Z = impedance (16.151)

CHAPTER

1

Introduction

OUTLINE

1.1. Fluid Mechanics   2

1.2. Units of Measurement   3

1.3. Solids, Liquids, and Gases   3

1.4. Continuum Hypothesis   5

1.5. Molecular Transport Phenomena   5

1.6. Surface Tension   8

1.7. Fluid Statics   9

1.8. Classical Thermodynamics   12

1.9. Perfect Gas   16

1.10. Stability of Stratified Fluid Media   18

1.11. Dimensional Analysis   21

Exercises   30

Literature Cited   36

Supplemental Reading   37

CHAPTER OBJECTIVES

- To properly introduce the subject of fluid mechanics and its importance

- To state the assumptions upon which the subject is based

- To review the basic background science of liquids and gases

- To present the relevant features of fluid statics

- To establish dimensional analysis as an intellectual tool for use in the remainder of the text

1.1. FLUID MECHANICS

Fluid mechanics is the branch of science concerned with moving and stationary fluids. Given that the vast majority of the observable mass in the universe exists in a fluid state, that life as we know it is not possible without fluids, and that the atmosphere and oceans covering this planet are fluids, fluid mechanics has unquestioned scientific and practical importance. Its allure crosses disciplinary boundaries, in part because it is described by a nonlinear field theory and also because it is readily observed. Mathematicians, physicists, biologists, geologists, oceanographers, atmospheric scientists, engineers of many types, and even artists have been drawn to study, harness, and exploit fluid mechanics to develop and test formal and computational techniques, to better understand the natural world, and to attempt to improve the human condition. The importance of fluid mechanics cannot be overstated for applications involving transportation, power generation and conversion, materials processing and manufacturing, food production, and civil infrastructure. For example, in the twentieth century, life expectancy in the United States approximately doubled. About half of this increase can be traced to advances in medical practice, particularly antibiotic therapies. The other half largely resulted from a steep decline in childhood mortality from water-borne diseases, a decline that occurred because of widespread delivery of clean water to nearly the entire population—a fluids-engineering and public-works achievement. Yet, the pursuits of mathematicians, scientists, and engineers are interconnected: Engineers need to understand natural phenomena to be successful, scientists strive to provide this understanding, and mathematicians pursue the formal and computational tools that support these efforts.

Advances in fluid mechanics, like any other branch of physical science, may arise from mathematical analyses, computer simulations, or experiments. Analytical approaches are often successful for finding solutions to idealized and simplified problems and such solutions can be of immense value for developing insight and understanding, and for comparisons with numerical and experimental results. Thus, some fluency in mathematics, especially multivariable calculus, is helpful in the study of fluid mechanics. In practice, drastic simplifications are frequently necessary to find analytical solutions because of the complexity of real fluid flow phenomena. Furthermore, it is probably fair to say that some of the greatest theoretical contributions have come from people who depended rather strongly on their physical intuition. Ludwig Prandtl, one of the founders of modern fluid mechanics, first conceived the idea of a boundary layer based solely on physical intuition. His knowledge of mathematics was rather limited, as his famous student Theodore von Karman (1954, page 50) testifies. Interestingly, the boundary layer concept has since been expanded into a general method in applied mathematics.

As in other scientific fields, mankind's mathematical abilities are often too limited to tackle the full complexity of real fluid flows. Therefore, whether we are primarily interested in understanding flow physics or in developing fluid-flow applications, we often must depend on observations, computer simulations, or experimental measurements to test hypotheses and analyses, and develop insights into the phenomena under study. This book is an introduction to fluid mechanics that should appeal to anyone pursuing fluid mechanical inquiry. Its emphasis is on fully presenting fundamental concepts and illustrating them with examples drawn from various scientific and engineering fields. Given its finite size, this book provides—at best—an incomplete description of the subject. However, the purpose of this book will be fulfilled if the reader becomes more curious and interested in fluid mechanics as a result of its perusal.

1.2. UNITS OF MEASUREMENT

For mechanical systems, the units of all physical variables can be expressed in terms of the units of four basic variables, namely, *length, mass, time,* and *temperature*. In this book, the international system of units (Système international d'unités) commonly referred to as SI (or MKS) units, is preferred. The basic units of this system are *meter* for length, *kilogram* for mass, *second* for time, and *Kelvin* for temperature. The units for other variables can be derived from these basic units. Some of the common variables used in fluid mechanics, and their SI units, are listed in Table 1.1. Some useful conversion factors between different systems of units are listed in Appendix A. To avoid very large or very small numerical values, prefixes are used to indicate multiples of the units given in Table 1.1. Some of the common prefixes are listed in Table 1.2.

Strict adherence to the SI system is sometimes cumbersome and will be abandoned occasionally for simplicity. For example, temperatures will be frequently quoted in degrees Celsius (°C), which is related to Kelvin (K) by the relation °C = K − 273.15. However, the English system of units (foot, pound, °F) will not be used, even though this unit system remains in use in some places in the world.

1.3. SOLIDS, LIQUIDS, AND GASES

The various forms of matter may be broadly categorized as being fluid or solid. A fluid is a substance that deforms continuously under an applied shear stress or, equivalently, one that does not have a preferred shape. A solid is one that does not deform continuously under an applied shear stress, and does have a preferred shape to which it relaxes when external forces on it are withdrawn. Consider a rectangular element of a solid ABCD (Figure 1.1a).

TABLE 1.1 SI Units

Quantity	Name of unit	Symbol	Equivalent
Length	Meter	m	
Mass	Kilogram	kg	
Time	Second	s	
Temperature	Kelvin	K	
Frequency	Hertz	Hz	s^{-1}
Force	Newton	N	$kg\,ms^{-2}$
Pressure	Pascal	Pa	$N\,m^{-2}$
Energy	Joule	J	$N\,m$
Power	Watt	W	$J\,s^{-1}$

TABLE 1.2   Common Prefixes

Prefix	Symbol	Multiple
Mega	M	10^6
Kilo	k	10^3
Deci	d	10^{-1}
Centi	c	10^{-2}
Milli	m	10^{-3}
Micro	μ	10^{-6}

Under the action of a shear force F the element assumes the shape ABC′D′. If the solid is perfectly elastic, it returns to its preferred shape ABCD when F is withdrawn. In contrast, a fluid deforms *continuously* under the action of a shear force, *however small*. Thus, the element of the fluid ABCD confined between parallel plates (Figure 1.1b) successively deforms to shapes such as ABC′D′ and ABC″D″, and keeps deforming, as long as the force F is maintained on the upper plate. When F is withdrawn, the fluid element's final shape is retained; it does not return to a prior shape. Therefore, we say that a fluid flows.

The qualification "however small" in the description of a fluid is significant. This is because some solids also deform continuously if the shear stress exceeds a certain limiting value, corresponding to the *yield point* of the solid. A solid in such a state is known as *plastic*, and plastic deformation changes the solid object's unloaded shape. Interestingly, the distinction between solids and fluids may not be well defined. Substances like paints, jelly, pitch, putty, polymer solutions, and biological substances (for example, egg whites) may simultaneously display both solid and fluid properties. If we say that an elastic solid has a perfect memory of its preferred shape (because it always springs back to its preferred shape when unloaded) and that an ordinary viscous fluid has zero memory (because it never springs back when unloaded), then substances like egg whites can be called *viscoelastic* because they partially rebound when unloaded.

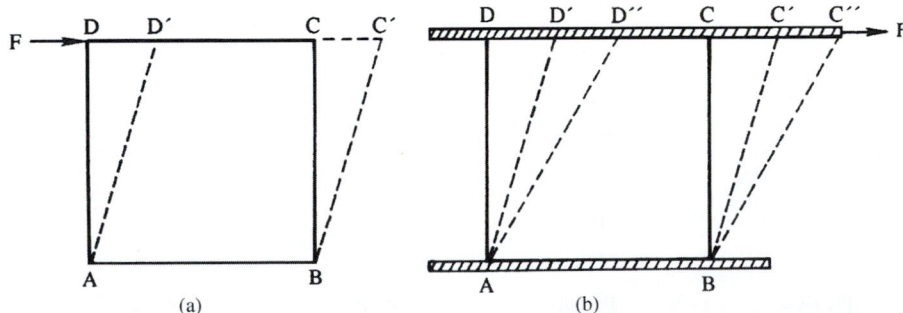

FIGURE 1.1   Deformation of solid and fluid elements under a constant externally applied shear force. (a) Solid; here the element deflects until its internal stress balances the externally applied force. (b) Fluid; here the element deforms continuously as long as the shear force is applied.

Although solids and fluids behave very differently when subjected to shear stresses, they behave similarly under the action of compressive normal stresses. However, tensile normal stresses again lead to differences in fluid and solid behavior. Solids can support both tensile and compressive normal stresses, while fluids typically expand or change phase (i.e., boil) when subjected to tensile stresses. Some liquids can support a small amount of tensile stress, the amount depending on the degree of molecular cohesion and the duration of the tensile stress.

Fluids generally fall into two classes, liquids and gases. A gas always expands to fill the entire volume of its container. In contrast, the volume of a liquid changes little, so that it cannot completely fill a large container; in a gravitational field, a free surface forms that separates a liquid from its vapor.

1.4. CONTINUUM HYPOTHESIS

A fluid is composed of a large number of molecules in constant motion undergoing collisions with each other, and is therefore discontinuous or discrete at the most microscopic scales. In principle, it is possible to study the mechanics of a fluid by studying the motion of the molecules themselves, as is done in kinetic theory or statistical mechanics. However, we are generally interested in the *average manifestation* of the molecular motion. For example, forces are exerted on the boundaries of a fluid's container due to the constant bombardment of the fluid molecules; the statistical average of these collision forces per unit area is called *pressure*, a macroscopic property. So long as we are not interested in the molecular mechanics of the origin of pressure, we can ignore the molecular motion and think of pressure as simply the average force per unit area exerted by the fluid.

When the molecular density of the fluid and the size of the region of interest are large enough, such average properties are sufficient for the explanation of macroscopic phenomena and the discrete molecular structure of matter may be ignored and replaced with a continuous distribution, called a *continuum*. In a continuum, fluid properties like temperature, density, or velocity are defined at every point in space, and these properties are known to be appropriate averages of molecular characteristics in a small region surrounding the point of interest. The continuum approximation is valid when the Knudsen number, $Kn = l/L$ where l is the mean free path of the molecules and L is the length scale of interest (a body length, a pore diameter, a turning radius, etc.), is much less than unity. For most terrestrial situations, this is not a great restriction since $l \approx 50$ *nm* for air at room temperature and pressure, and l is more than two orders of magnitude smaller for water under the same conditions. However, a molecular-kinetic-theory approach may be necessary for analyzing flows over very small objects or in very narrow flow paths, or in the tenuous gases at the upper reaches of the atmosphere.

1.5. MOLECULAR TRANSPORT PHENOMENA

Although the details of molecular motions may be locally averaged to compute temperature, density, or velocity, random molecular motions still lead to diffusive transport of molecular species, temperature, or momentum that impact fluid properties at macroscopic scales.

Consider a surface area AB within a mixture of two gases, say, nitrogen and oxygen (Figure 1.2), and assume that the nitrogen mass fraction Y varies across AB. Here the mass of nitrogen per unit volume is ρY (sometimes known as the nitrogen *concentration* or *density*), where ρ is the overall density of the gas mixture. Random migration of molecules across AB in both directions will result in a *net* flux of nitrogen across AB, from the region of higher Y toward the region of lower Y. To a good approximation, the flux of one constituent in a mixture is proportional to its gradient:

$$\mathbf{J}_m = -\rho \kappa_m \nabla Y. \tag{1.1}$$

Here the vector $\mathbf{J}_m$ is the mass flux (kg m^{-2} s^{-1}) of the constituent, ∇Y is the mass-fraction gradient of that constituent, and κ_m is a (positive) constant of proportionality that depends on the particular pair of constituents in the mixture and the local thermodynamic state. For example, κ_m for diffusion of nitrogen in a mixture with oxygen is different than κ_m for diffusion of nitrogen in a mixture with carbon dioxide. The linear relation (1.1) for mass diffusion is generally known as *Fick's law*, and the minus sign reflects the fact that species diffuse from higher to lower concentrations. Relations like this are based on empirical evidence, and

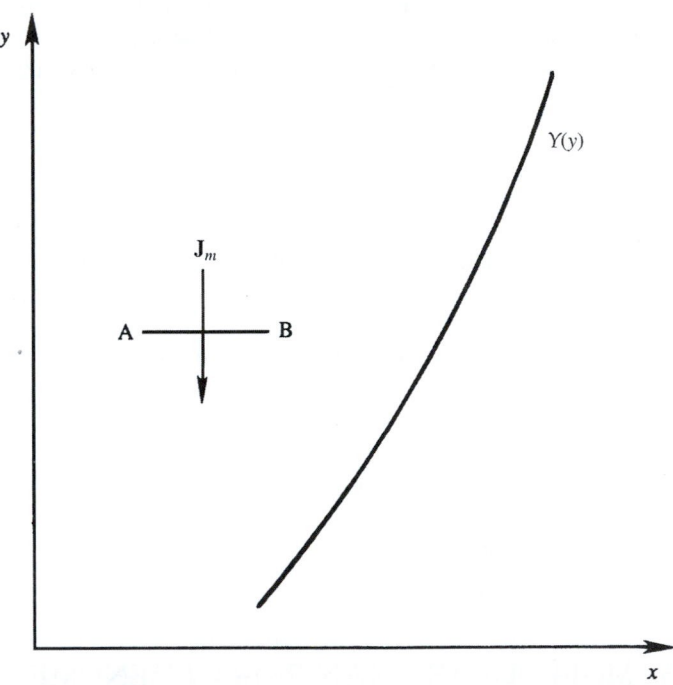

FIGURE 1.2   Mass flux $\mathbf{J}_m$ due to variation in the mass fraction $Y(y)$. Here the mass fraction profile increases with increasing Y, so Fick's law of diffusion states that the diffusive mass flux that acts to smooth out mass-fraction differences is downward across AB.

are called *phenomenological laws*. Statistical mechanics can sometimes be used to derive such laws, but only for simple situations.

The analogous relation for heat transport via a temperature gradient ∇T is *Fourier's law*,

$$\mathbf{q} = -k\nabla T, \tag{1.2}$$

where $\mathbf{q}$ is the heat flux ($J\ m^{-2}\ s^{-1}$), and k is the material's thermal conductivity.

The analogous relationship for momentum transport via a velocity gradient is qualitatively similar to (1.1) and (1.2) but is more complicated because momentum and velocity are vectors. So as a first step, consider the effect of a vertical gradient, du/dy, in the horizontal velocity u (Figure 1.3). Molecular motion and collisions cause the faster fluid above AB to pull the fluid underneath AB forward, thereby speeding it up. Molecular motion and collisions also cause the slower fluid below AB to pull the upper faster fluid backward, thereby slowing it down. Thus, without an external influence to maintain du/dy, the flow profile shown by the solid curve will evolve toward a profile shown by the dashed curve. This is analogous to saying that u, the horizontal momentum per unit mass (a momentum *concentration*), *diffuses* downward. Here, the resulting momentum flux, from high to low u, is equivalent to a shear stress, τ, existing in the fluid. Experiments show that the magnitude of τ along a surface such as AB is, to a good approximation, proportional to the local velocity gradient,

$$\tau = \mu(du/dy), \tag{1.3}$$

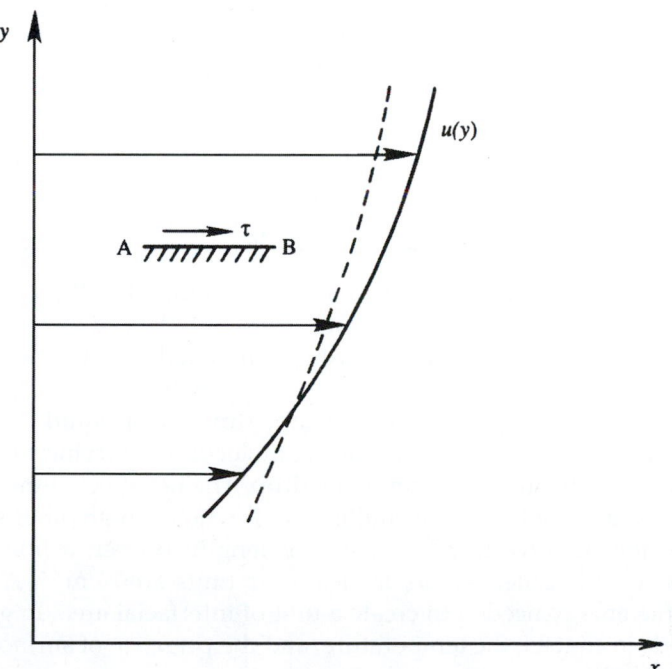

FIGURE 1.3 Shear stress τ on surface AB. The diffusive action of fluid viscosity tends to decrease velocity gradients, so that the continuous line tends toward the dashed line.

where the constant of proportionality μ (with units of kg m^{-1} s^{-1}) is known as the *dynamic viscosity*. This is *Newton's law* of friction. It is analogous to (1.1) and (1.2) for the simple unidirectional shear flow depicted in Figure 1.3. However, it is an incomplete scalar statement of molecular momentum transport when compared to the more complete vector relationships (1.1) and (1.2) for species and thermal molecular transport. A more general tensor form of (1.3) that accounts for three velocity components and three possible orientations of the surface AB is presented in Chapter 4 after the mathematical and kinematical developments in Chapters 2 and 3. For gases and liquids, μ depends on the local temperature T. In ideal gases, the random thermal speed is roughly proportional to $T^{1/2}$, so molecular momentum transport, and consequently μ, also vary approximately as $T^{1/2}$. For liquids, shear stress is caused more by the intermolecular cohesive forces than by the thermal motion of the molecules. These cohesive forces decrease with increasing T so μ for a liquid decreases with increasing T.

Although the shear stress is proportional to μ, we will see in Chapter 4 that the tendency of a fluid to transport velocity gradients is determined by the quantity

$$\nu \equiv \mu/\rho, \tag{1.4}$$

where ρ is the density (kg m^{-3}) of the fluid. The units of ν (m^2 s^{-1}) do not involve the mass, so ν is frequently called the *kinematic viscosity*.

Two points should be noticed about the transport laws (1.1), (1.2), and (1.3). First, only *first* derivatives appear on the right side in each case. This is because molecular transport is carried out by a nearly uncountable number of molecular interactions at length scales that are too small to be influenced by higher derivatives of the species mass fractions, temperature, or velocity profiles. Second, nonlinear terms involving higher powers of the first derivatives, for example $|\nabla u|^2$, do not appear. Although this is only expected for small first-derivative magnitudes, experiments show that the linear relations are accurate enough for most practical situations involving mass fraction, temperature, or velocity gradients.

1.6. SURFACE TENSION

A density discontinuity may exist whenever two immiscible fluids are in contact, for example at the interface between water and air. Here unbalanced attractive intermolecular forces cause the interface to behave as if it were a stretched membrane under tension, like the surface of a balloon or soap bubble. This is why small drops of liquid in air or small gas bubbles in water tend to be spherical in shape. Imagine a liquid drop surrounded by an insoluble gas. Near the interface, all the liquid molecules are trying to pull the molecules on the interface inward toward the center of the drop. The net effect of these attractive forces is for the interface area to contract until equilibrium is reached with other surface forces. The magnitude of the tensile force that acts per unit length to open a line segment lying in the surface like a seam is called *surface tension* σ; its units are N m^{-1}. Alternatively, σ can be thought of as the energy needed to create a unit of interfacial area. In general, σ depends on the pair of fluids in contact, the temperature, and the presence of surface-active chemicals (surfactants) or impurities, even at very low concentrations.

An important consequence of surface tension is that it causes a pressure difference across curved interfaces. Consider a spherical interface having a radius of curvature R (Figure 1.4a).

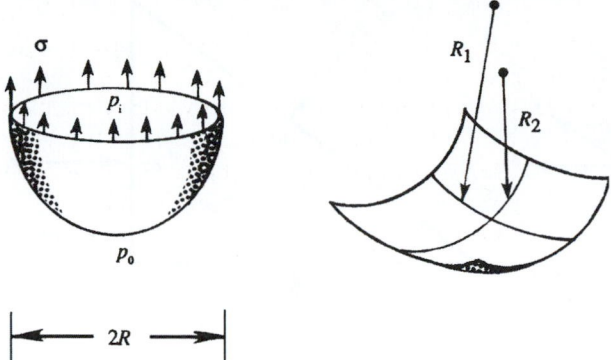

FIGURE 1.4 (a) Section of a spherical droplet, showing surface tension forces. (b) An interface with radii of curvatures R_1 and R_2 along two orthogonal directions.

If p_i and p_o are the pressures on the inner and outer sides of the interface, respectively, then a static force balance gives

$$\sigma(2\pi R) = (p_i - p_o)\pi R^2,$$

from which the pressure jump is found to be

$$p_i - p_o = 2\sigma/R,$$

showing that the pressure on the concave side (the inside) is higher.

The curvature of a general surface can be specified by the radii of curvature along two orthogonal directions, say, R_1 and R_2 (Figure 1.4b). A similar analysis shows that the pressure difference across the interface is given by

$$p_i - p_o = \sigma\left(\frac{1}{R_1} + \frac{1}{R_2}\right), \tag{1.5}$$

which agrees with the spherical interface result when $R_1 = R_2$. This pressure difference is called the Laplace pressure.

It is well known that the free surface of a liquid in a narrow tube rises above the surrounding level due to the influence of surface tension. This is demonstrated in Example 1.1. Narrow tubes are called *capillary tubes* (from Latin *capillus*, meaning hair). Because of this, the range of phenomena that arise from surface tension effects is called *capillarity*. A more complete discussion of surface tension is presented at the end of Chapter 4 as part of the section on boundary conditions.

1.7. FLUID STATICS

The magnitude of the force per unit area in a static fluid is called the *pressure*; pressure in a moving medium will be defined in Chapter 4. Sometimes the ordinary pressure is called the

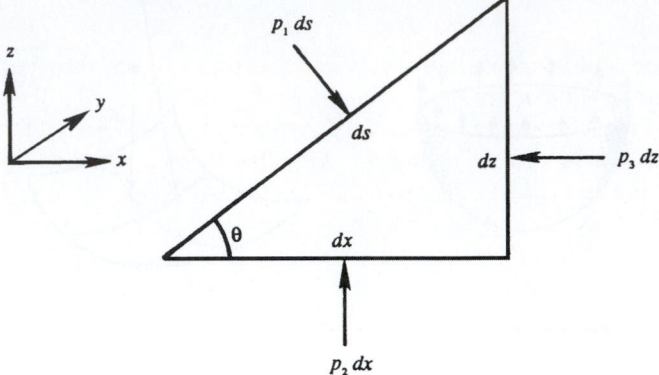

FIGURE 1.5 Demonstration that $p_1 = p_2 = p_3$ in a static fluid. Here the vector sum of the three arrows is zero when the volume of the element shrinks to zero.

absolute pressure, in order to distinguish it from the *gauge pressure,* which is defined as the absolute pressure minus the atmospheric pressure:

$$p_{\text{gauge}} = p - p_{\text{atm}}.$$

The standard value for atmospheric pressure p_{atm} is 101.3 kPa = 1.013 bar where 1 bar = 10^5 Pa. An absolute pressure of zero implies vacuum while a gauge pressure of zero implies atmospheric pressure.

In a fluid at rest, tangential viscous stresses are absent and the only force between adjacent surfaces is normal to the surface. We shall now demonstrate that in such a case the surface force per unit area (or pressure) is equal in all directions. Consider a small volume of fluid with a triangular cross section (Figure 1.5) of unit thickness normal to the paper, and let p_1, p_2, and p_3 be the pressures on the three faces. The z-axis is taken vertically upward. The only forces acting on the element are the pressure forces normal to the faces and the weight of the element. Because there is no acceleration of the element in the x direction, a balance of forces in that direction gives

$$(p_1 \, ds)\sin \theta - p_3 \, dz = 0.$$

Because $dz = \sin\theta \, ds$, the foregoing gives $p_1 = p_3$. A balance of forces in the vertical direction gives

$$-(p_1 \, ds)\cos \theta + p_2 \, dx - (1/2)\rho g \, dx \, dz = 0.$$

As $\cos\theta \, ds = dx$, this gives

$$p_2 - p_1 - (1/2)\rho g \, dz = 0.$$

As the triangular element is shrunk to a point, that is, $dz \rightarrow 0$ with $\theta = $ constant, the gravity force term drops out, giving $p_1 = p_2$. Thus, at a point in a static fluid, we have

$$p_1 = p_2 = p_3, \qquad (1.6)$$

so that the force per unit area is independent of the angular orientation of the surface. The pressure is therefore a scalar quantity.

We now proceed to determine the *spatial distribution* of pressure in a static fluid. Consider an infinitesimal cube of sides dx, dy, and dz, with the z-axis vertically upward (Figure 1.6). A balance of forces in the x direction shows that the pressures on the two sides perpendicular to the x-axis are equal. A similar result holds in the y direction, so that

$$\partial p / \partial x = \partial p / \partial y = 0. \qquad (1.7)$$

This fact is expressed by *Pascal's law*, which states that all points in a resting fluid medium (and connected by the *same* fluid) are at the same pressure if they are at the same depth. For example, the pressure at points F and G in Figure 1.7 are the same.

Vertical equilibrium of the element in Figure 1.6 requires that

$$p \, dx \, dy - (p + dp) \, dx \, dy - \rho g \, dx \, dy \, dz = 0,$$

which simplifies to

$$dp/dz = -\rho g. \qquad (1.8)$$

This shows that the pressure in a static fluid subject to a constant gravitational field decreases with height. For a fluid of uniform density, (1.8) can be integrated to give

$$p = p_0 - \rho g z, \qquad (1.9)$$

where p_0 is the pressure at $z = 0$. Equation (1.9) is the well-known result of *hydrostatics*, and shows that the pressure in a liquid decreases *linearly* with increasing height. It implies that

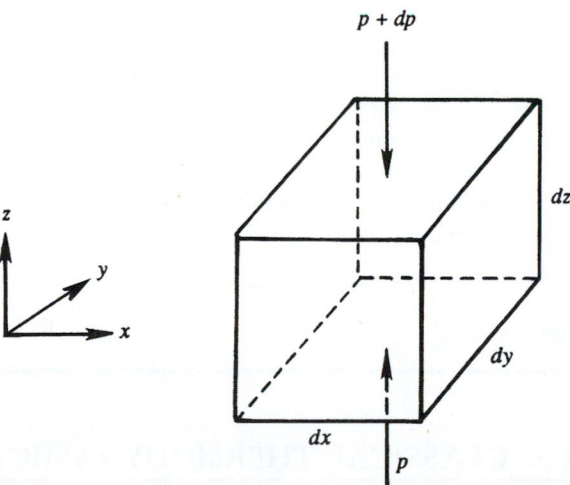

FIGURE 1.6 Fluid element at rest. Here the pressure difference between the top and bottom of the element balances the element's weight.

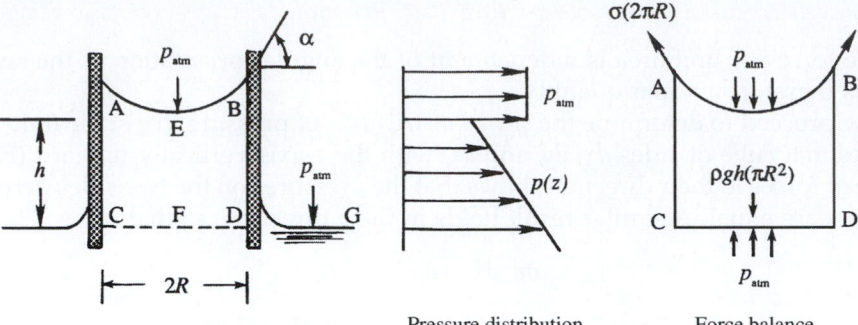

Pressure distribution   Force balance

FIGURE 1.7   Rise of a liquid in a narrow tube (Example 1.1) because of the action of surface tension. The curvature of the surface and the surface tension cause a pressure difference to occur across the surface.

the pressure rise at a depth h below the free surface of a liquid is equal to ρgh, which is the weight of a column of liquid of height h and unit cross section.

EXAMPLE 1.1

Using Figure 1.7, show that the rise of a liquid in a narrow tube of radius R is given by

$$h = \frac{2\sigma \sin\alpha}{\rho gR},$$

where σ is the surface tension and α is the *contact angle* between the fluid and the tube's inner surface.

Solution

Since the free surface is concave upward and exposed to the atmosphere, the pressure just below the interface at point E is below atmospheric. The pressure then increases linearly along EF. At F the pressure again equals the atmospheric pressure, since F is at the same level as G where the pressure is atmospheric. The pressure forces on faces AB and CD therefore balance each other. Vertical equilibrium of the element ABCD then requires that the weight of the element balances the vertical component of the surface tension force, so that

$$\sigma(2\pi R)\sin\alpha = \rho gh(\pi R^2),$$

which gives the required result.

1.8. CLASSICAL THERMODYNAMICS

Classical thermodynamics is the study of equilibrium states of matter, in which the properties are assumed uniform in space and time. Here, the reader is assumed to be familiar with the basic thermodynamic concepts, so this section merely reviews the main ideas and the most commonly used relations in this book.

A thermodynamic *system* is a quantity of matter that exchanges heat and work, but no mass, with its surroundings. A system in equilibrium is free of fluctuations, such as those generated during heat or work input from, or output to, its surroundings. After any such thermodynamic change, fluctuations die out or *relax*, a new equilibrium is reached, and once again the system's properties, such as pressure and temperature, are well defined. Here, the system's *relaxation time* is defined as the time taken by the system to adjust to a new thermodynamic state.

This thermodynamic system concept is obviously not directly applicable to a macroscopic volume of a moving fluid in which pressure and temperature may vary considerably. However, experiments show that classical thermodynamics does apply to small fluid volumes commonly called *fluid particles*. A fluid particle is a small deforming volume carried by the flow that: 1) always contains the same fluid molecules, 2) is large enough so that its thermodynamic properties are well defined when it is at equilibrium, but 3) is small enough so that its *relaxation time* is short compared to the time scales of fluid-motion-induced thermodynamic changes. Under ordinary conditions (the emphasis in this text), molecular densities, speeds, and collision rates are high enough so that the conditions for the existence of fluid particles are met, and classical thermodynamics can be directly applied to flowing fluids. However, there are circumstances involving rarified gases, shock waves, and high-frequency acoustic waves where one or more of the fluid particle requirements are not met and molecular-kinetic and quantum theories are needed.

The basic laws of classical thermodynamics are empirical, and cannot be derived from anything *more* fundamental. These laws essentially establish *definitions*, upon which the subject is built. The first law of thermodynamics can be regarded as a principle that defines the *internal energy* of a system, and the second law can be regarded as the principle that defines the *entropy* of a system.

First Law of Thermodynamics

The first law of thermodynamics states that the energy of a system is conserved;

$$\delta q + \delta w = \Delta e, \tag{1.10}$$

where δq is the heat added to the system, δw is the work done on the system, and Δe is the increase of the system's *internal energy*. All quantities in (1.10) are normalized by the mass of the system and therefore have units of J kg^{-1} and appear as lowercase letters. When (1.10) is written with capital letters, $\delta Q + \delta W = \Delta E$, it portrays the same thermodynamic law without normalization by the system mass. The internal energy (aka, thermal energy) is a manifestation of the random molecular motion of the system's constituents. In fluid flows, the kinetic energy of the fluid particles' macroscopic motion has to be included in the e-term in (1.10) in order that the principle of conservation of energy is satisfied. For developing the relations of classical thermodynamics, however, we shall only include the thermal energy in the term e.

It is important to realize the difference between heat and internal energy. Heat and work are forms of *energy in transition*, which appear at the *boundary* of the system and are *not contained* within the matter. In contrast, the internal energy resides within the matter. If two equilibrium states 1 and 2 of a system are known, then Q and W depend on the *process* or *path* followed by the system in going from state 1 to state 2. The change $\Delta e = e_2 - e_1$, in contrast, does not depend on the path. In short, e is a thermodynamic property and is a function of the thermodynamic state of the system. Thermodynamic properties are called *state functions*, in contrast to heat and work, which are *path functions*.

Frictionless quasi-static processes, carried out at an extremely slow rate so that the system is at all times in equilibrium with the surroundings, are called *reversible processes*. For a compressible fluid, the most common type of reversible work is by the expansion or contraction of the boundaries of the fluid particle. Let $v = 1/\rho$ be the *specific volume*, that is, the volume per unit mass. The work done per unit mass by a fluid particle in an infinitesimal reversible process is $-p\,dv$, where dv is the increase of v. The first law (1.10) for a reversible process then becomes

$$de = dq - p\,dv, \tag{1.11}$$

provided that q is also reversible. Note that irreversible forms of work, such as those done against frictional stresses, are excluded from (1.11).

Equations of State

A relation defining one state function in terms of two or more others is called an *equation of state*. For a simple compressible substance composed of a single component (the applicable model for nearly all pure fluids), the specification of two independent thermodynamic properties completely determines the state of the system. We can write relations such as the *thermal* and *caloric equations of state*:

$$p = p(v, T) \text{ or } e = e(p, T), \tag{1.12}$$

respectively. For more complicated systems composed of more than one component, the specification of additional properties is needed to completely determine the state. For example, seawater contains dissolved salt so its density is a function of temperature, pressure, and salinity.

Specific Heats

Before we define the specific heats of a substance, we define the thermodynamic property *enthalpy* as

$$h \equiv e + pv. \tag{1.13}$$

It is the sum of the thermal energy and the pressure-volume potential energy, and arises naturally in the study of compressible fluid flows.

For single-component systems, the specific heat capacities at constant pressure and constant volume are defined as

$$C_p \equiv (\partial h/\partial T)_p \text{ and } C_v \equiv (\partial e/\partial T)_v, \qquad (1.14, 1.15)$$

respectively. Here, (1.14) means that we regard h as a function of p and T, and find the partial derivative of h with respect to T, keeping p constant. Equation (1.15) has an analogous interpretation. The specific heats as defined are thermodynamic properties because they are defined in terms of other properties of the system. That is, C_p and C_v can be determined when two other system properties (say, p and T) are known.

For certain processes common in fluid flows, the heat exchange can be related to the specific heats. Consider a reversible process in which the work done is given by pdv, so that the first law of thermodynamics has the form of (1.11). Dividing by the change of temperature, it follows that the heat transferred per unit mass per unit temperature change in a constant volume process is

$$(\partial Q/\partial T)_v = (\partial e/\partial T)_v = C_v.$$

This shows that $C_v \, dT$ represents the heat transfer per unit mass in a reversible constant-volume process, in which the only type of work done is of the pdv type. It is misleading to define $C_v = (dQ/dT)_v$ without any restrictions imposed, as the temperature of a constant-volume system can increase without heat transfer, such as by vigorous stirring.

Similarly, the heat transferred at constant pressure during a reversible process is given by

$$(\partial Q/\partial T)_p = (\partial h/\partial T)_p = C_p.$$

Second Law of Thermodynamics

The second law of thermodynamics restricts the direction in which real processes can proceed as time increases. Its implications are discussed in Chapter 4. Some consequences of this law are the following:

(i) There must exist a thermodynamic property s, known as *entropy*, whose change between states 1 and 2 is given by

$$s_2 - s_1 = \int_1^2 \frac{dq_{\text{rev}}}{T}, \qquad (1.16)$$

where the integral is taken along any reversible process between the two states.

(ii) For an *arbitrary* process between states 1 and 2, the entropy change is

$$s_2 - s_1 \geq \int_1^2 \frac{dq_{\text{rev}}}{T} \text{ (Clausius-Duhem)},$$

which states that the entropy of an isolated system ($dQ = 0$) can only increase. Such increases are caused by friction, mixing, and other irreversible phenomena.

(iii) Molecular transport coefficients such as viscosity μ and thermal conductivity k must be positive. Otherwise, spontaneous unmixing or momentum separation would occur and lead to a decrease of entropy of an isolated system.

Property Relations

Two common relations are useful in calculating entropy changes during a process. For a reversible process, the entropy change is given by

$$Tds = dq. \tag{1.17}$$

On substituting into (1.11) and using (1.13), we obtain

$$Tds = de + pdv, \text{ or } Tds = dh - vdp. \text{ (Gibbs)} \tag{1.18}$$

It is interesting that these relations (1.18) are also valid for irreversible (frictional) processes, although the relations (1.11) and (1.17), from which equations (1.18) are derived, are true for reversible processes only. This is because (1.18) are relations between thermodynamic *state functions* alone and are therefore true for *any* process. The association of Tds with heat and $-pdv$ with work does not hold for irreversible processes. Consider stirring work done at constant volume that raises a fluid element's temperature; here $de = Tds$ is the increment of stirring work done.

Speed of Sound

In a compressible fluid, infinitesimal isentropic changes in density and pressure propagate through the medium at a finite speed, c. In Chapter 15, we shall prove that the square of this speed is given by

$$c^2 = (\partial p / \partial \rho)_s, \tag{1.19}$$

where the subscript s signifies that the derivative is taken at constant entropy. This is the speed of sound. For incompressible fluids, $\partial \rho / \partial p \to 0$ under all conditions so $c \to \infty$.

Thermal Expansion Coefficient

When fluid density is a function of temperature, we define the thermal expansion coefficient

$$\alpha \equiv -\frac{1}{\rho}\left(\frac{\partial \rho}{\partial T}\right)_p, \tag{1.20}$$

where the subscript p signifies that the partial derivative is taken at constant pressure. This expansion coefficient appears frequently in the study of nonisothermal systems.

1.9. PERFECT GAS

A basic result from kinetic theory and statistical mechanics for the thermal equation of state for n identical noninteracting gas molecules confined within a container having volume V is

$$pV = nk_BT, \tag{1.21}$$

where p is the average pressure on the inside surfaces of the container, $k_B = 1.381 \times 10^{-23} JK^{-1}$ is Boltzmann's constant, and T is the absolute temperature. Equation (1.21) is the molecule-based version of the perfect gas law. It is valid when attractive forces between the molecules are negligible and when V/n is much larger than the (average) volume of an individual molecule. When used with the continuum approximation, (1.21) is commonly rearranged by noting that $\rho = mn/V$, where m is the (average) mass of one gas molecule. Here m is calculated (in SI units) as M_w/A_o where M_w is the (average) molecular weight in kg (kg-mole)$^{-1}$ of the gas molecules, and A_o is the kilogram-based version of Avogadro's number, 6.023×10^{26}(kg-mole)$^{-1}$. With these replacements, (1.21) becomes

$$p = \frac{n}{V} k_B T = \frac{nm}{V}\left(\frac{k_B}{m}\right) T = \rho\left(\frac{k_B A_o}{M_w}\right) T = \rho\left(\frac{R_u}{M_w}\right) T = \rho R T, \qquad (1.22)$$

where the product $k_B A_o = R_u = 8314$ J kmol^{-1} K^{-1} is the *universal gas constant*, and $R = R_u/M_w$ is the *gas constant* for the gas under consideration. A perfect gas is one that obeys (1.22), even if it is a mixture of several different molecular species. For example, the average molecular weight of dry air is 28.966 kg kmol^{-1}, for which (1.22) gives $R = 287$ J kg^{-1} K^{-1}. At ordinary temperatures and pressures most gases can be treated as perfect gases.

The gas constant for a particular gas is related to the specific heats of the gas through the relation

$$R = C_p - C_v, \qquad (1.23)$$

where C_p and C_v are the specific heat capacities at constant pressure and volume, respectively. In general, C_p and C_v increase with temperature. The ratio of specific heats

$$\gamma \equiv C_p/C_v \qquad (1.24)$$

is important in compressible fluid dynamics. For air at ordinary temperatures, $\gamma = 1.40$ and $C_p = 1004$ J kg^{-1} K^{-1}. It can be shown that (1.21) or (1.22) is equivalent to $e = e(T)$ and $h = h(T)$, and conversely, so that the internal energy and enthalpy of a perfect gas are only functions of temperature (Exercise 1.10).

A process is called *adiabatic* if it takes place without the addition of heat. A process is called *isentropic* if it is adiabatic and frictionless, for then the entropy of the fluid does not change. From (1.18) it can be shown (Exercise 1.11) that isentropic flow of a perfect gas with constant specific heats obeys

$$p/\rho^\gamma = const. \qquad (1.25)$$

Using (1.22) and (1.25), the temperature and density changes during an isentropic process from a reference state (subscript 0) to a current state (no subscript) are

$$T/T_0 = (p/p_0)^{(\gamma-1)/\gamma} \text{ and } \rho/\rho_0 = (p/p_0)^{1/\gamma} \qquad (1.26)$$

(see Exercise 1.8). In addition, simple expressions can be found for the speed of sound c and the thermal expansion coefficient α for a perfect gas:

$$c = \sqrt{\gamma RT} \text{ and } \alpha = 1/T. \qquad (1.27, 1.28)$$

1.10. STABILITY OF STRATIFIED FLUID MEDIA

In a static fluid environment subject to a gravitational field, p, ρ, and T may vary with height z, but (1.8) and (1.12) provide two constraints so the p, ρ, and T variations cannot be arbitrary. Furthermore, these constraints imply that the specification of the vertical profile of any one thermodynamic variable allows the profiles of the others to be determined. In addition, our experience suggests that the fluid medium will be stable if $\rho(z)$ decreases with increasing z. Interestingly, the rate at which the density decreases also plays a role in the stability of the fluid medium when the fluid is compressible, as in a planetary atmosphere.

To assess the stability of a static fluid medium, consider a fluid particle with density $\rho(z_0)$ in an atmosphere (or ocean) at equilibrium at height z_0 that is displaced upward a small distance ζ via a frictionless adiabatic process and then released from rest. At its new height, $z_0 + \zeta$, the fluid particle will have a different density, $\rho(z_0) + (d\rho_a/dz)\zeta + ...$, where $d\rho_a/dz$ is the isentropic density gradient for the displaced particle at height z_0 (see Figure 1.8). The density of the fluid particles already at height $z_0 + \zeta$ is $\rho(z_0 + \zeta) = \rho(z_0) + (d\rho/dz)\zeta + ...$, where $d\rho/dz$ is the equilibrium density gradient at height z_0 in the fluid medium. A vertical-direction application of Newton's second law including weight and buoyancy for the displaced element leads to

$$\frac{d^2\zeta}{dt^2} - \frac{g}{\rho(z_0)}\left(\frac{d\rho}{dz} - \frac{d\rho_a}{dz}\right)\zeta = 0$$

when first-order terms in ζ are retained (see Exercise 1.13). The coefficient of ζ in the second term is the square of the *Brunt-Väisälä frequency*, N,

$$N^2 = -\frac{g}{\rho(z_0)}\left(\frac{d\rho}{dz} - \frac{d\rho_a}{dz}\right). \tag{1.29}$$

When N^2 is positive, the fluid medium is *stable*; the displaced fluid particle will accelerate back toward z_0 after release and the action of viscous forces and thermal conduction will arrest any oscillatory motion. Thus, a stable atmosphere (or ocean) is one in which the density decreases with height *faster* than in an isentropic atmosphere (or ocean). When N^2 is negative, the fluid medium is *unstable*; the displaced fluid element will accelerate away from z_0 after

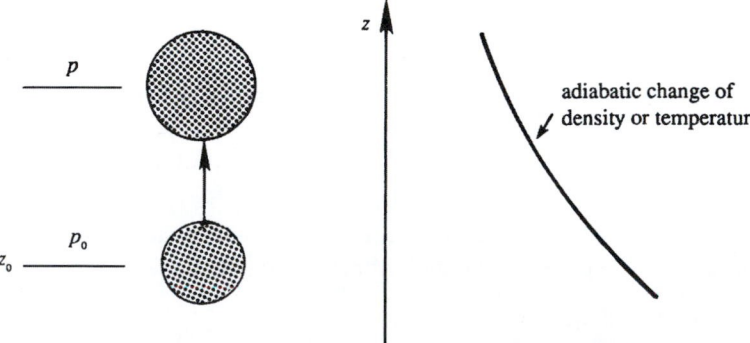

FIGURE 1.8 Adiabatic expansion of a fluid particle displaced upward in a compressible medium. In a static pressure field, if the fluid particle rises it encounters a lower pressure and may expand adiabatically.

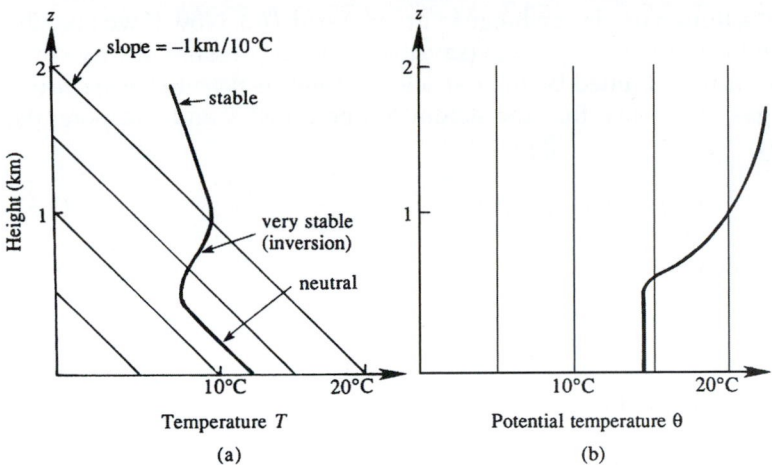

FIGURE 1.9 Vertical variation of the (a) actual and (b) potential temperature in the atmosphere. Thin straight lines represent temperatures for a neutral atmosphere. Slopes less than the neutral atmosphere lines lead to atmospheric instability. Slopes greater than the neutral atmosphere lines indicate a stable atmosphere.

release and further increase its displacement ζ. When N^2 is zero, the fluid medium is *neutrally stable* and the element will not move if released from rest; it will have zero vertical acceleration. There are two ways to achieve neutral stability: 1) the fluid density may be independent of the vertical coordinate so that $d\rho/dz = d\rho_a/dz = 0$, or 2) the equilibrium density gradient in the fluid medium may equal the isentropic density gradient, $d\rho/dz = d\rho_a/dz$. The former case implies that constant-density fluid media are neutrally stable. The latter case requires a *neutrally stable atmosphere* to be one where p, ρ, and T decrease with increasing height in such a way that the entropy is constant.

In atmospheric science, $\Gamma \equiv dT/dz$ is the atmospheric temperature gradient or *lapse rate*. The rate of temperature decrease in an isentropic atmosphere Γ_a is

$$dT_a/dz \equiv \Gamma_a = -g\alpha T/C_p, \qquad (1.30)$$

(see Exercise 1.14) and is called the *adiabatic temperature gradient* or *adiabatic lapse rate*. It is the steepest rate at which the temperature can decrease with increasing height without causing instability. In the earth's atmosphere, the adiabatic lapse rate is approximately $-10°C\ km^{-1}$.

Figure 1.9a shows a typical distribution of temperature in the earth's atmosphere. The lower part has been drawn with a slope nearly equal to the adiabatic temperature gradient because mixing processes near the ground tend to form a neutral (constant entropy) atmosphere. Observations show that the neutral atmosphere ends a layer where the temperature increases with height, a very stable situation. Meteorologists call this an *inversion*, because the temperature gradient changes signs here. Atmospheric turbulence and mixing processes below such an inversion typically cannot penetrate above it. Above this inversion layer the temperature decreases again, but less rapidly than near the ground, which again corresponds to stability. An isothermal atmosphere (a vertical line in Figure 1.9a) is quite stable.

Potential Temperature and Density

The foregoing discussion of static stability of a compressible atmosphere can be expressed in terms of the concept of *potential temperature*, which is generally denoted by θ. Suppose the pressure and temperature of a fluid particle at a height z are $p(z)$ and $T(z)$. Now if we take the particle *adiabatically* to a standard pressure $p_o = p(0)$ (say, the sea level pressure, nearly equal to 100 kPa), then the temperature θ attained by the particle is called its *potential temperature*. Using (1.26) for a perfect gas, it follows that the actual temperature T and the potential temperature θ are related by

$$T(z) = \theta(z)\big(p(z)/p_o\big)^{(\gamma-1)/\gamma}. \tag{1.31}$$

Taking the logarithm and differentiating, we obtain

$$\frac{1}{T}\frac{dT}{dz} = \frac{1}{\theta}\frac{d\theta}{dz} + \frac{(\gamma-1)}{\gamma p}\frac{dp}{dz}.$$

Substituting $dp/dz = -\rho g$, $p = \rho RT$, and $\alpha = 1/T$ produces

$$\frac{T}{\theta}\frac{d\theta}{dz} = \frac{dT}{dz} + \frac{g}{C_p} = \frac{d}{dz}(T - T_a) = \Gamma - \Gamma_a. \tag{1.32}$$

If the temperature decreases at a rate $\Gamma = \Gamma_a$, then the potential temperature θ (and therefore the entropy) is uniform with height. It follows that an atmosphere is stable, neutral, or unstable depending upon whether $d\theta/dz$ is positive, zero, or negative, respectively. This is illustrated in Figure 1.9b. It is the gradient of *potential* temperature that determines the stability of a column of gas, not the gradient of the actual temperature. However, this difference is negligible for laboratory-scale phenomena. For example, a 1.0 m vertical change may result in an air temperature decrease of only $1.0 \text{ m} \times (10°\text{C km}^{-1}) = 10^{-2}°\text{C}$.

Similarly, *potential density* ρ_θ is the density attained by a fluid particle if taken via an isentropic process to a standard pressure p_o. Using (1.26), the actual density $\rho(z)$ and potential density are related by

$$\rho(z) = \rho_\theta(z)\big(p(z)/p_o\big)^{1/\gamma}. \tag{1.33}$$

Multiplying (1.31) and (1.33), and using $p = \rho RT$, we obtain $\theta \rho_\theta = p_o/R = \text{const}$. Taking the logarithm and differentiating, we obtain

$$-\frac{1}{\rho_\theta}\frac{d\rho_\theta}{dz} = \frac{1}{\theta}\frac{d\theta}{dz}. \tag{1.34}$$

Thus, an atmosphere is stable, neutral, or unstable depending upon whether $d\rho_\theta/dz$ is negative, zero, or positive, respectively.

Interestingly, compressibility effects are also important in the deep ocean where saltwater density depends not only on the temperature and pressure, but also on the *salinity* (*S*) defined as kilograms of salt per kilogram of water. The average salinity of seawater is approximately 3.5%. Here, the potential density is defined as the density attained if a fluid particle is taken to a reference pressure via an isentropic process *and* at constant salinity. The potential density thus defined must decrease with height for stable water column conditions. Oceanographers automatically account for the compressibility of seawater by converting their density measurements at any depth to the sea level pressure, which serves as the reference pressure.

Because depth change—induced density changes are relatively small in percentage terms (~0.5% for a 1.0 km change in depth) for seawater, the static stability of the *ocean* is readily determined from (1.29). In particular, the vertical isentropic density gradient in (1.29) may be rewritten using $dp_a/dz = -\rho_a g$ and the definition of the sound speed c (1.19) to find

$$\frac{d\rho_a}{dz} = \left(\frac{\partial \rho_a}{\partial p}\right)_{s,S} \frac{dp_a}{dz} = -\left(\frac{\partial \rho_a}{\partial p}\right)_{s,S} \rho_a g = -\frac{\rho_a g}{c^2} \cong -\frac{\rho g}{c^2},$$

where the approximation $\rho_a \cong \rho$ produces the final result. Thus, (1.29) and its ensuing discussion imply that the ocean is stable, neutral, or unstable depending upon whether

$$\frac{d\rho_\theta}{dz} = \frac{d\rho}{dz} - \frac{d\rho_a}{dz} \cong \frac{d\rho}{dz} + \frac{\rho g}{c^2} \tag{1.35}$$

is negative, zero, or positive, respectively.

Scale Height of the Atmosphere

Approximate expressions for the pressure distribution and the thickness or *scale height* of the atmosphere can be obtained by assuming isothermal conditions. This is a reasonable assumption in the lower 70 km of the atmosphere, where the absolute temperature generally remains within 15% of 250 K. The hydrostatic distribution (1.8) and perfect gas law (1.22) require

$$dp/dz = -\rho g = -pg/RT.$$

When g, R, and T are constants, integration gives

$$p(z) = p_0 \, e^{-gz/RT},$$

where p_0 is the pressure at $z = 0$. The pressure therefore falls to e^{-1} of its surface value in a height $H = RT/g$. Thus, the quantity RT/g is called the *scale height* of the atmosphere, and it provides a reasonable quantitative measure of the thickness of the atmosphere. For an average atmospheric temperature of $T = 250$ K, the scale height is $RT/g = 7.3$ km.

1.11. DIMENSIONAL ANALYSIS

Interestingly, a physical quantity's units may be exploited to learn about its relationship to other physical quantities. This possibility exists because the natural realm does not need mankind's units of measurement to function. Natural laws are independent of any unit system imposed on them by human beings. Consider Newton's second law, generically stated as *force* = (*mass*) × (*acceleration*); it is true whether a scientist or engineer uses cgs (centimeter, gram, second), MKS (meter, kilogram, second), or even English (inch or foot, pound, second) units in its application. Because nature is independent of our systems of units, we can draw two important conclusions: 1) all correct physical relationships can be stated in dimensionless form, and 2) in any comparison, the units of the items being compared must be the same for the comparison to be valid. The first conclusion leads to the problem-simplification or scaling-law-development technique known as *dimensional analysis*. The second conclusion is known as the principle of *dimensional homogeneity*. It requires all terms in an equation to have the same dimension(s) and thereby provides an effective means for error catching within derivations and in derived answers. If terms in an equation do not have the same dimension(s) then the equation is not correct and a mistake has been made.

Dimensional analysis is a broadly applicable technique for developing scaling laws, interpreting experimental data, and simplifying problems. Occasionally it can even be used to solve problems. Dimensional analysis has utility throughout the physical sciences and it is routinely taught to students of fluid mechanics. Thus, it is presented here for subsequent use in this chapter's exercises and in the remaining chapters of this text.

Of the various formal methods of dimensional analysis, the description here is based on Buckingham's method from 1914. Let $q_1, q_2, ..., q_n$ be n variables and parameters involved in a particular problem or situation, so that there must exist a functional relationship of the form

$$f(q_1, q_2, ..., q_n) = 0. \tag{1.36}$$

Buckingham's theorem states that the n variables can always be combined to form exactly $(n - r)$ independent dimensionless parameter groups, where r is the number of independent dimensions. Each dimensionless parameter group is commonly called a *Π-group* or a *dimensionless group*. Thus, (1.36) can be written as a functional relationship

$$\phi(\Pi_1, \Pi_2, ..., \Pi_{n-r}) = 0 \text{ or } \Pi_1 = \varphi(\Pi_2, \Pi_3, ..., \Pi_{n-r}). \tag{1.37}$$

The dimensionless groups are not unique, but $(n - r)$ of them are *independent* and form a *complete set* that spans the parametric solution space of (1.37). The power of dimensional analysis is most apparent when n and r are single-digit numbers of comparable size so (1.37), which involves $n - r$ dimensionless groups, represents a significant simplification of (1.36), which has n parameters. The process of dimensional analysis is presented here as a series of six steps that should be followed by a seventh whenever possible. Each step is described in the following paragraphs and illustrated via the example of determining the functional dependence of the pressure difference Δp between two locations in a round pipe carrying a flowing viscous fluid.

Step 1. Select Variables and Parameters

Creating the list of variables and parameters to include in a dimensional analysis effort is the most important step. The parameter list should usually contain only one unknown variable, the *solution variable*. The rest of the variables and parameters should come from the problem's geometry, boundary conditions, initial conditions, and material parameters. Physical constants and other fundamental limits may also be included. However, shorter parameter lists tend to produce the most powerful dimensional analysis results; expansive lists commonly produce less useful results.

For the round-pipe pressure drop example, select Δp as the solution variable, and then choose as additional parameters: the distance Δx between the pressure measurement locations, the inside diameter d of the pipe, the average height ε of the pipe's wall roughness, the average flow velocity U, the fluid density ρ, and the fluid viscosity μ. The resulting functional dependence between these seven parameters can be stated as

$$f(\Delta p, \Delta x, d, \varepsilon, U, \rho, \mu) = 0. \tag{1.38}$$

Note, (1.38) does not include the fluid's thermal conductivity, heat capacities, thermal expansion coefficient, or speed of sound, so this dimensional analysis example will not account for the thermal or compressible flow effects embodied by these missing parameters.

Step 2. Create the Dimensional Matrix

Fluid flow problems without electromagnetic forces and chemical reactions involve only mechanical variables (such as velocity and density) and thermal variables (such as temperature and specific heat). The dimensions of all these variables can be expressed in terms of four basic dimensions—mass M, length L, time T, and temperature θ. We shall denote the dimension of a variable q by $[q]$. For example, the dimension of the velocity u is $[u] = L/T$, that of pressure is $[p] = [\text{force}]/[\text{area}] = MLT^{-2}/L^2 = M/LT^2$, and that of specific heat is $[C_p] = [\text{energy}]/[\text{mass}][\text{temperature}] = MLT^{-2}L/M\theta = L^2/\theta T^2$. When thermal effects are not considered, all variables can be expressed in terms of three fundamental dimensions, namely, M, L, and T. If temperature is considered only in combination with Boltzmann's constant $(k_B\theta)$, a gas constant $(R\theta)$, or a specific heat $(C_p\theta)$, then the units of the combination are simply L^2/T^2, and only the three dimensions M, L, and T are required.

The dimensional matrix is created by listing the powers of M, L, T, and θ in a column for each parameter selected. For the pipe-flow pressure difference example, the selected variables and their dimensions produce the following dimensional matrix:

	Δp	Δx	d	ε	U	ρ	μ
M	1	0	0	0	0	1	1
L	−1	1	1	1	1	−3	−1
T	−2	0	0	0	−1	0	−1

$$\tag{1.39}$$

where the seven variables have been written above the matrix entries and the three units have been written in a column to the left of the matrix. The matrix in (1.39) portrays $[\Delta p] = ML^{-1}T^{-2}$ via the first column of numeric entries.

Step 3. Determine the Rank of the Dimensional Matrix

The *rank* r of any matrix is defined to be the size of the largest square submatrix that has a nonzero determinant. Testing the determinant of the first three rows and columns of (1.39), we obtain

$$\begin{vmatrix} 1 & 0 & 0 \\ -1 & 1 & 1 \\ -2 & 0 & 0 \end{vmatrix} = 0.$$

However, (1.39) does include a nonzero third-order determinant, for example, the one formed by the last three columns:

$$\begin{vmatrix} 0 & 1 & 1 \\ 1 & -3 & -1 \\ -1 & 0 & -1 \end{vmatrix} = -1.$$

Thus, the rank of the dimensional matrix (1.39) is $r = 3$. If *all* possible third-order determinants were zero, we would have concluded that $r < 3$ and proceeded to testing second-order determinants.

For dimensional matrices, the rank is less than the number of rows only when one of the rows can be obtained by a linear combination of the other rows. For example, the matrix (not from 1.39)

$$\begin{vmatrix} 0 & 1 & 0 & 1 \\ -1 & 2 & 1 & -2 \\ -1 & 4 & 1 & 0 \end{vmatrix}$$

has $r = 2$, as the last row can be obtained by adding the second row to twice the first row. A rank of less than 3 commonly occurs in statics problems, in which mass or density is really not relevant but the dimensions of the variables (such as force) involve M. In most fluid mechanics problems without thermal effects, $r = 3$.

Step 4. Determine the Number of Dimensionless Groups

The number of dimensionless groups is $n - r$ where n is the number of variables and parameters, and r is the rank of the dimensional matrix. In the pipe-flow pressure difference example, the number of dimensionless groups is $4 = 7 - 3$.

Step 5. Construct the Dimensionless Groups

This can be done by exponent algebra or by inspection. The latter is preferred because it commonly produces dimensionless groups that are easier to interpret, but the former is sometimes required. Examples of both techniques follow here. Whatever the method, the best approach is usually to create the first dimensionless group with the solution variable appearing to the first power.

When using exponent algebra, select r parameters from the dimensional matrix as *repeating parameters* that will be found in all the subsequently constructed dimensionless groups. These repeating parameters must span the appropriate r-dimensional dimension space of M, L, and/or T, that is, the determinant of the dimensional matrix formed from these r parameters must be nonzero. For many fluid-flow problems, a characteristic velocity, a characteristic length, and a fluid property involving mass are ideal repeating parameters.

To form dimensionless groups for the pipe-flow problem, choose U, d, and ρ as the repeating parameters. The determinant of the dimensional matrix formed by these three parameters is nonzero. Other repeating parameter choices will result in a different set of dimensionless groups, but any such alternative set will still span the solution space of the problem. Thus, any satisfactory choice of the repeating parameters is equvialent to any other, so choices that simplify the work are most appropriate. Each dimensionless group is formed by combining the three repeating parameters, raised to unknown powers, with one of the nonrepeating variables or parameters from the list constructed for the first step. Here we ensure that the first dimensionless group involves the solution variable raised to the first power:

$$\Pi_1 = \Delta p U^a d^b \rho^c.$$

The exponents a, b, and c are obtained from the requirement that Π_1 is dimensionless. Replicating this equation in terms of dimensions produces

$$M^0 L^0 T^0 = \left[\Pi_1\right] = [\Delta p U^a d^b \rho^c] = (ML^{-1}T^{-2})(LT^{-1})^a (L)^b (ML^{-3})^c = M^{c+1} L^{a+b-3c-1} T^{-a-2}.$$

Equating exponents between the two extreme ends of this extended equality produces three algebraic equations that are readily solved to find $a = -2$, $b = 0$, $c = -1$, so

$$\Pi_1 = \Delta p / \rho U^2.$$

A similar procedure with Δp replaced by the other unused variables (Δx, ε, μ) produces

$$\Pi_2 = \Delta x / d, \ \Pi_3 = \varepsilon / d, \text{ and } \Pi_4 = \mu / \rho U d.$$

The inspection method proceeds directly from the dimensional matrix, and may be less tedious than exponent algebra. It involves selecting individual parameters from the dimensional matrix and sequentially eliminating their M, L, T, and θ units by forming ratios with other parameters. For the pipe-flow pressure difference example we again start with the solution variable $[\Delta p] = ML^{-1}T^{-2}$ and notice that the next entry in (1.32) that includes units of mass is $[\rho] = ML^{-3}$. To eliminate M from a combination of Δp and ρ, we form the ratio $[\Delta p / \rho] = L^2 T^{-2} = [\text{velocity}^2]$. An examination of (1.39) shows that U has units of velocity, LT^{-1}. Thus, $\Delta p / \rho$ can be made dimensionless if it is divided by U^2 to find: $[\Delta p / \rho U^2] =$ dimensionless. Here we have the good fortune to eliminate L and T in the same step. Therefore, the first dimensionless group is $\Pi_1 = \Delta p / \rho U^2$. To find the second dimensionless group Π_2, start with Δx, the left most unused parameter in (1.32), and note $[\Delta x] = L$. The first unused parameter to the right of Δx involving only length is d. Thus, $[\Delta x / d] =$ dimensionless so $\Pi_2 = \Delta x / d$. The third dimensionless group is obtained by starting with the next unused parameter, ε, to find $\Pi_3 = \varepsilon / d$. The final dimensionless group must include the last unused parameter $[\mu] = ML^{-1}T^{-1}$. Here it is better to eliminate the mass dimension with the density since reusing Δp would place the solution variable in two places in the final scaling law, an unnecessary complication. Therefore, form the ratio μ / ρ which has units $[\mu / \rho] = L^2 T^{-1}$. These can be eliminated with d and U, $[\mu / \rho U d] =$ dimensionless, so $\Pi_4 = \mu / \rho U d$.

Forming the dimensionless groups by inspection becomes easier with experience. For example, since there are three length scales Δx, d, and ε in (1.39), the dimensionless groups $\Delta x / d$ and ε / d can be formed immediately. Furthermore, Bernoulli equations (see Section 4.9, "Bernoulli Equations") tell us that ρU^2 has the same units as p so $\Delta p / \rho U^2$ is readily identified as a dimensionless group. Similarly, the dimensionless group that describes viscous effects in the fluid mechanical equations of motion is found to be $\mu/\rho U d$ when these equations are cast in dimensionless form (see Section 4.11).

Other dimensionless groups can be obtained by combining esblished groups. For the pipe-flow example, the group $\Delta p d^2 \rho/\mu^2$ can be formed from Π_1/Π_4^2, and the group $\varepsilon/\Delta x$ can be formed as Π_3/Π_2. However, only four dimensionless groups will be independent in the pipe-flow example.

Step 6. State the Dimensionless Relationship

This step merely involves placing the $(n - r)$ Π-groups in one of the forms in (1.37). For the pipe-flow example, this dimensionless relationship is

$$\frac{\Delta p}{\rho U^2} = \varphi\left(\frac{\Delta x}{d}, \frac{\varepsilon}{d}, \frac{\mu}{\rho U d}\right), \tag{1.40}$$

where φ is an undetermined function. This relationship involves only four dimensionless groups, and is therefore a clear simplification of (1.36) which lists seven independent parameters. The four dimensionless groups in (1.40) have familiar physical interpretations and have even been given special names. For example, $\Delta x/d$ is the pipe's aspect ratio, and ε/d is the pipe's roughness ratio. Common dimensionless groups in fluid mechanics are presented and discussed in Section 4.11.

Step 7. Use Physical Reasoning or Additional Knowledge to Simplify the Dimensionless Relationship

Sometimes there are only two extensive thermodynamic variables involved and these must be proportional in the final scaling law. An overall conservation law can be applied that restricts one or more parametric dependencies, or a phenomena may be known to be linear, quadratic, etc., in one of the parameters and this dependence must be reflected in the final scaling law. This seventh step may not always be possible, but when it is, significant and powerful results may be achieved from dimensional analysis.

EXAMPLE 1.2

Use dimensional analysis to find the parametric dependence of the scale height H in a static isothermal atmosphere at temperature T_o composed of a perfect gas with average molecular weight M_w when the gravitational acceleration is g.

Solution

Follow the six steps just described.

1. The parameter list must include H, T_o, M_w, and g. Here there is no velocity parameter, and there is no need for a second specification of a thermodynamic variable since a static pressure gradient prevails. However, the universal gas constant R_u must be included to help relate the thermal variable T_o to the mechanical ones.

2. The dimensional matrix is:

	H	T_o	M_w	g	R_u
M	0	0	1	0	1
L	1	0	0	1	2
T	0	0	0	−2	−2
θ	0	1	0	0	−1

Note that the $kmole^{-1}$ specification of M_w and R_u is lost in the matrix above since a $kmole$ is a pure number.

3. The rank of this matrix is four, so $r = 4$.
4. The number of dimensionless groups is: $n - r = 5 - 4 = 1$.
5. Use H as the solution parameter, and the others as the repeating parameters. Proceed with exponent algebra to find the dimensionless group:

$$M^0 L^0 T^0 \theta^0 = [\Pi_1] = \left[H T_o^a M_w^b g^c R_u^d \right] = (L)(\theta)^a (M)^b (LT^{-2})^c (ML^2 T^{-2}\theta^{-1})^d$$
$$= M^{b+d} L^{1+c+2d} T^{-2c-2d} \theta^{a-d}.$$

Equating exponents yields four linear algebraic equations:

$$b + d = 0, \ 1 + c + 2d = 0, \ -2c - 2d = 0, \ \text{and} \ a - d = 0,$$

which are solved by: $a = -1$, $b = 1$, $c = 1$, and $d = -1$. Thus, the lone dimensionless group is: $\Pi_1 = HgM_w/R_uT_o$.

6. Because there is only a single dimensionless group, its most general behavior is to equal a constant, so $HgM_w/R_uT_o = \varphi(\ldots) = const.$, or $H = const. (R_uT_o/gM_w)$. Based on the finding at the end of the previous section and $R = R_u/M_w$ from (1.22), this parametric dependence is correct and the constant is unity in this case.

EXAMPLE 1.3

Use dimensional analysis and Figure 1.10 to prove the Pythagorean theorem based on a right triangle's area a, the radian measure β of its most acute angle, and the length C of its longest side (Barenblatt, 1979).

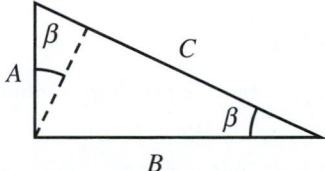

FIGURE 1.10 A right triangle with area a, smallest acute angle β, and hypotenuse C. The dashed line is perpendicular to side C.

Solution

Follow the six steps given earlier and then consider similarity between the main triangle and two sub-triangles.

1. The parameter list (a, β, C) is given in the problem statement so $n = 3$.
2. The dimensional matrix is:

	a	β	C
M	0	0	0
L	2	0	1
T	0	0	0

3. With no M or T units, the rank of this matrix is one, so $r = 1$.
4. The number of dimensionless groups is: $n - r = 3 - 1 = 2$.
5. Let the triangle's area a be the solution parameter. By inspection, $\Pi_1 = a/C^2$, and $\Pi_2 = \beta$.
6. Therefore, the dimensionless relationship is: $a/C^2 = \varphi(\beta)$ or $a = C^2\varphi(\beta)$.
7. When the dashed line is perpendicular to side C, then the large triangle is divided into two smaller ones that are similar to the larger one. These sub-triangles have A and B as their longest sides and both have the same acute angle as the large triangle. Therefore, the sub-triangle areas can be written as $A^2\varphi(\beta)$ and $B^2\varphi(\beta)$. Summing the sub-triangle areas produces: $A^2\varphi(\beta) + B^2\varphi(\beta) = C^2\varphi(\beta)$ or $A^2 + B^2 = C^2$ when $\varphi(\beta) \neq 0$.

EXAMPLE 1.4

Use dimensional analysis to determine the energy E released in an intense point blast if the blast-wave propagation distance D into an undisturbed atmosphere of density ρ is known as a function of time t following the energy release (Taylor, 1950; see Figure 1.11).

FIGURE 1.11   In an atmosphere with undisturbed density ρ, a point release of energy E produces a hemispherical blast wave that travels a distance D in time t.

Solution

Again follow the six steps given earlier.

1. The parameter list (E, D, ρ, t) is given in the problem statement so $n = 4$.
2. The dimensional matrix is:

	E	D	ρ	t
M	1	0	1	0
L	2	1	-3	0
T	-2	0	0	1

3. The rank of this matrix is three, so $r = 3$.
4. The number of dimensionless groups is: $n - r = 4 - 3 = 1$.
5. Let the point-blast energy be the solution parameter and construct the lone dimensionless group by inspection. First use E and ρ to eliminate M: $[E/\rho] = L^5 T^{-2}$. Next use D to eliminate L: $[E/\rho D^5] = T^{-2}$. Then use t to eliminate T: $[Et^2/\rho D^5] =$ dimensionless, so $\Pi_1 = Et^2/\rho D^5$.
6. Here there is only a single dimensionless group, so it must be a constant (K). This produces: $Et^2/\rho D^5 = \varphi(\ldots) = K$ which implies: $E = K\rho D^5/t^2$, where K is not determined by dimensional analysis.
7. The famous fluid mechanician G. I. Taylor was able to estimate the yield of the first atomic-bomb test conducted on the White Sands Proving Grounds in New Mexico in July 1945 using: 1) the dimensional analysis shown above, 2) a declassified movie made by J. E. Mack, and 3) timed photographs supplied by the Los Alamos National Laboratory and the Ministry of Supply. He determined the fireball radius as a function of time and then estimated E using a nominal atmospheric value for ρ. His estimate of $E = 17$ kilotons of TNT was very close to the actual yield (20 kilotons of TNT) in part because the undetermined constant K is close to unity in this case. At the time, the movie and the photographs were not classified but the yield of the bomb was entirely secret.

EXAMPLE 1.5

Use dimensional analysis to determine how the average light intensity S (Watts/m^2) scattered from an isolated particle depends on the incident light intensity I (Watts/m^2), the wavelength of the light λ (m), the volume of the particle V (m^3), the index of refraction of the particle n_s (dimensionless), and the distance d (m) from the particle to the observation point. Can the resulting dimensionless relationship be simplified to better determine parametric effects when $\lambda \gg V^{1/3}$?

Solution

Again follow the six steps given earlier, knowing that the seventh step will likely be necessary to produce a useful final relationship.

1. The parameter list (S, I, λ, V, n_s, d) is given in the problem statement so $n = 6$.
2. The dimensional matrix is:

	S	I	λ	V	n_s	d
M	1	1	0	0	0	0
L	0	0	1	3	0	1
T	-3	-3	0	0	0	0

3. The rank of this matrix is 2 because all the dimensions are either intensity or length, so $r = 2$.
4. The number of dimensionless groups is: $n - r = 6 - 2 = 4$.
5. Let scattered light intensity S be the solution parameter. By inspection the four dimensionless groups are:

$$\Pi_1 = S/I, \ \Pi_2 = d/\lambda, \ \Pi_3 = V/\lambda^3, \ \text{and} \ \Pi_4 = n_s.$$

6. Therefore, the dimensionless relationship is: $S/I = \varphi_1(d/\lambda, V/\lambda^3, n_s)$.
7. There are two physical features of this problem that allow refinement of this dimensional analysis result. First, light scattering from the particle must conserve energy and this implies: $4\pi d^2 S = const.$ so $S \propto 1/d^2$. Therefore, the result in step 6 must simplify to: $S/I = (\lambda/d)^2 \varphi_2$ $(V/\lambda^3, n_s)$. Second, when λ is large compared to the size of the scatterer, the scattered field amplitude will be produced from the dipole moment induced in the scatterer by the incident field, and this scattered field amplitude will be proportional to V. Thus, S, which is proportional to field amplitude squared, will be proportional to V^2. These deductions allow a further simplification of the dimensional analysis result to:

$$\frac{S}{I} = \left(\frac{\lambda}{d}\right)^2 \left(\frac{V}{\lambda^3}\right)^2 \varphi_3(n_s) = \frac{V^2}{d^2 \lambda^4} \varphi_3(n_s).$$

This is Lord Rayleigh's celebrated small-particle scattering law. He derived it in the 1870s while investigating light scattering from small scatterers to understand why the cloudless daytime sky was blue while the sun appeared orange or red at dawn and sunset. At the time, he imagined that the scatterers were smoke, dust, mist, aerosols, etc. However, the atmospheric abundance of these are insufficient to entirely explain the color change phenomena but the molecules that compose the atmosphere can accomplish enough scattering to explain the observations.

EXERCISES

1.1. [1]Many centuries ago, a mariner poured 100 cm^3 of water into the ocean. As time passed, the action of currents, tides, and weather mixed the liquid uniformly throughout the earth's oceans, lakes, and rivers. Ignoring salinity, estimate the probability that the next cup of water you drink will contain at least one water molecule that was dumped by the mariner. Assess your chances of ever drinking truly pristine water. (*Consider the following facts*: M_w for water is 18.0 kg per kg-mole, the radius of the earth is 6370 km, and the mean depth of the oceans is approximately 3.8 km, and they cover 71% of the surface of the earth. One cup is $\sim$240 ml.)

1.2. [1]An adult human expels approximately 500 mL of air with each breath during ordinary breathing. Imagining that two people exchanged greetings (one breath each) many centuries ago and that their breath subsequently has been mixed uniformly throughout the atmosphere, estimate the probability that the next breath you take will contain at least one air molecule from that age-old verbal exchange. Assess your chances of ever getting a truly fresh breath of air. For this problem, assume that air is composed of identical molecules having $M_w = 29.0$ kg per kg-mole and that the average atmospheric pressure on the surface of the earth is 100 kPa. Use 6370 km for the radius of the earth and 1.20 kg/m^3 for the density of air at room temperature and pressure.

[1]Based on a homework problem posed by Professor P. E. Dimotakis

1.3. In Cartesian coordinates, the Maxwell probability distribution, $f(\mathbf{u}) = f(u_1, u_2, u_3)$, of molecular velocities in a gas flow with average velocity $\mathbf{U} = (U_1, U_2, U_3)$ is

$$f(\mathbf{u}) = \left(\frac{m}{2\pi k_B T}\right)^{3/2} \exp\left\{-\frac{m}{2k_B T}|\mathbf{u} - \mathbf{U}|^2\right\},$$

where n is the number of gas molecules in volume V, m is the molecular mass, k_B is Boltzmann's constant and T is the absolute temperature.

a) Verify that $\mathbf{U}$ is the average molecular velocity, and determine the standard deviations (σ_1, σ_2, σ_3) of each component of $\mathbf{U}$ using $\sigma_i = [\iiint_{all\mathbf{u}} (u_i - U_i)^2 f(\mathbf{u}) d^3 u]^{1/2}$ for $i = 1, 2$, and 3.

b) Using (1.21), the molecular version of the perfect gas law, determine n/V at room temperature $T = 295$ K and atmospheric pressure $p = 101.3$ kPa.

c) Determine n for volumes $V = (10\ \mu m)^3$, $1\ \mu m^3$, and $(0.1\ \mu m)^3$.

d) For the ith velocity component, the standard deviation of the average, $\sigma_{a,i}$, over n molecules is $\sigma_{a,i} = \sigma_i / \sqrt{n}$ when $n \gg 1$. For an airflow at $\mathbf{U} = (1.0\ ms^{-1}, 0, 0)$, compute the relative uncertainty, $2\sigma_{a,1}/U_1$, at the 95% confidence level for the average velocity for the three volumes listed in part c).

e) For the conditions specified in parts b) and d), what is the smallest volume of gas that ensures a relative uncertainty in U of one percent or less?

1.4. Using the Maxwell molecular velocity distribution given in Exercise 1.3 with $\mathbf{U} = 0$, determine the average molecular speed $= \bar{v} = [\iiint_{all\mathbf{u}} |\mathbf{u}|^2 f(\mathbf{u}) d^3 u]^{1/2}$ and compare it with $c = $ speed of sound in a perfect gas under the same conditions.

1.5. By considering the volume swept out by a moving molecule, estimate how the mean-free path, l, depends on the average molecular cross section dimension $\bar{d}$ and the molecular number density $\tilde{n}$ for nominally spherical molecules. Find a formula for $l\tilde{n}^{1/3}$ (the ratio of the mean-free path to the mean intermolecular spacing) in terms of the *molecular volume* ($\bar{d}^3$) and the available *volume per molecule* ($1/\tilde{n}$). Is this ratio typically bigger or smaller than one?

1.6. In a gas, the molecular momentum flux (MF_{ij}) in the j-coordinate direction that crosses a flat surface of unit area with coordinate normal direction i is:

$$MF_{ij} = \frac{n}{V} \iiint_{all\mathbf{u}} m u_i u_j f(\mathbf{u}) d^3 u$$ where $f(\mathbf{u})$ is the Maxwell distribution given in Exercise

1.3, and n is the number of molecules in volume V. For a perfect gas that is not moving on average (i.e., $\mathbf{U} = 0$), show that $MF_{ij} = p$ (the pressure), when $i = j$, and that $MF_{ij} = 0$, when $i \neq j$.

1.7. Consider the viscous flow in a channel of width $2b$. The channel is aligned in the x-direction, and the velocity u in the x-direction at a distance y from the channel centerline is given by the parabolic distribution $u(y) = U_0[1 - (y/b)^2]$. Calculate the shear stress τ as a function y, μ, b, and U_0. What is the shear stress at $y = 0$?

1.8. Estimate the height to which water at 20°C will rise in a capillary glass tube 3 mm in diameter that is exposed to the atmosphere. For water in contact with glass the wetting angle is nearly 90°. At 20°C, the surface tension of a water-air interface is $\sigma = 0.073$ N/m.

1.9. A *manometer* is a U-shaped tube containing mercury of density ρ_m. Manometers are used as pressure-measuring devices. If the fluid in tank A has a pressure p and density ρ, then show that the gauge pressure in the tank is: $p - p_{atm} = \rho_m gh - \rho ga$. Note that the last term on the right side is negligible if $\rho \ll \rho_m$. (*Hint*: Equate the pressures at X and Y.)

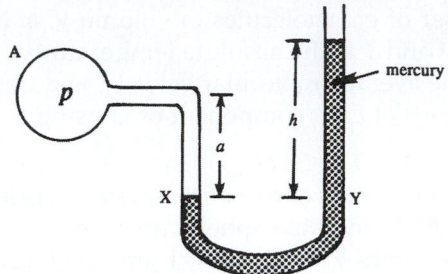

1.10. Prove that if $e(T, v) = e(T)$ only and if $h(T, p) = h(T)$ only, then the (thermal) equation of state is (1.22) or $pv = kT$, where k is a constant.

1.11. Starting from the property relationships (1.18) prove (1.25) and (1.26) for a reversible adiabatic process when the specific heats C_p and C_v are constant.

1.12. A cylinder contains 2 kg of air at 50°C and a pressure of 3 bars. The air is compressed until its pressure rises to 8 bars. What is the initial volume? Find the final volume for both isothermal compression and isentropic compression.

1.13. Derive (1.29) starting from the arguments provided at the beginning of Section 1.10 and Figure 1.8.

1.14. Starting with the hydrostatic pressure law (1.8), prove (1.30) without using perfect gas relationships.

1.15. Assume that the temperature of the atmosphere varies with height z as $T = T_0 + Kz$ where K is a constant. Show that the pressure varies with height as

$$p = p_0 \left[\frac{T_0}{T_0 + Kz}\right]^{g/KR}, \text{ where } g \text{ is the acceleration of gravity and } R \text{ is the gas constant for}$$

the atmospheric gas.

1.16. Suppose the atmospheric temperature varies according to: $T = 15 - 0.001z$, where T is in degrees Celsius and height z is in meters. Is this atmosphere stable?

1.17. Consider the case of a pure gas planet where the hydrostatic law is: $dp/dz = -\rho(z)Gm(z)/z^2$. Here G is the gravitational constant, and $m(z) = 4\pi \int_0^z \rho(\zeta)\zeta^2 d\zeta$ is the planetary mass up to distance z from the center of the planet. If the planetary gas is perfect with gas constant R, determine $\rho(z)$ and $p(z)$ if this atmosphere is isothermal at temperature T. Are these vertical profiles of ρ and p valid as z increases without bound?

1.18. Consider a heat-insulated enclosure that is separated into two compartments of volumes V_1 and V_2, containing perfect gases with pressures of p_1 and p_2, and temperatures of T_1 and T_2, respectively. The compartments are separated by an impermeable membrane that conducts heat (but not mass). Calculate the final steady-state temperature assuming each gas has constant specific heats.

1.19. Consider the initial state of an enclosure with two compartments as described in Exercise 1.18. At $t = 0$, the membrane is broken and the gases are mixed. Calculate the final temperature.

1.20. A heavy piston of weight W is dropped onto a thermally insulated cylinder of cross-sectional area A containing a perfect gas of constant specific heats, and initially having the external pressure p_1, temperature T_1, and volume V_1. After some oscillations, the piston reaches an equilibrium position L meters below the equilibrium position of a weightless piston. Find L. Is there an entropy increase?

1.21. [2]A gas of noninteracting particles of mass m at temperature T has density ρ, and internal energy per unit volume ε.

    a) Using dimensional analysis, determine how ε must depend on ρ, T, and m. In your formulation use k_B = Boltzmann's constant, h = Plank's constant, and c = speed of light to include possible quantum and relativistic effects.

    b) Consider the limit of slow-moving particles without quantum effects by requiring c and h to drop out of your dimensionless formulation. How does ε depend on ρ and T? What type of gas follows this thermodynamic law?

    c) Consider the limit of massless particles (i.e., photons) by requiring m and ρ to drop out of your dimensionless formulation of part a). How does ε depend on T in this case? What is the name of this radiation law?

1.22. Many flying and swimming animals—as well as human-engineered vehicles—rely on some type of repetitive motion for propulsion through air or water. For this problem, assume the average travel speed U depends on the repetition frequency f, the characteristic length scale of the animal or vehicle L, the acceleration of gravity g, the density of the animal or vehicle ρ_o, the density of the fluid ρ, and the viscosity of the fluid μ.

    a) Formulate a dimensionless scaling law for U involving all the other parameters.

    b) Simplify your answer for part a) for turbulent flow where μ is no longer a parameter.

    c) Fish and animals that swim at or near a water surface generate waves that move and propagate because of gravity, so g clearly plays a role in determining U. However, if fluctuations in the propulsive thrust are small, then f may not be important. Thus, eliminate f from your answer for part b) while retaining L, and determine how U depends on L. Are successful competitive human swimmers likely to be shorter or taller than the average person?

    d) When the propulsive fluctuations of a surface swimmer are large, the characteristic length scale may be U/f instead of L. Therefore, drop L from your answer for part b). In this case, will higher speeds be achieved at lower or higher frequencies?

    e) While traveling submerged, fish, marine mammals, and submarines are usually neutrally buoyant ($\rho_o \approx \rho$) or very nearly so. Thus, simplify your answer for part b) so that g drops out. For this situation, how does the speed U depend on the repetition frequency f?

    f) Although fully submerged, aircraft and birds are far from neutrally buoyant in air, so their travel speed is predominately set by balancing lift and weight. Ignoring

---

[2]Drawn from thermodynamics lectures of Prof. H. W. Liepmann

frequency and viscosity, use the remaining parameters to construct dimensionally accurate surrogates for lift and weight to determine how U depends on ρ_0/ρ, L, and g.

1.23. The acoustic power W generated by a large industrial blower depends on its volume flow rate Q, the pressure rise ΔP it works against, the air density ρ, and the speed of sound c. If hired as an acoustic consultant to quiet this blower by changing its operating conditions, what is your first suggestion?

1.24. A machine that fills peanut-butter jars must be reset to accommodate larger jars. The new jars are twice as large as the old ones but they must be filled in the same amount of time by the same machine. Fortunately, the viscosity of peanut butter decreases with increasing temperature, and this property of peanut butter can be exploited to achieve the desired results since the existing machine allows for temperature control.

  a) Write a dimensionless law for the jar-filling time t_f based on: the density of peanut butter ρ, the jar volume V, the viscosity of peanut butter μ, the driving pressure that forces peanut butter out of the machine P, and the diameter of the peanut butter–delivery tube d.

  b) Assuming that the peanut butter flow is dominated by viscous forces, modify the relationship you have written for part a) to eliminate the effects of fluid inertia.

  c) Make a reasonable assumption concerning the relationship between t_f and V when the other variables are fixed so that you can determine the viscosity ratio μ_{new}/μ_{old} necessary for proper operation of the old machine with the new jars.

1.25. As an idealization of fuel injection in a diesel engine, consider a stream of high-speed fluid (called a *jet*) that emerges into a quiescent air reservoir at $t = 0$ from a small hole in an infinite plate to form a *plume* where the fuel and air mix.

  a) Develop a scaling law via dimensional analysis for the penetration distance D of the plume as a function of: Δp the pressure difference across the orifice that drives the jet, d_o the diameter of the jet orifice, ρ_o the density of the fuel, μ_∞ and ρ_∞ the viscosity and density of the air, and t the time since the jet was turned on.

  b) Simplify this scaling law for turbulent flow where air viscosity is no longer a parameter.

  c) For turbulent flow and $D \ll d_o$, d_o and ρ_∞ are not parameters. Recreate the dimensionless law for D.

  d) For turbulent flow and $D \gg d_o$, only the momentum flux of the jet matters, so Δp and d_o are replaced by the single parameter J_o = jet momentum flux (J_o has the units of force and is approximately equal to $\Delta p d_o^2$). Recreate the dimensionless law for D using the new parameter J_o.

1.26. [3]One of the simplest types of gasoline carburetors is a tube with a small port for transverse injection of fuel. It is desirable to have the fuel uniformly mixed in the passing airstream as quickly as possible. A prediction of the mixing length L is sought. The parameters of this problem are: ρ = density of the flowing air, d = diameter of the tube, μ = viscosity of the flowing air, U = mean axial velocity of the flowing air, and J = momentum flux of the fuel stream.

---

[3]Developed from research discussions with Professor R. Breidenthal

a) Write a dimensionless law for L.

b) Simplify your result from part a) for turbulent flow where μ must drop out of your dimensional analysis.

c) When this flow is turbulent, it is observed that mixing is essentially complete after one rotation of the counter-rotating vortices driven by the injected-fuel momentum (see the downstream view of the drawing for this problem), and that the vortex rotation rate is directly proportional to J. Based on this information, assume that $L \propto$ (rotation time)(U) to eliminate the arbitrary function in the result of part b). The final formula for L should contain an undetermined dimensionless constant.

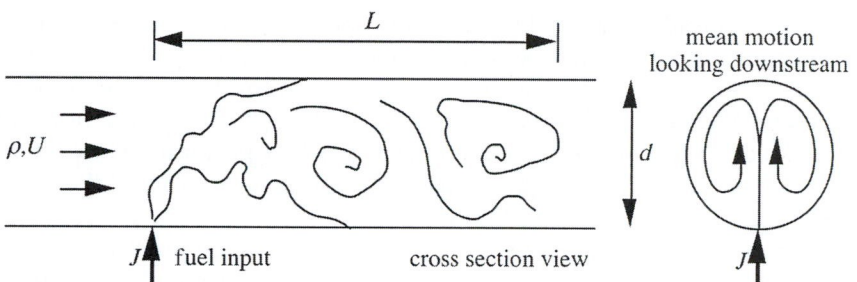

1.27. Consider dune formation in a large horizontal desert of deep sand.

a) Develop a scaling relationship that describes how the height h of the dunes depends on the average wind speed U, the length of time the wind has been blowing Δt, the average weight and diameter of a sand grain w and d, and the air's density ρ and kinematic viscosity ν.

b) Simplify the result of part a) when the sand-air interface is fully rough and ν is no longer a parameter.

c) If the sand dune height is determined to be proportional to the density of the air, how do you expect it to depend on the weight of a sand grain?

1.28. An isolated nominally spherical bubble with radius R undergoes shape oscillations at frequency f. It is filled with air having density ρ_a and resides in water with density ρ_w and surface tension σ. What frequency ratio should be expected between two isolated bubbles with 2 cm and 4 cm diameters undergoing geometrically similar shape oscillations? If a soluble surfactant is added to the water that lowers σ by a factor of two, by what factor should air bubble oscillation frequencies increase or decrease?

1.29. In general, boundary layer skin friction, τ_w, depends on the fluid velocity U above the boundary layer, the fluid density ρ, the fluid viscosity μ, the nominal boundary layer thickness δ, and the surface roughness length scale ε.

a) Generate a dimensionless scaling law for boundary layer skin friction.

b) For laminar boundary layers, the skin friction is proportional to μ. When this is true, how must τ_w depend on U and ρ?

c) For turbulent boundary layers, the dominant mechanisms for momentum exchange within the flow do not directly involve the viscosity μ. Reformulate your dimensional analysis without it. How must τ_w depend on U and ρ when μ is not a parameter?

d) For turbulent boundary layers on smooth surfaces, the skin friction on a solid wall occurs in a viscous sublayer that is very thin compared to δ. In fact, because the boundary layer provides a buffer between the outer flow and this viscous sub-layer, the viscous sublayer thickness l_v does not depend directly on U or δ. Determine how l_v depends on the remaining parameters.

e) Now consider nontrivial roughness. When ε is larger than l_v a surface can no longer be considered fluid-dynamically smooth. Thus, based on the results from parts a) through d) and anything you may know about the relative friction levels in laminar and turbulent boundary layers, are high- or low-speed boundary layer flows more likely to be influenced by surface roughness?

1.30. Turbulent boundary layer skin friction is one of the fluid phenomena that limit the travel speed of aircraft and ships. One means for reducing the skin friction of liquid boundary layers is to inject a gas (typically air) from the surface on which the boundary layer forms. The shear stress, τ_w, that is felt a distance L downstream of such an air injector depends on: the volumetric gas flux per unit span q (in m^2/s), the free stream flow speed U, the liquid density ρ, the liquid viscosity μ, the surface tension σ, and the gravitational acceleration g.

a) Formulate a dimensionless law for τ_w in terms of the other parameters.

b) Experimental studies of air injection into liquid turbulent boundary layers on flat plates has found that the bubbles may coalesce to form an air film that provides near perfect lubrication, $\tau_w \rightarrow 0$ for $L > 0$, when q is high enough and gravity tends to push the injected gas toward the plate surface. Reformulate your answer to part a) by dropping τ_w and L to determine a dimensionless law for the minimum air injection rate, q_c, necessary to form an air layer.

c) Simplify the result of part b) when surface tension can be neglected.

d) Experimental studies (Elbing et al., 2008) find that q_c is proportional to U^2. Using this information, determine a scaling law for q_c involving the other parameters. Would an increase in g cause q_c to increase or decrease?

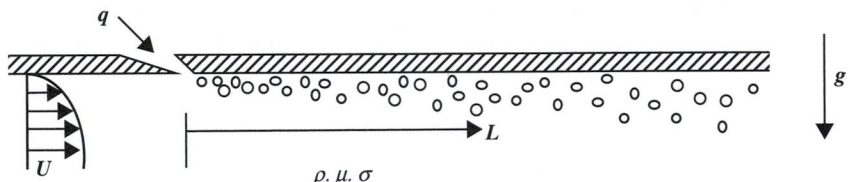

Literature Cited

Barenblatt, G. I. (1979). *"Similarity, Self-Similarity, and Intermediate Asymptotics"* (translated from Russian by Norman Stein). New York: Consultants Bureau.

Elbing, B. R., Winkel, E. S., Lay, K. A., Ceccio, S. L., Dowling, D. R., and Perlin, M. (2008). "Bubble-induced skin friction drag reduction and the abrupt transition to air-layer drag reduction." *Journal of Fluid Mechanics, 612,* 201–236.

Taylor, G. I. (1950). The formation of blast wave by a very intense explosion. Parts I and II. *Proceedings of the Royal Society of London, Series A, 201*, 159–174, and 175–186.

von Karman, T. (1954). *Aerodynamics*. New York: McGraw-Hill.

Supplemental Reading

Batchelor, G. K. (1967). *An Introduction to Fluid Dynamics*. London: Cambridge University Press. (A detailed discussion of classical thermodynamics, kinetic theory of gases, surface tension effects, and transport phenomena is given.)

Bridgeman, P. W. (1963). *Dimensional Analysis*. New Haven: Yale University Press.

Çengel, Y. A., and Cimbala, J. M. (2010). *Fluid Mechanics: Fundamentals and Applications* (2nd ed.). New York: McGraw-Hill. (A fine undergraduate fluid mechanics textbook. Chapter 7 covers dimensional analysis.)

Fox, R. W., Pritchard, P. J., and MacDonald, A. T. (2009). *Introduction to Fluid Mechanics* (7th ed.). New York: John Wiley. (A fine undergraduate fluid mechanics textbook. Chapter 7 covers dimensional analysis.)

Hatsopoulos, G. N., and Keenan, J. H. (1981). *Principles of General Thermodynamics*. Melbourne, FL: Krieger Publishing Co. (This is a good text on thermodynamics.)

Prandtl, L., and Tietjens, O. G. (1934). *Fundamentals of Hydro- and Aeromechanics*. New York: Dover Publications. (A clear and simple discussion of potential and adiabatic temperature gradients is given.)

Taylor, G. I. (1974). "The interaction between experiment and theory in fluid mechanics." *Annual Review of Fluid Mechanics, 6*, 1–16.

Tritton, D. J. (1988). *Physical Fluid Dynamics* (2nd ed.). Oxford: Clarendon Press. (A well-written book with broad coverage of fluid mechanics.)

Vincenti, W. G., and Kruger, C. H. (1965). *Introduction of Physical Gas Dynamics*. Malabar, Florida: Krieger. (A fine reference for kinetic theory and statistical mechanics.)

Cartesian Tensors

OUTLINE

2.1. Scalars, Vectors, Tensors, Notation   39

2.2. Rotation of Axes: Formal Definition of a Vector   42

2.3. Multiplication of Matrices   44

2.4. Second-Order Tensors   45

2.5. Contraction and Multiplication   47

2.6. Force on a Surface   48

2.7. Kronecker Delta and Alternating Tensor   50

2.8. Vector, Dot, and Cross Products   51

2.9. Gradient, Divergence, and Curl   52

2.10. Symmetric and Antisymmetric Tensors   55

2.11. Eigenvalues and Eigenvectors of a Symmetric Tensor   56

2.12. Gauss' Theorem   58

2.13. Stokes' Theorem   60

2.14. Comma Notation   62

Exercises   62

Literature Cited   64

Supplemental Reading   64

CHAPTER OBJECTIVES

- To define the notation used in this text for scalars, vectors, and tensors

- To review the basic algebraic manipulations of vectors and matrices

- To present how vector differentiation is applied to scalars, vectors, and tensors

- To review the fundamental theorems of vector field theory

2.1. SCALARS, VECTORS, TENSORS, NOTATION

The physical quantities in fluid mechanics vary in their complexity, and may involve multiple spatial directions. Their proper specification in terms of *scalars*, *vectors*, and (second order) *tensors* is the subject of this chapter. Here, three independent spatial dimensions are assumed to exist. The reader can readily simplify, or extend, the various results presented here for fewer, or more, independent spatial dimensions.

Scalars or zero-order tensors may be defined with a single magnitude and appropriate units, may vary with spatial location, but are independent of coordinate directions. Scalars are typically denoted herein by italicized symbols. For example, common scalars in fluid mechanics are pressure p, temperature T, and density ρ.

Vectors or first-order tensors have both a magnitude and a direction. A vector can be completely described by its components along three orthogonal coordinate directions. Thus, the components of a vector may change with a change in coordinate system. A vector is usually denoted herein by a boldface symbol. For example, common vectors in fluid mechanics are position $\mathbf{x}$, fluid velocity $\mathbf{u}$, and gravitational acceleration $\mathbf{g}$. In a Cartesian coordinate system with unit vectors $\mathbf{e}_1$, $\mathbf{e}_2$, and $\mathbf{e}_3$, in the three mutually perpendicular directions, the position vector $\mathbf{x}$, OP in Figure 2.1, may be written

$$\mathbf{x} = \mathbf{e}_1 x_1 + \mathbf{e}_2 x_2 + \mathbf{e}_3 x_3, \tag{2.1}$$

where x_1, x_2, and x_3 are the components of $\mathbf{x}$ along each Cartesian axis. Here, the subscripts of $\mathbf{e}$ do *not* denote vector components but rather reference the coordinate axes 1, 2, and 3; hence, the $\mathbf{e}$s are vectors themselves. Sometimes, to save writing, the components of a vector are denoted with an italic symbol having one index—such as i, j, or k—that implicitly is known to take on three possible values: 1, 2, or 3. For example, the components of $\mathbf{x}$ can be denoted by x_i or x_j (or x_k, etc.). For algebraic manipulation, a vector is written as a column matrix; thus, (2.1) is consistent with the following vector specifications:

$$\mathbf{x} = \begin{bmatrix} x_1 \\ x_2 \\ x_3 \end{bmatrix} \quad \text{where} \quad \mathbf{e}_1 = \begin{bmatrix} 1 \\ 0 \\ 0 \end{bmatrix}, \quad \mathbf{e}_2 = \begin{bmatrix} 0 \\ 1 \\ 0 \end{bmatrix}, \quad \text{and} \quad \mathbf{e}_3 = \begin{bmatrix} 0 \\ 0 \\ 1 \end{bmatrix}.$$

The *transpose* of the matrix (denoted by a superscript T) is obtained by interchanging rows and columns, so the transpose of the column matrix $\mathbf{x}$ is the row matrix:

$$\mathbf{x}^T = \begin{bmatrix} x_1 & x_2 & x_3 \end{bmatrix}.$$

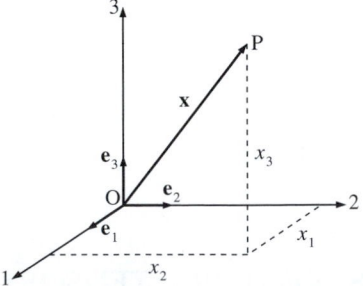

FIGURE 2.1   Position vector OP and its three Cartesian components (x_1, x_2, x_3). The three unit vectors for the coordinate directions are $\mathbf{e}_1$, $\mathbf{e}_2$, and $\mathbf{e}_3$. Once the coordinate system is chosen, the vector $\mathbf{x}$ is completely defined by its components, x_i where $i = 1, 2,$ or 3.

However, to save space in the text, the square-bracket notation for vectors shown here is typically replaced by triplets (or doublets) of values separated by commas and placed inside ordinary parentheses, for example, $\mathbf{x} = (x_1, x_2, x_3)$.

Second-order tensors have a component for each *pair* of coordinate directions and therefore may have as many as $3 \times 3 = 9$ separate components. A second-order tensor is sometimes denoted by a boldface symbol. For example, a common second-order tensor in fluid mechanics is the stress τ. Like vector components, second-order tensor components change with a change in coordinate system. Once a coordinate system is chosen, the nine components of a second-order tensor can be represented by a 3×3 matrix, or by an italic symbol having two indices, such as τ_{ij} for the stress tensor. Here again the indices i and j are known implicitly to separately take on the values 1, 2, or 3. Second-order tensors are further discussed in Section 2.4.

A second implicit feature of *index-based* or *indicial* notation is the implied sum over a repeated index in terms involving multiple indices. This notational convention can be stated as follows: *Whenever an index is repeated in a term, a summation over this index is implied, even though no summation sign is explicitly written.* This notational convention saves writing and increases mathematical precision when dealing with products of first- and higher-order tensors. It was introduced by Albert Einstein and is sometimes referred to as the *Einstein summation convention*. It can be illustrated by a simple example involving the ordinary dot product of two vectors $\mathbf{a}$ and $\mathbf{b}$ having components a_i and b_j, respectively. Their dot product is the sum of component products,

$$\mathbf{a} \cdot \mathbf{b} = a_1 b_1 + a_2 b_2 + a_3 b_3 = \sum_{i=1}^{3} a_i b_i \equiv a_i b_i, \tag{2.2}$$

where the final three-line *definition* equality ($\equiv$) follows from the repeated-index implied-sum convention. Since this notational convention is unlikely to be comfortable to the reader after a single exposure, it is repeatedly illustrated via definition equalities in this chapter before being adopted in the remainder of this text wherever indicial notation is used.

Both boldface (aka, *vector* or *dyadic*) and indicial (aka, *tensor*) notations are used throughout this text. With boldface notation the physical meaning of terms is generally clearer, and there are no subscripts to consider. Unfortunately, algebraic manipulations may be difficult and not distinct in boldface notation since the product $\mathbf{ab}$ may not be well defined nor equal to $\mathbf{ba}$ when $\mathbf{a}$ and $\mathbf{b}$ are second-order tensors. Boldface notation has other problems too; for example, the order or rank of a tensor is not clear if one simply calls it $\mathbf{a}$.

Indicial notation avoids these problems because it deals only with tensor *components*, which are *scalars*. Algebraic manipulations are simpler and better defined, and special attention to the ordering of terms is unnecessary (unless differentiation is involved). In addition, the number of indices or subscripts clearly specifies the order of a tensor. However, the physical structure and meaning of terms written with index notation only become apparent after an examination of the indices. Hence, indices must be clearly written to prevent mistakes and to promote proper understanding of the terms they help define. In addition, the cross product involves the possibly cumbersome alternating tensor ε_{ijk} as described in Sections 2.7 and 2.9.

2.2. ROTATION OF AXES: FORMAL DEFINITION OF A VECTOR

A vector can be formally defined as any quantity whose components change similarly to the components of the position vector under rotation of the coordinate system. Let O123 be the original coordinate system, and O1'2'3' be the rotated system that shares the same origin O (see Figure 2.2). The position vector $\mathbf{x}$ can be written in either coordinate system,

$$\mathbf{x} = x_1\mathbf{e}_1 + x_2\mathbf{e}_2 + x_3\mathbf{e}_3 \quad \text{or} \quad \mathbf{x} = x_1'\mathbf{e}_1' + x_2'\mathbf{e}_2' + x_3'\mathbf{e}_3', \tag{2.1, 2.3}$$

where the components of $\mathbf{x}$ in O123 and O1'2'3' are x_i and x_j', respectively, and the $\mathbf{e}_j'$ are the unit vectors in O1'2'3'. Forming a dot product of $\mathbf{x}$ with $\mathbf{e}_1'$, and using both (2.1) and (2.3) produces

$$\mathbf{x} \cdot \mathbf{e}_1' = x_1\mathbf{e}_1 \cdot \mathbf{e}_1' + x_2\mathbf{e}_2 \cdot \mathbf{e}_1' + x_3\mathbf{e}_3 \cdot \mathbf{e}_1' = x_1', \tag{2.4}$$

where we recognize the dot products between unit vectors as direction cosines; $\mathbf{e}_1 \cdot \mathbf{e}_1'$ is the cosine of the angle between the 1 and 1' axes, $\mathbf{e}_2 \cdot \mathbf{e}_1'$ is the cosine of the angle between the 2 and 1' axes, and $\mathbf{e}_3 \cdot \mathbf{e}_1'$ is the cosine of the angle between the 3 and 1' axes. Forming the dot products $\mathbf{x} \cdot \mathbf{e}_2' = x_2'$ and $\mathbf{x} \cdot \mathbf{e}_3' = x_3'$, and then combining these results with (2.3) produces

$$x_j' = x_1 C_{1j} + x_2 C_{2j} + x_3 C_{3j} = \sum_{i=1}^{3} x_i C_{ij} \equiv x_i C_{ij}, \tag{2.5}$$

where $C_{ij} = \mathbf{e}_i \cdot \mathbf{e}_j'$ is a 3×3 matrix of direction cosines and the definition equality follows from the summation convention. In (2.5) the *free* or not-summed-over index is j, while the

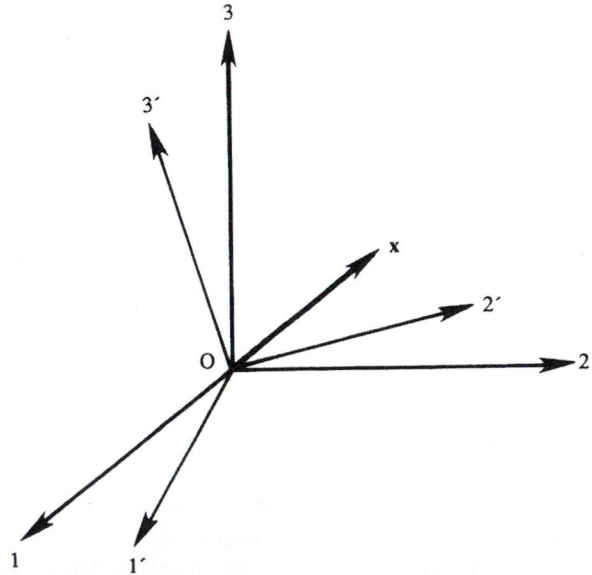

FIGURE 2.2　A rotation of the original Cartesian coordinate system O123 to a new system O1'2'3'. Here the $\mathbf{x}$ vector is unchanged, but its components in the original system x_i and in the rotated system x_i' will not be the same.

repeated or summed-over index can be any letter other than j. Thus, the rightmost term in (2.5) could equally well have been written $x_k C_{kj}$ or $x_m C_{mj}$. Similarly, any letter can also be used for the free index, as long as the same free index is used on *both* sides of the equation. For example, denoting the free index by i and the summed index by k allows (2.5) to be written with indicial notation as

$$x_i' = x_k C_{ki}. \tag{2.6}$$

This index-choice flexibility exists because the three algebraic equations represented by (2.5), corresponding to the three values of j, are the same as those represented by (2.6) for the three values of i.

It can be shown (see Exercise 2.2) that the components of $\mathbf{x}$ in O123 are related to those in O1'2'3' by

$$x_j = \sum_{i=1}^{3} x_i' C_{ji} \equiv x_i' C_{ji}. \tag{2.7}$$

The indicial positions on the right side of this relation are different from those in (2.5), because the first index of C_{ij} is summed in (2.5), whereas the second index of C_{ij} is summed in (2.7).

We can now formally define a Cartesian vector as any quantity that transforms like the position vector under rotation of the coordinate system. Therefore, by analogy with (2.5), $\mathbf{u}$ is a vector if its components transform as

$$u_j' = \sum_{i=1}^{3} u_i C_{ij} \equiv u_i C_{ij}. \tag{2.8}$$

EXAMPLE 2.1

Convert the two-dimensional vector $\mathbf{u} = (u_1, u_2)$ from Cartesian (x_1, x_2) to polar (r, θ) coordinates (see Figure 3.3a).

Solution

Clearly $\mathbf{u}$ can be represented in either coordinate system: $\mathbf{u} = u_1 \mathbf{e}_1 + u_2 \mathbf{e}_2 = u_r \mathbf{e}_r + u_\theta \mathbf{e}_\theta$, where u_r and u_θ are the components in polar coordinates, and $\mathbf{e}_r$ and $\mathbf{e}_\theta$ are the unit vectors in polar coordinates. Here the polar coordinate system is rotated compared to the Cartesian system, as illustrated in Figure 2.3. Forming the dot product of the above equation with $\mathbf{e}_r$ and $\mathbf{e}_\theta$ produces two algebraic equations that are equivalent to (2.5)

$$u_r = u_1 \mathbf{e}_1 \cdot \mathbf{e}_r + u_2 \mathbf{e}_2 \cdot \mathbf{e}_r$$
$$u_\theta = u_1 \mathbf{e}_1 \cdot \mathbf{e}_\theta + u_2 \mathbf{e}_2 \cdot \mathbf{e}_\theta,$$

with subscripts r and θ replacing $j = 1$ and 2 in (2.5). Evaluation of the unit vector dot products leads to

$$u_r = u_1 \cos\theta + u_2 \cos\left(\frac{\pi}{2} - \theta\right) = u_1 \cos\theta + u_2 \sin\theta, \text{ and}$$
$$u_\theta = u_1 \cos\left(\theta + \frac{\pi}{2}\right) + u_2 \cos\theta = -u_1 \sin\theta + u_2 \cos\theta.$$

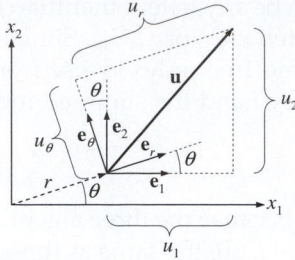

FIGURE 2.3   Resolution of a two-dimensional vector **u** in (x_1, x_2)-Cartesian and (r, θ)-polar coordinates. The angle between the $\mathbf{e}_1$ and $\mathbf{e}_r$ unit vectors, and the $\mathbf{e}_2$ and $\mathbf{e}_\theta$ unit vectors, is θ. The angle between the $\mathbf{e}_r$ and $\mathbf{e}_2$ unit vectors is $\pi/2 - \theta$, and the angle between the $\mathbf{e}_1$ and $\mathbf{e}_\theta$ unit vectors is $\pi/2 + \theta$. Here **u** does not emerge from the origin of coordinates (as in Figure 2.2) but it may be well defined in either coordinate system even though its components are not the same in the (x_1, x_2)- and (r, θ)-coordinates.

Thus, in this case:

$$C_{ij} = \begin{bmatrix} \mathbf{e}_1 \cdot \mathbf{e}_r & \mathbf{e}_1 \cdot \mathbf{e}_\theta \\ \mathbf{e}_2 \cdot \mathbf{e}_r & \mathbf{e}_2 \cdot \mathbf{e}_\theta \end{bmatrix} = \begin{bmatrix} \cos\theta & -\sin\theta \\ \sin\theta & \cos\theta \end{bmatrix}.$$

2.3. MULTIPLICATION OF MATRICES

Let **A** and **B** be two 3×3 matrices. The inner product of **A** and **B** is defined as the matrix **P** whose elements are related to those of **A** and **B** by

$$P_{ij} = \sum_{k=1}^{3} A_{ik}B_{kj} \equiv A_{ik}B_{kj}, \quad \text{or} \quad \mathbf{P} = \mathbf{A} \cdot \mathbf{B}, \qquad (2.9,\ 2.10)$$

where the definition equality in (2.9) follows from the summation convention, and the single dot between **A** and **B** in (2.10) signifies that a single index is summed to find **P**. An important feature of (2.9) is that the elements are summed over the inner or *adjacent* index k. It is sometimes useful to write (2.9) as

$$P_{ij} = A_{ik}B_{kj} = (\mathbf{A} \cdot \mathbf{B})_{ij},$$

where the last term is to be read as, "the *ij*-element of the inner product of matrices **A** and **B**." In explicit form, (2.9) is written as

$$\begin{bmatrix} P_{11} & P_{12} & P_{13} \\ P_{21} & P_{22} & P_{23} \\ P_{31} & P_{32} & P_{33} \end{bmatrix} = \begin{bmatrix} A_{11} & A_{12} & A_{13} \\ A_{21} & A_{22} & A_{23} \\ A_{31} & A_{32} & A_{33} \end{bmatrix} \begin{bmatrix} B_{11} & B_{12} & B_{13} \\ B_{21} & B_{22} & B_{23} \\ B_{31} & B_{32} & B_{33} \end{bmatrix}. \qquad (2.11)$$

This equation signifies that the *ij*-element of **P** is determined by multiplying the elements in the *i*-row of **A** and the *j*-column of **B**, and summing. For example,

$$P_{12} = A_{11}B_{12} + A_{12}B_{22} + A_{13}B_{32},$$

as indicated by the dashed-line boxes in (2.11). Naturally, the inner product $\mathbf{A} \cdot \mathbf{B}$ is only defined if the number of columns of $\mathbf{A}$ equals the number of rows of $\mathbf{B}$.

Equation (2.9) also applies to the inner product of a 3×3 matrix and a column vector. For example, (2.6) can be written as $x_i' = C_{ik}^T x_k$, which is now of the form of (2.9) because the summed index k is adjacent. In matrix form, (2.6) can therefore be written as

$$\begin{bmatrix} x_1' \\ x_2' \\ x_3' \end{bmatrix} = \begin{bmatrix} C_{11} & C_{12} & C_{13} \\ C_{21} & C_{22} & C_{23} \\ C_{31} & C_{32} & C_{33} \end{bmatrix}^T \begin{bmatrix} x_1 \\ x_2 \\ x_3 \end{bmatrix}.$$

Symbolically, the preceding is $\mathbf{x}' = \mathbf{C}^T \cdot \mathbf{x}$, whereas (2.7) is $\mathbf{x} = \mathbf{C} \cdot \mathbf{x}'$.

2.4. SECOND-ORDER TENSORS

A simple-to-complicated hierarchical description of physically meaningful quantities starts with scalars, proceeds to vectors, and then continues to second- and higher-order tensors. A scalar can be represented by a single value. A vector can be represented by three components, one for each of three orthogonal spatial directions denoted by a single free index. A second-order tensor can be represented by nine components, one for each pair of directions, and denoted by two free indices. Nearly all the tensors considered in Newtonian fluid mechanics are zero-, first-, or second-order tensors.

To better understand the structure of second-order tensors, consider the stress tensor $\boldsymbol{\tau}$ or τ_{ij}. Its two free indices specify two directions; the first indicates the orientation of the *surface* on which the stress is applied while the second indicates the component of the *force per unit area* on that surface. In particular, the first (i) index of τ_{ij} denotes the direction of the surface normal, and the second (j) index denotes the force component direction. This situation is illustrated in Figure 2.4, which shows the normal and shear stresses on an infinitesimal cube having surfaces parallel to the coordinate planes. The stresses are positive if they are directed as shown in this figure. The sign convention is that, on a surface whose outward normal points in the positive direction of a coordinate axis, the normal and shear stresses are positive if they point in the positive directions of the other axes. For example, on the surface ABCD, whose outward normal points in the positive x_2 direction, the positive stresses τ_{21}, τ_{22}, and τ_{23} point in the x_1, x_2, and x_3 directions, respectively. Normal stresses are positive if they are tensile and negative if they are compressive. On the opposite face EFGH the stress components have the same value as on ABCD, but their directions are reversed. This is because Figure 2.4 represents stresses *at a point*. The cube shown is intended to be vanishingly small, so that the faces ABCD and EFGH are just opposite sides of a plane perpendicular to the x_2-axis. Thus, stresses on the opposite faces are equal and opposite, and satisfy Newton's third law.

A vector $\mathbf{u}$ is completely specified by the three components u_i (where $i = 1, 2, 3$) because the components of $\mathbf{u}$ in any direction other than the original axes can be found from (2.8). Similarly, the state of stress at a point can be completely specified by the nine components τ_{ij} (where $i, j = 1, 2, 3$) that can be written as the matrix

FIGURE 2.4 Illustration of the stress field at a point via stress components on a cubic volume element. Here each surface may experience one normal and two shear components of stress. The directions of positive normal and shear stresses are shown. For clarity, the stresses on faces FBCG and CDHG are not labeled.

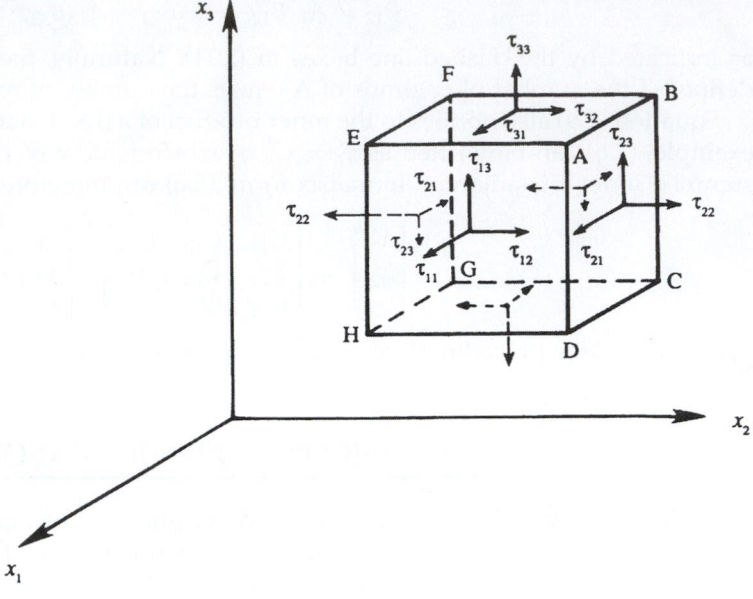

$$\tau = \begin{bmatrix} \tau_{11} & \tau_{12} & \tau_{13} \\ \tau_{21} & \tau_{22} & \tau_{23} \\ \tau_{31} & \tau_{32} & \tau_{33} \end{bmatrix}.$$

The specification of these nine stress components on surfaces perpendicular to the coordinate axes completely determines the state of stress at a point because the stresses on any arbitrary plane can be determined from them. To find the stresses on any arbitrary surface, we can consider a rotated coordinate system $O1'2'3'$ having one axis perpendicular to the given surface. It can be shown by a force balance on a tetrahedron element (see, e.g., Sommerfeld, 1964, page 59) that the components of τ in the rotated coordinate system are

$$\tau'_{mn} = \sum_{i=1}^{3} \sum_{j=1}^{3} C_{im} C_{jn} \tau_{ij} \equiv C_{im} C_{jn} \tau_{ij}, \tag{2.12}$$

where the definition equality follows from the summation convention. This equation may also be written as: $\tau'_{mn} = C^{\mathrm{T}}_{mi} \tau_{ij} C_{jn}$ or $\tau' = \mathbf{C}^{\mathrm{T}} \cdot \tau \cdot \mathbf{C}$. Note the similarity between the vector transformation rule (2.8) and (2.12). In (2.8) the first index of $\mathbf{C}$ is summed, while its second index is free. Equation (2.12) is identical, except that $\mathbf{C}$ is used twice. A quantity that obeys (2.12) is called a *second-order tensor*.

Tensor and matrix concepts are not quite the same. A matrix is any *arrangement* of elements, written as an array. The elements of a matrix represent the components of a second-order tensor only if they obey (2.12). In general, tensors can be of any order and the number of free indices corresponds to the order of the tensor. For example, $\mathbf{A}$ is a fourth-order tensor if it has four free indices, and the associated $3^4 = 81$ components change under a rotation of the coordinate system according to

$$A'_{mnpq} = \sum_{i=1}^{3} \sum_{j=1}^{3} \sum_{k=1}^{3} \sum_{l=1}^{3} C_{im} C_{jn} C_{kp} C_{lq} A_{ijkl} \equiv C_{im} C_{jn} C_{kp} C_{lq} A_{ijkl}, \tag{2.13}$$

where again the definition equality follows from the summation convention. Tensors of various orders arise in fluid mechanics. Common second-order tensors are the stress tensor τ_{ij} and the velocity-gradient tensor $\partial u_i/\partial x_j$. The nine products $u_i v_j$ formed from the components of the two vectors $\mathbf{u}$ and $\mathbf{v}$ also transform according to (2.12), and therefore form a second-order tensor. In addition, the *Kronecker-delta* and *alternating tensors* are also frequently used; these are defined and discussed in Section 2.7.

2.5. CONTRACTION AND MULTIPLICATION

When the two indices of a tensor are equated, and a summation is performed over this repeated index, the process is called *contraction*. An example is

$$\sum_{j=1}^{3} A_{jj} \equiv A_{jj} = A_{11} + A_{22} + A_{33},$$

which is the sum of the diagonal terms of A_{ij}. Clearly, A_{jj} is a scalar and therefore independent of the coordinate system. In other words, A_{jj} is an *invariant*. (There are three independent invariants of a second-order tensor, and A_{jj} is one of them; see Exercise 2.9.)

Higher-order tensors can be formed by multiplying lower-order tensors. If $\mathbf{A}$ and $\mathbf{B}$ are two second-order tensors, then the 81 numbers defined by $P_{ijkl} \equiv A_{ij} B_{kl}$ transform according to (2.13), and therefore form a fourth-order tensor.

Lower-order tensors can be obtained by performing a contraction within a multiplied form. The four contractions of $A_{ij} B_{kl}$ are

$$\sum_{i=1}^{3} A_{ij} B_{ki} \equiv A_{ij} B_{ki} = B_{ki} A_{ij} = (\mathbf{B} \cdot \mathbf{A})_{kj},$$

$$\sum_{i=1}^{3} A_{ij} B_{ik} \equiv A_{ij} B_{ik} = A_{ji}^{T} B_{ik} = (\mathbf{A}^{T} \cdot \mathbf{B})_{jk},$$

$$\sum_{i=1}^{3} A_{ij} B_{kj} \equiv A_{ij} B_{kj} = A_{ij} B_{jk}^{T} = (\mathbf{A} \cdot \mathbf{B}^{T})_{ik}, \tag{2.14}$$

$$\sum_{i=1}^{3} A_{ij} B_{jk} \equiv A_{ij} B_{jk} = (\mathbf{A} \cdot \mathbf{B})_{ik},$$

where all the definition equalities follow from the summation convention. All four products in (2.14) are second-order tensors. Note also in (2.14) how the terms have been rearranged until the summed index is adjacent; at this point they can be written as a product of matrices.

The contracted product of a second-order tensor $\mathbf{A}$ and a vector $\mathbf{u}$ is a vector. The two possibilities are

$$\sum_{i=1}^{3} A_{ij} u_j \equiv A_{ij} u_j = (\mathbf{A} \cdot \mathbf{u})_i, \text{ and}$$

$$\sum_{i=1}^{3} A_{ij} u_i \equiv A_{ij} u_i = A_{ji}^{\mathrm{T}} u_i = (\mathbf{A}^{\mathrm{T}} \cdot \mathbf{u})_j,$$

where again the definition equalities follow from the summation convention. The doubly contracted product of two second-order tensors $\mathbf{A}$ and $\mathbf{B}$ is a scalar. Using all three notations, the two possibilities are

$$\sum_{i=1}^{3} \sum_{j=1}^{3} A_{ij} B_{ji} \equiv A_{ij} B_{ji} (= \mathbf{A} : \mathbf{B}) \text{ and } \sum_{i=1}^{3} \sum_{j=1}^{3} A_{ij} B_{ij} \equiv A_{ij} B_{ij} (= \mathbf{A} : \mathbf{B}^{\mathrm{T}}),$$

where the bold colon (:) implies a *double* contraction or double dot product.

2.6. FORCE ON A SURFACE

A surface area element has a size (or magnitude) and an orientation, so it can be treated as a vector $d\mathbf{A}$. If dA is the surface element's size, and $\mathbf{n}$ is its normal unit vector, then $d\mathbf{A} = \mathbf{n}dA$.

Suppose the nine components, τ_{ij}, of the stress tensor with respect to a given set of Cartesian coordinates O123 are given, and we want to find the force per unit area, $\mathbf{f}(\mathbf{n})$ with components f_i, on an arbitrarily oriented surface element with normal $\mathbf{n}$ (see Figure 2.5). One way of completing this task is to switch to a rotated coordinate system, and use (2.12) to find the normal and shear stresses on the surface element. An alternative method is described here. Consider the tetrahedral element shown in Figure 2.5. The net force f_1 on the element in the first direction produced by the stresses τ_{ij} is

$$f_1 dA = \tau_{11} dA_1 + \tau_{21} dA_2 + \tau_{31} dA_3.$$

The geometry of the tetrahedron requires: $dA_i = n_i dA$, where n_i are the components of the surface normal vector $\mathbf{n}$. Thus, the net force equation can be rewritten as

$$f_1 dA = \tau_{11} n_1 dA + \tau_{21} n_2 dA + \tau_{31} n_3 dA.$$

Dividing by dA then produces $f_1 = \tau_{j1} n_j$ (with summation implied), or for any component of $\mathbf{f}$,

$$f_i = \sum_{j=1}^{3} \tau_{ji} n_j \equiv \tau_{ji} n_j \quad \text{or} \quad \mathbf{f} = \mathbf{n} \cdot \boldsymbol{\tau}, \tag{2.15}$$

where the boldface-only version of (2.15) follows when $\tau_{ij} = \tau_{ji}$, a claim that is proved in Chapter 4. Therefore, the contracted or inner product of the stress tensor $\boldsymbol{\tau}$ and the unit

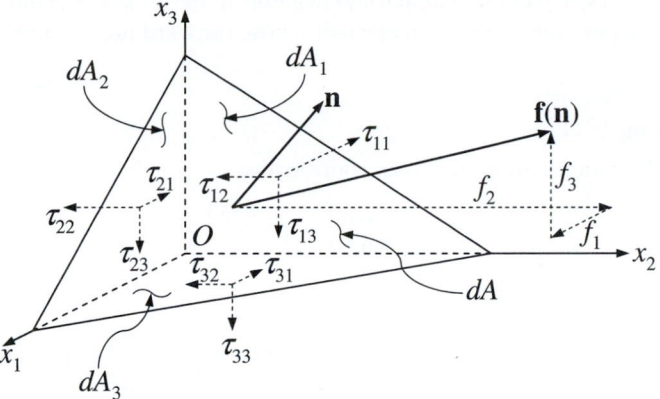

FIGURE 2.5   Force **f** per unit area on a surface element whose outward normal is **n**. The areas of the tetrahedron's faces that are perpendicular to the ith coordinate axis are dA_i. The area of the largest tetrahedron face is dA. As in Figure 2.4, the directions of positive normal and shear stresses are shown.

normal vector **n** gives the force per unit area on a surface perpendicular to **n**. This result is analogous to $u_n = \mathbf{u} \cdot \mathbf{n}$, where u_n is the component of the vector **u** along **n**; however, whereas u_n is a scalar, **f** in (2.15) is a vector.

EXAMPLE 2.2

In two spatial dimensions, x_1 and x_2, consider parallel flow through a channel (see Figure 2.6). Choose x_1 parallel to the flow direction. The viscous stress tensor at a point in the flow has the form

$$\tau = \begin{bmatrix} 0 & a \\ a & 0 \end{bmatrix},$$

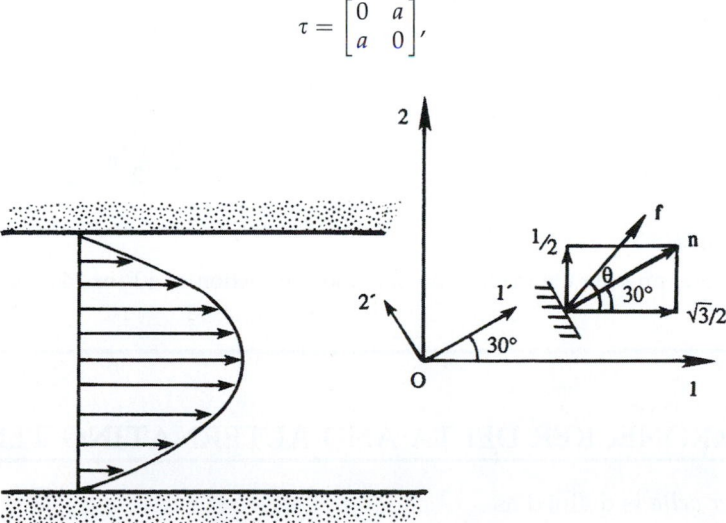

FIGURE 2.6   Determination of the force per unit area on a small area element with a normal vector rotated 30° from the flow direction in a simple unidirectional shear flow parallel to the x_1-axis.

where a is positive in one half of the channel, and negative in the other half. Find the magnitude and direction of the force per unit area $\mathbf{f}$ on an element whose outward normal points $\phi = 30°$ from the flow direction.

Solution by Using (2.15)

Start with the definition of $\mathbf{n}$ in the given coordinates:

$$\mathbf{n} = \begin{bmatrix} \cos\phi \\ \sin\phi \end{bmatrix} = \begin{bmatrix} \sqrt{3}/2 \\ 1/2 \end{bmatrix}.$$

The force per unit area is therefore

$$\mathbf{f} = \tau_{ji}n_j = \tau_{ij}n_j = \begin{bmatrix} 0 & a \\ a & 0 \end{bmatrix}\begin{bmatrix} \cos\phi \\ \sin\phi \end{bmatrix} = \begin{bmatrix} a\sin\phi \\ a\cos\phi \end{bmatrix} = \begin{bmatrix} a/2 \\ a\sqrt{3}/2 \end{bmatrix} = \begin{bmatrix} f_1 \\ f_2 \end{bmatrix}.$$

The magnitude of $\mathbf{f}$ is

$$f = |\mathbf{f}| = \left(f_1^2 + f_2^2\right)^{1/2} = |a|.$$

If θ is the angle of $\mathbf{f}$ with respect to the x_1 axis, then

$$\sin\theta = f_2/f = \left(\sqrt{3}/2\right)(a/|a|) \quad \text{and} \quad \cos\theta = f_1/f = (1/2)(a/|a|).$$

Thus $\theta = 60°$ if a is positive (in which case both $\sin\theta$ and $\cos\theta$ are positive), and $\theta = 240°$ if a is negative (in which case both $\sin\theta$ and $\cos\theta$ are negative).

Solution by Using (2.12)

Consider a rotated coordinate system $O1'2'$ with the x_1'-axis coinciding with $\mathbf{n}$ as shown in Figure 2.6. Using (2.12), the components of the stress tensor in the rotated frame are

$$\tau_{11}' = C_{11}C_{21}\tau_{12} + C_{21}C_{11}\tau_{21} = (\cos\phi\sin\phi)a + (\sin\phi\cos\phi)a = \frac{\sqrt{3}}{2}\frac{1}{2}a + \frac{1}{2}\frac{\sqrt{3}}{2}a = \frac{\sqrt{3}}{2}a \text{ and } [?],$$

$$\tau_{12}' = C_{11}C_{22}\tau_{12} + C_{21}C_{12}\tau_{21} = (\cos\phi)^2 a - (\sin\phi)^2 a = \frac{\sqrt{3}}{2}\frac{\sqrt{3}}{2}a - \frac{1}{2}\frac{1}{2}a = \frac{1}{2}a,$$

where $C_{ij} = \begin{bmatrix} \cos\phi & -\sin\phi \\ \sin\phi & \cos\phi \end{bmatrix}$. The normal stress is therefore $\sqrt{3}a/2$, and the shear stress is $a/2$.

These results again provide the magnitude of a and a direction of 60° or 240° depending on the sign of a.

2.7. KRONECKER DELTA AND ALTERNATING TENSOR

The *Kronecker delta* is defined as

$$\delta_{ij} = \left\{ \begin{array}{l} 1 \text{ if } i = j \\ 0 \text{ if } i \neq j \end{array} \right\}. \tag{2.16}$$

In three spatial dimensions it is the 3×3 identity matrix:

$$\delta = \begin{bmatrix} 1 & 0 & 0 \\ 0 & 1 & 0 \\ 0 & 0 & 1 \end{bmatrix}.$$

In matrix multiplication operations involving the Kronecker delta, it simply replaces its summed-over index by its other index. Consider

$$\sum_{j=1}^{3} \delta_{ij} u_j \equiv \delta_{ij} u_j = \delta_{i1} u_1 + \delta_{i2} u_2 + \delta_{i3} u_3;$$

the right-hand side is u_1 when $i = 1$, u_2 when $i = 2$, and u_3 when $i = 3$; thus

$$\delta_{ij} u_j = u_i. \tag{2.17}$$

From its definition it is clear that δ_{ij} is an *isotropic tensor* in the sense that its components are unchanged by a rotation of the frame of reference, that is, $\delta'_{ij} = \delta_{ij}$. Isotropic tensors can be of various orders. There is no isotropic tensor of first order, and δ_{ij} is the only isotropic tensor of second order. There is also only one isotropic tensor of third order. It is called the *alternating tensor* or *permutation symbol*, and is defined as

$$\varepsilon_{ijk} = \left\{ \begin{array}{ll} 1 & \text{if } ijk = 123, 231, \text{ or } 312 \text{ (cyclic order)}, \\ 0 & \text{if any two indices are equal}, \\ -1 & \text{if } ijk = 321, 213, \text{ or } 132 \text{ (anti-cyclic order)} \end{array} \right\}. \tag{2.18}$$

From this definition, it is clear that *an index on ε_{ijk} can be moved two places (either to the right or to the left) without changing its value*. For example, $\varepsilon_{ijk} = \varepsilon_{jki}$ where i has been moved two places to the right, and $\varepsilon_{ijk} = \varepsilon_{kij}$ where k has been moved two places to the left. For a movement of one place, however, the sign is reversed. For example, $\varepsilon_{ijk} = -\varepsilon_{ikj}$ where j has been moved one place to the right.

A very frequently used relation is the *epsilon delta relation*:

$$\sum_{k=1}^{3} \varepsilon_{ijk} \varepsilon_{klm} \equiv \varepsilon_{ijk} \varepsilon_{klm} = \delta_{il} \delta_{jm} - \delta_{im} \delta_{jl}. \tag{2.19}$$

The reader can verify the validity of this relationship by choosing some values for the indices $ijlm$. This relationship can be remembered by noting the following two points: 1) The adjacent index k is summed; and 2) the first two indices on the right side, namely, i and l, are the first index of ε_{ijk} and the first *free* index of ε_{klm}. The remaining indices on the right side then follow immediately.

2.8. VECTOR, DOT, AND CROSS PRODUCTS

The dot product of two vectors $\mathbf{u}$ and $\mathbf{v}$ is defined as the scalar

$$\mathbf{u} \cdot \mathbf{v} = \mathbf{v} \cdot \mathbf{u} = u_1 v_1 + u_2 v_2 + u_3 v_3 = \sum_{i=1}^{3} u_i v_i \equiv u_i v_i.$$

It is easy to show that $\mathbf{u} \cdot \mathbf{v} = uv \cos \theta$, where u and v are the vectors' magnitudes and θ is the angle between the vectors (see Exercises 2.12 and 2.13). The dot product is therefore the magnitude of one vector times the component of the other in the direction of the first. The dot product $\mathbf{u} \cdot \mathbf{v}$ is equal to the sum of the diagonal terms of the tensor $u_i v_j$.

The cross product between two vectors $\mathbf{u}$ and $\mathbf{v}$ is defined as the vector $\mathbf{w}$ whose magnitude is $uv \sin \theta$ where θ is the angle between $\mathbf{u}$ and $\mathbf{v}$, and whose direction is perpendicular to the plane of $\mathbf{u}$ and $\mathbf{v}$ such that $\mathbf{u}$, $\mathbf{v}$, and $\mathbf{w}$ form a right-handed system. Clearly, $\mathbf{u} \times \mathbf{v} = -\mathbf{v} \times \mathbf{u}$. Furthermore, unit vectors in right-handed coordinate systems obey the cyclic rule $\mathbf{e}_1 \times \mathbf{e}_2 = \mathbf{e}_3$. From these requirements it can be shown that

$$\mathbf{u} \times \mathbf{v} = (u_2 v_3 - u_3 v_2)\mathbf{e}_1 + (u_3 v_1 - u_1 v_3)\mathbf{e}_2 + (u_1 v_2 - u_2 v_1)\mathbf{e}_3 \tag{2.20}$$

(see Exercise 2.14). Equation (2.20) can be written as the determinant of a matrix

$$\mathbf{u} \times \mathbf{v} = \det \begin{bmatrix} \mathbf{e}_1 & \mathbf{e}_2 & \mathbf{e}_3 \\ u_1 & u_2 & u_3 \\ v_1 & v_2 & v_3 \end{bmatrix}.$$

In indicial notation, the k-component of $\mathbf{u} \times \mathbf{v}$ can be written as

$$(\mathbf{u} \times \mathbf{v})_k = \sum_{i=1}^{3} \sum_{j=1}^{3} \varepsilon_{ijk} u_i v_j \equiv \varepsilon_{ijk} u_i v_j = \varepsilon_{kij} u_i v_j. \tag{2.21}$$

As a check, for $k = 1$ the nonzero terms in the double sum in (2.21) result from $i = 2, j = 3$, and from $i = 3, j = 2$. This follows from the definition (2.18) that the permutation symbol is zero if any two indices are equal. Thus, (2.21) gives

$$(\mathbf{u} \times \mathbf{v})_1 = \varepsilon_{ij1} u_i v_j = \varepsilon_{231} u_2 v_3 + \varepsilon_{321} u_3 v_2 = u_2 v_3 - u_3 v_2,$$

which agrees with (2.20). Note that the third form of (2.21) is obtained from the second by moving the index k two places to the left; see the remark following (2.18).

2.9. GRADIENT, DIVERGENCE, AND CURL

The vector-differentiation operator "del"[i] is defined symbolically by

$$\nabla = \mathbf{e}_1 \frac{\partial}{\partial x_1} + \mathbf{e}_2 \frac{\partial}{\partial x_2} + \mathbf{e}_3 \frac{\partial}{\partial x_3} = \sum_{i=1}^{3} \mathbf{e}_i \frac{\partial}{\partial x_i} \equiv \mathbf{e}_i \frac{\partial}{\partial x_i}. \tag{2.22}$$

When operating on a scalar function of position ϕ, it generates the vector

$$\nabla \phi = \sum_{i=1}^{3} \mathbf{e}_i \frac{\partial \phi}{\partial x_i} \equiv \mathbf{e}_i \frac{\partial \phi}{\partial x_i},$$

[i]The inverted Greek delta is called a "nabla" ($\nu\alpha\beta\lambda\alpha$). The word originates from the Hebrew word for lyre, an ancient harp-like stringed instrument. It was on his instrument that the boy, David, entertained King Saul (Samuel II) and it is mentioned repeatedly in Psalms as a musical instrument to use in the praise of God.

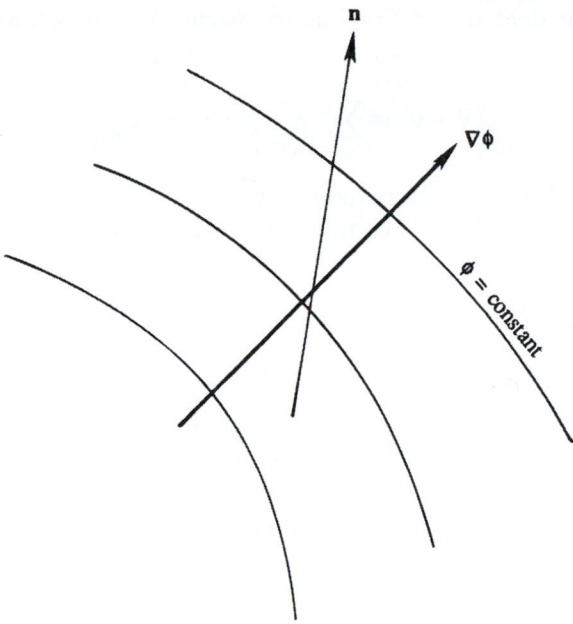

FIGURE 2.7 An illustration of the gradient, $\nabla\phi$, of a scalar function ϕ. The curves of constant ϕ and $\nabla\phi$ are perpendicular, and the spatial derivative of ϕ in the direction $\mathbf{n}$ is given by $\mathbf{n}\cdot\nabla\phi$. The most rapid change in ϕ is found when $\mathbf{n}$ and $\nabla\phi$ are parallel.

whose i-component is $(\nabla\phi)_i = \partial\phi/\partial x_i$. The vector $\nabla\phi$ is called the *gradient* of ϕ, and $\nabla\phi$ is perpendicular to surfaces defined by $\phi = $ constant. In addition, it specifies the magnitude and direction of the *maximum* spatial rate of change of ϕ (Figure 2.7). The spatial rate of change of ϕ in any other direction $\mathbf{n}$ is given by

$$\partial\phi/\partial n = \nabla\phi\cdot\mathbf{n}.$$

In Cartesian coordinates, the *divergence* of a vector field $\mathbf{u}$ is defined as the scalar

$$\nabla\cdot\mathbf{u} = \frac{\partial u_1}{\partial x_1} + \frac{\partial u_2}{\partial x_2} + \frac{\partial u_3}{\partial x_3} = \sum_{i=1}^{3} \frac{\partial u_i}{\partial x_i} \equiv \frac{\partial u_i}{\partial x_i}. \tag{2.23}$$

So far, we have defined the operations of the gradient of a scalar and the divergence of a vector. We can, however, generalize these operations. For example, the divergence of a second-order tensor τ can be defined as the vector whose i-component is

$$(\nabla\cdot\tau)_i = \sum_{j=1}^{3} \frac{\partial \tau_{ij}}{\partial x_j} \equiv \frac{\partial \tau_{ij}}{\partial x_j}.$$

It is evident that the divergence operation *decreases* the order of the tensor by one. In contrast, the gradient operation *increases* the order of a tensor by one, changing a zero-order tensor to a first-order tensor, and a first-order tensor to a second-order tensor, i.e., $\partial u_i/\partial x_j$.

The *curl* of a vector field **u** is defined as the vector $\nabla \times \mathbf{u}$, whose *i*-component can be written as

$$(\nabla \times \mathbf{u})_i = \sum_{j=1}^{3} \sum_{k=1}^{3} \varepsilon_{ijk} \frac{\partial u_k}{\partial x_j} \equiv \varepsilon_{ijk} \frac{\partial u_k}{\partial x_j} \tag{2.24}$$

using (2.21) and (2.22). The three components of the vector $\nabla \times \mathbf{u}$ can easily be found from the right-hand side of (2.24). For the $i = 1$ component, the nonzero terms in the double sum result from $j = 2, k = 3$, and from $j = 3, k = 2$. The three components of $\nabla \times \mathbf{u}$ are finally found as

$$(\nabla \times \mathbf{u})_1 = \frac{\partial u_3}{\partial x_2} - \frac{\partial u_2}{\partial x_3}, \quad (\nabla \times \mathbf{u})_2 = \frac{\partial u_1}{\partial x_3} - \frac{\partial u_3}{\partial x_1}, \quad \text{and} \quad (\nabla \times \mathbf{u})_3 = \frac{\partial u_2}{\partial x_1} - \frac{\partial u_1}{\partial x_2}. \tag{2.25}$$

A vector field **u** is called *solenoidal* or *divergence free* if $\nabla \cdot \mathbf{u} = 0$, and *irrotational* or *curl free* if $\nabla \times \mathbf{u} = 0$. The word solenoidal refers to the fact that the divergence of the magnetic induction is always zero because of the absence of magnetic monopoles. The reason for the word irrotational is made clear in Chapter 3.

EXAMPLE 2.3

If a is a positive constant and **b** is a constant vector, determine the divergence and the curl of a vector field that diverges from the origin of coordinates, $\mathbf{u} = a\mathbf{x}$, and a vector field indicative of solid body rotation about a fixed axis, $\mathbf{u} = \mathbf{b} \times \mathbf{x}$.

Solution

Using $\mathbf{u} = a\mathbf{x} = ax_1\mathbf{e}_1 + ax_2\mathbf{e}_2 + ax_2\mathbf{e}_2$ in (2.23) and (2.25) produces:

$$\nabla \cdot \mathbf{u} = \frac{\partial ax_1}{\partial x_1} + \frac{\partial ax_2}{\partial x_2} + \frac{\partial ax_3}{\partial x_3} = a + a + a = 3a,$$

$$(\nabla \times \mathbf{u})_1 = \frac{\partial ax_3}{\partial x_2} - \frac{\partial ax_2}{\partial x_3} = 0, \quad (\nabla \times \mathbf{u})_2 = \frac{\partial ax_1}{\partial x_3} - \frac{\partial ax_3}{\partial x_1} = 0, \quad \text{and}$$

$$(\nabla \times \mathbf{u})_3 = \frac{\partial ax_2}{\partial x_1} - \frac{\partial ax_1}{\partial x_2} = 0.$$

Thus, $\mathbf{u} = a\mathbf{x}$ has a constant nonzero divergence and is irrotational. Using $\mathbf{u} = (b_2 x_3 - b_3 x_2)\mathbf{e}_1 + (b_3 x_1 - b_1 x_3)\mathbf{e}_2 + (b_1 x_2 - b_2 x_1)\mathbf{e}_3$ in (2.23) and (2.25) produces:

$$\nabla \cdot \mathbf{u} = \frac{\partial (b_2 x_3 - b_3 x_2)}{\partial x_1} + \frac{\partial (b_3 x_1 - b_1 x_3)}{\partial x_2} + \frac{\partial (b_1 x_2 - b_2 x_1)}{\partial x_3} = 0,$$

$$(\nabla \times \mathbf{u})_1 = \frac{\partial (b_1 x_2 - b_2 x_1)}{\partial x_2} - \frac{\partial (b_3 x_1 - b_1 x_3)}{\partial x_3} = 2b_1,$$

$$(\nabla \times \mathbf{u})_2 = \frac{\partial (b_2 x_3 - b_3 x_2)}{\partial x_3} - \frac{\partial (b_1 x_2 - b_2 x_1)}{\partial x_1} = 2b_2, \quad \text{and}$$

$$(\nabla \times \mathbf{u})_3 = \frac{\partial (b_3 x_1 - b_1 x_3)}{\partial x_1} - \frac{\partial (b_2 x_3 - b_3 x_2)}{\partial x_2} = 2b_3.$$

Thus, $\mathbf{u} = \mathbf{b} \times \mathbf{x}$ is divergence free and rotational.

2.10. SYMMETRIC AND ANTISYMMETRIC TENSORS

A tensor **B** is called *symmetric* in the indices i and j if the components do not change when i and j are interchanged, that is, if $B_{ij} = B_{ji}$. Thus, the matrix of a symmetric second-order tensor is made up of only six distinct components (the three on the diagonal where $i = j$, and the three above or below the diagonal where $i \neq j$). On the other hand, a tensor is called *antisymmetric* if $B_{ij} = -B_{ji}$. An antisymmetric tensor must have zero diagonal components, and is made up of only three distinct components (the three above or below the diagonal). Any tensor can be represented as the sum of a symmetric part and an antisymmetric part. For if we write

$$B_{ij} = \frac{1}{2}(B_{ij} + B_{ji}) + \frac{1}{2}(B_{ij} - B_{ji}) = S_{ij} + A_{ij},$$

then the operation of interchanging i and j does not change the first term, but changes the sign of the second term. Therefore, $(B_{ij} + B_{ji})/2 \equiv S_{ij}$ is called the symmetric part of B_{ij}, and $(B_{ij} - B_{ji})/2 \equiv A_{ij}$ is called the antisymmetric part of B_{ij}.

Every vector can be associated with an antisymmetric tensor, and vice versa. For example, we can associate the vector **ω** having components ω_i, with an antisymmetric tensor:

$$\mathbf{R} = \begin{bmatrix} 0 & -\omega_3 & \omega_2 \\ \omega_3 & 0 & -\omega_1 \\ -\omega_2 & \omega_1 & 0 \end{bmatrix}. \tag{2.26}$$

The two are related via

$$R_{ij} = \sum_{k=1}^{3} -\varepsilon_{ijk}\omega_k \equiv -\varepsilon_{ijk}\omega_k, \quad \text{and} \quad \omega_k = \sum_{i-1}^{3}\sum_{j=1}^{3} -\frac{1}{2}\varepsilon_{ijk}R_{ij} \equiv -\frac{1}{2}\varepsilon_{ijk}R_{ij}. \tag{2.27}$$

As a check, (2.27) gives $R_{11} = 0$ and $R_{12} = -\varepsilon_{123}\omega_3 = -\omega_3$, in agreement with (2.26). (In Chapter 3, **R** is recognized as the *rotation* tensor corresponding to the *vorticity* vector **ω**.)

A commonly occurring operation is the doubly contracted product, P, of a *symmetric* tensor τ and another tensor **B**:

$$P = \sum_{k=1}^{3}\sum_{l=1}^{3} \tau_{kl}B_{kl} \equiv \tau_{kl}B_{kl} = \tau_{kl}(S_{kl} + A_{kl}) = \tau_{kl}S_{kl} + \tau_{kl}A_{kl} = \tau_{ij}S_{ij} + \tau_{ij}A_{ij}, \tag{2.28}$$

where **S** and **A** are the symmetric and antisymmetric parts of **B** (see above). The final equality follows from the index-summation convention; sums are completed over both k and l, so these indices can be replaced by any two distinct indices. Exchanging the indices of **A** in the final term of (2.28) produces $P = \tau_{ij}S_{ij} - \tau_{ij}A_{ji}$, but this can also be written $P = \tau_{ji}S_{ji} - \tau_{ji}A_{ji}$ because S_{ij} and τ_{ij} are symmetric. Now, replace the index j by k and the index i by l to find:

$$P = \tau_{kl}S_{kl} - \tau_{kl}A_{kl}. \tag{2.29}$$

This relationship and the fourth part of the extended equality in (2.28) require that $\tau_{ij}A_{ij} = \tau_{kl}A_{kl} = 0$, and

$$\tau_{ij}B_{ij} = \tau_{ij}S_{ij} = \frac{1}{2}\tau_{ij}\left(B_{ij} + B_{ji}\right).$$

Thus, the doubly contracted product of a symmetric tensor τ with any tensor $\mathbf{B}$ equals τ doubly contracted with the symmetric part of $\mathbf{B}$, and the doubly contracted product of a symmetric tensor and an antisymmetric tensor is zero. The latter result is analogous to the fact that the definite integral over an even (symmetric) interval of the product of a symmetric and an antisymmetric function is zero.

2.11. EIGENVALUES AND EIGENVECTORS OF A SYMMETRIC TENSOR

The reader is assumed to be familiar with the concepts of eigenvalues and eigenvectors of a matrix, so only a brief review of the main results is provided here. Suppose τ is a symmetric tensor with real elements, for example, the stress tensor. Then the following facts can be proved:

(1) There are three real eigenvalues λ^k ($k = 1, 2, 3$), which may or may not all be distinct. (Here, the superscript k is not an exponent, and λ^k does not denote the k-component of a vector.) These eigenvalues (λ^1, λ^2, and λ^3) are the roots or solutions of the third-degree polynomial

$$\det \left| \tau_{ij} - \lambda \delta_{ij} \right| = 0.$$

(2) The three eigenvectors $\mathbf{b}^k$ corresponding to distinct eigenvalues λ^k are mutually orthogonal. These eigenvectors define the directions of the *principal axes* of τ. Each $\mathbf{b}$ is found by solving three algebraic equations,

$$\left(\tau_{ij} - \lambda \delta_{ij} \right) b_j = 0$$

($i = 1, 2,$ or 3), where the superscript k on λ and $\mathbf{b}$ has been omitted for clarity because there is no sum over k.

(3) If the coordinate system is rotated so that its unit vectors coincide with the eigenvectors, then τ is diagonal with elements λ^k in this rotated coordinate system,

$$\tau' = \begin{bmatrix} \lambda^1 & 0 & 0 \\ 0 & \lambda^2 & 0 \\ 0 & 0 & \lambda^3 \end{bmatrix}.$$

(4) Although the elements τ_{ij} change as the coordinate system is rotated, they cannot be larger than the largest λ or smaller than the smallest λ; the λ^k represent the extreme values of τ_{ij}.

EXAMPLE 2.4

The strain rate tensor $\mathbf{S}$ is related to the velocity vector $\mathbf{u}$ by

$$S_{ij} = \frac{1}{2} \left(\frac{\partial u_i}{\partial x_j} + \frac{\partial u_j}{\partial x_i} \right).$$

For a two-dimensional flow parallel to the 1-direction,

$$\mathbf{u} = \begin{bmatrix} u_1(x_2) \\ 0 \end{bmatrix},$$

show how S is diagonalized in a frame of reference rotated to coincide with the principal axes.

Solution

For the given velocity profile $u_1(x_2)$, it is evident that $S_{11} = S_{22} = 0$, and $2S_{12} = 2S_{21} = du_1/dx_2 = \Gamma$. The strain rate tensor in the original coordinate system is therefore

$$\mathbf{S} = \begin{bmatrix} 0 & \Gamma \\ \Gamma & 0 \end{bmatrix}.$$

The eigenvalues are determined from

$$\det|E_{ij} - \lambda\delta_{ij}|\begin{vmatrix} -\lambda & \Gamma \\ \Gamma & -\lambda \end{vmatrix} = 0,$$

which has solutions $\lambda^1 = \Gamma$ and $\lambda^2 = -\Gamma$. The first eigenvector b^1 is given by

$$\begin{bmatrix} 0 & \Gamma \\ \Gamma & 0 \end{bmatrix}\begin{bmatrix} b_1^1 \\ b_2^1 \end{bmatrix} = \lambda^1 \begin{bmatrix} b_1^1 \\ b_2^1 \end{bmatrix},$$

which has solution $b_1^1 = b_2^1 = 1/\sqrt{2}$, when b^1 is normalized to have magnitude unity. The second eigenvector is similarly found so that

$$\mathbf{b}^1 = \begin{bmatrix} 1/\sqrt{2} \\ 1/\sqrt{2} \end{bmatrix}, \quad \text{and} \quad \mathbf{b}^2 = \begin{bmatrix} -1/\sqrt{2} \\ 1/\sqrt{2} \end{bmatrix}.$$

These eigenvectors are shown in Figure 2.8. The direction cosine matrix of the original and the rotated coordinate system is therefore

$$\mathbf{C} = \begin{bmatrix} \dfrac{1}{\sqrt{2}} & -\dfrac{1}{\sqrt{2}} \\ \dfrac{1}{\sqrt{2}} & \dfrac{1}{\sqrt{2}} \end{bmatrix},$$

which represents rotation of the coordinate system by 45°. Using the transformation rule (2.12), the components of S in the rotated system are found as follows:

$$S_{12}' = C_{i1}C_{j2}S_{ij} = C_{11}C_{22}S_{12} + C_{21}C_{12}S_{21} = \frac{1}{\sqrt{2}}\frac{1}{\sqrt{2}}\Gamma - \frac{1}{\sqrt{2}}\frac{1}{\sqrt{2}}\Gamma = 0,$$

$$S_{11}' = C_{i1}C_{j1}S_{ij} = C_{11}C_{21}S_{12} + C_{21}C_{11}S_{21} = \Gamma, \quad \text{and}$$

$$S_{22}' = C_{i2}C_{j2}S_{ij} = C_{12}C_{22}S_{12} + C_{22}C_{12}S_{21} = -\Gamma.$$

(Instead of using (2.12), all the components of S in the rotated system can be found by carrying out the matrix product $\mathbf{C}^T \cdot \mathbf{S} \cdot \mathbf{C}$.) The matrix of S in the rotated frame is therefore:

$$\mathbf{S}' = \begin{bmatrix} \Gamma & 0 \\ 0 & -\Gamma \end{bmatrix}.$$

FIGURE 2.8 Original coordinate system Ox_1x_2 and the rotated coordinate system $Ox_1'x_2'$ having unit vectors that coincide with the eigenvectors of the strain-rate tensor in Example 2.4. Here the strain rate is determined from a unidirectional flow having only cross-stream variation, and the angle of rotation is determined to be 45°.

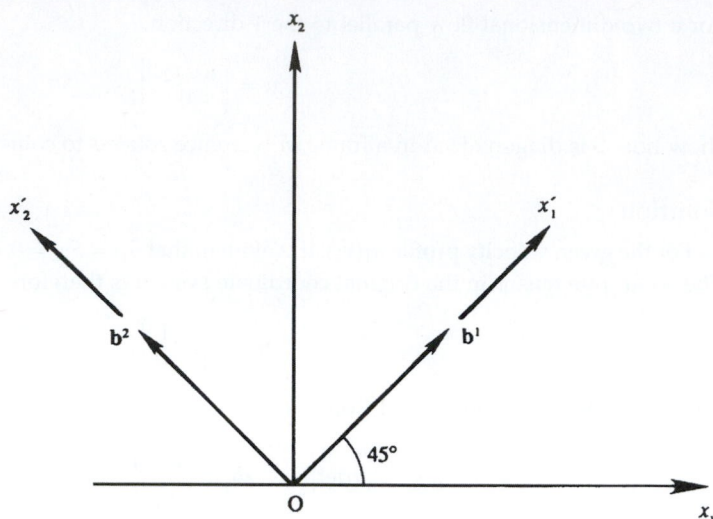

The foregoing matrix contains only diagonal terms. For positive Γ, it will be shown in the next chapter that it represents a linear stretching at a rate Γ along one principal axis, and a linear compression at a rate $-\Gamma$ along the other; the shear strains are zero in the principal-axis coordinate system of the strain rate tensor.

2.12. GAUSS' THEOREM

This very useful theorem relates volume and surface integrals. Let V be a volume bounded by a closed surface A. Consider an infinitesimal surface element dA having outward unit normal $\mathbf{n}$ with components n_i (Figure 2.9), and let $Q(\mathbf{x})$ be a scalar, vector, or tensor field of any order. Gauss' theorem states that

$$\iiint_V \frac{\partial Q}{\partial x_i} dV = \iint_A n_i Q dA. \qquad (2.30)$$

The most common form of Gauss' theorem is when $\mathbf{Q}$ is a vector, in which case the theorem is

$$\iiint_V \sum_{i=1}^3 \frac{\partial Q_i}{\partial x_i} dV \equiv \iiint_V \frac{\partial Q_i}{\partial x_i} dV = \iint_A \sum_{i=1}^3 n_i Q_i \, dA \equiv \iint_A n_i Q_i \, dA, \quad \text{or} \quad \iiint_V \nabla \cdot \mathbf{Q} dV$$

$$= \iint_A \mathbf{n} \cdot \mathbf{Q} dA,$$

which is commonly called the *divergence theorem*. In words, the theorem states that the volume integral of the divergence of $\mathbf{Q}$ is equal to the surface integral of the outflux of $\mathbf{Q}$.

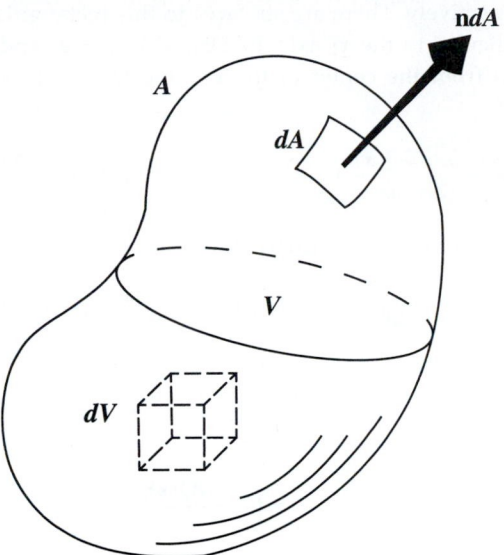

FIGURE 2.9 Illustration of Gauss' theorem for a volume V enclosed by surface area A. A small volume element, dV, and a small area element, dA, with outward normal $\mathbf{n}$ are shown.

Alternatively, (2.30) defines a generalized field derivative, denoted by $\mathcal{D}$, of Q when considered in its limiting form for a vanishingly small volume,

$$\mathcal{D}Q = \lim_{V \to 0} \frac{1}{V} \iint_A n_i Q dA. \tag{2.31}$$

Interestingly, this form is readily specialized to the gradient, divergence, and curl of any scalar, vector, or tensor Q. Moreover, by regarding (2.31) as a definition, the recipes for the computation of vector field derivatives may be obtained in any coordinate system. As stated, (2.31) defines the gradient of a tensor Q of any order. For a tensor of order one or higher, the divergence and curl are defined by including a dot (scalar) product or a cross (vector) product, respectively, under the integral:

$$\nabla \cdot \mathbf{Q} = \lim_{V \to 0} \frac{1}{V} \iint_A \mathbf{n} \cdot \mathbf{Q} dA, \quad \text{and} \quad \nabla \times \mathbf{Q} = \lim_{V \to 0} \frac{1}{V} \iint_A \mathbf{n} \times \mathbf{Q} dA. \tag{2.32, 2.33}$$

EXAMPLE 2.5

Obtain the recipe for the divergence of a vector $\mathbf{Q}(\mathbf{x})$ in Cartesian coordinates from the integral definition (2.32).

Solution

Consider an elemental rectangular volume centered on $\mathbf{x}$ with faces perpendicular to the coordinate axes (see Figure 2.4). Denote the lengths of the sides parallel to each coordinate axis by Δx_1, Δx_2, and Δx_3, respectively. There are six faces to this rectangular volume. First consider the two that are perpendicular to the x_1 axis, EADH with $\mathbf{n} = \mathbf{e}_1$ and FBCG with $\mathbf{n} = -\mathbf{e}_1$. A Taylor expansion of $\mathbf{Q}(\mathbf{x})$ from the center of the volume to the center of each of these sides produces

$$[\mathbf{Q}]_{EADH} = \mathbf{Q}(\mathbf{x}) + \frac{\Delta x_1}{2} \frac{\partial \mathbf{Q}(\mathbf{x})}{\partial x_1} + \dots \quad \text{and} \quad [\mathbf{Q}]_{FBCG} = \mathbf{Q}(\mathbf{x}) - \frac{\Delta x_1}{2} \frac{\partial \mathbf{Q}(\mathbf{x})}{\partial x_1} + \dots,$$

so that the x-direction contribution to the surface integral in (2.32) is

$$([\mathbf{n} \cdot \mathbf{Q}]_{EADH} + [\mathbf{n} \cdot \mathbf{Q}]_{FBCG}) dA = \left(\left[\mathbf{e}_1 \cdot \mathbf{Q}(\mathbf{x}) + \mathbf{e}_1 \cdot \frac{\Delta x_1}{2} \frac{\partial \mathbf{Q}(\mathbf{x})}{\partial x_1} + \dots \right] \right.$$

$$\left. + \left[-\mathbf{e}_1 \cdot \mathbf{Q}(\mathbf{x}) + \mathbf{e}_1 \cdot \frac{\Delta x_1}{2} \frac{\partial \mathbf{Q}(\mathbf{x})}{\partial x_1} + \dots \right] \right) \Delta x_2 \Delta x_3$$

$$= \left(\mathbf{e}_1 \cdot \frac{\partial \mathbf{Q}(\mathbf{x})}{\partial x_1} + \dots \right) \Delta x_1 \Delta x_2 \Delta x_3.$$

Similarly for the other two directions:

$$([\mathbf{n} \cdot \mathbf{Q}]_{ABCD} + [\mathbf{n} \cdot \mathbf{Q}]_{EFGH}) dA = \left(\mathbf{e}_2 \cdot \frac{\partial \mathbf{Q}(\mathbf{x})}{\partial x_2} + \dots \right) \Delta x_1 \Delta x_2 \Delta x_3$$

$$([\mathbf{n} \cdot \mathbf{Q}]_{ABFE} + [\mathbf{n} \cdot \mathbf{Q}]_{DCGH}) dA = \left(\mathbf{e}_3 \cdot \frac{\partial \mathbf{Q}(\mathbf{x})}{\partial x_3} + \dots \right) \Delta x_1 \Delta x_2 \Delta x_3$$

Assembling the contributions from all six faces (or all three directions) to evaluate (2.32) produces

$$\nabla \cdot \mathbf{Q} = \lim_{V \to 0} \frac{1}{V} \iint_A \mathbf{n} \cdot \mathbf{Q} dA$$

$$= \lim_{\substack{\Delta x_1 \to 0 \\ \Delta x_2 \to 0 \\ \Delta x_3 \to 0}} \frac{1}{\Delta x_1 \Delta x_2 \Delta x_3} \left(\mathbf{e}_1 \cdot \frac{\partial \mathbf{Q}(\mathbf{x})}{\partial x_1} + \mathbf{e}_2 \cdot \frac{\partial \mathbf{Q}(\mathbf{x})}{\partial x_2} + \mathbf{e}_3 \cdot \frac{\partial \mathbf{Q}(\mathbf{x})}{\partial x_3} + \dots \right) \Delta x_1 \Delta x_2 \Delta x_3,$$

and when the limit is taken, the expected Cartesian-coordinate form of the divergence emerges:

$$\nabla \cdot \mathbf{Q} = \mathbf{e}_1 \cdot \frac{\partial \mathbf{Q}(\mathbf{x})}{\partial x_1} + \mathbf{e}_2 \cdot \frac{\partial \mathbf{Q}(\mathbf{x})}{\partial x_2} + \mathbf{e}_3 \cdot \frac{\partial \mathbf{Q}(\mathbf{x})}{\partial x_3}.$$

2.13. STOKES' THEOREM

Stokes' theorem relates the integral over an open surface A to the line integral around the surface's bounding curve C. Here, unlike Gauss' theorem, the inside and outside of A are not well defined so an arbitrary choice must be made for the direction of the outward normal $\mathbf{n}$. Once this choice is made, the unit tangent vector to C, $\mathbf{t}$, points in the counterclockwise direction when looking at the outside of A; it is defined as $\mathbf{t} = \mathbf{n}_c \times \mathbf{n}$, where

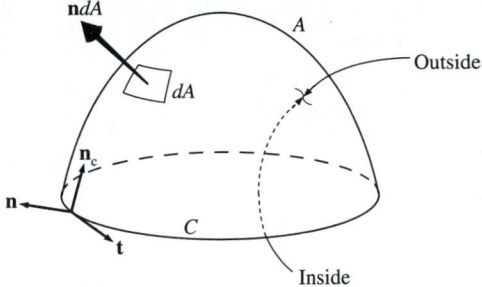

FIGURE 2.10 Illustration of Stokes' theorem for surface A bounded by the closed curve C. For the purposes of defining unit vectors, the *inside* and *outside* of A must be chosen, and one such choice is illustrated here. The unit vector $\mathbf{n}_c$ is perpendicular to C but is locally tangent to the surface A. The unit vector $\mathbf{n}$ is perpendicular to A and originates from the outside of A. The unit vector $\mathbf{t}$ is locally tangent to the curve C. The unit vectors $\mathbf{n}_c$, $\mathbf{n}$, and $\mathbf{t}$ define a right-handed triad of directions, $\mathbf{n}_c \times \mathbf{n} = \mathbf{t}$.

$\mathbf{n}_c$ is the unit normal to C that is locally tangent to A (Figure 2.10). For this geometry, Stokes' theorem states

$$\iint_A (\nabla \times \mathbf{u}) \cdot \mathbf{n} \, dA = \int_C \mathbf{u} \cdot \mathbf{t} \, ds, \tag{2.34}$$

where s is the arc length of the closed curve C. This theorem signifies that the surface integral of the curl of a vector field $\mathbf{u}$ is equal to the line integral of $\mathbf{u}$ along the bounding curve of the surface. In fluid mechanics, the right side of (2.34) is called the *circulation* of $\mathbf{u}$ about C. In addition, (2.34) can be used to define the curl of a vector through the limit of the circulation about an infinitesimal surface as

$$\mathbf{n} \cdot (\nabla \times \mathbf{u}) = \lim_{A \to 0} \frac{1}{A} \int_C \mathbf{u} \cdot \mathbf{t} \, ds. \tag{2.35}$$

The advantage of integral definitions of field derivatives is that such definitions do not depend on the coordinate system.

EXAMPLE 2.6

Obtain the recipe for the curl of a vector $\mathbf{u}(\mathbf{x})$ in Cartesian coordinates from the integral definition given by (2.35).

Solution

This is obtained by considering rectangular contours in three perpendicular planes intersecting at the point (x,y,z). First, consider the elemental rectangle in the $x = const.$ plane. The central point in this plane is (x,y,z) and the element's area is $\Delta y \Delta z$. It may be shown by careful integration of a Taylor expansion of the integrand that the integral along each line segment may be represented by the product of the integrand at the center of the segment and the length of the segment with attention paid to the direction of integration ds. Thus we obtain

$$(\nabla \times \mathbf{u})_x = \lim_{\Delta y \to 0} \lim_{\Delta z \to 0} \left\{ \begin{array}{l} \dfrac{1}{\Delta y \Delta z}\left[u_z\left(x, y + \dfrac{\Delta y}{2}, z\right) - u_z\left(x, y - \dfrac{\Delta y}{2}, z\right)\right]\Delta z \\[2ex] +\dfrac{1}{\Delta y \Delta z}\left[u_z\left(x, y, z - \dfrac{\Delta z}{2}\right) - u_z\left(x, y, z + \dfrac{\Delta z}{2}\right)\right]\Delta y \end{array} \right\}.$$

Taking the limits produces

$$(\nabla \times \mathbf{u})_x = \frac{\partial u_z}{\partial y} - \frac{\partial u_y}{\partial z}.$$

Similarly, integrating around the elemental rectangles in the other two planes leads to

$$(\nabla \times \mathbf{u})_y = \frac{\partial u_x}{\partial z} - \frac{\partial u_z}{\partial x}, \quad \text{and} \quad (\nabla \times \mathbf{u})_z = \frac{\partial u_y}{\partial x} - \frac{\partial u_x}{\partial y}.$$

2.14. COMMA NOTATION

Sometimes it is convenient to use an even more compact notation for partial derivatives

$$A,_i \equiv \partial A / \partial x_i, \tag{2.36}$$

where A is a tensor of any order. Here, the comma after the A indicates a spatial derivative in the direction of the following index or indices. Thus, as last illustrations of the implied-sum-over-repeated-index notation and additional examples of the comma notation, consider the divergence and curl of a vector $\mathbf{u}$ written in vector, ordinary, indicial, and comma notations:

$$\nabla \cdot \mathbf{u} = \sum_{i=1}^{3} \frac{\partial u_i}{\partial x_i} \equiv \frac{\partial u_i}{\partial x_i} \equiv u_{i,i} \quad \text{and} \quad (\nabla \times \mathbf{u})_i = \sum_{j=1}^{3}\sum_{k=1}^{3} \varepsilon_{ijk}\frac{\partial u_k}{\partial x_j} \equiv \varepsilon_{ijk}\frac{\partial u_k}{\partial x_j} \equiv \varepsilon_{ijk}\, u_{k,j}.$$

The comma notation has two advantages compared to the others. It is compact and allows all subscripts to be written on one line so that both indices of second-order tensors like $u_{i,j}$ are easily identified. Its disadvantages arise from its compactness. An imperfectly attentive reader may overlook a comma in a subscript listing. Plus, the comma must be written clearly in order to avoid confusion with other indices. The comma notation is adopted in Section 5.6 where the extent of the expressions is otherwise too cumbersome.

EXERCISES

2.1. For three spatial dimensions, rewrite the following expressions in index notation and evaluate or simplify them using the values or parameters given, and the definitions of δ_{ij} and ε_{ijk} wherever possible. In parts b) through e), $\mathbf{x}$ is the position vector, with components x_i.
   a) $\mathbf{b} \cdot \mathbf{c}$, where $\mathbf{b} = (1, 4, 17)$ and $\mathbf{c} = (-4, -3, 1)$.
   b) $(\mathbf{u} \cdot \nabla)\mathbf{x}$, where $\mathbf{u} =$ a vector with components u_i.

c) $\nabla\phi$, where $\phi = \mathbf{h} \cdot \mathbf{x}$ and $\mathbf{h}$ is a constant vector with components h_i.

d) $\nabla \times \mathbf{u}$, where $\mathbf{u} = \mathbf{\Omega} \times \mathbf{x}$ and $\mathbf{\Omega}$ is a constant vector with components Ω_i.

e) $\mathbf{C} \cdot \mathbf{x}$, where

$$\mathbf{C} = \left\{ \begin{array}{ccc} 1 & 2 & 3 \\ 0 & 1 & 2 \\ 0 & 0 & 1 \end{array} \right\}.$$

2.2. Starting from (2.1) and (2.3), prove (2.7).

2.3. Using Cartesian coordinates where the position vector is $\mathbf{x} = (x_1, x_2, x_3)$ and the fluid velocity is $\mathbf{u} = (u_1, u_2, u_3)$, write out the three components of the vector: $(\mathbf{u} \cdot \nabla)\mathbf{u} = u_i(\partial u_j/\partial x_i)$.

2.4. Convert $\nabla \times \nabla\rho$ to indicial notation and show that it is zero in Cartesian coordinates for any twice-differentiable scalar function ρ.

2.5. Using indicial notation, show that $\mathbf{a} \times (\mathbf{b} \times \mathbf{c}) = (\mathbf{a} \cdot \mathbf{c})\mathbf{b} - (\mathbf{a} \cdot \mathbf{b})\mathbf{c}$. [*Hint*: Call $\mathbf{d} \equiv \mathbf{b} \times \mathbf{c}$. Then $(\mathbf{a} \times \mathbf{d})_m = \varepsilon_{pqm}a_p d_q = \varepsilon_{pqm}a_p \varepsilon_{ijq}b_i c_j$. Using (2.19), show that $(\mathbf{a} \times \mathbf{d})_m = (\mathbf{a} \cdot \mathbf{c})b_m - (\mathbf{a} \cdot \mathbf{b})c_m$.]

2.6. Show that the condition for the vectors $\mathbf{a}$, $\mathbf{b}$, and $\mathbf{c}$ to be coplanar is $\varepsilon_{ijk}a_i b_j c_k = 0$.

2.7. Prove the following relationships: $\delta_{ij}\delta_{ij} = 3$, $\varepsilon_{pqr}\varepsilon_{pqr} = 6$, and $\varepsilon_{pqi}\varepsilon_{pqj} = 2\delta_{ij}$.

2.8. Show that $\mathbf{C} \cdot \mathbf{C}^T = \mathbf{C}^T \cdot \mathbf{C} = \delta$, where $\mathbf{C}$ is the direction cosine matrix and δ is the matrix of the Kronecker delta. Any matrix obeying such a relationship is called an *orthogonal matrix* because it represents transformation of one set of orthogonal axes into another.

2.9. Show that for a second-order tensor $\mathbf{A}$, the following quantities are invariant under the rotation of axes:

$$I_1 = A_{ii}$$
$$I_2 = \begin{vmatrix} A_{11} & A_{12} \\ A_{21} & A_{22} \end{vmatrix} + \begin{vmatrix} A_{22} & A_{23} \\ A_{32} & A_{33} \end{vmatrix} + \begin{vmatrix} A_{11} & A_{13} \\ A_{31} & A_{33} \end{vmatrix}$$
$$I_3 = \det(A_{ij}).$$

[*Hint*: Use the result of Exercise 2.8 and the transformation rule (2.12) to show that $I'_1 = A'_{ii} = A_{ii} = I_1$. Then show that $A_{ij}A_{ji}$ and $A_{ij}A_{jk}A_{ki}$ are also invariants. In fact, *all* contracted scalars of the form $A_{ij}A_{jk} \cdots A_{mi}$ are invariants. Finally, verify that

$$I_2 = \frac{1}{2}\left[I_1^2 - A_{ij}A_{ji}\right]$$

$$I_3 = \frac{1}{3}\left[A_{ij}A_{jk}A_{ki} - I_1 A_{ij}A_{ji} + I_2 A_{ii}\right].$$

Because the right-hand sides are invariant, so are I_2 and I_3.]

2.10. If $\mathbf{u}$ and $\mathbf{v}$ are vectors, show that the products $u_i v_j$ obey the transformation rule (2.12), and therefore represent a second-order tensor.

2.11. Show that δ_{ij} is an isotropic tensor. That is, show that $\delta'_{ij} = \delta_{ij}$ under rotation of the coordinate system. [*Hint*: Use the transformation rule (2.12) and the results of Exercise 2.8.]

2.12. If $\mathbf{u}$ and $\mathbf{v}$ are arbitrary vectors resolved in three-dimensional Cartesian coordinates, show that $\mathbf{u} \cdot \mathbf{v} = 0$ when $\mathbf{u}$ and $\mathbf{v}$ are perpendicular.

2.13. If $\mathbf{u}$ and $\mathbf{v}$ are vectors with magnitudes u and v, use the finding of Exercise 2.12 to show that $\mathbf{u} \cdot \mathbf{v} = uv\cos\theta$ where θ is the angle between $\mathbf{u}$ and $\mathbf{v}$.

2.14. Determine the components of the vector $\mathbf{w}$ in three-dimensional Cartesian coordinates when $\mathbf{w}$ is defined by: $\mathbf{u} \cdot \mathbf{w} = 0$, $\mathbf{v} \cdot \mathbf{w} = 0$, and $\mathbf{w} \cdot \mathbf{w} = u^2 v^2 \sin^2\theta$, where $\mathbf{u}$ and $\mathbf{v}$ are known vectors with components u_i and v_i and magnitudes u and v, respectively, and θ is the angle between $\mathbf{u}$ and $\mathbf{v}$. Choose the sign(s) of the components of $\mathbf{w}$ so that $\mathbf{w} = \mathbf{e}_3$ when $\mathbf{u} = \mathbf{e}_1$ and $\mathbf{v} = \mathbf{e}_2$.

2.15. If a is a positive constant and $\mathbf{b}$ is a constant vector, determine the divergence and the curl of $\mathbf{u} = a\mathbf{x}/x^3$ and $\mathbf{u} = \mathbf{b} \times (\mathbf{x}/x^2)$ where $x = \sqrt{x_1^2 + x_2^2 + x_3^2} \equiv \sqrt{x_i x_i}$ is the length of $\mathbf{x}$.

2.16. Obtain the recipe for the gradient of a scalar function in cylindrical polar coordinates from the integral definition (2.32).

2.17. Obtain the recipe for the divergence of a vector function in cylindrical polar coordinates from the integral definition (2.32).

2.18. Obtain the recipe for the divergence of a vector function in spherical polar coordinates from the integral definition (2.32).

2.19. Use the vector integral theorems to prove that $\nabla \cdot (\nabla \times \mathbf{u}) = 0$ for any twice-differentiable vector function $\mathbf{u}$ regardless of the coordinate system.

2.20. Use Stokes' theorem to prove that $\nabla \times (\nabla\phi) = 0$ for any single-valued twice-differentiable scalar ϕ regardless of the coordinate system.

Literature Cited

Sommerfeld, A. (1964). *Mechanics of Deformable Bodies*. New York: Academic Press. (Chapter 1 contains brief but useful coverage of Cartesian tensors.)

Supplemental Reading

Aris, R. (1962). *Vectors, Tensors, and the Basic Equations of Fluid Mechanics*. Englewood Cliffs, NJ: Prentice-Hall. (This book gives a clear and easy treatment of tensors in Cartesian and non-Cartesian coordinates, with applications to fluid mechanics.)

Prager, W. (1961). *Introduction to Mechanics of Continua*. New York: Dover Publications. (Chapters 1 and 2 contain brief but useful coverage of Cartesian tensors.)

O U T L I N E

3.1. Introduction and Coordinate Systems   65

3.2. Particle and Field Descriptions of Fluid Motion   67

3.3. Flow Lines, Fluid Acceleration, and Galilean Transformation   71

3.4. Strain and Rotation Rates   76

3.5. Kinematics of Simple Plane Flows   82

3.6. Reynolds Transport Theorem   85

Exercises   89

Literature Cited   93

Supplemental Reading   93

CHAPTER OBJECTIVES

- To review the basic Cartesian and curvilinear coordinates systems
- To link fluid flow kinematics with the particle kinematics
- To define the various flow lines in unsteady fluid velocity fields

- To present fluid acceleration in the Eulerian flow-field formulation
- To establish the fundamental meaning of the strain rate and rotation tensors
- To present the means for time differentiating general three-dimensional volume integrations

3.1. INTRODUCTION AND COORDINATE SYSTEMS

Kinematics is the study of motion without reference to the forces or stresses that produce the motion. In this chapter, fluid kinematics is presented in two and three dimensions starting with simple fluid-particle-path concepts and then proceeding to topics of greater complexity.

These include: particle- and field-based descriptions for the time-dependent position, velocity, and acceleration of fluid particles; the relationship between the fluid velocity gradient tensor and the deformation and rotation of fluid elements; and the general mathematical relationships that govern arbitrary volumes that move and deform within flow fields. The forces and stresses that cause fluid motion are considered in subsequent chapters covering the *dynamics* or *kinetics* of the fluid motion.

In general, three independent spatial dimensions and time are needed to fully describe fluid motion. When a flow does not depend on time, it is called *steady*; when it does depend on time it is called *unsteady*. In addition, fluid motion is studied in fewer than three dimensions whenever possible because the necessary analysis is usually simpler and relevant phenomena are more easily understood and visualized.

A truly *one-dimensional flow* is one in which the flow's characteristics can be entirely described with one independent spatial variable. Few real flows are strictly one dimensional, although flows in long, straight constant-cross-section conduits come close. Here, the independent coordinate may be aligned with the flow direction, as in the case of low-frequency pulsations in a pipe as shown in Figure 3.1a, where z is the independent coordinate and darker gray indicates higher gas density. Alternatively, the independent coordinate may be aligned in the cross-stream direction, as in the case of viscous flow in a round tube where the radial distance, R, from the tube's centerline is the independent coordinate (Figure 3.1b). In addition, higher dimensional flows are sometimes analyzed in one dimension by averaging the properties of the higher dimensional flow over an appropriate distance or area (Figure 3.1c and d).

A *two-dimensional,* or *plane,* flow is one in which the variation of flow characteristics can be described by two spatial coordinates. The flow of an ideal fluid past a circular cylinder of infinite length having its axis perpendicular to the primary flow direction is an example of a plane flow (see Figure 3.2a). (Here we should note that the word *cylinder* may also be used in this context for any body having a cross-sectional shape that is invariant along its length even if this shape is not circular.) This definition of two-dimensional flow officially includes the flow around bodies of revolution where flow characteristics are identical in any plane that contains the body's axis (see Figure 6.27). However, such flows are customarily called *three-dimensional axisymmetric flows.*

A *three-dimensional* flow is one that can only be properly described with three independent spatial coordinates and is the most general case considered in this text. Sometimes curvilinear coordinates that match flow-field boundaries or symmetries simplify the analysis and description of flow fields. Thus, several different coordinate systems are used in this text (see Figure 3.3). Two-dimensional (plane) Cartesian and polar coordinates for an arbitrary point P (Figure 3.3a) may be denoted by the coordinate pairs (x, y), (x_1, x_2), or (r, θ) with the corresponding velocity components (u, v), (u_1, u_2), or (u_r, u_θ). In three dimensions, Cartesian coordinates (Figures 2.1 and 3.3b) may be used to locate a point P via the coordinate triplets (x, y, z) or (x_1, x_2, x_3) with corresponding velocity components (u, v, w) or (u_1, u_2, u_3). Cylindrical polar coordinates for P (Figure 3.3c) are denoted by (R, φ, z) with corresponding velocity components (u_R, u_φ, u_z). In addition, they will occasionally be denoted (r, θ, z) or (r, θ, x). Spherical polar coordinates for P (Figure 3.3d) are denoted by (r, θ, φ) with the corresponding velocity components $(u_r, u_\theta, u_\varphi)$. In all cases, unit vectors are denoted by **e** with an appropriate

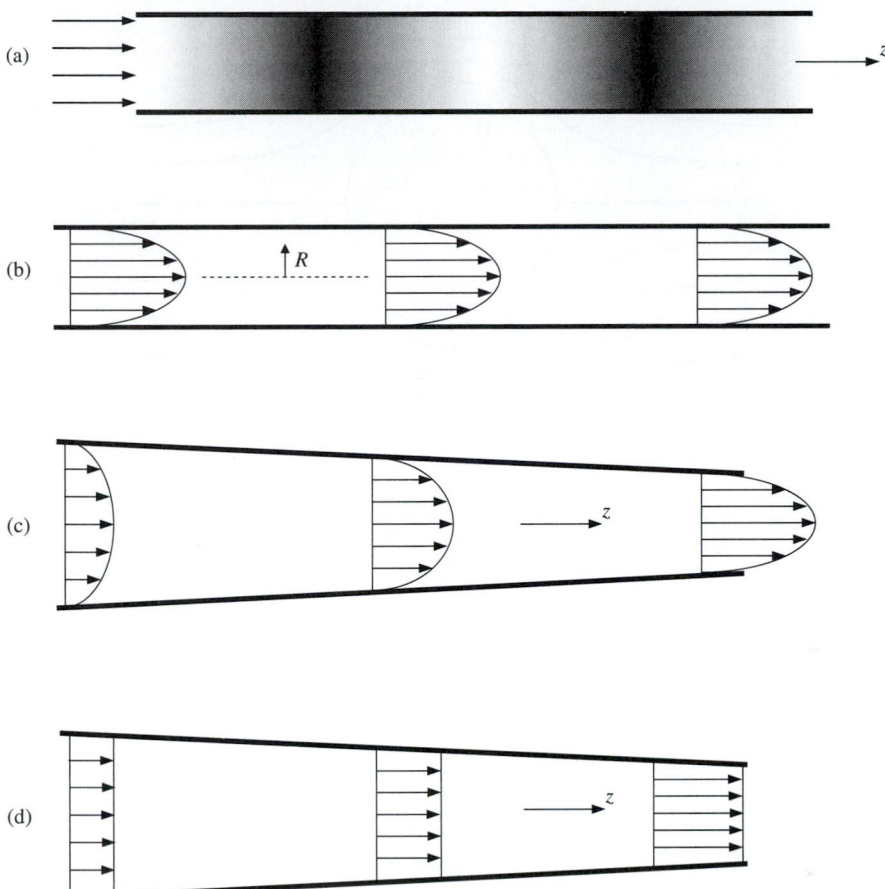

FIGURE 3.1 (a) Example of a one-dimensional fluid flow in which the gas density, shown by the grayscale, varies in the stream-wise z direction but not in the cross-stream direction. (b) Example of a one-dimensional fluid flow in which the fluid velocity varies in the cross-stream R direction but not in the stream-wise direction. (c) Example of a two-dimensional fluid flow where the fluid velocity varies in the cross-stream and stream-wise directions. (d) The one-dimensional approximation to the flow show in part (c). Here the approximate flow field varies only in the stream-wise z direction.

subscript as in (2.1) and Figure 2.1. More information about these coordinate systems is provided in Appendix B.

3.2. PARTICLE AND FIELD DESCRIPTIONS OF FLUID MOTION

There are two ways to describe fluid motion. In the *Lagrangian* description, fluid particles are followed as they move through a flow field. In the *Eulerian* description, a flow field's characteristics are monitored at fixed locations or in stationary regions of space. In fluid

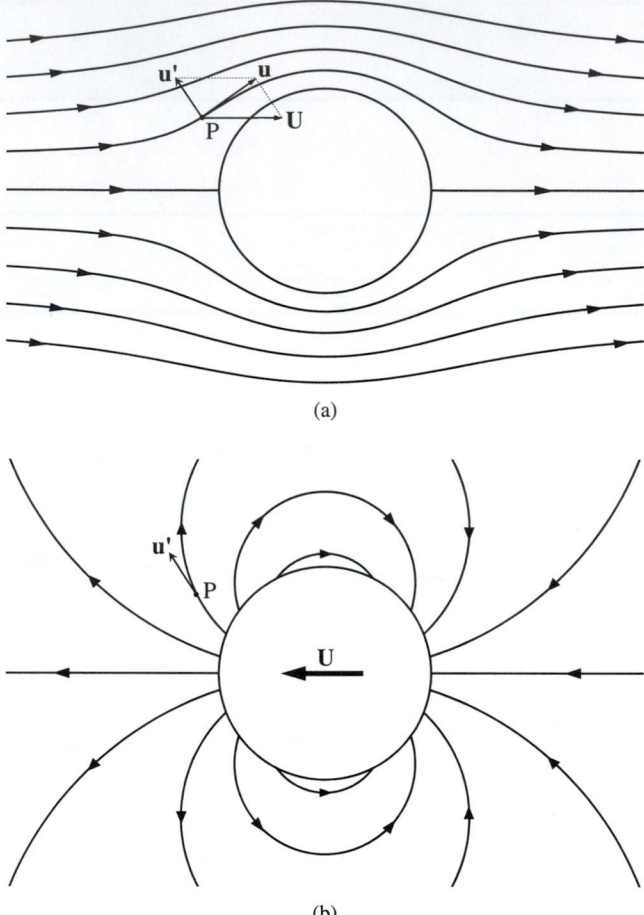

FIGURE 3.2   Sample flow fields where two spatial coordinates are needed. (a) Steady flow of an ideal incompressible fluid past a long stationary circular cylinder with its axis perpendicular to the flow. Here the total fluid velocity **u** at point P can be considered a sum of the flow velocity far from the cylinder **U**, and a velocity component **u′** caused by the presence of the cylinder. (b) Unsteady flow of a nominally quiescent ideal incompressible fluid around a moving long circular cylinder with its axis perpendicular to the page. Here the cylinder velocity **U** is shown inside the cylinder, and the fluid velocity **u′** at point P is caused by the presence of the moving cylinder alone. Although the two fields look very different, they only differ by a Galilean transformation. The streamlines in (a) can be changed to those in (b) by switching to a frame of reference where the fluid far from the cylinder is motionless.

mechanics, an understanding of both descriptions is necessary because the acceleration following a fluid particle is needed for the application of Newton's second law to fluid motion while observations, measurements, and simulations of fluid flows are commonly made at fixed locations or in stationary spatial regions with the fluid moving past the locations or through the regions of interest.

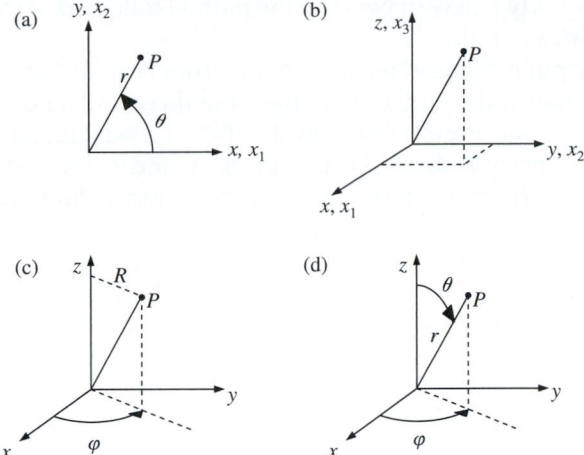

FIGURE 3.3 Coordinate systems commonly used in this text. In each case P is an arbitrary point away from the origin. (a) Plane Cartesian or polar coordinates where P is located by the coordinate pairs (x, y), (x_1, x_2), or (r, θ). (b) Three-dimensional Cartesian coordinates where P is located by the coordinate triplets (x, y, z) or (x_1, x_2, x_3). (c) Cylindrical polar coordinates where P is located by the coordinate triplet (R, φ, z). (d) Spherical polar coordinates where P is located by the coordinate triplet (r, θ, φ).

The Lagrangian description is based on the motion of fluid particles. It is the direct extension of single particle kinematics (e.g., see Meriam & Kraige, 2007) to a whole field of fluid particles that are labeled by their location, $\mathbf{r}_o$, at a reference time, $t = t_o$. The subsequent position $\mathbf{r}$ of each fluid particle as a function of time, $\mathbf{r}(t;\mathbf{r}_o,t_o)$, specifies the flow field. Here, $\mathbf{r}_o$ and t_o are boundary or initial condition parameters that label fluid particles, and are not independent variables. Thus, the current velocity $\mathbf{u}$ and acceleration $\mathbf{a}$ of the fluid particle that was located at $\mathbf{r}_o$ at time t_o are obtained from the first and second temporal derivatives of particle position $\mathbf{r}(t;\mathbf{r}_o,t_o)$:

$$\mathbf{u} = d\mathbf{r}(t;\mathbf{r}_o, t_o)/dt \quad \text{and} \quad \mathbf{a} = d^2\mathbf{r}(t;\mathbf{r}_o, t_o)/dt^2. \tag{3.1}$$

These values for $\mathbf{u}$ and $\mathbf{a}$ are valid for the fluid particle as it moves along its trajectory through the flow field (Figure 3.4). In this particle-based Lagrangian description of fluid motion, fluid particle kinematics are identical to that in ordinary particle mechanics, and any scalar, vector,

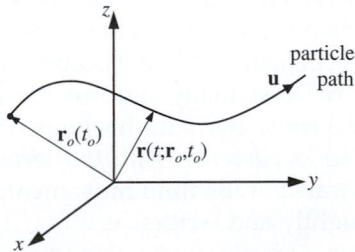

FIGURE 3.4 Lagrangian description of the motion of a fluid particle that started at location $\mathbf{r}_o$ at time t_o. The particle path or particle trajectory $\mathbf{r}(t;\mathbf{r}_o,t_o)$ specifies the location of the fluid particle at later times.

or tensor flow-field property F may depend on the path(s) followed of the relevant fluid particle(s) and time: $F = F[\mathbf{r}(t;\mathbf{r}_o,t_o), t]$.

The Eulerian description focuses on flow field properties at locations or in regions of interest, and involves four independent variables: the three spatial coordinates represented by the position vector $\mathbf{x}$, and time t. Thus, in this field-based Eulerian description of fluid motion, a flow-field property F depends directly on $\mathbf{x}$ and t: $F = F(\mathbf{x}, t)$. Even though this description complicates the calculation of $\mathbf{a}$, because individual fluid particles are not followed, it is the favored description of fluid motion.

Kinematic relationships between the two descriptions can be determined by requiring equality of flow-field properties when $\mathbf{r}$ and $\mathbf{x}$ define the same point in space, both are resolved in the same coordinate system, and a common clock is used to determine the time t:

$$F[\mathbf{r}(t;\mathbf{r}_o,t_o), t] = F(\mathbf{x},t) \quad \text{when} \quad \mathbf{x} = \mathbf{r}(t;\mathbf{r}_o,t_o). \tag{3.2}$$

Here the second equation specifies the trajectory followed by a fluid particle. This compatibility requirement forms the basis for determining and interpreting time derivatives in the Eulerian description of fluid motion. Applying a total time derivative to the first equation in (3.2) produces

$$\frac{d}{dt}F[\mathbf{r}(t;\mathbf{r}_o,t_o), t] = \frac{\partial F}{\partial r_1}\frac{dr_1}{dt} + \frac{\partial F}{\partial r_2}\frac{dr_2}{dt} + \frac{\partial F}{\partial r_3}\frac{dr_3}{dt} + \frac{\partial F}{\partial t} = \frac{d}{dt}F(\mathbf{x},t) \quad \text{when} \quad \mathbf{x} = \mathbf{r}(t;\mathbf{r}_o,t_o), \tag{3.3}$$

where the components of $\mathbf{r}$ are r_i. In (3.3), the time derivatives of r_i are the components u_i of the fluid particle's velocity $\mathbf{u}$ from (3.1). In addition, $\partial F/\partial r_i = \partial F/\partial x_i$ when $\mathbf{x} = \mathbf{r}$, so (3.3) becomes

$$\frac{d}{dt}F[\mathbf{r}(t;\mathbf{r}_o,t_o), t] = \frac{\partial F}{\partial x_1}u_1 + \frac{\partial F}{\partial x_2}u_2 + \frac{\partial F}{\partial x_3}u_3 + \frac{\partial F}{\partial t} = (\nabla F) \cdot \mathbf{u} + \frac{\partial F}{\partial t} \equiv \frac{D}{Dt}F(\mathbf{x},t), \tag{3.4}$$

where the final equality defines D/Dt as the total time derivative in the Eulerian description of fluid motion. It is the equivalent of the total time derivative d/dt in the Lagrangian description and is known as the *material derivative*, *substantial derivative*, or *particle derivative*, where the final attribution emphasizes the fact that it provides time derivative information following a fluid particle.

The material derivative D/Dt defined in (3.4) is composed of unsteady and advective acceleration terms. (1) The *unsteady* part of DF/Dt, $\partial F/\partial t$, is the *local* temporal rate of change of F at the location $\mathbf{x}$. It is zero when F is independent of time. (2) The *advective* (or *convective*) part of DF/Dt, $\mathbf{u} \cdot \nabla F$, is the rate of change of F that occurs as fluid particles move from one location to another. It is zero where F is spatially uniform, the fluid is not moving, or $\mathbf{u}$ and ∇F are perpendicular. For clarity and consistency in this book, the movement of fluid particles from place to place is referred to as *advection* with the term *convection* being reserved for the special circumstance of heat transport by fluid movement. In vector and index notations, (3.4) is commonly rearranged slightly and written as

$$\frac{DF}{Dt} \equiv \frac{\partial F}{\partial t} + \mathbf{u} \cdot \nabla F, \quad \text{or} \quad \frac{DF}{Dt} \equiv \frac{\partial F}{\partial t} + u_i\frac{\partial F}{\partial x_i}. \tag{3.5}$$

The scalar product $\mathbf{u} \cdot \nabla F$ is the magnitude of $\mathbf{u}$ times the component of ∇F in the direction of $\mathbf{u}$ so (3.5) can then be written in scalar notation as

$$\frac{DF}{Dt} \equiv \frac{\partial F}{\partial t} + |\mathbf{u}|\frac{\partial}{\partial s},$$

(3.6)

where s is a path-length coordinate on the fluid particle trajectory $\mathbf{x} = \mathbf{r}(t; \mathbf{r}_o, t_o)$, that is, $d\mathbf{r} = \mathbf{e_u}ds$ with $\mathbf{e_u} = \mathbf{u}/|\mathbf{u}|$.

3.3. FLOW LINES, FLUID ACCELERATION, AND GALILEAN TRANSFORMATION

In the Eulerian description, three types of curves are commonly used to describe fluid motion—streamlines, path lines, and streak lines. These are defined and described here assuming that the fluid velocity vector, $\mathbf{u}$, is known at every point of space and instant of time throughout the region of interest. Streamlines, path lines, and streak lines all coincide when the flow is steady. These curves are often valuable for understanding fluid motion and form the basis for experimental techniques that track seed particles or dye filaments. Pictorial and photographic examples of flow lines can be found in specialty volumes devoted to flow visualization (Van Dyke, 1982; Samimy et al., 2003).

A *streamline* is a curve that is instantaneously tangent to the fluid velocity throughout the flow field. In unsteady flows the streamline pattern changes with time. In Cartesian coordinates, if $d\mathbf{s} = (dx, dy, dz)$ is an element of arc length along a streamline (Figure 3.5) and $\mathbf{u} = (u, v, w)$ is the local fluid velocity vector, then the tangency requirement on $d\mathbf{s}$ and $\mathbf{u}$ leads to

$$dx/u = dy/v = dz/w$$

(3.7)

(see Exercise 3.3), and $\mathbf{u} \times d\mathbf{s} = 0$ because $d\mathbf{s}$ and $\mathbf{u}$ are locally parallel. Integrating (3.7) in both the upstream and downstream directions from a variety of reference locations allows streamlines to be determined throughout the flow field. If these reference

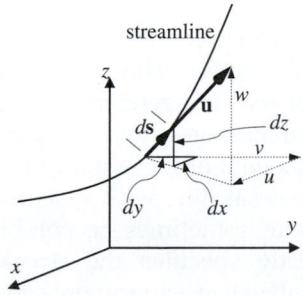

FIGURE 3.5 Streamline geometry. The arc-length element of a streamline, $d\mathbf{s}$, is locally tangent to the fluid velocity $\mathbf{u}$ so its components and the components of the velocity must follow (3.7).

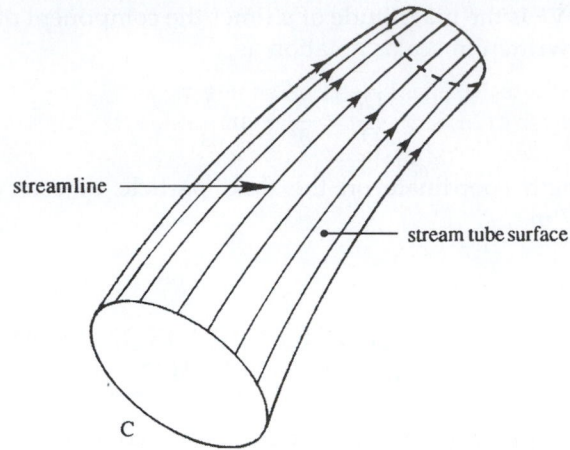

FIGURE 3.6 Stream tube geometry for the closed curve C.

locations lie on a closed curve C, the resulting stream surface is called a *stream tube* (Figure 3.6). No fluid crosses a stream tube's surface because the fluid velocity vector is everywhere tangent to it. Streamlines are useful in the depiction of flow fields and important for calculations involving simplifications (Bernoulli equations) of the full equations of fluid motion. In experiments, streamlines may be visualized by particle streak photography or by integrating (3.7) using measured velocity fields.

A *path line* is the trajectory of a fluid particle of fixed identity. It is defined in (3.2) and (3.3) as $\mathbf{x} = \mathbf{r}(t;\mathbf{r}_o,t_o)$. The equation of the path line for the fluid particle launched from $\mathbf{r}_o$ at t_o is obtained from the fluid velocity $\mathbf{u}$ by integrating

$$d\mathbf{r}/dt = [\mathbf{u}(\mathbf{x},t)]_{\mathbf{x}=\mathbf{r}} = \mathbf{u}(\mathbf{r},t) \qquad (3.8)$$

subject to the requirement $\mathbf{r}(t_o) = \mathbf{r}_o$. Other path lines are obtained by integrating (3.8) from different values of $\mathbf{r}_o$ or t_o. A discretized version of (3.8) is the basis for particle image velocimetry (PIV), a popular and powerful flow field measurement technique (Raffel et al., 1998).

A *streak line* is the curve obtained by connecting all the fluid particles that will pass or have passed through a fixed point in space. The streak line through the point $\mathbf{x}_o$ at time t is found by integrating (3.8) for all relevant reference times, t_o, subject to the requirement $\mathbf{r}(t_o) = \mathbf{x}_o$. When completed, this integration provides a path line, $\mathbf{x} = \mathbf{r}(t;\mathbf{x}_o,t_o)$, for each value of t_o. At a fixed time t, the components of these path-line equations, $x_i = r_i(t;\mathbf{x}_o,t_o)$, provide a parametric specification of the streak line with t_o as the parameter. Alternatively, these path-line component equations can sometimes be combined to eliminate t_o and thereby produce an equation that directly specifies the streak line through the point $\mathbf{x}_o$ at time t. Streak lines may be visualized in experiments by injecting a passive marker, like dye or smoke, from a small port and observing were it goes as it is carried through the flow field by the moving fluid.

EXAMPLE 3.1

In two-dimensional Cartesian coordinates, determine the streamline, path line, and streak line that pass through the origin of coordinates at $t = t'$ in the unsteady near-surface flow field typical of long-wavelength water waves with amplitude ξ_o: $u = \omega\xi_o\cos(\omega t)$ and $v = \omega\xi_o\sin(\omega t)$.

Streamline Solution

Utilize the first equality in (3.7) to find:

$$\frac{dy}{dx} = \frac{v}{u} = \frac{\omega\xi_o \sin(\omega t')}{\omega\xi_o \cos(\omega t')} = \tan\left(\omega t'\right).$$

Integrating once produces: $y = x\tan(\omega t') + const$. For the streamline to pass through the origin $(x = y = 0)$, the constant must equal zero, so the streamline equation is: $y = x\tan(\omega t')$.

Path-line Solution

Set $\mathbf{r} = [(x(t), y(t)]$, and use both components of (3.8) to find:

$$dx/dt = u = \omega\xi_o \cos(\omega t), \quad \text{and} \quad dy/dt = v = \omega\xi_o \sin(\omega t).$$

Integrate each of these equations once to find: $x = \xi_o \sin(\omega t) + x_o$, and $y = -\xi_o \cos(\omega t) + y_o$, where x_o and y_o are integration constants. The path-line requirement at $x = y = 0$ and $t = t'$ implies $x_o = -\xi_o \sin(\omega t')$, and $y_o = \xi_o \cos(\omega t')$, so the path-line component equations are:

$$x = \xi_o[\sin(\omega t) - \sin(\omega t')] \quad \text{and} \quad y = \xi_o[-\cos(\omega t) + \cos(\omega t')].$$

Here, the time variable t can be eliminated via a little algebra to find

$$\left(x + \xi_o \sin(\omega t')\right)^2 + \left(y - \xi_o \cos(\omega t')\right)^2 = \xi_o^2,$$

which is the equation of a circle of radius ξ_o centered on the location $[-\xi_o \sin(\omega t'), \xi_o \cos(\omega t')]$.

Streak-line Solution

To determine the streak line that passes through the origin of coordinates at $t = t'$, the location of the fluid particle that passed through $x = y = 0$ at $t = t_o$ must be found. Use the path-line results above but evaluate at t_o instead of t' to find different constants. Thus the parametric streak-line component equations are:

$$x = \xi_o[\sin(\omega t) - \sin(\omega t_o)] \quad \text{and} \quad y = \xi_o[-\cos(\omega t) + \cos(\omega t_o)].$$

Combine these equations to eliminate t_o and evaluate the result at $t = t'$ to find the required streak line:

$$\left(x - \xi_o \sin(\omega t')\right)^2 + \left(y + \xi_o \cos(\omega t')\right)^2 = \xi_o^2.$$

This is the equation of a circle of radius ξ_o centered on the location $[\xi_o \sin(\omega t'), -\xi_o \cos(\omega t')]$. The three flow lines in this example are shown in Figure 3.7. In this case, the streamline, path line, and streak line are all tangent to each other at the origin of coordinates.

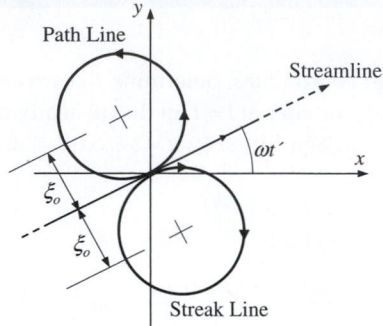

FIGURE 3.7 Streamline, path line, and streak line for Example 3.1. All three are distinct because the flow field is unsteady.

From this example it should be clear that streamlines, path lines, and streak lines differ in an unsteady flow field. This situation is also illustrated in Figure 3.2, which shows streamlines when there is relative motion of a circular cylinder and an ideal fluid. Figure 3.2a shows streamlines for a stationary cylinder with the fluid moving past it, a steady flow. Here, fluid particles that approach the cylinder are forced to move up or down to go around it. Figure 3.2b shows streamlines for a moving cylinder in a nominally quiescent fluid, an unsteady flow. Here, streamlines originate on the left side of the advancing cylinder where fluid particles are pushed to the left to make room for the cylinder. These streamlines curve backward and fluid particles move rightward at the cylinder's widest point. These streamlines terminate on the right side of the cylinder where fluid particles again move to the left to fill in the region behind the moving cylinder. Although their streamline patterns appear dissimilar, these flow fields only differ by a Galilean transformation. Consider the fluid velocity at a point P that lies at the same location relative to the cylinder in both fields. If $\mathbf{u}'$ is the fluid velocity at P in Figure 3.2b where the cylinder is moving at speed $\mathbf{U}$, then the fluid velocity $\mathbf{u}$ at P in Figure 3.2a is $\mathbf{u} = \mathbf{U} + \mathbf{u}'$. If $\mathbf{U}$ is constant, the fluid acceleration in both fields must be the same at the same location relative to the cylinder.

This expectation can be verified in general using (3.5) with F replaced by the fluid velocity observed in different coordinate frames. Consider a Cartesian coordinate system $O'x'y'z'$ that moves at a constant velocity $\mathbf{U}$ with respect to a stationary system $Oxyz$ having parallel axes (Figure 3.8). The fluid velocity $\mathbf{u}'(\mathbf{x}', t')$ observed in $O'x'y'z'$ will be related to the fluid velocity $\mathbf{u}(\mathbf{x},t)$ observed in $Oxyz$ by $\mathbf{u}(\mathbf{x},t) = \mathbf{U} + \mathbf{u}'(\mathbf{x}', t')$ when $t = t'$ and $\mathbf{x} = \mathbf{x}' + \mathbf{U}t + \mathbf{x}'_0$, where $\mathbf{x}'_0$ is the vector distance from O to O' at $t = 0$. Under these conditions it can be shown that

$$\frac{\partial \mathbf{u}}{\partial t} + (\mathbf{u} \cdot \nabla)\mathbf{u} = \left(\frac{D\mathbf{u}}{Dt}\right)_{in\ Oxyz} = \left(\frac{D\mathbf{u}'}{Dt'}\right)_{in\ O'x'y'z'} = \frac{\partial \mathbf{u}'}{\partial t'} + (\mathbf{u}' \cdot \nabla')\mathbf{u}' \qquad (3.9)$$

(Exercise 3.12) where ∇' operates on the primed coordinates. The first and second terms of the leftmost part of (3.9) are the unsteady and advective acceleration terms in $Oxyz$. The unsteady acceleration term, $\partial \mathbf{u}/\partial t$, is nonzero at $\mathbf{x}$ when $\mathbf{u}$ varies with time at $\mathbf{x}$. It is zero everywhere when the flow is steady. The advective acceleration term, $(\mathbf{u} \cdot \nabla)\mathbf{u}$, is nonzero when fluid particles move between locations where the fluid velocity is different. It is zero when the fluid

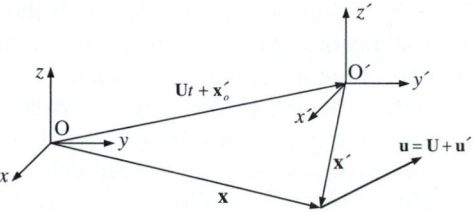

FIGURE 3.8 Geometry for showing that the fluid particle acceleration as determined by (3.9) is independent of the frame of reference when the frames differ by a Galilean transformation. Here $Oxyz$ is stationary and $O'x'y'z'$ moves with respect to it at a constant speed $\mathbf{U}$, the axes of the two frames are parallel, and $\mathbf{x}$ and $\mathbf{x}'$ represent the same location. The fluid velocity observed at $\mathbf{x}$ in frame $Oxyz$ is $\mathbf{u}$. The fluid velocity observed at $\mathbf{x}'$ in frame $O'x'y'z'$ is $\mathbf{u}'$.

velocity is zero, the fluid velocity is uniform in space, or when the fluid velocity only varies in the cross-stream direction. In addition, the unsteady term is linear in $\mathbf{u}$ while the advective term is nonlinear (quadratic) in $\mathbf{u}$. This nonlinearity is a primary feature of fluid mechanics. When $\mathbf{u}$ is small enough for this nonlinearity to be ignored, fluid mechanics reduces to acoustics or, when $\mathbf{u} = 0$, to fluid statics.

When examined together, the sample flow fields in Figure 3.2 and the Galilean invariance of the Eulerian fluid acceleration, (3.9), show that the relative importance of the steady and advective fluid-acceleration terms depends on the frame of reference of the observer. Figure 3.2a depicts a steady flow where the streamlines do not depend on time. Thus, the unsteady acceleration term, $\partial \mathbf{u}/\partial t$, is zero. However, the streamlines do bend in the vicinity of the cylinder so fluid particles must feel some acceleration because the absence of fluid-particle acceleration in a flow field corresponds to constant fluid-particle velocity and straight streamlines. Therefore, the advective acceleration term, $(\mathbf{u} \cdot \nabla)\mathbf{u}$, is nonzero for the flow in Figure 3.2a. In Figure 3.2b, the flow is unsteady and the streamlines are curved, so both acceleration terms in the rightmost part of (3.9) are nonzero. These observations imply that a Galilean transformation can alter the relative importance of the unsteady and advective fluid acceleration terms without changing the overall fluid-particle acceleration. Thus, an astutely chosen, steadily moving coordinate system can be used to enhance (or reduce) the relative importance of either the unsteady or advective fluid-acceleration term.

Additional insights into the character of the unsteady and advective acceleration terms might also be obtained from the reader's observations and experiences. For example, a nonzero unsteady acceleration is readily observed at any street intersection regulated by a traffic light with the moving or stationary vehicles taking the place of fluid particles. Here, a change in the traffic light may halt east-west vehicle flow and allow north-south vehicle flow to begin, thereby producing a time-dependent 90° rotation of the traffic-flow streamlines at the intersection location. Similarly, a nonzero advective acceleration is readily observed or experienced by roller-coaster riders when an analogy is made between the roller-coaster track and a streamline. While stationary and waiting in line, soon-to-be roller-coaster riders can observe that the track's shape involves hills, curves, and bends, and that this shape does not depend on time. This situation is analogous to the stationary observer of a nontrivial steady fluid flow—like that depicted in Figure 3.2a—who readily notes that streamlines curve and bend but do not depend on time. Thus, the unsteady acceleration term is zero for both the roller coaster and a steady flow because both the roller-coaster cars and fluid particles travel through space on fixed-shape trajectories and achieve consistent (time-independent) velocities at any point along the track or streamline. However, anyone who has ever ridden a roller coaster will know that significant acceleration is possible while following a roller-coaster's fixed-shape track because a roller-coaster car's velocity varies as it traverses the track. These velocity variations result from the advective acceleration, and fluid particles that follow curved fixed-shape streamlines experience it as well. Within this roller coaster-streamline analogy a nonzero unsteady acceleration would correspond to roller-coaster cars and fluid particles following time-dependent paths. Such a possibility is certainly unusual for roller-coaster riders; roller-coaster tracks are nearly rigid, seldom fall down (thankfully), and are typically designed to produce consistent car velocities at each point along the track.

3.4. STRAIN AND ROTATION RATES

Given the definition of a fluid as a material that deforms continuously under the action of a shear stress, the basic constitutive law for fluids relates fluid element *deformation rates* to the stresses (surface forces per unit area) applied to a fluid element. This section describes fluid-element deformation and rotation rates in terms of the fluid *velocity gradient tensor*, $\partial u_i / \partial x_j$. The constitutive law for Newtonian fluids is covered in the next chapter. The various illustrations and interpretations provided here are analogous to their counterparts in solid mechanics when the fluid-appropriate *strain rate* (based on velocity **u**) is replaced by the solid-appropriate *strain* (based on displacement **u**).

The relative motion between two neighboring points can be written as the sum of the motion due to local rotation and deformation. Consider the situation depicted in Figure 3.9, and let **u**(**x**,*t*) be the velocity at point O (position vector **x**), and let **u** + *d***u** be the velocity at the same time at a nearby neighboring point P (position vector **x** + *d***x**). A three-dimensional first-order Taylor expansion of **u** about **x** leads to the following relationship between the components of *d***u** and *d***x**:

$$du_i = (\partial u_i / \partial x_j) dx_j. \tag{3.10}$$

The term in parentheses in (3.10), $\partial u_i / \partial x_j$, is the velocity gradient tensor, and it can be decomposed into symmetric, S_{ij}, and antisymmetric, R_{ij}, tensors:

$$\frac{\partial u_i}{\partial x_j} = S_{ij} + \frac{1}{2}R_{ij}, \quad \text{where} \quad S_{ij} = \frac{1}{2}\left(\frac{\partial u_i}{\partial x_j} + \frac{\partial u_j}{\partial x_i}\right), \quad \text{and} \quad R_{ij} = \frac{\partial u_i}{\partial x_j} - \frac{\partial u_j}{\partial x_i}. \tag{3.11, 3.12, 3.13}$$

Here, S_{ij} is the *strain rate tensor*, and R_{ij} is the *rotation tensor*. The decomposition of $\partial u_i / \partial x_j$ provided by (3.11) is important when formulating the conservation equations for fluid motion because S_{ij}, which embodies fluid element deformation, is related to the stress field in a moving fluid while R_{ij}, which embodies fluid element rotation, is not.

The strain rate tensor has on- and off-diagonal terms. The diagonal terms of S_{ij} represent elongation and contraction per unit length in the various coordinate directions, and are sometimes called *linear* strain rates. A geometrical interpretation of S_{ij}'s first component, S_{11}, is

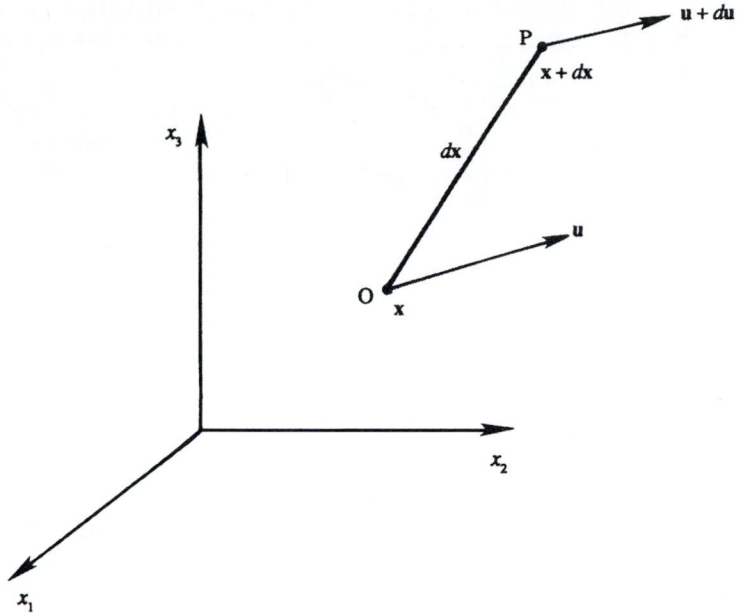

FIGURE 3.9 Velocity vectors **u** and **u** + *d***u** at two neighboring points O and P, respectively, that are separated by the short distance *d***x**.

provided in Figure 3.10. The rate of change of fluid element length in the x_1-direction per unit length in this direction is

$$\frac{1}{\delta x_1}\frac{D}{Dt}(\delta x_1) = \lim_{dt \to 0}\frac{1}{dt}\left(\frac{A'B' - AB}{AB}\right) = \lim_{dt \to 0}\frac{1}{\delta x_1 dt}\left(\delta x_1 + \frac{\partial u_1}{\partial x_1}\delta x_1 dt - \delta x_1\right) = \frac{\partial u_1}{\partial x_1},$$

where D/Dt indicates that the fluid element is followed as extension takes place. This simple construction is readily extended to the other two Cartesian directions, and in general the linear strain rate in the η direction is $\partial u_\eta / \partial x_\eta$ where *no summation* over the repeated η-index is implied. (Greek subscripts are commonly used when the summation convention is not followed.)

FIGURE 3.10 Illustration of positive linear strain rate in the first coordinate direction. Here $A'B' = AB + BB' - AA'$, and a positive $S_{11} = \partial u_1/\partial x_1$ corresponds to a lengthening of the fluid element.

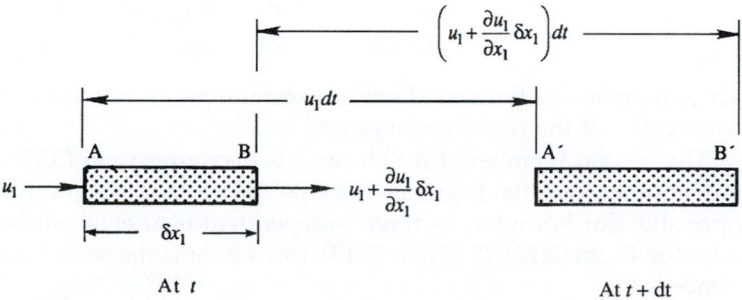

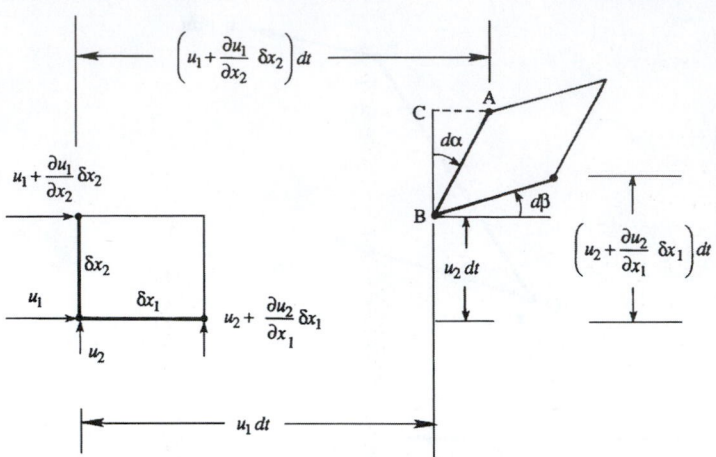

FIGURE 3.11 Illustration of positive deformation of a fluid element in the plane defined by the first and second coordinate directions. Here, both $\partial u_1/\partial x_2$ and $\partial u_2/\partial x_1$ are shown as positive, so $S_{12} = S_{21}$ from (3.12) is also positive. The deformation angle $d\alpha = \angle CBA$ is proportional to $\partial u_1/\partial x_2$ while $d\beta$ is proportional to $\partial u_2/\partial x_1$.

The off-diagonal terms of S_{ij} represent shear deformations that change the relative orientations of line segments initially parallel to the i- and j-directions in the flow. A geometrical interpretation of S_{ij}'s first off-diagonal component, $S_{12} = S_{21}$, is provided in Figure 3.11. The average rate at which the initially perpendicular segments δx_1 and δx_2 rotate toward each other is

$$\frac{1}{2}\frac{D(\alpha + \beta)}{Dt} = \lim_{dt \to 0} \frac{1}{2dt}\left(\frac{1}{\delta x_2}\left(\frac{\partial u_1}{\partial x_2}\delta x_2 dt\right) + \frac{1}{\delta x_1}\left(\frac{\partial u_2}{\partial x_1}\delta x_1 dt\right)\right) = \frac{1}{2}\left(\frac{\partial u_1}{\partial x_2} + \frac{\partial u_2}{\partial x_1}\right) = S_{12} = S_{21},$$

where again D/Dt indicates that the fluid element is followed as shear deformation takes place, and again this simple construction is readily extended to the other two Cartesian direction pairs. Thus, the off-diagonal terms of S_{ij} represent the average rate at which line segments initially parallel to the i- and j-directions rotate *toward* each other.

Here we also note that S_{ij} is zero for any rigid body motion composed of translation at a spatially uniform velocity $\mathbf{U}$ and rotation at a constant rate $\mathbf{\Omega}$ (see Exercise 3.17). Thus, S_{ij} is independent of the frame of reference in which it is observed, even if $\mathbf{U}$ depends on time and the frame of reference is rotating.

The first invariant of S_{ij} (the sum of its diagonal terms) is the *volumetric strain rate* or *bulk strain rate*. For a small volume $\delta V = \delta x_1 \delta x_2 \delta x_3$, it can be shown (Exercise 3.18) that

$$\frac{1}{\delta V}\frac{D}{Dt}(\delta V) = \frac{\partial u_1}{\partial x_1} + \frac{\partial u_2}{\partial x_2} + \frac{\partial u_3}{\partial x_3} = \frac{\partial u_i}{\partial x_i} = S_{ii}. \tag{3.14}$$

Thus, S_{ii} specifies the rate of volume change per unit volume and it does not depend on the orientation of the coordinate system.

The second member of the strain-rate decomposition (3.11) is the rotation tensor, R_{ij}. It is antisymmetric so its diagonal elements are zero and its off-diagonal elements are equal and opposite. Furthermore, its three independent elements can be put in correspondence with a vector. From (2.26), (2.27), or (3.13), this vector is the *vorticity*, $\mathbf{\omega} = \nabla \times \mathbf{u}$, and the correspondence is

$$R_{ij} = -\varepsilon_{ijk}(\nabla \times \mathbf{u})_k = -\varepsilon_{ijk}\omega_k = \begin{bmatrix} 0 & -\omega_3 & \omega_2 \\ \omega_3 & 0 & -\omega_1 \\ -\omega_2 & \omega_1 & 0 \end{bmatrix}, \qquad (2.26, 2.27, 3.15)$$

where

$$\omega_1 = \frac{\partial u_3}{\partial x_2} - \frac{\partial u_2}{\partial x_3}, \quad \omega_2 = \frac{\partial u_1}{\partial x_3} - \frac{\partial u_3}{\partial x_1}, \quad \text{and} \quad \omega_3 = \frac{\partial u_2}{\partial x_1} - \frac{\partial u_1}{\partial x_2}. \qquad (2.25, 3.16)$$

Figure 3.11 illustrates the motion of an initially square fluid element in the (x_1, x_2)-plane when $\partial u_1/\partial x_2$ and $\partial u_2/\partial x_1$ are nonzero and unequal so that $-\omega_3 = R_{12} = -R_{21} \neq 0$. In this situation, the fluid element translates and deforms in the (x_1, x_2)-plane, and rotates about the third coordinate axis. The average rotation rate is

$$\frac{1}{2}\frac{D(-\alpha + \beta)}{Dt} = \lim_{dt \to 0} \frac{1}{2dt}\left(-\frac{1}{\delta x_2}\left(\frac{\partial u_1}{\partial x_2}\delta x_2 dt\right) + \frac{1}{\delta x_1}\left(\frac{\partial u_2}{\partial x_1}\delta x_1 dt\right)\right) = \frac{1}{2}\left(-\frac{\partial u_1}{\partial x_2} + \frac{\partial u_2}{\partial x_1}\right)$$

$$= -\frac{R_{12}}{2} = \frac{R_{21}}{2},$$

where again D/Dt indicates that the fluid element is followed as rotation takes place, and again this simple construction is readily extended to the other two Cartesian direction pairs. Thus, $\boldsymbol{\omega}$ and R_{ij} represent twice the fluid element rotation rate (see also Exercise 2.1). This means that $\boldsymbol{\omega}$ and R_{ij} depend on the frame of reference in which they are determined since it is possible to choose a frame of reference that rotates with the fluid particle of interest at the time of interest. In such a co-rotating frame, $\boldsymbol{\omega}$ and R_{ij} will be zero but they will be nonzero if they are determined in a frame of reference that rotates at a different rate (see Exercise 3.19).

Interestingly, the presence or absence of fluid rotation often determines the character of a flow, and this dependence leads to two additional kinematic concepts related to fluid rotation. First, fluid motion is called *irrotational* if

$$\boldsymbol{\omega} = 0, \quad \text{or equivalently} \quad R_{ij} = \partial u_i/\partial x_j - \partial u_j/\partial x_i = 0. \qquad (3.17)$$

When (3.17) is true, the fluid velocity $\mathbf{u}$ can be written as the gradient of a scalar function $\phi(\mathbf{x}, t)$ because $u_i = \partial\phi/\partial x_i$ satisfies the condition of irrotationality (see Exercises 2.4 and 2.20). Although this may seem to be an unnecessary mathematical complication, finding a scalar function $\phi(\mathbf{x}, t)$ such that $\nabla\phi$ solves the irrotational equations of fluid motion is sometimes easier than solving these equations directly for the vector velocity $\mathbf{u}(\mathbf{x}, t)$ in the same circumstance.

The second concept related to fluid rotation is the extension of the vorticity, twice the fluid rotation rate at a point, to the *circulation* Γ, the amount of fluid rotation within a closed contour (or *circuit*) C. Here the circulation Γ is defined by

$$\Gamma \equiv \oint_C \mathbf{u} \cdot d\mathbf{s} = \int_A \boldsymbol{\omega} \cdot \mathbf{n}dA, \qquad (3.18)$$

where $d\mathbf{s}$ is an element of C, and the geometry is shown in Figure 3.12. The loop through the first integral sign signifies that C is a closed circuit and is often omitted. The second equality in (3.18) follows from Stokes' theorem (Section 2.13) and the definition of the vorticity

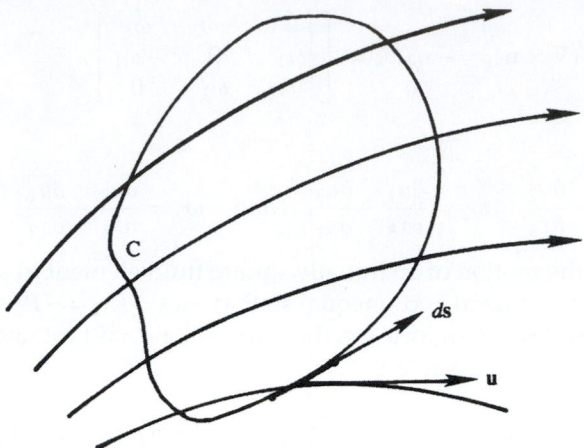

FIGURE 3.12 The circulation around the closed contour C is the line integral of the dot product of the velocity $\mathbf{u}$ and the contour element $d\mathbf{s}$.

$\boldsymbol{\omega} = \nabla \times \mathbf{u}$. The second equality requires the line integral of $\mathbf{u}$ around a closed curve C to be equal to the *flux of vorticity* through the arbitrary surface A bounded by C. Here, and elsewhere in this text, the term *flux* is used for the integral of a vector field normal to a surface. Equation (3.18) allows $\boldsymbol{\omega}$ to be identified as *the circulation per unit area*. This identification also follows directly from the definition of the curl as the limit of the circulation integral (see (2.35)).

Returning to the situation in Figure 3.9, equations (3.11) through (3.14) allow (3.10) to be rewritten as

$$du_i = \left(S_{ij} - \frac{1}{2}\varepsilon_{ijk}\omega_k \right) dx_j = S_{ij}dx_j + \frac{1}{2}(\boldsymbol{\omega} \times d\mathbf{x})_i, \tag{3.19}$$

where $\varepsilon_{ijk}\omega_k dx_j$ is the i-component of the cross product $-\boldsymbol{\omega} \times d\mathbf{x}$ (see (2.21)). Thus, the meaning of the second term in (3.19) can be deduced as follows. The velocity at a distance $\mathbf{x}$ from the axis of rotation of a rigid body rotating at angular velocity $\boldsymbol{\Omega}$ is $\boldsymbol{\Omega} \times \mathbf{x}$. The second term in (3.19) therefore represents the velocity of point P relative to point O because of an angular velocity of $\boldsymbol{\omega}/2$.

The first term in (3.19) is the relative velocity between point P and point O caused by deformation of the fluid element defined by $d\mathbf{x}$. This deformation becomes particularly simple in a coordinate system coinciding with the principal axes of the strain-rate tensor. The components S_{ij} change as the coordinate system is rotated, and for one particular orientation of the coordinate system, a symmetric tensor has only diagonal components; these are called the *principal axes* of the tensor (see Section 2.12 and Example 2.4). Denoting the variables in this principal coordinate system by an over bar (Figure 3.13), the *first* part of (3.19) can be written as:

$$d\bar{\mathbf{u}} = \bar{\mathbf{S}} \cdot d\bar{\mathbf{x}} = \begin{bmatrix} \bar{S}_{11} & 0 & 0 \\ 0 & \bar{S}_{22} & 0 \\ 0 & 0 & \bar{S}_{33} \end{bmatrix} \begin{bmatrix} d\bar{x}_1 \\ d\bar{x}_2 \\ d\bar{x}_3 \end{bmatrix}. \tag{3.20}$$

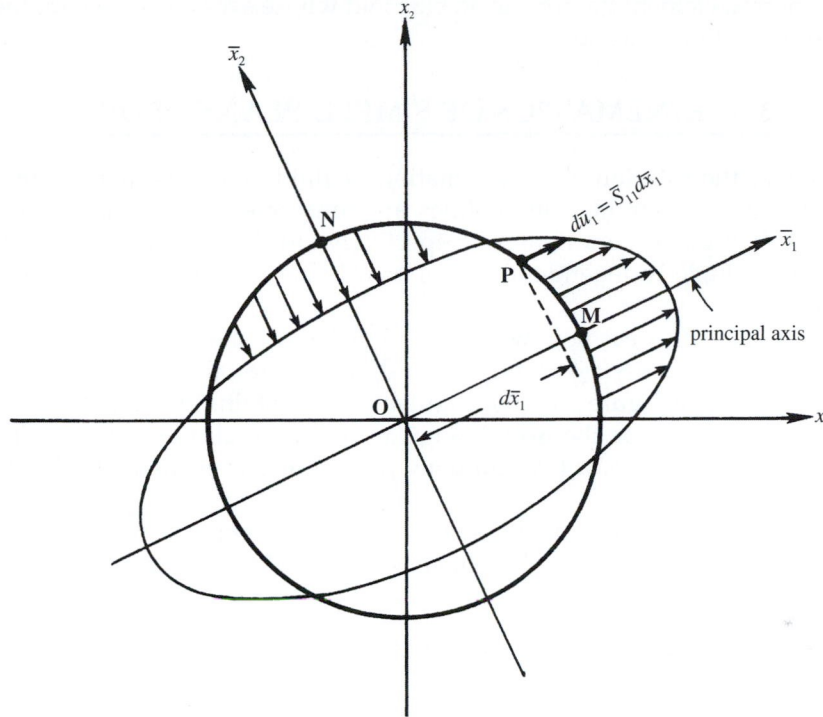

FIGURE 3.13 Deformation of a spherical fluid element into an ellipsoid. Here only the intersection of the element with the plane defined by the first and second coordinate directions is shown.

Here, $\bar{S}_{11}$, $\bar{S}_{22}$, and $\bar{S}_{33}$ are the diagonal components of $\mathbf{S}$ in the principal-axis coordinate system and are called the *eigenvalues* of $\mathbf{S}$. The three components of (3.20) can be written as

$$d\bar{u}_1 = \bar{S}_{11}d\bar{x}_1, \quad d\bar{u}_2 = \bar{S}_{22}d\bar{x}_2, \quad \text{and} \quad d\bar{u}_3 = \bar{S}_{33}d\bar{x}_3. \tag{3.21}$$

Consider the significance of $d\bar{u}_1 = \bar{S}_{11}d\bar{x}_1$ when $\bar{S}_{11}$ is positive. This equation implies that point P in Figure 3.9 is moving *away* from point O in the $\bar{x}_1$-direction at a rate proportional to the distance $d\bar{x}_1$. Considering all points on the surface of a sphere centered on O and having radius $|d\mathbf{x}|$ (see Figure 3.13), the movement of P in the $\bar{x}_1$ direction is maximum when P coincides with point M (where $d\bar{x}_1 = |d\mathbf{x}|$) and is zero when P coincides with point N (where $d\bar{x}_1 = 0$). Figure 3.13 illustrates the intersection of this sphere with the $(\bar{x}_1, \bar{x}_2)$-plane for the case where $\bar{S}_{11} > 0$ and $\bar{S}_{22} < 0$; the deformation in the x_3 direction is not shown in this figure. In a small interval of time, *a spherical fluid element around* O *therefore becomes an ellipsoid whose axes are the principal axes of the strain-rate tensor* $\mathbf{S}$.

Summary

The relative velocity in the neighborhood of a point can be divided into two parts. One part comes from rotation of the element, and the other part comes from deformation of the element. A spherical element deforms to an ellipsoid whose axes coincide with the principal axes of the local strain-rate tensor.

3.5. KINEMATICS OF SIMPLE PLANE FLOWS

In this section, the rotation and deformation of fluid elements in two simple steady flows with straight and circular streamlines are considered in two-dimensional (x_1,x_2)-Cartesian and (r,θ)-polar coordinates, respectively. In both cases, the flows can be described with a single independent spatial coordinate that increases perpendicular to the flow direction.

First consider parallel shear flow where $\mathbf{u} = (u_1(x_2), 0)$ as shown in Figure 3.14. The lone nonzero velocity gradient is $\gamma(x_2) \equiv du_1/dx_2$, and, from (3.16), the only nonzero component of vorticity is $\omega_3 = -\gamma$. In Figure 3.14, the angular velocity of line element AB is $-\gamma$, and that of BC is zero, giving $-\gamma/2$ as the overall angular velocity (half the vorticity). The average value does not depend on *which* two mutually perpendicular elements in the (x_1,x_2)-plane are chosen to compute it.

In contrast, the components of the strain rate do depend on the orientation of the element. From (3.11), S_{ij} for a fluid element such as ABCD, with sides parallel to the x_1, x_2-axes, is

$$S_{ij} = \begin{bmatrix} 0 & \gamma/2 \\ \gamma/2 & 0 \end{bmatrix},$$

which shows that there are only off-diagonal elements of **S**. Therefore, the element ABCD undergoes shear, but no normal strain. As discussed in Section 2.11 and Example 2.4, a symmetric tensor with zero diagonal elements can be diagonalized by rotating the coordinate system through 45°. It is shown there that, along these *principal axes* (denoted by an overbar in Figure 3.14), the strain rate tensor is

$$\overline{S}_{ij} = \begin{bmatrix} \gamma/2 & 0 \\ 0 & -\gamma/2 \end{bmatrix},$$

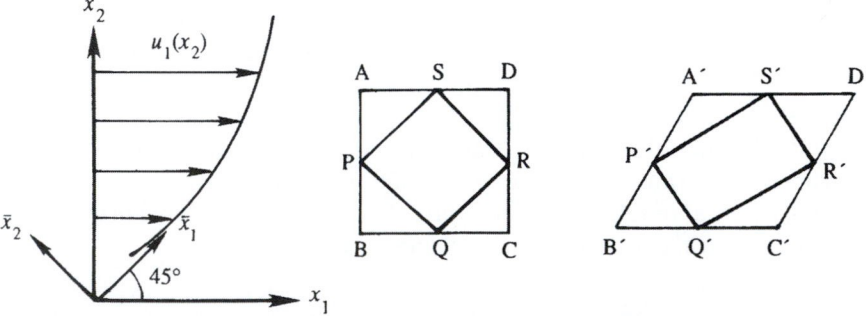

FIGURE 3.14   Deformation of elements in a parallel shear flow. The element is stretched along the principal axis $\bar{x}_1$ and compressed along the principal axis $\bar{x}_2$. The lengths of the sides of ADCB remain unchanged while the corner angles of SRQP remain unchanged.

so that along the first principle axis there is a linear extension rate of $\gamma/2$, along the second principle axis there is a linear compression rate of $-\gamma/2$, and there is no shear. This can be seen geometrically in Figure 3.14 by examining the deformation of an element PQRS oriented at 45°, which deforms to P′Q′R′S′. It is clear that the side PS elongates and the side PQ contracts, but the angles between the sides of the element remain at 90°. In a small time interval, a small spherical element in this flow would become an ellipsoid oriented at 45° to the x_1, x_2-coordinate system.

In summary, the element ABCD in a parallel shear flow deforms via shear without normal strain, whereas the element PQRS deforms via normal strain without shear strain. However, both elements rotate at the same angular velocity.

Now consider two steady vortex flows having circular streamlines. In (r,θ)-polar coordinates, both flows are defined by $u_r = 0$ and $u_\theta = u_\theta(r)$, with the first one being *solid-body rotation*,

$$u_r = 0 \quad \text{and} \quad u_\theta = \omega_0 r, \tag{3.22}$$

where ω_0 is a constant equal to the angular velocity of each particle about the origin (Figure 3.15). Such a flow can be generated by steadily rotating a cylindrical tank containing a viscous fluid about its axis and waiting until the transients die out. From Appendix B, the vorticity component in the z-direction perpendicular to the (r,θ)-plane is

$$\omega_z = \frac{1}{r}\frac{\partial}{\partial r}(r u_\theta) - \frac{1}{r}\frac{\partial u_r}{\partial \theta} = 2\omega_0, \tag{3.23}$$

which is independent of location. Thus, each fluid element is rotating about its own center at the same rate that it rotates about the origin of coordinates. This is evident in Figure 3.15, which shows the location of element ABCD at two successive times. The two mutually perpendicular fluid lines AD and AB both rotate counterclockwise (about the center of the

FIGURE 3.15 Solid-body rotation. The streamlines are circular and fluid elements spin about their own centers at the same rate that they revolve around the origin. There is no deformation of the elements, only rotation.

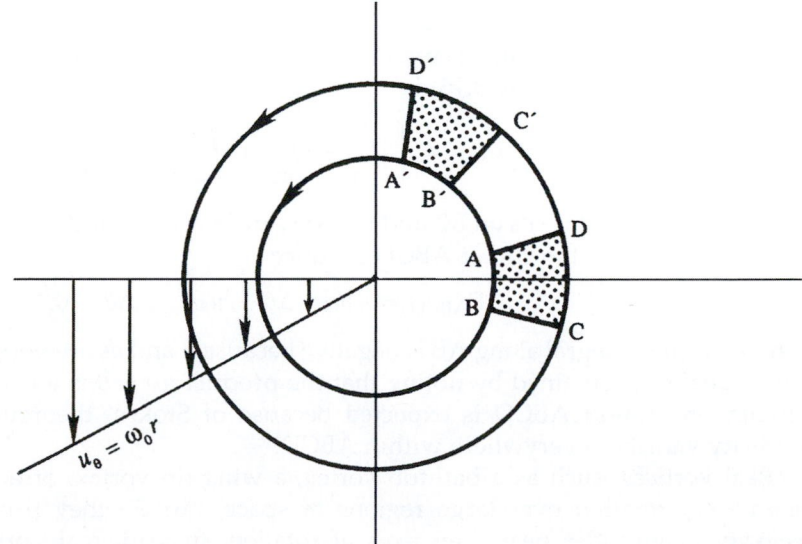

element) with speed ω_0. The time period for one *rotation* of the particle about its own center equals the time period for one *revolution* around the origin of coordinates. In addition, $\mathbf{S} = 0$ for this flow so fluid elements do not deform and each retains its location relative to other elements, as is expected for solid-body rotation.

The circulation around a circuit of radius r in this flow is

$$\Gamma = \oint_C \mathbf{u} \cdot d\mathbf{s} = \int_0^{2\pi} u_\theta r d\theta = 2\pi r u_\theta = 2\pi r^2 \omega_0, \tag{3.24}$$

which shows that circulation equals the vorticity, $2\omega_0$, times the area contained by C. This result is true for *any* circuit C, regardless of whether or not it contains the origin (see Exercise 3.23).

Another flow with circular streamlines is that from an ideal vortex line oriented perpendicular to the (r,θ)-plane. Here, the θ-component of fluid velocity is inversely proportional to the radius of the streamline and the radial velocity is again zero:

$$u_r = 0 \quad \text{and} \quad u_\theta = B/r, \tag{3.25}$$

where B is constant. From (3.23), the vorticity in this flow at any point away from the origin is $\omega_z = 0$, but the circulation around a circuit of radius r centered on the origin is a nonzero constant,

$$\Gamma = \int_0^{2\pi} u_\theta r d\theta = 2\pi r u_\theta = 2\pi B, \tag{3.26}$$

independent of r. Thus, considering vorticity to be the circulation per unit area, as in (3.18) when $\mathbf{n} = \mathbf{e}_z$, then (3.26) implies that the flow specified by (3.25) is *irrotational everywhere except at $r = 0$ where the vorticity is infinite with a finite area integral*:

$$[\omega_z]_{r \to 0} = \lim_{r \to 0} \frac{1}{A} \int_A \omega_z dA = \lim_{r \to 0} \frac{1}{\pi r^2} \oint_C \mathbf{u} \cdot d\mathbf{s} = \lim_{r \to 0} \frac{2B}{r^2}. \tag{3.27}$$

Although the circulation around a circuit containing the origin in an irrotational vortex flow is nonzero, that around a circuit *not* containing the origin is zero. The circulation around the contour ABCD (Figure 3.16) is

$$\Gamma_{ABCD} = \left\{ \int_{AB} + \int_{BC} + \int_{CD} + \int_{DA} \right\} \mathbf{u} \cdot d\mathbf{s}.$$

The line integrals of $\mathbf{u} \cdot d\mathbf{s}$ on BC and DA are zero because $\mathbf{u}$ and $d\mathbf{s}$ are perpendicular, and the remaining parts of the circuit ABCD produce

$$\Gamma_{ABCD} = -[u_\theta r]_r \Delta\theta + [u_\theta r]_{r+\Delta r} \Delta\theta = 0,$$

where the line integral along AB is negative because $\mathbf{u}$ and $d\mathbf{s}$ are oppositely directed, and the final equality is obtained by noting that the product $u_\theta r = B$ is a constant. In addition, zero circulation around ABCD is expected because of Stokes' theorem and the fact that the vorticity vanishes everywhere within ABCD.

Real vortices, such as a bathtub vortex, a wing-tip vortex, or a tornado, do not mimic solid-body rotation over large regions of space, nor do they produce unbounded fluid velocity magnitudes near their axes of rotation. Instead, real vortices combine elements

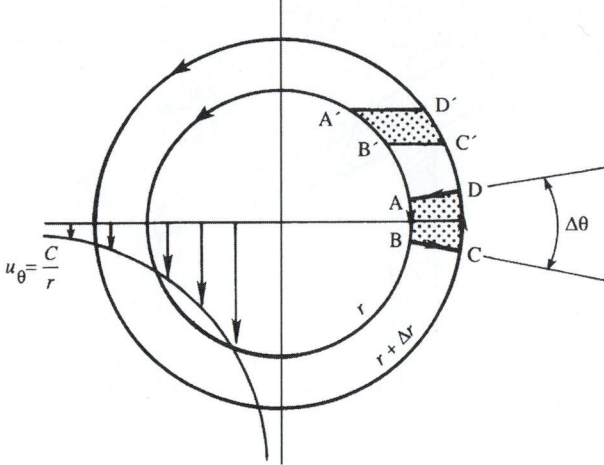

FIGURE 3.16 Irrotational vortex. The streamlines are circular, as for solid-body rotation, but the fluid velocity varies with distance from the origin so that fluid elements only deform; they do not spin. The vorticity of fluid elements is zero everywhere, except at the origin where it is infinite.

of the ideal vortex flows described by (3.22) and (3.25). Near the center of rotation, a real vortex's core flow is nearly solid-body rotation, but far from this core, real-vortex-induced flow is nearly irrotational. Two common idealizations of this behavior are the Rankine vortex defined by

$$\omega_z(r) = \begin{Bmatrix} \Gamma/\pi\sigma^2 = const. & \text{for } r \leq \sigma \\ 0 & \text{for } r > \sigma \end{Bmatrix} \quad \text{and} \quad u_\theta(r) = \begin{Bmatrix} (\Gamma/2\pi\sigma^2)r & \text{for } r \leq \sigma \\ \Gamma/2\pi r & \text{for } r > \sigma \end{Bmatrix}, \quad (3.28)$$

and the Gaussian vortex defined by

$$\omega_z(r) = \frac{\Gamma}{\pi\sigma^2} \exp\left(-r^2/\sigma^2\right) \quad \text{and} \quad u_\theta(r) = \frac{\Gamma}{2\pi r}\left(1 - \exp(-r^2/\sigma^2)\right). \quad (3.29)$$

In both cases, σ is a core-size parameter that determines the radial distance where real vortex behavior transitions from solid-body rotation to irrotational vortex flow. For the Rankine vortex, this transition is abrupt and occurs at $r = \sigma$ where u_θ reaches its maximum. For the Gaussian vortex, this transition is gradual and the maximum value of u_θ is reached at $r \approx 1.12091\sigma$ (see Exercise 3.26).

3.6. REYNOLDS TRANSPORT THEOREM

The final kinematic result needed for developing the differential and the control-volume versions of the conservation equations for fluid motion is the Reynolds transport theorem for time differentiation of integrals over arbitrarily moving and deforming volumes. Reynolds transport theorem is the three-dimensional extension of *Leibniz's theorem* for differentiating a single-variable integral having a time-dependent integrand and time-dependent limits (see Riley et al., 1998).

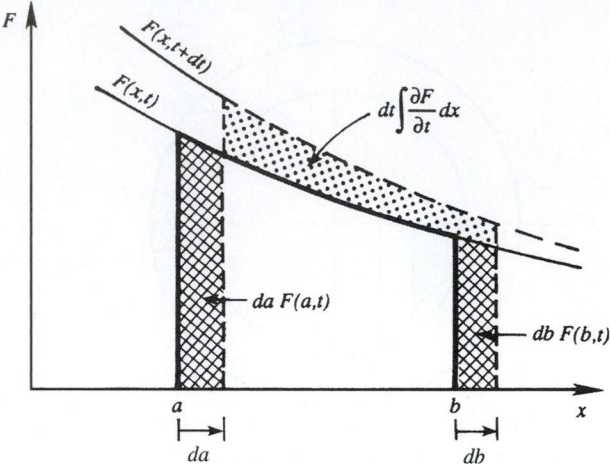

FIGURE 3.17   Graphical illustration of the Liebniz theorem. The three marked areas correspond to the three contributions shown on the right in (3.30). Here da, db, and $\partial F/\partial t$ are all shown as positive.

Consider a function F that depends on one independent spatial variable, x, and time t. In addition assume that the time derivative of its integral is of interest when the limits of integration, a and b, are themselves functions of time. Leibniz's theorem states the time derivative of the integral of $F(x,t)$ between $x=a(t)$ and $x=b(t)$ is

$$\frac{d}{dt}\int_{x=a(t)}^{x=b(t)} F(x,t)\,dx = \int_{a}^{b} \frac{\partial F}{\partial t}\,dx + \frac{db}{dt}F(b,t) - \frac{da}{dt}F(a,t), \qquad (3.30)$$

where a, b, F, and their derivatives appearing on the right side of (3.30) are all evaluated at time t. This situation is depicted in Figure 3.17, where the three contributions are shown by dots and cross-hatches. The continuous line shows the integral $\int F dx$ at time t, and the dashed line shows the integral at time $t + dt$. The first term on the right side of (3.30) is the integral of $\partial F/\partial t$ between $x=a$ and b, the second term is the gain of F at the upper limit which is moving at rate db/dt, and the third term is the loss of F at the lower limit which is moving at rate da/dt. The essential features of (3.30) are the total time derivative on the left, an integral over the partial time derivative of the integrand on the right, and terms that account for the time-dependence of the limits of integration on the right. These features persist when (3.30) is generalized to three dimensions.

A largely geometrical development of this generalization is presented here using notation drawn from Thompson (1972). Consider a moving volume $V^*(t)$ having a (closed) surface $A^*(t)$ with outward normal $\mathbf{n}$ and let $\mathbf{b}$ denote the local velocity of A^* (Figure 3.18). The volume V^* and its surface A^* are commonly called a *control volume* and its *control surface*, respectively. The situation is quite general. The volume and its surface need not coincide with any particular boundary, interface, or surface. The velocity $\mathbf{b}$ need not be steady or uniform over $A^*(t)$. No specific coordinate system or origin of coordinates is needed. The goal of this effort is to determine the time derivative of the integral of a single-valued

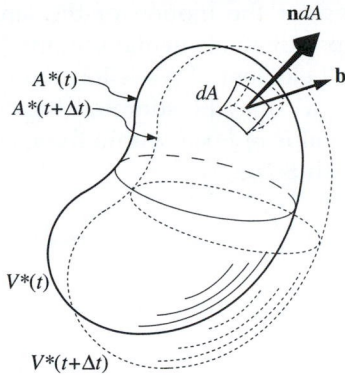

FIGURE 3.18 Geometrical depiction of a control volume $V^*(t)$ having a surface $A^*(t)$ that moves at a nonuniform velocity **b** during a small time increment Δt. When Δt is small enough, the volume increment $\Delta V = V^*(t + \Delta t) - V^*(t)$ will lie very near $A^*(t)$, so the volume-increment element adjacent to dA will be $(\mathbf{b}\Delta t) \cdot \mathbf{n}dA$ where **n** is the outward normal on $A^*(t)$.

continuous function $F(\mathbf{x}, t)$ in the volume $V^*(t)$. The starting point for this effort is the definition of a time derivative:

$$\frac{d}{dt}\int_{V^*(t)} F(\mathbf{x},t)dV = \lim_{\Delta t \to 0}\frac{1}{\Delta t}\left\{ \int_{V^*(t+\Delta t)} F(\mathbf{x},t+\Delta t)dV - \int_{V^*(t)} F(\mathbf{x},t)dV \right\}. \tag{3.31}$$

The geometry for the two integrals inside the {,}-braces is shown in Figure 3.18 where solid lines are for time t while the dashed lines are for time $t + \Delta t$. The time derivative of the integral on the left is properly written as a total time derivative since the volume integration subsumes the possible spatial dependence of F. The first term inside the {,}-braces can be expanded to four terms by defining the volume increment $\Delta V \equiv V^*(t + \Delta t) - V^*(t)$ and Taylor expanding the integrand function $F(\mathbf{x}, t + \Delta t) \cong F(\mathbf{x}, t) + \Delta t(\partial F/\partial t)$ for $\Delta t \to 0$:

$$\int_{V^*(t+\Delta t)} F(\mathbf{x},t+\Delta t)dV \cong \int_{V^*(t)} F(\mathbf{x},t)dV + \int_{V^*(t)} \Delta t\frac{\partial F(\mathbf{x},t)}{\partial t}dV + \int_{\Delta V} F(\mathbf{x},t)dV$$
$$+ \int_{\Delta V} \Delta t\frac{\partial F(\mathbf{x},t)}{\partial t}dV. \tag{3.32}$$

The first term on the right in (3.32) will cancel with the final term in (3.31), and, when the limit in (3.31) is taken, both Δt and ΔV go to zero so the final term in (3.32) will not contribute because it is second order. Thus, when (3.32) is substituted into (3.31), the result is

$$\frac{d}{dt}\int_{V^*(t)} F(\mathbf{x},t)dV = \lim_{\Delta t \to 0}\frac{1}{\Delta t}\left\{ \int_{V^*(t)} \Delta t\frac{\partial F(\mathbf{x},t)}{\partial t}dV + \int_{\Delta V} F(\mathbf{x},t)dV \right\}, \tag{3.33}$$

and this limit may be taken once the relationship between ΔV and Δt is known.

To find this relationship consider the motion of the small area element dA shown in Figure 3.18. In time Δt, dA sweeps out an elemental volume $(\mathbf{b}\Delta t) \cdot \mathbf{n}dA$ of the volume increment ΔV. Furthermore, this small element of ΔV is located adjacent to the surface $A^*(t)$. All these elemental contributions to ΔV may be summed together via a surface integral, and, as Δt goes to zero, the integrand value of $F(\mathbf{x},t)$ within these elemental volumes may be taken as that of F on the surface $A^*(t)$, thus

$$\int_{\Delta V} F(\mathbf{x},t)dV \cong \int_{A^*(t)} F(\mathbf{x},t)(\mathbf{b}\Delta t \cdot \mathbf{n})dA \quad \text{as} \quad \Delta t \to 0. \tag{3.34}$$

Substituting (3.34) into (3.33), and taking the limit, produces the following statement of Reynolds transport theorem:

$$\frac{d}{dt}\int_{V^*(t)} F(\mathbf{x},t)dV = \int_{V^*(t)} \frac{\partial F(\mathbf{x},t)}{\partial t}dV + \int_{A^*(t)} F(\mathbf{x},t)\mathbf{b} \cdot \mathbf{n}dA. \tag{3.35}$$

This final result follows the pattern set by Liebniz's theorem that the total time derivative of an integral with time-dependent limits equals the integral of the partial time derivative of the integrand plus a term that accounts for the motion of the integration boundary. In (3.35), both inflows and outflows of F are accounted for through the dot product in the surface-integral term that monitors whether $A^*(t)$ is locally advancing ($\mathbf{b} \cdot \mathbf{n} > 0$) or retreating ($\mathbf{b} \cdot \mathbf{n} < 0$) along $\mathbf{n}$, so separate terms as in (3.30) are unnecessary. In addition, the $(\mathbf{x},t)$-space-time dependence of the control volume's surface velocity $\mathbf{b}$ and unit normal $\mathbf{n}$ are not explicitly shown in (3.35) because $\mathbf{b}$ and $\mathbf{n}$ are only defined on $A^*(t)$; neither is a field quantity like $F(\mathbf{x},t)$. Equation (3.35) is an entirely kinematic result, and it shows that d/dt may be moved inside a volume integral and replaced by $\partial/\partial t$ only when the integration volume, $V^*(t)$, is fixed in space so that $\mathbf{b} = 0$.

There are two physical interpretations of (3.35). The first, obtained when $F = 1$, is that volume is conserved as $V^*(t)$ moves through three-dimensional space, and under these conditions (3.35) is equivalent to (3.14) for small volumes (see Exercise 3.28). The second is that (3.35) is the extension of (3.5) to finite-size volumes (see Exercise 3.30). Nevertheless, (3.35) and judicious choices of F and $\mathbf{b}$ are the starting points in the next chapter for deriving the field equations of fluid motion from the principles of mass, momentum, and energy conservation.

EXAMPLE 3.2

The base radius r of a fixed-height right circular cone is increasing at the rate $\dot{r}$. Use Reynolds transport theorem to determine the rate at which the cone's volume is increasing when the cone's base radius is r_0 if its height is h.

Solution

At any time, the volume V of the right circular cone is: $V = 1/3 \, \pi h r^2$, which can be differentiated directly and evaluated at $r = r_0$ to find $dV/dt = 2/3 \, \pi h r_0 \dot{r}$. However, the task is to obtain this answer using (3.35). Choose V^* to perfectly enclose the cone so that $V^* = V$, and set $F = 1$ in (3.35) so that the time derivative of the cone's volume appears on the left:

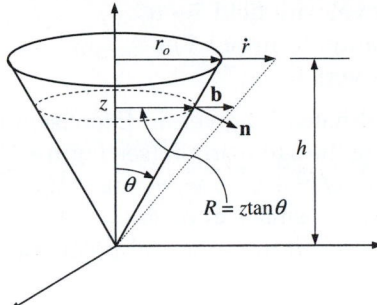

FIGURE 3.19 Conical geometry for Example 3.2. The cone's height is fixed but the radius of its circular surface (base) is increasing.

$$dV/dt = \int_{A^*(t)} \mathbf{b} \cdot \mathbf{n} dA.$$

Use the cylindrical coordinate system shown in Figure 3.19 with the cone's apex at the origin. Here, $\mathbf{b}=0$ on the cone's base while $\mathbf{b} = (z/h)\dot{r}\mathbf{e}_R$ on its conical sides. The normal vector on the cone's sides is $\mathbf{n} = \mathbf{e}_R\cos\theta - \mathbf{e}_z\sin\theta$ where $r_o/h = \tan\theta$. Here, at the height z, the cone's surface area element is $dA = z\tan\theta d\varphi(dz/\cos\theta)$, where φ is the azimuthal angle, and the extra cosine factor enters because the conical surface is sloped. Thus, the volumetric rate of change becomes

$$\frac{dV}{dt} = \int_{z=0}^{h}\int_{\varphi=0}^{2\pi}\frac{z}{h}\dot{r}\mathbf{e}_R \cdot (\mathbf{e}_R\cos\theta - \mathbf{e}_z\sin\theta)\, z[?]\tan\theta d\varphi\left(\frac{dz}{\cos\theta}\right)$$

$$= 2\pi\frac{\dot{r}\tan\theta}{h}\int_{z=0}^{h}z^2 dz = \frac{2}{3}\pi h^2\dot{r}\tan\theta = \frac{2}{3}\pi h r_o\dot{r},$$

which recovers the answer obtained by direct differentiation.

EXERCISES

3.1. The gradient operator in Cartesian coordinates (x, y, z) is:
$\nabla = \mathbf{e}_x(\partial/\partial x) + \mathbf{e}_y(\partial/\partial y) + \mathbf{e}_z(\partial/\partial z)$ where $\mathbf{e}_x$, $\mathbf{e}_y$, and $\mathbf{e}_z$ are the unit vectors. In cylindrical polar coordinates (R, φ, z) having the same origin (see Figure 3.3b), coordinates and unit vectors are related by: $R = \sqrt{x^2+y^2}$, $\varphi = \tan^{-1}(y/x)$, and $z=z$; and $\mathbf{e}_R = \mathbf{e}_x\cos\varphi + \mathbf{e}_y\sin\varphi$, $\mathbf{e}_\varphi = -\mathbf{e}_x\sin\varphi + \mathbf{e}_y\cos\varphi$, and $\mathbf{e}_z = \mathbf{e}_z$. Determine the following in the cylindrical polar coordinate system.
a) $\partial\mathbf{e}_R/\partial\varphi$ and $\partial\mathbf{e}_\varphi/\partial\varphi$
b) the gradient operator ∇
c) the Laplacian operator $\nabla \cdot \nabla \equiv \nabla^2$

d) the divergence of the velocity field $\nabla \cdot \mathbf{u}$
e) the advective acceleration term $(\mathbf{u} \cdot \nabla)\mathbf{u}$
[See Appendix B for answers.]

3.2. Consider Cartesian coordinates (as given in Exercise 3.1) and spherical polar coordinates (r, θ, φ) having the same origin (see Figure 3.3c). Here coordinates and unit vectors are related by: $r = \sqrt{x^2 + y^2 + z^2}$, $\theta = \tan^{-1}(\sqrt{x^2 + y^2}/z)$, and $\varphi = \tan^{-1}(y/x)$; and $\mathbf{e}_r = \mathbf{e}_x \cos\varphi\sin\theta + \mathbf{e}_y \sin\varphi\sin\theta + \mathbf{e}_z \cos\theta$, $\mathbf{e}_\theta = \mathbf{e}_x \cos\varphi\cos\theta + \mathbf{e}_y \sin\varphi\cos\theta - \mathbf{e}_z \sin\theta$, and $\mathbf{e}_\varphi = -\mathbf{e}_x \sin\varphi + \mathbf{e}_y \cos\varphi$. In the spherical polar coordinate system, determine the following items.
a) $\partial\mathbf{e}_r/\partial\theta$, $\partial\mathbf{e}_r/\partial\varphi$, $\partial\mathbf{e}_\theta/\partial\theta$, $\partial\mathbf{e}_\theta/\partial\varphi$, and $\partial\mathbf{e}_\varphi/\partial\varphi$
b) the gradient operator ∇
c) the Laplacian $\nabla \cdot \nabla \equiv \nabla^2$
d) the divergence of the velocity field $\nabla \cdot \mathbf{u}$
e) the advective acceleration term $(\mathbf{u} \cdot \nabla)\mathbf{u}$
[See Appendix B for answers.]

3.3. If $d\mathbf{s} = (dx, dy, dz)$ is an element of arc length along a streamline (Figure 3.5) and $\mathbf{u} = (u, v, w)$ is the local fluid velocity vector, show that if $d\mathbf{s}$ is everywhere tangent to $\mathbf{u}$ then $dx/u = dy/v = dz/w$.

3.4. For the two-dimensional steady flow having velocity components $u = Sy$ and $v = Sx$, determine the following when S is a positive real constant having units of inverse time.
a) equations for the streamlines with a sketch of the flow pattern
b) the components of the strain-rate tensor
c) the components of the rotation tensor
d) the coordinate rotation that diagonalizes the strain-rate tensor, and the principal strain rates

3.5. Repeat Exercise 3.4 when $u = -Sx$ and $v = +Sy$. How are the two flows related?

3.6. At the instant shown in Figure 3.2b, the (u,v)-velocity field in Cartesian coordinates is $u = A(y^2 - x^2)/(x^2 + y^2)^2$, and $v = -2Axy/(x^2 + y^2)^2$ where A is a positive constant. Determine the equations for the streamlines by rearranging the first equality in (3.7) to read $udy - vdx = 0 = (\partial\psi/\partial y)dy + (\partial\psi/\partial x)dx$ and then looking for a solution in the form $\psi(x, y) = const$.

3.7. Determine the equivalent of the first equality in (3.7) for two-dimensional (r,θ)-polar coordinates, and then find the equation for the streamline that passes through (r_o, θ_o) when $\mathbf{u} = (u_r, u_\theta) = (A/r, B/r)$ where A and B are constants.

3.8. Determine the streamline, path line, and streak line that pass through the origin of coordinates at $t = t'$ when $u = U_o + \omega\xi_o\cos(\omega t)$ and $v = \omega\xi_o\sin(\omega t)$ in two-dimensional Cartesian coordinates where U_o is a constant horizontal velocity. Compare your results to those in Example 3.1 for $U_o \rightarrow 0$.

3.9. Compute and compare the streamline, path line, and streak line that pass through $(1,1,0)$ at $t = 0$ for the following Cartesian velocity field $\mathbf{u} = (x, -yt, 0)$.

3.10. Consider a time-dependent flow field in two-dimensional Cartesian coordinates where $u = \ell\tau/t^2$, $v = xy/\ell\tau$, and ℓ and τ are constant length and time scales, respectively.
a) Use dimensional analysis to determine the functional form of the streamline through $\mathbf{x}'$ at time t'.

b) Find the equation for the streamline through $\mathbf{x}'$ at time t' and put your answer in dimensionless form.

c) Repeat part b) for the path line through $\mathbf{x}'$ at time t'.

d) Repeat part b) for the streak line through $\mathbf{x}'$ at time t'.

3.11. The velocity components in an unsteady plane flow are given by $u = x/(1+t)$ and $v = 2y/(2+t)$. Determine equations for the streamlines and path lines subject to $\mathbf{x} = \mathbf{x}_0$ at $t = 0$.

3.12. Using the geometry and notation of Figure 3.8, prove (3.9).

3.13. Determine the unsteady, $\partial\mathbf{u}/\partial t$, and advective, $(\mathbf{u}\cdot\nabla)\mathbf{u}$, fluid acceleration terms for the following flow fields specified in Cartesian coordinates.

a) $\mathbf{u} = (u(y,z,t),0,0)$

b) $\mathbf{u} = \mathbf{\Omega}\times\mathbf{x}$ where $\mathbf{\Omega} = (0,0,\Omega_z(t))$

c) $\mathbf{u} = A(t)(x,-y,0)$

d) $\mathbf{u} = (U_o + u_o\sin(kx - \Omega t), 0, 0)$ where U_o, u_o, k, and Ω are positive constants

3.14. Consider the following Cartesian velocity field $\mathbf{u} = A(t)(f(x),g(y),h(z))$ where A, f, g, and h are nonconstant functions of only one independent variable.

a) Determine $\partial\mathbf{u}/\partial t$ and $(\mathbf{u}\cdot\nabla)\mathbf{u}$ in terms of A, f, g, and h, and their derivatives.

b) Determine A, f, g, and h when $D\mathbf{u}/Dt = 0$, $\mathbf{u} = 0$ at $\mathbf{x} = 0$, and $\mathbf{u}$ is finite for $t > 0$.

c) For the conditions in part b), determine the equation for the path line that passes through $\mathbf{x}_o$ at time t_o, and show directly that the acceleration $\mathbf{a}$ of the fluid particle that follows this path is zero.

3.15. If a velocity field is given by $u = ay$ and $v = 0$, compute the circulation around a circle of radius r_o that is centered on the origin. Check the result by using Stokes' theorem.

3.16. Consider a plane Couette flow of a viscous fluid confined between two flat plates a distance b apart. At steady state the velocity distribution is $u = Uy/b$ and $v = w = 0$, where the upper plate at $y = b$ is moving parallel to itself at speed U, and the lower plate is held stationary. Find the rates of linear strain, the rate of shear strain, and vorticity in this flow.

3.17. For the flow field $\mathbf{u} = \mathbf{U} + \mathbf{\Omega}\times\mathbf{x}$, where $\mathbf{U}$ and $\mathbf{\Omega}$ are constant linear- and angular-velocity vectors, use Cartesian coordinates to a) show that S_{ij} is zero, and b) determine R_{ij}.

3.18. Starting with a small rectangular volume element $\delta V = \delta x_1\delta x_2\delta x_3$, prove (3.14).

3.19. Let $Oxyz$ be a stationary frame of reference, and let the z-axis be parallel with fluid vorticity vector in the vicinity of O so that $\boldsymbol{\omega} = \nabla\times\mathbf{u} = \omega_z\mathbf{e}_z$ in this frame of reference. Now consider a second rotating frame of reference $Ox'y'z'$ having the same origin that rotates about the z-axis at angular rate $\Omega\mathbf{e}_z$. Starting from the kinematic relationship, $\mathbf{u} = (\Omega\mathbf{e}_z)\times\mathbf{x} + \mathbf{u}'$, show that in the vicinity of O the vorticity $\boldsymbol{\omega}' = \nabla'\times\mathbf{u}'$ in the rotating frame of reference can only be zero when $2\Omega = \omega_z$, where ∇' is the gradient operator in the primed coordinates. The following unit vector transformation rules may be of use: $\mathbf{e}'_x = \mathbf{e}_x\cos(\Omega t) + \mathbf{e}_y\sin(\Omega t)$, $\mathbf{e}'_y = -\mathbf{e}_x\sin(\Omega t) + \mathbf{e}_y\cos(\Omega t)$, and $\mathbf{e}'_z = \mathbf{e}_z$.

3.20. Consider a plane-polar area element having dimensions dr and $rd\theta$. For two-dimensional flow in this plane, evaluate the right-hand side of Stokes' theorem

$\int \boldsymbol{\omega} \cdot \mathbf{n} dA = \int \mathbf{u} \cdot d\mathbf{s}$ and thereby show that the expression for vorticity in plane-polar coordinates is: $\omega_z = \frac{1}{r}\frac{\partial}{\partial r}(ru_\theta) - \frac{1}{r}\frac{\partial u_r}{\partial \theta}$.

3.21. The velocity field of a certain flow is given by $u = 2xy^2 + 2xz^2$, $v = x^2 y$, and $w = x^2 z$. Consider the fluid region inside a spherical volume $x^2 + y^2 + z^2 = a^2$. Verify the validity of Gauss' theorem $\iiint\limits_{V} \nabla \cdot \mathbf{u} dV = \iint\limits_{A} \mathbf{u} \cdot \mathbf{n} dA$ by integrating over the sphere.

3.22. A flow field on the xy-plane has the velocity components $u = 3x + y$ and $v = 2x - 3y$. Show that the circulation around the circle $(x-1)^2 + (y-6)^2 = 4$ is 4π.

3.23. Consider solid-body rotation about the origin in two dimensions: $u_r = 0$ and $u_\theta = \omega_0 r$. Use a polar-coordinate element of dimension $rd\theta$ and dr, and verify that the circulation is vorticity times area. (In Section 3.5 this was verified for a circular element surrounding the origin.)

3.24. Consider the following steady Cartesian velocity field $\mathbf{u} = \left(\frac{-Ay}{(x^2+y^2)^\beta}, \frac{+Ax}{(x^2+y^2)^\beta}, 0\right)$.

   a) Determine the streamline that passes through $\mathbf{x} = (x_o, y_o, 0)$.
   b) Compute R_{ij} for this velocity field.
   c) For $A > 0$, explain the sense of rotation (i.e., clockwise or counterclockwise) for fluid elements for $\beta < 1$, $\beta = 1$, and $\beta > 1$.

3.25. Using indicial notation (and no vector identities), show that the acceleration $\mathbf{a}$ of a fluid particle is given by $\mathbf{a} = \partial \mathbf{u}/\partial t + \nabla\left(\frac{1}{2}|\mathbf{u}|^2\right) + \boldsymbol{\omega} \times \mathbf{u}$, where $\boldsymbol{\omega}$ is the vorticity.

3.26. Starting from (3.29), show that the maximum u_θ in a Gaussian vortex occurs when $1 + 2(r^2/\sigma^2) = \exp(r^2/\sigma^2)$. Verify that this implies $r \approx 1.12091\sigma$.

3.27. [1]For the following time-dependent volumes $V^*(t)$ and smooth single-valued integrand functions F, choose an appropriate coordinate system and show that $(d/dt)\int_{V^*(t)} F dV$ obtained from (3.30) is equal to that obtained from (3.35).

   a) $V^*(t) = L_1(t)L_2 L_3$ is a rectangular solid defined by $0 \leq x_i \leq L_i$, where L_1 depends on time while L_2 and L_3 are constants, and the integrand function $F(x_1,t)$ depends only on the first coordinate and time.
   b) $V^*(t) = (\pi/4)d^2(t)L$ is a cylinder defined by $0 \leq R \leq d(t)/2$ and $0 \leq z \leq L$, where the cylinder's diameter d depends on time while its length L is constant, and the integrand function $F(R,t)$ depends only on the distance from the cylinder's axis and time.
   c) $V^*(t) = (\pi/6)D^3(t)$ is a sphere defined by $0 \leq r \leq D(t)/2$ where the sphere's diameter D depends on time, and the integrand function $F(r,t)$ depends only on the radial distance from the center of the sphere and time.

3.28. Starting from (3.35), set $F = 1$ and derive (3.14) when $\mathbf{b} = \mathbf{u}$ and $V^*(t) = \delta V \to 0$.

3.29. For a smooth, single-valued function $F(\mathbf{x})$ that only depends on space and an arbitrarily shaped control volume that moves with velocity $\mathbf{b}(t)$ that only depends on time, show that $(d/dt)\int_{V^*(t)} F(\mathbf{x})dV = \mathbf{b} \cdot \left(\int_{V^*(t)} \nabla F(\mathbf{x})dV\right)$.

3.30. Show that (3.35) reduces to (3.5) when $V^*(t) = \delta V \to 0$ and the control surface velocity $\mathbf{b}$ is equal to the fluid velocity $\mathbf{u}(\mathbf{x},t)$.

---

[1]Developed from Problem 1.9 on page 48 in Thompson (1972)

Literature Cited

Meriam, J. L., and Kraige, L. G. (2007). *Engineering Mechanics, Dynamics* (6th ed). Hoboken, NJ: John Wiley and Sons, Inc.

Raffel, M., Willert, C., and Kompenhans, J. (1998). *Particle Image Velocimetry.* Berlin: Springer.

Riley, K. F., Hobson, M. P., and Bence, S. J. (1998). *Mathematical Methods for Physics and Engineering* (3rd ed.). Cambridge, UK: Cambridge University Press.

Samimy, M., Breuer, K. S., Leal, L. G., and Steen, P. H. (2003). *A Gallery of Fluid Motion.* Cambridge, UK: Cambridge University Press.

Thompson, P. A. (1972). *Compressible-Fluid Dynamics.* New York: McGraw-Hill.

Van Dyke, M. (1982). *An Album of Fluid Motion.* Stanford, California: Parabolic Press.

Supplemental Reading

Aris, R. (1962). *Vectors, Tensors, and the Basic Equations of Fluid Mechanics.* Englewood Cliffs, NJ: Prentice-Hall. (The distinctions among streamlines, path lines, and streak lines in unsteady flows are explained, with examples.)

Prandtl, L., and Tietjens, O. C. (1934). *Fundamentals of Hydro- and Aeromechanics.* New York: Dover Publications. (Chapter V contains a simple but useful treatment of kinematics.)

Prandtl, L., and Tietjens, O. G. (1934). *Applied Hydro- and Aeromechanics.* New York: Dover Publications. (This volume contains classic photographs from Prandtl's laboratory.)

OUTLINE

4.1. Introduction   96

4.2. Conservation of Mass   96

4.3. Stream Functions   99

4.4. Conservation of Momentum   101

4.5. Constitutive Equation for a Newtonian Fluid   111

4.6. Navier-Stokes Momentum Equation   114

4.7. Noninertial Frame of Reference   116

4.8. Conservation of Energy   121

4.9. Special Forms of the Equations   125

4.10. Boundary Conditions   137

4.11. Dimensionless Forms of the Equations and Dynamic Similarity   143

Exercises   151

Literature Cited   168

Supplemental Reading   168

CHAPTER OBJECTIVES

- To present a derivation of the governing equations for moving fluids starting from the principles of mass, momentum, and energy conservation for a material volume.

- To illustrate the application of the integral forms of the mass and momentum conservation equations to stationary, steadily moving, and accelerating control volumes.

- To develop the constitutive equation for a Newtonian fluid and provide the Navier-Stokes differential momentum equation.

- To show how the differential momentum equation is modified in noninertial frames of reference.

- To develop the differential energy equation and highlight its internal coupling between mechanical and thermal energies.

- To present several common extensions and simplified forms of the equations of motion.

- To derive and describe the dimensionless numbers that appear naturally when the equations of motion are put in dimensionless form.

4.1. INTRODUCTION

The governing principles in fluid mechanics are the conservation laws for mass, momentum, and energy. These laws are presented in this order in this chapter and can be stated in *integral* form, applicable to an extended region, or in *differential* form, applicable at a point or to a fluid particle. Both forms are equally valid and may be derived from each other. The integral forms of the equations of motion are stated in terms of the evolution of a control volume and the fluxes of mass, momentum, and energy that cross its control surface. The integral forms are typically useful when the spatial extent of potentially complicated flow details are small enough for them to be neglected and an average or integral flow property, such as a mass flux, a surface pressure force, or an overall velocity or acceleration, is sought. The integral forms are commonly taught in first courses on fluid mechanics where they are specialized to a variety of different control volume conditions (stationary, steadily moving, accelerating, deforming, etc.). Nevertheless, the integral forms of the equations are developed here for completeness and to unify the various control volume concepts.

The differential forms of the equations of motion are coupled nonlinear partial differential equations for the dependent flow-field variables of density, velocity, pressure, temperature, etc. Thus, the differential forms are often more appropriate for detailed analysis when field information is needed instead of average or integrated quantities. However, both approaches can be used for either scenario when appropriately refined for the task at hand. In the development of the differential equations of fluid motion, attention is given to determining when a solvable system of equations has been found by comparing the number of equations with the number of unknown dependent field variables. At the outset of this monitoring effort, the fluid's thermodynamic characteristics are assumed to provide as many as two equations, the thermal and caloric equations of state (1.12).

The development of the integral and differential equations of fluid motion presented in this chapter is not unique, and alternatives are readily found in other references. The version presented here is primarily based on that in Thompson (1972).

4.2. CONSERVATION OF MASS

Setting aside nuclear reactions and relativistic effects, mass is neither created nor destroyed. Thus, individual mass elements—molecules, grains, fluid particles, etc.—may be tracked within a flow field because they will not disappear and new elements will not spontaneously appear. The equations representing conservation of mass in a flowing fluid are based on the principle that the mass of a specific collection of neighboring fluid particles is constant. The volume occupied by a specific collection of fluid particles is called a *material volume* $V(t)$. Such a volume moves and deforms within a fluid flow so that it always contains the same mass elements; none enter the volume and none leave it. This implies that a material volume's surface $A(t)$, a material surface, must move at the local fluid velocity $\mathbf{u}$ so that fluid particles inside $V(t)$ remain inside and fluid particles outside $V(t)$ remain outside. Thus, a statement of conservation of mass for a material volume in a flowing fluid is:

$$\frac{d}{dt} \int_{V(t)} \rho(\mathbf{x},t)dV = 0. \tag{4.1}$$

where ρ is the fluid density. Figure 3.18 depicts a material volume when the control surface velocity $\mathbf{b}$ is equal to $\mathbf{u}$. The primary concept here is equivalent to an infinitely flexible, perfectly sealed thin-walled balloon containing fluid. The balloon's contents play the role of the material volume $V(t)$ with the balloon itself defining the material surface $A(t)$. And, because the balloon is sealed, the total mass of fluid inside the balloon remains constant as the balloon moves, expands, contracts, or deforms.

Based on (4.1), the principle of mass conservation clearly constrains the fluid density. The implications of (4.1) for the fluid velocity field may be better displayed by using Reynolds transport theorem (3.35) with $F = \rho$ and $\mathbf{b} = \mathbf{u}$ to expand the time derivative in (4.1):

$$\int_{V(t)} \frac{\partial \rho(\mathbf{x},t)}{\partial t}dV + \int_{A(t)} \rho(\mathbf{x},t)\mathbf{u}(\mathbf{x},t)\cdot\mathbf{n}dA = 0. \tag{4.2}$$

This is a mass-balance statement between integrated density changes within $V(t)$ and integrated motion of its surface $A(t)$. Although general and correct, (4.2) may be hard to utilize in practice because the motion and evolution of $V(t)$ and $A(t)$ are determined by the flow, which may be unknown.

To develop the integral equation that represents mass conservation for an *arbitrarily moving* control volume $V^*(t)$ with surface $A^*(t)$, (4.2) must be modified to involve integrations over $V^*(t)$ and $A^*(t)$. This modification is motivated by the frequent need to conserve mass within a volume that is not a material volume, for example a stationary control volume. The first step in this modification is to set $F = \rho$ in (3.35) to obtain

$$\frac{d}{dt} \int_{V^*(t)} \rho(\mathbf{x},t)dV - \int_{V^*(t)} \frac{\partial \rho(\mathbf{x},t)}{\partial t}dV - \int_{A^*(t)} \rho(\mathbf{x},t)\mathbf{b}\cdot\mathbf{n}dA = 0. \tag{4.3}$$

The second step is to choose the arbitrary control volume $V^*(t)$ to be instantaneously coincident with material volume $V(t)$ so that *at the moment of interest* $V(t) = V^*(t)$ and $A(t) = A^*(t)$. At this coincidence moment, the $(d/dt)\int \rho dV$-terms in (4.1) and (4.3) are not equal; however, the volume integration of $\partial\rho/\partial t$ in (4.2) is equal to that in (4.3) and the surface integral of $\rho\mathbf{u}\cdot\mathbf{n}$ over $A(t)$ is equal to that over $A^*(t)$:

$$\int_{V^*(t)} \frac{\partial \rho(\mathbf{x},t)}{\partial t}dV = \int_{V(t)} \frac{\partial \rho(\mathbf{x},t)}{\partial t}dV = -\int_{A(t)} \rho(\mathbf{x},t)\mathbf{u}(\mathbf{x},t)\cdot\mathbf{n}dA = -\int_{A^*(t)} \rho(\mathbf{x},t)\mathbf{u}(\mathbf{x},t)\cdot\mathbf{n}dA$$

$$\tag{4.4}$$

where the middle equality follows from (4.2). The two ends of (4.4) allow the central volume-integral term in (4.3) to be replaced by a surface integral to find:

$$\frac{d}{dt} \int_{V^*(t)} \rho(\mathbf{x}, t) dV + \int_{A^*(t)} \rho(\mathbf{x}, t)(\mathbf{u}(\mathbf{x}, t) - \mathbf{b}) \cdot \mathbf{n} dA = 0, \tag{4.5}$$

where $\mathbf{u}$ and $\mathbf{b}$ must both be observed in the same frame of reference; they are not otherwise restricted. This is the general integral statement of conservation of mass for an arbitrarily moving control volume. It can be specialized to stationary, steadily moving, accelerating, or deforming control volumes by appropriate choice of $\mathbf{b}$. In particular, when $\mathbf{b} = \mathbf{u}$, the arbitrary control volume becomes a material volume and (4.5) reduces to (4.1).

The differential equation that represents mass conservation is obtained by applying Gauss' divergence theorem (2.30) to the surface integration in (4.2):

$$\int_{V(t)} \frac{\partial \rho(\mathbf{x}, t)}{\partial t} dV + \int_{A(t)} \rho(\mathbf{x}, t)\mathbf{u}(\mathbf{x}, t) \cdot \mathbf{n} dA = \int_{V(t)} \left\{ \frac{\partial \rho(\mathbf{x}, t)}{\partial t} + \nabla \cdot \left(\rho(\mathbf{x}, t)\mathbf{u}(\mathbf{x}, t) \right) \right\} dV = 0. \tag{4.6}$$

The final equality can only be possible if the integrand vanishes at every point in space. If the integrand did not vanish at every point in space, then integrating (4.6) in a small volume around a point where the integrand is nonzero would produce a nonzero integral. Thus, (4.6) requires:

$$\frac{\partial \rho(\mathbf{x}, t)}{\partial t} + \nabla \cdot (\rho(\mathbf{x}, t)\mathbf{u}(\mathbf{x}, t)) = 0, \text{ or, in index notation } \frac{\partial \rho}{\partial t} + \frac{\partial}{\partial x_i}(\rho u_i) = 0. \tag{4.7}$$

This relationship is called the *continuity equation*. It expresses the principle of conservation of mass in differential form, but is insufficient for fully determining flow fields because it is a single equation that involves two field quantities, ρ and $\mathbf{u}$, and $\mathbf{u}$ is a vector with three components.

The second term in (4.7) is the divergence of the mass-density flux $\rho\mathbf{u}$. Such *flux divergence* terms frequently arise in conservation statements and can be interpreted as the net loss at a point due to divergence of a flux. For example, the local ρ will increase with time if $\nabla \cdot (\rho\mathbf{u})$ is negative. Flux divergence terms are also called *transport* terms because they transfer quantities from one region to another without making a net contribution over the entire field. When integrated over the entire domain of interest, their contribution vanishes if there are no sources at the boundaries.

The continuity equation may alternatively be written using (3.5) the definition of D/Dt and $\partial(\rho u_i)/\partial x_i = u_i \partial \rho/\partial x_i + \rho \partial u_i/\partial x_i$ [see (B3.6)]:

$$\frac{1}{\rho(\mathbf{x}, t)} \frac{D}{Dt}\rho(\mathbf{x}, t) + \nabla \cdot \mathbf{u}(\mathbf{x}, t) = 0. \tag{4.8}$$

The derivative $D\rho/Dt$ is the time rate of change of fluid density following a fluid particle. It will be zero for *constant density* flow where ρ = constant throughout the flow field, and for *incompressible* flow where the density of fluid particles does not change but different fluid particles may have different density:

$$\frac{D\rho}{Dt} \equiv \frac{\partial \rho}{\partial t} + \mathbf{u} \cdot \nabla \rho = 0. \tag{4.9}$$

Taken together, (4.8) and (4.9) imply:

$$\nabla \cdot \mathbf{u} = 0 \tag{4.10}$$

for incompressible flows. Constant density flows are a subset of incompressible flows; ρ = constant is a solution of (4.9) but it is not a general solution. A fluid is usually called *incompressible* if its density does not change with *pressure*. Liquids are almost incompressible. Gases are compressible, but for flow speeds less than ~100 m/s (that is, for Mach numbers < 0.3) the fractional change of absolute pressure in an air flow is small. In this and several other situations, density changes in the flow are also small and (4.9) and (4.10) are valid.

The general form of the continuity equation (4.7) is typically required when the derivative $D\rho/Dt$ is nonzero because of changes in the pressure, temperature, or molecular composition of fluid particles.

4.3. STREAM FUNCTIONS

Consider the steady form of the continuity equation (4.7),

$$\nabla \cdot (\rho \mathbf{u}) = 0. \tag{4.11}$$

The divergence of the curl of any vector field is identically zero (see Exercise 2.19), so $\rho \mathbf{u}$ will satisfy (4.11) when written as the curl of a vector potential $\mathbf{\Psi}$,

$$\rho \mathbf{u} = \nabla \times \mathbf{\Psi}, \tag{4.12}$$

which can be specified in terms of two scalar functions: $\mathbf{\Psi} = \chi \nabla \psi$. Putting this specification for $\mathbf{\Psi}$ into (4.12) produces $\rho \mathbf{u} = \nabla \chi \times \nabla \psi$, because the curl of any gradient is identically zero (see Exercise 2.20). Furthermore, $\nabla \chi$ is perpendicular to surfaces of constant χ, and $\nabla \psi$ is perpendicular to surfaces of constant ψ, so the mass flux $\rho \mathbf{u} = \nabla \chi \times \nabla \psi$ will be parallel to surfaces of constant χ and constant ψ. Therefore, three-dimensional streamlines are the intersections of the two stream surfaces, or stream functions in a three-dimensional flow.

The situation is illustrated in Figure 4.1. Consider two members of each of the families of the two stream functions $\chi = a$, $\chi = b$, $\psi = c$, $\psi = d$. The intersections shown as darkened lines in Figure 4.1 are the streamlines. The mass flux $\dot{m}$ through the surface A bounded by the four stream functions (shown in gray in Figure 4.1) is calculated with area element dA having $\mathbf{n}$ as shown and Stokes' theorem.

Defining the mass flux $\dot{m}$ through A, and using Stokes' theorem produces

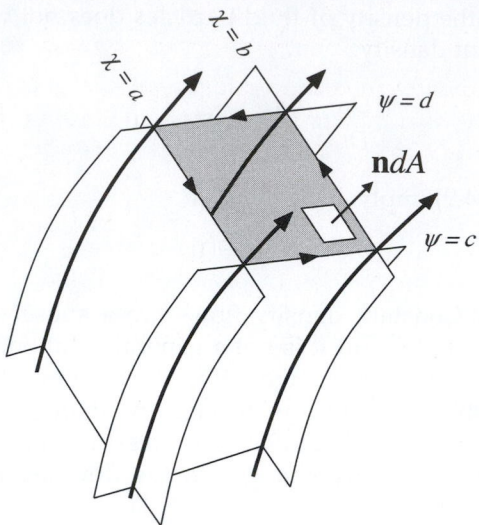

FIGURE 4.1 Isometric view of two members from each family of stream surfaces. The solid curves are streamlines and these lie at the intersections of the surfaces. The unit vector **n** points in the stream direction and is perpendicular to the gray surface that is bordered by the nearly rectangular curve C made up of segments defined by $\chi = a$, $\chi = b$, $\psi = c$, and $\psi = d$. The arrows on this border indicate the integration direction for Stokes' theorem.

$$\dot{m} = \int_A \rho \mathbf{u} \cdot \mathbf{n} dA = \int_A (\boldsymbol{\nabla} \times \boldsymbol{\Psi}) \cdot \mathbf{n} dA = \int_C \boldsymbol{\Psi} \cdot d\mathbf{s} = \int_C \chi \boldsymbol{\nabla} \psi \cdot d\mathbf{s} = \int_C \chi d\psi$$
$$= b(d - c) + a(c - d) = (b - a)(d - c).$$

Here we have used the vector identity $\boldsymbol{\nabla} \psi \cdot d\mathbf{s} = d\psi$. The mass flow rate of the stream tube defined by adjacent members of the two families of stream functions is just the product of the differences of the numerical values of the respective stream functions.

As a special case, consider two-dimensional flow in (x, y)-Cartesian coordinates where all the streamlines lie in $z = $ constant planes. In this situation, z is one of the three-dimensional stream functions, so we can set $\chi = -z$, where the sign is chosen to obey the usual convention. This produces $\boldsymbol{\nabla} \chi = -\mathbf{e}_z$, so $\rho \mathbf{u} = -\mathbf{e}_z \times \boldsymbol{\nabla} \psi$, or

$$\rho u = \partial \psi / \partial y, \quad \text{and} \quad \rho v = -\partial \psi / \partial x$$

in conformity with Exercise 4.7.

Similarly, for axisymmetric three-dimensional flow in cylindrical polar coordinates (Figure 3.3c), all the streamlines lie in $\varphi = $ constant planes that contain the z-axis so $\chi = -\varphi$ is one of the stream functions. This produces $\boldsymbol{\nabla} \chi = -R^{-1}\mathbf{e}_\varphi$ and $\rho \mathbf{u} = \rho(u_R, u_z) = -R^{-1}\mathbf{e}_\varphi \times \boldsymbol{\nabla} \psi$, or

$$\rho u_R = -R^{-1}(\partial \psi / \partial z), \quad \text{and} \quad \rho u_z = R^{-1}(\partial \psi / \partial R).$$

We note here that if the density is constant, mass conservation reduces to $\boldsymbol{\nabla} \cdot \mathbf{u} = 0$ (steady or not) and the entire preceding discussion follows for **u** rather than $\rho \mathbf{u}$ with the interpretation of stream function values in terms of volumetric flux rather than mass flux.

4.4. CONSERVATION OF MOMENTUM

In this section, the momentum-conservation equivalent of (4.5) is developed from Newton's second law, the fundamental principle governing fluid momentum. When applied to a material volume $V(t)$ with surface area $A(t)$, Newton's second law can be stated directly as:

$$\frac{d}{dt} \int_{V(t)} \rho(\mathbf{x}, t)\mathbf{u}(\mathbf{x}, t)dV = \int_{V(t)} \rho(\mathbf{x}, t)\mathbf{g}dV + \int_{A(t)} \mathbf{f}(\mathbf{n}, \mathbf{x}, t)dA, \qquad (4.13)$$

where $\rho\mathbf{u}$ is the momentum per unit volume of the flowing fluid, $\mathbf{g}$ is the body force per unit mass acting on the fluid within $V(t)$, $\mathbf{f}$ is the surface force per unit area acting on $A(t)$, and $\mathbf{n}$ is the outward normal on $A(t)$. The implications of (4.13) are better displayed when the time derivative is expanded using Reynolds transport theorem (3.35) with $F = \rho\mathbf{u}$ and $\mathbf{b} = \mathbf{u}$:

$$\int_{V(t)} \frac{\partial}{\partial t}(\rho(\mathbf{x}, t)\mathbf{u}(\mathbf{x}, t))dV + \int_{A(t)} \rho(\mathbf{x}, t)\mathbf{u}(\mathbf{x}, t)(\mathbf{u}(\mathbf{x}, t) \cdot \mathbf{n})dA$$

$$= \int_{V(t)} \rho(\mathbf{x}, t)\mathbf{g}dV + \int_{A(t)} \mathbf{f}(\mathbf{n}, \mathbf{x}, t)dA. \qquad (4.14)$$

This is a momentum-balance statement between integrated momentum changes within $V(t)$, integrated momentum contributions from the motion of $A(t)$, and integrated volume and surface forces. It is the momentum conservation equivalent of (4.2).

To develop an integral equation that represents momentum conservation for an arbitrarily moving control volume $V^*(t)$ with surface $A^*(t)$, (4.14) must be modified to involve integrations over $V^*(t)$ and $A^*(t)$. The steps in this process are entirely analogous to those taken between (4.2) and (4.5) for conservation of mass. First set $F = \rho u$ in (3.35) and rearrange it to obtain:

$$\int_{V^*(t)} \frac{\partial}{\partial t}(\rho(\mathbf{x}, t)\mathbf{u}(\mathbf{x}, t))dV = \frac{d}{dt}\int_{V^*(t)} \rho(\mathbf{x}, t)\mathbf{u}(\mathbf{x}, t)dV - \int_{A^*(t)} \rho(\mathbf{x}, t)\mathbf{u}(\mathbf{x}, t)\mathbf{b} \cdot \mathbf{n}dA = 0. \quad (4.15)$$

Then choose $V^*(t)$ to be instantaneously coincident with $V(t)$ so that at the moment of interest:

$$\int_{V(t)} \frac{\partial}{\partial t}(\rho(\mathbf{x}, t)\mathbf{u}(\mathbf{x}, t))dV = \int_{V^*(t)} \frac{\partial}{\partial t}(\rho(\mathbf{x}, t)\mathbf{u}(\mathbf{x}, t))dV,$$

$$\int_{A(t)} \rho(\mathbf{x}, t)\mathbf{u}(\mathbf{x}, t)(\mathbf{u}(\mathbf{x}, t) \cdot \mathbf{n})dA = \int_{A^*(t)} \rho(\mathbf{x}, t)\mathbf{u}(\mathbf{x}, t)(\mathbf{u}(\mathbf{x}, t) \cdot \mathbf{n})dA,$$

$$\int_{V(t)} \rho(\mathbf{x}, t)\mathbf{g}dV = \int_{V^*(t)} \rho(\mathbf{x}, t)\mathbf{g}dV, \quad \text{and} \quad \int_{A(t)} \mathbf{f}(\mathbf{n}, \mathbf{x}, t)dA = \int_{A^*(t)} \mathbf{f}(\mathbf{n}, \mathbf{x}, t)dA.$$

$$(4.16a, 4.16b, 4.16c, 4.16d)$$

Now substitute (4.16a) into (4.15) and use this result plus (4.16b, 4.16c, 4.16d) to convert (4.14) to:

$$\frac{d}{dt} \int\limits_{V^*(t)} \rho(\mathbf{x},t)\mathbf{u}(\mathbf{x},t)dV + \int\limits_{A^*(t)} \rho(\mathbf{x},t)\mathbf{u}(\mathbf{x},t)(\mathbf{u}(\mathbf{x},t) - \mathbf{b})\cdot\mathbf{n}dA$$

$$= \int\limits_{V^*(t)} \rho(\mathbf{x},t)\mathbf{g}dV + \int\limits_{A^*(t)} \mathbf{f}(\mathbf{n},\mathbf{x},t)dA. \qquad (4.17)$$

This is the general integral statement of momentum conservation for an arbitrarily moving control volume. Just like (4.5), it can be specialized to stationary, steadily moving, accelerating, or deforming control volumes by appropriate choice of $\mathbf{b}$. For example, when $\mathbf{b} = \mathbf{u}$, the arbitrary control volume becomes a material volume and (4.17) reduces to (4.13).

At this point, the forces in (4.13), (4.14), and (4.17) merit some additional description that facilitates the derivation of the differential equation representing momentum conservation and allows its simplification under certain circumstances.

The body force, $\rho\mathbf{g}dV$, acting on the fluid element dV does so without physical contact. Body forces commonly arise from gravitational, magnetic, electrostatic, or electromagnetic *force fields*. In addition, in accelerating or rotating frames of reference, *fictitious* body forces arise from the frame's noninertial motion (see Section 4.7). By definition body forces are distributed through the fluid and are proportional to mass (or electric charge, electric current, etc.). In this book, body forces are specified per unit mass and carry the units of acceleration.

Body forces may be conservative or nonconservative. *Conservative body forces* are those that can be expressed as the gradient of a potential function:

$$\mathbf{g} = -\nabla\Phi \quad \text{or} \quad g_j = -\partial\Phi/\partial x_j, \qquad (4.18)$$

where Φ is called the *force potential*; it has units of energy per unit mass. When the z-axis points vertically upward, the force potential for gravity is $\Phi = gz$, where g is the acceleration of gravity, and (4.18) produces $\mathbf{g} = -g\mathbf{e}_z$. Forces satisfying (4.18) are called *conservative* because the work done by conservative forces is independent of the path, and the sum of fluid-particle kinetic and potential energies is conserved when friction is absent.

Surface forces, $\mathbf{f}$, act on fluid elements through direct contact with the surface of the element. They are proportional to the contact area and carry units of stress (force per unit area). Surface forces are commonly resolved into components normal and tangential to the contact area. Consider an arbitrarily oriented element of area dA in a fluid (Figure 2.5). If $\mathbf{n}$ is the surface normal with components n_i, then from (2.15) the components f_j of the surface force per unit area $\mathbf{f}(\mathbf{n}, \mathbf{x}, t)$ on this element are $f_j = n_i\tau_{ij}$, where τ_{ij} is the stress tensor. Thus, the normal component of $\mathbf{f}$ is $\mathbf{n}\cdot\mathbf{f} = n_if_i$, while the tangential component is the vector $\mathbf{f} - (\mathbf{n}\cdot\mathbf{f})\mathbf{n}$ which has components $f_k - (n_if_i)n_k$.

Other forces that influence fluid motion are surface- and interfacial-tension forces that act on lines or curves embedded within interfaces between liquids and gases or between immiscible liquids (see Figure 1.4). Although these forces are commonly important in flows with such interfaces, they do not appear directly in the equations of motion, entering instead through the boundary conditions.

Before proceeding to the differential equation representing momentum conservation, the use of (4.5) and (4.17) for stationary, moving, and accelerating control volumes having a variety of sizes and shapes is illustrated through a few examples. In all four examples, equations representing mass and momentum conservation must be solved simultaneously.

EXAMPLE 4.1

A long bar with constant cross section is held perpendicular to a uniform horizontal flow of speed U_∞, as shown in Figure 4.2. The flowing fluid has density ρ and viscosity μ (both constant). The bar's cross section has characteristic transverse dimension d, and the span of the bar is l with $l \gg d$. The average horizontal velocity profile measured downstream of the bar is $U(y)$, which is less than U_∞ due to the presence of the bar. Determine the required force per unit span, $-F_D/l$, applied to the ends of the bar to hold it in place. Assume the flow is steady and two dimensional in the plane shown. Ignore body forces.

Solution

Before beginning, it is important to explain the sign convention for fluid dynamic drag forces. The drag force on an inanimate object is *the force applied to the object by the fluid*. Thus, for stationary objects, drag forces are positive in the downstream direction, the direction the object would accelerate if released. However, the control volume laws are written for *forces applied to the contents of the volume*. Thus, from Newton's third law, a positive drag force on an object implies a negative force on the fluid. Therefore, the F_D appearing in Figure 4.2 is a positive number and this will be borne out by the final results. Here we also note that since the horizontal velocity downstream of the bar, the wake velocity $U(y)$, is less than U_∞, the fluid has been decelerated inside the control volume and this is consistent with a force from the body opposing the motion of the fluid as shown.

The basic strategy is to select a stationary control volume, and then use (4.5) and (4.17) to determine the force F_D that the body exerts on the fluid per unit span. The first quantitative step in the solution is to select a rectangular control volume with flat control surfaces aligned with the coordinate directions. The inlet, outlet, and top and bottom sides of such a control volume are shown in Figure 4.2. The vertical sides parallel to the x-y plane are not shown. However, the flow does not vary in the third direction and is everywhere parallel to these surfaces so these merely need be selected a comfortable distance l apart. The inlet control surface should be far enough upstream of the bar so that the inlet fluid velocity is $U_\infty \mathbf{e}_x$, the pressure is p_∞, and both are uniform.

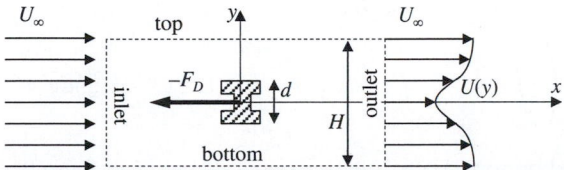

FIGURE 4.2 Momentum and mass balance for flow past long bar of constant cross section placed perpendicular to the flow. The intersection of the recommended stationary control volume with the x-y plane is shown with dashed lines. The force $-F_D$ holds the bar in place and slows the fluid that enters the control volume.

The top and bottom control surfaces should be separated by a distance H that is large enough so that these boundaries are free from shear stresses, and the horizontal velocity and pressure are so close to U_∞ and p_∞ that any difference can be ignored. And finally, the outlet surface should be far enough downstream so that streamlines are nearly horizontal there and the pressure can again be treated as equal to p_∞.

For steady flow and the chosen stationary volume, the control surface velocity is $\mathbf{b} = 0$ and the time derivative terms in (4.5) and (4.17) are both zero. In addition, the surface force integral contributes $-F_D\mathbf{e}_x$ where the beam crosses the control volume's vertical sides parallel to the x-y plane. The remainder of the surface force integral contains only pressure terms since the shear stress is zero on the control surface boundaries. After setting the pressure to p_∞ on all control surfaces, (4.5) and (4.17) simplify to:

$$\int_{A^*(t)} \rho\mathbf{u}(\mathbf{x})\cdot\mathbf{n}dA = 0, \quad \text{and} \quad \int_{A^*} \rho\mathbf{u}(\mathbf{x})\mathbf{u}(\mathbf{x})\cdot\mathbf{n}dA = -\int_{A^*} p_\infty\mathbf{n}dA - F_D\mathbf{e}_x.$$

In this case the pressure integral may be evaluated immediately by using Gauss' divergence theorem:

$$\int_{A^*} p_\infty\mathbf{n}dA = \int_{V^*} \nabla p_\infty dV = 0,$$

with the final value (zero) occurring because p_∞ is a constant. After this simplification, denote the fluid velocity components by $(u,v) = \mathbf{u}$, and evaluate the mass and x-momentum conservation equations:

$$-\int_{inlet} \rho U_\infty l dy + \int_{top} \rho v l dx - \int_{bottom} \rho v l dx + \int_{outlet} \rho U(y) l dy = 0, \quad \text{and}$$

$$-\int_{inlet} \rho U_\infty^2 l dy + \int_{top} \rho U_\infty v l dx - \int_{bottom} \rho U_\infty v l dx + \int_{outlet} \rho U^2(y) l dy = -F_D$$

where $\mathbf{u}\cdot\mathbf{n}dA$ is: $-U_\infty l dy$ on the inlet surface, $+v l dx$ on the top surface, $-v l dx$ on the bottom surface, and $+U(y)l dy$ on the outlet surface where l is the span of the flow into the page. Dividing both equations by ρl, and combining like integrals produces:

$$\int_{top} v dx - \int_{bottom} v dx = \int_{-H/2}^{+H/2} (U_\infty - U(y))dy, \quad \text{and}$$

$$U_\infty \left(\int_{top} v dx - \int_{bottom} v dx \right) + \int_{-H/2}^{+H/2} (U^2(y) - U_\infty^2)dy = -F_D/\rho l.$$

Eliminating the top and bottom control surface integrals between these two equations leads to:

$$F_D/l = \rho \int_{-H/2}^{+H/2} U(y)(U_\infty - U(y))dy,$$

which produces a positive value of F_D when $U(y)$ is less than U_∞. An essential feature of this analysis is that there are nonzero mass fluxes through the top and bottom control surfaces. The final formula here is genuinely useful in experimental fluid mechanics since it allows F_D/l to be determined from single-component velocity measurements made in the wake of an object.

EXAMPLE 4.2

Using a stream-tube control volume of differential length ds, derive the Bernoulli equation, $(\frac{1}{2})\rho U^2 + gz + p/\rho =$ constant along a streamline, for steady, inviscid, constant density flow where U is the local flow speed.

Solution

The basic strategy is to use a stationary stream-tube-element control volume, (4.5), and (4.17) to determine a simple differential relationship that can be integrated along a streamline. The geometry is shown in Figure 4.3. For steady inviscid flow and a stationary control volume, the control surface velocity $\mathbf{b} = 0$, the surface friction forces are zero, and the time derivative terms in (4.5) and (4.17) are both zero. Thus, these two equations simplify to:

$$\int_{A^*(t)} \rho \mathbf{u}(\mathbf{x}) \cdot \mathbf{n} dA = 0 \quad \text{and} \quad \int_{A^*(t)} \rho \mathbf{u}(\mathbf{x})\mathbf{u}(\mathbf{x}) \cdot \mathbf{n} dA = \int_{V^*(t)} \rho \mathbf{g} dV - \int_{A^*(t)} p \mathbf{n} dA.$$

The geometry of the volume plays an important role here. The nearly conical curved surface is tangent to the velocity while the inlet and outlet areas are perpendicular to it. Thus, $\mathbf{u} \cdot \mathbf{n} dA$ is: $-U dA$ on the inlet surface, zero on the nearly conical curved surface, and $+[U + (\partial U/\partial s)ds]dA$ on the outlet surface. Therefore, conservation of mass with constant density leads to

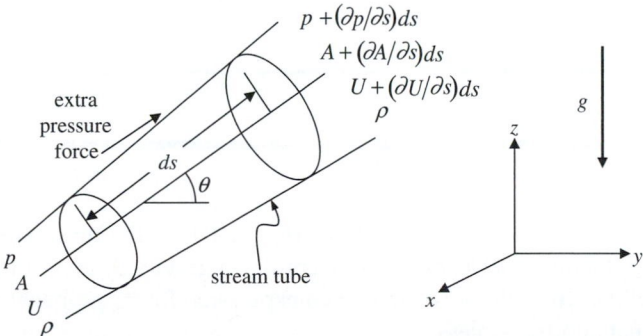

FIGURE 4.3 Momentum and mass balance for a short segment of a stream tube in steady inviscid constant density flow. Here, the inlet and outlet areas are perpendicular to the flow direction, and they are small enough so that only first-order corrections in the stream direction need to be considered. The alignment of gravity and stream tube leads to a vertical change of $\sin\theta\, ds = dz$ between its two ends. The area difference between the two ends of the stream tube leads to an extra pressure force.

$$-\rho UA + \rho\left(U + \frac{\partial U}{\partial s}ds\right)\left(A + \frac{\partial A}{\partial s}ds\right) = 0,$$

where first-order variations in U and A in the stream-wise direction are accounted for. Now consider the stream-wise component of the momentum equation recalling that $\mathbf{u} = U\mathbf{e}_u$ and setting $\mathbf{g} = -g\mathbf{e}_z$. For inviscid flow, the only surface force is pressure, so the simplified version of (4.17) becomes

$$-\rho U^2 A + \rho\left(U + \frac{\partial U}{\partial s}ds\right)^2\left(A + \frac{\partial A}{\partial s}ds\right)$$

$$= -\rho g \sin\theta\left(A + \frac{\partial A}{\partial s}\frac{ds}{2}\right)ds + pA + \left(p + \frac{\partial p}{\partial s}\frac{ds}{2}\right)\frac{\partial A}{\partial s}ds - \left(p + \frac{\partial p}{\partial s}ds\right)\left(A + \frac{\partial A}{\partial s}ds\right).$$

Here, the middle pressure term comes from the extra pressure force on the nearly conical surface of the stream tube.

To reach the final equation, use the conservation of mass result to simplify the flux terms on the left side of the stream-wise momentum equation. Then, simplify the pressure contributions by canceling common terms, and note that $\sin\theta\, ds = dz$ to find

$$-\rho U^2 A + \rho U\left(U + \frac{\partial U}{\partial s}ds\right)A = \rho UA\frac{\partial U}{\partial s}ds$$

$$= -\rho g\left(A + \frac{\partial A}{\partial s}\frac{ds}{2}\right)dz + \frac{\partial p}{\partial s}\frac{\partial A}{\partial s}\frac{(ds)^2}{2} - A\frac{\partial p}{\partial s}ds - \frac{\partial p}{\partial s}\frac{\partial A}{\partial s}(ds)^2.$$

Continue by dropping the second-order terms that contain $(ds)^2$ or $dsdz$, and divide by ρA to reach:

$$U\frac{\partial U}{\partial s}ds = -gdz - \frac{1}{\rho}\frac{\partial p}{\partial s}ds, \quad \text{or} \quad \left[d(U^2/2) + gdz + (1/\rho)dp = 0\right]_{\text{along a streamline}}.$$

Integrate the final differential expression along the streamline to find:

$$\frac{1}{2}U^2 + gz + p/\rho = \text{a constant along a streamline.} \tag{4.19}$$

EXAMPLE 4.3

Consider a small solitary wave that moves from right to left on the surface of a water channel of undisturbed depth h (Figure 4.4). Denote the acceleration of gravity by g. Assuming a small change in the surface elevation across the wave, derive an expression for its propagation speed, U, when the channel bed is flat and frictionless.

Solution

Before starting the control volume part of this problem, a little dimensional analysis goes a *long* way toward determining the final solution. The statement of the problem has only three parameters,

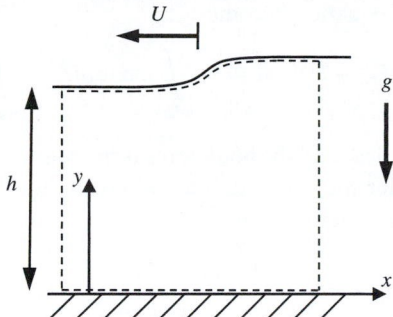

FIGURE 4.4 Momentum and mass balance for a small amplitude water wave moving into quiescent water of depth h. The recommended moving control volume is shown with dashed lines. The wave is driven by the imbalance of static pressure forces on the vertical inlet (left) and outlet (right) control surfaces.

U, g, and h, and there are two independent units (length and time). Thus, there is only one dimensionless group, U^2/gh, so it must be a constant. Therefore, the final answer must be in the form: $U = const \cdot \sqrt{gh}$, so the value of the following control volume analysis lies merely in determining the constant.

Choose the control volume shown and assume it is moving at speed $\mathbf{b} = -U\mathbf{e}_x$. Here we assume that the upper and lower control surfaces coincide with the water surface and the channel's frictionless bed. They are shown close to these boundaries in Figure 4.4 for clarity. Apply the integral conservation laws for mass and momentum.

$$\frac{d}{dt} \int_{V^*(t)} \rho dV + \int_{A^*(t)} \rho(\mathbf{u} - \mathbf{b}) \cdot \mathbf{n} \, dA = 0, \quad \frac{d}{dt} \int_{V^*(t)} \rho \mathbf{u} dV + \int_{A^*(t)} \rho \mathbf{u}(\mathbf{u} - \mathbf{b}) \cdot \mathbf{n} dA$$

$$= \int_{V^*(t)} \rho \mathbf{g} dV + \int_{A^*(t)} \mathbf{f} dA.$$

With this choice of a moving control volume, its contents are constant so both the d/dt terms are zero; thus,

$$\int_{A^*} \rho(\mathbf{u} + U\mathbf{e}_x) \cdot \mathbf{n} dA = 0, \quad \text{and} \quad \int_{A^*(t)} \rho \mathbf{u}(\mathbf{u} + U\mathbf{e}_x) \cdot \mathbf{n} dA = \int_{V^*(t)} \rho \mathbf{g} dV + \int_{A^*(t)} \mathbf{f} dA.$$

Here, all velocities are referred to a stationary coordinate frame, so that $\mathbf{u} = 0$ on the inlet side of the control volume in the undisturbed fluid layer. In addition, label the inlet (left) and outlet (right) water depths as h_{in} and h_{out}, respectively, and save consideration of the simplifications that occur when $(h_{out} - h_{in}) \ll (h_{out} + h_{in})/2$ for the end of the analysis. Let U_{out} be the horizontal flow speed on the outlet side of the control volume and assume its profile is uniform. Therefore $(\mathbf{u} + U\mathbf{e}_x) \cdot \mathbf{n} dA$ is $-Uldy$ on the inlet surface, and $+(U_{out} + U)ldy$ on the outlet surface, where l is (again) the width of the flow into the page. With these replacements, the conservation of mass equation becomes:

$$-\rho U h_{in} l + \rho(U_{out} + U)h_{out} l = 0, \quad \text{or} \quad U h_{in} = (U_{out} + U)h_{out},$$

and the horizontal momentum equation becomes:

$$-\rho(0)(0 + U)h_{in}l + \rho U_{out}(U_{out} + U)h_{out}l = -\int_{inlet} p\mathbf{n}\cdot\mathbf{e}_x dA - \int_{outlet} p\mathbf{n}\cdot\mathbf{e}_x dA - \int_{top} p\mathbf{n}\cdot\mathbf{e}_x dA.$$

Here, no friction terms are included, and the body force term does not appear because it has no horizontal component. First consider the pressure integral on the top of the control volume, and let $y = h(x)$ define the shape of the water surface:

$$-p_o \int \mathbf{n}\cdot\mathbf{e}_x dA = -p_o \int \frac{(-dh/dx, 1)}{\sqrt{1 + (dh/dx)^2}}\cdot(1,0)l\sqrt{1 + (dh/dx)^2}dx$$

$$= -p_o \int \left(-\frac{dh}{dx}\right)dx = p_o \int_{h_{in}}^{h_{out}} dh = p_o(h_{out} - h_{in})$$

where the various square-root factors arise from the surface geometry; p_o is the (constant) atmospheric pressure on the water surface. The pressure on the inlet and outlet sides of the control volume is hydrostatic. Using the coordinate system shown, integrating (1.8), and evaluating the constant on the water surface produces $p = p_o + \rho g(h - y)$. Thus, the integrated inlet and outlet pressure forces are:

$$\int_{inlet} pdA - \int_{outlet} pdA - \int_{top} p_o\mathbf{n}\cdot\mathbf{e}_x dA$$

$$= \int_0^{h_{in}} (p_o + \rho g(h_{in} - y))ldy - \int_0^{h_{out}} (p_o + \rho g(h_{out} - y))ldy + p_o(h_{out} - h_{in})l$$

$$= \int_0^{h_{in}} \rho g(h_{in} - y)ldy - \int_0^{h_{out}} \rho g(h_{out} - y)ldy = \rho g\left(\frac{h_{in}^2}{2} - \frac{h_{out}^2}{2}\right)l$$

where the signs of the inlet and outlet integrals have been determined by evaluating the dot products and we again note that the constant reference pressure p_o does not contribute to the net pressure force. Substituting this pressure force result into the horizontal momentum equation produces:

$$-\rho(0)(0 + U)h_{in}l + \rho U_{out}(U_{out} + U)h_{out}l = \frac{\rho g}{2}(h_{in}^2 - h_{out}^2)l.$$

Dividing by the common factors of ρ and l,

$$U_{out}(U_{out} + U)h_{out} = \frac{g}{2}(h_{in}^2 - h_{out}^2),$$

and eliminating U_{out} via the conservation of mass relationship, $U_{out} = (h_{in} - h_{out})U/h_{out}$, leads to:

$$U\frac{(h_{in} - h_{out})}{h_{out}}\left(U\frac{(h_{in} - h_{out})}{h_{out}} + U\right)h_{out} = \frac{g}{2}(h_{in}^2 - h_{out}^2).$$

Dividing by the common factor of $(h_{in} - h_{out})$ and simplifying the left side of the equation produces:

$$u^2\frac{h_{in}}{h_{out}} = \frac{g}{2}(h_{in} + h_{out}), \quad \text{or} \quad U = \sqrt{\frac{gh_{out}}{2h_{in}}(h_{in} + h_{out})} \approx \sqrt{gh},$$

where the final approximate equality holds when the inlet and outlet heights differ by only a small amount with both nearly equal to h.

EXAMPLE 4.4

Derive the differential equation for the vertical motion for a simple rocket having nozzle area A_e that points downward, exhaust discharge speed V_e, and exhaust density ρ_e, without considering the internal flow within the rocket (Figure 4.5). Denote the mass of the rocket by $M(t)$ and assume the discharge flow is uniform.

Solution

Select a control volume (not shown) that contains the rocket and travels with it. This will be an accelerating control volume and its velocity $\mathbf{b} = b(t)\mathbf{e}_z$ will be the rocket's vertical velocity. In addition, the discharge velocity is specified with respect to the rocket, so in a stationary frame of reference, the absolute velocity of the rocket's exhaust is $\mathbf{u} = u_z\mathbf{e}_z = (-V_e + b)\mathbf{e}_z$.

The conservation of mass and vertical-momentum equations are:

$$\frac{d}{dt}\int_{V^*(t)} \rho dV + \int_{A^*(t)} \rho(\mathbf{u} - \mathbf{b})\cdot\mathbf{n}\, dA = 0, \quad \frac{d}{dt}\int_{V^*(t)} \rho u_z dV + \int_{A^*(t)} \rho u_z(\mathbf{u} - \mathbf{b})\cdot\mathbf{n}\, dA$$

$$= -g\int_{V^*(t)} \rho dV + \int_{A^*(t)} f_z dA.$$

Here we recognize the first term in each equation as the time derivative of the rocket's mass M, and the rocket's vertical momentum Mb, respectively. (The second of these identifications is altered when the rocket's internal flows are considered; see Thompson, 1972, pp. 43–47.) For ordinary rocketry, the

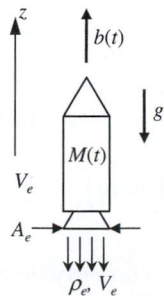

FIGURE 4.5 Geometry and parameters for a simple rocket having mass $M(t)$ that is moving vertically at speed $b(t)$. The rocket's exhaust area, density, and velocity (or specific impulse) are A_e, ρ_e, and V_e, respectively.

rocket exhaust exit will be the only place that mass and momentum cross the control volume boundary and here $\mathbf{n} = -\mathbf{e}_z$; thus $(\mathbf{u} - \mathbf{b})\cdot\mathbf{n}\,dA = (-V_e\mathbf{e}_z)\cdot(-\mathbf{e}_z)dA = V_e dA$ over the nozzle exit. In addition, we will denote the integral of vertical surface stresses by F_S, a force that includes the aerodynamic drag on the rocket and the pressure thrust produced when the rocket nozzle's outlet pressure exceeds the local ambient pressure. With these replacements, the above equations become:

$$\frac{dM}{dt} + \rho_e V_e A_e = 0, \quad \frac{d}{dt}(Mb) + \rho_e(-V_e + b)V_e A_e = -Mg + F_S.$$

Eliminating $\rho_e V_e A_e$ between the two equations produces:

$$\frac{d}{dt}(Mb) + (-V_e + b)\left(-\frac{dM}{dt}\right) = -Mg + F_S,$$

which reduces to:

$$M\frac{d^2 z_R}{dt^2} = -V_e \frac{dM}{dt} - Mg + F_S,$$

where z_R is the rocket's vertical location and $dz_R/dt = b$. From this equation it is clear that negative dM/dt (mass loss) may produce upward acceleration of the rocket when its exhaust discharge velocity V_e is high enough. In fact, V_e is the crucial figure of merit in rocket propulsion and is commonly referred to as the *specific impulse*, the thrust produced per unit rate of mass discharged.

Returning now to the development of the equations of motion, the differential equation that represents momentum conservation is obtained from (4.14) after collecting all four terms into the same volume integration. The first step is to convert the two surface integrals in (4.14) to volume integrals using Gauss' theorem (2.30):

$$\int_{A(t)} \rho(\mathbf{x},t)\mathbf{u}(\mathbf{x},t)(\mathbf{u}(\mathbf{x},t)\cdot\mathbf{n})dA = \int_{V(t)} \nabla\cdot(\rho(\mathbf{x},t)\mathbf{u}(\mathbf{x},t)\mathbf{u}(\mathbf{x},t))dV = \int_{V(t)} \frac{\partial}{\partial x_i}\left(\rho u_i u_j\right)dV, \quad \text{and}$$

$$\int_{A(t)} \mathbf{f}(\mathbf{n},\mathbf{x},t)dA = \int_{A(t)} n_i \tau_{ij} dA = \int_{V(t)} \frac{\partial}{\partial x_i}\left(\tau_{ij}\right)dV,$$

$$(4.20a, 4.20b)$$

where the explicit listing of the independent variables has been dropped upon moving to index notation. Substituting (4.20a, 4.20b) into (4.14) and collecting all the terms on one side of the equation into the same volume integration produces:

$$\int_{V(t)} \left\{ \frac{\partial}{\partial t}\left(\rho u_j\right) + \frac{\partial}{\partial x_i}\left(\rho u_i u_j\right) - \rho g_j - \frac{\partial}{\partial x_i}\left(\tau_{ij}\right) \right\}dV = 0. \tag{4.21}$$

Similarly to (4.6), the integral in (4.21) can only be zero for any material volume if the integrand vanishes at every point in space; thus (4.21) requires:

$$\frac{\partial}{\partial t}\left(\rho u_j\right) + \frac{\partial}{\partial x_i}\left(\rho u_i u_j\right) = \rho g_j + \frac{\partial}{\partial x_i}\left(\tau_{ij}\right). \tag{4.22}$$

This equation can be put into a more standard form by expanding the leading two terms,

$$\frac{\partial}{\partial t}\left(\rho u_j\right) + \frac{\partial}{\partial x_i}\left(\rho u_i u_j\right) = \rho\frac{\partial u_j}{\partial t} + u_j\left[\frac{\partial \rho}{\partial t} + \frac{\partial}{\partial x_i}(\rho u_i)\right] + \rho u_i\frac{\partial u_j}{\partial x_i} = \rho\frac{Du_j}{Dt}, \qquad (4.23)$$

recognizing that the contents of the [,]-brackets are zero because of (4.7), and using the definition of D/Dt from (3.5). The final result is:

$$\rho\frac{Du_j}{Dt} = \rho g_j + \frac{\partial}{\partial x_i}\left(\tau_{ij}\right), \qquad (4.24)$$

which is sometimes called *Cauchy's equation of motion*. It relates fluid-particle acceleration to the net body (ρg_i) and surface force ($\partial \tau_{ij}/\partial x_j$) on the particle. It is true in any continuum, solid or fluid, no matter how the stress tensor τ_{ij} is related to the velocity field. However, (4.24) does not provide a complete description of fluid dynamics, even when combined with (4.7) because the number of dependent field variables is greater than the number of equations. Taken together, (4.7), (4.24), and two thermodynamic equations provide at most $1 + 3 + 2 = 6$ scalar equations but (4.7) and (4.24) contain ρ, u_j, and τ_{ij} for a total of $1 + 3 + 9 = 13$ unknowns. Thus, the number of unknowns must be decreased to produce a solvable system. The fluid's stress-strain rate relationship(s) or constitutive equation provides much of the requisite reduction.

4.5. CONSTITUTIVE EQUATION FOR A NEWTONIAN FLUID

As previously described in Section 2.4, the stress at a point can be completely specified by the nine components of the stress tensor τ; these components are illustrated in Figures 2.4 and 2.5, which show the directions of *positive* stresses on the various faces of small cubical and tetrahedral fluid elements. The first index of τ_{ij} indicates the direction of the normal to the surface on which the stress is considered, and the second index indicates the direction in which the stress acts. The diagonal elements τ_{11}, τ_{22}, and τ_{33} of the stress matrix are the normal stresses, and the off-diagonal elements are the tangential or shear stresses. Although finite size elements are shown in these figures, the stresses apply on the various planes when the elements shrink to a point and the elements have vanishingly small mass. Denoting the cubical volume in Figure 2.4 by $dV = dx_1 dx_2 dx_3$ and considering the torque produced on it by the various stresses' components, it can be shown that the stress tensor is symmetric,

$$\tau_{ij} = \tau_{ji}, \qquad (4.25)$$

by considering the element's rotational dynamics in the limit $dV \to 0$ (see Exercise 4.30). Therefore, the stress tensor has only six independent components. However, this symmetry is violated if there are body-force couples proportional to the mass of the fluid element, such as those exerted by an electric field on polarized fluid molecules. Antisymmetric stresses must be included in such circumstances.

The relationship between the stress and deformation in a continuum is called a *constitutive equation*, and a linear constitutive equation between stress τ_{ij} and $\partial u_i/\partial x_j$ is examined here. A fluid that follows the simplest possible linear constitutive equation is known as a *Newtonian* fluid.

In a fluid at rest, there are only normal components of stress on a surface, and the stress does not depend on the orientation of the surface; the stress is *isotropic*. The only second-order isotropic tensor is the Kronecker delta, δ_{ij}, from (2.16). Therefore, the stress in a static fluid must be of the form

$$\tau_{ij} = -p\delta_{ij}, \tag{4.26}$$

where p is the *thermodynamic pressure* related to ρ and T by an equation of state such as that for a perfect gas $p = \rho RT$ (1.22). The negative sign in (4.26) occurs because the normal components of τ are regarded as positive if they indicate tension rather than compression (see Figure 2.4).

A moving fluid develops additional stress components, σ_{ij}, because of viscosity, and these stress components appear as both diagonal and off-diagonal components within τ. A simple extension of (4.26) that captures this phenomenon and reduces to (4.26) when fluid motion ceases is:

$$\tau_{ij} = -p\delta_{ij} + \sigma_{ij}. \tag{4.27}$$

This decomposition of the stress into fluid-static (p) and fluid-dynamic (σ_{ij}) contributions is approximate, because p is only well defined for equilibrium conditions. However, molecular densities, speeds, and collision rates are typically high enough, so that fluid particles (as defined in Section 1.8) reach local thermodynamic equilibrium conditions in nearly all fluid flows so that p in (4.27) is still the thermodynamic pressure.

The fluid-dynamic contribution, σ_{ij}, to the stress tensor is called the *deviatoric stress tensor*. For it to be invariant under Galilean transformations, it cannot depend on the absolute fluid velocity so it must depend on the velocity gradient tensor $\partial u_i / \partial x_j$. However, by definition, stresses only develop in fluid elements that change shape. Therefore, only the symmetric part of $\partial u_i / \partial x_j$, S_{ij} from (3.12), should be considered in the fluid constitutive equation because the antisymmetric part of $\partial u_i / \partial x_j$, R_{ij} from (3.13), corresponds to pure rotation of fluid elements. The most general linear relationship between σ_{ij} and S_{ij} that produces $\sigma_{ij} = 0$ when $S_{ij} = 0$ is

$$\sigma_{ij} = K_{ijmn} S_{mn}, \tag{4.28}$$

where K_{ijmn} is a fourth-order tensor having 81 components that may depend on the local thermodynamic state of the fluid. Equation (4.28) allows *each* of the nine components of σ_{ij} to be linearly related to *all* nine components of S_{ij}. However, this level of generality is unnecessary when the stress tensor is symmetric, and the fluid is isotropic.

In an isotropic fluid medium, the stress–strain rate relationship is independent of the orientation of the coordinate system. This is only possible if K_{ijmn} is an isotropic tensor. All fourth-order isotropic tensors must be of the form:

$$K_{ijmn} = \lambda \delta_{ij} \delta_{mn} + \mu \delta_{im} \delta_{jn} + \gamma \delta_{in} \delta_{jm} \tag{4.29}$$

(see Aris, 1962, pp. 30–33), where λ, μ, and γ are scalars that depend on the local thermodynamic state. In addition, σ_{ij} is symmetric in i and j, so (4.28) requires that K_{ijmn} also be symmetric in i and j, too. This requirement is consistent with (4.29) only if

$$\gamma = \mu. \tag{4.30}$$

Therefore, only two constants, μ and λ, of the original 81, remain after the imposition of material-isotropy and stress-symmetry restrictions. Substitution of (4.29) into the constitutive equation (4.28) yields

$$\sigma_{ij} = 2\mu S_{ij} + \lambda S_{mm}\,\delta_{ij},$$

where $S_{mm} = \nabla \cdot \mathbf{u}$ is the volumetric strain rate (see Section 3.6). The complete stress tensor (4.27) then becomes

$$\tau_{ij} = -p\delta_{ij} + 2\mu S_{ij} + \lambda S_{mm}\,\delta_{ij}, \tag{4.31}$$

and this is the appropriate multi-dimensional extension of (1.3).

The two scalar constants μ and λ can be further related as follows. Setting $i = j$, summing over the repeated index, and noting that $\delta_{ii} = 3$, we obtain

$$\tau_{ii} = -3p + (2\mu + 3\lambda)S_{mm},$$

from which the pressure is found to be

$$p = -\frac{1}{3}\tau_{ii} + \left(\frac{2}{3}\mu + \lambda\right)\nabla \cdot \mathbf{u}. \tag{4.32}$$

The diagonal terms of S_{ij} in a flow may be unequal. In such a case the stress tensor τ_{ij} can have unequal diagonal terms because of the presence of the term proportional to μ in (4.31). We can therefore take the average of the diagonal terms of τ and define a *mean pressure* (as opposed to thermodynamic pressure p) as

$$\bar{p} \equiv -\frac{1}{3}\tau_{ii}. \tag{4.33}$$

Substitution into (4.32) gives

$$p - \bar{p} = \left(\frac{2}{3}\mu + \lambda\right)\nabla \cdot \mathbf{u}. \tag{4.34}$$

For a completely incompressible fluid we can only define a mechanical or mean pressure, because there is no equation of state to determine a thermodynamic pressure. (In fact, the *absolute pressure in an incompressible fluid is indeterminate,* and only its *gradients* can be determined from the equations of motion.) The λ-term in the constitutive equation (4.31) drops out when $S_{mm} = \nabla \cdot \mathbf{u} = 0$, and no consideration of (4.34) is necessary. So, for *incompressible fluids*, the constitutive equation (4.31) takes the simple form:

$$\tau_{ij} = -p\delta_{ij} + 2\mu S_{ij} \quad \text{(incompressible)}, \tag{4.35}$$

where p can only be interpreted as the mean pressure experienced by a fluid particle. For a compressible fluid, on the other hand, a thermodynamic pressure can be defined, and it seems that p and $\bar{p}$ can be different. In fact, equation (4.34) relates this difference to the rate of expansion through the proportionality constant $\mu_v = \lambda + 2\mu/3$, which is called the *coefficient of bulk viscosity.* It has an appreciable effect on sound absorption and shock-wave structure. It is generally found to be nonzero in polyatomic gases because of relaxation effects associated with molecular rotation. However, the *Stokes assumption,*

$$\lambda + \frac{2}{3}\mu = 0, \tag{4.36}$$

is found to be accurate in many situations because either the fluid's μ_v or the flow's dilatation rate is small. Interesting historical aspects of the Stokes assumption can be found in Truesdell (1952).

Without using (4.36), the stress tensor (4.31) is:

$$\tau_{ij} = -p\delta_{ij} + 2\mu\left(S_{ij} - \frac{1}{3}S_{mm}\delta_{ij}\right) + \mu_v S_{mm}\,\delta_{ij}. \tag{4.37}$$

This linear relation between τ and $\mathbf{S}$ is consistent with Newton's definition of the viscosity coefficient μ in a simple parallel flow $u(y)$, for which (4.37) gives a shear stress of $\tau = \mu(du/dy)$. Consequently, a fluid obeying equation (4.37) is called a *Newtonian fluid* where μ and μ_v may only depend on the local thermodynamic state. The off-diagonal terms of (4.37) are of the type

$$\tau_{12} = \mu\left(\frac{\partial u_1}{\partial x_2} + \frac{\partial u_2}{\partial x_1}\right),$$

and directly relate the shear stress to shear strain rate via the viscosity μ. The diagonal terms of (4.37) combine pressure and viscous effects. For example, the first diagonal component of (4.37) is

$$\tau_{11} = -p + 2\mu\left(\frac{\partial u_1}{\partial x_1}\right) + \left(\mu_v - \frac{2}{3}\mu\right)\frac{\partial u_m}{\partial x_m},$$

which means that the normal viscous stress on a plane normal to the x_1-axis is proportional to the extension rate in the x_1 direction and the average expansion rate at the point.

The linear Newtonian friction law (4.37) might only be expected to hold for small strain rates since it is essentially a first-order expansion of the stress in terms of S_{ij} around $\tau_{ij} = 0$. However, the linear relationship is surprisingly accurate for many common fluids such as air, water, gasoline, and oils. Yet, other liquids display non-Newtonian behavior at moderate rates of strain. These include solutions containing long-chain polymer molecules, concentrated soaps, melted plastics, emulsions and slurries containing suspended particles, and many liquids of biological origin. These liquids may violate Newtonian behavior in several ways. For example, shear stress may be a *nonlinear* function of the local strain rate, which is the case for many liquid plastics that are *shear thinning*; their viscosity drops with increasing strain rate. Alternatively, the stress on a non-Newtonian fluid particle may depend on the local strain rate and on its *history*. Such memory effects give the fluid some elastic properties that may allow it to mimic solid behavior over short periods of time. In fact there is a whole class of *viscoelastic* substances that are neither fully fluid nor fully solid. Non-Newtonian fluid mechanics is beyond the scope of this text but its fundamentals are well covered elsewhere (see Bird et al., 1987).

4.6. NAVIER-STOKES MOMENTUM EQUATION

The momentum conservation equation for a Newtonian fluid is obtained by substituting (4.37) into Cauchy's equation (4.24) to obtain:

$$\rho\left(\frac{\partial u_j}{\partial t} + u_i\frac{\partial u_j}{\partial x_i}\right) = -\frac{\partial p}{\partial x_j} + \rho g_j + \frac{\partial}{\partial x_i}\left[\mu\left(\frac{\partial u_j}{\partial x_i} + \frac{\partial u_i}{\partial x_j}\right) + \left(\mu_v - \frac{2}{3}\mu\right)\frac{\partial u_m}{\partial x_m}\delta_{ij}\right], \qquad (4.38)$$

where we have used $(\partial p/\partial x_i)\delta_{ij} = \partial p/\partial x_j$, (3.5) with $F = u_j$, and (3.12). This is the *Navier-Stokes momentum equation*. The viscosities, μ and μ_v, in this equation can depend on the thermodynamic state and indeed μ, for most fluids, displays a rather strong dependence on temperature, decreasing with T for liquids and increasing with T for gases. Together, (4.7) and (4.38) provide $1 + 3 = 4$ scalar equations, and they contain ρ, p, and u_j for $1 + 1 + 3 = 5$ dependent variables. Therefore, when combined with suitable boundary conditions, (4.7) and (4.38) provide a complete description of fluid dynamics when ρ is constant or when a single (known) relationship exists between p and ρ. In the later case, the fluid or the flow is said to be *barotropic*. When the relationship between p and ρ also includes the temperature T, the internal (or thermal) energy e of the fluid must also be considered. These additions allow a caloric equation of state to be added to the equation listing, but introduces two more dependent variables, T and e. Thus, in general, a third field equation representing conservation of energy is needed to fully describe fluid dynamics.

When temperature differences are small within the flow, μ and μ_v can be taken outside the spatial derivative operating on the contents of the [,]-brackets in (4.38), which then reduces to

$$\rho\frac{Du_j}{Dt} = -\frac{\partial p}{\partial x_j} + \rho g_j + \mu\frac{\partial^2 u_j}{\partial x_i^2} + \left(\mu_v + \frac{1}{3}\mu\right)\frac{\partial}{\partial x_j}\frac{\partial u_m}{\partial x_m} \text{ (compressible)}. \qquad (4.39a)$$

For incompressible fluids $\nabla \cdot \mathbf{u} = \partial u_m/\partial x_m = 0$, so (4.39a) in vector notation reduces to:

$$\rho\frac{D\mathbf{u}}{Dt} = -\nabla p + \rho\mathbf{g} + \mu\nabla^2\mathbf{u} \text{ (incompressible)}. \qquad (4.39b)$$

Interestingly, the net viscous force per unit volume in incompressible flow, the last term on the right in this equation, can be obtained from the divergence of the strain rate tensor or from the curl of the vorticity (see Exercise 4.38):

$$(\mu\nabla^2\mathbf{u})_j = \mu\frac{\partial^2 u_j}{\partial x_i^2} = 2\mu\frac{\partial S_{ij}}{\partial x_i} = \mu\frac{\partial}{\partial x_i}\left(\frac{\partial u_j}{\partial x_i} + \frac{\partial u_i}{\partial x_j}\right) = -\mu\varepsilon_{jik}\frac{\partial \omega_k}{\partial x_i} = -\mu(\nabla \times \boldsymbol{\omega})_j. \qquad (4.40)$$

This result would seem to pose a paradox since it shows that the net viscous force depends on the vorticity even though rotation of fluid elements was explicitly excluded from entering (4.37), the precursor of (4.40). This paradox is resolved by realizing that the net viscous force is given by either a spatial *derivative* of the vorticity or a spatial *derivative* of the deformation rate. The net viscous force vanishes when $\boldsymbol{\omega}$ is uniform in space (as in solid-body rotation), in which case the incompressibility condition requires that the deformation rate is zero everywhere as well.

If viscous effects are negligible, which is commonly true away from the boundaries of the flow field, (4.39) further simplifies to the *Euler equation*

$$\rho\frac{D\mathbf{u}}{Dt} = -\nabla p + \rho\mathbf{g}. \qquad (4.41)$$

4.7. NONINERTIAL FRAME OF REFERENCE

The equations of fluid motion in a noninertial frame of reference are developed in this section. The equations of motion given in Sections 4.4 through 4.6 are valid in an inertial frame of reference, one that is stationary or that is moving at a constant speed with respect to a stationary frame of reference. Although a stationary frame of reference cannot be defined precisely, a frame of reference that is stationary with respect to distant stars is adequate for our purposes. Thus, noninertial-frame effects may be found in other frames of reference known to undergo nonuniform translation and rotation. For example, the fluid mechanics of rotating machinery is often best analyzed in a rotating frame of reference, and modern life commonly places us in the noninertial frame of reference of a moving and maneuvering vehicle. Fortunately, in many laboratory situations, the relevant distances and time scales are short enough so that a frame of reference attached to the earth (sometimes referred to as the *laboratory frame* of reference) is a suitable inertial frame of reference. However, in atmospheric, oceanic, or geophysical studies where time and length scales are much larger, the earth's rotation may play an important role, so an earth-fixed frame of reference must often be treated as a noninertial frame of reference.

In a noninertial frame of reference, the continuity equation (4.7) is unchanged but the momentum equation (4.38) must be modified. Consider a frame of reference $O'1'2'3'$ that translates at velocity $d\mathbf{X}(t)/dt = \mathbf{U}(t)$ and rotates at angular velocity $\mathbf{\Omega}(t)$ with respect to a stationary frame of reference $O123$ (see Figure 4.6). The vectors $\mathbf{U}$ and $\mathbf{\Omega}$ may be resolved in either frame. The same clock is used in both frames so $t = t'$. A fluid particle P can be located in the rotating frame $\mathbf{x}' = (x'_1, x'_2, x'_3)$ or in the stationary frame $\mathbf{x} = (x_1, x_2, x_3)$, and these distances are simply related via vector addition: $\mathbf{x} = \mathbf{X} + \mathbf{x}'$. The velocity $\mathbf{u}$ of the fluid particle is obtained by time differentiation:

$$\mathbf{u} = \frac{d\mathbf{x}}{dt} = \frac{d\mathbf{X}}{dt} + \frac{d\mathbf{x}'}{dt} = \mathbf{U} + \frac{d}{dt}(x'_1\mathbf{e}'_1 + x'_2\mathbf{e}'_2 + x'_3\mathbf{e}'_3)$$

$$= \mathbf{U} + \frac{dx'_1}{dt}\mathbf{e}'_1 + \frac{dx'_2}{dt}\mathbf{e}'_2 + \frac{dx'_3}{dt}\mathbf{e}'_3 + x'_1\frac{d\mathbf{e}'_1}{dt} + x'_2\frac{d\mathbf{e}'_2}{dt} + x'_3\frac{d\mathbf{e}'_3}{dt} = \mathbf{U} + \mathbf{u}' + \mathbf{\Omega} \times \mathbf{x}', \qquad (4.42)$$

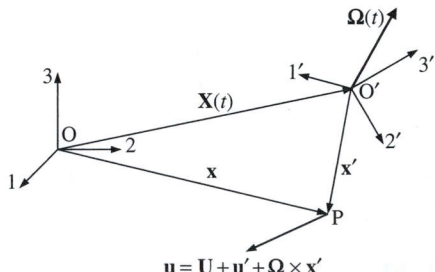

$$\mathbf{u} = \mathbf{U} + \mathbf{u}' + \mathbf{\Omega} \times \mathbf{x}'$$

FIGURE 4.6 Geometry showing the relationship between a stationary coordinate system O123 and a noninertial coordinate system $O'1'2'3'$ that is moving, accelerating, and rotating with respect to O123. In particular, the vector connecting O and O' is $\mathbf{X}(t)$ and the rotational velocity of $O'1'2'3'$ is $\mathbf{\Omega}(t)$. The vector velocity $\mathbf{u}$ at point P in O123 is shown. The vector velocity $\mathbf{u}'$ at point P in $O'1'2'3'$ differs from u because of the motion of $O'1'2'3'$.

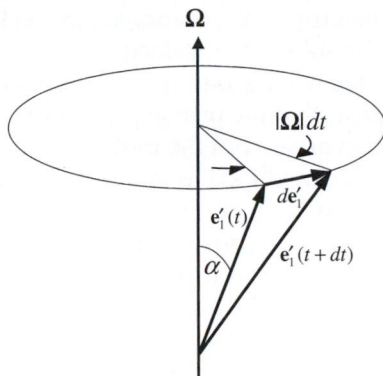

FIGURE 4.7 Geometry showing the relationship between Ω, the rotational velocity vector of $O'1'2'3'$, and the first coordinate unit vector $\mathbf{e}'_1$ in $O'1'2'3'$. Here, the increment $d\mathbf{e}'_1$ is perpendicular to Ω and $\mathbf{e}'_1$.

where the final equality is based on the geometric construction of the cross product shown in Figure 4.7 for $\mathbf{e}'_1$, one of the unit vectors in the rotating frame. In a small time dt, the rotation of $O'1'2'3'$ causes $\mathbf{e}'_1$ to trace a small portion of a cone with radius $\sin\alpha$ as shown. The magnitude of the change in $\mathbf{e}'_1$ is $|\mathbf{e}'_1| = (\sin\alpha)|\Omega|dt$, so $d|\mathbf{e}'_1|/dt = (\sin\alpha)|\Omega|$, which is equal to the magnitude of $\Omega \times \mathbf{e}'_1$. The direction of the rate of change of $\mathbf{e}'_1$ is perpendicular to Ω and $\mathbf{e}'_1$, which is the direction of $\Omega \times \mathbf{e}'_1$. Thus, by geometric construction, $d\mathbf{e}'_1/dt = \Omega \times \mathbf{e}'_1$, and by direct extension to the other unit vectors, $d\mathbf{e}'_i/dt = \Omega \times \mathbf{e}'_i$ (in mixed notation).

To find the acceleration $\mathbf{a}$ of a fluid particle at P, take the time derivative of the final version of (4.42) to find:

$$\mathbf{a} = \frac{d\mathbf{u}}{dt} = \frac{d}{dt}(\mathbf{U} + \mathbf{u}' + \Omega \times \mathbf{x}') = \frac{d\mathbf{U}}{dt} + \mathbf{a}' + 2\Omega \times \mathbf{u}' + \frac{d\Omega}{dt} \times \mathbf{x}' + \Omega \times (\Omega \times \mathbf{x}'). \quad (4.43)$$

(see Exercise 4.42) where $d\mathbf{U}/dt$ is the acceleration of O' with respect to O, $\mathbf{a}'$ is the fluid particle acceleration viewed in the noninertial frame, $2\Omega \times \mathbf{u}'$ is the *Coriolis* acceleration, $(d\Omega/dt) \times \mathbf{x}'$ is the acceleration caused by angular acceleration of the noninertial frame, and the final term is the *centripetal* acceleration.

In fluid mechanics, the acceleration $\mathbf{a}$ of fluid particles is denoted $D\mathbf{u}/Dt$, so (4.43) is rewritten:

$$\left(\frac{D\mathbf{u}}{Dt}\right)_{O123} = \left(\frac{D'\mathbf{u}'}{Dt}\right)_{O'1'2'3'} + \frac{d\mathbf{U}}{dt} + 2\Omega \times \mathbf{u}' + \frac{d\Omega}{dt} \times \mathbf{x}' + \Omega \times (\Omega \times \mathbf{x}'). \quad (4.44)$$

This equation states that fluid particle acceleration in an inertial frame is equal to the sum of: the particle's acceleration in the noninertial frame, the acceleration of the noninertial frame, the Coriolis acceleration, the particle's apparent acceleration from the noninertial frame's angular acceleration, and the particle's centripetal acceleration. Substituting (4.44) into (4.39), produces:

$$\rho\left(\frac{D'\mathbf{u}'}{Dt}\right)_{O'1'2'3'} = -\nabla'p + \rho\left[\mathbf{g} - \frac{d\mathbf{U}}{dt} - 2\Omega \times \mathbf{u}' - \frac{d\Omega}{dt} \times \mathbf{x}' - \Omega \times (\Omega \times \mathbf{x}')\right] + \mu\nabla'^2\mathbf{u}' \quad (4.45)$$

as the incompressible-flow momentum conservation equation in a noninertial frame of reference where the primes denote differentiation, velocity, and position in the noninertial frame. Thermodynamic variables and the net viscous stress are independent of the frame of reference. Equation (4.45) makes it clear that the primary effect of a noninertial frame is the addition of extra body force terms that arise from the motion of the noninertial frame. The terms in [,]-brackets reduce to $\mathbf{g}$ alone when O'1'2'3' is an inertial frame ($\mathbf{U}$ = constant and $\mathbf{\Omega}$ = 0).

The four new terms in (4.45) may each be significant. The first new term $d\mathbf{U}/dt$ accounts for the acceleration of O' relative to O. It provides the apparent force that pushes occupants back into their seats or makes them tighten their grip on a handrail when a vehicle accelerates. An aircraft that is flown on a parabolic trajectory produces weightlessness in its interior when its acceleration $d\mathbf{U}/dt$ equals $\mathbf{g}$.

The second new term, the Coriolis term, depends on the fluid particle's velocity, not on its position. Thus, even at the earth's rotation rate of one cycle per day, it has important consequences for the accuracy of artillery and for navigation during air and sea travel. The earth's angular velocity vector $\mathbf{\Omega}$ points out of the ground in the northern hemisphere. The Coriolis acceleration $-2\mathbf{\Omega} \times \mathbf{u}$ therefore tends to deflect a particle to the right of its direction of travel in the northern hemisphere and to the left in the southern hemisphere. Imagine a low-drag projectile shot horizontally from the north pole with speed u (Figure 4.8). The Coriolis acceleration $2\Omega u$ constantly acts perpendicular to its path and therefore does not change the speed u of the projectile. The forward distance traveled in time t is ut, and the deflection is $\Omega u t^2$. The angular deflection is $\Omega u t^2 / ut = \Omega t$, which is the earth's rotation in time t. This demonstrates that the projectile in fact travels in a straight line if observed from outer space (an inertial frame); its apparent deflection is merely due to the rotation of the earth underneath it. Observers on earth need an imaginary force to account for this deflection. A clear physical explanation of the Coriolis force, with applications to mechanics, is given by Stommel and Moore (1989).

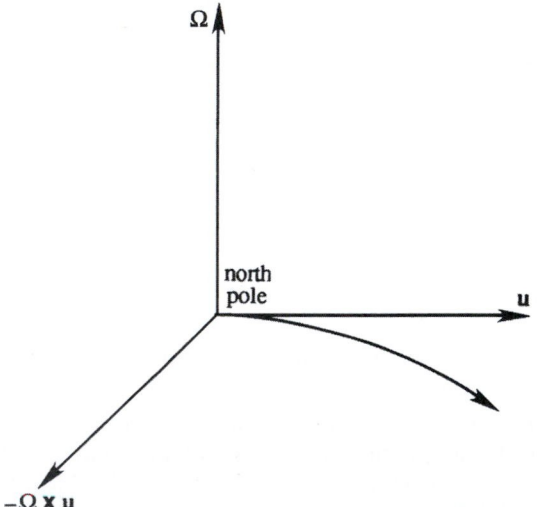

FIGURE 4.8   Particle trajectory deflection caused by the Coriolis acceleration when observed in a rotating frame of reference. If observed from a stationary frame of reference, the particle trajectory would be straight.

In the atmosphere, the Coriolis acceleration is responsible for wind circulation patterns around centers of high and low pressure in the earth's atmosphere. In an inertial frame, a nonzero pressure gradient accelerates fluid from regions of higher pressure to regions of lower pressure, as the first term on the right of (4.38) and (4.45) indicates. Imagine a cylindrical polar coordinate system (Figure 3.3c), with the z-axis normal to the earth's surface and the origin at the center of a high- or low-pressure region in the atmosphere. If it is a high pressure zone, u_R would be outward (positive) away from the z-axis in the absence of rotation since fluid will leave a center of high pressure. In this situation when there is rotation, the Coriolis acceleration $-2\Omega \times \mathbf{u} = -2\Omega_z u_R e_\varphi$ is in the $-\varphi$ direction (in the Northern hemisphere), or clockwise as viewed from above. On the other hand, if the flow is inward toward the center of a low-pressure zone, which reverses the direction of u_R, the Coriolis acceleration is counterclockwise. In the southern hemisphere, the direction of Ω_z is reversed so that the circulation patterns described above are reversed. Although the effects of a rotating frame will be commented on occasionally in this and subsequent chapters, most of the discussions involving Coriolis forces are given in Chapter 13, which covers geophysical fluid dynamics.

The third new acceleration term in [,]-brackets in (4.45) is caused by changes in the rotation rate of the frame of reference so it is of little importance for geophysical flows or for flows in machinery that rotate at a constant rate about a fixed axis. However, it does play a role when rotation speed or the direction of rotation vary with time.

The final new acceleration term in (4.45), the centrifugal acceleration, depends strongly on the rotation rate and the distance of the fluid particle from the axis of rotation. If the rotation rate is steady and the axis of rotation coincides with the z-axis of a cylindrical polar

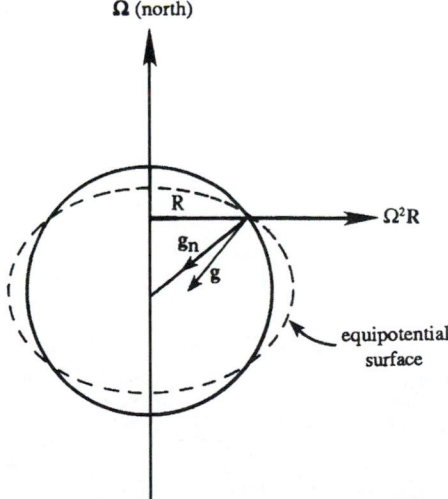

FIGURE 4.9 The earth's rotation causes it to budge near the equator and this leads to a mild distortion of equipotential surfaces from perfect spherical symmetry. The total gravitational acceleration is a sum of a centrally directed acceleration $\mathbf{g_n}$ (the Newtonian gravitation) and a rotational correction $\Omega^2 R$ that points away from the axis of rotation.

coordinate system so that $\mathbf{\Omega} = (0, 0, \Omega)$ and $\mathbf{x'} = (R, \varphi, z)$, then $-\mathbf{\Omega} \times (\mathbf{\Omega} \times \mathbf{x'}) = +\Omega^2 R\mathbf{e}_R$. This additional apparent acceleration can be added to the gravitational acceleration $\mathbf{g}$ to define an *effective gravity* $\mathbf{g}_e = \mathbf{g} + \Omega^2 R\mathbf{e}_R$ (Figure 4.9). Interestingly, a body-force potential for the new term can be found, but its impact might only be felt for relatively large atmospheric- or oceanic-scale flows (Exercise 4.43). The effective gravity is not precisely directed at the center of the earth and its acceleration value varies slightly over the surface of the earth. The equipotential surfaces (shown by the dashed lines in Figure 4.9) are perpendicular to the effective gravity, and the average sea level is one of these equipotential surfaces. Thus, at least locally on the earth's surface, we can write $\Phi_e = gz$, where·z is measured perpendicular to an equipotential surface, and g is the local acceleration caused by the effective gravity. Use of the locally correct acceleration and direction for the earth's gravitational body force in the equations of fluid motion accounts for the centrifugal acceleration and the fact that the earth is really an ellipsoid with equatorial diameter 42 km larger than the polar diameter.

EXAMPLE 4.5

Find the radial, angular, and axial fluid momentum equations for viscous flow in the gaps between plates of a von Karman viscous impeller pump (see Figure 4.10) that rotates at a constant angular speed Ω_z. Assume steady constant-density constant-viscosity flow, neglect the body force for simplicity, and use cylindrical coordinates (Figure 3.3c).

Solution

First a little background: A von Karman viscous impeller pump uses rotating plates to pump viscous fluids via a combination of viscous and centrifugal forces. Although such pumps may be inefficient, they are wear-tolerant and may be used to pump abrasive fluids that would damage the

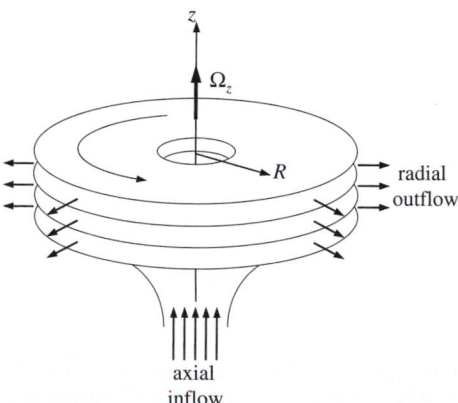

FIGURE 4.10   Schematic drawing of the impeller of a von Karman pump (Example·4.5).

vanes or blades of other pumps. Plus, their pumping action is entirely steady so they are exceptionally quiet, a feature occasionally exploited for air-moving applications in interior spaces occupied by human beings.

For steady, constant-density, constant-viscosity flow without a body force in a steadily rotating frame of reference, the momentum equation is a simplified version of (4.45):

$$\rho(\mathbf{u}'\cdot\nabla')\mathbf{u}' = -\nabla'p + \rho[-2\Omega\times\mathbf{u}' - \Omega\times(\Omega\times\mathbf{x}')] + \mu\nabla'^2\mathbf{u}'.$$

Here we are not concerned with the axial inflow or the flow beyond the outer edges of the disks. Now choose the z-axis of the coordinate system to be coincident with the axis of rotation. For this choice, the flow between the disks should be axisymmetric, so we can presume that u'_R, u'_φ, u'_z, and p only depend on R and z. To further simplify the momentum equation, drop the primes, evaluate the cross products,

$$\Omega\times\mathbf{u} = \Omega_z\mathbf{e}_z\times(u_R\mathbf{e}_R + u_\varphi\mathbf{e}_\varphi) = +\Omega_z u_R\mathbf{e}_\varphi - \Omega_z u_\varphi\mathbf{e}_R, \quad\text{and}\quad \Omega\times(\Omega\times\mathbf{x}') = -\Omega_z^2 R\mathbf{e}_R,$$

and separate the radial, angular, and axial components to find:

$$\rho\left(u_R\frac{\partial u_R}{\partial R} + u_z\frac{\partial u_R}{\partial z} - \frac{u_\varphi^2}{R}\right) = -\frac{\partial p}{\partial R} + \rho[2\Omega_z u_\varphi + \Omega_z^2 R] + \mu\left(\frac{1}{R}\frac{\partial}{\partial R}\left(R\frac{\partial u_R}{\partial R}\right) + \frac{\partial^2 u_R}{\partial z^2} - \frac{u_R}{R^2}\right)$$

$$\rho\left(u_R\frac{\partial u_\varphi}{\partial R} + u_z\frac{\partial u_\varphi}{\partial z} + \frac{u_R u_\varphi}{R}\right) = \rho[-2\Omega_z u_R] + \mu\left(\frac{1}{R}\frac{\partial}{\partial R}\left(R\frac{\partial u_\varphi}{\partial R}\right) + \frac{\partial^2 u_\varphi}{\partial z^2} - \frac{u_\varphi}{R^2}\right)$$

$$\rho\left(u_R\frac{\partial u_z}{\partial R} + u_z\frac{\partial u_z}{\partial z}\right) = -\frac{\partial p}{\partial z} + \mu\left(\frac{1}{R}\frac{\partial}{\partial R}\left(R\frac{\partial u_z}{\partial R}\right) + \frac{\partial^2 u_z}{\partial z^2}\right).$$

Here we have used the results found in the Appendix B for cylindrical coordinates. In the first two momentum equations, the terms in [,]-brackets result from rotation of the coordinate system.

4.8. CONSERVATION OF ENERGY

In this section, the integral energy-conservation equivalent of (4.5) and (4.17) is developed from a mathematical statement of conservation of energy for a fluid particle in an inertial frame of reference. The subsequent steps that lead to a differential energy-conservation equivalent of (4.7) and (4.24) follow the pattern set in Sections 4.2 and 4.5. For clarity and conciseness, the explicit listing of independent variables is dropped from the equations in this section.

When applied to a material volume $V(t)$ with surface area $A(t)$, conservation of internal energy per unit mass e and the kinetic energy per unit mass $(\frac{1}{2})|\mathbf{u}|^2$ can be stated:

$$\frac{d}{dt}\int_{V(t)}\rho\left(e+\frac{1}{2}|\mathbf{u}|^2\right)dV = \int_{V(t)}\rho\mathbf{g}\cdot\mathbf{u}dV + \int_{A(t)}\mathbf{f}\cdot\mathbf{u}dA - \int_{A(t)}\mathbf{q}\cdot\mathbf{n}dA, \quad (4.46)$$

where the terms on the right are: work done on the fluid in $V(t)$ by body forces, work done on the fluid in $V(t)$ by surface forces, and heat transferred out of $V(t)$. Here, $\mathbf{q}$ is the heat flux vector and in general includes thermal conduction and radiation. The final term in (4.46) has a negative sign because the energy in $V(t)$ decreases when heat leaves $V(t)$ and this occurs when $\mathbf{q} \cdot \mathbf{n}$ is positive. Again, the implications of (4.46) are better displayed when the time derivative is expanded using Reynolds transport theorem (3.35),

$$
\int_{V(t)} \frac{\partial}{\partial t}\left(\rho e + \frac{\rho}{2}|\mathbf{u}|^2\right)dV + \int_{A(t)} \left(\rho e + \frac{\rho}{2}|\mathbf{u}|^2\right)(\mathbf{u} \cdot \mathbf{n})dA
$$
$$
= \int_{V(t)} \rho \mathbf{g} \cdot \mathbf{u}\, dV + \int_{A(t)} \mathbf{f} \cdot \mathbf{u}\, dA - \int_{A(t)} \mathbf{q} \cdot \mathbf{n}\, dA.
\tag{4.47}
$$

Similar to the prior developments for mass and momentum conservation, this result can be generalized to an arbitrarily moving control volume $V^*(t)$ with surface $A^*(t)$:

$$
\frac{d}{dt} \int_{V^*(t)} \rho\left(e + \frac{1}{2}|\mathbf{u}|^2\right)dV + \int_{A^*(t)} \left(\rho e + \frac{\rho}{2}|\mathbf{u}|^2\right)(\mathbf{u} - \mathbf{b}) \cdot \mathbf{n}\, dA
$$
$$
= \int_{V^*(t)} \rho \mathbf{g} \cdot \mathbf{u}\, dV + \int_{A^*(t)} \mathbf{f} \cdot \mathbf{u}\, dA - \int_{A^*(t)} \mathbf{q} \cdot \mathbf{n}\, dA,
\tag{4.48}
$$

when $V^*(t)$ is instantaneously coincident with $V(t)$. And, just like (4.5) and (4.17), (4.48) can be specialized to stationary, steadily moving, accelerating, or deforming control volumes by appropriate choice of the control surface velocity $\mathbf{b}$.

The differential equation that represents energy conservation is obtained from (4.47) after collecting all four terms under the same volume integration. The first step is to convert the three surface integrals in (4.47) to volume integrals using Gauss' theorem (2.30):

$$
\int_{A(t)} \left(\rho e + \frac{\rho}{2}|\mathbf{u}|^2\right)(\mathbf{u} \cdot \mathbf{n})dA = \int_{V(t)} \nabla \cdot \left(\rho e \mathbf{u} + \frac{\rho}{2}|\mathbf{u}|^2 \mathbf{u}\right)dV
$$
$$
= \int_{V(t)} \frac{\partial}{\partial x_i}\left(\rho\left(e + \frac{1}{2}u_j^2\right)u_i\right)dV,
\tag{4.49}
$$

$$
\int_{A(t)} \mathbf{f} \cdot \mathbf{u}\, dA = \int_{A(t)} n_i \tau_{ij} u_j\, dA = \int_{V(t)} \frac{\partial}{\partial x_i}(\tau_{ij} u_j)dV,
\tag{4.50}
$$

and

$$
\int_{A(t)} \mathbf{q} \cdot \mathbf{n}\, dA = \int_{A(t)} q_i n_i\, dA = \int_{V(t)} \nabla \cdot \mathbf{q}\, dA = \int_{V(t)} \frac{\partial}{\partial x_i} q_i\, dA,
\tag{4.51}
$$

where in (4.49) $u_j^2 = u_1^2 + u_2^2 + u_3^2$ because the summation index j is implicitly repeated. Substituting (4.49) through (4.51) into (4.47) and putting all the terms together into the same volume integration produces:

$$\int\limits_{V(t)} \left\{ \frac{\partial}{\partial t}\left(\rho\left[e + \frac{1}{2}u_j^2 \right] \right) + \frac{\partial}{\partial x_i}\left(\rho\left[e + \frac{1}{2}u_j^2 \right]u_i \right) - \rho g_i u_i - \frac{\partial}{\partial x_i}\left(\tau_{ij} u_j \right) + \frac{\partial q_i}{\partial x_i} \right\} dV = 0. \quad (4.52)$$

Similar to (4.6) and (4.21), the integral in (4.52) can only be zero for any material volume if its integrand vanishes at every point in space; thus (4.52) requires:

$$\frac{\partial}{\partial t}\left(\rho\left[e + \frac{1}{2}u_j^2 \right] \right) + \frac{\partial}{\partial x_i}\left(\rho\left[e + \frac{1}{2}u_j^2 \right]u_i \right) = \rho g_i u_i + \frac{\partial}{\partial x_i}\left(\tau_{ij} u_j \right) - \frac{\partial q_i}{\partial x_i}. \quad (4.53)$$

This differential equation is a statement of conservation of energy containing terms for fluid particle internal energy, fluid particle kinetic energy, work, energy exchange, and heat transfer. It is commonly revised and simplified so that its terms are more readily interpreted. The second term on the right side of (4.53) represents the total rate of work done on a fluid particle by surface stresses. By performing the differentiation, and then using (4.27) to separate out pressure and viscous surface-stress terms, it can be decomposed as follows:

$$\frac{\partial}{\partial x_i}\left(\tau_{ij} u_j \right) = \tau_{ij}\frac{\partial u_j}{\partial x_i} + u_j\frac{\partial \tau_{ij}}{\partial x_i} = \left(-p\frac{\partial u_j}{\partial x_j} + \sigma_{ij}\frac{\partial u_j}{\partial x_i} \right) + \left(-u_j\frac{\partial p}{\partial x_j} + u_j\frac{\partial \sigma_{ij}}{\partial x_i} \right). \quad (4.54)$$

In the final equality, the terms in the first set of (,)-parentheses are the pressure and viscous-stress work terms that lead to the deformation of fluid particles while the terms in the second set of (,)-parentheses are the product of the local fluid velocity with the net pressure force and the net viscous force that lead to either an increase or decrease in the fluid particle's kinetic energy. (Recall from (4.24) that $\partial \tau_{ij}/\partial x_j$ represents the net surface force.) Substituting (4.54) into (4.53), expanding the differentiations on the left in (4.53), and using the continuity equation (4.7) to drop terms produces:

$$\rho\frac{D}{Dt}\left(e + \frac{1}{2}u_j^2 \right) = \rho g_i u_i + \left(-p\frac{\partial u_j}{\partial x_j} + \sigma_{ij}\frac{\partial u_j}{\partial x_i} \right) + \left(-u_j\frac{\partial p}{\partial x_j} + u_j\frac{\partial \sigma_{ij}}{\partial x_i} \right) - \frac{\partial q_i}{\partial x_i} \quad (4.55)$$

(see Exercise 4.45). This equation contains both mechanical and thermal energy terms. A separate equation for the mechanical energy can be constructed by multiplying (4.22) by u_j and summing over j. After some manipulation, the result is:

$$\rho\frac{D}{Dt}\left(\frac{1}{2}u_j^2 \right) = \rho g_j u_j - u_j\frac{\partial p}{\partial x_j} + u_j\frac{\partial}{\partial x_i}\left(\sigma_{ij} \right) \quad (4.56)$$

(see Exercise 4.46), where (4.27) has been used for τ_{ij}. Subtracting (4.56) from (4.55), dividing by $\rho = 1/v$, and using (4.8) produces:

$$\frac{De}{Dt} = -p\frac{Dv}{Dt} + \frac{1}{\rho}\sigma_{ij}S_{ij} - \frac{1}{\rho}\frac{\partial q_i}{\partial x_i}, \quad (4.57)$$

where the fact that σ_{ij} is symmetric has been exploited so $\sigma_{ij}(\partial u_j/\partial x_i) = \sigma_{ij}(S_{ji} + R_{ji}) = \sigma_{ij}S_{ij}$ with S_{ij} given by (3.12). Equation (4.57) is entirely equivalent to the first law of thermodynamics (1.10)—the change in energy of a system equals the work put into the system minus the heat lost by the system. The difference is that in (4.57), all the terms have units of power per unit mass instead of energy. The first two terms on the right in (4.57) are the pressure and viscous work done on a fluid particle while the final term represents heat transfer from the fluid particle. The pressure work and heat transfer terms may have either sign.

The viscous work term in (4.57) is the kinetic energy dissipation rate per unit mass, and it is commonly denoted by $\varepsilon = (1/\rho)\sigma_{ij}S_{ij}$. It is the product of the viscous stress acting on a fluid element and the deformation rate of a fluid element, and represents the viscous work put into fluid element deformation. This work is irreversible because deformed fluid elements do not return to their prior shape when a viscous stress is relieved. Thus, ε represents the irreversible conversion of mechanical energy to thermal energy through the action of viscosity. It is always positive and can be written in terms of the viscosities and squares of velocity field derivatives (see Exercise 4.47):

$$\varepsilon \equiv \frac{1}{\rho}\sigma_{ij}S_{ij} = \frac{1}{\rho}\left(2\mu S_{ij} + \left(\mu_v - \frac{2}{3}\mu\right)\frac{\partial u_m}{\partial x_m}\delta_{ij}\right)S_{ij} = 2\nu\left(S_{ij} - \frac{1}{3}\frac{\partial u_m}{\partial x_m}\delta_{ij}\right)^2 + \frac{\mu_v}{\rho}\left(\frac{\partial u_m}{\partial x_m}\right)^2, \quad (4.58)$$

where $\nu \equiv \mu/\rho$ is the kinematic viscosity, (1.4), and

$$\sigma_{ij} = +\mu\left(\frac{\partial u_i}{\partial x_j} + \frac{\partial u_j}{\partial x_i}\right) + \left(\mu_v - \frac{2}{3}\mu\right)\frac{\partial u_m}{\partial x_m}\delta_{ij} \quad (4.59)$$

for a Newtonian fluid. Here we note that only shear deformations contribute to ε when $\mu_v = 0$ or when the flow is in incompressible. As described in Chapter 12, ε plays an important role in the physics and description of turbulent flow. It is proportional to μ (and μ_v) and the square of velocity gradients, so it is more important in regions of high shear. The internal energy increase resulting from high ε could appear as a hot lubricant in a bearing, or as burning of the surface of a spacecraft on reentry into the atmosphere.

The final energy-equation manipulation is to express q_i in terms of the other dependent field variables. For nearly all the circumstances considered in this text, heat transfer is caused by thermal conduction alone, so using (4.58) and Fourier's law of heat conduction (1.2), (4.57) can be rewritten:

$$\rho\frac{De}{Dt} = -p\frac{\partial u_m}{\partial x_m} + 2\mu\left(S_{ij} - \frac{1}{3}\frac{\partial u_m}{\partial x_m}\delta_{ij}\right)^2 + \mu_v\left(\frac{\partial u_m}{\partial x_m}\right)^2 + \frac{\partial}{\partial x_i}\left(k\frac{\partial T}{\partial x_i}\right), \quad (4.60)$$

where k is the fluid's thermal conductivity. It is presumed to only depend on thermodynamic conditions, as is the case for μ and μ_v.

At this point the development of the differential equations of fluid motion is complete. The field equations (4.7), (4.38), and (4.60) are general for a Newtonian fluid that follows Fourier's law of heat conduction. These field equations and two thermodynamic equations provide: $1 + 3 + 1 + 2 = 7$ scalar equations. The dependent variables in these equations are ρ, e, p, T, and u_j, a total of $1 + 1 + 1 + 1 + 3 = 7$ unknowns. The number of equations is equal to the number of unknown field variables; therefore, solutions are in principle possible for suitable boundary conditions.

Interestingly, the evolution of the entropy s in fluid flows can be deduced from (4.57) by using Gibb's property relation (1.18) for the internal energy $de = Tds - pd(1/\rho)$. When made specific to time variations following a fluid particle, it becomes:

$$\frac{De}{Dt} = T\frac{Ds}{Dt} - p\frac{D(1/\rho)}{Dt}. \tag{4.61}$$

Combining (4.57), (4.58), and (4.61) produces:

$$\frac{Ds}{Dt} = -\frac{1}{\rho T}\frac{\partial q_i}{\partial x_i} + \frac{\varepsilon}{T} = -\frac{1}{\rho}\frac{\partial}{\partial x_i}\left(\frac{q_i}{T}\right) - \frac{q_i}{\rho T^2}\left(\frac{\partial T}{\partial x_i}\right) + \frac{\varepsilon}{T}, \tag{4.62}$$

and using Fourier's law of heat conduction, this becomes:

$$\frac{Ds}{Dt} = +\frac{1}{\rho}\frac{\partial}{\partial x_i}\left(\frac{k}{T}\frac{\partial T}{\partial x_i}\right) + \frac{k}{\rho T^2}\left(\frac{\partial T}{\partial x_i}\right)^2 + \frac{\varepsilon}{T}. \tag{4.63}$$

The first term on the right side is the entropy gain or loss from heat conduction. The last two terms, which are proportional to the square of temperature and velocity gradients (see (4.58)), represent the *entropy production* caused by heat conduction and viscous generation of heat. The second law of thermodynamics requires that the entropy production due to irreversible phenomena should be positive, so that $\mu, \kappa, k > 0$. Thus, explicit appeal to the second law of thermodynamics is not required in most analyses of fluid flows because it has already been satisfied by taking positive values for the viscosities and the thermal conductivity. In addition (4.63) requires that fluid particle entropy be preserved along particle trajectories when the flow is inviscid and non-heat-conducting, i.e., when $Ds/Dt = 0$.

4.9. SPECIAL FORMS OF THE EQUATIONS

The general equations of motion for a fluid may be put into a variety of special forms when certain symmetries or approximations are valid. Several special forms are presented in this section. The first applies to the integral form of the momentum equation and corresponds to the classical mechanics principle of conservation of angular momentum. The second through fifth special forms arise from manipulations of the differential equations to generate Bernoulli equations. The sixth special form applies when the flow has constant density and the gravitational body force and hydrostatic pressure cancel. The final special form for the equations of motion presented here, known as the Boussinesq approximation, is for low-speed incompressible flows with constant transport coefficients and small changes in density.

Angular Momentum Principle for a Stationary Control Volume

In the mechanics of solids bodies it is shown that

$$d\mathbf{H}/dt = \mathbf{M}, \tag{4.64}$$

where $\mathbf{M}$ is the torque of all external forces on the body about any chosen axis, and $d\mathbf{H}/dt$ is the rate of change of angular momentum of the body about the same axis. For the fluid in a material control volume, the angular momentum is

FIGURE   4.11 Definition sketch for the angular momentum theorem where $dm = \rho dV$. Here the chosen axis points out of the page, and elemental contributions to the angular momentum about this axis are $\mathbf{r} \times \rho \mathbf{u} dV$.

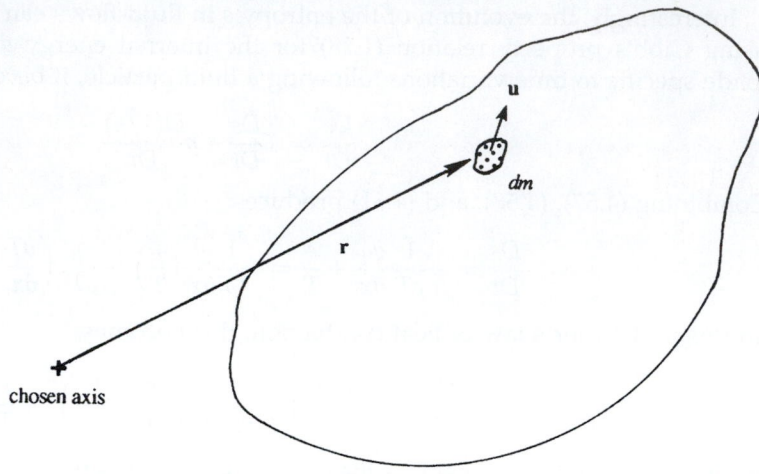

$$\mathbf{H} = \int_{V(t)} (\mathbf{r} \times \rho \mathbf{u}) dV,$$

where $\mathbf{r}$ is the position vector from the chosen axis (Figure 4.11). Inserting this in (4.64) produces:

$$\frac{d}{dt} \int_{V(t)} (\mathbf{r} \times \rho \mathbf{u}) dV = \int_{V(t)} (\mathbf{r} \times \rho \mathbf{g}) dV + \int_{A(t)} (\mathbf{r} \times \mathbf{f}) dA,$$

where the two terms on the right are the torque produced by body forces and surface stresses, respectively. As before, the left-hand term can be expanded via Reynolds transport theorem to find:

$$\frac{d}{dt} \int_{V(t)} (\mathbf{r} \times \rho \mathbf{u}) dV = \int_{V(t)} \frac{\partial}{\partial t}(\mathbf{r} \times \rho \mathbf{u}) dV + \int_{A(t)} (\mathbf{r} \times \rho \mathbf{u})(\mathbf{u} \cdot \mathbf{n}) dA$$

$$= \int_{V_o} \frac{\partial}{\partial t}(\mathbf{r} \times \rho \mathbf{u}) dV + \int_{A_o} (\mathbf{r} \times \rho \mathbf{u})(\mathbf{u} \cdot \mathbf{n}) dA$$

$$= \frac{d}{dt} \int_{V_o} (\mathbf{r} \times \rho \mathbf{u}) dV + \int_{A_o} (\mathbf{r} \times \rho \mathbf{u})(\mathbf{u} \cdot \mathbf{n}) dA,$$

where V_o and A_o are the volume and surface of a stationary control volume that is instantaneously coincident with the material volume, and the final equality holds because V_o does not vary with time. Thus, the stationary volume angular momentum principle is:

$$\frac{d}{dt} \int_{V_o} (\mathbf{r} \times \rho \mathbf{u}) dV + \int_{A_o} (\mathbf{r} \times \rho \mathbf{u})(\mathbf{u} \cdot \mathbf{n}) dA = \int_{V_o} (\mathbf{r} \times \rho \mathbf{g}) dV + \int_{A_o} (\mathbf{r} \times \mathbf{f}) dA. \qquad (4.65)$$

The angular momentum principle (4.65) is analogous to the linear momentum principle (4.17) when $\mathbf{b} = 0$, and is very useful in investigating rotating fluid systems such as turbomachines, fluid couplings, dishwashing-machine spray rotors, and even lawn sprinklers.

EXAMPLE 4.6

Consider a lawn sprinkler as shown in Figure 4.12. The area of each nozzle exit is A, and the jet velocity is U. Find the torque required to hold the rotor stationary.

Solution

Select a stationary volume V_0 with area A_0 as shown by the dashed lines. Pressure everywhere on the control surface is atmospheric, and there is no net moment due to the pressure forces. The control surface cuts through the vertical support and the torque M exerted by the support on the sprinkler arm is the only torque acting on V_0. Apply the angular momentum balance

$$\int_{A_o} (\mathbf{r} \times \rho\mathbf{u})(\mathbf{u}\cdot\mathbf{n})dA = \int_{A_o} (\mathbf{r} \times \mathbf{f})dA = M,$$

where the time derivative term must be zero for a stationary rotor. Evaluating the surface flux terms produces:

$$\int_{A_o} (\mathbf{r} \times \rho\mathbf{u})(\mathbf{u}\cdot\mathbf{n})dA = (a\rho U \cos \alpha)UA + (a\rho U \cos \alpha)UA = 2a\rho AU^2 \cos \alpha.$$

Therefore, the torque required to hold the rotor stationary is $M = 2a\rho AU^2 \cos \alpha$. When the sprinkler is rotating at a steady state, this torque is balanced by both air resistance and mechanical friction.

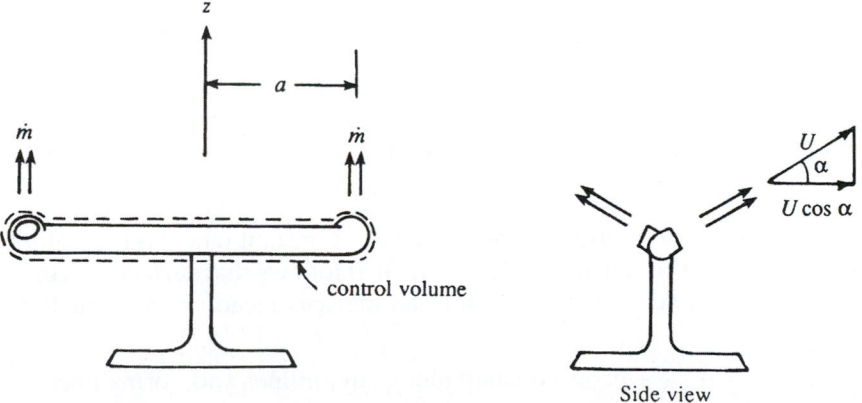

FIGURE 4.12   Lawn sprinkler.

Bernoulli Equations

Various conservation laws for mass, momentum, energy, and entropy were presented in the preceding sections. Bernoulli equations are not separate laws, but are instead derived from the Navier-Stokes momentum equation (4.38) and the energy equation (4.60) under various sets of conditions.

First consider inviscid flow ($\mu = \mu_v = 0$) where gravity is the only body force so that (4.38) reduces to the Euler equation (4.41):

$$\frac{\partial u_j}{\partial t} + u_i \frac{\partial u_j}{\partial x_i} = -\frac{1}{\rho} \frac{\partial p}{\partial x_j} - \frac{\partial}{\partial x_j} \Phi, \tag{4.66}$$

where $\Phi = gz$ is the body force potential, g is the acceleration of gravity, and the z-axis is vertical. If the flow is also barotropic, then $\rho = \rho(p)$, and

$$\frac{1}{\rho} \frac{\partial p}{\partial x_j} = \frac{\partial}{\partial x_j} \int_{p_o}^{p} \frac{dp'}{\rho(p')}, \tag{4.67}$$

where dp/ρ is a perfect differential, p_o is a reference pressure, and p' is the integration variable. In this case the integral depends only on its endpoints, and not on the path of integration. Constant density, isothermal, and isentropic flows are barotropic. In addition, the advective acceleration in (4.66) may be rewritten in terms of the velocity-vorticity cross product, and the gradient of the kinetic energy per unit mass:

$$u_i \frac{\partial u_j}{\partial x_i} = -(\mathbf{u} \times \boldsymbol{\omega})_j + \frac{\partial}{\partial x_j} \left(\frac{1}{2} u_i^2 \right) \tag{4.68}$$

(see Exercise 4.50). Substituting (4.67) and (4.68) into (4.66) produces:

$$\frac{\partial u_j}{\partial t} + \frac{\partial}{\partial x_j} \left[\frac{1}{2} u_i^2 + \int_{p_o}^{p} \frac{dp'}{\rho(p')} + gz \right] = (\mathbf{u} \times \boldsymbol{\omega})_j, \tag{4.69}$$

where all the gradient terms have been collected together to form the Bernoulli function $B =$ the contents of the [,]-brackets.

Equation (4.69) can be used to deduce the evolution of the Bernoulli function in inviscid barotropic flow. First consider steady flow ($\partial u_j / \partial t = 0$) so that (4.69) reduces to

$$\nabla B = \mathbf{u} \times \boldsymbol{\omega}. \tag{4.70}$$

The left-hand side is a vector normal to the surface $B =$ constant whereas the right-hand side is a vector perpendicular to both $\mathbf{u}$ and $\boldsymbol{\omega}$ (Figure 4.13). It follows that surfaces of constant B must contain the streamlines and vortex lines. Thus, an inviscid, steady, barotropic flow satisfies

$$\frac{1}{2} u_i^2 + \int_{p_o}^{p} \frac{dp'}{\rho(p')} + gz = \text{constant along streamlines and vortex lines.} \tag{4.71}$$

This is the first of several possible *Bernoulli equations*. If, in addition, the flow is irrotational ($\boldsymbol{\omega} = 0$), then (4.70) implies that

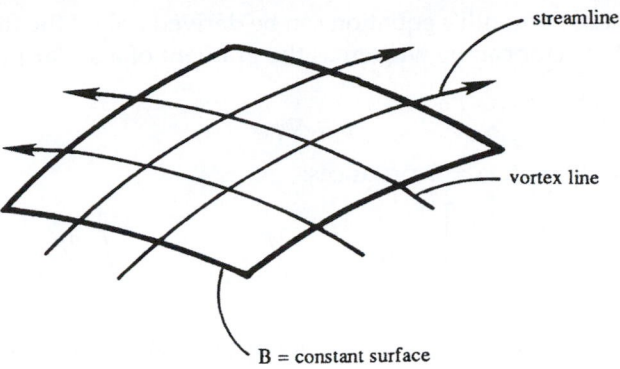

FIGURE 4.13 A surface defined by streamlines and vortex lines. Within this surface the Bernoulli function defined as the contents of the [,]-brackets in (4.69) is constant in steady flow. Note that the streamlines and vortex lines can be at an arbitrary angle.

$$\frac{1}{2}u_i^2 + \int_{p_0}^{p} \frac{dp'}{\rho(p')} + gz = \text{constant everywhere.} \tag{4.72}$$

It may be shown that a sufficient condition for the existence of the surfaces containing streamlines and vortex lines is that the flow be barotropic. Incidentally, these are called *Lamb surfaces* in honor of the distinguished English applied mathematician and hydrodynamicist, Horace Lamb. In a general nonbarotropic flow, a path composed of streamline and vortex line segments can be drawn between any two points in a flow field. Then (4.71) is valid with the proviso that the integral be evaluated on the specific path chosen. As written, (4.71) requires that the flow be steady, inviscid, and have only gravity (or other conservative) body forces acting upon it. Irrotational flow is presented in Chapter 6. We shall note only the important point here that, in a nonrotating frame of reference, barotropic irrotational flows remain irrotational if viscous effects are negligible. Consider the flow around a solid object, say an airfoil (Figure 4.14). The flow is irrotational at all points outside the thin viscous layer close to the surface of the body. This is because a particle P on a streamline outside the viscous layer started from some point S, where the flow is uniform and consequently irrotational. The Bernoulli equation (4.72) is therefore satisfied everywhere outside the viscous layer in this example.

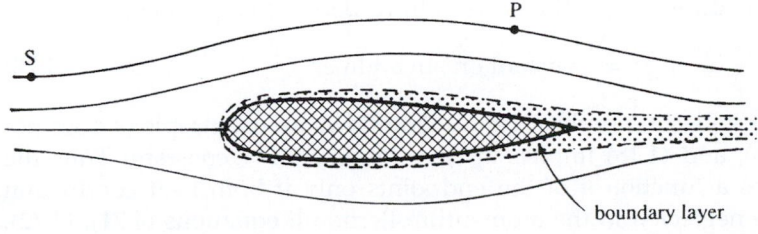

FIGURE 4.14 Flow over a solid object. Viscous shear stresses are usually confined to a thin layer near the body called a *boundary layer*. Flow outside the boundary layer is irrotational, so if a fluid particle at S is initially irrotational it will remain irrotational at P because the streamline it travels on does not enter the boundary layer.

An unsteady form of Bernoulli's equation can be derived only if the flow is irrotational. In this case, the velocity vector can be written as the gradient of a scalar potential ϕ (called the *velocity potential*):

$$\mathbf{u} \equiv \nabla\phi. \tag{4.73}$$

Putting (4.73) into (4.69) with $\boldsymbol{\omega} = 0$ produces:

$$\nabla\left[\frac{\partial\phi}{\partial t} + \frac{1}{2}|\nabla\phi|^2 + \int_{p_o}^{p}\frac{dp'}{\rho(p')} + gz\right] = 0, \quad \text{or} \quad \frac{\partial\phi}{\partial t} + \frac{1}{2}|\nabla\phi|^2 + \int_{p_o}^{p}\frac{dp'}{\rho(p')} + gz = B(t), \tag{4.74}$$

where the integration function $B(t)$ is independent of location. Here ϕ can be redefined to include B,

$$\phi \to \phi + \int_{t_o}^{t} B(t')dt',$$

without changing its use in (4.73); then the second part of (4.74) provides a second Bernoulli equation for unsteady, inviscid, irrotational, barotropic flow:

$$\frac{\partial\phi}{\partial t} + \frac{1}{2}|\nabla\phi|^2 + \int_{p_o}^{p}\frac{dp'}{\rho(p')} + gz = \text{constant.} \tag{4.75}$$

This form of the Bernoulli equation will be used in studying irrotational wave motions in Chapter 7.

A third Bernoulli equation can be obtained for steady flow ($\partial/\partial t = 0$) from the energy equation (4.55) in the absence of viscous stresses and heat transfer ($\sigma_{ij} = q_i = 0$):

$$\rho u_i \frac{\partial}{\partial x_i}\left(e + \frac{1}{2}u_j^2\right) = \rho u_i g_i - \frac{\partial}{\partial x_j}\left(\rho u_j p/\rho\right). \tag{4.76}$$

When the body force is conservative with potential gz, and the steady continuity equation, $\partial(\rho u_i)/\partial x_i = 0$, is used to simplify (4.76), it becomes:

$$\rho u_i \frac{\partial}{\partial x_i}\left(e + \frac{p}{\rho} + \frac{1}{2}u_j^2 + gz\right) = 0. \tag{4.77}$$

From (1.13) $h = e + p/\rho$, so (4.77) states that gradients of the sum $h + |\mathbf{u}|^2/2 + gz$ must be normal to the local streamline direction u_i. Therefore, a third Bernoulli equation is:

$$h + \frac{1}{2}|\mathbf{u}|^2 + gz = \text{constant on streamlines.} \tag{4.78}$$

Equation (4.63) requires that inviscid, non-heat-conducting flows are isentropic (s does not change along particle paths), and (1.18) implies $dp/\rho = dh$ when $s = $ constant. Thus the path integral $\int dp/\rho$ becomes a function h of the endpoints only if both heat conduction and viscous stresses may be neglected in the momentum Bernoulli equations (4.71), (4.72), and (4.75). Equation (4.78) is very useful for high-speed gas flows where there is significant interplay between kinetic and thermal energies along a streamline. It is nearly the same as

(4.71), but does not include the other barotropic and vortex-line-evaluation possibilities allowed by (4.71).

Interestingly, there is also a Bernoulli equation for constant-viscosity constant-density irrotational flow. It can be obtained by starting from (4.39), using (4.68) for the advective acceleration, and noting from (4.40) that $\nabla^2 \mathbf{u} = -\nabla \times \boldsymbol{\omega}$ in incompressible flow:

$$\rho \frac{D\mathbf{u}}{Dt} = \rho \frac{\partial \mathbf{u}}{\partial t} + \rho \nabla \left(\frac{1}{2} |\mathbf{u}|^2 \right) - \rho \mathbf{u} \times \boldsymbol{\omega} = -\nabla p + \rho \mathbf{g} + \mu \nabla^2 \mathbf{u} = -\nabla p + \rho \mathbf{g} - \mu \nabla \times \boldsymbol{\omega}. \quad (4.79)$$

When $\rho = $ constant, $\mathbf{g} = -\nabla(gz)$, and $\boldsymbol{\omega} = 0$, the second and final parts of this extended equality require:

$$\rho \frac{\partial \mathbf{u}}{\partial t} + \nabla \left(\frac{1}{2} \rho |\mathbf{u}|^2 + \rho gz + p \right) = 0. \quad (4.80)$$

Now, form the dot product of this equation with the arc-length element $\mathbf{e}_u ds = d\mathbf{s}$ directed along a chosen streamline, integrate from location 1 to location 2 along this streamline, and recognize that $\mathbf{e}_u \cdot \nabla = \partial/\partial s$ to find:

$$\rho \int_1^2 \frac{\partial \mathbf{u}}{\partial t} \cdot d\mathbf{s} + \int_1^2 \mathbf{e}_u \cdot \nabla \left(\frac{1}{2} \rho |\mathbf{u}|^2 + \rho gz + p \right) ds = \rho \int_1^2 \frac{\partial \mathbf{u}}{\partial t} \cdot d\mathbf{s} + \int_1^2 \frac{\partial}{\partial s} \left(\frac{1}{2} \rho |\mathbf{u}|^2 + \rho gz + p \right) ds = 0.$$

$$(4.81)$$

The integration in the second term is elementary, so a fourth Bernoulli equation for constant-viscosity constant-density irrotational flow is:

$$\int_1^2 \frac{\partial \mathbf{u}}{\partial t} \cdot d\mathbf{s} + \left(\frac{1}{2} |\mathbf{u}|^2 + gz + \frac{p}{\rho} \right)_2 = \left(\frac{1}{2} |\mathbf{u}|^2 + gz + \frac{p}{\rho} \right)_1, \quad (4.82)$$

where 1 and 2 denote upstream and downstream locations on the same streamline at a single instant in time. Alternatively, (4.80) can be written using (4.73) as

$$\frac{\partial \phi}{\partial t} + \frac{1}{2} |\nabla \phi|^2 + gz + \frac{p}{\rho} = \text{constant}. \quad (4.83)$$

To summarize, there are (at least) four Bernoulli equations: (4.71) is for inviscid, steady, barotropic flow; (4.75) is for inviscid, irrotational, unsteady, barotropic flow; (4.78) is for inviscid, isentropic, steady flow; and (4.82) or (4.83) is for constant-viscosity, irrotational, unsteady, constant density flow. Perhaps the simplest form of these is (4.19).

There are many useful and important applications of Bernoulli equations. A few of these are described in the following paragraphs.

Consider first a simple device to measure the local velocity in a fluid stream by inserting a narrow bent tube (Figure 4.15), called a *pitot tube* after the French mathematician Henri Pitot (1695–1771), who used a bent glass tube to measure the velocity of the river Seine. Consider two points (1 and 2) at the same level, point 1 being away from the tube and point 2 being immediately in front of the open end where the fluid velocity $\mathbf{u}_2$ is zero. If the flow is steady

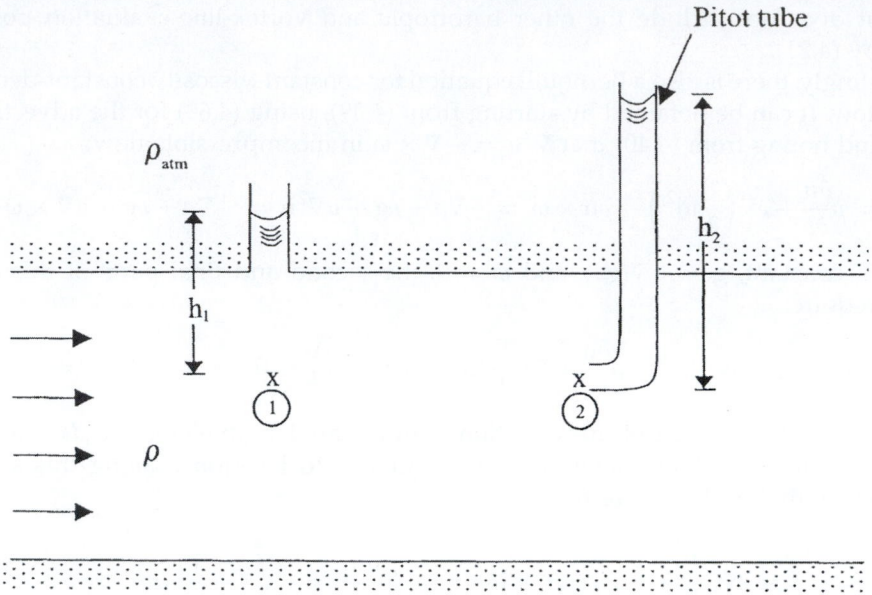

FIGURE 4.15 Pitot tube for measuring velocity in a duct. The first port measures the static pressure while the second port measures the static and dynamic pressure. Using the steady Bernoulli equation for incompressible flow, the height difference $h_2 - h_1$ can be related to the flow speed.

and irrotational with constant density along the streamline that connects 1 and 2, then (4.19) gives

$$\frac{p_1}{\rho} + \frac{1}{2}|\mathbf{u}|_1^2 = \frac{p_2}{\rho} + \frac{1}{2}|\mathbf{u}|_2^2 = \frac{p_2}{\rho},$$

from which the magnitude of $\mathbf{u}_1$ is found to be

$$|\mathbf{u}|_1 = \sqrt{2(p_2 - p_1)/\rho}.$$

Pressures at the two points are found from the hydrostatic balance

$$p_1 = \rho g h_1 \quad \text{and} \quad p_2 = \rho g h_2,$$

so that the magnitude of $\mathbf{u}_1$ can be found from

$$|\mathbf{u}|_1 = \sqrt{2g(h_2 - h_1)}.$$

Because it is assumed that the fluid density is very much greater than that of the atmosphere to which the tubes are exposed, the pressures at the tops of the two fluid columns are assumed to be the same. They will actually differ by $\rho_{atm} g(h_2 - h_1)$. Use of the hydrostatic

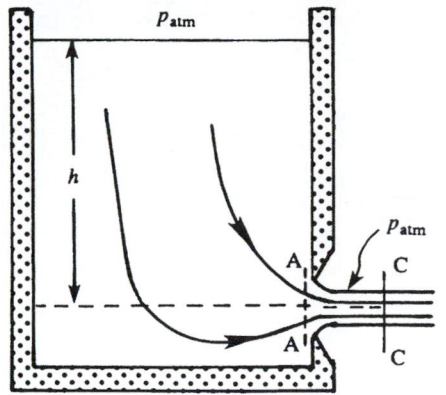

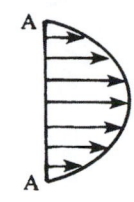

Distribution of
$(p - p_{atm})$ at orifice

FIGURE 4.16 Flow through a sharp-edged orifice. Pressure has the atmospheric value everywhere across section CC; its distribution across orifice AA is indicated. The basic finding here is that the width of the fluid jet that emerges from the tank at AA is larger than the width of the jet that crosses CC.

approximation above station 1 is valid when the streamlines are straight and parallel between station 1 and the upper wall.

The pressure p_2 measured by a pitot tube is called *stagnation pressure* or *total pressure*, which is larger than the local static pressure. Even when there is no pitot tube to measure the stagnation pressure, it is customary to refer to the local value of the quantity $(p + \rho|\boldsymbol{u}|^2/2)$ as the local *stagnation pressure*, defined as the pressure that would be reached if the local flow is *imagined* to slow down to zero velocity frictionlessly. The quantity $\rho u^2/2$ is sometimes called the *dynamic pressure*; stagnation pressure is the sum of static and dynamic pressures.

As another application of Bernoulli's equation, consider the flow through an orifice or opening in a tank (Figure 4.16). The flow is slightly unsteady due to lowering of the water level in the tank, but this effect is small if the tank area is large compared to the orifice area. Viscous effects are negligible everywhere away from the walls of the tank. All streamlines can be traced back to the free surface in the tank, where they have the same value of the Bernoulli constant $B = |\boldsymbol{u}|^2/2 + p/\rho + gz$. It follows that the flow is irrotational, and B is constant *throughout* the flow.

We want to apply a Bernoulli equation between a point at the free surface in the tank and a point in the jet. However, the conditions right at the opening (section A in Figure 4.16) are not simple because the pressure is *not* uniform across the jet. Although pressure has the atmospheric value everywhere on the free surface of the jet (neglecting small surface tension effects), it is not equal to the atmospheric pressure *inside* the jet at this section. The streamlines at the orifice are curved, which requires that pressure must vary across the width of the jet in order to balance the centrifugal force. The pressure distribution across the orifice (section A) is shown in Figure 4.16. However, the streamlines in the jet become parallel a short distance away from the orifice (section C in Figure 4.16), where the jet area is smaller than the orifice area. The pressure across section C is uniform and equal to the atmospheric value (p_{atm}) because it has that value at the surface of the jet.

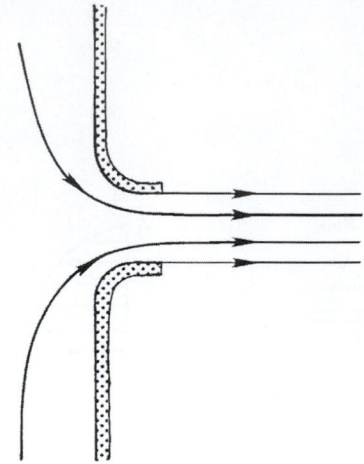

FIGURE 4.17 Flow through a rounded orifice. Here the pressure and velocity can achieve parallel outflow inside the tank, so the width of the jet does not change outside the tank.

Application of the Bernoulli equation (4.19) for steady constant-density flow between a point on the free surface in the tank and a point at C gives

$$\frac{p_{atm}}{\rho} + gh = \frac{p_{atm}}{\rho} + \frac{u^2}{2},$$

from which the average jet velocity magnitude u is found as

$$u = \sqrt{2gh},$$

which simply states that the loss of potential energy equals the gain of kinetic energy. The mass flow rate is approximately

$$\dot{m} = \rho A_c u = \rho A_c \sqrt{2gh},$$

where A_c is the area of the jet at C. For orifices having a sharp edge, A_c has been found to be $\approx 62\%$ of the orifice area because the jet contracts downstream of the orifice opening.

If the orifice has a well-rounded opening (Figure 4.17), then the jet does not contract, the streamlines right at the exit are then parallel, and the pressure at the exit is uniform and equal to the atmospheric pressure. Consequently the mass flow rate is simply $\rho A \sqrt{2gh}$, where A equals the orifice area.

Neglect of Gravity in Constant Density Flows

When the flow velocity is zero, the Navier-Stokes momentum equation for incompressible flow (4.39b) reduces to a balance between the hydrostatic pressure p_s, and the steady body force acting on the hydrostatic density ρ_s,

$$0 = -\nabla p_s + \rho_s \mathbf{g},$$

which is equivalent to (1.8). When this hydrostatic balance is subtracted from (4.39b), the pressure difference from hydrostatic, $p' = p - p_s$, and the density difference from hydrostatic, $\rho' = \rho - \rho_s$, appear:

$$\rho \frac{D\mathbf{u}}{Dt} = -\nabla p' + \rho' \mathbf{g} + \mu \nabla^2 \mathbf{u}. \tag{4.84}$$

When the fluid density is constant, $\rho' = 0$ and the gravitational-body-force term disappears leaving:

$$\rho \frac{D\mathbf{u}}{Dt} = -\nabla p' + \mu \nabla^2 \mathbf{u}. \tag{4.85}$$

Because of this, steady body forces (like gravity) in constant density flow are commonly omitted from the momentum equation, and pressure is measured relative to its local hydrostatic value. Furthermore, the prime on p in (4.85) is typically dropped in this situation. However, when the flow includes a free surface, a fluid-fluid interface across which the density changes, or other variations in density, the gravitational-body-force term should reappear.

The Boussinesq Approximation

For flows satisfying certain conditions, Boussinesq in 1903 suggested that density changes in the fluid can be neglected except where ρ is multiplied by g. This approximation also treats the other properties of the fluid (such as μ, k, C_p) as constants. It is commonly useful for analyzing oceanic and atmospheric flows. Here we shall discuss the basis of the approximation in a somewhat intuitive manner and examine the resulting simplifications of the equations of motion. A formal justification, and the conditions under which the Boussinesq approximation holds, is given in Spiegel and Veronis (1960).

The Boussinesq approximation replaces the full continuity equation (4.7) by its incompressible form (4.10), $\nabla \cdot \mathbf{u} = 0$, to indicate that the relative density changes following a fluid particle, $\rho^{-1}(D\rho/Dt)$, are small compared to the velocity gradients that compose $\nabla \cdot \mathbf{u}$. Thus, the Boussinesq approximation cannot be applied to high-speed gas flows where density variations induced by velocity divergence cannot be neglected (see Section 4.11). Similarly, it cannot be applied when the vertical scale of the flow is so large that hydrostatic pressure variations cause significant changes in density. In a hydrostatic field, the vertical distance over which the density changes become important is of order $c^2/g \sim 10$ km for air where c is the speed of sound. (This vertical distance estimate is consistent with the scale height of the atmosphere; see Section 1.10.) The Boussinesq approximation therefore requires that the vertical scale of the flow be $L \ll c^2/g$.

In both cases just mentioned, density variations are caused by pressure variations. Now suppose that such pressure-compressibility effects are small and that density changes are caused by temperature variations alone, as in a thermal convection problem. In this case, the Boussinesq approximation applies when the temperature variations in the flow are small. Assume that ρ changes with T according to $\delta\rho/\rho = -\alpha\delta T$, where $\alpha = -\rho^{-1}(\partial\rho/\partial T)_p$ is the thermal expansion coefficient (1.20). For a perfect gas at room temperature $\alpha = 1/T \sim 3 \times 10^{-3}\,\mathrm{K}^{-1}$ but for typical liquids $\alpha \sim 5 \times 10^{-4}\,\mathrm{K}^{-1}$. Thus, for a temperature difference in the fluid of 10°C, density variations can be at most a few percent and it turns out that $\rho^{-1}(D\rho/Dt)$ can also be no larger than a few percent of the velocity gradients in $\nabla \cdot \mathbf{u}$. To see this, assume that the flow field is characterized by a length scale L, a velocity scale U, and a temperature scale δT. By this we mean that the velocity varies by U and the temperature varies by δT between locations separated by a distance of order L. The ratio of the magnitudes of the two terms in the continuity equation is

$$\frac{(1/\rho)(D\rho/Dt)}{\nabla \cdot \mathbf{u}} \sim \frac{(1/\rho)u(\partial\rho/\partial x)}{\partial u/\partial x} \sim \frac{(U/\rho)(\delta\rho/L)}{U/L} = \frac{\delta\rho}{\rho} = \alpha\delta T \ll 1,$$

which allows (4.7) to be replaced by its incompressible form (4.10).

The Boussinesq approximation for the momentum equation is based on its form for incompressible flow, and proceeds from (4.84) divided by ρ_s:

$$\frac{\rho}{\rho_s}\frac{D\mathbf{u}}{Dt} = -\frac{1}{\rho_s}\nabla p' + \frac{\rho'}{\rho_s}\mathbf{g} + \frac{\mu}{\rho_s}\nabla^2\mathbf{u}.$$

When the density fluctuations are small $\rho/\rho_s \cong 1$ and $\mu/\rho_s \cong \nu$ (= the kinematic viscosity), so this equation implies:

$$\frac{D\mathbf{u}}{Dt} = -\frac{1}{\rho_0}\nabla p' + \frac{\rho'}{\rho_0}\mathbf{g} + \nu\nabla^2\mathbf{u}, \tag{4.86}$$

where ρ_0 is a constant reference value of ρ_s. This equation states that density changes are negligible when conserving momentum, except when ρ' is multiplied by $\mathbf{g}$. In flows involving buoyant convection, the magnitude of $\rho'g/\rho_s$ is of the same order as the vertical acceleration $\partial w/\partial t$ or the viscous term $\nu\nabla^2 w$.

The Boussinesq approximation to the energy equation starts from (4.60), written in vector notation,

$$\rho\frac{De}{Dt} = -p\nabla \cdot \mathbf{u} + \rho\varepsilon - \nabla \cdot \mathbf{q}, \tag{4.87}$$

where (4.58) has been used to insert ε, the kinetic energy dissipation rate per unit mass. Although the continuity equation is approximately $\nabla \cdot \mathbf{u} = 0$, an important point is that the volume expansion term $p(\nabla \cdot \mathbf{u})$ is *not* negligible compared to other dominant terms of equation (4.87); only for incompressible liquids is $p(\nabla \cdot \mathbf{u})$ negligible in (4.87). We have

$$-p\nabla \cdot \mathbf{u} = \frac{p}{\rho}\frac{D\rho}{Dt}; \frac{p}{\rho}\left(\frac{\partial\rho}{\partial T}\right)_{\mathrm{p}}\frac{DT}{Dt} = -p\alpha\frac{DT}{Dt}.$$

Assuming a perfect gas, for which $p = \rho RT$, $C_{\mathrm{p}} - C_{\mathrm{v}} = R$, and $\alpha = 1/T$, the foregoing estimate becomes

$$-p\nabla \cdot \mathbf{u} = -\rho RT\alpha\frac{DT}{Dt} = -\rho(C_{\mathrm{p}} - C_{\mathrm{v}})\frac{DT}{Dt}.$$

Equation (4.87) then becomes

$$\rho C_{\mathrm{p}} \frac{DT}{Dt} = \rho\varepsilon - \nabla \cdot \mathbf{q}, \tag{4.88}$$

where $e = C_v T$ for a perfect gas. Note that C_v (instead of C_p) would have appeared on the left side of (4.88) if $\nabla \cdot \mathbf{u}$ had been dropped from (4.87).

The heating due to viscous dissipation of energy is negligible under the restrictions underlying the Boussinesq approximation. Comparing the magnitude of $\rho\varepsilon$ with the left-hand side of (4.88), we obtain

$$\frac{\rho\varepsilon}{\rho C_{\mathrm{p}}(DT/Dt)} \sim \frac{2\mu S_{ij}S_{ij}}{\rho C_{\mathrm{p}} u_i(\partial T/\partial x_i)} \sim \frac{\mu U^2/L^2}{\rho C_{\mathrm{p}} U(\delta T/L)} = \frac{\nu U}{(C_{\mathrm{p}}\delta T)L}.$$

In typical situations this is extremely small ($\sim 10^{-7}$). Neglecting $\rho\varepsilon$, and assuming Fourier's law of heat conduction (1.2) with constant k, (4.88) finally reduces to

$$\frac{DT}{Dt} = \kappa\nabla^2 T, \tag{4.89}$$

where $\kappa \equiv k/\rho C_{\mathrm{p}}$ is the thermal diffusivity.

Summary

The Boussinesq approximation applies if the Mach number of the flow is small, propagation of sound or shock waves is not considered, the vertical scale of the flow is not too large, and the temperature differences in the fluid are small. Then the density can be treated as a constant in both the continuity and the momentum equations, except in the gravity term. Properties of the fluid such as μ, k, and C_p are also assumed constant. Omitting Coriolis accelerations, the set of equations corresponding to the Boussinesq approximation is: (4.9) and/or (4.10), (4.86) with $\mathbf{g} = -g\mathbf{e}_z$, (4.89), and $\rho = \rho_0[1 - \alpha(T - T_0)]$, where the z-axis points upward. The constant ρ_0 is a reference density corresponding to a reference temperature T_0, which can be taken to be the mean temperature in the flow or the temperature at an appropriate boundary. Applications of the Boussinesq set can be found in several places in this book, for example, in the analysis of wave propagation in a density-stratified medium, thermal instability, turbulence in a stratified medium, and geophysical fluid dynamics.

4.10. BOUNDARY CONDITIONS

The differential equations for the conservation laws require boundary conditions for proper solution. Specifically, the Navier-Stokes momentum equation (4.38) requires the specification of the velocity vector on all surfaces bounding the flow domain. For an external flow, one that is not contained by walls or surfaces at specified locations, the fluid's velocity vector and the thermodynamic state must be specified on a closed distant surface.

On a solid boundary or at the interface between two immiscible fluids, conditions may be derived from the three basic conservation laws as follows. In Figure 4.18, a small cylindrical control volume is drawn through the interface separating medium 1 (fluid) from medium 2 (solid or liquid immiscible with fluid 1). Here $+\mathbf{n}dA$ and $-\mathbf{n}dA$ are the end face-directed area

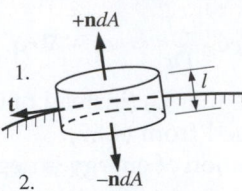

FIGURE 4.18   Interface between two media for evaluation of boundary conditions. Here medium 1 is a fluid, and medium 2 is a solid or a second fluid that is immiscible with the first fluid. Boundary conditions can be determined by evaluating the equations of motion in the small cylindrical control volume shown and then letting l go to zero with the volume straddling the interface.

elements in medium 1 and medium 2, respectively. The circular surfaces are locally tangent to the interface, and separated from each other by a distance l. Now apply the conservation laws to the volume defined by the cylindrical volume. Next, let $l \rightarrow 0$, keeping the two round area elements in the two different media. As $l \rightarrow 0$, all volume integrals $\rightarrow 0$ and the integral over the side area, which is proportional to l, tends to zero as well. The unit vector $\mathbf{n}$ is normal to the interface and points into medium 1. Mass conservation gives $\rho_1 \mathbf{u}_1 \cdot \mathbf{n} = \rho_2 \mathbf{u}_2 \cdot \mathbf{n}$ at each point on the interface as the end face area becomes small. (Here we assume that the coordinates are fixed to the interface, that is, the interface is at rest. Later in this section we show the modifications necessary when the interface is moving.)

If medium 2 is a solid, then $\mathbf{u}_2 = 0$ there. If medium 1 and medium 2 are immiscible liquids, no mass flows across the boundary surface. In either case, $\mathbf{u}_1 \cdot \mathbf{n} = 0$ on the boundary. The same procedure applied to the integral form of the momentum equation (4.17) gives the result that the force/area on the surface, $n_i \tau_{ij}$, is continuous across the interface if surface tension is neglected. If surface tension is included, a jump in pressure in the direction normal to the interface must be added; see Section 1.6 and the discussion later in this section.

Applying the integral form of energy conservation (4.48) to a small cylindrical control volume of infinitesimal height l that straddles the interface gives the result that $n_i q_i$ is continuous across the interface, or explicity, $k_1(\partial T_1 / \partial n) = k_2(\partial T_2 / \partial n)$ at the interface surface. The heat flux must be continuous at the interface; it cannot store heat.

Two more boundary conditions are required to completely specify a problem and these are not consequences of any conservation law. These boundary conditions are: no slip of a viscous fluid is permitted at a solid boundary $\mathbf{u}_1 \cdot \mathbf{t} = 0$; and no temperature jump is permitted at the boundary $T_1 = T_2$. Here $\mathbf{t}$ is a unit vector tangent to the boundary.

Known violations of the no-slip boundary condition occur for superfluid helium at or below 2.17 K, which has an immeasurably small (essentially zero) viscosity. On the other hand, the *appearance* of slip is created when water or water-based fluids flow over finely textured *super-hydrophobic* (strongly water repellent) coated surfaces. This is described by Gogte et al. (2005). Surface textures must be much smaller than the capillary length for water and were typically about $10 \mu m$ in this case. The fluid did not slip on the protrusions but did not penetrate the valleys because of the surface tension, giving the appearance of slip. Both slip and temperature jump are known to occur in highly rarefied gases, where the mean distances between intermolecular collisions become of the order of the length scales of interest in the problem. The details are closely related to the manner of gas-surface interaction of momentum and energy. A review of these topics is provided by McCormick (2005).

Moving and Deforming Boundaries

Consider a surface in space that may be moving or deforming in some arbitrary way. Examples may be flexible solid boundaries, the interface between two immiscible liquids, or a moving shock wave, which is described in Chapter 15. The first two examples do not permit mass flow across the interface, whereas the third does. Such a surface can be defined and its motion described in inertial coordinates by $\eta(x, y, z, t) = 0$. We often must treat problems in which boundary conditions must be satisfied on such a moving, deforming interface. Let the velocity of a point that remains on the interface, where $\eta = 0$, be $\mathbf{u}_s$. An observer riding on that point sees:

$$d\eta/dt = \partial\eta/\partial t + (\mathbf{u}_s \cdot \nabla)\eta = 0 \quad \text{on} \quad \eta = 0. \tag{4.90}$$

A fluid particle has velocity $\mathbf{u}$. If no fluid flows across $\eta = 0$, then $\mathbf{u} \cdot \nabla \eta = \mathbf{u}_s \cdot \nabla \eta = -\partial\eta/\partial t$. Thus the condition that there be no mass flow across the surface becomes

$$\partial\eta/\partial t + (\mathbf{u} \cdot \nabla)\eta \equiv D\eta/Dt = 0 \quad \text{on} \quad \eta = 0. \tag{4.91}$$

If there is mass flow across the surface, it is proportional to the relative velocity between the fluid and the surface, $(u_{rel})_n = \mathbf{u} \cdot \mathbf{n} - \mathbf{u}_s \cdot \mathbf{n}$, where $\mathbf{n} = \nabla\eta/|\nabla\eta|$.

$$(u_{rel})_n = (\mathbf{u} \cdot \nabla\eta + \partial\eta/\partial t)/|\nabla\eta| = (1/|\nabla\eta|)D\eta/Dt. \tag{4.92}$$

Thus the mass flow rate across the surface (per unit surface area) is represented by

$$(\rho/|\nabla\eta|)D\eta/Dt \quad \text{on} \quad \eta = 0. \tag{4.93}$$

Again, if no mass flows across the surface, the requirement is $D\eta/Dt = 0$ on $\eta = 0$.

Surface Tension Revisited

As discussed in Section 1.6, attractive intermolecular forces dominate in a liquid, whereas in a gas repulsive forces are larger. However, as a liquid-gas phase boundary is approached from the liquid side, these attractive forces are not felt equally because there are many fewer liquid-phase molecules near the phase boundary. Thus there tends to be an unbalanced attraction to the interior of the liquid of the molecules on the phase boundary. This unbalanced attraction is called *surface tension* and its manifestation is a pressure increment across a curved interface. A somewhat more detailed description is provided in texts on physico-chemical hydrodynamics. Two excellent sources are Probstein (1994, Chapter 10) and Levich (1962, Chapter VII).

Lamb, in *Hydrodynamics* (1945, 6th Edition, p. 456), writes, "Since the condition of stable equilibrium is that the free energy be a minimum, the surface tends to contract as much as is consistent with the other conditions of the problem." Thus we are led to introduce the Helmholtz free energy (per unit mass) f via

$$f = e - Ts, \tag{4.94}$$

where the notation is consistent with that used in Section 1.8. If the free energy is a minimum, then the system is in a state of stable equilibrium, and F is called the *thermodynamic potential at constant volume* (Fermi, 1956, *Thermodynamics*, p. 80). For a reversible, isothermal change, the work done on the system is the gain in total free energy F,

$$df = de - Tds - sdT, \qquad (4.95)$$

where the last term is zero for an isothermal change. Then, from (1.18), $dF = -pdv =$ work done on the system. (These relations suggest that surface tension decreases with increasing temperature.)

For an interface of area $= A$, separating two fluids of densities ρ_1 and ρ_2, with volumes V_1 and V_2, respectively, and with a surface tension coefficient σ (corresponding to free energy per unit area), the total (Helmholtz) free energy F of the system can be written as

$$F = \rho_1 V_1 f_1 + \rho_2 V_2 f_2 + A\sigma. \qquad (4.96)$$

If $\sigma > 0$, then the two media (fluids) are immiscible and A will reach a local minimum value at equilibrium. On the other hand, if $\sigma < 0$, corresponding to surface compression, then the two fluids mix freely since the minimum free energy will occur when A has expanded to the point that the spacing between its folds reaches molecular dimensions and the two-fluid system has uniform composition.

When $\sigma > 0$, it and the curvature of the fluid interface determine the pressure difference across the interface. Here, we shall assume that $\sigma = $ const. Flows driven by surface tension gradients are called *Marangoni flows* and are not discussed here. Consider the situation depicted in Figure 4.19 where the pressure above a curved interface is higher than that below it by an increment Δp, and the shape of the fluid interface is given by $\eta(x, y, z) = z - h(x,y) = 0$. The origin of coordinates and the direction of the z-axis are chosen so that h, $\partial h/\partial x$, and $\partial h/\partial y$ are all zero at $\mathbf{x} = (0, 0, 0)$. Plus, the directions of the x- and y-axes are chosen so that the surface's principal radii of curvature, R_1 and R_2, are found in the x-z and y-z planes, respectively. Thus, the surface's shape is given by

$$\eta(x,y,z) = z - (x^2/2R_1) - (y^2/2R_2) = 0$$

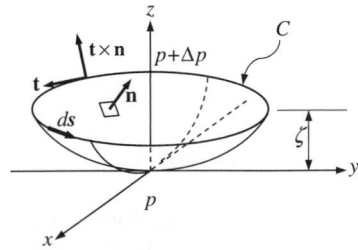

FIGURE 4.19   The curved surface shown is tangent to the x-y plane at the origin of coordinates. The pressure above the surface is Δp higher than the pressure below the surface, creating a downward force. Surface tension forces pull in the local direction of $\mathbf{t} \times \mathbf{n}$, which is slightly upward, all around the curve C and thereby balances the downward pressure force.

in the vicinity of the origin. A closed curve C is defined by the intersection of the curved surface and the plane $z = \zeta$. The goal here is to determine how the pressure increment Δp depends on R_1 and R_2 when pressure and surface tension forces are balanced as the area enclosed by C approaches zero.

First determine the net pressure force $\mathbf{F}_p$ on the surface A bounded by C. The unit normal $\mathbf{n}$ to the surface η is

$$\mathbf{n} = \frac{\nabla \eta}{|\nabla \eta|} = \frac{(-x/R_1, -y/R_2, 1)}{\sqrt{(x/R_1)^2 + (y/R_2)^2 + 1}},$$

and the area element is

$$dA = \sqrt{1 + (\partial \eta/\partial x)^2 + (\partial \eta/\partial y)^2}\, dxdy = \sqrt{1 + (x/R_1)^2 + (y/R_2)^2}\, dxdy,$$

so

$$\mathbf{F}_p = -\iint_A \Delta p\, \mathbf{n}\, dA = -\Delta p \int_{-\sqrt{2R_1\zeta}}^{+\sqrt{2R_1\zeta}} \left[\int_{-\sqrt{2R_2\zeta - x^2 R_2/R_1}}^{+\sqrt{2R_2\zeta - x^2 R_2/R_1}} (-x/R_1, -y/R_2, 1)dy \right] dx. \tag{4.97}$$

The minus sign appears here because greater pressure above the surface (positive Δp) must lead to a downward force and the vertical component of $\mathbf{n}$ is positive. The x- and y-components of $\mathbf{F}_p$ are zero because of the symmetry of the situation (odd integrand with even limits). The remaining double integration for the z-component of $\mathbf{F}_p$ produces:

$$(\mathbf{F}_p)_z = \mathbf{e}_z \cdot \mathbf{F}_p = -\pi \Delta p \sqrt{2R_1\zeta}\sqrt{2R_2\zeta}.$$

The net surface tension force $\mathbf{F}_{st}$ on C can be determined from the integral,

$$\mathbf{F}_{st} = \sigma \oint_C \mathbf{t} \times \mathbf{n}\, ds, \tag{4.98}$$

where $ds = dx\sqrt{1 + (dy/dx)^2}$ is an arc length element of curve C, and $\mathbf{t}$ is the unit tangent to C so

$$\mathbf{t} = \frac{(1, dy/dx, 0)}{\sqrt{1 + (dy/dx)^2}} = \frac{(-y/R_2, x/R_1, 0)}{\sqrt{(y/R_2)^2 + (x/R_1)^2}},$$

and dy/dx is found by differentiating the equation for C, $\zeta = (x^2/2R_1) - (y^2/2R_2)$, with ζ regarded as constant. On each element of C, the surface tension force acts perpendicular to $\mathbf{t}$ and tangent to A. This direction is given by $\mathbf{t} \times \mathbf{n}$ so the integrand in (4.98) is

$$\mathbf{t} \times \mathbf{n}\, ds = \frac{(R_2/y)dx}{\sqrt{1 + (x/R_1)^2 + (y/R_2)^2}} \left(\frac{x}{R_1}, \frac{y}{R_2}, \frac{x^2}{R_1^2} + \frac{y^2}{R_2^2} \right) \cong \frac{R_2}{y}\left(\frac{x}{R_1}, \frac{y}{R_2}, \frac{x^2}{R_1^2} + \frac{y^2}{R_2^2} \right)dx,$$

where the approximate equality holds when x/R_1 and $y/R_2 \ll 1$ and the area enclosed by C approaches zero. The symmetry of the integration path will cause the x- and y-components of $\mathbf{F}_{st}$ to be zero, leaving

$$(\mathbf{F}_{st})_z = \mathbf{e}_z \cdot \mathbf{F}_{st} = 4\sigma \int_0^{\sqrt{2R_1\zeta}} \frac{R_2}{\sqrt{2R_2\zeta - (R_2/R_1)x^2}} \left[\frac{2\zeta}{R_2} + \frac{x^2}{R_1}\left(\frac{1}{R_1} - \frac{1}{R_2}\right)\right] dx,$$

where y has been eliminated from the integrand using the equation for C, and the factor of four appears because the integral shown only covers one-quarter of the path defined by C. An integration variable substitution in the form $\sin \xi = x/\sqrt{2R_1\zeta}$ allows the integral to be evaluated:

$$(\mathbf{F}_{st})_z = \mathbf{e}_z \cdot \mathbf{F}_{st} = \pi\sigma\sqrt{2R_1\zeta}\sqrt{2R_2\zeta}\left(\frac{1}{R_1} + \frac{1}{R_2}\right).$$

For static equilibrium, $\mathbf{F}_p + \mathbf{F}_{st} = 0$, so the evaluated results of (4.97) and (4.98) require:

$$\Delta p = \sigma(1/R_1 + 1/R_2), \tag{1.5}$$

where the pressure is greater on the side of the surface with the centers of curvature of the interface. Batchelor (1967, p. 64) writes,

> *An unbounded surface with a constant sum of the principal curvatures is spherical, and this must be the equilibrium shape of the surface. This result also follows from the fact that in a state of (stable) equilibrium the energy of the surface must be a minimum consistent with a given value of the volume of the drop or bubble, and the sphere is the shape which has the least surface area for a given volume.*

The original source of this analysis is Lord Rayleigh's (1890) "On the Theory of Surface Forces."

For an air bubble in water, gravity is an important factor for bubbles of millimeter size, as we shall see here. The hydrostatic pressure in a liquid is obtained from $p_L + \rho gz = $ const., where z is measured positively upwards from the free surface and gravity acts downwards. Thus for a gas bubble beneath the free surface,

$$p_G = p_L + \sigma\left(1/R_1 + 1/R_2\right) = \text{const.} - \rho gz + \sigma\left(1/R_1 + 1/R_2\right).$$

Gravity and surface tension forces are of the same order over a length scale $(\sigma/\rho g)^{1/2}$. For an air bubble in water at 288 K, this scale $= [7.35 \times 10^{-2}\ \text{N/m}/(9.81\ \text{m/s}^2 \times 10^3\ \text{kg/m}^3)]^{1/2} = 2.74 \times 10^{-3}$ m.

EXAMPLE 4.7

Calculation of the shape of the free surface of a liquid adjoining an infinite vertical plane wall. Here let $z = \zeta(x)$ define the free surface shape. With reference to Figure 4.20 where the y-axis points into the page, $1/R_1 = [\partial^2\zeta/\partial x^2][1 + (\partial\zeta/\partial x)^2]^{-3/2}$, and $1/R_2 = [\partial^2\zeta/\partial y^2][1 + (\partial\zeta/\partial y)^2]^{-3/2} = 0$. At the free surface, $\rho g\zeta - \sigma/R_1 = const$. As $x \to \infty$, $\zeta \to 0$, and $R_2 \to \infty$, so $const. = 0$. Then $\rho g\zeta/\sigma - \zeta''/(1 + \zeta'^2)^{3/2} = 0$.

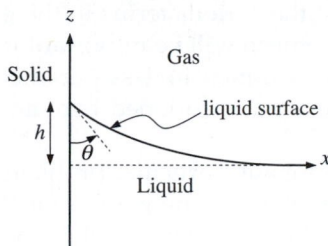

FIGURE 4.20 Free surface of a liquid adjoining a vertical plane wall. Here the contact angle is θ and the liquid rises to $z = h$ at the solid wall.

Multiply by the integrating factor ζ' and integrate. We obtain $(\rho g/2\sigma)\zeta^2 + (1 + \zeta'^2)^{-1/2} = C$. Evaluate C as $x \rightarrow \infty$, $\zeta \rightarrow 0$, $\zeta' \rightarrow 0$. Then $C = 1$. We look at $x = 0$, $z = \zeta(0) = h$ to find h. The slope at the wall, $\zeta' = \tan(\theta + \pi/2) = -\cot\theta$. Then $1 + \zeta'^2 = 1 + \cot^2\theta = \csc^2\theta$. Thus we now have $(\rho g/2\sigma)h^2 = 1 - 1/\csc\theta = 1 - \sin\theta$, so that $h^2 = (2\sigma/\rho g)(1 - \sin\theta)$. Finally we seek to integrate to obtain the shape of the interface. Squaring and rearranging the result above, the differential equation we must solve may be written as $1 + (d\zeta/dx)^2 = [1 - (\rho g/2\sigma)\zeta^2]^{-2}$. Solving for the slope and taking the negative square root (since the slope is negative for positive x),

$$d\zeta/dx = -\left\{1 - [1 - (\rho g/2\sigma)\zeta^2]^2\right\}^{1/2}[1 - (\rho g/2\sigma)\zeta^2]^{-1}.$$

Define $\sigma/\rho g = \delta^2$, $\zeta/\delta = \gamma$. Rewriting the equation in terms of x/δ and γ, and separating variables:

$$2(1 - \gamma^2/2)\gamma^{-1}(4 - \gamma^2)^{-1/2}d\gamma = d\left(x/\delta\right).$$

The integrand on the left is simplified by partial fractions and the constant of integration is evaluated at $x = 0$ when $\eta = h/\delta$. Finally,

$$\cosh^{-1}(2\delta/\zeta) - (4 - \zeta^2/\delta^2)^{1/2} - \cosh^{-1}(2\delta/h) + (4 - h^2/\delta^2)^{1/2} = x/\delta$$

gives the shape of the interface in terms of $x(\zeta)$.

Analysis of surface tension effects results in the appearance of additional dimensionless parameters in which surface tension is compared with other effects such as viscous stresses, body forces such as gravity, and inertia. These are defined in the next section.

4.11. DIMENSIONLESS FORMS OF THE EQUATIONS AND DYNAMIC SIMILARITY

For a properly specified fluid flow problem or situation, the differential equations of fluid motion, the fluid's constitutive and thermodynamic properties, and the boundary conditions may be used to determine the dimensionless parameters that govern the situation of interest even before a solution of the equations is attempted. The dimensionless parameters so determined set the importance of the various terms in the governing differential equations, and thereby indicate which phenomena will be important in the resulting flow. This section describes and presents the primary dimensionless parameters or numbers required in the remainder of the text. Many others not mentioned here are defined and used in the broad realm of fluid mechanics.

The dimensionless parameters for any particular problem can be determined in two ways. They can be deduced directly from the governing differential equations if these equations are known; this method is illustrated here. However, if the governing differential equations are unknown or the parameter of interest does not appear in the known equations, dimensionless parameters can be determined from dimensional analysis (see Section 1.11). The advantage of adopting the former strategy is that dimensionless parameters determined from the equations of motion are more readily interpreted and linked to the physical phenomena occurring in the flow. Thus, knowledge of the relevant dimensionless parameters frequently aids the solution process, especially when assumptions and approximations are necessary to reach a solution.

In addition, the dimensionless parameters obtained from the equations of fluid motion set the conditions under which scale model testing with small models will prove useful for predicting the performance of larger devices. In particular, two flow fields are considered to be dynamically similar when their dimensionless parameters match, and their geometries are scale similar; that is, any length scale in the first flow field may be mapped to its counterpart in the second flow field by multiplication with a single scale ratio. When two flows are dynamically similar, analysis, simulations, or measurements from one flow field are directly applicable to the other when the scale ratio is accounted for. Moreover, use of standard dimensionless parameters typically reduces the parameters that must be varied in an experiment or calculation, and greatly facilitates the comparison of measured or computed results with prior work conducted under potentially different conditions.

To illustrate these advantages, consider the drag force F_D on a sphere, of diameter d moving at a speed U through a fluid of density ρ and viscosity μ. Dimensional analysis (Section 1.11) using these five parameters produces the following possible dimensionless scaling laws:

$$\frac{F_D}{\rho U^2 d^2} = \Psi\left(\frac{\rho U d}{\mu}\right), \quad \text{or} \quad \frac{F_D \rho}{\mu^2} = \Phi\left(\frac{\mu}{\rho U d}\right). \tag{4.99}$$

Both are valid, but the first is preferred because it contains dimensionless groups that either come from the equations of motion or are traditionally defined in the study of fluid dynamic drag. If dimensionless groups were not used, experiments would have to be conducted to determine F_D as a function of d, keeping U, ρ, and μ fixed. Then, experiments would have to be conducted to determine F_D as a function of U, keeping d, ρ, and μ fixed, and so on. However, such a duplication of effort is unnecessary when dimensionless groups are used. In fact, use of the first dimensionless law above allows experimental results from a wide range of conditions to be simply plotted with two axes (see Figure 4.21) even though the full complement of experiments may have involved variations in all five dimensional parameters.

The idea behind dimensional analysis is intimately associated with the concept of similarity. In fact, a collapse of all the data on a single graph such as the one in Figure 4.21 is

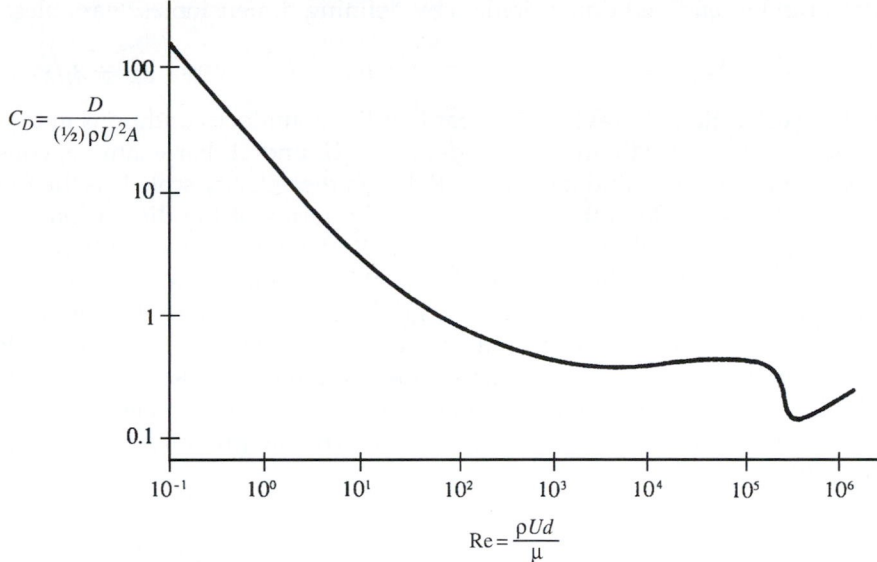

FIGURE 4.21   Coefficient of drag C_D for a sphere vs. the Reynolds number Re based on sphere diameter. At low Reynolds number $C_D \sim 1/\text{Re}$, and above Re $\sim 10^3$, $C_D \sim$ constant (except for the dip between Re $= 10^5$ and 10^6). These behaviors (except for the dip) can be explained by simple dimensional reasoning. The reason for the dip is the transition of the laminar boundary layer to a turbulent one, as explained in Chapter 9.

possible only because in this problem all flows having the same value of the dimensionless group known as the *Reynolds number* Re $= \rho U d/\mu$ are dynamically similar. This dynamic similarity is assured because the Reynolds number appears when the equations of motion are cast in dimensionless form.

The use of dimensionless parameters pervades fluid mechanics to such a degree that this chapter and this text would be considered incomplete without this section, even though this topic is typically covered in first-course fluid mechanics texts. For clarity, the discussion begins with the momentum equation, and then proceeds to the continuity and energy equations.

Consider the flow of a fluid having nominal density ρ and viscosity μ through a flow field characterized by a length scale l, a velocity scale U, and a rotation or oscillation frequency Ω. The situation here is intended to be general so that the dimensional parameters obtained from this effort will be broadly applicable. Particular situations that would involve all five parameters include pulsating flow through a tube, flow past an undulating self-propelled body, or flow through a turbomachine.

The starting point is the Navier-Stokes momentum equation (4.39) simplified for incompressible flow. (The effect of compressibility is deduced from the continuity equation in the next subsection.)

$$\rho\left(\frac{\partial \mathbf{u}}{\partial t} + (\mathbf{u}\cdot\nabla)\mathbf{u}\right) = -\nabla p + \rho\mathbf{g} + \mu\nabla^2\mathbf{u} \tag{4.39b}$$

This equation can be rendered dimensionless by defining dimensionless variables:

$$x_i^* = x_i/l, \quad t^* = \Omega t, \quad u_j^* = u_j/U, \quad p^* = (p - p_\infty)/\rho U^2, \quad \text{and} \quad g_j^* = g_j/g, \qquad (4.100)$$

where g is the acceleration of gravity. It is clear that the boundary conditions in terms of the dimensionless variables (4.100) are independent of l, U, and Ω. For example, consider the viscous flow over a circular cylinder of radius R. When the velocity scale U is the free-stream velocity and the length scale is the radius R, then, in terms of the dimensionless velocity $u^* = u/U$ and the dimensionless coordinate $r^* = r/R$, the boundary condition at infinity is $u^* \to 1$ as $r^* \to \infty$, and the condition at the surface of the cylinder is $u^* = 0$ at $r^* = 1$. In addition, because pressure enters (4.39b) only as a gradient, the pressure itself is not of consequence; only pressure differences are important. The conventional practice is to render $p - p_\infty$ dimensionless, where p_∞ is a suitably chosen reference pressure. Depending on the nature of the flow, $p - p_\infty$ could be made dimensionless in terms of a generic viscous stress $\mu U/l$, a hydrostatic pressure $\rho g l$, or, as in (4.100), a dynamic pressure ρU^2. Substitution of (4.100) into (4.39) produces:

$$\left[\frac{\Omega l}{U}\right] \frac{\partial \mathbf{u}^*}{\partial t^*} + (\mathbf{u}^* \cdot \nabla^*)\mathbf{u}^* = -\nabla^* p^* + \left[\frac{gl}{U^2}\right] \mathbf{g}^* + \left[\frac{\mu}{\rho U l}\right] \nabla^{*2} \mathbf{u}^*, \qquad (4.101)$$

where $\nabla^* = l\nabla$. It is apparent that two flows (having different values of Ω, U, l, g, or μ) will obey the same dimensionless differential equation if the values of the dimensionless groups $\Omega l/U$, gl/U^2, and $\mu/\rho U l$ are identical. Because the dimensionless boundary conditions are also identical in the two flows, it follows that *they will have the same dimensionless solutions*. Products of these dimensionless groups appear as coefficients in front of different terms when the pressure is presumed to have alternative scalings (see Exercise 4.59).

The parameter groupings shown in [,]-brackets in (4.100) have the following names and interpretations:

$$\text{St} = \text{Strouhal number} \equiv \frac{\text{unsteady acceleration}}{\text{advective acceleration}} \propto \frac{\partial u/\partial t}{u(\partial u/\partial x)} \propto \frac{\Omega U}{U^2/l} = \frac{\Omega l}{U}, \qquad (4.102)$$

$$\text{Re} = \text{Reynolds number} \equiv \frac{\text{inertia force}}{\text{viscous force}} \propto \frac{\rho u(\partial u/\partial x)}{\mu(\partial^2 u/\partial x^2)} \propto \frac{\rho U^2/l}{\mu U/l^2} = \frac{\rho U l}{\mu}, \text{ and} \qquad (4.103)$$

$$\text{Fr} = \text{Froude number} \equiv \left[\frac{\text{inertia force}}{\text{gravity force}}\right]^{1/2} \propto \left[\frac{\rho u(\partial u/\partial x)}{\rho g}\right]^{1/2} \propto \left[\frac{\rho U^2/l}{\rho g}\right]^{1/2} = \frac{U}{\sqrt{gl}}. \qquad (4.104)$$

The Strouhal number sets the importance of unsteady fluid acceleration in flows with oscillations. It is relevant when flow unsteadiness arises naturally or because of an imposed frequency. The Reynolds number is the most common dimensionless number in fluid mechanics. Low Re flows involve small sizes, low speeds, and high kinematic viscosity such as bacteria swimming through mucous. High Re flows involve large sizes, high speeds, and low kinematic viscosity such as an ocean liner steaming at full speed.

St, Re, and Fr have to be equal for dynamic similarity of two flows in which unsteadiness and viscous and gravitational effects are important. Note that the mere presence of gravity does not make the gravitational effects dynamically important. For flow around an object in a homogeneous fluid, gravity is important only if surface waves are generated. Otherwise, the effect of gravity is simply to add a hydrostatic pressure to the entire system that can be combined with the pressure gradient (see "Neglect of Gravity in Constant Density Flows" earlier in this chapter).

Interestingly, in a density-stratified fluid, gravity can play a significant role without the presence of a free surface. The effective gravity force in a two-layer situation is the buoyancy force per unit volume $(\rho_2 - \rho_1)g$, where ρ_1 and ρ_2 are fluid densities in the two layers. In such a case, an internal Froude number is defined as:

$$\text{Fr}' \equiv \left[\frac{\text{inertia force}}{\text{buoyancy force}}\right]^{1/2} \propto \left[\frac{\rho_1 U^2/l}{(\rho_2 - \rho_1)g}\right]^{1/2} = \frac{U}{\sqrt{g'l}}, \tag{4.105}$$

where $g' \equiv g\,(\rho_2 - \rho_1)/\rho_1$ is the *reduced gravity*. For a continuously stratified fluid having a maximum buoyancy frequency N (see 1.29), the equivalent of (4.104) is $\text{Fr}' \equiv U/Nl$. Alternatively, the internal Froude number may be replaced by the Richardson Number $= \text{Ri} \equiv 1/\text{Fr}'^2 = g'l/U^2$, which can also be refined to a gradient Richardson number $\equiv N^2(z)/(dU/dz)^2$ that is important in studies of instability and turbulence in stratified fluids.

Under dynamic similarity, all the dimensionless numbers match and there is one dimensionless solution. The dimensional consistency of the equations of motion ensures that all flow quantities may be set in dimensionless form. For example, the local pressure at point $\mathbf{x} = (x, y, z)$ can be made dimensionless in the form

$$\frac{p(\mathbf{x}, t) - p_\infty}{(1/2)\rho U^2} \equiv C_p = \Psi\left(\text{St}, \text{Fr}, \text{Re}; \frac{\mathbf{x}}{l}, \Omega t\right), \tag{4.106}$$

where $C_p = (p - p_\infty)/(1/2)\rho U^2$ is called the *pressure coefficient* (or the Euler number), and Ψ represents the dimensionless solution for the pressure coefficient in terms of dimensionless parameters and variables. The factor of ½ in (4.106) is conventional but not necessary. Similar relations also hold for any other dimensionless flow variables such as velocity $\mathbf{u}/U$. It follows that in dynamically similar flows, dimensionless flow variables are identical at *corresponding points and times* (that is, for identical values of $\mathbf{x}/l$, and Ωt). Of course there are many instances where the flow geometry may require two or more length scales: $l, l_1, l_2, \ldots l_n$. When this is the case, the aspect ratios $l_1/l, l_2/l, \ldots l_n/l$ provide a dimensionless description of the geometry, and would also appear as arguments of the function Ψ in a relationship like (4.106). Here a difference between relations (4.99) and (4.106) should be noted. Equation (4.99) is a relation between *overall* flow parameters, whereas (4.106) holds *locally* at a point.

In the foregoing analysis we have assumed that the imposed unsteadiness in boundary conditions is important. However, time may also be made dimensionless via $t^* = Ut/l$, as would be appropriate for a flow with steady boundary conditions. In this case, the time derivative in (4.39) should still be retained because the resulting flow may still be naturally unsteady since flow oscillations can arise spontaneously even if the boundary conditions are steady. But, we know from dimensional considerations, such unsteadiness must have a time scale proportional to l/U.

In the foregoing analysis we have also assumed that an imposed velocity U is relevant. Consider now a situation in which the imposed boundary conditions are purely unsteady.

To be specific, consider an object having a characteristic length scale l oscillating with a frequency Ω in a fluid at rest at infinity. This is a problem having an imposed length scale and an *imposed time scale* $1/\Omega$. In such a case a velocity scale $U = l\Omega$ can be constructed. The preceding analysis then goes through, leading to the conclusion that $St = 1$, $Re = Ul/\nu = \Omega l^2/\nu$, and $Fr = U/(gl)^{1/2} = \Omega(l/g)^{1/2}$ have to be duplicated for dynamic similarity.

All dimensionless quantities are identical in dynamically similar flows. For flow around an immersed body, like a sphere, we can define the (dimensionless) drag and lift coefficients,

$$C_D \equiv \frac{F_D}{(1/2)\rho U^2 A} \quad \text{and} \quad C_L \equiv \frac{F_L}{(1/2)\rho U^2 A}, \qquad (4.107,\ 4.108)$$

where A is a reference area, and F_D and F_L are the drag and lift forces, respectively, experienced by the body; as in (4.106) the factor of $1/2$ in (4.107) and (4.108) is conventional but not necessary. For blunt bodies such as spheres and cylinders, A is taken to be the maximum cross section perpendicular to the flow. Therefore, $A = \pi d^2/4$ for a sphere of diameter d, and $A = bd$ for a cylinder of diameter d and length b, with its axis perpendicular to the flow. For flows over flat plates, and airfoils, on the other hand, A is taken to be the *planform area*, that is, $A = sl$; here, l is the average length of the plate or chord of the airfoil in the direction of flow and s is the width perpendicular to the flow, sometimes called the *span*.

The values of the drag and lift coefficients are identical for dynamically similar flows. For flow about a steadily moving ship, the drag is caused both by gravitational and viscous effects so we must have a functional relation of the form $C_D = C_D(Fr, Re)$. However, in many flows gravitational effects are unimportant. An example is flow around a body that is far from a free surface and does not generate gravity waves. In this case, Fr is irrelevant, so $C_D = C_D(Re)$ is all that is needed when the effects of compressibility are unimportant. This is the situation portrayed by the first member of (4.99) and illustrated in Figure 4.21.

A dimensionless form of the continuity equation should indicate when flow-induced pressure differences induce significant departures from incompressible flow. However, the simplest possible scaling fails to provide any insights because the continuity equation itself does not contain the pressure. Thus, a more fruitful starting point for determining the relative size of $\nabla \cdot \mathbf{u}$ is (4.9),

$$\nabla \cdot \mathbf{u} = -\frac{1}{\rho}\frac{D\rho}{Dt} = -\frac{1}{\rho c^2}\frac{Dp}{Dt}, \qquad (4.9)$$

along with the assumption that pressure-induced density changes will be isentropic, $dp = c^2 d\rho$ where c is the sound speed (1.19). Using the following dimensionless variables,

$$x_i^* = x_i/l, \quad t^* = Ut/l, \quad u_j^* = u_j/U, \quad p^* = (p - p_\infty)/\rho_o U^2, \quad \text{and} \quad \rho^* = \rho/\rho_o, \qquad (4.109)$$

where ρ_o is a reference density, the outside members of (4.9) can be rewritten:

$$\nabla^* \cdot \mathbf{u}^* = -\left[\frac{U^2}{c^2}\right]\frac{1}{\rho^*}\frac{Dp^*}{Dt^*}, \qquad (4.110)$$

which specifically shows that the square of

$$M = \text{Mach number} \equiv \left[\frac{\text{inertia force}}{\text{compressibility force}}\right]^{1/2} \propto \left[\frac{\rho U^2/l}{\rho c^2/l}\right]^{1/2} = \frac{U}{c} \qquad (4.111)$$

sets the size of isentropic departures from incompressible flow. In engineering practice, gas flows are considered incompressible when $M < 0.3$, and from (4.110) this corresponds to ~10% departure from ideal incompressible behavior when $(1/\rho^*)(Dp^*/Dt^*)$ is unity. Of course, there may be nonisentropic changes in density too and these are considered in Thompson (1972, pp. 137−146). Flows in which $M < 1$ are called *subsonic*, whereas flows in which $M > 1$ are called *supersonic*. At high subsonic and supersonic speeds, matching Mach number between flows is required for dynamic similarity.

There are many possible thermal boundary conditions for the energy equation, so a fully general scaling of (4.60) is not possible. Instead, a simple scaling is provided here based on constant specific heats, neglect of μ_v, and constant free-stream and wall temperatures, T_o and T_w, respectively. In addition, for simplicity, an imposed flow oscillation frequency is not considered. The starting point of the scaling provided here is a mild revision of (4.60) that involves the enthalpy h per unit mass,

$$\rho\frac{Dh}{Dt} = \frac{Dp}{Dt} + \rho\varepsilon + \frac{\partial}{\partial x_i}\left(k\frac{\partial T}{\partial x_i}\right), \qquad (4.112)$$

where ε is given by (4.58). Using $dh \cong C_p dT$, the following dimensionless variables:

$$\varepsilon^* = \rho_o l^2\varepsilon/\mu_o U^2, \quad \mu^* = \mu/\mu_o, \quad k^* = k/k_o, \quad T^* = (T - T_o)/(T_w - T_o), \qquad (4.113)$$

and those defined in (4.106), (4.107) becomes:

$$\rho^*\frac{DT^*}{Dt^*} = \left[\frac{U^2}{C_p(T_w - T_o)}\right]\frac{Dp^*}{Dt^*} + \left[\frac{U^2}{C_p(T_w - T_o)}\frac{\mu_o}{\rho_o U l}\right]\rho^*\varepsilon^* + \left[\frac{k_o}{C_p\mu_o}\frac{\mu_o}{\rho_o U l}\right]\nabla^*(k^*\nabla^*T^*). \qquad (4.114)$$

Here the relevant dimensionless parameters are:

$$Ec = \text{Eckert number} \equiv \frac{\text{kinetic energy}}{\text{thermal energy}} = \frac{U^2}{C_p(T_w - T_o)}, \qquad (4.115)$$

$$Pr = \text{Prandtl number} \equiv \frac{\text{momentum diffusivity}}{\text{thermal diffusivity}} = \frac{\nu}{\kappa} = \frac{\mu_o/\rho_o}{k_o/\rho_o C_p} = \frac{\mu_o C_p}{k_o}, \qquad (4.116)$$

and we recognize $\rho_o U l/\mu_o$ as the Reynolds number in (4.114) as well. In low-speed flows, where the Eckert number is small the middle terms drop out of (4.114), and the full energy equation (4.112) reduces to (4.89). Thus, low Ec is needed for the Boussinesq approximation.

The Prandtl number is a ratio of two molecular transport properties. It is therefore a fluid property and independent of flow geometry. For air at ordinary temperatures and pressures, $Pr = 0.72$, which is close to the value of 0.67 predicted from a simplified kinetic theory model assuming hard spheres and monatomic molecules (Hirschfelder, Curtiss, & Bird, 1954, pp. 9−16). For water at 20 °C, $Pr = 7.1$. Dynamic similarity between flows involving thermal effects requires equality of the Eckert, Prandtl, and Reynolds numbers.

And finally, for flows involving surface tension σ, there are several relevant dimensionless numbers:

$$We = \text{Weber number} \equiv \frac{\text{inertia force}}{\text{surface tension force}} \propto \frac{\rho U^2 l^2}{\sigma l} = \frac{\rho U^2 l}{\sigma}, \qquad (4.117)$$

$$Bo = \text{Bond number} \equiv \frac{\text{gravity force}}{\text{surface tension force}} \propto \frac{\rho l^3 g}{\sigma l} = \frac{\rho l^2 g}{\sigma}, \qquad (4.118)$$

$$Ca = \text{Capillary number} \equiv \frac{\text{viscous stress}}{\text{surface tension stress}} \propto \frac{\mu U/l}{\sigma/l} = \frac{\mu U}{\sigma}. \qquad (4.119)$$

Here, for the Weber and Bond numbers, the ratio is constructed based on a ratio of forces as in (4.107) and (4.108), and not forces per unit volume as in (4.103), (4.104), and (4.111). At high Weber number, droplets and bubbles are easily deformed by fluid acceleration or deceleration, for example during impact with a solid surface. At high Bond numbers surface tension effects are relatively unimportant compared to gravity, as is the case for long-wavelength, ocean surface waves. At high capillary numbers viscous forces dominate those from surface tension; this is commonly the case in machinery lubrication flows. However, for slow bubbly flow through porous media or narrow tubes (low Ca) the opposite is true.

EXAMPLE 4.8

A ship 100 m long is expected to cruise at 10 m/s. It has a submerged surface of 300 m². Find the model speed for a 1/25 scale model, neglecting frictional effects. The drag is measured to be 60 N when the model is tested in a towing tank at the model speed. Estimate the full scale drag when the skin-friction drag coefficient for the model is 0.003 and that for the full-scale prototype is half of that.

Solution

Estimate the model speed neglecting frictional effects. Then the nondimensional drag force depends only on the Froude number:

$$D/\rho U^2 l^2 = f\left(U/\sqrt{gl}\right).$$

Equating Froude numbers for the model (denoted by subscript "m") and full-size prototype (denoted by subscript "p"), we get

$$U_m = U_p \sqrt{g_m l_m / g_p l_p} = 10\sqrt{1/25} = 2 \text{ m/s}.$$

The total drag on the model was measured to be 60 N at this model speed. Of the total measured drag, a part was due to frictional effects. The frictional drag can be estimated by treating the surface of the hull as a flat plate, for which the drag coefficient C_D is a function of the Reynolds number. Using a value of $v = 10^{-6}$ m²/s for water, we get

$$Ul/\nu \ (\text{model}) = [2(100/25)]/10^{-6} = 8 \times 10^6,$$

$$Ul/\nu \ (\text{prototype}) = 10 \ (100)/10^{-6} = 10^9.$$

The problem statement sets the frictional drag coefficients as

$$C_D(\text{model}) = 0.003,$$
$$C_D(\text{prototype}) = 0.0015.$$

and these are consistent with Figure 9.11. Using a value of $\rho = 1000 \ \text{kg/m}^3$ for water, we estimate:

Frictional drag on model $= \frac{1}{2}C_D\rho U^2 A = (0.5) \ (0.003) \ (1000) \ (2)^2 \ (300/25^2) = 2.88 \ \text{N}.$
Out of the total model drag of 60 N, the wave drag is therefore $60 - 2.88 = 57.12 \ \text{N}.$

Now the *wave drag* still obeys the scaling law above, which means that $D/\rho U^2 l^2$ for the two flows are identical, where D represents wave drag alone. Therefore,

Wave drag on prototype $= (\text{Wave drag on model}) \ (\rho_p/\rho_m) \ (l_p/l_m)^2 \ (U_p/U_m)^2$
$$= 57.12 \ (1) \ (25)^2 \ (10/2)^2 = 8.92 \times 10^5 \ \text{N}.$$

Having estimated the wave drag on the prototype, we proceed to determine its frictional drag.

Frictional drag on prototype $= \frac{1}{2}C_D\rho U^2 A = (0.5) \ (0.0015) \ (1000) \ (10)^2 \ (300) = 0.225 \times 10^5 \ \text{N}.$

Therefore, total drag on prototype $= (8.92 + 0.225) \times 10^5 = 9.14 \times 10^5 \ \text{N}.$
If we did not correct for the frictional effects, and assumed that the measured model drag was all due to wave effects, then we would have found a prototype drag of

$$D_p = D_m(\rho_p/\rho_m) \ (l_p/l_m)^2 \ (U_p/U_m)^2 = 60 \ (1) \ (25)^2 \ (10/2)^2 = 9.37 \times 10^5 \ \text{N}.$$

EXERCISES

4.1. Let a one-dimensional velocity field be $u = u(x, t)$, with $v = 0$ and $w = 0$. The density varies as $\rho = \rho_0(2 - \cos \omega t)$. Find an expression for $u(x, t)$ if $u(0, t) = U$.

4.2. Consider the one-dimensional Cartesian velocity field: $\mathbf{u} = (\alpha x/t, 0, 0)$ where α is a constant. Find a spatially uniform, time-dependent density field, $\rho = \rho(t)$, that renders this flow field mass conserving when $\rho = \rho_0$ at $t = t_0$.

4.3. Find a nonzero density field $\rho(x,y,z,t)$ that renders the following Cartesian velocity fields mass conserving. Comment on the physical significance and uniqueness of your solutions.

    a) $\mathbf{u} = (U \sin(\Omega t - kx), 0, 0)$ where U, ω, k are positive constants
    [*Hint*: exchange the independent variables x,t for a single independent variable $\xi = \omega t - kx$]

b) $\mathbf{u} = (-\Omega y, +\Omega x, 0)$ with Ω = constant [*Hint*: switch to cylindrical coordinates]

c) $\mathbf{u} = (A/x, B/y, C/z)$ where A, B, C are constants

4.4. A proposed conservation law for ξ, a new fluid property, takes the following form:

$$\frac{d}{dt} \int_{V(t)} \rho\xi dV + \int_{A(t)} \Theta \cdot \mathbf{n} dS = 0,$$ where $V(t)$ is a material volume that moves with the

fluid velocity $\mathbf{u}$, $A(t)$ is the surface of $V(t)$, ρ is the fluid density, and $\Theta = -\rho\gamma\nabla\xi$.

a) What partial differential equation is implied by the above conservation statement?

b) Use the part a) result and the continuity equation to show: $\dfrac{\partial\xi}{\partial t} + \mathbf{u}\cdot\nabla\xi = \dfrac{1}{\rho}\nabla\cdot(\rho\gamma\nabla\xi)$.

4.5. The components of a mass flow vector $\rho\mathbf{u}$ are $\rho u = 4x^2 y$, $\rho v = xyz$, $\rho w = yz^2$.

a) Compute the net mass outflow through the closed surface formed by the planes $x = 0$, $x = 1$, $y = 0$, $y = 1$, $z = 0$, $z = 1$.

b) Compute $\nabla\cdot(\rho\mathbf{u})$ and integrate over the volume bounded by the surface defined in part a).

c) Explain why the results for parts a) and b) should be equal or unequal.

4.6. Consider a simple fluid mechanical model for the atmosphere of an ideal spherical star that has a surface gas density of ρ_o and a radius r_o. The escape velocity from the surface of the star is v_e. Assume that a tenuous gas leaves the star's surface radially at speed v_o uniformly over the star's surface. Use the steady continuity equation for the gas density ρ and fluid velocity $\mathbf{u} = (u_r, u_\theta, u_\varphi)$ in spherical coordinates

$$\frac{1}{r^2}\frac{\partial}{\partial r}(r^2 \rho u_r) + \frac{1}{r\sin\theta}\frac{\partial}{\partial\theta}(\rho u_\theta \sin\theta) + \frac{1}{r\sin\theta}\frac{\partial}{\partial\varphi}(\rho u_\varphi) = 0$$

to determine the following.

a) Determine ρ when $v_o \geq v_e$ so that $\mathbf{u} = (u_r, u_\theta, u_\varphi) = (v_o\sqrt{1 - (v_e^2/v_o^2)(1 - (r_o/r))}, 0, 0)$.

b) Simplify the result from part a) when $v_o \gg v_e$ so that: $\mathbf{u} = (u_r, u_\theta, u_\varphi) = (v_o, 0, 0)$.

c) Simplify the result from part a) when $v_o = v_e$.

d) Use words, sketches, or equations to describe what happens when $v_o < v_e$. State any assumptions that you make.

4.7. The definition of the stream function for two-dimensional, constant-density flow in the x-y plane is: $\mathbf{u} = -\mathbf{e}_z \times \nabla\psi$, where $\mathbf{e}_z$ is the unit vector perpendicular to the x-y plane that determines a right-handed coordinate system.

a) Verify that this vector definition is equivalent to $u = \partial\psi/\partial y$ and $v = -\partial\psi/\partial x$ in Cartesian coordinates.

b) Determine the velocity components in r-θ polar coordinates in terms of r-θ derivatives of ψ.

c) Determine an equation for the z-component of the vorticity in terms of ψ.

4.8. A curve of $\psi(x, y) = C_1$ (= a constant) specifies a streamline in steady two-dimensional, constant-density flow. If a neighboring streamline is specified by $\psi(x, y) = C_2$, show that the volume flux per unit depth into the page between the streamlines equals $C_2 - C_1$ when $C_2 > C_1$.

4.9. The well-known undergraduate fluid mechanics textbook by Fox et al. (2009) provides the following statement of conservation of momentum for a constant-shape (nonrotating) control volume moving at a non-constant velocity $\mathbf{U} = \mathbf{U}(t)$:

$$\frac{d}{dt} \int_{V^*(t)} \rho \mathbf{u}_{rel} dV + \int_{A^*(t)} \rho \mathbf{u}_{rel}(\mathbf{u}_{rel} \cdot \mathbf{n}) dA = \int_{V^*(t)} \rho \mathbf{g} dV + \int_{A^*(t)} \mathbf{f} dA - \int_{V^*(t)} \rho \frac{d\mathbf{U}}{dt} dV.$$

Here $\mathbf{u}_{rel} = \mathbf{u} - \mathbf{U}(t)$ is the fluid velocity observed in a frame of reference moving with the control volume while $\mathbf{u}$ and $\mathbf{U}$ are observed in a nonmoving frame. Meanwhile, equation (4.17) states this law as

$$\frac{d}{dt} \int_{V^*(t)} \rho \mathbf{u} dV + \int_{A^*(t)} \rho \mathbf{u}(\mathbf{u} - \mathbf{U}) \cdot \mathbf{n} dA = \int_{V^*(t)} \rho \mathbf{g} dV + \int_{A^*(t)} \mathbf{f} dA$$

where the replacement $\mathbf{b} = \mathbf{U}$ has been made for the velocity of the accelerating control surface $A^*(t)$. Given that the two equations above are not identical, determine if these two statements of conservation of fluid momentum are contradictory or consistent.

4.10. A jet of water with a diameter of 8 cm and a speed of 25 m/s impinges normally on a large stationary flat plate. Find the force required to hold the plate stationary. Compare the average pressure on the plate with the stagnation pressure if the plate is 20 times the area of the jet.

4.11. Show that the thrust developed by a stationary rocket motor is $F = \rho A U^2 + A(p - p_{atm})$, where p_{atm} is the atmospheric pressure, and p, ρ, A, and U are, respectively, the pressure, density, area, and velocity of the fluid at the nozzle exit.

4.12. Consider the propeller of an airplane moving with a velocity U_1. Take a reference frame in which the air is moving and the propeller [disk] is stationary. Then the effect of the propeller is to accelerate the fluid from the upstream value U_1 to the downstream value $U_2 > U_1$. Assuming incompressibility, show that the thrust developed by the propeller is given by $F = \rho A(U_2^2 - U_1^2)/2$, where A is the projected area of the propeller and ρ is the density (assumed constant). Show also that the velocity of the fluid at the plane of the propeller is the average value $U = (U_1 + U_2)/2$. [*Hint:* The flow can be idealized by a pressure jump of magnitude $\Delta p = F/A$ right at the location of the propeller. Also apply Bernoulli's equation between a section far upstream and a section immediately upstream of the propeller. Also apply the Bernoulli equation between a section immediately downstream of the propeller and a section far downstream. This will show that $\Delta p = \rho(U_2^2 - U_1^2)/2$.]

4.13. Generalize the control volume analysis of Example 4.1 by considering the control volume geometry shown for steady two-dimensional flow past an arbitrary body in the absence of body forces. Show that the force the fluid exerts on the body is given by the Euler momentum integral: $F_j = -\int_{A_1} (\rho u_i u_j - \tau_{ij}) n_i dA$, and $0 = \int_{A_1} \rho u_i n_i dA$.

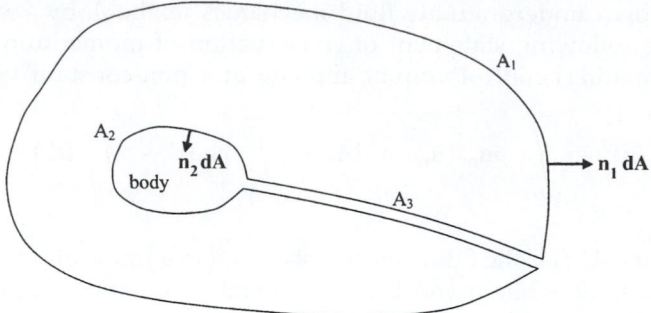

4.14. The pressure rise $\Delta p = p_2 - p_1$ that occurs for flow through a sudden pipe-cross-sectional-area expansion can depend on the average upstream flow speed U_{ave}, the upstream pipe diameter d_1, the downstream pipe diameter d_2, and the fluid density ρ and viscosity μ. Here p_2 is the pressure downstream of the expansion where the flow is first fully adjusted to the larger pipe diameter.

a) Find a dimensionless scaling law for Δp in terms of U_{ave}, d_1, d_2, ρ, and μ.

b) Simplify the result of part a) for high-Reynolds-number turbulent flow where μ does not matter.

c) Use a control volume analysis to determine Δp in terms of U_{ave}, d_1, d_2, and ρ for the high Reynolds number limit. [*Hints:* 1) a streamline drawing might help in determining or estimating the pressure on the vertical surfaces of the area transition, and 2) assume uniform flow profiles wherever possible.]

d) Compute the ideal flow value for Δp using the Bernoulli equation (4.19) and compare this to the result from part c) for a diameter ratio of $d_1/d_2 = \frac{1}{2}$. What fraction of the maximum possible pressure rise does the sudden expansion achieve?

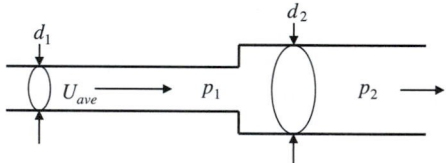

4.15. Consider how pressure gradients and skin friction develop in an empty wind tunnel or water tunnel test section when the flow is incompressible. Here the fluid has viscosity μ and density ρ, and flows into a horizontal cylindrical pipe of length L with radius R at a uniform horizontal velocity U_o. The inlet of the pipe lies at $x = 0$. Boundary layer growth on the pipe's walls induces the horizontal velocity on the pipe's centerline to be U_L at $x = L$; however, the pipe-wall boundary layer thickness remains much smaller than R. Here, L/R is of order 10, and $\rho U_o R/\mu \gg 1$. The radial coordinate from the pipe centerline is r.

a) Determine the displacement thickness, δ_L^*, of the boundary layer at $x = L$ in terms of U_o, U_L, and R. Assume that the boundary layer displacement thickness is zero at $x = 0$. [The boundary layer displacement thickness, δ^*, is the thickness of the zero-flow-speed layer that displaces the outer flow by the same amount as the actual boundary layer. For a boundary layer velocity profile $u(y)$ with y = wall-normal coordinate and U = outer flow velocity, δ^* is defined by: $\delta^* = \int_0^\infty (1 - (u/U))dy$.]

b) Determine the pressure difference, $\Delta P = P_L - P_o$, between the ends of the pipe in terms of ρ, U_o, and U_L.

c) Assume the horizontal velocity profile at the outlet of the pipe can be approximated by: $u(r, x = L) = U_L(1 - (r/R)^n)$ and estimate average skin friction, $\overline{\tau}_w$, on the inside of the pipe between $x = 0$ and $x = L$ in terms of ρ, U_o, U_L, R, L, and n.

d) Calculate the skin friction coefficient, $c_f = \overline{\tau}_w/(1/2)\rho U_o^2$, when $U_o = 20.0$ m/s, $U_L = 20.5$ m/s, $R = 1.5$ m, $L = 12$ m, $n = 80$, and the fluid is water, i.e., $\rho = 10^3$ kg/m^3.

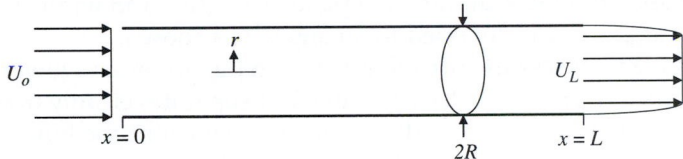

4.16. Consider the situation depicted below. Wind strikes the side of a simple residential structure and is deflected up over the top of the structure. Assume the following: two-dimensional steady inviscid constant-density flow, uniform upstream velocity profile, linear gradient in the downstream velocity profile (velocity U at the upper boundary and zero velocity at the lower boundary as shown), no flow through the upper boundary of the control volume, and constant pressure on the upper boundary of the control volume. Using the control volume shown:

a) Determine h_2 in terms of U and h_1.

b) Determine the *direction* and *magnitude* of the horizontal force on the house per unit depth into the page in terms of the fluid density ρ, the upstream velocity U, and the height of the house h_1.

c) Evaluate the magnitude of the force for a house that is 10 m tall and 20 m long in wind of 22 m/sec (approximately 50 miles per hour).

4.17. A large wind turbine with diameter D extracts a fraction η of the kinetic energy from the airstream (density $= \rho =$ constant) that impinges on it with velocity U.

a) What is the diameter of the wake zone, E, downstream of the windmill?

b) Determine the magnitude and direction of the force on the windmill in terms of ρ, U, D, and η.

c) Does your answer approach reasonable limits as $\eta \to 0$ and $\eta \to 1$?

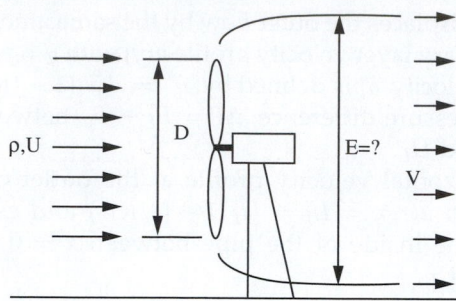

4.18. An incompressible fluid of density ρ flows through a horizontal rectangular duct of height h and width b. A uniform flat plate of length L and width b attached to the top of the duct at point A is deflected to an angle θ as shown.

    a) Estimate the pressure difference between the upper and lower sides of the plate in terms of x, ρ, U_o, h, L, and θ when the flow separates cleanly from the tip of the plate.

    b) If the plate has mass M and can rotate freely about the hinge at A, determine a formula for the angle θ in terms of the other parameters. You may leave your answer in terms of an integral.

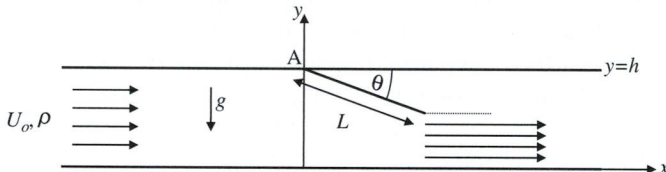

4.19. A pipe of length L and cross sectional area A is to be used as a fluid-distribution manifold that expels a steady uniform volume flux per unit length of an incompressible liquid from $x = 0$ to $x = L$. The liquid has density ρ, and is to be expelled from the pipe through a slot of varying width, $w(x)$. The goal of this problem is to determine $w(x)$ in terms of the other parameters of the problem. The pipe-inlet pressure and liquid velocity at $x = 0$ are P_o and U_o, respectively, and the pressure outside the pipe is P_e. If $P(x)$ denotes the pressure on the inside of the pipe, then the liquid velocity through the slot U_e is determined from: $P(x) - P_e = (1/2)\rho U_e^2$. For this problem assume that the expelled liquid exits the pipe perpendicular to the pipe's axis, and note that $wU_e = \text{const.} = U_o A/L$, even though w and U_e both depend on x.

    a) Formulate a dimensionless scaling law for w in terms of x, L, A, ρ, U_o, P_o, and P_e.

    b) Ignore the effects of viscosity, assume all profiles through the cross section of the pipe are uniform, and use a suitable differential-control-volume analysis to show that:

$$A\frac{dU}{dx} + wU_e = 0, \quad \text{and} \quad \rho\frac{d}{dx}U^2 = -\frac{dP}{dx}.$$

c) Use these equations and the relationships stated above to determine $w(x)$ in terms of x, L, A, ρ, U_o, P_o, and P_e. Is the slot wider at $x = 0$ or at $x = L$?

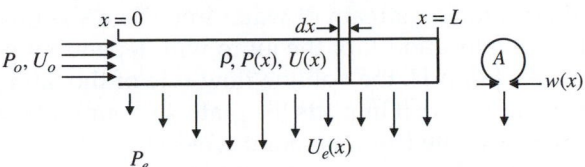

4.20. The take-off mass of a Boeing 747-400 may be as high as 400,000 kg. An Airbus A380 may be even heavier. Using a control volume (CV) that comfortably encloses the aircraft, explain why such large aircraft do not crush houses or people when they fly low overhead. Of course, the aircraft's wings generate lift but they are entirely contained within the CV and do not coincide with any of the CV's surfaces; thus merely stating the lift balances weight is not a satisfactory explanation. Given that the CV's vertical body-force term, $-g \int_{CV} \rho dV$, will exceed 4×10^6 N when the airplane and air in the CV's interior are included, your answer should instead specify which of the CV's surface forces or surface fluxes carries the signature of a large aircraft's impressive weight.

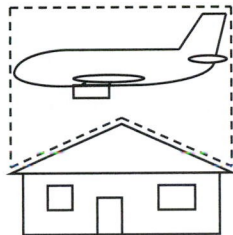

4.21. [1]An inviscid incompressible liquid with density ρ flows in a wide conduit of height H and width B into the page. The inlet stream travels at a uniform speed U and fills the conduit. The depth of the outlet stream is less than H. Air with negligible density fills the gap above the outlet stream. Gravity acts downward with acceleration g. Assume the flow is steady for the following items.

a) Find a dimensionless scaling law for U in terms of ρ, H, and g.

b) Denote the outlet stream depth and speed by h and u, respectively, and write down a set of equations that will allow U, u, and h to be determined in terms of ρ, H, and g.

c) Solve for U, u, and h in terms of ρ, H, and g. [*Hint:* solve for h first.]

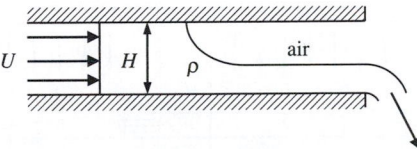

[1]Based on a lecture example of Professor P. E. Dimotakis.

4.22. A hydraulic jump is the shallow-water-wave equivalent of a gas-dynamic shock wave. A steady radial hydraulic jump can be observed safely in one's kitchen, bathroom, or backyard where a falling stream of water impacts a shallow pool of water on a flat surface. The radial location R of the jump will depend on gravity g, the depth of the water behind the jump H, the volume flow rate of the falling stream Q, and the stream's speed, U, where it impacts the plate. In your work, assume $\sqrt{2gh} \ll U$ where r is the radial coordinate from the point where the falling stream impacts the surface.

a) Formulate a dimensionless law for R in terms of the other parameters.

b) Use the Bernoulli equation and a control volume with narrow angular and negligible radial extents that contains the hydraulic jump to show that:

$$R \cong \frac{Q}{2\pi U H^2}\left(\frac{2U^2}{g} - H\right).$$

c) Rewrite the result of part b) in terms of the dimensionless parameters found for part a).

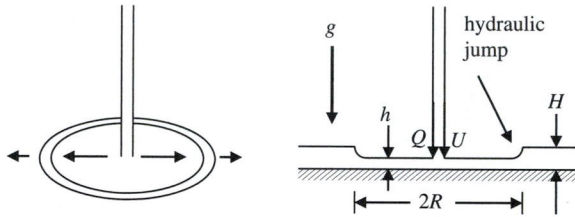

4.23. A fine uniform mist of an inviscid incompressible fluid with density ρ settles steadily at a total volume flow rate (per unit depth into the page) of $2q$ onto a flat horizontal surface of width $2s$ to form a liquid layer of thickness $h(x)$ as shown. The geometry is two dimensional.

a) Formulate a dimensionless scaling law for h in terms of x, s, q, ρ, and g.

b) Use a suitable control volume analysis, assuming $u(x)$ does not depend on y, to find a single cubic equation for $h(x)$ in terms of $h(0)$, s, q, x, and g.

c) Determine $h(0)$.

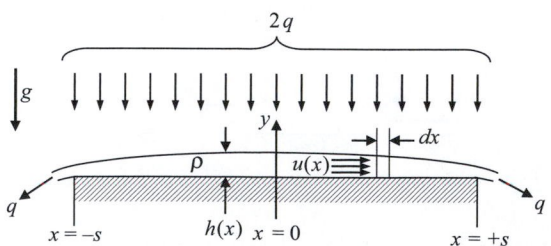

4.24. A thin-walled pipe of mass m_o, length L, and cross-sectional area A is free to swing in the x-y plane from a frictionless pivot at point O. Water with density ρ fills the pipe, flows into it at O perpendicular to the x-y plane, and is expelled at a right angle from the pipe's end as shown. The pipe's opening also has area A and gravity g acts downward. For a steady mass flow rate of $\dot{m}$, the pipe hangs at angle θ with respect to the vertical as shown. Ignore fluid viscosity.

  a) Develop a dimensionless scaling law for θ in terms of m_o, L, A, ρ, $\dot{m}$, and g.
  b) Use a control volume analysis to determine the force components, F_x and F_y, applied to the pipe at the pivot point in terms of θ, m_o, L, A, ρ, $\dot{m}$, and g.
  c) Determine θ in terms of m_o, L, A, ρ, $\dot{m}$, and g.
  d) Above what value of $\dot{m}$ will the pipe rotate without stopping?

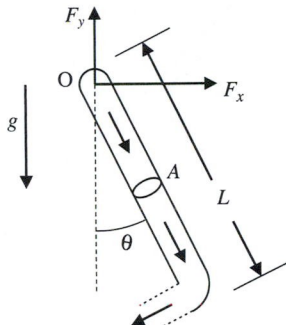

4.25. Construct a house of cards, or light a candle. Get the cardboard tube from the center of a roll of paper towels and back away from the cards or candle a meter or two so that by blowing you cannot knock down the cards or blow out the candle unaided. Now use the tube in two slightly different configurations. First, place the tube snugly against your face encircling your mouth, and try to blow down the house of cards or blow out the candle. Repeat the experiment while moving closer until the cards are knocked down or the candle is blown out (you may need to get closer to your target than might be expected; do not hyperventilate; do not start the cardboard tube on fire). Note the distance between the end of the tube and the card house or candle at this point. Rebuild the card house or relight the candle and repeat the experiment, except this time hold the tube a few centimeters away from your face and mouth, and blow through it. Again, determine the greatest distance from which you can knock down the cards or blow out the candle.

  a) Which configuration is more effective at knocking the cards down or blowing the candle out?
  b) Explain your findings with a suitable control-volume analysis.
  c) List some practical applications where this effect might be useful.

4.26. [2]Attach a drinking straw to a 15-cm-diameter cardboard disk with a hole at the center using tape or glue. Loosely fold the corners of a standard piece of paper upward so that

---

[2]Based on a demonstration done for the 3rd author by Professor G. Tryggvason

the paper mildly cups the cardboard disk (see drawing). Place the cardboard disk in the central section of the folded paper. Attempt to lift the loosely folded paper off a flat surface by blowing or sucking air through the straw.

a) Experimentally determine if blowing or suction is more effective in lifting the folded paper.

b) Explain your findings with a control volume analysis.

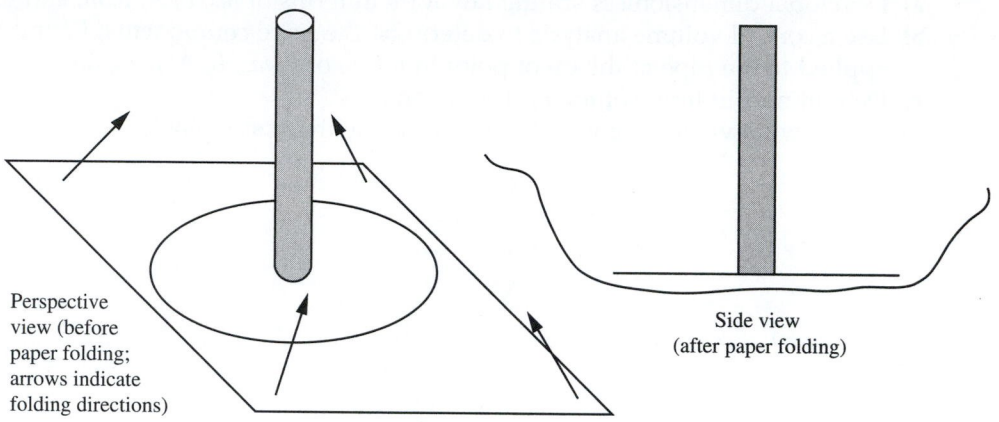

Perspective view (before paper folding; arrows indicate folding directions)

Side view (after paper folding)

4.27. A rectangular tank is placed on wheels and is given a constant horizontal acceleration a. Show that, at steady state, the angle made by the free surface with the horizontal is given by $\tan \theta = a/g$.

4.28. Starting from rest at $t = 0$, an airliner of mass M accelerates at a constant rate $\mathbf{a} = a\mathbf{e}_x$ into a headwind, $\mathbf{u} = -u_i\mathbf{e}_x$. For the following items, assume that: 1) the x-component of the fluid velocity is $-u_i$ on the front, sides, and back upper half of the control volume (CV), 2) the x-component of the fluid velocity is $-u_o$ on the back lower half of the CV, 3) changes in M can be neglected, 4) changes of air momentum inside the CV can be neglected, and 5) the airliner has frictionless wheels. In addition, assume constant air density ρ and uniform flow conditions exist on the various control surfaces. In your work, denote the CV's front and back area by A. (This approximate model is appropriate for real commercial airliners that have the engines hung under the wings.)

a) Find a dimensionless scaling law for u_o at $t = 0$ in terms of u_i, ρ, a, M, and A.

b) Using a CV that encloses the airliner (as shown) determine a formula for $u_o(t)$, the time-dependent air velocity on the lower half of the CV's back surface.

c) Evaluate u_o at $t = 0$, when $M = 4 \times 10^5$ kg, $a = 2.0$ m/s^2, $u_i = 5$ m/s, $\rho = 1.2$ kg/m^3, and $A = 1200$ m^2. Would you be able to walk comfortably behind the airliner?

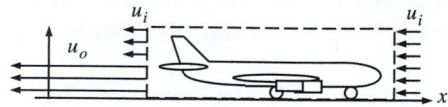

4.29. [3]A cart that can roll freely in the x direction deflects a horizontal water jet into its tank with a vane inclined to the vertical at an angle θ. The jet issues steadily at velocity U with density ρ, and has cross-sectional area A. The cart is initially at rest with a mass of m_0. Ignore the effects of surface tension, the cart's rolling friction, and wind resistance in your answers.

  a) Formulate dimensionless law for the mass, $m(t)$, in the cart at time t in terms of t, θ, U, ρ, A, and m_0.

  b) Formulate a differential equation for $m(t)$.

  c) Find a solution for $m(t)$ and put it in dimensionless form.

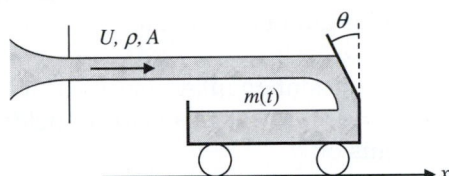

4.30. Prove that the stress tensor is symmetric by considering first-order changes in surface forces on a vanishingly small cube in rotational equilibrium. Work with rotation about the number 3 coordinate axis to show $\tau_{12} = \tau_{21}$. Cyclic permutation of the indices will suffice for showing the symmetry of the other two shear stresses.

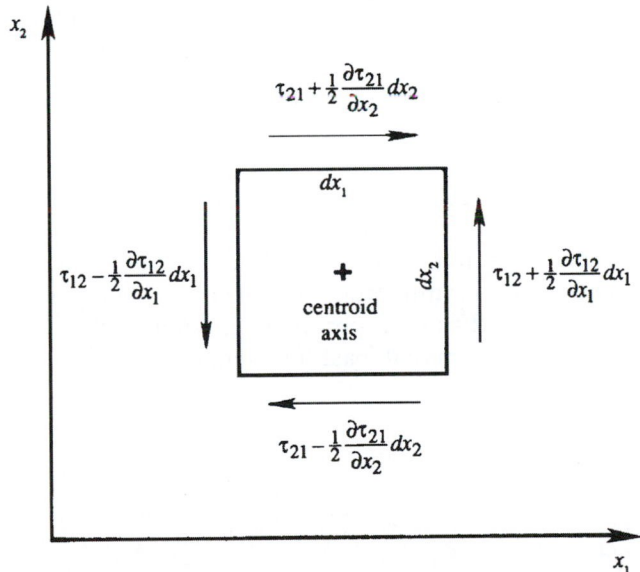

4.31. Obtain an empty plastic milk jug with a cap *that seals tightly*, and a frying pan. Fill both the pan and jug with water to a depth of approximately 1 cm. Place the jug in the pan with the cap off. Place the pan on a stove and turn up the heat until the water in the

[3]Similar to problem 4.170 on page 157 in Fox et al. (2009)

frying pan boils vigorously for a few minutes. Turn the stove off, and quickly put the cap tightly on the jug. *Avoid spilling or splashing hot water on yourself.* Remove the capped jug from the frying pan and let it cool to room temperature. Does anything interesting happen? If something does happen, explain your observations in terms of surface forces. What is the origin of these surface forces? Can you make any quantitative predictions about what happens?

4.32. In cylindrical coordinates (R,φ,z), two components of a steady incompressible viscous flow field are known: $u_\varphi = 0$, and $u_z = -Az$ where A is a constant, and body force is zero.

   a) Determine u_R so that the flow field is smooth and conserves mass.

   b) If the pressure, p, at the origin of coordinates is P_o, determine $p(R,\varphi,z)$ when the density is constant.

4.33. Solid body rotation with a constant angular velocity, Ω, is described by the following Cartesian velocity field: $\mathbf{u} = \Omega \times \mathbf{x}$. For this velocity field:

   a) Compute the components of:

$$\tau_{ij} = -p\delta_{ij} + \mu\left[\left(\frac{\partial u_i}{\partial x_j} + \frac{\partial u_j}{\partial x_i}\right) - \frac{2}{3}\delta_{ij}\frac{\partial u_k}{\partial x_k}\right] + \mu_v\delta_{ij}\frac{\partial u_k}{\partial x_k}.$$

   b) Consider the case of $\Omega_1 = \Omega_2 = 0$, $\Omega_3 \neq 0$, with $p = p_o$ at $x_1 = x_2 = 0$. Use the differential momentum equation in Cartesian coordinates to determine $p(r)$, where $r^2 = x_1^2 + x_2^2$, when there is no body force and $\rho = $ constant. Does your answer make sense? Can you check it with a simple experiment?

4.34. Using only (4.7), (4.22), (4.37), and (3.12) show that $\rho\dfrac{D\mathbf{u}}{Dt} + \nabla p = \rho\mathbf{g} + \mu\nabla^2\mathbf{u} + \left(\mu_v + \dfrac{1}{3}\mu\right)\nabla(\nabla\cdot\mathbf{u})$ when the dynamic (μ) and bulk (μ_v) viscosities are constants.

4.35. [4]Air, water, and petroleum products are important engineering fluids and can usually be treated as Newtonian fluids. Consider the following materials and try to classify them as: Newtonian fluid, non-Newtonian fluid, or solid. State the reasons for your choices and note the temperature range where you believe your answers are correct. Simple impact, tensile, and shear experiments in your kitchen or bathroom are recommended. Test and discuss at least five items.

   a) toothpaste

   b) peanut butter

   c) shampoo

   d) glass

   e) honey

   f) mozzarella cheese

   g) hot oatmeal

   h) creamy salad dressing

   i) ice cream

   j) silly putty

[4]Based on a suggestion from Professor W. W. Schultz

4.36. The equations for conservation of mass and momentum for a viscous Newtonian fluid are (4.7) and (4.39a) when the viscosities are constant.

    a) Simplify these equations and write them out in primitive form for steady constant-density flow in two dimensions where $u_i = (u_1(x_1, x_2), u_2(x_1, x_2), 0)$, $p = p(x_1, x_2)$, and $g_j = 0$.

    b) Determine $p = p(x_1, x_2)$ when $u_1 = Cx_1$ and $u_2 = -Cx_2$, where C is a positive constant.

4.37. Starting from (4.7) and (4.39b), derive a Poisson equation for the pressure, p, by taking the divergence of the constant-density momentum equation. [In other words, find an equation where $\partial^2 p / \partial x_j^2$ appears by itself on the left side and other terms not involving p appear on the right side.] What role does the viscosity μ play in determining the pressure in constant density flow?

4.38. Prove the equality of the two ends of (4.40) without leaving index notation or using vector identities.

4.39. The viscous compressible fluid conservation equations for mass and momentum are (4.7) and (4.38). Simplify these equations for constant-density, constant-viscosity flow and where the body force has a potential, $g_j = -\partial\Phi/\partial x_j$. Assume the velocity field can be found from $u_j = \partial\phi/\partial x_j$, where the scalar function ϕ depends on space and time. What are the simplified conservation of mass and momentum equations for ϕ?

4.40. The viscous compressible fluid conservation equations for mass and momentum are (4.7) and (4.38).

    a) In Cartesian coordinates (x, y, z) with $\mathbf{g} = (g_x, 0, 0)$, simplify these equations for un-steady one-dimensional unidirectional flow where: $\rho = \rho(x, t)$ and $\mathbf{u} = (u(x, t), 0, 0)$.

    b) If the flow is also incompressible, show that the fluid velocity depends only on time, i.e., $u(x, t) = U(t)$, and show that the equations found for part a) reduce to

$$\frac{\partial \rho}{\partial t} + u\frac{\partial \rho}{\partial x} = 0, \quad \text{and} \quad \rho\frac{\partial u}{\partial t} = -\frac{\partial p}{\partial x} + \rho g_x.$$

    c) If $\rho = \rho_o(x)$ at $t = 0$, and $u = U(0) = U_o$ at $t = 0$, determine implicit or explicit solutions for $\rho = \rho(x, t)$ and $U(t)$ in terms of x, t, $\rho_o(x)$, U_o, $\partial p/\partial x$, and g_x.

4.41. [5]**a)** Derive the following equation for the velocity potential for irrotational inviscid compressible flow in the absence of a body force:

$$\frac{\partial^2 \phi}{\partial t^2} + \frac{\partial}{\partial t}\left(|\nabla\phi|^2\right) + \frac{1}{2}\nabla\phi\cdot\nabla\left(|\nabla\phi|^2\right) - c^2\nabla^2\phi = 0$$

where $\nabla\phi = \mathbf{u}$. Start from the Euler equation (4.41), use the continuity equation, assume that the flow is isentropic so that p depends only on ρ, and denote $(\partial p/\partial\rho)_s = c^2$.

    b) What limit does $c \to \infty$ imply?

    c) What limit does $|\nabla\phi| \to 0$ imply?

---

[5]Obtained from Professor Paul Dimotakis

4.42. Derive (4.43) from (4.42).

4.43. For steady constant-density inviscid flow with body force per unit mass $\mathbf{g} = -\nabla\Phi$, it is possible to derive the following Bernoulli equation: $p + \frac{1}{2}\rho|\mathbf{u}|^2 + \rho\Phi =$ constant along a streamline.

   a) What is the equivalent form of the Bernoulli equation for constant-density inviscid flow that appears steady when viewed in a frame of reference that rotates at a constant rate about the z-axis, i.e., when $\mathbf{\Omega} = (0, 0, \Omega_z)$ with Ω_z constant?

   b) If the extra term found in the Bernoulli equation is considered a pressure correction: Where on the surface of the earth (i.e., at what latitude) will this pressure correction be the largest? What is the absolute size of the maximum pressure correction when changes in R on a streamline are 1 m, 1 km, and 10^3 km?

4.44. For many atmospheric flows, rotation of the earth is important. The momentum equation for inviscid flow in a frame of reference rotating at a constant rate $\mathbf{\Omega}$ is:

$$\partial\mathbf{u}/\partial t + (\mathbf{u}\cdot\nabla)\mathbf{u} = -\nabla\Phi - (1/\rho)\nabla p - 2\mathbf{\Omega}\times\mathbf{u} - \mathbf{\Omega}\times(\mathbf{\Omega}\times\mathbf{x})$$

For steady two-dimensional horizontal flow, $\mathbf{u} = (u, v, 0)$, with $\Phi = gz$ and $\rho = \rho(z)$, show that the streamlines are parallel to constant pressure lines when the fluid particle acceleration is dominated by the Coriolis acceleration $|(\mathbf{u}\cdot\nabla)\mathbf{u}| \ll |2\mathbf{\Omega}\times\mathbf{u}|$, and when the local pressure gradient dominates the centripetal acceleration $|\mathbf{\Omega}\times(\mathbf{\Omega}\times\mathbf{x})| \ll |\nabla p|/\rho$. [This seemingly strange result governs just about all large-scale weather phenomena like hurricanes and other storms, and it allows weather forecasts to be made based on surface pressure measurements alone.]

   Hints:

   1. If $Y(x)$ defines a streamline contour, then $dY/dx = v/u$ is the streamline slope.

   2. Write out all three components of the momentum equation and build the ratio v/u.

   3. Using hint 1, the pressure increment along a streamline is: $dp = (\partial p/\partial x)dx + (\partial p/\partial y)dY$.

4.45. Show that (4.55) can be derived from (4.7), (4.53), and (4.54).

4.46. Multiply (4.22) by u_j and sum over j to derive (4.56).

4.47. Starting from $\varepsilon = (1/\rho)\sigma_{ij}S_{ij}$, derive the rightmost expression in (4.58).

4.48. For many gases and liquids (and solids too!), the following equations are valid:

$\mathbf{q} = -k\nabla T$ (Fourier's law of heat conduction, $k =$ thermal conductivity, $T =$ temperature),

$e = e_o + C_vT$ ($e =$ internal energy per unit mass, $C_v =$ specific heat at constant volume), and

$h = h_o + C_pT$ ($h =$ enthalpy per unit mass, $C_p =$ specific heat at constant pressure), where e_o and h_o are constants, and C_v and C_p are also constants. Start with the energy equation

$$\rho\frac{\partial e}{\partial t} + \rho u_i\frac{\partial e}{\partial x_i} = -p\frac{\partial u_i}{\partial x_i} + \sigma_{ij}S_{ij} - \frac{\partial q_i}{\partial x_i}$$

for each of the following items.

a) Derive an equation for T involving u_j, k, ρ, and C_v for incompressible flow when $\sigma_{ij} = 0$.

b) Derive an equation for T involving u_j, k, ρ, and C_p for flow with $p = const.$ and $\sigma_{ij} = 0$.

c) Provide a physical explanation as to why the answers to parts a) and b) are different.

4.49. Derive the following alternative form of (4.60): $\rho C_p \dfrac{DT}{Dt} = \alpha T \dfrac{Dp}{Dt} + \rho\varepsilon + \dfrac{\partial}{\partial x_i}\left(k\dfrac{\partial T}{\partial x_i}\right)$,

where ε is given by (4.58) and α is the thermal expansion coefficient defined in (1.20). [Hint: $dh = (\partial h/\partial T)_p dT + (\partial h/\partial p)_T dp$]

4.50. Show that (4.68) is true without abandoning index notation or using vector identities.

4.51. Consider an incompressible planar Couette flow, which is the flow between two parallel plates separated by a distance b. The upper plate is moving parallel to itself at speed U, and the lower plate is stationary. Let the x-axis lie on the lower plate. The pressure and velocity fields are independent of x, and the fluid is uniform with constant viscosity.

a) Show that the pressure distribution is hydrostatic and that the solution of the Navier-Stokes equation is $u(y) = Uy/b$.

b) Write the expressions for the stress and strain rate tensors, and show that the viscous kinetic-energy dissipation per unit volume is $\mu U^2/b^2$.

c) Using a rectangular control volume for which the two horizontal surfaces coincide with the walls and the two vertical surfaces are perpendicular to the flow: evaluate the kinetic energy equation (4.56) within this control volume, and show that the balance is between the viscous dissipation and the work done in moving the upper surface.

4.52. Determine the outlet speed, U_2, of a chimney in terms of ρ_0, ρ_2, g, H, A_1, and A_2. For simplicity, assume the fire merely decreases the density of the air from ρ_0 to ρ_2 ($\rho_0 > \rho_2$) and does not add any mass to the airflow. (This mass flow assumption isn't true, but it serves to keep the algebra under control in this problem.) The relevant parameters are shown in the figure. Use the steady Bernoulli equation into the inlet and from the outlet of the fire, but perform a control volume analysis across the fire. Ignore the vertical extent of A_1 compared to H and the effects of viscosity.

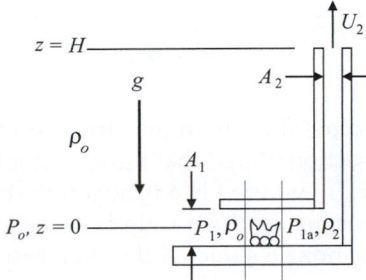

4.53. A hemispherical vessel of radius R containing an inviscid constant-density liquid has a small rounded orifice of area A at the bottom. Show that the time required to lower the level from h_1 to h_2 is given by

$$t = \frac{2\pi}{A\sqrt{2g}}\left[\frac{2}{3}R\left(h_1^{3/2} - h_2^{3/2}\right) - \frac{1}{5}\left(h_1^{5/2} - h_2^{5/2}\right)\right].$$

4.54. Water flows through a pipe in a gravitational field as shown in the accompanying figure. Neglect the effects of viscosity and surface tension. Solve the appropriate conservation equations for the variation of the cross-sectional area of the fluid column $A(z)$ after the water has left the pipe at $z = 0$. The velocity of the fluid at $z = 0$ is uniform at V_0 and the cross-sectional area is A_0.

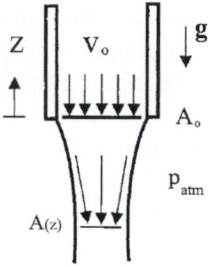

4.55. Redo the solution for the orifice-in-a-tank problem allowing for the fact that in Figure 4.16, $h = h(t)$, but ignoring fluid acceleration. Estimate how long it takes for the tank to empty.

4.56. A circular plate is forced down at a steady velocity U_o against a flat surface. Frictionless incompressible fluid of density ρ fills the gap $h(t)$. Assume that $h \ll r_o$, the plate radius, and that the radial velocity $u_r(r,t)$ is constant across the gap.
   a) Obtain a formula for $u_r(r,t)$ in terms of r, U_o, and h.
   b) Determine $\partial u_r(r,t)/\partial t$.
   c) Calculate the pressure distribution under the plate assuming that $p(r = r_o) = 0$.

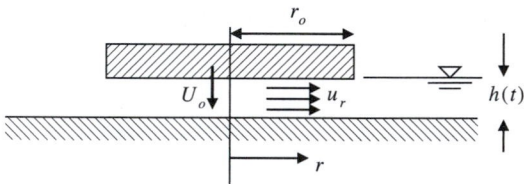

4.57. A frictionless, incompressible fluid with density ρ resides in a horizontal nozzle of length L having a cross-sectional area that varies smoothly between A_i and A_o via: $A(x) = A_i + (A_o - A_i)f(x/L)$, where f is a function that goes from 0 to 1 as x/L goes from 0 to 1. Here the x-axis lies on the nozzle's centerline, and $x = 0$ and $x = L$ are the horizontal locations of the nozzle's inlet and outlet, respectively. At $t = 0$, the pressure at the inlet of the nozzle is raised to $p_i > p_o$, where p_o is the (atmospheric) outlet pressure of the nozzle, and the fluid begins to flow horizontally through the nozzle.

a) Derive the following equation for the time-dependent volume flow rate $Q(t)$ through the nozzle from the unsteady Bernoulli equation and an appropriate conservation-of-mass relationship.

$$\frac{\dot{Q}(t)}{A_i} \int\limits_{x=0}^{x=L} \frac{A_i}{A(x)} \, dx + \frac{Q^2(t)}{2}\left(\frac{1}{A_o^2} - \frac{1}{A_i^2}\right) = \left(\frac{p_i - p_o}{\rho}\right)$$

b) Solve the equation of part a) when $f(x/L) = x/L$.

c) If $\rho = 10^3$ kg/m^3, $L = 25$ cm, $A_i = 100$ cm^2, $A_o = 30$ cm^2, and $p_i - p_o = 100$ kPa for $t \geq 0$, how long does it take for the flow rate to reach 99% of its steady-state value?

4.58. Using the small slope version of the surface curvature $1/R_1 \approx d^2\zeta/dx^2$, redo Example 4.7 to find h and $\zeta(x)$ in terms of x, σ, ρ, g, and θ. Show that the two answers are consistent when θ approaches $\pi/2$.

4.59. Redo the dimensionless scaling leading to (4.101) by choosing a generic viscous stress, $\mu U/l$, and then a generic hydrostatic pressure, $\rho g l$, to make $p - p_\infty$ dimensionless. Interpret the revised dimensionless coefficients that appear in the scaled momentum equation, and relate them to St, Re, and Fr.

4.60. From Figure 4.21, it can be seen that $C_D \propto 1/\text{Re}$ at small Reynolds numbers and that C_D is approximately constant at large Reynolds numbers. Redo the dimensional analysis leading to (4.99) to verify these observations when:

a) Re is low and fluid inertia is unimportant so ρ is no longer a parameter.

b) Re is high and the drag force is dominated by fore-aft pressure differences on the sphere and μ is no longer a parameter.

4.61. Suppose that the power to drive a propeller of an airplane depends on d (diameter of the propeller), U (free-stream velocity), ω (angular velocity of the propeller), c (velocity of sound), ρ (density of fluid), and μ (viscosity). Find the dimensionless groups. In your opinion, which of these are the most important and should be duplicated in model testing?

4.62. A 1/25 scale model of a submarine is being tested in a wind tunnel in which $p = 200$ kPa and $T = 300$ K. If the prototype speed is 30 km/hr, what should be the free-stream velocity in the wind tunnel? What is the drag ratio? Assume that the submarine would not operate near the free surface of the ocean.

4.63. A set of small-scale tank-draining experiments are performed to predict the liquid depth, h, as a function of time t for the draining process of a large cylindrical tank that is geometrically similar to the small-experiment tanks. The relevant parameters are gravity g, initial tank depth H, tank diameter D, orifice diameter d, and the density and viscosity of the liquid, ρ and μ, respectively.

a) Determine a general relationship between h and the other parameters.

b) Using the following small-scale experiment results, determine whether or not the liquid's viscosity is an important parameter.

| H = 8 cm, D = 24 cm, d = 8 mm | | H = 16 cm, D = 48 cm, d = 1.6 cm | |
h (cm)	t (s)	h (cm)	t (s)
8.0	0.00	16.0	0.00
6.8	1.00	13.3	1.50
5.0	2.00	9.5	3.00
3.0	3.00	5.3	4.50
1.2	4.00	1.8	6.00
0.0	5.30	0.0	7.50

c) Using the small-scale-experiment results above, predict how long it takes to completely drain the liquid from a large tank having $H = 10$ m, $D = 30$ m, and $d = 1.0$ m.

Literature Cited

Aris, R. (1962). *Vectors, Tensors, and the Basic Equations of Fluid Mechanics*. Englewood Cliffs, NJ: Prentice-Hall. (The basic equations of motion and the various forms of the Reynolds transport theorem are derived and discussed.)

Batchelor, G. K. (1967). *An Introduction to Fluid Dynamics*. London: Cambridge University Press. (This contains an excellent and authoritative treatment of the basic equations.)

Bird, R. B., Armstrong, R. C., & Hassager, O. (1987). *Dynamics of Polymeric Liquids, Vol. 1. Fluid Mechanics* (2nd ed.). New York: John Wiley & Sons.

Fermi, E. (1956). *Thermodynamics*. New York: Dover Publications, Inc.

Gogte, S. P., Vorobieff, R., Truesdell, A., Mammoli, F., van Swol, P. Shah, & Brinker, C. J. (2005). Effective slip on textured superhydrophobic surfaces. *Phys. Fluids*, 17, 051701.

Hirschfelder, J. O., Curtiss, C. F., & Bird, R. B. (1954). *Molecular Theory of Gases and Liquids*. New York: John Wiley and Sons.

Lamb, H. (1945). *Hydrodynamics* (6th ed.). New York: Dover Publications, Inc.

Levich, V. G. (1962). *Physicochemical Hydrodynamics* (2nd ed.). Englewood Cliffs, NJ: Prentice-Hall.

McCormick, N. J. (2005). Gas-surface accomodation coefficients from viscous slip and temperature jump coefficients. *Phys. Fluids*, 17, 107104.

Probstein, R. F. (1994). *Physicochemical Hydrodynamics* (2nd ed.). New York: John Wiley & Sons.

Rayleigh, Lord, & Strutt, J. W. (1890). On the theory of surface forces. *Phil. Mag. (Ser. 5)*, 30, 285–298, 456–475.

Spiegel, E. A., & Veronis, G. (1960). On the Boussinesq approximation for a compressible fluid. *Astrophysical Journal*, 131, 442–447.

Stommel, H. M., & Moore, D. W. (1989). *An Introduction to the Coriolis Force*. New York: Columbia University Press.

Thompson, P. A. (1972). *Compressible-Fluid Dynamics*. New York: McGraw-Hill.

Truesdell, C. A. (1952). Stokes' principle of viscosity. *Journal of Rational Mechanics and Analysis*, 1, 228–231.

Supplemental Reading

Çengel, Y. A., & Cimbala, J. M. (2010). *Fluid Mechanics: Fundamentals and Applications* (2nd ed.). New York: McGraw-Hill. (A fine undergraduate-level fluid mechanics textbook.)

Chandrasekhar, S. (1961). *Hydrodynamic and Hydromagnetic Stability*. London: Oxford University Press. (This is a good source to learn the basic equations in a brief and simple way.)

Dussan, V. E. B. (1979). On the spreading of liquids on solid surfaces: Static and dynamic contact lines. *Annual Rev. of Fluid Mech*, 11, 371–400.

Fox, R. W., Pritchard, P. J., & MacDonald, A. T. (2009). *Introduction to Fluid Mechanics* (7th ed.). New York: John Wiley. (Another fine undergraduate-level fluid mechanics textbook.)

Levich, V. G., & Krylov, V. S. (1969). Surface tension driven phenomena. *Annual Rev. of Fluid Mech*, 1, 293–316.

Pedlosky, J. (1987). *Geophysical Fluid Dynamics*. New York: Springer-Verlag.

Sabersky, R. H., Acosta, A. J., Hauptmann, E. G., & Gates, E. M. (1999). *Fluid Flow: A First Course in Fluid Mechanics* (4th ed.). New Jersey: Prentice Hall. (Another fine undergraduate-level fluid mechanics textbook.)

White, F. M. (2008). *Fluid Mechanics* (6th ed.). New York: McGraw-Hill. (Another fine undergraduate-level fluid mechanics textbook.)

CHAPTER

5

Vorticity Dynamics

OUTLINE

5.1. Introduction   171

5.2. Kelvin's Circulation Theorem   176

5.3. Helmholtz's Vortex Theorems   179

5.4. Vorticity Equation in a Nonrotating Frame   180

5.5. Velocity Induced by a Vortex Filament: Law of Biot and Savart   181

5.6. Vorticity Equation in a Rotating Frame   183

5.7. Interaction of Vortices   187

5.8. Vortex Sheet   191

Exercises   192

Literature Cited   195

Supplemental Reading   196

CHAPTER OBJECTIVES

- To introduce the basic concepts and phenomena associated with vortex lines, tubes, and sheets in viscous and inviscid flows.

- To derive and state classical theorems and equations for vorticity production and transport in inertial and rotating frames of reference.

- To develop the relationship that describes how vorticity at one location induces fluid velocity at another.

- To present some of the intriguing phenomena of vortex dynamics.

5.1. INTRODUCTION

Vorticity is a vector field that is twice the angular velocity of a fluid particle. A concentration of codirectional or nearly codirectional vorticity is called a *vortex*. Fluid motion leading to circular or nearly circular streamlines is called *vortex motion*. In two dimensions (r,θ), a uniform distribution of plane-normal vorticity with magnitude ω produces solid body rotation,

$$u_\theta = \omega r/2, \tag{5.1}$$

while a perfect concentration of plane-normal vorticity located at $r = 0$ with circulation Γ produces irrotational flow for $r > 0$,

$$u_\theta = \Gamma/2\pi r. \tag{5.2}$$

Both of these flow fields are steady and both produce closed (circular) streamlines. However, in the first, fluid particles rotate, but in the second, for $r \neq 0$, they do not. In the second flow field, the vorticity is infinite on a line perpendicular to the r-θ plane that intersects it at $r = 0$, but is zero elsewhere. Thus, such an *ideal line vortex* is also known as an *irrotational vortex*. It is a useful idealization that will be exploited in this chapter, in Chapter 6, and in Chapter 14.

   In general, vorticity in a flowing fluid is neither unidirectional nor steady. In fact, we can commonly think of vorticity as being embedded in fluid elements so that an element's vorticity may be reoriented or concentrated or diffused depending on the motion and deformation of that fluid element and on the torques applied to it by surrounding fluid elements. This conjecture is based on the fact that the dynamics of three-dimensional time-dependent vorticity fields can often be interpreted in terms of a few fundamental principles. This chapter presents these principles and some aspects of flows with vorticity, starting with fundamental vortex concepts.

   A *vortex line* is a curve in the fluid that is everywhere tangent to the local vorticity vector. Here, of course, we recognize that a vortex line is not strictly linear; it may be curved just as a streamline may be curved. A vortex line is related to the vorticity vector the same way a streamline is related to the velocity vector. Thus, if ω_x, ω_y and ω_z are the Cartesian components of the vorticity vector $\boldsymbol{\omega}$, then the components of an element $d\mathbf{s} = (dx, dy, dz)$ of a vortex line satisfy

$$dx/\omega_x = dy/\omega_y = dz/\omega_z, \tag{5.3}$$

which is analogous to (3.7) for a streamline. As a further similarity, vortex lines do not exist in irrotational flow just as streamlines do not exist in stationary fluid. Elementary examples of vortex lines are supplied by the flow fields (5.1) and (5.2). For solid-body rotation (5.1), all lines perpendicular to the r-θ plane are vortex lines, while in the flow field of an irrotational vortex (5.2) the lone vortex line is perpendicular to the r-θ plane and passes through it at $r = 0$.

   In a region of flow with nontrivial vorticity, the vortex lines passing through any closed curve form a tubular surface called a *vortex tube* (Figure 5.1), which is akin to a stream tube (Figure 3.6). The circulation around a narrow vortex tube is $d\Gamma = \boldsymbol{\omega} \cdot \mathbf{n} dA$ just as the volume flow rate in a narrow stream tube is $dQ = \mathbf{u} \cdot \mathbf{n} dA$. The *strength of a vortex tube* is defined as the circulation computed on a closed circuit lying on the surface of the tube that encircles it just once. From Stokes' theorem it follows that the strength of a vortex tube, Γ, is equal to the vorticity in the tube integrated over its cross-sectional area. Thus, when Gauss' theorem is applied to the volume V defined by a section of a vortex tube, such as that shown in Figure 5.1, we find that

$$\int_V \nabla \cdot \boldsymbol{\omega} \, dV = \int_A \boldsymbol{\omega} \cdot \mathbf{n} \, dA = \left\{ \int_{lower\ end} + \int_{curved\ side} + \int_{upper\ end} \right\} \boldsymbol{\omega} \cdot \mathbf{n} \, dA$$

$$= -\Gamma_{lower\ end} + \Gamma_{upper\ end} = 0, \tag{5.4}$$

where $\boldsymbol{\omega} \cdot \mathbf{n}$ is zero on the curved sides of the tube, and the final equality follows from $\nabla \cdot \boldsymbol{\omega} = \nabla \cdot (\nabla \times \mathbf{u}) = 0$. Equation (5.4) states that a vortex tube's strength is independent of where it is measured, $\Gamma_{lower\ end} = \Gamma_{upper\ end}$, and this implies that vortex tubes cannot end within the fluid, a concept that can be extended to vortex lines in the limit as a vortex tube's cross-sectional area goes to zero. However, vortex lines and tubes can terminate on solid surfaces or free surfaces, or they can form loops. This kinematic constraint is often useful for determining the topology of vortical flows.

As we will see in this and other chapters, fluid viscosity plays an essential role in the diffusion of vorticity, and in the reconnection of vortex lines. However, before considering these effects, the role of viscosity in the two basic vortex flows (5.1) and (5.2) is examined. Assuming incompressible flow, we shall see that in one of these flows the viscous terms in the momentum equation drop out, although the viscous stress and dissipation of energy are nonzero.

As discussed in Chapter 3, fluid elements undergoing solid-body rotation (5.1) do not deform ($S_{ij} = 0$), so the Newtonian viscous stress tensor (4.37) reduces to $\tau_{ij} = -p\delta_{ij}$, and Cauchy's equation (4.24) reduces to Euler's equation (4.41). When the solid-body rotation field, $u_r = 0$ and $u_\theta = \omega r / 2$, is substituted into (4.41), it simplifies to:

$$-\rho u_\theta^2 / r = -\partial p / \partial r, \quad \text{and} \quad 0 = -\partial p / \partial z - \rho g. \tag{5.5a, 5.5b}$$

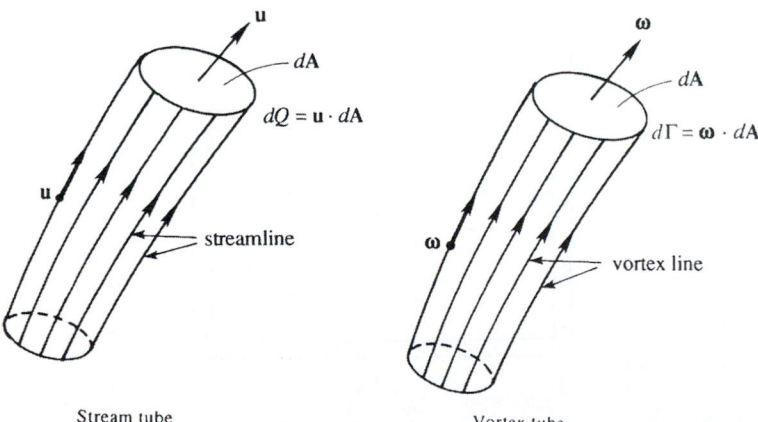

FIGURE 5.1   Analogy between stream tubes and vortex tubes. The lateral sides of stream and vortex tubes are locally tangent to the flow's velocity and vorticity fields, respectively. Stream and vortex tubes with cross-sectional area dA carry constant volume flux $\mathbf{u} \cdot dA$ and constant circulation $\boldsymbol{\omega} \cdot dA$, respectively.

Integrating (5.5a) produces $p(r,z) = \rho\omega^2 r^2/8 + f(z)$, where f is an undetermined function. Integrating (5.5b) produces $p(r,z) = -\rho g z + g(r)$, where g is an undetermined function. These two equations are consistent when:

$$p(r,z) - p_0 = \frac{1}{8}\rho\omega^2 r^2 - \rho g z \qquad (5.6)$$

where p_0 is the pressure at $r=0$ and $z=0$. To determine the shape of constant pressure surfaces, solve (5.6) for z to find:

$$z = \frac{\omega^2 r^2}{8g} - \frac{p(r,z) - p_0}{\rho g}.$$

Hence, surfaces of constant pressure are paraboloids of revolution (Figure 5.2).

The important point to note is that viscous stresses are absent in steady solid-body rotation. (The viscous stresses, however, are important during the transient period of *initiating* solid body rotation, say by steadily rotating a tank containing a viscous fluid initially at rest.) In terms of velocity, (5.6) can be written as

$$-\frac{1}{2}u_\theta^2 + gz + \frac{p(r,z)}{\rho} = const.,$$

and, when compared to (4.19), this shows that the Bernoulli function $B = u_\theta^2/2 + gz + p/\rho$ is *not* constant for points on different streamlines. This outcome is expected because the flow is rotational.

For the flow induced by an irrotational vortex (5.2), the viscous stress is

$$\sigma_{r\theta} = \mu\left[\frac{1}{r}\frac{\partial u_r}{\partial\theta} + r\frac{\partial}{\partial r}\left(\frac{u_\theta}{r}\right)\right] = -\frac{\mu\Gamma}{\pi r^2},$$

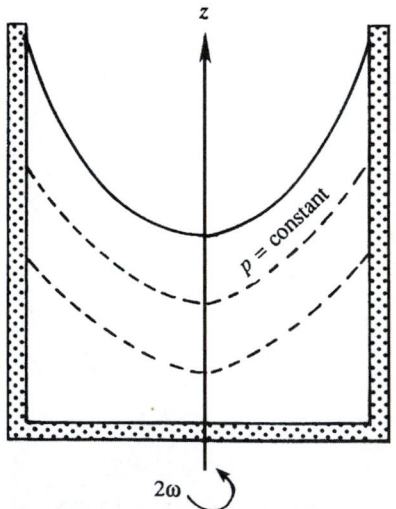

FIGURE 5.2   The steady flow field of a viscous liquid in a steadily rotating tank is solid body rotation. When the axis of rotation is parallel to the (downward) gravitational acceleration, surfaces of constant pressure in the liquid are paraboloids of revolution.

which is nonzero everywhere because fluid elements deform (see Figure 3.16). However, the interesting point is that the *net viscous force* on an element is zero for $r > 0$ (see Exercise 5.4) because the viscous forces on the surfaces of an element cancel out, leaving a zero resultant. Thus, momentum conservation is again represented by the Euler equation. Substitution of (5.2) into (5.5), followed by integration, yields

$$p(r,z) - p_\infty = -\frac{\rho \Gamma^2}{8\pi^2 r^2} - \rho g z, \tag{5.7}$$

where p_∞ is the pressure far from the line vortex at $z = 0$. This can be rewritten:

$$z = -\frac{\Gamma^2}{8\pi^2 r^2 g} - \frac{p(r,z) - p_\infty}{\rho g},$$

which shows that surfaces of constant pressure are hyperboloids of revolution of the second degree (Figure 5.3). Equation (5.7) can also be rewritten:

$$\frac{1}{2}u_\theta^2 + gz + \frac{p(r,z)}{\rho} = const.,$$

which shows that Bernoulli's equation is applicable between any two points in the flow field, as is expected for irrotational flow.

One way of generating the flow field from an irrotational vortex is by rotating a solid circular cylinder with radius a in an infinite viscous fluid (see Figure 8.7). It is shown in Section 8.2 that the steady solution of the Navier-Stokes equations satisfying the no-slip boundary condition ($u_\theta = \omega a/2$ at $r = a$) is

$$u_\theta = \omega a^2/2r \text{ for } r \geq a,$$

FIGURE 5.3 Surfaces of constant pressure in the flow induced by an ideal linear vortex that coincides with the z-axis and is parallel to the (downward) gravitational acceleration.

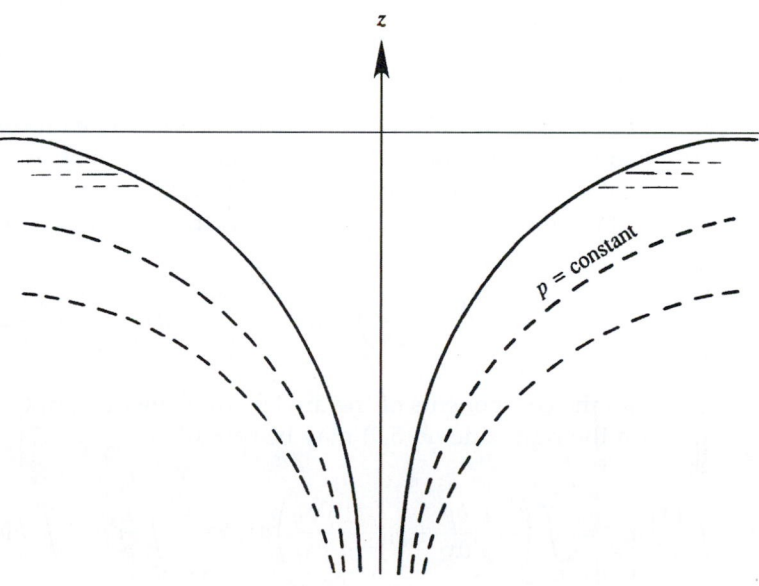

where $\omega/2$ is the cylinder's constant rotation rate; see (8.11). When the motions inside and outside the cylinder are considered, this flow field precisely matches that of a Rankine vortex with core size a; see (3.28) with $\Gamma = \pi a^2 \omega$. The presence of the nonzero-radius cylinder leads to a flow field without a singularity that is irrotational for $r > a$. Viscous stresses are present, and the resulting viscous dissipation of kinetic energy is exactly compensated by the work done at the surface of the cylinder. However, there is no *net* viscous force at any point in the steady state. Interestingly, the application of the moment of momentum principle (see Section 4.9) to a large-radius cylindrical control volume centered on the rotating solid cylinder shows that the torque that rotates the solid cylinder is transmitted to an arbitrarily large distance from the axis of rotation. Thus, any attempt to produce this flow in a stationary container would require the application of counteracting torque on the container.

These examples suggest that *irrotationality does not imply the absence of viscous stresses.* Instead, it implies the *absence of net viscous forces.* Viscous stresses will be present whenever fluid elements deform. Yet, when $\boldsymbol{\omega}$ is uniform and nonzero (solid body rotation), there is no viscous stress at all. However, solid-body rotation is unique in this regard, and this uniqueness is built into the Newtonian-fluid viscous stress tensor (4.59). In general, fluid element rotation is accomplished and accompanied by viscous effects. Indeed, viscosity is a primary agent for vorticity generation and diffusion.

5.2. KELVIN'S CIRCULATION THEOREM

By considering the analogy with electrodynamics, Helmholtz published several theorems for vortex motion in an inviscid fluid in 1858. Ten years later, Kelvin introduced the idea of circulation and proved the following theorem: *In an inviscid, barotropic flow with conservative body forces, the circulation around a closed curve moving with the fluid remains constant with time*, if the motion is observed from a nonrotating frame. This theorem can be stated mathematically as

$$D\Gamma/Dt = 0 \tag{5.8}$$

where D/Dt is defined by (3.5) and represents the total time rate of change following the fluid elements that define the closed curve, C (a material contour), used to compute the circulation Γ. Such a material contour is shown in Figure 5.4.

Kelvin's theorem can be proved by time differentiating the definition of the circulation (3.18):

$$\frac{D\Gamma}{Dt} = \frac{D}{Dt}\int_C u_i dx_i = \int_C \frac{Du_i}{Dt}dx_i + \int_C u_i \frac{D}{Dt}(dx_i), \tag{5.9}$$

where dx_i are the components of the arc length element $d\mathbf{x}$ of C. Using (4.39) and (4.59), the first term on the right side of (5.9) may be rewritten:

$$\int_C \frac{Du_i}{Dt}dx_i = \int_C \left(-\frac{1}{\rho}\frac{\partial p}{\partial x_i} + g_i + \frac{1}{\rho}\frac{\partial \sigma_{ij}}{\partial x_j}\right)dx_i = -\int_C \frac{1}{\rho}dp - \int_C d\Phi + \int_C \left(\frac{1}{\rho}\frac{\partial \sigma_{ij}}{\partial x_j}\right)dx_i, \tag{5.10}$$

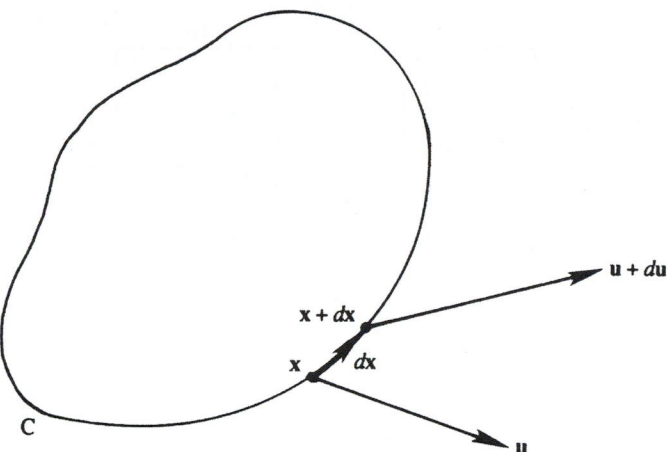

FIGURE 5.4 Contour geometry for the Proof of Kelvin's circulation theorem. Here the short segment dx of the contour C moves with the fluid so that $D(dx)/Dt = d\mathbf{u}$.

where the replacements $(\partial p / \partial x_i)dx_i = dp$ and $(\partial \Phi / \partial x_i)dx_i = d\Phi$ have been made, and Φ is the body force potential (4.18). For a barotropic fluid, the first term on the right side of (5.10) is zero because C is a closed contour, and ρ and p are single valued at each point in space. Similarly, the second integral on the right side of (5.10) is zero since Φ is also single valued at each point in space.

Now consider the second term on the right side of (5.9). The velocity at point $\mathbf{x} + d\mathbf{x}$ on C is:

$$\mathbf{u} + d\mathbf{u} = \frac{D}{Dt}(\mathbf{x} + d\mathbf{x}) = \frac{D\mathbf{x}}{Dt} + \frac{D}{Dt}(d\mathbf{x}), \text{ so } du_i = \frac{D}{Dt}(dx_i).$$

Thus, the last term in (5.9) then becomes

$$\int_C u_i \frac{D}{Dt}(dx_i) = \int_C u_i du_i = \int_C d\left(\frac{1}{2}u_i^2\right) = 0,$$

where the final equality again follows because C is a closed contour and $\mathbf{u}$ is a single-valued vector function. Hence, (5.9) simplifies to:

$$\frac{D\Gamma}{Dt} = \int_C \left(\frac{1}{\rho}\frac{\partial \sigma_{ij}}{\partial x_j}\right)dx_i, \tag{5.11}$$

and Kelvin's theorem (5.8) is proved when the fluid is inviscid ($\mu = \mu_v = 0$) or when the net viscous force ($\partial \sigma_{ij}/\partial x_j$) is zero along C. This latter condition occurs when C lies entirely in irrotational fluid.

From this short proof we see that the three ways to create or destroy vorticity in a flow are: nonconservative body forces, a nonbarotropic pressure-density relationship, and nonzero net viscous forces. Examples of each follow. The Coriolis acceleration is a nonconservative body force that occurs in rotating frames of reference, and it generates a drain or bathtub

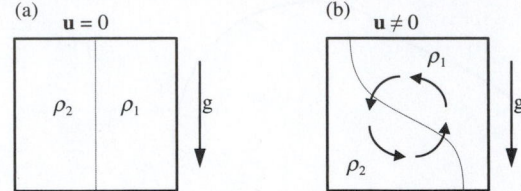

FIGURE 5.5   Schematic drawings of two fluids with differing density that are initially stationary and separated within a rectangular container. Gravity acts downward as shown. Here the density difference is baroclinic because it depends on fluid composition and pressure, not on pressure alone. (a) This drawing shows the initial condition immediately before the barrier between the two fluids is removed. (b) This drawing shows the resulting fluid motion a short time after barrier removal. The deflection of the fluid interface clearly indicates that vorticity has been created.

vortex when a filled tank, initially at rest on the earth's surface, is drained. Nonbarotropic effects can lead to vorticity generation when a vertical barrier is removed between two side-by-side initially motionless fluids having different densities in the same container and subject to a gravitational field. The two fluids will tumble as the heavier one slumps to the container's bottom and the lighter one surges to the container's top (see Figure 5.5 and Exercise 5.5). Nonzero net viscous forces create vorticity at solid boundaries where the no-slip condition is maintained. A short distance away from a solid boundary, the velocity parallel to the boundary may be large. Vorticity is created when such near-wall velocity gradients arise.

Kelvin's theorem implies that irrotational flow will remain irrotational if the following four restrictions are satisfied.

(1) There are no net viscous forces along C. If C moves into regions where there are net viscous forces such as within a boundary layer that forms on a solid surface, then the circulation changes. The presence of viscous effects causes *diffusion* of vorticity into or out of a fluid circuit and consequently changes the circulation.

(2) The body forces are conservative. Conservative body forces such as gravity act through the center of mass of a fluid particle and therefore do not generate torques that cause fluid particle rotation.

(3) The fluid density must depend on pressure only (barotropic flow). A flow will be barotropic if the fluid is homogeneous and one of the two independent thermodynamic variables is constant. Isentropic, isothermal, and constant density conditions lead to barotropic flow. Flows that are not barotropic are called *baroclinic*. Here fluid density depends on the pressure *and* the temperature, composition, salinity, and/or concentration of dissolved constituents. Consider fluid elements in barotropic and baroclinic flows (Figure 5.6). For the barotropic element, lines of constant p are parallel to lines of constant ρ, which implies that the resultant pressure forces pass through the center of mass of the element. For the baroclinic element, the lines of constant p and ρ are not parallel. The net pressure force does not pass through the center of mass, and the resulting torque changes the vorticity and circulation. As described above, Figure 5.6 depicts a situation where vorticity is generated in a baroclinic flow.

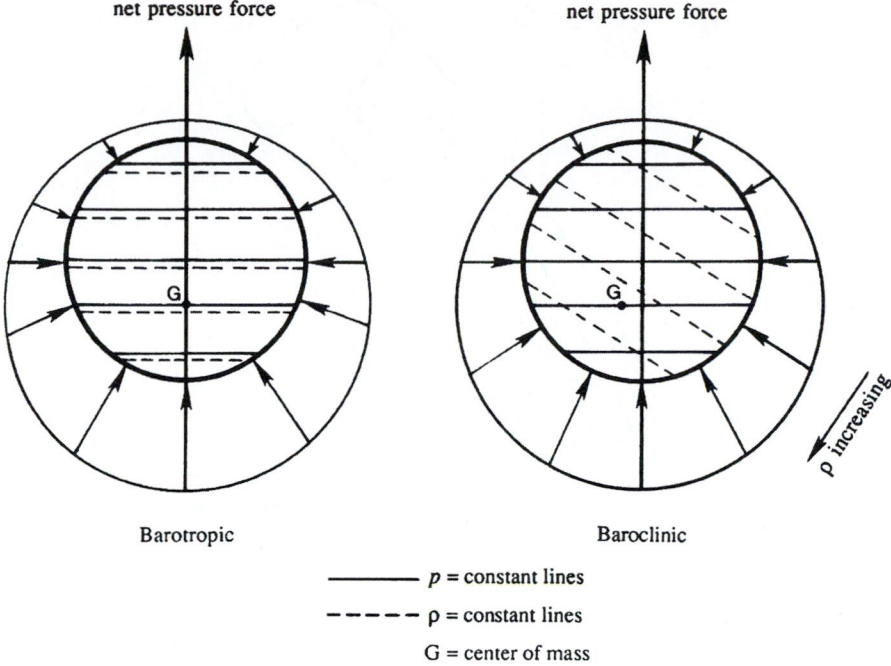

net pressure force   net pressure force

G   G

Barotropic   Baroclinic

————— p = constant lines

– – – – ρ = constant lines

G = center of mass

FIGURE 5.6 Mechanism of vorticity generation in baroclinic flow, showing that the net pressure force does not pass through the center of mass G of the fluid element. The radially inward arrows indicate pressure forces on an element.

(4) The frame of reference must be an inertial frame. As described in Section 4.7, the conservation of momentum equation includes extra terms when the frame of reference rotates and accelerates, and these extra terms were not considered in the short proof given above.

5.3. HELMHOLTZ'S VORTEX THEOREMS

Under the same four restrictions, Helmholtz proved the following theorems for vortex motion:

(1) Vortex lines move with the fluid.
(2) The strength of a vortex tube (its circulation) is constant along its length.
(3) A vortex tube cannot end within the fluid. It must either end at a solid boundary or form a closed loop—a *vortex ring* or *loop*.
(4) The strength of a vortex tube remains constant in time.

Here, we only highlight the proof of the first theorem, which essentially says that fluid particles that at any time are part of a vortex line always belong to the same vortex line. To prove this result, consider an area S, bounded by a curve, lying on the surface of a vortex

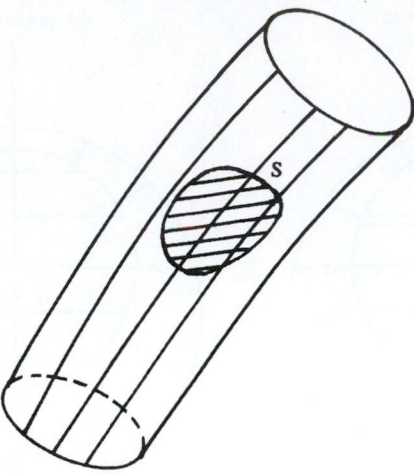

FIGURE 5.7　Vortex tube and surface geometry for Helmholtz's first vortex theorem. The surface S lies within a closed contour on the surface of a vortex tube.

tube without embracing it (Figure 5.7). Since the vorticity vectors are everywhere lying parallel to S (none are normal to S), it follows that the circulation around the edge of S is zero. After an interval of time, the same fluid particles form a new surface, S'. According to Kelvin's theorem, the circulation around S' must also be zero. As this is true for any S, the component of vorticity normal to every element of S' must vanish, demonstrating that S' must lie on the surface of the vortex tube. Thus, vortex tubes move with the fluid, a result we will also be able to attain from the field equation for vorticity. Applying this result to an infinitesimally thin vortex tube, we get the Helmholtz vortex theorem that vortex lines move with the fluid. A different proof may be found in Sommerfeld (1964, pp. 130–132).

5.4. VORTICITY EQUATION IN A NONROTATING FRAME

An equation governing the vorticity in an inertial frame of reference is derived in this section. The *fluid density is assumed to be constant*, so that the flow is barotropic. Viscous effects are retained but the viscosity is assumed to be constant. Baroclinic effects and a rotating frame of reference are considered in Section 5.6. The derivation given here uses vector notation and several vector identities. In Section 5.6, the derivation is completed in tensor notation.

Vorticity $\boldsymbol{\omega}$ is the curl of the velocity, so, as previously noted, $\nabla \cdot \boldsymbol{\omega} = \nabla \cdot (\nabla \times \mathbf{u}) = 0$. An equation for the vorticity can be obtained from the curl of the momentum conservation equation (4.39b)

$$\nabla \times \left\{ \frac{D\mathbf{u}}{Dt} = -\frac{1}{\rho}\nabla p + \mathbf{g} + \nu\nabla^2\mathbf{u} \right\}. \tag{5.12}$$

When $\mathbf{g}$ is conservative and (4.18) applies, the curl of the first two terms on the right side of (5.12) will be zero because they are gradients of scalar functions. The acceleration term on the left side of (5.12) becomes:

$$\nabla \times \left\{ \frac{\partial \mathbf{u}}{\partial t} + (\mathbf{u} \cdot \nabla)\mathbf{u} \right\} = \frac{\partial \boldsymbol{\omega}}{\partial t} + \nabla \times \{(\mathbf{u} \cdot \nabla)\mathbf{u}\} = \frac{\partial \boldsymbol{\omega}}{\partial t} + \nabla \times \{\nabla(\mathbf{u} \cdot \mathbf{u}) + \boldsymbol{\omega} \times \mathbf{u}\}$$

$$= \frac{\partial \boldsymbol{\omega}}{\partial t} + \nabla \times (\boldsymbol{\omega} \times \mathbf{u}),$$

so (5.12) reduces to

$$\frac{\partial \boldsymbol{\omega}}{\partial t} + \nabla \times (\boldsymbol{\omega} \times \mathbf{u}) = \nu \nabla^2 \boldsymbol{\omega},$$

where we have also used the identity $\nabla \times \nabla^2 \mathbf{u} = \nabla^2 (\nabla \times \mathbf{u})$ in rewriting the viscous term. The second term in the above equation can be written as

$$\nabla \times (\boldsymbol{\omega} \times \mathbf{u}) = (\mathbf{u} \cdot \nabla)\boldsymbol{\omega} - (\boldsymbol{\omega} \cdot \nabla)\mathbf{u},$$

based on the vector identity (B.3.10), and the fact that $\nabla \cdot \mathbf{u} = 0$ and $\nabla \cdot \boldsymbol{\omega} = 0$. Thus, (5.12) becomes

$$\frac{D\boldsymbol{\omega}}{Dt} = (\boldsymbol{\omega} \cdot \nabla)\mathbf{u} + \nu \nabla^2 \boldsymbol{\omega}. \tag{5.13}$$

This is the field equation governing vorticity in a fluid with constant ρ and conservative body forces. The term $\nu \nabla^2 \boldsymbol{\omega}$ represents the rate of change of $\boldsymbol{\omega}$ caused by diffusion of vorticity in the same way that $\nu \nabla^2 \mathbf{u}$ represents acceleration caused by diffusion of momentum. The term $(\boldsymbol{\omega} \cdot \nabla)\mathbf{u}$ represents the rate of change of vorticity caused by the stretching and tilting of vortex lines. This important mechanism of vorticity alteration is discussed further in Section 5.6. Note that pressure and gravity terms do not appear in (5.13) since these forces act through the center of mass of an element and therefore generate no torque. In addition, note that (5.13) might appear upon first glance to be a linear equation for $\boldsymbol{\omega}$. However, the vorticity is the curl of the velocity so both the advective part of the $D\boldsymbol{\omega}/Dt$ term and the $(\boldsymbol{\omega} \cdot \nabla)\mathbf{u}$ term represent nonlinearities.

5.5. VELOCITY INDUCED BY A VORTEX FILAMENT: LAW OF BIOT AND SAVART

For a variety of applications in aero- and hydrodynamics, the flow induced by a concentrated distribution of vorticity (a vortex) with arbitrary orientation must be calculated. Here we consider the simple case of incompressible flow where $\nabla \cdot \mathbf{u} = 0$. Taking the curl of the vorticity produces:

$$\nabla \times \boldsymbol{\omega} = \nabla \times (\nabla \times \mathbf{u}) = \nabla(\nabla \cdot \mathbf{u}) - \nabla^2 \mathbf{u} = -\nabla^2 \mathbf{u},$$

where the second equality follows from an identity of vector calculus (B.3.13). The two ends of this extended equality form a Poisson equation, and its solution is the vorticity-induced portion of the fluid velocity:

$$\mathbf{u}(\mathbf{x}, t) = -\frac{1}{4\pi} \int_{V'} \frac{1}{|\mathbf{x} - \mathbf{x}'|} (\nabla' \times \boldsymbol{\omega}(\mathbf{x}', t)) d^3 x', \tag{5.14}$$

where V' encloses the vorticity of interest and ∇' operates on the $\mathbf{x}'$ coordinates (see Exercise 5.8). This result can be further simplified by rewriting the integrand in (5.14):

$$\frac{1}{|\mathbf{x}-\mathbf{x}'|}(\nabla' \times \omega(\mathbf{x}',t)) = \nabla' \times \left(\frac{\omega(\mathbf{x}',t)}{|\mathbf{x}-\mathbf{x}'|}\right) - \nabla'\left(\frac{1}{|\mathbf{x}-\mathbf{x}'|}\right) \times \omega(\mathbf{x}',t)$$

$$= \nabla' \times \left(\frac{\omega(\mathbf{x}',t)}{|\mathbf{x}-\mathbf{x}'|}\right) + \left(\frac{\mathbf{x}-\mathbf{x}'}{|\mathbf{x}-\mathbf{x}'|^3}\right) \times \omega(\mathbf{x}',t),$$

to obtain:

$$\mathbf{u}(\mathbf{x},t) = -\frac{1}{4\pi}\int_{V'} \nabla' \times \left(\frac{\omega(\mathbf{x}',t)}{|\mathbf{x}-\mathbf{x}'|}\right)d^3x' + \frac{1}{4\pi}\int_{V'} \frac{\omega(\mathbf{x}',t) \times (\mathbf{x}-\mathbf{x}')}{|\mathbf{x}-\mathbf{x}'|^3}d^3x'.$$

Here the first integral is zero when V' is chosen to capture a segment of the vortex, but it takes several steps to deduce this. First, rewrite the curl operation in index notation and apply Gauss' divergence theorem:

$$\int_{V'} \nabla' \times \left(\frac{\omega(\mathbf{x}',t)}{|\mathbf{x}-\mathbf{x}'|}\right)d^3x' = \int_{V'} \varepsilon_{kij}\frac{\partial}{\partial x_i'}\left(\frac{\omega_j(\mathbf{x}',t)}{|\mathbf{x}-\mathbf{x}'|}\right)d^3x' = \int_{A'} \varepsilon_{kij}\left(\frac{\omega_j(\mathbf{x}',t)}{|\mathbf{x}-\mathbf{x}'|}\right)n_i d^2x'$$

$$= \int_{A'} \frac{\mathbf{n} \times \omega(\mathbf{x}',t)}{|\mathbf{x}-\mathbf{x}'|}d^2x', \tag{5.15}$$

where A' is the surface of V' and $\mathbf{n}$ is the outward normal on A'. Now choose V' to be a volume aligned so that its end surfaces are locally normal to $\omega(\mathbf{x}',t)$ while its curved lateral surface lies outside the concentration of vorticity as shown in Figure 5.8. For this volume, the final integral in (5.15) is zero because $\mathbf{n} \times \omega = 0$ on its end surfaces since $\omega(\mathbf{x}',t)$ and $\mathbf{n}$ are parallel there, and because $\omega(\mathbf{x}',t) = 0$ on its lateral surface. Thus, (5.14) reduces to:

$$\mathbf{u}(\mathbf{x},t) = \frac{1}{4\pi}\int_{V'} \frac{\omega(\mathbf{x}',t) \times (\mathbf{x}-\mathbf{x}')}{|\mathbf{x}-\mathbf{x}'|^3}d^3x'. \tag{5.16}$$

If an elemental vortex segment of length dl is considered so that $V' = \Delta A' dl$, and the observation location, $\mathbf{x}$, is sufficiently distant from the vorticity concentration location $\mathbf{x}'$ so that $(\mathbf{x}-\mathbf{x}')/|\mathbf{x}-\mathbf{x}'|^3$ is effectively constant over the vorticity concentration, then (5.16) may be simplified to:

$$d\mathbf{u}(\mathbf{x},t) \cong \frac{1}{4\pi}\int_{\Delta A'} |\omega(\mathbf{x}',t)|\mathbf{e}_\omega d^2x' \times \frac{(\mathbf{x}-\mathbf{x}')}{|\mathbf{x}-\mathbf{x}'|^3}dl = \frac{\Gamma dl}{4\pi}\mathbf{e}_\omega \times \frac{(\mathbf{x}-\mathbf{x}')}{|\mathbf{x}-\mathbf{x}'|^3}, \tag{5.17}$$

where $d\mathbf{u}$ is the velocity induced by the vortex segment, and Γ and $\mathbf{e}_\omega$ are the strength and direction of the vortex segment at $\mathbf{x}'$, respectively. This is an expression of the Biot-Savart vortex induction law.

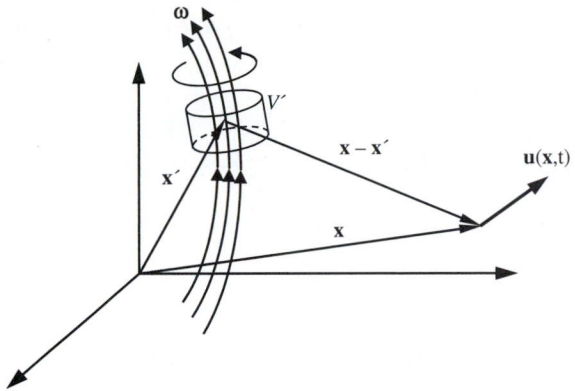

FIGURE 5.8 Geometry for derivation of Law of Biot and Savart. The location of the vorticity concentration or vortex is $\mathbf{x}'$. The location of the vortex-induced velocity $\mathbf{u}$ is $\mathbf{x}$. The volume V' contains a segment of the vortex. Its flat ends are perpendicular to the vorticity in the vortex, while its curved lateral sides lie outside the vortex.

5.6. VORTICITY EQUATION IN A ROTATING FRAME

A vorticity equation was derived in Section 5.4 for a fluid of uniform density observed from an inertial frame of reference. Here, this equation is generalized to a rotating frame of reference and a nonbarotropic fluid. The flow, however, will be assumed nearly incompressible in the Boussinesq sense, so that the continuity equation is approximately $\nabla \cdot \mathbf{u} = 0$. And, for conciseness, the comma notation for spatial derivatives (Section 2.14) is adopted.

The first step is to show that $\nabla \cdot \boldsymbol{\omega} = \omega_{i,i}$ is zero. From the definition $\boldsymbol{\omega} = \nabla \times \mathbf{u}$, we obtain

$$\omega_{i,i} = (\varepsilon_{inq} u_{q,n})_{,i} = \varepsilon_{inq} u_{q,ni}.$$

In the last term, ε_{inq} is antisymmetric in i and n, whereas the derivative $u_{q,ni}$ is symmetric in i and n. As the contracted product of a symmetric and an antisymmetric tensor is zero, it follows that

$$\omega_{i,i} = 0 \quad \text{or} \quad \nabla \cdot \boldsymbol{\omega} = 0 . \tag{5.18}$$

Hence, the vorticity field is divergence free (solenoidal), even for compressible and unsteady flows.

The continuity and momentum equations for a nearly incompressible flow in a steadily rotating coordinate system are

$$u_{i,i} = 0, \text{ and } \frac{\partial u_i}{\partial t} + u_j u_{i,j} + 2\varepsilon_{ijk}\Omega_j u_k = -\frac{1}{\rho}p_{,i} + g_i + \nu u_{i,jj}, \tag{5.19, 5.20}$$

where $\boldsymbol{\Omega}$ is the angular velocity of the coordinate system and g_i is the effective gravity (including centrifugal acceleration); see Section 4.7. The advective acceleration can be written as

$$u_j u_{i,j} = u_j(u_{i,j} - u_{j,i}) + u_j u_{j,i} = -u_j \varepsilon_{ijk}\omega_k + \frac{1}{2}(u_j u_j)_{,i} = -(\mathbf{u}\times\boldsymbol{\omega})_i + \frac{1}{2}(u_j^2)_{,i}, \quad (5.21)$$

where we have used the relation

$$\varepsilon_{ijk}\omega_k = \varepsilon_{ijk}\varepsilon_{kmn}u_{n,m} = (\delta_{im}\delta_{jn} - \delta_{in}\delta_{jm})u_{n,m} = u_{j,i} - u_{i,j}. \quad (5.22)$$

The viscous diffusion term can be written as

$$\nu u_{i,jj} = \nu(u_{i,j} - u_{j,i})_{,j} + \nu u_{j,ij} = -\nu\varepsilon_{ijk}\omega_{k,j}, \quad (5.23)$$

where we have used (5.22) and the fact that $u_{j,ij} = 0$ because of (5.19). Equation (5.23) says that $\nu\nabla^2\mathbf{u} = -\nu\nabla\times\boldsymbol{\omega}$, which we have used several times before (e.g., see (4.40)). Because $\boldsymbol{\Omega}\times\mathbf{u} = -\mathbf{u}\times\boldsymbol{\Omega}$, the Coriolis acceleration term in (5.20) can be rewritten

$$2\varepsilon_{ijk}\Omega_j u_k = -2\varepsilon_{ijk}\Omega_k u_j. \quad (5.24)$$

Substituting (5.21), (5.23), and (5.24) into (5.20), we obtain

$$\partial u_i/\partial t + \left(\frac{1}{2}u_j^2 + \Phi\right)_{,i} - \varepsilon_{ijk}u_j(\omega_k + 2\Omega_k) = -(1/\rho)p_{,i} - \nu\varepsilon_{ijk}\omega_{k,j}, \quad (5.25)$$

where we have also set $\mathbf{g} = -\nabla\Phi$; see (4.18).

Equation (5.25) is another form of the Navier-Stokes momentum equation, so the rotating-frame-of-reference vorticity equation is obtained by taking its curl. Since $\omega_n = \varepsilon_{nqi}u_{i,q}$, we need to operate on (5.25) by $\varepsilon_{nqi}(\ )_{,q}$ which produces:

$$\frac{\partial}{\partial t}\left(\varepsilon_{nqi}u_{i,q}\right) + \varepsilon_{nqi}\left(\frac{1}{2}u_j^2 + \Phi\right)_{,iq} - \varepsilon_{nqi}\varepsilon_{ijk}\left[u_j(\omega_k + 2\Omega_k)\right]_{,q} = -\varepsilon_{nqi}\left(\frac{1}{\rho}p_{,i}\right)_{,q} - \nu\varepsilon_{nqi}\varepsilon_{ijk}\omega_{k,jq}. \quad (5.26)$$

The second term on the left side vanishes on noticing that ε_{nqi} is antisymmetric in q and i, whereas the derivative $(u_j^2/2 + \Pi)_{,iq}$ is symmetric in q and i. The third term on the left side of (5.26) can be written as

$$-\varepsilon_{nqi}\varepsilon_{ijk}\left[u_j\left(\omega_k + 2\Omega_k\right)\right]_{,q} = -(\delta_{nj}\delta_{qk} - \delta_{nk}\delta_{qj})\left[u_j(\omega_k + 2\Omega_k)\right]_{,q}$$

$$= -\left[u_n(\omega_k + 2\Omega_k)\right]_{,k} + \left[u_j(\omega_n + 2\Omega_n)\right]_{,j}$$

$$= -u_n(\omega_{k,k} + 2\Omega_{k,k}) - u_{n,k}(\omega_k + 2\Omega_k) + u_j(\omega_n + 2\Omega_n)_{,j}$$

$$= -u_n(0 + 0) - u_{n,k}(\omega_k + 2\Omega_k) + u_j(\omega_n + 2\Omega_n)_{,j}$$

$$= -u_{n,j}(\omega_j + 2\Omega_j) + u_j\,\omega_{n,j}, \quad (5.27)$$

where we have used $u_{i,i} = 0$, $\omega_{i,i} = 0$ and the fact that the derivatives of Ω are zero.

The first term on the right-hand side of (5.26) can be written as

$$-\varepsilon_{nqi}\left(\frac{1}{\rho}p_{,i}\right)_{,q} = -\frac{1}{\rho}\varepsilon_{nqi}\,p_{,iq} + \frac{1}{\rho^2}\varepsilon_{nqi}\rho_{,q}p_{,i}$$

$$= 0 + \frac{1}{\rho^2}[\nabla\rho \times \nabla p]_n, \tag{5.28}$$

which involves the n-component of the vector $\nabla\rho \times \nabla p$. The viscous term in (5.26) can be written as

$$-\nu\varepsilon_{nqi}\varepsilon_{ijk}\omega_{k,jq} = -\nu(\delta_{nj}\delta_{qk} - \delta_{nk}\delta_{qj})\omega_{k,jq}$$

$$= -\nu\omega_{k,nk} + \nu\omega_{n,jj} = \nu\omega_{n,jj}. \tag{5.29}$$

If we use (5.27) through (5.29), then (5.26) becomes

$$\frac{\partial\omega_n}{\partial t} = u_{n,j}(\omega_j + 2\Omega_j) - u_j\omega_{n,j} + \frac{1}{\rho^2}[\nabla\rho \times \nabla p]_n + \nu\omega_{n,jj}.$$

Changing the free index from n to i produces

$$\frac{D\omega_i}{Dt} = (\omega_j + 2\Omega_j)u_{i,j} + \frac{1}{\rho^2}[\nabla\rho \times \nabla p]_i + \nu\omega_{i,jj}.$$

In vector notation this can be written:

$$\frac{D\mathbf{\omega}}{Dt} = (\mathbf{\omega} + 2\mathbf{\Omega})\cdot\nabla\mathbf{u} + \frac{1}{\rho^2}\nabla\rho \times \nabla p + \nu\nabla^2\mathbf{\omega}. \tag{5.30}$$

This is the *vorticity equation* for a nearly incompressible (i.e., Boussinesq) fluid observed from a frame of reference rotating at a constant rate Ω. Here $\mathbf{u}$ and $\mathbf{\omega}$ are, respectively, the velocity and vorticity observed in this rotating frame of reference. As vorticity is defined as twice the angular velocity, 2Ω is the *planetary vorticity* and $(\mathbf{\omega} + 2\mathbf{\Omega})$ is the *absolute vorticity* of the fluid, measured in an inertial frame. In a nonrotating frame, the vorticity equation is obtained from (5.30) by setting Ω to zero and interpreting $\mathbf{u}$ and $\mathbf{\omega}$ as the absolute velocity and vorticity, respectively.

The left side of (5.30) represents the rate of change of vorticity following a fluid particle. The last term $\nu\nabla^2\mathbf{\omega}$ represents the rate of change of $\mathbf{\omega}$ due to molecular diffusion of vorticity, in the same way that $\nu\nabla^2\mathbf{u}$ represents acceleration due to diffusion of velocity. The second term on the right-hand side is the rate of generation of vorticity due to baroclinicity of the flow, as discussed in Section 5.2. In a barotropic flow, density is a function of pressure alone, so $\nabla\rho$ and ∇p are parallel vectors. The first term on the right side of (5.30) represents vortex stretching and plays a crucial role in the dynamics of vorticity even when $\Omega = 0$.

To better understand the vortex-stretching term, consider the natural coordinate system where s is the arc length along a vortex line, n points away from the center of vortex-line curvature, and m lies along the second normal to s (Figure 5.9). Then,

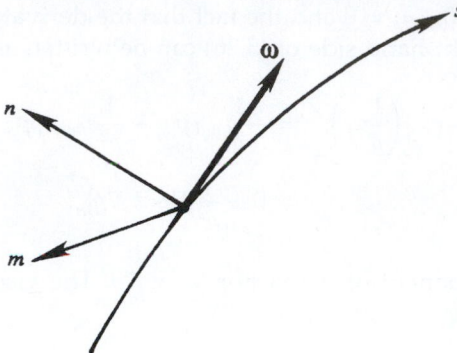

FIGURE 5.9   Coordinate system aligned with the vorticity vector.

$$(\boldsymbol{\omega}\cdot\nabla)\mathbf{u} \;=\; \left[\boldsymbol{\omega}\cdot\left(\mathbf{e}_s\frac{\partial}{\partial s}+\mathbf{e}_n\frac{\partial}{\partial n}+\mathbf{e}_m\frac{\partial}{\partial m}\right)\right]\mathbf{u} \;=\; \omega\frac{\partial}{\partial s}, \tag{5.31}$$

where we have used $\boldsymbol{\omega}\cdot\mathbf{e}_n=\boldsymbol{\omega}\cdot\mathbf{e}_m=0$, and $\boldsymbol{\omega}\cdot\mathbf{e}_s=\omega=|\boldsymbol{\omega}|$. Equation (5.31) shows that $(\boldsymbol{\omega}\cdot\nabla)\mathbf{u}$ equals the magnitude of $\boldsymbol{\omega}$ times the derivative of $\mathbf{u}$ in the direction of $\boldsymbol{\omega}$. The quantity $\omega(\partial\mathbf{u}/\partial s)$ is a vector and has the components $\omega(\partial u_s/\partial s)$, $\omega(\partial u_n/\partial s)$, and $\omega(\partial u_m/\partial s)$. Among these, $\partial u_s/\partial s$ represents the increase of u_s along the vortex line s, that is, the stretching of vortex lines. On the other hand, $\partial u_n/\partial s$ and $\partial u_m/\partial s$ represent the change of the normal velocity components along s and, therefore, the rate of turning or tilting of vortex lines about the m and n axes, respectively.

To see the effect of these terms more clearly, write out the components of (5.30) for barotropic inviscid flow observed in an inertial frame of reference:

$$\frac{D\omega_s}{Dt}=\omega\frac{\partial u_s}{\partial s}, \quad \frac{D\omega_n}{Dt}=\omega\frac{\partial u_n}{\partial s}, \quad \text{and} \quad \frac{D\omega_m}{Dt}=\omega\frac{\partial u_m}{\partial s}. \tag{5.32}$$

The first equation of (5.32) shows that the vorticity along s changes due to stretching of vortex lines, reflecting the principle of conservation of angular momentum. Stretching decreases the moment of inertia of fluid elements that constitute a vortex line, resulting in an increase of their angular rotation speed. Vortex stretching plays an especially crucial role in the dynamics of turbulent and geophysical flows. The second and third equations of (5.32) show how vorticity along n and m is created by the tilting of vortex lines. For example, in Figure 5.9, the turning of the vorticity vector $\boldsymbol{\omega}$ toward the n-axis will generate a vorticity component along n. *The vortex stretching and tilting term $(\boldsymbol{\omega}\cdot\nabla)\mathbf{u}$ is absent in two-dimensional flows, in which $\boldsymbol{\omega}$ is perpendicular to the plane of flow.*

To better understand how frame rotation influences vorticity, consider $\boldsymbol{\Omega}=\Omega\mathbf{e}_z$ so that $2(\boldsymbol{\Omega}\cdot\nabla)\mathbf{u}=2\Omega(\partial\mathbf{u}/\partial z)$ and suppress all other terms on the right side of (5.30) to obtain the component equations:

$$\frac{D\omega_z}{Dt}=2\Omega\frac{\partial w}{\partial z}, \quad \frac{D\omega_x}{Dt}=2\Omega\frac{\partial u}{\partial z}, \quad \text{and} \quad \frac{D\omega_y}{Dt}=2\Omega\frac{\partial v}{\partial z}.$$

FIGURE 5.10 Generation of relative vorticity due to stretching of fluid columns parallel to the planetary vorticity 2Ω. A fluid column acquires ω_z (in the same sense as Ω) by moving from location A to location B.

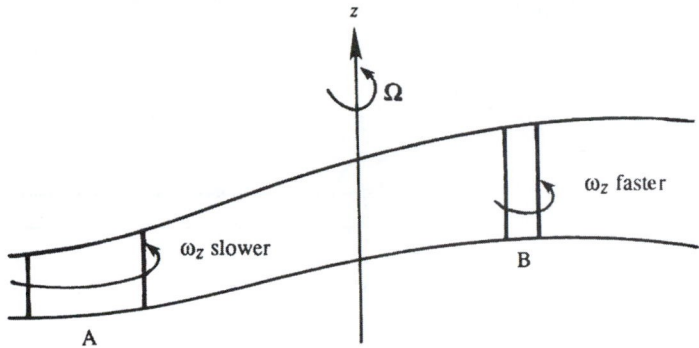

This shows that stretching of fluid lines in the z direction increases ω_z, whereas a tilting of vertical lines changes the relative vorticity along the x and y directions. Note that merely stretching or turning of vertical *fluid lines* is required for this mechanism to operate, in contrast to $(\boldsymbol{\omega} \cdot \nabla) \mathbf{u}$ where a stretching or turning of *vortex lines* is needed. This is because vertical fluid lines contain the planetary vorticity 2Ω. A vertically stretching fluid column tends to acquire positive ω_z, and a vertically shrinking fluid column tends to acquire negative ω_z (Figure 5.10). For this reason large-scale geophysical flows are almost always full of vorticity, and the change of $\boldsymbol{\omega}$ due to the presence of planetary vorticity 2Ω is a central feature of geophysical fluid dynamics.

Kelvin's circulation theorem for inviscid flow in a rotating frame of reference is modified to

$$\frac{D\Gamma_a}{Dt} = 0 \quad \text{where} \quad \Gamma_a \equiv \int_A (\boldsymbol{\omega} + 2\Omega) \cdot \mathbf{n} dA = \Gamma + 2\int_A \Omega \cdot \mathbf{n} dA \qquad (5.33)$$

(see Exercise 5.11). Here, Γ_a is circulation due to the absolute vorticity $(\boldsymbol{\omega} + 2\Omega)$ and differs from Γ by the amount of planetary vorticity intersected by the area A.

5.7. INTERACTION OF VORTICES

Vortices placed close to one another can mutually interact through their induced velocities and generate interesting motions. To examine such interactions, consider ideal concentrated-line vortices. A real vortex, with a core within which vorticity is distributed, can be idealized by a concentrated vortex line with circulation equal to the average vorticity in the core times the core area. Motion outside the vortex core is assumed irrotational, and therefore inviscid. It will be shown in the next chapter that irrotational motion of a constant density fluid is governed by the linear Laplace equation so the principle of superposition applies, and the velocity at a point can be obtained by adding the contribution of all vortices in the field. To determine the mutual interaction of line vortices, the important principle to keep in mind is the first Helmholtz vortex theorem—vortex lines move with the flow.

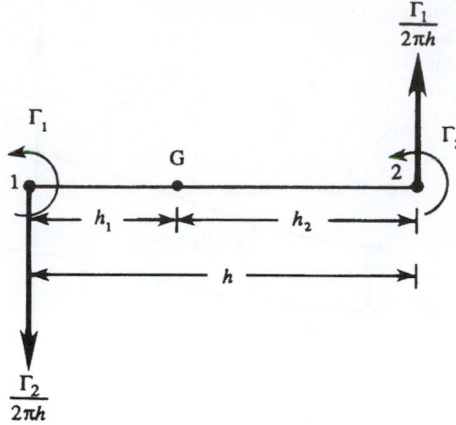

FIGURE 5.11   Interaction of two line vortices of the same sign. Here the induced velocities are in opposite directions and perpendicular to the line connecting the vortices. Thus, if free to move, the two vortices will travel on circular paths centered on the point G where the combined velocity induced by the two vortices is zero.

Consider the interaction of two ideal line vortices of strengths Γ_1 and Γ_2, where both Γ_1 and Γ_2 are positive (i.e., counterclockwise vorticity). Let $h = h_1 + h_2$ be the distance between the vortices (Figure 5.11). Then the velocity at point 2 due to vortex Γ_1 is directed upward and equals

$$V_1 = \Gamma_1/2\pi h.$$

Similarly, the velocity at point 1 due to vortex Γ_2 is downward and equals

$$V_2 = \Gamma_2/2\pi h.$$

The vortex pair therefore rotates counterclockwise around this center of vorticity G, which remains stationary.

Now suppose that the two vortices have the same circulation of magnitude Γ, but an opposite sense of rotation (Figure 5.12). Then the velocity of each vortex at the location of the other is $\Gamma/(2\pi h)$ so the dual-vortex system translates at a speed $\Gamma/(2\pi h)$ relative to the fluid. A pair of counter-rotating vortices can be set up by stroking the paddle of a boat, or by briefly moving the blade of a knife in a bucket of water (Figure 5.13). After the paddle or knife is withdrawn, the vortices do not remain stationary but continue to move.

The behavior of a single vortex near a wall can be found by superposing two vortices of equal and opposite strength. The technique involved is called the *method of images* and has wide application in irrotational flow, heat conduction, acoustics, and electromagnetism. It is clear that the inviscid flow pattern due to vortex A at distance h from a wall can be obtained by eliminating the wall and introducing instead a vortex of equal and opposite strength at the image point B (Figure 5.14). The velocity at any point P on the wall, made up of V_A due to the real vortex and V_B due to the image vortex, is parallel to the wall. The wall is therefore a streamline, and the inviscid boundary condition of zero normal velocity across a solid wall is satisfied. Because of the flow induced by the image vortex, vortex A moves with speed

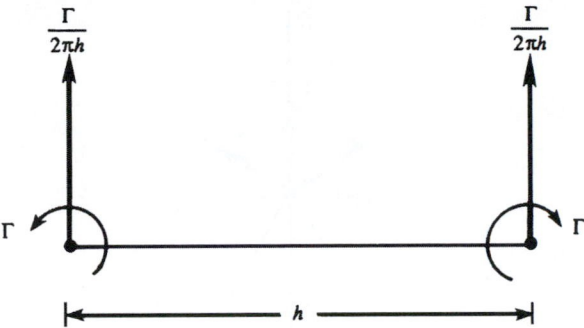

FIGURE 5.12 Interaction of line vortices of opposite spin, but of the same magnitude. Here Γ refers to the *magnitude* of circulation, and the induced velocities are in same direction and perpendicular to the line connecting the vortices. Thus, if free to move, the two vortices will travel along straight lines in the direction shown at speed $\Gamma/2\pi h$.

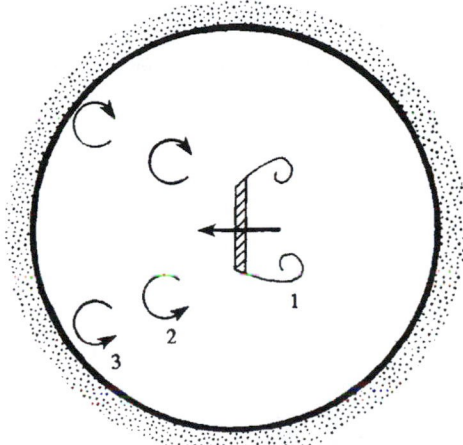

FIGURE 5.13 Top view of a vortex pair generated by moving the blade of a knife in a bucket of water. Positions at three instances of time 1, 2, and 3 are shown. (*After Lighthill, 1986.*)

$\Gamma/(4\pi h)$ parallel to the wall. For this reason, vortices in the example of Figure 5.13 move apart along the boundary on reaching the side of the vessel.

Now consider the interaction of two doughnut-shaped vortex rings (such as smoke rings) of equal and opposite circulation (Figure 5.15a). According to the method of images, the flow field for a single ring near a wall is identical to the flow of two rings of opposite circulations. The translational motion of each element of the ring is caused by the induced velocity from each element of the same ring, plus the induced velocity from each element of the other vortex ring. In the figure, the motion at A is the resultant of V_B, V_C, and V_D, and this resultant has components parallel to and toward the wall. Consequently, the vortex ring increases in diameter and moves toward the wall with a speed that decreases monotonically (Figure 5.15b).

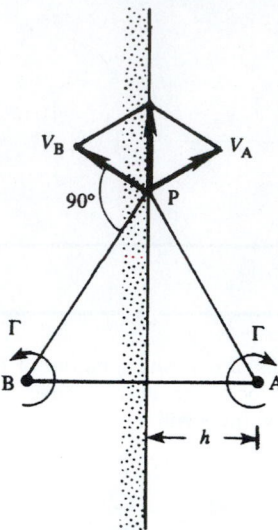

FIGURE 5.14   Line vortex A near a wall and its image B. The sum of the induced velocities is parallel to the wall at all points P on the wall when the two vortices have equal and opposite strengths and they are equidistant from the wall.

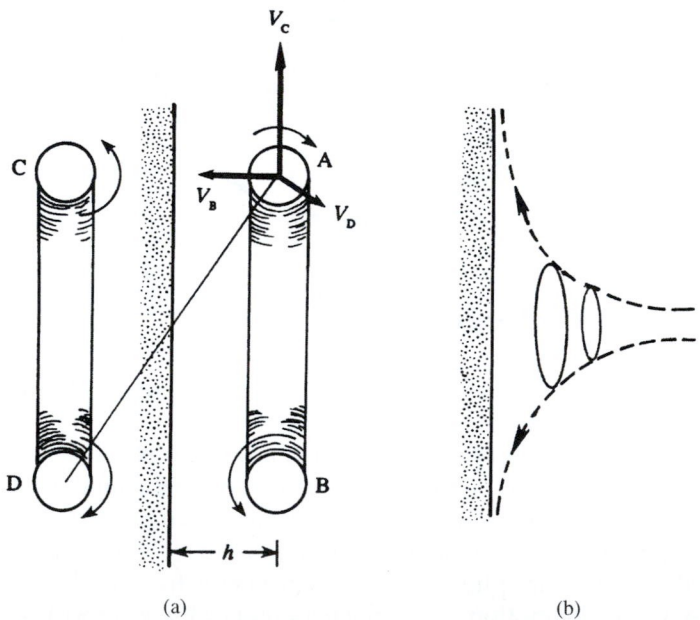

(a)   (b)

FIGURE 5.15   (a) Torus or doughnut-shaped vortex ring near a wall and its image. A section through the middle of the ring is shown along with primary induced velocities at A from the vortex segments located at B, C, and D. (b) Trajectory of a vortex ring, showing that it widens while its translational velocity toward the wall decreases.

Finally, consider the interaction of two vortex rings of equal magnitude and similar sense of rotation. It is left to the reader (Exercise 5.15) to show that they should both translate in the same direction, but the one in front increases in radius and therefore slows down in its translational speed, while the rear vortex contracts and translates faster. This continues until the smaller ring passes through the larger one, at which point the roles of the two vortices are reversed. The two vortices can pass through each other forever in an ideal fluid. Further discussion of this intriguing problem can be found in Sommerfeld (1964, p. 161).

5.8. VORTEX SHEET

Consider an infinite number of ideal line vortices, placed side by side on a surface AB (Figure 5.16). Such a surface is called a *vortex sheet*. If the vortex filaments all rotate clockwise, then the tangential velocity immediately above AB is to the right, while that immediately below AB is to the left. Thus, a discontinuity of tangential velocity exists across a vortex sheet. If the vortex filaments are not infinitesimally thin, then the vortex sheet has a finite thickness, and the velocity change is spread out.

In Figure 5.16, consider the circulation around a circuit of dimensions dn and ds. The normal velocity component v is continuous across the sheet ($v = 0$ if the sheet does not move normal to itself), while the tangential component u experiences a sudden jump. If u_1 and u_2 are the tangential velocities on the two sides, then

$$d\Gamma = u_2\,ds + v\,dn - u_1\,ds - v\,dn = (u_2 - u_1)\,ds.$$

Therefore the circulation per unit length, called the *strength of a vortex sheet*, equals the jump in tangential velocity:

$$\Gamma \equiv \frac{d\Gamma}{ds} = u_2 - u_1.$$

The concept of a vortex sheet is especially useful in discussing the flow over aircraft wings (Chapter 14).

FIGURE 5.16 A vortex sheet produces a change in the velocity that is tangent to it. Vortex sheets may be formed by placing many parallel ideal line vortices next to each other. The strength of a vortex sheet, $d\Gamma/ds = u_1 - u_2$, can be determined by computing the circulation on the rectangular contour shown and this strength may depend on the sheet-tangent coordinate.

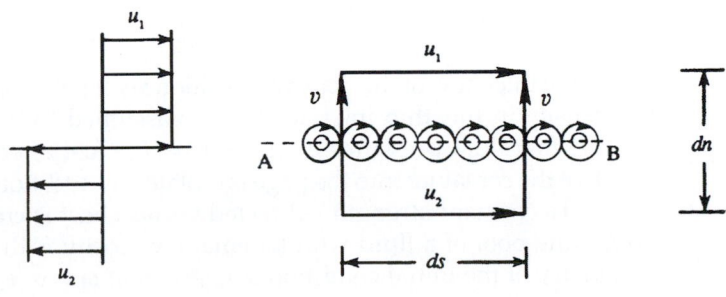

EXERCISES

5.1. A closed cylindrical tank 4 m high and 2 m in diameter contains water to a depth of 3 m. When the cylinder is rotated at a constant angular velocity of 40 rad/s, show that nearly 0.71 m^2 of the bottom surface of the tank is uncovered. [*Hint:* The free surface is in the form of a paraboloid. For a point on the free surface $p = p_o$, let h be the height above the (imaginary) vertex of the paraboloid and r be the local radius of the paraboloid. From Section 5.1 we have $h = \omega_0^2 r^2 / 2g$, where ω_0 is the angular velocity of the tank. Apply this equation to the two points where the paraboloid cuts the top and bottom surfaces of the tank.]

5.2. A tornado can be idealized as a Rankine vortex with a core of diameter 30 m. The gauge pressure at a radius of 15 m is -2000 N/m^2 (i.e., the absolute pressure is 2000 N/m^2 below atmospheric).

   (a) Show that the circulation around any circuit surrounding the core is 5485 m^2/s. [*Hint:* Apply the Bernoulli equation between infinity and the edge of the core.]

   (b) Such a tornado is moving at a linear speed of 25 m/s relative to the ground. Find the time required for the gauge pressure to drop from -500 to -2000 N/m^2. Neglect compressibility effects and assume an air temperature of 25°C. (Note that the tornado causes a sudden decrease of the local atmospheric pressure. The damage to structures is often caused by the resulting excess pressure on the inside of the walls, which can cause a house to explode.)

5.3. The velocity field of a flow in cylindrical coordinates (R, φ, z) is $\mathbf{u} = (u_R, u_\varphi, u_z) = (0, aRz, 0)$ where a is a constant.

   (a) Show that the vorticity components are $\boldsymbol{\omega} = (\omega_R, \omega_\varphi, \omega_z) = (-aR, 0, 2az)$.

   (b) Verify that $\nabla \cdot \boldsymbol{\omega} = 0$.

   (c) Sketch the streamlines and vortex lines in an Rz plane. Show that the vortex lines are given by $zR^2 = $ constant.

5.4. Starting from the flow field of an ideal vortex (5.2), compute the viscous stresses σ_{rr}, $\sigma_{r\theta}$, and $\sigma_{\theta\theta}$, and show that the net viscous force on a fluid element, $(\partial \sigma_{ij}/\partial x_i)$, is zero.

5.5. Consider the situation depicted in Figure 5.5. Use a Cartesian coordinate system with a horizontal x-axis that puts the barrier at $x = 0$, a vertical y-axis that puts the bottom of the container at $y = 0$ and the top of the container at $y = H$, and a z-axis that points out of the page. Show that, at the instant the barrier is removed, the rate of baroclinic vorticity production at the interface between the two fluids is:

$$\frac{D\omega_z}{Dt} = \frac{2(\rho_2 - \rho_1)g}{(\rho_2 + \rho_1)\delta},$$

where the thickness of the density transition layer just after barrier removal is $\delta \ll H$, and the density in this thin interface layer is assumed to be $(\rho_1 + \rho_2)/2$. If necessary, also assume that fluid pressures match at $y = H/2$ just after barrier removal, and that the width of the container into the page is b. State any additional assumptions that you make.

5.6. At $t = 0$ a constant-strength z-directed vortex sheet is created in an x-z plane ($y = 0$) in an infinite pool of a fluid with kinematic viscosity ν, that is, $\boldsymbol{\omega}(y,0) = \mathbf{e}_z \gamma \delta(y)$. The symmetry of the initial condition suggests that $\boldsymbol{\omega} = \omega_z \mathbf{e}_z$ and that ω_z will only depend on y and t. Determine $\boldsymbol{\omega}(y,t)$ for $t > 0$ via the following steps.

a) Determine a dimensionless scaling law for ω_z in terms of γ, ν, y, and t.

b) Simplify the general vorticity equation (5.13) to a linear field equation for ω_z for this situation.

c) Based on the fact that the field equation is linear, simplify the result of part a) by requiring ω_z to be proportional to γ, plug the simplified dimensionless scaling law into the equation determined for part b), and solve this equation to find the undetermined function to reach:

$$\omega_z(y,t) = \frac{\gamma}{2\sqrt{\pi\nu t}} \exp\left\{-\frac{y^2}{4\nu t}\right\}$$

5.7. [1]a) Starting from the continuity and Euler equations for an inviscid compressible fluid, $\partial\rho/\partial t + \nabla\cdot(\rho\mathbf{u}) = 0$ and $\rho(D\mathbf{u}/Dt) = -\nabla p + \rho\mathbf{g}$, derive the Vazsonyi equation:

$$\frac{D}{Dt}\left(\frac{\boldsymbol{\omega}}{\rho}\right) = \left(\frac{\boldsymbol{\omega}}{\rho}\right)\cdot\nabla u + \frac{1}{\rho^3}\nabla\rho \times \nabla p,$$

when the body force is conservative: $\mathbf{g} = -\nabla\Phi$. This equation shows that $\boldsymbol{\omega}/\rho$ in a compressible flow plays nearly the same dynamic role as $\boldsymbol{\omega}$ in an incompressible flow [see (5.30) with $\boldsymbol{\Omega} = 0$ and $\nu = 0$].

b) Show that the final term in the Vazsonyi equation may also be written: $(1/\rho)\nabla T \times \nabla s$.

c) Simplify the Vazsonyi equation for barotropic flow.

5.8. Starting from the unsteady momentum equation for a compressible fluid with constant viscosities:

$$\rho\frac{D\mathbf{u}}{Dt} + \nabla p = \rho\mathbf{g} + \mu\nabla^2\mathbf{u} + \left(\mu_v + \frac{1}{3}\mu\right)\nabla(\nabla\cdot\mathbf{u}),$$

show that

$$\frac{\partial\mathbf{u}}{\partial t} + \boldsymbol{\omega} \times \mathbf{u} = T\nabla s - \nabla\left(h + \frac{1}{2}|\mathbf{u}|^2 + \Phi\right) - \frac{\mu}{\rho}\nabla \times \boldsymbol{\omega} + \frac{1}{\rho}\left(\mu_v + \frac{4}{3}\mu\right)\nabla(\nabla\cdot\mathbf{u})$$

where T = temperature, h = enthalpy per unit mass, s = entropy per unit mass, and the body force is conservative: $\mathbf{g} = -\nabla\Phi$. This is the viscous Crocco-Vazsonyi equation. Simplify this equation for steady inviscid non-heat-conducting flow to find the Bernoulli equation (4.78), $h + \frac{1}{2}|\mathbf{u}|^2 + \Phi$ = constant along a streamline, which is valid when the flow is rotational and nonisothermal.

5.9. a) Solve $\nabla^2 G(\mathbf{x},\mathbf{x}') = \delta(\mathbf{x} - \mathbf{x}')$ for $G(\mathbf{x},\mathbf{x}')$ in a uniform, unbounded three-dimensional domain, where $\delta(\mathbf{x} - \mathbf{x}') = \delta(x - x')\delta(y - y')\delta(z - z')$ is the three-dimensional Dirac delta function.

b) Use the result of part a) to show that: $\phi(\mathbf{x}) = -\frac{1}{4\pi}\int\limits_{all\ \mathbf{x}'} \frac{q(\mathbf{x}')}{|\mathbf{x} - \mathbf{x}'|}d^3x'$ is the solution of the Poisson equation $\nabla^2\phi(\mathbf{x}) = q(\mathbf{x})$ in a uniform, unbounded three-dimensional domain.

[1]Obtained from Professor Paul Dimotakis

5.10. Start with the equations of motion in the rotating steadily coordinates, and prove Kelvin's circulation theorem $\frac{D}{Dt}(\Gamma_a) = 0$ where $\Gamma_a = \int (\omega + 2\Omega) \cdot d\mathbf{A}$. Assume that Ω is constant, the flow is inviscid and barotropic and that the body forces are conservative. Explain the result physically.

5.11. In (R,φ,z) cylindrical coordinates, consider the radial velocity $u_R = -R^{-1}(\partial\psi/\partial z)$, and the axial velocity $u_z = R^{-1}(\partial\psi/\partial R)$ determined from the axisymmetric stream function

$$\psi(R,z) = \frac{Aa^4}{10}\left(\frac{R^2}{a^2}\right)\left(1 - \frac{R^2}{a^2} - \frac{z^2}{a^2}\right) \text{ where } A \text{ is a constant. This flow is known as } Hill's$$

spherical vortex.

  a) For $R^2 + z^2 \le a^2$, sketch the streamlines of this flow in a plane that contains the z-axis. What does a represent?

  b) Determine $\mathbf{u} = u_R(R,z)\mathbf{e}_R + u_z(R,z)\mathbf{e}_z$.

  c) Given $\omega_\varphi = (\partial u_R/\partial z) - (\partial u_z/\partial R)$, show that $\boldsymbol{\omega} = AR\mathbf{e}_\varphi$ in this flow and that this vorticity field is a solution of the vorticity equation (5.13).

  d) Does this flow include stretching of vortex lines?

5.12. In (R,φ,z) cylindrical coordinates, consider the flow field $u_R = -\alpha R/2$, $u_\phi = 0$, and $u_z = \alpha z$.

  a) Compute the strain rate components S_{RR}, S_{zz}, and S_{Rz}. What sign of α causes fluid elements to elongate in the z direction? Is this flow incompressible?

  b) Show that it is possible for a steady vortex (a Burgers' vortex) to exist in this flow field by adding $u_\varphi = (\Gamma/2\pi R)[1 - \exp(-\alpha R^2/4\nu)]$ to u_R and u_z from part a) and then determining a pressure field $p(R,z)$ that together with $\mathbf{u} = (u_R, u_\varphi, u_z)$ solves the Navier-Stokes momentum equation for a fluid with constant density ρ and kinematic viscosity ν.

  c) Determine the vorticity in the Burgers' vortex flow of part b).

  d) Explain how the vorticity distribution can be steady when $\alpha \ne 0$ and fluid elements are stretched or compressed.

  e) Interpret what is happening in this flow when $\alpha > 0$ and when $\alpha < 0$.

5.13. An ideal line vortex parallel to the z-axis of strength Γ intersects the x-y plane at $x = 0$ and $y = h$. Two solid walls are located at $y = 0$ and $y = H > 0$. Use the method of images for the following.

  a) Based on symmetry arguments, determine the horizontal velocity u of the vortex when $h = H/2$.

  b) Show that for $0 < h < H$ the horizontal velocity of the vortex is:

$$u(0,h) = \frac{\Gamma}{4\pi h}\left(1 - 2\sum_{n=1}^{\infty}\frac{1}{(nH/h)^2 - 1}\right),$$

and evaluate the sum when $h = H/2$ to verify your answer to part a).

5.14. The axis of an infinite solid circular cylinder with radius a coincides with the z-axis. The cylinder is stationary and immersed in an incompressible inviscid fluid, and the net circulation around it is zero. An ideal line vortex parallel to the cylinder with circulation Γ passes through the x-y plane at $x = L > a$ and $y = 0$. Here two image vortices are needed to satisfy the boundary condition on the cylinder's surface. If one of these is located at $x = y = 0$ and has strength Γ, determine the strength and location of the second image vortex.

5.15. Consider the interaction of two vortex rings of equal strength and similar sense of rotation. Argue that they go through each other, as described near the end of Section 5.7.

5.16. A constant-density irrotational flow in a rectangular torus has a circulation Γ and volumetric flow rate Q. The inner radius is r_1, the outer radius is r_2, and the height is h. Compute the total kinetic energy of this flow in terms of only ρ, Γ, and Q.

5.17. Consider a cylindrical tank of radius R filled with a viscous fluid spinning steadily about its axis with constant angular velocity $\mathbf{\Omega}$. Assume that the flow is in a steady state.

  (a) Find $\int_A \boldsymbol{\omega} \cdot dA$ where A is a horizontal plane surface through the fluid normal to the axis of rotation and bounded by the wall of the tank.

  (b) The tank then stops spinning. Find again the value of $\int_A \boldsymbol{\omega} \cdot dA$.

5.18. In Figure 5.11, locate point G.

5.19. Consider two-dimensional steady flow in the x-y plane outside of a long circular cylinder of radius a that is centered on and rotating about the z-axis at a constant angular rate of Ω_z. Show that the fluid velocity on the x-axis is $\mathbf{u}(x,0) = (\Omega_z a^2/x)\mathbf{e}_y$ for $x > a$ when the cylinder is replaced by:

  a) A circular vortex sheet of radius a with strength $\gamma = \Omega_z a$

  b) A circular region of uniform vorticity $\boldsymbol{\omega} = 2\Omega_z \mathbf{e}_z$ with radius a.

  c) Describe the flow for $x^2 + y^2 < a^2$ for parts a) and b).

5.20. An ideal line vortex in a half space filled with an inviscid constant-density fluid has circulation Γ, lies parallel to the z-axis, and passes through the x-y plane at $x = 0$ and $y = h$. The plane defined by $y = 0$ is a solid surface.

  a) Use the method of images to find $\mathbf{u}(x,y)$ for $y > 0$ and show that the fluid velocity on $y = 0$ is $\mathbf{u}(x,0) = \Gamma h \mathbf{e}_x / [\pi(x^2 + h^2)]$.

  b) Show that $\mathbf{u}(0,y)$ is unchanged for $y > 0$ if the image vortex is replaced by a vortex sheet of strength $\gamma(x) = -u(x,0)\mathbf{e}_z$ on $y = 0$.

  c) (If you have the patience) Repeat part b) for $\mathbf{u}(x,y)$ when $y > 0$.

Literature Cited

Lighthill, M. J. (1986). *An Informal Introduction to Theoretical Fluid Mechanics.* Oxford, England: Clarendon Press.
Sommerfeld, A. (1964). *Mechanics of Deformable Bodies.* New York: Academic Press.

Supplemental Reading

Batchelor, G. K. (1967). *An Introduction to Fluid Dynamics*. London: Cambridge University Press.

Pedlosky, J. (1987). *Geophysical Fluid Dynamics*. New York: Springer-Verlag. (This book discusses the vorticity dynamics in rotating coordinates, with application to geophysical systems.)

Prandtl, L., & Tietjens, O. G. (1934). *Fundamentals of Hydro- and Aeromechanics*. New York: Dover Publications. (This book contains a good discussion of the interaction of vortices.)

Saffman, P. G. (1992). *Vortex Dynamics*. Cambridge: Cambridge University Press. (This book presents a wide variety of vortex topics from an applied mathematics perspective.)

CHAPTER

6

Ideal Flow

OUTLINE

6.1. Relevance of Irrotational Constant-Density Flow Theory   198

6.2. Two-Dimensional Stream Function and Velocity Potential   200

6.3. Construction of Elementary Flows in Two Dimensions   203

6.4. Complex Potential   216

6.5. Forces on a Two-Dimensional Body   219

6.6. Conformal Mapping   222

6.7. Numerical Solution Techniques in Two Dimensions   225

6.8. Axisymmetric Ideal Flow   231

6.9. Three-Dimensional Potential Flow and Apparent Mass   236

6.10. Concluding Remarks   240

Exercises   241

Literature Cited   251

Supplemental Reading   251

CHAPTER OBJECTIVES

- To describe the formulation and limitations of ideal flow theory.

- To illustrate the use of the stream function and the velocity potential in two-dimensional, axisymmetric, and three-dimensional flows.

- To derive and present classical ideal flow results for flows past simple objects.

6.1. RELEVANCE OF IRROTATIONAL CONSTANT-DENSITY FLOW THEORY

When a constant-density fluid flows without rotation, and pressure is measured relative to its local hydrostatic value (see Section 4.9), the equations of fluid motion in an inertial frame of reference, (4.7) and (4.38), simplify to:

$$\nabla \cdot \mathbf{u} = 0 \text{ and } \rho(D\mathbf{u}/Dt) = -\nabla p, \qquad\qquad (4.10, 6.1)$$

even though the fluid's viscosity μ may be nonzero. These are the equations of *ideal* flow. They are useful for developing a first-cut understanding of nearly any macroscopic fluid flow, and are directly applicable to low-Mach-number irrotational flows of homogeneous fluids away from solid boundaries. Ideal flow theory has abundant applications in the exterior aero- and hydrodynamics of moderate- to large-scale objects at nontrivial subsonic speeds. Here, moderate size (L) and nontrivial speed (U) are determined jointly by the requirement that the Reynolds number, $\text{Re} = \rho UL/\mu$ (4.103), be large enough (typically $\text{Re} \sim 10^3$ or greater) so that the combined influence of fluid viscosity and fluid element rotation is confined to thin layers on solid surfaces, commonly known as *boundary layers*.

The conditions necessary for the application of ideal flow theory are commonly present on the upstream side of many ordinary objects, and may even persist to the downstream side of some. Ideal flow analysis can predict fluid velocity away from solid surfaces, surface-normal pressure forces (when the boundary layer is thin and attached), acoustic streamlines, flow patterns that minimize form drag, and unsteady-flow, fluid-inertia effects. Ideal flow theory does not predict viscous effects like skin friction or energy dissipation, so it is not directly applicable to interior flows in pipes and ducts, to boundary layer flows, or to any rotational flow region. This final specification excludes low-Re flows and regions of turbulence.

Because (6.1) involves only first-order spatial derivatives, ideal flows only satisfy the no-through-flow boundary condition on solid surfaces. The no-slip boundary condition is not applied in ideal flows, so nonzero tangential velocity at a solid surface may exist

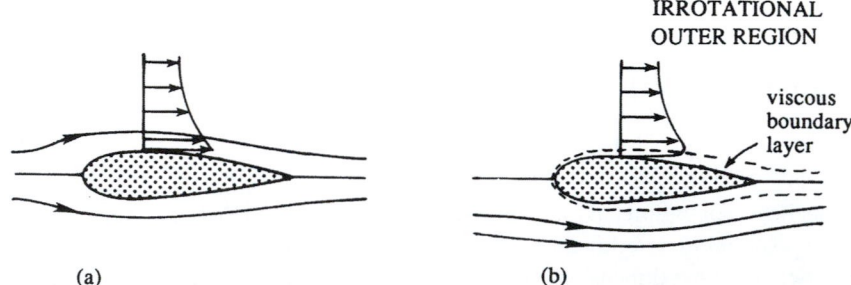

(a)   (b)

FIGURE 6.1 Comparison of a completely irrotational constant-density (ideal) flow (a) and a high Reynolds number flow (b). In both cases the no-through-flow boundary condition is applied. However, the ideal flow is effectively inviscid and the fluid velocity is tangent to and nonzero on the body surface. The high-Re flow includes thin boundary layers where fluid rotation and viscous effects are prevalent, and the non-slip boundary condition is enforced, but the velocity above the thin boundary layer is similar to that in the ideal flow.

(Figure 6.1a). In contrast, a real fluid with a nonzero shear viscosity must satisfy a no-slip boundary condition because (4.38) contains second-order spatial derivatives. At sufficiently high Re, there are two primary differences between ideal and real flows over the same object. First, viscous boundary layers containing rotational fluid form on solid surfaces in the real flow, and the thickness of such boundary layers, within which viscous diffusion of vorticity is important, approaches zero as Re $\rightarrow \infty$ (Figure 6.1b). The second difference is the possible formation in the real flow of *separated flow* or *wake* regions that occur when boundary layers leave the surface on which they have developed to create a wider zone of rotational flow (Figure 6.2). Ideal flow theory is not directly applicable to such layers or regions of rotational flow. However, rotational flow regions may be easy to anticipate or identify, and may represent a small fraction of a total flow field so that predictions from ideal flow theory may remain worthwhile even when viscous flow phenomena are present. Further discussion of viscous-flow phenomena is provided in Chapters 8 and 9.

For (4.10) and (6.1) to apply, fluid density ρ must be constant and the flow must be irrotational. If the flow is merely incompressible and contains baroclinic density variations, (4.10) will still be satisfied but (6.1) will not; it will need a body-force term like that in (4.84) and the reference pressure would have to be redefined. If the fluid is a homogeneous compressible gas with sound speed c, the constant density requirement will be satisfied when the Mach number, $M = U/c$ (4.111), of the flow is much less than unity. The irrotationality condition is satisfied when fluid elements enter the flow field of interest without rotation and do not acquire any while they reside in it. Based on Kelvin's circulation theorem (5.11) for constant density flow, this is possible when the body force is conservative and the net viscous force on a fluid element is zero. Thus, a fluid element that is initially irrotational is likely to stay that way unless it enters a boundary layer, wake, or separated flow region where it acquires rotation via viscous diffusion. So, when initially irrotational fluid flows over a solid object, ideal flow theory most readily applies to the *outer* region of the flow away from the object's surface(s) where the flow is irrotational. Viscous flow theory is needed in the *inner* region where viscous diffusion of vorticity is important. Often, at high Re, the outer flow can be approximately predicted by ignoring the existence of viscous boundary layers. With this outer flow prediction, viscous flow equations can be solved for the boundary-layer flow and, under the right conditions, the two solutions can be adjusted until they match in a suitable region of overlap. This approach works well for objects like thin airfoils at low angles of

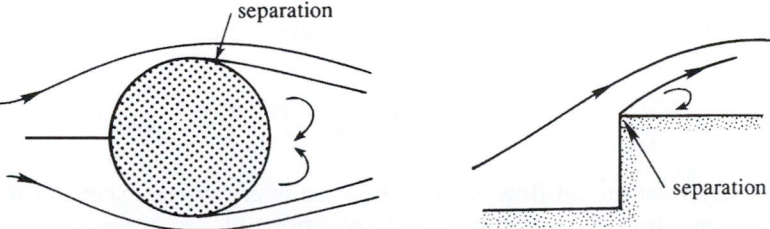

FIGURE 6.2   Schematic drawings of flows with boundary layer separation. (a) Real flow past a cylinder where the boundary layers on the top and bottom of the cylinder leave the surface near its widest point. (b) Real flow past a sharp-cornered obstacle where the boundary layer leaves the surface at the corner. Upstream of the point of separation, ideal flow theory is usually a good approximation of the real flow.

attack when boundary layers remain thin and stay attached all the way to the foil's trailing edge (Figure 6.1b). However, it is not satisfactory when the solid object has such a shape that one or more boundary layers separate from its surface before reaching its downstream edge (Figure 6.2), giving rise to a rotational wake flow or region of separated flow (sometimes called a *bubble*) that is not necessarily thin, no matter how high the Reynolds number. In this case, the limit of a real flow as $\mu \rightarrow 0$ does not approach that of an ideal flow ($\mu = 0$). Yet, upstream of boundary layer separation, ideal flow theory may still provide a good approximation of the real flow.

In summary, the theory presented here does not apply to inhomogeneous fluids, high subsonic or supersonic flow speeds, boundary layer flows, wake flows, interior flows, or any flow region where fluid elements rotate. However, the remaining flow possibilities are abundant, and include those that are commercially valuable (flight), naturally important (water waves), or readily encountered in our everyday lives (flow around vehicles; also see Chapters 7 and 14 for further examples).

Steady and unsteady irrotational constant-density flow fields around simple objects and through simple geometries in two and three dimensions are the subjects of this chapter. All the coordinate systems presented in Figure 3.3 are utilized herein.

6.2. TWO-DIMENSIONAL STREAM FUNCTION AND VELOCITY POTENTIAL

The two-dimensional incompressible continuity equation,

$$\partial u/\partial x + \partial v/\partial y = 0, \tag{6.2}$$

is identically satisfied when u,v-velocity components are determined from a single scalar function ψ:

$$u \equiv \partial\psi/\partial y, \text{ and } v \equiv -\partial\psi/\partial x. \tag{6.3}$$

The function $\psi(x,y)$ is the stream function in two dimensions. Along a curve of $\psi = $ constant, $d\psi = 0$, and this implies

$$0 = d\psi = \frac{\partial\psi}{\partial x}dx + \frac{\partial\psi}{\partial y}dy = -vdx + udy, \text{ or } \left(\frac{dy}{dx}\right)_{\psi=const} = \frac{v}{u},$$

which is the definition of a streamline in two dimensions. The vorticity ω_z in a flow described by ψ is:

$$\frac{\partial v}{\partial x} - \frac{\partial u}{\partial y} = \omega_z = \frac{\partial}{\partial x}\left(-\frac{\partial\psi}{\partial x}\right) - \frac{\partial}{\partial y}\left(\frac{\partial\psi}{\partial y}\right) = -\nabla^2\psi. \tag{6.4}$$

In constant-density irrotational flow, ω_z will be zero everywhere except at the locations of irrotational vortices. Thus, we are interested in solutions of

$$\nabla^2\psi = 0, \text{ and } \nabla^2\psi = -\Gamma\delta(x - x')\delta(y - y'), \tag{6.5, 6.6}$$

where δ is the Dirac delta-function (see Appendix B.4), and $\mathbf{x}' = (x', y')$ is the location of an ideal irrotational vortex of strength Γ.

In an unbounded domain, the most elementary nontrivial solutions of (6.5) and (6.6) are

$$\psi = -Vx + Uy, \text{ and } \psi = -\frac{\Gamma}{2\pi}\ln\sqrt{(x-x')^2+(y-y')^2}, \qquad (6.7, 6.8)$$

respectively. These correspond to uniform fluid velocity with horizontal component U and vertical component V, and to the flow induced by an irrotational vortex located at $\mathbf{x}'$ (see Exercise 6.1).

These two-dimensional stream function results have been obtained by considering incompressibility first, and irrotationality second. An equivalent formulation of two-dimensional ideal flow that leads to a different scalar function is possible when incompressibility and irrotationality are considered in the other order. The condition of irrotationality in two dimensions is

$$\partial v/\partial x - \partial u/\partial y = 0, \qquad (6.9)$$

and it is identically satisfied when u,v-velocity components are determined from a single scalar function ϕ:

$$u \equiv \partial\phi/\partial x, \text{ and } v \equiv \partial\phi/\partial y. \qquad (6.10)$$

The function $\phi(x,y)$ is known as the *velocity potential* in two dimensions because (6.10) implies $\nabla\phi = \mathbf{u}$. In fact, a velocity potential must exist in all irrotational flows, so such flows are frequently called *potential flows*. Curves of $\phi = $ constant are defined by

$$0 = d\phi = \frac{\partial\phi}{\partial x}dx + \frac{\partial\phi}{\partial y}dy = udx + vdy \text{ or } \left(\frac{dy}{dx}\right)_{\phi=const} = -\frac{u}{v},$$

and are perpendicular to streamlines. When using $\phi(x,y)$, the condition for incompressibility becomes:

$$\frac{\partial u}{\partial x} + \frac{\partial v}{\partial y} = \frac{\partial}{\partial x}\left(\frac{\partial\phi}{\partial x}\right) + \frac{\partial}{\partial y}\left(\frac{\partial}{\partial y}\right)\phi = \nabla^2\phi = q(x,y), \qquad (6.11)$$

where $q(x,y)$ is the spatial distribution of the source strength in the flow field. Of course, in real incompressible flows, $q(x,y) = 0$; however, ideal point sources and sinks of fluid are useful idealizations that allow the flow around objects of various shapes to be determined. These point sources and sinks are the ϕ-field equivalents of positive- and negative-circulation ideal vortices in flow fields described by ψ. Thus, we are interested in solutions of

$$\nabla^2\phi = 0, \text{ and } \nabla^2\phi = m\delta(x-x')\delta(y-y'), \qquad (6.12, 6.13)$$

where m is a constant that sets the strength of the singularity at $\mathbf{x}'$.

In an unbounded domain, the most elementary solutions of (6.12) and (6.13) are

$$\phi = Ux + Vy, \text{ and } \phi = \frac{m}{2\pi}\ln\sqrt{(x-x')^2+(y-y')^2}, \qquad (6.14, 6.15)$$

respectively. These correspond to uniform fluid velocity with horizontal component U and vertical component V, and to the flow induced by an ideal point source of strength m located at $\mathbf{x}'$ (see Exercise 6.3). Here, m is the source's volume flow rate per unit depth perpendicular to the plane of the flow.

Either ψ or ϕ can provide a complete description of a two-dimensional ideal flow, and they can be combined to form a complex potential that follows the theory of *harmonic* functions (see Section 6.4). In addition, ψ is readily extended to rotational flows while ϕ is readily extended to unsteady and three-dimensional flows.

Boundary conditions must be considered to extend the elementary ideal flow solutions (6.7), (6.8), (6.14), and (6.15) to more interesting geometries. The boundary conditions normally encountered in irrotational flows are as follows.

(1) *No flow through a solid surface.* The component of fluid velocity normal to a solid surface must equal the velocity of the boundary normal to itself. This can be stated as $\mathbf{n} \cdot \mathbf{U_s} = (\mathbf{n} \cdot \mathbf{u})_{on\ the\ surface}$, where $\mathbf{n}$ is the surface's normal and $\mathbf{U_s}$ is the velocity of the surface at the point of interest. For a stationary body, this condition reduces to $(\mathbf{n} \cdot \mathbf{u})_{on\ the\ surface} = 0$, which implies:

$$\partial\phi/\partial n = 0 \text{ or } \partial\psi/\partial s = 0 \text{ on the surface,} \tag{6.16}$$

where s is the arc-length along the surface, and n is the surface-normal coordinate. However, $\partial\psi/\partial s$ is also zero along a streamline. Thus, a stationary solid boundary in an ideal flow must also be a streamline. Therefore, if any ideal-flow streamline is replaced by a stationary solid boundary having the same shape, then the remainder of the flow is not changed.

(2) *Recovery of conditions at infinity.* For the typical case of a body immersed in a uniform fluid flowing in the x direction with speed U, the condition far from the body is

$$\partial\phi/\partial x = U, \text{ or } \partial\psi/\partial y = U. \tag{6.17}$$

When $U = 0$, the fluid far from the body is said to be quiescent.

Solving the Laplace equation, (6.5) or (6.12), with complicated-geometry boundary conditions like (6.16) and (6.17) requires numerical techniques. Historically, irrotational flow theory was developed by finding functions that satisfy the Laplace equation and then determining the boundary conditions met by those functions. Since the Laplace equation is linear, any superposition of known solutions provides another solution, but the superposition of two or more solutions may satisfy different boundary conditions than any of the constituents of the superposition. Thus, through collection and combination, a rich variety of interesting ideal-flow solutions has emerged. This solution-construction approach to ideal flow theory is adopted in this chapter, except in Sections 6.7 and 6.8 where numerical methods for solving (6.5) or (6.12) subject to (6.16) or (6.17) are presented.

After a solution of the Laplace equation has been obtained, the velocity components are determined by taking derivatives of ϕ or ψ. Then, the conservation of momentum equation (6.1) is satisfied for steady flow by determining pressure from the Bernoulli equation:

$$p + \frac{1}{2}\rho|\mathbf{u}|^2 = p + \frac{1}{2}\rho(u^2 + v^2) = p + \frac{1}{2}\rho|\nabla\phi|^2 = p + \frac{1}{2}\rho|\nabla\psi|^2 = const. \tag{6.18}$$

For unsteady flow, the term $\rho(\partial\phi/\partial t)$ must be added (see Exercise 6.4). With this procedure, solutions of (4.10) and (6.1) for $\mathbf{u}$ and p are obtained for ideal flows in a simple manner even though (6.1) is nonlinear.

For quick reference, the important equations in planar polar coordinates are:

$$\frac{1}{r}\frac{\partial}{\partial r}(ru_r) + \frac{1}{r}\frac{\partial u_\theta}{\partial \theta} = 0 \qquad \text{(continuity)}, \tag{6.19}$$

$$\frac{1}{r}\frac{\partial}{\partial r}(ru_\theta) - \frac{1}{r}\frac{\partial u_r}{\partial \theta} = 0 \quad \text{(irrotationality)}, \tag{6.20}$$

$$u_r = \frac{\partial \phi}{\partial r} = \frac{1}{r}\frac{\partial \psi}{\partial \theta}, \tag{6.21}$$

$$u_\theta = \frac{1}{r}\frac{\partial \phi}{\partial \theta} = -\frac{\partial \psi}{\partial r}, \tag{6.22}$$

$$\nabla^2 \psi = \frac{1}{r}\frac{\partial}{\partial r}\left(r\frac{\partial \psi}{\partial r}\right) + \frac{1}{r^2}\frac{\partial^2 \psi}{\partial \theta^2} = 0, \text{ and } \nabla^2 \phi = \frac{1}{r}\frac{\partial}{\partial r}\left(r\frac{\partial \phi}{\partial r}\right) + \frac{1}{r^2}\frac{\partial^2 \phi}{\partial \theta^2} = 0. \tag{6.23a, 6.23b}$$

6.3. CONSTRUCTION OF ELEMENTARY FLOWS IN TWO DIMENSIONS

In this section, elementary solutions of the Laplace equation are developed and then superimposed to produce a variety of geometrically simple ideal flows in two dimensions.

First consider polynomial solutions of the Laplace equation in Cartesian coordinates. A zero-order polynomial, ψ or ϕ = constant, is not interesting since with either field function the result is $\mathbf{u} = 0$. First-order polynomial solutions are given by (6.7) and (6.14), and these solutions represent spatially uniform velocity fields, $\mathbf{u} = (U,V)$. Quadratic functions in x and y are the next possibilities, and there are two of these:

$$\psi = 2Axy \text{ or } \phi = 2Axy, \text{ and } \psi = A(x^2 - y^2) \text{ or } \phi = A(x^2 - y^2), \tag{6.24--6.27}$$

where A is a constant. (The reason for the two in (6.24) and (6.25) will be clear in the next section.) Here we will only construct the flow fields for $2Axy$. Production of flow-field results for ψ and $\phi = A(x^2 - y^2)$ is left as an exercise.

Examine $\psi = 2Axy$ first, and by direct differentiation find $u = 2Ax$, and $v = -2Ay$. Thus, for $A > 0$, the flow is toward the origin along the y-axis, away from it along the x-axis, and the streamlines are hyperbolae given by $xy = \psi/2A$ (Figure 6.3). Considering the first quadrant only, this is flow in a 90° corner. Now consider $\phi = 2Axy$, and by direct differentiation find $u = 2Ay$, and $v = 2Ax$. The equipotential lines are hyperbolae given by $xy = \phi/2A$. The flow is away from the origin along the line $y = x$ and toward it along the line $y = -x$. Thus, $\phi = 2Axy$ produces a flow that is equivalent to that of $\psi = 2Axy$ after a 45° rotation. Interestingly, higher-order polynomial solutions lead to flows in smaller-angle corners, while fractional powers lead to flows in larger-angle corners (see Section 6.4 and Exercises 6.6 and 6.7).

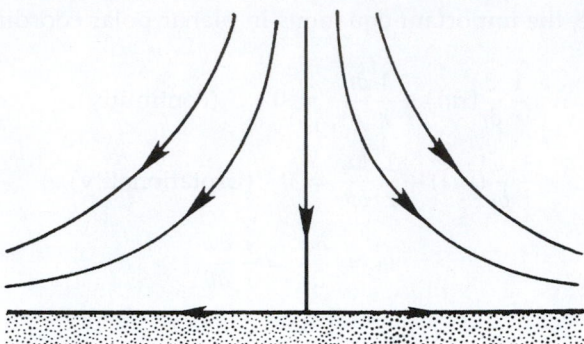

FIGURE 6.3   Stagnation point flow represented by $\psi = 2Axy$. Here the flow impinges on the flat surface from above. The stagnation point is located where the single vertical streamline touches the horizontal surface.

The next set of solutions to consider are (6.8) and (6.15) with $x' = y' = 0$. In this case, curves of $\psi = -(\Gamma/2\pi)\ln\sqrt{x^2 + y^2} = const.$ are circles centered on the origin of coordinates (Figure 6.4), and direct differentiation of (6.8) produces:

$$u = \frac{\partial}{\partial y}\left(-\frac{\Gamma}{2\pi}\ln\sqrt{x^2 + y^2}\right) = -\frac{\Gamma}{2\pi}\frac{y}{x^2 + y^2} = -\frac{\Gamma}{2\pi r}\sin\theta \text{ , and}$$

$$v = -\frac{\partial}{\partial x}\left(-\frac{\Gamma}{2\pi}\ln\sqrt{x^2 + y^2}\right) = +\frac{\Gamma}{2\pi}\frac{x}{x^2 + y^2} = \frac{\Gamma}{2\pi r}\cos\theta,$$

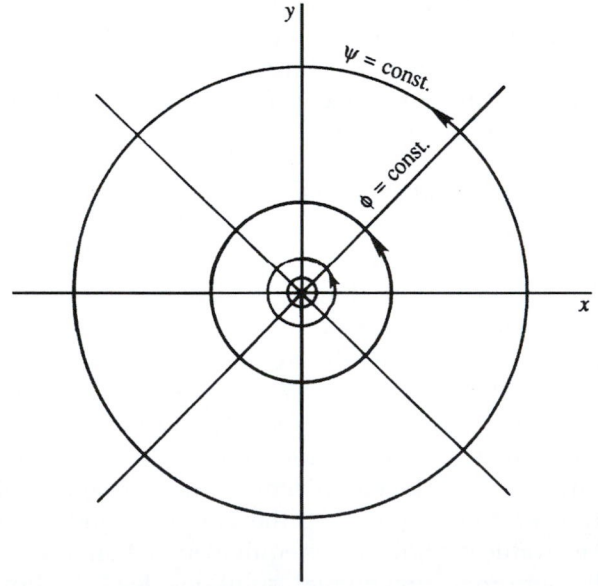

FIGURE 6.4   The flow field of an ideal vortex located at the origin of coordinates. The streamlines are circles and the potential lines are radials. Here, the vortex line is perpendicular to the x-y plane.

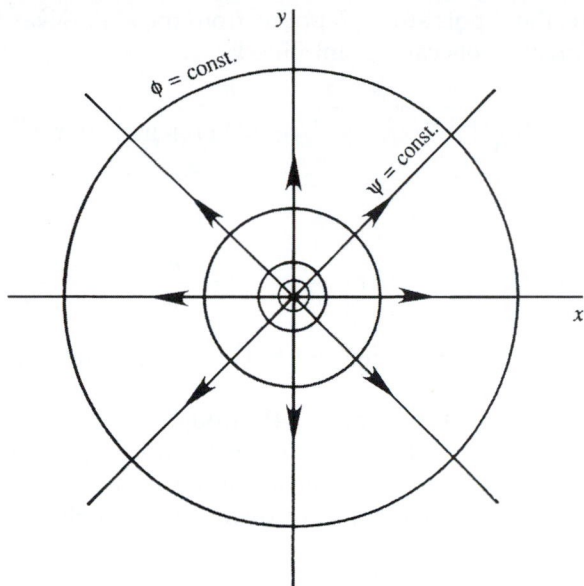

FIGURE 6.5   The flow field of an ideal source located at the origin of coordinates in two dimensions. The streamlines are radials and the potential lines are circles.

where r and θ are defined in Figure 3.3a. These results may be rewritten using the outcome of Example 2.1 as $u_r = 0$ and $u_\theta = \Gamma/2\pi r$, which is the flow field of the ideal irrotational vortex (5.2). Similarly, curves of $\phi = (m/2\pi)\ln\sqrt{x^2 + y^2} = const.$ are circles centered on the origin of coordinates (Figure 6.5), and direct differentiation of (6.8) produces:

$$u = \frac{\partial}{\partial x}\left(\frac{m}{2\pi}\ln\sqrt{x^2 + y^2}\right) = \frac{m}{2\pi}\frac{x}{x^2 + y^2} = \frac{m}{2\pi r}\cos\theta, \text{ and}$$

$$v = \frac{\partial}{\partial y}\left(\frac{m}{2\pi}\ln\sqrt{x^2 + y^2}\right) = \frac{m}{2\pi}\frac{y}{x^2 + y^2} = \frac{m}{2\pi r}\sin\theta.$$

These results may be rewritten as $u_r = m/2\pi r$ and $u_\theta = 0$, which is purely radial flow away from the origin (Figure 6.5). Here, $\nabla \cdot \mathbf{u}$ is zero everywhere except at the origin. Thus, this potential represents flow from an ideal incompressible point source for $m > 0$, or sink for $m < 0$, that is located at $r = 0$ in two dimensions.

A source of strength $+m$ at $(-\varepsilon, 0)$ and sink of strength $-m$ at $(+\varepsilon, 0)$, can be considered together

$$\phi = \frac{m}{2\pi}\ln\sqrt{(x + \varepsilon)^2 + y^2} - \frac{m}{2\pi}\ln\sqrt{(x - \varepsilon)^2 + y^2}$$

to obtain the potential for a *doublet* in the limit that $\varepsilon \to 0$ and $m \to \infty$, so that the dipole strength vector

$$\mathbf{d} = \sum_{sources} \mathbf{x}_i m_i = -\varepsilon \mathbf{e}_x m + \varepsilon \mathbf{e}_x(-m) = -2m\varepsilon \mathbf{e}_x \qquad (6.28)$$

remains constant. Here, the dipole strength points from the sink toward the source. As $\varepsilon \to 0$, the logarithm of the square roots can be simplified:

$$\ln\sqrt{(x \pm \varepsilon)^2 + y^2} = \ln r + \ln\sqrt{1 \pm 2\varepsilon x/r^2 + \varepsilon^2/r^2} = \ln r + \ln(1 \pm \varepsilon x/r^2 + \ldots) \cong \ln(r) \pm \varepsilon x/r^2,$$

where $r^2 = x^2 + y^2$, so

$$\lim_{\varepsilon \to 0} \lim_{m \to \infty} \phi \cong \frac{m}{2\pi}\left(\ln r + \frac{\varepsilon x}{r^2} + \ldots - \ln r + \frac{\varepsilon x}{r^2} + \ldots\right) = \frac{m\varepsilon}{\pi}\frac{x}{r^2} = -\frac{\mathbf{d} \cdot \mathbf{x}}{2\pi r^2} = \frac{|\mathbf{d}|}{2\pi}\frac{\cos\theta}{r}. \tag{6.29}$$

The doublet flow field is illustrated in Figure 6.6. The stream function for the doublet can be derived from (6.29) (Exercise 6.9).

The flows described by (6.7), (6.14), and (6.24) through (6.27) are solutions of the Laplace equation. The flows described by (6.8), (6.15), and (6.29) are singular at the origin and satisfy the Laplace equation for $r > 0$. Perhaps the most common and useful superposition of these solutions involves the combining of a uniform stream parallel to the x-axis, $\psi = Uy$ or $\phi = Ux$,

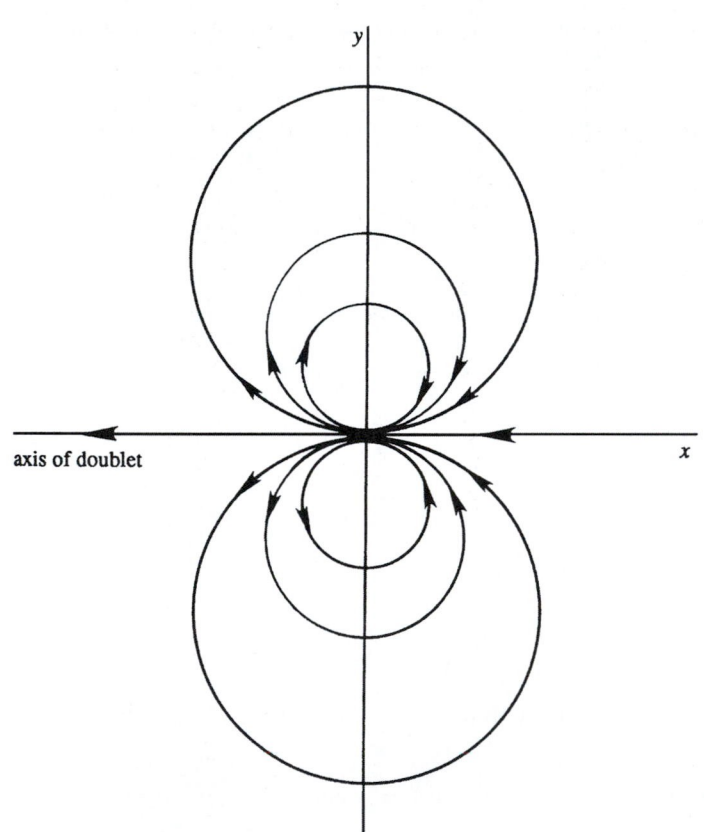

axis of doublet

FIGURE 6.6 The flow field of an ideal doublet that points along the negative x-axis. The net source strength is zero so all streamlines begin and end at the origin. In this flow, the streamlines are circles tangent to the x-axis at the origin.

and one or more of the singular solutions. The simplest example is the combination of a source and a uniform stream, which can be written in Cartesian and polar coordinates as:

$$\phi = Ux + \frac{m}{2\pi}\ln\sqrt{x^2+y^2} = Ur\cos\theta + \frac{m}{2\pi}\ln r, \text{ or} \tag{6.30}$$

$$\psi = Uy + \frac{m}{2\pi}\tan^{-1}\left(\frac{y}{x}\right) = Ur\sin\theta + \frac{m}{2\pi}\theta. \tag{6.31}$$

Here the velocity field is:

$$u = U + \frac{m}{2\pi}\frac{x}{x^2+y^2} \quad \text{and} \quad v = \frac{m}{2\pi}\frac{y}{x^2+y^2},$$

and streamlines are shown in Figure 6.7. The stagnation point is located at $x=-a=-m/2\pi U$, and $y=0$, and the value of the stream function on the stagnation streamline is $\psi = m/2$.

The streamlines that emerge vertically from the stagnation point (the darker curves in Figure 6.7) form a semi-infinite body with a smooth nose, generally called a *half-body*. These stagnation streamlines divide the field into regions external and internal to the half-body. The internal flow consists entirely of fluid emanating from the source, and the external region contains fluid from upstream of the source. The half-body resembles several practical shapes, such as the leading edge of an airfoil or the front part of a bridge pier; the upper half of the flow resembles the flow over a cliff or a side contraction in a wide channel. The half-width of the body, h, can be found from (6.31) with $\psi = m/2$:

$$h = m(\pi - \theta)/2\pi U.$$

Far downstream ($\theta \to 0$), the half-width tends to $h_{max} = m/2U$ (Figure 6.7).

The pressure distribution on the half-body can be found from Bernoulli's equation, (6.18) with $const. = p_\infty + \rho U^2/2$, and is commonly reported as a dimensionless excess pressure via the *pressure coefficient* C_p or Euler number (4.106):

$$C_p = \frac{p - p_\infty}{\frac{1}{2}\rho U^2} = 1 - \frac{|\mathbf{u}|^2}{U^2}. \tag{6.32}$$

FIGURE 6.7 Ideal flow past a two-dimensional half-body formed from a horizontal free stream and a point source at the origin. The boundary streamline, shown as a darker curve, is given by $\psi = m/2$.

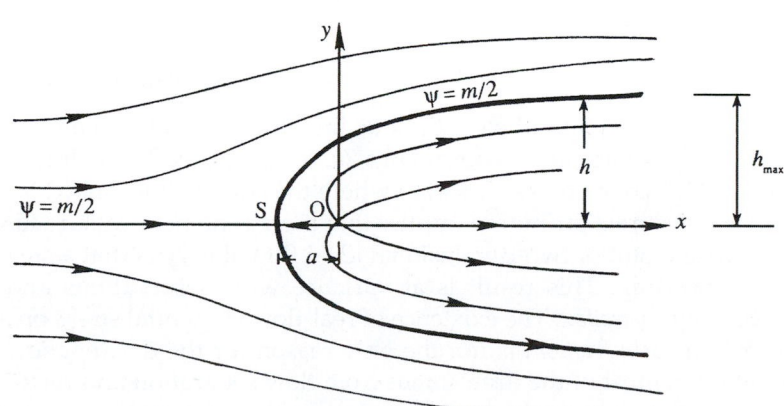

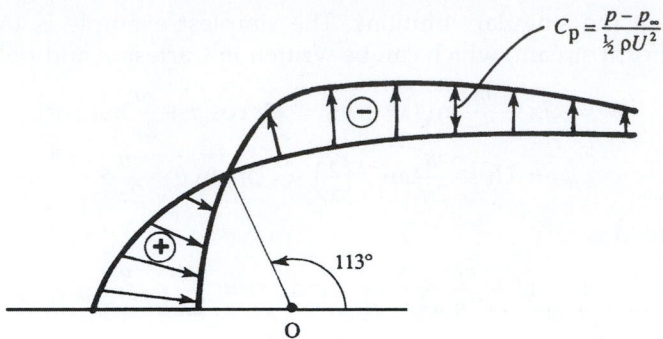

FIGURE 6.8   Pressure distribution in ideal flow over the half-body shown in Figure 6.7. Pressure excess near the nose is indicated by the circled "+" and pressure deficit elsewhere is indicated by the circled "−".

A plot of C_p on the surface of the half-body is given in Figure 6.8, which shows that there is pressure excess near the nose of the body and a pressure deficit beyond it. It is easy to show by integrating p over the surface that the net pressure force is zero (Exercise 6.13).

As a second example of flow construction via superposition, consider a horizontal free stream U and a doublet with strength $\mathbf{d} = -2\pi U a^2 \mathbf{e}_x$:

$$\phi = Ux + \frac{Ua^2 x}{x^2 + y^2} = U\left(r + \frac{a^2}{r}\right)\cos\theta, \text{ or } \psi = Uy - \frac{Ua^2 y}{x^2 + y^2} = U\left(r - \frac{a^2}{r}\right)\sin\theta. \quad (6.33)$$

Here, $\psi = 0$ at $r = a$ for all values of θ, showing that the streamline $\psi = 0$ represents a circular cylinder of radius a. The streamline pattern is shown in Figure 6.9 (and Figure 3.2a). In this flow, the net source strength is zero, so the cylindrical body is closed and does not extend downstream. The velocity field is:

$$u_r = U\left(1 - \frac{a^2}{r^2}\right)\cos\theta, \text{ and } u_\theta = -U\left(1 + \frac{a^2}{r^2}\right)\sin\theta. \quad (6.34)$$

The velocity components on the surface of the cylinder are $u_r = 0$ and $u_\theta = -2U\sin\theta$, so the cylinder-surface pressure coefficient is:

$$C_p(r = a, \theta) = 1 - 4\sin^2\theta, \quad (6.35)$$

and this is shown by the continuous line in Figure 6.10. There are stagnation points on the cylinder's surface at r-θ coordinates, $(a, 0)$ and (a, π). The cylinder-surface pressure minima occur at r-θ coordinates $(a, \pm\pi/2)$ where the surface flow speed is maximum. The symmetry of the pressure distribution implies that there is no net pressure force on the cylinder. In fact, a general result of two-dimensional ideal flow theory is that a steadily moving body experiences no drag. This result is at variance with observations and is sometimes known as *d'Alembert's paradox*. The existence of real-flow tangential stress on a solid surface, commonly known as *skin friction*, is not the only reason for the discrepancy. For blunt bodies such as a cylinder, most of the drag comes from flow separation and the formation of a wake, which is likely to be unsteady or even turbulent. When a wake is present, the flow loses fore-aft

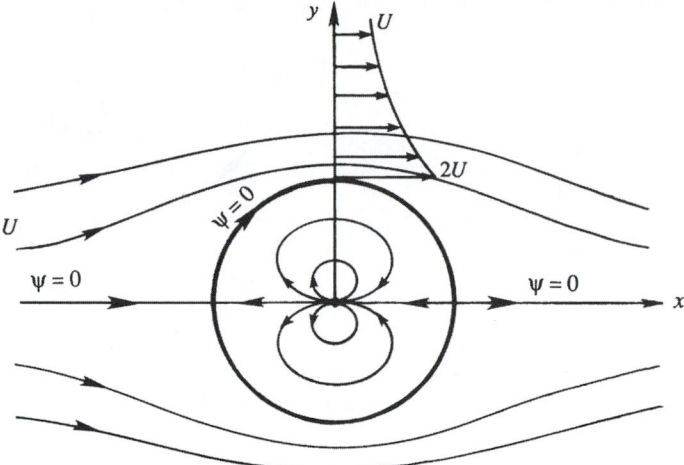

FIGURE 6.9 Idea flow past a circular cylinder without circulation. This flow field is formed by combining a horizontal uniform stream flowing in the $+x$ direction with a doublet pointing in the $-x$ direction. The streamline that passes through the two stagnation points and forms the body surface is given by $\psi = 0$.

symmetry, and the surface pressure on the downstream side of the object is smaller than that predicted by ideal flow theory (Figure 6.10), resulting in pressure drag. These facts will be discussed further in Chapter 9.

As discussed in Section 3.3, the flow due to a cylinder moving steadily through a fluid appears unsteady to an observer at rest with respect to the fluid at infinity. This flow is shown in Figure 3.3b and can be obtained by superposing a uniform stream in the negative x direction with the flow shown in Figure 6.9. The resulting instantaneous streamline pattern is simply that of a doublet, as is clear from the decomposition shown in Figure 6.11.

FIGURE 6.10 Comparison of irrotational and observed pressure distributions over a circular cylinder. Here $0°$ is the most upstream point of the cylinder and $180°$ is the most downstream point. The observed distribution changes with the Reynolds number; a typical behavior at high Re is indicated by the dashed line.

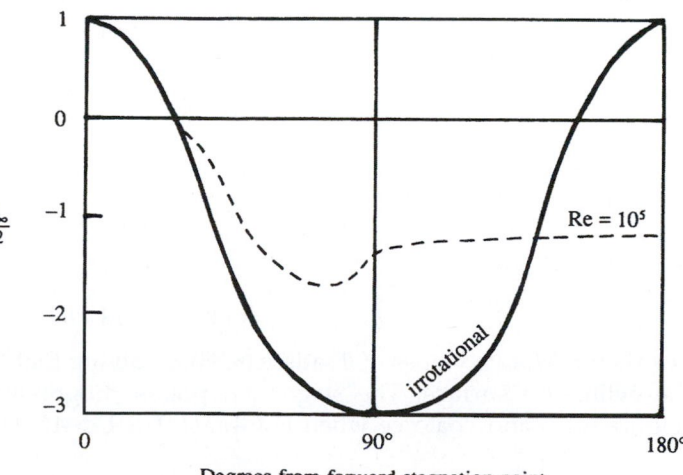

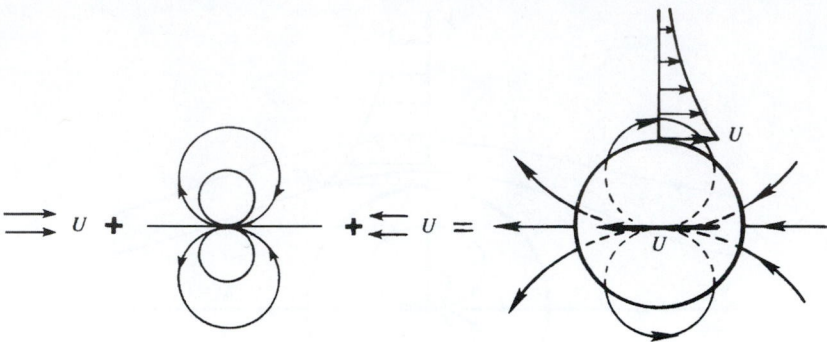

FIGURE 6.11 Decomposition of the irrotational flow pattern due to a moving cylinder. Here a horizontal free stream of $+U$ and doublet form a cylinder. When a uniform stream of $-U$ is added, the flow field of a moving cylinder is obtained.

Although there is no net drag force on a circular cylinder in steady irrotational flow, there may be a lateral or lift force perpendicular to the free stream when circulation is added. Consider the flow field (6.33) with the addition of a point vortex of circulation $-\Gamma$[1] at the origin that induces a *clockwise* velocity:

$$\psi = U\left(r - \frac{a^2}{r}\right)\sin\theta + \frac{\Gamma}{2\pi}\ln\left(\frac{r}{a}\right). \qquad (6.36)$$

Here, a has been added to the logarithm's argument to make it dimensionless.

Figure 6.12 shows the resulting streamline pattern for various values of Γ. The close streamline spacing and higher velocity on top of the cylinder is due to the addition of velocities from the clockwise vortex and the uniform stream. In contrast, the smaller velocities at the bottom of the cylinder are a result of the vortex field counteracting the uniform stream. Bernoulli's equation consequently implies a higher pressure below the cylinder than above it, and this pressure difference leads to an upward lift force on the cylinder.

The tangential velocity component at any point in the flow is

$$u_\theta = -\frac{\partial\psi}{\partial r} = -U\left(1 + \frac{a^2}{r^2}\right)\sin\theta - \frac{\Gamma}{2\pi r}.$$

At the surface of the cylinder, the fluid velocity is entirely tangential and is given by

$$u_\theta(r = a, \theta) = -2U\sin\theta - \Gamma/2\pi a, \qquad (6.37)$$

which vanishes if

$$\sin\theta = -\Gamma/4\pi a U. \qquad (6.38)$$

For $\Gamma < 4\pi a U$, two values of θ satisfy (6.38), implying that there are two stagnation points on the cylinder's surface. The stagnation points progressively move down as Γ increases (Figure 6.12) and coalesce when $\Gamma = 4\pi a U$. For $\Gamma > 4\pi a U$, the stagnation point moves out

---

[1]This minus sign is necessary to achieve the usual fluid dynamic result given by (6.40).

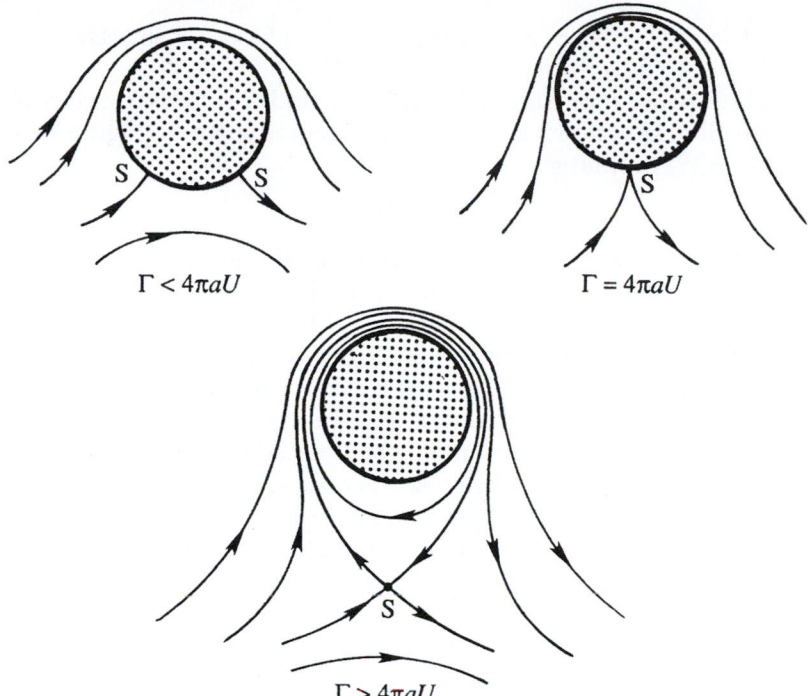

$\Gamma < 4\pi a U$ $\Gamma = 4\pi a U$

$\Gamma > 4\pi a U$

FIGURE 6.12 Irrotational flow past a circular cylinder for different circulation values. Here S represents the stagnation point(s) in the flow. (a) At low values of the circulation, there are two stagnation points on the surface of the cylinder. (b) When the circulation is equal to $4\pi a U$, there is one stagnation point on the surface of the cylinder. (c) When the circulation is even greater, there is one stagnation point below the cylinder.

into the flow along the negative y-axis. The radial distance of the stagnation point in this case is found from

$$u_\theta(r, \theta = -\pi/2) = U\left(1 + \frac{a^2}{r^2}\right) - \frac{\Gamma}{2\pi r} = 0, \text{ or } r = \frac{1}{4\pi U}\left[\Gamma \pm \sqrt{\Gamma^2 - (4\pi a U)^2}\right],$$

one root of which has $r > a$; the other root corresponds to a stagnation point inside the cylinder.

The cylinder surface pressure is found from (6.18) with $const. = p_\infty + \rho U^2/2$, and (6.37) to be

$$p(r = a, \theta) = p_\infty + \frac{1}{2}\rho\left[U^2 - \left(-2U\sin\theta - \frac{\Gamma}{2\pi a}\right)^2\right]. \tag{6.39}$$

The upstream-downstream symmetry of the flow implies that the pressure force on the cylinder has no stream-wise component. The lateral pressure force (per unit length perpendicular to the flow plane) is

$$L = -\int_0^{2\pi} p(r = a, \theta)\mathbf{n}dl\cdot\mathbf{e}_y = -\int_0^{2\pi} p(r = a, \theta)\sin\theta a d\theta,$$

where $\mathbf{n} = \mathbf{e}_r$ is the outward normal from the cylinder, and $dl = ad\theta$ is a surface element of the cylinder's cross section; L is known as the *lift* force in aerodynamics (Figure 6.13). Evaluating the integral using (6.39) produces:

$$L = \rho U\Gamma. \tag{6.40}$$

It is shown in Section 6.5 that (6.40) holds for irrotational flow around *any* two-dimensional object; it is not just for circular cylinders. The result that L is proportional to Γ is of fundamental importance in aerodynamics. Equation (6.40) was proved independently by the German mathematician Wilhelm Kutta and the Russian aerodynamist Nikolai Zhukhovsky just after 1900; it is called the *Kutta-Zhukhovsky lift theorem*. (Older Western texts transliterated Zhukhovsky's name as Joukowsky.) The interesting question of how certain two-dimensional shapes, such as an airfoil, develop circulation when placed in a moving fluid is discussed in Chapter 14. It is shown there that fluid viscosity is responsible for the development of circulation. The magnitude of circulation, however, is independent of viscosity but does depend on the flow speed U, and the shape and orientation of the object.

For a circular cylinder, the only way to develop circulation is by rotating it. Although viscous effects are important in this case, the observed flow pattern for *large* values of cylinder rotation displays a striking similarity to the ideal flow pattern for $\Gamma > 4\pi aU$; see Figure 3.25 in the book by Prandtl (1952). For lower rates of cylinder rotation, the retarded flow in the boundary layer is not able to overcome the adverse pressure gradient behind the cylinder, leading to separation; the real flow is therefore rather unlike the irrotational pattern. However, even in the presence of separation, observed flow speeds are higher on the upper surface of the cylinder, implying a lift force.

A second reason for the presence of lift on a rotating cylinder is the flow asymmetry generated by a delay of boundary layer separation on the upper surface of the cylinder. The contribution of this mechanism is small for two-dimensional objects such as the circular cylinder,

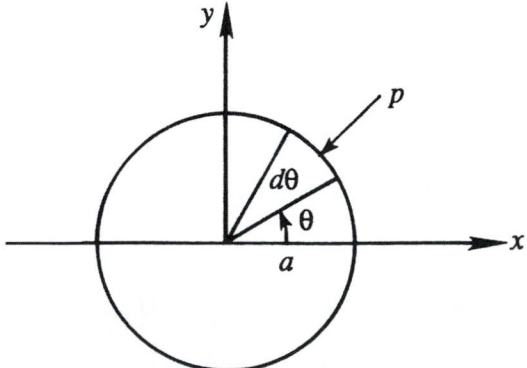

FIGURE 6.13 Calculation of pressure force on a circular cylinder. Surface pressure forces on top of the cylinder, where **n** has a positive vertical component, push the cylinder down. Thus the surface integral for the lift force applied by the fluid to the cylinder contains a minus sign.

but it is the only mechanism for side forces experienced by spinning three-dimensional objects like sports balls. The interesting question of why spinning balls follow curved paths is discussed in Section 9.9. The lateral force experienced by rotating bodies is called the *Magnus effect*.

Two-dimensional ideal flow solutions are commonly not unique and the topology of the flow domain determines uniqueness. Simply stated, a two-dimensional ideal flow solution is unique when any closed contour lying entirely within the fluid can be reduced to a point by continuous deformation without ever cutting through a flow-field boundary. Such fluid domains are *singly connected*. Thus, fluid domains, like those shown in Figures 6.9 and 6.12, that entirely encircle an object may not provide unique ideal flow solutions based on boundary conditions alone. In particular, consider the ideal flow (6.36) depicted in Figure 6.12 for various values of Γ. All satisfy the *same* boundary condition on the solid surface ($u_r = 0$) and at infinity ($\mathbf{u} = U\mathbf{e}_x$). The ambiguity occurs in these domains because there exist closed contours lying entirely within the fluid that cannot be reduced to a point, and on these contours a nonzero circulation can be computed. Fortunately, this ambiguity may often be resolved by considering real-flow effects. For example, the circulation strength that should be assigned to a streamlined object in two-dimensional ideal flow can be determined by applying the viscous flow–based *Kutta* condition at the object's trailing edge. This point is further explained in Chapter 14.

Another important consequence of the superposition principle for ideal flow is that it allows boundaries to be built into ideal flows through the method of images. For example, if the flow of interest in an unbounded domain is the solution of $\nabla^2 \psi_1 = -\omega_1(x,y)$, then $\nabla^2 \psi_2 = -\omega_1(x,y) + \omega_1(x, -y)$ will determine the solution for the same vorticity distribution with a solid wall along the x-axis. Here, $\psi_2 = \psi_1(x,y) - \psi_1(x, -y)$, so that the *zero* streamline, $\psi_2 = 0$, occurs on $y = 0$ (Figure 6.14). Similarly, if the flow of interest in an unbounded domain is the solution of $\nabla^2 \phi_1 = q_1(x,y)$, then $\nabla^2 \phi_2 = q_1(x,y) + q_1(x, -y)$ will determine the solution for the same source distribution with a solid wall along the x-axis. Here, $\phi_2 = \phi_1(x,y) + \phi_1(x, -y)$ so that $v = \partial \phi_2/\partial y = 0$ on $y = 0$ (Figure 6.15).

As an example of the method of images, consider the flow induced by an ideal source of strength m a distance a from a straight vertical wall (Figure 6.16). Here an image source of the same strength and sign is needed a distance a on the other side of the wall. The stream function and potential for this flow are:

$$\psi = \frac{m}{2\pi}\left[\tan^{-1}\left(\frac{y}{x+a}\right) + \tan^{-1}\left(\frac{y}{x-a}\right)\right] \text{ and } \phi = \frac{m}{2\pi}\left[\ln\sqrt{(x+a)^2+y^2} + \ln\sqrt{(x-a)^2+y^2}\right],$$

$$(6.41)$$

FIGURE 6.14 Illustration of the method of images for the flow near a horizontal wall generated by a vorticity distribution. An image distribution of equal strength and opposite sign mimics the effect of the solid wall.

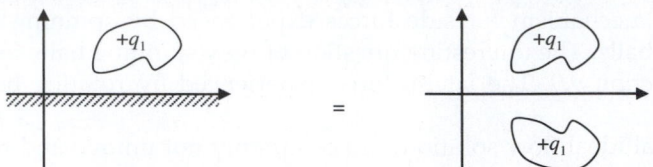

FIGURE 6.15   Illustration of the method of images for the flow near a horizontal wall generated by a source distribution. An image distribution of equal strength mimics the effect of the solid wall.

respectfully. After some rearranging and use of the two-angle formula for the tangent function, the equation for the streamlines may be found:

$$x^2 - y^2 - 2xy \cot(2\pi\psi/m) = a^2.$$

The x and y axes form part of the streamline pattern, with the origin as a stagnation point. This flow represents three interesting situations: flow from two equal sources (all of Figure 6.16), flow from a source near a flat vertical wall (right half of Figure 6.16), and flow through a narrow slit at $x = a$ into a right-angled corner (first quadrant of Figure 6.16).

   The method of images can commonly be extended to circular boundaries by allowing more than one image vortex or source (Exercises 5.14, 6.26), and to unsteady flows as long as the image distributions of vorticity or source strength move appropriately. Unsteady two-dimensional ideal flow merely involves the inclusion of time as an independent variable in ψ or ϕ, and the addition of $\rho(\partial\phi/\partial t)$ in the Bernoulli equation (see Exercise 6.4). The following example, which validates the free-vortex results stated in Chapter 5.7, illustrates these changes.

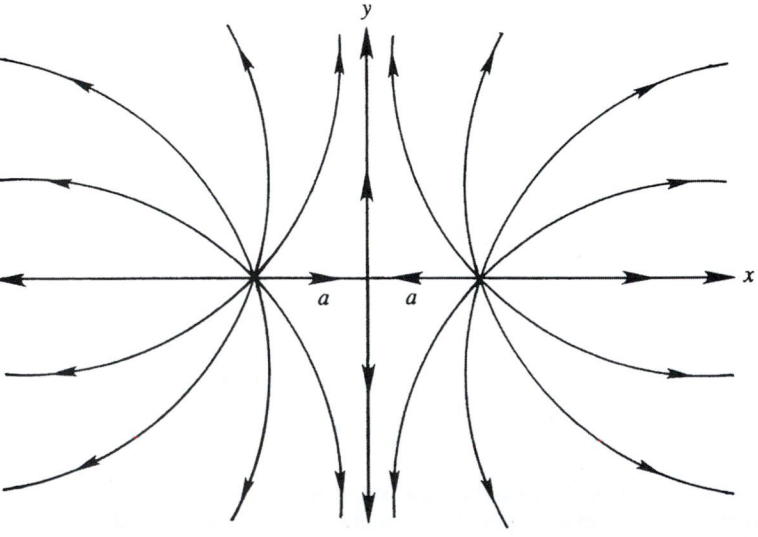

FIGURE 6.16 Ideal flow from two equal sources placed at $x = \pm a$. The origin is a stagnation point. The vertical axis is a streamline and may be replaced by a solid surface. This flow field further illustrates the method of images.

EXAMPLE 6.1

At $t = 0$, an ideal free vortex with strength $-\Gamma$ is located at point A near a flat solid vertical wall as shown in Figure 5.14. If the x,y-coordinates of A are $(h,0)$ and the fluid far from the vortex is quiescent at pressure p_∞, determine the trajectory $\xi(t) = (\xi_x, \xi_y)$ of the vortex, and the pressure at the origin of coordinates as a function of time.

Solution

From the method of images, the stream function for this flow field will be:

$$\psi(x,y,t) = \frac{\Gamma}{2\pi}\left[-\ln\sqrt{\left(x + \xi_x(t)\right)^2 + \left(y - \xi_y(t)\right)^2} + \ln\sqrt{\left(x - \xi_x(t)\right)^2 + \left(y - \xi_y(t)\right)^2}\right].$$

The first term is for the image vortex and the second term is for the original vortex. The horizontal and vertical components of the induced velocity for both vortices are:

$$u(x,y,t) = \frac{\partial\psi}{\partial y} = \frac{\Gamma}{2\pi}\left[-\frac{y - \xi_y(t)}{\left(x + \xi_x(t)\right)^2 + \left(y - \xi_y(t)\right)^2} + \frac{y - \xi_y(t)}{\left(x - \xi_x(t)\right)^2 + \left(y - \xi_y(t)\right)^2}\right] = \frac{\partial\phi}{\partial x},$$

$$v(x,y,t) = -\frac{\partial\psi}{\partial x} = -\frac{\Gamma}{2\pi}\left[-\frac{x + \xi_x(t)}{\left(x + \xi_x(t)\right)^2 + \left(y - \xi_y(t)\right)^2} + \frac{x - \xi_x(t)}{\left(x - \xi_x(t)\right)^2 + \left(y - \xi_y(t)\right)^2}\right] = \frac{\partial\phi}{\partial y}.$$

As expected, the use of an opposite sign image vortex produces $u(0,y) = 0$. Free vortices move with fluid elements and follow path lines, thus:

$$\frac{d\xi_x(t)}{dt} = \lim_{x\to\xi}(u(x,y,t)), \text{ and } \frac{d\xi_y(t)}{dt} = \lim_{x\to\xi}(v(x,y,t)).$$

The equations for the velocity components given above include contributions from the image and original vortices. The limit of the image vortex's induced velocity at the location of the original vortex is well defined. The velocity induced on the original vortex by itself is not well defined, but this is a mathematical artifact of ideal vortices. Any real vortex has a finite core size, and the self-induced velocity is well defined and equal to zero on the vortex axis when the core is axisymmetric. Thus, the self-induced velocity of an ideal vortex is taken to be zero, and the path-line equations above become:

$$\frac{d\xi_x}{dt} = 0, \text{ and } \frac{d\xi_y}{dt} = \frac{\Gamma}{2\pi}\frac{1}{2\xi_x}.$$

The solution of the first path-line equation is $\xi_x = \text{const.} = h$ where the second equality follows from the initial condition. The solution of the second equation is: $\xi_y = \Gamma t/4\pi h$, where the initial condition requires the constant of integration to be zero. Therefore, the vortex trajectory is: $\xi(t) = (h, \Gamma t/4\pi h)$.

To determine the pressure, integrate the velocity components to determine the potential:

$$\phi(x,y,t) = +\frac{\Gamma}{2\pi}\tan^{-1}\left(\frac{y - \Gamma t/4\pi h}{x + h}\right) - \frac{\Gamma}{2\pi}\tan^{-1}\left(\frac{y - \Gamma t/4\pi h}{x - h}\right).$$

The fluid far from the vortex is quiescent so the appropriate Bernoulli equation is:

$$\frac{\partial \phi}{\partial t} + \frac{1}{2}|\nabla \phi|^2 + \frac{p}{\rho} = const. = \frac{p_\infty}{\rho}$$

[(4.83) with $g = 0$], where the second equality follows from evaluating the constant far from the vortex. Thus, at $x = y = 0$:

$$\frac{p(0,0,t) - p_\infty}{\rho} = -\left(\frac{\partial \phi}{\partial t} + \frac{1}{2}(u^2 + v^2)\right)_{x=y=0}$$

Here,

$$\left(\frac{\partial \phi}{\partial t}\right)_{x=y=0} = \frac{-\Gamma^2}{4\pi^2}\left(\frac{1}{h^2 + (\Gamma t/4\pi h)^2}\right), \quad u(0.0,t) = 0, \quad \text{and} \quad v(0,0,t) = \frac{\Gamma h}{\pi}\left(\frac{1}{h^2 + (\Gamma t/4\pi h)^2}\right),$$

so

$$\frac{p(0,0,t) - p_\infty}{\rho} = \frac{\Gamma^2}{4\pi^2}\left(\frac{1}{h^2 + (\Gamma t/4\pi h)^2}\right) - \frac{\Gamma^2 h^2}{2\pi^2}\left(\frac{1}{h^2 + (\Gamma t/4\pi h)^2}\right)^2 = \frac{\Gamma^2}{4\pi^2}\frac{(\Gamma t/4\pi h)^2 - h^2}{\left((\Gamma t/4\pi h)^2 + h^2\right)^2}.$$

6.4. COMPLEX POTENTIAL

Using complex variables and complex functions, the developments for ψ and ϕ provided in the prior two sections can be recast in terms of a single complex potential $w(z)$,

$$w \equiv \phi + i\psi, \tag{6.42}$$

where z is a complex variable:

$$z \equiv x + iy = re^{i\theta}, \tag{6.43}$$

$i = \sqrt{-1}$ is the imaginary root, (x,y) are plane Cartesian coordinates, and (r,θ) are plane polar coordinates. There are many fine texts, such as Churchill et al. (1974) or Carrier et al. (1966), that cover the relevant mathematics of complex analysis, so it is merely alluded to here. In its Cartesian form, the complex number z represents a point in the x,y-plane with x increasing on the real axis and y increasing on the imaginary axis (Figure 6.17). In its polar form, z represents the position vector Oz, with magnitude $r = (x^2 + y^2)^{1/2}$ and angle with respect to the x-axis of $\theta = \tan^{-1}(y/x)$.

The complex function $w(z)$ is *analytic* and has a unique derivative dw/dz independent of the direction of differentiation within the complex z-plane. This condition leads to the *Cauchy-Riemann conditions*:

$$\frac{\partial \phi}{\partial x} = \frac{\partial \psi}{\partial y}, \quad \text{and} \quad \frac{\partial \phi}{\partial y} = -\frac{\partial \psi}{\partial x} \tag{6.44}$$

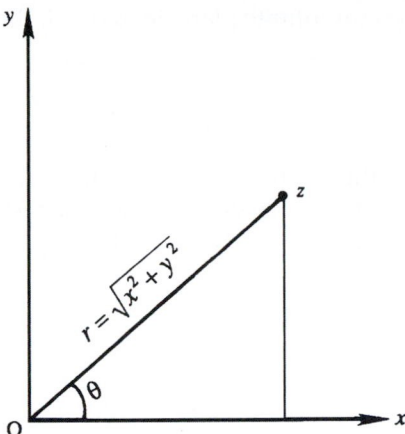

FIGURE 6.17 The complex plane where $z = x + iy = re^{i\theta}$ is the independent complex variable, $i = \sqrt{-1}$, $r = (x^2 + y^2)^{1/2}$, and $\tan\theta = y/x$.

where derivatives of $w(z)$ in the x- and iy-directions are computed separately and then equated. These equations imply that lines of constant ϕ and ψ are orthogonal. Points in the z-plane where w or dw/dz is zero or infinite are called *singularities* and at these points this orthogonality is lost.

If ϕ is interpreted as the velocity potential and ψ as the stream function, then w is the *complex potential* for the flow and (6.43) ensures the equality of the u,v-velocity components. Here, the *complex velocity* can be determined from

$$dw/dz = u - iv. \tag{6.45}$$

Applying the *Cauchy-Riemann conditions* to the complex velocity in (6.45) leads to the conditions for incompressible (6.2) and irrotational (6.9) flow, and Laplace equations for ϕ and ψ, (6.5) and (6.12), respectively. Thus, any twice-differentiable complex function of $z = x + iy$ produces solutions to Laplace's equation in the (x,y)-plane, a genuinely remarkable result! In general, a function of the two variables (x,y) may be written as $f(z, z^*)$ where $z^* = x - iy$ is the complex conjugate of z. Thus, it is the special case when $f(z, z^*) = w(z)$ alone that is considered here.

With these formal mathematical results, the correspondence between $w(z)$ and the prior results for ψ and ϕ are summarized in this and the following short paragraphs. The complex potential for flow in a corner of angle $\alpha = \pi/n$ is obtained from a power law in z,

$$w(z) = Az^n = A(re^{i\theta})^n = Ar^n(\cos n\theta + i \sin n\theta) \text{ for } n \geq 1/2, \tag{6.46}$$

where A is a real constant. When $n = 2$, the streamline pattern, $\psi = \text{Im}\{w\} = Ar^2\sin 2\theta$, represents flow in a region bounded by perpendicular walls, and (6.24) and (6.27) are readily recovered from (6.46). By including the field within the second quadrant of the z-plane, it is clear that $n = 2$ also represents the flow impinging against a flat wall (Figure 6.3). The streamlines and equipotential lines are all rectangular hyperbolas. This is called a *stagnation flow* because it includes a stagnation point. For comparison, the streamline pattern for $n = 1/2$

corresponds to flow around a semi-infinite plate. In general, the complex velocity computed from (6.46) is:

$$dw/dz = nAz^{n-1} = (A\pi/\alpha)z^{(\pi-\alpha)/\alpha},$$

which shows that $dw/dz = 0$ at the origin for $\alpha < \pi$ while $dw/dz \to \infty$ at the origin for $\alpha > \pi$. Thus, in this flow the origin is a stagnation point for flow in a wall angle smaller than 180°; in contrast, it is a point of infinite velocity for wall angles larger than 180°. In both cases it is a singular point.

The complex potential for an irrotational vortex of strength Γ at (x', y'), the equivalent of (6.8), is:

$$w(z) = -\frac{i\Gamma}{2\pi}\ln(z - z') = \frac{\Gamma}{2\pi}\theta' - i\frac{\Gamma}{2\pi}\ln r', \tag{6.47}$$

where $z' = x' + iy'$, $r' = \sqrt{(x - x')^2 + (y - y')^2}$, and $\theta' = \tan^{-1}((y - y')/(x - x'))$.

The complex potential for a source or sink of volume flow rate m per unit depth located at (x', y'), the equivalent of (6.15), is:

$$w(z) = \frac{m}{2\pi}\ln(z - z') = \frac{m}{2\pi}\ln(r'e^{i\theta'}) = \frac{m}{2\pi}\ln r' + i\frac{m\theta'}{2\pi}. \tag{6.48}$$

The complex potential for a doublet with dipole strength $-d\mathbf{e}_x$ located at (x', y'), the equivalent of (6.29), is:

$$w = \frac{d}{2\pi(z - z')}. \tag{6.49}$$

The complex potential for uniform flow at speed U past a half body—see (6.29), (6.30), and Figure 6.7—is the combination of a source of strength m at the origin and a uniform horizontal stream,

$$w(z) = Uz + \frac{m}{2\pi}\ln z. \tag{6.50}$$

The complex potential for uniform flow at speed U past a circular cylinder, see (6.33) and Figure 6.9, is the combination of a doublet with dipole strength $\mathbf{d} = -2\pi Ua^2\mathbf{e}_x$ and a uniform stream:

$$w = U\left(z + \frac{a^2}{z}\right). \tag{6.51}$$

When *clockwise* circulation Γ is added to the cylinder, the complex potential becomes

$$w = U\left(z + \frac{a^2}{z}\right) + \frac{i\Gamma}{2\pi}\ln(z/a), \tag{6.52}$$

the flow field is altered (see Figure 6.12), and the cylinder experiences a lift force. Here, the imaginary part of (6.52) reproduces (6.36).

The method of images also applies to the complex potential. For example, the complex potential for the flow described by (6.41) is:

$$w = \frac{m}{2\pi}\ln\left(\frac{z-a}{a}\right) + \frac{m}{2\pi}\ln\left(\frac{z+a}{a}\right) = \frac{m}{2\pi}\ln\left(\frac{z^2 - a^2}{a^2}\right) = \frac{m}{2\pi}\ln\left(\frac{x^2 - y^2 - a^2 + 2ixy}{a^2}\right). \tag{6.53}$$

The complex variable description of ideal flow also allows some very general results to be obtained for pressure forces (per unit depth perpendicular to the plane of the flow) that act on two-dimensional bodies.

6.5. FORCES ON A TWO-DIMENSIONAL BODY

In Section 3 we demonstrated that the drag on a circular cylinder in steady flow is zero while the lift equals $L = \rho U\Gamma$ when the circulation is clockwise. These results are also valid for any object with an *arbitrary* noncircular cross section that does not vary in the z-direction.

Blasius Theorem

Consider a stationary object of this type with extent B perpendicular to the plane of the flow, and let D (drag) be the stream-wise (x) force component and L (lift) be cross-stream or lateral (y) force (per unit depth) exerted on the object by the surrounding fluid. Thus, from Newton's third law, the total force applied to the fluid by the object is $\mathbf{F} = -B(D\mathbf{e}_x + L\mathbf{e}_y)$. For steady irrotational constant-density flow, conservation of momentum (4.17) within a stationary control volume implies:

$$\int_{A^*} \rho\mathbf{u}(\mathbf{u}\cdot\mathbf{n})dA = -\int_{A^*} p\mathbf{n}dA + \mathbf{F}. \tag{6.54}$$

If the control surface A^* is chosen to coincide with the body surface and the body is not moving, then $\mathbf{u}\cdot\mathbf{n} = 0$ and the flux integral on the left in (6.54) is zero, so

$$D\mathbf{e}_x + L\mathbf{e}_y = -\frac{1}{B}\int_{A*} p\mathbf{n}dA. \tag{6.55}$$

If C is the contour of the body's cross section, then $dA = Bds$ where $d\mathbf{s} = \mathbf{e}_x dx + \mathbf{e}_y dy$ is an element of C and $ds = [(dx)^2 + (dy)^2]^{1/2}$. By definition, $\mathbf{n}$ must have unit magnitude, must be perpendicular to $d\mathbf{s}$, and must point outward from the control volume, so $\mathbf{n} = (\mathbf{e}_x dy - \mathbf{e}_y dx)/ds$. Using these relationships for $\mathbf{n}$ and dA, (6.55) can be separated into force components,

$$D\mathbf{e}_x + L\mathbf{e}_y = -\frac{1}{B}\oint_C p\frac{(\mathbf{e}_x dy - \mathbf{e}_y dx)}{ds}Bds = \left(-\oint_C pdy\right)\mathbf{e}_x + \left(\oint_C pdx\right)\mathbf{e}_y, \tag{6.56}$$

to identify the contour integrals leading to D and L. Here, C must be traversed in the counterclockwise direction.

Now switch from the physical domain to the complex z-plane to make use of the complex potential. This switch is accomplished here by replacing ds with $dz = dx + idy$ and exploiting

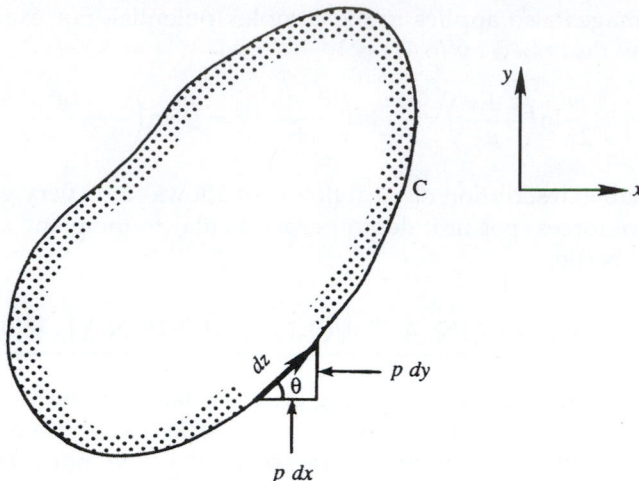

FIGURE 6.18   Elemental forces in a plane on two-dimensional object. Here the elemental horizontal and vertical force components (per unit depth) are $-pdy$ and $+pdx$, respectively.

the dichotomy between real and imaginary parts to keep track of horizontal and vertical components (see Figure 6.18). To achieve the desired final result, construct the complex force,

$$D - iL = \left(-\oint_C pdy\right) - i\left(\oint_C pdx\right) = -i\oint_C p(dx - idy) = -i\oint_C p\,dz^*, \qquad (6.57)$$

where * denotes a complex conjugate. The pressure p is found from the Bernoulli equation (6.18),

$$p_\infty + \tfrac{1}{2}\rho U^2 = p + \tfrac{1}{2}\rho(u^2 + v^2) = p + \tfrac{1}{2}\rho(u - iv)(u + iv),$$

where p_∞ and U are the pressure and horizontal flow speed far from the body. Inserting this into (6.57) produces:

$$D - iL = -i\oint_C \left[p_\infty + \tfrac{1}{2}\rho U^2 - \tfrac{1}{2}\rho(u - iv)(u + iv)\right]dz^*. \qquad (6.58)$$

The integral of the constant terms, $p_\infty + \rho U^2/2$, around a closed contour is zero. The body-surface velocity vector and the surface element $dz = |dz|e^{i\theta}$ are parallel, so $(u + iv)dz^*$ can be rewritten

$$(u + iv)dz^* = \left[u^2 + v^2\right]^{1/2}e^{i\theta}|dz|e^{-i\theta} = \left[u^2 + v^2\right]^{1/2}e^{-i\theta}|dz|e^{i\theta} = (u - iv)dz = (dw/dz)dz,$$
$$(6.59)$$

where (6.45) has been used for the final equality. Thus, (6.58) reduces to

$$D - iL = \frac{i\rho}{2}\oint_C \left(\frac{dw}{dz}\right)^2 dz, \qquad (6.60)$$

a result known as the *Blasius theorem*. It applies to any plane steady ideal flow. Interestingly, the integral need not be carried out along the contour of the body because the theory of complex variables allows *any contour surrounding the body to be chosen* provided there are no singularities in $(dw/dz)^2$ between the body and the contour chosen.

Kutta–Zhukhovsky Lift Theorem

The Blasius theorem can be readily applied to an arbitrary cross-section object around which there is circulation $-\Gamma$. The flow can be considered a superposition of a uniform stream and a set of singularities such as vortex, doublet, source, and sink.

As there are no singularities outside the body, we shall take the contour C in the Blasius theorem at a very large distance from the body. From large distances, all singularities appear to be located near the origin $z = 0$, so the complex potential on the contour C will be of the form:

$$w = Uz + \frac{m}{2\pi}\ln z + \frac{i\Gamma}{2\pi}\ln z + \frac{d}{2\pi z} + \dots$$

When U, m, Γ, and d are positive and real, the first term represents a uniform flow in the x-direction, the second term represents a net source of fluid, the third term represents a clockwise vortex, and the fourth term represents a doublet. Because the body contour is closed, there can be no net flux of fluid into the domain. The sinks must scavenge all the flow introduced by the sources, so $m = 0$. The Blasius theorem, (6.60), then becomes

$$D - iL = \frac{i\rho}{2}\oint_C \left(U + \frac{i\Gamma}{2\pi z} - \frac{d}{2\pi z^2} + \dots\right)^2 dz = \frac{i\rho}{2}\oint_C \left(U^2 + \frac{iU\Gamma}{\pi}\frac{1}{z} + \left(\frac{Ud}{\pi} - \frac{\Gamma^2}{4\pi^2}\right)\frac{1}{z^2} + \dots\right)^2 dz.$$

$$(6.61)$$

To evaluate the contour integral in (6.61), we simply have to find the coefficient of the term proportional to $1/z$ in the integrand. This coefficient is known as the *residue* at $z = 0$ and the residue theorem of complex variable theory states that the value of a contour integral like (6.61) is $2\pi i$ times the sum of the residues at all singularities inside C. Here, the only singularity is at $z = 0$, and its residue is $iU\Gamma/\pi$, so

$$D - iL = \frac{i\rho}{2}2\pi i\left(\frac{iU\Gamma}{\pi}\right) = -i\rho U\Gamma \text{ or } D = 0 \text{ and } L = \rho U\Gamma. \quad (6.62)$$

Thus, there is no drag on an arbitrary-cross-section object in steady two-dimensional, irrotational constant-density flow, a more general statement of d'Alembert's paradox. Given that nonzero drag forces are an omnipresent fact of everyday life, this might seem to eliminate any practical utility for ideal flow. However, there are at least three reasons to avoid this presumption. First of all, ideal flow streamlines indicate what a real flow should look like to achieve minimum pressure drag. Lower drag on real objects is often realized when object-geometry changes are made or boundary-layer, separation-control strategies are implemented that allow real-flow streamlines to better match their ideal-flow counterparts. Second, the predicted circulation-dependent force on the object perpendicular to the oncoming stream—the lift force, $L = \rho U\Gamma$—is basically correct. The result (6.62) is called the *Kutta-Zhukhovsky lift theorem*, and it plays a fundamental role in aero- and hydrody-namics. As described in Chapter 14, the circulation developed by an air- or hydrofoil is nearly proportional to U, so L is nearly proportional to U^2. And third, the influence of viscosity in real fluid flows takes some time to develop, so impulsively started flows and rapidly oscil-lating flows (i.e., acoustic fluctuations) often follow ideal flow streamlines.

6.6. CONFORMAL MAPPING

We shall now introduce a method by which complex flow patterns can be transformed into simple ones using a technique known as *conformal mapping* in complex variable theory. Consider the functional relationship $w = f(z)$, which maps a point in the w-plane to a point in the z-plane, and vice versa. We shall prove that infinitesimal figures in the two planes preserve their geometric similarity if $w = f(z)$ is analytic. Let lines C_z and C_z' in the z-plane be transforma-tions of the curves C_w and C_w' in the w-plane, respectively (Figure 6.19). Let δz, $\delta'z$, δw, and $\delta'w$ be infinitesimal elements along the curves as shown. The four elements are related by

$$\delta w = \frac{dw}{dz}\delta z, \tag{6.63}$$

$$\delta'w = \frac{dw}{dz}\delta'z. \tag{6.64}$$

If $w = f(z)$ is analytic, then dw/dz is independent of orientation of the elements, and therefore has the same value in (6.63) and (6.64). These two equations then imply that the elements δz and $\delta'z$ are rotated by the *same amount* (equal to the argument of dw/dz) to obtain the elements δw and $\delta'w$. It follows that

$$\alpha = \beta,$$

which demonstrates that infinitesimal figures in the two planes are geometrically similar. The demonstration fails at singular points at which dw/dz is either zero or infinite. Because dw/dz is a function of z, the amount of magnification and rotation that an element δz undergoes during transformation from the z-plane to the w-plane varies. Consequently, *large* figures become distorted during the transformation.

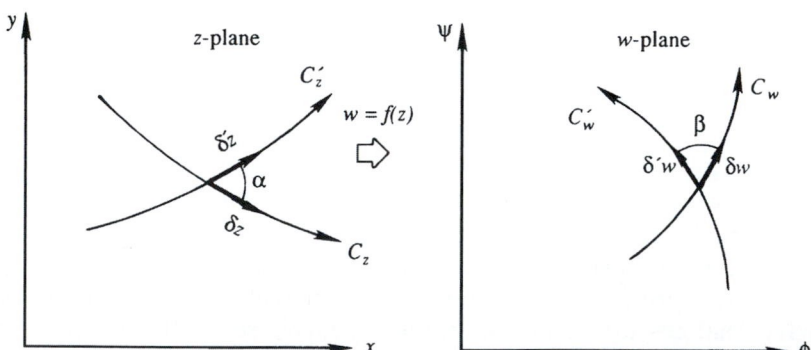

FIGURE 6.19 Preservation of geometric similarity of small elements in conformal mapping between the complex z- and w-planes.

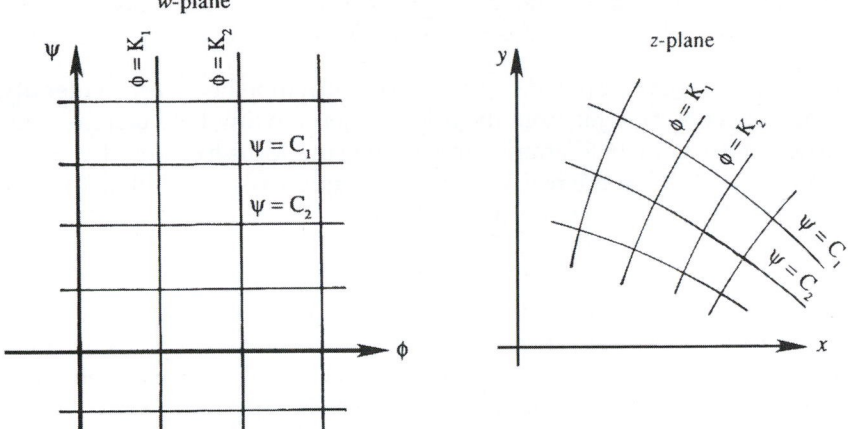

FIGURE 6.20 Flow patterns in the complex w-plane and the z-plane. The w-plane represents uniform flow with straight potential lines and streamlines. In the z-plane these lines curve to represent the flow of interest.

In application of conformal mapping, we always choose a rectangular grid in the w-plane consisting of constant ϕ and ψ lines (Figure 6.20). In other words, we define ϕ and ψ to be the real and imaginary parts of w:

$$w = \phi + i\psi.$$

The rectangular net in the w-plane represents a uniform flow in this plane. The constant ϕ and ψ lines are transformed into certain curves in the z-plane through the transformation $w = f(z)$ or its inverse $f^{-1}(w) = z$. *The pattern in the z-plane is the physical pattern under investigation*, and the images of constant ϕ and ψ lines in the z-plane form the equipotential lines and stream-lines, respectively, of the desired flow. We say that $w = f(z)$ transforms a uniform flow in the w-plane into the desired flow in the z-plane. In fact, all the preceding flow patterns studied through the transformation $w = f(z)$ can be interpreted this way.

If the physical pattern under investigation is too complicated, we may introduce intermediate transformations in going from the w-plane to the z-plane. For example, the transformation $w = \ln(\sin z)$ can be broken into

$$w = \ln \zeta \qquad \zeta = \sin z.$$

Velocity components in the z-plane are given by

$$u - iv = \frac{dw}{dz} = \frac{dw}{d\zeta}\frac{d\zeta}{dz} = \frac{1}{\zeta}\cos z = \cot z.$$

As a simple example of conformal mapping, consider the transformation, $w = \phi + i\psi = z^2 = x^2 + y^2 + 2ixy$. Streamlines are given by $\psi = const = 2xy$, rectangular hyperbolae (see Figure 6.5). Here uniform flow in the w-plane has been mapped onto flow in a 90° corner in the z-plane by this transformation. A more involved example follows. Additional applications are discussed in Chapter 14.

The Zhukhovsky transformation relates two complex variables z and ζ, and has important applications in airfoil theory,

$$z = \zeta + \frac{b^2}{\zeta}. \tag{6.65}$$

When $|\zeta|$ or $|z|$ is very large compared to b, this transformation becomes an identity, so it does not change the flow condition far from the origin when moving between the z and ζ planes. However, close to the origin, (6.65) transforms a circle of radius b centered at the origin of the ζ-plane into a line segment on the real axis of the z-plane. To establish this, let $\zeta = b\exp(i\theta)$ on the circle (Figure 6.21) so that (6.65) provides the corresponding point in the z-plane as:

$$z = be^{i\theta} + be^{-i\theta} = 2b\cos\theta.$$

As θ varies from 0 to π, z goes along the x-axis from $2b$ to $-2b$. As θ varies from π to 2π, z goes from $-2b$ to $2b$. The circle of radius b in the ζ-plane is thus transformed into a line segment of length $4b$ in the z-plane. The region *outside* the circle in the ζ-plane is mapped into the *entire z-*plane. It can be shown that the region inside the circle is also transformed into the entire z-plane. This, however, is of no concern to us because we shall not consider the interior of the circle in the ζ-plane.

Now consider a circle of radius $a > b$ in the ζ-plane (Figure 6.21). A point $\zeta = a\exp(i\theta)$ on this circle is transformed to

$$z = ae^{i\theta} + \frac{b^2}{a}e^{-i\theta}, \tag{6.66}$$

which traces out an ellipse for various values of θ; the geometry becomes clear by separating real and imaginary parts of (6.66) and eliminating θ:

$$\frac{x^2}{(a + b^2/a)^2} + \frac{y^2}{(a - b^2/a)^2} = 1. \tag{6.67}$$

For various values of $a > b$, (6.67) represents a family of ellipses in the z-plane, with foci at $x = \pm 2b$.

The flow around one of these ellipses (in the z-plane) can be determined by first finding the flow around a circle of radius a in the ζ-plane, and then using (6.65) to go to the z-plane.

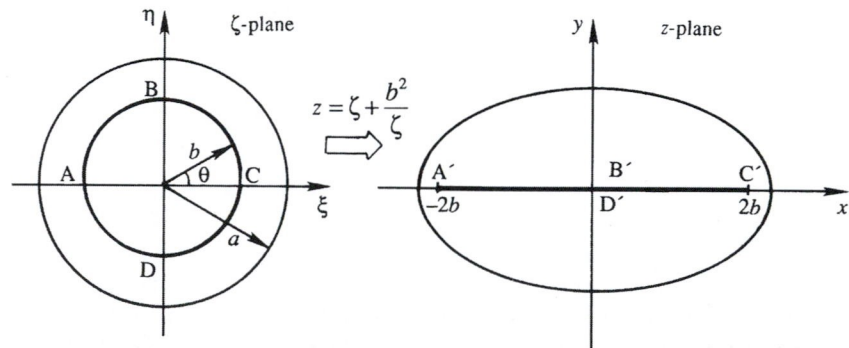

FIGURE 6.21 Transformation of a circle of radius a in the ζ-plane into an ellipse in the z-plane by means of the Zhukhovsky transformation $z = \zeta + b^2/\zeta$. A circle of radius b in the ζ-plane transforms into a line segment between $z = \pm 2b$ in the z-plane.

To be specific, suppose the desired flow in the z-plane is that of flow around an elliptic cylinder with clockwise circulation Γ placed in a stream moving at U. The corresponding flow in the ζ-plane is that of flow with the same circulation around a circular cylinder of radius a placed in a horizontal stream of speed U. The complex potential for this flow is (6.52) with z replaced by ζ:

$$w = U\left(\zeta + \frac{a^2}{\zeta}\right) + \frac{i\Gamma}{2\pi}\ln(\zeta/a). \tag{6.68}$$

The complex potential $w(z)$ in the z-plane can be found by substituting the inverse of (6.65),

$$\zeta = \frac{1}{2}z + \frac{1}{2}(z^2 - 4b^2)^{1/2}, \tag{6.69}$$

into (6.68). Here, the negative root, which falls inside the cylinder, has been excluded from (6.69). Instead of finding the complex velocity in the z-plane by directly differentiating $w(z)$, it is easier to find it as

$$u - iv = \frac{dw}{dz} = \frac{dw}{d\zeta}\frac{d\zeta}{dz}.$$

The resulting flow around an elliptic cylinder with circulation is qualitatively quite similar to that around a circular cylinder as shown in Figure 6.12.

6.7. NUMERICAL SOLUTION TECHNIQUES IN TWO DIMENSIONS

Exact solutions can be obtained only for flows with relatively simple geometries, so approximate methods of solution become necessary for complicated geometries. One of these approximate methods is that of building up a flow by superposing a distribution of sources and sinks; this method is illustrated in Section 6.8 for axisymmetric flows. Another method is to apply perturbation techniques by assuming that the body is thin. A third method is to solve the Laplace equation numerically. In this section we shall illustrate the numerical method in its simplest form without worrying about computational efficiency. It is hoped that the reader will have an opportunity to learn numerical methods that are becoming increasingly important in the applied sciences in a separate study. Introductory material on several important techniques of computational fluid dynamics is provided in Chapter 10.

Numerical techniques for solving the Laplace equation typically rely on discretizing the spatial domain. When using finite difference techniques, the flow field is discretized into a system of *grid points*, and field derivatives are computed by taking differences between field values at adjacent grid points. Let the coordinates of a point be represented by

$$x = i\Delta x(i = 0, 1, 2, \ldots), \text{ and } y = j\Delta y(j = 0, 1, 2, \ldots).$$

Here, Δx and Δy are the dimensions of a grid cell, and the integers i and j are the indices associated with a grid point (Figure 6.22). The stream function $\psi(x,y)$ can be represented at these discrete locations by

$$\psi(x, y) = \psi\left(i\,\Delta x, j\,\Delta y\right) \equiv \psi_{i,j},$$

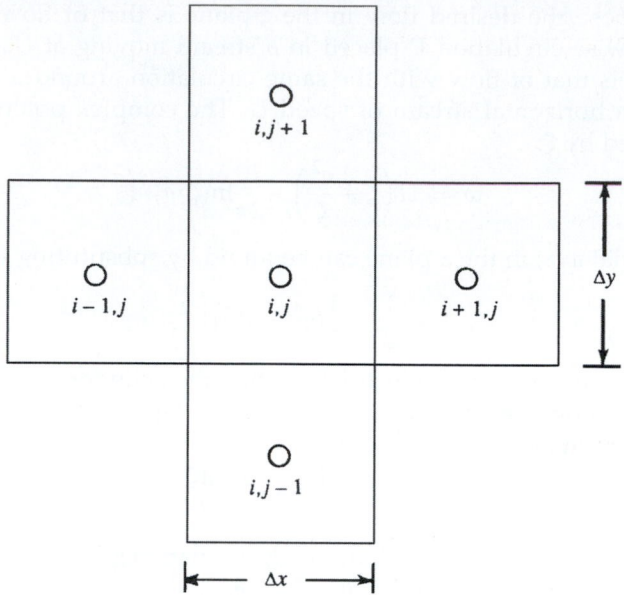

FIGURE 6.22   Adjacent grid boxes in a numerical calculation showing how the indices i and j change.

where $\psi_{i,j}$ is the value of ψ at the grid point (i, j) (the comma notation for derivatives is not applied here). In finite difference form, the first derivatives of ψ_i are approximated as first-order central differences:

$$\left(\frac{\partial \psi}{\partial x}\right)_{i,j} \simeq \frac{1}{\Delta x}\left(\psi_{i+\frac{1}{2},j} - \psi_{i-\frac{1}{2},j}\right), \text{ and } \left(\frac{\partial \psi}{\partial y}\right)_{i,j} \simeq \frac{1}{\Delta y}\left(\psi_{i,j+\frac{1}{2}} - \psi_{i,j-\frac{1}{2}}\right).$$

The quantities on the right side of each equation (such as $\psi_{i+1/2,j}$) are halfway between the grid points and therefore undefined. However, this potential difficulty is avoided here because the Laplace equation does not involve first derivatives. Applying the same approach to the second-order x-derivative produces:

$$\begin{aligned}
\left(\frac{\partial^2 \psi}{\partial x^2}\right)_{i,j} &\simeq \frac{1}{\Delta x}\left[\left(\frac{\partial \psi}{\partial x}\right)_{i+\frac{1}{2},j} - \left(\frac{\partial \psi}{\partial x}\right)_{i-\frac{1}{2},j}\right], \\
&\simeq \frac{1}{\Delta x}\left[\frac{1}{\Delta x}\left(\psi_{i+1,j} - \psi_{i,j}\right) - \frac{1}{\Delta x}\left(\psi_{i,j} - \psi_{i-1,j}\right)\right], \\
&= \frac{1}{\Delta x^2}\left[\psi_{i+1,j} - 2\psi_{i,j} + \psi_{i-1,j}\right].
\end{aligned} \tag{6.70}$$

Similarly,

$$\left(\frac{\partial^2 \psi}{\partial y^2}\right)_{i,j} \simeq \frac{1}{\Delta y^2}\left[\psi_{i,j+1} - 2\psi_{i,j} + \psi_{i,j-1}\right]. \tag{6.71}$$

Using (6.70) and (6.71), the Laplace equation (6.5) for the stream function in a plane two-dimensional flow has a finite difference representation:

$$\frac{1}{\Delta x^2}\left[\psi_{i+1,j} - 2\psi_{i,j} + \psi_{i-1,j}\right] + \frac{1}{\Delta y^2}\left[\psi_{i,j+1} - 2\psi_{i,j} + \psi_{i,j-1}\right] = 0.$$

Taking $\Delta x = \Delta y$, for simplicity, this reduces to

$$\psi_{i,j} = \frac{1}{4}\left[\psi_{i-1,j} + \psi_{i+1,j} + \psi_{i,j-1} + \psi_{i,j+1}\right], \tag{6.72}$$

which shows that ψ satisfies the Laplace equation if its value at a grid point equals the average of the values at the four surrounding points.

Equation (6.72) can be solved by a simple iteration technique when the values of ψ are given on the boundary. First consider a readily countable number of grid points covering the rectangular region of Figure 6.23 where the flow field is discretized with 16 grid points with $1 \le i, j \le 4$. Of these, the values of ψ are presumed known at the 12 boundary points indicated by open circles. The values of ψ at the four interior points indicated by solid circles are unknown. For these interior points, (6.72) gives

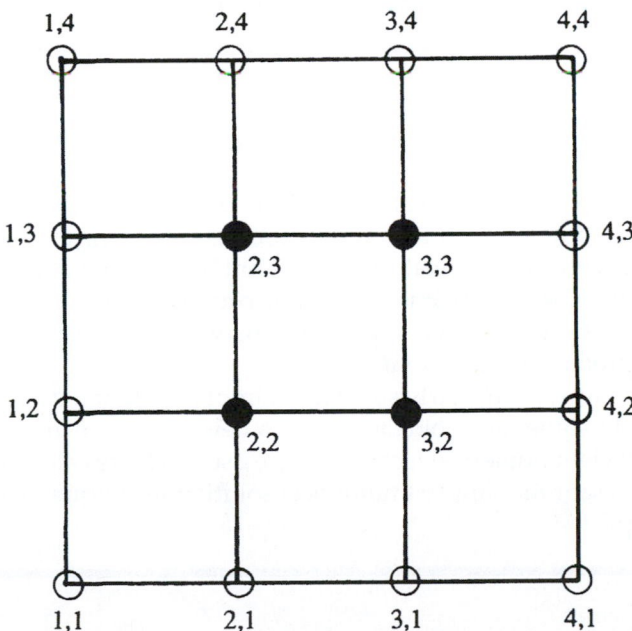

FIGURE 6.23 Network of sixteen grid points arrayed in a rectangular grid. The i,j-values at each point are listed. Boundary points with known values are indicated by open circles. The four interior points with unknown values are indicated by solid circles.

$$\psi_{2,2} = \frac{1}{4}\Big[\psi_{1,2}^{B} + \psi_{3,2} + \psi_{2,1}^{B} + \psi_{2,3}\Big],$$

$$\psi_{3,2} = \frac{1}{4}\Big[\psi_{2,2} + \psi_{4,2}^{B} + \psi_{3,1}^{B} + \psi_{3,3}\Big],$$

(6.73)

$$\psi_{2,3} = \frac{1}{4}\Big[\psi_{1,3}^{B} + \psi_{3,3} + \psi_{2,2} + \psi_{2,4}^{B}\Big],$$

$$\psi_{3,3} = \frac{1}{4}\Big[\psi_{2,3} + \psi_{4,3}^{B} + \psi_{3,2} + \psi_{3,4}^{B}\Big].$$

Here, the known boundary values have been indicated by a superscript "B." The equation set (6.73) represents four linear algebraic equations in four unknowns and is therefore solvable.

In practice, however, a much larger number of grid points will likely be used to represent the flow field, and a computer will determine a numerical solution of the resulting large number of simultaneous algebraic equations. One of the simplest techniques of solving such a large equation set is the *iteration method* where an initial solution guess is gradually improved and updated until (6.73) is satisfied at every point. For example, suppose the initial guesses for ψ at the four unknown points of Figure 6.23 are all zero. From (6.73), the first estimate of $\psi_{2,2}$ can be computed as

$$\psi_{2,2} = \frac{1}{4}\Big[\psi_{1,2}^{B} + 0 + \psi_{2,1}^{B} + 0\Big].$$

The initial zero value for $\psi_{2,2}$ is now replaced by this *updated* value. The first estimate for the next grid point is then obtained as

$$\psi_{3,2} = \frac{1}{4}\Big[\psi_{2,2} + \psi_{4,2}^{B} + \psi_{3,1}^{B} + 0\Big],$$

where the updated value of $\psi_{2,2}$ has been used on the right-hand side. In this manner, we can sweep over the entire flow domain in a systematic manner, *always using the latest available value at each point*. Once the first estimate at every point has been obtained, the domain sweep can be repeated to obtain second estimates and this process can be repeated again and again until the values of $\psi_{i,j}$ do not change appreciably between two successive sweeps. At this point, the iteration process has *converged*.

The foregoing scheme is particularly suitable for implementation using a computer, where it is easy to replace old values at a point as soon as a new value is available. In practice, more sophisticated and efficient numerical techniques are used in large calculations. However, the purpose here is to present the simplest numerical solution technique, which is illustrated in the following example.

EXAMPLE 6.2

Figure 6.24 shows a contraction in a channel through which the flow rate per unit depth is 5 m^2/s. The velocity is uniform and parallel across the inlet and outlet sections. Find the stream function values within this flow field.

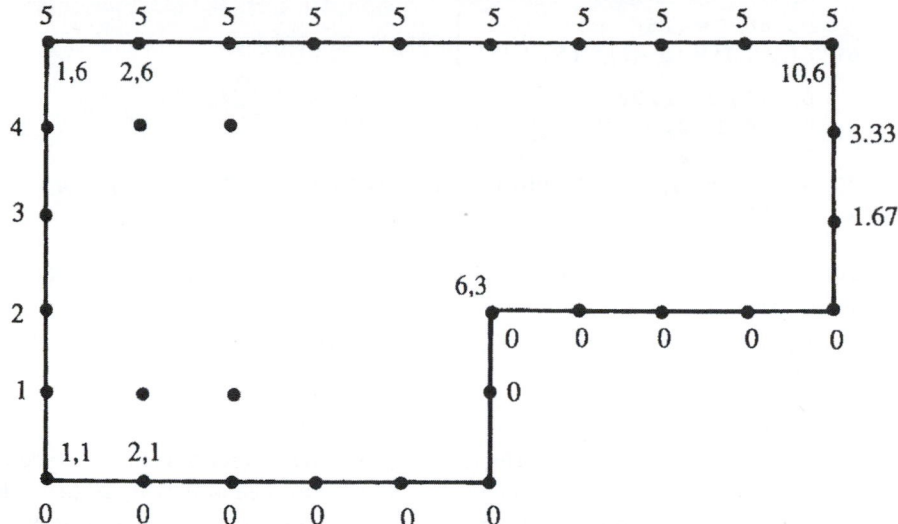

FIGURE 6.24 Grid pattern for irrotational flow through an abrupt, sharp-cornered contraction (Example 6.2). The flow enters on the left and exits on the right. The boundary values of ψ are indicated on the outside. The values of i, j for some grid points are indicated on the inside.

Solution

The region of flow is two-dimensional but is singly connected because the flow field *is interior* to the boundaries and every fluid-only circuit can be reduced to a point. Therefore, the problem has a unique solution that can be found numerically.

The difference in ψ values is equal to the flow rate between two streamlines. Thus, setting $\psi = 0$ at the bottom wall requires $\psi = 5$ m^2/s at the top wall. We divide the field into the system of grid points shown, with $\Delta x = \Delta y = 1$ m. Because $\Delta\psi/\Delta y$ ($= u$) is given to be uniform across the inlet and the outlet, we must have $\Delta\psi = 1$ m^2 at the inlet and $\Delta\psi = 5/3 = 1.67$ m^2/s at the outlet. The resulting values of ψ at the boundary points are indicated in Figure 6.25. A FORTRAN code for solving the problem is as follows:

```
DIMENSION S(10, 6)

      DO 10 I = 1, 6
10 S(I, 1) = 0.
      DO 20 I = 2, 3
20 S(6, J) = 0.                  Set ψ on top and bottom walls
      DO 30 I = 7, 10
30 S(I, 3) = 0.
      DO 40 I = 1, 10
40 S(I, 6) = 5.

      DO 50 J = 2, 6
50 S(1, J) = J - 1.              Set ψ at inlet
```

```
        DO 60 J = 4, 6                      ⎫
     60 S(10, J) = (J - 3) * (5. / 3.)      ⎬ Set ψ at outlet
                                            ⎭
        DO 100 N = 1, 20
        DO 70 I = 2, 5
        DO 70 J = 2, 5
     70 S(I, J) = (S(I, J + 1) + S(I, J-1) + S(I + 1, J) + S(I - 1, J)) / 4.
        DO 80 I = 6, 9
        DO 80 J = 4, 5
     80 S(I, J) = (S(I, J + 1) + S(I, J - 1) + S(I + 1, J) + S(I - 1, J)) / 4.
    100 CONTINUE
        PRINT 1, ((S(I, J), I = 1, 10), J = 1, 6)
      1 FORMAT (' ', 10 E 12.4')
        END
```

Here, S denotes the stream function ψ. The code first sets the boundary values. The iteration is performed in the N loop. In practice, iterations will not be performed arbitrarily 20 times. Instead the convergence of the iteration process will be checked, and the process is continued until some reasonable criterion (such as less than 1% change at every point) is met. However, some caution is appropriate. To be sure a numerical solution has been obtained, all the terms in the field equation must be calculated and the satisfaction of the field equation by the numerical solution must be verified. Such improvements are easy to implement, so the code above is left in its simplest form. The values of ψ at the grid points after 50 iterations, and the corresponding streamlines, are shown in Figure 6.25.

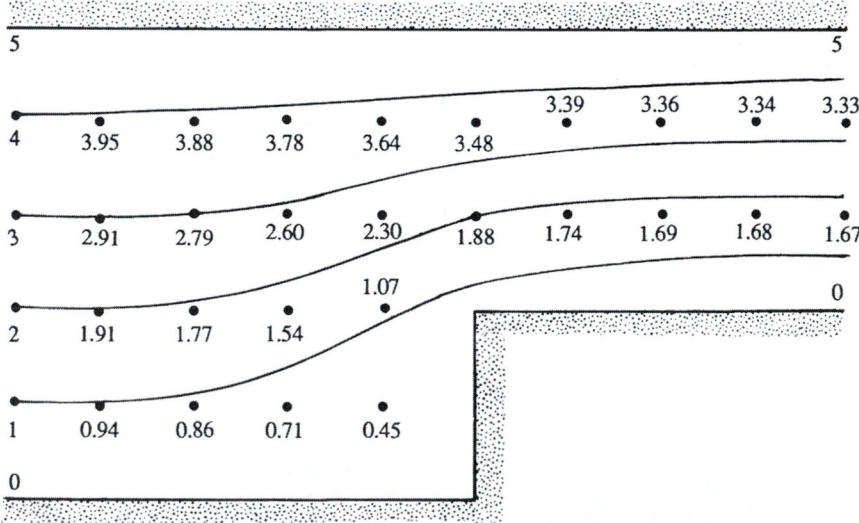

FIGURE 6.25 Numerical solution of Example 6.2 for the boundary conditions shown in Figure 6.24. The numerical values listed are for the corresponding grid points shown as black dots. Note that there is some streamline curvature after the abrupt contraction.

6.8. AXISYMMETRIC IDEAL FLOW

Two stream functions are required to describe a fully three-dimensional flow (Section 4.3). However, when the flow is axisymmetric, a single stream function can again be used. Thus, the development presented here parallels that in Section 6.2 for plane ideal flow.

In (R, φ, z) cylindrical coordinates, the axisymmetric incompressible continuity equation is:

$$\frac{1}{R}\frac{\partial}{\partial R}(Ru_R) + \frac{\partial u_z}{\partial z} = 0. \tag{6.74}$$

As discussed near the end of Section 4.3, this equation may be solved by choosing the first three-dimensional stream function $\chi = -\varphi$, so that $\mathbf{u} = (u_R, 0, u_z) = \nabla\chi \times \nabla\psi = -(1/R)\mathbf{e}_\varphi \times \nabla\psi$, which implies

$$u_R = -\frac{1}{R}\frac{\partial\psi}{\partial z}, \text{ and } u_z = \frac{1}{R}\frac{\partial\psi}{\partial R}. \tag{6.75}$$

Substituting these into the equation for the φ-component of vorticity,

$$\omega_\varphi = \frac{\partial u_R}{\partial z} - \frac{\partial u_z}{\partial R}, \tag{6.76}$$

produces the field equation for the axisymmetric stream function in irrotational flow:

$$\frac{\partial}{\partial R}\left(\frac{1}{R}\frac{\partial\psi}{\partial R}\right) + \frac{1}{R}\frac{\partial^2\psi}{\partial z^2} = -\omega_\varphi = 0. \tag{6.77}$$

This *is not* the two-dimensional Laplace equation. Therefore, the complex variable formulation for plane ideal flows does not apply to axisymmetric ideal flow.

The axisymmetric stream function is sometimes called the *Stokes stream function*. It has units of m^3/s, in contrast to the plane-flow stream function, which has units of m^2/s. Surfaces of $\psi =$ constant in axisymmetric flow are surfaces of revolution. The volume flow rate dQ between two axisymmetric stream surfaces described by constant values ψ and $\psi + d\psi$ (Figure 6.26) is

$$dQ = 2\pi R(\mathbf{u}\cdot\mathbf{n})ds = 2\pi R(-u_R dz + u_z dR) = 2\pi\left(\frac{\partial\psi}{\partial z}dz + \frac{\partial\psi}{\partial R}dR\right) = 2\pi d\psi, \tag{6.78}$$

where $\mathbf{u} = u_R\mathbf{e}_R + u_z\mathbf{e}_z$, $\mathbf{n} = (-\mathbf{e}_R dz + \mathbf{e}_z dR)/ds$, and (6.75) has been used. Note that as drawn in Figure 6.26 dz is negative. The form $d\psi = dQ/2\pi$ shows that the difference in ψ values is the flow rate between two concentric stream surfaces per unit radian angle around the axis. This is consistent with the discussion of stream functions in Section 4.3. The factor of 2π is absent in plane flows, where $d\psi = dQ$ is the flow rate per unit depth perpendicular to the plane of the flow.

Here also an axisymmetric potential function ϕ can be defined via $\mathbf{u} = \nabla\phi$ or

$$u_R = \partial\phi/\partial R \text{ and } u_z = \partial\phi/\partial z, \tag{6.79}$$

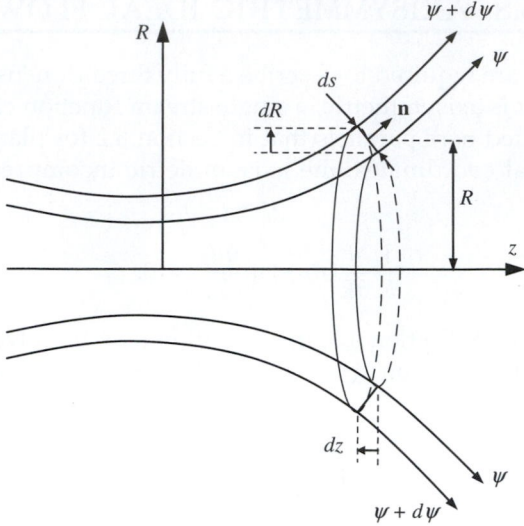

FIGURE 6.26 Geometry for calculating the volume flow rate between axisymmetric-flow stream surfaces with values of ψ and $\psi + d\psi$. The z-axis is the axis of symmetry, R is the radial distance from the z-axis, $ds = \mathbf{e}_R dR + \mathbf{e}_z dz$ is the distance between the two surfaces, and $\mathbf{n}$ is a unit vector perpendicular to ds. The volume flow rate between the two surfaces is $2\pi d\psi$.

so that the flow identically satisfies $\omega_\varphi = 0$. Substituting (6.79) into the incompressible continuity equation produces the field equation for axisymmetric potential function:

$$\frac{1}{R}\frac{\partial}{\partial R}\left(R\frac{\partial \phi}{\partial R}\right) + \frac{\partial^2 \phi}{\partial z^2} = 0. \tag{6.80}$$

While this *is* the axisymmetric Laplace equation, it is not the same as the two-dimensional Cartesian version.

In axisymmetric flow problems, both (R, φ, z)-cylindrical and (r, θ, φ)-spherical polar coordinates are commonly used. These are illustrated in Figure 3.3 with the z-axis and polar-axis vertical. The angle φ is the same in both systems. Axisymmetric flows are independent φ, and their velocity component, u_φ, in the φ direction is zero. In this section, we will commonly point the z-axis horizontal. Note that R is the radial distance from the axis of symmetry (the z-axis or polar axis) in cylindrical coordinates, whereas r is the distance from the origin in spherical coordinates. Important expressions for these curvilinear coordinates are listed in Appendix B. Several relevant expressions are provided here for easy reference.

Cylindrical	Spherical	
$x = R\cos\varphi$	$x = r\sin\theta\cos\varphi$	
$y = R\sin\varphi$	$y = r\sin\theta\sin\varphi$	(6.81)
$z = z$	$z = r\cos\theta$	

Cylindrical	Spherical	
Continuity equations (6.74)	$\dfrac{1}{r}\dfrac{\partial}{\partial r}(r^2 u_r) + \dfrac{1}{\sin\theta}\dfrac{\partial}{\partial\theta}(u_\theta \sin\theta) = 0$	(6.82)
Velocity components (6.75), (6.79)	$u_r = \dfrac{1}{r^2 \sin\theta}\dfrac{\partial\psi}{\partial\theta} = \dfrac{\partial\phi}{\partial r}, \; u_\theta = -\dfrac{1}{r\sin\theta}\dfrac{\partial\psi}{\partial r} = \dfrac{1}{r}\dfrac{\partial\phi}{\partial\theta}$	(6.83)
Vorticity (6.76)	$\omega_\varphi = \dfrac{1}{r}\left[\dfrac{\partial}{\partial r}(r u_\theta) - \dfrac{\partial u_r}{\partial\theta}\right]$	(6.84)
Laplace equation (6.80)	$\dfrac{1}{r^2}\dfrac{\partial}{\partial r}\left(r^2\dfrac{\partial\phi}{\partial r}\right) + \dfrac{1}{r^2 \sin\theta}\dfrac{\partial}{\partial\theta}\left(\sin\theta\dfrac{\partial\phi}{\partial\theta}\right) = 0$	(6.85)

Some simple examples of axisymmetric irrotational flows around bodies of revolution, such as spheres and airships, are provided in the rest of this section.

Axisymmetric ideal flows can be constructed from elementary solutions in the same manner as plane flows, except that complex variables cannot be used. Several elementary flows are tabulated here

Cylindrical	Spherical	
Uniform flow in the z direction $\phi = Uz, \; \psi = \dfrac{1}{2}UR^2$	$\phi = Ur\cos\theta, \; \psi = \dfrac{1}{2}Ur^2\sin^2\theta$	(6.86)
Point source of strength Q(m³/s) at the origin of coordinates $\phi = \dfrac{-Q}{4\pi\sqrt{R^2 + z^2}}, \; \psi = \dfrac{-Qz}{4\pi\sqrt{R^2 + z^2}}$	$\phi = -\dfrac{Q}{4\pi r}, \; \psi = -\dfrac{Q}{4\pi}\cos\theta$	(6.87)
Doublet with dipole strength −de_z at the origin of coordinates $\phi = \dfrac{d}{4\pi}\dfrac{z}{(R^2 + z^2)^{3/2}}, \; \psi = -\dfrac{d}{4\pi}\dfrac{R^2}{(R^2 + z^2)^{3/2}}$	$\phi = \dfrac{d}{4\pi r^2}\cos\theta, \; \psi = -\dfrac{d}{4\pi r}\sin^2\theta$	(6.88)

For these three flows, streamlines in any plane containing the axis of symmetry will be qualitatively similar to those of their two-dimensional counterparts.

Potential flow around a sphere can be generated by the superposition of a uniform stream Ue_z and an axisymmetric doublet opposing the stream of strength $d = 2\pi a^3 U$. In spherical coordinates, the stream and potential functions are:

$$\psi = \frac{1}{2}Ur^2\sin^2\theta - \frac{d}{4\pi r}\sin^2\theta = \frac{1}{2}Ur^2\left(1 - \frac{a^3}{r^3}\right)\sin^2\theta;$$

$$\phi = Ur\cos\theta + \frac{d}{4\pi r^2}\cos\theta = Ur\left(1 + \frac{a^3}{2r^3}\right)\cos\theta. \tag{6.89}$$

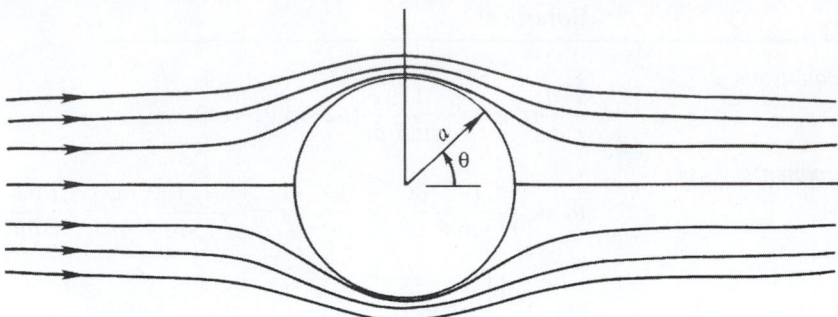

FIGURE 6.27 Axisymmetric streamlines for ideal flow past a sphere in a plane containing the axis of symmetry. The flow is fastest and the streamlines are closest together at $\theta = 90°$. The streamlines upstream and downstream of the sphere are the same, so there is no drag on the sphere.

This shows that $\psi = 0$ for $\theta = 0$ or π (for any $r \neq 0$), or for $r = a$ (for any θ). Thus the entire z-axis and the spherical surface of radius a form the stream surface $\psi = 0$. Streamlines for this flow are shown in Figure 6.27. The velocity components are:

$$u_r = \frac{1}{r^2 \sin\theta} \frac{\partial\psi}{\partial\theta} = U\left[1 - \left(\frac{a}{r}\right)^3\right]\cos\theta,$$

$$u_\theta = -\frac{1}{r\sin\theta}\frac{\partial\psi}{\partial r} = -U\left[1 + \frac{1}{2}\left(\frac{a}{r}\right)^3\right]\sin\theta. \tag{6.90}$$

The pressure coefficient on the sphere's surface is

$$C_p = \frac{p - p_\infty}{\frac{1}{2}\rho U^2} = 1 - \left(\frac{u_\theta}{U}\right)^2 = 1 - \frac{9}{4}\sin^2\theta, \tag{6.91}$$

which is fore-aft symmetrical, again demonstrating zero drag in steady ideal flow.

Interestingly, the potential for this flow can be rewritten to eliminate the dependence on the coordinate system. Start from the first equality for the potential in (6.89) and use $\mathbf{x} = r\mathbf{e}_r$, $|\mathbf{x}| = r$, $\cos\theta = \mathbf{e}_z \cdot \mathbf{e}_r$, $\mathbf{U} = U\mathbf{e}_z$, and $\mathbf{d} = -d\mathbf{e}_z$, to find:

$$\phi = Ur\cos\theta + \frac{d}{4\pi r^2}\cos\theta = U\mathbf{e}_z\cdot r\mathbf{e}_r - \frac{\mathbf{d}}{4\pi r^3}\cdot r\mathbf{e}_r$$

$$= \mathbf{U}\cdot\mathbf{x} - \frac{\mathbf{d}}{4\pi|\mathbf{x}|^3}\cdot\mathbf{x} = \left(\mathbf{U} - \frac{\mathbf{d}}{4\pi|\mathbf{x}|^3}\right)\cdot\mathbf{x}, \tag{6.92}$$

a result that will be useful in the next section.

As in plane flows, the motion around a closed body of revolution can be generated by superposition of a uniform stream and a collection of sources and sinks whose net strength is zero. The closed surface becomes *streamlined* (that is, has a gradually tapering tail) if, for example, the sinks are *distributed* over a finite length. Consider Figure 6.28, where there is a point source Q (m^3/s) at the origin O, and a continuously distributed line sink on the z-axis from O to A (distance $= a$). Let the volume absorbed per unit length of the line sink

FIGURE 6.28 Ideal flow past an axisymmetric streamlined body generated by a point source at O and a distributed line sink from O to A. The upper half of the figure shows the streamlines induced by the source and the line-segment sink alone. The lower half of the figure shows streamlines when a uniform stream along the axis of symmetry is added to the flow in the upper half of the figure.

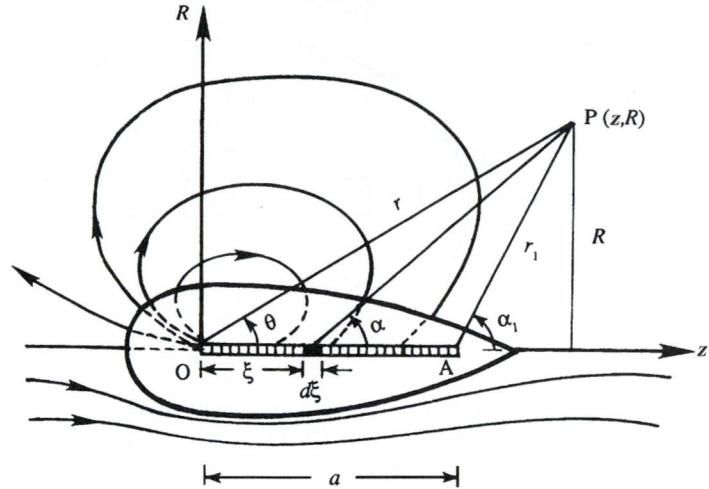

be k (m^3/s per m). An elemental length $d\xi$ of the sink can be regarded as a point sink of strength $kd\xi$, for which the stream function at any point P is

$$d\psi_{sink} = \frac{k\,d\xi}{4\pi}\cos\alpha$$

[see (6.87)]. The total stream function at P due to the entire line sink from O to A is

$$\psi_{sink} = \frac{k}{4\pi}\int_0^a \cos\alpha\,d\xi. \tag{6.93}$$

The integral can be evaluated by noting that $z - \xi = R\cot\alpha$. This gives $d\xi = Rd\alpha/\sin^2\alpha$ because z and R remain constant as we go along the sink. The stream function of the line sink is therefore

$$\psi_{sink} = \frac{k}{4\pi}\int_\theta^{\alpha_1}\cos\alpha\frac{R}{\sin^2\alpha}d\alpha = \frac{kR}{4\pi}\int_\theta^{\alpha_1}\frac{d(\sin\alpha)}{\sin^2\alpha}, = \frac{kR}{4\pi}\left[\frac{1}{\sin\theta} - \frac{1}{\sin\alpha_1}\right] = \frac{k}{4\pi}(r - r_1). \tag{6.94}$$

To obtain a closed body, we must adjust the strengths so that the volume flow from the source (Q) is absorbed by the sink, that is, $Q = ak$. Then the stream function at any point P due to the superposition of a point source of strength Q, a distributed line sink of strength $k = Q/a$, and a uniform stream of velocity U along the z-axis, is

$$\psi = -\frac{Q}{4\pi}\cos\theta + \frac{Q}{4\pi a}(r - r_1) + \frac{1}{2}Ur^2\sin^2\theta. \tag{6.95}$$

A plot of the steady streamline pattern is shown in the bottom half of Figure 6.28, in which the top half shows instantaneous streamlines in a frame of reference at rest with respect to the fluid at infinity.

Here we have assumed that the strength of the line sink is uniform along its length. Other interesting streamline patterns can be generated by assuming that the strength $k(\xi)$ is nonuniform.

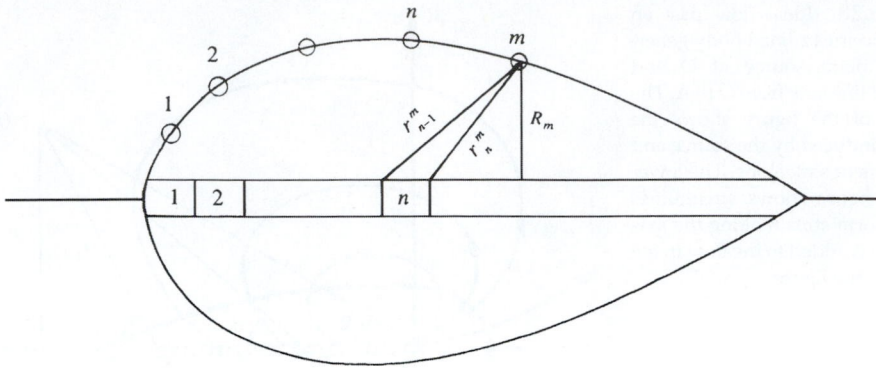

FIGURE 6.29 Flow around an arbitrary axisymmetric shape generated by superposition of a series of line-segment sources distributed along the axis of symmetry.

So far, we have assumed certain distributions of singularities, and then determined the resulting body shape when the distribution is superposed on a uniform stream. The flow around a body with a given shape can be simulated by superposing a uniform stream on a series of sources and sinks of unknown strength distributed on a line coinciding with the axis of the body. The strengths of the sources and sinks are then adjusted so that, when combined with a given uniform flow, a closed stream surface coincides with the given body. Such a calculation is typically done numerically using a computer.

Let the body length L be divided into N equal segments of length $\Delta\xi$, and let k_n be the strength (m^2/s) of one of these line segments, which may be positive or negative (Figure 6.29). The stream function at any body surface point m due to the line-segment source n is, using (6.94),

$$\psi_{mn} = -\frac{k_n}{4\pi}\left(r^m_{n-1} - r^m_n\right),$$

where the negative sign is introduced because (6.94) is for a sink. When combined with a uniform stream, the stream function at point m in Figure 6.29 due to all N line sources is

$$\psi_m = -\sum_{n=1}^{N}\frac{k_n}{4\pi}\left(r^m_{n-1} - r^m_n\right) + \tfrac{1}{2}U\,R^2_m.$$

Setting $\psi_m = 0$ for all N values of m, we obtain a set of N linear algebraic equations in N unknowns $k_n(n = 1, 2, \ldots, N)$, which can be solved by the iteration technique described in Section 6.7 or a matrix inversion routine.

6.9. THREE-DIMENSIONAL POTENTIAL FLOW AND APPARENT MASS

In three dimensions, ideal flow concepts can be used effectively for a variety of problems in aerodynamics and hydrodynamics. However, d'Alembert's paradox persists and it can be shown that steady ideal flow in three dimensions cannot predict fluid mechanical drag on

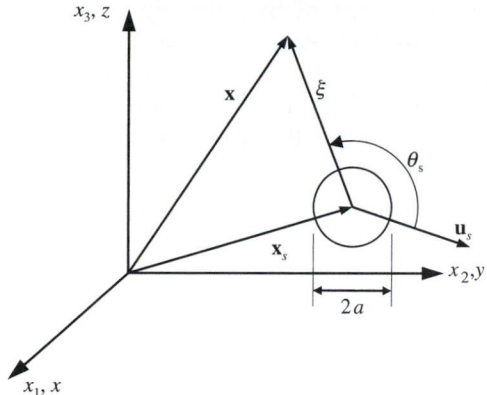

FIGURE 6.30 Three-dimensional geometry for calculating the fluid dynamic force on an arbitrarily moving submerged sphere of radius a centered at $\mathbf{x}_s$. The angle between the sphere's velocity, $\mathbf{u}_s$, and the observation point $\mathbf{x}$ is θ_s. The distance from the center of the sphere to the observation point is ξ.

closed bodies (Exercise 6.39). However, nonzero drag forces can be predicted on submerged three-dimensional objects when the flow is unsteady or some vorticity is present. This section concentrates on the former while leaving the latter to Chapter 14. The objective here is to establish the origin of the *apparent mass* or *added mass* of an accelerating object immersed in a fluid. In general terms, apparent mass is the enhanced and/or altered inertia of an object that is caused by motion of the fluid around the object. Knowledge of apparent mass is essential for predicting the performance of underwater vehicles, lighter-than-air airships, and ultra-light aircraft. It is important for describing and understanding the maneuverability of fish, the dynamics of kites and bubbles, and the differences between sailboat and motorboat motions on the surface of a wavy sea. However, to minimize complexity and to emphasize the core concepts, the focus here is on the simplest possible three-dimensional object, a sphere.

In general, the velocity potential ϕ is extended to the three dimensions merely by considering all three components of its definition $\mathbf{u} \equiv \nabla\phi$ to be nonzero. In Cartesian coordinates, this means augmenting (6.10) to include $w \equiv \partial\phi/\partial z$, where (in this section) w is the z-component of the fluid velocity and z is the third spatial coordinate.

The situation of interest is depicted in Figure 6.30, where a sphere with radius a moves in a quiescent fluid with undisturbed pressure p_∞ via an external force, $\mathbf{F}_E$, that acts only on the sphere.[2] The location $\mathbf{x}_s(t)$, velocity $\mathbf{u}_s(t) = d\mathbf{x}_s/dt$, and acceleration $d\mathbf{u}_s/dt$ of the sphere are presumed known, and the fluid dynamic force $\mathbf{F}_s$ on the sphere is to be determined. This is an idealization of the situation for a maneuvering fish, submarine, or airship.

The potential for an arbitrarily moving sphere is a modified version of (6.92) with the sphere centered at $\mathbf{x}_s(t)$ and the fluid far from the sphere at rest. These changes are implemented by replacing $\mathbf{x}$ in (6.92) with $\mathbf{x} - \mathbf{x}_s(t)$, and by setting $\mathbf{U} = 0$, which leaves:

$$\phi = -\frac{1}{4\pi|\mathbf{x} - \mathbf{x}_s(t)|^3}\mathbf{d}\cdot(\mathbf{x} - \mathbf{x}_s(t)). \tag{6.96}$$

[2]The development provided here is based on a lecture given by Professor P. Dimotakis in 1984.

For this potential to represent a constant-size moving sphere at each instant in time, the dipole strength must continuously change direction and strength to point into the flow impinging on the sphere. If the sphere's velocity is $\mathbf{u}_s(t)$, then, to an observer on the sphere, the oncoming flow velocity is $-\mathbf{u}_s(t)$. Thus, at any instant in time the dipole strength must be $\mathbf{d}(t) = -2\pi a^3[-\mathbf{u}_s(t)]$, a direct extension of the steady flow result. Substitution of this $\mathbf{d}(t)$ into (6.96) produces:

$$\phi(\mathbf{x}, \mathbf{x}_s, \mathbf{u}_s) = -\frac{a^3}{2|\mathbf{x} - \mathbf{x}_s|^3}\mathbf{u}_s \cdot (\mathbf{x} - \mathbf{x}_s) = -\frac{a^3}{2|\xi|^3}\mathbf{u}_s \cdot \xi. \tag{6.97}$$

Here, explicit listing of the time argument of $\mathbf{x}_s$ and $\mathbf{u}_s$ has been dropped for clarity, and $\xi = \mathbf{x} - \mathbf{x}_s$ is the vector distance from the center of the sphere to the location $\mathbf{x}$.

For ideal flow, an integral of the pressure forces over the surface of the sphere determines $\mathbf{F}_s$:

$$\mathbf{F}_s = -\int_{\text{sphere's surface}} (p - p_\infty)\mathbf{n}dA. \tag{6.98}$$

This is the three-dimensional equivalent of (6.55) since $\int p_\infty \mathbf{n}dA = 0$ for a closed surface and constant p_∞. The pressure difference in (6.98) can be obtained from the unsteady Bernoulli equation evaluated on the sphere's surface and far from the sphere where the pressure is p_∞, $\mathbf{u} = 0$, and $\partial\phi/\partial t = 0$:

$$\left[\frac{\partial\phi}{\partial t} + \frac{1}{2}|\nabla\phi|^2 + \frac{p}{\rho}\right]_{\text{sphere's surface}} = \frac{p_\infty}{\rho}. \tag{6.99}$$

For the geometry shown in Figure 6.30 the sphere's surface is defined by $|\mathbf{x} - \mathbf{x}_s| = |\xi| = a$, so for notational convenience the subscript "a" will denote quantities evaluated on the sphere's surface. Thus, (6.99) can be rewritten:

$$\frac{p_a - p_\infty}{\rho} = -\left(\frac{\partial\phi}{\partial t}\right)_a - \frac{1}{2}|\nabla\phi|_a^2. \tag{6.100}$$

The time derivative of ϕ can be evaluated as follows:

$$\frac{\partial}{\partial t}\phi(\mathbf{x}, \mathbf{x}_s, \mathbf{u}_s) = \frac{\partial\phi}{\partial(x_s)_i}\frac{d(x_s)_i}{dt} + \frac{\partial\phi}{\partial(u_s)_i}\frac{d(u_s)_i}{dt} = -\mathbf{u}\cdot\mathbf{u}_s - \frac{a^3}{2|\xi|^3}\xi\cdot\frac{d\mathbf{u}_s}{dt}, \tag{6.101}$$

where the middle of this extended equality presents a temporary switch to index notation. The final form in (6.101) is obtained from the definition $d(x_s)_i/dt = \mathbf{u}_s$, and the fact that $\partial\phi/\partial(x_s)_i = -\partial\phi/\partial x_i = -\mathbf{u}$ for the potential (6.97) since it only depends on $\mathbf{x} - \mathbf{x}_s$. When evaluated on the sphere's surface, this becomes:

$$\left(\frac{\partial\phi}{\partial t}\right)_a = -\mathbf{u}_a\cdot\mathbf{u}_s - \frac{a}{2}\mathbf{e}_\xi\cdot\frac{d\mathbf{u}_s}{dt}, \tag{6.102}$$

where $\mathbf{e}_\xi = \xi/|\xi|$. The independent spatial variable $\mathbf{x}$ appears twice in (6.97) so the gradient of ϕ involves two terms,

$$\nabla\phi = -\frac{a^3}{2}\left[-\frac{3(\mathbf{x}-\mathbf{x}_s)}{|\mathbf{x}-\mathbf{x}_s|^5}\mathbf{u}_s\cdot(\mathbf{x}-\mathbf{x}_s)+\frac{1}{|\mathbf{x}-\mathbf{x}_s|^3}\mathbf{u}_s\right],\tag{6.103}$$

which are readily evaluated on the surface of the sphere where $(\mathbf{x}-\mathbf{x}_s)_a = a\mathbf{e}_\xi$:

$$\mathbf{u}_a = (\nabla\phi)_a = -\frac{a^3}{2}\left[-\frac{3a\mathbf{e}_\xi}{a^5}\mathbf{u}_s\cdot a\mathbf{e}_\xi - \frac{1}{a^3}\mathbf{u}_s\right] = \frac{3}{2}(\mathbf{u}_s\cdot\mathbf{e}_\xi)\,\mathbf{e}_\xi - \frac{1}{2}\mathbf{u}_s.\tag{6.104}$$

Combining (6.100), (6.102), and (6.104) produces a final relationship for the surface pressure p_a on the sphere in terms of the orientation $\mathbf{e}_\xi$, and the sphere's velocity $\mathbf{u}_s$ and acceleration $d\mathbf{u}_s/dt$:

$$\frac{p_a - p_\infty}{\rho} = \left(\frac{3}{2}(\mathbf{u}_s\cdot\mathbf{e}_\xi)\,\mathbf{e}_\xi - \frac{1}{2}\mathbf{u}_s\right)\cdot\mathbf{u}_s + \frac{a}{2}\mathbf{e}_\xi\cdot\frac{d\mathbf{u}_s}{dt} - \frac{1}{2}\left|\frac{3}{2}(\mathbf{u}_s\cdot\mathbf{e}_\xi)\,\mathbf{e}_\xi - \frac{1}{2}\mathbf{u}_s\right|^2$$

$$= \left(\frac{3}{2}(\mathbf{u}_s\cdot\mathbf{e}_\xi)^2 - \frac{1}{2}|\mathbf{u}_s|^2\right) + \frac{a}{2}\mathbf{e}_\xi\cdot\frac{d\mathbf{u}_s}{dt} - \frac{1}{8}\left(9(\mathbf{u}_s\cdot\mathbf{e}_\xi)^2 - 6(\mathbf{u}_s\cdot\mathbf{e}_\xi)^2 + |\mathbf{u}_s|^2\right)$$

$$= \frac{1}{2}|\mathbf{u}_s|^2\left(\frac{9}{4}\frac{(\mathbf{u}_s\cdot\mathbf{e}_\xi)^2}{|\mathbf{u}_s|^2} - \frac{5}{4}\right) + \frac{a}{2}\mathbf{e}_\xi\cdot\frac{d\mathbf{u}_s}{dt}\tag{6.105}$$

When the sphere is not accelerating and θ_s is the angle between $\mathbf{e}_\xi$ and $\mathbf{u}_s$, then

$$\left(\frac{p_a - p_\infty}{\frac{1}{2}\rho|\mathbf{u}_s|^2}\right)_{steady} = \frac{9}{4}\cos^2\theta_s - \frac{5}{4} = 1 - \frac{9}{4}\sin^2\theta_s,\tag{6.106}$$

which is identical to (6.91). Thus, as expected from the Galilean invariance of Newtonian mechanics, steady flow past a stationary sphere and steady motion of a sphere through an otherwise quiescent fluid lead to the same pressure distribution on the sphere. And, once again, no drag on the sphere is predicted.

However, (6.105) includes a second term that depends on the direction and magnitude of the sphere's acceleration. To understand the effects of this term, reorient the coordinate system in Figure 6.30 so that at the time of interest the sphere is at the origin of coordinates and its acceleration is parallel to the polar z- or x_3-axis: $d\mathbf{u}_s/dt = |d\mathbf{u}_s/dt|\mathbf{e}_z$. With this revised geometry, $\mathbf{e}_\xi\cdot d\mathbf{u}_s/dt = |d\mathbf{u}_s/dt|\cos\theta$, and the fluid dynamic force on the sphere can be obtained from (6.98) in spherical polar coordinates:

$$\mathbf{F}_s = -\rho\frac{a}{2}\left|\frac{d\mathbf{u}_s}{dt}\right|\int_{\theta=0}^{\pi}\int_{\varphi=0}^{2\pi}\cos\theta(\mathbf{e}_x\sin\theta\cos\varphi + \mathbf{e}_y\sin\theta\sin\varphi + \mathbf{e}_z\cos\theta)a^2\sin\theta d\varphi d\theta.\tag{6.107}$$

The φ-integration causes the x- and y-force components to be zero, leaving:

$$\mathbf{F}_s = -\pi\rho a^3\left|\frac{d\mathbf{u}_s}{dt}\right|\mathbf{e}_z\int_{\theta=0}^{\pi}\cos^2\theta\,\sin\theta d\varphi d\theta = -\frac{2}{3}\pi\rho a^3\frac{d\mathbf{u}_s}{dt} = -M\frac{d\mathbf{u}_s}{dt},\tag{6.108}$$

where $M = 2\pi a^3 \rho / 3$ is the *apparent* or *added mass* of the sphere. Thus, the ideal-flow fluid-dynamic force on an accelerating sphere opposes the acceleration, and its magnitude is proportional to the sphere's acceleration and one-half of the mass of the fluid displaced by the sphere.

This fluid-inertia-based loading is the apparent mass or added mass of the sphere. It occurs because fluid must move more rapidly out of the way, in front of, and more rapidly fill in behind, an accelerating sphere. To illustrate its influence, consider an elementary dynamics problem involving a rigid sphere of mass m and radius a that is subject to an external force $\mathbf{F}_E$ while submerged in a large bath of nominally quiescent inviscid fluid with density ρ. In this case, Newton's second law (sum of forces = mass × acceleration) implies:

$$\mathbf{F}_E + \mathbf{F}_s = \mathbf{F}_E - M\frac{d\mathbf{u}_s}{dt} = m\frac{d\mathbf{u}_s}{dt}, \text{ or } \mathbf{F}_E = \left(m + \frac{2\pi}{3}\rho a^3\right)\frac{d\mathbf{u}_s}{dt}. \qquad (6.109)$$

Thus, a submerged sphere will behave as if its inertia is larger by one-half of the mass of the fluid it displaces compared to its behavior in vacuum. For a sphere, the apparent mass is a scalar because of its rotational symmetry. In general, apparent mass is a tensor and the final equality in (6.108) is properly stated $(F_s)_i = M_{ij}\, d(u_s)_j / dt$.

6.10. CONCLUDING REMARKS

The theory of irrotational constant-density (ideal) flow has reached a highly developed stage during the last 250 years because of the efforts of theoretical physicists such as Euler, Bernoulli, d'Alembert, Lagrange, Stokes, Helmholtz, Kirchhoff, and Kelvin. The special interest in the subject has resulted from the applicability of potential theory to other fields such as heat conduction, elasticity, and electromagnetism. When applied to fluid flows, however, the theory predicts zero fluid dynamic drag on a moving body, a result that is at variance with observations. Meanwhile, the theory of viscous flow was developed during the middle of the nineteenth century, after the Navier-Stokes equations were formulated. The viscous solutions generally applied either to very slow flows where the nonlinear advection terms in the equations of motion were negligible, or to flows in which the advective terms were identically zero (such as the viscous flow through a straight pipe). The viscous solutions were highly rotational, and it was not clear where the irrotational flow theory was applicable and why. This was left for Prandtl to explain (see Chapter 9).

It is probably fair to say that ideal flow theory does not occupy center stage in fluid mechanics any longer, although it did so in the past. However, the subject is still quite useful in several fields, especially in aerodynamics and hydrodynamics. We shall see in Chapter 9 that the pressure distribution around streamlined bodies can still be predicted with a fair degree of accuracy from the ideal flow theory. In Chapter 14 we shall see that the lift of an airfoil is due to the development of circulation around it, and the magnitude of the lift agrees with the Kutta-Zhukhovsky lift theorem. The technique of conformal mapping will also be useful in our study of flow around airfoil shapes.

EXERCISES

6.1. a) Show that (6.7) solves (6.5) and leads to $\mathbf{u} = (U,V)$.

   b) Integrate (6.6) within a circular area centered on (x', y') of radius

   $r' = \sqrt{(x - x')^2 + (y - y')^2}$ to show that (6.8) is a solution of (6.6).

6.2. For two-dimensional ideal flow, show separately that:

   a) $\nabla\psi \cdot \nabla\phi = 0$

   b) $-\nabla\psi \times \nabla\phi = |\mathbf{u}|^2 \mathbf{e}_z$

   c) $|\nabla\psi|^2 = |\nabla\phi|^2$

   d) $\nabla\phi = -\mathbf{e}_z \times \nabla\psi$

6.3. a) Show that (6.14) solves (6.12) and leads to $\mathbf{u} = (U,V)$.

   b) Integrate (6.13) within circular area centered on (x', y') of radius

   $r' = \sqrt{(x - x')^2 + (y - y')^2}$ to show that (6.15) is a solution of (6.13).

   c) For the flow described by (6.15), show that the volume flux (per unit depth into the page) $= \oint_C \mathbf{u} \cdot \mathbf{n} ds$ computed from a closed contour C that encircles the point (x', y') is m. Here $\mathbf{n}$ is the outward normal on C and ds is a differential element of C.

6.4. Show that (6.1) reduces to $\frac{\partial\phi}{\partial t} + \frac{1}{2}|\nabla\phi|^2 + \frac{p}{\rho} = const.$ when the flow is described by the velocity potential ϕ.

6.5. Determine u and v, and sketch streamlines for:

   a) $\psi = A(x^2 - y^2)$

   b) $\phi = A(x^2 - y^2)$

6.6. Assume $\psi = ax^3 + bx^2y + cxy^2 + dy^3$ where a, b, c, and d are constants; and determine two independent solutions to the Laplace equation. Sketch the streamlines for both flow fields.

6.7. Repeat Exercise 6.6 for $\psi = ax^4 + bx^3y + cx^2y^2 + dxy^3 + ey^4$ where a, b, c, d, and e are constants.

6.8. Without using complex variables, determine:

   a) The potential ϕ for an ideal vortex of strength Γ starting from (6.8).

   b) The stream function for an ideal point source of strength Q starting from (6.15).

   c) Is there any ambiguity in your answers to parts a) and b)? If so, does this ambiguity influence the fluid velocity?

6.9. Determine the stream function of a doublet starting from (6.29) and show that the streamlines are circles having centers on the y-axis that are tangent to the x-axis at the origin.

6.10. Consider steady horizontal flow at speed U past a stationary source of strength m located at the origin of coordinates in two dimensions, (6.30) or (6.31). To hold it in place, an external force per unit depth into the page, $\mathbf{F}$, is applied to the source.

   a) Develop a dimensionless scaling law for $F = |\mathbf{F}|$.

   b) Use a cylindrical control volume centered on the source with radius R and having depth B into the page, the steady ideal-flow momentum conservation equation for a control volume,

$$\int_{A^*} \rho\mathbf{u}(\mathbf{u} \cdot \mathbf{n})dA = -\int_{A^*} p\mathbf{n}dA + \mathbf{F},$$

and an appropriate Bernoulli equation to determine the magnitude and direction of $\mathbf{F}$.

c) Is the direction of **F** unusual in any way? Explain it physically.

6.11. Repeat all three parts of Exercise 6.10 for steady ideal flow past a stationary irrotational vortex located at the origin when the control volume is centered on the vortex. The stream function for this flow is: $\psi = Ur \sin\theta - (\Gamma/2\pi)\ln(r)$.

6.12. Use the principle of conservation of mass (4.5) and an appropriate control volume to show that maximum half thickness of the half-body described by (6.30) or (6.31) is $h_{max} = m/2U$.

6.13. By integrating the surface pressure, show that the drag on a plane half-body (Figure 6.7) is zero.

6.14. Ideal flow past a cylinder (6.33) is perturbed by adding a small vertical velocity without changing the orientation of the doublet:

$$\psi = -U\gamma x + Uy - \frac{Ua^2 y}{x^2 + y^2} = -U\gamma r \cos\theta + U\left(r - \frac{a^2}{r}\right)\sin\theta.$$

a) Show that the stagnation point locations are $r_s = a$ and $\theta_s = \gamma/2, \pi + \gamma/2$ when $\gamma \ll 1$.
b) Does this flow include a closed body?

6.15. For the following flow fields (b, U, Q, and Γ are positive real constants), sketch streamlines.

a) $\psi = b\sqrt{r}\cos(\theta/2)$ for $|\theta| < 180°$

b) $\psi = Uy + (\Gamma/2\pi)\left[\ln(\sqrt{x^2 + (y-b)^2}) - \ln(\sqrt{x^2 + (y+b)^2})\right]$

c) $\phi = \sum_{n=-\infty}^{n=+\infty}(Q/2\pi)\ln\left(\sqrt{x^2 + (y-2na)^2}\right)$ for $|y| < a$

6.16. [1]Take a standard sheet of paper and cut it in half. Make a simple airfoil with one half and a cylinder with the other half that are approximately the same size as shown.

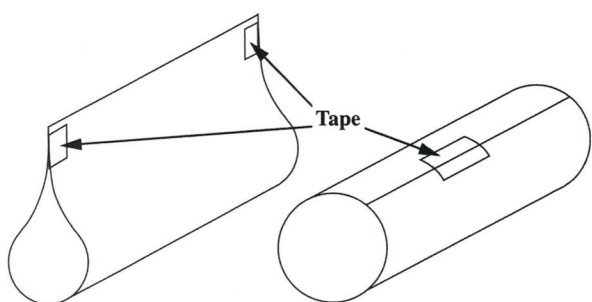

a) If the cylinder and the airfoil are dropped from the same height at the same time with the airfoil pointed toward the ground in its most streamlined configuration, predict which one reaches the ground first.

b) Stand on a chair and perform this experiment. What happens? Are your results repeatable?

c) Can you explain what you observe?

[1]Based on a suggestion from Professor William Schultz

6.17. Consider the following two-dimensional stream function composed of a uniform horizontal stream of speed U and two vortices of equal and opposite strength in (x,y)-Cartesian coordinates.

$$\psi(x,y) = Uy + (\Gamma/2\pi)\ln\sqrt{x^2 + (y-b)^2} - (\Gamma/2\pi)\ln\sqrt{x^2 + (y+b)^2}$$

a) Simplify this stream function for the combined limit of $b \to 0$ and $\Gamma \to \infty$ when $2b\Gamma = C = \text{constant}$ to find: $\psi(x,y) = Uy(1 - (C/2\pi U)(x^2 + y^2)^{-1})$.
b) Switch to (r,θ)-polar coordinates and find both components of the velocity using the simplified stream function.
c) For the simplified stream function, determine where $u_r = 0$.
d) Sketch the streamlines for the simplified stream function, and describe this flow.

6.18. Graphically generate the streamline pattern for a plane half-body in the following manner. Take a source of strength $m = 200 \text{ m}^2/\text{s}$ and a uniform stream $U = 10 \text{ m/s}$. Draw radial streamlines from the source at equal intervals of $\Delta\theta = \pi/10$, with the corresponding stream function interval

$$\Delta\psi_{\text{source}} = \frac{m}{2\pi}\Delta\theta = 10 \text{ m}^2/\text{s}.$$

Now draw streamlines of the uniform flow with the same interval, that is,

$$\Delta\psi_{\text{stream}} = U\Delta y = 10 \text{ m}^2/\text{s}.$$

This requires $\Delta y = 1$ m, which you can plot assuming a linear scale of 1 cm $= 1$ m. Now connect points of equal $\psi = \psi_{\text{source}} + \psi_{\text{stream}}$.

6.19. Consider the two-dimensional steady flow formed by combining a uniform stream of speed U in the positive x direction, a source of strength $m > 0$ at $(x, y) = (-a, 0)$, and a sink of strength m at $(x, y) = (+a, 0)$ where $a > 0$. The pressure far upstream of the origin is p_∞.
a) Write down the velocity potential and the stream function for this flow field.
b) What are the coordinates of the stagnation points?
c) Determine the pressure in this flow field along the y-axis.

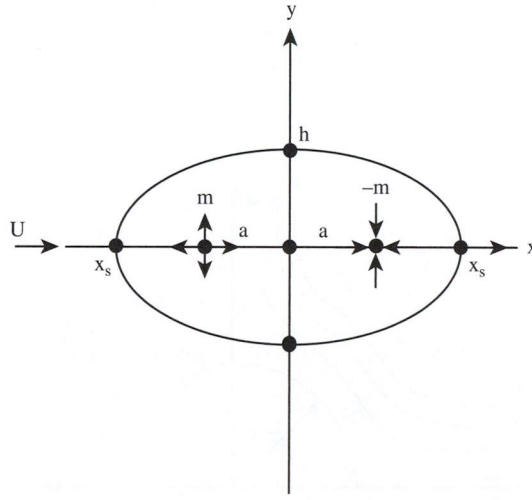

d) There is a closed streamline in this flow that defines a Rankine body. Obtain a transcendental algebraic equation for this streamline, and show that the half-width, h, of the body in the y direction is given by: $\dfrac{h}{a} = \cot\left(\dfrac{\pi U h}{m}\right)$.

(The introduction of angles may be useful here.)

6.20. A stationary ideal two-dimensional vortex with clockwise circulation Γ is located at point $(0, a)$, above a flat plate. The plate coincides with the x-axis. A uniform stream U directed along the x-axis flows over the vortex. Sketch the flow pattern and show that it represents the flow over an oval-shaped body when $\Gamma/\pi a > U$. [*Hint*: Introduce the image vortex and locate the two stagnation points on the x-axis.]

If the pressure at $x = \pm\infty$ is P_∞, and that *below* the plate is also P_∞, then show that the pressure at any point on the plate is given by: $p_\infty - p = \dfrac{\rho\Gamma^2 a^2}{2\pi^2(x^2 + a^2)^2} - \dfrac{\rho U \Gamma a}{\pi(x^2 + a^2)}$.

Show that the total upward force per unit depth on the plate is: $F = \dfrac{\rho\Gamma^2}{4\pi a} - \rho U \Gamma$.

6.21. Consider plane flow around a circular cylinder. Use the complex potential and Blasius theorem (6.60) to show that the drag is zero and the lift is $L = \rho U \Gamma$. (In Section 6.3, these results were obtained by integrating the surface pressure distribution.)

6.22. For the doublet flow described by (6.29) and sketched in Figure 6.6, show $u < 0$ for $y < x$ and $u > 0$ for $y > x$. Also, show that $v < 0$ in the first quadrant and $v > 0$ in the second quadrant.

6.23. Hurricane winds blow over a *Quonset hut*, that is, a long half-circular cylindrical cross-section building, 6 m in diameter. If the velocity far upstream is $U_\infty = 40$ m/s and $p_\infty = 1.003 \times 10^5$ N/m, $\rho_\infty = 1.23$ kg/m³, find the force per unit depth on the building, assuming the pressure inside the hut is a) p_∞, and b) stagnation pressure, $p_\infty + \dfrac{1}{2}\rho_\infty U_\infty^2$.

6.24. In a two-dimensional ideal flow, a source of strength m is located a meters above an infinite plane. Find the velocity on the plane, the pressure on the plane, and the reaction force on the plane assuming constant pressure p_∞ below the plane.

6.25. Consider a two-dimensional ideal flow over a circular cylinder of radius $r = a$ with axis coincident with a right-angle corner, as shown in the figure below. Assuming that $\psi = Axy$ (with $A =$ constant) when the cylinder is absent, solve for the stream function and velocity components.

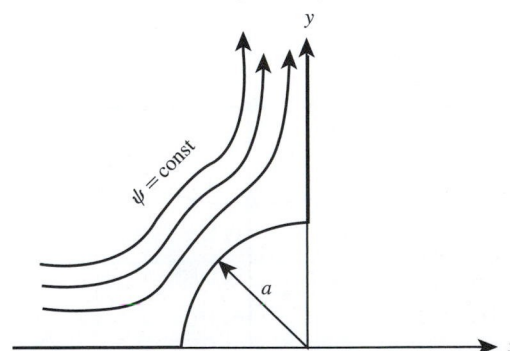

6.26. Consider the following two-dimensional velocity potential consisting of two sources and one sink, all of strength m:

$$\phi(x,y) = (m/2\pi)\left(\ln\sqrt{(x-b)^2+y^2} + \ln\sqrt{(x-a^2/b)^2+y^2} - \ln\sqrt{x^2+y^2} \right).$$

Here a and b are positive constants and $b > a$.
a) Determine the locations of the two stagnation points in this flow field.
b) Sketch the streamlines in this flow field.
c) Show that the closed streamline in this flow is given by $x^2 + y^2 = a^2$ by finding the radial location where $u_r = 0$.

6.27. Without using complex variables, derive the results of the Kutta-Zhukhovsky lift theorem (6.62) for steady two-dimensional irrotational constant-density flow past an arbitrary-cross-section object by considering the *clam-shell* control volume (shown as a dashed line) in the limit as $r \to \infty$. Here A_1 is a large circular contour, A_2 follows the object's cross-section contour, and A_3 connects A_1 and A_2. Let p_∞ and Ue_x be the pressure and flow velocity far from the origin of coordinates, and denote the flow extent perpendicular to the x-y plane by B.

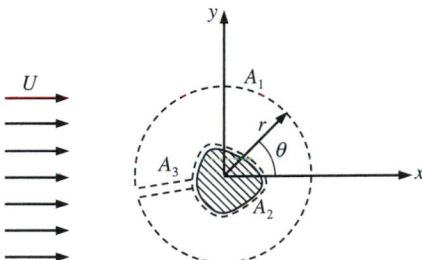

6.28. Pressure fluctuations in wall-bounded turbulent flows are a common source of flow noise. Such fluctuations are caused by turbulent eddies as they move over the bounding surface. A simple ideal-flow model that captures some of the important phenomena involves a two-dimensional vortex that moves above a flat surface in a fluid of density ρ. Thus, for the following items, use the potential:

$$\phi(x,y,t) = -\frac{\Gamma}{2\pi}\tan^{-1}\left(\frac{y-h}{x-Ut}\right) + \frac{\Gamma}{2\pi}\tan^{-1}\left(\frac{y+h}{x-Ut}\right),$$

where h is the distance of the vortex above the flat surface, Γ is the vortex strength, and U is the convection speed of the vortex.
a) Compute the horizontal u and vertical v velocity components and verify that $v = 0$ on $y = 0$.
b) Determine the pressure at $x = y = 0$ in terms of ρ, t, Γ, h, and U.
c) Based on your results from part b), is it possible for a fast-moving, high-strength vortex far from the surface to have the same pressure signature as a slow-moving, low-strength vortex closer to the surface?

6.29. A pair of equal strength ideal line vortices having axes perpendicular to the x-y plane are located at $\mathbf{x}_a(t) = (x_a(t), y_a(t))$, and $\mathbf{x}_b(t) = (x_b(t), y_b(t))$, and move in their mutually induced velocity fields. The stream function for this flow is given by:

$$\psi(x, y, t) = -\frac{\Gamma}{2\pi}(\ln|\mathbf{x} - \mathbf{x}_a(t)| + \ln|\mathbf{x} - \mathbf{x}_b(t)|).$$ Explicitly determine $\mathbf{x}_a(t)$ and $\mathbf{x}_b(t)$,

given $\mathbf{x}_a(0) = (-r_o, 0)$ and $\mathbf{x}_b(0) = (r_o, 0)$. Switching to polar coordinates at some point in your solution may be useful.

6.30. Consider the unsteady potential flow of two ideal sinks located at $\mathbf{x}_a(t) = (x_a(t), 0)$ and $\mathbf{x}_b(t) = (x_b(t), 0)$ that are free to move along the x-axis in an ideal fluid that is stationary far from the origin. Assume that each sink will move in the velocity field induced by the other.

$$\phi(x, y, t) = -\frac{Q}{2\pi}\left[\ln\sqrt{(x - x_a(t))^2 + y^2} + \ln\sqrt{(x - x_b(t))^2 + y^2}\right], \text{ with } Q > 0.$$

a) Determine $x_a(t)$ and $x_b(t)$ when $\mathbf{x}_a(0) = (-L, 0)$ and $\mathbf{x}_b(0) = (+L, 0)$.

b) If the pressure far from the origin is p_∞ and the fluid density is ρ, determine the pressure p at $x = y = 0$ as a function of p_∞, ρ, Q, and $x_a(t)$.

6.31. Consider the unsteady potential flow of an ideal source and sink located at $\mathbf{x}_1(t) = (x_1(t), 0)$ and $\mathbf{x}_2(t) = (x_2(t), 0)$ that are free to move along the x-axis in an ideal fluid that is stationary far from the origin. Assume that the source and sink will move in the velocity field induced by the other.

$$\phi(x, y, t) = \frac{m}{2\pi}\left[\ln\sqrt{(x - x_1(t))^2 + y^2} - \ln\sqrt{(x - x_2(t))^2 + y^2}\right], \text{ with } m > 0.$$

a) Determine $x_1(t)$ and $x_2(t)$ when $\mathbf{x}_1(0) = (-\lambda, 0)$ and $\mathbf{x}_2(0) = (+\lambda, 0)$.

b) If the pressure far from the origin is p_∞ and the fluid density is ρ, determine the pressure p at $x = y = 0$ as a function of p_∞, ρ, m, λ, and t.

6.32. Consider a free ideal line vortex oriented parallel to the z-axis in a $90°$ corner defined by the solid walls $\theta = 0$ and $\theta = 90°$. If the vortex passes through the plane of the flow at (x, y), show that the vortex path is given by: $\frac{1}{x^2} + \frac{1}{y^2} =$ constant. [*Hint*: Three image vortices are needed at points $(-x, -y)$, $(-x, y)$, and $(x, -y)$. Carefully choose the directions of rotation of these image vortices, show that $dy/dx = v/u = -y^3/x^3$, and integrate to produce the desired result.]

6.33. In ideal flow, streamlines are defined by $d\psi = 0$, and potential lines are defined by $d\phi = 0$. Starting from these relationships, show that streamlines and potential lines are perpendicular:

a) in plane flow where x and y are the independent spatial coordinates, and

b) in axisymmetric flow where R and z are the independent spatial coordinates.

[*Hint*: For any two independent coordinates x_1 and x_2, the unit tangent to the curve $x_2 = f(x_1)$ is $\mathbf{t} = (\mathbf{e}_1 + (df/dx_1)\mathbf{e}_2)/\sqrt{1 + (df/dx_1)^2}$; thus, for parts a) and b) it is sufficient to show $(\mathbf{t})_{\psi=const} \cdot (\mathbf{t})_{\phi=const} = 0$.]

6.34. Consider a three-dimensional point source of strength Q (m³/s). Use a spherical control volume and the principle of conservation of mass to argue that the velocity components in spherical coordinates are $u_\theta = 0$ and $u_r = Q/4\pi r^2$ and that the velocity potential and stream function must be of the form $\phi = \phi(r)$ and $\psi = \psi(\theta)$. Integrate the velocity, to show that $\phi = -Q/4\pi r$ and $\psi = -Q\cos\theta/4\pi$.

6.35. Solve the Poisson equation $\nabla^2\phi = Q\delta(\mathbf{x} - \mathbf{x}')$ in a uniform, unbounded three-dimensional domain to obtain the velocity potential $\phi = -Q/4\pi|\mathbf{x} - \mathbf{x}'|$ for an ideal point source located at $\mathbf{x}'$.

6.36. Using (R, φ, z)-cylindrical coordinates, consider steady three-dimensional potential flow for a point source of strength Q at the origin in a free stream flowing along the z-axis at speed U:

$$\phi(R, \varphi, z) = Uz - \frac{Q}{4\pi\sqrt{R^2 + z^2}}.$$

a) Sketch the streamlines for this flow in any R-z half-plane.
b) Find the coordinates of the stagnation point that occurs in this flow.
c) Determine the pressure gradient, ∇p, at the stagnation point found in part b).
d) If $R = a(z)$ defines the stream surface that encloses the fluid that emerges from the source, determine $a(z)$ for $z \to +\infty$.
e) Use Stokes' stream function to determine an equation for $a(z)$ that is valid for any value of z.
f) Use the control-volume momentum equation, $\int_S \rho\mathbf{u}(\mathbf{u}\cdot\mathbf{n})dS = -\int_S p\mathbf{n}dS + \mathbf{F}$ where $\mathbf{n}$ is the outward normal from the control volume, to determine the force $\mathbf{F}$ applied to the point source to hold it stationary.
g) If the fluid expelled from the source is replaced by a solid body having the same shape, what is the drag on the front of this body?

6.37. In (R, φ, z) cylindrical coordinates, the three-dimensional potential for a point source at $(0,0,s)$ is given by: $\phi = -(Q/4\pi)[R^2 + (z - s)^2]^{-1/2}$.
a) By combining a source of strength $+Q$ at $(0,0,-b)$, a sink of strength $-Q$ at $(0, 0, +b)$, and a uniform stream with velocity $U\mathbf{e}_z$, derive the potential (6.89) for flow around a sphere of radius a by taking the limit as $Q \to \infty$ and $b \to 0$, such that $\mathbf{d} = -2bQ\mathbf{e}_z = -2\pi a^3 U\mathbf{e}_z = $ constant. Put your final answer in spherical coordinates in terms of U, r, θ, and a.
b) Repeat part a) for the Stokes stream function starting from $\psi = -(Q/4\pi)(z - s)[R^2 + (z - s)^2]^{-1/2}$.

6.38. a) Determine the locus of points in uniform ideal flow past a circular cylinder of radius a without circulation where the velocity perturbation produced by the presence of the cylinder is 1% of the free-stream value.
b) Repeat for uniform ideal flow past a sphere.
c) Explain the physical reason(s) for the differences between the answers for parts a) and b).

6.39. Using the figure for Exercise 6.27 with $A_3 \rightarrow 0$ and $r \rightarrow \infty$, expand the three-dimensional potential for a stationary arbitrary-shape closed body in inverse powers of the distance r and prove that ideal flow theory predicts zero drag on the body.

6.40. Consider steady ideal flow over a hemisphere of constant radius a lying on the y-z plane. For the spherical coordinate system shown, the potential for this flow is:

$$\phi(r,\theta,\varphi) = Ur(1 + a^3/2r^3)\cos\theta,$$

where U is the flow velocity far from the hemisphere. Assume gravity acts downward along the x-axis. Ignore fluid viscosity in this problem.

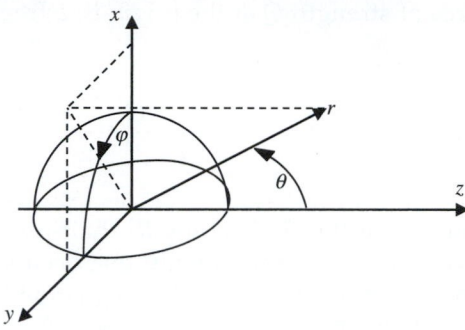

a) Determine all three components of the fluid velocity on the surface of the hemisphere, $r = a$, in spherical polar coordinates: $(u_r, u_\theta, u_\varphi) = \nabla\phi = \left(\frac{\partial\phi}{\partial r}, \frac{1}{r}\frac{\partial\phi}{\partial\theta}, \frac{1}{r\sin\theta}\frac{\partial\phi}{\partial\varphi}\right)$.

b) Determine the pressure, p, on $r = a$.

c) Determine the hydrodynamic force, R_x, on the hemisphere assuming stagnation pressure is felt everywhere underneath the hemisphere. [*Hints*: $\mathbf{e}_r \cdot \mathbf{e}_x = \sin\theta\cos\varphi$, $\int_0^\pi \sin^2\theta \, d\theta = \pi/2$, and $\int_0^\pi \sin^4\theta \, d\theta = 3\pi/8$.]

d) For the conditions of part c) what density ρ_h must the hemisphere have to remain on the surface.

6.41. The flow-field produced by suction flow into a round vacuum cleaner nozzle held above a large flat surface can be easily investigated with a simple experiment, and analyzed via potential flow in (R, φ, z)-cylindrical coordinates with the method of images.

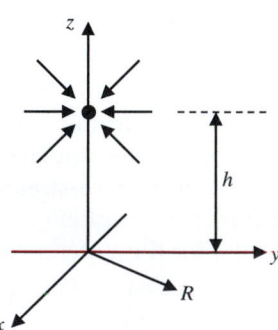

a) Do the experiment first. Obtain a vacuum cleaner that has a hose for attachments. Remove any cleaning attachments (brush, wand, etc.) or unplug the hose from the cleaning head, and attach an extension hose or something with a round opening (~4 cm diameter is recommended). Find a smooth, dry, flat horizontal surface that is a ~0.5 meter or more in diameter. Sprinkle the central third of the surface with a light granular material that is easy to see (granulated sugar, dry coffee grounds, salt, flour, talcum powder, etc., should work well). The grains should be 0.5 to 1 mm apart on average. Turn on the vacuum cleaner and lower the vacuum hose opening from ~0.25 meter above the surface toward the surface with the vacuum opening facing toward the surface. When the hose gets to within about one opening diameter of the surface or so, the granular material should start to move. Once the granular material starts moving, hold the hose opening at the same height or lift the hose slightly so that grains are not sucked into it. If many grains are vacuumed up, distribute new ones in the bare spot(s) and start over. Once the correct hose-opening-to-surface distance is achieved, hold the hose steady and let the suction airflow of the vacuum cleaner scour a pattern into the distributed granular material. Describe the shape of the final pattern, and measure any relevant dimensions.

Now see if ideal flow theory can explain the pattern observed in part a). As a first approximation, the flow field near the hose inlet can be modeled as a sink (a source with strength $-Q$) above an infinite flat boundary since the vacuum cleaner outlet (a source with strength $+Q$) is likely to be far enough away to be ignored. Denote the fluid density by ρ, the pressure far away by p_∞, and the pressure on the flat surface by $p(R)$. The potential for this flow field will be the sum of two terms:

$$\phi(R,z) = \frac{+Q}{4\pi\sqrt{R^2 + (z-h)^2}} + K(R,z).$$

b) Sketch the streamlines in the y-z plane for $z > 0$.
c) Determine $K(R,z)$.
d) Use dimensional analysis to determine how $p(R) - p_\infty$ must depend on ρ, Q, R, and h.
e) Compute $p(R) - p_\infty$ from the steady Bernoulli equation. Is this pressure distribution consistent with the results of part a)? Where is the lowest pressure? (This is also the location of the highest speed surface flow.) Is a grain at the origin of coordinates the one most likely to be picked up by the vacuum cleaner?

6.42. There is a point source of strength Q (m^3/s) at the origin, and a uniform line sink of strength $k = Q/a$ extending from $z = 0$ to $z = a$. The two are combined with a uniform stream U parallel to the z-axis. Show that the combination represents the flow past a closed surface of revolution of airship shape, whose total length is the difference of the roots of:

$$\frac{z^2}{a^2}\left(\frac{z}{a} \pm 1\right) = \frac{Q}{4\pi U a^2}.$$

6.43. Using a computer, determine the surface contour of an axisymmetric half-body formed by a line source of strength k (m^2/s) distributed uniformly along the z-axis from $z = 0$

to $z = a$ and a uniform stream. The nose of this body is more pointed than that formed by the combination of a point source and a uniform stream. From a mass balance, show that far downstream the radius of the half-body is $r = \sqrt{ak/\pi U}$.

6.44. [2]Consider the radial flow induced by the collapse of a spherical cavitation bubble of radius $R(t)$ in a large quiescent bath of incompressible inviscid fluid of density ρ. The pressure far from the bubble is p_∞. Ignore gravity.

a) Determine the velocity potential $\phi(r,t)$ for the radial flow outside the bubble.

b) Determine the pressure $p(R(t), t)$ on the surface of the bubble.

c) Suppose that at $t = 0$ the pressure on the surface of the bubble is p_∞, the bubble radius is R_0, and its initial velocity is $-\dot{R}_0$ (i.e., the bubble is shrinking); how long will it take for the bubble to completely collapse if its surface pressure remains constant?

6.45. Derive the apparent mass per unit depth into the page of a cylinder of radius a that travels at speed $U_c(t) = dx_c/dt$ along the x-axis in a large reservoir of an ideal quiescent fluid with density ρ. Use an appropriate Bernoulli equation and the following time-dependent two-dimensional potential: $\phi(x,y,t) = -\dfrac{a^2 U_c(x - x_c)}{(x - x_c)^2 + y^2}$,

where $x_c(t)$ is location of the center of the cylinder, and the Cartesian coordinates are x and y. [*Hint:* Steady cylinder motion does *not* contribute to the cylinder's apparent mass; keep only the term (or terms) from the Bernoulli equation necessary to determine apparent mass.]

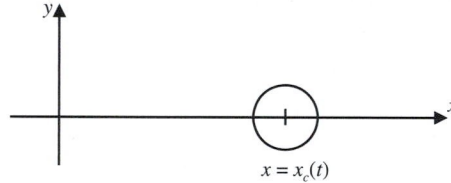

6.46. A stationary sphere of radius a and mass m resides in inviscid fluid with constant density ρ.

a) Determine the buoyancy force on the sphere when gravity g acts downward.

b) At $t = 0$, the sphere is released from rest. What is its initial acceleration?

c) What is the sphere's initial acceleration if it is a bubble in a heavy fluid (i.e., when $m \to 0$)?

6.47. A sphere of mass m and volume V is attached to the end of a light, thin, flexible cable of length L. In a vacuum, with gravity g acting, the natural frequencies for small longitudinal (bouncing) and transverse (pendulum) oscillations of the sphere are ω_b and ω_p. Ignore the effects of viscosity and estimate these natural frequencies when the same sphere and cable are submerged in water with density ρ_w. What is ω_p when $m \ll \rho_w V$?

6.48. Determine the ideal-flow force on a stationary sphere for the following unsteady flow conditions.

[2]Based on problem 5.7 in Currie (1993)

a) The free stream of velocity Ue_z is constant but the sphere's radius $a(t)$ varies.
b) The free stream velocity magnitude changes, $U(t)e_z$, but the sphere's radius a is constant.
c) The free stream velocity changes direction $U(e_x\cos\Omega t + e_y\sin\Omega t)$, but its magnitude U and the sphere's radius a are constant.

6.49. In three dimensions, consider a solid object moving with velocity **U** near the origin of coordinates in an unbounded quiescent bath of inviscid incompressible fluid with density ρ. The kinetic energy of the moving fluid in this situation is:

$$KE = \frac{1}{2}\rho \int_V |\nabla\phi|^2 dV,$$

where ϕ is the velocity potential and V is a control volume that contains all of the moving fluid but excludes the object. (Such a control volume is shown in the figure for Exercise 6.27 when $A_3 \to 0$ and $U = 0$.)

a) Show that $KE = -\frac{1}{2}\rho \int_A \phi(\nabla\phi \cdot \mathbf{n}) dA$ where A encloses the body and is coincident with its surface, and **n** is the outward normal on A.

b) The apparent mass, M, of the moving body may be defined by $KE = \frac{1}{2}M|\mathbf{U}|^2$. Using this definition, the result of part a), and (6.97) with $\mathbf{x}_s = 0$, show that $M = 2\pi a^3 \rho/3$ for a sphere.

Literature Cited

Carrier, G. F., Krook, M., and Pearson, C. E. (1966). *Functions of a Complex Variable*. New York: McGraw-Hill.
Churchill, R. V., Brown, J. W., and Verhey, R. F. (1974). *Complex Variables and Applications* (3rd ed.). New York: McGraw-Hill.
Currie, I. G. (1993). *Fundamentals of Fluid Mechanics* (2nd ed.). New York: McGraw Hill.
Prandtl, L. (1952). *Essentials of Fluid Dynamics*. New York: Hafner Publishing.

Supplemental Reading

Batchelor, G. K. (1967). *An Introduction to Fluid Dynamics*. London: Cambridge University Press.
Milne-Thompson, L. M. (1962). *Theoretical Hydrodynamics*. London: Macmillan Press.
Shames, I. H. (1962). *Mechanics of Fluids*. New York: McGraw-Hill.
Vallentine, H. R. (1967). *Applied Hydrodynamics*. New York: Plenum Press.

Gravity Waves

O U T L I N E

7.1. Introduction   254

7.2. Linear Liquid-Surface Gravity
Waves   256

7.3. Influence of Surface Tension   269

7.4. Standing Waves   271

7.5. Group Velocity, Energy Flux,
and Dispersion   273

7.6. Nonlinear Waves in Shallow
and Deep Water   279

7.7. Waves on a Density Interface   286

7.8. Internal Waves in a Continuously
Stratified Fluid   293

Exercises   304

Literature Cited   307

CHAPTER OBJECTIVES

- To develop the equations and boundary conditions for surface, interface, and internal waves.

- To derive linear gravity-capillary wave propagation speed(s), pressure fluctuations, dispersion, particle motion, and energy flux for surface waves on a liquid layer of arbitrary but constant depth.

- To describe and highlight wave refraction and nonlinear gravity wave results in shallow and deep water.

- To determine linear density-interface wave characteristics with and without an additional free surface.

- To present the characteristics of gravity waves on a density gradient with constant buoyancy frequency.

7.1. INTRODUCTION

There are three types of waves commonly considered in the study of fluid mechanics: interface waves, internal waves, and compression and expansion waves. In all cases, the waves are traveling fluid oscillations, impulses, or pressure changes sustained by the interplay of fluid inertia and a restoring force or a pressure imbalance. For interface waves the restoring forces are gravity and surface tension. For internal waves, the restoring force is gravity. For expansion and compression waves, the restoring force comes directly from the compressibility of the fluid. The basic elements of linear and nonlinear compression and expansion waves are presented in Chapter 15, which covers compressible fluid dynamics. This chapter covers interface and internal waves with an emphasis on gravity as the restoring force. The approach and results from the prior chapter will be exploited here since the wave physics and wave phenomena presented in this chapter primarily involve irrotational flow.

Perhaps the simplest and most readily observed fluid waves are those that form and travel on the density discontinuity provided by an air-water interface. Such *surface gravity-capillary waves*, sometimes simply called *water waves*, involve fluid particle motions parallel and perpendicular to the direction of wave propagation. Thus, the waves are neither longitudinal nor transverse. When generalized to internal waves that propagate in a fluid medium having a continuous density gradient, the situation may be even more complicated. This chapter presents some basic features of wave motion and illustrates them with water waves because water wave phenomena are readily observed and this aids comprehension. Throughout this chapter, the wave frequency will be assumed much higher than the Coriolis frequency so the wave motion is unaffected by the earth's rotation. Waves affected by planetary rotation are considered in Chapter 13. And, unless specified otherwise, wave amplitudes are assumed small enough so that the governing equations and boundary conditions are linear.

For such linear waves, Fourier superposition of sinusoidal waves allows arbitrary waveforms to be constructed and sinusoidal waveforms arise naturally from the linearized equations for water waves (see Exercise 7.3). Consequently, a simple sinusoidal traveling wave of the form

$$\eta(x,t) = a \cos\left[\frac{2\pi}{\lambda}(x - ct)\right] \tag{7.1}$$

is a foundational element for what follows. In Cartesian coordinates with x horizontal and z vertical, $z = \eta(x,t)$ specifies the *waveform* or surface shape where a is the wave *amplitude*, λ is the *wavelength*, c is the *phase speed*, and $2\pi(x - ct)/\lambda$ is the *phase*. In addition, the spatial frequency $k \equiv 2\pi/\lambda$, with units of rad./m, is known as the *wave number*. If (7.1) describes the vertical deflection of an air-water interface, then the height of wave crests is $+a$ and the depth of the wave troughs is $-a$ compared to the undisturbed water-surface location $y = 0$. At any instant in time, the distance between successive wave crests is λ. At any fixed x-location, the time between passage of successive wave crests is the *period*, $T = 2\pi/kc = \lambda/c$. Thus, the wave's *cyclic frequency* is $\nu = 1/T$ with units of Hz, and its *radian frequency* is $\omega = 2\pi\nu$ with units of rad./s. In terms of k and ω, (7.1) can be written:

$$\eta(x,t) = a \cos[kx - \omega t]. \tag{7.2}$$

The wave propagation speed is readily deduced from (7.1) or (7.2) by determining the travel speed of wave crests. This means setting the phase in (7.1) or (7.2) so that the cosine function is unity and $\eta = +a$. This occurs when the phase is $2n\pi$ where n is an integer,

$$\frac{2\pi}{\lambda}(x_{crest} - ct) = 2n\pi = kx_{crest} - \omega t, \tag{7.3}$$

and x_{crest} is the time-dependent location where $\eta = +a$. Solving for the crest location produces:

$$x_{crest} = (\omega/k)t + 2n\pi/k.$$

Therefore, in a time increment Δt, a wave crest moves a distance $\Delta x_{crest} = (\omega/k)\Delta t$. Thus,

$$c = \omega/k = \lambda\nu \tag{7.4}$$

is known as the phase speed because it specifies the travel speed of constant-phase wave features, like wave crests or troughs.

Although instructive, (7.1) and (7.2) are limited to propagation in the positive x direction only. In general, waves may propagate in any direction. A useful three-dimensional generalization of (7.2) is:

$$\eta = a\cos(kx + ly + mz - \omega t) = a\cos(\mathbf{K}\cdot\mathbf{x} - \omega t), \tag{7.5}$$

where $\mathbf{K} = (k, l, m)$ is a vector, called the *wave number vector*, whose magnitude is given by

$$K^2 = k^2 + l^2 + m^2. \tag{7.6}$$

The wavelength derived from (7.5) is

$$\lambda = 2\pi/K, \tag{7.7}$$

which is illustrated in Figure 7.1 in two dimensions. The magnitude of the phase velocity is $c = \omega/K$, and the direction of propagation is parallel to $\mathbf{K}$, so the phase velocity vector is:

$$\mathbf{c} = (\omega/K)\mathbf{e}_K, \tag{7.8}$$

where $\mathbf{e}_K = \mathbf{K}/K$.

From Figure 7.1, it is also clear that $c_x = \omega/k$, $c_y = \omega/l$, and $c_z = \omega/m$ are each larger than the resultant $c = \omega/K$, because k, l, and m are individually smaller than K when all three are nonzero, as required by (7.6). Thus, c_x, c_y, and c_z are not vector components of the phase velocity in the usual sense, but they do reflect the fact that constant-phase surfaces appear to travel faster along directions not coinciding with the direction of propagation, the x and y directions in Figure 7.1 for example. Any of the three axis-specific phase speeds is sometimes called the *trace velocity* along its associated axis.

If sinusoidal fluid waves exist in a fluid moving with uniform speed $\mathbf{U}$, then the observed phase speed is $\mathbf{c}_0 = \mathbf{c} + \mathbf{U}$. Forming a dot product of $\mathbf{c}_0$ with $\mathbf{K}$, and using (7.8), produces

$$\omega_0 = \omega + \mathbf{U}\cdot\mathbf{K}, \tag{7.9}$$

where ω_0 is the *observed frequency* at a fixed point, and ω is the *intrinsic frequency* measured by an observer moving with the flow. It is apparent that the frequency of

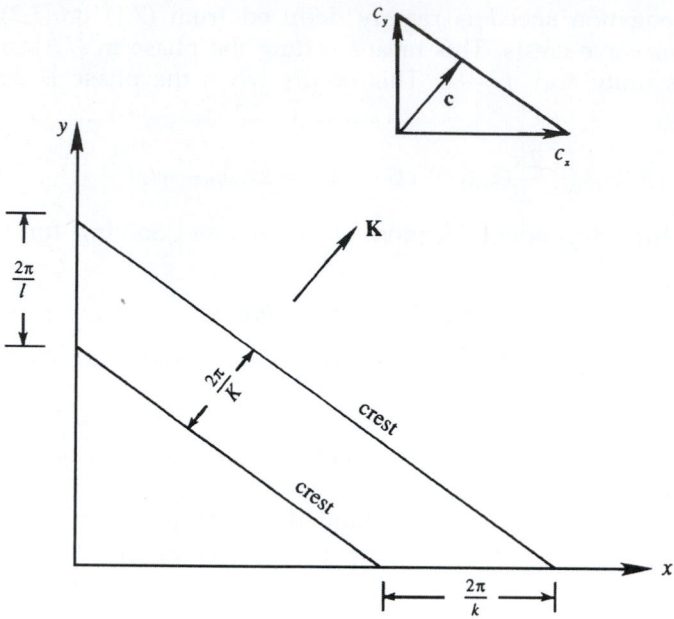

FIGURE 7.1   Wave crests propagating in the x-y plane. The crest spacing along the coordinate axes is larger than the wavelength $\lambda = 2\pi/K$. The inset shows how the trace velocities c_x and c_y are combined to give the phase velocity vector **c**.

a wave is *Doppler shifted* by an amount **U·K** in nonzero flow. Equation (7.9) may be understood by considering a situation in which the intrinsic frequency ω is zero, but the flow pattern has a periodicity in the x direction of wavelength $2\pi/k$. If this sinusoidal pattern is translated in the x direction at speed U, then the observed frequency at a fixed point is $\omega_0 = Uk$. The effects of uniform flow on frequency will not be considered further, and all frequencies in the remainder of this chapter should be interpreted as intrinsic frequencies.

7.2. LINEAR LIQUID-SURFACE GRAVITY WAVES

Starting from the equations for ideal flow, this section develops the properties of small-slope, small-amplitude gravity waves on the free surface of a constant-density liquid layer of uniform depth H, which may be large or small compared to the wavelength λ. The limitation to waves with small slopes and amplitudes implies $a/\lambda \ll 1$ and $a/H \ll 1$, respectively. These two conditions allow the problem to be linearized. In this first assessment of wave motion, surface tension is neglected for simplicity; in water its effect is limited to wavelengths less than 5 to 10 centimeters, as discussed in Section 7.3. In addition, the air above the liquid is ignored, and the liquid's motion is presumed to be irrotational and entirely caused by the surface waves.

FIGURE 7.2 Geometry for determining the properties of linear gravity waves on the surface of a liquid layer of depth H. Gravity points downward along the z-axis. The undisturbed liquid surface location is $z = 0$ so the bottom is located at $z = -H$. The surface's vertical deflection or waveform is $\eta(x,t)$. When η is sinusoidal, its peak deflection from $z = 0$ is the sinusoid's amplitude a.

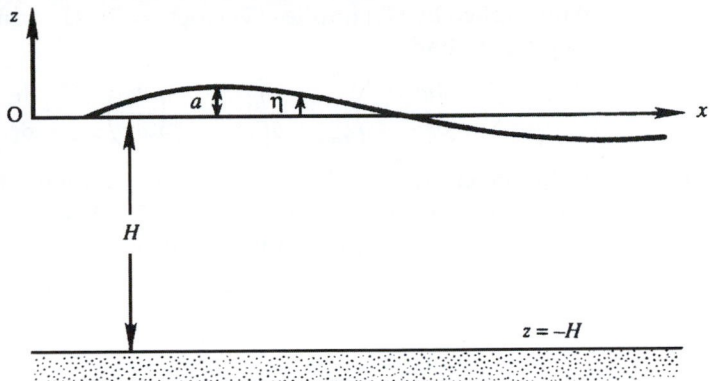

To get started, choose the x-axis in the direction of wave propagation with the z-axis vertical so that the motion is two dimensional in the x-z plane (Figure 7.2). Let $\eta(x,t)$ denote the vertical liquid-surface displacement from its undisturbed location $z = 0$. Because the liquid's motion is irrotational, a velocity potential $\phi(x, z, t)$ can be defined such that

$$u = \partial\phi/\partial x, \quad \text{and} \quad w = \partial\phi/\partial z, \tag{7.10}$$

so the incompressible continuity equation $\partial u/\partial x + \partial w/\partial z = 0$ implies

$$\partial^2\phi/\partial x^2 + \partial^2\phi/\partial z^2 = 0. \tag{7.11}$$

There are three boundary conditions. The condition at the bottom of the liquid layer is zero normal velocity, that is,

$$w = \partial\phi/\partial z = 0 \quad on \quad z = -H. \tag{7.12}$$

At the free surface, a *kinematic boundary condition* is applied that requires the fluid-particle velocity normal to the surface, $\mathbf{u} \cdot \mathbf{n}$, and on the surface be the same as the velocity of the surface $\mathbf{U}_s$ normal to itself:

$$(\mathbf{n} \cdot \mathbf{u})_{z=\eta} = \mathbf{n} \cdot \mathbf{U}_s, \tag{7.13}$$

where $\mathbf{n}$ is the surface normal. This ensures that the liquid elements that define the interface do not become separated from the interface while still allowing these interface elements to move along the interface.

For the current situation, the equation for the surface may be written $f(x, z, t) = z - \eta(x, t) = 0$, so the surface normal $\mathbf{n}$, which points upward out of the liquid will be:

$$\mathbf{n} = \nabla f/|\nabla f| = (-(\partial\eta/\partial x)\mathbf{e}_x + \mathbf{e}_z)/\sqrt{(\partial\eta/\partial x)^2 + 1}. \tag{7.14}$$

The velocity of the surface $\mathbf{U}_s$ at any location x can be considered purely vertical:

$$\mathbf{U}_s = (\partial\eta/\partial t)\mathbf{e}_z. \tag{7.15}$$

Thus, (7.13) multiplied by $|\nabla f|$ implies $(\nabla f \cdot \mathbf{u})_{z=\eta} = \nabla f \cdot \mathbf{U}_s$, which can be evaluated using (7.14) and $\mathbf{u} = u\mathbf{e}_x + w\mathbf{e}_z$ to find:

$$\left(-u\frac{\partial \eta}{\partial x} + w\right)_{z=\eta} = \frac{\partial \eta}{\partial t}, \quad \text{or} \quad \left(\frac{\partial \phi}{\partial z}\right)_{z=\eta} = \frac{\partial \eta}{\partial t} + \frac{\partial \eta}{\partial x}\left(\frac{\partial \phi}{\partial x}\right)_{z=\eta}, \tag{7.16}$$

where (7.10) has been used for the fluid velocity components to achieve the second form of (7.16). For small-slope waves, the final term in (7.16) is small compared to the other two, so the kinematic boundary condition can be approximated:

$$\left(\frac{\partial \phi}{\partial z}\right)_{z=\eta} \cong \frac{\partial \eta}{\partial t}. \tag{7.17}$$

For consistency, the left side of (7.17) must also be approximated for small wave slopes, and this is readily accomplished via a Taylor series expansion around $z = 0$:

$$\left(\frac{\partial \phi}{\partial z}\right)_{z=\eta} = \left(\frac{\partial \phi}{\partial z}\right)_{z=0} + \eta\left(\frac{\partial^2 \phi}{\partial z^2}\right)_{z=0} + \dots \cong \frac{\partial \eta}{\partial t}.$$

Thus, when a/λ is small enough, the simplest version of (7.13) is

$$\left(\frac{\partial \phi}{\partial z}\right)_{z=0} \cong \frac{\partial \eta}{\partial t}. \tag{7.18}$$

These simplifications of the kinematic boundary are justified when $ka = 2\pi a/\lambda \ll 1$ (see Exercise 7.2).

In addition to the kinematic condition at the surface, there is a *dynamic condition* that the pressure just below the liquid surface be equal to the ambient pressure, with surface tension neglected. Taking the ambient air pressure above the liquid to be a constant atmospheric pressure, the dynamic surface condition can be stated,

$$(p)_{z=\eta} = 0, \tag{7.19}$$

where p in (7.19) is the gauge pressure. Equation (7.19) follows from the boundary condition on $\boldsymbol{\tau} \cdot \mathbf{n}$, which is continuous across an interface as established in Section 4.10. Equation (7.19) and the neglect of any shear stresses on $z = \eta$ define a stress-free boundary. Thus, the water surface in this ideal case is commonly called a *free surface*. For consistency, this condition should also be simplified for small-slope waves by dropping the nonlinear term $|\nabla \phi|^2$ in the relevant Bernoulli equation (4.83):

$$\frac{\partial \phi}{\partial t} + \frac{p}{\rho} + gz \cong 0, \tag{7.20}$$

where the Bernoulli constant has been evaluated on the undisturbed liquid surface far from the surface wave. Evaluating (7.20) on $z = \eta$ and applying (7.19) produces:

$$\left(\frac{\partial \phi}{\partial t}\right)_{z=\eta} \cong \left(\frac{\partial \phi}{\partial t}\right)_{z=0} \cong -g\eta. \tag{7.21}$$

The first approximate equality follows because $(\partial \phi/\partial t)_{z=0}$ is the first term in a Taylor series expansion of $(\partial \phi/\partial t)_{z=\eta}$ in powers of η about $\eta = 0$. This approximation is consistent with (7.18).

Interestingly, even with the specification of the field equation (7.11) and the three boundary conditions, (7.12), (7.18), and (7.21), the overall linear surface-wave problem is not fully defined without initial condition for the surface shape (Exercise 7.3). For simplicity, we chose $\eta(x, t = 0) = a\cos(kx)$, since it is satisfied by the simple sinusoidal wave (7.2), which now becomes a foundational part of the solution. To produce a cosine dependence for η on the phase $(kx - \omega t)$ in (7.2), conditions (7.18) and (7.21) require ϕ to be a sine function of $(kx - \omega t)$. Consequently, a solution is sought for ϕ in the form

$$\phi(x, z, t) = f(z)\sin(kx - \omega(k)t), \tag{7.22}$$

where $f(z)$ and $\omega = \omega(k)$ are to be determined. Substitution of (7.22) into the Laplace equation (7.11) gives

$$d^2f/dz^2 - k^2f = 0,$$

which has the general solution $f(z) = Ae^{kz} + Be^{-kz}$, where A and B are constants. Thus, (7.22) implies

$$\phi = (Ae^{kz} + Be^{-kz})\sin(kx - \omega t). \tag{7.23}$$

The constants A and B can be determined by substituting (7.23) into (7.12):

$$k(Ae^{-kH} - Be^{+kH})\sin(kx - \omega t) = 0 \text{ or } B = Ae^{-2kH}, \tag{7.24}$$

and by substituting (7.2) and (7.23) into (7.18),

$$k(A - B)\sin(kx - \omega t) = \omega a\sin(kx - \omega t) \text{ or } k(A - B) = \omega a. \tag{7.25}$$

Solving (7.24) and (7.25) for A and B produces:

$$A = \frac{a\omega}{k(1 - e^{-2kH})} \qquad B = \frac{a\omega\, e^{-2kH}}{k(1 - e^{-2kH})}.$$

The velocity potential (7.23) then becomes:

$$\phi = \frac{a\omega}{k}\frac{\cosh(k(z + H))}{\sinh(kH)}\sin(kx - \omega t), \tag{7.26}$$

from which the fluid velocity components are found as:

$$u = a\omega\frac{\cosh(k(z + H))}{\sinh(kH)}\cos(kx - \omega t), \text{ and } w = a\omega\frac{\sinh(k(z + H))}{\sinh(kH)}\sin(kx - \omega t). \tag{7.27}$$

This solution of the Laplace equation has been found using kinematic boundary conditions alone, and this is typical of irrotational constant-density flows where fluid pressure is determined through a Bernoulli equation after the velocity field has been found. Here the dynamic surface boundary condition (7.21) enforces $p = 0$ on the liquid surface, and substitution of (7.2) and (7.26) into (7.21) produces:

$$\left(\frac{\partial\phi}{\partial t}\right)_{z=0} = -\frac{a\omega^2}{k}\frac{\cosh(kH)}{\sinh(kH)}\cos(kx - \omega t) \cong -g\eta = -ag\cos(kx - \omega t),$$

which simplifies to a relation between ω and k (or equivalently, between the wave period T and the wave length λ):

$$\omega = \sqrt{gk \tanh(kH)} \text{ or } T = \sqrt{\frac{2\pi\lambda}{g} \coth\left(\frac{2\pi H}{\lambda}\right)}. \tag{7.28}$$

The first part of (7.28) specifies how temporal and spatial frequencies of the surface waves are related, and it is known as a *dispersion relation*. The phase speed c of these surface waves is given by:

$$c = \frac{\omega}{k} = \sqrt{\frac{g}{k} \tanh(kH)} = \sqrt{\frac{g\lambda}{2\pi} \tanh\left(\frac{2\pi H}{\lambda}\right)}. \tag{7.29}$$

This result is of fundamental importance for water waves. It shows that surface waves are *dispersive* because their propagation speed depends on wave number, with lower k (longer wavelength) waves traveling faster. (*Dispersion* is a term borrowed from optics, where it signifies separation of different colors due to the speed of light in a medium depending on the wavelength.) Thus, a concentrated wave packet made up of many different wavelengths (or frequencies) will not maintain a constant waveform or shape. Instead, it will disperse or spread out as it travels. The longer wavelength components will travel faster than the shorter wavelength ones so that an initial impulse evolves into a wide wave train. This is precisely what happens when an object is dropped onto the surface of a quiescent pool, pond, or lake. The radial extent of the circular waves increases with time, and the longest wavelengths appear farthest from the point of impact while the shortest wavelengths are seen closest to the point of impact.

The rest of this section covers some implications of the linear surface-wave solution (7.26) and the dispersion relation (7.28). Given the ease with which it can be measured, the pressure below the liquid surface is considered first. In particular, the time-dependent perturbation pressure,

$$p' \equiv p + \rho g z, \tag{7.30}$$

produced by surface waves is of interest. Using this and (7.26) in the linearized Bernoulli equation (7.20) leads to

$$p' = -\rho\frac{\partial\phi}{\partial t} = \rho\frac{a\omega^2}{k}\frac{\cosh(k(z+H))}{\sinh(kH)}\cos(kx - \omega t) = \rho g a \frac{\cosh(k(z+H))}{\cosh(kH)}\cos(kx - \omega t), \tag{7.31}$$

where the second equality follows when (7.28) is used to eliminate ω^2. The perturbation pressure therefore decreases with increasing depth, and the extent of this decrease depends on the wavelength through k.

Another interesting feature of linear surface waves is the fact that they travel and cause fluid elements to move, but they do not cause fluid elements to travel. To ascertain what happens when a linear surface wave passes, consider the fluid element that follows a path $\mathbf{x}_p(t) = x_p(t)\mathbf{e}_x + z_p(t)\mathbf{e}_z$. The path-line equations (3.8) for this fluid element are

$$\frac{dx_p(t)}{dt} = u(x_p, z_p, t), \text{ and } \frac{dz_p(t)}{dt} = w(x_p, z_p, t), \tag{7.32}$$

which imply:

$$\frac{dx_p}{dt} = a\omega \frac{\cosh\left(k(z_p + H)\right)}{\sinh(kH)} \cos(kx_p - \omega t), \text{ and } \frac{dz_p}{dt} = a\omega \frac{\sinh\left(k(z_p + H)\right)}{\sinh(kH)} \sin(kx_p - \omega t), \quad (7.33)$$

when combined with (7.27). To be consistent with the small amplitude approximation, these equations can be linearized by setting $x_p(t) = x_0 + \xi(t)$ and $z_p(t) = z_0 + \zeta(t)$, where (x_0, z_0) is the average fluid element location and the element excursion vector (ξ, ζ) (see Figure 7.3) is assumed to be small compared to the wavelength. Thus, the linearized versions of (7.33) are obtained by evaluating the right side of each equation at (x_0, z_0):

$$\frac{d\xi}{dt} \cong a\omega \frac{\cosh(k(z_0 + H))}{\sinh(kH)} \cos(kx_0 - \omega t), \text{ and } \frac{d\zeta}{dt} \cong a\omega \frac{\sinh(k(z_0 + H))}{\sinh(kH)} \sin(kx_0 - \omega t),$$

$$(7.34a, 7.34b)$$

where x_0 and z_0 have been assumed independent of time. This linearization is valid when the velocity of the fluid element along its path is nearly equal to the fluid velocity at (x_0, z_0) at that instant. It is accurate when $a \ll \lambda$. The equations (7.34a, 7.34b) are reminiscent of those in Example 3.1, and are readily time-integrated:

$$\xi \cong -a \frac{\cosh(k(z_0 + H))}{\sinh(kH)} \sin(kx_0 - \omega t), \text{ and } \zeta \cong a \frac{\sinh(k(z_0 + H))}{\sinh(kH)} \cos(kx_0 - \omega t). \quad (7.35a, 7.35b)$$

Here we note that $\xi(t)$ and $\zeta(t)$ are entirely oscillatory. Neither contains a term that increases with time so the assumption that x_0 and z_0 are time independent is self consistent when $a \ll \lambda$. Elimination of the phase $(kx_0 - \omega t)$ from (7.35a, 7.35b) gives:

$$\xi^2 \Big/ \left[a \frac{\cosh(k(z_0 + H))}{\sinh(kH)} \right]^2 + \zeta^2 \Big/ \left[a \frac{\sinh(k(z_0 + H))}{\sinh(kH)} \right]^2 = 1, \quad (7.36)$$

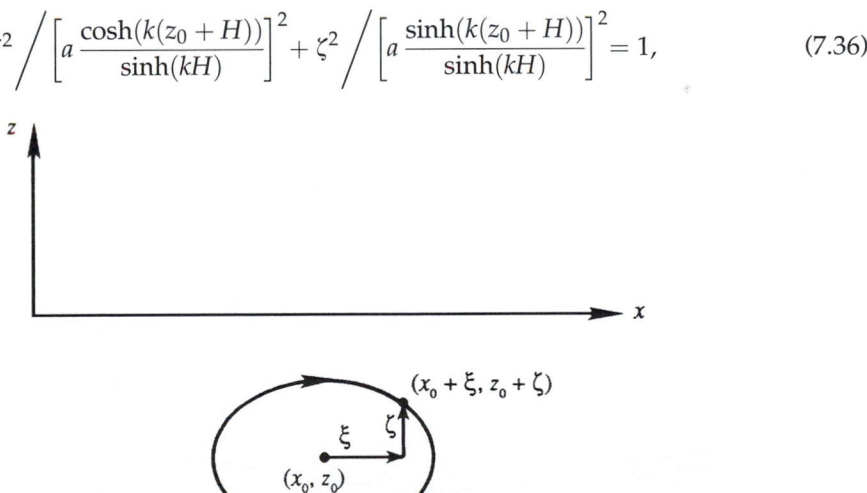

FIGURE 7.3   Orbit of a fluid particle below a linear surface wave. The average position of the particle is (x_0, z_0), and $\xi(t)$ and $\zeta(t)$ are small time-dependent displacements in the horizontal and vertical directions, respectively. When the surface wave is sinusoidal, travels to the right, and has small amplitude, the fluid particles below the surface traverse closed elliptical orbits in the clockwise direction.

which represents an ellipse. Both the semi-major axis, $a\cosh[k(z_0 + H)]/\sinh(kH)$, and the semi-minor axis, $a\sinh[k(z_0 + H)]/\sinh(kH)$, decrease with depth, with the minor axis vanishing at $z_0 = -H$ (Figure 7.4b). The distance between foci remains constant with depth. Equations (7.35a, 7.35b) show that the phase of the motion is independent of z_0, so fluid elements in any vertical column move in phase. That is, if one of them is at the top of its orbit, then all elements at the same x_0 are at the top of their orbits.

Streamlines may be found from the stream function ψ, which can be determined by integrating the velocity component equations $\partial\psi/\partial z = u$ and $-\partial\psi/\partial x = w$ when u and w are given by (7.27):

$$\psi = \frac{a\omega}{k}\frac{\sinh(k(z + H))}{\sinh(kH)}\cos(kx - \omega t) \tag{7.37}$$

(Exercise 7.4). To understand the streamline structure, consider a particular time, $t = 0$, when

$$\psi \propto \sin k(z + H)\cos kx.$$

It is clear that $\psi = 0$ at $z = -H$, so that the bottom wall is a part of the $\psi = 0$ streamline. However, ψ is also zero at $kx = \pm\pi/2, \pm3\pi/2, \ldots$ for any z. At $t = 0$ and at these values of kx, η from (7.2) vanishes. The resulting streamline pattern is shown in Figure 7.5. It is seen that the *velocity is in the direction of propagation (and horizontal) at all depths below the crests, and opposite to the direction of propagation at all depths below troughs.*

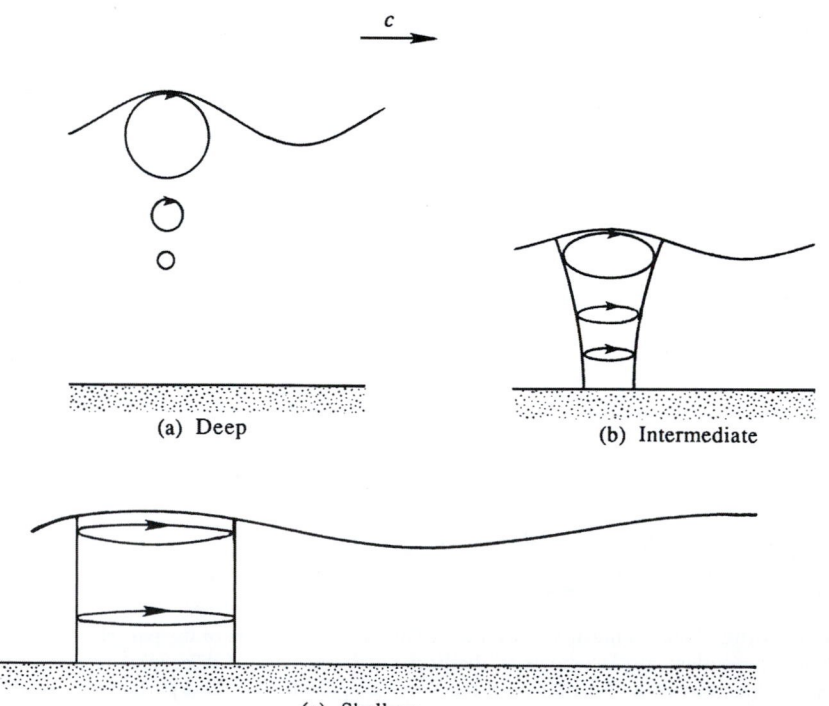

(a) Deep

(b) Intermediate

(c) Shallow

FIGURE 7.4 Fluid particle orbits caused by a linear sinusoidal surface wave traveling to the right for three liquid depths. (a) When the liquid is deep and $\tanh(kH) \approx 1$, then particle orbits are circular and decrease in size with increasing depth. (b) At intermediate depths, the particle orbits are broad ellipses that narrow and contract with increasing depth. (c) When the water is shallow and $\tanh(kH) \approx \sinh(kH) \approx kH$, the orbits are thin ellipses that become thinner with increasing depth.

FIGURE 7.5 Instantaneous streamline pattern for a sinusoidal surface wave propagating to the right. Here the $\psi = 0$ streamline follows the bottom and jumps up to contact the surface where $\eta = 0$. The remaining streamlines start and end on the liquid surface with purely horizontal motion found in the $+x$ direction below a wave crest and in $-x$ direction below a wave trough.

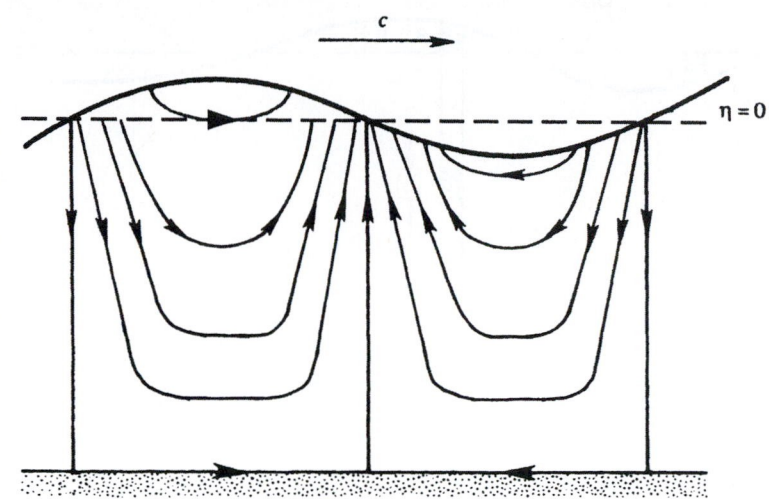

Surface gravity waves possess kinetic energy in the motion of the fluid and potential energy in the vertical deformation of the free surface. The kinetic energy per unit horizontal area, E_k, is found by integrating over the depth and averaging over a wavelength:

$$E_k = \frac{\rho}{2\lambda} \int_0^\lambda \int_{-H}^0 (u^2 + w^2)\, dz\, dx.$$

Here the z-integral is taken from $z = -H$ to $z = 0$, consistent with the linearization performed to reach (7.26); integrating from $z = -H$ to $z = \eta$ merely introduces a higher-order term. Substitution of the velocity components from (7.27) gives:

$$E_k = \frac{\rho\omega^2}{2\sinh^2 kH} \left[\frac{1}{\lambda} \int_0^\lambda a^2 \cos^2(kx - \omega t)\, dx \int_{-H}^0 \cosh^2 k(z + H)\, dz \right.$$
$$\left. + \frac{1}{\lambda} \int_0^\lambda a^2 \sin^2(kx - \omega t)\, dx \int_{-H}^0 \sinh^2 k(z + H)\, dz \right].$$

(7.38)

In terms of free-surface displacement η, the x-integrals in (7.38) can be written as

$$\frac{1}{\lambda} \int_0^\lambda a^2 \cos^2(kx - \omega t)\, dx = \frac{1}{\lambda} \int_0^\lambda a^2 \sin^2(kx - \omega t)\, dx = \frac{1}{\lambda} \int_0^\lambda \eta^2 dx = \overline{\eta^2},$$

where $\overline{\eta^2}$ is the mean-square vertical surface displacement. The z-integrals in (7.38) are easy to evaluate by expressing the hyperbolic functions in terms of exponentials. Using the dispersion relation (7.28), (7.38) finally becomes

$$E_k = \frac{1}{2}\rho g \overline{\eta^2},$$

(7.39)

which is the kinetic energy of the wave motion per unit horizontal area.

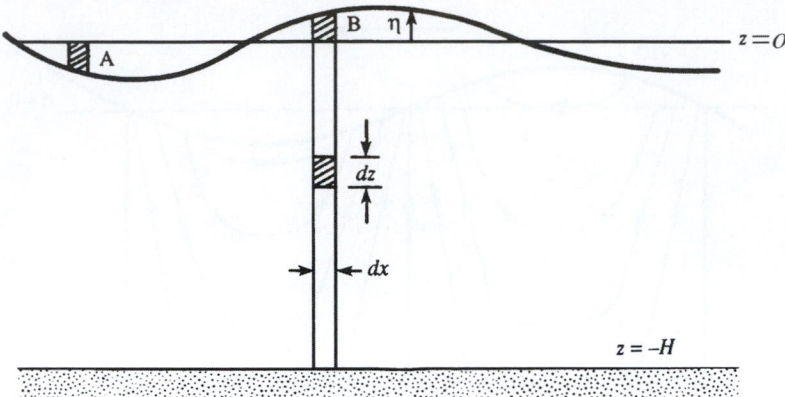

FIGURE 7.6 Calculation of potential energy of a fluid column. Here work must be done to push the liquid surface down below $z = 0$ (A), and lift the liquid surface up above $z = 0$ (B).

The *potential energy* E_p of the wave system is defined as the work done per unit area to deform a horizontal free surface into the disturbed state. It is therefore equal to the *difference* of potential energies of the system in the disturbed and undisturbed states. As the potential energy of an element in the fluid (per unit length in y) is $\rho g z \, dx \, dz$ (Figure 7.6), E_p can be calculated as

$$E_p = \frac{\rho g}{\lambda} \int_0^\lambda \int_{-H}^\eta z \, dz \, dx - \frac{\rho g}{\lambda} \int_0^\lambda \int_{-H}^0 z \, dz \, dx, \; = \frac{\rho g}{\lambda} \int_0^\lambda \int_0^\eta z \, dz \, dx = \frac{\rho g}{2\lambda} \int_0^\lambda \eta^2 dx. \qquad (7.40)$$

(An easier way to arrive at the expression for E_p is to note that the potential energy increase due to wave motion equals the work done in raising column A in Figure 7.6 to the location of column B, and integrating over *half the wavelength*. This is because an interchange of A and B over half a wavelength automatically forms a complete wavelength of the deformed surface. The mass (per unit length in y) of column A is $\rho \eta dx$, and the center of gravity is raised by η when A is taken to B. This agrees with the last form in (7.40).) Equation (7.40) can also be written in terms of the mean square displacement as

$$E_p = \frac{1}{2} \rho g \overline{\eta^2}. \qquad (7.41)$$

Thus, the average kinetic and potential energies are equal. This is called the *principle of equipartition of energy* and is valid in conservative dynamical systems undergoing small oscillations that are unaffected by planetary rotation. However, it is not valid when Coriolis forces are included, as described in Chapter 13. The total wave energy in the water column per unit horizontal area is

$$E = E_p + E_k = \rho g \overline{\eta^2} = \frac{1}{2} \rho g a^2, \qquad (7.42)$$

where the last form in terms of the amplitude a is valid if η is assumed sinusoidal, since the average over a wavelength of the square of a sinusoid is ½.

Next, consider the rate of transmission of energy due to a single sinusoidal component of wave number k. The *energy flux* across the vertical plane $x = 0$ is the pressure work done by the fluid in the region $x < 0$ on the fluid in the region $x > 0$. The time average energy flux F per unit length of crest is (writing p as the sum of a perturbation p' and a background pressure $-\rho g z$):

$$F = \left\langle \int_{-H}^{0} pu \, dz \right\rangle = \left\langle \int_{-H}^{0} p'u \, dz \right\rangle - \rho g \langle u \rangle \int_{-H}^{0} z \, dz = \left\langle \int_{-H}^{0} p'u \, dz \right\rangle, \qquad (7.43)$$

where $\langle \ \rangle$ denotes an average over a wave period, and we have used the fact that $\langle u \rangle = 0$. Substituting for p' from (7.31) and u from (7.28), (7.43) becomes

$$F = \langle \cos^2 (kx - \omega t) \rangle \frac{\rho a^2 \omega^3}{k \sinh^2 kH} \int_{-H}^{0} \cosh^2 k(z + H) \, dz.$$

The time average of $\cos^2(kx - \omega t)$ is ½, and the z-integral can be carried out by writing it in terms of exponentials, thus

$$F = \left[\frac{1}{2} \rho g a^2 \right] \left[\frac{c}{2} \left(1 + \frac{2kH}{\sinh 2kH} \right) \right]. \qquad (7.44)$$

The first factor is the wave energy per unit area given in (7.42). Therefore, the second factor must be the speed of propagation of the wave energy of component k. This energy propagation speed is called the *group speed*, and is further discussed in Section 7.5.

Approximations for Deep and Shallow Water

The preceding analysis is applicable for any value of H/λ. However, interesting simplifications are provided in the next few paragraphs for deep water, $H/\lambda \gg 1$, and shallow water, $H/\lambda \ll 1$.

Consider deep water first. The general expression for phase speed is (7.29), but we know that $\tanh(x) \to 1$ for $x \to \infty$ (Figure 7.7). However, x need not be very large for this approximation to be valid, because $\tanh(x) = 0.96403$ for $x = 2.0$. It follows that, with 2% accuracy, (7.29) can be approximated by

$$c = \sqrt{g/k} = \sqrt{g\lambda/2\pi} \qquad (7.45)$$

for $H > 0.32\lambda$ (corresponding to $kH > 2.0$). Surface waves are therefore classified as *deep-water waves* if the depth is more than one-third of the wavelength. Here, it is clear that deep-water waves are dispersive since their phase speed depends on wavelength.

A dominant period of wind-generated surface gravity waves in the ocean is ~10 s, which, via the dispersion relation (7.28), corresponds to a wavelength of 150 m. The water depth on a typical continental shelf is ~100 m, and in the open ocean it is ~4 km. Thus, the dominant wind waves in the ocean, even over the continental shelf, act as deep-water waves and do not feel the effects of the ocean bottom until they arrive near a coastline. This is not true of the very long wavelength gravity waves or tsunamis generated by tidal forces or earthquakes. Such waves may have wavelengths of hundreds of kilometers.

In deep water, the semi-major and semi-minor axes of particle orbits produced by small-amplitude gravity waves are nearly equal to ae^{kz} since

$$\frac{\cosh(k(z+H))}{\sinh(kH)} \approx \frac{\sinh(k(z+H))}{\sinh(kH)} \approx e^{kz}$$

for $kH > 2.0$, so the deep-water wave-induced fluid particle motions are:

$$\xi \cong -ae^{kz_0}\sin(kx_0 - \omega t), \text{ and } \zeta \cong ae^{kz_0}\cos(kx_0 - \omega t). \qquad (7.46)$$

The orbits are circles (Figure 7.4a). At the surface, their radius is a, the amplitude of the wave. The fluid velocity components for deep-water waves are

$$u = a\omega e^{kz}\cos(kx - \omega t), \text{ and } w = a\omega e^{kz}\sin(kx - \omega t). \qquad (7.47)$$

At a fixed spatial location, the velocity vector rotates clockwise (for a wave traveling in the positive x direction) at frequency ω, while its magnitude remains constant at $a\omega e^{kz}$.

For deep-water waves, the perturbation pressure from (7.31) simplifies to

$$p' = \rho g a e^{kz}\cos(kx - \omega t), \qquad (7.48)$$

which shows the wave-induced pressure change decays exponentially with depth, reaching 4% of its surface magnitude at a depth of $\lambda/2$. Thus, a bottom-mounted sensor used to record

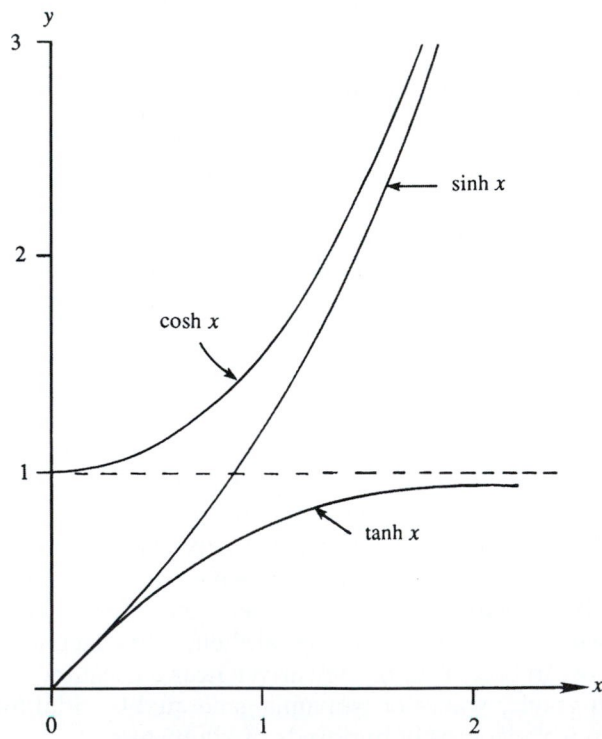

FIGURE 7.7   Behavior of hyperbolic functions $\cosh(x)$, $\sinh(x)$, and $\tanh(x)$ vs. x. For small x, $\cosh(x) \approx 1$ and $\sinh(x) \approx \tanh(x) \approx x$. For large x, $\cosh(x) \approx \sinh(x)$ and $\tanh(x) \approx 1$.

wave-induced pressure fluctuations will respond as a low-pass filter. Its signal will favor long waves while rejecting short ones.

The shallow water limit is also important and interesting. We know that $\tanh(x) \approx x$ as $x \to 0$ (Figure 7.7), so for $H/\lambda \ll 1$:

$$\tanh(2\pi H/\lambda) \approx 2\pi H/\lambda,$$

in which case the phase speed from (7.29) simplifies to

$$c = \sqrt{gH}, \tag{7.49}$$

and this matches the control volume result from Example 4.3. The approximation gives better than 3% accuracy if $H < 0.07\lambda$. Therefore, surface waves are regarded as *shallow-water waves* only if they are 14 times longer than the water depth. For these waves, (7.49) shows that the wave speed increases with water depth, and that it is independent of wavelength, so shallow-water waves are *nondispersive*.

To determine the approximate form of particle orbits for shallow-water waves, substitute the following approximations into (7.35):

$$\cosh(k(z+H)) \cong 1, \quad \sinh(k(z+H)) \cong k(z+H), \quad \text{and} \quad \sinh(kH) \cong kH.$$

The particle excursions then become

$$\xi \cong -\frac{a}{kH} \sin(kx_0 - \omega t), \quad \text{and} \quad \zeta \cong a\left(1 + \frac{z}{H}\right)\cos(kx_0 - \omega t). \tag{7.50}$$

These represent thin ellipses (Figure 7.4c), with a depth-independent semi-major axis a/kH and a semi-minor axis $a(1 + z/H)$ that linearly decreases to zero at the bottom wall.

From (7.27), the velocity field is

$$u = \frac{a\omega}{kH} \cos(kx - \omega t), \quad \text{and} \quad w = a\omega\left(1 + \frac{z}{H}\right)\sin(kx - \omega t), \tag{7.51}$$

which shows that the vertical component is much smaller than the horizontal component.

The pressure change from the undisturbed state is found from (7.31) to be

$$p' = \rho g a \cos(kx - \omega t) = \rho g \eta, \tag{7.52}$$

where (7.2) has been used to express the pressure change in terms of η. This shows that the pressure change at any point is independent of depth, and equals the hydrostatic increase of pressure due to the surface elevation change η. *The pressure field is therefore completely hydrostatic in shallow-water waves.* Vertical accelerations are negligible because of the small w-field. For this reason, shallow water waves are also called *hydrostatic waves*. Any worthwhile pressure sensor mounted on the bottom will sense these waves.

The depth-dependent wave speed (7.49) in shallow water leads to the phenomenon of shallow-water wave *refraction* observed at coastlines around the world. Consider a sloping beach, with depth contours parallel to the coastline (Figure 7.8). Assume that waves are propagating toward the coast from the deep ocean, with their crests at an angle to the coastline. Sufficiently near the coastline they begin to feel the effect of the bottom and finally become shallow-water waves. Their frequency does not change along the path, but their speed of propagation $c = (gH)^{1/2}$ and their wavelength λ become smaller. Consequently, the crest

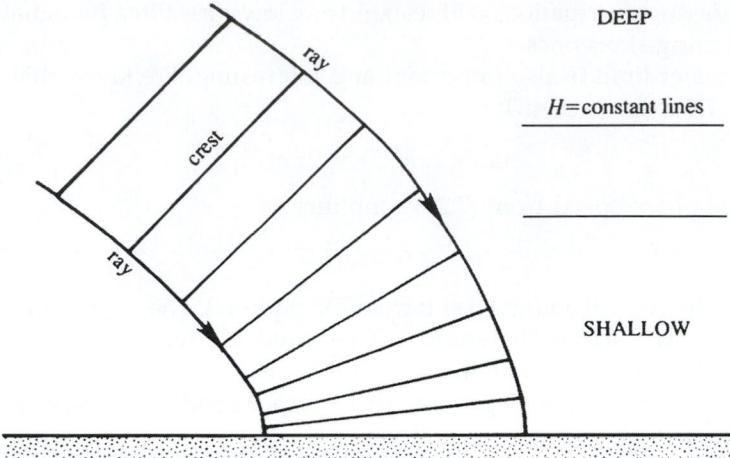

FIGURE 7.8 Refraction of a surface gravity wave approaching a sloping beach caused by changes in depth. In deep water, wave crests are commonly misaligned with isobaths. However, as a wave approaches the shore from any angle, the portion of the wave in shallower water will be slowed compared to that in deeper water. Thus, the wave crests will rotate and tend to become parallel to the shore as they approach it.

lines, which are perpendicular to the local direction of c, tend to become parallel to the coast. This is why the waves coming toward a gradually sloping beach always seem to have their crests parallel to the coastline.

An interesting example of wave refraction occurs when a deep-water wave train with straight crests approaches an island (Figure 7.9). Assume that the water gradually becomes shallower as the island is approached, and that the constant depth contours are circles

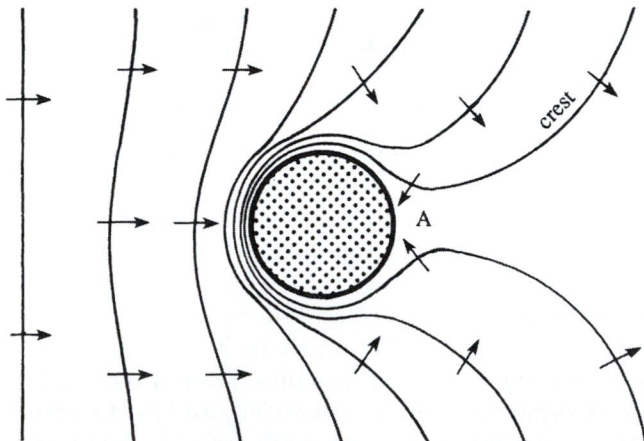

FIGURE 7.9 Refraction of surface gravity waves approaching a circular island with a gradually sloping beach. Crest lines are shown and are observed to travel toward the island, even on its shadow side A. *Reprinted with the permission of Mrs. Dorothy Kinsman Brown: B. Kinsman,* Wind Waves, *Prentice-Hall, Englewood Cliffs, NJ, 1965.*

concentric with the island. Figure 7.9 shows that the waves always come in *toward* the island, even on the shadowed-side marked A.

The bending of wave paths in an inhomogeneous medium is called *wave refraction*. In this case the source of inhomogeneity is the spatial dependence of H. The analogous phenomenon in optics is the bending of light due to density changes in its path.

7.3. INFLUENCE OF SURFACE TENSION

As described in Section 1.6, the interface between two immiscible fluids is in a state of tension. The tension acts as another restoring force on surface deformation, enabling the interface to support waves in a manner analogous to waves on a stretched membrane or string. Waves due to the presence of surface tension are called *capillary waves*. Although gravity is not needed to support these waves, the existence of surface tension alone without gravity is uncommon in terrestrial environments. Thus, the preceding results for pure gravity waves are modified to include surface tension in this section.

As shown in Section 4.10, there is a pressure difference Δp across a curved interface with nonzero surface tension σ. When the surface's principal radii of curvature are R_1 and R_2, this pressure difference is

$$\Delta p = \sigma(1/R_1 + 1/R_2), \tag{1.5}$$

where the pressure is greater on the side of the surface with the centers of curvature of the interface. This pressure difference modifies the free-surface boundary condition (7.19).

For straight-crested surface waves that produce fluid motion in the x-z plane, there is no variation in the y direction, so one of the radii of curvature is infinite, and the other, denoted R, lies in the x-z plane. Thus, if the pressure above the liquid is atmospheric, p_a, then pressure p in the liquid at the surface $z = \eta$ can be found from (1.5):

$$p_a - (p)_{z=\eta} = \sigma \frac{1}{R} = \sigma \frac{\partial^2 \eta / \partial x^2}{\left[1 + (\partial \eta / \partial x)^2\right]^{3/2}} \cong \sigma \frac{\partial^2 \eta}{\partial x^2}, \tag{7.53}$$

where the second equality follows from the definition of the curvature $1/R$ and the final approximate equality holds when the liquid surface slope $\partial \eta / \partial x$ is small. As before we can choose p to be a gauge pressure and this means setting $p_a = 0$ in (7.53), which leaves

$$(p)_{z=\eta} = -\sigma \frac{\partial^2 \eta}{\partial x^2}, \tag{7.54}$$

as the pressure-matching boundary condition at the liquid surface for small-slope surface waves. As before, this can be combined with the linearized unsteady Bernoulli equation (7.20) and evaluated on $z = 0$ for small-slope surface waves:

$$\left(\frac{\partial \phi}{\partial t}\right)_{z=0} = \frac{\sigma}{\rho} \frac{\partial^2 \eta}{\partial x^2} - g\eta. \tag{7.55}$$

The linear capillary-gravity, surface-wave solution now proceeds in an identical manner to that for pure gravity waves, except that the pressure boundary condition (7.21) is replaced by

(7.55). This modification only influences the dispersion relation $\omega(k)$, which is found by substitution of (7.2) and (7.26) into (7.55), to give

$$\omega = \sqrt{k\left(g + \frac{\sigma k^2}{\rho}\right)\tanh(kH)}, \tag{7.56}$$

so the phase velocity is

$$c = \sqrt{\left(\frac{g}{k} + \frac{\sigma k}{\rho}\right)\tanh kH} = \sqrt{\left(\frac{g\lambda}{2\pi} + \frac{2\pi\sigma}{\rho\lambda}\right)\tanh\frac{2\pi H}{\lambda}}. \tag{7.57}$$

A plot of (7.57) is shown in Figure 7.10. The primary effect of surface tension is to increase c above its value for pure gravity waves at all wavelengths. This increase occurs because there are two restoring forces that act together on the surface, instead of just one. However, the effect of surface tension is only appreciable for small wavelengths. The nominal size of these wavelengths is obtained by noting that there is a minimum phase speed at $\lambda = \lambda_m$, and surface tension dominates for $\lambda < \lambda_m$ (Figure 7.10). Setting $dc/d\lambda = 0$ in (7.57), and assuming deep water, $H > 0.32\lambda$ so $\tanh(2\pi H/\lambda) \approx 1$, produces:

$$c_{min} = \left[\frac{4g\sigma}{\rho}\right]^{1/4} \quad \text{at} \quad \lambda_m = 2\pi\sqrt{\frac{\sigma}{\rho g}}. \tag{7.58}$$

For an air-water interface at 20°C, the surface tension is $\sigma = 0.073$ N/m, giving

$$c_{min} = 23.1 \text{ cm/s} \quad \text{at} \quad \lambda_m = 1.71 \text{ cm.} \tag{7.59}$$

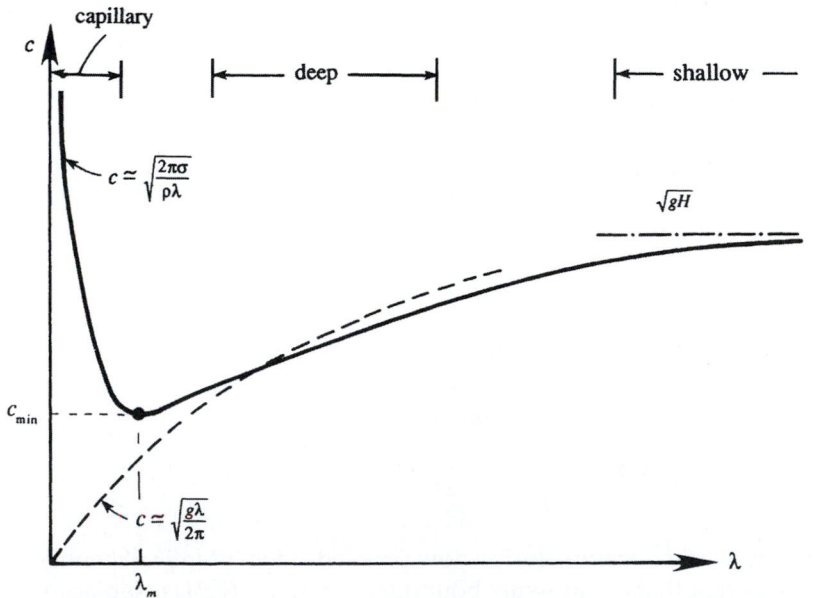

FIGURE 7.10 Generic sketch of the phase velocity c vs. wavelength λ for waves on the surface of liquid layer of depth H. The phase speed of the shortest waves is set by the liquid's surface tension σ and density ρ. The phase speed of the longest waves is set by gravity g and depth H. In between these limits, the phase speed has a minimum that typically occurs when the effects of surface tension and gravity are both important.

Therefore, only short-wavelength waves ($\lambda < \sim 7$ cm for an air-water interface), called *ripples*, are affected by surface tension. The waves specified by (7.59) are readily observed as the wave rings closest to the point of impact after an object is dropped onto the surface of a quiescent pool, pond, or lake of clean water. Surfactants and surface contaminants may lower σ or even introduce additional surface properties like surface viscosity or elasticity. Water surface wavelengths below 4 mm are dominated by surface tension and are essentially unaffected by gravity. From (7.57), the phase speed of *pure capillary waves* is

$$c = \sqrt{\frac{2\pi\sigma}{\rho\lambda}}, \tag{7.60}$$

where again $\tanh(2\pi H/\lambda) \approx 1$ has been assumed.

7.4. STANDING WAVES

The wave motion results presented so far are for one propagation direction ($+x$) as specified by (7.2). However, a small-amplitude sinusoidal wave with phase ($kx + \omega t$) is an equally valid solution of (7.11). Such a waveform,

$$\eta(x,t) = a\cos[kx + \omega t], \tag{7.61}$$

only differs from (7.2) in its direction of propagation. Its wave crests move in the $-x$ direction with increasing time. Interestingly, nonpropagating waves can be generated by superposing two waves with the same amplitude and wavelength that move in opposite directions. The resulting surface displacement is

$$\eta = a\cos(kx - \omega t) + a\cos(kx + \omega t) = 2a\cos kx \cos \omega t.$$

Here it follows that $\eta = 0$ at $kx = \pm\pi/2, \pm 3\pi/2$, etc., for all time. Such locations of zero surface displacement are called *nodes*. In this case, deflections of the liquid surface do not travel. The surface simply oscillates up and down at frequency ω with a spatially varying amplitude, keeping the nodal points fixed. Such waves are called *standing waves*. The corresponding stream function, a direct extension of (7.37), includes both the $\cos(kx - \omega t)$ and $\cos(kx + \omega t)$ components:

$$\psi = \frac{a\omega}{k}\frac{\sinh k(z+H)}{\sinh kH}[\cos(kx - \omega t) - \cos(kx + \omega t)] = \frac{2a\omega}{k}\frac{\sinh k(z+H)}{\sinh kH}\sin kx \sin \omega t. \tag{7.62}$$

The instantaneous streamline pattern shown in Figure 7.11 should be compared with the streamline pattern for a propagating wave (Figure 7.5).

Standing waves may form in a limited body of water such as a tank, pool, or lake when traveling waves reflect from its walls, sides, or shores. A standing-wave oscillation in a lake is called a *seiche* (pronounced "saysh"), in which only certain wavelengths and frequencies ω (eigenvalues) are allowed by the system. Consider a lake of length L with uniform depth H and vertical shores (walls), and assume that the waves are invariant along y. The possible wavelengths are found by setting $u = 0$ at the two walls. Here, $u = \partial\psi/\partial z$, so (7.62) gives

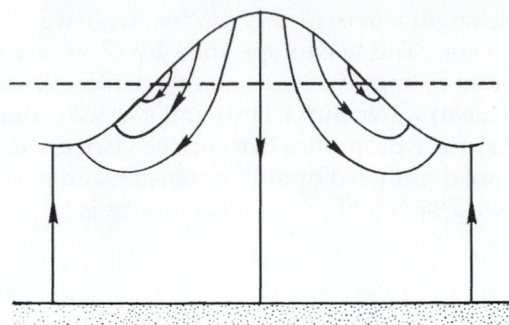

FIGURE 7.11 Instantaneous streamline pattern in a standing surface gravity wave. Here the $\psi = 0$ streamline follows the bottom and jumps up to contact the surface at wave crests and troughs where the horizontal velocity is zero. If this standing wave represents the $n = 0$ mode of a reservoir of length L with vertical walls, then $L = \lambda/2$ is the distance between a crest and a trough. If it represents the $n = 1$ mode, then $L = \lambda$ is the distance between successive crests or successive troughs.

$$u = 2a\omega \frac{\cosh k(z + H)}{\sinh kH} \sin kx \sin \omega t. \tag{7.63}$$

Taking the walls at $x = 0$ and L, the condition of no flow through the walls requires $\sin(kL) = 0$, that is,

$$kL = (n + 1)\pi \quad n = 0, 1, 2, \ldots,$$

which gives the allowable wavelengths as

$$\lambda = \frac{2L}{n + 1}. \tag{7.64}$$

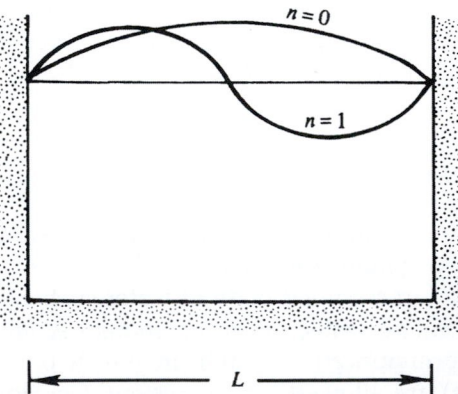

FIGURE 7.12 Distributions of horizontal velocity u for the first two normal modes in a lake or reservoir with vertical sides. Here the boundary conditions require $u = 0$ on the vertical sides. These distributions are consistent with the streamline pattern of Figure 7.11.

The largest possible wavelength is $2L$ and the next smaller is L (Figure 7.12). The allowed frequencies can be found from the dispersion relation (7.28), giving

$$\omega = \sqrt{\frac{\pi g(n+1)}{L} \tanh\left[\frac{(n+1)\pi H}{L}\right]},\qquad (7.65)$$

which are the natural frequencies of the lake.

7.5. GROUP VELOCITY, ENERGY FLUX, AND DISPERSION

A variety of interesting phenomena take place when waves are dispersive and their phase speed depends on wavelength. Such wavelength-dependent propagation is common for waves that travel on interfaces between different materials (Graff, 1975). Examples are Rayleigh waves (vacuum and a solid), Stonely waves (a solid and another material), or interface waves (two different immiscible liquids). Here we consider only air-water interface waves and emphasize deep-water gravity waves for which c is proportional to $\sqrt{\lambda}$.

In a dispersive system, the energy of a wave component does not propagate at the phase velocity $c = \omega/k$, but at the *group velocity* defined as $c_g = d\omega/dk$. To understand this, consider the superposition of two sinusoidal wave components of equal amplitude but slightly different wave number (and consequently slightly different frequency because $\omega = \omega(k)$). The waveform of the combination is

$$\eta = a\cos(k_1 x - \omega_1 t) + a\cos(k_2 x - \omega_2 t).$$

Applying the trigonometric identity for the sum of cosines of different arguments, we obtain

$$\eta = 2a\cos\left(\frac{1}{2}\Delta kx - \frac{1}{2}\Delta\omega x\right)\cos(kx - \omega t),\qquad (7.66)$$

where $\Delta k = k_2 - k_1$ and $\Delta\omega = \omega_2 - \omega_1$, $k = (k_1 + k_2)/2$, and $\omega = (\omega_1 + \omega_2)/2$. Here, $\cos(kx - \omega t)$ is a progressive wave with a phase speed of $c = \omega/k$. However, its amplitude $2a$ is modulated by a *slowly varying* function $\cos[\Delta kx/2 - \Delta\omega t/2]$, which has a large wavelength $4\pi/\Delta k$, a long period $4\pi/\Delta\omega$, and propagates at a speed (wavelength/period) of

$$c_g = \Delta\omega/\Delta k = d\omega/dk,\qquad (7.67)$$

where the second equality holds in the limit as Δk and $\Delta\omega \to 0$. Multiplication of a rapidly varying sinusoid and a slowly varying sinusoid, as in (7.66), generates repeating wave groups (Figure 7.13). The individual wave crests (and troughs) propagate with the speed $c = \omega/k$, but the envelope of the wave groups travels with the speed c_g, which is therefore called the *group velocity*. If $c_g < c$, then individual wave crests appear spontaneously at a nodal point, proceed forward through the wave group, and disappear at the next nodal point. If, on the other hand, $c_g > c$, then individual wave crests emerge from a forward nodal point and vanish at a backward nodal point.

Equation (7.67) shows that the group speed of waves of a certain wave number k is given by the slope of the *tangent* to the dispersion curve $\omega(k)$. In contrast, the phase velocity is given by the slope of the radius or distance vector on the same plot (Figure 7.14).

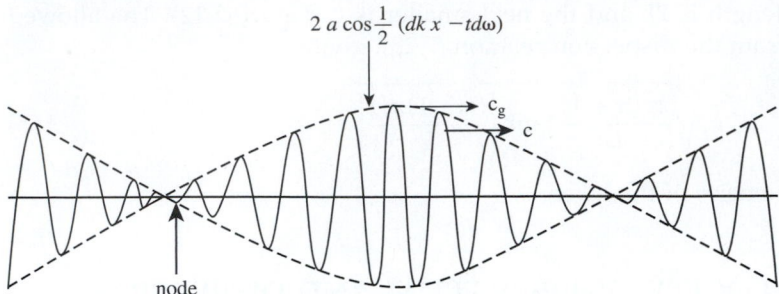

$$2\,a\,\cos\frac{1}{2}\,(dk\,x - td\omega)$$

node

FIGURE 7.13 Linear combination of two equal amplitude sinusoids of nearly the same frequency that form a modulated wave train. Individual wave crests or troughs travel at the phase speed. However, the nodal locations which partition the wave train into groups, travel at the group speed.

A particularly illuminating example of the idea of group velocity is provided by the concept of a *wave packet*, formed by combining all wave numbers in a certain narrow band δk around a central value k. In physical space, the wave appears nearly sinusoidal with wavelength $2\pi/k$, but the amplitude *dies away* over a distance proportional to $1/\delta k$ (Figure 7.15). If the spectral width δk is narrow, then decay of the wave amplitude in physical space is slow. The concept of such a wave packet is more realistic than the one in Figure 7.13, which is rather unphysical because the wave groups repeat themselves. Suppose that, at some initial time, the wave group is represented by

$$\eta = a(x)\cos kx.$$

It can be shown (see, for example, Phillips, 1977, p. 25) that for small times, the subsequent evolution of the wave profile is approximately described by

$$\eta = a(x - c_g t)\cos(kx - \omega t), \tag{7.68}$$

where $c_g = d\omega/dk$. This shows that the *amplitude of a wave packet travels with the group speed*. It follows that c_g must equal the speed of propagation of *energy* of a certain wavelength. The fact

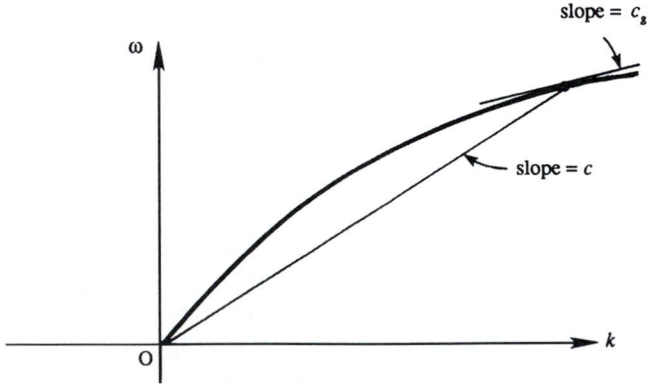

FIGURE 7.14 Graphical depiction of the phase speed, c, and group speed, c_g, on a generic plot of a gravity wave dispersion relation, $\omega(k)$ vs. k. If a sinusoidal wave has frequency ω and wave number k, then the phase speed c is the slope of the straight line through the points $(0, 0)$ and (k, ω), while the group speed c_g is the tangent to the dispersion relation at the point (k, ω). For the dispersion relation depicted here, c_g is less than c.

FIGURE 7.15 A wave packet composed of a wave number lying in a confined bandwidth δk. The length of the wave packet in physical space is proportional to $1/\delta k$. Thus, narrowband packets are longer than broadband packets.

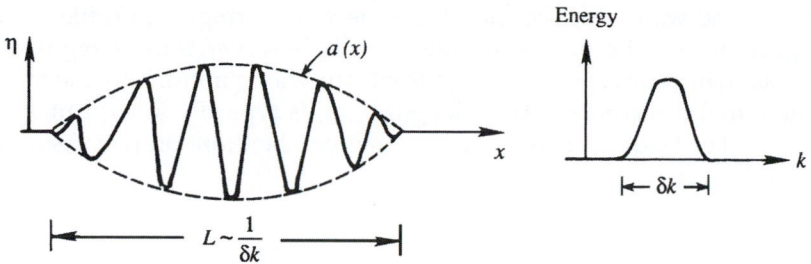

that c_g is the speed of energy propagation is also evident in Figure 7.13 because the nodal points travel at c_g and no energy crosses nodal points.

For surface gravity waves having the dispersion relation (7.29), the group velocity is found to be

$$c_g = \frac{c}{2}\left[1 + \frac{2kH}{\sinh(2kH)}\right], \tag{7.69}$$

which has two limiting cases:

$$c_g = c/2 \,(\text{deep water}), \text{ and } c_g = c \,(\text{shallow water}). \tag{7.70}$$

The group velocity of deep-water gravity waves is half the deep-water phase speed while shallow-water waves are nondispersive with $c = c_g$. For a linear nondispersive system, any waveform preserves its shape as it travels because all the wavelengths that make up the waveform travel at the same speed. For a pure capillary wave, the group velocity is $c_g = 3c/2$ (Exercise 7.9).

The rate of energy transmission for gravity waves is given by (7.44), namely

$$F = E\frac{c}{2}\left[1 + \frac{2kH}{\sinh(2kH)}\right] = Ec_g, \tag{7.71}$$

where $E = \rho g a^2/2$ is the average energy in the water column per unit horizontal area. This signifies that the *rate of transmission of energy of a sinusoidal wave component is wave energy times the group velocity,* and reinforces the interpretation of the group velocity as the speed of propagation of wave energy.

In three dimensions, the dispersion relation $\omega = \omega(k, l, m)$ may depend on all three components of the wave number vector $\mathbf{K} = (k, l, m)$. Here, using index notation, the group velocity vector is given by

$$c_{gi} = \frac{\partial \omega}{\partial K_i},$$

so the group velocity vector is the gradient of ω in the wave number space.

As mentioned in connection with (7.29) and (7.59), deep-water wave dispersion readily explains the evolution of the surface disturbance generated by dropping a stone into a quiescent pool, pond, or lake. Here, the initial disturbance can be thought of as being composed of a great many wavelengths, but the longer ones will travel faster. A short time after impact, at

$t = t_1$, the water surface may have the rather irregular profile shown in Figure 7.16. The appearance of the surface at a later time t_2, however, is more regular, with the longer components (which travel faster) out in front. The waves in front are the longest waves produced by the initial disturbance. Their length, λ_{max}, is typically a few times larger than the dropped object. The leading edge of the wave system therefore propagates at the group speed of these wavelengths:

$$c_{g\ max} = \frac{1}{2}\sqrt{\frac{g\lambda_{max}}{2\pi}}.$$

Of course, pure capillary waves can propagate faster than this speed, but they may have small amplitudes and are dissipated quickly. Interestingly, the region of the impact becomes calm because there is a minimum group velocity of water waves due to the influence of surface tension, namely 17.8 cm/s (Exercise 7.10). The trailing edge of the wave system therefore travels at speed

$$c_{g\ min} = 17.8 \ \text{cm/s}.$$

With $c_{gmax} > 17.8$ cm/s for ordinary hand-size stones, the length of the disturbed region gets larger, as shown in Figure 7.16. The wave heights become correspondingly smaller because there is a fixed amount of energy in the wave system. (Wave dispersion, therefore, makes the linearity assumptions more accurate.) The smoothing of the waveform and the spreading of the region of disturbance continue until the amplitudes become imperceptible or the waves are damped by viscous dissipation (Exercise 7.11). It is clear that the *initial superposition of various wavelengths, running for some time, will sort themselves from slowest to fastest traveling components* since the different sinusoidal components, differing widely in their wave numbers, become spatially *separated*, with the slow ones close to the point of impact and

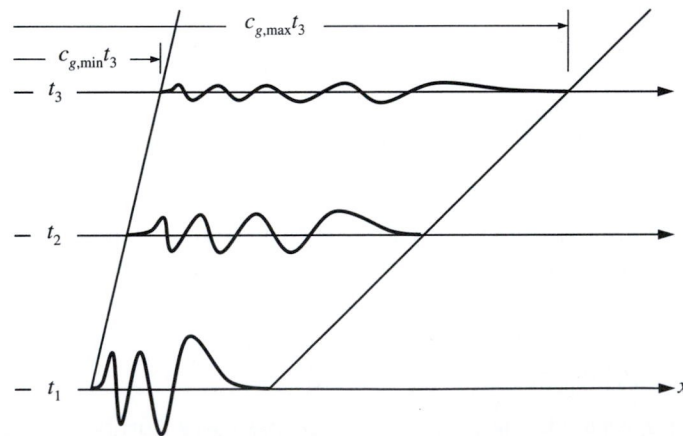

FIGURE 7.16 Generic surface profiles at three successive times of the wave train produced by dropping a stone into a deep quiescent pool. As time increases, the initial disturbance's long-wave (low-frequency) components travel faster than its short-wave (high-frequency) components. Thus, the wave train lengthens, the number of crests and troughs increases, and amplitudes fall (to conserve energy).

the fast ones further away. This is a basic feature of the behavior of dispersive wave propagation.

In the case of deep-water surface waves described here, the wave group as a whole travels slower than individual crests. Therefore, if we try to follow the last crest at the rear of the train, quite soon it is the second one from the rear; a new crest has appeared behind it. In fact, new crests are constantly appearing at the rear of the train, propagating through the train, and finally disappearing at the front of the train. This is because, by following a particular crest, we are traveling at roughly twice the speed at which the wave energy is traveling. Consequently, *we do not see a wave of fixed wavelength if we follow a particular crest.* In fact, an individual wave constantly becomes longer as it propagates through the train. When its length becomes equal to the longest wave generated initially, it cannot evolve anymore and dies out. Clearly, the waves at the front of the train are the longest Fourier components present in the initial disturbance. In addition, the temporal frequencies of the highest and lowest speed wave components of the wave group are typically different enough so that the number of wave crests in the train increases with time.

Another way to understand the group velocity is to consider the k or λ determined by an observer traveling at speed c_g with a slowly varying wave train described by

$$\eta = a(x,t)\cos[\theta(x,t)], \tag{7.72}$$

in an otherwise quiescent pool of water with constant depth H. Here $a(x, t)$ is a slowly varying amplitude and $\theta(x, t)$ is the local phase. For a specific wave number k and frequency ω, the phase is $\theta = kx - \omega t$. For a slowly varying wave train, define the *local* wave number $k(x, t)$ and the *local* frequency $\omega(x, t)$ as the rate of change of phase in space and time, respectively,

$$k(x,t) \equiv (\partial/\partial x)\theta(x,t) \text{ and } \omega(x,t) \equiv -(\partial/\partial t)\theta(x,t). \tag{7.73}$$

Cross differentiation leads to

$$\partial k/\partial t + \partial \omega/\partial x = 0, \tag{7.74}$$

but when there is a dispersion relationship $\omega = \omega(k)$, the spatial derivative of ω can be rewritten using the chain rule, $\partial\omega/\partial x = (d\omega/dk)\partial k/\partial x$, so that (7.74) becomes

$$\frac{\partial k}{\partial t} + c_g\frac{\partial k}{\partial x} = 0, \tag{7.75}$$

where $c_g = d\omega/dk$. The left-hand side of (7.75) is similar to the material derivative and gives the rate of change of k as seen by an observer traveling at speed c_g, which in this case is zero. Therefore, such an observer will always see the same wavelength. The *group velocity is therefore the speed at which wave numbers are advected.* This is shown in the xt-diagram of Figure 7.17, where wave crests follow lines with $dx/dt = c$ and wavelengths are preserved along the lines $dx/dt = c_g$. Note that the width of the disturbed region, bounded by the first and last thick lines in Figure 7.17, increases with time, and that the crests constantly appear at the back of the group and vanish at the front.

Now consider the same traveling observer, but allow there to be smooth variations in the water depth $H(x)$. Such depth variation creates an inhomogeneous medium when the waves are long enough to feel the presence of the bottom. Here, the dispersion relationship will be

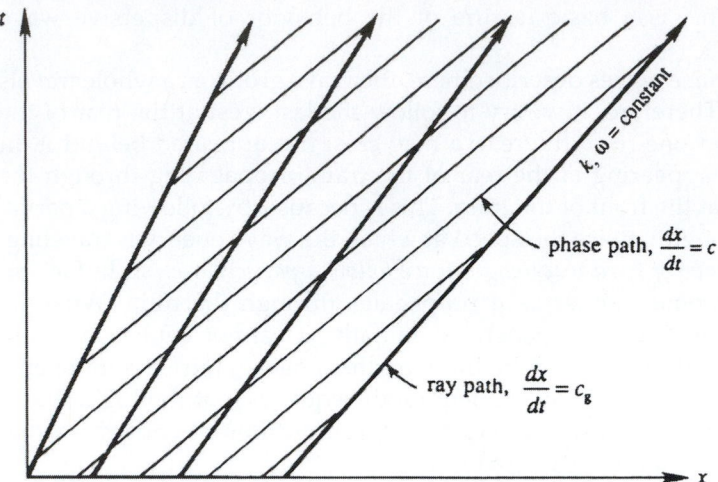

FIGURE 7.17 Propagation of a wave group in a homogeneous medium, represented on an *x-t* plot. Thin lines indicate paths taken by wave crests, and thick lines represent paths along which k and ω are constant. *M. J. Lighthill*, Waves in Fluids, *1978, reprinted with the permission of Cambridge University Press, London.*

$$\omega = \sqrt{gk \tanh \left[kH(x)\right]},$$

which is of the form

$$\omega = \omega(k, x). \tag{7.76}$$

Thus, a local value of the group velocity can be defined:

$$\partial \omega(k, x)/\partial k = c_g, \tag{7.77}$$

which on multiplication by $\partial k/\partial t$ gives

$$c_g \frac{\partial k}{\partial t} = \frac{\partial \omega}{\partial k}\frac{\partial k}{\partial t} = \frac{\partial \omega}{\partial t}. \tag{7.78}$$

Multiplying (7.74) by c_g and using (7.78) we obtain

$$\frac{\partial \omega}{\partial t} + c_g \frac{\partial \omega}{\partial x} = 0. \tag{7.79}$$

In three dimensions, this implies

$$\partial \omega/\partial t + \mathbf{c_g} \cdot \nabla \omega = 0,$$

which shows that ω, the frequency of the wave, remains constant to an observer traveling with the group velocity in an inhomogeneous medium.

Summarizing, an observer traveling at c_g in a homogeneous medium sees constant values of k, $\omega(k)$, c, and $c_g(k)$. Consequently, ray paths describing group velocity in the *x-t* plane are straight lines (Figure 7.17). In an inhomogeneous medium ω remains constant along the lines $dx/dt = c_g$, but k, c, and c_g can change. Consequently, ray paths are not straight in this case (Figure 7.18).

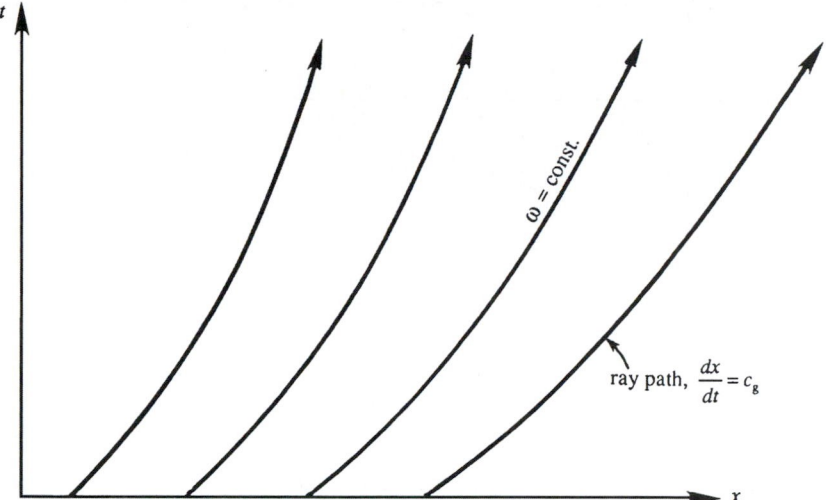

FIGURE 7.18 Propagation of a wave group in an inhomogeneous medium represented on an *x-t* plot. Only ray paths along which ω is constant are shown. *M. J. Lighthill,* Waves in Fluids, *1978, reprinted with the permission of Cambridge University Press, London.*

7.6. NONLINEAR WAVES IN SHALLOW AND DEEP WATER

In the first five sections of this chapter, the wave slope has been assumed to be small enough so that neglect of higher-order terms in the Bernoulli equation and application of the boundary conditions at $z = 0$ instead of at the free surface $z = \eta$ are acceptable approximations. One consequence of such linear analysis has been that shallow-water waves of arbitrary shape propagate unchanged in form. The unchanging form results from the fact that all wavelengths composing the initial waveform propagate at the same speed, $c = (gH)^{1/2}$, provided all the sinusoidal components satisfy the shallow-water approximation $kH \ll 1$. Such waveform invariance no longer occurs if *finite amplitude* effects are considered. This and several other nonlinear effects will also be discussed in this section.

Finite amplitude effects can be formally treated by the *method of characteristics*; this is discussed, for example, in Liepmann and Roshko (1957) and Lighthill (1978). Instead, a qualitative approach is adopted here. Consider a finite amplitude surface displacement consisting of a wave crest and trough, propagating in shallow-water of undisturbed depth H (Figure 7.19). Let a little wavelet be superposed on the crest at point x', at which the water depth is H' and the fluid velocity due to the wave motion is $u(x')$. Relative to an observer moving with the fluid velocity u, the wavelet propagates at the local shallow-water speed $c' = \sqrt{gH'}$. The speed of the wavelet relative to a frame of reference fixed in the undisturbed fluid is therefore $c = c' + u$. It is apparent that the local wave speed c is no longer constant because $c'(x)$ and $u(x)$ are variables. This is in contrast to the linearized theory in which u is negligible and c' is constant because $H' \approx H$.

Let us now examine the effect of variable phase speed on the wave profile. The value of c' is larger for points near the wave crest than for points in the wave trough. From Figure 7.5 we

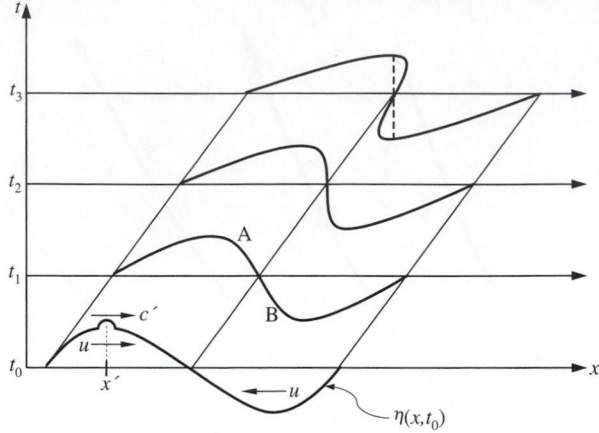

FIGURE 7.19 Finite-amplitude surface wave profiles at four successive times. When the wave amplitude is large enough, the fluid velocity below a crest or trough may be an appreciable fraction of the phase speed. This will cause wave crests to overtake wave troughs and will steepen the compressive portion of the wave (section A-B at time t_1). As this steepening continues, the wave-compression surface slope may become very large (t_2), or the wave may overturn and become a plunging breaker (t_3). Depending on the dynamics of the actual wave, the conditions shown at t_2 and t_3 may or may not occur since additional nonlinear processes (not described here) may contribute to the wave's evolution after t_1. If the waves were longitudinal (as in one-dimensional gas dynamics), the waveform at t_2 would represent a nascent shockwave, while the waveform at t_3 would represent a fully formed shockwave and would follow the dashed line to produce a single-valued profile.

also know that the fluid velocity u is positive (i.e., in the direction of wave propagation) under a wave crest and negative under a trough. It follows that wave speed c is larger for points on the crest than for points on the trough, so that the waveform deforms as it propagates, the crest region tending to overtake the trough region (Figure 7.19).

We shall call the front face AB a *compression region* because the surface here is rising with time and this implies an increase in pressure at any depth within the liquid. Figure 7.19 shows that the net effect of nonlinearity is a steepening of the compression region. For finite amplitude waves in a nondispersive medium like shallow water, therefore, there is an important distinction between compression and expansion regions. A compression region tends to steepen with time, while an expansion region tends to flatten out. This eventually would lead to the wave shape shown at the top of Figure 7.19, where there are three values of surface elevation at a point. While this situation is certainly possible and is readily observed as plunging breakers develop in the surf zone along ocean coastlines, the actual wave dynamics of such a situation lie beyond the scope of this discussion. However, even before the formation of a plunging breaker, the wave slope becomes infinite (profile at t_2 in Figure 7.19), so that additional physical processes including wave breaking, air entrainment, and foaming become important, and the current ideal flow analysis becomes inapplicable. Once the wave has broken, it takes the form of a front that propagates into still fluid at a constant speed that lies between $\sqrt{gH_1}$ and $\sqrt{gH_2}$, where H_1 and H_2 are the water depths on the two sides of the front (Figure 7.20). Such a wave is called a *hydraulic jump*, and it is similar to a *shockwave* in a compressible flow. Here it should be noted that the t_3 wave profile shown in Figure 7.19 is

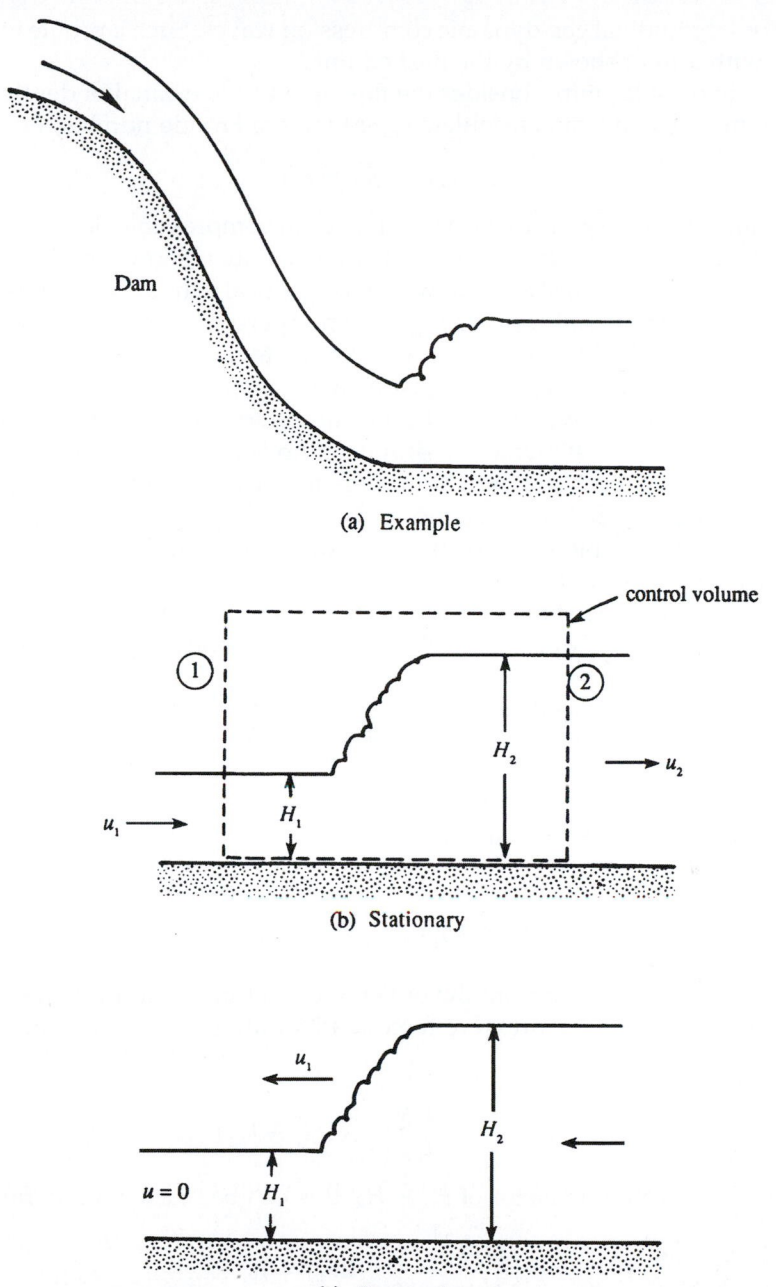

(a) Example

(b) Stationary

(c) Propagating

FIGURE 7.20   Schematic cross-section drawings of hydraulic jumps. (a) A stationary hydraulic jump formed at the bottom of a damn's spillway. (b) A stationary hydraulic jump and a stationary rectangular control volume with vertical inlet surface (1) and vertical outlet surface (2). (c) A hydraulic jump moving into a quiescent fluid layer of depth H_1. The flow speed behind the jump is nonzero.

not possible for longitudinal gas-dynamic compression waves. Such a profile instead leads to a shockwave with a front shown by the dashed line.

To analyze a hydraulic jump, consider the flow in a shallow canal of depth H. If the flow speed is u, we may define a dimensionless speed via the Froude number, Fr:

$$\text{Fr} \equiv u/\sqrt{gH} = u/c. \tag{4.104}$$

The Froude number is analogous to the *Mach number* in compressible flow. The flow is called *supercritical* if $\text{Fr} > 1$, and *subcritical* if $\text{Fr} < 1$. For the situation shown in Figure 7.20b, where the jump is stationary, the upstream flow is supercritical while the downstream flow is subcritical, just as a compressible flow changes from supersonic to subsonic by going through a shockwave (see Chapter 15). The depth of flow is greater downstream of a hydraulic jump, just as the gas pressure is greater downstream of a shockwave. However, dissipative processes act within shockwaves and hydraulic jumps so that mechanical energy is lost in both cases. An example of a stationary hydraulic jump is found at the foot of a dam, where the flow almost always reaches a supercritical state because of the freefall (Figure 7.20a). A tidal bore propagating into a river mouth is an example of a propagating hydraulic jump. A circular hydraulic jump can be made by directing a vertically falling water stream onto a flat horizontal surface (Exercise 4.22).

The planar hydraulic jump shown in cross section in Figure 7.20b can be analyzed by using the dashed control volume shown, the goal being to determine how the depth ratio depends on the upstream Froude number. As shown, the depth rises from H_1 to H_2 and the velocity falls from u_1 to u_2. If the velocities are uniform through the depth and Q is the volume flow rate per unit width normal to the plane of the paper, then mass conservation requires

$$Q = u_1 H_1 = u_2 H_2.$$

Conserving momentum with the same control volume via (4.17) with $d/dt = 0$ and $\mathbf{b} = 0$ produces

$$\rho Q(u_2 - u_1) = \frac{1}{2}\rho g(H_1^2 - H_2^2),$$

where the left-hand terms come from the outlet and inlet momentum fluxes, and the right-hand terms are the hydrostatic pressure forces. Substituting $u_1 = Q/H_1$ and $u_2 = Q/H_2$ on the right side yields:

$$Q^2\left(\frac{1}{H_2} - \frac{1}{H_1}\right) = \frac{1}{2}g(H_1^2 - H_2^2). \tag{7.80}$$

After canceling out a common factor of $H_1 - H_2$, this can be rearranged to find:

$$\left(\frac{H_2}{H_1}\right)^2 + \frac{H_2}{H_1} - 2\text{Fr}_1^2 = 0, .$$

where $\text{Fr}_1^2 = Q^2/gH_1^3 = u_1^2/gH_1$. The physically meaningful solution is

$$\frac{H_2}{H_1} = \frac{1}{2}\left(-1 + \sqrt{1 + 8\text{Fr}_1^2}\right). \tag{7.81}$$

For supercritical flows $Fr_1 > 1$, for which (7.81) shows that $H_2 > H_1$, and this verifies that water depth increases through a hydraulic jump.

Although a solution with $H_2 < H_1$ for $Fr_1 < 1$ is mathematically allowed, such a solution violates the second law of thermodynamics, because it implies an increase of mechanical energy through the jump. To see this, consider the mechanical energy of a fluid particle at the surface, $E = u^2/2 + gH = Q^2/2H^2 + gH$. Eliminating Q by using (7.80) we obtain, after some algebra,

$$E_2 - E_1 = -(H_2 - H_1)\frac{g(H_2 - H_1)^2}{4H_1 H_2}.$$

This shows that $H_2 < H_1$ implies $E_2 > E_1$, which violates the second law of thermodynamics. The mechanical energy, in fact, *decreases* in a hydraulic jump because of the action of viscosity.

Hydraulic jumps are not limited to air-water interfaces and may also appear at density interfaces in a stratified fluid, in the laboratory as well as in the atmosphere and the ocean. (For example, see Turner, 1973, Figure 3.11, for a photograph of an internal hydraulic jump on the lee side of a mountain.)

In a nondispersive medium, nonlinear effects may continually accumulate until they become large changes. Such an accumulation is prevented in a dispersive medium because the different Fourier components propagate at different speeds and tend to separate from each other. In a dispersive system, then, nonlinear steepening could cancel out the dispersive spreading, resulting in finite amplitude waves of constant form. This is indeed the case. A brief description of the phenomenon is given here; further discussion can be found in Whitham (1974), Lighthill (1978), and LeBlond and Mysak (1978).

In 1847 Stokes showed that periodic waves of finite amplitude are possible in deep water. In terms of a power series in the amplitude a, he showed that the surface deflection of irrotational waves in deep water is given by

$$\eta = a \cos k(x - ct) + \frac{1}{2}ka^2 \cos 2k(x - ct) + \frac{3}{8}k^2 a^3 \cos 3k(x - ct) + \ldots, \qquad (7.82)$$

where the speed of propagation is

$$c = \sqrt{\frac{g}{k}\left(1 + k^2 a^2 + \ldots\right)}. \qquad (7.83)$$

Equation (7.82) shows the first three terms in a Fourier series for the waveform η. The addition of Fourier components of different wavelengths in (7.82) shows that the wave profile η is no longer exactly sinusoidal. The arguments in the cosine terms show that all the Fourier components propagate at the same speed c, so that the wave profile propagates unchanged in time. It has now been established that the existence of periodic wave trains of unchanging form is a typical feature of nonlinear dispersive systems. Another important result, generally valid for nonlinear systems, is that the wave speed depends on the amplitude, as in (7.83).

Periodic finite-amplitude irrotational waves in deep water are frequently called *Stokes waves*. They have a flattened trough and a peaked crest (Figure 7.21). The maximum possible amplitude is $a_{max} = 0.07\lambda$, at which point the crest becomes a sharp 120° angle. Attempts at generating waves of larger amplitude result in the appearance of foam (*white caps*) at these sharp crests.

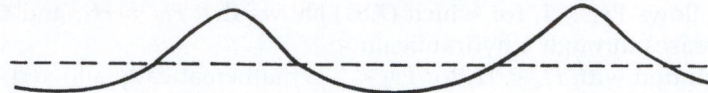

FIGURE 7.21　The waveform of a Stokes wave. Stokes waves are finite-amplitude, periodic irrotational waves in deep water with crests that are more pointed and troughs that are broader than sinusoidal waves.

When finite amplitude waves are present, fluid particles no longer trace closed orbits, but undergo a slow drift in the direction of wave propagation. This is called *Stokes drift*. It is a second-order or finite-amplitude effect that causes fluid particle orbits to no longer close and instead take a shape like that shown in Figure 7.22. The mean velocity of a fluid particle is therefore not zero, although the mean velocity at a fixed point in space must be zero if the wave motion is periodic. The drift occurs because the particle moves forward faster when at the top of its trajectory than it does backward when at the bottom of its trajectory.

To find an expression for the Stokes drift, start from the path-line equations (7.32) for the fluid particle trajectory $\mathbf{x}_p(t) = x_p(t)\mathbf{e}_x + z_p(t)\mathbf{e}_z$, but this time include first-order variations in the u and w fluid velocities via a first-order Taylor series in $\xi = x_p - x_0$, and $\zeta = z_p - z_0$:

$$\frac{dx_p(t)}{dt} = u(x_p, z_p, t) = u(x_0, z_0, t) + \xi\left(\frac{\partial u}{\partial x}\right)_{x_0, z_0} + \zeta\left(\frac{\partial u}{\partial z}\right)_{x_0, z_0} + \ldots, \quad (7.84a)$$

and

$$\frac{dz_p(t)}{dt} = w(x_p, z_p, t) = w(x_0, z_0, t) + \xi\left(\frac{\partial w}{\partial x}\right)_{x_0, z_0} + \zeta\left(\frac{\partial w}{\partial z}\right)_{x_0, z_0} + \ldots, \quad (7.84b)$$

where (x_0, z_0) is the fluid element location in the absence of wave motion. The Stokes drift is the time average of (7.84a). However, the time average of $u(x_0, z_0, t)$ is zero; thus, the Stokes drift is given by the time average of the next two terms of (7.84a). These terms were neglected in the fluid particle trajectory analysis in Section 7.2, and the result was closed orbits.

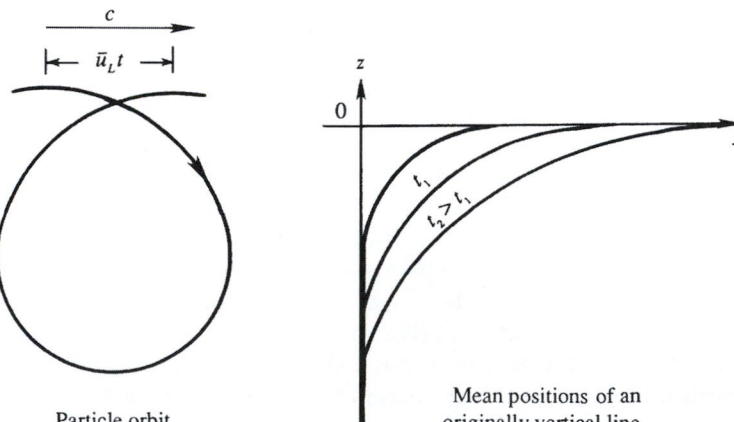

Particle orbit

Mean positions of an originally vertical line

FIGURE 7.22　The Stokes drift. The drift velocity $\bar{u}_L$ is a finite-amplitude effect and occurs because near-surface fluid particle paths are no longer closed orbits. The mean position of an initially vertical line of fluid particles extending downward from the liquid surface will increasingly bend in the direction of wave propagation with increasing time.

For deep-water gravity waves, the Stokes drift speed $\bar{u}_L$ can be estimated by evaluating the time average of (7.84a) using (7.46) and (7.47) to produce:

$$\bar{u}_L = a^2 \omega k e^{2kz_0}, \tag{7.85}$$

which is the Stokes drift speed in deep water. Its surface value is $a^2 \omega k$, and the vertical decay rate is twice that for the fluid velocity components. It is therefore confined very close to the sea surface. For arbitrary water depth, (7.85) may be generalized to

$$\bar{u}_L = a^2 \omega k \frac{\cosh(2k(z_0 + H))}{2 \sinh^2(kH)} \tag{7.86}$$

(Exercise 7.14). As might be expected, the vertical component of the Stokes drift is zero.

The Stokes drift causes mass transport in the fluid so it is also called the *mass transport velocity*. A vertical column of fluid elements marked by some dye gradually bends near the surface (Figure 7.22). In spite of this mass transport, the mean fluid velocity at any point that resides within the liquid for the entire wave period is exactly zero (to any order of accuracy), if the flow is irrotational. This follows from the condition of irrotationality $\partial u / \partial z = \partial w / \partial x$, a vertical integral which gives

$$u = u|_{z=-H} + \int_{-H}^{z} \frac{\partial w}{\partial x} dz,$$

showing that the mean of u is proportional to the mean of $\partial w / \partial x$ over a wavelength, which is zero for periodic flows.

There are also a variety of wave analyses for specialized circumstances that involve dispersion, nonlinearity, and viscosity to varying degrees. So, before moving on to internal waves, one of the classical examples of this type of specialization is presented here for nonlinear waves that are slightly dispersive. In 1895 Korteweg and de Vries showed that waves with λ / H in the range between 10 and 20 satisfy:

$$\frac{\partial \eta}{\partial t} + c_0 \frac{\partial \eta}{\partial x} + \frac{3}{2} c_0 \frac{\eta}{H} \frac{\partial \eta}{\partial x} + \frac{1}{6} c_0 H^2 \frac{\partial^3 \eta}{\partial x^3} = 0, \tag{7.87}$$

where $c_0 = \sqrt{gH}$. This is the *Korteweg–de Vries equation*. The first two terms are linear and nondispersive. The third term is nonlinear and represents finite amplitude effects. The fourth term is linear and results from weak dispersion due to the water depth not being shallow enough. If the nonlinear term in (7.87) is neglected, then setting $\eta = a \cos(kx - \omega t)$ leads to the dispersion relation $c = c_0 (1 - (1/6)k^2 H^2)$. This agrees with the first two terms in the Taylor series expansion of the dispersion relation $c = \sqrt{(g/k)\tanh kH}$ for small kH, verifying that weak dispersive effects are indeed properly accounted for by the last term in (7.87).

The ratio of nonlinear and dispersion terms in (7.88) is

$$\frac{\eta}{H} \frac{\partial \eta}{\partial x} \bigg/ H^2 \frac{\partial^3 \eta}{\partial x^3} \sim \frac{a\lambda^2}{H^3}.$$

When $a\lambda^2/H^3$ is larger than ~16, nonlinear effects sharpen the forward face of the wave, leading to a hydraulic jump, as discussed earlier in this section. For lower values of

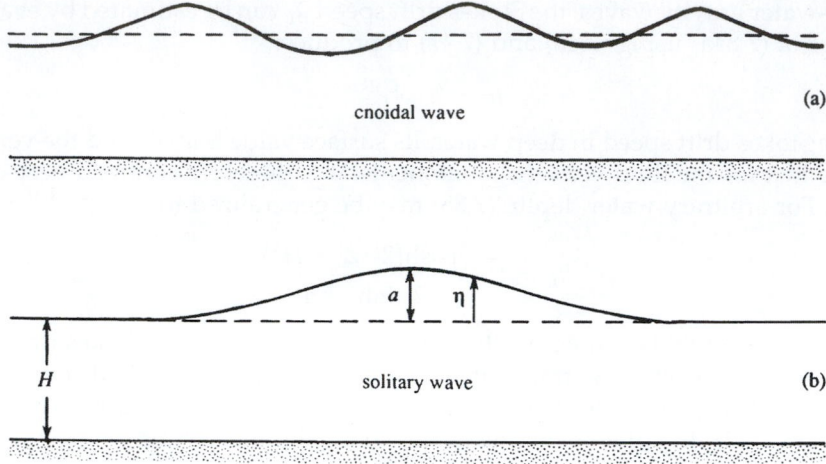

FIGURE 7.23 Finite-amplitude waves of unchanging form: (a) cnoidal waves and (b) a solitary wave. In both cases, the processes of nonlinear steepening and dispersive spreading balance so that the waveform is unchanged.

$a\lambda^2/H^3$, a balance can be achieved between nonlinear steepening and dispersive spreading, and waves of unchanging form become possible.

Analysis of the Korteweg–de Vries equation shows that two types of solutions are then possible—a periodic solution and a solitary wave solution. The periodic solution is called a *cnoidal wave*, because it is expressed in terms of elliptic functions denoted by $cn(x)$. Its waveform is shown in Figure 7.23. The other possible solution of the Korteweg–de Vries equation involves only a single wave crest and is called a *solitary wave* or *soliton*. Its profile is given by

$$\eta = a \operatorname{sech}^2\left[\left(\frac{3a}{4H^3}\right)^{1/2}(x - ct)\right], \tag{7.88}$$

where the speed of propagation is

$$c = c_0\left(1 + \frac{a}{2H}\right),$$

showing that the propagation velocity increases with amplitude. The validity of (7.88) can be checked by substitution into (7.87) (Exercise 7.15). The waveform of the solitary wave is shown in Figure 7.23.

An isolated hump propagating at constant speed with unchanging form and in fairly shallow water was first observed experimentally by S. Russell in 1844. Solitons have been observed to exist not only as surface waves, but also as internal waves in stratified fluids, in the laboratory as well as in the ocean (see Turner, 1973, Figure 3.3).

7.7. WAVES ON A DENSITY INTERFACE

To this point, waves at the surface of a liquid have been considered without regard to the gas (or liquid) above the surface. Yet, gravity and capillary waves can also exist at the

FIGURE 7.24 Internal wave at a density interface between two infinitely deep fluids. Here the horizontal velocity is equal and opposite above and below the interface, so there is a time-dependent vortex sheet at the interface.

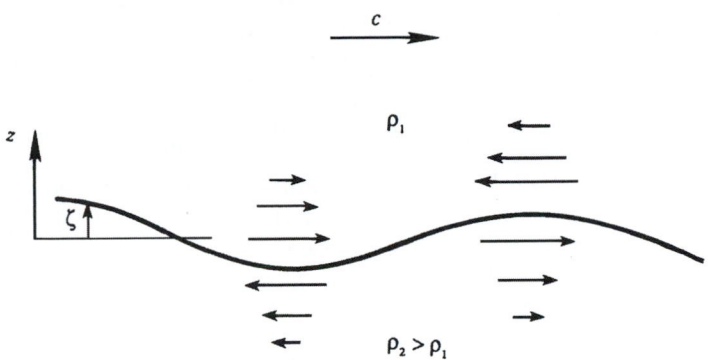

interface between two immiscible liquids of different densities. A sharp-density gradient can be readily generated in the laboratory (at least temporarily) between gases with different densities, and between oil and water. In the ocean sharp-density gradients may be generated by solar heating of the upper layer, or in an estuary (that is, a river mouth) or a fjord into which fresh (less saline) river water flows over oceanic water, which is more saline and consequently heavier. The basic situation can be idealized by considering a lighter fluid of density ρ_1 lying over a heavier fluid of density ρ_2 (Figure 7.24).

For simplicity ignore interfacial (surface) tension, and assume that only small-slope linear waves exist on the interface and that both fluids are infinitely deep, so that only those solutions that decay exponentially from the interface are allowed. In this section and in the rest of this chapter, *complex notation* will be used to ease the algebraic and trigonometric effort. This means that (7.2) will be replaced by

$$\zeta(x,t) = \text{Re}\{a \exp[i(kx - \omega t)]\},$$

where Re{} is the operator that extracts the real part of the complex function in {}-braces, and $i = \sqrt{-1}$ as usual. When using complex numbers and variables in linear mathematical analyses it is customary to drop Re{} and simply write

$$\zeta(x,t) = a \exp[i(kx - \omega t)] \qquad (7.89)$$

until reporting the final results when Re{} commonly reappears. Any analysis done with (7.89) includes an imaginary part, sometimes denoted Im{}, that winds up being of no consequence in the final results.

To determine wave properties in this situation, the Laplace equation for the velocity potential must be solved in both fluids subject to the continuity of p and w at the interface. The equations are

$$\frac{\partial^2 \phi_1}{\partial x^2} + \frac{\partial^2 \phi_1}{\partial z^2} = 0 \quad \text{and} \quad \frac{\partial^2 \phi_2}{\partial x^2} + \frac{\partial^2 \phi_2}{\partial z^2} = 0, \qquad (7.90)$$

subject to

$$\phi_1 \to 0 \quad \text{as} \quad z \to \infty, \qquad (7.91)$$
$$\phi_2 \to 0 \quad \text{as} \quad z \to -\infty, \qquad (7.92)$$

$$\frac{\partial \phi_1}{\partial z} = \frac{\partial \phi_2}{\partial z} = \frac{\partial \zeta}{\partial t} \quad \text{at } z = 0, \text{ and} \tag{7.93}$$

$$\rho_1 \frac{\partial \phi_1}{\partial t} + \rho_1 g \zeta = \rho_2 \frac{\partial \phi_2}{\partial t} + \rho_2 g \zeta \quad \text{at } z = 0. \tag{7.94}$$

Equation (7.93) follows from equating the vertical velocity of the fluid on both sides of the interface to the rate of rise of the interface. Equation (7.94) follows from the continuity of pressure across the interface in the absence of interfacial (surface) tension, $\sigma = 0$. As in the case of surface waves, the boundary conditions are linearized and applied at $z = 0$ instead of at $z = \zeta$. Conditions (7.91) and (7.92) require that the solutions of (7.90) must be of the form

$$\phi_1 = A \, e^{-kz} e^{i(kx - \omega t)} \text{ and}$$
$$\phi_2 = B \, e^{kz} e^{i(kx - \omega t)},$$

because a solution proportional to e^{kz} is not allowed in the upper fluid, and a solution proportional to e^{-kz} is not allowed in the lower fluid. Here the amplitudes A and B can be complex. As in Section 7.2, the constants are determined from the kinematic boundary conditions (7.93), giving

$$A = -B = i\omega a / k.$$

The dynamic boundary condition (7.94) then leads to the dispersion relation

$$\omega = \sqrt{gk \left(\frac{\rho_2 - \rho_1}{\rho_2 + \rho_1} \right)} = \varepsilon \sqrt{gk}, \tag{7.95}$$

where $\varepsilon^2 \equiv (\rho_2 - \rho_1)/(\rho_2 + \rho_1)$ is a small number if the density difference between the two liquids is small. The case of small density difference is relevant in geophysical situations; for example, a 10°C temperature change causes the density of an upper layer of the ocean to decrease by 0.3%. Equation (7.95) shows that waves at the interface between two liquids of infinite thickness travel like deep-water surface waves, with ω proportional to $\sqrt{gk}$, but at a frequency that is lower by the factor ε. In general, *internal waves have a lower frequency and slower phase speed than surface waves with the same wave number*. As expected, (7.95) recovers (7.45) as $\varepsilon \to 1$ when $\rho_1/\rho_2 \to 0$.

The kinetic energy E_k per unit area of interface of the field can be found by integrating $\rho(u^2 + w^2)/2$ over the range $z = \pm \infty$ (Exercise 7.16):

$$E_k = \frac{1}{4} (\rho_2 - \rho_1) g a^2.$$

The potential energy can be calculated by finding the rate of work done in deforming a flat interface to the wave shape. In Figure 7.25, this involves a transfer of column A of density ρ_2 to location B, a simultaneous transfer of column B of density ρ_1 to location A, and integrating the work over *half the wavelength*, since the resulting exchange forms a complete wavelength; see the previous discussion of Figure 7.6. The potential energy per unit horizontal area is therefore

$$E_p = \frac{1}{\lambda} \int_0^{\lambda/2} \rho_2 g \zeta^2 \, dx - \frac{1}{\lambda} \int_0^{\lambda/2} \rho_1 g \zeta^2 \, dx = \frac{g(\rho_2 - \rho_1)}{2\lambda} \int_0^{\lambda/2} \zeta^2 \, dx = \frac{1}{4} (\rho_2 - \rho_1) g a^2.$$

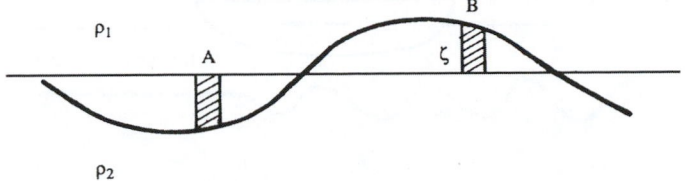

FIGURE 7.25 Geometry for calculating the potential energy of a sinusoidal displacement of the interface between two incompressible fluids with different densities. The work done in transferring element A to the vertical location of element B equals the weight of A times the vertical displacement of its center of gravity.

The total wave energy per unit horizontal area is

$$E = E_k + E_p = \frac{1}{2}(\rho_2 - \rho_1)ga^2. \tag{7.96}$$

In a comparison with (7.42), it follows that the amplitude of ocean internal waves is usually much larger than those of surface waves for the same amount of energy per unit interface area when $(\rho_2 - \rho_1) \ll \rho_2$.

The horizontal velocity components in the two layers are

$$u_1 = \frac{\partial \phi_1}{\partial x} = -\omega a e^{-kz} e^{i(kx-\omega t)} \text{ and}$$

$$u_2 = \frac{\partial \phi_2}{\partial x} = \omega a e^{kz} e^{i(kx-\omega t)},$$

and are oppositely directed (Figure 7.24). The interface is therefore a time-dependent *vortex sheet* and the tangential velocity is discontinuous across it. It can be expected that a continuously stratified medium, in which the density varies continuously as a function of z, will support internal waves whose vorticity is distributed throughout the flow. Consequently, *internal waves in a continuously stratified fluid are not irrotational and do not satisfy the Laplace equation.*

The existence of internal waves at a density discontinuity has explained an interesting phenomenon observed in Norwegian fjords (Gill, 1982). It was known for a long time that ships experienced unusually high drags on entering these fjords. The phenomenon was a mystery (and was attributed to "dead water") until Bjerknes, a Norwegian oceanographer, explained it as due to the internal waves at the interface generated by the motion of the ship (Figure 7.26). (Note that the product of the drag times the speed of the ship gives the rate of generation of wave energy, with other sources of resistance neglected.)

As a second example of an internal wave at a density discontinuity, consider the case in which the upper layer is not infinitely thick but has a finite thickness; the lower layer is initially assumed to be infinitely thick. The case of two infinitely deep liquids, treated in the preceding section, is then a special case of the present situation. Whereas only waves at the interface were allowed in the preceding section, the presence of a free surface now allows surface waves to enter the problem. It is clear that the present configuration will allow two modes of oscillation where the free-surface and interface waves are in or out of phase.

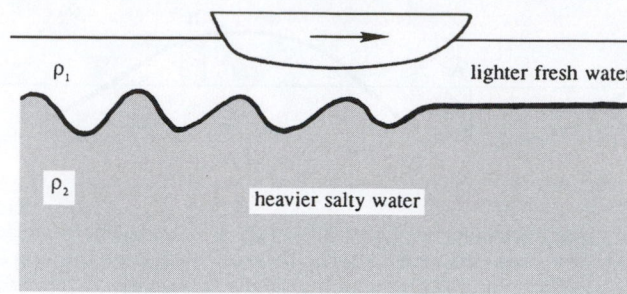

FIGURE 7.26 Schematic explanation for the phenomenon of *dead water* in Norwegian fjords. The ship on the ocean surface may produce waves on the ocean surface and on an interface between lighter, fresher water and cooler, saltier water. Wave production leads to drag on the ship and both types of waves are generated under certain conditions.

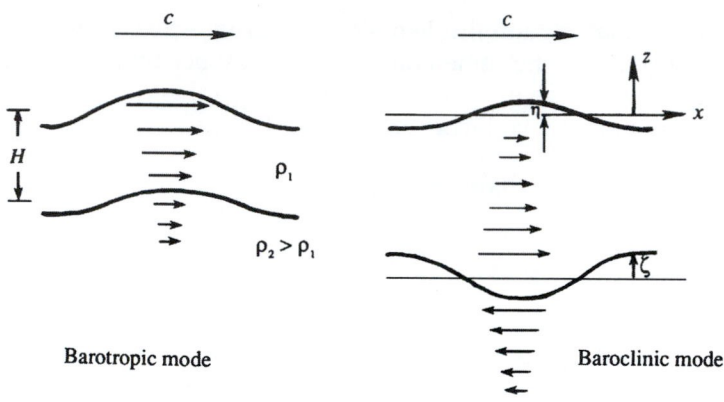

FIGURE 7.27 The two modes of motion of a layer of fluid overlying an infinitely deep fluid. The barotropic mode is an extension of the surface wave motion discussed in the first six sections of this chapter. The baroclinc mode is the extension of the interface wave motion described earlier in this section and shown in Figure 7.24. The baroclinic mode includes vorticity at the density interface; the barotropic mode does not.

To analyze this situation, let H be the thickness of the upper layer, and let the origin be placed at the mean position of the free surface (Figure 7.27). The field equations are (7.90) and the boundary conditions are:

$$\phi_2 \to 0 \quad \text{at } z \to -\infty, \tag{7.97}$$

$$\frac{\partial \phi_1}{dz} = \frac{\partial \eta}{\partial t} \quad \text{at } z = 0, \tag{7.98}$$

$$\frac{\partial \phi_1}{\partial t} + g\eta = 0 \quad \text{at } z = 0, \tag{7.99}$$

$$\frac{\partial \phi_1}{\partial z} = \frac{\partial \phi_2}{\partial z} = \frac{\partial \zeta}{\partial t} \quad \text{at } z = -H, \text{ and} \tag{7.100}$$

$$\rho_1 \frac{\partial \phi_1}{\partial t} + \rho_1 g\zeta = \rho_2 \frac{\partial \phi_2}{\partial t} + \rho_2 g\zeta \quad \text{at } z = -H. \tag{7.101}$$

In addition, assume a free-surface displacement of the form

$$\eta = ae^{i(kx-\omega t)}, \tag{7.102}$$

and an interface displacement of the form

$$\zeta = b e^{i(kx-\omega t)}. \tag{7.103}$$

Without losing generality, we can regard a as real, which means that we are considering a surface wave of the form $\eta = a \cos(kx - \omega t)$. The constant b should be left complex since ζ and η may not be in phase, and the solution of the problem should determine such phase differences.

The velocity potentials in the layers must be of the form

$$\phi_1 = (A\, e^{kz} + B\, e^{-kz})\, e^{i(kx-\omega t)}, \tag{7.104}$$

$$\phi_2 = C\, e^{kz}\, e^{i(kz-\omega t)}. \tag{7.105}$$

The form (7.105) satisfies (7.97). Conditions (7.98) through (7.100) allow a solution for the constants in terms of a, ω, k, g, and H:

$$A = -\frac{ia}{2}\left(\frac{\omega}{k} + \frac{g}{\omega}\right), \tag{7.106}$$

$$B = \frac{ia}{2}\left(\frac{\omega}{k} - \frac{g}{\omega}\right), \tag{7.107}$$

$$C = -\frac{ia}{2}\left(\frac{\omega}{k} + \frac{g}{\omega}\right) - \frac{ia}{2}\left(\frac{\omega}{k} - \frac{g}{\omega}\right)e^{2kH}, \text{ and} \tag{7.108}$$

$$b = \frac{a}{2}\left(1 + \frac{gk}{\omega^2}\right)e^{-kH} + \frac{a}{2}\left(1 - \frac{gk}{\omega^2}\right)e^{kH}. \tag{7.109}$$

Substitution into (7.101) leads to the dispersion relation $\omega(k)$. After some algebraic manipulations, the result can be written as (see Exercise 7.19):

$$\left(\frac{\omega^2}{gk} - 1\right)\left\{\frac{\omega^2}{gk}[\rho_1 \sinh kH + \rho_2 \cosh kH] - (\rho_2 - \rho_1)\sinh kH\right\} = 0. \tag{7.110}$$

One possible root of (7.110) is

$$\omega^2 = gk, \tag{7.111}$$

which is the same as that for a deep-water gravity wave. Substituting (7.111) into (7.109) leads to

$$b = ae^{-kH}, \tag{7.112}$$

which implies that the interface waves are in phase with the surface waves but are reduced in amplitude by the factor e^{-kH}. This mode is similar to a gravity wave propagating on the free surface of the upper liquid, in which the motion decays as e^{-kz} from the free surface. It is called the *barotropic mode*, because the surfaces of constant pressure and density coincide.

The other root of (7.110) is

$$\omega^2 = \frac{gk(\rho_2 - \rho_1)\sinh kH}{\rho_2 \cosh kH + \rho_1 \sinh kH}, \tag{7.113}$$

which reduces to (7.95) when $kH \to \infty$. Substituting (7.113) into (7.109) leads to

$$\eta = -\zeta \left(\frac{\rho_2 - \rho_1}{\rho_1} \right) e^{-kH}, \qquad (7.114)$$

which demonstrates that η and ζ have opposite signs and that the interface displacement (ζ) is much larger than the surface displacement (η) if the density difference is small. This is the *baroclinic* or *internal mode* because the surfaces of constant pressure and density do not coincide. Here the horizontal velocity u changes sign across the interface. The existence of a density difference has therefore generated a motion that is quite different from the barotropic mode, (7.111) and (7.112). The case described at the beginning of this section, where the fluids have infinite depth and no free surface, has only a baroclinic mode and no barotropic mode.

A very common simplification, frequently made in geophysical situations, involves assuming that the wavelengths are large compared to the upper layer depth. For example, the depth of the oceanic upper layer, below which there is a sharp-density gradient, could be ≈ 50 m thick, but interfacial waves much longer than this may be of interest. The relevant approximation in this case, $kH \ll 1$, is called the *shallow-water* or *long-wave approximation* and is implemented via:

$$\sinh(kH) \cong kH \quad \text{and} \quad \cosh(kH) \cong 1,$$

so the dispersion relation (7.113) corresponding to the baroclinic mode reduces to

$$\omega^2 = kg \left(\frac{\rho_2 - \rho_1}{\rho_2} \right) kH \qquad (7.115)$$

to lowest order in the small parameter kH. The phase velocity of waves at the interface is

$$c = [g'H]^{1/2}, \quad \text{where } g' = g \left(\frac{\rho_2 - \rho_1}{\rho_2} \right) \qquad (7.116, 7.117)$$

is the reduced gravity. Equation (7.116) is similar to the corresponding expression for *surface* waves in a shallow homogeneous layer of thickness H, namely, $c = \sqrt{gH}$, except that the speed is reduced by the factor $\sqrt{(\rho_2 - \rho_1)/\rho_2}$. This agrees with the previous conclusion that internal waves propagate slower than surface waves. Under the shallow-water approximation, (7.114) reduces to

$$\eta = -\zeta \left(\frac{\rho_2 - \rho_1}{\rho_1} \right). \qquad (7.118)$$

In Section 7.2, the shallow-water approximation for surface waves is found equivalent to a hydrostatic approximation and results in a depth-independent horizontal velocity. This conclusion also holds for interfacial waves. The fact that u_1 is independent of z follows from (7.104) on noting that $e^{kz} \approx e^{-kz} \approx 1$. To see that pressure is hydrostatic, the perturbation pressure p' in the upper layer determined from (7.104) is

$$p' = -\rho_1 \frac{\partial \phi_1}{\partial t} = i\rho_1 \omega \left(A + B \right) e^{i(kx - \omega t)} = \rho_1 g \eta, \qquad (7.119)$$

FIGURE 7.28 The two modes of motion in a shallow-water, two-layer system in the Boussinesq limit. These profiles are the limiting case of those in Figure 7.27 when the lower fluid layer depth is shallow. As before, the baroclinic mode includes vorticity at the density interface; the barotropic mode does not.

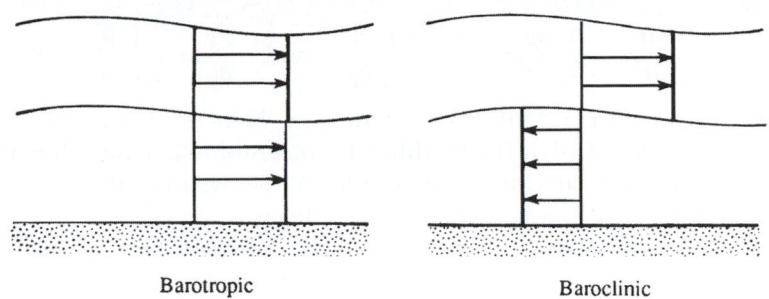

Barotropic   Baroclinic

where the constants given in (7.106) and (7.107) have been used. This shows that p' is independent of z and equals the hydrostatic pressure change due to the free-surface displacement.

So far, the lower fluid has been assumed to be infinitely deep, resulting in an exponential decay of the flow field from the interface into the lower layer, with a decay scale of the order of the wavelength. If the lower layer is now considered thin compared to the wavelength, then the horizontal velocity will be depth independent, and the flow hydrostatic, in the lower layer. If *both* layers are considered thin compared to the wavelength, then the flow is hydrostatic (and the horizontal velocity field is depth independent) in *both* layers. This is the *shallow-water* or *long-wave approximation* for a two-layer fluid. In such a case the horizontal velocity field in the barotropic mode has a discontinuity at the interface, which vanishes in the Boussinesq limit $(\rho_2 - \rho_1)/\rho_1 \ll 1$. Under these conditions the two modes of a two-layer system have a simple structure (Figure 7.28): a barotropic mode in which the horizontal velocity is depth independent across the entire water column; and a baroclinic mode in which the horizontal velocity is directed in opposite directions in the two layers (but is depth independent in each layer).

7.8. INTERNAL WAVES IN A CONTINUOUSLY STRATIFIED FLUID

Waves may also exist in the interior of a pool, reservoir, lake, or ocean when the fluid's density in a quiescent state is a continuous function of the vertical coordinate z. The equations of motion for internal waves in such a stratified medium presented here are simplifications of the Boussinesq set specified at the end of Section 4.9. The Boussinesq approximation treats the density as constant, except in the vertical momentum equation. For simplicity, we shall also assume that: 1) the wave motion is effectively inviscid because the velocity gradients are small and the Reynolds number is large, 2) the wave amplitudes are small enough so that the nonlinear advection terms can be neglected, and 3) the frequency of wave motion is much larger than the Coriolis frequency so it does not affect the wave motion. Effects of the earth's rotation are considered in Chapter 13. The Boussinesq set then simplifies to:

$$\frac{D\rho}{Dt} = 0, \quad \frac{\partial u}{\partial x} + \frac{\partial v}{\partial y} + \frac{\partial w}{\partial z} = 0, \qquad (4.9, 4.10)$$

$$\frac{\partial u}{\partial t} = -\frac{1}{\rho_0}\frac{\partial p}{\partial x}, \quad \frac{\partial v}{\partial t} = -\frac{1}{\rho_0}\frac{\partial p}{\partial y}, \quad \text{and} \quad \frac{\partial w}{\partial t} = -\frac{1}{\rho_0}\frac{\partial p}{\partial z} - \frac{\rho g}{\rho_0}, \qquad (7.120, 7.121, 7.122)$$

where ρ_0 is a constant reference density. Here, (4.9) expresses constancy of fluid-particle density while (4.10) is the condition for incompressible flow. If temperature is the only agency that changes the density, then $D\rho/Dt = 0$ follows from the heat equation in the nondiffusive form $DT/Dt = 0$ and a temperature-only equation of state, in the form $\delta\rho/\rho = -\alpha\delta T$, where α is the coefficient of thermal expansion. If the density changes are due to changes in the concentration S of a constituent (e.g., salinity in the ocean or water vapor in the atmosphere), then $D\rho/Dt = 0$ follows from $DS/Dt = 0$ (the nondiffusive form of the constituent conservation equation) and a concentration-only equation of state, $\rho = \rho(S)$, in the form of $\delta\rho/\rho = \beta\delta S$, where β is the coefficient describing how the density changes due to concentration of the constituent. In both cases, the principle underlying $D\rho/Dt = 0$ is an equation of state that does not include pressure. In terms of common usage, this equation is frequently called the *density equation*, as opposed to the *continuity equation* (4.10).

The five equations (4.9), (4.10), and (7.120) through (7.122) contain five unknowns (u, v, w, p, ρ). Before considering wave motions, first define the quiescent density $\bar{\rho}\,(z)$ and pressure $\bar{p}\,(z)$ profiles in the medium as those that satisfy a hydrostatic balance:

$$0 = -\frac{1}{\rho_0}\frac{d\bar{p}}{dz} - \frac{\bar{\rho}g}{\rho_0}. \qquad (7.123)$$

When the motion develops, the pressure and density will change relative to their quiescent values:

$$p = \bar{p}\,(z) + p', \quad \rho = \bar{\rho}(z) + \rho'. \qquad (7.124)$$

The density equation (4.9) then becomes

$$\frac{\partial}{\partial t}(\bar{\rho} + \rho') + u\frac{\partial}{\partial x}(\bar{\rho} + \rho') + v\frac{\partial}{\partial y}(\bar{\rho} + \rho') + w\frac{\partial}{\partial z}(\bar{\rho} + \rho') = 0. \qquad (7.125)$$

Here, $\partial\bar{\rho}/\partial t = \partial\bar{\rho}/\partial x = \partial\bar{\rho}/\partial y = 0$. The nonlinear terms (namely, $u\partial\rho'/\partial x$, $v\partial\rho'/\partial y$, and $w\partial\rho'/\partial z$) are also negligible for small-amplitude motions. The *linear* part of the fourth term, $w\,d\bar{\rho}/dz$, must be retained, so the linearized version of (4.9) is

$$\frac{\partial\rho'}{\partial t} + w\frac{d\bar{\rho}}{dz} = 0, \qquad (7.126)$$

which states that the density perturbation at a point is generated only by the vertical advection of the *background* density distribution. We now introduce the *Brunt–Väisälä frequency*, or *buoyancy frequency*:

$$N^2 \equiv -\frac{g}{\rho_0}\frac{d\bar{\rho}}{dz}. \qquad (7.127)$$

This is (1.29) when the adiabatic density gradient is zero. As described in Section 1.10, $N(z)$ has units of rad./s and is the oscillation frequency of a vertically displaced fluid particle released from rest in the absence of fluid friction. Using (7.123) and (7.127) in (7.120) through (7.122) and (7.126) produces

$$\frac{\partial u}{\partial t} = -\frac{1}{\rho_0}\frac{\partial p'}{\partial x}, \quad \frac{\partial v}{\partial t} = -\frac{1}{\rho_0}\frac{\partial p'}{\partial y}, \quad \frac{\partial w}{\partial t} = -\frac{1}{\rho_0}\frac{\partial p'}{\partial z} - \frac{\rho' g}{\rho_0}, \qquad (7.128, 7.129, 7.130)$$

and

$$\frac{\partial \rho'}{\partial t} - \frac{N^2 \rho_0}{g} w = 0. \qquad (7.131)$$

Comparing (7.120) through (7.122) and (7.128) through (7.130), we see that the only difference is the replacement of the total density ρ and pressure p with the perturbation density ρ' and pressure p'.

The full set of equations for linear wave motion in a stratified fluid are (4.10) and (7.128) through (7.131), where ρ may be a function of temperature T and concentration S of a constituent, but not of pressure. At first this does not seem to be a good assumption. The compressibility effects in the atmosphere are certainly not negligible; even in the ocean the density changes due to the huge changes in the background pressure are as much as 4%, which is ≈ 10 times the density changes due to the variations of the salinity and temperature. The effects of compressibility, however, can be handled within the Boussinesq approximation if we regard $\bar{\rho}$ in the definition of N as the background *potential density*, that is, the density distribution from which the adiabatic changes of density, due to the changes of pressure, have been subtracted out. The concept of potential density is explained in Chapter 1. Oceanographers account for compressibility effects by converting all their density measurements to the standard atmospheric pressure; thus, when they report variations in density (what they call "sigma tee") they are generally reporting variations due only to changes in temperature and salinity.

A useful condensation of the above equations involving only w can be obtained by taking the time derivative of (4.10) and using the horizontal momentum equations (7.128) and (7.129) to eliminate u and v. The result is

$$\frac{1}{\rho_0}\nabla_H^2 p' = \frac{\partial^2 w}{\partial z\,\partial t}, \qquad (7.132)$$

where $\nabla_H^2 \equiv \partial^2/\partial x^2 + \partial^2/\partial y^2$ is the *horizontal* Laplacian operator. Elimination of ρ' from (7.130) and (7.131) gives

$$\frac{1}{\rho_0}\frac{\partial^2 p'}{\partial t\,\partial z} = -\frac{\partial^2 w}{\partial t^2} - N^2 w. \qquad (7.133)$$

Finally, p' can be eliminated by taking ∇_H^2 (7.133), and inserting the result in (7.132) to find:

$$\frac{\partial^2}{\partial t\,\partial z}\left(\frac{\partial^2 w}{\partial t\,\partial z}\right) = -\nabla_H^2\left(\frac{\partial^2 w}{\partial t^2} + N^2 w\right),$$

which can be written as

$$\frac{\partial^2}{\partial t^2}\nabla^2 w + N^2 \nabla_H^2 w = 0, \qquad (7.134)$$

where $\nabla^2 \equiv \partial^2/\partial x^2 + \partial^2/\partial y^2 + \partial^2/\partial z^2 = \nabla_H^2 + \partial^2/\partial z^2$ is the three-dimensional Laplacian operator. This equation for the vertical velocity w can be used to derive the dispersion relation for internal gravity waves.

Internal Waves in a Stratified Fluid

The situation embodied in (7.134) is fundamentally different from that of interface waves because there is no obvious direction of propagation. For interface waves constrained to follow a horizontal surface with the x-axis chosen along the direction of wave propagation, a dispersion relation $\omega(k)$ was obtained that is independent of the wave direction. Furthermore, wave crests and wave groups propagate in the same direction, although at different speeds. However, in the current situation, the fluid is *continuously* stratified and internal waves might propagate in any direction and at any angle to the vertical. In such a case the *direction* of the wave number vector $\mathbf{K} = (k, l, m)$ becomes important and the dispersion relationship is *anisotropic* and depends on the wave number components:

$$\omega = \omega(k, l, m) = \omega(\mathbf{K}). \tag{7.135}$$

Consequently, the wave number, phase velocity, and group velocity are no longer scalars and the prototype sinusoidal wave (7.2) must be replaced with its three-dimensional extension (7.5). However, (7.135) must still be isotropic in k and l, the wave number components in the two horizontal directions.

The propagation of internal waves is a baroclinic process, in which the surfaces of constant pressure do not coincide with the surfaces of constant density. It was shown in Section 5.4, in connection with Kelvin's circulation theorem, that baroclinic processes generate vorticity. Internal waves in a continuously stratified fluid are therefore rotational. Waves at a density interface constitute a limiting case in which all the vorticity is concentrated in the form of a velocity discontinuity at the interface. The Laplace equation can therefore be used to describe the flow field within each layer. However, internal waves in a continuously stratified fluid cannot be described by the Laplace equation.

To reveal the structure of the situation described by (7.134) and (7.135), consider the complex version of (7.5) for the vertical velocity

$$w = w_0 e^{i(kx+ly+mz-\omega t)} = w_0 e^{i(\mathbf{K}\cdot\mathbf{x}-\omega t)} \tag{7.136}$$

in a fluid medium having a constant buoyancy frequency. Substituting (7.136) into (7.134) with constant N leads to the dispersion relation:

$$\omega^2 = \frac{k^2 + l^2}{k^2 + l^2 + m^2} N^2. \tag{7.137}$$

For simplicity choose the x-z plane so it contains $\mathbf{K}$ and $l = 0$. No generality is lost through this choice because the medium is horizontally isotropic, but k now represents the entire horizontal wave number and (7.137) can be written:

$$\omega = \frac{kN}{\sqrt{k^2 + m^2}} = \frac{kN}{K}. \tag{7.138}$$

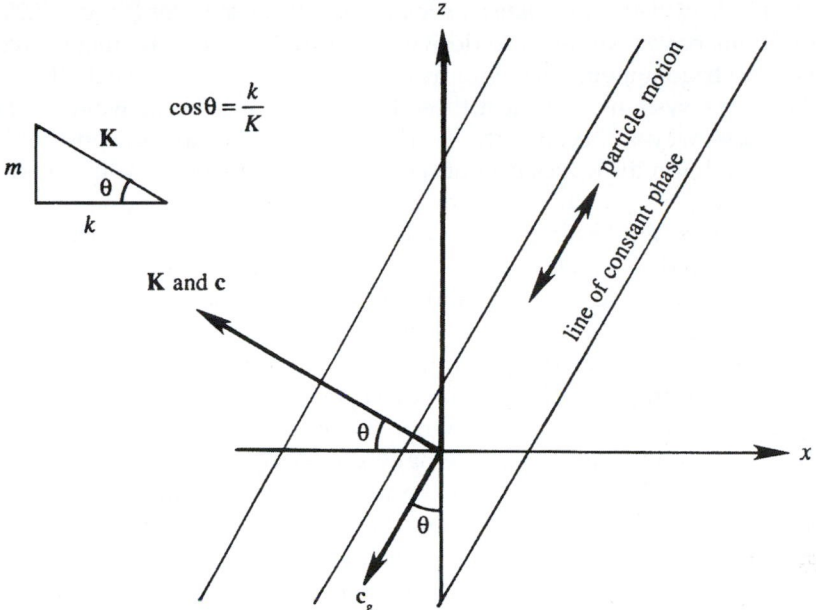

FIGURE 7.29   Geometric parameters for internal waves. Here z is vertical and x is horizontal. Note that $\mathbf{c}$ and $\mathbf{c}_g$ are at right angles and have opposite vertical components while $\mathbf{u}$ is parallel to the group velocity. Thus, internal wave packets slide along their crests.

This is the dispersion relation for internal gravity waves and can also be written as

$$\omega = N \cos \theta, \qquad (7.139)$$

where $\theta = \tan^{-1}(m/k)$ is the angle between the phase velocity vector $\mathbf{c}$ (and therefore $\mathbf{K}$) and the horizontal direction (Figure 7.29). Interestingly, (7.139) states that the frequency of an internal wave in a stratified fluid depends only on the *direction* of the wave number vector and not on its magnitude. This is in sharp contrast with surface and interfacial gravity waves, for which frequency depends only on the magnitude. In addition, the wave frequency lies in the range $0 < \omega < N$, and this indicates that *N is the maximum possible frequency of internal waves in a stratified fluid.*

Before further investigation of the dispersion relation, consider particle motion in an incompressible internal wave. For consistency with (7.136), the horizontal fluid velocity is written as

$$u = u_0\, e^{i(kx+ly+mz-\omega t)}, \qquad (7.140)$$

plus two similar expressions for v and w. Differentiating produces:

$$\frac{\partial u}{\partial x} = iku_0\, e^{i(kx+ly+mz-\omega t)} = iku.$$

Thus, (4.10) then requires that $ku + lv + mw = 0$, that is,

$$\mathbf{K} \cdot \mathbf{u} = 0, \qquad (7.141)$$

showing that *particle motion is perpendicular to the wave number vector* (Figure 7.29). Note that only two conditions have been used to derive this result, namely the incompressible continuity equation and trigonometric behavior in *all* spatial directions. As such, the result is valid for many other wave systems that meet these two conditions. These waves are called *shear waves* (or transverse waves) because the fluid moves parallel to the constant phase lines. Surface or interfacial gravity waves do not have this property because the field varies *exponentially* in the vertical.

We can now interpret θ in the dispersion relation (7.139) as the angle between the particle motion and the *vertical* direction (Figure 7.29). The maximum frequency $\omega = N$ occurs when $\theta = 0$, that is, when the particles move up and down vertically. This case corresponds to $m = 0$ (see (7.138)), showing that the motion is independent of the z-coordinate. The resulting motion consists of a series of vertical columns, all oscillating at the buoyancy frequency N, with the flow field varying in the horizontal direction only.

At the opposite extreme we have $\omega = 0$ when $\theta = \pi/2$, that is, when the particle motion is completely horizontal. In this limit our internal wave solution (7.138) would seem to require $k = 0$, that is, horizontal independence of the motion. However, such a conclusion is not valid; pure horizontal motion is not a limiting case of internal waves, and it is necessary to examine the basic equations to draw any conclusion for this case. An examination of the governing set, (4.10) and (7.128) through (7.131), shows that a possible steady solution is $w = p' = \rho' = 0$, with u and v and *any* functions of x and y satisfying

$$\frac{\partial u}{\partial x} + \frac{\partial v}{\partial y} = 0. \tag{7.142}$$

The z-dependence of u and v is arbitrary. The motion is therefore two dimensional in the horizontal plane, with the motion in the various horizontal planes decoupled from each other. This is why clouds in the upper atmosphere seem to move in flat horizontal sheets, as often observed in airplane flights (Gill, 1982). For a similar reason a cloud pattern pierced by a mountain peak sometimes shows *Karman vortex streets*, a two-dimensional feature; see the striking photograph in Figure 9.19. A restriction of strong stratification is necessary for such almost horizontal flows, because (7.131) suggests that the vertical motion is small if N is large.

The foregoing discussion leads to the interesting phenomenon of *blocking* in a strongly stratified fluid. Consider a two-dimensional body placed in such a fluid, with its axis

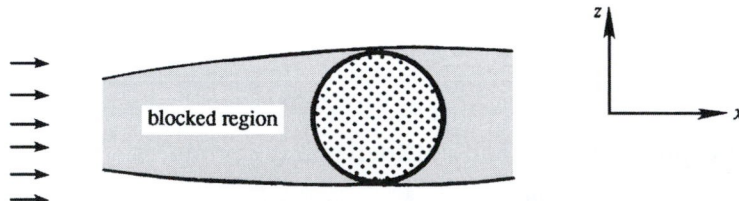

FIGURE 7.30 Blocking in strongly stratified flow. The circular region represents a two-dimensional body with its axis along the y direction (perpendicular to the page). Horizontal flow in the shaded region is blocked by the body when the stratification is strong enough to prevent fluid in the blocked layer from going over or under the body.

horizontal (Figure 7.30). The two dimensionality of the body requires $\partial v/\partial y = 0$, so that the continuity equation (7.142) reduces to $\partial u/\partial x = 0$. A horizontal layer of fluid ahead of the body, bounded by tangents above and below it, is therefore blocked and held motionless. (For photographic evidence see Figure 3.18 in the book by Turner (1973).) This happens because the strong stratification suppresses the w field and prevents the fluid from going below or over the body.

Dispersion of Internal Waves in a Stratified Fluid

The dispersion relationship (7.138) for linear internal waves with constant buoyancy frequency contains a few genuine surprises that challenge our imaginations and violate the intuition acquired by observing surface or interface waves. One of these surprises involves the phase, $\mathbf{c}$, and group, $\mathbf{c_g}$, velocity vectors. In multiple dimensions, these are defined by

$$\mathbf{c} = (\omega/K)\mathbf{e}_K \quad \text{and} \quad \mathbf{c_g} = \mathbf{e}_x\frac{\partial\omega}{\partial k} + \mathbf{e}_y\frac{\partial\omega}{\partial l} + \mathbf{e}_z\frac{\partial\omega}{\partial m}, \qquad (7.8,\ 7.143)$$

where $\mathbf{e}_K = \mathbf{K}/K$. Interface waves $\mathbf{c}$ and $\mathbf{c_g}$ are in the same direction, although their magnitudes can be different. For internal waves, (7.138), (7.8), and (7.143) can be used to determine:

$$\mathbf{c} = \frac{\omega}{K^2}(k\mathbf{e}_x + m\mathbf{e}_z), \quad \text{and} \quad \mathbf{c_g} = \frac{Nm}{K^3}(m\mathbf{e}_x - k\mathbf{e}_z). \qquad (7.144,\ 7.145)$$

Forming the dot product of these two equations produces:

$$\mathbf{c_g}\cdot\mathbf{c} = 0! \qquad (7.146)$$

Thus, the *phase and group velocity vectors are perpendicular* as shown on Figure 7.29. Equations (7.144) and (7.145) do place the horizontal components of $\mathbf{c}$ and $\mathbf{c_g}$ in the same direction, but their vertical components are equal and opposite. In fact, $\mathbf{c}$ and $\mathbf{c_g}$ form two sides of a right triangle whose hypotenuse is horizontal (Figure 7.31). Consequently, the phase velocity has an upward component when the group velocity has a downward component, and vice versa. Equations (7.141) and (7.146) are consistent because $\mathbf{c}$ and $\mathbf{K}$ are parallel and $\mathbf{c_g}$ and $\mathbf{u}$ are

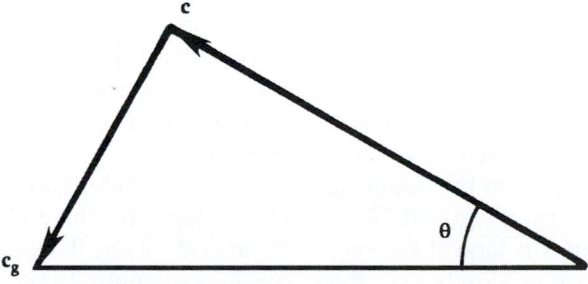

FIGURE 7.31 Orientation of phase and group velocity for internal waves. The vertical components of the phase and group velocities are equal and opposite.

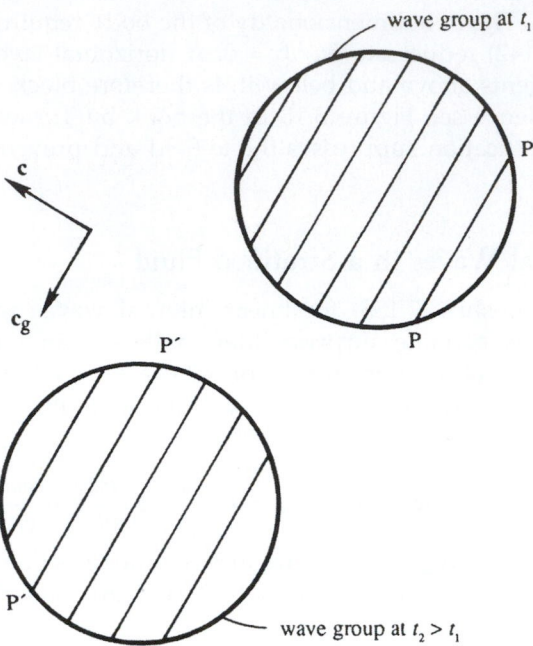

FIGURE 7.32 Illustration of phase and group propagation in a circular internal-wave packet. Positions of the wave packet at two times are shown. The constant-phase line PP (a crest perhaps) at time t_1 propagates to P'P' at t_2.

parallel. The fact that **c** and c_g are perpendicular, and have opposite vertical components, is illustrated in Figure 7.32. It shows that the phase lines are propagating toward the left and upward, whereas the wave group is propagating to the left and downward. Wave crests are constantly appearing at one edge of the group, propagating through the group, and vanishing at the other edge.

The group velocity here has the usual significance of being the velocity of propagation of energy of a certain sinusoidal component. Suppose a source is oscillating at frequency ω. Then its energy will only be transmitted outward along four beams oriented at an angle θ with the vertical, where $\cos \theta = \omega/N$. This has been verified in a laboratory experiment (Figure 7.33). The source in this case was a vertically oscillating cylinder with its axis perpendicular to the plane of paper. The frequency was $\omega < N$. The light and dark lines in the photograph are lines of constant density, made visible by an optical technique. The experiment showed that the energy radiated along four beams that became more vertical as the frequency was increased, which agrees with $\cos \theta = \omega/N$.

These results were obtained by assuming that N is depth independent, an assumption that may seem unrealistic at first. Figure 13.2 shows N vs. depth for the deep ocean, and $N < 0.01$ rad./s everywhere, but N is largest between ~200 m and ~2 km. These results can be considered *locally* valid if N varies slowly over the vertical wavelength $2\pi/m$ of the motion. The so-called *WKB approximation* for internal waves, in which such a slow variation of $N(z)$ is not neglected, is discussed in Chapter 13.

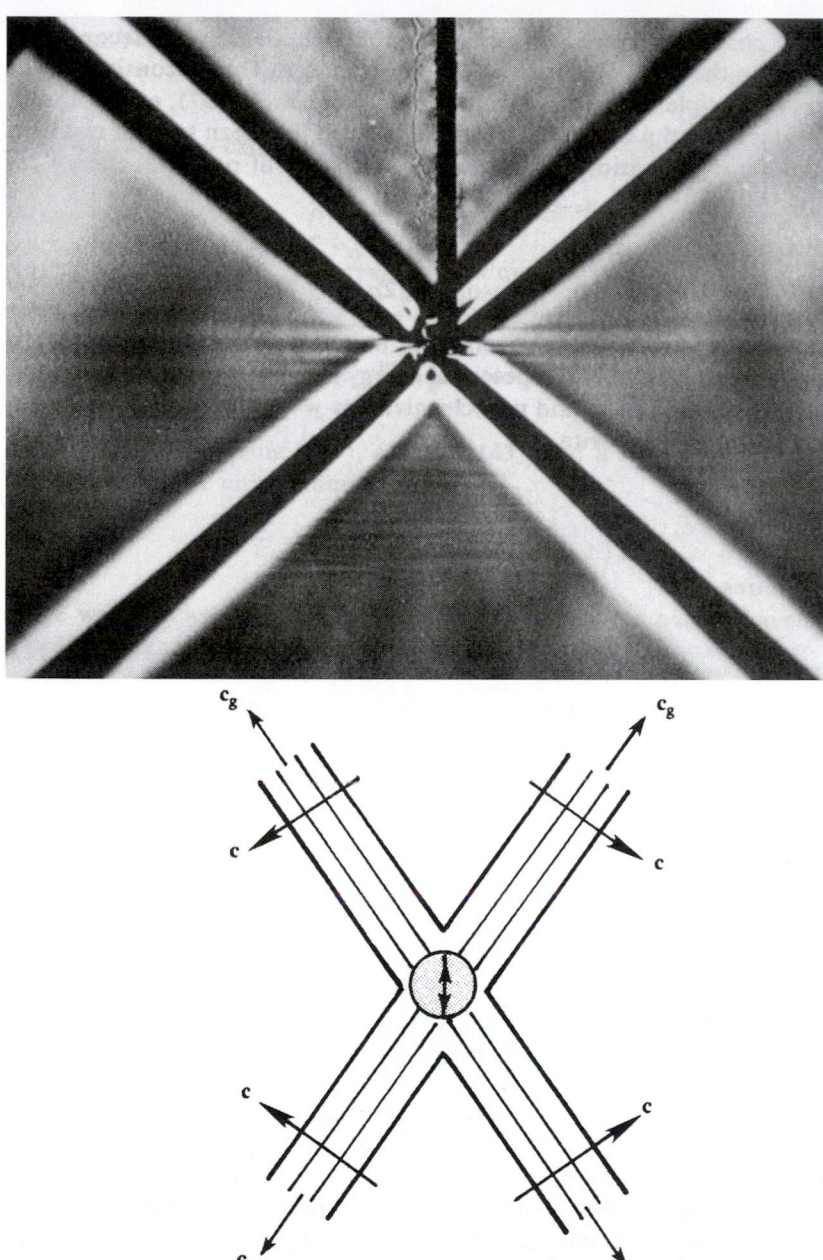

FIGURE 7.33 Waves generated in a stratified fluid of uniform buoyancy frequency $N = 1$ rad./s. The forcing agency is a horizontal cylinder, with its axis perpendicular to the plane of the paper, oscillating vertically at frequency $\omega = 0.71$ rad./s. With $\omega/N = 0.71 = \cos\theta$, this agrees with the observed angle of $\theta = 45°$ made by the beams with the horizontal direction. The vertical dark line in the upper half of the photograph is the cylinder support and should be ignored. The light and dark radial lines represent contours of constant ρ' and are therefore constant phase lines. The schematic diagram below the photograph shows the directions of $\mathbf{c}$ and $\mathbf{c}_g$ for the four beams. *Reprinted with the permission of Dr. T. Neil Stevenson, University of Manchester.*

Energy Considerations for Internal Waves in a Stratified Fluid

The energy carried by an internal wave travels in the direction and at the speed of the group velocity. To show this is the case, construct a mechanical energy equation from (7.128) through (7.130) by multiplying the first equation by $\rho_0 u$, the second by $\rho_0 v$, the third by $\rho_0 w$, and summing the results to find:

$$\frac{\partial}{\partial t}\left[\frac{1}{2}\rho_0(u^2 + v^2 + w^2)\right] + g\rho'w + \nabla\cdot(p'\mathbf{u}) = 0. \tag{7.147}$$

Here the continuity equation has been used to write $u(\partial p'/\partial x) + v(\partial p'/\partial y) + w(\partial p'/\partial z) = \nabla \cdot (p'\mathbf{u})$, which represents the net work done by pressure forces. Another interpretation is that $\nabla\cdot(p'\mathbf{u})$ is the divergence of the *energy flux* $p'\mathbf{u}$, which must change the wave energy at a point. As the first term in (7.147) is the rate of change of kinetic energy, we can anticipate that the second term $g\rho'w$ must be the rate of change of potential energy. This is consistent with the energy principle derived in Chapter 4 (see (4.56)), except that ρ' and p' replace ρ and p because we have subtracted the mean state of rest here. Using the density equation (7.131), the rate of change of potential energy can be written as

$$\frac{\partial E_P}{\partial t} = g\rho'w = \frac{\partial}{\partial t}\left[\frac{g^2\rho'^2}{2\rho_0 N^2}\right], \tag{7.148}$$

which shows that the potential energy per unit volume must be the positive quantity $E_P = g^2\rho'^2/2\rho_0 N^2$. The potential energy can also be expressed in terms of the displacement ζ of a fluid particle, given by $w = \partial\zeta/\partial t$. Using the density equation (7.131), we can write

$$\frac{\partial\rho'}{\partial t} = \frac{N^2\rho_0}{g}\frac{\partial\zeta}{\partial t}, \quad \text{which requires that } \rho' = \frac{N^2\rho_0\zeta}{g}. \tag{7.149}$$

The potential energy *per unit volume* is therefore

$$E_P = \frac{g^2\rho'^2}{2\rho_0 N^2} = \frac{1}{2}N^2\rho_0\zeta^2. \tag{7.150}$$

This expression is consistent with our previous result from (7.96) for two infinitely deep fluids, for which the average potential energy of the entire water column *per unit horizontal area* was shown to be

$$\frac{1}{4}(\rho_2 - \rho_1)ga^2, \tag{7.151}$$

where the interface displacement is of the form $\zeta = a\cos(kx - \omega t)$ and $(\rho_2 - \rho_1)$ is the density discontinuity. To see the consistency, we shall symbolically represent the buoyancy frequency of a density discontinuity at $z = 0$ as

$$N^2 = -\frac{g}{\rho_0}\frac{d\bar{\rho}}{dz} = \frac{g}{\rho_0}(\rho_2 - \rho_1)\delta(z), \tag{7.152}$$

where $\delta(z)$ is the Dirac delta function (see Appendix B.4). (As with other relations involving the delta function, equation (7.152) is valid in the *integral* sense, that is, the integral (across the origin) of the last two terms is equal because $\int \delta(z)\, dz = 1$.) Using (7.152), a vertical integral of (7.150), coupled with horizontal averaging over a wavelength, gives the expression (7.151). Note that for surface or interfacial waves, E_k and E_p represent kinetic and potential energies of the entire water column, per unit horizontal area. In a continuously stratified fluid, they represent energies per unit volume.

We shall now demonstrate that the average kinetic and potential energies are equal for internal wave motion. Assume periodic solutions

$$\left[u, w, p', \rho'\right] = \left[\widehat{u}, \widehat{w}, \widehat{p}, \widehat{\rho}\right] e^{i(kx+mz-\omega t)}.$$

Then all variables can be expressed in terms of w:

$$p' = -\frac{\omega m \rho_0}{k^2}\widehat{w}\, e^{i(kx+mz-\omega t)}, \rho' = \frac{iN^2\rho_0}{\omega g}\widehat{w}\, e^{i(kx+mz-\omega t)}, u = -\frac{m}{k}\widehat{w}\, e^{i(kx+mz-\omega t)}, \qquad (7.153)$$

where p' is derived from (7.132), ρ' from (7.131), and u from (7.128). The average kinetic energy per unit volume is therefore

$$E_k = \frac{1}{2}\rho_0\overline{(u^2 + w^2)} = \frac{1}{4}\rho_0\left(\frac{m^2}{k^2}+1\right)\widehat{w}^2, \qquad (7.154)$$

where we have taken real parts of the various expressions in (7.153) before computing quadratic quantities and used the fact that the average of $\cos^2()$ over a wavelength is ½. The average potential energy per unit volume is

$$E_p = \frac{g^2\overline{\rho'^2}}{2\rho_0 N^2} = \frac{N^2\rho_0}{4\omega^2}\widehat{w}^2, \qquad (7.155)$$

where we have used $\overline{\rho'^2} = \widehat{w}^2 N^4 \rho_0^2/2\omega^2 g^2$, found from (7.153) after taking its real part. Use of the dispersion relation $\omega^2 = k^2 N^2/(k^2 + m^2)$ shows that

$$E_k = E_p, \qquad (7.156)$$

which is a general result for small oscillations of a conservative system without Coriolis forces. The total wave energy is

$$E = E_k + E_p = \frac{1}{2}\rho_0\left(\frac{m^2}{k^2}+1\right)\widehat{w}^2. \qquad (7.157)$$

Last, we shall show that $\mathbf{c_g}$ times the wave energy equals the energy flux. The average energy flux $\mathbf{F}$ across a unit area can be found from (7.153):

$$\mathbf{F} = \overline{p'\mathbf{u}} = \mathbf{e}_x\overline{p'u} + \mathbf{e}_z\overline{p'w} = \frac{\rho_0\omega m\widehat{w}^2}{2k^2}\left(\mathbf{e}_x\frac{m}{k} - \mathbf{e}_z\right). \qquad (7.158)$$

Using (7.145) and (7.157), group velocity times wave energy is

$$\mathbf{c_g}E = \frac{Nm}{K^3}(m\mathbf{e}_x - k\mathbf{e}_z)\left[\frac{\rho_0}{2}\left(\frac{m^2}{k^2}+1\right)\widehat{w}^2\right],$$

which reduces to (7.158) on using the dispersion relation (7.138), so it follows that

$$F = c_g E.$$ (7.159)

This result also holds for surface or interfacial gravity waves. However, in that case F represents the flux per unit width perpendicular to the propagation direction (integrated over the entire depth), and E represents the energy per unit horizontal area. In (7.159), on the other hand, F is the flux per unit area, and E is the energy per unit volume.

EXERCISES

7.1. Starting from (7.5) and working in (x, y, z) Cartesian coordinates, determine an equation that specifies the locus of points that defines a wave crest. Verify that the travel speed of the crests in the direction of $\mathbf{K} = (k, l, m)$ is $c = \omega/|\mathbf{K}|$. Can anything be determined about the wave crest travel speed in other directions?

7.2. For $ka \ll 1$, use the potential for linear deep-water waves $\phi(z, x, t) = a(\omega/k)e^{kz}\sin(kx - \omega t)$ and the waveform $\eta(x, t) = a\cos(kx - \omega t) + \alpha ka^2 \cos[2(kx - \omega t)]$ to show that:
   a) With an appropriate choice of the constant α, the kinematic boundary condition (7.16) can be satisfied for terms proportional to $(ka)^0$ and $(ka)^1$ once the common factor of $a\omega$ has been divided out.
   b) With an appropriate choice of the constant γ, the dynamic boundary condition (7.19) can be satisfied for terms proportional to $(ka)^0$, $(ka)^1$, and $(ka)^2$ when $\omega^2 = gk(1 + \gamma k^2 a^2)$ once the common factor of ag has been divided out.

7.3. The field equation for surface waves on a deep fluid layer in two dimensions (x, z) is: $\dfrac{\partial^2 \phi}{\partial x^2} + \dfrac{\partial^2 \phi}{\partial z^2} = 0$, where ϕ is the velocity potential, $\nabla\phi = (u, w)$. The linearized free-surface boundary conditions and the bottom boundary condition are:

$$(\partial\phi/\partial z)_{z=0} \cong \partial\eta/\partial t, \quad (\partial\phi/\partial t)_{z=0} + g\eta \cong 0, \quad \text{and} \quad (\partial\phi/\partial z)_{z\to-\infty} = 0,$$

where $z = \eta(x,t)$ defines the free surface, gravity g points downward along the z-axis, and the undisturbed free surface lies at $z = 0$. The goal of this problem is to develop the general solution for these equations without assuming a sinusoidal form for the free surface as was done in Sections 7.1 and 7.2.

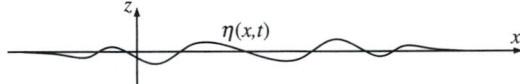

   a) Assume $\phi(x, z, t) = \Lambda(x, t)Z(z)$, and use the field equation and bottom boundary condition to show that $\phi(x, z, t) = \Lambda(x, t)\exp(+kz)$, where k is a positive real constant.
   b) Use the results of part a) and the remaining boundary conditions to show:

$$\frac{\partial^2 \Lambda}{\partial t^2} + gk\Lambda = 0 \quad \text{and} \quad \frac{\partial^2 \Lambda}{\partial x^2} + k^2\Lambda = 0.$$

c) For a fixed value of k, find $\Lambda(x, t)$ in terms of four unknown amplitudes $A, B, C,$ and D.

d) For the initial conditions: $\eta = h(x)$ and $\partial\eta/\partial t = \dot{h}(x)$ at $t = 0$, determine the general form of $\phi(x, z, t)$.

7.4. Derive (7.37) from (7.27).

7.5. Consider stationary surface gravity waves in a rectangular container of length L and breadth b, containing water of undisturbed depth H. Show that the velocity potential $\phi = A \cos(m\pi x/L)\cos(n\pi y/b)\cosh k(z + H) e^{-i\omega t}$ satisfies the Laplace equation and the wall boundary conditions, if $(m\pi/L)^2 + (n\pi/b)^2 = k^2$. Here m and n are integers. To satisfy the linearized free-surface boundary condition, show that the allowable frequencies must be $\omega^2 = gk \tanh kH$. [*Hint*: combine the two boundary conditions (7.18) and (7.21) into a single equation $\partial^2\phi/\partial t^2 = -g\,\partial\phi/\partial z$ at $z = 0$.]

7.6. A lake has the following dimensions: $L = 30$ km, $b = 2$ km, and $H = 100$ m. If the wind sets up the mode $m = 1$ and $n = 0$, show that the period of the oscillation is 31.7 min.

7.7. Fill a square or rectangular cake pan half way with water. Do the same for a round frying pan of about the same size. Agitate the water by carrying the two pans while walking briskly (outside) at a consistent pace on a horizontal surface.

a) Which shape lends itself better to spilling?

b) At what portion of the perimeter of the rectangular pan does spilling occur most readily?

c) Explain your observations in terms of standing wave modes.

7.8. Using the approach of (7.43), show that the time-average energy flux F per unit length of crest is zero for the standing wave described by (7.62).

7.9. Show that the group velocity of pure capillary waves in deep water, for which the gravitational effects are negligible, is $c_g = \dfrac{3}{2}c$.

7.10. Plot the group velocity of surface gravity waves, including surface tension σ, as a function of λ.

a) Assuming deep water, show that the group velocity is

$$c_g = \frac{1}{2}\sqrt{\frac{g}{k}\frac{1 + 3\sigma k^2/\rho g}{\sqrt{1 + \sigma k^2/\rho g}}}.$$

b) Show that this becomes minimum at a wave number given by

$$\frac{\sigma k^2}{\rho g} = \frac{2}{\sqrt{3}} - 1.$$

c) For cool water ($\rho = 1000$ kg/m^3 and $\sigma = 0.074$ N/m), verify that $c_{g\,min} = 17.8$ cm/s.

7.11. The effect of viscosity on the energy of linear deep-water surface waves can be determined from the wave motion's velocity components and the viscous dissipation (4.58).

a) For incompressible flow, the viscous dissipation of energy per unit mass of fluid is $\varepsilon = 2(\mu/\rho)S_{ij}^2$, where S_{ij} is the strain-rate tensor and μ is the fluid's viscosity. Determine ε using (7.47).

b) The total wave energy per unit surface area, E, for a linear sinusoidal water wave with amplitude a is given by (7.42). Assume that a is a function of time, set $dE/dt = -\varepsilon$, and show that $a(t) = a_0\exp[-2(\mu/\rho)k^2 t]$, where a_0 is the wave amplitude at $t = 0$.

c) Using a nominal value of $\mu/\rho = 10^{-6}$ m^2/s for water, determine the time necessary for an amplitude reduction of 50% for water-surface waves having $\lambda = 1$ mm, 1 cm, 10 cm, 1 m, 10 m, and 100 m.

d) Convert the times calculated in part c) to travel distances by multiplication with an appropriate group speed. Remember to include surface tension. Can a typhoon located near New Zealand produce increased surf on the west coast of North America? [The circumference of the earth is approximately 40,000 km.]

7.12. Consider a deep-water wave train with a Gaussian envelope that resides near $x = 0$ at $t = 0$ and travels in the positive x direction. The surface shape at any time is a Fourier superposition of waves with all possible wave numbers:

$$\eta(x,t) = \int_{-\infty}^{+\infty} \tilde{\eta}(k)\exp\left[i\left(kx - (g|k|)^{1/2}t\right)\right]dk, \qquad (\dagger)$$

where $\tilde{\eta}(k)$ is the amplitude of the wave component with wave number k, and the dispersion relation is $\omega = (gk)^{1/2}$. For the following items assume the surface shape at $t = 0$ is:

$$\eta(x,0) = \frac{1}{\sqrt{2\pi}\alpha}\exp\left\{-\frac{x^2}{2\alpha^2} + ik_d x\right\}.$$

Here, $k_d > 0$ is the dominant wave number, and α sets the initial horizontal extent of the wave train, with larger α producing a longer wave train.

a) Plot Re$\{\eta(x, 0)\}$ for $|x| \le 40$ m when $\alpha = 10$ m and $k_d = 2\pi/\lambda_d = 2\pi/10$ m^{-1}.

b) Use the inverse Fourier transform at $t = 0$, $\tilde{\eta}(k) = (1/2\pi)\int_{-\infty}^{+\infty} \eta(x,0)\exp[-ikx]dx$ to find the wave amplitude distribution: $\tilde{\eta}(k) = (1/2\pi)\exp\left\{-\frac{1}{2}(k - k_d)^2\alpha^2\right\}$, and plot this function for $0 < k < 2k_d$ using the numerical values from part a). Does the dominant contribution to the wave activity come from wave numbers near k_d for the part a) values?

c) For large x and t, the integrand of ($\dagger$) will be highly oscillatory unless the phase $\Phi \equiv kx - (g|k|)^{1/2}t$ happens to be constant. Thus, for any x and t, the primary contribution to η will come from the region where the phase in ($\dagger$) does not depend on k. Thus, set $d\Phi/dk = 0$, and solve for k_s (= the wave number where the phase is independent of k) in terms of x, t, and g.

d) Based on the result of part b), set $k_s = k_d$ to find the x-location where the dominant portion of the wave activity occurs at time t. At this location, the ratio x/t is the propagation speed of the dominant portion of the wave activity. Is this propagation speed the phase speed, the group speed, or another speed altogether?

7.13. Show that the vertical component of the Stokes drift is zero starting from (7.85) and using (7.47) and (7.48).

7.14. Extend the deep water Stokes drift result (7.85) to arbitrary depth to derive (7.86).

7.15. Explicitly show through substitution and differentiation that (7.88) is a solution of (7.87).

7.16. A *thermocline* is a thin layer in the upper ocean across which water temperature and, consequently, water density change rapidly. Suppose the thermocline in a very deep ocean is at a depth of 100 m from the ocean surface, and that the temperature drops across it from 30°C to 20°C. Show that the reduced gravity is $g' = 0.025$ m/s^2. Neglecting Coriolis effects, show that the speed of propagation of long gravity waves on such a thermocline is 1.58 m/s.

7.17. Consider internal waves in a continuously stratified fluid of buoyancy frequency $N = 0.02$ s^{-1} and average density 800 kg/m^3. What is the direction of ray paths if the frequency of oscillation is $\omega = 0.01$ s^{-1}? Find the energy flux per unit area if the amplitude of the vertical velocity is $\hat{w} = 1$ cm/s and the horizontal wavelength is π meters.

7.18. Consider internal waves at a density interface between two infinitely deep fluids, and show that the average kinetic energy per unit horizontal area is $E_k = (\rho_2 - \rho_1)ga^2/4$.

7.19. Consider waves in a finite layer overlying an infinitely deep fluid. Using the constants given in equations (7.106) through (7.109), prove the dispersion relation (7.110).

7.20. A simple model of oceanic internal waves involves two ideal incompressible fluids ($\rho_2 > \rho_1$) trapped between two horizontal surfaces at $z = h_1$ and $z = -h_2$, and having an average interface location of $z = 0$. For traveling waves on the interface, assume that the interface deflection from $z = 0$ is $\xi = \xi_o$ Re$\{\exp(i(\omega t - kx))\}$. The phase speed of the waves is $c = \omega/k$.

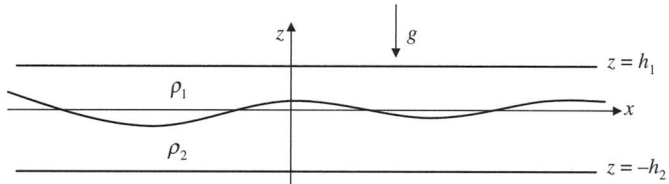

a) Show that the dispersion relationship is $\omega^2 = \dfrac{gk(\rho_2 - \rho_1)}{\rho_2 \coth(kh_2) + \rho_1 \coth(kh_1)}$, where g is the acceleration of gravity.

b) Determine the limiting form of c for short (i.e., unconfined) waves, kh_1 and $kh_2 \to \infty$.

c) Determine the limiting form of c for long (i.e., confined) waves, kh_1 and $kh_2 \to 0$.

d) At fixed wavelength λ (or fixed $k = 2\pi/\lambda$), do confined waves go faster or slower than unconfined waves?

e) At a fixed frequency, what happens to the wavelength and phase speed as $\rho_2 - \rho_1 \to 0$?

f) What happens if $\rho_2 < \rho_1$?

Literature Cited

Gill, A. (1982). *Atmosphere–Ocean Dynamics*. New York: Academic Press.

Graff, K. A. (1975). *Wave Motion in Elastic Solids*. Oxford: Oxford University Press.

Kinsman, B. (1965). *Wind Waves*. Englewood Cliffs, New Jersey: Prentice-Hall.

LeBlond, P. H., & Mysak, L. A. (1978). *Waves in the Ocean*. Amsterdam: Elsevier Scientific Publishing.

Liepmann, H. W., & Roshko, A. (1957). *Elements of Gasdynamics*. New York: Wiley.

Lighthill, M. J. (1978). *Waves in Fluids*. London: Cambridge University Press.

Phillips, O. M. (1977). *The Dynamics of the Upper Ocean*. London: Cambridge University Press.

Turner, J. S. (1973). *Buoyancy Effects in Fluids*. London: Cambridge University Press.

Whitham, G. B. (1974). *Linear and Nonlinear Waves*. New York: Wiley.

OUTLINE

8.1. Introduction   309

8.2. Exact Solutions for Steady
     Incompressible Viscous Flow   312

8.3. Elementary Lubrication Theory   318

8.4. Similarity Solutions for Unsteady
     Incompressible Viscous Flow   326

8.5. Flow Due to an Oscillating Plate 337

8.6. Low Reynolds Number Viscous
     Flow Past a Sphere   338

8.7. Final Remarks   347

Exercises   347

Literature Cited   359

Supplemental Reading   359

CHAPTER OBJECTIVES

• To present a variety of exact and approximate solutions to the viscous equations of fluid motion in confined and unconfined geometries.

• To introduce lubrication theory and indicate its utility.

• To define and present similarity solutions to exact and approximate viscous flow field equations.

• To develop the equations for creeping flow and illustrate their use.

8.1. INTRODUCTION

Chapters 6 and 7 covered flows in which the viscous terms in the Navier-Stokes equations were dropped because the flow was ideal (irrotational and constant density) or the effects of viscosity were small. For these situations, the underlying assumptions were either that 1) viscous forces and rotational flow were spatially confined to a small portion of the flow domain (thin boundary layers near solid surfaces), or that 2) fluid particle accelerations caused by fluid inertia $\sim U^2/L$ were much larger than those caused by viscosity $\sim \mu U/\rho L^2$, where U is a characteristic velocity, L is a characteristic length, ρ is the fluid's density, and μ is the fluid's kinematic viscosity. Both of these assumptions are valid if the Reynolds number is large and boundary layers stay attached to the surface on which they have formed.

However, for low values of the Reynolds number, the *entire flow* may be influenced by viscosity, and inviscid flow theory is no longer even approximately correct. The purpose of this chapter is to present some exact and approximate solutions of the Navier-Stokes equations for simple geometries and situations, retaining the viscous terms in (4.38) everywhere in the flow and applying the no-slip boundary condition at solid surfaces (see Section 4.10).

Viscous flows generically fall into two categories, *laminar* and *turbulent*, but the boundary between them is imperfectly defined. The basic difference between the two categories is phenomenological and was dramatically demonstrated in 1883 by Reynolds, who injected a thin stream of dye into the flow of water through a tube (Figure 8.1). At low flow rates, the dye stream was observed to follow a well-defined straight path, indicating that the fluid moved in parallel layers (laminae) with no unsteady macroscopic mixing or overturning motion of the layers. Such smooth orderly flow is called *laminar*. However, if the flow rate

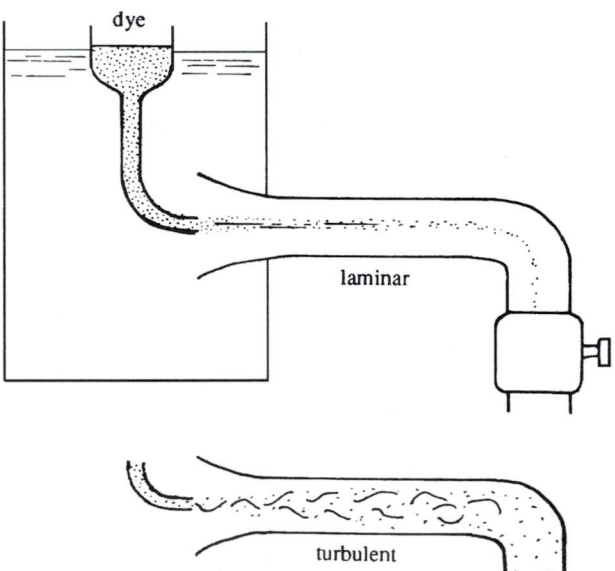

FIGURE 8.1   Reynolds's experiment to distinguish between laminar and turbulent flows. At low flow rates (the upper drawing), the pipe flow was laminar and the dye filament moved smoothly through the pipe. At high flow rates (the lower drawing), the flow became turbulent and the dye filament was mixed throughout the cross section of the pipe.

was increased beyond a certain critical value, the dye streak broke up into irregular filaments and spread throughout the cross section of the tube, indicating the presence of unsteady, apparently chaotic three-dimensional macroscopic mixing motions. Such irregular disorderly flow is called a *turbulent*. Reynolds demonstrated that the transition from laminar to turbulent flow always occurred at a fixed value of the ratio that bears his name, the Reynolds number, $\text{Re} = Ud/\nu \sim 2000$ to 3000 where U is the velocity averaged over the tube's cross section, d is the tube diameter, and $\nu = \mu/\rho$ is the kinematic viscosity.

As will be further verified in Section 8.4, the fluid's kinematic viscosity specifies the propensity for vorticity to diffuse through a fluid. Consider (5.13) for the z-component of vorticity in a two-dimensional flow confined to the x-y plane so that $\boldsymbol{\omega} \cdot \nabla \mathbf{u} = 0$:

$$D\omega_z/Dt = \nu\nabla^2\omega_z.$$

This equation states that the rate of change of ω_z following a fluid particle is caused by diffusion of vorticity. Clearly, for the same initial vorticity distribution, a fluid with larger ν will produce a larger diffusion term, $\nu\nabla^2\omega$, and more rapid changes in the vorticity. This equation is similar to the Boussinesq heat equation,

$$DT/Dt = \kappa\nabla^2 T, \tag{4.89}$$

where $\kappa \equiv k/\rho C_p$ is the *thermal diffusivity*, and this similarity suggests that vorticity diffuses in a manner analogous to heat. At a coarse level, this suggestion is correct since both ν and κ arise from molecular processes in real fluids and both have the same units (length^2/time). The similarity emphasizes that the diffusive effects are controlled by ν and κ, and not by μ (viscosity) and k (thermal conductivity). In fact, the constant-density, constant-viscosity momentum equation,

$$D\mathbf{u}/Dt = -(1/\rho)\nabla p + \nu\nabla^2\mathbf{u}, \tag{4.85, 8.1}$$

also shows that the acceleration due to viscous diffusion is proportional to ν. Thus, at room temperature and pressure, air ($\nu = 15 \times 10^{-6}$ m^2/s) is 15 times more diffusive than water ($\nu = 1 \times 10^{-6}$ m^2/s), although μ for water is larger. Both ν and κ have the units of m^2/s; thus, the kinematic viscosity ν is sometimes called the *momentum diffusivity*, in analogy with κ, the *thermal diffusivity*. However, velocity cannot be simply regarded as being diffused and advected in a flow because of the presence of the pressure gradient in (8.1).

Laminar flows in which viscous effects are important throughout the flow are the subject of the present chapter. The primary field equations will be $\nabla \cdot \mathbf{u} = 0$ (4.10) and (8.1) or the version that includes a body force (4.39b). The velocity boundary conditions on a solid surface are:

$$\mathbf{n} \cdot \mathbf{U}_s = (\mathbf{n} \cdot \mathbf{u})_{\text{on the surface}} \text{ and } \mathbf{t} \cdot \mathbf{U}_s = (\mathbf{t} \cdot \mathbf{u})_{\text{on the surface}}, \tag{8.2, 8.3}$$

where $\mathbf{U}_s$ is the velocity of the surface, $\mathbf{n}$ is the normal to the surface, and $\mathbf{t}$ is the tangent to the surface. Here fluid density will be assumed constant, and the frame of reference will be inertial. Thus, gravity can be dropped from the momentum equation as long as no free surface is present (see Section 4.9 "Neglect of Gravity in Constant Density Flows"). Laminar flows in which frictional effects are confined to boundary layers near solid surfaces are discussed in the next chapter. Chapter 11 considers the stability of laminar flows and their transition to turbulence; fully turbulent flows are discussed in Chapter 12. Some viscous flow solutions in rotating coordinates, such as the Ekman layers, are presented in Chapter 13.

8.2. EXACT SOLUTIONS FOR STEADY INCOMPRESSIBLE VISCOUS FLOW

Because of the presence of the nonlinear acceleration term $\mathbf{u} \cdot \nabla \mathbf{u}$ in (8.1), very few exact solutions of the Navier-Stokes equations are known in closed form. An example of an exact solution is that for steady laminar flow between infinite parallel plates (Figure 8.2). Such a flow is said to be *fully developed* when its velocity profile $u(x,y)$ becomes independent of the downstream coordinate x so that $u = u(y)$ alone. The entrance length of the flow, where the velocity profile depends on the downstream distance, may be several or even many times longer than the spacing between the plates. Within the entrance length, the derivative $\partial u / \partial x$ is not zero so the continuity equation $\partial u / \partial x + \partial v / \partial y = 0$ requires that $v \neq 0$, so that the flow is *not* parallel to the walls within the entrance length. Laminar flow development is the subject of Section 8.4, and the next chapter. Here we are interested in steady, fully developed flows.

Steady Flow between Parallel Plates

Consider the situation depicted in Figure 8.3 where a viscous fluid flows between plates parallel to the x-axis with lower and upper plates at $y = 0$ and $y = h$, respectively. The flow

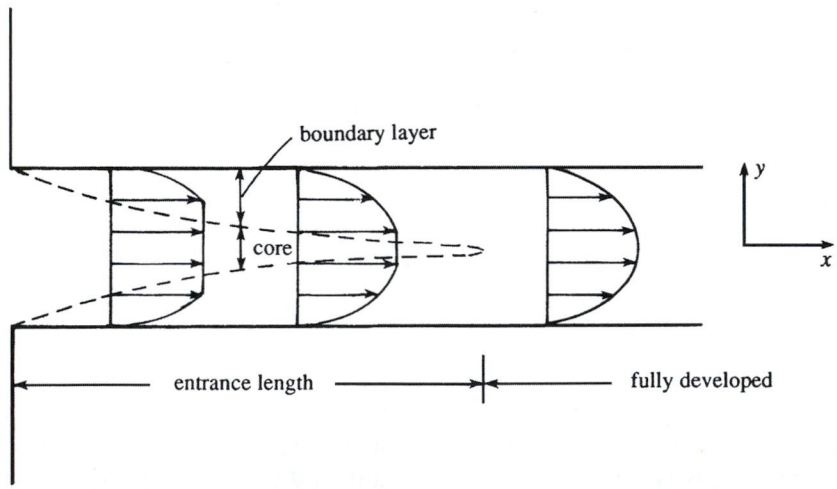

FIGURE 8.2 Developing and fully developed flows in a channel. Within the entrance length, the viscous boundary layers on the upper and lower walls are separate and the flow profile $u(x,y)$ depends on both spatial coordinates. Downstream of the point where the boundary layers merge, the flow is fully developed and its profile $u(y)$ is independent of the stream-wise coordinate x.

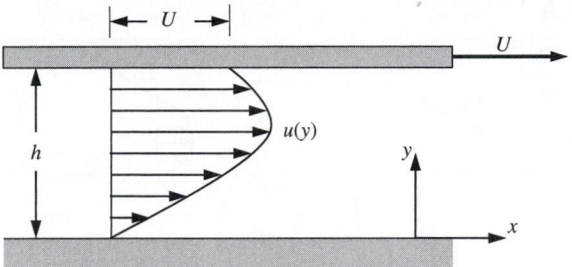

FIGURE 8.3   Flow between parallel plates when the lower plate at $y = 0$ is stationary, the upper plate at $y = h$ is moving in the positive-x direction at speed U, and a nonzero $dp/dx < 0$ leads to velocity profile curvature.

is sustained by an externally applied pressure gradient ($\partial p/\partial x \neq 0$) in the x-direction, and horizontal motion of the upper plate at speed U in the x-direction. For this situation, the flow should be independent of the z-direction so $w = 0$ and $\partial/\partial z = 0$ can be used in the equations of motion. A steady, fully developed flow will have a horizontal velocity $u(y)$ that does not depend on x so $\partial u/\partial x = 0$. Thus, the continuity equation, $\partial u/\partial x + \partial v/\partial y = 0$, requires $\partial v/\partial y = 0$, and since $v = 0$ at $y = 0$ and h, it follows that $v = 0$ everywhere, which reflects the fact that the flow is parallel to the walls. Under these circumstances, $\mathbf{u} = (u(y), 0, 0)$, and the x- and y-momentum equations reduce to:

$$0 = -\frac{1}{\rho}\frac{\partial p}{\partial x} + \nu\frac{d^2u}{dy^2}, \quad \text{and} \quad 0 = -\frac{1}{\rho}\frac{\partial p}{\partial y}. \tag{8.4a,b}$$

The y-momentum equation shows that p is not a function of y, so $p = p(x)$. Thus, the first term in the x-momentum equation must be a function of x alone, while the second term must be a function of y alone. The only way the equation can be satisfied throughout x-y space is if both terms are constant. The *pressure gradient is therefore a constant*, which implies that the pressure varies linearly along the channel. Integrating the x-momentum equation twice, we obtain

$$0 = -\frac{y^2}{2}\frac{dp}{dx} + \mu u + Ay + B,$$

where A and B are constants and dp/dx has replaced $\partial p/\partial x$ because p is a function of x alone. The constants are determined from the boundary conditions: $u = 0$ at $y = 0$, and $u = U$ at $y = h$. The results are $B = 0$ and $A = (h/2)(dp/dx) - \mu U/h$, so the velocity profile becomes

$$u(y) = \frac{U}{h}y - \frac{1}{2\mu}\frac{dp}{dx}y(h - y), \tag{8.5}$$

which is illustrated in Figure 8.4 for various cases. The volume flow rate Q per unit width of the channel is

$$Q = \int_0^h u\,dy = U\frac{h}{2}\left[1 - \frac{h^2}{6\mu U}\frac{dp}{dx}\right],$$

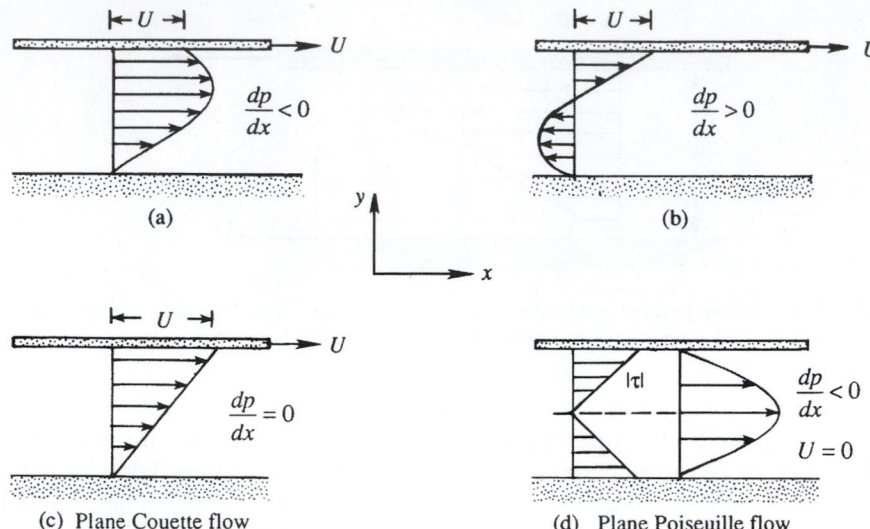

FIGURE 8.4 Various cases of parallel flow in a channel: (a) positive U and favorable $dp/dx < 0$, (b) positive U and adverse $dp/dx > 0$, (c) positive U and $dp/dx = 0$, and (d) $U = 0$ and favorable $dp/dx < 0$.

so that the average velocity is

$$V \equiv \frac{Q}{h} = \int_0^h u\, dy = \frac{U}{2}\left[1 - \frac{h^2}{6\mu U}\frac{dp}{dx}\right].$$

Here, negative and positive pressure gradients increase and decrease the flow rate, respectively.

When the flow is driven by motion of the upper plate alone, without any externally imposed pressure gradient, it is called a *plane Couette flow*. In this case (8.5) reduces to $u(y) = Uy/h$, and the magnitude of the shear stress is $\tau = \mu(du/dy) = \mu U/h$, which is uniform across the channel.

When the flow is driven by an externally imposed pressure gradient without motion of either plate, it is called a *plane Poiseuille flow*. In this case (8.5) reduces to the parabolic profile (Figure 8.4d):

$$u(y) = -\frac{1}{2\mu}\frac{dp}{dx}y(h - y).$$

The shear stress is

$$\tau = \mu\frac{du}{dy} = -\left(\frac{h}{2} - y\right)\frac{dp}{dx},$$

which is linear with a magnitude of $(h/2)(dp/dx)$ at the walls (Figure 8.4d).

Interestingly, the constancy of the pressure gradient and the linearity of the shear stress distribution are general results for a fully developed channel flow and persist for appropriate averages of these quantities when the flow is turbulent.

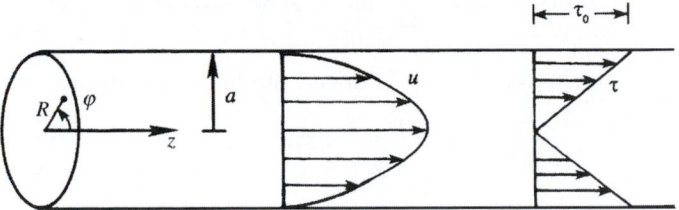

FIGURE 8.5 Laminar flow through a round tube. The flow profile is parabolic, similar to pressure-driven flow between stationary parallel plates (Figure 8.4d).

Steady Flow in a Round Tube

A second geometry for which there is an exact solution of (4.10) and (8.1) is steady, fully developed laminar flow through a round tube of constant radius a, frequently called *circular Poiseuille flow*. We employ cylindrical coordinates (R, φ, z), with the z-axis coinciding with the axis of the tube (Figure 8.5). The equations of motion in cylindrical coordinates are given in Appendix B. The only nonzero component of velocity is the axial velocity $u_z(R)$, and $\mathbf{u} = (0, 0, u_z(R))$ automatically satisfies the continuity equation. The radial and angular equations of motion reduce to

$$0 = \partial p/\partial \varphi \text{ and } 0 = \partial p/\partial R,$$

so p is a function of z alone. The z-momentum equation gives

$$0 = -\frac{dp}{dz} + \frac{\mu}{R}\frac{d}{dR}\left(R\frac{du_z}{dR}\right).$$

As for flow between parallel plates, the first term must be a function of the stream-wise coordinate, z, alone and the second term must be a function of the cross-stream coordinate, R, alone, so both terms must be constant. The pressure therefore falls linearly along the length of the tube. Integrating the stream-wise momentum equation twice produces

$$u_z(R) = \frac{R^2}{4\mu}\frac{dp}{dz} + A \ln R + B.$$

To keep u_z bounded at $R = 0$, the constant A must be zero. The no-slip condition $u_z = 0$ at $R = a$ gives $B = -(a^2/4\mu)(dp/dz)$. The velocity distribution therefore takes the parabolic shape:

$$u_z(R) = \frac{R^2 - a^2}{4\mu}\frac{dp}{dz}. \tag{8.6}$$

From Appendix B, the shear stress at any point is

$$\tau_{zR} = \mu\left(\frac{\partial u_R}{\partial z} + \frac{\partial u_z}{\partial R}\right).$$

In this case the radial velocity u_R is zero. Dropping the subscripts on τ and differentiating (8.6) yields

$$\tau = \mu \frac{\partial u_z}{\partial R} = \frac{R}{2} \frac{dp}{dz}, \tag{8.7}$$

which shows that the stress distribution is linear, having a maximum value at the wall of

$$\tau_0 = \frac{a}{2} \frac{dp}{dz}. \tag{8.8}$$

Here again, (8.8) is also valid for appropriate averages of τ_0 and p for turbulent flow in a round pipe.

The volume flow rate in the tube is:

$$Q = \int_0^a u(R) 2\pi R dR = -\frac{\pi a^4}{8\mu} \frac{dp}{dz},$$

where the negative sign offsets the negative value of dp/dz. The average velocity over the cross section is

$$V = \frac{Q}{\pi a^2} = -\frac{a^2}{8\mu} \frac{dp}{dz}.$$

Steady Flow between Concentric Rotating Cylinders

A third example in which the nonlinear advection terms drop out of the equations of motion is steady flow between two concentric, rotating cylinders, also know as *circular Couette flow*. Let the radius and angular velocity of the inner cylinder be R_1 and Ω_1 and those for the outer cylinder be R_2 and Ω_2 (Figure 8.6). Using cylindrical coordinates and assuming that $\mathbf{u} = (0, u_\varphi(R), 0)$, the continuity equation is automatically satisfied, and the momentum equations for the radial and tangential directions are

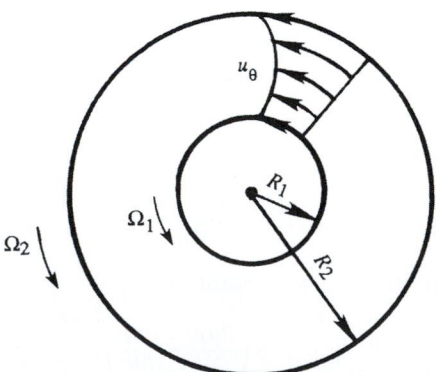

FIGURE 8.6   Circular Couette flow. The viscous fluid flows in the gap between an inner cylinder with radius R_1 that rotates at angular speed Ω_1 and an outer cylinder with radius R_2 that rotates at angular speed Ω_2.

$$-\frac{u_\varphi^2}{R} = -\frac{1}{\rho}\frac{dp}{dR}, \quad \text{and} \quad 0 = \mu\frac{d}{dR}\left[\frac{1}{R}\frac{d}{dR}(Ru_\varphi)\right].$$

The R-momentum equation shows that the pressure increases radially outward due to the centrifugal acceleration. The pressure distribution can therefore be determined once $u_\varphi(R)$ has been found. Integrating the φ-momentum equation twice produces

$$u_\varphi(R) = AR + B/R. \tag{8.9}$$

Using the boundary conditions $u_\varphi = \Omega_1 R_1$ at $R = R_1$, and $u_\varphi = \Omega_2 R_2$ at $R = R_2$, A and B are found to be

$$A = \frac{\Omega_2 R_2^2 - \Omega_1 R_1^2}{R_2^2 - R_1^2}, \quad \text{and} \quad B = -\frac{(\Omega_2 - \Omega_1)R_1^2 R_2^2}{R_2^2 - R_1^2}.$$

Substitution of these into (8.9) produces the velocity distribution,

$$u_\varphi(R) = \frac{1}{R_2^2 - R_1^2}\left\{[\Omega_2 R_2^2 - \Omega_1 R_1^2]R - [\Omega_2 - \Omega_1]\frac{R_1^2 R_2^2}{R}\right\}, \tag{8.10}$$

which has interesting limiting cases when $R_2 \to \infty$ with $\Omega_2 = 0$, and when $R_1 \to 0$ with $\Omega_1 = 0$.

The first limiting case produces the flow outside a long circular cylinder with radius R_1 rotating with angular velocity Ω_1 in an infinite bath of viscous fluid (Figure 8.7). By direct simplification of (8.10), the velocity distribution is

$$u_\varphi(R) = \frac{\Omega_1 R_1^2}{R}, \tag{8.11}$$

FIGURE 8.7 Rotation of a solid cylinder of radius R_1 in an infinite body of viscous fluid. If gravity points downward along the cylinder's axis, the shape of a free surface pierced by the cylinder is also indicated. The flow field is viscous but irrotational.

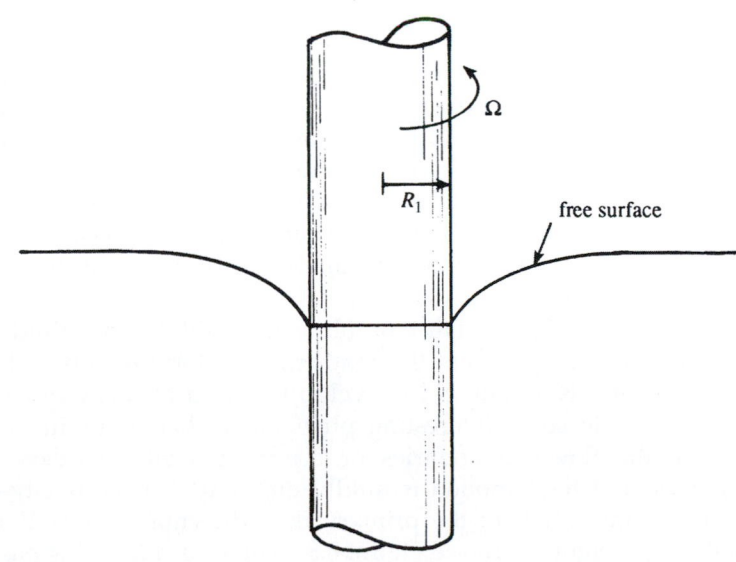

which is identical to that of an ideal vortex, see (5.2), for $R > R_1$ when $\Gamma = 2\pi\Omega_1 R_1^2$. This is the only example of a viscous solution that is completely irrotational. As described in Section 5.1, shear stresses do exist in this flow, but there is no *net* viscous force on a fluid element. The viscous shear stress at any point is given by

$$\sigma_{R\varphi} = \mu\left[\frac{1}{R}\frac{\partial u_R}{\partial \varphi} + R\frac{\partial}{\partial R}\left(\frac{u_\varphi}{R}\right)\right] = -\frac{2\mu\Omega_1 R_1^2}{R^2}.$$

The mechanical power supplied to the fluid (per unit length of cylinder) is $(2\pi R_1)\tau_{R\varphi}u_\varphi$, and it can be shown that this power equals the integrated viscous dissipation of the flow field (Exercise 8.12).

The second limiting case of (8.10) produces steady viscous flow within a cylindrical tank of radius R_2 rotating at rate Ω_2. Setting R_1 and Ω_1 equal to zero in (8.10) leads to

$$u\phi(R) = \Omega_2 R, \tag{8.12}$$

which is the velocity field of solid body rotation, see (5.1) and Section 5.1.

The three exact solutions of the incompressible viscous flow equations (4.10) and (8.1) described in this section are all known as internal or confined flows. In each case, the velocity field was confined between solid walls and the symmetry of each situation eliminated the nonlinear advective acceleration term from the equations. Other exact solutions of the incompressible viscous flow equations for confined and unconfined flows are described in other fine texts (Sherman, 1990; White, 2006), in Section 8.4, and in the Exercises of this chapter. However, before proceeding to these, a short diversion into elementary lubrication theory is provided in the next section.

8.3. ELEMENTARY LUBRICATION THEORY

The exact viscous flow solutions for ideal geometries presented in the prior section indicate that a linear or simply varying velocity profile is a robust solution for flow within a confined space. This observation has been developed into the theory of lubrication, which provides approximate solutions to the viscous flow equations when the geometry is not ideal but at least one transverse flow dimension is small. The elementary features of lubrication theory are presented here because of its connection to the exact solutions described in Section 8.2, especially the Couette and Poiseuille flow solutions. Plus, the development of approximate equations in this section parallels that necessary for the boundary layer approximation (see Section 9.1).

The economic importance of lubrication with viscous fluids is hard to overestimate, and lubrication theory covers the mathematical formulation and analysis of such flows. The purpose of this section is to develop the most elementary equations of lubrication theory and illustrate some interesting phenomena that occur in viscous constant-density flows where the flow's boundaries or confining walls are close together, but not precisely parallel, and their motion is mildly unsteady. For simplicity consider two spatial dimensions, x and y, where the primary flow direction, x, lies along the narrow flow passage with gap height $h(x,t)$ (see Figure 8.8). The length L of this passage is presumed to be large

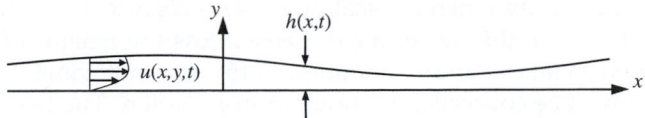

FIGURE 8.8 Nearly parallel flow of a viscous fluid having a film thickness of $h(x,t)$ above a flat stationary surface.

compared to h so that viscous and pressure forces are the primary terms in any fluid-momentum balance. If the passage is curved, this will not influence the analysis as long as the radius of curvature is much larger than the gap height h. The field equations are (4.10) and (8.1) for the horizontal u, and vertical v velocity components, and the pressure p in the fluid:

$$\frac{\partial u}{\partial x} + \frac{\partial v}{\partial y} = 0, \tag{6.2}$$

$$\frac{\partial u}{\partial t} + u\frac{\partial u}{\partial x} + v\frac{\partial u}{\partial y} = -\frac{1}{\rho}\frac{\partial p}{\partial x} + \frac{\mu}{\rho}\left(\frac{\partial^2 u}{\partial x^2} + \frac{\partial^2 u}{\partial y^2}\right), \text{ and } \frac{\partial v}{\partial t} + u\frac{\partial v}{\partial x} + v\frac{\partial v}{\partial y} = -\frac{1}{\rho}\frac{\partial p}{\partial x} + \frac{\mu}{\rho}\left(\frac{\partial^2 v}{\partial x^2} + \frac{\partial^2 v}{\partial y^2}\right).$$

$$\tag{8.13a, 8.13b}$$

Here, the boundary conditions are $u = U_0(t)$ on $y = 0$ and $u = U_h(t)$ on $y = h(x, t)$, and the pressure is presumed to be time dependent as well.

To determine which terms are important and which may be neglected when the passage is narrow, recast these equations in terms of dimensionless variables:

$$x^* = x/L, \ y^* = y/h = y/\varepsilon L, \ t^* = Ut/L, \ u^* = u/U, \ v^* = v/\varepsilon U, \text{ and } p^* = p/P_a, \tag{8.14}$$

where U is a characteristic velocity of the flow, P_a is atmospheric pressure, and $\varepsilon = h/L$ is the passage's *fineness ratio* (the inverse of its aspect ratio). The goal of this effort is to find a set of approximate equations that are valid for common lubrication geometries where $\varepsilon \ll 1$ and the flow is unsteady. Because of the passage geometry, the magnitude of v is expected to be much less than the magnitude of u and gradients along the passage, $\partial/\partial x \sim 1/L$, are expected to be much smaller than gradients across it, $\partial/\partial y \sim 1/h$. These expectations have been incorporated into the dimensionless scaling (8.14). Combining (6.2), (8.13), and (8.14) leads to the following dimensionless equations:

$$\frac{\partial u^*}{\partial x^*} + \frac{\partial v^*}{\partial y^*} = 0, \tag{8.15}$$

$$\varepsilon^2 \, \text{Re}_L\left(\frac{\partial u^*}{\partial t^*} + u^*\frac{\partial u^*}{\partial x^*} + v^*\frac{\partial u^*}{\partial y^*}\right) = -\frac{1}{\Lambda}\frac{\partial p^*}{\partial x^*} + \varepsilon^2\frac{\partial^2 u^*}{\partial x^{*2}} + \frac{\partial^2 u^*}{\partial y^{*2}}, \text{ and }$$

$$\tag{8.16a, 8.16b}$$

$$\varepsilon^4 \, \text{Re}_L\left(\frac{\partial v^*}{\partial t^*} + u^*\frac{\partial v^*}{\partial x^*} + v^*\frac{\partial v^*}{\partial y^*}\right) = -\frac{1}{\Lambda}\frac{\partial p^*}{\partial y^*} + \varepsilon^4\frac{\partial^2 v^*}{\partial x^{*2}} + \varepsilon^2\frac{\partial^2 v^*}{\partial y^{*2}},$$

where $\text{Re}_L = \rho UL/\mu$, and $\Lambda = \mu UL/P_a h^2$ is the ratio of the viscous and pressure forces on a fluid element; it is sometimes called the *bearing number*. All the dimensionless derivative terms should be of order unity when the scaling (8.14) is correct. Thus, possible simplifying approximations are based on the size of the dimensionless coefficients of the various terms. The scaled two-dimensional continuity equation (8.15) does not contain any dimensionless coefficients so mass must be conserved without approximation. The two-scaled momentum equations (8.16) contain ε, Re_L, and Λ. For the present purposes, Λ must be considered to be near unity, Re_L must be finite, and $\varepsilon \ll 1$. When $\varepsilon^2 Re_L \ll 1$, the left side and the middle term on the right side of (8.16a) may be ignored. In (8.16b) the pressure derivative is the only term not multiplied by ε. Therefore, the momentum equations can be approximately simplified to:

$$0 \cong -\frac{1}{\rho}\frac{\partial p}{\partial x} + \frac{\partial^2 u}{\partial y^2} \quad \text{and} \quad 0 \cong -\frac{1}{\rho}\frac{\partial p}{\partial y}, \tag{8.17a, 8.17b}$$

when $\varepsilon^2 Re_L \to 0$. As a numerical example of this approximation, $\varepsilon^2 Re_L = 0.001$ for room temperature flow of common 30-weight oil with $\nu \approx 4 \times 10^{-4}$ m^2/s within a 0.1 mm gap between two 25-cm-long surfaces moving with a differential speed of 10 m/s. When combined with a statement of conservation of mass, the equations (8.17) are the simplest form of the lubrication approximation (the *zeroth*-order approximation), and these equations are readily extended to two-dimensional gap-thickness variations (see Exercise 8.19). Interestingly, the approximations leading to (8.17) eliminated both the unsteady and the advective fluid acceleration terms from (8.16); a steady-flow approximation was not made. Therefore, time is still an independent variable in (8.17) even though it does not explicitly appear.

A generic solution to (8.17) is readily produced by following the steps used to solve (8.4a, 8.4b). Equation (8.17b) implies that p is not a function of y, so (8.16a) can be integrated twice to produce:

$$u(x,y,t) \cong \frac{1}{\mu}\frac{\partial p(x,t)}{\partial x}\frac{y^2}{2} + Ay + B, \tag{8.18}$$

where A and B might be functions of x and t but not y. Applying the boundary conditions mentioned earlier allows A and B to be evaluated, and the fluid velocity within the gap is found to be:

$$u(x,y,t) \cong -\frac{h^2(x,t)}{2\mu}\frac{\partial p(x,t)}{\partial x}\frac{y}{h(x,t)}\left(1 - \frac{y}{h(x,t)}\right) + U_h(t)\frac{y}{h(x,t)} + U_0(t). \tag{8.19}$$

The basic result here is that balancing viscous and pressure forces leads to a velocity profile that is parabolic in the cross-stream direction. While (8.19) represents a significant simplification of the two momentum equations (8.13), it is not a complete solution because the pressure $p(x, t)$ within the gap has not yet been determined. The complete solution to an elementary lubrication flow problem is typically obtained by combining (8.19), or an appropriate equivalent, with a differential or integral form of (4.10) or (6.2), and pressure boundary conditions. Such solutions are illustrated in the following examples.

EXAMPLE 8.1

A sloped bearing pad of width B into the page moves horizontally at a steady speed U on a thin layer of oil with density ρ and viscosity μ. The gap between the bearing pad and a stationary hard, flat surface located at $y = 0$ is $h(x) = h_0(1 + \alpha x/L)$ where $\alpha \ll 1$. If p_e is the exterior pressure and $p(x)$ is the pressure in the oil under the bearing pad, determine the load W (per unit width into the page) that the bearing can support.

Solution

The solution plan is to conserve mass exactly using (4.5), a control volume (CV) that is attached to the bearing pad, and the generic velocity profile (8.19). Then, pressure boundary conditions at the ends of the bearing pad should allow the pressure distribution under the pad to be found. Finally, W can be determined by integrating this pressure distribution.

Use the fixed-shape, but moving CV shown in Figure 8.9 that lies between x_1 and x_2 at the moment of interest. The mass of fluid in the CV is constant so the unsteady term in (4.5) is zero, and the control surface velocity is $\mathbf{b} = U\mathbf{e}_x$. Denote the fluid velocity as $\mathbf{u} = u(x, y)\mathbf{e}_x$, and recognize $\mathbf{n}dA = -\mathbf{e}_x B dy$ on the vertical CV surface at x_1 and $\mathbf{n}dA = +\mathbf{e}_x B dy$ on the vertical CV surface at x_2. Thus, (4.5) simplifies to:

$$\rho B \left[-\int_0^{h(x_1)} (u(x_1, y) - U)dy + \int_0^{h(x_2)} (u(x_2, y) - U)dy \right] = 0.$$

Dividing this equation by $\rho B(x_2 - x_1)$ and taking the limit as $(x_2 - x_1) \to 0$ produces:

$$\frac{d}{dx}\left[\int_0^{h(x)} (u(x, y) - U)dy \right] = 0, \text{ or } \int_0^{h(x)} (u(x, y) - U)dy = C_1,$$

where C_1 is a constant. For the flow geometry and situation in Figure 8.9, (8.19) simplifies to:

$$u(x, y, t) \cong -\frac{h^2(x)}{2\mu}\frac{dp(x)}{dx}\frac{y}{h(x)}\left(1 - \frac{y}{h(x)}\right) + U\frac{y}{h(x)},$$

which can substituted into with the conservation of mass result and integrated to determine C_1 in terms of dp/dx and $h(x)$:

$$C_1 = -\frac{h^3(x)}{12\mu}\frac{dp(x)}{dx} - \frac{Uh(x)}{2}, \text{ or } \frac{dp(x)}{dx} = -\frac{12\mu}{h^3(x)}C_1 - \frac{6\mu U}{h^2(x)}.$$

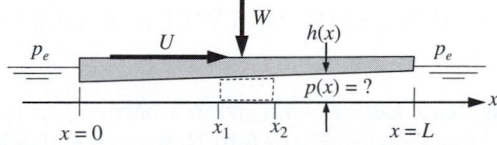

FIGURE 8.9 Schematic drawing of a bearing pad with load W moving above a stationary flat surface coated with viscous oil. The gap below the pad has a mild slope and the pressure ahead, behind, and on top of the bearing pad is p_e.

The second equation is just an algebraic rearrangement of the first, and is a simple first-order differential equation for the pressure that can be integrated using the known $h(x)$ from the problem statement:

$$p(x) = -\frac{12\mu C_1}{h_o^3}\int \frac{dx}{(1 - \alpha x/L)^3} - \frac{6\mu U}{h_o^2}\int \frac{dx}{(1 - \alpha x/L)^2} + C_2 = \frac{6\mu L}{h_o^2 \alpha}\left[\frac{C_1/h_o}{(1 - \alpha x/L)^2} + \frac{U}{1 - \alpha x/L}\right] + C_2.$$

Using the two ends of this extended equality and the pressure conditions $p(x = 0) = p(x = L) = p_e$ produces two algebraic equations that can be solved simultaneously for the constants C_1 and C_2:

$$C_1 = -\left(\frac{1 + \alpha}{2 + \alpha}\right)Uh_o, \text{ and } C_2 = p_e - \frac{6\mu LU}{h_o^2 \alpha}\left(\frac{1}{2 + \alpha}\right).$$

Thus, after some algebra the following pressure distribution is found:

$$p(x) - p_e = \frac{6\mu LU}{h_o^2}\left[\frac{\alpha(x/L)(1 - x/L)}{(2 + \alpha)(1 + \alpha x/L)}\right].$$

However, this distribution may contain some superfluous dependence on α and x, because no approximations have been made regarding the size of α while (8.19) is only valid when $\alpha \ll 1$. Thus, keeping only the linear term in α produces:

$$p(x) - p_e \cong \frac{3\alpha\mu LU}{h_o^2}\left(\frac{x}{L}\right)\left(1 - \frac{x}{L}\right) \text{ for } \alpha \ll 1.$$

The bearing load per unit depth into the page is

$$W = \int_0^L (p(x) - p_e)\,dx = \frac{3\alpha\mu LU}{h_o^2}\int_0^L \left(\frac{x}{L}\right)\left(1 - \frac{x}{L}\right)dx = \frac{\alpha\mu L^2 U}{2h_o^2}.$$

This result shows that larger loads may be carried when the bearing slope, the fluid viscosity, the bearing size, and/or the bearing speed are larger, or the oil passage is smaller. Thus, the lubrication action of this bearing pad as described in this example is stable to load perturbations when the other parameters are held constant; an increase in load will lead to a smaller h_o where the bearing's load-carrying capacity is higher. However, the load-carrying capacity of this bearing goes to zero when α, μ, L, or U go to zero, and this bearing design fails (i.e., W becomes negative so the pad and surface are drawn into contact) when either α or U are negative. Thus, the bearing only works when it moves in the correct direction. A more detailed analysis of this bearing flow is provided in Sherman (1990).

EXAMPLE 8.2: VISCOUS FLOW BETWEEN PARALLEL PLATES (HELE-SHAW 1898)

A viscous fluid flows with velocity $\mathbf{u} = (u, v, w)$ in a narrow gap between stationary parallel plates lying at $z = 0$ and $z = h$ as shown in Figure 8.10. Nonzero x- and y-directed pressure gradients are maintained at the plates' edges, and obstacles or objects of various sizes may be placed between the plates. Using the continuity equation (4.10) and the two horizontal (x, y) and one vertical (z) momentum equations (deduced in Exercise 8.19),

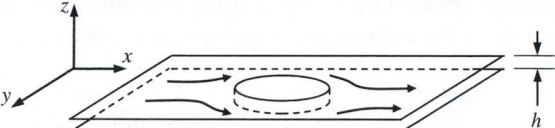

FIGURE 8.10   Pressure-driven viscous flow between parallel plates that trap an obstacle. The gap height h is small compared to the extent of the plates and the extent of the obstacle, shown here as a round disk.

$$0 \cong -\frac{1}{\rho}\frac{\partial p}{\partial x}+\frac{\partial^2 u}{\partial z^2}, \quad 0 \cong -\frac{1}{\rho}\frac{\partial p}{\partial y}+\frac{\partial^2 v}{\partial z^2}, \text{ and } 0 \cong -\frac{1}{\rho}\frac{\partial p}{\partial z},$$

show that the two in-plane velocity components parallel to the plates, u and v, can be determined from the equations for two-dimensional potential flow:

$$u = \frac{\partial \phi}{\partial x} \text{ and } v = \frac{\partial \phi}{\partial y} \text{ with } \frac{\partial^2 \phi}{\partial x^2}+\frac{\partial^2 \phi}{\partial y^2}=0, \qquad\qquad (6.10, 6.12)$$

for an appropriate choice of ϕ.

Solution

The solution plan is to use the two horizontal momemtum equations given above to determine the functional forms of u and v. Then ϕ can be determined via integration of (6.10). Combining these results into (4.10) should produce (6.12), the two-dimensional Laplace equation for ϕ. Integrating the two horizontal momemtum equations twice in the z-direction produces:

$$u \cong \frac{1}{\mu}\frac{\partial p}{\partial x}\frac{z^2}{2}+Az+B, \quad \text{and} \quad v \cong \frac{1}{\mu}\frac{\partial p}{\partial y}\frac{z^2}{2}+Cz+D,$$

where A, B, C, and D are constants that can be determined from the boundary conditions on $y = 0$, $u = v = 0$ which produces $B = D = 0$, and on $y = h$, $u = v = 0$, which produces $A = -(h/2\mu)(\partial p/\partial x)$ and $C = -(h/2\mu)(\partial p/\partial y)$. Thus, the two in-plane velocity components are:

$$u \cong -\frac{1}{2\mu}\frac{\partial p}{\partial x}z(h-z)=\frac{\partial \phi}{\partial x} \quad \text{and} \quad v \cong -\frac{1}{2\mu}\frac{\partial p}{\partial y}z(h-z)=\frac{\partial \phi}{\partial y}.$$

Integrating the second equality in each case produces:

$$\phi = -\frac{z(h-z)}{2\mu}p+E(y) \quad \text{and} \quad \phi = -\frac{z(h-z)}{2\mu}p+F(x).$$

These equations are consistent when $E = F = const.$, and this constant can be set to zero without loss of generality because it does not influence u and v, which are determined from derivatives of ϕ. Therefore, the velocity field requires a potential ϕ of the form:

$$\phi = -\frac{z(h-z)}{2\mu}p.$$

To determine the equation satisfied by p or ϕ, place the results for u and v into the continuity equation (4.10) and integrate in the z-direction from $z = 0$ to h to find:

$$\int_0^h \left(\frac{\partial u}{\partial x} + \frac{\partial v}{\partial y} \right) dz = - \int_0^h \frac{\partial w}{\partial z} dz = -(w)_{z=0}^{z=h} = 0 \rightarrow -\frac{1}{2\mu} \left(\frac{\partial^2 p}{\partial x^2} + \frac{\partial^2 p}{\partial y^2} \right) \int_0^h z(h-z) dz = 0,$$

where the no-through-flow boundary condition ensures $w = 0$ on $y = 0$ and h. The vertical momemtum equation given above requires p to be independent of z, that is, $p = p(x, y, t)$, so p may be taken outside the z integration. The integral of $z(h - z)$ from $z = 0$ to h is not zero, so it and $-1/2\mu$ can be divided out of the last equation to achieve:

$$\frac{\partial^2 p}{\partial x^2} + \frac{\partial^2 p}{\partial y^2} = 0, \text{ or } \frac{\partial^2 \phi}{\partial x^2} + \frac{\partial^2 \phi}{\partial y^2} = 0,$$

where the final equation follows from the form of ϕ determined from the velocity field.

This is a rather unusual and unexpected result because it requires viscous flow between closely spaced parallel plates to produce the same potential-line and streamline patterns as two-dimensional ideal flow. Interestingly, this suggestion is correct, except in thin layers having a thickness of order h near the surface of obstacles where the no-slip boundary condition on the obstacle prevents the tangential-flow slip that occurs in ideal flow. (Hele-Shaw flow near the surface of an obstacle is considered in Exercise 8.34). Thus, two-dimensional, ideal-flow streamlines past an object or obstacle may be visualized by injecting dye into pressure-driven viscous flow between closely spaced glass plates that trap a cross-sectional slice of the object or obstacle. Hele-Shaw flow has practical applications, too. Much, if not all, of the manufacturing design analysis done to create molds and tooling for plastic-forming operations is based on Hele-Shaw flow.

The basic balance of pressure and viscous stresses underlying lubrication theory can be extended to gravity-driven viscous flows by appropriately revising the meaning of the pressure gradient and evaluating the constants A and B in (8.18) for different boundary conditions. Such an extension is illustrated in the next example in two dimensions for gravity-driven flow of magma, paint, or viscous oil over a horizontal surface. Gravity re-enters the formulation here because there is a large density change across the free surface of the viscous fluid (see Section 4.9 "Neglect of Gravity in Constant Density Flows").

EXAMPLE 8.3

A two-dimensional bead of a viscous fluid with density ρ and viscosity μ spreads slowly on a smooth horizontal surface under the action of gravity. Ignoring surface tension and fluid acceleration, determine a differential equation for the thickness $h(x, t)$ of the spreading bead as a function of time.

Solution

The solution plan is to conserve mass exactly using (4.5), a stationary control volume (CV) of thickness dx, height h, and unit depth into the page (see Figure 8.11), and the generic velocity profile (8.18) when the pressure gradient is recast in terms of the thickness gradient $\partial h/\partial x$. The constants A and B in (8.18) can be determined from the no-slip condition at $y = 0$, and a stress-free condition on

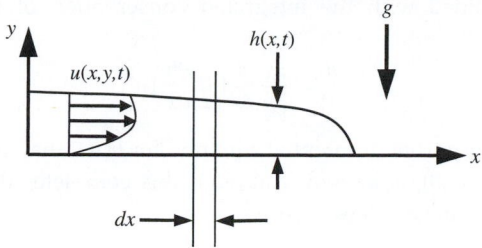

FIGURE 8.11 Gravity-driven spreading of a two-dimensional drop or bead on a flat, stationary surface. The fluid is not confined from above. Hydrostatic pressure forces cause the fluid to move but it is impeded by the viscous shear stress at $y = 0$. The flow is assumed to be symmetric about $x = 0$ so only half of it is shown.

$y = h$. When this refined version of (8.18) is put into the conservation of mass statement, the result is the differential equation that is sought.

By conserving mass between the two vertical lines in Figure 8.11, (4.5) becomes:

$$\rho \frac{\partial h}{\partial t} dx - \int_0^{h(x,t)} \rho u(x,y,t)\, dy + \int_0^{h(x+dx,t)} \rho u(x+dx,y,t)\, dy = 0.$$

When rearranged and the limit $dx \to 0$ is taken, this becomes:

$$\frac{\partial h}{\partial t} + \frac{\partial}{\partial x}\left(\int_0^{h(x,t)} u(x,y,t)\, dy \right) = 0.$$

At any location within the spreading bead, the pressure p is hydrostatic: $p(x, y, t) = \rho g(h(x, t) - y)$ when fluid acceleration is ignored. Thus, the horizontal pressure gradient in the viscous fluid is

$$\frac{\partial p}{\partial x} = \frac{\partial}{\partial x}\left[\rho g(h(x,t) - y) \right] = \rho g \frac{\partial h}{\partial x},$$

which is independent of y, so (8.18) becomes:

$$u(x,y,t) \cong \frac{\rho g}{2\mu} \frac{\partial h(x,t)}{\partial x} y^2 + Ay + B.$$

The no-slip condition at $y = 0$ implies that $B = 0$, and the no-stress condition at $y = h$ implies:

$$0 = \mu \left(\frac{\partial u}{\partial y} \right)_{y=h(x,t)} = \rho g \frac{\partial h(x,t)}{\partial x} h(x,t) + \mu A, \text{ so } A = -\frac{\rho g}{\mu} h \frac{\partial h}{\partial x}.$$

So, the velocity profile within the bead is:

$$u \cong -\frac{\rho g}{2\mu} \frac{\partial h}{\partial x} y(2h - y),$$

and its integral is:

$$\int_0^h u(x,y,t)\, dy \cong -\frac{\rho g}{2\mu} \frac{\partial h}{\partial x} \int_0^h y(2h - y)\, dy = -\frac{\rho g}{3\mu} h^3 \frac{\partial h}{\partial x}.$$

When this result is combined with the integrated conservation of mass statement, the final equation is:

$$\frac{\partial h}{\partial t} = \frac{\rho g}{3\mu} \frac{\partial}{\partial x}\left(h^3 \frac{\partial h}{\partial x}\right).$$

This is a single nonlinear partial differential equation for $h(x,t)$ that in principle can be solved if a bead's initial thickness, $h(x,0)$, is known. Although this completes the effort for this example, a *similarity solution* to this equation does exist.

Similarity solutions to partial differential equations are possible when a variable transformation exists that allows the partial differential equation to be rewritten as an ordinary differential equation, and several such solutions for the Navier-Stokes equations are presented in the next section.

8.4. SIMILARITY SOLUTIONS FOR UNSTEADY INCOMPRESSIBLE VISCOUS FLOW

So far, we have considered steady flows with parallel, or nearly parallel, streamlines. In this situation, the nonlinear advective acceleration is zero, or small, and the stream-wise velocity reduces to a function of one spatial coordinate, and time. When a viscous flow with parallel or nearly parallel streamlines is impulsively started from rest, the flow depends on the spatial coordinate and time. For such unsteady flows, exact solutions still exist because the nonlinear advective acceleration drops out again (see Exercise 8.31). In this section, several simple and physically revealing unsteady flow problems are presented and solved. The first is the flow due to impulsive motion of a flat plate parallel to itself, commonly known as *Stokes' first problem*. (The flow is sometimes unfairly associated with the name of Rayleigh, who used Stokes' solution to predict the thickness of a developing boundary layer on a semi-infinite plate.)

A similarity solution is one of several ways to solve Stokes' first problem. The geometry of this problem is shown in Figure 8.12. An infinite flat plate lies along $y = 0$, surrounded by an initially quiescent fluid (with constant ρ and μ) for $y > 0$. The plate is impulsively given a velocity U at $t = 0$ and constant pressure is maintained at $x = \pm\infty$. At first, only the fluid near the plate will be drawn into motion, but as time progresses the thickness of this moving region will increase. Since the resulting flow at any time is invariant in the x direction ($\partial/\partial x = 0$), the continuity equation $\partial u/\partial x + \partial v/\partial y = 0$ requires $\partial v/\partial y = 0$. Thus, it follows that $v = 0$ everywhere since it is zero at $y = 0$. Therefore, the simplified horizontal and vertical momentum equations are:

$$\rho \frac{\partial u}{\partial t} = -\frac{\partial p}{\partial x} + \mu \frac{\partial^2 u}{\partial y^2}, \quad \text{and} \quad 0 = -\frac{\partial p}{\partial y}.$$

Just before $t = 0$, all the fluid is at rest so $p = $ constant. For $t > 0$, the vertical momentum equation only allows the fluid pressure to depend on x and t. However, at any finite time, there will be a vertical distance from the plate where the fluid velocity is still zero, and, at this

FIGURE 8.12 Laminar flow due to a flat plate that starts moving parallel to itself at speed U at $t = 0$. Before $t = 0$, the entire fluid half-space ($y > 0$) was quiescent. As time progresses, more and more of the viscous fluid above the plate is drawn into motion. Thus, the flow profile with greater vertical extent corresponds to the later time.

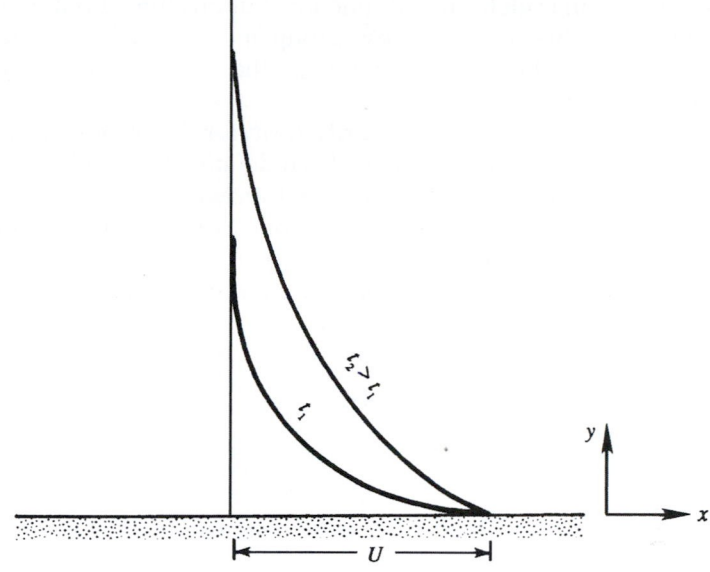

vertical distance from the plate, $\partial p/\partial x$ is zero. However, if $\partial p/\partial x = 0$ far from the plate, then $\partial p/\partial x = 0$ on the plate because $\partial p/\partial y = 0$. Thus, the horizontal momentum equation reduces to

$$\frac{\partial u}{\partial t} = \nu \frac{\partial^2 u}{\partial y^2},\tag{8.20}$$

subject to the boundary and initial conditions:

$$u(y, t = 0) = 0, u(y = 0, t) = \left\{ \begin{array}{l} 0 \text{ for } t < 0 \\ U \text{ for } t \geq 0 \end{array} \right\}, \text{ and } u(y \to \infty, t) = 0. \tag{8.21, 8.22, 8.23}$$

The problem is well posed because (8.22) and (8.23) are conditions at two values of y, and (8.21) is a condition at one value of t; this is consistent with (8.20), which involves a first derivative in t and a second derivative in y.

The partial differential equation (8.20) can be transformed into an ordinary differential equation by switching to a similarity variable. The reason for this is the absence of enough other parameters in this problem to render y and t dimensionless without combining them. Based on dimensional analysis (see Section 1.11), the functional form of the solution to (8.20) can be written:

$$u/U = f(y/\sqrt{\nu t}, y/Ut).\tag{8.24}$$

where f is an undetermined function. However, (8.20) is a linear equation, so u must be proportional to U. This means that the final dimensionless group in (8.24) must be dropped, leaving:

$$u/U = F(y/\sqrt{\nu t}) \equiv F(\eta),\tag{8.25}$$

where F is an undetermined function, but this time it is a function of only one dimensionless group and this dimensionless group $\eta = y/(\nu t)^{1/2}$ combines both independent variables. This reduces the dimensionality of the solution space from two to one, an enormous simplification.

Equation (8.25) is the similarity form for the fluid velocity in Stokes's first problem. The similarity variable η could have been defined differently, such as $\nu t/y^2$, but different choices for η merely change F, not the final answer. The chosen η allows F to be interpreted as a velocity profile function with y appearing to the first power in the numerator of η. At any fixed $t > 0$, y and η are proportional.

Using (8.25) to form the derivatives in (8.20) leads to

$$\frac{\partial u}{\partial t} = U\frac{dF}{d\eta}\frac{\partial \eta}{\partial t} = -\frac{Uy}{2\sqrt{\nu t^3}}\frac{dF}{d\eta} = -\frac{U\eta}{2t}\frac{dF}{d\eta} \quad \text{and}$$

$$U\frac{\partial^2 F}{\partial y^2} = U\frac{\partial}{\partial y}\left(\frac{dF}{d\eta}\frac{\partial \eta}{\partial y}\right) = U\frac{\partial}{\partial y}\left(\frac{1}{\sqrt{\nu t}}\frac{dF}{d\eta}\right) = \frac{U}{\sqrt{\nu t}}\frac{d}{d\eta}\left(\frac{dF}{d\eta}\right)\frac{\partial \eta}{\partial y} = \frac{U}{\nu t}\frac{d}{d\eta}\left(\frac{dF}{d\eta}\right),$$

and these can be combined to provide the equivalent of (8.20) in similarity form:

$$-\frac{\eta}{2}\frac{dF}{d\eta} = \frac{d}{d\eta}\left(\frac{dF}{d\eta}\right). \tag{8.26}$$

The initial and boundary conditions (8.21) through (8.23) for F reduce to

$$F(\eta = 0) = 1, \text{ and } F(\eta \to \infty = 0), \tag{8.27, 8.28}$$

because (8.21) and (8.23) reduce to the same condition in terms of η. This reduction is expected because (8.20) was a partial differential equation and needed two conditions in y and one condition in t. In contrast, (8.26) is a second-order ordinary differential equation and needs only two boundary conditions.

Equation (8.26) is readily separated:

$$-\frac{\eta}{2}d\eta = \frac{d(dF/d\eta)}{dF/d\eta},$$

and integrated:

$$-\frac{\eta^2}{4} = \ln(dF/d\eta) + const.$$

Exponentiating produces:

$$dF/d\eta = A\exp(-\eta^2/4),$$

where A is a constant. Integrating again leads to:

$$F(\eta) = A\int_0^\eta \exp(-\xi^2/4)d\xi + B, \tag{8.29}$$

where ξ is just an integration variable and B is another constant. The condition (8.27) sets $B = 1$, while condition (8.28) gives

$$0 = A \int_0^\infty \exp(-\xi^2/4)d\xi + 1, \text{ or } 0 = 2A \int_0^\infty \exp(-\zeta^2)d\zeta + 1, \text{ so}$$

$$0 = 2A(\sqrt{\pi}/2) + 1, \text{ thus } A = -1/\sqrt{\pi},$$

where the tabulated integral $\int_{-\infty}^{+\infty} \exp(-\zeta^2)d\zeta = \sqrt{\pi}$ has been used. The final solution for u then becomes:

$$\frac{u(y,t)}{U} = 1 - \mathrm{erf}\left(\frac{y}{2\sqrt{vt}}\right), \text{ where } \mathrm{erf}(\zeta) = \frac{2}{\sqrt{\pi}} \int_0^\zeta \exp(-\xi^2)d\xi \qquad (8.30)$$

is the error function and again ξ is just an integration variable. The error function is a standard tabulated function (see Abramowitz & Stegun, 1972). It is apparent that *the solutions at different times all collapse into a single curve of u/U vs η, as shown in Figure 8.13.*

The nature of the variation of u/U with y for various values of t is sketched in Figure 8.12. The solution clearly has a diffusive nature. At $t = 0$, a vortex sheet (that is, a velocity discontinuity) is created at the plate surface. The initial vorticity is in the form of a delta function, which is infinite at the plate surface and zero elsewhere. The integral $\int_0^\infty \omega dy = \int_0^\infty (-\partial u/\partial y)dy = -U$ is independent of time, so *no new vorticity is generated after the initial time.* The flow given by (8.30) occurs as the initial vorticity diffuses away from the wall. The situation is analogous to a heat conduction problem in a semi-infinite solid extending from $y = 0$ to $y = \infty$. Initially, the solid has a uniform temperature, and at $t = 0$ the face at $y = 0$ is suddenly brought to a different temperature. The temperature distribution for this heat conduction problem is given by an equation similar to (8.30).

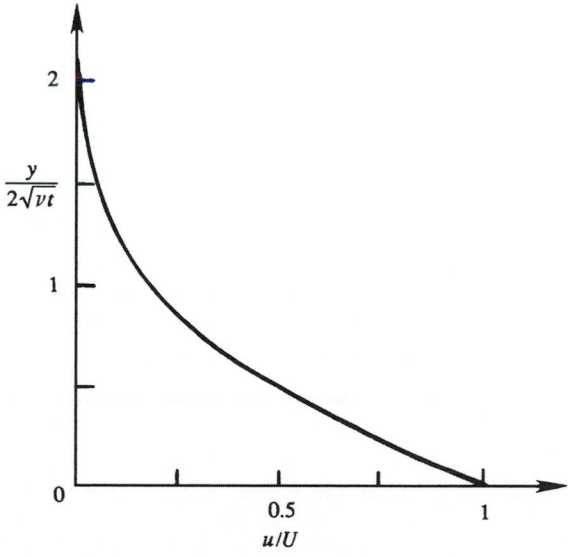

FIGURE 8.13 Similarity solution of laminar flow due to an impulsively started flat plate. Using these scaled coordinates, all flow profiles like those shown on Figure 8.12 will collapse to the same curve. The factor of two in the scaling of the vertical axis follows from (8.30).

We may arbitrarily define the thickness of the diffusive layer as the distance at which u falls to 1% of U. From Figure 8.13, $u/U = 0.01$ corresponds to $y/(\nu t)^{1/2} = 3.64$. Therefore, in time t the diffusive effects propagate to a distance of

$$\delta_{99} \sim 3.64\sqrt{\nu t}, \tag{8.31}$$

which defines the 99% thickness of the layer of moving fluid and this layer's thickness increases as $t^{1/2}$. Obviously, the factor of 3.64 is somewhat arbitrary and can be changed by choosing a different ratio of u/U as the definition for the edge of the diffusive layer. However, 99% thicknesses are commonly considered in boundary layer theory (see Chapter 9).

Stokes' first problem illustrates an important class of fluid mechanical problems that have *similarity solutions*. Because of the absence of suitable scales to render the independent variables dimensionless, the only possibility was a combination of variables that resulted in a reduction in the number of independent variables required to describe the problem. In this case the reduction was from two (y, t) to one (η) so that the formulation reduced a partial differential equation in y and t to an ordinary differential equation in η.

The solution (8.30) for $u(y, t)$ is *self-similar* in the sense that at different times $t_1, t_2, t_3, \ldots$ the various velocity profiles $u(y, t_1), u(y, t_2), u(y, t_3), \ldots$ all collapse into a single curve if the velocity is scaled by U and y is scaled by the thickness $(\nu t)^{1/2}$. Moreover, such a collapse will occur for different values of U and for fluids having different ν.

Similarity solutions arise in situations in which there are no imposed length or time scales provided by the initial or boundary conditions (or the field equation). A similarity solution would not be possible if, for example, the boundary conditions were changed after a certain time t_1 since this introduces a time scale into the problem (see Exercise 8.30). Likewise, if the flow in Stokes' first problem was bounded above by a second parallel plate, there could be no similarity solution because the distance to the second plate introduces a length scale into the problem.

Similarity solutions are often ideal for developing an understanding of flow phenomena, so they are sought wherever possible. A method for finding similarity solutions starts from a presumed form for the solution:

$$\gamma = At^{-n}F(\xi/\delta(t)) \equiv At^{-n}F(\eta) \text{ or } \gamma = A\xi^{-n}F(\xi/\delta(t)) \equiv A\xi^{-n}F(\eta), \tag{8.32a,b}$$

where γ is the dependent field variable of interest, A is a constant (units $= [\gamma] \times [\text{time}]^n$ or $[\gamma] \times [\text{length}]^n$), ξ is the independent spatial coordinate, t is time, and $\eta = \xi/\delta$ is the similarity variable, and $\delta(t)$ is a time-dependent length scale. The factor of At^{-n} or $A\xi^{-n}$ that multiplies F in (8.32) is sometimes needed for similarity solutions that are infinite (or zero) at $t = 0$ or $\xi = 0$. Use of (8.32) is illustrated in the following examples.

EXAMPLE 8.4

Use (8.32a) to find the similarity solution to Stokes' first problem.

Solution

The solution plan is to populate (8.32a) with the appropriate variables, substitute it into the field equation (8.20), and then require that the coefficients all have the same time dependence. For Stokes' first problem $\gamma = u/U$, and the independent spatial variable is y. For this flow, the coefficient At^{-n} is not needed since $u/U = 1$ at $\eta = 0$ for all $t > 0$ and this can only happen when $A = 1$ and $n = 0$. Thus, the dimensional analysis result (8.25) may be replaced by (8.32a) with $A = 1$, $n = 0$, and $\xi = y$:

$$u/U = F(y/\delta(t)) \equiv F(\eta).$$

A time derivative produces:

$$\frac{\partial u}{\partial t} = U\frac{dF}{d\eta}\left(-\frac{y}{\delta^2}\right)\frac{d\delta}{dt},$$

while two y-derivatives produce:

$$\frac{\partial^2 u}{\partial y^2} = U\frac{d^2F}{d\eta^2}\frac{1}{\delta^2}.$$

Reconstructing (8.20) with these replacements yields:

$$\frac{\partial u}{\partial t} = \nu\frac{\partial^2 u}{\partial y^2} \rightarrow U\frac{dF}{d\eta}\left(-\frac{y}{\delta^2}\right)\frac{d\delta}{dt} = \nu U\frac{d^2F}{d\eta^2}\frac{1}{\delta^2},$$

which can be rearranged to find:

$$-\left[\frac{1}{\delta}\frac{d\delta}{dt}\right]\eta\frac{dF}{d\eta} = \left[\frac{\nu}{\delta^2}\right]\frac{d^2F}{d\eta^2}.$$

For this equation to be in similarity form, the coefficients in [,]-brackets must both have the same time dependence so that division by this common time dependence will leave an ordinary differential equation for $F(\eta)$ and t will no longer appear. Thus, we require the two coefficients to be proportional:

$$\frac{1}{\delta}\frac{d\delta}{dt} = C_1\frac{\nu}{\delta^2},$$

where C_1 is the constant of proportionality. This is a simple differential equation for $\delta(t)$ that is readily rearranged and solved:

$$\delta\frac{d\delta}{dt} = C_1\nu \rightarrow \frac{\delta^2}{2} = C_1\nu t + C_2 \rightarrow \delta = \sqrt{2C_1\nu t},$$

where the condition $\delta(0) = 0$ has been used to determine that the constant of integration $C_2 = 0$. When $C_1 = \frac{1}{2}$, the prior definition of η in (8.25) is recovered, and the solution for u proceeds as before (see (8.26) through (8.30)).

EXAMPLE 8.5

At $t = 0$ an infinitely thin vortex sheet in a fluid with density ρ and viscosity μ coincides with the plane defined by $y = 0$, so that the fluid velocity is U for $y > 0$ and $-U$ for $y < 0$. The coordinate axes are aligned so that only the z-component of vorticity is nonzero. Determine the similarity solution for $\omega_z(y, t)$ for $t > 0$.

Solution

The solution plan is the same as for Example 8.4, except here the coefficient At^{-n} must be included. In this circumstance, there will be only one component of the fluid velocity, $\mathbf{u} = u(y, t)\mathbf{e}_x$, so $\omega_z(y, t) = -\partial u/\partial y$. The independent coordinate y does not appear in the initial condition, so (8.32a) is the preferred choice. Its appropriate form is:

$$\omega_z(y, t) = At^{-n}F(y/\delta(t)) \equiv At^{-n}F(\eta),$$

and the field equation,

$$\frac{\partial \omega_z}{\partial t} = \nu \frac{\partial^2 \omega_z}{\partial y^2},$$

is obtained by applying $\partial/\partial y$ to (8.20). Here, the derivatives of the similarity solution are:

$$\frac{\partial \omega_z}{\partial t} = -nAt^{-n-1}F(\eta) + At^{-n}\frac{dF}{d\eta}\left(-\frac{y}{\delta^2}\right)\frac{d\delta}{dt'}, \text{ and } \frac{\partial^2 \omega_z}{\partial y^2} = At^{-n}\frac{d^2F}{d\eta^2}\frac{1}{\delta^2}.$$

Reassembling the field equation and canceling common factors produces:

$$-\left[\frac{n}{t}\right]F(\eta) - \left[\frac{1}{\delta}\frac{d\delta}{dt}\right]\eta\frac{dF}{d\eta} = \left[\frac{\nu}{\delta^2}\right]\frac{d^2F}{d\eta^2}.$$

From Example 8.4, we know that requiring the second and third coefficients in [,]-brackets to be proportional with a proportionality constant of ½ produces $\delta = (\nu t)^{1/2}$. With this choice for δ, each of the coefficients in [,]-brackets is proportional to $1/t$ so, the similarity equation becomes:

$$-nF(\eta) - \frac{1}{2}\eta\frac{dF}{d\eta} = \frac{d^2F}{d\eta^2}.$$

The boundary conditions are: 1) at any finite time the vorticity must go to zero infinitely far from the initial location of the vortex sheet, $F(\eta) \rightarrow 0$ for $\eta \rightarrow \infty$, and 2) the velocity difference across the diffusing vortex sheet is constant and equal to $2U$:

$$-\int_{-\infty}^{+\infty} \omega_z \, dy = \int_{-\infty}^{+\infty} \frac{\partial u}{\partial y} \, dy = [u(y, t)]_{-\infty}^{+\infty} = U - (-U) = 2U.$$

Substituting the similarity solution into this second requirement leads to:

$$-\int_{-\infty}^{+\infty} \omega_z \, dy = -\int_{-\infty}^{+\infty} At^{-n}F(\eta) \, dy = -At^{-n}\delta\int_{-\infty}^{+\infty} F(\eta)d(y/\delta) = -At^{-n}\delta\int_{-\infty}^{+\infty} F(\eta)d\eta = 2U.$$

The final integral is just a number so $t^{-n}\delta(t)$ must be constant, and this implies $n = ½$ so the similarity equation may be rewritten, and integrated:

$$-\frac{1}{2}\left(F(\eta) + \eta\frac{dF}{d\eta}\right) = -\frac{1}{2}\frac{d}{d\eta}(\eta F) = \frac{d}{d\eta}\left(\frac{dF}{d\eta}\right) \rightarrow \frac{dF}{d\eta} + \frac{1}{2}\eta F = C.$$

The first boundary condition implies that both F and $dF/d\eta \to 0$ when η is large enough. Therefore, assume that $\eta F \to 0$ when $\eta \to \infty$ so that the constant of integration C can be set to zero (this assumption can be checked once F is found). When $C = 0$, the last equation can be separated and integrated to find:

$$F(\eta) = D \exp(-\eta^2/4),$$

where D is a constant, and the assumed limit, $\eta F \to 0$ when $\eta \to \infty$, is verified so C is indeed zero. The velocity-difference constraint and the tabulated integral used to reach (8.30) allow the product AD to be evaluated. Thus, the similarity solutions for the vorticity $\omega_z = -\partial u/\partial y$ and velocity u are:

$$\omega_z(y,t) = -\frac{U}{\sqrt{\pi \nu t}} \exp\left\{-\frac{y^2}{4\nu t}\right\}, \text{ and } u(y,t) = U \text{erf}\left\{\frac{y}{2\sqrt{\nu t}}\right\}.$$

Schematic plots of the vorticity and velocity distributions are shown in Figure 8.14. If we define the width of the velocity transition layer as the distance between the points where $u = \pm 0.95U$, then the corresponding values of η are ± 2.76 and consequently the width of the transition layer is $5.54(\nu t)^{1/2}$.

The results of this example are closely related to Stokes' first problem, and to the laminar boundary layer flows discussed in the next chapter, for several reasons. First of all, this flow is essentially the same as that in Stokes' first problem. The velocity field in the upper half of Figure 8.14 is identical to that in Figure 8.13 after a Galilean transformation to a coordinate system

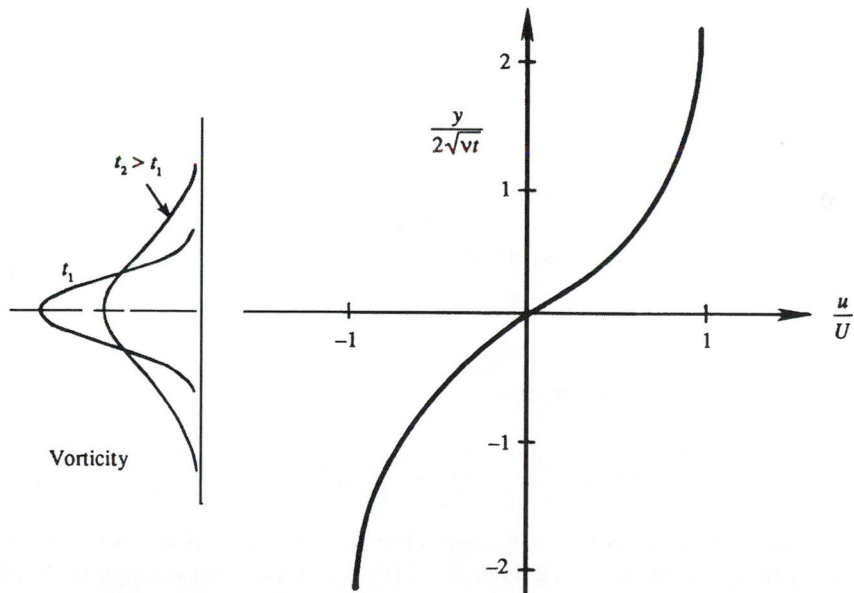

FIGURE 8.14 Viscous thickening of a vortex sheet. The left panel indicates the vorticity distribution at two times, while the right panel shows the velocity field solution in similarity coordinates. The upper half of this flow is equivalent to a temporally developing boundary layer.

moving at speed $+U$ followed by a sign change. In addition, the flow for $y > 0$ represents a *temporally developing* boundary layer that begins at $t = 0$. The velocity far from the surface is irrotational and uniform at speed U while the no-slip condition ($u = 0$) is satisfied at $y = 0$. Here, the wall shear stress and skin friction coefficient C_f are time dependent:

$$\tau_w = \mu\left(\frac{\partial u}{\partial y}\right)_{y=0} = \frac{\mu U}{\sqrt{\pi\nu t}}, \text{ or } C_f = \frac{\tau_w}{\frac{1}{2}\rho U^2} = \frac{2}{\sqrt{\pi}}\sqrt{\frac{\nu}{U^2 t}}.$$

When Ut is interpreted as a surrogate for the downstream distance, x, in a *spatially developing* boundary layer, the last square-root factor above becomes $(\nu/Ux)^{1/2} = \text{Re}_x^{-1/2}$, and this is the correct parametric dependence for C_f in a laminar boundary layer that develops on a smooth, flat surface below a steady uniform flow.

EXAMPLE 8.6

A thin, rapidly spinning cylinder produces the two-dimensional flow field, $u_\theta = \Gamma/2\pi r$, of an ideal vortex of strength Γ located at $r = 0$. At $t = 0$, the cylinder stops spinning. Use (8.32) to determine $u_\theta(r, t)$ for $t > 0$.

Solution

Follow the approach specified for the Example 8.4 but this time use (8.32b) because r appears in the initial condition. Here u_θ is the dependent field variable and r is the independent spatial variable, so the appropriate form of (8.32b) is:

$$u_\theta(r,t) = Ar^{-n}F(r/\delta(t)) \equiv Ar^{-n}F(\eta).$$

The initial and boundary conditions are: $u_\theta(r, 0) = \Gamma/2\pi r = u_\theta(r \to \infty, t)$, and $u_\theta(0, t) = 0$ for $t > 0$, which are simplified to $F(\eta \to \infty) = 1$ and $F(0) = 0$ when Ar^{-n} is set equal to $\Gamma/2\pi r$. In this case, the field equation for u_θ (see Appendix B) is

$$\frac{\partial u_\theta}{\partial t} = \nu\frac{\partial}{\partial r}\left(\frac{1}{r}\frac{\partial}{\partial r}(ru_\theta)\right).$$

Inserting $u_\theta = (\Gamma/2\pi r)F(\eta)$ produces:

$$-\frac{\Gamma}{2\pi r}\left(\frac{1}{\delta}\frac{d\delta}{dt}\right)\eta\frac{dF}{d\eta} = \nu\frac{\Gamma}{2\pi}\left(-\frac{1}{\delta r^2}\frac{dF}{d\eta} + \frac{1}{\delta^2 r}\frac{d^2F}{d\eta^2}\right) \to -\left[\frac{r^2}{\nu\delta}\frac{d\delta}{dt}\right]\eta\frac{dF}{d\eta} = -\eta\frac{dF}{d\eta} + \eta^2\frac{d^2F}{d\eta^2}.$$

For a similarity solution, the coefficient in [,]-brackets must depend on η alone, not on r or t. Here, this coefficient reduces to $\eta^2/2$ when $\delta = (\nu t)^{1/2}$ (as in the prior examples). With this replacement, the similarity equation can be integrated twice:

$$\left(\frac{1}{\eta} - \frac{\eta}{2}\right)\frac{dF}{d\eta} = \frac{d}{d\eta}\left(\frac{dF}{d\eta}\right) \to \ln\eta - \frac{\eta^2}{4} + const. = \ln\left(\frac{dF}{d\eta}\right) \to C\int\eta\exp\{-\eta^2/4\}d\eta + D = F(\eta).$$

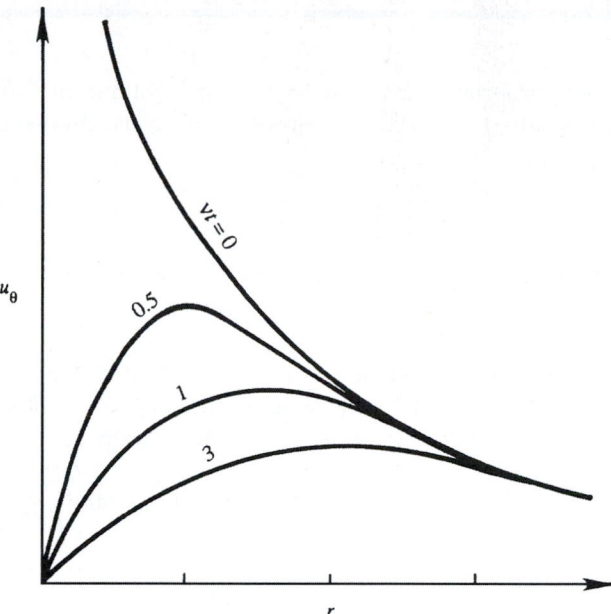

FIGURE 8.15 Viscous decay of a line vortex showing the tangential velocity u_θ at different times. The velocity field nearest to the axis of rotation changes the most quickly. At large radii, flow alterations occur more slowly.

The remaining integral is elementary, and the boundary conditions given above for F allow the constants C and D to be evaluated. The final result is $F(\eta) = 1 - \exp\{-\eta^2/4\}$, so the velocity distribution is:

$$u_\theta(r,t) = \frac{\Gamma}{2\pi r}\left[1 - \exp\left\{-\frac{r^2}{4\nu t}\right\}\right],$$

which is identical to the Gaussian vortex of (3.29) when $\sigma^2 = 4\nu t$. A sketch of the velocity distribution for various values of t is given in Figure 8.15. Near the center, $r \ll (\nu t)^{1/2}$, the flow has the form of rigid-body rotation, while in the outer region, $r \gg (\nu t)^{1/2}$, the motion has the form of an ideal vortex.

The foregoing presentation applies to the *decay* of a line vortex. The case where a line vortex is suddenly *introduced* into a fluid at rest leads to the velocity distribution,

$$u_\theta(r,t) = \frac{\Gamma}{2\pi r}\exp\left\{-\frac{r^2}{4\nu t}\right\}$$

(see Exercise 8.26). This situation is equivalent to the impulsive rotational start of an infinitely thin and quickly rotating cylinder.

EXAMPLE 8.7

Use (8.32a) and an appropriate constraint on the total volume of fluid to determine the form of the similarity solution to the two-dimensional, viscous, drop-spreading equation of Example 8.3.

Solution

The solution plan is to populate (8.32a) with the appropriate variables,

$$h = At^{-n}F(x/\delta(t)) \equiv At^{-n}F(\eta),$$

substitute it into the equation from Example 8.3, and require that: 1) the coefficients all have the same time dependence, and 2) the total fluid volume per unit depth into the page, $\int_{-\infty}^{+\infty} h(x,t)dx$, is independent of time. The starting point is the evaluation of derivatives:

$$\frac{\partial h}{\partial t} = -nAt^{-n-1}F(\eta) + At^{-n}\frac{dF}{d\eta}\left(-\frac{x}{\delta^2}\right)\frac{d\delta}{dt}, \text{ and } \frac{\partial h}{\partial x} = At^{-n}\frac{dF}{d\eta}\left(\frac{1}{\delta}\right),$$

which, when inserted in the final equation of Example 8.3, produces:

$$\frac{\partial h}{\partial t} = -\left[nAt^{-n-1}\right]F(\eta) - \left[At^{-n}\frac{1}{\delta}\frac{d\delta}{dt}\right]\eta\frac{dF}{d\eta} = \left[\frac{\rho g}{3\mu}A^4t^{-4n}\frac{1}{\delta^2}\right]\left(3F^2\left(\frac{dF}{d\eta}\right)^2 + F^3\frac{d^2F}{d\eta^2}\right) = \frac{\rho g}{3\mu}\frac{\partial}{\partial x}\left(h^3\frac{\partial h}{\partial x}\right).$$

Requiring proportionality between the first two coefficients in [,]-brackets with C as the constant of proportionality yields:

$$CnAt^{-n-1} = At^{-n}\frac{1}{\delta}\frac{d\delta}{dt} \rightarrow C\frac{n}{t} = \frac{1}{\delta}\frac{d\delta}{dt}.$$

The second equation is satisfied when $\delta = Dt^m$ where D is another constant and $m = Cn$. Requiring proportionality between the second and third coefficients and using $\delta = Dt^m$ produces:

$$EAt^{-n}\frac{1}{\delta}\frac{d\delta}{dt} = \frac{\rho g}{3\mu}A^4t^{-4n}\frac{1}{\delta^2} \rightarrow -E\frac{m}{t} = \frac{\rho g}{3\mu}A^3t^{-3n}D^2t^{-2m} \rightarrow -1 = -3n - 2m,$$

where E is another constant of proportionality; the final equation for the exponents follows from equating powers of t in the second equation. These results set the form of $\delta(t)$ and specify one relationship between n and m. A second relationship between m and n comes from conserving the volume per unit depth into the page:

$$\int\limits_{-\infty}^{+\infty} h(x,t)dx = \int\limits_{-\infty}^{+\infty} At^{-n}F(\eta)dx = At^{-n}Dt^m\int\limits_{-\infty}^{+\infty} F(\eta)d\eta = const.$$

The final integral is just a number so the exponents of t outside this integral must sum to zero for the volume to be constant. This implies: $-n + m = 0$. Taken together, the two equations for m and n imply: $n = m = \frac{1}{5}$. Thus, the form of the similarity solution of the final equation of Example 8.2 is:

$$h(x,t) = At^{-1/5}F(x/Dt^{1/5}).$$

Determining the constants A and D requires solution of the equation for F and knowledge of the bead's volume per unit depth, and is beyond the scope of this example.

After reviewing these examples, it should be clear that diffusive length scales in unsteady viscous flow are proportional to $(vt)^{1/2}$. The viscous bead-spreading example produces a length scale with a different power, but this is not a diffusion time scale. Instead it is an advection time scale that specifies how far fluid elements travel in the direction of the flow.

8.5. FLOW DUE TO AN OSCILLATING PLATE

The unsteady flows discussed in the preceding sections have similarity solutions, because there were no imposed or specified length or time scales. The flow discussed here is an unsteady viscous flow that includes an imposed time scale.

Consider an infinite flat plate lying at $y = 0$ that executes sinusoidal oscillations parallel to itself. (This is sometimes called *Stokes' second problem*.) Here, only the steady periodic solution after the starting transients have died out is considered, thus there are no initial conditions to satisfy. The governing equation (8.20) is the same as that for Stokes' first problem. The boundary conditions are:

$$u(y = 0, t) = U\cos(\omega t), \text{ and } u(y \to \infty, t) = \text{bounded}, \qquad (8.33, 8.34)$$

where ω is the oscillation frequency (rad./s). In the steady state, the flow variables must have a periodicity equal to the periodicity of the boundary motion. Consequently, a complex separable solution of the form,

$$u(y, t) = \text{Re}\{e^{i\omega t}f(y)\}, \qquad (8.35)$$

is used here, and the specification of the real part is dropped until the final equation for u is reached. Substitution of (8.33) into (8.20) produces:

$$i\omega f = v(d^2f/dy^2), \qquad (8.36)$$

which is an ordinary differential equation with constant coefficients. It has exponential solutions of the form: $f = \exp(ky)$ where $k = (i\omega/v)^{1/2} = \pm(i + 1)(\omega/2v)^{1/2}$. Thus, the solution of (8.36) is

$$f(y) = A \exp\{-(i+1)y\sqrt{\omega/2v}\} + B \exp\{+(i+1)y\sqrt{\omega/2v}\}. \qquad (8.37)$$

The condition (8.34) requires that the solution must remain bounded as $y \to \infty$, so $B = 0$, and the complex solution only involves the first term in (8.37). The surface boundary condition (8.33) requires $A = U$. Thus, after taking the real part of (8.37), the final velocity distribution for Stokes' second problem is

$$u(y, t) = U \exp\left\{-y\sqrt{\frac{\omega}{2v}}\right\}\cos\left(\omega t - y\sqrt{\frac{\omega}{2v}}\right). \qquad (8.38)$$

The cosine factor in (8.38) represents a dispersive wave traveling in the positive-y direction, while the exponential term represents amplitude decay with increasing y. The flow therefore

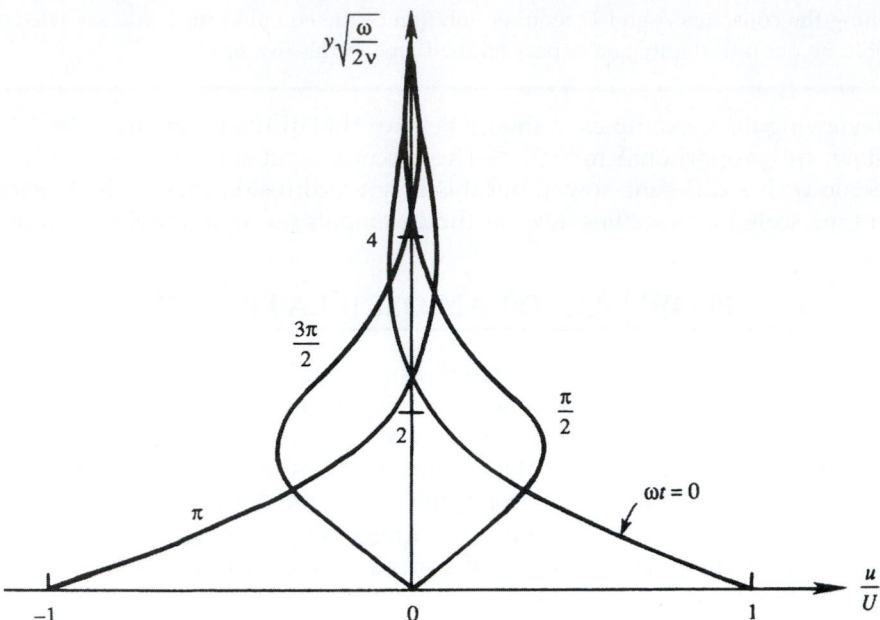

FIGURE 8.16 Velocity distribution in laminar flow near an oscillating plate. The distributions at $\omega t = 0$, $\pi/2$, π, and $3\pi/2$ are shown. The diffusive distance is of order $\delta \sim 4(\nu/\omega)^{1/2}$.

resembles a highly damped transverse wave (Figure 8.16). However, this is a diffusion problem and *not* a wave-propagation problem because there are no restoring forces involved here. The apparent propagation is merely a result of the oscillating boundary condition. For $y = 4(\nu/\omega)^{1/2}$, the amplitude of u is $U\exp\{-4/\sqrt{2}\} = 0.06U$, which means that the influence of the wall is confined within a distance of order $\delta \sim 4(\nu/\omega)^{1/2}$, which decreases with increasing frequency.

The solution (8.38) has several interesting features. First of all, it cannot be represented by a single curve in terms of dimensionless variables. A dimensional analysis of Stokes' second problem produces three dimensionless groups: u/U, ωt, and $y(\omega/\nu)^{1/2}$. Here the independent spatial variable y can be fully separated from the independent time variable t. Self-similar solutions exist only when the independent spatial and temporal variables must be combined in the absence of imposed time or length scales. However, the fundamental concept associated with viscous diffusion holds true, the spatial extent of the solution is parameterized by $(\nu/\omega)^{1/2}$, the square root of the product of the kinematic viscosity, and the imposed time scale $1/\omega$. In addition, (8.38) can be used to predict the weak absorption of sound at solid flat surfaces.

8.6. LOW REYNOLDS NUMBER VISCOUS FLOW PAST A SPHERE

Many physical problems can be described by the behavior of a system when a certain parameter is either very small or very large. Consider the problem of steady constant-density flow of a viscous fluid at speed U around an object of size L. The governing equations will be (4.10) and the steady flow version of (8.1):

$$\rho \mathbf{u} \cdot \nabla \mathbf{u} + \nabla p = \mu \nabla^2 \mathbf{u}. \tag{8.39}$$

As described in Chapter 4.11, this equation can be scaled to determine which terms are most important. The purpose of such a scaling is to generate dimensionless terms that are of order unity in the flow field. For example, when the flow speeds are high and the viscosity is small, the pressure and inertia forces dominate the momentum balance, showing that pressure changes are of order ρU^2. Consequently, for high Reynolds number, the scaling (4.100) is appropriate for nondimensionalizing (8.39) to obtain

$$\mathbf{u}^* \cdot \nabla^* \mathbf{u}^* + \nabla^* p^* = \frac{1}{\text{Re}} \nabla^{*2} \mathbf{u}^*, \tag{8.40}$$

where $\text{Re} = \rho U L / \mu$ is the Reynolds number. For $\text{Re} \gg 1$, (8.40) may be solved by treating $1/\text{Re}$ as a small parameter, and as a first approximation, $1/\text{Re}$ may be set to zero everywhere in the flow, which reduces (8.40) to the inviscid Euler equation without a body force.

However, viscous effects may still be felt at high Re because a single length scale is typically inadequate to describe all regions of high-Re flows. For example, complete omission of the viscous term cannot be valid near a solid surface because the inviscid flow cannot satisfy the no-slip condition at the body surface. Viscous forces are important near solid surfaces because of the high shear in the boundary layer near the body surface. The scaling (4.100), which assumes that velocity gradients are proportional to U/L, is invalid in such boundary layers. Thus, there is a region of *nonuniformity* near the body where a perturbation expansion in terms of the small parameter $1/\text{Re}$ becomes *singular*. The proper scaling in the *boundary layer* and a procedure for analyzing wall-bounded high Reynolds number flows will be discussed in Chapter 9. A hint of what is to come is provided by the scaling (8.14), which leads to the lubrication approximation and involves different-length scales for the stream-wise and cross-stream flow directions.

Now consider flows in the opposite limit of very low Reynolds numbers, $\text{Re} \to 0$. Clearly such flows should have negligible inertia forces, with pressure and viscous forces providing the dominant balance. Therefore, multiply (8.40) by Re to obtain

$$\text{Re}(\mathbf{u}^* \cdot \nabla^* \mathbf{u}^* + \nabla^* p^*) = \nabla^{*2} \mathbf{u}^*. \tag{8.41}$$

Although this equation does have negligible inertia terms as $\text{Re} \to 0$, it does not lead to a balance of pressure and viscous forces as $\text{Re} \to 0$ since it reduces to $0 = \mu \nabla^2 \mathbf{u}$, which is not the proper governing equation for low Reynolds number flows. The source of the inadequacy is the scaling of the pressure term specified by (4.100). For low Reynolds number flows, pressure is *not* of order ρU^2. Instead, at low Re, pressure differences should be scaled with a generic viscous stress such as $\mu \partial u / \partial y \sim \mu U / L$. Thus, the pressure scaling $p^* = (p - p_\infty)/\rho U^2$ in (4.100) should be replaced by $p^* = (p - p_\infty)L/\mu U$, and this leads to a correctly revised version of (8.41)

$$\text{Re}(\mathbf{u}^* \cdot \nabla^* \mathbf{u}^*) = -\nabla^* p^* + \nabla^{*2} \mathbf{u}^*, \tag{8.42}$$

which does exhibit the proper balance of terms as $\text{Re} \to 0$ and becomes the linear (dimensional) equation

$$\nabla p = \mu \nabla^2 \mathbf{u}, \tag{8.43}$$

when this limit is taken.

Flows with Re $\to$ 0 are called *creeping* flows, and they occur at low flow speeds of viscous fluids past small objects or through narrow passages. Examples of such flows are the motion of a thin film of oil in the bearing of a shaft, settling of sediment particles in nominally quiescent water, the fall of mist droplets in the atmosphere, or the flow of molten plastic during a molding process. A variety of other creeping flow examples are presented in Sherman (1990).

From this discussion of scaling, we conclude that *the proper length and time scales depend on the nature and the region of the flow, and are obtained by balancing the terms that are most important in the region of the flow field under consideration.* Identifying the proper length and time scales is commonly the goal of experimental and numerical investigations of viscous flows, so that the most appropriate simplified versions of the full equations for fluid motion may be analyzed. The remainder of this section presents a solution for the creeping flow past a sphere, first given by Stokes in 1851. This is a flow where different field equations should be used in regions close to and far from the sphere.

We begin by considering the near-field flow around a stationary sphere of radius a placed in a uniform stream of speed U (Figure 8.17) with Re $\to$ 0. The problem is axisymmetric, that is, the flow patterns are identical in all planes parallel to U and passing through the center of

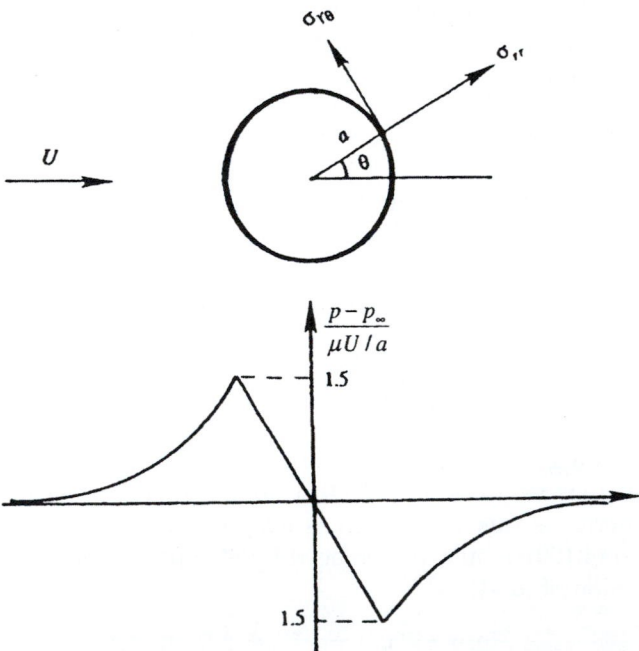

FIGURE 8.17 Creeping flow over a sphere. The upper panel shows the viscous stress components at the surface. The lower panel shows the pressure distribution in an axial ($\varphi =$ const.) plane.

the sphere. Since Re $\to$ 0, as a first approximation, neglect the inertia forces altogether and seek a solution to (8.43). Taking the curl of (8.43) produces an equation for the vorticity alone

$$\nabla^2 \boldsymbol{\omega} = 0,{}^1$$

because $\nabla \times \nabla p = 0$ and the order of the operators curl and ∇^2 can be interchanged. (The reader may verify this using indicial notation.) The only component of vorticity in this axisymmetric problem is ω_φ, the component perpendicular to $\varphi = $ constant planes in Figure 8.17, and it is given by

$$\omega_\varphi = \frac{1}{r}\left[\frac{\partial(ru_\theta)}{\partial r} - \frac{\partial u_r}{\partial \theta}\right].$$

This is an axisymmetric flow, so the r and θ velocity components can be found from an axisymmetric stream function ψ:

$$u_r = \frac{1}{r^2 \sin \theta}\frac{\partial \psi}{\partial \theta}, \text{ and } u_\theta = -\frac{1}{r \sin \theta}\frac{\partial \psi}{\partial r}. \tag{6.83}$$

In terms of this stream function, the vorticity becomes

$$\omega_\varphi = -\frac{1}{r}\left[\frac{1}{\sin \theta}\frac{\partial^2 \psi}{\partial r^2} + \frac{1}{r^2}\frac{\partial}{\partial \theta}\left(\frac{1}{\sin\theta}\frac{\partial \psi}{\partial \theta}\right)\right],$$

which is governed by:

$$\nabla^2 \omega_\varphi = 0.{}^1$$

Combining the last two equations, we obtain

$$\left[\frac{\partial^2}{\partial r^2} + \frac{\sin \theta}{r^2}\frac{\partial}{\partial \theta}\left(\frac{1}{\sin \theta}\frac{\partial}{\partial \theta}\right)\right]^2 \psi = 0.{}^2 \tag{8.44}$$

The boundary conditions on the preceding equation are

$$\psi(r = a, \theta) = 0 \ [u_r = 0 \text{ at surface}], \tag{8.45}$$

$$\partial\psi(r = a, \theta)/\partial r = 0 \ [u_\theta = 0 \text{ at surface}], \text{ and} \tag{8.46}$$

$$\psi(r \to \infty, \theta) = \frac{1}{2}Ur^2 \sin^2 \theta \ [\text{uniform flow far from the sphere}]. \tag{8.47}$$

The last condition follows from the fact that the stream function for a uniform flow is ½$Ur^2\sin^2\theta$ in spherical coordinates (see (6.86)).

The far-field condition (8.47) suggests a separable solution of the form

$$\psi(r, \theta) = f(r) \sin^2 \theta.$$

[1]In spherical polar coordinates, the operator in the footnoted equations is actually $-\nabla \times \nabla \times$ (−curl curl_), which is different from the Laplace operator defined in Appendix B.

[2]Equation (8.44) is the square of the operator, and not the biharmonic.

Substitution of this into the governing equation (8.44) gives

$$f^{iv} - \frac{4f''}{r^2} + \frac{8f'}{r^3} - \frac{8f}{r^4} = 0,$$

whose solution is

$$f = Ar^4 + Br^2 + Cr + D/r.$$

The far-field boundary condition (8.47) requires that $A = 0$ and $B = U/2$. The surface boundary conditions then give $C = -3Ua/4$ and $D = Ua^3/4$. The solution then reduces to

$$\psi = Ur^2 \sin^2 \theta \left(\frac{1}{2} - \frac{3a}{4r} + \frac{a^3}{4r^3} \right). \tag{8.48}$$

The velocity components are found from (8.48) using (6.83):

$$u_r = U \cos \theta \left(1 - \frac{3a}{2r} + \frac{a^3}{2r^3} \right), \text{ and } u_\theta = - U \sin \theta \left(1 - \frac{3a}{4r} - \frac{a^3}{4r^3} \right). \tag{8.49}$$

The pressure is found by integrating the momentum equation $\nabla p = \mu \nabla^2 \mathbf{u}$. The result is

$$p - p_\infty = -\frac{3\mu a U \cos \theta}{2r^2}, \tag{8.50}$$

which is sketched in Figure 8.17. The maximum $p - p_\infty = 3\mu U/2a$ occurs at the forward stagnation point $(\theta = \pi)$, while the minimum $p - p_\infty = 3\mu U/2a$ occurs at the rear stagnation point $(\theta = 0)$.

The drag force D on the sphere can be determined by integrating its surface pressure and shear stress distributions (see Exercise 8.35) to find:

$$D = 6\pi\mu a U, \tag{8.51}$$

of which one-third is pressure drag and two-thirds is skin friction drag. It follows that drag in a creeping flow is proportional to the velocity; this is known as *Stokes' law of resistance*.

In a well-known experiment to measure the charge of an electron, Millikan (1911) used (8.51) to estimate the radius of an oil droplet falling through air. Suppose ρ' is the density of a spherical falling particle and ρ is the density of the surrounding fluid. Then the effective weight of the sphere is $4\pi a^3 g(\rho' - \rho)/3$, which is the weight of the sphere minus the weight of the displaced fluid. The falling body reaches its terminal velocity when it no longer accelerates, at which point the viscous drag equals the effective weight. Then,

$$(4/3)\pi a^3 g(\rho' - \rho) = 6\pi\mu a U,$$

from which the radius a can be estimated.

Millikan (1911) was able to deduce the charge on an electron (and win a Nobel prize) making use of Stokes' drag formula by the following experiment. Two horizontal parallel plates can be charged by a battery (see Figure 8.18). Oil is sprayed through a very fine hole in the upper plate and develops static charge (+) by losing a few (n) electrons in passing through the small hole. If the plates are charged, then an electric force neE will act on each of the drops. Now n is not known but $E = -V_b/L$, where V_b is the battery voltage and L is the gap between the plates, provided that the charge density in the gap is very low. With the

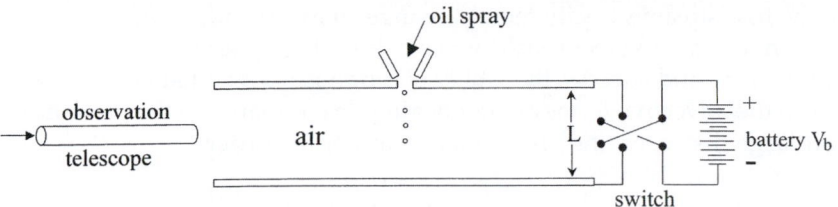

FIGURE 8.18 Simplified schematic of the Millikan oil drop experiment where observations of charged droplet motion and Stokes' drag law were used to determine the charge on an electron.

plates uncharged, measurement of the downward terminal velocity allowed the radius of a drop to be calculated assuming that the viscosity of the drop is much larger than the viscosity of the air. The switch is thrown to charge the upper plate negatively. The same droplet then reverses direction and is forced upward. It quickly achieves its terminal velocity U_u by virtue of the balance of upward forces (electric + buoyancy) and downward forces (weight + drag). This gives

$$6\pi\mu U_u a + (4/3)\pi a^3 g(\rho' - \rho) = neE,$$

where U_u is measured by the observation telescope and the radius of the particle is now known. The data then allow for the calculation of ne. As n must be an integer, data from many droplets may be differenced to identify the minimum difference that must be e, the charge of a single electron.

The drag coefficient, C_D, defined by (4.107) with $A = \pi a^2$, for Stokes' sphere is

$$C_D = \frac{D}{\frac{1}{2}\rho U^2 \pi a^2} = \frac{24}{Re'} \tag{8.52}$$

where $Re = 2aU/\nu$ is the Reynolds number based on the diameter of the sphere. This dependence on the Reynolds number can be predicted from dimensional analysis when fluid inertia, represented by ρ, is not a parameter (see Exercise 4.60). Without fluid density, the drag force on a slowly moving sphere may only depend on the other parameters of the problem:

$$D = f(\mu, U, a).$$

Here there are four variables and the three basic dimensions of mass, length, and time. Therefore, only one dimensionless parameter, $D/\mu Ua$, can be formed. Hence, it must be a constant, and this leads to $C_D \propto 1/Re$.

The flow pattern in a reference frame fixed to the fluid at infinity can be found by superposing a uniform velocity U to the left. This cancels out the first term in (8.48), giving

$$\psi = Ur^2 \sin^2 \theta \left(-\frac{3a}{4r} + \frac{a^3}{4r^3}\right),$$

which gives the streamline pattern for a sphere moving from right to left in front of an observer (Figure 8.19). The pattern is symmetric between the upstream and the downstream directions, which is a result of the linearity of the governing equation (9.63); reversing the direction of the free-stream velocity merely changes $\mathbf{u}$ to $-\mathbf{u}$ and $p - p_\infty$ to $-p + p_\infty$. The flow therefore does not leave a velocity-field wake behind the sphere.

In spite of its fame and success, the Stokes solution is not valid at large distances from the sphere because the advective terms are not negligible compared to the viscous terms at these distances. At large distances, the viscous terms are of the order

$$\text{viscous force/volume} = \text{stress gradient} \sim \frac{\mu U a}{r^3} \text{ as } r \to \infty,$$

while from (8.49), the largest inertia term is:

$$\text{inertia force/volume} \sim \rho u_r \frac{\partial u_\theta}{\partial r} \sim \frac{\rho U^2 a}{r^2} \text{ as } r \to \infty;$$

therefore,

$$\text{inertia force/viscous force} \sim \frac{\rho U a}{\mu} \frac{r}{a} \sim \text{Re} \frac{r}{a} \text{ as } r \to \infty,$$

which shows that the inertia forces are not negligible for distances larger than $r/a \sim 1/\text{Re}$.

Solutions of problems involving a small parameter can be developed in terms of a perturbation series in which the higher-order terms act as corrections on the lower-order terms. If

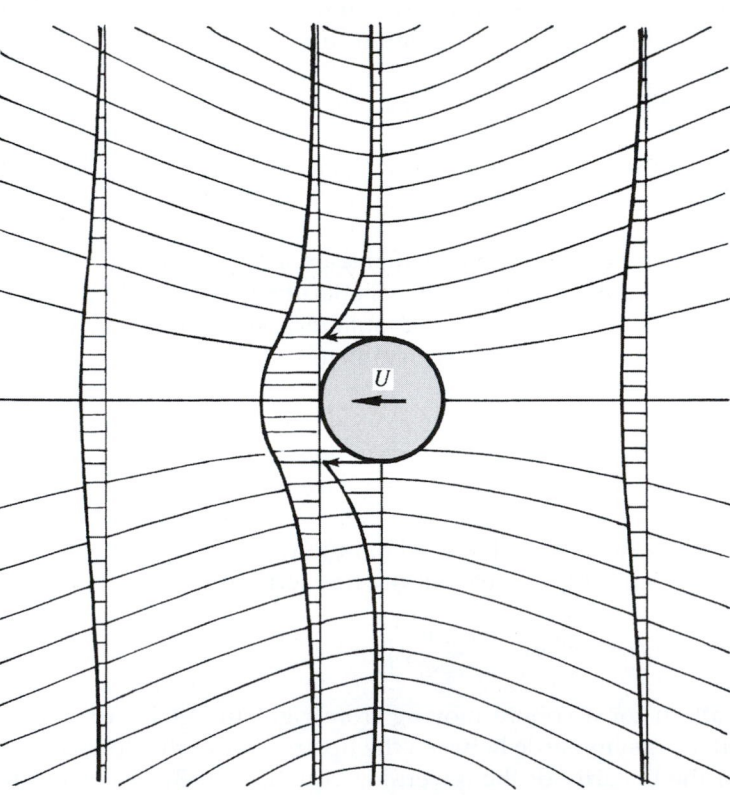

FIGURE 8.19 Streamlines and velocity distributions in Stokes' solution of creeping flow due to a moving sphere. Note the upstream and downstream symmetry, which is a result of complete neglect of nonlinearity.

we regard the Stokes solution as the first term of a series expansion in the small parameter Re, then the expansion is not uniformly valid because it breaks down as $r \to \infty$. If we tried to calculate the next term (to order Re) of the perturbation series, we would find that the velocity corresponding to the higher-order term becomes unbounded compared to that of the first term as $r \to \infty$.

The situation becomes worse for two-dimensional objects such as the circular cylinder. In this case, the Stokes balance, $\nabla p = \mu \nabla^2 \mathbf{u}$, has *no solution at all* that can satisfy the uniform-flow boundary condition at infinity. From this, Stokes concluded that steady, slow flows around cylinders cannot exist in nature. It has now been realized that the nonexistence of a first approximation of the Stokes flow around a cylinder is due to the *singular* nature of low Reynolds number flows in which there is a region of *nonuniformity* for $r \to \infty$. The nonexistence of the second approximation for flow around a sphere is due to the same reason. In a different (and more familiar) class of singular perturbation problems, the region of nonuniformity is a thin layer (the boundary layer) near the surface of an object. This is the class of flows with Re $\to \infty$, that are discussed in the next chapter. For these high Reynolds number flows the small parameter 1/Re multiplies the *highest*-order derivative in the governing equations, so that the solution with 1/Re identically set to zero cannot satisfy all the boundary conditions. In low Reynolds number flows this classic symptom of the loss of the highest derivative is absent, but it is a singular perturbation problem nevertheless.

Oseen (1910) provided an improvement to Stokes' solution by partly accounting for the inertia terms at large distances. He made the substitutions,

$$u = U + u', v = v', \text{ and } w = w',$$

where u', v', and w' are the Cartesian components of the perturbation velocity, and are small at large distances. Substituting these, the advection term of the x-momentum equation becomes

$$u\frac{\partial u}{\partial x} + v\frac{\partial u}{\partial y} + w\frac{\partial u}{\partial z} = U\frac{\partial u'}{\partial x} + \left[u'\frac{\partial u'}{\partial x} + v'\frac{\partial u'}{\partial y} + w'\frac{\partial u'}{\partial z}\right].$$

Neglecting the quadratic terms, a revised version of the equation of motion (8.43) becomes

$$\rho U\frac{\partial u'_i}{\partial x} = \frac{\partial p}{\partial x_i} + \mu \nabla^2 u'_i,$$

where u'_i represents u', v', or w'. This is called *Oseen's equation*, and the approximation involved is called *Oseen's approximation*. In essence, the Oseen approximation linearizes the advective acceleration term $\mathbf{u} \cdot \nabla \mathbf{u}$ to $U(\partial \mathbf{u}/\partial x)$, whereas the Stokes approximation drops advection altogether. Near the body both approximations have the same order of accuracy. However, the Oseen approximation is better in the far field where the velocity is only slightly different from U. The Oseen equations provide a lowest-order solution that is uniformly valid everywhere in the flow field.

The boundary conditions for a stationary sphere with the fluid moving past it at velocity $U\mathbf{e}_x$ are:

$$u', v', w' \to 0 \text{ as } r \to \infty, \text{ and } u' = -U \text{ and } v', w' = 0$$

on the sphere's surface. The solution found by Oseen is:

$$\frac{\psi}{Ua^2} = \left[\frac{r^2}{2a^2} + \frac{a}{4r}\right] \sin^2\theta - \frac{3}{Re}(1 + \cos\theta)\left\{1 - \exp\left[-\frac{Re}{4}\frac{r}{a}(1 - \cos\theta)\right]\right\}, \quad (8.53)$$

where $Re = 2aU/\nu$. Near the surface $r/a \approx 1$, a series expansion of the exponential term shows that Oseen's solution is identical to the Stokes solution (9.68) to the lowest order. The Oseen approximation predicts that the drag coefficient is

$$C_D = \frac{24}{Re}\left(1 + \frac{3}{16}Re\right),$$

which should be compared with the Stokes formula (8.52). Experimental results show that the Oseen and the Stokes formulas for C_D are both fairly accurate for $Re < 5$ (experimental results fall between them), an impressive range of validity for a theory developed for $Re \to 0$.

The streamlines corresponding to the Oseen solution (8.53) are shown in Figure 8.20, where a uniform flow of U is added to the left to generate the pattern of flow due to a sphere moving in front of a stationary observer. It is seen that the flow is no longer symmetric, but has a wake where the streamlines are closer together than in the Stokes flow. The velocities in

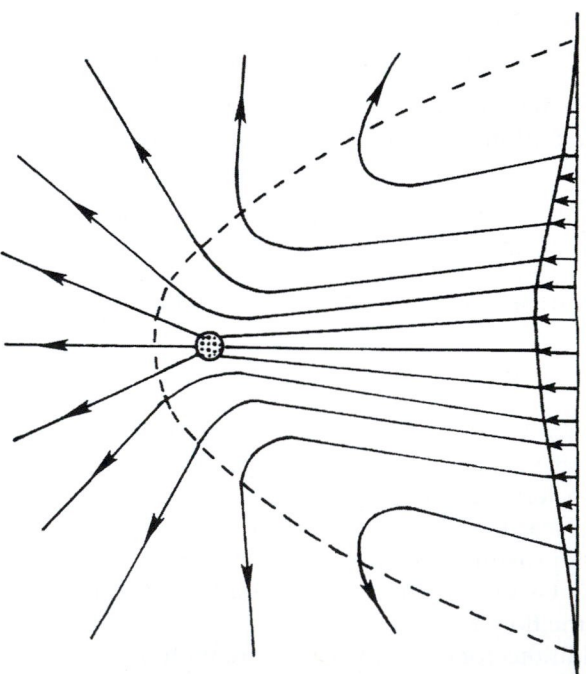

FIGURE 8.20 Streamlines and velocity distribution in Oseen's solution of creeping flow due to a moving sphere. Note the upstream and downstream asymmetry, which is a result of partial accounting for advection in the far field.

the wake are larger than in front of the sphere. Relative to the sphere, the flow is slower in the wake than in front of the sphere.

In 1957, Oseen's correction to Stokes' solution was rationalized independently by Kaplun (1957), and Proudman and Pearson (1957) in terms of matched asymptotic expansions. Higher-order corrections were obtained by Chester and Breach (1969).

8.7. FINAL REMARKS

As in other fields, analytical methods in fluid flow problems are useful in understanding the physics of fluid flows and in making generalizations. However, it is probably fair to say that most of the analytically tractable problems in ordinary laminar flow have already been solved, and approximate methods are now necessary for further advancing our knowledge. One of these approximate techniques is the perturbation method, where the flow is assumed to deviate slightly from a basic linear state. Another method that is playing an increasingly important role is that of solving the Navier-Stokes equations numerically using a computer. A proper application of such techniques requires considerable care and familiarity with various iterative techniques and their limitations. It is hoped that the reader will have the opportunity to learn numerical methods in a separate study. However, for completeness, Chapter 10 introduces several basic methods for computational fluid dynamics.

EXERCISES

8.1. **a)** Write out the three components of (8.1) in x-y-z Cartesian coordinates.

   b) Set $\mathbf{u} = (u(y), 0, 0)$, and show that the x- and y-momentum equations reduce to:

$$0 = -\frac{1}{\rho}\frac{\partial p}{\partial x} + \nu\frac{d^2 u}{dy^2}, \text{ and } 0 = -\frac{1}{\rho}\frac{\partial p}{\partial y}.$$

8.2. For steady pressure-driven flow between parallel plates (see Figure 8.3), there are 7 parameters: $u(y)$, U, y, h, ρ, μ, and dp/dx. Determine a dimensionless scaling law for $u(y)$, and rewrite the flow-field solution (8.5) in dimensionless form.

8.3. An incompressible viscous liquid with density ρ fills the gap between two large, smooth parallel walls that are both stationary. The upper and lower walls are located at $x_2 = \pm h$, respectively. An additive in the liquid causes its viscosity to vary in the x_2 direction. Here the flow is driven by a constant nonzero pressure gradient: $\partial p/\partial x_1 = const.$

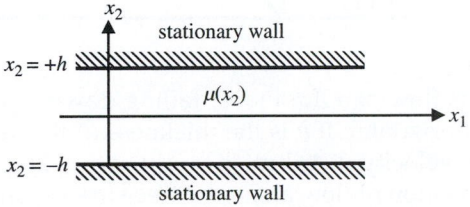

a) Assume steady flow, ignore the body force, set $\mathbf{u} = (u_1(x_2),0,0)$ and use

$$\frac{\partial \rho}{\partial t} + \frac{\partial}{\partial x_i}(\rho u_i) = 0, \quad \rho\frac{\partial u_j}{\partial t} + \rho u_i\frac{\partial u_j}{\partial x_i}$$

$$= -\frac{\partial p}{\partial x_j} + \rho g_j + \frac{\partial}{\partial x_i}\left[\mu\left(\frac{\partial u_i}{\partial x_j} + \frac{\partial u_j}{\partial x_i}\right)\right] + \frac{\partial}{\partial x_j}\left[\left(\mu_v - \frac{2}{3}\mu\right)\frac{\partial u_i}{\partial x_i}\right]$$

to determine $u_1(x_2)$ when $\mu = \mu_o(1 + \gamma(x_2/h)^2)$

b) What shear stress is felt on the lower wall?

c) What is the volume flow rate (per unit depth into the page) in the gap when $\gamma = 0$?

d) If $-1 < \gamma < 0$, will the volume flux be higher or lower than the case when $\gamma = 0$?

8.4. An incompressible viscous liquid with density ρ fills the gap between two large, smooth parallel plates. The upper plate at $x_2 = h$ moves in the positive x_1-direction at speed U. The lower plate at $x_2 = 0$ is stationary. An additive in the liquid causes its viscosity to vary in the x_2 direction.

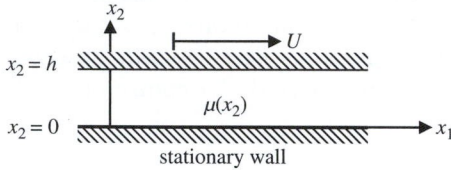

a) Assume steady flow, ignore the body force, set $\mathbf{u} = (u_1(x_2),0,0)$ and $\partial p/\partial x_1 = 0$, and use the equations specified in Exercise 8.3 to determine $u_1(x_2)$ when $\mu = \mu_o(1 + \gamma x_2/h)$.

b) What shear stress is felt on the lower plate?

c) Are there any physical limits on γ? If, so specify them.

8.5. Planar Couette flow is generated by placing a viscous fluid between two infinite parallel plates and moving one plate (say, the upper one) at a velocity U with respect to the other one. The plates are a distance h apart. Two immiscible viscous liquids are placed between the plates as shown in the diagram. The lower fluid layer has thickness d. Solve for the velocity distributions in the two fluids.

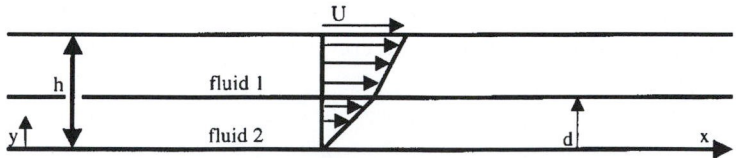

8.6. Consider the laminar flow of a fluid layer falling down a plane inclined at an angle θ with respect to the horizontal. If h is the thickness of the layer in the fully developed stage, show that the velocity distribution is $u(y) = (g/2\nu)(h^2 - y^2)\sin\theta$, where the x-axis points in the direction of flow along the free surface, and the y-axis points toward the plane. Show that the volume flow rate per unit width is $Q = (gh^3/3\nu)\sin\theta$, and that the frictional stress on the wall is $\tau_0 = \rho g h \sin\theta$.

8.7. Room temperature water drains through a round vertical tube with diameter d. The length of the tube is L. The pressure at the tube's inlet and outlet is atmospheric, the flow is steady, and $L \gg d$.

    a) Using dimensional analysis, write a physical law for the mass flow rate $\dot{m}$ through the tube.

    b) Assume that the velocity profile in the tube is independent of the vertical coordinate, determine a formula for $\dot{m}$, and put it in dimensionless form.

    c) What is the change in $\dot{m}$ if the temperature is raised and the water's viscosity drops by a factor of two?

8.8. Consider steady laminar flow through the annular space formed by two coaxial tubes aligned with the z-axis. The flow is along the axis of the tubes and is maintained by a pressure gradient dp/dz. Show that the axial velocity at any radius R is

$$u_z(R) = \frac{1}{4\mu}\frac{dp}{dz}\left[R^2 - a^2 - \frac{b^2-a^2}{\ln(b/a)}\ln\frac{R}{a}\right],$$

where a is the radius of the inner tube and b is the radius of the outer tube. Find the radius at which the maximum velocity is reached, the volume flow rate, and the stress distribution.

8.9. A long, round wire with radius a is pulled at a steady speed U, along the axis of a long round tube of radius b that is filled with a viscous fluid. Assuming laminar, fully developed axial flow with $\partial p/\partial z = 0$ in cylindrical coordinates (R, φ, z) with $\mathbf{u} = (0, 0, w(R))$, determine $w(R)$ assuming constant fluid density ρ and viscosity μ with no body force.

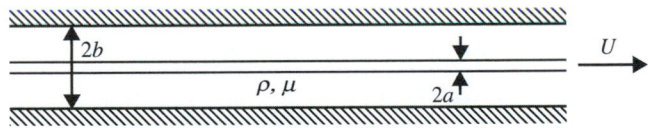

    a) What force per unit length of the wire is needed to maintain the motion of the wire?

    b) Explain what happens to $w(R)$ when $b \to \infty$. Is this situation physically meaningful? What additional term(s) from the equations of motion need to be retained to correct this situation?

8.10. [1]Consider steady unidirectional incompressible viscous flow in Cartesian coordinates, $u = v = 0$ with $w = w(x,y)$ without body forces.

    a) Starting from the steady version of (8.1), derive a single equation for w assuming that $\partial p/\partial z$ is nonzero and constant.

    b) Guess $w(x,y)$ for a tube with elliptical cross section $(x/a)^2+(y/b)^2 = 1$.

---

[1]Inspired by problem 2 on page 383 in Yih (1979).

c) Determine $w(x,y)$ for a tube of rectangular cross section: $-a/2 \leq x \leq +a/2, -b/2 \leq y \leq +b/2$. [*Hint*: find particular (a polynomial) and homogeneous (a Fourier series) solutions for w.]

8.11. A long vertical cylinder of radius b rotates with angular velocity Ω concentrically outside a smaller stationary cylinder of radius a. The annular space is filled with fluid of viscosity μ. Show that the steady velocity distribution is: $u_\theta = \dfrac{r^2 - a^2}{b^2 - a^2}\dfrac{b^2\Omega}{r}$, and that the torque exerted on either cylinder, per unit length, equals $4\pi\mu\Omega a^2 b^2/(b^2 - a^2)$.

8.12. Consider a solid cylinder of radius a, steadily rotating at angular speed Ω in an infinite viscous fluid. The steady solution is irrotational: $u_\theta = \Omega a^2/R$. Show that the work done by the external agent in maintaining the flow (namely, the value of $2\pi R u_\theta \tau_{r\theta}$ at $R = a$) equals the viscous dissipation rate of fluid kinetic energy in the flow field.

8.13. For lubrication flow under the sloped bearing of Example 8.1, the assumed velocity profile was $u(x,y) = -(1/2\mu)(dP/dx)y(h(x) - y)) + Uy/h(x)$, the derived pressure was $P(x) = P_e + (3\mu U a/h_0^2 L)x(L - x)$, and the load (per unit depth) carried by the bearing was $W = \mu U a L^2/2h_0^2$. Use these equations to determine the frictional force (per unit depth), F_f, applied to the lower (flat) stationary surface in terms of $W, h_0/L$, and α. What is the spatially averaged coefficient of friction under the bearing?

8.14. A bearing pad of total length $2L$ moves to the right at constant speed U above a thin film of incompressible oil with viscosity μ and density ρ. There is a step change in the gap thickness (from h_1 to h_2) below the bearing as shown. Assume that the oil flow under the bearing pad follows: $u(y) = -\dfrac{y(h_j - y)}{2\mu}\dfrac{dP(x)}{dx} + \dfrac{Uy}{h_j}$, where $j = 1$ or 2. The pad is instantaneously aligned above the coordinate system shown. The pressure in the oil ahead and behind the bearing is P_e.

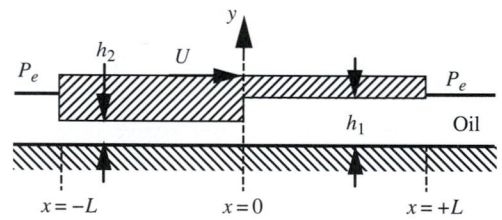

a) By conserving mass for the oil flow, find a relationship between $\mu, U, h_j, dP/dx$, and an unknown constant C.

b) Use the result of part a) and continuity of the pressure at $x = 0$, to determine

$$P(0) - P_e = \frac{6\mu U L(h_1 - h_2)}{h_2^3 + h_1^3}.$$

c) Can this bearing support an externally applied downward load when $h_1 < h_2$?

8.15. A flat disk of radius a rotates above a solid boundary at a steady rotational speed of Ω. The gap, h ($\ll a$), between the disk and the boundary is filled with an incompressible Newtonian fluid with viscosity μ and density ρ. The pressure at the edge of the disk is $p(a)$.

a) Using cylindrical coordinates and assuming that the only nonzero velocity component is $u_\varphi(R,z)$, determine the torque necessary to keep the disk turning.

b) If $p(a)$ acts on the exposed (upper) surface of the disk, will the pressure distribution on the disk's wetted surface tend to pull the disk *toward* or *away from* the solid boundary?

c) If the gap is increased, eventually the assumption of part a) breaks down. What happens? Explain why and where u_R and u_z might be nonzero when the gap is no longer narrow.

8.16. A circular block with radius a and weight W is released at $t = 0$ on a thin layer of an incompressible fluid with viscosity μ that is supported by a smooth horizontal motionless surface. The fluid layer's initial thickness is h_o. Assume that flow in the gap between the block and the surface is quasi-steady with a parabolic velocity profile:

$$u_R(R,z,t) = -(dP(R)/dR)z(h(t) - z))/2\mu,$$

where R is the distance from the center of the block, $P(R)$ is the pressure at R, z is the vertical coordinate from the smooth surface, $h(t)$ is the gap thickness, and t is time.

a) By considering conservation of mass, show that: $dh/dt = (h^3/6\mu R)(dP/dR)$.

b) If W is known, determine $h(t)$ and note how long it takes for $h(t)$ to reach zero.

8.17. Consider the inverse of the previous exercise. A block and a smooth surface are separated by a thin layer of a viscous fluid with thickness h_o. At $t = 0$, a force, F, is applied to separate them. If h_o is arbitrarily small, can the block and plate be separated easily? Perform some tests in your kitchen. Use maple syrup, creamy peanut butter, liquid soap, pudding, etc. for the viscous liquid. The flat top side of a metal jar lid or the flat bottom of a drinking glass makes a good circular block. (Lids with raised edges and cups and glasses with ridges or sloped bottoms do not work well). A flat countertop or the flat portion of a dinner plate can be the motionless smooth surface. Can the item used for the block be more easily separated from the surface when tilted relative to the surface? Describe your experiments and try to explain your results.

8.18. A rectangular slab of width $2L$ (and depth B into the page) moves *vertically* on a thin layer of oil that flows horizontally as shown. Assume $u(y,t) = -(h^2/2\mu)(dP/dx)(y/h)(1 - y/h)$, where $h(x,t)$ is the instantaneous gap between the slab and the surface, μ is the oil's viscosity, and $P(x,t)$ is the pressure in the oil below the slab. The slab is slightly misaligned with the surface so that $h(x,t) = h_o(1 + \alpha x/L) + \dot{h}_o t$ where $\alpha \ll 1$ and $\dot{h}_o$ is the vertical velocity of the slab. The pressure in the oil outside the slab is P_o. Consider the instant $t = 0$ in your work below.

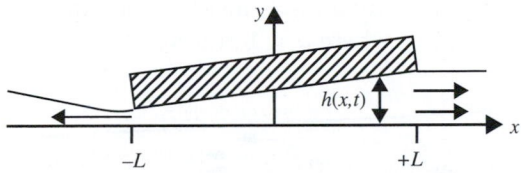

a) Conserve mass in an appropriate CV to show that: $\dfrac{\partial h}{\partial t} + \dfrac{\partial}{\partial x}\left(\displaystyle\int_0^{h(x,t)} u(y,t)dy \right) = 0.$

b) Keeping only linear terms in α, and noting that C and D are constants, show that:

$$P(x, t = 0) = \frac{12\mu}{h_o^3}\left(\frac{\dot{h}_o x^2}{2}\left(1 - \frac{2\alpha x}{L}\right) + Cx\left(1 - \frac{3\alpha x}{2L}\right)\right) + D.$$

c) State the boundary conditions necessary to evaluate the constants C and D.

d) Evaluate the constants to show that the pressure distribution below the slab is:

$$P(x, t) - P_o = -\frac{6\mu\dot{h}_o L^2}{h_o^3(t)}(1 - (x/L)^2)\left(1 - 2\alpha\frac{x}{L}\right).$$

e) Does this pressure distribution act to increase or decrease alignment between the slab and surface when the slab is moving downward? Answer the same question for upward slab motion.

8.19. Show that the lubrication approximation can be extended to viscous flow within narrow gaps $h(x,y,t)$ that depend on two spatial coordinates. Start from (4.10) and (8.1), and use Cartesian coordinates oriented so that x-y plane is locally tangent to the center-plane of the gap. Scale the equations using a direct extension of (8.14):

$$x^* = \frac{x}{L}, \ y^* = \frac{y}{L}, \ z^* = \frac{z}{h} = \frac{y}{\varepsilon L}, \ t^* = \frac{Ut}{L}, \ u^* = \frac{u}{U}, \ v^* = \frac{v}{U}, \ w^* = \frac{w}{\varepsilon U}, \text{ and } p^* = \frac{p}{P_a},$$

where L is the characteristic distance for the gap thickness to change in either the x or y direction, and $\varepsilon = h/L$. Simplify these equation when $\varepsilon^2 Re_L \rightarrow 0$, but $\mu UL/P_a h^2$ remains of order unity to find:

$$0 \cong -\frac{1}{\rho}\frac{\partial p}{\partial x} + \frac{\partial^2 u}{\partial z^2}, \ 0 \cong -\frac{1}{\rho}\frac{\partial p}{\partial y} + \frac{\partial^2 v}{\partial z^2}, \text{ and } 0 \cong -\frac{1}{\rho}\frac{\partial p}{\partial z}.$$

8.20. A thin film of viscous fluid is bounded below by a flat stationary plate at $z = 0$. If the in-plane velocity at the upper film surface, $z = h(x, y, t)$, is $\mathbf{U} = U(x, y, t)\mathbf{e}_x + V(x, y, t)\mathbf{e}_y$ use the equations derived in Exercise 8.19 to produce the Reynolds equation for constant-density, thin-film lubrication:

$$\nabla\cdot\left[\left(\frac{h^3}{\mu}\right)\nabla p\right] = 12\frac{\partial h}{\partial t} + 6\nabla\cdot(h\mathbf{U}).$$

where $\nabla = \mathbf{e}_x(\partial/\partial x) + \mathbf{e}_y(\partial/\partial y)$ merely involves the two in-plane dimensions.

8.21. Fluid of density ρ and viscosity μ flows inside a long tapered tube of length L and radius $R(x) = (1 - \alpha x/L)R_o$, where $\alpha < 1$ and $R_o \ll L$.

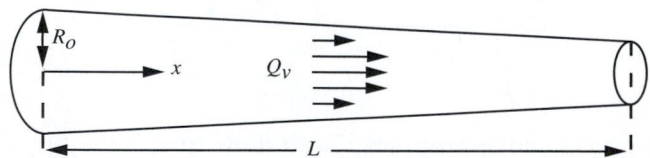

a) Estimate the volume discharge rate Q_v through the tube, for a given pressure difference Δp sustained between the inlet and the outlet.

b) Discuss the range of validity of your solution in terms of the parameters of the problem.

8.22. A circular lubricated bearing of radius a holds a stationary round shaft. The bearing hub rotates at angular rate Ω as shown. A load per unit depth on the shaft, **W**, causes the center of the shaft to be displaced from the center of the rotating hub by a distance εh_o, where h_o is the average gap thickness and $h_o \ll a$. The gap is filled with an incompressible oil of viscosity μ.

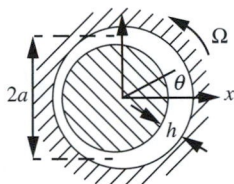

a) Determine a dimensionless scaling law for $|\mathbf{W}|$.

b) Determine **W** by assuming a lubrication flow profile in the gap and $h(\theta) = h_o(1 + \varepsilon\cos\theta)$ with $\varepsilon \ll 1$.

c) If **W** is increased a little bit, is the lubrication action stabilizing?

8.23. As a simple model of small-artery blood flow, consider slowly varying viscous flow through a round flexible tube with inlet at $z = 0$ and outlet at $z = L$. At $z = 0$, the volume flux entering the tube is $Q_o(t)$. At $z = L$, the pressure equals the exterior pressure p_e. The radius of the tube, $a(z,t)$, expands and contracts in proportion to pressure variations within the tube so that: 1) $a - a_e = \gamma(p - p_e)$, where a_e is the tube radius when the pressure, $p(z,t)$, in the tube is equal to p_e, and γ is a positive constant. Assume the local volume flux, $Q(z,t)$, is related to $\partial p / \partial z$ by 2) $Q = -(\pi a^4 / 8\mu)(\partial p / \partial z)$.

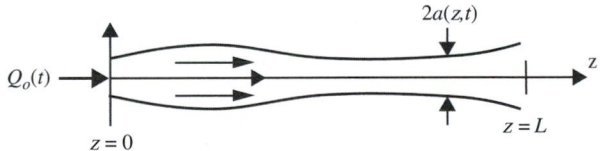

a) By conserving mass, find a partial differential equation that relates Q and a.

b) Combine 1), 2), and the result of part a) into one partial differential equation for $a(z,t)$.

c) Determine $a(z)$ when Q_o is a constant and the flow is perfectly steady.

8.24. Consider a simple model of flow from a tube of toothpaste. A liquid with viscosity μ and density ρ is squeezed out of a round horizontal tube having radius $a(t)$. In your work, assume that a is decreasing and use cylindrical coordinates with the z-axis coincident with the centerline of the tube. The tube is closed at $z = 0$, but is open to the atmosphere at $z = L$. Ignore gravity.

a) If w is the fluid velocity along the z-axis, show that: $za\dfrac{da}{dt} + \displaystyle\int_0^a w(z, R, t)R\,dR = 0$.

b) Determine the pressure distribution, $p(z) - p(L)$, by assuming the flow in the tube can be treated within the lubrication approximation by setting $w(z, R, t) = -\frac{1}{4\mu}\frac{dp}{dz}(a^2(t) - R^2)$.

c) Find the cross-section-average flow velocity $w_{ave}(z, t)$ in terms of z, a, and da/dt.

d) If the pressure difference between $z = 0$ and $z = L$ is ΔP, what is the volume flux exiting the tube as a function of time. Does this answer partially explain why fully emptying a toothpaste tube by squeezing it is essentially impossible?

8.25. A large flat plate below an infinite stationary incompressible viscous fluid is set in motion with a constant acceleration, $\dot{u}$, at $t = 0$. A prediction for the subsequent fluid motion, $u(y,t)$, is sought.

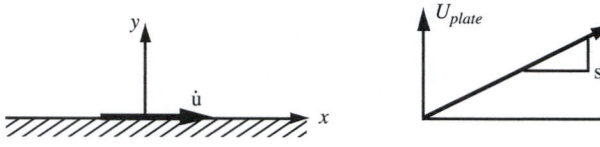

a) Use dimensional analysis to write a physical law for $u(y,t)$ in this flow.

b) Starting from the x-component of (8.1) determine a linear partial differential equation for $u(y,t)$.

c) The linearity of the equation obtained for part c) suggests that $u(y,t)$ must be directly proportional to $\dot{u}$. Simplify your dimensional analysis to incorporate this requirement.

d) Let $\eta = y/(vt)^{1/2}$ be the independent variable, and derive a second-order ordinary linear differential equation for the unknown function $f(\eta)$ left from the dimensional analysis.

e) From an analogy between fluid acceleration in this problem and fluid velocity in Stokes' first problem, deduce the solution $u(y, t) = \dot{u}\int_0^t[1 - erf(y/2\sqrt{vt'})]dt'$ and show that it solves the equation of part b).

f) Determine $f(\eta)$ and—if your patience holds out—show that it solves the equation found in part d).

g) Sketch the expected velocity profile shapes for several different times. Note the direction of increasing time on your sketch.

8.26. Suppose a line vortex of circulation Γ is suddenly introduced into a fluid at rest at $t = 0$. Show that the solution is $u_\theta(r,t) = (\Gamma/2\pi r)\exp\{-r^2/4vt\}$. Sketch the velocity distribution at different times. Calculate and plot the vorticity, and observe how it diffuses outward.

8.27. Obtain several liquids of differing viscosity (water, cooking oil, pancake syrup, shampoo, etc.). Using an eyedropper, a small spoon, or your finger, place a drop of each on a smooth vertical surface (a bathroom mirror perhaps) and measure how far the drops have moved or extended in a known period of time (perhaps a minute or two). Try to make the mass of all the drops equal. Using dimensional analysis, determine how the drop-sliding distance depends on the other parameters. Does this match your experimental results?

8.28. A drop of an incompressible viscous liquid is allowed to spread on a flat horizontal surface under the action of gravity. Assume the drop spreads in an axisymmetric fashion and use cylindrical coordinates (R, φ, z). Ignore the effects of surface tension.

a) Show that conservation of mass implies: $\dfrac{\partial h}{\partial t} + \dfrac{1}{R}\dfrac{\partial}{\partial R}\left(R\int_0^h u\,dz\right) = 0$, where $u = u(R, z, t)$ is the horizontal velocity within the drop, and $h = h(R, t)$ is the thickness of the spreading drop.

b) Assume that the lubrication approximation applies to the horizontal velocity profile, that is, $u(R, z, t) = a(R, t) + b(R, t)z + c(R, t)z^2$, apply the appropriate boundary conditions on the upper and lower drop surfaces, and require a pressure and shear-stress force balance within a differential control volume $h(R,t)R\,dR\,d\theta$ to show that: $u(R, z, t) = -\dfrac{g}{2\nu}\dfrac{\partial h}{\partial R}z(2h - z)$.

c) Combine the results of a) and b) to find $\dfrac{\partial h}{\partial t} = \dfrac{g}{3\nu R}\dfrac{\partial}{\partial R}\left(Rh^3\dfrac{\partial h}{\partial R}\right)$.

d) Assume a similarity solution: $h(R, t) = \dfrac{A}{t^n}f(\eta)$ with $\eta = \dfrac{BR}{t^m}$, use the result of part c) and $2\pi \int_0^{R_{max}(t)} h(R, t)R\,dR = V$, where $R_{max}(t)$ is the radius of the spreading drop and V is the initial volume of the drop to determine $m = 1/8$, $n = 1/4$, and a single nonlinear ordinary differential equation for $f(\eta)$ involving only $A, B, g/\nu$, and η. You need not solve this equation for f. [Given that $f \to 0$ as $\eta \to \infty$, there will be a finite value of η for which f is effectively zero. If this value of η is η_{max} then the radius of the spreading drop, $R(t)$, will be: $R_{max}(t) = \eta_{max}t^m/B$.]

8.29. Obtain a clean, flat glass plate, a watch, a ruler, and some nonvolatile oil that is more viscous than water. The plate and oil should be at room temperature. Dip the tip of one of your fingers in the oil and smear it over the center of the plate so that a thin bubble-free oil film covers a circular area ~10 to 15 cm in diameter. Set the plate on a horizontal surface and place a single drop of oil at the center of the oil-film area and observe how the drop spreads. Measure the spreading drop's diameter 1, 10, 10^2, 10^3, and 10^4 seconds after the drop is placed on the plate. Plot your results and determine if the spreading drop diameter grows as $t^{1/8}$ (the predicted drop-diameter time dependence from the prior problem) to within experimental error.

8.30. An infinite flat plate located at $y = 0$ is stationary until $t = 0$ when it begins moving horizontally in the positive x-direction at a constant speed U. This motion continues until $t = T$ when the plate suddenly stops moving.

a) Determine the fluid velocity field, $u(y, t)$ for $t > T$. At what height above the plate does the peak velocity occur for $t > T$? [Hint: the governing equation is linear so superposition of solutions is possible.]

b) Determine the mechanical impulse I (per unit depth and length) imparted to the fluid while the plate is moving: $I = \int_0^T \tau_w dt$.

c) As $t \to \infty$, the fluid slows down and eventually stops moving. How and where was the mechanical impulse dissipated? What is t/T when 99% of the initial impulse has been lost?

8.31. Consider the development from rest of plane Couette flow. The flow is bounded by two rigid boundaries at $y = 0$ and $y = h$, and the motion is started from rest by suddenly accelerating the lower plate to a steady velocity U. The upper plate is held stationary. Here a similarity solution cannot exist because of the appearance of the parameter h. Show that the velocity distribution is given by

$$u(y,t) = U\left(1 - \frac{y}{h}\right) - \frac{2U}{\pi}\sum_{n=1}^{\infty}\frac{1}{n}\exp\left(-n^2\pi^2\frac{\nu t}{h^2}\right)\sin\left(\frac{n\pi y}{h}\right).$$

Sketch the flow pattern at various times, and observe how the velocity reaches the linear distribution for large times.

8.32. [2]Two-dimensional flow between flat nonparallel plates can be formulated in terms of a normalized angular coordinate, $\eta = \theta/\alpha$, where α is the half angle between the plates, and a normalized radial velocity, $u_r(r,\theta) = u_{max}(r)f(\eta)$, where $\eta = \theta/\alpha$ for $|\theta| \le \alpha$. Here, $u_\theta = 0$, the Reynolds number is $Re = u_{max}r\alpha/\nu$, and Q is the volume flux (per unit width perpendicular to the page).

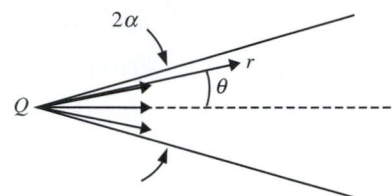

a) Using the appropriate versions of (4.10) and (8.1), show that $f'' + Re\ \alpha f^2 + 4\alpha^2 f = const.$

b) Find $f(\eta)$ for symmetric creeping flow, that is, $Re = 0 = f(+1) = f(-1)$, and $f(0) = 1$.

c) Above what value of the channel half-angle will backflow always occur?

8.33. Consider steady viscous flow inside a cone of constant angle θ_o. The flow has constant volume flux $= Q$, and the fluid has constant density $= \rho$ and constant kinematic viscosity $= \nu$. Use spherical coordinates, and assume that the flow only has a radial component, $\mathbf{u} = (u_r(r,\theta),0,0)$, which is independent of the azimuthal angle φ, so that the equations of motion are:

Conservation of mass,

$$\frac{1}{r^2}\frac{\partial}{\partial r}(r^2 u_r) = 0,$$

Conservation of radial momentum,

$$u_r\frac{\partial u_r}{\partial r} = -\frac{1}{\rho}\frac{\partial p}{\partial r} + \nu\left(\frac{1}{r^2}\frac{\partial}{\partial r}\left(r^2\frac{\partial u_r}{\partial r}\right) + \frac{1}{r^2\sin\theta}\frac{\partial}{\partial\theta}\left(\sin\theta\frac{\partial u_r}{\partial\theta}\right) - \frac{2}{r^2}u_r\right), \text{ and}$$

Conservation of θ-momerntum,

$$0 = -\frac{1}{\rho r}\frac{\partial p}{\partial\theta} + \nu\left(\frac{2}{r^2}\frac{\partial u_r}{\partial\theta}\right).$$

[2]Rephrased from White (2006) p. 211, problem 3.32.

For the following items, assume the radial velocity can be determined using: $u_r(r,\theta) = QR(r)\Theta(\theta)$. Define the Reynolds number of this flow as: $Re = Q/(\pi v r)$.

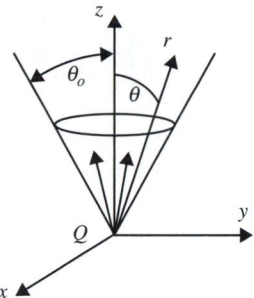

a) Use the continuity equation to determine $R(r)$.

b) Integrate the θ-momentum equation, assume the constant of integration is zero, and combine the result with the radial momentum equation to determine a single differential equation for $\Theta(\theta)$ in terms of θ and Re.

c) State the matching and/or boundary conditions that $\Theta(\theta)$ must satisfy.

8.34. The boundary conditions on obstacles in Hele-Shaw flow were not considered in Example 8.2. Therefore, consider them here by examining Hele-Shaw flow parallel to a flat obstacle surface at $y = 0$. The Hele-Shaw potential in this case is:

$$\phi = Ux\frac{z}{h}\left(1 - \frac{z}{h}\right),$$

where (x, y, z) are Cartesian coordinates and the flow is confined to $0 < z < h$ and $y > 0$.

a) Show that this potential leads to a slip velocity of $u(x, y \to 0) = U(z/h)(1 - z/h)$, and determine the pressure distribution implied by this potential.

b) Since this is a viscous flow, the slip velocity must be corrected to match the genuine no-slip condition on the obstacle's surface at $y = 0$. The analysis of Example (8.2) did not contain the correct scaling for this situation near $y = 0$. Therefore, rescale the x-component of (8.1) using:

$$x^* = x/L, \quad y^* = y/h = y/\varepsilon L, \quad z^* = z/h = z/\varepsilon L, \quad t^* = Ut/L, \quad u^* = u/U, \quad v^* = v/\varepsilon U,$$
$$w^* = w/\varepsilon U, \text{ and } p^* = p/Pa,$$

and then take the limit as $\varepsilon^2 Re_L \to 0$, with $\mu UL/P_a h^2$ remaining of order unity, to simplify the resulting dimensionless equation that has

$$0 \cong \frac{dp}{dx} + \mu\left(\frac{\partial^2 u}{\partial y^2} + \frac{\partial^2 u}{\partial z^2}\right)$$

as its dimensional counterpart.

c) Using boundary conditions of $u = 0$ on $y = 0$, and $u = U(z/h)(1 - z/h)$ for $y \gg h$. Show that

$$u(x,y,z) = U\frac{z}{h}\left(1 - \frac{z}{h}\right) + \sum_{n=1}^{\infty} A_n \sin\left(\frac{n\pi}{h}z\right) \exp\left(-\frac{n\pi}{h}y\right), \text{ where}$$

$$A_n = -\frac{2U}{h}\int_0^h \frac{z}{h}\left(1 - \frac{z}{h}\right)\sin\left(\frac{n\pi}{h}z\right)dz.$$

[The results here are directly applicable to the surfaces of curved obstacles in Hele-Shaw flow when the obstacle's radius of curvature is much greater than h.]

8.35. Using the velocity field (8.49), determine the drag on Stokes' sphere from the surface pressure and the viscous surface stresses σ_{rr} and $\sigma_{r\theta}$.

8.36. Calculate the drag on a spherical droplet of radius $r = a$, density ρ' and viscosity μ' moving with velocity U in an infinite fluid of density ρ and viscosity μ. Assume $Re = \rho Ua/\mu \ll 1$. Neglect surface tension.

8.37. Consider a very low Reynolds number flow over a circular cylinder of radius $r = a$. For $r/a = O(1)$ in the $Re = Ua/\nu \rightarrow 0$ limit, find the equation governing the stream function $\psi(r, \theta)$ and solve for ψ with the least singular behavior for large r. There will be one remaining constant of integration to be determined by asymptotic matching with the large r solution (which is not part of this problem). Find the domain of validity of your solution.

8.38. A small, neutrally buoyant sphere is centered at the origin of coordinates in a deep bath of a quiescent viscous fluid with density ρ and viscosity μ. The sphere has radius a and is initially at rest. It begins rotating about the z-axis with a constant angular velocity Ω at $t = 0$. The relevant equations for the fluid velocity, $\mathbf{u} = (u_r, u_\theta, u_\varphi)$, in spherical coordinates (r, θ, φ) are:

$$\frac{1}{r^2}\frac{\partial}{\partial r}(r^2 u_r) + \frac{1}{r\sin\theta}\frac{\partial}{\partial\theta}(u_\theta\sin\theta) + \frac{1}{r\sin\theta}\frac{\partial}{\partial\varphi}(u_\varphi) = 0, \text{ and}$$

$$\frac{\partial u_\varphi}{\partial t} + u_r\frac{\partial u_\varphi}{\partial r} + \frac{u_\theta}{r}\frac{\partial u_\varphi}{\partial\theta} + \frac{u_\varphi}{r\sin\theta}\frac{\partial u_\varphi}{\partial\varphi} + \frac{1}{r}(u_r u_\varphi + u_\theta u_\varphi\cot\theta)$$

$$= -\frac{1}{\rho r\sin\theta}\frac{\partial p}{\partial\varphi} + \nu\left(\frac{1}{r^2}\frac{\partial}{\partial r}\left(r^2\frac{\partial u_\varphi}{\partial r}\right) + \frac{1}{r^2\sin\theta}\frac{\partial}{\partial\theta}\left(\sin\theta\frac{\partial u_\varphi}{\partial\theta}\right)\right.$$

$$\left. + \frac{1}{r^2\sin^2\theta}\frac{\partial^2 u_\varphi}{\partial\varphi^2} - \frac{u_\varphi}{r^2\sin^2\theta} + \frac{2}{r^2\sin^2\theta}\frac{\partial u_r}{\partial\varphi} + \frac{2\cos\theta}{r^2\sin^2\theta}\frac{\partial u_\theta}{\partial\varphi}\right).$$

a) Assume $\mathbf{u} = (0, 0, u_\varphi)$ and reduce these equations to:

$$\frac{\partial u_\varphi}{\partial t} = -\frac{1}{\rho r\sin\theta}\frac{\partial p}{\partial\varphi} + \nu\left(\frac{1}{r^2}\frac{\partial}{\partial r}\left(r^2\frac{\partial u_\varphi}{\partial r}\right) + \frac{1}{r^2\sin\theta}\frac{\partial}{\partial\theta}\left(\sin\theta\frac{\partial u_\varphi}{\partial\theta}\right) - \frac{u_\varphi}{r^2\sin^2\theta}\right)$$

b) Set $u_\varphi(r, \theta, t) = \Omega a F(r, t) \sin \theta$, make an appropriate assumption about the pressure field, and derive the following equation for F: $\dfrac{\partial F}{\partial t} = \nu \left(\dfrac{1}{r^2} \dfrac{\partial}{\partial r} \left(r^2 \dfrac{\partial F}{\partial r} \right) - 2 \dfrac{F}{r^2} \right)$.

c) Determine F for $t \to \infty$ for boundary conditions $F = 1$ at $r = a$, and $F \to 0$ as $r \to \infty$.

d) Find the surface shear stress and torque on the sphere.

Literature Cited

Abramowitz, M., & Stegun, I. A. (1972). *Handbook of Mathematical Functions*. Washington, DC: U.S. Department of Commerce, National Bureau of Standards.

Chester, W., & Breach, D. R. (with I. Proudman) (1969). On the flow past a sphere at low Reynolds number. *J. Fluid Mech.*, 37, 751–760.

Hele-Shaw, H. S. (1898). Investigations of the Nature of Surface Resistance of Water and of Stream Line Motion Under Certain Experimental Conditions. *Trans. Roy. Inst. Naval Arch.*, 40, 21–46.

Kaplun, S. (1957). Low Reynolds number flow past a circular cylinder. *J. Math. Mech.*, 6, 585–603.

Millikan, R. A. (1911). The isolation of an ion, a precision measurement of its charge, and the correction of Stokes' law. *Phys. Rev.*, 32, 349–397.

Oseen, C. W. (1910). Über die Stokes'sche Formel, und über eine verwandte Aufgabe in der Hydrodynamik. *Ark Math. Astrom. Fys.*, 6(No. 29).

Proudman, I., & Pearson, J. R. A. (1957). Expansions at small Reynolds numbers for the flow past a sphere and a circular cylinder. *J. Fluid Mech.*, 2, 237–262.

Sherman, F. S. (1990). *Viscous Flow*. New York: McGraw-Hill.

White, F. M. (2006). *Viscous Fluid Flow*. New York: McGraw-Hill.

Yih, C.-S. (1979). *Fluid Mechanics*. Ann Arbor: West River Press.

Supplemental Reading

Batchelor, G. K. (1967). *An Introduction to Fluid Dynamics*. London: Cambridge University Press.

Lighthill, M. J. (1986). *An Informal Introduction to Theoretical Fluid Mechanics*. Oxford, England: Clarendon Press.

Schlichting, H. (1979). *Boundary Layer Theory*. New York: McGraw-Hill.

CHAPTER

9

Boundary Layers and Related Topics

OUTLINE

9.1. Introduction 362

9.2. Boundary-Layer Thickness Definitions 367

9.3. Boundary Layer on a Flat Plate: Blasius Solution 369

9.4. Falkner-Skan Similarity Solutions of the Laminar Boundary-Layer Equations 373

9.5. Von Karman Momentum Integral Equation 375

9.6. Thwaites' Method 377

9.7. Transition, Pressure Gradients, and Boundary-Layer Separation 382

9.8. Flow Past a Circular Cylinder 388

9.9. Flow Past a Sphere and the Dynamics of Sports Balls 395

9.10. Two-Dimensional Jets 399

9.11. Secondary Flows 407

Exercises 408

Literature Cited 418

Supplemental Reading 419

CHAPTER OBJECTIVES

- To describe the boundary-layer concept and the mathematical simplifications it allows in the complete equations of motion for a viscous fluid.

- To present the equations of fluid motion for attached laminar boundary layers.

- To provide a variety of exact and approximate steady laminar boundary-layer solutions.

- To describe the basic phenomenology of boundary-layer transition and separation.

- To discuss the Reynolds number dependent phenomena associated with flow past bluff bodies.

- To illustrate the use of a boundary-layer solution methodology for free and wall-bounded jet flows.

9.1. INTRODUCTION

Until the beginning of the twentieth century, analytical solutions of steady fluid flows were generally known for two typical situations. One of these was that of parallel viscous flows and low Reynolds number flows, in which the nonlinear advective terms were zero and the balance of forces was that between pressure and viscous forces. The second type of solution was that of inviscid flows around bodies of various shapes, in which the balance of forces was that between inertia and pressure forces. Although the equations of motion are nonlinear in this second situation, the velocity field can be determined by solving the linear Laplace equation. These irrotational solutions predicted pressure forces on a streamlined body that agreed surprisingly well with experimental data for flow of fluids of small viscosity. However, these solutions also predicted zero drag force and a nonzero tangential velocity at the body surface, features that did not agree with the experiments.

In 1905 Ludwig Prandtl, an engineer by profession and therefore motivated to find realistic fields near bodies of various shapes, first hypothesized that, for small viscosity, the viscous forces are negligible everywhere except close to solid boundaries where the no-slip condition has to be satisfied. The thickness of these boundary layers approaches zero as the viscosity goes to zero. Prandtl's hypothesis reconciled two rather contradictory facts. It supported the intuitive idea that the effects of viscosity are indeed negligible in most of the flow field if ν is small, but it also accounted for drag by insisting that the no-slip condition must be satisfied at a solid surface, no matter how small the viscosity. This reconciliation was Prandtl's aim, which he achieved brilliantly, and in such a simple way that it now seems strange that nobody before him thought of it. Prandtl also showed how the equations of motion within the boundary layer can be simplified. Since the time of Prandtl, the concept of the boundary layer has been generalized, and the mathematical techniques involved have been formalized, extended, and applied in other branches of physical science (see van Dyke, 1975; Bender & Orszag, 1978; Kevorkian & Cole 1981; Nayfeh, 1981). The concept of the boundary layer is considered a cornerstone in the intellectual foundation of fluid mechanics.

This chapter presents the boundary-layer hypothesis and examines its consequences. The equations of fluid motion within the boundary layer can be simplified because of the layer's thinness, and exact or approximate solutions can be obtained in many cases. In addition, boundary-layer phenomena provide explanations for the lift and drag characteristics of bodies of various shapes in high Reynolds number flows, including turbulent flows. In particular, the fluid mechanics of curved sports-ball trajectories is described here.

The fundamental assumption of boundary-layer theory is that the layer is thin compared to other length scales such as the length of the surface or its local radius of curvature. Across this thin layer, which can exist only in high Reynolds number flows, the velocity varies rapidly enough for viscous effects to be important. This is shown in Figure 9.1, where the boundary-layer thickness is greatly exaggerated. (On a typical airplane wing, which may have a chord of several meters, the boundary-layer thickness is of order one centimeter.) However, thin viscous layers exist not only next to solid walls but also in the form of jets, wakes, and shear layers if the Reynolds number is sufficiently high. So, to be specific, we shall first consider the boundary layer contiguous to a solid surface, adopting a curvilinear

FIGURE 9.1 A boundary layer forms when a viscous fluid moves over a solid surface. Only the boundary layer on the top surface of the foil is depicted in the figure and its thickness, δ, is greatly exaggerated. Here, U_∞ is the oncoming free-stream velocity and U_e is the velocity at the edge of the boundary layer. The usual boundary-layer coordinate system allows the x-axis to coincide with a mildly curved surface so that the y-axis always lies in the surface-normal direction.

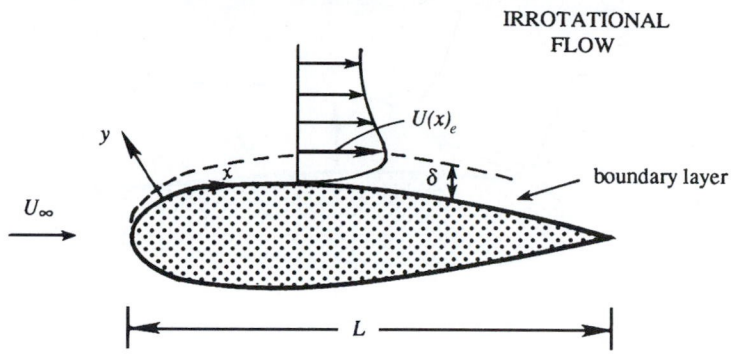

coordinate system that conforms to the surface where x increases along the surface and y increases normal to it. Here the surface may be curved but the radius of curvature of the surface must be much larger than the boundary-layer thickness. We shall refer to the solution of the irrotational flow outside the boundary layer as the *outer* problem and that of the boundary-layer flow as the *inner* problem.

For a thin boundary layer that is contiguous to the solid surface on which it has formed, the full equations of motion for a constant-density constant-viscosity fluid, (4.10) and (8.1), may be simplified. Let $\bar{\delta}(x)$ be the average thickness of the boundary layer at downstream location x on the surface of a body having a local radius of curvature R. The steady-flow momentum equation for the surface-parallel velocity component, u, is

$$u\frac{\partial u}{\partial x} + v\frac{\partial u}{\partial y} = -\frac{1}{\rho}\frac{\partial p}{\partial x} + \nu\left(\frac{\partial^2 u}{\partial x^2} + \frac{\partial^2 u}{\partial y^2}\right), \tag{9.1}$$

which is valid when $\bar{\delta}/R \ll 1$. The more general curvilinear form for arbitrary $R(x)$ is given in Goldstein (1938) and Schlichting (1979), but the essential features of viscous boundary layers can be illustrated via (9.1) without additional complications.

Let a characteristic magnitude of u in be U_∞, the velocity at a large distance upstream of the body, and let L be the stream-wise distance over which u changes appreciably. The longitudinal length of the body can serve as L, because u within the boundary layer does change in the stream-wise direction by a large fraction of U_∞ over a distance L (Figure 9.2). A measure of $\partial u/\partial x$ is therefore U_∞/L, so that the approximate size of the first advective term in (9.1) is

$$u(\partial u/\partial x) \sim U_\infty^2/L, \tag{9.2}$$

where $\sim$ is to be interpreted as "of order." We shall see shortly that the other advective term in (9.1) is of the same order. The approximate size of the viscous term in (9.1) is

$$\nu(\partial^2 u/\partial y^2) \sim \nu U_\infty/\bar{\delta}^2. \tag{9.3}$$

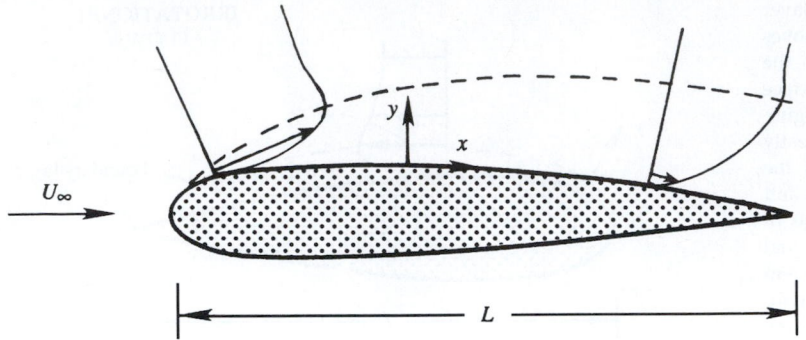

FIGURE 9.2 Velocity profiles at two positions within the boundary layer. Here again, the boundary-layer thickness is greatly exaggerated. The two velocity vectors are drawn at the same distance y from the surface, showing that the variation of u over a distance $x \sim L$ is of the order of the free-stream velocity U_∞.

The magnitude of $\bar{\delta}$ can now be estimated by noting that the advective and viscous terms should be of the same order within the boundary layer. Equating advective and viscous terms in (9.2) and (9.3) leads to:

$$\bar{\delta} \sim \sqrt{\nu L/U_\infty} \quad \text{or} \quad \bar{\delta}/L \sim 1/\sqrt{\mathrm{Re}}. \tag{9.4}$$

This estimate of $\bar{\delta}$ can also be obtained by noting that viscous effects diffuse to a distance of order $[\nu t]^{1/2}$ in time t and that the time-of-flight for a fluid element along a body of length L is of order L/U_∞. Substituting L/U_∞ for t in $[\nu t]^{1/2}$ suggests the viscous layer's diffusive thickness at $x = L$ is of order $[\nu L/U_\infty]^{1/2}$.

A formal simplification of the equations of motion within the boundary layer can now be performed. The basic idea is that variations across the boundary layer occur over a much shorter length scale than variations along the layer, that is:

$$\partial/\partial x \sim 1/L \quad \text{and} \quad \partial/\partial y \sim 1/\bar{\delta}, \tag{9.5}$$

where $\bar{\delta} \ll L$ when $\mathrm{Re} \gg 1$ from (9.4). When applied to the continuity equation, $\partial u/\partial x + \partial v/\partial y = 0$, this derivative scaling requires $U_\infty/L \sim v/\bar{\delta}$, so the proper velocity scale for v is $\bar{\delta} U_\infty/L = U_\infty \mathrm{Re}^{-1/2}$. At high Re, experimental data show that the pressure distribution on the body is nearly that in an irrotational flow around the body, implying that the pressure variations scale with the fluid inertia: $p - p_\infty \sim \rho U_\infty^2$. Thus, the proper dimensionless variables for boundary-layer flow are:

$$x^* = x/L, \; y^* = y/\bar{\delta} = (y/L)\mathrm{Re}^{1/2}, \; u^* = u/U_\infty, \; v^* = (v/U_\infty)\mathrm{Re}^{1/2}, \; \text{and } p^* = (p - p_\infty)/\rho U_\infty^2. \tag{9.6}$$

For the coordinates and the velocities, this scaling is similar to that of (8.14) with $\varepsilon = \mathrm{Re}^{-1/2}$. The primary effect of (9.6) is a magnification of the surface-normal coordinate y and velocity v by a factor of $\mathrm{Re}^{1/2}$ compared to the stream-wise coordinate x and velocity u. In terms of these dimensionless variables, the steady two-dimensional equations of motion are:

$$\frac{\partial u^*}{\partial x^*} + \frac{\partial v^*}{\partial y^*} = 0, \tag{8.15}$$

$$u^* \frac{\partial u^*}{\partial x^*} + v^* \frac{\partial u^*}{\partial y^*} = -\frac{\partial p^*}{\partial x^*} + \frac{1}{Re} \frac{\partial^2 u^*}{\partial x^*} + \frac{\partial^2 u^*}{\partial y^*}, \quad \text{and} \tag{9.7}$$

$$\frac{1}{Re} \left(u^* \frac{\partial v^*}{\partial x^*} + v^* \frac{\partial v^*}{\partial y^*} \right) = -\frac{\partial p^*}{\partial y^*} + \frac{1}{Re^2} \frac{\partial^2 v^*}{\partial x^{*2}} + \frac{1}{Re} \frac{\partial^2 v^*}{\partial y^{*2}}, \tag{9.8}$$

where $Re \equiv U_\infty L/\nu$ is an overall Reynolds number. In these equations, each of the dimensionless variables and their derivatives should be of order unity when the scaling assumptions embodied in (9.6) are valid. Thus, it follows that the importance of each term in (8.15), (9.7), and (9.8) is determined by its coefficient. So, as $Re \to \infty$, the terms with coefficients of $1/Re$ or $1/Re^2$ drop out asymptotically. Thus, the relevant equations for laminar boundary-layer flow, in dimensional form, are:

$$\frac{\partial u}{\partial x} + \frac{\partial v}{\partial y} = 0, \tag{6.2}$$

$$u \frac{\partial u}{\partial x} + v \frac{\partial u}{\partial y} = -\frac{1}{\rho} \frac{\partial p}{\partial x} + \nu \frac{\partial^2 u}{\partial y^2}, \quad \text{and} \quad 0 = -\frac{\partial p}{\partial y}. \tag{9.9, 9.10}$$

This scaling exercise has shown which terms must be kept and which terms may be dropped under the boundary-layer assumption. It differs from the scalings that produced (4.101) and (8.42) because the x and y directions are scaled differently in (9.6) which causes a second derivative term to be retained in (9.9).

Equation (9.10) implies that the pressure is approximately uniform across the boundary layer, an important result. The pressure at the surface is therefore equal to that at the edge of the boundary layer, so it can be found from an ideal outer-flow solution for flow above the surface. Thus, the outer flow imposes the pressure on the boundary layer. This justifies the experimental fact that the observed surface pressures underneath attached boundary layers are approximately equal to that calculated from ideal flow theory. A vanishing $\partial p/\partial y$, however, is not valid if the boundary layer separates from the surface or if the radius of curvature of the surface is not large compared with the boundary-layer thickness.

Although the steady boundary-layer equations (6.2), (9.9), and (9.10) do represent a significant simplification of the full equations, they are still nonlinear second-order partial differential equations that can only be solved when appropriate boundary and matching conditions are specified. If the exterior flow is presumed to be known and irrotational (and the fluid density is constant), the pressure gradient at the edge of the boundary layer can be found by differentiating the steady constant-density Bernoulli equation (without the body force term), $p + \frac{1}{2}\rho U_e^2 = const.$, to find:

$$-\frac{1}{\rho} \frac{dp}{dx} = U_e \frac{dU_e}{dx}, \tag{9.11}$$

where $U_e(x)$ is the velocity at the *edge* of the boundary layer. Equation (9.11) represents a matching condition between the outer solution and the inner boundary-layer solution in the region where both solutions must be valid. The (usual) remaining boundary conditions on the fluid velocities of the inner solution are:

$$u(x,0) = 0 \quad \text{(no slip at the wall)}, \tag{9.12}$$

$$v(x, 0) = 0 \quad \text{(no flow through the wall)}, \tag{9.13}$$

$$u(x, y \to \infty) = U_e(x) \quad \text{(matching of inner and outer solutions), and} \tag{9.14}$$

$$u(x_0, y) = u_{in}(y) \quad \text{(inlet boundary condition at } x_0\text{)}. \tag{9.15}$$

For two-dimensional flow, (6.2), (9.9), and the conditions (9.11) through (9.15), completely specify the inner solution as long as the boundary layer remains thin and contiguous to the surface on which it develops. Condition (9.14) merely means that the boundary layer must join smoothly with the outer flow; for the inner solution, points outside the boundary layer are represented by $y \to \infty$, although we mean this strictly in terms of the dimensionless distance $y/\bar{\delta} = (y/L)\text{Re}^{1/2} \to \infty$. Condition (9.15) implies that an initial velocity profile $u_{in}(y)$ at some location x_0 is required for solving the problem. Such a condition is needed because the terms $u\partial u/\partial x$ and $v\partial^2 u/\partial y^2$ give the boundary-layer equations a parabolic character, with x playing the role of a time-like variable. Recall the Stokes problem of a suddenly accelerated plate, discussed in the preceding chapter, where the simplified field equation is $\partial u/\partial t = v\partial^2 u/\partial y^2$. In such problems governed by parabolic equations, the field at a certain time or place depends only on its *past* or *upstream history*. Boundary layers therefore transfer viscous effects only in the *downstream* direction. In contrast, the complete Navier-Stokes equations are elliptic and thus require boundary conditions on the velocity (or its derivative normal to the boundary) upstream, downstream, and on the top and bottom boundaries, that is, all around. (The upstream influence of the downstream boundary condition is a common concern in fluid dynamic computations.)

Considering two dimensions, an ideal outer flow solution from (6.5) or (6.12) and (6.18), and a viscous inner flow solution as described here would seem to fully solve the problem of uniform flow of a viscous fluid past a solid object. The solution procedure could be a two-step process. First, the outer flow is determined from (6.5) or (6.12) and (6.18), neglecting the existence of the boundary layer, an error that gets smaller when the boundary layer becomes thinner. Then, (9.11) could be used to determine the surface pressure, and (6.2) and (9.9) could be solved for the boundary-layer flow using the surface-pressure gradient determined from the outer flow solution. If necessary this process might even be iterated to achieve higher accuracy by re-solving for the outer flow with the first-pass-solution boundary-layer characteristics included, and then proceeding to a second solution of the boundary-layer equations using the corrected outer-flow solution. In practice, such an approach can be successful and it converges when the boundary layer stays thin and attached. However, it does not converge when the boundary layer thickens or departs (separates) from the surface on which it has developed. Boundary-layer separation occurs when the surface shear stress, τ_0, produced by the boundary layer vanishes and reverse (or upstream-directed) flow occurs near the surface. Boundary-layer separation is discussed further in Sections 9.6 and 9.7. Here it is sufficient to point out that ideal flow around nonslender or *bluff* bodies typically produces surface pressure gradients that lead to boundary-layer separation.

In summary, the simplifications of the boundary-layer assumption are as follows. First, diffusion in the stream-wise direction is negligible compared to that in the wall normal direction. Second, the pressure field can be found from the outer flow, so that it is regarded as a known quantity within the boundary layer. Here, the boundary layer is so thin that the pressure does not change across it. Furthermore, a crude estimate of τ_0, the wall shear stress, can be made from the various scalings employed earlier: $\tau_0 \sim \mu U / \bar{\delta} \sim (\mu U / L) \mathrm{Re}^{1/2}$. This implies a skin friction coefficient of

$$\frac{\tau_0}{\frac{1}{2}\rho U^2} \sim \frac{(\mu U/L)\sqrt{\mathrm{Re}}}{\frac{1}{2}\rho U^2} = \frac{2}{\sqrt{\mathrm{Re}}}.$$

This expression provides the correct order of magnitude and parametric dependence on Reynolds number. Only the numerical factor differs between different laminar boundary-layer flows.

9.2. BOUNDARY-LAYER THICKNESS DEFINITIONS

Since the fluid velocity in the boundary layer smoothly joins that of the outer flow, there is no obvious demarcation of the boundary layer's edge. Thus, a variety of thickness definitions are used to define a boundary layer's character. The three most common thickness definitions are described here.

The first, δ_{99}, is an overall boundary-layer thickness that specifies the distance from the wall where the stream-wise velocity in the boundary layer is $0.99U_e$, where U_e is the local free-stream speed. For a known boundary-layer stream-wise velocity profile, $u(x, y)$, at downstream distance x, this thickness is defined by: $u(x, \delta_{99}) = 0.99U_e(x)$. This thickness primarily plays a conceptual role in boundary-layer research. In practice it is difficult to measure accurately, and its physical importance is subjective since the choice of 99% instead of 95%, 98%, 99.5%, or another percentage is arbitrary.

A second measure of the boundary-layer thickness, and one in which there is no arbitrariness, is the *displacement thickness*, which is commonly denoted δ^* or δ_1. It is defined as the thickness of a layer of zero-velocity fluid that has the same velocity deficit as the actual boundary layer. The velocity deficit in a boundary layer is $U_e - u$, so this definition implies

$$\int_{y=0}^{h} (U_e - u)dy = \int_{y=0}^{\delta^*} (U_e - 0)dy = U_e\delta^*, \quad \text{or} \quad \delta^* = \int_{y=0}^{\infty} \left(1 - \frac{u}{U_e}\right)dy, \qquad (9.16)$$

where h is a surface-normal distance that lies far outside the boundary layer (Figure 9.3). Here the extension $h \to \infty$ in the upper limit in the last integration is not problematic because $U_e - u \to 0$ exponentially fast as $y \to \infty$. Alternatively, the displacement thickness is the distance by which the wall would have to be displaced outward in a hypothetical frictionless flow to maintain the same mass flux as that in the actual flow. This means that the displacement thickness can be interpreted as the distance by which streamlines outside the boundary layer are displaced due to the presence of the boundary layer. Figure 9.4 shows the displacement of streamlines over a flat plate. Equating mass flux across two sections A and B, we obtain

$$U_e h = \int_{y=0}^{h+\delta^*} u\,dy = \int_{y=0}^{h} u\,dy + U_e\delta^*, \quad \text{or} \quad U_e\delta^* = \int_{y=0}^{h} (U_e - u)dy,$$

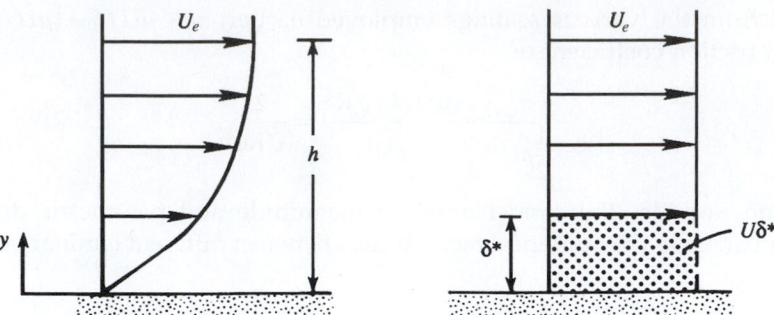

FIGURE 9.3 Schematic depiction of the displacement thickness. The panel on the left shows a typical laminar boundary-layer profile. The panel on the right shows an equivalent ideal-flow velocity profile with a zero-velocity layer having the same volume-flux deficit as the actual boundary layer. The thickness of this zero-velocity layer is the displacement thickness δ^*.

where h is the wall-normal distance defined above. Here again, it can be replaced by ∞ without changing the integral in the final equation, which then reduces to (9.16).

The displacement thickness is used in the design of airfoils, ducts, nozzles, intakes of air-breathing engines, wind tunnels, etc. by first assuming a frictionless flow and then revising the device's geometry to produce the desired flow condition with the boundary layer present. Here, the method for the geometric revisions involves using δ^* to correct the outer flow solution for the presence of the boundary layer. As mentioned in Section 9.1, the first approximation is to neglect the existence of the boundary layer, and calculate the ideal-flow dp/dx over the surface of interest. A solution of the boundary-layer equations gives $u(x,y)$, which can be integrated using (9.16) to find $\delta^*(x)$, the displacement thickness. The flow device's surface is then displaced outward by this amount and a next approximation of dp/dx is found from a solution for flow over the mildly revised geometry (see Exercise 9.22). Thus, $\delta^*(x)$ is

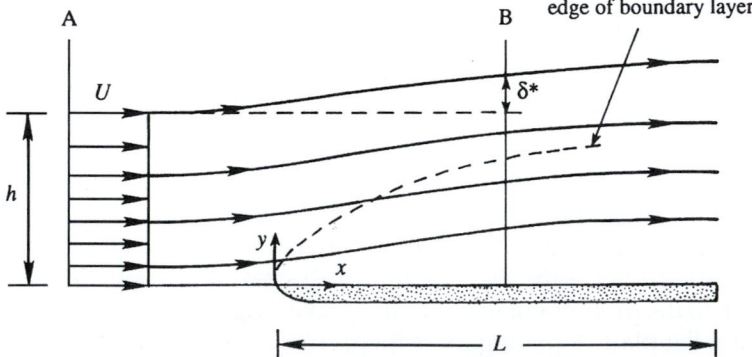

FIGURE 9.4 Displacement thickness and streamline displacement. Within the boundary layer, fluid motion in the downstream direction is retarded, that is, $\partial u/\partial x$ is negative. Thus, the continuity equation (6.2) requires $\partial v/\partial y$ to be positive, so the boundary layer produces a surface-normal velocity that deflects streamlines away from the surface. Above the boundary layer, the extent of this deflection is the displacement thickness δ^*.

a critical ingredient in such an iterative solution procedure that alternates between the outer- and inner-flow solutions.

A third measure of the boundary-layer thickness is the momentum thickness θ or δ_2. It is defined such that $\rho U^2 \theta$ is the momentum loss in the actual flow because of the presence of the boundary layer. A control volume calculation (see Exercise 9.6) leads to the following definition:

$$\theta = \int_{y=0}^{\infty} \frac{u}{U_e}\left(1 - \frac{u}{U_e}\right) dy. \tag{9.17}$$

The momentum thickness embodies the integrated signature of the wall shear stress from the beginning of the boundary layer to the stream-wise location of interest.

9.3. BOUNDARY LAYER ON A FLAT PLATE: BLASIUS SOLUTION

The simplest possible boundary layer forms on a semi-infinite flat plate with a constant free-stream flow speed, $U_e = U = \text{constant}$. In this case, the boundary-layer equations simplify to:

$$\frac{\partial u}{\partial x} + \frac{\partial v}{\partial y} = 0 \quad \text{and} \quad u\frac{\partial u}{\partial x} + v\frac{\partial u}{\partial y} = \nu\frac{\partial^2 u}{\partial y^2}, \tag{6.2, 9.18}$$

where (9.11) requires $dp/dx = 0$ because $dU_e/dx = 0$. Here, the independent variables are x and y, and the dependent field quantities are u and v. The flow is incompressible but rotational, so a guaranteed solution of (6.2) may be sought in terms of a stream function, ψ, with the two velocity components determined via derivatives of ψ (see (6.3)). Here, the flow is steady and there is no imposed length scale, so a similarity solution for ψ can be proposed based on (8.32) and (9.6):

$$\psi = U\delta(x)f(\eta) \quad \text{where } \eta = y/\delta(x), \tag{9.19}$$

where x is the time-like independent variable, f is an unknown dimensionless function, and $\delta(x)$ is a boundary-layer thickness that is to be determined as part of the solution (it is not a Dirac delta-function). Here the coefficient $U\delta$ in (9.19) has replaced $UA\xi^{-n}$ in (8.32) based on dimensional considerations; the stream function must have dimensions of length^2/time. A more general form of (9.19) that uses UAx^{-n} as the coefficient of $f(\eta)$ produces the same results when combined with (9.18).

The solution to (6.2) and (9.18) should be valid for $x > 0$, so the boundary conditions are:

$$u = v = 0 \quad \text{on } y = 0, \tag{9.20}$$

$$u(x, y) \to U \quad \text{as } y/\delta \to \infty, \text{ and} \tag{9.21}$$

$$\delta \to 0 \quad \text{as } x \to 0. \tag{9.22}$$

Here we note that the boundary-layer approximation will not be valid near $x = 0$ (the leading edge of the plate) where the high Reynolds number approximation, $\text{Re}_x = Ux/\nu \gg 1$, used to reach (9.18) is not valid. Ideally, the exact equations of motion would be solved from $x = 0$ up some location, x_0, where $Ux_0/\nu \gg 1$. Then, the stream-wise velocity profile at this location would be used in the inlet boundary condition (9.15), and (9.18) could be solved for $x > x_0$ to determine the boundary-layer flow. However, for this similarity solution, we are effectively assuming that the distance x_0 is small compared to x and can be ignored. Thus, the boundary condition (9.22), which replaces (9.15), is really an assumption that must be shown to produce self-consistent results when $Re_x \gg 1$.

The prior discussion touches on the question of a boundary layer's downstream dependence on, or memory of, its initial state. If the external stream $U_e(x)$ admits a similarity solution, is the initial condition forgotten? And, if so, how soon? Serrin (1967) and Peletier (1972) showed that for $U_e dU_e/dx > 0$ (*favorable* pressure gradients) and allowing similarity solutions, the initial condition is forgotten and the larger the free-stream acceleration the sooner similarity is achieved. However, a decelerating flow will accentuate details of the boundary layer's initial state and similarity will never be found even if it is mathematically possible. This is consistent with the experimental findings of Gallo et al. (1970). Interestingly, a flat plate for which $U_e(x) = U = \text{const.}$ is the borderline case; similarity is eventually achieved. Thus, a solution in the form (9.19) is pursued here.

The first solution's steps involve performing derivatives of ψ to find u and v:

$$u = \frac{\partial \psi}{\partial y} = U\delta \frac{df}{d\eta} \frac{1}{\delta} = Uf',$$

$$v = -\frac{\partial \psi}{\partial x} = -U\left(\frac{d\delta}{dx}f + \delta\frac{df}{d\eta}\left(-\frac{y}{\delta^2}\right)\frac{d\delta}{dx}\right) = U\delta'(-f + \eta f'),$$

(9.23, 9.24)

where a prime denotes differentiation of a function with respect to its argument. When substituted into (9.18), these produce:

$$Uf'Uf''\left(-\frac{y}{\delta^2}\right)\delta' + U\delta'(-f + \eta f')\frac{U}{\delta}f'' = \nu\frac{U}{\delta^2}f''', \text{ or}$$

$$\left[-\frac{U^2}{\delta}\delta'\right]\eta f'f'' - \left[\frac{U^2}{\delta}\delta'\right]ff'' + \left[\frac{U^2}{\delta}\delta'\right]\eta f'f'' = \left[\nu\frac{U}{\delta^2}\right]f'''.$$

The first and third terms on the left are equal and opposite, so (9.18) finally reduces to:

$$-\left[\frac{U^2}{\delta}\delta'\right]ff'' = \left[\nu\frac{U}{\delta^2}\right]f'''.$$

(9.25)

For a similarity solution, the coefficients in [,]-braces must be proportional:

$$C\frac{U^2}{\delta}\delta' = \nu\frac{U}{\delta^2} \to C\delta\frac{d\delta}{dx} = \frac{\nu}{U} \to C\frac{\delta^2}{2} = \frac{\nu x}{U} + D,$$

where C and D are constants. Here (9.22) requires $D = 0$, and C can be chosen equal to 2 to simplify the resulting expression for δ:

$$\delta(x) = [\nu x/U]^{1/2}.$$

(9.26)

As described above, this result will be imperfect as $x \to 0$ since it is based on equations that are only valid when $Re_x \gg 1$. However, it is self-consistent since it produces a boundary

FIGURE 9.5   The Blasius similarity solution of velocity distribution in a laminar boundary layer on a flat plate with zero-pressure gradient, $U_e = U = $ constant.   The finite slope at $\eta = 0$ implies a nonzero wall shear stress τ_0. The boundary layer's velocity profile smoothly asymptotes to U as $\eta \to \infty$. The momentum θ and displacement δ^* thicknesses are indicated by arrows on the horizontal axis.

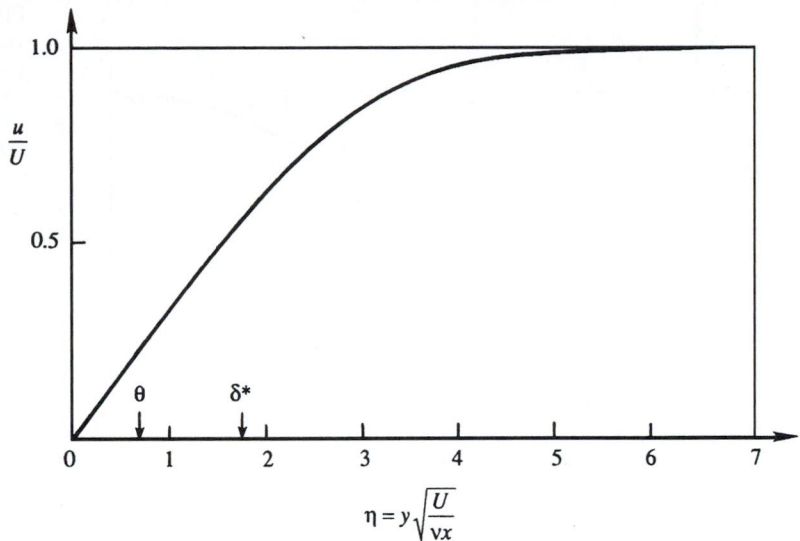

layer that thins with decreasing distance so that $u \to U$ as $x \to 0$. When (9.26) is substituted into (9.25), the final equation for f is found:

$$\frac{d^3f}{d\eta^3} + \frac{1}{2}f\frac{d^2f}{d\eta^2} = 0, \quad \text{or} \quad f''' + \frac{1}{2}ff'' = 0. \tag{9.27}$$

The boundary conditions for (9.27) are:

$$df/d\eta = 0 \quad \text{and} \quad f = 0 \quad \text{at } \eta = 0, \text{ and} \tag{9.28}$$

$$df/d\eta \to 1 \quad \text{as } \eta \to \infty, \tag{9.29}$$

which replace (9.20) and (9.21), respectively.

A series solution of (9.27), subject to (9.28) and (9.29), was given by Blasius; today it is much easier to numerically determine $f(\eta)$. The resulting profile of $u/U = f'(\eta)$ is shown in Figure 9.5. The solution makes the profiles at various downstream distances collapse into a single curve of u/U vs. $y[U/\nu x]^{1/2}$, and is in excellent agreement with experimental data for laminar flow at high Reynolds numbers. The profile has a point of inflection (i.e., zero curvature) at the wall, where $\partial^2 u/\partial y^2 = 0$. This is a result of the absence of a pressure gradient in the flow (see Section 9.7).

The Blasius boundary-layer profile has a variety of noteworthy properties. First of all, an asymptotic analysis of the solution to (9.27) shows that $(df/d\eta - 1) \sim (1/\eta)\exp(-\eta^2/4)$ as $\eta \to \infty$ so u approaches U very smoothly with increasing wall-normal distance. Second, the wall-normal velocity is:

$$v = -\frac{\partial \psi}{\partial x} = \frac{1}{2}\sqrt{\frac{\nu U}{x}}\left(-f + \eta\frac{df}{d\eta}\right), \quad \text{or} \quad \frac{v}{U} = \frac{1}{2\mathrm{Re}_x^{1/2}}\left(-f + \eta\frac{df}{d\eta}\right) \sim \frac{0.86}{\mathrm{Re}_x^{1/2}} \quad \text{as } \eta \to \infty,$$

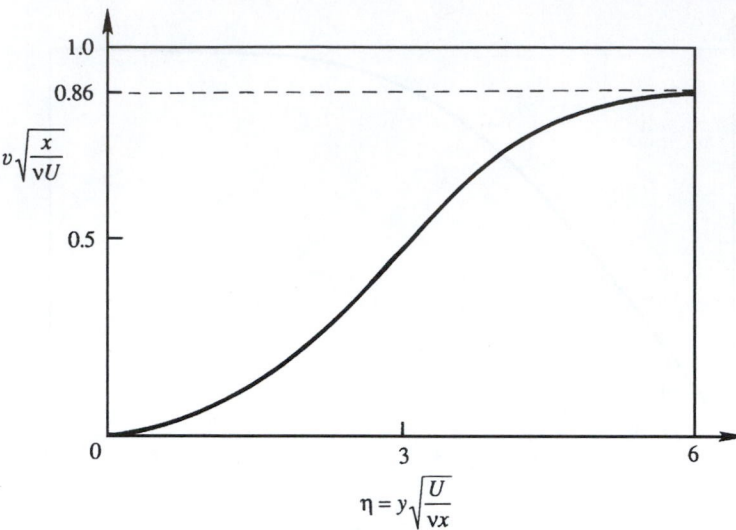

FIGURE 9.6 Surface-normal velocity component, v, in a laminar boundary layer on a flat plate with constant free-stream speed U. Here the scaling on the vertical axis, $(v/U)\sqrt{Re_x}$, causes it to be expanded compared to that in Figure 9.5.

a plot of which is shown in Figure 9.6. The wall-normal velocity increases from zero at the wall to a maximum value at the edge of the boundary layer, a pattern that is in agreement with the streamline shapes sketched in Figure 9.4.

The various thicknesses for the Blasius boundary layer are as follows. From Figure 9.5, the distance where $u = 0.99U$ is $\eta = 4.93$, so

$$\delta_{99} = 4.93\sqrt{\nu x/U} \quad \text{or} \quad \delta_{99}/x = 4.93/Re_x^{1/2}. \tag{9.30}$$

The half-power boundary-layer thickness dependence, $\delta \sim x^{1/2}$, is in good agreement with experiments. For air at ordinary temperatures flowing at $U = 1$ m/s, the Reynolds number at a distance of 1 m from leading edge of a flat plate is $Re_x = 6 \times 10^4$, and (9.30) gives $\delta_{99} = 2$ cm, showing that the boundary layer is indeed thin. The displacement and momentum thicknesses, (9.16) and (9.17), of the Blasius boundary layer are

$$\delta^* = 1.72\sqrt{\nu x/U}, \quad \text{and} \quad \theta = 0.664\sqrt{\nu x/U}.$$

These thicknesses are indicated along the abscissa of Figure 9.5.

The local wall shear stress is $\tau_0 = \mu(du/dy)_{y=0} = (\mu U/\delta)(d^2 f/d\eta^2)_{\eta=0}$, or

$$\tau_0 = 0.332\rho U^2/\sqrt{Re_x}. \tag{9.31}$$

The wall shear stress therefore decreases as $x^{-1/2}$, a result of the thickening of the boundary layer and the associated decrease of the velocity gradient at the surface. Note that the wall shear stress at the plate's leading edge has an integrable singularity. This is a manifestation of the fact that boundary-layer theory breaks down near the leading edge where the assumptions $Re_x \gg 1$ and $\partial/\partial x \ll \partial/\partial y$ are invalid. The wall shear stress is commonly expressed in terms of the dimensionless *skin friction coefficient*:

$$C_f \equiv \frac{\tau_0}{\frac{1}{2}\rho U^2} = \frac{0.664}{\sqrt{Re_x}}. \tag{9.32}$$

The drag force per unit width on one side of a plate of length L is

$$F_D = \int_0^L \tau_0 dx = \frac{0.664\rho U^2 L}{\sqrt{Re_L}},$$

where $Re_L \equiv UL/v$ is the Reynolds number based on the plate length. This equation shows that the drag force is proportional to the 3/2 power of velocity. This should be compared with small Reynolds number flows, where the drag is proportional to the first power of velocity. We shall see later in this chapter that the drag on a *blunt* body in a high Reynolds number flow is proportional to the *square* of velocity.

The overall *drag coefficient* for one side of the plate, defined in the usual manner, is

$$C_D = \frac{F_D}{\frac{1}{2}\rho U^2 L} = \frac{1.33}{\sqrt{Re_L}}. \tag{9.33}$$

It is clear from (9.32) and (9.33) that

$$C_D = \frac{1}{L}\int_0^L C_f dx,$$

which says that the overall drag coefficient is the spatial average of the local skin friction coefficient. However, carrying out an integration from $x = 0$ may be of questionable validity because the equations and solutions are valid only for $Re_x \gg 1$. Nevertheless, (9.33) is found to be in good agreement with laminar flow experiments for $Re_L > 10^3$.

9.4. FALKNER-SKAN SIMILARITY SOLUTIONS OF THE LAMINAR BOUNDARY-LAYER EQUATIONS

The Blasius boundary-layer solution is one of a whole class of similarity solutions to the boundary-layer equations that were investigated by Falkner and Skan (1931). In particular, similarity solutions of (6.2), (9.9), and (9.10) are possible when $U_e(x) = ax^n$, and $Re_x = ax^{(n+1)}/v$ is sufficiently large so that the boundary-layer approximation is valid and any dependence on an initial velocity profile has been forgotten. In this case, the initial location x_0 again disappears from the problem and a similarity solution may be sought in the form:

$$\psi(x, y) = \sqrt{vxU_e(x)}f(\eta), \quad \text{where } \eta = \frac{y}{\delta(x)} = \frac{y}{x}\sqrt{Re_x} = y\sqrt{\frac{a}{v}}x^{(n-1)/2}. \tag{9.34}$$

This is a direct extension of (9.19) to boundary-layer flow with a spatially varying free-stream speed $U_e(x)$. Here, $u/U_e = f'(\eta)$ as in the Blasius solution, but now the pressure gradient is nontrivial:

$$-dp/dx = U_e(dU_e/dx) = na^2 x^{2n-1}, \tag{9.35}$$

and the generic boundary-layer thickness is

$$\delta(x) = \sqrt{\nu x / U_e(x)} = \sqrt{\nu x^{1-n}/a},$$

which increases in size when $n < 1$ and decreases in size when $n > 1$ as x increases. When $n = 1$, then $\delta(x)$ is constant. Substituting (9.34) and (9.35) into (9.9) allows it to be reduced to the similarity form:

$$\frac{d^3f}{d\eta^3} + \frac{n+1}{2}f\frac{d^2f}{d\eta^2} - n\left(\frac{df}{d\eta}\right)^2 + n = 0, \quad \text{or} \quad f''' + \frac{n+1}{2}ff'' - nf'^2 + n = 0, \tag{9.36}$$

where f is subject to the boundary conditions (9.28) and (9.29). The Blasius equation (9.27) is a special case of (9.36) for $n = 0$, that is, $U_e(x) = U = \text{constant}$.

Solutions to (9.36) are displayed in Figure 5.9.1 of Batchelor (1967) and are reproduced here in Figure 9.7. They are parameterized by the power law exponent, n, which also sets the pressure gradient. The shapes of the various profiles can be understood by comparing them to the stream-wise velocity profiles obtained for flow between parallel plates when the upper plate moves with a positive horizontal velocity. They show a monotonically increasing shear stress [$f''(0)$] as n increases. When n is positive, the flow accelerates as it moves to higher x, the pressure gradient is *favorable* ($dp/dx < 0$), the wall shear stress is nonzero and positive, and $(\partial^2 u/\partial y^2)_{y\,=\,0} < 0$. Thus, the profiles for $n > 0$ in Figure 9.7 are similar to the lower half of the profiles shown in Figures 8.4a and 8.4d. When $n = 0$, there is no flow acceleration or pressure gradient and $(\partial^2 u/\partial y^2)_{y\,=\,0} = 0$. This case corresponds

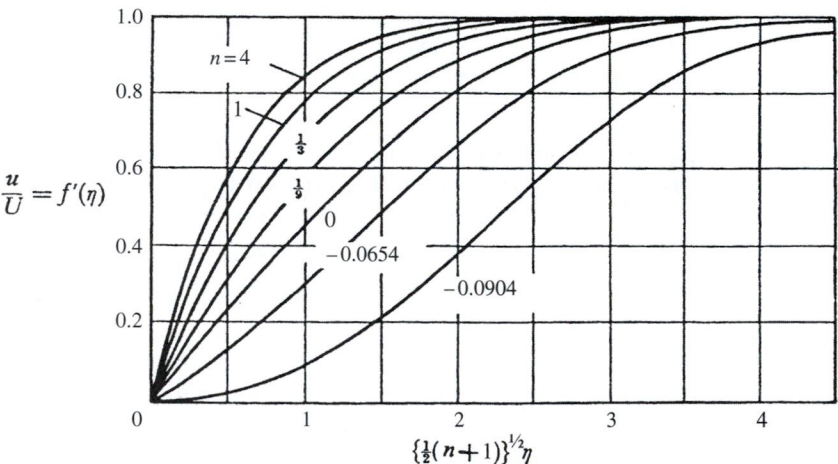

FIGURE 9.7 Falkner-Skan profiles of stream-wise velocity in a laminar boundary layer when the external stream is $U_e = ax^n$. The horizontal axis is the scaled surface-normal coordinate. The various curves are labeled by their associated value of n. When $n > 0$, the free-stream speed increases with increasing x, and $\partial^2 u/\partial y^2$ is negative throughout the boundary layer. When $n = 0$ (the Blasius boundary layer), the free-stream speed is constant, and $\partial^2 u/\partial y^2 = 0$ at the wall and is negative throughout the boundary layer. When $n < 0$, the free-stream speed decreases with increasing x, and $\partial^2 u/\partial y^2$ is positive near the wall but negative higher up in the boundary layer so there is an inflection point in the stream-wise velocity profile at a finite distance from the surface. *Reprinted with the permission of Cambridge University Press, from: G. K. Batchelor,* An Introduction to Fluid Dynamics, *1st ed. (1967).*

to Figure 8.4c. When n is negative, the flow decelerates as it moves downstream, the pressure gradient is *adverse* ($dp/dx > 0$), the wall shear stress may approach zero, and $(\partial^2 u/\partial y^2)_{y\,=\,0} < 0$. Thus, the profiles for $n < 0$ in Figure 9.7 approach that shown in Figure 8.4b. For $n = -0.0904$, $f''(0) = 0$, so $\tau_0 = 0$, and boundary-layer separation is imminent all along the surface. Solutions of (9.36) exist for $n < -0.0904$ but these solutions involve reverse flow, like that shown in Figure 8.4b, and do not necessarily represent boundary layers because the stream-wise velocity scaling in (9.6) used to reach (9.9), $u \sim U$, is invalid when $u = 0$ away from the wall.

In many real flows, boundary or initial conditions prevent similarity solutions from being directly applicable. However, after a variety of empirical and analytical advances made in the middle of the twentieth century, useful approximate methods were found to predict the properties of laminar boundary layers. These approximate techniques are based on the von Karman boundary-layer integral equation, which is derived in the next section. Then, in Section 9.6, the Thwaites method for estimating the surface shear stress, the displacement thickness, and the momentum thickness for attached laminar boundary layers is presented. In the most general cases or when greater accuracy is required, the boundary-layer equations with boundary and initial conditions as written in (6.2) and (9.9) through (9.15) must be solved by procedures that are discussed in more detail in the next chapter.

9.5. VON KARMAN MOMENTUM INTEGRAL EQUATION

Exact solutions of the boundary-layer equations are possible only in simple cases. In more complicated problems, approximate methods satisfy only an *integral* of the boundary-layer equations across the layer thickness. When this integration is performed, the resulting ordinary differential equation involves the boundary layer's displacement and momentum thicknesses, and its wall shear stress. This simple differential equation was derived by von Karman in 1921 and applied to several situations by Pohlhausen (1921).

The common emphasis of an integral formulation is to obtain critical information with minimum effort. The important results of boundary-layer calculations are the wall shear stress, displacement thickness, momentum thickness, and separation point (when one exists). The von Karman boundary-layer momentum integral equation explicitly links the first three of these, and can be used to estimate, or at least determine the existence of, the fourth. The starting points are (6.2) and (9.9), with the pressure gradient specified in terms of $U_e(x)$ from (9.11) and the shear stress $\tau = \mu(\partial u/\partial y)$:

$$u\frac{\partial u}{\partial x} + v\frac{\partial u}{\partial y} = U_e\frac{dU_e}{dx} + \frac{1}{\rho}\frac{\partial \tau}{\partial y}. \tag{9.37}$$

Multiply (6.2) by u and add it to the left side of this equation:

$$u\left(\frac{\partial u}{\partial x} + \frac{\partial v}{\partial y}\right) + u\frac{\partial u}{\partial x} + v\frac{\partial u}{\partial y} = \frac{\partial(u^2)}{\partial x} + \frac{\partial(vu)}{\partial y} = U_e\frac{dU_e}{dx} + \frac{1}{\rho}\frac{\partial \tau}{\partial y}. \tag{9.38}$$

Move the term involving U_e to the other side of the last equality, and integrate (6.2) and (9.38) from $y = 0$ where $u = v = 0$ to $y = \infty$ where $u = U_e$ and $v = v_\infty$:

$$\int_0^\infty \left(\frac{\partial u}{\partial x} + \frac{\partial v}{\partial y} \right) dy = 0 \rightarrow \int_0^\infty \frac{\partial u}{\partial x} dy = -\int_0^\infty \frac{\partial v}{\partial y} dy = -[v]_0^\infty = -v_\infty, \tag{9.39}$$

$$\int_0^\infty \left(\frac{\partial(u^2)}{\partial x} + \frac{\partial(vu)}{\partial y} - U_e \frac{dU_e}{dx} \right) dy$$

$$= +\frac{1}{\rho} \int_0^\infty \frac{\partial \tau}{\partial y} dy \rightarrow \int_0^\infty \left(\frac{\partial(u^2)}{\partial x} - U_e \frac{dU_e}{dx} \right) dy + U_e v_\infty = -\frac{1}{\rho} \tau_0, \tag{9.40}$$

where τ_0 is the shear stress at $y = 0$ and $\tau = 0$ at $y = \infty$. Use the final form of (9.39) to eliminate v_∞ from (9.40), and exchange the order of integration and differentiation in the first term of (9.40):

$$\frac{d}{dx} \int_0^\infty u^2 dy - \int_0^\infty U_e \frac{dU_e}{dx} dy - U_e \int_0^\infty \frac{\partial u}{\partial x} dy = -\frac{1}{\rho} \tau_0. \tag{9.41}$$

Now, note that

$$U_e \int_0^\infty \frac{\partial u}{\partial x} dy = U_e \frac{d}{dx} \int_0^\infty u \, dy = \frac{d}{dx} \left(U_e \int_0^\infty u \, dy \right) - \frac{dU_e}{dx} \int_0^\infty u \, dy,$$

and use this to rewrite the third term on the left side of (9.41) to find:

$$\frac{d}{dx} \int_0^\infty \left(u^2 - U_e u \right) dy + \frac{dU_e}{dx} \int_0^\infty (u - U_e) \, dy = -\frac{1}{\rho} \tau_0. \tag{9.42}$$

A few final algebraic rearrangements produce:

$$\frac{1}{\rho} \tau_0 = \frac{d}{dx} \left[U_e^2 \int_0^\infty \frac{u}{U_e} \left(1 - \frac{u}{U_e} \right) dy \right] + \frac{dU_e}{dx} U_e \int_0^\infty \left(1 - \frac{u}{U_e} \right) dy, \text{ or}$$

$$\frac{1}{\rho} \tau_0 = \frac{d}{dx} [U_e^2 \theta] + U_e \delta^* \frac{dU_e}{dx}. \tag{9.43}$$

Throughout these manipulations, U_e and dU_e/dx may be moved inside or taken outside the vertical-direction integrations because they only depend on x.

Equation (9.43) is known as the *von Karman boundary-layer momentum integral equation*, and it is valid for steady laminar boundary layers and for time-averaged flow in turbulent boundary layers. It is a single ordinary differential equation that relates three unknowns θ, δ^*, and τ_0, so additional assumptions must be made or correlations provided to obtain solutions for these parameters. The search for appropriate assumptions and empirical correlations was actively pursued by many researchers in the middle of the twentieth century starting with Pohlhausen (1921) and ending with Thwaites (1949) who combined analysis and inspired guesswork with the laminar boundary-layer measurements and equation solutions known at that time to develop the approximate empirical laminar-boundary-layer solution procedure for (9.43) described in the next section.

9.6. THWAITES' METHOD

To solve (9.43) at least two additional equations are needed. Using the correlation parameter,

$$\lambda \equiv \frac{\theta^2}{\nu} \frac{dU_e}{dx}, \tag{9.44}$$

introduced by Holstein and Bohlen (1940), Thwaites (1949) developed an approximate solution to (9.43) that involves two empirical dimensionless functions $l(\lambda)$ and $H(\lambda)$,

$$\tau_0 \equiv \mu \frac{U_e}{\theta} l(\lambda) \quad \text{and} \quad \frac{\delta^*}{\theta} \equiv H(\lambda), \tag{9.45, 9.46}$$

that are listed in Table 9.1. This tabulation is identical to Thwaites' original for $\lambda \geq -0.060$ but includes the improvements recommended by Curle and Skan a few years later (see Curle, 1962) for $\lambda < -0.060$. The function $l(\lambda)$ is sometimes known as the shear correlation while $H(\lambda)$ is commonly called the shape factor.

Thwaites' method is developed from (9.43) by multiplying it with $\rho\theta/\mu U_e$:

$$\frac{\theta\tau_0}{\mu U_e} = \frac{\rho\theta}{\mu U_e} \frac{d}{dx}(U_e^2\theta) + \frac{\rho\theta}{\mu U_e}U_e\delta^*\frac{dU_e}{dx}, \quad \text{or} \quad \frac{\theta\tau_0}{\mu U_e} = 2\frac{\theta^2}{\nu}\frac{dU_e}{dx} + \frac{U_e\theta}{\nu}\frac{d\theta}{dx} + \frac{\theta^2}{\nu}\frac{\delta^*}{\theta}\frac{dU_e}{dx}. \tag{9.47}$$

The definitions of l and H allow the second equation of (9.47) to be simplified:

$$l(\lambda) = (2 + H(\lambda))\frac{\theta^2}{\nu}\frac{dU_e}{dx} + \frac{U_e}{2}\frac{d}{dx}\left(\frac{\theta^2}{\nu}\right).$$

The momentum thickness θ can be eliminated from this equation using (9.44), to find:

$$U_e\frac{d}{dx}\left(\frac{\lambda}{dU_e/dx}\right) = 2l(\lambda) - 2(2 + H(\lambda))\lambda \equiv L(\lambda). \tag{9.48}$$

Fortunately, $L(\lambda) \approx 0.45 - 6.0\lambda = 0.45 + 6.0m$ is approximately linear as shown in Figure 9.8 which is taken from Thwaites' (1949) original paper where $m = -\lambda$. With this linear fit, (9.48) can be integrated:

$$U_e\frac{d}{dx}\left(\frac{\lambda}{dU_e/dx}\right) = 0.45 - 6.0\lambda \rightarrow \frac{d}{dx}\left(\frac{\theta^2}{\nu}\right) + \frac{6.0}{U_e}\frac{dU_e}{dx}\frac{\theta^2}{\nu} = \frac{0.45}{U_e}. \tag{9.49}$$

The second version of (9.49) is a first-order linear inhomogeneous differential equation for θ^2/ν; its integrating factor is U_e^6. The resulting solution for θ^2 involves a simple

TABLE 9.1   Universal Functions for Thwaites' Method

λ	$l(\lambda)$	$H(\lambda)$
0.25	0.500	2.00
0.20	0.463	2.07
0.14	0.404	2.18
0.12	0.382	2.23
0.10	0.359	2.28
0.08	0.333	2.34
0.064	0.313	2.39
0.048	0.291	2.44
0.032	0.268	2.49
0.016	0.244	2.55
0	0.220	2.61
−0.008	0.208	2.64
−0.016	0.195	2.67
−0.024	0.182	2.71
−0.032	0.168	2.75
−0.040	0.153	2.81
−0.048	0.138	2.87
−0.052	0.130	2.90
−0.056	0.122	2.94
−0.060	0.113	2.99
−0.064	0.104	3.04
−0.068	0.095	3.09
−0.072	0.085	3.15
−0.076	0.072	3.22
−0.080	0.056	3.30
−0.084	0.038	3.39
−0.086	0.027	3.44
−0.088	0.015	3.49
−0.090	0	3.55
	(Separation)	

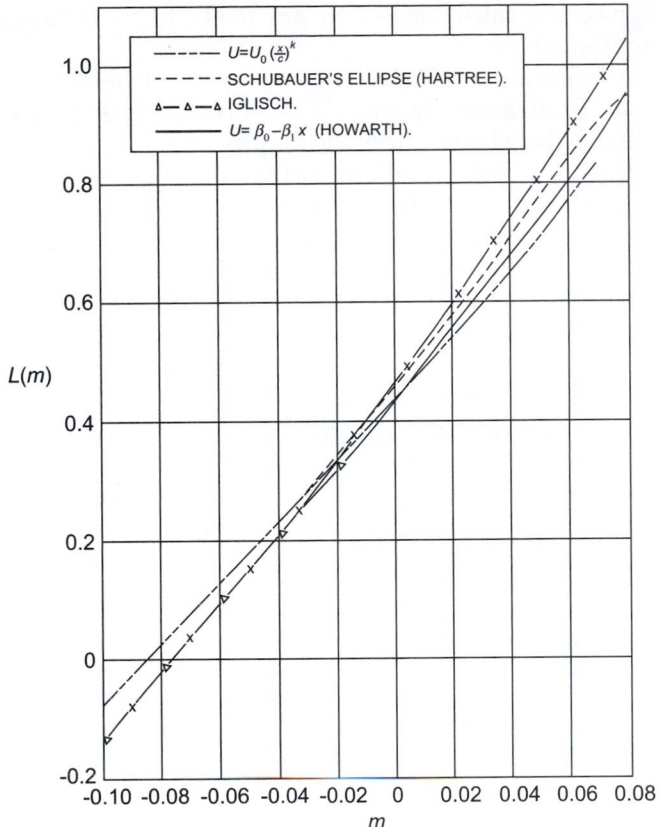

FIGURE 9.8 Plot of $L(m)$ from (9.48) vs. $m = -\lambda$ from Thwaites' 1949 paper. Here a suitable empirical fit to the four sources of laminar boundary-layer data is provided by $L(m) = 0.45 + 6.0m = 0.45 - 6.0\lambda$. *Reprinted with the permission of The Royal Aeronautical Society.*

integral of the fifth power of the free-stream velocity at the edge of the boundary layer:

$$\frac{\theta^2 U_e^6(x)}{\nu} = 0.45 \int_0^x U_e^5(x')\, dx' + \frac{\theta_0^2 U_0^6}{\nu} \quad \text{or} \quad \theta^2 = \frac{0.45\nu}{U_e^6(x)} \int_0^x U_e^5(x')\, dx' + \frac{\theta_0^2 U_0^6}{U_e^6(x)}, \qquad (9.50)$$

where x' is an integration variable, $\theta = \theta_0$, and $U_e = U_0$ at $x = 0$. If $x = 0$ is a stagnation point ($U_e = 0$), then $\theta_0 = 0$. Once the integration specified by (9.50) is complete, the surface shear stress and displacement thickness can be recovered by computing λ and then using (9.45), (9.46), and Table 9.1. Overall, the accuracy of Thwaites' method is $\pm 3\%$ or so for favorable pressure gradients, and $\pm 10\%$ for adverse pressure gradients but perhaps slightly worse near boundary-layer separation. The great strength of Thwaites' method is that it involves only one parameter (λ) and requires only a single integration. This simplicity makes it ideal for preliminary engineering calculations that are likely to be followed by more formal computations or experiments.

Before proceeding to example calculations, an important limitation of boundary-layer calculations that start from a steady presumed surface pressure distribution (such as Thwaites' method) must be stated. Such techniques can only predict the existence of boundary-layer separation; they do not reliably predict the location of boundary-layer separation. As will be further discussed in the next section, once a boundary layer separates from the surface on which it has formed, the fluid mechanics of the situation are entirely changed. First of all, the boundary-layer approximation is invalid downstream of the separation point because the layer is no longer thin and contiguous to the surface; thus, the scaling (9.6) is no longer valid. Second, separation commonly leads to unsteadiness because separated boundary layers are unstable and may produce fluctuations even if all boundary conditions are steady. And third, a separated boundary layer commonly has an enormous flow-displacement effect that drastically changes the outer flow so that it no longer imposes the presumed attached boundary-layer surface pressure distribution. Thus, any boundary-layer calculation that starts from a presumed surface pressure distribution should be abandoned once that calculation predicts the occurrence of boundary-layer separation.

The following two examples illustrate the use of Thwaites' method with and without a prediction of boundary-layer separation.

EXAMPLE 9.1

Use Thwaites' method to estimate the momentum thickness, displacement thickness, and wall shear stress of the Blasius boundary layer with $\theta_0 = 0$ at $x = 0$.

Solution

The solution plan is to use (9.50) to obtain θ. Then, because $dU_e/dx = 0$ for the Blasius boundary layer, $\lambda = 0$ at all downstream locations and the remaining boundary-layer parameters can be determined from the θ results, (9.45), (9.46), and Table 9.1. The first step is setting $U_e = U = \text{constant}$ in (9.50) with $\theta_0 = 0$:

$$\theta^2 = \frac{0.45\nu}{U^6} \int_0^x U^5 dx = \frac{0.45\nu}{U}x, \quad \text{or} \quad \theta = 0.671\sqrt{\frac{\nu x}{U}}.$$

This approximate answer is only 1% higher than the Blasius-solution value. For $\lambda = 0$, the tabulated shape factor is $H(0) = 2.61$, so

$$\delta^* = \theta\left(\frac{\delta^*}{\theta}\right) = \theta H(0) = 0.671\sqrt{\frac{\nu x}{U}}(2.61) = 1.75\sqrt{\frac{\nu x}{U}}.$$

This approximate answer is less than 2% higher than the Blasius-solution value. For $\lambda = 0$, the shear correlation value is $l(0) = 0.220$, so

$$\tau_0 = \mu\frac{U}{\theta}l(0) = \frac{\mu U}{0.671\sqrt{\nu x/U}}(0.220) = \frac{1}{2}\rho U^2(0.656)\sqrt{\frac{\nu}{Ux}},$$

which implies a skin friction coefficient of

$$C_f = \frac{\tau_0}{\frac{1}{2}\rho U^2} = \frac{0.656}{\sqrt{Re_x}},$$

which is only 1.2% below the Blasius-solution value.

EXAMPLE 9.2

A shallow-angle, two-dimensional diffuser of length L is designed for installation downstream of a blower in a ventilation system to slow the blower-outlet airflow via an increase in duct cross-sectional area (see Figure 9.9). If the diffuser should reduce the flow speed by half by doubling the flow area and the boundary layer is laminar, is boundary-layer separation likely to occur in this diffuser?

Solution

The first step is to determine the outer flow $U_e(x)$ by assuming uniform (ideal) flow within the diffuser. Then, (9.49) can be used to estimate $\theta^2(x)$ and $\lambda(x)$. Boundary-layer separation will occur if λ falls below -0.090.

For uniform incompressible flow within the diffuser: $U_1 A_1 = U_e(x)A(x)$, where (1) denotes the diffuser inlet, $U_e(x)$ is the flow speed, and $A(x)$ is the cross-sectional area. For flat diffuser sides, a doubling of the flow area in a distance L requires $A(x) = A_1(1 + x/L)$, so the ideal outer flow velocity is $U_e(x) = U_1(1 + x/L)^{-1}$. With this exterior velocity the Thwaites' integral becomes:

$$\theta^2 = \frac{0.45\nu}{U_e^6(x)} \int_0^x U_e^5(x')\, dx' + \frac{\theta_0^2 U_0^6}{U_e^6(x)} = \frac{0.45\nu}{U_1}\left(1 + \frac{x}{L}\right)^6 \int_0^x \left(1 + \frac{x}{L}\right)^{-5} dx + \theta_0^2 \left(1 + \frac{x}{L}\right)^6,$$

where $U_0 = U_1$ in this case. The 0-to-x integration is readily completed and this produces:

$$\theta^2 = \frac{0.45\nu}{U_1}\left(1 + \frac{x}{L}\right)^6 \frac{L}{4}\left[1 - \left(1 + \frac{x}{L}\right)^{-4}\right] + \theta_0^2 \left(1 + \frac{x}{L}\right)^6.$$

From this equation it is clear that θ grows with increasing x. This relationship can be converted to λ by multiplying it with $(1/\nu)dU_e/dx = -(U_1/\nu L)(1 + x/L)^{-2}$:

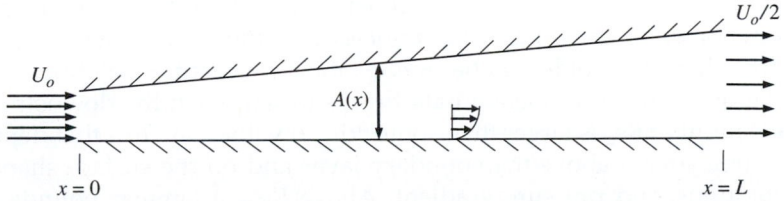

FIGURE 9.9 A simple two-dimensional diffuser of length L intended to slow the incoming flow to half its speed by doubling the flow area. The resulting adverse pressure gradient in the diffuser influences the character of the boundary layers that develop on the diffuser's inner surfaces, especially when these boundary layers are laminar.

$$\lambda = \frac{\theta^2}{\nu}\frac{dU_e}{dx} = -\frac{0.45}{4}\left[\left(1+\frac{x}{L}\right)^4 - 1\right] - \frac{\theta_0^2 U_1}{\nu L}\left(1+\frac{x}{L}\right)^4.$$

Here, we can see that even when $\theta_0 = 0$, λ will (at best) start at zero and become increasingly negative with increasing x. At this point, a determination of whether or not boundary-layer separation will occur involves calculating λ as a function of x/L. The following table comes from evaluating the last equation with $\theta_0 = 0$.

x/L	λ
0	0
0.05	−0.02424
0.1	−0.05221
0.15	−0.08426
0.2	−0.12078

Here, Thwaites' method predicts that boundary-layer separation will occur, since λ will fall below -0.090 at $x/L \approx 0.16$, a location that is far short of the end of the diffuser at $x = L$. While it is tempting to consider this a prediction of the location of boundary-layer separation, such a temptation should be avoided. In addition, if θ_0 was nonzero, then λ would decrease even more quickly than shown in the table, making the positive prediction of boundary-layer separation even more firm. (In reality, diffusers in duct work and flow systems are common but they typically operate with turbulent boundary layers that more effectively resist separation.)

9.7. TRANSITION, PRESSURE GRADIENTS, AND BOUNDARY-LAYER SEPARATION

The analytical and empirical results provided in the prior sections are altered when a boundary layer transitions from laminar to turbulent flow, and when a boundary layer separates from the surface on which it has developed. Both of these phenomena, especially the second, are influenced by the pressure gradient felt by the boundary layer.

The process of changing from laminar to turbulent flow is called *transition*, and it occurs in a wide variety of flows as the Reynolds number increases. For the present purposes, the complicated phenomenon known as *boundary-layer transition* is described in general terms. Interestingly for a high Reynolds number theory, the agreement of solutions to the laminar boundary equations with experimental data breaks down when the downstream-distance-based Reynolds number Re_x is larger than some critical value, say Re_{cr}, that depends on fluctuations in the free stream above the boundary layer and on the surface shape, curvature, roughness, vibrations, and pressure gradient. Above Re_{cr}, a laminar boundary-layer flow becomes unstable and transitions to turbulence. Typically, the critical Reynolds number decreases when the surface roughness or free-stream fluctuation levels increase. In general, Re_{cr} varies greatly and detailed predictions of transition are often a difficult task or research endeavor. Within a factor of five or so, the transition Reynolds number for a smooth, flat-plate boundary layer is found to be

$$\text{Re}_{cr} \sim 10^6 \text{ (flat plate)}.$$

Figure 9.10 schematically depicts the flow regimes on a semi-infinite flat plate (with the vertical direction greatly exaggerated). In the leading-edge region, $\text{Re}_x = Ux/\nu \sim 1$, the full Navier-Stokes equations are required to properly describe the flow. As Re_x increases toward the downstream limit of the leading-edge region, we can locate x_0 as the maximal upstream location where the laminar boundary-layer equations are valid. For some distance $x > x_0$, the boundary layer's condition at $x = x_0$ is remembered. Eventually, the influence of the initial condition may be neglected and the solution becomes of similarity form. For somewhat larger Re_x, a bit farther downstream, an initial instability appears and fluctuations of a specific wavelength or frequency may be amplified. With increasing downstream distance, a wider spatial or temporal frequency range of fluctuations may be amplified and these fluctuations interact with each other nonlinearly through the advective acceleration terms in the momentum equation. As Re_x increases further, the fluctuations may increase in strength and the flow becomes increasingly chaotic and irregular with increasing downstream distance. When the fluctuations cease their rapid growth, the flow is said to be fully turbulent and transition is complete.

Laminar and turbulent boundary layers differ in many important ways. A fully turbulent boundary layer produces significantly more average surface shear stress τ_0 than an

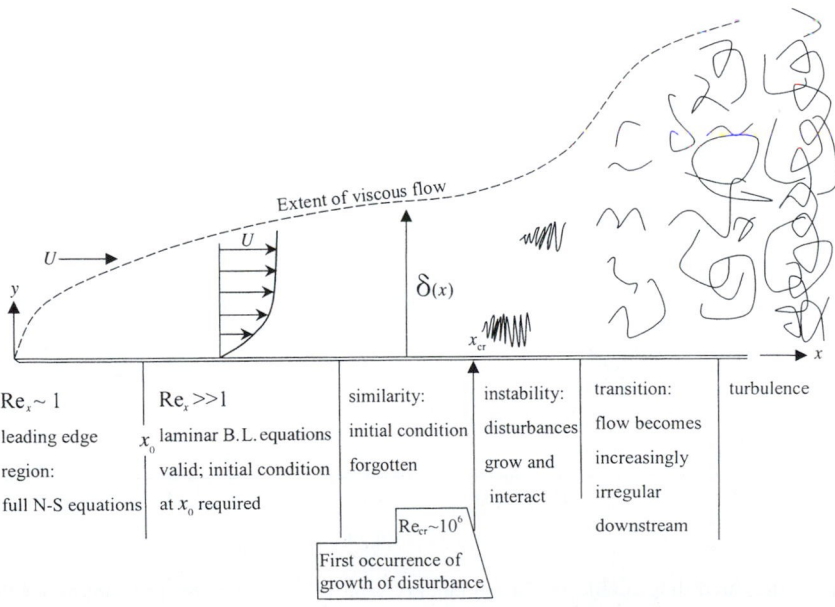

FIGURE 9.10   Schematic depiction of flow over a semi-infinite flat plate. Here, increasing x is synonymous with increasing Reynolds number.

equivalent laminar boundary layer, and a fully turbulent boundary-layer velocity profile has a different shape and different parametric dependencies than an equivalent laminar one. For example, the thickness of a zero-pressure-gradient turbulent boundary layer grows faster than $x^{1/2}$ (Figure 9.10), and the wall shear stress increases faster with U than in a laminar boundary layer where $\tau_0 \propto U^{3/2}$. This increase in friction occurs because turbulent fluctuations produce more wall-normal transport of momentum than that possible from steady viscous diffusion alone. However, both types of boundary layers respond similarly to pressure gradients but with different sensitivities.

Figure 9.11 sketches the nature of the observed variation of the drag coefficient in a flow over a flat plate, as a function of the Reynolds number. The lower curve applies if the boundary layer is laminar over the entire length of the plate, and the upper curve applies if the boundary layer is turbulent over the entire length. The curve joining the two applies to a boundary layer that is laminar over the initial part of the plate, begins transition at $Re_L \sim 5 \times 10^5$, and is fully turbulent for $Re_L > 10^7$. The exact point at which the observed drag deviates from the wholly laminar behavior depends on flow conditions, flow geometry, and surface conditions.

Although surface pressure gradients do affect transition, it may be argued that their most important influence is on boundary-layer separation. A fundamental discussion of boundary-layer separation begins with the steady stream-wise, boundary-layer flow momentum equation, (9.9), where the pressure gradient is found from the external velocity field via (9.11) and with x taken in the stream-wise direction along the surface of interest. At the surface, both velocity components are zero so (9.9) reduces to

$$\mu \left(\frac{\partial^2 u}{\partial y^2} \right)_{wall} = \frac{dp}{dx}.$$

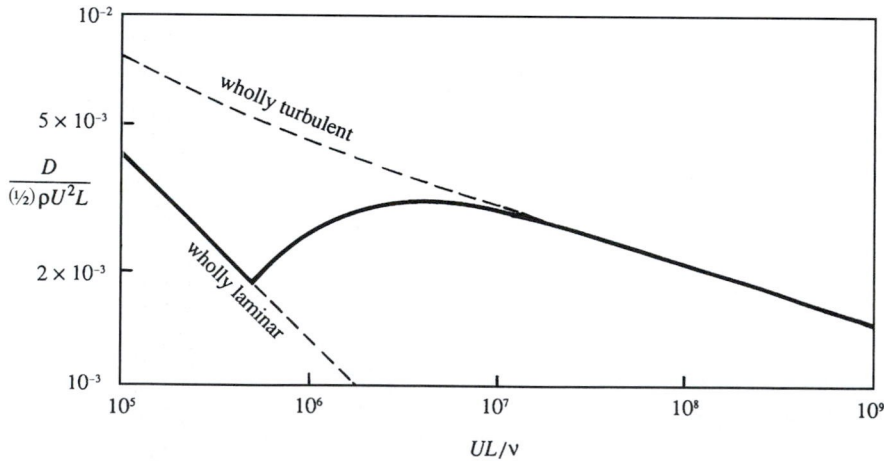

FIGURE 9.11   Measured drag coefficient for a boundary layer over a flat plate. The continuous line shows the drag coefficient for a plate on which the flow is partly laminar and partly turbulent, with the transition taking place at a position where the local Reynolds number is 5×10^5. The dashed lines show the behavior if the boundary layer was either completely laminar or completely turbulent over the entire length of the plate.

In an accelerating stream $dp/dx < 0$, so

$$\left(\frac{\partial^2 u}{\partial y^2}\right)_{wall} < 0 \quad \text{(accelerating)}. \tag{9.51}$$

Given that the velocity profile has to blend smoothly with the external profile, the gradient $\partial u/\partial y$ slightly below the edge of the boundary layer decreases with increasing y from a positive value to zero; therefore, $\partial^2 u/\partial y^2$ slightly below the boundary-layer edge is negative. Equation (9.51) then shows that $\partial^2 u/\partial y^2$ has the same sign at the wall and at the boundary-layer edge, and presumably throughout the boundary layer. In contrast, for a decelerating external stream, $dp/dx > 0$, the curvature of the velocity profile at the wall is

$$\left(\frac{\partial^2 u}{\partial y^2}\right)_{wall} > 0 \quad \text{(decelerating)}, \tag{9.52}$$

so that the profile curvature changes sign somewhere within the boundary layer. In other words, the boundary-layer profile in a decelerating flow has a *point of inflection* where $\partial^2 u/\partial y^2 = 0$, an important fact for boundary-layer stability and transition (see Chapter 11). In the special case of the Blasius boundary layer, the profile's inflection point is at the wall.

The shape of the velocity profiles in Figure 9.12 suggests that a decelerating exterior flow tends to increase the thickness of the boundary layer. This can also be seen from the continuity equation:

$$v(y) = -\int_0^y (\partial u/\partial x)dy.$$

Compared to flow over a flat plate, a decelerating external stream causes a larger $-(\partial u/\partial x)$ within the boundary layer because the deceleration of the outer flow adds to the viscous

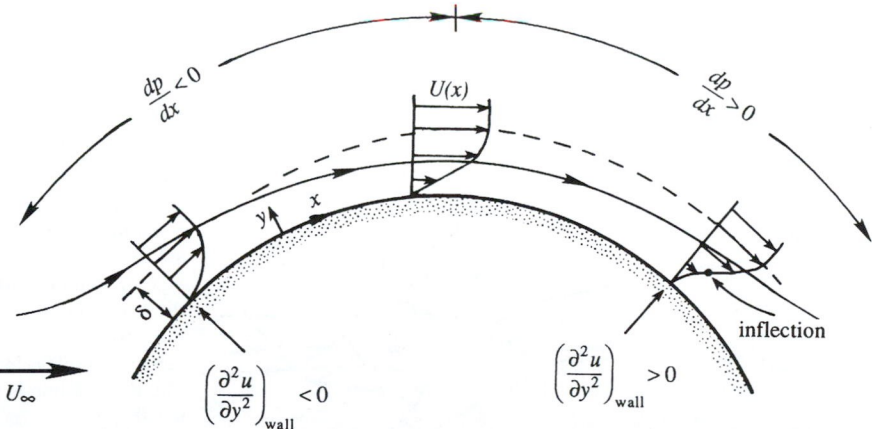

FIGURE 9.12   Velocity profiles across boundary layers with favorable $(dp/dx < 0)$ and adverse $(dp/dx > 0)$ pressure gradients, as indicated above the flow. The surface shear stress and stream-wise fluid velocity near the surface are highest and lowest in the favorable and adverse pressure gradients, respectively, with the $dp/dx = 0$ case falling between these limits.

deceleration within the boundary layer. It follows from the foregoing equation that the v-field, directed away from the surface, is larger for a decelerating flow. The boundary layer therefore thickens not only by viscous diffusion but also by advection away from the surface, resulting in a more rapid increase in the boundary-layer thickness with x than when the exterior flow is constant or accelerating.

If p falls with increasing x, $dp/dx < 0$, the pressure gradient is said to be *favorable*. If the p rises with increasing x, $dp/dx > 0$, the pressure gradient is said to be *adverse*. In an adverse pressure gradient, the boundary-layer flow decelerates, thickens, and develops a point of inflection. When the adverse pressure gradient is strong enough or acts over a long enough distance, the flow next to the wall reverses direction (Figure 9.13). The point S at which the reverse flow meets the forward flow is a local stagnation point and is known as the *separation point*. Fluid elements approach S (from either side) and are then transported away from the wall. Thus, a separation streamline emerges from the surface at S. Furthermore, the surface shear stress changes sign across S because the surface flow changes direction. Thus, the surface shear stress at S is zero, which implies

$$\left(\frac{\partial u}{\partial y}\right)_{wall} = 0 \quad \text{(separation).}$$

Once a boundary layer separates from the surface on which it has formed, the surface displacement effect produced by divergence of the body contour and the separation streamline may be enormous. Additionally, at high Reynolds numbers, a separated boundary layer commonly acquires the properties of a vortex sheet and may rapidly become unstable and transition to a thick zone of turbulence. Thus, boundary-layer separation typically requires the presumed geometry-based, inner-outer and rotational-irrotational flow dichotomies to be reconsidered or even abandoned. In such cases, recourse to experiments or multidimensional numerical solutions may be the only choices for flow investigation.

At Reynolds numbers that are not too large, flow separation may not lead to unsteadiness. For flow past a circular cylinder for $4 < \text{Re} < 40$ the reversed flow downstream of a separation point may form part of a steady vortex behind the cylinder (see Figure 9.16 in Section 9.8). At

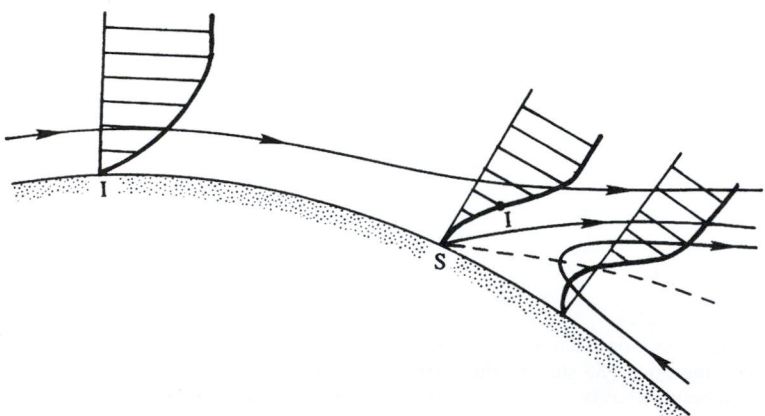

FIGURE 9.13 Streamlines and velocity profiles near a separation point S where a streamline emerges from the surface. The usual boundary-layer equations are not valid downstream of S. The inflection point in the stream-wise velocity profile is indicated by I. The dashed line is the locus of $u = 0$.

higher Reynolds numbers, when the flow on the upstream side of the cylinder develops genuine boundary-layer characteristics, the flow downstream of separation is unsteady and frequently turbulent.

The adverse-pressure gradient strength that a boundary layer can withstand without separating depends on the geometry of the flow and whether the boundary layer is laminar or turbulent. However, a severe adverse-pressure gradient, such as that behind a blunt body, invariably leads to a separation. In contrast, the boundary layer on the trailing surface of a slender body may overcome the weak pressure gradients involved. Therefore, to avoid separation and the resulting form drag penalty, the trailing section of a submerged body should be *gradually* reduced in size, giving it a *streamlined* (or teardrop) shape.

Experimental evidence indicates that the point of separation is relatively insensitive to the Reynolds number as long as the boundary layer is laminar. However, a *transition to turbulence delays boundary-layer separation*; that is, a turbulent boundary layer is more capable of withstanding an adverse pressure gradient. This is because the velocity profile in a turbulent boundary layer places more high-speed fluid near the surface (Figure 9.14). For example, the laminar boundary layer over a circular cylinder separates at ~82° from the forward stagnation point, whereas a turbulent layer over the same body separates at 125° (shown later in Figure 9.16). Experiments show that the surface pressure remains fairly uniform downstream of separation and has a lower value than the pressures on the forward face of the body. The resulting drag due to such fore-aft pressure differences is called *form drag*, as it depends crucially on the shape of the body (and the location of boundary-layer separation). For a blunt body like a sphere, the form drag is larger than the skin friction drag because of the occurrence of separation. For a streamlined body like a rowing shell for crew races, skin friction is generally larger than the form drag. As long as the separation point is located at the same place on the body, the drag coefficient of a blunt body is nearly constant at high Reynolds numbers. However, the drag coefficient drops suddenly when the boundary layer

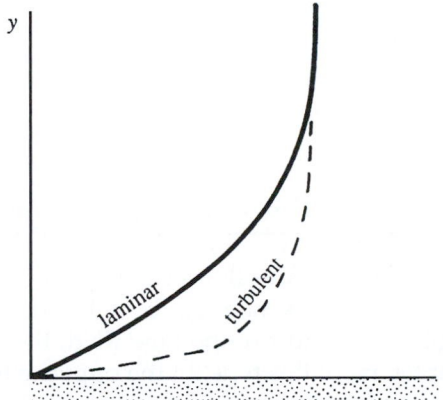

FIGURE 9.14 Nominal comparison of laminar and mean-turbulent, stream-wise velocity profiles in a boundary layer. Here the primary differences are the presence of higher speed fluid closer to the surface and greater surface shear stress in the turbulent layer.

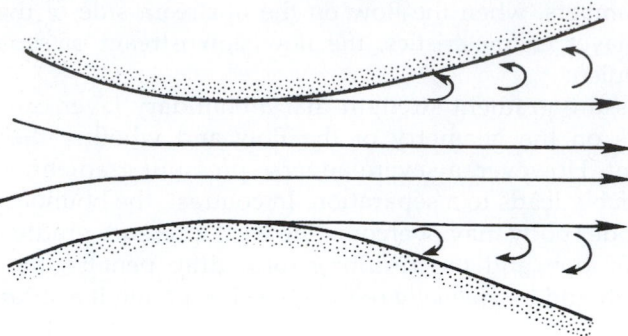

FIGURE 9.15 Separation of flow in a divergent channel. Here, an adverse pressure gradient has led to boundary-layer separation just downstream of the narrowest part of the channel. Such separated flows are instable and are exceedingly likely to be unsteady, even if all the boundary conditions are time independent.

undergoes transition to turbulence, the separation point moves aft, and the body's wake becomes narrower (see Figure 9.21 in Section 9.8).

Boundary-layer separation may take place in internal as well as external flows. An example is a divergent channel or diffuser (Example 9.2, Figure 9.15). Downstream of a narrow point in a ducted flow, an adverse-pressure gradient can cause separation. Elbows, tees, and valves in pipes and tubes commonly lead to regions of internal flow separation, too.

Again it must be emphasized that the boundary-layer equations are valid only as far downstream as the point of separation, if it is known. Beyond separation, the basic underlying assumptions of boundary-layer theory become invalid. Moreover, the parabolic character of the boundary-layer equations requires that a numerical integration is possible only in the direction of advection (along which information is propagated). In a region of reversed flow, this integration direction is *upstream*. Thus, a forward (downstream) integration of the boundary-layer equations breaks down after separation. Furthermore, potential theory may not be used to determine the pressure in a separated flow region, since the flow there is rotational and the effective boundary between irrotational and rotational flow regions is no longer the solid surface but some unknown shape encompassing part of the body's contour, the separation streamline, and, possibly, a wake-zone contour.

9.8. FLOW PAST A CIRCULAR CYLINDER

In general, analytical solutions of viscous flows can be found (possibly in terms of perturbation series) only in two limiting cases, namely Re ≪ 1 and Re ≫ 1. In the Re ≪ 1 limit the inertia forces are negligible over most of the flow field; the Stokes-Oseen solutions discussed in the preceding chapter are of this type. In the opposite limit of Re ≫ 1, the viscous forces are negligible everywhere except close to the surface, and a solution may be attempted by matching an irrotational outer flow with a boundary layer near the surface. In the intermediate range of Reynolds numbers, finding analytical solutions becomes almost an impossible task, and one has to depend on experimentation and numerical solutions. Some of these experimental flow patterns are described in this section, taking the flow over a circular cylinder as an example. Instead of discussing only the intermediate Reynolds number range, we shall describe the experimental data for the entire range from small to very high Reynolds numbers.

Low Reynolds Numbers

Consider creeping flow around a circular cylinder, characterized by $Re = U_\infty d/\nu < 1$, where U_∞ is the upstream flow speed and d is the cylinder's diameter. Vorticity is generated close to the surface because of the no-slip boundary condition. In the Stokes approximation this vorticity is simply diffused, not advected, which results in a fore and aft symmetry. The Oseen approximation partially takes into account the advection of vorticity, and results in an asymmetric velocity distribution *far* from the body (which was shown in Figure 8.20). The vorticity distribution is qualitatively analogous to the dye distribution caused by a source of colored fluid at the position of the body. The color diffuses symmetrically in very slow flows, but at higher flow speeds the dye source is confined behind a parabolic boundary with the dye source at the parabola's focus.

For increasing Re above unity, the Oseen approximation breaks down, and the vorticity is increasingly confined behind the cylinder because of advection. For Re > 4, two small steady eddies appear behind the cylinder and form a closed separation zone contained with a separation streamline. This zone is sometimes called a *separation bubble*. The cylinder's wake is completely laminar and the vortices rotate in a manner that is consistent with the exterior flow (Figure 9.16). These eddies grow in length and width as Re increases.

Moderate Reynolds Numbers

A very interesting sequence of events begins to develop when Re reaches 40, the point at which the wake behind the cylinder becomes unstable. Experiments show that for Re ~ 10^2 the wake develops a slow oscillation in which the velocity is periodic in time and downstream distance, with the amplitude of the oscillation increasing downstream. The oscillating wake rolls up into two staggered rows of vortices with opposite sense of rotation (Figure 9.17). Von Karman investigated the phenomenon as a problem of superposition of irrotational vortices; he concluded that a nonstaggered row of vortices is unstable, and a staggered row is stable only if the ratio of lateral distance between the vortices to their longitudinal distance is 0.28. Because of the similarity of the wake with footprints in a street, the staggered row of vortices behind a blunt body is called a *von Karman vortex street*. The vortices move downstream at a speed smaller than U_∞. This means that the vortex pattern slowly follows the cylinder if it is pulled through a stationary fluid.

In the range 40 < Re < 80, the vortex street does not interact with the pair of attached vortices. As Re increases above 80, the vortex street forms closer to the cylinder, and the attached eddies (whose downstream length has now grown to be about twice the diameter of the cylinder) themselves begin to oscillate. Finally the attached eddies periodically break off alternately from the two sides of the cylinder. While an eddy on one side is shed, that on the other side forms, resulting in an unsteady flow near the cylinder. As vortices of opposite circulations are shed off alternately from the two sides, the circulation around the cylinder

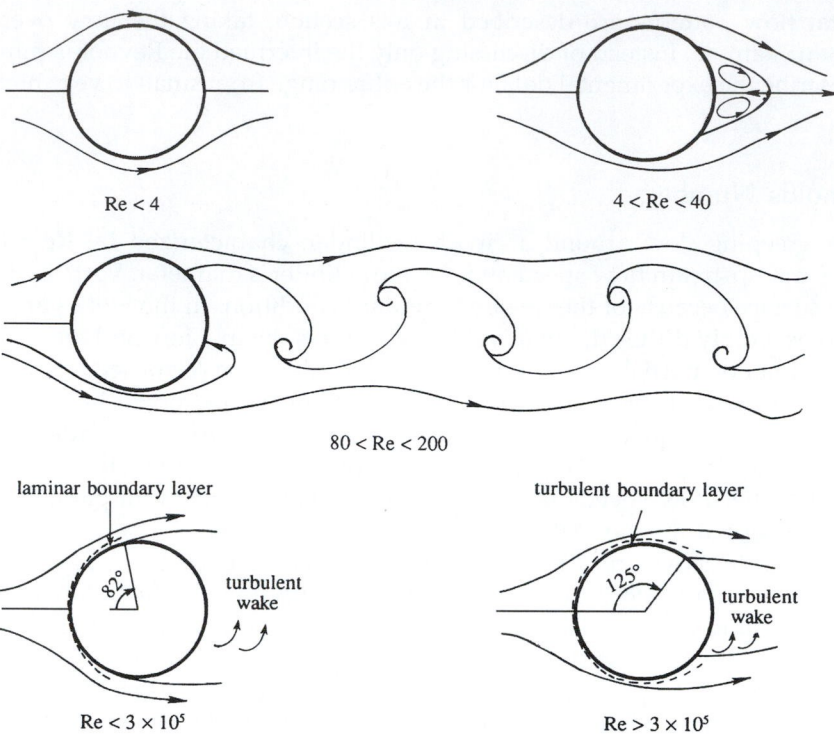

FIGURE 9.16 Depiction of some of the flow regimes for a circular cylinder in a steady uniform cross flow. Here $Re = U_\infty d/\nu$ is the Reynolds number based on free-stream speed U_∞ and cylinder diameter d. At the lowest Re, the streamlines approach perfect fore-aft symmetry. As Re increases, asymmetry increases and steady wake vortices form. With further increase in Re, the wake becomes unsteady and forms the alternating-vortex von Karman vortex street. For Re up to $Re_{cr} \sim 3 \times 10^5$, the laminar boundary layer separates approximately 82° from the forward separation point. Above this Re value, the boundary-layer transitions to turbulence, and separation is delayed to 125° from the forward separation point.

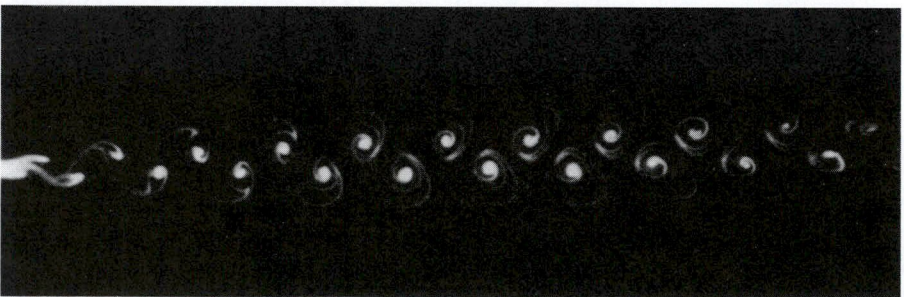

FIGURE 9.17 Von Karman vortex street downstream of a circular cylinder at $Re = 55$. *Flow visualized by condensed milk. S. Taneda, Jour. Phys. Soc., Japan* **20**: *1714–1721, 1965, and reprinted with the permission of The Physical Society of Japan and Dr. Sadatoshi Taneda.*

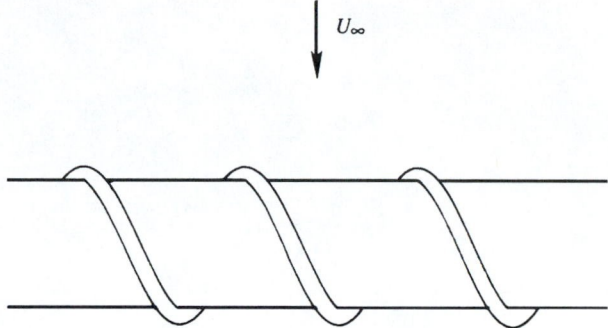

FIGURE 9.18 Spiral blades used for breaking up the span-wise coherence of vortex shedding from a cylindrical rod. Coherent vortex shedding can produce tonal noise and potentially large (and undesired) structural loads on engineered devices that encounter wind or water currents.

changes sign, resulting in an oscillating lift or lateral force perpendicular to the upstream flow direction. If the frequency of vortex shedding is close to the natural frequency of some structural mode of vibration of the cylinder and its supports, then an appreciable lateral vibration may be observed. Engineered structures such as suspension bridges, oil drilling platforms, and even automobile components are designed to prevent coherent shedding of vortices from cylindrical structures. This is done by including spiral blades protruding out of the cylinder's surface, which break up the span-wise coherence of vortex shedding, forcing the vortices to detach at different times along the length of these structures (Figure 9.18).

The passage of regular vortices causes velocity measurements in the cylinder's wake to have a dominant periodicity, and this frequency Ω is commonly expressed as a *Strouhal number* (4.102), $St = \Omega d / U_\infty$. Experiments show that for a circular cylinder the value of S remains close to 0.2 for a large range of Reynolds numbers. For small values of cylinder diameter and moderate values of U_∞, the resulting frequencies of the vortex shedding and oscillating lift lie in the acoustic range. For example, at $U_\infty = 10$ m/s and a wire diameter of 2 mm, the frequency corresponding to a Strouhal number of 0.2 is $n = 1000$ cycles per second. The *singing* of telephone and electrical transmission lines and automobile radio antennae have been attributed to this phenomenon. The value of the St given here is that observed in three-dimensional flows with nominally two-dimensional boundary conditions. Moving soap-film experiments and calculations suggest a somewhat higher value of $St = 0.24$ in perfectly two-dimensional flow (see Wen & Lin, 2001).

Below Re = 200, the vortices in the wake are laminar and continue to be so for very large distances downstream. Above 200, the vortex street becomes unstable and irregular, and the flow within the vortices themselves becomes chaotic. However, the flow in the wake continues to have a strong frequency component corresponding to a Strouhal number of $S = 0.2$. However, above a Reynolds number of several thousand, periodicity in the wake is only perceptible near the cylinder, and the wake may be described as fully turbulent beyond several cylinder diameters downstream.

Striking examples of vortex streets have also been observed in stratified atmospheric flows. Figure 9.19 shows a satellite photograph of the wake behind several isolated mountain

FIGURE 9.19   A von Karman vortex street downstream of mountain peaks in a strongly stratified atmosphere. There are several mountain peaks along the linear, light-colored feature running diagonally in the upper left-hand corner of the photograph. North is upward, and the wind is blowing toward the southeast. *R. E. Thomson and J. F. R. Gower,* Monthly Weather Review *105: 873–884, 1977; reprinted with the permission of the American Meteorological Society.*

peaks through which the wind is blowing toward the southeast. The mountains pierce through the cloud level, and the flow pattern becomes visible by the cloud pattern. The wakes behind at least two mountain peaks display the characteristics of a von Karman vortex street. The strong density stratification in this flow has prevented vertical motions, giving the flow the two-dimensional character necessary for the formation of vortex streets.

High Reynolds Numbers

At high Reynolds numbers the frictional effects upstream of separation are confined near the surface of the cylinder, and the boundary-layer approximation is valid as far downstream

FIGURE 9.20 Surface pressure distribution around a circular cylinder at subcritical and super-critical Reynolds numbers. Note that the pressure is nearly constant within the wake and that the wake is narrower for flow at supercritical Re. The change in the top- and bottom-side, boundary-layer separation points near Re_{cr} is responsible for the change in C_p shown.

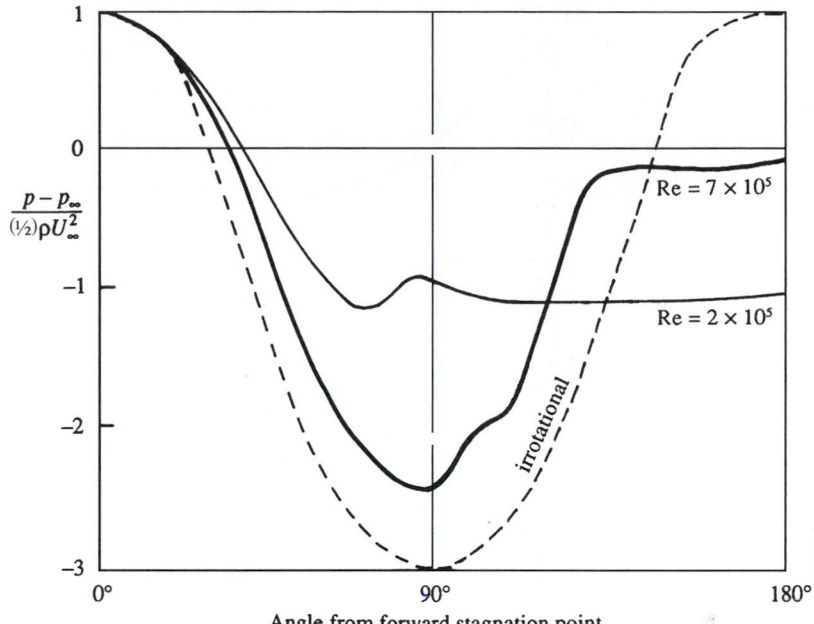

as the point of separation. For a smooth cylinder up to $Re < 3 \times 10^5$, the boundary layer remains laminar, although the wake formed behind the cylinder may be completely turbulent. The laminar boundary layer separates at $\approx 82°$ from the forward stagnation point (Figure 9.16). The pressure in the wake downstream of the point of separation is nearly constant and lower than the upstream pressure (Figure 9.20). The drag on the cylinder in this Re range is primarily due to the asymmetry in the pressure distribution caused by boundary-layer separation, and, since the point of separation remains fairly stationary in this Re range, the cylinder's drag coefficient C_D also stays constant at a value near unity (see Figure 9.21).

Important changes take place beyond the critical Reynolds number of $Re_{cr} \sim 3 \times 10^5$. When $3 \times 10^5 < Re < 3 \times 10^6$, the laminar boundary layer becomes unstable and transitions to turbulence. Because of its greater average near-surface flow speed, a turbulent boundary layer is able to overcome a larger adverse-pressure gradient. In the case of a circular cylinder the turbulent boundary layer separates at $125°$ from the forward stagnation point, resulting in a thinner wake and a pressure distribution more similar to that of potential flow. Figure 9.20 compares the pressure distributions around the cylinder for two values of Re, one with a laminar and the other with a turbulent boundary layer. It is apparent that the pressures within the wake are higher when the boundary layer is turbulent, resulting in a drop in the drag coefficient from 1.2 to 0.33 at the point of transition. For values of $Re > 3 \times 10^6$, the separation point slowly moves upstream as the Reynolds number increases, resulting in a mild increase of the drag coefficient (Figure 9.21).

It should be noted that the critical Reynolds number at which the boundary layer undergoes transition is strongly affected by two factors, namely the intensity of fluctuations

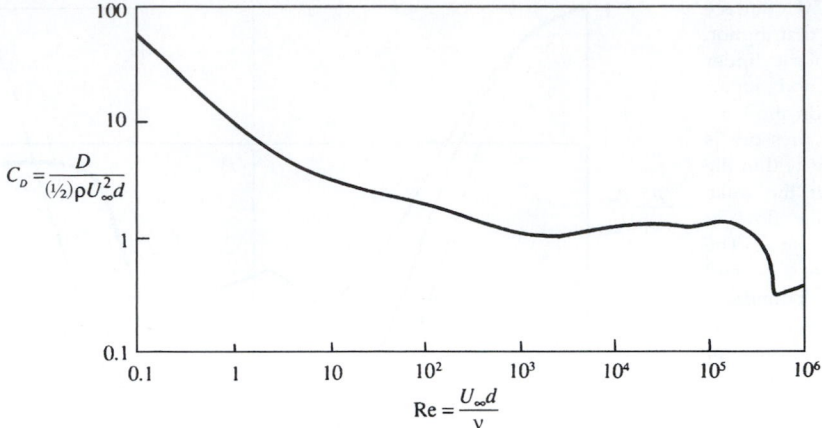

FIGURE 9.21   Measured drag coefficient, C_D, of a smooth circular cylinder vs. $Re = U_\infty d/\nu$. The sharp dip in C_D near Re_{cr} is due to the transition of the boundary layer to turbulence, and the consequent downstream movement of the point of separation and change in the cylinder's surface pressure distribution.

existing in the approaching stream and the roughness of the surface, an increase in either of which decreases Re_{cr}. The value of 3×10^5 is found to be valid for a smooth circular cylinder at low levels of fluctuation of the oncoming stream.

We close this section by noting that this flow illustrates three instances where the solution is counterintuitive. First, small causes can have large effects. If we solve for the flow of a fluid with zero viscosity around a circular cylinder, we obtain the results of Section 6.3. The inviscid flow has fore-aft symmetry and the cylinder experiences zero drag. The bottom two panels of Figure 9.16 illustrate the flow for small viscosity. In the limit as viscosity tends to zero, the flow must look like the last panel in which there is substantial fore-aft asymmetry, a significant wake, and significant drag. This is because of the necessity of a boundary layer and the satisfaction of the no-slip boundary condition on the surface so long as viscosity is not exactly zero. When viscosity is exactly zero, there is no boundary layer and there is slip at the surface. Thus, the resolution of d'Alembert's paradox lies in the existence of, and an understanding of, the boundary layer.

The second instance of counterintuitivity is that symmetric problems can have nonsymmetric solutions. This is evident in the intermediate Reynolds number middle panel of Figure 9.16. Beyond a Reynolds number of ≈ 40, the symmetric wake becomes unstable and a pattern of alternating vortices called a *von Karman vortex street* is established. Yet the equations and boundary conditions are symmetric about a central plane in the flow. If one were to solve only a half problem, assuming symmetry, a solution would be obtained, but it would be unstable to infinitesimal disturbances and unlikely to be observed in a laboratory.

The third instance of counterintuitivity is that there is a range of Reynolds numbers where roughening the surface of the body can reduce its drag. This is true for all blunt bodies. In this range of Reynolds numbers, the boundary layer on the surface of a blunt body is laminar, but sensitive to disturbances such as surface roughness, which would cause earlier transition of the boundary layer to turbulence than would occur on a smooth body. Although the skin

FIGURE 9.22 Measured drag coefficient, C_D, of a smooth sphere vs. $Re = U_\infty d/\nu$. The Stokes solution is $C_D = 24/Re$, and the Oseen solution is $C_D = (24/Re)\ (1 + 3Re/16)$; these two solutions are discussed at the end of Chapter 8. The increase of drag coefficient in the range A–B has relevance in explaining why the flight paths of sports balls bend in the air.

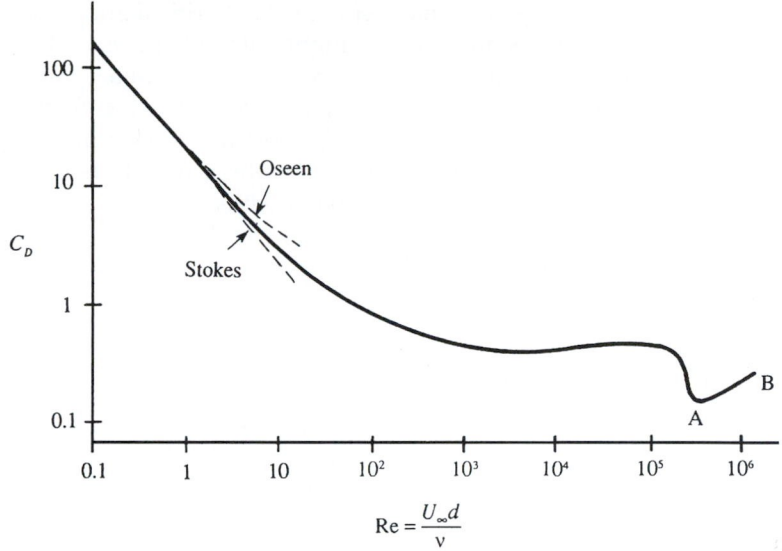

friction of a turbulent boundary layer is much larger than that of a laminar boundary layer, most of the drag on a bluff body is caused by incomplete pressure recovery on its downstream side as shown in Figure 9.20, rather than by skin friction. In fact, it is because the skin friction of a turbulent boundary layer is much larger—as a result of a larger velocity gradient at the surface—that a turbulent boundary layer can remain attached farther on the downstream side of a blunt body, leading to a narrower wake, more complete pressure recovery, and reduced drag. The drag reduction attributed to the turbulent boundary layer is shown in Figure 9.21 for a circular cylinder and Figure 9.22 for a sphere.

9.9. FLOW PAST A SPHERE AND THE DYNAMICS OF SPORTS BALLS

Several features of the description of flow over a circular cylinder qualitatively apply to flows over other two-dimensional blunt bodies. For example, a vortex street is observed in a flow perpendicular to a finite flat plate. The flow over a three-dimensional body, however, has one fundamental difference in that a regular vortex street is absent. For flow around a sphere at low Reynolds numbers, there is an attached eddy in the form of a doughnut-shaped ring; in fact, an axial section of the flow looks similar to that shown in Figure 9.16 for the range $4 < Re < 40$. For $Re > 130$ the ring-eddy oscillates, and some of it breaks off periodically in the form of distorted vortex loops.

The behavior of the boundary layer around a sphere is similar to that around a circular cylinder. In particular it undergoes transition to turbulence at a critical Reynolds number of $Re_{cr} \sim 5 \times 10^5$, which corresponds to a sudden dip of the drag coefficient (Figure 9.22). As in the case of a circular cylinder, the *separation point slowly moves upstream for postcritical Reynolds numbers*, accompanied by a rise in the drag coefficient. The behavior of the separation point for flow around a sphere at subcritical and supercritical Reynolds numbers is responsible for the bending in the flight paths of sports balls.

In many sports (tennis, cricket, soccer, ping-pong, baseball, golf, etc.), the trajectory of a moving ball may bend in potentially unexpected ways. Such bending may be known as *curve, swing, hook, swerve, slice,* etc. The problem has been investigated by wind-tunnel tests and by stroboscopic photographs of flight paths in field tests, a summary of which was given by Mehta (1985). Evidence indicates that the mechanics of trajectory bending is different for spinning and nonspinning balls. The following discussion gives a qualitative explanation of the mechanics of sports-ball trajectory bending. (Readers not interested in sports may omit the rest of this section!)

Cricket Ball Dynamics

A cricket ball has a prominent (1-mm high) seam, and tests show that the orientation of the seam is responsible for the bending of the ball's flight path. It is known to bend when thrown at speeds of around 30 m/s, which is equivalent to a Reynolds number of Re $= U_\infty d/\nu \sim 10^5$, U_∞ is the speed of the ball, and d is its diameter. This Re is somewhat less than the critical value of $Re_{cr} = 5 \times 10^5$ necessary for transition of the boundary layer on a smooth sphere into turbulence. However, the presence of the seam is able to trip the laminar boundary layer into turbulence on one side of the ball (the lower side in Figure 9.23), while the boundary layer on the other side remains laminar. This transition asymmetry leads to boundary-layer separation asymmetry. Typically, the boundary layer on the laminar side separates at $\approx 85°$, whereas that on the turbulent side separates at $120°$. Compared to region B, the surface pressure near region A is therefore closer to that given by the potential flow theory (which predicts a suction pressure of $(p_{\min} - p_\infty)/(1/2)\rho U_\infty^2 = -5/4$; see (6.91)). In other words, the pressures are lower on side A, resulting in a downward force on the ball. (Note that Figure 9.23 is a view of the flow pattern looking downward on the ball, so that it corresponds to a ball that bends to the left in its flight. The flight of a cricket ball oriented as in Figure 9.23 is called an *outswinger* in cricket literature, in contrast to an *inswinger* for which the seam is oriented in the opposite direction so as to generate an upward force in Figure 9.23.)

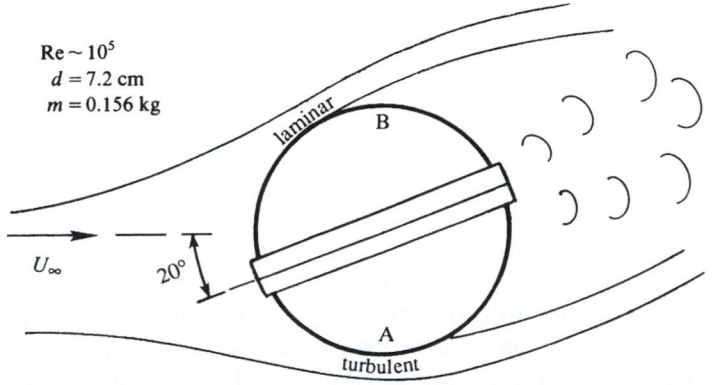

$Re \sim 10^5$
$d = 7.2$ cm
$m = 0.156$ kg

laminar

B

U_∞

$20°$

A

turbulent

FIGURE 9.23 The swing (or curve) of a cricket ball. The seam is oriented in such a way that a difference in boundary-layer separation points on the top and bottom sides of the ball lead to a downward lateral force in the figure; the surface pressure at A is less than the surface pressure at B.

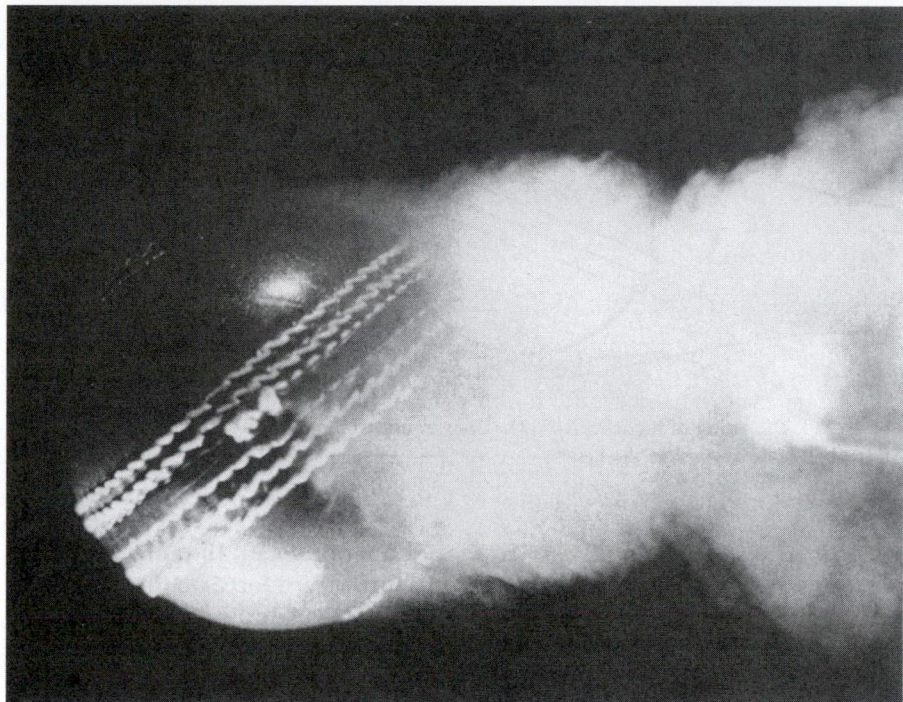

FIGURE 9.24 Smoke photograph of flow over a cricket ball in the same orientation and flow condition as that depicted in Figure 9.23. The flow is from left to right, the seam angle is 40°, the flow speed is 17 m/s, and Re = 0.85×10^5. R. Mehta, Ann. Rev Fluid Mech. *17*: 151–189, 1985. *Photograph reproduced with permission from the Annual Review of Fluid Mechanics, Vol. 17 © 1985 by Annual Reviews, www.AnnualReviews.org.*

Figure 9.24, a photograph of a cricket ball in a wind-tunnel experiment, clearly shows the delayed separation on the seam side. Note that the wake has been deflected upward by the presence of the ball, implying that an upward force has been exerted by the ball on the fluid. It follows that a downward force has been exerted by the fluid on the ball.

In practice some spin is invariably imparted to the ball. The ball is held along the seam and, because of the round arm action of the bowler, some backspin is always imparted *along the seam*. This has the important effect of stabilizing the orientation of the ball and preventing it from wobbling. A typical cricket ball can generate side forces amounting to almost 40% of its weight. A constant lateral force oriented in the same direction causes a deflection proportional to the square of time. The ball therefore travels in a parabolic path that can bend as much as 0.8 m by the time it reaches the batsman.

It is known that the trajectory of the cricket ball does not bend if the ball is thrown too slow or too fast. In the former case even the presence of the seam is not enough to trip the boundary layer into turbulence, and in the latter case the boundary layer on both sides could be turbulent; in both cases an asymmetric flow is prevented. It is also clear why only a new, shiny ball is able to swing, because the rough surface of an old ball causes the boundary layer to become turbulent on both sides. Fast bowlers in cricket maintain one hemisphere of the ball in a smooth state by constant polishing. It therefore seems that most of the known facts about the swing of a cricket ball have been adequately explained by scientific research. The feature that has not been explained is the universally observed fact that a cricket ball swings more in humid conditions. The changes in density and viscosity due to changes in humidity can change the Reynolds number by only 2%, which cannot explain this phenomenon.

Tennis Ball Dynamics

Unlike the cricket ball, the path of the tennis ball bends because of spin. A ball hit with topspin curves downward, whereas a ball hit with underspin (backspin) travels in a much flatter trajectory. The direction of the lateral force is therefore in the same sense as that of the Magnus effect experienced by a circular cylinder in potential flow with circulation (see Section 6.3). The mechanics, however, are different. The potential flow argument (involving the Bernoulli equation) offered to account for the lateral force around a circular cylinder cannot explain why a *negative* Magnus effect is universally observed at lower Reynolds numbers. (By a negative Magnus effect we mean a lateral force opposite to that experienced by a cylinder with a circulation of the same sense as the rotation of the sphere.) The correct argument seems to be the asymmetric boundary-layer separation caused by the spin. In fact, the phenomenon was not properly explained until the boundary-layer concepts were understood in the twentieth century. Some pioneering experimental work on the bending paths of spinning spheres was conducted by Robins (1742) over two hundred years ago; the deflection of rotating spheres is sometimes called the *Robins effect*.

Experimental data on nonrotating spheres (Figure 9.22) shows that the boundary layer on a sphere undergoes transition at $Re_{cr} = 5 \times 10^5$, as indicated by a sudden drop in the drag coefficient. This drop is due to the transition of the laminar boundary layer to turbulence. An important point for the present discussion is that for supercritical Reynolds numbers the separation point slowly moves upstream, as evidenced by the increase of the drag coefficient after the sudden drop shown in Figure 9.22.

With this background, we are now in a position to understand how a spinning ball generates a negative Magnus effect at $Re < Re_{cr}$ and a positive Magnus effect at $Re < Re_{cr}$. For a clockwise rotation of the ball, the fluid velocity *relative to the surface* is larger on the lower side (Figure 9.25). For the lower Reynolds number case (Figure 9.25a), this causes a transition of the boundary layer on the lower side, while the boundary layer on the upper side remains laminar. The result is a delayed separation and lower pressure on the bottom surface, and a consequent downward force on the ball. The force here is in a sense opposite to that of the Magnus effect.

The rough surface of a tennis ball lowers the critical Reynolds number, so that for a well-hit tennis ball the boundary layers on both sides of the ball have already undergone transition. Due to the higher relative velocity, the flow near the bottom has a higher Reynolds number, and is therefore farther along the Re-axis of Figure 9.22, in the range AB in which the separation point moves upstream with an increase of the Reynolds number. The separation therefore occurs *earlier* on the bottom side, resulting in a higher pressure there than on the top. This causes an upward lift force and a positive Magnus effect. Figure 9.25b shows that a tennis ball hit with underspin (backspin) generates an upward force; this overcomes

FIGURE 9.25 Curving flight of rotating spheres, in which F indicates the force exerted by the fluid: (a) negative Magnus effect; and (b) positive Magnus effect. A well-hit tennis ball with spin is likely to display the positive Magnus effect.

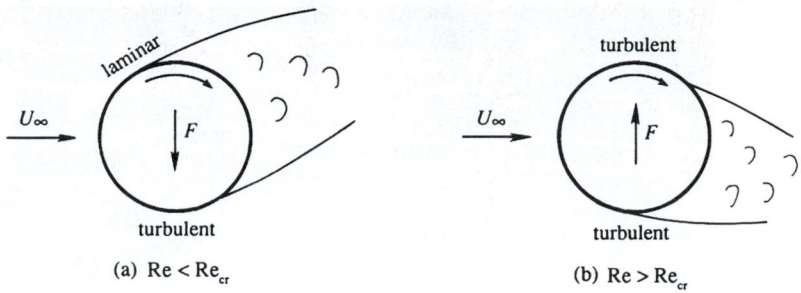

(a) $Re < Re_{cr}$

(b) $Re > Re_{cr}$

a large fraction of the weight of the ball, resulting in a much flatter trajectory than that of a tennis ball hit with topspin. A *slice serve*, in which the ball is hit tangentially on the right-hand side, curves to the left due to the same effect. Presumably soccer balls curve in the air due to similar dynamics.

Baseball Dynamics

A baseball pitcher uses different kinds of deliveries, a typical Reynolds number being 1.5×10^5. One type of delivery is called a *curveball*, caused by sidespin imparted by the pitcher to bend away from the side of the throwing arm. A *screwball* has the opposite spin and oppositely curved trajectory, when thrown correctly. The dynamics are similar to that of a spinning tennis ball (Figure 9.25b). Figure 9.26 is a photograph of the flow over a spinning baseball, showing an asymmetric separation, a crowding together of streamlines at the bottom, and an upward deflection of the wake that corresponds to a downward force on the ball.

The *knuckleball*, on the other hand, is released without any spin. In this case the path of the ball bends due to an asymmetric separation caused by the orientation of the seam, much like the cricket ball. However, the cricket ball is released with spin along the seam, which stabilizes the orientation and results in a predictable bending. The knuckleball, on the other hand, tumbles in its flight because of a lack of stabilizing spin, resulting in an irregular orientation of the seam and a consequent irregular trajectory.

9.10. TWO-DIMENSIONAL JETS

The previous nine sections have considered boundary layers over solid surfaces. The concept of a boundary layer, however, is more general, and the approximations involved are applicable whenever the vorticity in the flow is confined in thin layers, even in the absence of a solid surface. Such a layer can be in the form of a jet of fluid ejected from an orifice, a wake (where the velocity is lower than the upstream velocity) behind a solid object, or a thin shear layer (vortex sheet) between two uniform streams of different speeds. As an illustration of the method of analysis of these *free shear flows*, we shall consider the case of a laminar two-dimensional jet, which is an efflux of fluid from a long and narrow orifice that issues into a large quiescent reservoir of the same fluid. Downstream from the orifice,

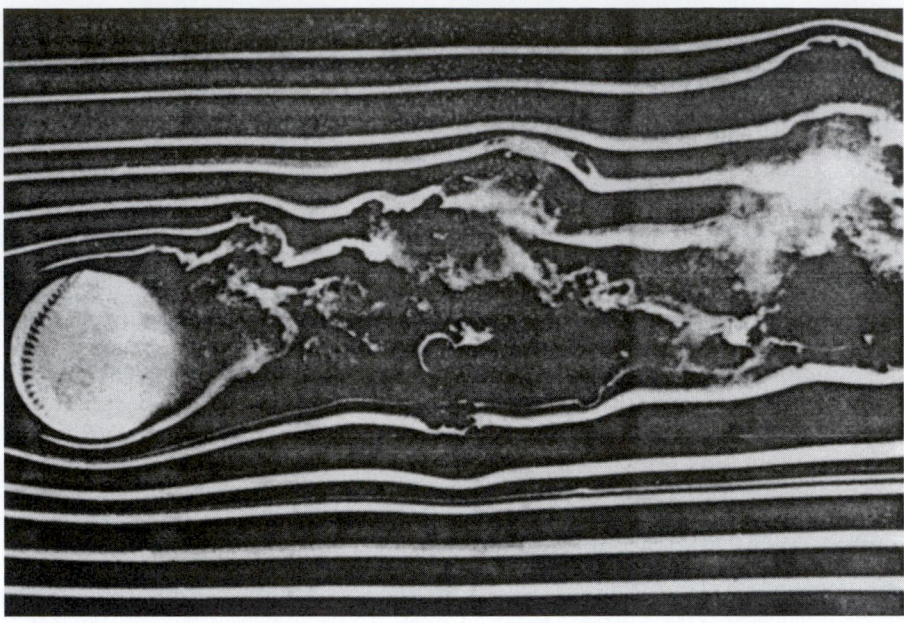

FIGURE 9.26   Smoke photograph of flow around a spinning baseball. Flow is from left to right, flow speed is 21 m/s, and the ball is spinning counterclockwise at 15 rev/s. *[Photograph by F. N. M. Brown, University of Notre Dame.] Photograph reproduced with permission from the* Annual Review of Fluid Mechanics, *Vol. 17 © 1985 by* Annual Reviews, www.AnnualReviews.org.

some of the ambient fluid is carried along with the moving jet fluid through viscous vorticity diffusion at the outer edge of the jet (Figure 9.27). The process of drawing reservoir fluid into the jet by is called *entrainment*.

The velocity distribution near the opening of the jet depends on the details of conditions upstream of the orifice exit. However, because of the absence of an externally imposed length

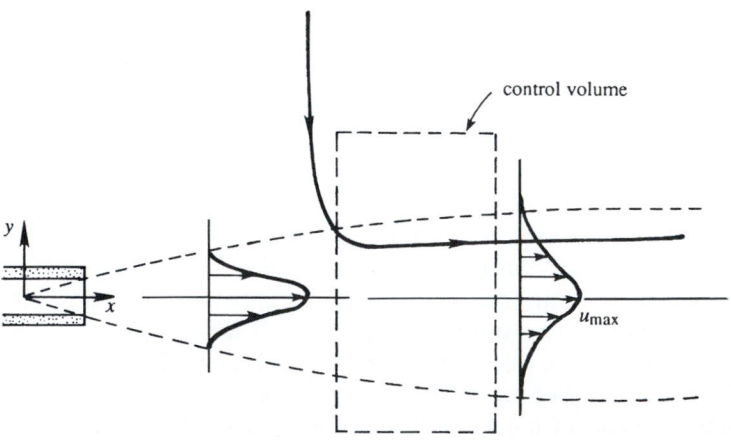

FIGURE 9.27 Simple laminar two-dimensional free jet. A narrow slot injects fluid horizontally with an initial momentum flux J into a nominally quiescent reservoir of the same fluid. The region of horizontally moving fluid slows and expands as x increases. A typical streamline showing entrainment of surrounding fluid is indicated.

scale in the downstream direction, the velocity profile in the jet approaches a self-similar shape not far from where it emerges into the reservoir, regardless of the velocity distribution at the orifice.

For large Reynolds numbers, the jet is narrow and the boundary-layer approximation can be applied. Consider a control volume with sides cutting across the jet axis at two sections (Figure 9.27); the other two sides of the control volume are taken at large distances from the jet axis. No external pressure gradient is maintained in the surrounding fluid so dp/dx is zero. According to the boundary-layer approximation, the same zero pressure gradient is also impressed upon the jet. There is, therefore, no net force acting on the surfaces of the control volume, and this requires the x-momentum flux at the two sections across the jet to be the same.

Let $u_0(x)$ be the stream-wise velocity on the x-axis and assume $\mathrm{Re} = u_0 x/\nu$ is sufficiently large for the boundary-layer equations to be valid. The flow is steady, two-dimensional (x, y), without body forces, and with constant properties (ρ, μ). Then $\partial/\partial y \gg \partial/\partial x$, $v \ll u$, $\partial p/\partial y = 0$, so the fluid equations of motion are the same as for the Blasius boundary layer: (6.2) and (9.18). However, the boundary conditions are different here:

$$u = 0 \quad \text{for } y \to \pm \infty \text{ and } x > 0, \tag{9.53}$$

$$v = 0 \quad \text{on } y = 0 \quad \text{for } x > 0, \text{ and} \tag{9.54}$$

$$u = \tilde{u}(y) \quad \text{on } x = x_0, \tag{9.55}$$

where $\tilde{u}$ is a (known) flow profile. Now partially follow the derivation of the von Karman boundary-layer integral equation. Multiply (6.2) by u and add it to the left side of (9.18) but this time integrate over all y to find:

$$\int_{-\infty}^{+\infty} 2u \frac{\partial u}{\partial x} dy + \int_{-\infty}^{+\infty} \left[u \frac{\partial v}{\partial y} + v \frac{\partial u}{\partial y} \right] dy = \int_{-\infty}^{+\infty} \frac{\partial \tau}{\partial y} dy, \quad \text{or} \quad \frac{d}{dx} \int_{-\infty}^{+\infty} u^2 dy + [uv]_{-\infty}^{+\infty} = [\tau]_{-\infty}^{+\infty}. \tag{9.56}$$

Since $u(y = \pm \infty) = 0$, all derivatives of u with respect to y must also be zero at $y = \pm \infty$. Thus, since $\tau = \mu(\partial u/\partial y)$, the second and third terms in the second equation of (9.56) are both zero. Hence, (9.56) reduces to

$$\frac{d}{dx} \int_{-\infty}^{+\infty} u^2 dy = 0, \tag{9.57}$$

a statement that the stream-wise momentum flux is conserved. Thus, when integrated, (9.57) becomes:

$$\int_{-\infty}^{+\infty} u^2 dy = const. = \int_{-\infty}^{+\infty} \tilde{u}^2(y) dy = J/\rho, \tag{9.58}$$

where the second equality follows from (9.55). Here, the constant is the momentum flux in the jet per unit span, J, divided by the fluid density, ρ.

A similarity solution is obtained far enough downstream so that the boundary-layer equations are valid and $\tilde{u}(y)$ has been forgotten. Thus, we can seek a solution in the form of (8.32) or (9.19):

$$\psi = u_0(x)\delta(x)f(\eta), \quad \text{where } \eta = y/\delta(x), \quad \delta(x) = [vx/u_0(x)]^{1/2}, \tag{9.59}$$

and $u_0(x)$ is the stream-wise velocity on $y = 0$. The stream-wise velocity throughout the field is obtained from differentiation:

$$u = \partial\psi/\partial y = [vxu_0(x)]^{1/2}(df/d\eta)[vx/u_0(x)]^{-1/2} = u_0(x)(df/d\eta). \tag{9.60}$$

The final equality here implies that $f' = 1$ on $\eta = 0$. When (9.60) is substituted into (9.58), the dependence of $u_0(x)$ on x is determined:

$$\frac{J}{\rho} = \int_{-\infty}^{+\infty} u^2 dy = u_0^2(x) \int_{-\infty}^{+\infty} f'^2(\eta)dy = u_0^2(x)\delta(x) \int_{-\infty}^{+\infty} f'^2(\eta)d\eta. \tag{9.61}$$

Since the integral is a dimensionless constant ($= C$), we must have

$$Cu_0^2(x)\delta(x) = Cu_0^{3/2}(x)\cdot(vx)^{1/2} = J/\rho,$$

so

$$u_0(x) = [J^2/C^2\rho^2 vx]^{1/3}, \quad \text{and} \quad \delta(x) = [C\rho v^2 x^2/J]^{1/3}. \tag{9.62, 9.63}$$

Thus, (9.59) becomes

$$\psi = [Jvx/C\rho]^{1/3}f(\eta) \quad \text{where } \eta = y/[C\rho v^2 x^2/J]^{1/3}. \tag{9.64}$$

In terms of the stream function, (9.18) becomes:

$$\frac{\partial\psi}{\partial y}\frac{\partial}{\partial x}\left(\frac{\partial\psi}{\partial y}\right) - \frac{\partial\psi}{\partial x}\frac{\partial}{\partial y}\left(\frac{\partial\psi}{\partial y}\right) = v\frac{\partial^2}{\partial y^2}\left(\frac{\partial\psi}{\partial y}\right). \tag{9.65}$$

Evaluating the derivatives using (9.64) and simplifying produces a differential equation for f:

$$3f''' + f''f + f'^2 = 0.$$

The boundary conditions for $x > 0$ are:

$$f' = 0 \quad \text{for} \quad \eta \to \pm\infty, \tag{9.66}$$

$$f' = 1 \quad \text{on } \eta = 0, \text{ and} \tag{9.67}$$

$$f = 0 \quad \text{on } \eta = 0. \tag{9.68}$$

Integrating the differential equation for f once produces:

$$3f'' + f'f = C_1.$$

Evaluating at $\eta = \pm\infty$ implies $C_1 = 0$ from (9.66) since $f' = 0$ implies $f'' = 0$ too. Integrating again yields:

$$3f' + f^2/2 = C_2. \tag{9.69}$$

Evaluating on $\eta = 0$ implies $C_2 = 3$ from (9.67) and (9.68). The independent and dependent variables in (9.69) can be separated and integrated:

$$3\frac{df}{d\eta} = 3 - \frac{f^2}{2} \quad \text{or} \quad \int \frac{df}{1 - f^2/6} = \int d\eta.$$

The integral on the left in the second equality can be evaluated via the variable substitution $f = \sqrt{6} \tanh \beta$, and leads to:

$$\tanh^{-1}\left(\frac{f}{\sqrt{6}}\right) = \frac{\eta}{\sqrt{6}} + C_3, \quad \text{or} \quad f = \sqrt{6} \tanh\left(\frac{\eta}{\sqrt{6}} + C_3\right). \tag{9.70}$$

Evaluating the final expression on $\eta = 0$ implies $C_3 = 0$ from (9.68). Thus, using (9.60), (9.62), and (9.70), the stream-wise velocity field is

$$u(x, y) = u_0(x)f'(\eta) = \left(\frac{J^2}{C^2\rho^2 vx}\right)^{1/3} \text{sech}^2\left(\frac{y}{\sqrt{6}}\left[\frac{J}{C\rho v^2 x^2}\right]^{1/3}\right), \tag{9.71}$$

and the dimensionless constant, C, is determined from

$$C = \int_{-\infty}^{+\infty} f'^2(\eta)\, d\eta = \int_{-\infty}^{+\infty} \text{sech}^4\left(\frac{\eta}{\sqrt{6}}\right) d\eta = \frac{4\sqrt{6}}{3}. \tag{9.72}$$

The mass flux of the jet per unit span is:

$$\dot{m} = \int_{-\infty}^{+\infty} \rho u_0(x)f'(\eta)dy = \rho u_0(x)\delta(x) \int_{-\infty}^{+\infty} f'(\eta)d\eta = \rho u_0(x)\delta(x)[f]_{-\infty}^{+\infty} = \rho u_0(x)\delta(x) \cdot 2\sqrt{6}.$$

Using (9.62), (9.63), and (9.72), this simplifies to:

$$\dot{m} = (36J\rho^2 vx)^{1/3}, \tag{9.73}$$

which shows that the jet's mass flux increases with increasing downstream distance as the jet entrains ambient reservoir fluid via the action of viscosity. The jet's entrainment induces flow toward the jet within the reservoir. The vertical velocity is:

$$v = -\frac{\partial \psi}{\partial x} = -\frac{1}{3}\left(\frac{Jv}{C\rho x^2}\right)^{1/3}[f - 2\eta f'], \quad \text{or} \quad \frac{v}{u_0(x)} = -\frac{[f - 2\eta f']}{3\sqrt{\text{Re}_x}} \quad \text{where Re}_x = \frac{xu_0(x)}{v}. \tag{9.74}$$

Here, $f(\eta) \to \pm\sqrt{6}$ and $2\eta f'(\eta) = 2\eta\, \text{sech}^2(\eta/\sqrt{6}) \to 0$ as $\eta \to \pm\infty$, so

$$\frac{v}{u_0(x)} \to \mp \frac{\sqrt{6}}{3\sqrt{\text{Re}_x}} \quad \text{as } \eta \to \pm\infty. \tag{9.75}$$

Thus, the jet's entrainment field is a flow of reservoir fluid toward the jet from above and below.

The jet spreads as it travels downstream, and this can be deduced from (9.71). Following the definition of δ_{99} in Section 9.2, the 99% half width of the jet, h_{99}, may be defined as the y-location where the horizontal velocity falls to 1% of its value at $y = 0$. Thus, from (9.71) we can determine:

$$\text{sech}^2\left(\frac{h_{99}}{\sqrt{6}}\left[\frac{J}{C\rho\nu^2 x^2}\right]^{1/3}\right) = 0.01 \rightarrow \frac{h_{99}}{\sqrt{6}}\left[\frac{J}{C\rho\nu^2 x^2}\right]^{1/3} \cong 2.2924 \rightarrow h_{99} \cong 5.6152\left[\frac{C\rho\nu^2 x^2}{J}\right]^{1/3},$$

(9.76)

which shows the jet width grows with increasing downstream distance like $x^{2/3}$. Viscosity increases the jet's thickness but higher momentum jets are thinner. The Reynolds numbers based on the stream-wise (x) and cross-stream (h_{99}) dimensions of the jet are:

$$\text{Re}_x = \frac{xu_0(x)}{\nu} = \left(\frac{3Jx}{4\sqrt{6}\rho\nu^2}\right)^{2/3} \quad \text{and} \quad \text{Re}_{h_{99}} = \frac{h_{99}u_0(x)}{\nu} = 5.6152\left(\frac{3Jx}{4\sqrt{6}\rho\nu^2}\right)^{1/3}.$$

Unfortunately, this steady-flow, two-dimensional laminar jet solution is not readily observable because the flow is unstable when $\text{Re} \gg 1$. The low critical Reynolds number for instability of a jet or wake is associated with the existence of one or more inflection points in the stream-wise velocity profile, as discussed in Chapter 11. Nevertheless, the laminar solution has revealed at least two significant phenomena—constancy of jet momentum flux and increase of jet mass flux through entrainment—that also apply to round jets and turbulent jets. However, the cross-stream spreading rate of a turbulent jet is found to be independent of Reynolds number and is faster than the laminar jet, being more like $h_{99} \propto x$ rather than $h_{99} \propto x^{2/3}$ (see Chapter 13).

A second example of a two-dimensional jet that also shares some boundary-layer characteristics is the *wall jet*. The solution here is due to Glauert (1956). We consider fluid exiting a narrow slot with its lower boundary being a planar wall taken along the x-axis (see Figure 9.28). Near the wall ($y = 0$) the flow behaves like a boundary layer, but far from the wall it behaves like a free jet. For large Re_x the jet is thin ($\delta/x \ll 1$) so $\partial p/\partial y \approx 0$ across it. The pressure is constant in the nearly stagnant outer fluid so $p \approx$ const. throughout the flow. Here again the fluid mechanical equations of motion are (6.2) and (9.18). This time the boundary conditions are:

$$u = v = 0 \quad \text{on } y = 0 \quad \text{for } x > 0, \text{ and} \qquad (9.77)$$

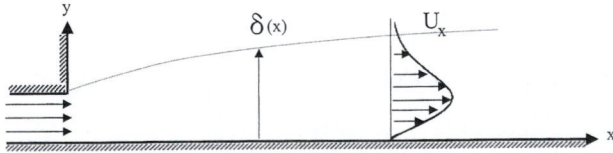

FIGURE 9.28   The laminar two-dimensional wall jet. A narrow slot injects fluid horizontally along a smooth flat wall. As for the free jet, the thickness of the region of horizontally moving fluid slows and expands as x increases but with different dependencies.

$$u(x, y) \rightarrow 0 \quad \text{as } y \rightarrow \infty. \tag{9.78}$$

Here again, a similarity solution valid for $\text{Re}_x \rightarrow \infty$ can be found under the assumption that the initial velocity distribution is forgotten by the flow. However, unlike the free jet, the momentum flux of the wall jet is not constant; it diminishes with increasing downstream distance because of the wall shear stress. To obtain the conserved property in the wall jet, integrate (9.18) from y to ∞:

$$\int_y^\infty u\frac{\partial u}{\partial x}dy + \int_y^\infty v\frac{\partial u}{\partial y}\, dy = v\int_y^\infty \frac{\partial^2 u}{\partial y^2}\, dy = v\left[\frac{\partial u}{\partial y}\right]_y^\infty = -v\frac{\partial u}{\partial y},$$

multiply this by u, and integrate from 0 to ∞:

$$\int_0^\infty \left(u\frac{\partial}{\partial x}\int_y^\infty \frac{u^2}{2}dy\right)dy + \int_0^\infty \left(u\int_y^\infty v\frac{\partial u}{\partial y}dy\right)dy = -\frac{v}{2}\int_0^\infty \frac{\partial u^2}{\partial y}\, dy = -\frac{v}{2}[u^2]_0^\infty = 0.$$

The final equality follows from the boundary conditions (9.77) and (9.78). Integrating the interior integral of the second term on the left by parts and using (6.2) yields a term equal to the first term and one that lacks any differentiation:

$$\int_0^\infty \left(u\frac{\partial}{\partial x}\int_y^\infty u^2 dy\right)dy - \int_0^\infty u^2 v dy = 0. \tag{9.79}$$

Now consider

$$\frac{d}{dx}\int_0^\infty \left(u\int_y^\infty u^2\, dy\right)dy = \int_0^\infty \left(\frac{\partial u}{\partial x}\int_y^\infty u^2\, dy\right)dy + \int_0^\infty \left(u\frac{\partial}{\partial x}\int_y^\infty u^2\, dy\right)dy,$$

use (6.2) in the first term on the right side, integrate by parts, and combine this with (9.79) to obtain

$$\frac{d}{dx}\int_0^\infty \left(u\int_y^\infty u^2 dy\right)dy = 0. \tag{9.80}$$

This says that the flux of exterior momentum flux remains constant with increasing downstream distance and is the necessary condition for obtaining similarity exponents.

As for the steady free laminar jet, the field equation is (9.65) and the solution is presumed to be in the similarity form specified by (9.59). Here $u_0(x)$ is to be determined and this similarity solution should be valid when $x \gg x_o$, where x_o is the location where the initial condition is specified, which we take to be the upstream extent of the validity of the boundary-layer momentum equation (9.18) or (9.65). Substituting $u = \partial \psi / \partial y = u_0(x) f'(\eta)$ from (9.59) into (9.80) produces:

$$\frac{d}{dx} \left[u_0^3(x) \cdot \frac{\nu x}{u_0(x)} \int_0^\infty \left(f' \int_\eta^\infty f'^2 d\eta \right) d\eta \right] = 0. \tag{9.81}$$

If the double integration is independent of x, then the factor outside the integral must be constant. Therefore, set $x u_0^2(x) = C^2$, which implies $u_0(x) = Cx^{-1/2}$ so (9.59) becomes:

$$\psi(x, y) = \left[\nu C x^{1/2} \right]^{1/2} f(\eta) \quad \text{where } \eta = y/\delta(x), \ \delta(x) = \left[\nu x^{3/2}/C \right]^{1/2}. \tag{9.82}$$

After appropriately differentiating (9.82), substituting into (9.65), and canceling common factors, (9.65) reduces to

$$f''' + ff'' + 2f'^2 = 0,$$

subject to the boundary conditions (9.77) and (9.78): $f(0) = 0; f'(0) = 0; f'(\infty) = 0$. This third-order equation can be integrated once after multiplying by the integrating factor f, to yield $4ff'' - 2f'^2 + f^2 f' = 0$, where the constant of integration has been evaluated at $\eta = 0$. Dividing by the integrating factor $4f^{3/2}$ allows another integration. The result is

$$f^{-1/2} f' + f^{3/2}/6 = C_1 \equiv f_\infty^{3/2}/6, \quad \text{where } f_\infty = f(\infty).$$

The final integration can be performed by separating variables and defining $g^2(\eta) = f/f_\infty$:

$$\int \frac{df}{f_\infty^{3/2} f - f^2} = \frac{1}{6} \int d\eta, \quad \text{or} \quad \int \frac{dg}{1 - g^3} = \frac{f_\infty}{12} \int d\eta.$$

The integration on the left may be performed via a partial fraction expansion using $1 - g^3 = (1 - g) \cdot (1 + g + g^2)$ with the final result in left-implicit form:

$$-\ln(1 - g) + \sqrt{3} \tan^{-1}\left(\frac{2g + 1}{\sqrt{3}} \right) + \ln\left(1 + g + g^2\right)^{1/2} = \frac{f_\infty}{4}\eta + \sqrt{3} \tan^{-1}\left(\frac{1}{\sqrt{3}} \right), \tag{9.83}$$

where the boundary condition $g(0) = 0$ was used to evaluate the constant of integration. The profiles of f and f' are plotted vs. η in Figure 9.29. We can verify easily that $f \to 0$ exponentially fast in η from this solution for $g(\eta)$. As $\eta \to \infty$, $g \to 1$, so for large η the solution for g reduces to $-\ln(1 - g) + \sqrt{3} \tan^{-1}\sqrt{3} + (1/2)\ln 3 \cong f_\infty \eta/4 + \sqrt{3} \tan^{-1}(1/\sqrt{3})$. The first term on each side of this equation dominates, leaving $1 - g \approx e^{-(f_\infty/4)\eta}$. Thus, for $\eta \to \infty$, we must have: $f' = 2f_\infty g g' \approx \frac{1}{2} f_\infty^2 \exp[-f_\infty \eta/4]$. The mass flow rate per unit span in the steady laminar wall jet is

$$\dot{m} = \int_0^\infty \rho u \, dy = \rho u_0(x) \delta(x) \int_0^\infty f'(\eta) d\eta = \rho \sqrt{\nu C} f_\infty x^{1/4}, \tag{9.84}$$

FIGURE 9.29 Variation of normalized mass flux (f) and normalized stream-wise velocity profile (f') with similarly variable η for the laminar two-dimensional wall jet. *Reprinted with the permission of Cambridge University Press.*

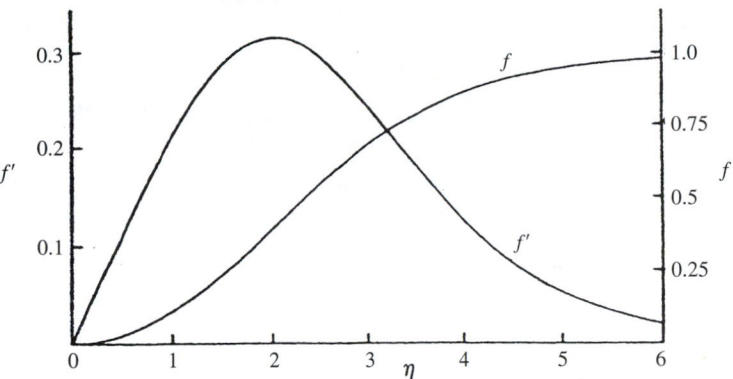

indicating that entrainment increases the mass flow rate in the jet with $x^{1/4}$. The two constants, C and f_∞, can be determined from the integrated form of (9.81) in terms of Ψ, the flux of the exterior momentum flux (a constant):

$$u_0^2(x)\nu x \int_0^\infty \left(f' \int_\eta^\infty f'^2 d\eta \right) d\eta = C^2 \nu \int_0^\infty \left(f' \int_\eta^\infty f'^2 d\eta \right) d\eta = \Psi, \qquad (9.85)$$

and knowledge of $\dot{m}$ at one downstream location.

The entrainment into the steady laminar wall jet is evident from the form of $v = -\partial\psi/\partial x = -\sqrt{\nu C}(f - 3\eta f')/4x^{3/4}$, which simplifies to $v \approx -\sqrt{\nu C}f_\infty/4x^{3/4}$ as $\eta \to \infty$, so, far above the jet, the flow is downward toward the jet.

9.11. SECONDARY FLOWS

Large Reynolds number flows with curved streamlines tend to generate additional velocity components because of the properties of boundary layers. These additional components are commonly called *secondary flows*. An example of such a flow is made dramatically visible by randomly dispersing finely crushed tea leaves into a cup of water, and then stirring vigorously in a circular motion. When the motion has ceased, all of the particles have collected in a mound at the center of the bottom of the cup (see Figure 9.30). An explanation of this phenomenon is given in terms of thin boundary layers. The stirring motion imparts a primary velocity, $u_\varphi(R)$ (see Appendix B.6 for coordinates), large enough for the Reynolds number to be large enough for the boundary layers on the cup's sidewalls and bottom to be thin. The two largest terms in the R-momentum equation are

$$\frac{\partial p}{\partial R} = \frac{\rho u_\varphi^2}{R}.$$

Away from the walls, the flow is inviscid. As the boundary layer on the bottom is thin, boundary-layer theory yields $\partial p/\partial z = 0$ from the axial momentum equation. Thus, the pressure in the bottom boundary layer is the same as for the inviscid flow just outside the

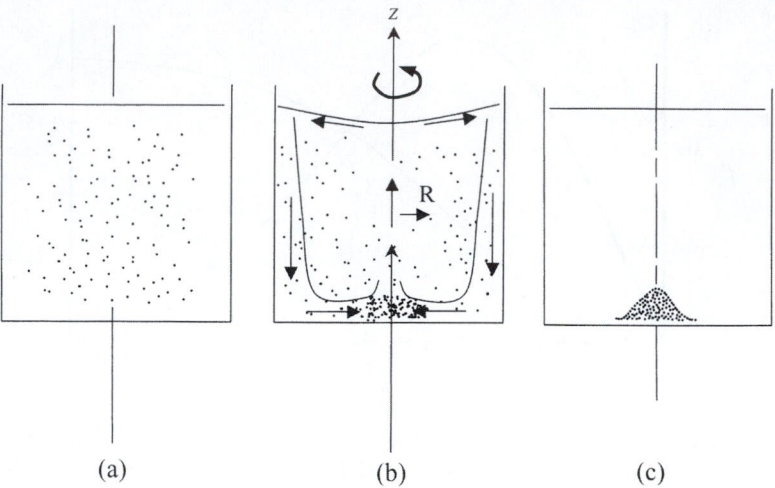

FIGURE 9.30 Secondary flow in a teacup. Tea leaf fragments are slightly denser than water. (a) Tea leaf fragments randomly dispersed—initial state; (b) stirred vigorously—transient motion; and (c) final state where all the tea leaf fragments are piled near the axis of rotation on the bottom of the cup.

(a)   (b)   (c)

boundary layer. However, within the boundary layer, u_ϕ is less than the inviscid value at the edge. Thus $p(R)$ is everywhere larger in the boundary layer than that required for circular streamlines inside the boundary layer, and this pressure difference pushes the streamlines inward toward the center of the cup. That is, the pressure gradient within the boundary layer generates an inwardly directed u_R. This motion induces a downwardly directed flow in the sidewall boundary layer and an outwardly directed flow on the top surface. This secondary flow is closed by an upward flow along the cup's centerline. The visualization is accomplished by crushed tea leaves which are slightly denser than water. They descend by gravity or are driven outward by centrifugal acceleration. If they enter the sidewall boundary layer, they are transported downward and thence to the center by the secondary flow. If the tea particles enter the bottom boundary layer from above, they are quickly swept to the center and dropped as the flow turns upward. All the particles collect at the center of the bottom of the teacup. A practical application of this effect, illustrated in Exercise 9.28, relates to sand and silt transport by the Mississippi River.

EXERCISES

9.1. A thin flat plate 2 m long and 1 m wide is placed at zero angle of attack in a low-speed wind tunnel in the two positions sketched below.

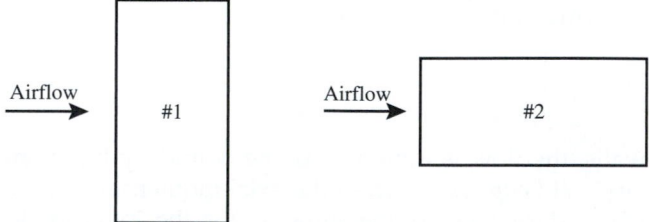

a) For steady airflow, what is the ratio: $\dfrac{\text{drag on the plate in position \#1}}{\text{drag on the plate in position \#2}}$?

b) For steady airflow at 10 m/sec, what is the total drag on the plate in position #1?

c) If the airflow is impulsively raised from zero to 10 m/sec at $t = 0$, will the initial drag on the plate in position #1 be greater or less than the steady-state drag value calculated for part b)?

d) Estimate how long it will take for drag on the plate in position #1 in the impulsively started flow to reach the steady-state drag value calculated for part b)?

9.2. Solve the Blasius equations (9.27) through (9.29) with a computer, using the Runge-Kutta scheme of numerical integration.

9.3. A flat plate 4 m wide and 1 m long (in the direction of flow) is immersed in kerosene at 20°C ($v = 2.29 \times 10^{-6}$ m²/s, $\rho = 800$ kg/m³), flowing with an undisturbed velocity of 0.5 m/s. Verify that the Reynolds number is less than critical everywhere, so that the flow is laminar. Show that the thickness of the boundary layer and the shear stress at the center of the plate are $\delta = 0.74$ cm and $\tau_0 = 0.2$ N/m², and those at the trailing edge are $\delta = 1.05$ cm and $\tau_0 = 0.14$ N/m². Show also that the total frictional drag on one side of the plate is 1.14 N. Assume that the similarity solution holds for the entire plate.

9.4. A simple realization of a temporal boundary layer involves the spinning fluid in a cylindrical container. Consider a viscous incompressible fluid (density $= \rho$, viscosity $= \mu$) in solid body rotation (rotational speed $= \Omega$) in a cylindrical container of diameter d. The mean depth of the fluid is h. An external stirring mechanism forces the fluid to maintain solid body rotation. At $t = 0$, the external stirring ceases. Denote the time for the fluid to spin-down (i.e., to stop rotating) by τ.

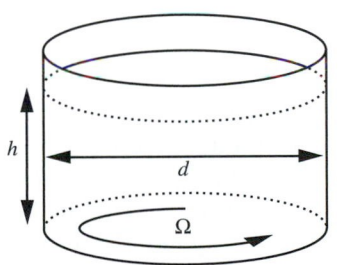

a) For $h \gg d$, write a simple laminar-flow scaling law for τ assuming that the velocity perturbation produced by the no-slip condition on the container's sidewall must travel inward a distance $d/2$ via diffusion.

b) For $h \ll d$, write a simple laminar-flow scaling law for τ assuming that the velocity perturbation produced by the no-slip condition on the container's bottom must travel upward a distance h via diffusion.

c) Using partially filled cylindrical containers of several different sizes (drinking glasses and pots and pans are suggested) with different amounts of water, test the validity of the above diffusion estimates. Use a spoon or a whirling motion of the container to bring the water into something approaching solid body rotation. You'll know when the fluid motion is close to solid body rotation because the fluid surface will be a paraboloid of revolution. Once you have this initial flow condition set up, cease the stirring or whirling and note how long it takes for the fluid to stop moving. Perform at least one test when d and h are several inches or more. Cookie or bread crumbs sprinkled on the water surface will help visualize surface motion. The judicious addition of a few drops of milk after the fluid starts slowing down may prove interesting.

d) Compute numbers from your scaling laws for parts a) and b) using the viscosity of water, the dimensions of the containers, and the experimental water depths. Are the scaling laws from parts a) and b) useful for predicting the experimental results? If not, explain why.

(The phenomena investigated here have some important practical consequences in atmospheric and oceanic flows and in IC engines where swirl and tumble are exploited to mix the fuel charge and increase combustion speeds.)

9.5. A square-duct wind tunnel of length $L = 1$ m is being designed to operate at room temperature and atmospheric conditions. A uniform airflow at $U = 1$ m/s enters through an opening of $D = 20$ cm. Due to the viscosity of air, it is necessary to design a variable cross-sectional area if a constant velocity is to be maintained in the middle part of the cross-section throughout the wind tunnel.

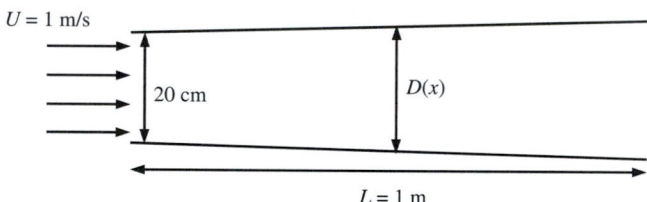

a) Determine the duct size, $D(x)$, as a function of x.
b) How will the result be affected if $U = 20$ m/s? At a given value of x, will $D(x)$ be larger or smaller (or the same) than the value obtained in part a)? Explain.
c) How will the result be affected if the wind tunnel is to be operated at 10 atm (and $U = 1$ m/s)? At a given value of x, will $D(x)$ be larger or smaller (or the same) than the value obtained in part a)? Explain. [*Hint*: the dynamic viscosity of air (μ) is largely unaffected by pressure.]
d) Does the airflow apply a net force to the wind tunnel? If so, indicate the direction of the force.

9.6. Use the control volume shown to derive the definition of the momentum thickness, θ, for flow over a flat plate:

$$\rho U^2 \theta = \rho U^2 \int_0^h \frac{u}{U}\left(1 - \frac{u}{U}\right) dy = \frac{\text{drag force on the plate from zero to } x}{\text{unit depth into the page}} = \int_0^x \tau_0 dx$$

The words in the figure describe the upper and lower control volume boundaries.

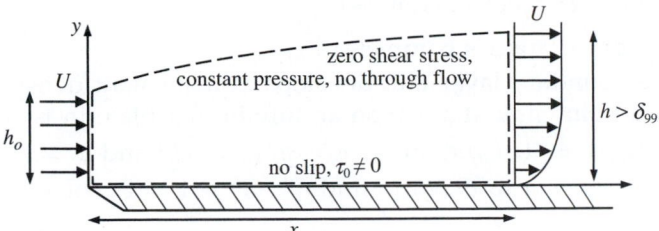

9.7. Estimate the 99% boundary-layer thickness on:
   a) A paper airplane wing (length $= 0.25$ m, $U = 1$ m/sec)
   b) The underside of a super tanker (length $= 300$ m, $U = 5$ m/sec)
   c) An airport runway on a blustery day (length $= 5$ km, $U = 10$ m/sec)
   d) Will these estimates be accurate in each case? Explain.

9.8. Air at 20°C and 100 kPa ($\rho = 1.167$ kg/m^3, $\nu = 1.5 \times 10^{-5}$ m^2/s) flows over a thin plate with a free-stream velocity of 6 m/s. At a point 15 cm from the leading edge, determine the value of y at which $u/U = 0.456$. Also calculate v and $\partial u/\partial y$ at this point. [*Answer:* $y = 0.857$ mm, $v = 0.39$ cm/s, $\partial u/\partial y = 3020$ s^{-1}. You may not be able to achieve this level of accuracy from Figure 9.5 alone.]

9.9. An incompressible fluid (density ρ, viscosity μ) flows steadily from a large reservoir into a long pipe with diameter D. Assume the pipe wall boundary-layer thickness is zero at $x = 0$. The Reynolds number based on D, Re$_D$, is greater than 10^4.

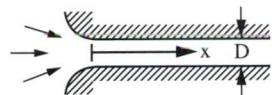

   a) Estimate the necessary pipe length for establishing a parabolic velocity profile in the pipe.
   b) Will the pressure drop in this entry length be larger or smaller than an equivalent pipe length in which the flow has a parabolic profile? Why?

9.10. [1]A variety of different dimensionless groups have been used to characterize the importance of a pressure gradient in boundary layer flows. Develop an expression for each of the following parameters for the Falkner-Skan boundary layer solutions in terms of the exponent n in $U_e(x) = ax^n$, Re$_x = U_e x/\nu$, integrals involving the profile function f', and $f''(0)$, the profile slope at $y = 0$. Here $u(x, y) = U_e(x)f'(y/\delta(x)) = U_e f'(\eta)$ and the wall shear stress $\tau_0 = \mu(\partial u/\partial y)_{y=0} = (\mu U_e/\delta(x))f''(0)$. What value does each parameter take in a Blasius boundary layer? What value does each parameter achieve at the separation condition?
   a) $(\nu/U_e^2)(dU_e/dx)$, an inverse Reynolds number
   b) $(\theta^2/\nu)(dU_e/dx)$, the Holstein and Bohlen correlation parameter

---

[1]Inspired by problem 4.10 on page 330 of White (2006)

c) $(\mu/\sqrt{\rho\tau_0^3})(dp/dx)$, Patel's parameter

d) $(\delta^*/\tau_0)(dp/dx)$, Clauser's parameter

9.11. Consider the boundary layer that develops as a constant density viscous fluid is drawn to a point sink at $x = 0$ on an infinite flat plate in two dimensions (x, y). Here $U_e(x) = -U_oL_o/x$, so set $\eta = y/\sqrt{vx/|U_e|}$ and $\psi = -\sqrt{vx|U_e|}f(\eta)$ and redo the steps leading to (9.36) to find $f''' - f'^2 + 1 = 0$. Solve this equation and utilize appropriate boundary conditions to find $f' = 3\left[\dfrac{1 - \alpha e^{-\sqrt{2}\eta}}{1 + \alpha e^{-\sqrt{2}\eta}}\right]^2 - 2$

where $\alpha = \dfrac{\sqrt{3} - \sqrt{2}}{\sqrt{3} + \sqrt{2}}$.

9.12. By completing the steps below, show that it is possible to derive von Karman's boundary-layer integral equation without integrating to infinity in the surface-normal direction using the three boundary-layer thicknesses commonly defined for laminar and turbulent boundary layers: 1) δ (or δ_{99}) = the full boundary-layer thickness that encompasses all (or 99%) of the region of viscous influence, 2) δ^* = the displacement thickness of the boundary layer, and 3) θ = momentum thickness of the boundary layer. Here, the definitions of the latter two involve the first:

$\delta^*(x) = \int\limits_{y=0}^{y=\delta}\left(1 - \dfrac{u(x, y)}{U_e(x)}\right)dy$ and $\theta(x) = \int\limits_{y=0}^{y=\delta}\dfrac{u(x, y)}{U_e(x)}\left(1 - \dfrac{u(x, y)}{U_e(x)}\right)dy$, where $U_e(x)$ is the

flow speed parallel to the wall outside the boundary layer, and δ is presumed to depend on x too.

a) Integrate the two-dimensional continuity equation from $y = 0$ to δ to show that the vertical velocity at the edge of the boundary layer is: $v(x, y = \delta) = \dfrac{d}{dx}(U_e(x)\delta^*(x))$ $-\delta\dfrac{dU_e}{dx}$.

b) Integrate the steady two-dimensional x-direction boundary-layer momentum equation from $y = 0$ to δ to show that: $\dfrac{\tau_0}{\rho} = \dfrac{d}{dx}(U_e^2(x)\theta(x)) + \dfrac{\delta^*(x)}{2}\dfrac{dU_e^2(x)}{dx}$.

[*Hint*: Use Leibniz's rule $\dfrac{d}{dx}\int\limits_{a(x)}^{b(x)}f(x, y)dy = \left[f(x, b)\dfrac{db}{dx}\right] - \left[f(x, a)\dfrac{da}{dx}\right] +$

$\int\limits_{a(x)}^{b(x)}\dfrac{\partial f(x, y)}{\partial x}dy$ to handle the fact that $\delta = \delta(x)$.]

9.13. Derive the von Karman boundary layer integral equation by conserving mass and momentum in a control volume (C.V.) of width dx and height h that moves at the exterior flow speed $U_e(x)$ as shown. Here h is a constant distance that is comfortably greater than the overall boundary layer thickness δ.

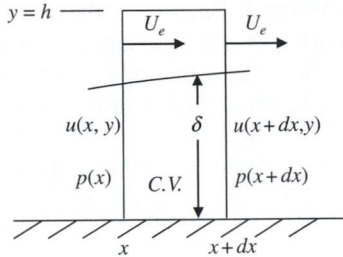

9.14. For the following approximate flat-plate boundary-layer profile:

$$\frac{u}{U} = \left\{ \begin{array}{ll} \sin(\pi y/2\delta) & \text{for } 0 \leq y \leq \delta \\ 1 & \text{for } y > \delta \end{array} \right\},$$

where δ is the generic boundary-layer thickness, determine:

a) The displacement thickness δ^*, the momentum thicknesses θ, and the shape factor $H = \delta^*/\theta$.

b) Use the zero-pressure gradient boundary-layer integral equation to find: $(\delta/x)Re_x^{1/2}$, $(\delta^*/x)Re_x^{1/2}$, $(\theta/x)Re_x^{1/2}$, $c_f Re_x^{1/2}$, and $C_D Re_L^{1/2}$ for the approximate profile.

c) Compare these results to their equivalent Blasius boundary-layer values.

9.15. An incompressible viscous fluid with kinematic viscosity ν flows steadily in a long two-dimensional horn with cross-sectional area $A(x) = A_o \exp\{\beta x\}$. At $x = 0$, the fluid velocity in the horn is uniform and equal to U_o. The boundary-layer momentum thickness is zero at $x = 0$.

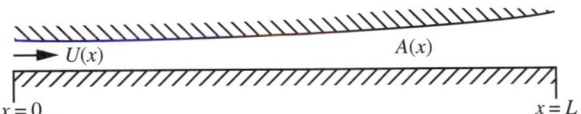

a) Assuming no separation, determine the boundary-layer momentum thickness, $\theta(x)$, on the lower horn boundary using Thwaites' method.

b) Determine the condition on β that makes the no-separation assumption valid for $0 < x < L$.

c) If $\theta(x = 0)$ was nonzero and positive, would the flow in the horn be more or less likely to separate than the $\theta(x = 0) = 0$ case with the same horn geometry?

9.16. The steady two-dimensional velocity potential for a source of strength m located a distance b above a large flat surface located at $y = 0$ is:

$$\phi(x, y) = \frac{m}{2\pi}\left(\ln\sqrt{x^2 + (y - b)^2} + \ln\sqrt{x^2 + (y + b)^2} \right)$$

a) Determine $U(x)$, the horizontal fluid velocity on $y = 0$.

b) Use this $U(x)$ and Thwaites' method to estimate the momentum thickness, $\theta(x)$, of the laminar boundary layer that develops on the flat surface when the initial momentum thickness θ_o is zero. [Potentially useful information:

$$\int_0^x \frac{\xi^5 d\xi}{(\xi^2 + b^2)^5} = \frac{x^6(x^2 + 4b^2)}{24b^4(x^2 + b^2)^4}]$$

c) Will boundary-layer separation occur in this flow? If so, at what value of x/b does Thwaites' method predict zero wall shear stress?

d) Using solid lines, sketch the streamlines for the ideal flow specified by the velocity potential given above. For comparison, on the same sketch, indicate with dashed lines the streamlines you expect for the flow of a real fluid in the same geometry at the same flow rate.

9.17. A fluid-mediated particle-deposition process requires a laminar boundary-layer flow with a *constant* shear stress, τ_w, on a smooth flat surface. The fluid has viscosity μ and density ρ (both constant). The flow is steady, incompressible, and two dimensional, and the flat surface extends from $0 < x < L$. The flow speed above the boundary layer is $U(x)$. Ignore body forces.

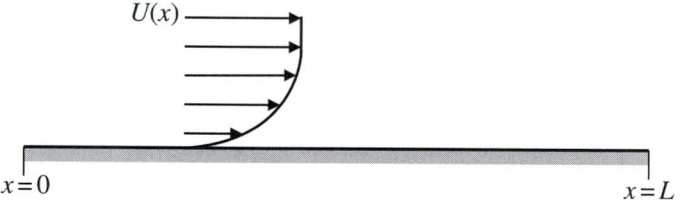

a) Assume the boundary-layer thickness is zero at $x = 0$, and use Thwaites' formulation for the shear stress, $\tau_w = (\mu U / \theta)l(\lambda)$ with $\lambda = (\theta^2/\nu)(dU/dx)$, to determine $\theta(x)$ and $U(x)$ in terms of λ, $\nu = \mu/\rho$, x, and $\tau_w/\mu = $ constant. [*Hint:* assume that $U/\theta = A$ and $l(\lambda)$ are both constants so that $\tau_w/\mu = Al(\lambda)$.]

b) Using the Thwaites integral (9.50) and the results of part a), determine λ.

c) Is boundary-layer separation a concern in this flow? Explain with words or equations.

9.18. The steady two-dimensional potential for incompressible flow at nominal horizontal speed U over a stationary but mildly wavy wall is: $\phi(x, y) = Ux - U\varepsilon \exp(-ky) \cos(kx)$, where $k\varepsilon \ll 1$. Here, ε is the amplitude of the waviness and $k = 2\pi/\lambda$, where λ is the wavelength of the waviness.

a) Use the potential to determine the horizontal velocity $u(x, y)$ on $y = 0$.

b) Assume that $u(x, 0)$ from part a) is the exterior velocity on the wavy wall and use Thwaites' method to approximately determine the momentum thickness, θ, of the

laminar boundary layer that develops on the wavy wall when the fluid viscosity is
μ, and $\theta = 0$ at $x = 0$. Keep only the linear terms in $k\varepsilon$ and ε/x to simplify your work.
c) Is the average wall shear stress higher for $\lambda/2 \leq x \leq 3\lambda/4$ or for $3\lambda/4 \leq x \leq \lambda$?
d) Does the boundary layer ever separate when $k\varepsilon = 0.01$?
e) In $0 \leq x \leq \lambda$, determine where the wall pressure is the highest and the lowest.
f) If the wavy surface were actually an air-water interface, would a steady wind tend
to increase or decrease water wave amplitudes? Explain.

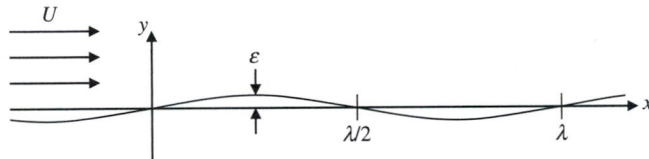

9.19. Consider the boundary layer that develops in stagnation point flow: $U_e(x) = U_o x/L$.
a) With $\theta = 0$ at $x = 0$, use Thwaites' method to determine $\delta^*(x)$, $\theta(x)$, and $c_f(x)$.
b) This flow also has an exact similarity solution of the full Navier-Stokes
equations. Numerical evaluation of the final nonlinear ordinary differential
equation produces: $c_f\sqrt{Re_x} = 2.4652$, where $Re_x = U_e x/\nu = U_o x^2/L\nu$. Assess
the accuracy of the predictions for $c_f(x)$ from the Thwaites' method for
this flow.
9.20. A laminar boundary layer develops on a large smooth flat surface under the influence
of an exterior flow velocity $U(x)$ that varies with downstream distance, x.

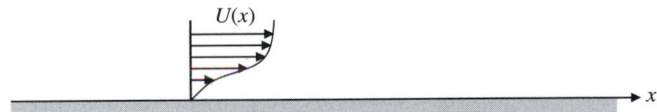

a) Using Thwaites' method, find a single integral-differential equation for $U(x)$ if the
boundary layer is to remain perpetually right on the verge of separation so that the
wall shear stress, τ_0, is zero. Assume that the boundary layer has zero thickness at
$x = 0$.
b) Assume $U(x) = U_o(x/L)^\gamma$ and use the result of part a) to find γ.
c) Compute the boundary-layer momentum thickness $\theta(x)$ for this situation.
d) Determine the extent to which the results of parts b) and c) satisfy the von Karman
boundary-layer integral equation, (9.43), when $\tau_0 = 0$ by computing the residual of
this equation. Interpret the meaning of your answer; is von Karman's equation well
satisfied, or is the residual of sufficient size to be problematic?
e) Can the $U(x)$ determined for part b) be produced in a duct with cross-sectional area
$A(x) = A_o(x/L)^{-\gamma}$? Explain your reasoning.

9.21. Consider the boundary layer that develops on a cylinder of radius a in a cross flow.

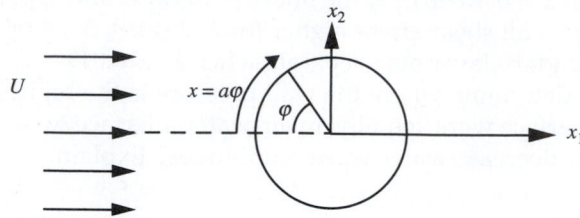

    a) Using Thwaites' method, determine the momentum thickness as a function of φ, the angle from the upstream stagnation point (see drawing).

    b) Make a sketch of c_f versus φ.

    c) At what angle does Thwaites method predict vanishing wall shear stress?

9.22. An incompressible viscous fluid flows steadily in a large duct with constant cross-sectional area A_o and interior perimeter b. A laminar boundary layer develops on the duct's sidewalls. At $x = 0$, the fluid velocity in the duct is uniform and equal to U_o, and the boundary-layer thickness is zero. Assume the thickness of the duct-wall boundary layer is small compared to A_o/b.

    a) Calculate the duct-wall boundary-layer momentum and displacement thicknesses, $\theta(x)$ and $\delta^*(x)$, respectively, using Thwaites' method when $U(x) = U_o$.

    b) Using the $\delta^*(x)$ found for part a), compute a more accurate version of $U(x)$ that includes boundary-layer displacement effects.

    c) Using the $U(x)$ found for part b), recompute $\theta(x)$ and compare to the results of part a). To simplify your work, linearize all the power-law expressions, i.e., $(1 - b\delta^*/A_o)^n \cong 1 - nb\delta^*/A_o$.

    d) If the duct area expanded as the flow moved downstream, would the correction for the presence of the sidewall boundary layers be more likely to move boundary-layer separation upstream or downstream? Explain.

9.23. Water flows over a flat plate 30 m long and 17 m wide with a free-stream velocity of 1 m/s. Verify that the Reynolds number at the end of the plate is larger than the critical value for transition to turbulence. Using the drag coefficient in Figure 9.11, estimate the drag on the plate.

9.24. A common means of assessing boundary-layer separation is to observe the surface streaks left by oil or paint drops that were smeared across a surface by the flow. Such investigations can be carried out in an elementary manner for cross-flow past a cylinder using a blow dryer, a cylinder 0.5 to 1 cm in diameter that is ~10 cm long (a common ball-point pen), and a suitable viscous liquid. Here, creamy salad dressing, shampoo, dish washing liquid, or molasses should work. And, for the best observations, the liquid should not be clear and the cylinder and liquid should be different colors. Dip your finger into the viscous liquid and wipe it over two thirds of the surface of the cylinder. The liquid layer should be thick enough so that you can easily tell where it is thick or thin. Use the remaining dry third of the cylinder to hold the cylinder horizontal. Now, turn on the blow dryer, leaving the heat off, and direct its outflow across the wetted portion of the horizontal cylinder to mimic the flow situation in the drawing for Exercise 9.21.

a) Hold the cylinder stationary, and observe how the viscous fluid moves on the surface of the cylinder and try to determine the angle φ_s at which boundary-layer separation occurs. To get consistent results you may have to experiment with different liquids, different initial liquid thicknesses, different blow-dryer fan settings, and different distances between cylinder and blow dryer. Estimate the cylinder-diameter-based Reynolds number of the flow you've studied.

b) If you have completed Exercise 9.21, do your boundary-layer separation observations match the calculations? Explain any discrepancies between your experiments and the calculations.

9.25. Find the diameter of a parachute required to provide a fall velocity no larger than that caused by jumping from a 2.5 m height, if the total load is 80 kg. Assume that the properties of air are $\rho = 1.167$ kg/m^3, $\nu = 1.5 \times 10^{-5}$ m^2/s, and treat the parachute as a hemispherical shell with $C_D = 2.3$. [*Answer:* 3.9 m]

9.26. Consider incompressible, slightly viscous flow over a semi-infinite flat plate with constant suction. The suction velocity $v(x, y = 0) = v_0 < 0$ is ordered by $O(\mathrm{Re}^{-1/2}) < v_0/U < O(1)$ where $\mathrm{Re} = Ux/\nu \to \infty$. The flow upstream is parallel to the plate with speed U. Solve for u, v in the boundary layer.

9.27. The boundary-layer approximation is sometimes applied to flows that do not have a bounding surface. Here the approximation is based on two conditions: downstream fluid motion dominates over the cross-stream flow, and any moving layer thickness defined in the transverse direction evolves slowly in the downstream direction. Consider a laminar jet of momentum flux J that emerges from a small orifice into a large pool of stationary viscous fluid at $z = 0$. Assume the jet is directed along the positive z-axis in a cylindrical coordinate system. In this case, the steady, incompressible, axisymmetric boundary-layer equations are:

$$\frac{1}{R}\frac{\partial(Ru_R)}{\partial R} + \frac{\partial w}{\partial z} = 0, \quad \text{and} \quad w\frac{\partial w}{\partial z} + u_R\frac{\partial w}{\partial R} = -\frac{1}{\rho}\frac{\partial p}{\partial z} + \frac{\nu}{R}\frac{\partial}{\partial R}\left(R\frac{\partial w}{\partial R}\right),$$

where w is the (axial) z-direction velocity component, and R is the radial coordinate. Let $r(z)$ denote the generic radius of the cone of jet flow.

a) Let $w(R, z) = (\nu/z)f(\eta)$ where $\eta = R/z$, and derive the following equation for f: $\eta f' + f \int^{\eta} \eta f d\eta = 0$.

b) Solve this equation by defining a new function $F = \int^{\eta} \eta f d\eta$. Determine constants from the boundary condition: $w \to 0$ as $\eta \to \infty$, and the requirement:

$$J = 2\pi\rho \int_{R=0}^{R=r(z)} w^2(R, z)RdR = const.$$

c) At fixed z, does $r(z)$ increase or decrease with increasing J?

[*Hints:* 1) The fact that the jet emerges into a pool of quiescent fluid should provide information about $\partial p/\partial z$, and 2) $f(\eta) \propto (1 + const \cdot \eta^2)^{-2}$, but try to obtain this result without using it.]

9.28. Mississippi River boatmen know that when rounding a bend in the river, they must stay close to the outer bank or else they will run aground. Explain in fluid mechanical terms the reason for the cross-sectional shape of the river at the bend.

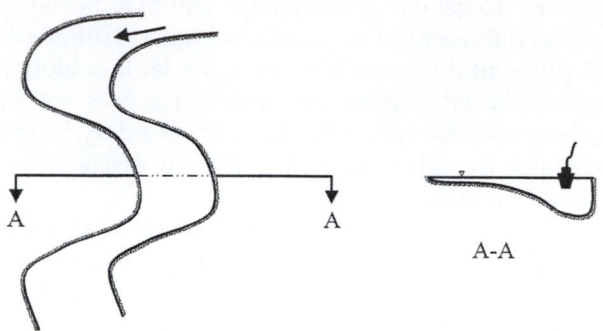

Literature Cited

Batchelor, G. K. (1967). *An Introduction to Fluid Dynamics*. London: Cambridge University Press.

Bender, C. M., & Orszag, S. A. (1978). *Advanced Mathematical Methods for Scientists and Engineers*. New York: McGraw-Hill.

Curle, N. (1962). *The Laminar Boundary Layer Equations*. Oxford: Clarendon Press.

Falkner, V. W., & Skan, S. W. (1931). Solutions of the boundary layer equations. *Phil. Mag. (Ser. 7), 12*, 865–896.

Gallo, W. F., Marvin, J. G., & Gnos, A. V. (1970). Nonsimilar nature of the laminar boundary layer. *AIAA J, 8*, 75–81.

Glauert, M. B. (1956). The Wall Jet. *J. Fluid Mech, 1*, 625–643.

Goldstein, S. (Ed.). (1938). *Modern Developments in Fluid Dynamics*. London: Oxford University Press. Reprinted by Dover, New York (1965).

Holstein, H., & Bohlen, T. (1940). Ein einfaches Verfahren zur Berechnung laminarer Reibungsschichten die dem Näherungsverfahren von K. Pohlhausen genügen. *Lilienthal-Bericht, S10*, 5–16.

Kevorkian, J., & Cole, J. D. (1981). *Perturbation Methods in Applied Mathematics*. New York: Springer-Verlag.

Mehta, R. (1985). Aerodynamics of sports balls. *Annual Review of Fluid Mechanics, 17*, 151–189.

Nayfeh, A. H. (1981). *Introduction to Perturbation Techniques*. New York: Wiley.

Peletier, L. A. (1972). On the asymptotic behavior of velocity profiles in laminar boundary layers. *Arch. for Rat. Mech. and Anal, 45*, 110–119.

Pohlhausen, K. (1921). Zur näherungsweisen Integration der Differentialgleichung der laminaren Grenzschicht. *Z. Angew. Math. Mech, 1*, 252–268.

Robbins, B. (1742). *New Principles of Gunnery: Containing the Determinations of the Force of Gun-powder and Investigations of the Difference in the Resisting Power of the Air to Swift and Slow Motions*. London: J. Nourse.

Schlichting, H. (1979). *Boundary Layer Theory* (7th ed.). New York: McGraw-Hill.

Serrin, J. (1967). Asymptotic behaviour of velocity profiles in the Prandtl boundary layer theory. *Proc. Roy. Soc. A, 299*, 491–507.

Taneda, S. (1965). Experimental investigation of vortex streets. *J. Phys. Soc. Japan, 20*, 1714–1721.

Thomson, R. E., & Gower, J. F. R. (1977). Vortex streets in the wake of the Aleutian Islands. *Monthly Weather Review, 105*, 873–884.

Thwaites, B. (1949). Approximate calculation of the laminar boundary layer. *Aero. Quart, 1*, 245–280.

van Dyke, M. (1975). *Perturbation Methods in Fluid Mechanics*. Stanford, CA: The Parabolic Press.

von Karman, T. (1921). Über laminare und turbulente Reibung. *Z. Angew. Math. Mech, 1*, 233–252.

Wen, C.-Y., & Lin, C.-Y. (2001). Two-dimensional vortex shedding of a circular cylinder. *Phys. Fluids, 13*, 557–560.

Supplemental Reading

Friedrichs, K. O. (1955). Asymptotic phenomena in mathematical physics. *Bull. Am. Math. Soc., 61*, 485–504.

Lagerstrom, P. A., & Casten, R. G. (1972). Basic concepts underlying singular perturbation techniques. *SIAM Review, 14*, 63–120.

Meksyn, D. (1961). *New Methods in Laminar Boundary Layer Theory.* New York: Pergamon Press.

Panton, R. L. (2005). *Incompressible Flow* (3rd ed.). New York: Wiley.

Sherman, F. S. (1990). *Viscous Flow.* New York: McGraw-Hill.

White, F. M. (2006). *Viscous Fluid Flow.* New York: McGraw-Hill.

Yih, C. S. (1977). *Fluid Mechanics: A Concise Introduction to the Theory.* Ann Arbor, MI: West River Press.

Computational Fluid Dynamics

Howard H. Hu

OUTLINE

10.1. Introduction   421

10.2. Finite-Difference Method   423

10.3. Finite-Element Method   429

10.4. Incompressible Viscous
Fluid Flow   436

10.5. Three Examples   449

10.6. Concluding Remarks   470

Exercises   470

Literature Cited   471

Supplemental Reading   472

CHAPTER OBJECTIVES

- To introduce the techniques for computational solutions of the fluid dynamic equations of motion
- To describe the finite-difference and finite-element formulations

- To specify finite-difference and finite-element equations for incompressible viscous flow
- To illustrate use of these equations via example calculations

10.1. INTRODUCTION

Computational fluid dynamics (CFD) is a science that, with the help of digital computers, produces quantitative predictions of fluid-flow phenomena based on the conservation laws (conservation of mass, momentum, and energy) governing fluid motion. These predictions normally occur under those conditions defined in terms of flow geometry, the physical properties of a fluid, and the boundary and initial conditions of a flow field. Such predictions generally concern sets of values of the flow variables, for example, velocity, pressure, or temperature at selected locations in the domain and for selected times. The predictions may also involve evaluations of overall flow behavior, such as the flow rate or the hydrodynamic force acting on an object in the flow.

During the past several decades different types of numerical methods have been developed to simulate fluid flows involving a wide range of applications. These methods include finite-difference, finite-element, finite-volume, and spectral methods. Some of them are discussed in this chapter.

As time has passed, CFD has increased in importance and in accuracy; however, its predictions are never completely exact. Because many potential sources of error may be involved, one has to be very careful when interpreting the results produced by CFD techniques. The most common sources of error are:

- *Discretization error.* This error is intrinsic to all numerical methods, and is incurred whenever a continuous system is approximated by a discrete one. For example, a finite number of locations in space (grid points) or instants of time may be used to resolve the flow field. Different numerical schemes may have different orders of magnitude of the discretization error. Even with the same method, the discretization error will be different depending upon the distribution of grid points used in a simulation. In most applications, one needs to properly select a numerical method and choose a grid to control discretization error.
- *Input data error.* This error arises from the fact that both the flow geometry and fluid properties may be known to only a certain level of precision, or possibly in only an approximate way.
- *Initial and boundary condition error.* It is common that the initial and boundary conditions of a flow field may represent the real situation with imperfect precision. For example, flow information is needed at locations where fluid enters and leaves the computational domain. Here, flow properties may not be known exactly and are thus approximated to complete a numerical calculation.
- *Modeling error.* More complicated flows may involve physical phenomena that are not perfectly described by current scientific theories. Models used to solve these problems certainly contain errors, for example, turbulence modeling, atmospheric modeling, polymeric-fluid constitutive modeling, multiphase flow modeling, and so on.

As a research and design tool, CFD normally complements experimental and theoretical fluid dynamics. However, CFD has a number of distinct advantages:

- It can be produced inexpensively and quickly, without an extraordinary amount of training, although interpreting results often requires experience. Yet, while the price of many commodities increases, computing costs are falling. According to Moore's law (Intel Corporation, 2003) based on the observation of the data for the last 40 years, computational power will double every two years into the foreseeable future.

- It generates complete information. Full-field CFD produces detailed and comprehensive information of all relevant variables throughout the domain of interest. This information can also be easily accessed.
- It allows easy parameter changes. CFD permits input parameters to be varied easily over wide ranges, thereby facilitating design optimization. Such variations are often either impossible or prohibitively expensive in experimental studies.
- It has the ability to simulate realistic conditions. CFD can simulate flows directly under practical conditions, unlike experiments, where a small- or a large-scale model may be needed, or analytical theories that may only be valid for limiting cases where one parameter or another is very large or small.
- It has the ability to simulate ideal conditions. CFD provides the convenience of switching off certain terms in the governing equations, which allows one to focus attention on a few essential parameters and eliminate all irrelevant features. Such parametric control is typically impossible in experiments.
- It permits investigation of unnatural or unwanted situations. CFD allows events to be studied so that every attempt is made to prevent, for example, conflagrations, explosions, or nuclear power plant failures.

The remainder of the chapter provides a self-contained survey of CFD techniques, so that the interested reader, who might further investigate CFD, will be aware of language and techniques when pursuing more detailed sources and current literature.

10.2. FINITE-DIFFERENCE METHOD

The key to various numerical methods is to convert the partial different equations that govern a physical phenomenon into a system of algebraic equations. Different techniques are available for this conversion. The finite-difference method is one of the most commonly used.

Approximation to Derivatives

Consider the one-dimensional transport equation,

$$\frac{\partial T}{\partial t} + u\frac{\partial T}{\partial x} = D\frac{\partial^2 T}{\partial x^2} \quad \text{for} \quad 0 \le x \le L. \tag{10.1}$$

This is the classic convection-diffusion problem for the scalar $T(x, t)$, where u is a convective velocity and D is a diffusion coefficient. For simplicity, assume that u and D are constants. This equation is written in dimensional form. The boundary conditions for this problem are

$$T(0, t) = g \quad \text{and} \quad \frac{\partial T}{\partial x}(L, t) = q, \tag{10.2}$$

where g and q are two constants. The initial condition is

$$T(x, 0) = T_0(x) \quad \text{for} \quad 0 \le x \le L, \tag{10.3}$$

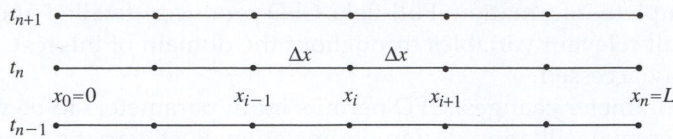

FIGURE 10.1   Uniform grid in space and time. The time epoch is listed at the left. The horizontal lines indicate the spatial domain. The small black dots indicate the spatial grid points where field quantities are determined.

where $T_0(x)$ is a given function that satisfies the boundary conditions (10.2).

Let us first discretize the transport equation (10.1) on a uniform grid with grid spacing Δx, as shown in Figure 10.1. Equation (10.1) is evaluated at spatial location $x = x_i$ and time $t = t_n$. Define $T(x_i, t_n)$ as the exact value of T at the location $x = x_i$ and time $t = t_n$, and let T_i^n be its approximation. Using the Taylor series expansion, we have:

$$T_{i+1}^n = T_i^n + \Delta x \left[\frac{\partial T}{\partial x}\right]_i^n + \frac{\Delta x^2}{2}\left[\frac{\partial^2 T}{\partial x^2}\right]_i^n + \frac{\Delta x^3}{6}\left[\frac{\partial^3 T}{\partial x^3}\right]_i^n + \frac{\Delta x^4}{24}\left[\frac{\partial^4 T}{\partial x^4}\right]_i^n + O(\Delta x^5), \qquad (10.4)$$

$$T_{i-1}^n = T_i^n - \Delta x \left[\frac{\partial T}{\partial x}\right]_i^n + \frac{\Delta x^2}{2}\left[\frac{\partial^2 T}{\partial x^2}\right]_i^n - \frac{\Delta x^3}{6}\left[\frac{\partial^3 T}{\partial x^3}\right]_i^n + \frac{\Delta x^4}{24}\left[\frac{\partial^4 T}{\partial x^4}\right]_i^n + O(\Delta x^5), \qquad (10.5)$$

where $O(\Delta x^5)$ means terms of the order of Δx^5. Therefore, the first spatial derivative may be approximated as

$$\begin{aligned}\left[\frac{\partial T}{\partial x}\right]_i^n &= \frac{T_{i+1}^n - T_i^n}{\Delta x} + O(\Delta x) \quad \text{(forward difference)}\\[2mm]
&= \frac{T_i^n - T_{i-1}^n}{\Delta x} + O(\Delta x) \quad \text{(backward difference)}\\[2mm]
&= \frac{T_{i+1}^n - T_{i-1}^n}{2\Delta x} + O(\Delta x^2) \quad \text{(centered difference)}\end{aligned} \qquad (10.6)$$

and the second-order derivative may be approximated as

$$\left[\frac{\partial^2 T}{\partial x^2}\right]_i^n = \frac{T_{i+1}^n - 2T_i^n + T_{i-1}^n}{\Delta x^2} + O(\Delta x^2). \qquad (10.7)$$

The orders of accuracy of the approximations (truncation errors) are also indicated in the expressions of (10.6) and (10.7). Higher order accuracy is generally desirable since the number of grid points is often a limitation for the size of a CFD computation. More accurate approximations generally require more values of the variable on the neighboring grid points. Similar expressions to (10.6) and (10.7) can be derived for nonuniform grids.

In the same fashion, the time derivative can be discretized as

$$\left[\frac{\partial T}{\partial t}\right]_i^n = \frac{T_i^{n+1} - T_i^n}{\Delta t} + O\left(\Delta t\right)$$

$$= \frac{T_i^n - T_i^{n-1}}{\Delta t} + O\left(\Delta t\right)$$

$$= \frac{T_i^{n+1} - T_i^{n-1}}{2\Delta t} + O\left(\Delta t^2\right), \tag{10.8}$$

where $\Delta t = t_{n+1} - t_n = t_n - t_{n-1}$ is the constant time step.

Discretization and Its Accuracy

A discretization of the transport equation (10.1) is obtained by evaluating the equation at fixed spatial and temporal grid points and using the approximations for the individual derivative terms listed in the preceding section. When the first expression in (10.8) is used, together with (10.7) and the centered difference in (10.6), (10.1) may be discretized by:

$$\frac{T_i^{n+1} - T_i^n}{\Delta t} + u\frac{T_{i+1}^n - T_{i-1}^n}{2\Delta x} = D\frac{T_{i+1}^n - 2T_i^n + T_{i-1}^n}{\Delta x^2} + O\left(\Delta t, \Delta x^2\right), \tag{10.9}$$

or

$$T_i^{n+1} \approx T_i^n - u\Delta t\frac{T_{i+1}^n - T_{i-1}^n}{2\Delta x} + D\Delta t\frac{T_{i+1}^n - 2T_i^n + T_{i-1}^n}{\Delta x^2}$$

$$= T_i^n - \alpha\left(T_{i+1}^n - T_{i-1}^n\right) + \beta\left(T_{i+1}^n - 2T_i^n + T_{i-1}^n\right), \tag{10.10}$$

where

$$\alpha = u\frac{\Delta t}{2\Delta x}, \quad \beta = D\frac{\Delta t}{\Delta x^2}. \tag{10.11}$$

Once the values of T_i^n are known, starting with the initial condition (10.3), the expression (10.10) simply updates the variable for the next time step $t = t_{n+1}$. This scheme is known as an explicit algorithm. The discretization (10.10) is first-order accurate in time and second-order accurate in space.

As another example, when the backward difference expression in (10.8) is used, we will have

$$\frac{T_i^n - T_i^{n-1}}{\Delta t} + u\frac{T_{i+1}^n - T_{i-1}^n}{2\Delta x} = D\frac{T_{i+1}^n - 2T_i^n + T_{i-1}^n}{\Delta x^2} + O\left(\Delta t, \Delta x^2\right), \tag{10.12}$$

or

$$T_i^n + \alpha \left(T_{i+1}^n - T_{i-1}^n\right) - \beta \left(T_{i+1}^n - 2T_i^n + T_{i-1}^n\right) \approx T_i^{n-1}. \tag{10.13}$$

At each time step $t = t_n$, a system of algebraic equations needs to be solved to advance the solution. This scheme is known as an *implicit algorithm*. Obviously, for the same accuracy, the explicit scheme (10.10) is much simpler than the implicit one (10.13). However, the explicit scheme has limitations.

Convergence, Consistency, and Stability

The result from the solution of the explicit scheme (10.10) or the implicit scheme (10.13) represents an approximate numerical solution to the original partial differential equation (10.1). One certainly hopes that the approximate solution will be close to the exact one. Thus we introduce the concepts of *convergence*, *consistency*, and *stability* of the numerical solution.

The approximate solution is said to be *convergent* if it approaches the exact solution as the grid spacings Δx and Δt tend to zero. We may define the solution error as the difference between the approximate solution and the exact solution,

$$e_i^n = T_i^n - T\left(x_i, t_n\right). \tag{10.14}$$

Thus, the approximate solution converges when $e_i^n \to 0$ as $\Delta x, \Delta t \to 0$. For a convergent solution, some measure of the solution error can be estimated as

$$\|e_i^n\| \le K \, \Delta x^a \, \Delta t^b, \tag{10.15}$$

where the measure may be the root mean square (rms) of the solution error on all the grid points; K is a constant independent of the grid spacing Δx and the time step Δt; and the indices a and b represent the convergence rates at which the solution error approaches zero.

One may reverse the discretization process and examine the limit of the discretized equations (10.10) and (10.13), as the grid spacing tends to zero. The discretized equation is said to be *consistent* if it recovers the original partial differential equation (10.1) in the limit of zero grid spacing.

Let us consider the explicit scheme (10.10). Substitution of the Taylor series expansions (10.4) and (10.5) into this scheme (10.10) produces

$$\left[\frac{\partial T}{\partial t}\right]_i^n + u\left[\frac{\partial T}{\partial x}\right]_i^n - D\left[\frac{\partial^2 T}{\partial x^2}\right]_i^n + E_i^n = 0, \tag{10.16}$$

where

$$E_i^n = \frac{\Delta t}{2}\left[\frac{\partial^2 T}{\partial t^2}\right]_i^n + u\frac{\Delta x^2}{6}\left[\frac{\partial^3 T}{\partial x^3}\right]_i^n - D\frac{\Delta x^2}{12}\left[\frac{\partial^4 T}{\partial x^4}\right]_i^n + O\left(\Delta t^2, \Delta x^4\right) \tag{10.17}$$

is the truncation error. Obviously, as the grid spacing $\Delta x, \Delta t \to 0$, this truncation error is of the order of $O(\Delta t, \Delta x^2)$ and tends to zero. Therefore, the explicit scheme (10.10) or expression

(10.16) recovers the original partial differential equation (10.1), so it is consistent. It is said to be first-order accurate in time and second-order accurate in space, according to the order of magnitude of the truncation error.

In addition to the truncation error introduced in the discretization process, other sources of error may be present in the approximate solution. Spontaneous disturbances (such as the round-off error produced from the finite precision arithmetic used by computers) may be introduced during either the evaluation or the numerical solution process. A numerical approximation is said to be *stable* if these disturbances decay and do not affect the solution.

The stability of the explicit scheme (10.10) may be examined using the von Neumann method in which numerical errors are represented via a Fourier decomposition. Let us consider the error at a grid point,

$$\xi_i^n = T_i^n - \overline{T}_i^n, \tag{10.18}$$

where T_i^n is the exact solution of the discretized system (10.10) and $\overline{T}_i^n$ is the approximate numerical solution of the same system. This error could be introduced due to the round-off error at each step of the computation. We need to monitor its decay/growth with time. It can be shown that the evolution of this error satisfies the same homogeneous algebraic system (10.10) or

$$\xi_i^{n+1} = (\alpha + \beta)\, \xi_{i-1}^n + (1 - 2\beta)\, \xi_i^n + (\beta - \alpha)\, \xi_{i+1}^n. \tag{10.19}$$

The error distributed along the grid line can always be decomposed in Fourier space as

$$\xi_i^n = \sum_{k=-\infty}^{\infty} g^n(k) e^{i\pi k x_i}, \tag{10.20}$$

where $i = \sqrt{-1}$, k is the wave number in Fourier space, and g^n represents the function g at time $t = t_n$. As the system is linear, we can examine one component of (10.20) at a time,

$$\xi_i^n = g^n(k) e^{i\pi k x_i}. \tag{10.21}$$

The component at the next time increment has a similar form:

$$\xi_i^{n+1} = g^{n+1}(k) e^{i\pi k x_i}. \tag{10.22}$$

Substituting the preceding two equations (10.21) and (10.22) into error equation (10.19), we obtain

$$g^{n+1} e^{i\pi k x_i} = g^n [(\alpha + \beta) e^{i\pi k x_{i-1}} + (1 - 2\beta) e^{i\pi k x_i} + (\beta - \alpha) e^{i\pi k x_{i+1}}] \tag{10.23}$$

or

$$\frac{g^{n+1}}{g^n} = [(\alpha + \beta) e^{-i\pi k \Delta x} + (1 - 2\beta) + (\beta - \alpha) e^{i\pi k \Delta x}]. \tag{10.24}$$

This ratio g^{n+1}/g^n is called the *amplification factor*. The condition for stability is that the magnitude of the error should decay with time, or

$$\left| \frac{g^{n+1}}{g^n} \right| \leq 1, \tag{10.25}$$

for any value of the wave number k. For this explicit scheme, the condition for stability (10.25) can be expressed as

$$\left(1 - 4\beta \sin^2\left(\frac{\theta}{2}\right)\right)^2 + (2\alpha \sin \theta)^2 \leq 1, \tag{10.26}$$

where $\theta = k\pi \, \Delta x$. The stability condition (10.26) also can be expressed as (Noye, 1983),

$$0 \leq 4\alpha^2 \leq 2\beta \leq 1. \tag{10.27}$$

For the pure diffusion problem ($u = 0$), the stability condition (10.27) for this explicit scheme requires that

$$0 \leq \beta \leq \frac{1}{2} \quad or \quad \Delta t \leq \frac{1}{2}\frac{\Delta x^2}{D}, \tag{10.28}$$

which limits the size of the time step. For the pure convection problem ($D = 0$), condition (10.27) will never be satisfied, which indicates that the scheme is always unstable and it means that any error introduced during the computation will explode with time. Thus, this explicit scheme is useless for pure convection problems. To improve the stability of the explicit scheme for the convection problem, one may use an upwind scheme to approximate the convective term,

$$T_i^{n+1} = T_i^n - 2\alpha \left(T_i^n - T_{i-1}^n\right), \tag{10.29}$$

where the stability condition requires that

$$u\frac{\Delta t}{\Delta x} \leq 1. \tag{10.30}$$

The condition (10.30) is known as the *Courant-Friedrichs-Lewy (CFL) condition*. This condition indicates that a fluid particle should not travel more than one spatial grid in one time step.

It can easily be shown that the implicit scheme (10.13) is also consistent and unconditionally stable.

It is normally difficult to show the convergence of an approximate solution theoretically. However, the *Lax Equivalence Theorem* (Richtmyer & Morton, 1967) states that: *for an approximation to a well-posed linear initial value problem, which satisfies the consistency condition, stability is a necessary and sufficient condition for the convergence of the solution.*

For convection-diffusion problems, the exact solution may change significantly in a narrow boundary layer. If the computational grid is not sufficiently fine to resolve the rapid variation of the solution in the boundary layer, the numerical solution may present unphysical oscillations adjacent to or in the boundary layer. To prevent the oscillatory solution, a condition on the cell Péclet number (or Reynolds number) is normally required (see Section 10.4),

$$R_{cell} = u\frac{\Delta x}{D} \leq 2. \tag{10.31}$$

10.3. FINITE-ELEMENT METHOD

The finite-element method was developed initially as an engineering procedure for stress and displacement calculations in structural analysis, where its success is impressive. The method was subsequently placed on a sound mathematical foundation with a variational interpretation of the potential energy of the system. For most fluid dynamics problems, finite-element applications have used the Galerkin finite-element formulation on which we will focus in this section.

Weak or Variational Form of Partial Differential Equations

Let us consider again the one-dimensional transport problem (10.1). The form of (10.1) with the boundary condition (10.2) and the initial conditions (10.3) is called the *strong* (or *classical*) form of the problem.

We first define a collection of trial solutions, which consists of all functions that have square-integrable first derivatives (H^1 functions, i.e., $\int_0^L (\partial T/\partial x)^2 dx < \infty$ if $T \in H^1$) and satisfy the Dirichlet type of boundary condition (where the value of the variable is specified) at $x = 0$. This is expressed as the trial functional space,

$$S = \{T | T \in H^1, T(0) = g\}. \tag{10.32}$$

The variational space of the trial solution is defined as

$$V = \{w | w \in H^1, w(0) = 0\}, \tag{10.33}$$

which requires a corresponding homogeneous boundary condition.

We next multiply the transport equation (10.1) by a function in the variational space ($w \in V$), and integrate the product over the domain where the problem is defined,

$$\int_0^L \left(\frac{\partial T}{\partial t} w\right) dx + u \int_0^L \left(\frac{\partial T}{\partial x} w\right) dx = D \int_0^L \left(\frac{\partial^2 T}{\partial x^2} w\right) dx. \tag{10.34}$$

Integrating the right-hand side of (10.34) by parts, we have:

$$\int_0^L \left(\frac{\partial T}{\partial t} w\right) dx + u \int_0^L \left(\frac{\partial T}{\partial x} w\right) dx + D \int_0^L \left(\frac{\partial T}{\partial x} \frac{\partial w}{\partial x}\right) dx = D \left[\frac{\partial T}{\partial x} w\right]_0^L = Dqw(L), \tag{10.35}$$

where the boundary conditions $\partial T/\partial x = q$ and $w(0) = 0$ are applied. The integral equation (10.35) is called the *weak* form of this problem. Therefore, the weak form can be stated as: Find $T \in S$ such that for all $w \in V$,

$$\int_0^L \left(\frac{\partial T}{\partial t} w\right) dx + u \int_0^L \left(\frac{\partial T}{\partial x} w\right) dx + D \int_0^L \left(\frac{\partial T}{\partial x} \frac{\partial w}{\partial x}\right) dx = Dqw(L). \tag{10.36}$$

It can be formally shown that the solution of the weak problem is identical to that of the strong problem, or that the strong and weak forms of the problem are *equivalent*. Obviously, if T is a solution of the strong problems (10.1) and (10.2), it must also be a solution of the weak problem (10.36) using the procedure for derivation of the weak formulation. However, let us assume that T is a solution of the weak problem (10.36). By reversing the order in deriving the weak formulation, we have:

$$\int_0^L \left(\frac{\partial T}{\partial t} + u\frac{\partial T}{\partial x} - D\frac{\partial^2 T}{\partial x^2} \right) wdx + D\left[\frac{\partial T}{\partial x}(L) - q \right] w\,(L) = 0. \tag{10.37}$$

Satisfying (10.37) for all possible functions of $w \in V$ requires that

$$\frac{\partial T}{\partial t} + u\frac{\partial T}{\partial x} - D\frac{\partial^2 T}{\partial x^2} = 0 \text{ for } x \in (0, L), \quad \text{and} \quad \frac{\partial T}{\partial x}(L) - q = 0, \tag{10.38}$$

which means that the solution T will be also a solution of the strong problem. It should be noted that the Dirichlet type of boundary condition (where the value of the variable is specified) is built into the trial functional space S, and is thus called an *essential* boundary condition. However, the Neumann type of boundary condition (where the derivative of the variable is imposed) is implied by the weak formulation as indicated in (10.38) and is referred to as a *natural* boundary condition.

Galerkin's Approximation and Finite-Element Interpolations

As shown earlier, the strong and weak forms of the problem are equivalent, and there is no approximation involved between these two formulations. Finite-element methods start with the weak formulation of the problem. Let us construct finite-dimensional approximations of S and V, which are denoted by S^h and V^h, respectively. The superscript refers to a discretization with a characteristic grid size h. The weak formulation (10.36) can be rewritten using these new spaces, as: Find $T^h \in S^h$ such that for all $w^h \in V^h$,

$$\int_0^L \left(\frac{\partial T^h}{\partial t} w^h \right) dx + u \int_0^L \left(\frac{\partial T^h}{\partial x} w^h \right) dx + D \int_0^L \left(\frac{\partial T^h}{\partial x} \frac{\partial w^h}{\partial x} \right) dx = Dqw^h\,(L). \tag{10.39}$$

Normally, S^h and V^h will be subsets of S and V, respectively. This means that if a function $\phi \in S^h$ then $\phi \in S$, and if another function $\psi \in V^h$ then $\psi \in V$. Therefore, (10.39) defines an approximate solution T^h to the exact weak form of the problem (10.36).

It should be noted that, up to the boundary condition $T(0) = g$, the function spaces S^h and V^h are composed of identical collections of functions. We may take out this boundary condition by defining a new function:

$$v^h\,(x, t) = T^h\,(x, t) - g^h\,(x), \tag{10.40}$$

where g^h is a specific function that satisfies the boundary condition $g^h\,(0) = g$. Thus, the functions v^h and w^h belong to the same space V^h. Equation (10.39) can be rewritten in terms of the new function v^h: Find $T^h = v^h + g^h$, where $v^h \in V^h$, such that for all $w^h \in V^h$,

$$\int_0^L \left(\frac{\partial v^h}{\partial t} w^h \right) dx + a\,(w^h, v^h) = Dqw^h\,(L) - a\,(w^h, g^h). \tag{10.41}$$

The operator $a\,(\cdot,\cdot)$ is defined as

$$a\,(w,v)\,=\,u\int_0^L \left(\frac{\partial v}{\partial x}w\right)dx + D\int_0^L \left(\frac{\partial v}{\partial x}\frac{\partial w}{\partial x}\right)dx. \tag{10.42}$$

The formulation (10.41) is called a *Galerkin formulation*, because the solution and the variational functions are in the same space. Again, the Galerkin formulation of the problem is an approximation to the weak formulation (10.36). Other classes of approximation methods, called *Petrov-Galerkin methods*, are those in which the solution function may be contained in a collection of functions other than V^h.

Next we need to explicitly construct the finite-dimensional variational space V^h. Let us assume that the dimension of the space is n and that the basis (shape or interpolation) functions for the space are

$$N_A\,(x),\ A\,=\,1, 2, ..., n. \tag{10.43}$$

Each shape function has to satisfy the boundary condition at $x = 0$,

$$N_A\,(0)\,=\,0,\ A\,=\,1, 2, ..., n, \tag{10.44}$$

which is required by the space V^h. The form of the shape functions will be discussed later. Any function $w^h \in V^h$ can be expressed as a linear combination of these shape functions:

$$w^h\,=\,\sum_{A=1}^n c_A N_A\,(x), \tag{10.45}$$

where the coefficients c_A are independent of x and uniquely define this function. We may introduce one additional function N_0 to specify the function g^h in (10.40) related to the essential boundary condition. This shape function has the property

$$N_0(0)\,=\,1. \tag{10.46}$$

Therefore, the function g^h can be expressed as

$$g^h(x)\,=\,gN_0\,(x),\quad\text{and}\quad g^h(0)\,=\,g. \tag{10.47}$$

With these definitions, the approximate solution can be written as

$$v^h(x,t)\,=\,\sum_{A=1}^n d_A\,(t)\,N_A\,(x), \tag{10.48}$$

and

$$T^h(x,t)\,=\,\sum_{A=1}^n d_A\,(t)\,N_A\,(x) + gN_0(x), \tag{10.49}$$

where d_A is a function of time only for time-dependent problems.

Matrix Equations, Comparison with Finite-Difference Method

With the construction of the finite-dimensional space V^h, the Galerkin formulation of the problem (10.41) leads to a coupled system of ordinary differential equations. Substitution of the expressions for the variational function (10.45) and for the approximate solution (10.48) into the Galerkin formulation (10.41) yields

$$\int_0^L \left(\sum_{B=1}^n \dot{d}_B N_B \sum_{A=1}^n c_A N_A \right) dx + a \left(\sum_{A=1}^n c_A N_A, \sum_{B=1}^n d_B N_B \right)$$

$$= Dq \sum_{A=1}^n c_A N_A(L) - a \left(\sum_{A=1}^n c_A N_A, g N_0 \right) \tag{10.50}$$

where $\dot{d}_B = d\,(d_B)/dt$. Rearranging the terms, (10.50) reduces to

$$\sum_{A=1}^n c_A G_A = 0, \tag{10.51}$$

where

$$G_A = \sum_{B=1}^n \dot{d}_B \int_0^L (N_A N_B)\, dx + \sum_{B=1}^n d_B a\,(N_A, N_B) - Dq N_A\,(L) + ga\,(N_A, N_0). \tag{10.52}$$

As the Galerkin formulation (10.41) should hold for all possible functions of $w^h \in V^h$, the coefficients, c_A, should be arbitrary. The necessary requirement for (10.51) to hold is that each G_A must be zero, that is,

$$\sum_{B=1}^n \dot{d}_B \int_0^L (N_B N_A)\, dx + \sum_{B=1}^n d_B a\,(N_A, N_B) = Dq N_A\,(L) - ga\,(N_A, N_0), \tag{10.53}$$

for $A = 1, 2, \ldots, n$. The system of equations (10.53) constitutes a system of n first-order ordinary differential equations (ODEs) for d_B. It can be put into a more concise matrix form. Let us define:

$$\mathbf{M} = [M_{AB}], \quad \mathbf{K} = [K_{AB}], \quad \mathbf{F} = \{F_A\}, \quad \mathbf{d} = \{d_B\}, \tag{10.54}$$

where

$$M_{AB} = \int_0^L (N_A N_B)\, dx, \tag{10.55}$$

$$K_{AB} = u \int_0^L (N_{B,x} N_A)\, dx + D \int_0^L (N_{B,x} N_{A,x})\, dx, \tag{10.56}$$

$$F_A = Dq N_A\,(L) - gu \int_0^L (N_{0,x} N_A)\, dx - gD \int_0^L (N_{0,x} N_{A,x})\, dx. \tag{10.57}$$

Equation (10.53) can then be written as

$$\mathbf{M}\dot{\mathbf{d}} + \mathbf{K}\mathbf{d} = \mathbf{F}. \tag{10.58}$$

The system of equations (10.58) is also termed the *matrix form* of the problem. Usually, $\mathbf{M}$ is called the *mass matrix*, $\mathbf{K}$ is the *stiffness matrix*, $\mathbf{F}$ is the *force vector*, and $\mathbf{d}$ is the *displacement vector*. This system of ODEs can be integrated by numerical methods, for example

Runge-Kutta methods, or discretized (in time) by finite-difference schemes as described in the previous section. The initial condition (10.3) will be used for integration. An alternative approach is to use a finite-difference approximation to the time-derivative term in the transport equation (10.1) at the beginning of the process, for example, by replacing $\partial T/\partial t$ with $(T^{n+1} - T^n)/\Delta t$, and then using the finite-element method to discretize the resulting equation.

Now let us consider the actual construction of the shape functions for the finite-dimensional variational space. The simplest example is to use piecewise-linear, finite-element space. We first partition the domain $[0, L]$ into n nonoverlapping subintervals (elements). A typical one is denoted as $[x_A, x_{A+1}]$. The shape functions associated with the interior nodes, $A = 1, 2, \ldots, n - 1$, are defined as

$$N_A(x) = \begin{cases} \dfrac{x - x_{A-1}}{x_A - x_{A-1}}, & x_{A-1} \le x < x_A, \\[2mm] \dfrac{x_{A+1} - x}{x_{A+1} - x_A}, & x_A \le x \le x_{A+1}, \\[2mm] 0, & \text{elsewhere.} \end{cases} \tag{10.59}$$

Further, for the boundary nodes, the shape functions are defined as

$$N_n(x) = \frac{x - x_{n-1}}{x_n - x_{n-1}}, x_{n-1} \le x \le x_n, \tag{10.60}$$

and

$$N_0(x) = \frac{x_1 - x}{x_1 - x_0}, x_0 \le x \le x_1. \tag{10.61}$$

These shape functions are graphically plotted in Figure 10.2. It should be noted that these shape functions have very compact (local) support and satisfy $N_A(x_B) = \delta_{AB}$, where δ_{AB} is the Kronecker delta (i.e., $\delta_{AB} = 1$ if $A = B$, whereas $\delta_{AB} = 0$ if $A \ne B$).

With the construction of the shape functions, the coefficient, d_A, in the expression for the approximate solution (10.49) represents the values of T^h at the nodes $x = x_A$ ($A = 1, 2, \ldots, n$), or

$$d_A = T^h(x_A) = T_A. \tag{10.62}$$

To compare the discretized equations generated from the finite-element method with those from finite-difference methods, we substitute (10.59) into (10.53) and evaluate the integrals. For an interior node x_A ($A = 1, 2, \ldots, n - 1$), we have:

$$\frac{d}{dt}\left(\frac{T_{A-1}}{6} + \frac{2T_A}{3} + \frac{T_{A+1}}{6}\right) + \frac{u}{2h}(T_{A+1} - T_{A-1}) - \frac{D}{h^2}(T_{A-1} - 2T_A + T_{A+1}) = 0, \tag{10.63}$$

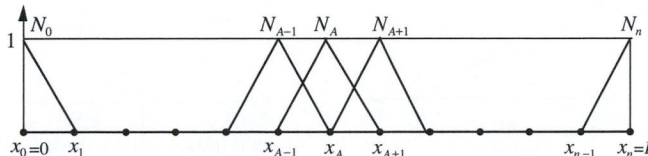

FIGURE 10.2 Piecewise-linear, finite-element space. These shape functions are compact.

where h is the uniform mesh size. The convective and diffusive terms in expression (10.63) have the same forms as those discretized using the standard second-order finite-difference method (centered difference) in (10.12). However, in the finite-element scheme, the time derivative term is presented with a three-point spatial average of the variable T, which differs from the finite-difference method. In general, the Galerkin finite-element formulation is equivalent to a finite-difference method. The advantage of the finite-element method lies in its flexibility to handle complex geometries.

Element Point of View of the Finite-Element Method

So far we have been using a global view of the finite-element method. The shape functions are defined on the global domain, as shown in Figure 10.2. However, it is also convenient to present the finite-element method using a local (or element) point of view. This viewpoint is useful for the evaluation of the integrals in (10.55) through (10.57) and the actual computer implementation of the finite-element method.

Figure 10.3 depicts the global and local descriptions of the eth element. The global description of the element e is just the "local" view of the full domain shown in Figure 10.2. Only two shape functions are nonzero within this element, N_{A-1} and N_A. Using the local coordinate in the standard element (parent domain) as shown on the right in Figure 10.3, we can write the standard shape functions as

$$N_1(\xi) = \frac{1}{2}(1 - \xi) \quad \text{and} \quad N_2(\xi) = \frac{1}{2}(1 + \xi). \tag{10.64}$$

Clearly, the standard shape function N_1 (or N_2) corresponds to the global shape function N_{A-1} (or N_A). The mapping between the domains of the global and local descriptions can easily be generated with the help of these shape functions,

$$x(\xi) = N_1(\xi)\, x_1^e + N_2(\xi)\, x_2^e = \frac{1}{2}\left[(x_A - x_{A-1})\,\xi + x_A + x_{A-1}\right], \tag{10.65}$$

with the notation that $x_1^e = x_{A-1}$ and $x_2^e = x_A$. One can also solve (10.65) for the inverse map:

$$\xi(x) = \frac{2x - x_A - x_{A-1}}{x_A - x_{A-1}}. \tag{10.66}$$

Within the element e, the derivative of the shape functions can be evaluated using the mapping equation (10.66):

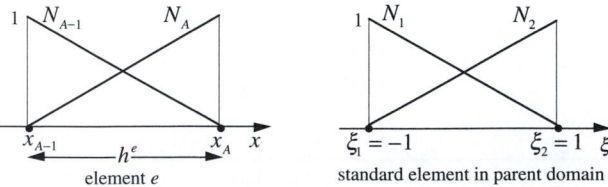

FIGURE 10.3　Global and local descriptions of an element.

$$\frac{dN_A}{dx} = \frac{dN_A}{d\xi}\frac{d\xi}{dx} = \frac{2}{x_A - x_{A-1}}\frac{dN_1}{d\xi} = \frac{-1}{x_A - x_{A-1}} \tag{10.67}$$

and

$$\frac{dN_{A+1}}{dx} = \frac{dN_{A+1}}{d\xi}\frac{d\xi}{dx} = \frac{2}{x_A - x_{A-1}}\frac{dN_2}{d\xi} = \frac{1}{x_A - x_{A-1}}. \tag{10.68}$$

The global mass matrix (10.55), the global stiffness matrix (10.56), and the global force vector (10.57) have been defined as the integrals over the global domain $[0, L]$. These integrals may be written as the summation of integrals over each element's domain. Thus

$$\mathbf{M} = \sum_{e=1}^{n_{el}}\mathbf{M}^e, \quad \mathbf{K} = \sum_{e=1}^{n_{el}}\mathbf{K}^e, \quad \mathbf{F} = \sum_{e=1}^{n_{el}}\mathbf{F}^e, \tag{10.69}$$

$$\mathbf{M}^e = [M^e_{AB}], \quad \mathbf{K}^e = [K^e_{AB}], \quad \mathbf{F}^e = \{F^e_A\} \tag{10.70}$$

where n_{el} is the total number of finite elements (in this case $n_{el} = n$), and

$$M^e_{AB} = \int_{\Omega^e}(N_A N_B)\, dx, \tag{10.71}$$

$$K^e_{AB} = u\int_{\Omega^e}(N_B,{}_x N_A)\, dx + D\int_{\Omega^e}(N_B,{}_x N_A,{}_x)\, dx, \tag{10.72}$$

$$F^e_A = Dq\delta_{en_{el}}\delta_{An} - gu\int_{\Omega^e}(N_0,{}_x N_A)\, dx - gD\int_{\Omega^e}(N_0,{}_x N_A,{}_x)\, dx, \tag{10.73}$$

and $\Omega^e = [x^e_1, x^e_2] = [x_{A-1}, x_A]$ is the domain of the eth element; and the first term on the right-hand side of (10.73) is nonzero only for $e = n_{el}$ and $A = n$.

Given the construction of the shape functions, most of the element matrices and force vectors in (10.71) through (10.73) will be zero. The nonzero ones require that $A = e$ or $e + 1$ and $B = e$ or $e + 1$. We may collect these nonzero terms and arrange them into the element mass matrix, stiffness matrix, and force vector as follows:

$$\mathbf{m}^e = [m^e_{ab}], \quad \mathbf{k}^e = [k^e_{ab}], \quad \mathbf{f}^e = \{f^e_a\}, \quad a, b = 1, 2, \tag{10.74}$$

where

$$m^e_{ab} = \int_{\Omega^e}(N_a N_b)\, dx, \tag{10.75}$$

$$k^e_{ab} = u\int_{\Omega^e}(N_b,{}_x N_a)\, dx + D\int_{\Omega^e}(N_b,{}_x N_a,{}_x)\, dx, \tag{10.76}$$

$$f^e_a = \begin{cases} -gk^e_{a1} & e = 1, \\ 0 & e = 2, 3, \dots, n_{el} - 1, \\ Dq\delta_{a2} & e = n_{el}, \end{cases} \tag{10.77}$$

Here, $\mathbf{m}^e$, $\mathbf{k}^e$, and $\mathbf{f}^e$ are defined with the local (element) ordering, and represent the nonzero terms in the corresponding $\mathbf{M}^e$, $\mathbf{K}^e$, and $\mathbf{F}^e$ with the global ordering. The terms in the local ordering need to be mapped back into the global ordering. For this example, the mapping is defined as

$$A = \begin{cases} e-1 & \text{if } a = 1 \\ e & \text{if } a = 2 \end{cases} \tag{10.78}$$

for element e.

Therefore, in the element viewpoint, the global matrices and the global vector can be constructed by summing the contributions of the element matrices and the element vector, respectively. The evaluation of the element matrices and the element vector can be performed on a standard element using the mapping between the global and local descriptions.

The finite-element methods for two- or three-dimensional problems will follow the same basic steps introduced in this section. However, the data structure and the forms of the elements or the shape functions will be more complicated. Refer to Hughes (1987) for a detailed discussion. Section 10.5 presents an example of a two-dimensional flow over a circular cylinder.

10.4. INCOMPRESSIBLE VISCOUS FLUID FLOW

This section discusses numerical schemes for solving incompressible viscous fluid flows. It focuses on techniques using the primitive variables (velocity and pressure). Other formulations using the stream function and vorticity are available in the literature (see Fletcher, 1988, Vol. II) and will not be discussed here since their extensions to three-dimensional flows are not straightforward. The schemes to be discussed normally apply to laminar flows. However, by incorporating additional appropriate turbulence models, these schemes will also be effective for turbulent flows.

For an incompressible Newtonian fluid, the fluid motion satisfies the constant-viscosity Navier-Stokes equation,

$$\rho\left(\frac{\partial \mathbf{u}}{\partial t} + (\mathbf{u}\cdot\nabla)\,\mathbf{u}\right) = \rho\mathbf{g} - \nabla p + \mu\nabla^2\mathbf{u}, \tag{10.79}$$

and the continuity equation,

$$\nabla\cdot\mathbf{u} = 0, \tag{10.80}$$

where (as in Chapter 4) $\mathbf{u}$ is the velocity vector, $\mathbf{g}$ is the body force per unit mass, which could be the gravitational acceleration, p is the pressure, and ρ and μ are the density and viscosity of the fluid, respectively. With the proper scaling, (10.79) can be written in the dimensionless form,

$$\frac{\partial \mathbf{u}}{\partial t} + (\mathbf{u}\cdot\nabla)\,\mathbf{u} = \mathbf{g} - \nabla p + \frac{1}{\text{Re}}\nabla^2\mathbf{u}, \tag{10.81}$$

where Re is the Reynolds number of the flow. In some approaches, the convective term is rewritten in conservative form,

$$(\mathbf{u} \cdot \nabla) \, \mathbf{u} = \nabla \cdot (\mathbf{u} \mathbf{u}),\tag{10.82}$$

because $\mathbf{u}$ is solenoidal.

In order to guarantee that a flow problem is well posed, appropriate initial and boundary conditions for the problem must be specified. For time-dependent flow problems, the initial condition for the velocity,

$$\mathbf{u} \, (\mathbf{x}, t = 0) = \mathbf{u}_0 \, (\mathbf{x}),\tag{10.83}$$

is required. The initial velocity field has to satisfy the continuity equation $\nabla \cdot \mathbf{u}_0 = 0$. At a solid surface, the fluid velocity should equal the surface velocity (no-slip condition). No boundary condition for the pressure is required at a solid surface. If the computational domain contains a section where the fluid enters the domain, the fluid velocity (and the pressure) at this inflow boundary should be specified. If the computational domain contains a section where the fluid leaves the domain (outflow section), appropriate outflow boundary conditions include zero tangential velocity and zero normal stress, or zero velocity derivatives, as further discussed in Gresho (1991). Because the conditions at the outflow boundary are artificial, it should be checked that the numerical results are not sensitive to the location of this boundary. In order to solve the Navier-Stokes equations, it is also appropriate to specify the value of the pressure at one reference point in the domain, because the pressure appears only as a gradient and can be determined up to a constant.

There are two major difficulties in solving the Navier-Stokes equations numerically. One is related to the unphysical oscillatory solution often found in a convection-dominated problem. The other is the treatment of the continuity equation that is a constraint on the flow to determine the pressure.

Convection-Dominated Problems

As mentioned in Section 10.2, the exact solution may change significantly in a thin boundary layer for convection-dominated transport problems. If the computational grid is not sufficiently fine to resolve the rapid variation of the solution in the boundary layer, the numerical solution may present unphysical oscillations adjacent the boundary. Let us examine the steady transport problem in one dimension,

$$u\frac{\partial T}{\partial x} = D\frac{\partial^2 T}{\partial x^2} \quad \text{for} \quad 0 \le x \le L,\tag{10.84}$$

with two boundary conditions:

$$T \, (0) = 0 \quad \text{and} \quad T \, (L) = 1.\tag{10.85}$$

The exact solution for this problem is

$$T = \frac{e^{Rx/L} - 1}{e^R - 1},\tag{10.86}$$

where

$$R = uL/D \tag{10.87}$$

is the global Péclet number. For large values of R, the solution (10.86) behaves as

$$T = e^{-R(1-x/L)}. \tag{10.88}$$

The essential feature of this solution is the existence of a boundary layer at $x = L$, and its thickness δ is of the order of

$$\frac{\delta}{L} = O\left(\frac{1}{|R|}\right). \tag{10.89}$$

At $1 - x/L = 1/R$, T is about 37% of the boundary value; while at $1 - x/L = 2/R$, T is about 13.5% of the boundary value.

If centered differences are used to discretize the steady transport equation (10.84) using the grid shown in Figure 10.1, the resulting finite-difference scheme is

$$\frac{u\Delta x}{2D}(T_{j+1} - T_{j-1}) = (T_{j+1} - 2T_j + T_{j-1}), \tag{10.90}$$

or

$$0.5 R_{cell}(T_{j+1} - T_{j-1}) = (T_{j+1} - 2T_j + T_{j-1}), \tag{10.91}$$

where the grid spacing $\Delta x = L/n$ and the cell Péclet number $R_{cell} = u\Delta x/D = R/n$. From the scaling of the boundary thickness (10.89) we know that it is of the order

$$\delta = O\left(\frac{L}{nR_{cell}}\right) = O\left(\frac{\Delta x}{R_{cell}}\right). \tag{10.92}$$

Physically, if T represents the temperature in the transport problem (10.84), the convective term brings the heat toward the boundary $x = L$, whereas the diffusive term conducts the heat away through the boundary. These two terms have to be balanced. The discretized equation (10.91) has the same physical meaning. Let us examine this balance for a node next to the boundary, $j = n - 1$. When the cell Péclet number $R_{cell} > 2$, according to (10.92) the thickness of the boundary layer is less than half the grid spacing, and the exact solution (10.86) indicates that the temperatures T_j and T_{j-1} are already outside the boundary layer and are essentially zero. Thus, the two sides of the discretized equation (10.91) cannot balance, or the conduction term is not strong enough to remove the heat convected to the boundary, assuming the solution is smooth. In order to force the heat balance, an unphysical oscillatory solution with $T_j < 0$ is generated to enhance the conduction term in the discretized problem (10.91). To prevent the oscillatory solution, the cell Péclet number is normally required to be less than two, which can be achieved by refining the grid to resolve the flow inside the boundary layer. In some respect, an oscillatory solution may be a virtue since it provides a warning that a physically important feature is not being properly resolved. To reduce the overall computational cost, nonuniform grids with local fine grid spacing inside the boundary layer will frequently be used to resolve the variables there.

Another common method to avoid the oscillatory solution is to use a first-order upwind scheme,

$$R_{cell} \, (T_j - T_{j-1}) = (T_{j+1} - 2T_j + T_{j-1}),$$ (10.93)

where a forward-difference scheme is used to discretize the convective term. It is easy to see that this scheme reduces the heat convected to the boundary and thus prevents the oscillatory solution. However, the upwind scheme is not very accurate (only first-order accurate). It can be easily shown that the upwind scheme (10.93) does not recover the original transport equation (10.84). Instead it is consistent with a slightly different transport equation (when the cell Péclet number is kept finite during the process),

$$u \frac{\partial T}{\partial x} = D \, (1 + 0.5 R_{cell}) \frac{\partial^2 T}{\partial x^2}.$$ (10.94)

Thus, another way to view the effect of the first-order upwind scheme (10.93) is that it introduces a numerical diffusivity of the value of $0.5 \, R_{cell} D$, which enhances the conduction of heat through the boundary. For an accurate solution, one normally requires that $0.5 \, R_{cell} \ll 1$, which is very restrictive and does not offer any advantage over the centered difference scheme (10.91).

Higher order upwind schemes may be introduced to obtain more accurate nonoscillatory solutions without excessive grid refinement. However, those schemes may be less robust. Refer to Fletcher (1988; see Vol. I, Chapter 9) for discussions.

Similarly, there are upwind schemes for finite-element methods to solve convection-dominated problems. Most of those are based on the Petrov-Galerkin approach that permits an effective upwind treatment of the convective term along local streamlines (Brooks & Hughes, 1982). Since then, stabilized finite-element methods have been developed where a least-squares term is added to the momentum balance equation to provide the necessary stability for convection-dominated flows (see Franca et al., 1992).

Incompressibility Condition

In solving the Navier-Stokes equations using the primitive variables (velocity and pressure), another numerical difficulty lies in the continuity equation: The continuity equation can be regarded either as a constraint on the flow field to determine the pressure, or the pressure plays the role of the Lagrange multiplier to satisfy the continuity equation.

In a flow field, the information (or disturbance) travels with both the flow and the speed of sound in the fluid. Since the speed of sound is infinite in an incompressible fluid, part of the information (pressure disturbance) is propagated instantaneously throughout the domain. In many numerical schemes the pressure is often obtained by solving a Poisson equation. The Poisson equation may occur in either continuous form or discrete form. Some of these schemes will be described here. In some of them, solving the pressure Poisson equation is the most costly step.

Another common technique to surmount the difficulty of the incompressible limit is to introduce an artificial compressibility (Chorin, 1967). This formulation is normally used for steady problems with a pseudotransient formulation. In the formulation, the continuity equation is replaced by

$$\frac{\partial p}{\partial t} + c^2 \nabla \cdot \mathbf{u} = 0, \tag{10.95}$$

where c is an arbitrary constant and could be the artificial speed of sound in a corresponding compressible fluid with the equation of state $p = c^2 \rho$. The formulation is called *pseudotransient* because (10.95) does not have any physical meaning before the steady state is reached. However, when c is large, (10.95) can be considered as an approximation to the unsteady solution of (10.80) in the incompressible Navier-Stokes problem.

Explicit MacCormack Scheme

Instead of using the artificial compressibility in (10.95), one may start with the exact compressible Navier-Stokes equations. In Cartesian coordinates, the component form of the continuity equation and compressible Navier-Stokes equation (with $\mu_v = 0$) in two dimensions can be explicitly written as:

$$\frac{\partial \rho}{\partial t} + \frac{\partial (\rho u)}{\partial x} + \frac{\partial (\rho v)}{\partial y} = 0, \tag{10.96}$$

$$\frac{\partial}{\partial t}(\rho u) + \frac{\partial}{\partial x}(\rho u^2) + \frac{\partial}{\partial y}(\rho v u) = \rho g_x - \frac{\partial p}{\partial x} + \mu \nabla^2 u + \frac{\mu}{3}\frac{\partial}{\partial x}\left(\frac{\partial u}{\partial x} + \frac{\partial v}{\partial y}\right), \tag{10.97}$$

$$\frac{\partial}{\partial t}(\rho v) + \frac{\partial}{\partial x}(\rho u v) + \frac{\partial}{\partial y}(\rho v^2) = \rho g_y - \frac{\partial p}{\partial y} + \mu \nabla^2 v + \frac{\mu}{3}\frac{\partial}{\partial y}\left(\frac{\partial u}{\partial x} + \frac{\partial v}{\partial y}\right), \tag{10.98}$$

with the equation of state,

$$p = c^2 \rho, \tag{10.99}$$

where c is the speed of sound in the medium. As long as the flows are limited to low Mach numbers and the conditions are almost isothermal, the solution to this set of equations should approximate the incompressible limit.

The explicit MacCormack scheme, after R. W. MacCormack (1969), is essentially a predictor-corrector scheme, similar to a second-order Runge-Kutta method commonly used to solve ordinary differential equations. For a system of equations of the form

$$\frac{\partial \mathbf{U}}{\partial t} + \frac{\partial \mathbf{E}\,(\mathbf{U})}{\partial x} + \frac{\partial \mathbf{F}\,(\mathbf{U})}{\partial y} = 0, \tag{10.100}$$

the explicit MacCormack scheme consists of two steps:

$$\text{predictor } \mathbf{U}^*_{i,j} = \mathbf{U}^n_{i,j} - \frac{\Delta t}{\Delta x}(\mathbf{E}^n_{i+1,j} - \mathbf{E}^n_{i,j}) - \frac{\Delta t}{\Delta y}(\mathbf{F}^n_{i,j+1} - \mathbf{F}^n_{i,j}), \tag{10.101}$$

$$\text{corrector } \mathbf{U}^{n+1}_{i,j} = \frac{1}{2}\left[\mathbf{U}^n_{i,j} + \mathbf{U}^*_{i,j} - \frac{\Delta t}{\Delta x}(\mathbf{E}^*_{i,j} - \mathbf{E}^*_{i-1,j}) - \frac{\Delta t}{\Delta y}(\mathbf{F}^*_{i,j} - \mathbf{F}^*_{i,j-1})\right] \tag{10.102}$$

Notice that the spatial derivatives in (10.100) are discretized with opposite one-sided finite differences in the predictor and corrector stages. The star variables are all evaluated at time level t_{n+1}. This scheme is second-order accurate in both time and space.

Applying the MacCormack scheme to the compressible Navier-Stokes equations (10.96) through (10.98) and replacing the pressure with (10.99), we have the predictor step:

$$\rho_{i,j}^* = \rho_{i,j}^n - c_1\left[(\rho u)_{i+1,j}^n - (\rho u)_{i,j}^n\right] - c_2\left[(\rho v)_{i,j+1}^n - (\rho v)_{i,j}^n\right] \tag{10.103}$$

$$\begin{aligned}
(\rho u)_{i,j}^* = {} & (\rho u)_{i,j}^n - c_1\left[\left(\rho u^2 + c^2\rho\right)_{i+1,j}^n - \left(\rho u^2 + c^2\rho\right)_{i,j}^n\right] \\
& - c_2\left[(\rho u v)_{i,j+1}^n - (\rho u v)_{i,j}^n\right] + \frac{4}{3}c_3\left(u_{i+1,j}^n - 2u_{i,j}^n + u_{i-1,j}^n\right) \\
& + c_4\left(u_{i,j+1}^n - 2u_{i,j}^n + u_{i,j-1}^n\right) \\
& + c_5\left(v_{i+1,j+1}^n + v_{i-1,j-1}^n - v_{i+1,j-1}^n - v_{i-1,j+1}^n\right)
\end{aligned} \tag{10.104}$$

$$\begin{aligned}
(\rho v)_{i,j}^* = {} & (\rho v)_{i,j}^n - c_1\left[(\rho u v)_{i+1,j}^n - (\rho u v)_{i,j}^n\right] \\
& - c_2\left[\left(\rho v^2 + c^2\rho\right)_{i,j+1}^n - \left(\rho v^2 + c^2\rho\right)_{i,j}^n\right] \\
& + c_3\left(v_{i+1,j}^n - 2v_{i,j}^n + v_{i-1,j}^n\right) + \frac{4}{3}c_4\left(v_{i,j+1}^n - 2v_{i,j}^n + v_{i,j-1}^n\right) \\
& + c_5\left(u_{i+1,j+1}^n + u_{i-1,j-1}^n - u_{i+1,j-1}^n - u_{i-1,j+1}^n\right).
\end{aligned} \tag{10.105}$$

Similarly, the corrector step is given by:

$$2\rho_{i,j}^{n+1} = \rho_{i,j}^n + \rho_{i,j}^* - c_1\left[(\rho u)_{i,j}^* - (\rho u)_{i-1,j}^*\right] - c_2\left[(\rho v)_{i,j}^* - (\rho v)_{i,j-1}^*\right] \tag{10.106}$$

$$\begin{aligned}
2(\rho u)_{i,j}^{n+1} = {} & (\rho u)_{i,j}^n + (\rho u)_{i,j}^* - c_1\left[\left(\rho u^2 + c^2\rho\right)_{i,j}^* - \left(\rho u^2 + c^2\rho\right)_{i-1,j}^*\right] \\
& - c_2\left[(\rho u v)_{i,j}^* - (\rho u v)_{i,j-1}^*\right] + \frac{4}{3}c_3\left(u_{i+1,j}^* - 2u_{i,j}^* + u_{i-1,j}^*\right) \\
& + c_4\left(u_{i,j+1}^* - 2u_{i,j}^* + u_{i,j-1}^*\right) \\
& + c_5\left(v_{i+1,j+1}^* + v_{i-1,j-1}^* - v_{i+1,j-1}^* - v_{i-1,j+1}^*\right)
\end{aligned} \tag{10.107}$$

$$\begin{aligned}
2\,(\rho v)_{i,j}^{n+1} = {} & (\rho v)_{i,j}^n + (\rho v)_{i,j}^* - c_1\left[(\rho u v)_{i,j}^* - (\rho u v)_{i-1,j}^*\right] \\
& - c_2\left[\left(\rho v^2 + c^2\rho\right)_{i,j}^* - \left(\rho v^2 + c^2\rho\right)_{i,j-1}^*\right] \\
& + c_3\left(v_{i+1,j}^* - 2v_{i,j}^* + v_{i-1,j}^*\right) + \frac{4}{3}c_4\left(v_{i,j+1}^* - 2v_{i,j}^* + v_{i,j-1}^*\right) \\
& + c_5\left(u_{i+1,j+1}^* + u_{i-1,j-1}^* - u_{i+1,j-1}^* - u_{i-1,j+1}^*\right)
\end{aligned} \tag{10.108}$$

The coefficients are defined as

$$c_1 = \frac{\Delta t}{\Delta x}, \quad c_2 = \frac{\Delta t}{\Delta y}, \quad c_3 = \frac{\mu\Delta t}{(\Delta x)^2}, \quad c_4 = \frac{\mu\Delta t}{(\Delta y)^2}, \quad c_5 = \frac{\mu\Delta t}{12\Delta x\Delta y}. \tag{10.109}$$

In both the predictor and corrector steps, the viscous terms (the second-order derivative terms) are all discretized with centered differences to maintain second-order accuracy. For brevity, body force terms in the momentum equations are neglected here.

During the predictor and corrector stages of the explicit MacCormack scheme (10.103) through (10.108), one-sided differences are arranged in the FF and BB fashion, respectively. Here, in the notation FF, the first F denotes the forward difference in the x-direction and the second F denotes the forward difference in the y-direction. Similarly, BB stands for backward differences in both x and y directions. We denote this arrangement as FF/BB. Similarly, one may get BB/FF, FB/BF, BF/FB arrangements. It is noted that some balanced cyclings of these arrangements generate better results than others.

Tannehill, Anderson, and Pletcher (1997) give the following semi-empirical stability criterion for the explicit MacCormack scheme:

$$\Delta t \leq \frac{\sigma}{(1 + 2/Re_\Delta)} \left[\frac{|u|}{\Delta x} + \frac{|v|}{\Delta y} + c\sqrt{\frac{1}{\Delta x^2} + \frac{1}{\Delta y^2}} \right]^{-1}, \tag{10.110}$$

where σ is a safety factor (≈ 0.9), and $Re_\Delta = \min (\rho|u|\Delta x/\mu, \rho|v|\Delta y/\mu)$ is the minimum mesh Reynolds number. This condition is quite conservative for flows with small-mesh Reynolds numbers.

One key issue for the explicit MacCormack scheme to work properly is the boundary conditions for density (thus pressure). We leave this issue to the next section where its implementation in two sample problems will be demonstrated.

MAC Scheme

Most numerical schemes developed for computational fluid dynamics problems can be characterized as operator-splitting algorithms. The operator-splitting algorithms divide each time step into several substeps. Each substep solves one part of the operator and thus decouples the numerical difficulties associated with each part of the operator. For example, consider a system,

$$\frac{d\phi}{dt} + A(\phi) = f, \tag{10.111}$$

with initial condition $\phi(0) = \phi_0$, where the operator A may be split into two operators

$$A(\phi) = A_1(\phi) + A_2(\phi). \tag{10.112}$$

Using a simple first-order accurate Marchuk-Yanenko fractional step scheme (Marchuk, 1975; Yanenko, 1971), the solution of the system at each time step $\phi^{n+1} = \phi((n+1)\Delta t)$ ($n = 0, 1, \ldots$) is approximated by solving the following two successive problems:

$$\frac{\phi^{n+1/2} - \phi^n}{\Delta t} + A_1(\phi^{n+1/2}) = f_1^{n+1}, \tag{10.113}$$

$$\frac{\phi^{n+1} - \phi^{n+1/2}}{\Delta t} + A_2(\phi^{n+1}) = f_2^{n+1}, \tag{10.114}$$

where $\phi^0 = \phi_0$, $\Delta t = t_{n+1} - t_n$, and $f_1^{n+1} + f_2^{n+1} = f^{n+1} = f((n+1)\Delta t)$. The time discretizations in (10.113) and (10.114) are implicit. Some schemes to be discussed in what follows actually use explicit discretizations. However, the stability conditions for those explicit schemes must be satisfied.

The MAC (marker-and-cell) method was first proposed by Harlow and Welch (1965) to solve flow problems with free surfaces. There are many variations of this method. It basically uses a finite-difference discretization for the Navier-Stokes equations and splits the equations into two operators:

$$\mathbf{A}_1(\mathbf{u}, p) = \begin{pmatrix} (\mathbf{u} \cdot \nabla)\mathbf{u} - \dfrac{1}{Re}\nabla^2 \mathbf{u} \\ 0 \end{pmatrix}, \quad \text{and} \quad \mathbf{A}_2(\mathbf{u}, p) = \begin{pmatrix} \nabla p \\ \nabla \cdot \mathbf{u} \end{pmatrix}. \tag{10.115}$$

Each time step is divided into two substeps as discussed in the Marchuk-Yanenko fractional-step scheme (10.113) and (10.114). The first step solves a convection and diffusion problem, which is discretized explicitly:

$$\frac{\mathbf{u}^{n+1/2} - \mathbf{u}^n}{\Delta t} + (\mathbf{u}^n \cdot \nabla)\mathbf{u}^n - \frac{1}{Re}\nabla^2 \mathbf{u}^n = \mathbf{g}^{n+1}. \tag{10.116}$$

In the second step, the pressure gradient operator is added (implicitly) and, at the same time, the incompressible condition is enforced:

$$\frac{\mathbf{u}^{n+1} - \mathbf{u}^{n+1/2}}{\Delta t} + \nabla p^{n+1} = 0, \tag{10.117}$$

and

$$\nabla \cdot \mathbf{u}^{n+1} = 0. \tag{10.118}$$

This step is also called a *projection step* to satisfy the incompressibility condition.

Normally, the MAC scheme is presented in a discretized form. A preferred feature of the MAC method is the use of the staggered grid. An example of a staggered grid in two dimensions is shown in Figure 10.4. On this staggered grid, pressure variables are defined at the centers of the cells and velocity components are defined at the cell faces.

Using the staggered grid, two components of the transport equation (10.116) can be written as

$$u_{i+1/2,j}^{n+1/2} = u_{i+1/2,j}^n - \Delta t \left(u\frac{\partial u}{\partial x} + v\frac{\partial u}{\partial y} - \frac{1}{Re}\nabla^2 u \right)_{i+1/2,j}^n + \Delta t\, f_{i+1/2,j}^{n+1}, \tag{10.119}$$

$$v_{i,j+1/2}^{n+1/2} = v_{i,j+1/2}^n - \Delta t \left(u\frac{\partial v}{\partial x} + v\frac{\partial v}{\partial y} - \frac{1}{Re}\nabla^2 v \right)_{i,j+1/2}^n + \Delta t\, g_{i,j+1/2}^{n+1}, \tag{10.120}$$

where $\mathbf{u} = (u, v)$, $\mathbf{g} = (f, g)$, $\left(u\frac{\partial u}{\partial x} + v\frac{\partial u}{\partial y} - \frac{1}{Re}\nabla^2 u \right)_{i+1/2,j}^n$, and $\left(u\frac{\partial v}{\partial x} + v\frac{\partial v}{\partial y} - \frac{1}{Re}\nabla^2 v \right)_{i,j+1/2}^n$ are the functions interpolated at the grid locations for the x-component of the velocity at $(i+1/2, j)$ and for the y-component of the velocity at $(i, j+1/2)$, respectively, and at the previous time $t = t_n$. The discretized form of (10.117) is

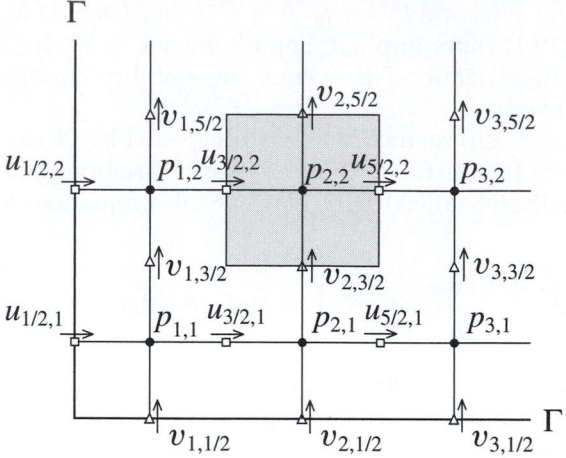

FIGURE 10.4 Staggered grid and a typical cell around $p_{2,2}$. Here velocity component values are computed halfway between the grid points for the pressure field.

$$u_{i+1/2,\,j}^{n+1} = u_{i+1/2,\,j}^{n+1/2} - \frac{\Delta t}{\Delta x}\left(p_{i+1,\,j}^{n+1} - p_{i,\,j}^{n+1}\right),$$ (10.121)

$$v_{i,\,j+1/2}^{n+1} = v_{i,\,j+1/2}^{n+1/2} - \frac{\Delta t}{\Delta y}\left(p_{i,\,j+1}^{n+1} - p_{i,\,j}^{n+1}\right),$$ (10.122)

where $\Delta x = x_{i+1} - x_i$ and $\Delta y = y_{j+1} - y_j$ are the uniform grid spacing in the x and y directions, respectively. The discretized continuity equation (10.118) can be written as:

$$\frac{u_{i+1/2,\,j}^{n+1} - u_{i-1/2,\,j}^{n+1}}{\Delta x} + \frac{v_{i,\,j+1/2}^{n+1} - v_{i,\,j-1/2}^{n+1}}{\Delta y} = 0.$$ (10.123)

Substitution of the two velocity components from (10.121) and (10.122) into the discretized continuity equation (10.123) generates a discrete Poisson equation for the pressure:

$$\nabla_d^2 p_{i,\,j}^{n+1} \equiv \frac{1}{\Delta x^2}\left(p_{i+1,\,j}^{n+1} - 2p_{i,\,j}^{n+1} + p_{i-1,\,j}^{n+1}\right) + \frac{1}{\Delta y^2}\left(p_{i,\,j+1}^{n+1} - 2p_{i,\,j}^{n+1} + p_{i,\,j-1}^{n+1}\right)$$

$$= \frac{1}{\Delta t}\left(\frac{u_{i+1/2,\,j}^{n+1/2} - u_{i-1/2,\,j}^{n+1/2}}{\Delta x} + \frac{v_{i,\,j+1/2}^{n+1/2} - v_{i,\,j-1/2}^{n+1/2}}{\Delta y}\right).$$ (10.124)

The major advantage of the staggered grid is that it prevents the appearance of oscillatory solutions. On a normal grid, the pressure gradient would have to be approximated using two alternate grid points (not the adjacent ones) when a central difference scheme is used, that is,

$$\left(\frac{\partial p}{\partial x}\right)_{i,\,j} = \frac{p_{i+1,\,j} - p_{i-1,\,j}}{2\Delta x} \quad \text{and} \quad \left(\frac{\partial p}{\partial y}\right)_{i,\,j} = \frac{p_{i,\,j+1} - p_{i,\,j-1}}{2\Delta y}.$$ (10.125)

Thus, a wavy pressure field (in a zigzag pattern) would be felt like a uniform one by the momentum equation. However, on a staggered grid, the pressure gradient is approximated by the difference of the pressures between two adjacent grid points. Consequently, a pressure field with a zigzag pattern would no longer be felt as a uniform pressure field and could not arise as a possible solution. It is also seen that the discretized continuity equation (10.123) contains the differences of the adjacent velocity components, which would prevent a wavy velocity field from satisfying the continuity equation.

Another advantage of the staggered grid is its accuracy. For example, the truncation error for (10.123) is $O\left(\Delta x^2, \Delta y^2\right)$ even though only four grid points are involved. The pressure gradient evaluated at the cell faces,

$$\left(\frac{\partial p}{\partial x}\right)_{i+1/2,\,j} = \frac{p_{i+1,\,j} - p_{i,\,j}}{\Delta x}, \quad \text{and} \quad \left(\frac{\partial p}{\partial y}\right)_{i,\,j+1/2} = \frac{p_{i,\,j+1} - p_{i,\,j}}{\Delta y}, \tag{10.126}$$

are all second-order accurate.

On the staggered grid, the MAC method does not require boundary conditions for the pressure equation (10.124). Let us examine a pressure node next to the boundary, for example, $p_{1,2}$ as shown in Figure 10.4. When the normal velocity is specified at the boundary, $u_{1/2,\,2}^{n+1}$ is known. In evaluating the discrete continuity equation (10.123) at the pressure node (1, 2), the velocity $u_{1/2,\,2}^{n+1}$ should not be expressed in terms of $u_{1/2,\,2}^{n+1/2}$ using (10.121). Therefore $p_{0,2}$ will not appear in equation (10.120), and no boundary condition for the pressure is needed. It should also be noted that (10.119) and (10.120) only update the velocity components for the interior grid points, and their values at the boundary grid points are not needed in the MAC scheme. Peyret and Taylor (1983) also noticed that the numerical solution in the MAC method is independent of the boundary values of $u^{n+1/2}$ and $v^{n+1/2}$, and a zero normal pressure gradient on the boundary would give satisfactory results. However, their explanation was more cumbersome.

In summary, for each time step in the MAC scheme, the intermediate velocity components, $u_{i+1/2,\,j}^{n+1/2}$ and $v_{i,\,j+1/2}^{n+1/2}$, in the interior of the domain are first evaluated using (10.119) and (10.120), respectively. Next, the discrete pressure Poisson equation (10.124) is solved. Finally, the velocity components at the new time step are obtained from (10.121) and (10.122). In the MAC scheme, the most costly step is the solution of the Poisson equation for the pressure (10.124).

Chorin (1968) and Temam (1969) independently presented a numerical scheme for the incompressible Navier-Stokes equations, termed the *projection method*. The projection method was initially proposed using the standard grid. However, when it is applied in an explicit fashion on the MAC staggered grid, it is identical to the MAC method as long as the boundary conditions are not considered, as shown in Peyret and Taylor (1983).

A physical interpretation of the MAC scheme or the projection method is that the explicit update of the velocity field does not generate a divergence-free velocity field in the first step. Thus an irrotational correction field, in the form of a velocity potential which is proportional to the pressure, is added to the nondivergence-free velocity field in the second step in order to enforce the incompressibility condition.

As the MAC method uses an explicit scheme in the convection-diffusion step, the stability conditions for this method are (Peyret & Taylor, 1983),

$$\frac{1}{2}(u^2 + v^2) \, \Delta t \, \text{Re} \leq 1, \tag{10.127}$$

and

$$\frac{4\Delta t}{\text{Re}\Delta x^2} \leq 1, \tag{10.128}$$

when $\Delta x = \Delta y$. The stability conditions (10.127) and (10.128) are quite restrictive on the size of the time step. These restrictions can be removed by using implicit schemes for the convection-diffusion step.

Θ-Scheme

The MAC algorithm described in the preceding section is only first-order accurate in time. In order to have a second-order accurate scheme for the Navier-Stokes equations, the Θ-scheme of Glowinski (1991) may be used. The Θ-scheme splits each time step symmetrically into three substeps, which are described here.

- Step 1:

$$\frac{\mathbf{u}^{n+\theta} - \mathbf{u}^n}{\theta \, \Delta t} - \frac{\alpha}{\text{Re}}\nabla^2 \mathbf{u}^{n+\theta} + \nabla p^{n+\theta} = \mathbf{g}^{n+\theta} + \frac{\beta}{\text{Re}}\nabla^2 \mathbf{u}^n - (\mathbf{u}^n \cdot \nabla)\mathbf{u}^n, \tag{10.129}$$

$$\nabla \cdot \mathbf{u}^{n+\theta} = 0. \tag{10.130}$$

- Step 2:

$$\frac{\mathbf{u}^{n+1-\theta} - \mathbf{u}^{n+\theta}}{(1-2\theta) \, \Delta t} - \frac{\beta}{\text{Re}}\nabla^2 \mathbf{u}^{n+1-\theta} + (\mathbf{u}^* \cdot \nabla)\mathbf{u}^{n+1-\theta} = \mathbf{g}^{n+1-\theta} + \frac{\alpha}{\text{Re}}\nabla^2 \mathbf{u}^{n+\theta} - \nabla p^{n+\theta}. \tag{10.131}$$

- Step 3:

$$\frac{\mathbf{u}^{n+1} - \mathbf{u}^{n+1-\theta}}{\theta \, \Delta t} - \frac{\alpha}{\text{Re}}\nabla^2 \mathbf{u}^{n+1} + \nabla p^{n+1} = \mathbf{g}^{n+1} + \frac{\beta}{\text{Re}}\nabla^2 \mathbf{u}^{n+1-\theta} - (\mathbf{u}^{n+1-\theta} \cdot \nabla)\mathbf{u}^{n+1-\theta}, \tag{10.132}$$

$$\nabla \cdot \mathbf{u}^{n+1} = 0. \tag{10.133}$$

It was shown that when $\theta = 1 - 1/\sqrt{2} = 0.29289...$, $\alpha + \beta = 1$, and $\beta = \theta/(1 - \theta)$, the scheme is second-order accurate. The first and third steps of the Θ-scheme are identical and are the Stokes flow problems. The second step, (10.131), represents a nonlinear convection-diffusion problem if $\mathbf{u}^* = \mathbf{u}^{n+1-\theta}$. However, it was concluded that there is practically no loss in accuracy and stability if $\mathbf{u}^* = \mathbf{u}^{n+\theta}$ is used. Numerical techniques for solving these substeps are discussed in Glowinski (1991).

Mixed Finite-Element Formulation

The weak formulation described in Section 10.3 can be directly applied to the Navier-Stokes equations (10.80) and (10.81), and it gives

$$\int_{\Omega} \left(\frac{\partial \mathbf{u}}{\partial t} + \mathbf{u} \cdot \nabla \mathbf{u} - \mathbf{g} \right) \cdot \tilde{\mathbf{u}} d\Omega + \frac{2}{\mathrm{Re}} \int_{\Omega} \mathbf{D}\,[\mathbf{u}] : \mathbf{D}\,[\tilde{\mathbf{u}}]\, d\Omega - \int_{\Omega} p\,(\nabla \cdot \tilde{\mathbf{u}})\, d\Omega = 0, \qquad (10.134)$$

$$\int_{\Omega} \tilde{p} \nabla \cdot \mathbf{u} d\Omega = 0, \qquad (10.135)$$

where $\tilde{\mathbf{u}}$ and $\tilde{p}$ are the variations of the velocity and pressure, respectively. The rate of strain tensor is given by

$$\mathbf{D}\,[\mathbf{u}] = \frac{1}{2}[\nabla \mathbf{u} + (\nabla \mathbf{u})^T]. \qquad (10.136)$$

The Galerkin finite-element formulation for the problem is identical to (10.134) and (10.135), except that all the functions are chosen from finite-dimensional subspaces and represented in the form of basis or interpolation functions.

The main difficulty with this finite-element formulation is the choice of the interpolation functions (or the types of the elements) for velocity and pressure. The finite-element approximations that use the same interpolation functions for velocity and pressure suffer from a highly oscillatory pressure field. As described in the previous section, a similar behavior in the finite-difference scheme is prevented by introducing the staggered grid. There are a number of options to overcome this problem with spurious pressure. One of them is the mixed finite-element formulation that uses different interpolation functions (or finite elements) for velocity and pressure. The requirement for the mixed finite-element approach is related to the so-called Babuska-Brezzi (or LBB) stability condition, or *inf-sup* condition. The detailed discussions for this condition can be found in Oden and Carey (1984). A common practice in the mixed finite-element formulation is to use a pressure interpolation function that is one order lower than a velocity interpolation function. As an example in two dimensions, a triangular element is shown in Figure 10.5a. On this mixed element, quadratic interpolation functions are used for the velocity components and are defined on all six nodes,

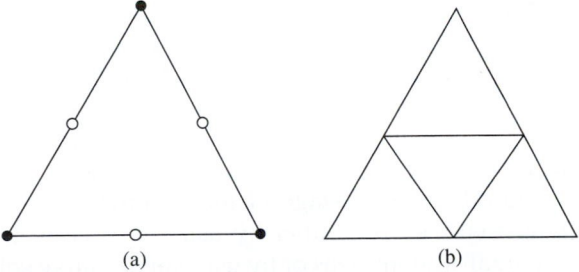

(a)　　　　　　　　(b)

FIGURE 10.5   Mixed finite elements. (a) Velocities are determined at all six points while pressure is only determined at the triangle's vertices (filled circles). (b) The velocity grid involves all four triangles while the pressure only involves the one larger composite triangle.

while linear interpolation functions are used for the pressure and are defined on three vertices only. A slightly different approach is to use a pressure grid that is twice coarser than the velocity one, and then use the same interpolation functions on both grids (Glowinski, 1991). For example, a piecewise-linear pressure is defined on the outside (coarser) triangle, while a piecewise-linear velocity is defined on all four subtriangles, as shown in Figure 10.5b.

Another option to prevent a spurious pressure field is to use the stabilized finite-element formulation while keeping the equal order interpolations for velocity and pressure. A general formulation in this approach is the Galerkin/least-squares (GLS) stabilization (Tezduyar, 1992). In the GLS stabilization, the stabilizing terms are obtained by minimizing the squared residual of the momentum equation integrated over each element domain. The choice of the stabilization parameter is discussed in Franca et al. (1992) and Franca and Frey (1992).

Comparing the mixed and the stabilized finite-element formulations, the mixed finite-element method is parameter free, as pointed out in Glowinski (1991). There is no need to adjust the stabilization parameters, which could be a delicate problem. More importantly, for a given flow problem the desired finite-element mesh size is generally determined based on the velocity behavior (e.g., it is defined by the boundary or shear layer thickness). Therefore, equal order interpolation will be more costly from the pressure point of view but without further gains in accuracy. However, the GLS-stabilized finite-element formulation has the additional benefit of preventing oscillatory solutions produced in the Galerkin finite-element method due to the large convective term in high Reynolds number flows.

Once the interpolation functions for the velocity and pressure in the mixed finite-element approximations are determined, the matrix form of equations (10.134) and (10.135) can be written as

$$\begin{pmatrix} \mathbf{M}\dot{\mathbf{u}} \\ \mathbf{0} \end{pmatrix} + \begin{pmatrix} \mathbf{A} & \mathbf{B} \\ \mathbf{B}^T & \mathbf{0} \end{pmatrix} \begin{pmatrix} \mathbf{u} \\ \mathbf{p} \end{pmatrix} = \begin{pmatrix} \mathbf{f}_u \\ \mathbf{f}_p \end{pmatrix}, \tag{10.137}$$

where $\mathbf{u}$ and $\mathbf{p}$ are the vectors containing all unknown values of the velocity components and pressure defined on the finite-element mesh, respectively; $\dot{\mathbf{u}}$ is the first-time derivative of $\mathbf{u}$; and $\mathbf{M}$ is the mass matrix corresponding to the time derivative term in equation (10.134). Matrix $\mathbf{A}$ depends on the value of $\mathbf{u}$ due to the nonlinear convective term in the momentum equation. The symmetry in the pressure terms in (10.134) and (10.135) results in the symmetric arrangement of $\mathbf{B}$ and $\mathbf{B}^T$ in the algebraic system (10.137). Vectors $\mathbf{f}_u$ and $\mathbf{f}_p$ come from the body-force term in the momentum equation and from the application of the boundary conditions.

The ordinary differential equation (10.137) can be further discretized in time with finite-difference methods. The resulting nonlinear system of equations is typically solved iteratively using Newton's method. At each stage of the nonlinear iteration, the sparse linear algebraic equations are normally solved either by using a direct solver such as the Gauss elimination procedure for small system sizes or by using an iterative solver such as the generalized minimum residual method (GMRES) for large systems. Other iterative solution methods for sparse nonsymmetric systems can be found in Saad (1996). An application of the mixed finite-element method is discussed as one of the examples in the next section.

10.5. THREE EXAMPLES

In this section, we will solve three sample problems. The first one is the classic driven-cavity flow problem. The second is flow around a square block confined between two parallel plates. These two problems will be solved by using the explicit MacCormack scheme, with details in Perrin and Hu (2006). The contribution by Andrew Perrin in preparing results for these two problems is greatly appreciated. The last problem is flow around a circular cylinder confined between two parallel plates. It will be solved by using a mixed finite-element formulation.

Explicit MacCormack Scheme for Driven-Cavity Flow Problem

The driven-cavity flow problem, in which a fluid-filled square box (or cavity) is swirled by a uniformly translating lid, as shown in Figure 10.6, is a classic problem in CFD. This problem is unambiguous with easily applied boundary conditions and has a wealth of documented analytical and computational results, for example Ghia et al. (1982). We will solve this flow using the explicit MacCormack scheme discussed in the previous section.

We may nondimensionalize the problem with the following scaling: length with D, velocity with U, time with D/U, density with a reference density ρ_0, and pressure with $\rho_0 U^2$. Using this scaling, the equation of state (10.99) becomes $p = \rho/M^2$, where $M = U/c$ is the Mach number. The Reynolds number is defined as $\text{Re} = \rho_0 U D/\mu$.

The boundary conditions for this problem are relatively simple. The velocity components on all four sides of the cavity are well defined. There are two singularities of velocity gradient at the two top corners where velocity u changes from 0 to U and from U to 0 directly underneath the sliding lid. However, these singularities will be smoothed out on a given grid, since the change of the velocity occurs linearly between two grid points. The boundary conditions for the density (hence the pressure) are more involved. Since the density is not specified on a solid surface, we need to generate an update scheme for values of density on all boundary points. A natural option is to derive it using the continuity equation.

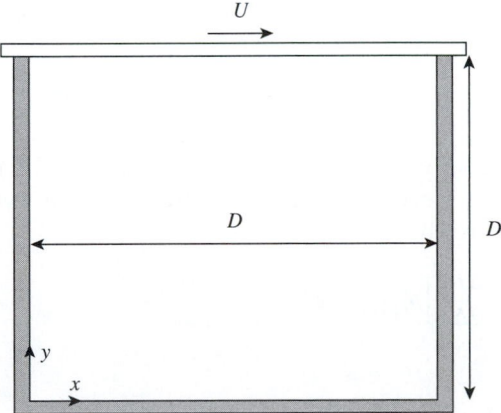

FIGURE 10.6 Driven-cavity flow problem. The cavity is filled with a fluid with the top lid sliding at a constant velocity U. The resulting flow field involves a large main vortex that is not symmetrical or centered in the box.

Consider the boundary on the left (at $x = 0$). Since $v = 0$ along the surface, the continuity equation (10.96) reduces to

$$\frac{\partial \rho}{\partial t} + \frac{\partial \rho u}{\partial x} = 0. \tag{10.138}$$

We may use a predictor-corrector scheme to update density on this surface with a one-sided, second-order accurate discretization for the spatial derivative:

$$\left(\frac{\partial f}{\partial x}\right)_i = \frac{1}{2\Delta x}(-f_{i+2} + 4f_{i+1} - 3f_i) + O\left(\Delta x^2\right)$$

$$\text{or} \quad \left(\frac{\partial f}{\partial x}\right)_i = \frac{-1}{2\Delta x}(-f_{i-2} + 4f_{i-1} - 3f_i) + O\left(\Delta x^2\right).$$

Therefore, on the surface of $x = 0$ (for $i = 0$, including two corner points on the left), we have the following update scheme for density:

$$\text{predictor} \quad \rho_{i,j}^* = \rho_{i,j}^n - \frac{\Delta t}{2\Delta x}\left[-(\rho u)_{i+2,j}^n + 4(\rho u)_{i+1,j}^n - 3(\rho u)_{i,j}^n\right], \tag{10.139}$$

$$\text{corrector} \quad 2\rho_{i,j}^{n+1} = \rho_{i,j}^n + \rho_{i,j}^* - \frac{\Delta t}{2\Delta x}\left[-(\rho u)_{i+2,j}^* + 4(\rho u)_{i+1,j}^* - 3(\rho u)_{i,j}^*\right]. \tag{10.140}$$

Similarly, on the right side of the cavity $x = D$ (for $i = n_x - 1$, where n_x is the number of grid points in the x-direction, including two corner points on the right), we have:

$$\text{predictor} \quad \rho_{i,j}^* = \rho_{i,j}^n + \frac{\Delta t}{2\Delta x}\left[-(\rho u)_{i-2,j}^n + 4(\rho u)_{i-1,j}^n - 3(\rho u)_{i,j}^n\right], \tag{10.141}$$

$$\text{corrector} \quad 2\rho_{i,j}^{n+1} = \rho_{i,j}^n + \rho_{i,j}^* + \frac{\Delta t}{2\Delta x}\left[-(\rho u)_{i-2,j}^* + 4(\rho u)_{i-1,j}^* - 3(\rho u)_{i,j}^*\right]. \tag{10.142}$$

On the bottom of the cavity $y = 0$ ($j = 0$),

$$\text{predictor} \quad \rho_{i,j}^* = \rho_{i,j}^n - \frac{\Delta t}{2\Delta y}\left[-(\rho v)_{i,j+2}^n + 4(\rho v)_{i,j+1}^n - 3(\rho v)_{i,j}^n\right], \tag{10.143}$$

$$\text{corrector} \quad 2\rho_{i,j}^{n+1} = \rho_{i,j}^n + \rho_{i,j}^* - \frac{\Delta t}{2\Delta y}\left[-(\rho v)_{i,j+2}^* + 4(\rho v)_{i,j+1}^* - 3(\rho v)_{i,j}^*\right]. \tag{10.144}$$

Finally, on the top of the cavity $y = D$ ($j = n_y - 1$) where n_y is the number of grid points in the y-direction), the density needs to be updated from slightly different expressions since $\partial \rho u / \partial x = U \partial \rho / \partial x$ is not zero there:

$$\text{predictor} \quad \rho_{i,j}^* = \rho_{i,j}^n - \frac{\Delta t U}{2\Delta x}\left[\rho_{i+1,j}^n - \rho_{i-1,j}^n\right] + \frac{\Delta t}{2\Delta y}\left[-(\rho v)_{i,j-2}^n + 4(\rho v)_{i,j-1}^n - 3(\rho v)_{i,j}^n\right], \tag{10.145}$$

$$\text{corrector} \quad 2\rho_{i,j}^{n+1} = \rho_{i,j}^n + \rho_{i,j}^* - \frac{\Delta t U}{2\Delta x}\left[\rho_{i+1,j}^* - \rho_{i-1,j}^*\right]$$
$$+ \frac{\Delta t}{2\Delta y}\left[-(\rho v)_{i,j-2}^* + 4(\rho v)_{i,j-1}^* - 3(\rho v)_{i,j}^*\right]. \tag{10.146}$$

In summary, we may organize the explicit MacCormack scheme at each time step (10.103) to (10.108) into the following six substeps.

Step 1: For $0 \le i < n_x$ and $0 \le j < n_y$ (all nodes):

$$u_{i,j} = (\rho u)^n_{i,j}/\rho^n_{i,j}, \quad v_{i,j} = (\rho v)^n_{i,j}/\rho^n_{i,j}.$$

Step 2: For $1 \le i < n_x - 1$ and $1 \le j < n_y - 1$ (all interior nodes):

$$\rho^*_{i,j} = \rho^n_{i,j} - a_1\left[(\rho u)^n_{i+1,j} - (\rho u)^n_{i,j}\right] - a_2\left[(\rho v)^n_{i,j+1} - (\rho v)^n_{i,j}\right],$$

$$(\rho u)^*_{i,j} = (\rho u)^n_{i,j} - a_3\left(\rho^n_{i+1,j} - \rho^n_{i,j}\right) - a_1\left[(\rho u^2)^n_{i+1,j} - (\rho u^2)^n_{i,j}\right]$$
$$- a_2\left[(\rho u v)^n_{i,j+1} - (\rho u v)^n_{i,j}\right] - a_{10}u_{i,j} + a_5(u_{i+1,j} + u_{i-1,j})$$
$$+ a_6(u_{i,j+1} + u_{i,j-1}) + a_9(v_{i+1,j+1} + v_{i-1,j-1} - v_{i+1,j-1} - v_{i-1,j+1}),$$

$$(\rho v)^*_{i,j} = (\rho v)^n_{i,j} - a_4\left(\rho^n_{i,j+1} - \rho^n_{i,j}\right) - a_1\left[(\rho u v)^n_{i+1,j} - (\rho u v)^n_{i,j}\right]$$
$$- a_2\left[(\rho v^2)^n_{i,j+1} - (\rho v^2)^n_{i,j}\right] - a_{11}v_{i,j} + a_7(v_{i+1,j} + v_{i-1,j})$$
$$+ a_8(v_{i,j+1} + v_{i,j-1}) + a_9(u_{i+1,j+1} + u_{i-1,j-1} - u_{i+1,j-1} - u_{i-1,j+1}).$$

Step 3: Impose boundary conditions (at time t_{n+1}) for $\rho^*_{i,j}$, $(\rho u)^*_{i,j}$, and $(\rho v)^*_{i,j}$.

Step 4: For $0 \le i < n_x$ and $0 \le j < n_y$ (all nodes):

$$u^*_{i,j} = (\rho u)^*_{i,j}/\rho^*_{i,j}, \quad v^*_{i,j} = (\rho v)^*_{i,j}/\rho^*_{i,j}.$$

Step 5: For $1 \le i < n_x - 1$ and $1 \le j < n_y - 1$ (all interior nodes):

$$2\rho^{n+1}_{i,j} = \left(\rho^n_{i,j} + \rho^*_{i,j}\right) - a_1 + \left[(\rho u)^*_{i,j} - (\rho u)^*_{i-1,j}\right] - a_2\left[(\rho v)^*_{i,j} - (\rho v)^*_{i,j-1}\right],$$

$$2(\rho u)^{n+1}_{i,j} = (\rho u)^n_{i,j} + (\rho u)^*_{i,j} - a_3\left(\rho^*_{i,j} - \rho^*_{i-1,j}\right) - a_1\left[(\rho u^2)^*_{i,j} - (\rho u^2)^*_{i-1,j}\right]$$
$$- a_2\left[(\rho u v)^*_{i,j} - (\rho u v)^*_{i,j-1}\right] - a_{10}u^*_{i,j} + a_5\left(u^*_{i+1,j} + u^*_{i-1,j}\right)$$
$$+ a_6\left(u^*_{i,j+1} + u^*_{i,j-1}\right) + a_9\left(v^*_{i+1,j+1} + v^*_{i-1,j-1} - v^*_{i+1,j-1} - v^*_{i-1,j+1}\right),$$

$$2(\rho v)^{n+1}_{i,j} = (\rho v)^n_{i,j} + (\rho v)^*_{i,j} - a_4\left(\rho^*_{i,j} - \rho^*_{i,j-1}\right) - a_1\left[(\rho u v)^*_{i,j} - (\rho u v)^*_{i-1,j}\right]$$
$$- a_2\left[(\rho v^2)^*_{i,j} - (\rho v^2)^*_{i,j-1}\right] - a_{11}v^*_{i,j} + a_7\left(v^*_{i+1,j} + v^*_{i-1,j}\right)$$
$$+ a_8\left(v^*_{i,j+1} + v^*_{i,j-1}\right) + a_9\left(u^*_{i+1,j+1} + u^*_{i-1,j-1} - u^*_{i+1,j-1} - u^*_{i-1,j+1}\right).$$

Step 6: Impose boundary conditions for $\rho_{i,j}^{n+1}$, $(\rho u)_{i,j}^{n+1}$, and $(\rho v)_{i,j}^{n+1}$. The coefficients are defined as:

$$a_1 = \frac{\Delta t}{\Delta x}, \quad a_2 = \frac{\Delta t}{\Delta y}, \quad a_3 = \frac{\Delta t}{\Delta x M^2}, \quad a_4 = \frac{\Delta t}{\Delta y M^2}, \quad a_5 = \frac{4\Delta t}{3Re(\Delta x)^2},$$

$$a_6 = \frac{\Delta t}{Re\,(\Delta y)^2}, \quad a_7 = \frac{\Delta t}{Re\,(\Delta x)^2}, \quad a_8 = \frac{4\Delta t}{3Re\,(\Delta y)^2}, \quad a_9 = \frac{\Delta t}{12Re\Delta x\Delta y},$$

$$a_{10} = 2\,(a_5 + a_6), \quad a_{11} = 2\,(a_7 + a_8).$$

For coding purposes, the variables $u_{i,j}$ $(v_{i,j})$ and $u_{i,j}^*$ $(v_{i,j}^*)$ can take the same storage space. At the end of each time step, the starting values of $\rho_{i,j}^n$, $(\rho u)_{i,j}^n$, and $(\rho v)_{i,j}^n$ will be replaced with the corresponding new values of $\rho_{i,j}^{n+1}$, $(\rho u)_{i,j}^{n+1}$, and $(\rho v)_{i,j}^{n+1}$.

Next we present some of the results and compare them with those in the paper by Hou et al. (1995) obtained by a lattice Boltzmann method. To keep the flow almost incompressible, the Mach number is chosen as $M = 0.1$. Flows with two Reynolds numbers, $Re = \rho_0 UD/\mu = 100$ and 400 are simulated. At these Reynolds numbers, the flow will eventually be steady. Thus calculations need to be run long enough to get to the steady state. A uniform grid of 256 by 256 was used for this example.

Figure 10.7 shows comparisons of the velocity field calculated by the explicit MacCormack scheme with the streamlines from Hou et al. (1995) at $Re = 100$ and 400. The agreement seems reasonable. It was also observed that the location of the center of the primary eddy agrees even better. When $Re = 100$, the center of the primary eddy is found at $(0.62 \pm 0.02, 0.74 \pm 0.02)$ from the MacCormack scheme in comparison with $(0.6196, 0.7373)$ from Hou. When $Re = 400$, the center of the primary eddy is found at $(0.57 \pm 0.02, 0.61 \pm 0.02)$ from the MacCormack scheme in comparison with $(0.5608, 0.6078)$ from Hou.

For a more quantitative comparison, Figure 10.8 plots the velocity profile along a vertical line cut through the center of the cavity $(x = 0.5D)$. The velocity profiles for two Reynolds numbers, $Re = 100$ and 400, are compared. The results from the explicit MacCormack scheme

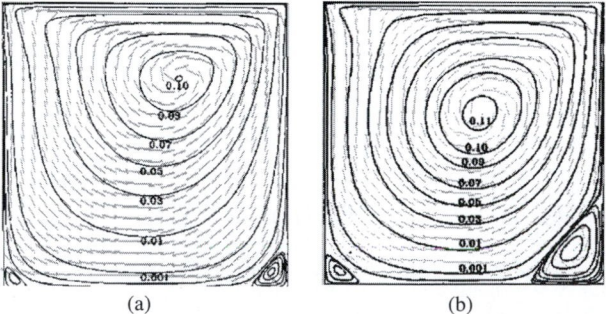

(a)   (b)

FIGURE 10.7   Comparisons of lid-driven square cavity results from the explicit MacCormack scheme (light gray, velocity vector field) and those from Hou et al. (1995) (dark solid streamlines) calculated using a lattice Boltzmann method. (a) Re = 100, (b) Re = 400.

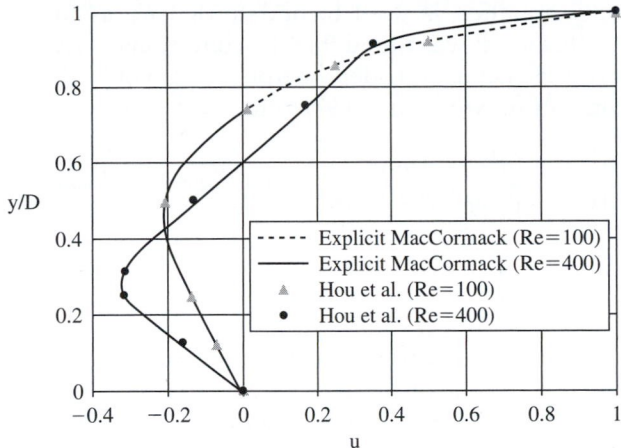

FIGURE 10.8 Comparison of velocity profiles along a line cut through the center of the cavity ($x = 0.5D$) at Re = 100 and 400. There is agreement between the results at the same Reynolds number from different computational schemes.

are shown in solid and dashed lines. The data points in symbols were directly converted from Hou's paper. The agreement is excellent.

Explicit MacCormack Scheme for Flow Over a Square Block

For the second example, we consider flow around a square block confined between two parallel plates. Fluid comes in from the left with a uniform velocity profile U, and the plates are sliding with the same velocity, as indicated in Figure 10.9. This flow corresponds to the block moving left with velocity U along the channel's centerline. In the calculation we set the channel width $H = 3D$ and the channel length $L = 35D$ with $15D$ ahead of the block and $19D$ behind. The Mach number is set at $M = 0.05$ to approximate the incompressible limit.

The velocity boundary conditions in this problem are specified as shown in Figure 10.9, except that at the outflow section, conditions $\partial \rho u / \partial x = 0$ and $\partial \rho v / \partial x = 0$ are used. The density (or pressure) boundary conditions are much more complicated, especially on the block surface. On all four sides of the outer boundary (top and bottom plates, inflow and

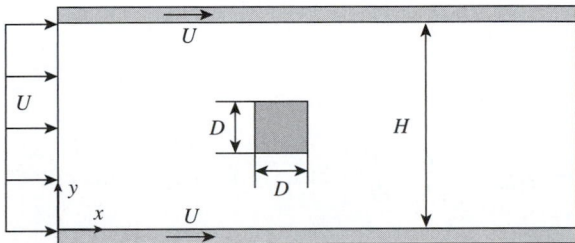

FIGURE 10.9 Computational domain for two-dimensional flow past a square obstruction (block) between two parallel plates.

outflow), the continuity equation is used to update density as in the previous example. However, on the block surface, it was found that the conditions derived from the momentum equations give better results. Let us consider the front section of the block, and evaluate the x-component of the momentum equation (10.97) with $u = v = 0$:

$$
\frac{\partial \rho}{\partial x} = M^2 \left[\frac{1}{Re} \left(\frac{4}{3} \frac{\partial^2 u}{\partial x^2} + \frac{1}{3} \frac{\partial^2 v}{\partial x \partial y} + \frac{\partial^2 u}{\partial y^2} \right) - \frac{\partial}{\partial x}(\rho u^2) - \frac{\partial}{\partial y}(\rho v u) - \frac{\partial}{\partial t}(\rho u) \right]_{\text{front suface}}
$$

$$
= \frac{M^2}{Re} \left(\frac{4}{3} \frac{\partial^2 u}{\partial x^2} + \frac{1}{3} \frac{\partial^2 v}{\partial x \partial y} \right). \tag{10.147}
$$

In (10.147), the variables are nondimensionalized with the same scaling as the driven-cavity flow problem except that the block size D is used for length. Furthermore, the density gradient may be approximated with a second-order backward finite-difference scheme:

$$
\left(\frac{\partial \rho}{\partial x} \right)_{i,j} = \frac{-1}{2\Delta x}(-\rho_{i-2,j} + 4\rho_{i-1,j} - 3\rho_{i,j}) + O(\Delta x^2). \tag{10.148}
$$

The second-order derivatives for the velocities are expressed as

$$
\left(\frac{\partial^2 u}{\partial x^2} \right)_{i,j} = \frac{1}{\Delta x^2}(2u_{i,j} - 5u_{i-1,j} + 4u_{i-2,j} - u_{i-3,j}) + O(\Delta x^2), \tag{10.149}
$$

and

$$
\left(\frac{\partial^2 v}{\partial x \partial y} \right)_{i,j} = \frac{-1}{4\Delta x \Delta y} \left[-(v_{i-2,j+1} - v_{i-2,j-1}) + 4(v_{i-1,j+1} - v_{i-1,j-1}) - 3(v_{i,j+1} - v_{i,j-1}) \right]
$$

$$
+ O(\Delta x^2, \Delta x \Delta y, \Delta y^2). \tag{10.150}
$$

Substituting (10.148) through (10.150) into (10.147), we have an expression for density at the front of the block:

$$
\rho_{i,j}|_{\text{front}} = \frac{1}{3}(4\rho_{i-1,j} - \rho_{i-2,j}) + \frac{8}{9\Delta x} \frac{M^2}{Re}(-5u_{i-1,j} + 4u_{i-2,j} - u_{i-3,j})
$$

$$
- \frac{1}{18\Delta y} \frac{M^2}{Re} \left[-(v_{i-2,j+1} - v_{i-2,j-1}) + 4(v_{i-1,j+1} - v_{i-1,j-1}) - 3(v_{i,j+1} - v_{i,j-1}) \right]. \tag{10.151}
$$

Similarly, at the back of the block:

$$
\rho_{i,j}|_{\text{back}} = \frac{1}{3}(4\rho_{i+1,j} - \rho_{i+2,j}) - \frac{8}{9\Delta x} \frac{M^2}{Re}(-5u_{i+1,j} + 4u_{i+2,j} - u_{i+3,j})
$$

$$
- \frac{1}{18\Delta y} \frac{M^2}{Re} \left[-(v_{i+2,j+1} - v_{i+2,j-1}) + 4(v_{i+1,j+1} - v_{i+1,j-1}) - 3(v_{i,j+1} - v_{i,j-1}) \right]. \tag{10.152}
$$

At the top of the block, the y-component of the momentum equation should be used, and it is easy to find:

$$\rho_{i,j}\big|_{top} = \frac{1}{3}(4\rho_{i,j+1} - \rho_{i,j+2}) - \frac{8}{9\Delta y}\frac{M^2}{Re}(-5v_{i,j+1} + 4v_{i,j+2} - v_{i,j+3})$$

$$-\frac{1}{18\Delta x}\frac{M^2}{Re}\Big[-(u_{i+1,j+2} - u_{i-1,j+2}) + 4(u_{i+1,j+1} - u_{i-1,j+1}) - 3(u_{i+1,j} - u_{i-1,j})\Big],$$

$$\text{(10.153)}$$

and, finally, at the bottom of the block:

$$\rho_{i,j}\big|_{bottom} = \frac{1}{3}(4\rho_{i,j-1} - \rho_{i,j-2}) + \frac{8}{9\Delta y}\frac{M^2}{Re}(-5v_{i,j-1} + 4v_{i,j-2} - v_{i,j-3})$$

$$-\frac{1}{18\Delta x}\frac{M^2}{Re}\Big[-(u_{i+1,j-2} - u_{i-1,j-2}) + 4(u_{i+1,j-1} - u_{i-1,j-1}) - 3(u_{i+1,j} - u_{i-1,j})\Big].$$

$$\text{(10.154)}$$

At the four corners of the block, the average values from the two corresponding sides may be used.

In computation, double precision numbers should be used: Otherwise cumulative round-off error may corrupt the simulation, especially for long runs. It is also helpful to introduce a new variable for density, $\rho' = \rho - 1$, such that only the density variation is computed. For this example, we may extend the *FF/BB* form of the explicit MacCormack scheme to have an *FB/BF* arrangement for one time step and a *BF/FB* arrangement for the subsequent time step. This cycling seems to generate better results.

We first plot the drag coefficient, C_D (4.107) and the lift coefficient, C_L (4.108) as functions of time for flows at two Reynolds numbers, Re = 20 and 100, in Figure 10.10. For Re = 20, after the initial messy transient (corresponding to sound waves bouncing around the block and reflecting at the outflow) the flow eventually settles into a steady state. The drag coefficient stabilizes at a constant value around $C_D = 6.94$ (obtained on a grid of 701×61). Calculation on a finer grid (1401×121) yields $C_D = 7.003$. This is in excellent agreement with the value of $C_D = 7.005$ obtained from an implicit finite-element calculation for incompressible flows (similar to the one used in the next example in this section) on a similar mesh to 1401×121. There is a small lift ($C_L = 0.014$) due to asymmetries in the numerical scheme. The lift reduces to $C_L = 0.003$ on the finer grid of 1401×121. For Re = 100, periodic vortex shedding occurs. Drag and lift coefficients are shown in Figure 10.10b. The mean value of the drag coefficient and the amplitude of the lift coefficient are $C_D = 3.35$ and $C_L = 0.77$, respectively. The finite-element results are $C_D = 3.32$ and $C_L = 0.72$ under similar conditions.

The flow field around the block at Re = 20 is shown in Figure 10.11. A steady wake is attached behind the block, and the circulation within the wake is clearly visible. Figure 10.12 displays a sequence of the flow field around the block during one cycle of vortex shedding at Re = 100.

Figure 10.13 shows the convergence of the drag coefficient as the grid spacing is reduced. Tests for two Reynolds numbers, Re = 20 and 100, are plotted. It seems that the solution with

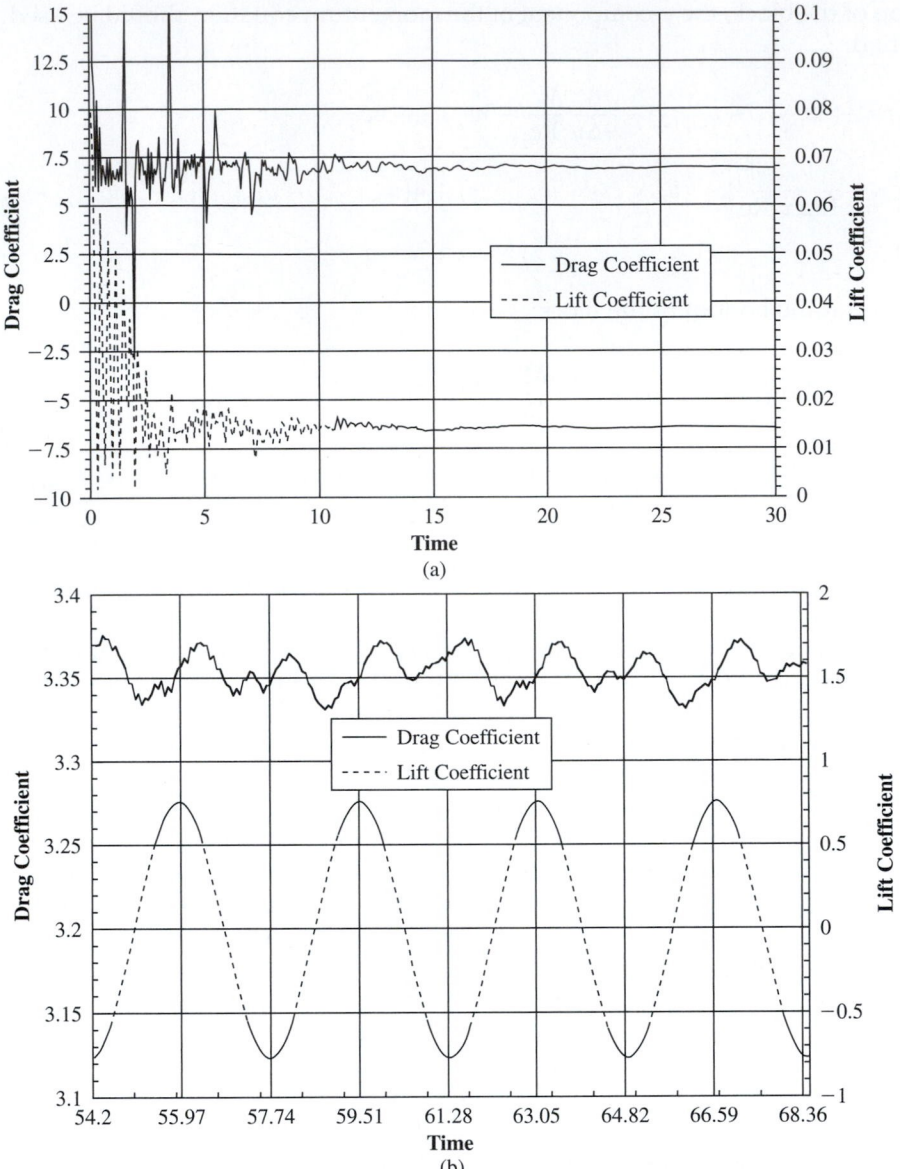

FIGURE 10.10   Drag and lift coefficients as functions of time for flow over a block. (a) Re = 20 on a grid of 701 ×
61, (b) Re = 100 on a grid of 1401 × 121. Clearly, this flow is oscillatory at the higher Reynolds number.

20 grid points across the block ($\Delta x = \Delta y = 0.05$) reasonably resolves the drag coefficient and
the singularity at the block corners does not affect this convergence very much.

The explicit MacCormack scheme can be quite efficient to compute flows at high Reynolds
numbers where small time steps are naturally needed to resolve high frequencies in the flow,

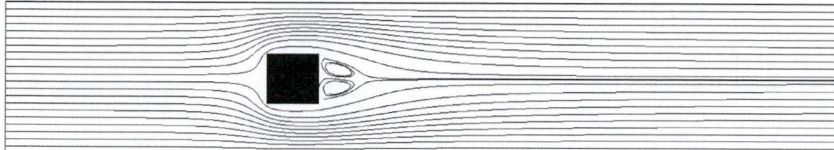

FIGURE 10.11 Streamlines for flow around a block at Re = 20. Although there is flow separation, the flow is steady.

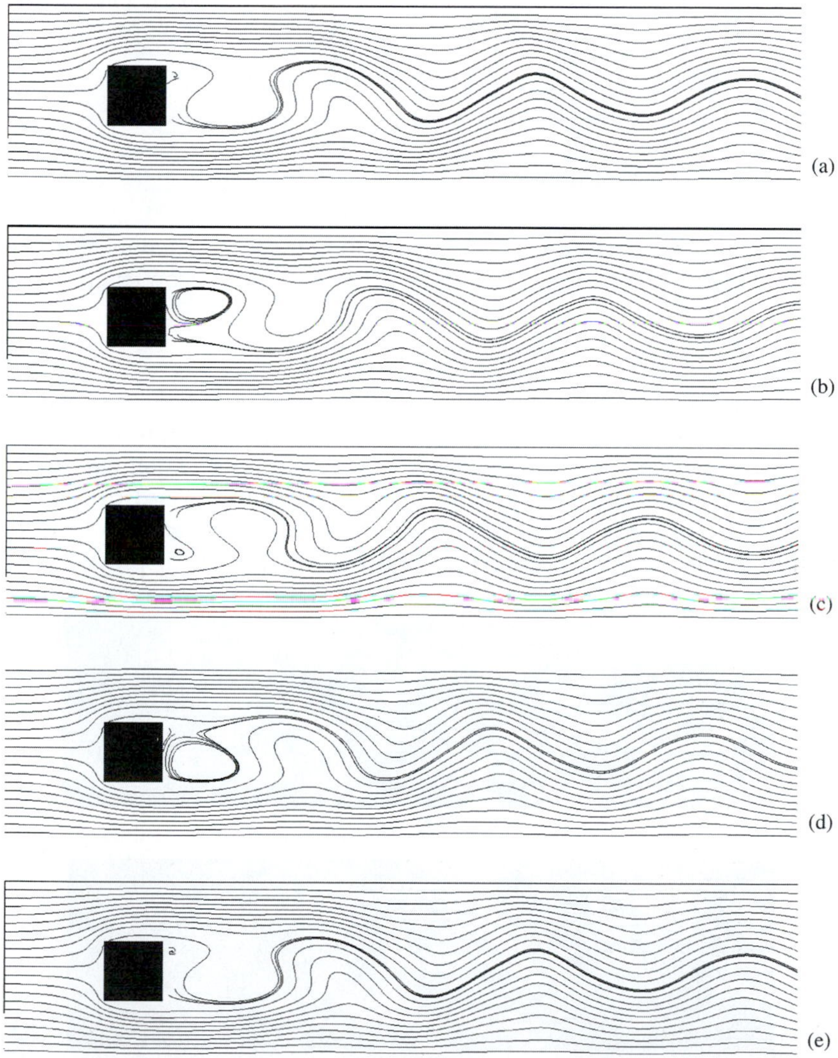

FIGURE 10.12 A sequence of flow fields around a block at Re = 100 during one period of vortex shedding. (a) $t = 40.53$, (b) $t = 41.50$, (c) $t = 42.48$, (d) $t = 43.45$, (e) $t = 44.17$. Here the unsteadiness fully washes away the attached near-wake vortices.

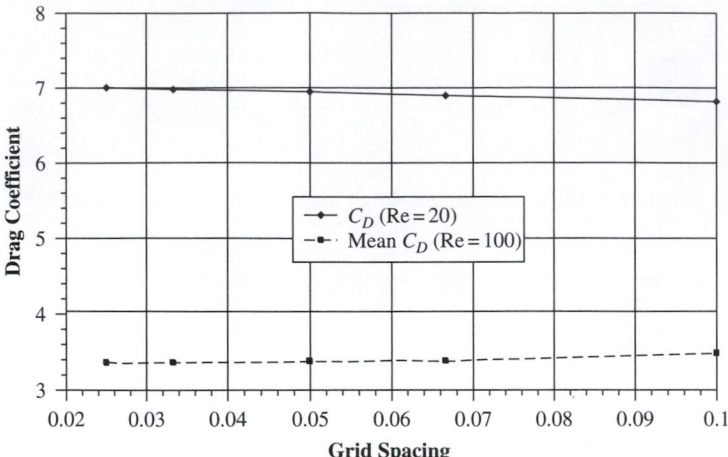

FIGURE 10.13 Convergence tests for the drag coefficient as the grid spacing decreases. The grid spacing is equal in both directions $\Delta x = \Delta y$, and time step Δt is determined by the stability condition. Confidence in computed results increases when such results are shown to be independent of the resolution of the computational grid for decreasing grid-point spacing.

and the stability condition for the time step is no longer too restrictive. With $\Delta x = \Delta y$ and large (grid) Reynolds numbers, the stability condition (10.110) becomes approximately:

$$\Delta t \leq \frac{\sigma}{\sqrt{2}} M \Delta x. \tag{10.155}$$

As a more complicated example, the flow around a circular cylinder confined between two parallel plates (the same geometry as the fourth example later in this section) is calculated at Re = 1000 using the explicit MacCormack scheme. For flow visualization, a smoke line is introduced at the inlet. Numerically, an additional convection-diffusion equation for smoke concentration is solved similarly, with an explicit scheme at each time step coupled with the computed flow field. Two snapshots of the flow field are displayed in Figure 10.14.

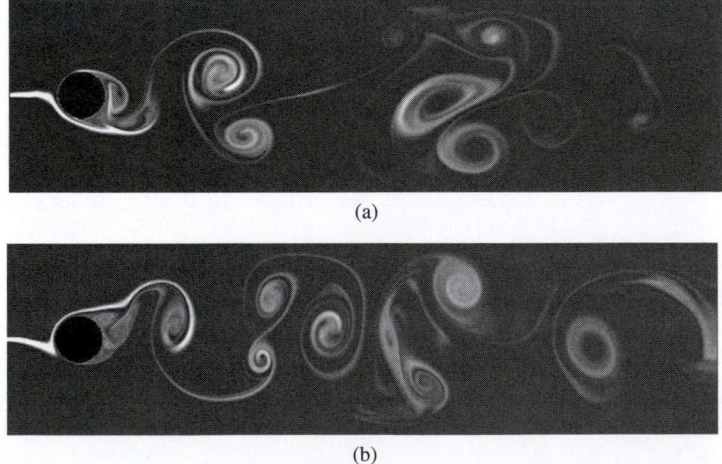

FIGURE 10.14   Smoke lines in flow around a circular cylinder between two parallel plates at Re = 1000. The flow geometry is the same as in the fourth example presented later in this section.

In this calculation, the flow Mach number is set at $M = 0.3$, and a uniform fine grid with 100 grid points across the cylinder diameter is used.

Finite-Element Formulation for Flow Over a Cylinder Confined in a Channel

We next consider the flow over a circular cylinder moving along the center of a channel. In the computation, we fix the cylinder and use the flow geometry as shown in Figure 10.15. The flow comes from the left with a uniform velocity U. Both plates of the channel are sliding to the right with the same velocity U. The diameter of the cylinder is d and the width of the channel is $W = 4d$. The boundary sections for the computational domain are indicated in the figure. The location of the inflow boundary Γ_1 is selected to be at $x_{min} = -7.5d$, and the location of the outflow boundary section Γ_2 is at $x_{max} = 15d$. They are both far away from the cylinder so as to minimize their influence on the flow field near the cylinder. In order to compute the flow at higher Reynolds numbers, we relax the assumptions that the flow is symmetric and steady. We will compute unsteady flow (with vortex shedding) in the full geometry and using the Cartesian coordinates shown in Figure 10.15.

The first step in the finite-element method is to discretize (mesh) the computational domain described in Figure 10.15. We cover the domain with triangular elements. A typical mesh is presented in Figure 10.16. The mesh size is distributed in a way that finer elements are used next to the cylinder surface to better resolve the local flow field. For this example, the mixed finite-element method will be used, such that each triangular element will have six nodes as shown Figure 10.5a. This element allows for curved sides that better capture the surface of the circular cylinder. The mesh in Figure 10.16 has 3320 elements, 6868 velocity nodes, and 1774 pressure nodes.

The weak formulation of the Navier-Stokes equations is given in (10.134) and (10.135). For this example the body-force term is zero, $\mathbf{g} = 0$. In Cartesian coordinates, the weak form of the momentum equation (10.134) can be written explicitly as

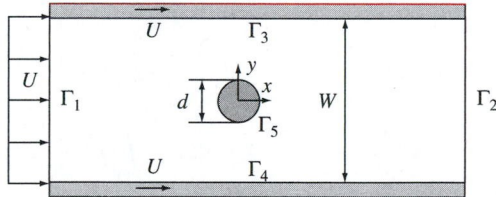

FIGURE 10.15 Computational domain for two-dimensional flow past a circular cylinder in a channel formed by parallel plates.

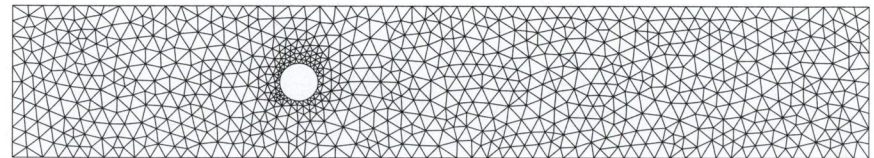

FIGURE 10.16 A finite-element mesh around a cylinder. Note that mesh spacing is finer near the cylinder to properly resolve the boundary layers that form on the cylinder.

$$
\int_{\Omega} \left(\frac{\partial \mathbf{u}}{\partial t} + u\frac{\partial \mathbf{u}}{\partial x} + v\frac{\partial \mathbf{u}}{\partial y} \right) \cdot \tilde{\mathbf{u}} d\Omega + \frac{2}{Re} \int_{\Omega} \left[\frac{\partial u}{\partial x} \frac{\partial \tilde{u}}{\partial x} + \frac{1}{2}\left(\frac{\partial u}{\partial y} + \frac{\partial v}{\partial x} \right)\left(\frac{\partial \tilde{u}}{\partial y} + \frac{\partial \tilde{v}}{\partial x} \right) + \frac{\partial v}{\partial y} \frac{\partial \tilde{v}}{\partial y} \right] d\Omega
$$

$$
- \int_{\Omega} p \left(\frac{\partial \tilde{u}}{\partial x} + \frac{\partial \tilde{v}}{\partial y} \right) d\Omega = 0, \tag{10.156}
$$

where Ω is the computational domain and $\tilde{\mathbf{u}} = (\tilde{u}, \tilde{v})$. Since the variational functions $\tilde{u}$ and $\tilde{v}$ are independent, the weak formulation (10.156) can be separated into two equations:

$$
\int_{\Omega} \left(\frac{\partial u}{\partial t} + u\frac{\partial u}{\partial x} + v\frac{\partial u}{\partial y} \right) \tilde{u} d\Omega - \int_{\Omega} p\frac{\partial \tilde{u}}{\partial x} d\Omega + \frac{1}{Re} \int_{\Omega} \left[2\frac{\partial u}{\partial x} \frac{\partial \tilde{u}}{\partial x} + \left(\frac{\partial u}{\partial y} + \frac{\partial v}{\partial x} \right) \frac{\partial \tilde{u}}{\partial y} \right] d\Omega = 0, \tag{10.157}
$$

$$
\int_{\Omega} \left(\frac{\partial v}{\partial t} + u\frac{\partial v}{\partial x} + v\frac{\partial v}{\partial y} \right) \tilde{v} d\Omega - \int_{\Omega} p\frac{\partial \tilde{v}}{\partial y} d\Omega + \frac{1}{Re} \int_{\Omega} \left[\left(\frac{\partial u}{\partial y} + \frac{\partial v}{\partial x} \right) \frac{\partial \tilde{v}}{\partial x} + 2\frac{\partial v}{\partial y} \frac{\partial \tilde{v}}{\partial y} \right] d\Omega = 0. \tag{10.158}
$$

The weak form of the continuity equation (10.135) is expressed as

$$
-\int_{\Omega} \left(\frac{\partial u}{\partial x} + \frac{\partial v}{\partial y} \right) \tilde{p} \, d\Omega = 0. \tag{10.159}
$$

Given a triangulation of the computational domain, for example, the mesh shown in Figure 10.16, the weak formulation of (10.157) through (10.159) can be approximated by the Galerkin finite-element formulation based on the finite-dimensional discretization of the flow variables. The Galerkin formulation can be written as:

$$
\int_{\Omega^h} \left(\frac{\partial u^h}{\partial t} + u^h\frac{\partial u^h}{\partial x} + v^h\frac{\partial u^h}{\partial y} \right) \tilde{u}^h d\Omega - \int_{\Omega^h} p^h\frac{\partial \tilde{u}^h}{\partial x} d\Omega
$$

$$
+ \frac{1}{Re}\int_{\Omega^h} \left[2\frac{\partial u^h}{\partial x} \frac{\partial \tilde{u}^h}{\partial x} + \left(\frac{\partial u^h}{\partial y} + \frac{\partial v^h}{\partial x} \right) \frac{\partial \tilde{u}^h}{\partial y} \right] d\Omega = 0, \tag{10.160}
$$

$$
\int_{\Omega^h} \left(\frac{\partial v^h}{\partial t} + u^h\frac{\partial v^h}{\partial x} + v^h\frac{\partial v^h}{\partial y} \right) \tilde{v}^h d\Omega - \int_{\Omega^h} p^h\frac{\partial \tilde{v}^h}{\partial y} d\Omega
$$

$$
+ \frac{1}{Re}\int_{\Omega^h} \left[\left(\frac{\partial u^h}{\partial y} + \frac{\partial v^h}{\partial x} \right) \frac{\partial \tilde{v}^h}{\partial x} + 2\frac{\partial v^h}{\partial y} \frac{\partial \tilde{v}^h}{\partial y} \right] d\Omega = 0, \tag{10.161}
$$

and

$$
-\int \Omega^h \left(\frac{\partial u^h}{\partial x} + \frac{\partial v^h}{\partial y} \right) \tilde{p}^h d\Omega = 0, \tag{10.162}
$$

where h indicates a given triangulation of the computational domain.

The time derivatives in (10.160) and (10.161) can be discretized by finite-difference methods. We first evaluate all the terms in (10.160) to (10.162) at a given time instant

$t = t_{n+1}$ (fully implicit discretization). Then the time derivative in (10.160) and (10.161) can be approximated as

$$\frac{\partial \mathbf{u}}{\partial t}(\mathbf{x}, t_{n+1}) \approx \alpha \frac{\mathbf{u}(\mathbf{x}, t_{n+1}) - \mathbf{u}(\mathbf{x}, t_n)}{\Delta t} - \beta \frac{\partial \mathbf{u}}{\partial t}(\mathbf{x}, t_n), \tag{10.163}$$

where $\Delta t = t_{n+1} - t_n$ is the time step. The approximation in (10.163) is first-order accurate in time when $\alpha = 1$ and $\beta = 0$. It can be improved to second-order accurate by selecting $\alpha = 2$ and $\beta = 1$ which is a variation of the well-known Crank-Nicolson scheme.

As (10.160) and (10.161) are nonlinear, iterative methods are often used for the solution. In Newton's method, the flow variables at the current time $t = t_{n+1}$ are often expressed as:

$$\begin{aligned} \mathbf{u}^h(\mathbf{x}, t_{n+1}) &= \mathbf{u}^*(\mathbf{x}, t_{n+1}) + \mathbf{u}'(\mathbf{x}, t_{n+1}), \\ p^h(\mathbf{x}, t_{n+1}) &= p^*(\mathbf{x}, t_{n+1}) + p'(\mathbf{x}, t_{n+1}), \end{aligned} \tag{10.164}$$

where $\mathbf{u}^*$ and p^* are the guesstimated values of velocity and pressure during the iteration; u' and p' are the corrections sought at each iteration.

Substituting (10.163) and (10.164) into Galerkin formulation, (10.160) through (10.162), and linearizing the equations with respect to the correction variables, we have:

$$\int_{\Omega^h} \left(\frac{\alpha}{\Delta t} u' + u^* \frac{\partial u'}{\partial x} + v^* \frac{\partial u'}{\partial y} + \frac{\partial u^*}{\partial x} u' + \frac{\partial u^*}{\partial y} v' \right) \tilde{u}^h \, d\Omega - \int_{\Omega^h} p' \frac{\partial \tilde{u}^h}{\partial x} \, d\Omega$$

$$+ \frac{1}{\mathrm{Re}} \int_{\Omega^h} \left[2 \frac{\partial u'}{\partial x} \frac{\partial \tilde{u}^h}{\partial x} + \left(\frac{\partial u'}{\partial y} + \frac{\partial v'}{\partial x} \right) \frac{\partial \tilde{u}^h}{\partial y} \right] d\Omega$$

$$= - \int_{\Omega^h} \left[\frac{\alpha}{\Delta t}(u^* - u(t_n)) - \beta \frac{\partial u}{\partial t}(t_n) + u^* \frac{\partial u^*}{\partial x} + v^* \frac{\partial u^*}{\partial y} \right] \tilde{u}^h \, d\Omega + \int_{\Omega^h} p^* \frac{\partial \tilde{u}^h}{\partial x} d\Omega$$

$$- \frac{1}{\mathrm{Re}} \int_{\Omega^h} \left[2 \frac{\partial u^*}{\partial x} \frac{\partial \tilde{u}^h}{\partial x} + \left(\frac{\partial u^*}{\partial y} + \frac{\partial v^*}{\partial x} \right) \frac{\partial \tilde{u}^h}{\partial y} \right] d\Omega, \tag{10.165}$$

$$\int_{\Omega^h} \left(\frac{\alpha}{\Delta t} v' + u^* \frac{\partial v'}{\partial x} + v^* \frac{\partial v'}{\partial y} + \frac{\partial v^*}{\partial x} u' + \frac{\partial v^*}{\partial y} v' \right) \tilde{v}^h \, d\Omega$$

$$- \int_{\Omega^h} p' \frac{\partial \tilde{v}^h}{\partial y} \, d\Omega + \frac{1}{\mathrm{Re}} \int_{\Omega^h} \left[\left(\frac{\partial u'}{\partial y} + \frac{\partial v'}{\partial x} \right) \frac{\partial \tilde{v}^h}{\partial x} + 2 \frac{\partial v'}{\partial y} \frac{\partial \tilde{v}^h}{\partial y} \right] d\Omega$$

$$= - \int_{\Omega^h} \left[\frac{\alpha}{\Delta t}(v^* - v(t_n)) - \beta \frac{\partial v^*}{\partial t}(t_n) + u^* \frac{\partial v^*}{\partial x} + v^* \frac{\partial v^*}{\partial y} \right] \tilde{v}^h \, d\Omega$$

$$+ \int_{\Omega^h} p^* \frac{\partial \tilde{v}^h}{\partial y} \, d\Omega - \frac{1}{\mathrm{Re}} \int_{\Omega^h} \left[\left(\frac{\partial u^*}{\partial y} + \frac{\partial v^*}{\partial x} \right) \frac{\partial \tilde{v}^h}{\partial x} + 2 \frac{\partial v^*}{\partial y} \frac{\partial \tilde{v}^h}{\partial y} \right] d\Omega, \tag{10.166}$$

and

$$-\int_{\Omega^h}\left(\frac{\partial u'}{\partial x}+\frac{\partial v'}{\partial y}\right)\tilde{p}^h\,d\Omega \;=\; \int_{\Omega^h}\left(\frac{\partial u^*}{\partial x}+\frac{\partial v^*}{\partial y}\right)\tilde{p}^h\,d\Omega.\tag{10.167}$$

As the functions in the integrals, unless specified otherwise, are all evaluated at the current time instant t_{n+1}, the temporal discretization in (10.165) and (10.166) is fully implicit and unconditionally stable. The terms on the right-hand side of (10.165) through (10.167) represent the residuals of the corresponding equations and can be used to monitor the convergence of the nonlinear iteration.

Similar to the one-dimensional case in Section 10.3, the finite-dimensional discretization of the low variables can be constructed using shape (or interpolation) functions:

$$u' = \sum_A u_A N_A^u(x,y),\quad v' = \sum_A v_A N_A^u(x,y),\quad p' = \sum_B p_B N_B^p(x,y),\tag{10.168}$$

where $N_A^u(x,y)$ and $N_B^p(x,y)$ are the shape functions for the velocity and the pressure, respectively. They are not necessarily the same. In order to satisfy the LBB stability condition, the shape function $N_A^u(x,y)$ in the mixed finite-element formulation should be one order higher than $N_B^p(x,y)$, as discussed in Section 10.4. The summation over A is through all the velocity nodes, while the summation over B runs through all the pressure nodes. The variational functions may be expressed in terms of the same shape functions:

$$\tilde{u}^h = \sum_A \tilde{u}_A N_A^u(x,y),\quad \tilde{v}^h = \sum_A \tilde{v}_A N_A^u(x,y),\quad \tilde{p}^h = \sum_B \tilde{p}_B N_B^p(x,y).\tag{10.169}$$

Since the Galerkin formulation (10.165) through (10.167) is valid for all possible choices of the variational functions, the coefficients in (10.169) should be arbitrary. In this way, the Galerkin formulation (10.165) through (10.167) reduces to a system of algebraic equations:

$$\sum_{A'} u_{A'}\int_{\Omega^h}\left[\left(\frac{\alpha}{\Delta t}N_{A'}^u + u^*\frac{\partial N_{A'}^u}{\partial x}+v^*\frac{\partial N_{A'}^u}{\partial y}+\frac{\partial u^*}{\partial x}N_{A'}^u\right)N_A^u + \frac{1}{Re}\left(2\frac{\partial N_{A'}^u}{\partial x}\frac{\partial N_A^u}{\partial x}+\frac{\partial N_{A'}^u}{\partial y}\frac{\partial N_A^u}{\partial y}\right)\right]d\Omega$$

$$+\sum_{A'} v_{A'}\int_{\Omega^h}\left(\frac{\partial u^*}{\partial y}N_{A'}^u N_A^u + \frac{1}{Re}\frac{\partial N_{A'}^u}{\partial x}\frac{\partial N_A^u}{\partial y}\right)d\Omega - \sum_{B'} p_{B'}\int_{\Omega^h}N_{B'}^{p'}\frac{\partial N_A^u}{\partial x}d\Omega$$

$$=-\int_{\Omega^h}\left[\frac{\alpha}{\Delta t}(u^* - u(t_n))-\beta\frac{\partial u}{\partial t}(t_n)+u^*\frac{\partial u^*}{\partial x}+v^*\frac{\partial u^*}{\partial y}\right]N_A^u d\Omega + \int_{\Omega^h}p^*\frac{\partial N_A^u}{\partial x}d\Omega$$

$$-\frac{1}{Re}\int_{\Omega^h}\left[2\frac{\partial u^*}{\partial x}\frac{\partial N_A^u}{\partial x}+\left(\frac{\partial u^*}{\partial y}+\frac{\partial v^*}{\partial x}\right)\frac{\partial N_A^u}{\partial y}\right]d\Omega,$$

$$\tag{10.170}$$

$$\sum_{A'} v_{A'} \int_{\Omega^h} \left[\left(\frac{\alpha}{\Delta t} N^u_{A'} + u^* \frac{\partial N^u_{A'}}{\partial x} + v^* \frac{\partial N^u_{A'}}{\partial y} + \frac{\partial u^*}{\partial y} N^u_{A'} \right) N^u_A + \frac{1}{Re} \left(\frac{\partial N^u_{A'}}{\partial x} \frac{\partial N^u_A}{\partial x} + 2 \frac{\partial N^u_{A'}}{\partial y} \frac{\partial N^u_A}{\partial y} \right) \right] d\Omega$$

$$+ \sum_{A'} u_{A'} \int_{\Omega^h} \left(\frac{\partial v^*}{\partial x} N^u_{A'} N^u_A + \frac{1}{Re} \frac{\partial N^u_{A'}}{\partial y} \frac{\partial N^u_A}{\partial x} \right) d\Omega - \sum_{B'} p_{B'} \int_{\Omega^h} N^p_{B'} \frac{\partial N^u_A}{\partial y} d\Omega$$

$$= - \int_{\Omega^h} \left[\frac{\alpha}{\Delta t} (v^* - v(t_n)) - \beta \frac{\partial v^*}{\partial t}(t_n) + u^* \frac{\partial v^*}{\partial x} + v^* \frac{\partial v^*}{\partial y} \right] N^u_A d\Omega + \int_{\Omega^h} p^* \frac{\partial N^u_A}{\partial y} d\Omega$$

$$- \frac{1}{Re} \int_{\Omega^h} \left[\left(\frac{\partial u^*}{\partial y} + \frac{\partial v^*}{\partial x} \right) \frac{\partial N^u_A}{\partial x} + 2 \frac{\partial v^*}{\partial y} \frac{\partial N^u_A}{\partial y} \right] d\Omega,$$

$$(10.171)$$

and

$$- \sum_{A'} u_{A'} \int_{\Omega^h} \left(\frac{\partial N^u_{A'}}{\partial x} N^p_B \right) d\Omega - \sum_{A'} u_{A'} \int_{\Omega^h} \left(\frac{\partial N^u_{A'}}{\partial y} N^p_B \right) d\Omega = \int_{\Omega^h} \left(\frac{\partial u^*}{\partial x} + \frac{\partial v^*}{\partial y} \right) N^p_B \, d\Omega, \quad (10.172)$$

for all the velocity nodes A and pressure nodes B. Equations (10.170) through (10.172) can be organized into matrix form:

$$\begin{pmatrix} \mathbf{A}_{uu} & \mathbf{A}_{uv} & \mathbf{B}_{up} \\ \mathbf{A}_{vu} & \mathbf{A}_{vv} & \mathbf{B}_{vp} \\ \mathbf{B}^T_{up} & \mathbf{B}^T_{vp} & 0 \end{pmatrix} \begin{pmatrix} \mathbf{u} \\ \mathbf{v} \\ \mathbf{p} \end{pmatrix} = \begin{pmatrix} \mathbf{f}_u \\ \mathbf{f}_v \\ \mathbf{f}_p \end{pmatrix}, \qquad (10.173)$$

where:

$$\mathbf{A}_{uu} = \left[A^{uu}_{AA'} \right], \quad \mathbf{A}_{uv} = \left[A^{uv}_{AA'} \right], \quad \mathbf{B}_{up} = \left[B^{up}_{AB'} \right],$$

$$\mathbf{A}_{vu} = \left[A^{vu}_{AA'} \right], \quad \mathbf{A}_{vv} = \left[A^{vv}_{AA'} \right], \quad \mathbf{B}_{vp} = \left[B^{vp}_{AB'} \right],$$

$$\mathbf{u} = \{ u_{A'} \}, \quad \mathbf{v} = \{ v_{A'} \}, \quad \mathbf{p} = \{ p_{B'} \},$$

$$\mathbf{f}_u = \{ f^u_A \}, \quad \mathbf{f}_v = \{ f^v_A \}, \quad \mathbf{f}_p = \{ f^p_B \},$$

$$(10.174)$$

and:

$$A^{uu}_{AA'} = \int_{\Omega^h} \left[\left(\frac{\alpha}{\Delta t} N^u_{A'} + u^* \frac{\partial N^u_{A'}}{\partial x} + v^* \frac{\partial N^u_{A'}}{\partial y} + \frac{\partial u^*}{\partial x} N^u_{A'} \right) N^u_A + \frac{1}{Re} \left(2 \frac{\partial N^u_{A'}}{\partial x} \frac{\partial N^u_A}{\partial x} + \frac{\partial N^u_{A'}}{\partial y} \frac{\partial N^u_A}{\partial y} \right) \right] d\Omega,$$

$$(10.175)$$

$$A^{uv}_{AA'} = \int_{\Omega^h} \left(\frac{\partial u^*}{\partial y} N^u_{A'} N^u_A + \frac{1}{Re} \frac{\partial N^u_{A'}}{\partial x} \frac{\partial N^u_A}{\partial y} \right) d\Omega, \qquad (10.176)$$

$$A^{vu}_{AA'} = \int_{\Omega^h} \left(\frac{\partial v^*}{\partial x} N^u_{A'} N^u_A + \frac{1}{Re} \frac{\partial N^u_{A'}}{\partial y} \frac{\partial N^u_A}{\partial x} \right) d\Omega, \qquad (10.177)$$

$$A^{vv}_{AA'} = \int_{\Omega^h} \left[\left(\frac{\alpha}{\Delta t} N^u_{A'} + u^* \frac{\partial N^u_{A'}}{\partial x} + v^* \frac{\partial N^u_{A'}}{\partial y} + \frac{\partial v^*}{\partial y} N^u_{A'} \right) N^u_A + \frac{1}{Re} \left(\frac{\partial N^u_{A'}}{\partial x} \frac{\partial N^u_A}{\partial x} + 2 \frac{\partial N^u_{A'}}{\partial y} \frac{\partial N^u_A}{\partial y} \right) \right] d\Omega,$$

$$(10.178)$$

$$B_{AB'}^{up} = -\int_{\Omega^h} N_{B'}^p \frac{\partial N_A^u}{\partial x} d\Omega, \tag{10.179}$$

$$B_{AB'}^{vp} = -\int_{\Omega^h} N_{B'}^p \frac{\partial N_A^u}{\partial y} d\Omega, \tag{10.180}$$

$$
\begin{aligned}
f_A^u = &-\int_{\Omega^h} \left[\frac{\alpha}{\Delta t}(u^* - u\,(t_n)) - \beta \frac{\partial u}{\partial t}(t_n) + u^* \frac{\partial u^*}{\partial x} + v^* \frac{\partial u^*}{\partial y} \right] N_A^u d\Omega \\
&+ \int_{\Omega^h} p^* \frac{\partial N_A^u}{\partial x} d\Omega - \frac{1}{Re} \int_{\Omega^h} \left[2\frac{\partial u^*}{\partial x} \frac{\partial N_A^u}{\partial x} + \left(\frac{\partial u^*}{\partial y} + \frac{\partial v^*}{\partial x} \right) \frac{\partial N_A^u}{\partial y} \right] d\Omega,
\end{aligned}
\tag{10.181}
$$

$$
\begin{aligned}
f_A^v = &-\int_{\Omega^h} \left[\frac{\alpha}{\Delta t}(v^* - v\,(t_n)) - \beta \frac{\partial v^*}{\partial t}(t_n) + u^* \frac{\partial v^*}{\partial x} + v^* \frac{\partial v^*}{\partial y} \right] N_A^u d\Omega \\
&+ \int_{\Omega^h} p^* \frac{\partial N_A^u}{\partial y} d\Omega - \frac{1}{Re} \int_{\Omega^h} \left[\left(\frac{\partial u^*}{\partial y} + \frac{\partial v^*}{\partial x} \right) \frac{\partial N_A^u}{\partial x} + 2\frac{\partial v^*}{\partial y} \frac{\partial N_A^u}{\partial y} \right] d\Omega,
\end{aligned}
\tag{10.182}
$$

$$f_B^p = \int_{\Omega^h} \left(\frac{\partial u^*}{\partial x} + \frac{\partial v^*}{\partial y} \right) N_B^p \, d\Omega. \tag{10.183}$$

The practical evaluation of the integrals in (10.175) through (10.183) is done element-wise. We need to construct the shape functions locally and transform these global integrals into local integrals over each element.

In the finite-element method, the global shape functions have very compact support. They are zero everywhere except in the neighborhood of the corresponding grid point in the mesh. It is convenient to cast the global formulation using the element point of view (Section 10.3). In this element view, the local shape functions are defined inside each element. The global shape functions are the assembly of the relevant local ones. For example, the global shape function corresponding to the grid point A in the finite-element mesh consists of the local shape functions of all the elements that share this grid point. An element in the physical space can be mapped into a standard element, as shown in Figure 10.17, and the local shape functions can be defined on this standard element. The mapping is given by

$$x\,(\xi, \eta) = \sum_{a=1}^{6} x_a^e \phi_a(\xi, \eta) \quad \text{and} \quad y\,(\xi, \eta) = \sum_{a=1}^{6} y_a^e \phi_a(\xi, \eta), \tag{10.184}$$

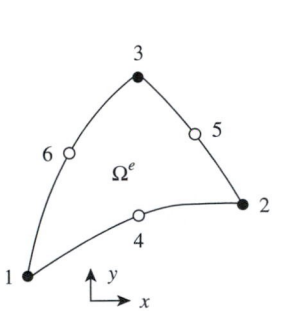

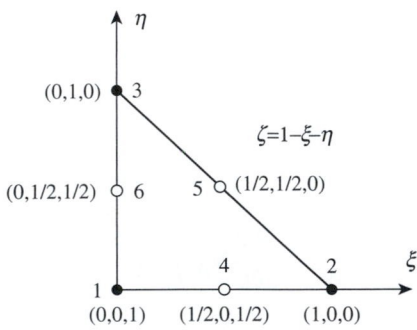

FIGURE 10.17 A quadratic triangular finite-element mapping into the standard element. The single-digit numbers indicate equivalent points on the left- and right-side elements.

where (x_a^e, y_a^e) are the coordinates of the nodes in the element e. The local shape functions are ϕ_a. For a quadratic triangular element they are defined as

$$\phi_1 = \zeta\,(2\zeta - 1), \quad \phi_2 = \xi\,(2\xi - 1), \quad \phi_3 = \eta\,(2\eta - 1), \quad \phi_4 = 4\xi\zeta, \quad \phi_5 = 4\xi\eta, \quad \phi_6 = 4\eta\zeta,$$

(10.185)

where $\zeta = 1 - \xi - \eta$. As shown in Figure 10.17, the mapping (10.184) is able to handle curved triangles. The variation of the flow variables within this element can also be expressed in terms of their values at the nodes of the element and the local shape functions,

$$u' = \sum_{a=1}^{6} u_a^e \phi_a(\xi, \eta), \quad v' = \sum_{a=1}^{6} v_a^e \phi_a(\xi, \eta), \quad v' = \sum_{b=1}^{3} p_b^e \psi_b(\xi, \eta).$$

(10.186)

Here the shape functions for velocities are quadratic and the same as the coordinates. The shape functions for the pressure are chosen to be linear, thus one order less than those for the velocities. They are given by

$$\psi_1 = \zeta, \quad \psi_2 = \xi, \quad \psi_3 = \eta.$$

(10.187)

Furthermore, the integration over the global computational domain can be written as the summation of the integrations over all the elements in the domain. As most of these integrations will be zero, the nonzero ones are grouped as element matrices and vectors:

$$\mathbf{A}_{uu}^e = \left[A_{aa'}^{euu}\right], \quad \mathbf{A}_{uv}^e = \left[A_{aa'}^{euv}\right], \quad \mathbf{B}_{up}^e = \left[B_{ab'}^{eup}\right],$$

$$\mathbf{A}_{vu}^e = \left[A_{aa'}^{evu}\right], \quad \mathbf{A}_{vv}^e = \left[A_{aa'}^{evv}\right], \quad \mathbf{B}_{vp}^e = \left[B_{ab'}^{evp}\right],$$

(10.188)

$$\mathbf{f}_u^e = \{f_a^{eu}\}, \quad \mathbf{f}_v^e = \{f_a^{ev}\}, \quad \mathbf{f}_p^e = \{f_b^{ep}\},$$

where:

$$A_{aa'}^{euu} = \int_{\Omega^e}\left[\left(\frac{\alpha}{\Delta t}\phi_{a'} + u^*\frac{\partial\phi_{a'}}{\partial x} + v^*\frac{\partial\phi_{a'}}{\partial y} + \frac{\partial u^*}{\partial x}\phi_{a'}\right)\phi_a + \frac{1}{Re}\left(2\frac{\partial\phi_{a'}}{\partial x}\frac{\partial\phi_a}{\partial x} + \frac{\partial\phi_{a'}}{\partial y}\frac{\partial\phi_a}{\partial y}\right)\right]d\Omega, \quad (10.189)$$

$$A_{aa'}^{euv} = \int_{\Omega^e}\left(\frac{\partial u^*}{\partial y}\phi_{a'}\phi_a + \frac{1}{Re}\frac{\partial\phi_{a'}}{\partial x}\frac{\partial\phi_a}{\partial y}\right)d\Omega, \quad (10.190)$$

$$A_{aa'}^{evu} = \int_{\Omega^e}\left(\frac{\partial v^*}{\partial x}\phi_{a'}\phi_a + \frac{1}{Re}\frac{\partial\phi_{a'}}{\partial y}\frac{\partial\phi_a}{\partial x}\right)d\Omega, \quad (10.191)$$

$$A_{aa'}^{evv} = \int_{\Omega^e}\left[\left(\frac{\alpha}{\Delta t}\phi_{a'} + u^*\frac{\partial\phi_{a'}}{\partial x} + v^*\frac{\partial\phi_{a'}}{\partial y} + \frac{\partial v^*}{\partial y}\phi_{a'}\right)\phi_a + \frac{1}{Re}\left(\frac{\partial\phi_{a'}}{\partial x}\frac{\partial\phi_a}{\partial x} + 2\frac{\partial\phi_{a'}}{\partial y}\frac{\partial\phi_a}{\partial y}\right)\right]d\Omega, \quad (10.192)$$

$$B_{ab'}^{eup} = -\int_{\Omega^e}\psi_{b'}\frac{\partial\phi_a}{\partial x}\,d\Omega, \quad (10.193)$$

$$B_{ab'}^{evp} = -\int_{\Omega^e}\psi_{b'}\frac{\partial\phi_a}{\partial y}\,d\Omega, \quad (10.194)$$

$$f_a^{eu} = -\int_{\Omega^e} \left[\frac{\alpha}{\Delta t}(u^* - u\,(t_n)) - \beta\frac{\partial u}{\partial t}(t_n) + u^*\frac{\partial u^*}{\partial x} + v^*\frac{\partial u^*}{\partial y}\right]\phi_a d\Omega + \int_{\Omega^e} p^*\frac{\partial \phi_a}{\partial x} d\Omega$$

$$- \frac{1}{Re}\int_{\Omega^e}\left[2\frac{\partial u^*}{\partial x}\frac{\partial \phi_a}{\partial x} + \left(\frac{\partial u^*}{\partial y} + \frac{\partial v^*}{\partial x}\right)\frac{\partial \phi_a}{\partial y}\right] d\Omega, \tag{10.195}$$

$$f_a^{ev} = -\int_{\Omega^e}\left[\frac{\alpha}{\Delta t}(v^* - v\,(t_n)) - \beta\frac{\partial v^*}{\partial t}(t_n) + u^*\frac{\partial v^*}{\partial x} + v^*\frac{\partial v^*}{\partial y}\right]\phi_a d\Omega + \int_{\Omega^e} p^*\frac{\partial \phi_a}{\partial y}\, d\Omega$$

$$- \frac{1}{Re}\int_{\Omega^e}\left[\left(\frac{\partial u^*}{\partial y} + \frac{\partial v^*}{\partial x}\right)\frac{\partial \phi_a}{\partial x} + 2\frac{\partial v^*}{\partial y}\frac{\partial \phi_a}{\partial y}\right] d\Omega, \tag{10.196}$$

$$f_b^{ep} = \int_{\Omega^e}\left(\frac{\partial u^*}{\partial x} + \frac{\partial v^*}{\partial y}\right)\psi_b d\Omega. \tag{10.197}$$

The indices a and a' run from 1 to 6, and b and b' run from 1 to 3.

The integrals in the above expressions can be evaluated by numerical integration rules,

$$\int_{\Omega^e} f(x, y)\, d\Omega = \int_0^1\int_0^{1-\eta} f\,(\xi, \eta)\, J\,(\xi, \eta)\, d\xi\, d\eta = \frac{1}{2}\sum_{l=1}^{N_{int}} f(\xi_l, \eta_l)\, J\,(\xi_l, \eta_l)\, W_l, \tag{10.198}$$

where the Jacobian of the mapping (10.184) is given by $J = x_\xi\, y_\eta - x_\eta\, y_\xi$. Here N_{int} is the number of numerical integration points and W_l is the weight of the lth integration point. For this example, a seven-point integration formula with degree of precision of five (see Hughes, 1987) was used.

The global matrices and vectors in (10.173) are the summations of the element matrices and vectors in (10.188) over all the elements. In the process of summation (assembly), a mapping of the local nodes in each element to the global node numbers is needed. This information is commonly available for any finite-element mesh.

Once the matrix equation (10.173) is generated, we may impose the essential boundary conditions for the velocities. One simple method is to use the equation of the boundary condition to replace the corresponding equation in the original matrix or one can multiply a large constant by the equation of the boundary condition and add this equation to the original system of equations in order to preserve the structure of the matrix. The resulting matrix equation may be solved using common direct or iterative solvers for a linear algebraic system of equations.

Figures 10.18 and 10.19 display the streamlines and vorticity lines around the cylinder at three Reynolds numbers: $Re = 1$, 10, and 40. For these Reynolds numbers, the flow is steady and should be symmetric above and below the cylinder. However, due to the imperfection in the mesh used for the calculation and as shown in Figure 10.16, the calculated flow field is not perfectly symmetric. From Figure 10.18 we observe the increase in the size of the wake behind the cylinder as the Reynolds number increases. In Figure 10.19, we see the effects of the Reynolds number in the vorticity build up in front of the cylinder, and in the convection of the vorticity by the flow.

We next compute the case with Reynolds number $Re = 100$. In this case, the flow is expected to be unsteady. Periodic vortex shedding occurs. In order to capture the details of the flow, we used a finer mesh than the one shown in Figure 10.16. The finer mesh has 9222 elements, 18,816 velocity nodes, and 4797 pressure nodes. In this calculation, the flow

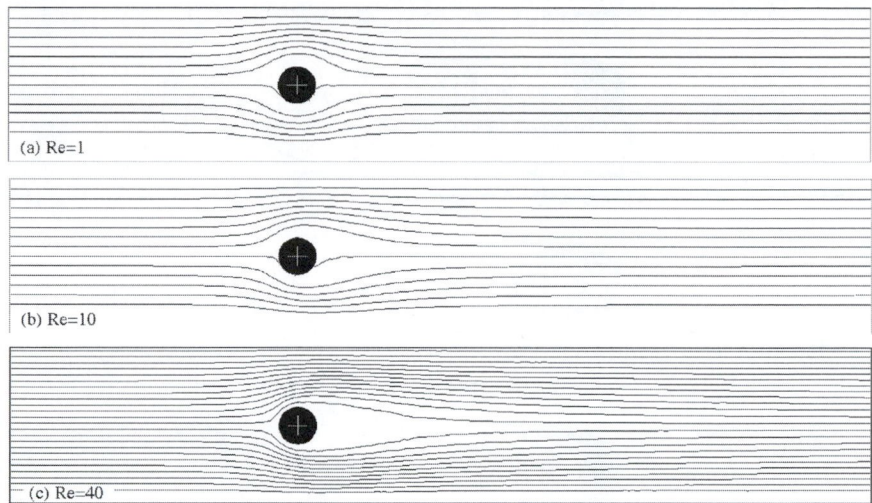

FIGURE 10.18 Streamlines for flow around a cylinder at three different Reynolds numbers: 1, 10, and 40. Here the streamlines are nearly fore-aft symmetric at the lowest Re, but the cylinder's wake grows with increasing Reynolds number.

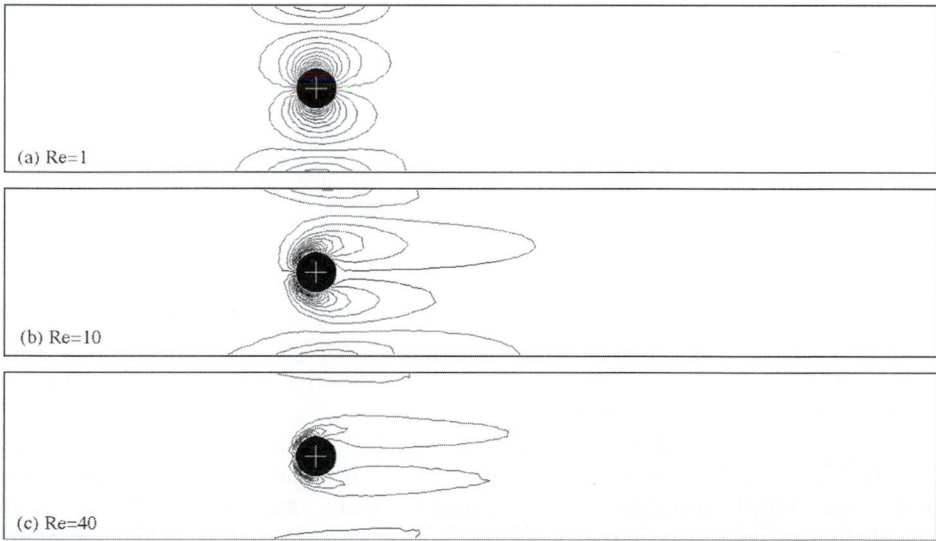

FIGURE 10.19 Vorticity contours for flow around a cylinder at the three different Reynolds numbers shown in Figure 10.18. Top-to-bottom asymmetry in the flow field arises because the mesh above and below the cylinder are not mirror images.

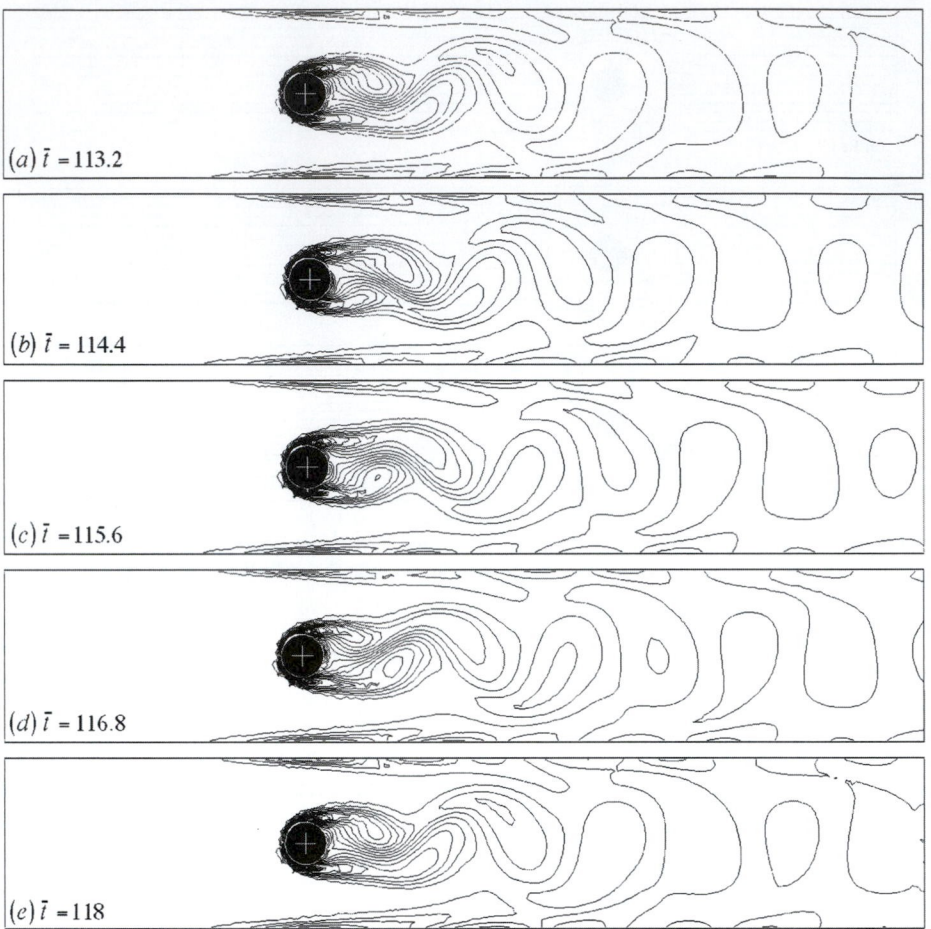

FIGURE 10.20 Vorticity lines for flow around a cylinder at Reynolds number Re = 100. $\bar{t} = tU/d$ is the dimensionless time. Here the formation of a von Karman vortex street is evident.

starts from rest. Initially, the flow is symmetric, and the wake behind the cylinder grows bigger and stronger. Then, the wake becomes unstable, undergoes a supercritical Hopf bifurcation, and sheds periodically away from the cylinder. The periodic vortex shedding forms the well-known von Karman vortex street. The vorticity lines are presented in Figure 10.20 for a complete cycle of vortex shedding.

For this case with Re = 100, we plot in Figure 10.21 the history of the forces and torque acting on the cylinder. The oscillations shown in the lift and torque plots are typical for the supercritical Hopf bifurcation. The nonzero mean value of the torque shown in Figure 10.21c is due to the asymmetry in the finite-element mesh. It is clear that the flow becomes fully periodic at the times shown in Figure 10.20. The period of the oscillation is measured as $\tau = 0.0475s$ or $\bar{\tau} = 4.75$ in the nondimensional units. This period corresponds to a nondimensional Strouhal number $S = nd/U = 0.21$, where n is the frequency of the

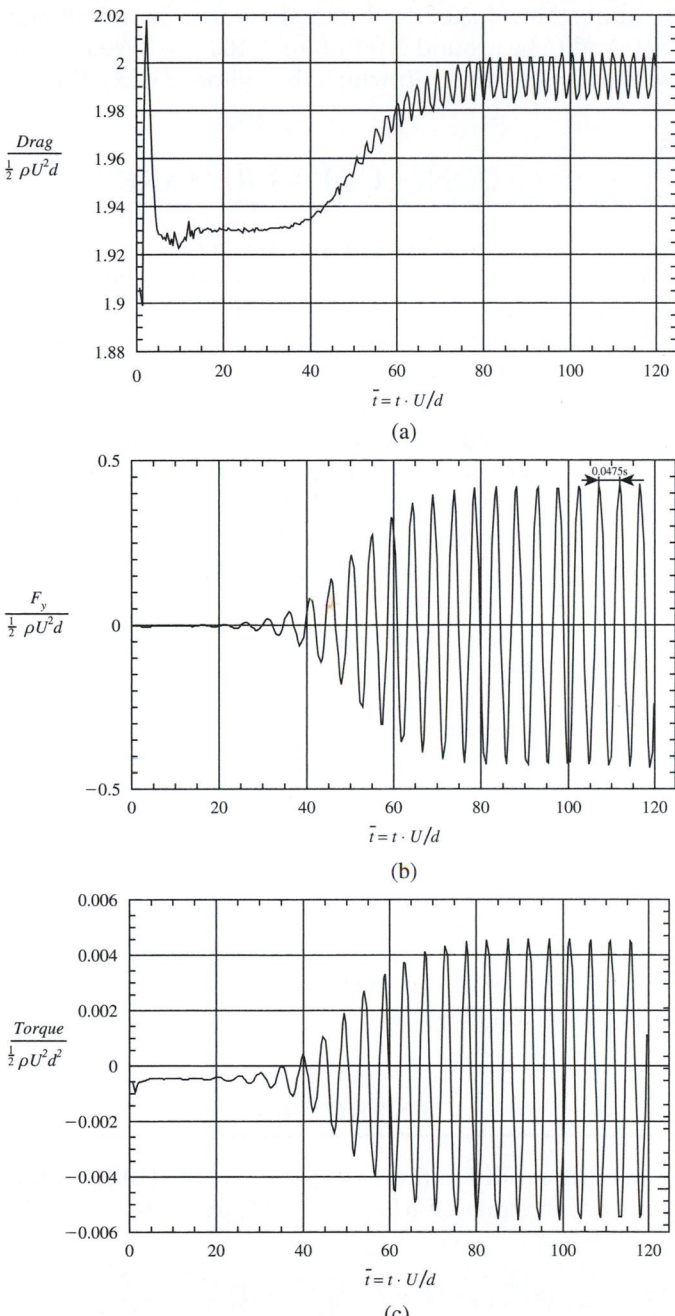

FIGURE 10.21 History of forces and torque acting on the cylinder at Re = 100: (a) drag coefficient; (b) lift coefficient; and (c) coefficient for the torque. Clearly the alternate shedding of vortices leads to oscillations in the fluid-dynamic loads on the cylinder.

shedding. In the literature, the value of the Strouhal number for an unbounded uniform flow around a cylinder is found to be around 0.167 at Re = 100 (e.g., Wen & Lin, 2001). The difference could be caused by the geometry in which the cylinder is confined in a channel.

10.6. CONCLUDING REMARKS

It should be strongly emphasized that CFD is merely a tool for analyzing fluid-flow problems. If it is used correctly, it can provide useful information cheaply and quickly. However, it can easily be misused or even abused. In today's computer age, there is a tendency for people to trust the output from a computer, especially when they do not know the equations that were solved or how the computations were performed. One certainly should be aware of the assumptions used in producing the results from a CFD model.

As we have previously discussed, CFD is never exact. There are uncertainties involved in CFD predictions. However, one is able to gain more confidence in CFD predictions by following a few steps. Tests on some benchmark problems with known solutions are often encouraged. A mesh-refinement test is normally a must in order to be sure that the numerical solution converges to something meaningful. A similar test with the time step for unsteady flow problems is often desired. If the boundary locations and conditions are in doubt, their effects on the CFD predictions should be minimized. Furthermore, the sensitivity of the CFD predictions to some key parameters in the problem should be investigated for practical design problems.

In this chapter, we have presented the basics of the finite-difference and finite-element methods and their applications in CFD. There are other kinds of numerical methods, for example, the spectral method and the spectral element method, which are often used in CFD. They share the common approach that discretizes the Navier-Stokes equations into a system of algebraic equations. However, a class of new numerical techniques including the lattice gas cellular automata, the lattice Boltzmann method, and dissipative particle dynamics does not start from the continuum Navier-Stokes equations. Unlike the conventional methods discussed in this chapter, they are based on simplified kinetic models that incorporate the essential physics of the microscopic or mesoscopic processes so that the macroscopic-averaged properties obey the desired macroscopic Navier-Stokes equations.

EXERCISES

10.1. Show that the stability condition for the explicit scheme (10.10) is the condition (10.26).

10.2. For the heat conduction equation $\partial T/\partial t - D(\partial^2 T/\partial x^2) = 0$, one of the discretized forms is

$$-sT_{j+1}^{n+1} + (1 + 2s)\, T_j^{n+1} - sT_{j-1}^{n+1} = T_j^n$$

where $s = D\, \Delta t/\Delta x^2$. Show that this implicit algorithm is always stable.

10.3. An insulated rod initially has a temperature of $T(x, 0) = 0°C$ $(0 \le x \le 1)$. At $t = 0$ hot reservoirs ($T = 100°C$) are brought into contact with the two ends, A ($x = 0$) and B

$(x = 1)$: $T(0, t) = T(1, t) = 100°C$. Numerically find the temperature $T(x, t)$ of any point in the rod. The governing equation of the problem is the heat conduction equation $\partial T/\partial t - D\,(\partial^2 T/\partial x^2) = 0$. The exact solution to this problem is

$$T^*(x_j, t_n) = 100 - \sum_{m=1}^{M} \frac{400}{(2m-1)\,\pi}\sin\left[\left(2m-1\right)\pi x_j\right]\exp\left[-D\,(2m-1)^2\pi^2 t_n\right],$$

(10.199)

where M is the number of terms used in the approximation.
   a) Try to solve the problem with the explicit forward time, centered space (FTCS) scheme. Use the parameter $s = D\,\Delta t/\Delta x^2 = 0.5$ and 0.6 to test the stability of the scheme.
   b) Solve the problem with a stable explicit or implicit scheme. Test the rate of convergence numerically using the error at $x = 0.5$.
10.4. Derive the weak form, Galerkin form, and the matrix form of the following strong problem:

> Given functions $D(x), f(x)$, and constants g, h, find $u(x)$ such that
> $[D(x)u, {}_x], {}_x + f(x) = 0$ on $\Omega = (0, 1)$,
> with $u(0) = g$ and $-u, {}_x(1) = h$.

Write a computer program solving this problem using piecewise-linear shape functions. You may set $D = 1$, $g = 1$, and $h = 1$. Check your numerical result with the exact solution.
10.5. Solve numerically the steady convective transport equation,

$$u\frac{\partial T}{\partial x} = D\frac{\partial^2 T}{\partial x^2}, \text{ for } 0 \le x \le 1,$$

with two boundary conditions $T(0) = 0$ and $T(1) = 1$, where u and D are two constants:
   a) using the centered finite-difference scheme in equation (10.91), and compare with the exact solution; and
   b) using the upwind scheme (10.93), and compare with the exact solution.
10.6. Code the explicit MacCormack scheme with the FF/BB arrangement for the driven-cavity flow problem as described in Section 10.5. Compute the flow field at Re $= 100$ and 400, and explore effects of Mach number and the stability condition (10.110).

Literature Cited

Brooks, A. N., & Hughes, T. J. R. (1982). Streamline-upwinding/Petrov-Galerkin formulation for convection dominated flows with particular emphasis on incompressible Navier-Stokes equation. *Comput. Methods Appl. Mech. Engrg., 30*, 199–259.

Chorin, A. J. (1967). A numerical method for solving incompressible viscous flow problems. *J. Comput. Phys., 2*, 12–26.

Chorin, A. J. (1968). Numerical solution of the Navier-Stokes equations. *Math. Comput., 22*, 745–762.

Fletcher, C. A. J. (1988). *Computational Techniques for Fluid Dynamics, I—Fundamental and General Techniques, and II—Special Techniques for Different Flow Categories*. New York: Springer-Verlag.

Franca, L. P., Frey, S. L., & Hughes, T. J. R. (1992). Stabilized finite element methods: I. Application to the advective-diffusive model. *Comput. Methods Appl. Mech. Engrg., 95*, 253—276.

Franca, L. P., & Frey, S. L. (1992). Stabilized finite element methods: II. The incompressible Navier-Stokes equations. *Comput. Methods Appl. Mech. Engrg., 99*, 209—233.

Ghia, U., Ghia, K. N., & Shin, C. T. (1982). High-Re solutions for incompressible flow using the Navier-Stokes equations and a multigrid method. *J. Comput. Phys., 48*, 387—411.

Glowinski, R. (1991). *Finite element methods for the numerical simulation of incompressible viscous flow, introduction to the control of the Navier-Stokes equations*. In: *Lectures in Applied Mathematics, Vol. 28*. Providence, R.I: American Mathematical Society. 219—301.

Gresho, P. M. (1991). Incompressible fluid dynamics: Some fundamental formulation issues. *Annu. Rev. Fluid Mech., 23*, 413—453.

Harlow, F. H., & Welch, J. E. (1965). Numerical calculation of time-dependent viscous incompressible flow of fluid with free surface. *Phys. Fluids, 8*, 2182—2189.

Hou, S., Zou, Q., Chen, S., Doolen, G. D., & Cogley, A. C. (1995). Simulation of cavity flow by the lattice Boltzmann method. *J. Comp. Phys., 118*, 329—347.

Hughes, T. J. R. (1987). *The Finite Element Method, Linear Static and Dynamic Finite Element Analysis*. Englewood Cliffs, NJ: Prentice-Hall.

Intel Corporation. (2003). Moore's Law and Intel Innovation: Advancing silicon technology. <http://www.intel.com/about/companyinfo/museum/exhibits/moore.htm>. Accessed 10/4/2011.

MacCormack, R. W. (1969). *The effect of viscosity in hypervelocity impact cratering*. Cincinnati, Ohio: AIAA Paper. 69—354.

Marchuk, G. I. (1975). *Methods of Numerical Mathematics*. New York: Springer-Verlag.

Noye, J. (Ed.), (1983). Chapter 2 in *Numerical Solution of Differential Equations*. Amsterdam: North-Holland.

Oden, J. T., & Carey, G. F. (1984). *Finite Elements: Mathematical Aspects, Vol. IV*. Englewood Cliffs, N.J: Prentice-Hall.

Perrin, A., & Hu, H. H. (2006). An explicit finite-difference scheme for simulation of moving particles. *J. Comput. Phys., 212*, 166—187.

Peyret, R., & Taylor, T. D. (1983). *Computational Methods for Fluid Flow*. New York: Springer-Verlag.

Richtmyer, R. D., & Morton, K. W. (1967). *Difference Methods for Initial-Value Problems*. New York: Interscience.

Saad, Y. (1996). *Iterative Methods for Sparse Linear Systems*. Boston: PWS Publishing Company.

Tannehill, J. C., Anderson, D. A., & Pletcher, R. H. (1997). *Computational Fluid Mechanics and Heat Transfer*. Washington, DC: Taylor & Francis.

Temam, R. (1969). Sur l'approximation des équations de Navier-Stokes par la méthode de pas fractionaires. *Archiv. Ration. Mech. Anal., 33*, 377—385.

Tezduyar, T. E. (1992). In J. W. Hutchinson, & T. Y. Wu (Eds.), *Advances in Applied Mechanics. Stabilized Finite Element Formulations for Incompressible Flow Computations, Vol. 28* (pp. 1—44). New York: Academic Press.

Wen, C. Y., & Lin, C. Y. (2001). Two dimensional vortex shedding of a circular cylinder. *Phys. Fluids, 13*, 557—560.

Yanenko, N. N. (1971). *The Method of Fractional Steps*. New York: Springer-Verlag.

Supplemental Reading

Dennis, S. C. R., & Chang, G. Z. (1970). Numerical solutions for steady flow past a circular cylinder at Reynolds numbers up to 100. *J. Fluid Mech, 42*, 471—489.

Sucker, D., & Brauer, H. (1975). Fluiddynamik bei der angeströmten Zylindern. *Wärme-Stoffübertrag, 8*, 149.

Takami, H., & Keller, H. B. (1969). Steady two-dimensional viscous flow of an incompressible fluid past a circular cylinder. *Phys. Fluids, 12*(Suppl.II), II-51—II-56.

Instability

OUTLINE

11.1. Introduction   474

11.2. Method of Normal Modes   475

11.3. Kelvin-Helmholtz Instability   477

11.4. Thermal Instability: The Bénard Problem   484

11.5. Double-Diffusive Instability   492

11.6. Centrifugal Instability: Taylor Problem   496

11.7. Instability of Continuously Stratified Parallel Flows   502

11.8. Squire's Theorem and the Orr-Sommerfeld Equation   508

11.9. Inviscid Stability of Parallel Flows   511

11.10. Results for Parallel and Nearly Parallel Viscous Flows   515

11.11. Experimental Verification of Boundary-Layer Instability   520

11.12. Comments on Nonlinear Effects   522

11.13. Transition   523

11.14. Deterministic Chaos   524

Exercises   532

Literature Cited   539

CHAPTER OBJECTIVES

- To present the mathematical theory of temporal flow instability

- To illustrate how this theory may be applied in a variety of confined and unconfined flows

- To present classic theoretical results for parallel flows

- To describe results for viscous flows

- To discuss nonlinearity and the possible role of chaos in the transition to turbulence

11.1. INTRODUCTION

Many phenomena that satisfy the conservation laws exactly are unobservable because they are unstable when subjected to the small disturbances that are invariably present in any real system. Consider the stability of two simple mechanical systems in a vertical gravitation field. A sharpened pencil may, in theory, be balanced on its point on a horizontal surface, but any small surface vibration or air pressure disturbance will knock it over. Thus, sharpened pencils on horizontal surfaces are commonly observed lying horizontally. Similarly, the position of a smooth ball resting on the inside surface of a hemispherical bowl is stable provided the bowl is concave upwards. However, the ball's position is unstable to small displacements if placed on the outer side of a hemispherical bowl when the bowl is concave downwards (Figure 11.1). In fluid flows, smooth laminar flows are stable to small disturbances only when certain conditions are satisfied. For example, in the flow of a homogeneous viscous fluid in a channel, the Reynolds number must be less than some critical value for the flow to be stable, and in a stratified shear flow, the Richardson number must be larger than a critical value for stability. When these conditions are not satisfied, infinitesimal disturbances may grow spontaneously and completely change the character of the original flow. Sometimes the disturbances can grow to finite amplitude and reach a new steady-state equilibrium. The new state may then become unstable to other types of disturbances, and may evolve to yet another steady state, and so on. As a limit of this situation, the

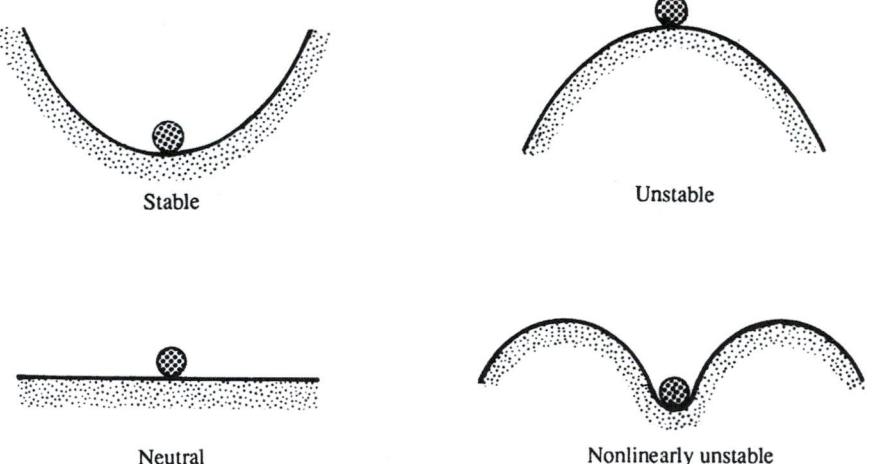

FIGURE 11.1 Stable and unstable mechanical systems. Here, gravity is presumed to act downward. In the upper left and lower right panels, a small displacement of the round object away from equilibrium will be opposed by the action of gravity and the object will move back toward its equilibrium location. These are linearly stable situations. In the upper right panel, a small displacement of the object will be enhanced by the action of gravity and the object will move away from its equilibrium location, an unstable situation. In the lower left, a small displacement of the object does not produce a new force, thus the situation is neutrally stable. In the lower right panel, a sufficiently large displacement of the object may place it beyond its region of stability; thus this situation is nonlinearly unstable.

flow becomes a superposition of a variety of interacting disturbances with nearly random phases, a state of chaotic or nearly chaotic fluctuations that is commonly described as *turbulence*. In fact, two primary motivations for studying fluid-flow stability are: 1) to understand the process of laminar to turbulent transition, and 2) to predict the onset of this transition. Finite-amplitude effects, including the development of chaotic solutions, are examined briefly later in this chapter.

The primary objective of this chapter, however, is the examination of stability of certain fluid flows with respect to infinitesimal disturbances. We shall introduce perturbations on a particular flow, and determine whether the equations of motion predict that the perturbations grow or decay. In this analysis, the perturbations are commonly assumed to be small enough so that linearization is possible through neglect of quadratic and higher order terms in the perturbation variables and their derivatives. While such linearization fruitfully allows the production of analytical results, it inherently limits the applicability of such results to the *initial* behavior of infinitesimal disturbances. The loss of stability does not in itself constitute a transition to turbulence since the linear theory can, at best, describe only the very beginning of the transition process. In addition, a real flow may be stable to infinitesimal disturbances (linearly stable), but may be unstable to sufficiently large disturbances (nonlinearly unstable); this is schematically represented in Figure 11.1.

In spite of these limitations, linear stability theory enjoys considerable success. There is excellent agreement between experimental results and the theoretical prediction of the onset of thermal convection in a layer of fluid, and of the onset of Tollmien-Schlichting waves in a viscous boundary layer. Taylor's experimental verification of his own theoretical prediction of the onset of secondary flow in a rotating Couette flow is so striking that it has led people to suggest that Taylor's work is the first rigorous confirmation of the Navier-Stokes equations on which the calculations are based.

This chapter describes the temporal instability of confined and unconfined flows where spatially extended perturbations decay, persist, or grow in time. The complimentary situation where spatially confined disturbances decay, persist, or grow while traveling in space is more complicated and is described elsewhere (see Huerre & Monkewitz, 1990). The primary analysis technique used here, the method of normal modes, is described in the next section. The third through eleventh sections of this chapter utilize this technique to illustrate basic flow physics and to present results for a variety of flows important in engineering applications and geophysical situations. The final few sections describe transition and the onset of turbulence. None of the flow situations discussed in this chapter contains Coriolis forces. *Baroclinic instability*, which does contain the Coriolis frequency, is discussed in Chapter 13. The books by Chandrasekhar (1961, 1981) and Drazin and Reid (1981) provide further information on flow instability. The review article by Bayly, Orszag, and Herbert (1988) is recommended as well.

11.2. METHOD OF NORMAL MODES

Basic linear stability analysis consists of presuming the existence of sinusoidal disturbances to a *basic state* (also called a background, initial, or equilibrium state), which is the flow whose stability is being investigated. For example, the velocity field of a basic state involving a flow parallel to the x-axis and varying along the z-axis is $\mathbf{U} = U(z)\mathbf{e}_x$. On this background flow we superpose a spatially extended disturbance of the form:

$$u(x,y,z,t) = \hat{u}(z)\,\exp\{ikx + imy + \sigma t\} = \hat{u}(z)\,\exp\{i|\mathbf{K}|(\mathbf{e}_K \cdot \mathbf{x} - ct)\} \qquad (11.1)$$

where $\hat{u}(z)$ is a complex amplitude, $i = \sqrt{-1}$ is the imaginary root, $\mathbf{K} = (k, m, 0)$ is the disturbance wave number, $\mathbf{e}_K = \mathbf{K}/|\mathbf{K}|$, $\mathbf{x} = (x, y, z)$, σ is the temporal growth rate, c is the complex phase speed of the disturbance, and the real part of (11.1) is taken to obtain physical quantities. The complex notation used here is explained in Section 7.7. The two forms of (11.1) are useful when the disturbance is stationary, and when it takes the form of a traveling wave, respectively. The reason solutions exponential in (x, y, t) are allowed in (11.1) is that, as we shall see, the coefficients of the differential equation governing the perturbation in this flow are independent of (x, y, t). The flow field is assumed to be unbounded in the x and y directions, hence the wave number components k and m can only be real (and $|\mathbf{K}|$ positive real) in order that the dependent variables remain bounded as x, $y \to \pm\infty$; however, $\sigma = \sigma_r + i\sigma_i$ and $c = c_r + ic_i$ are regarded as complex.

The behavior of the system for *all* possible wave numbers, $\mathbf{K}$, is examined in the analysis. If σ_r or c_i are positive for *any* value of the wave number, the system is unstable to disturbances of this wave number. If no such unstable state can be found, the system is stable. We say that

$$\sigma_r < 0 \quad \text{or} \quad c_i < 0 \text{ implies a } \textit{stable} \text{ flow,}$$

$$\sigma_r = 0 \quad \text{or} \quad c_i = 0 \text{ implies a } \textit{neutrally stable} \text{ flow, and}$$

$$\sigma_r > 0 \quad \text{or} \quad c_i > 0 \text{ implies an } \textit{unstable} \text{ flow.}$$

The method of analysis involving the examination of Fourier components such as (11.1) is called the *normal mode method*. An arbitrary disturbance can be decomposed into a complete set of normal modes. In this method the stability of each of the modes is examined separately, as the linearity of the problem implies that the various modes do not interact. The method leads to an eigenvalue problem.

The boundary between stability and instability is called the *marginal state*, for which $\sigma_r = c_i = 0$. There can be two types of marginal states, depending on whether σ_i or c_r is also zero or nonzero in this state. If $\sigma_i = c_r = 0$ in the marginal state, then (11.1) shows that the marginal state is characterized by a *stationary pattern* of motion; we shall see later that the instability here appears in the form of *cellular convection* or *secondary flow* (see Figure 11.18 later). If, on the other hand, $\sigma_i \neq 0$ or $c_r \neq 0$ in the marginal state, then the instability sets in as traveling oscillations of growing amplitude. Following Eddington, such a mode of instability is frequently called *overstability* because the restoring forces are so strong that the system overshoots its corresponding position on the other side of equilibrium. We prefer to avoid this term and instead call it the *oscillatory mode* of instability.

The difference between the *neutrally stable state* and the *marginal state* should be noted as both have $\sigma_r = c_i = 0$. However, the marginal state has the additional constraint that it lies at the *borderline* between stable and unstable solutions. That is, a slight change of parameters (such as the Reynolds number) from the marginal state can take the system into an unstable regime where $\sigma_r > 0$. In many cases we shall find the stability criterion by simply setting $\sigma_r = 0$ or $c_i = 0$, without formally demonstrating that these conditions define the borderline between unstable and stable states.

11.3. KELVIN-HELMHOLTZ INSTABILITY

Instability at the interface between two horizontal parallel fluid streams with different velocities and densities is called the *Kelvin-Helmholtz instability*. This is an inertial instability and it can be readily analyzed assuming ideal flow in each stream. The name is also commonly used to describe the instability of the more general case where the variations of velocity and density are continuous and occur over a finite thickness (see Section 11.7).

The Kelvin-Helmholtz instability leads to enhanced momentum, heat, and moisture transport in the atmosphere, plus it is routinely exploited in a variety of geometries for mixing two or more fluid streams in engineering applications. The simplest version is analyzed here in two dimensions (x, z), where x is the stream-wise coordinate and z is the vertical coordinate. Consider two fluid layers of infinite depth that meet at a zero-thickness interface located at $z = \zeta(x, t)$. Let U_1 and ρ_1 be the horizontal velocity and density of the basic state in the upper half-space, and U_2 and ρ_2 be those of the basic state in the lower half-space (Figure 11.2). From Kelvin's circulation theorem, the perturbed flow must be irrotational in each half-space because the motion develops from an irrotational basic state, uniform velocity in each half-space. Thus, the infinitesimally perturbed flow above (subscript 1) and below (subscript 2) the interface can be described by the velocity potentials:

$$\tilde{\phi}_1 = U_1 x + \phi_1, \quad \text{and} \quad \tilde{\phi}_2 = U_2 x + \phi_2, \tag{11.2}$$

where the U_1 and U_2 terms represent the basic state, and tildes ($\sim$) denote the total flow potentials (background plus perturbations), a notation used throughout this chapter. Here $\tilde{\phi}_1$ and $\tilde{\phi}_2$ must satisfy the Laplace equation, so the perturbation potentials, ϕ_1 and ϕ_2, must also satisfy Laplace equations:

$$\nabla^2 \phi_1 = 0 \quad \text{and} \quad \nabla^2 \phi_2 = 0. \tag{11.3}$$

There are a total of four boundary conditions:

$$\phi_1 \to 0 \quad \text{as } z \to +\infty, \quad \phi_2 \to 0 \quad \text{as } z \to -\infty, \tag{11.4, 11.5}$$

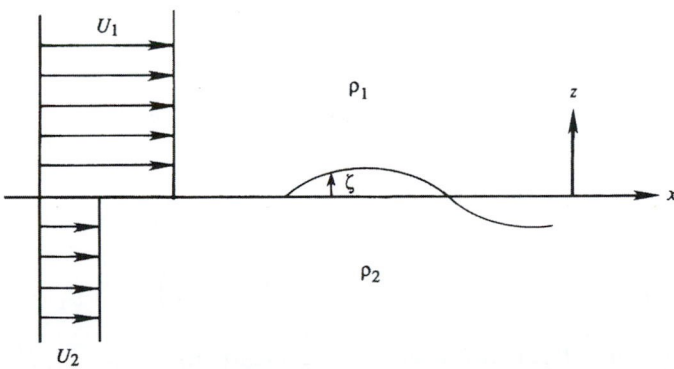

FIGURE 11.2 Basic flow configuration leading to the Kelvin-Helmholtz instability. Here the velocity and density profiles are discontinuous across an interface nominally located at $z = 0$. If the small vertical perturbation $\zeta(x,t)$ to this interface grows, then the flow is unstable.

$$\mathbf{n} \cdot \nabla \tilde{\phi}_1 = \mathbf{n} \cdot \mathbf{U}_s = \mathbf{n} \cdot \nabla \tilde{\phi}_2 \quad \text{on } z = \zeta, \text{ and} \tag{11.6}$$

$$p_1 = p_2 \quad \text{on } z = \zeta, \tag{11.7}$$

where $\mathbf{n}$ is the local normal to the interface, $\mathbf{U}_s$ is the velocity of the interface, and p_1 and p_2 are the pressures above and below the interface. Here, the kinematic and dynamic boundary conditions, (11.6) and (11.7), respectively, are conceptually similar to (7.14) and (7.20). The kinematic condition, (11.6), can be rewritten:

$$\mathbf{n} \cdot \left\{ \frac{\partial \tilde{\phi}_1}{\partial x} \mathbf{e}_x + \frac{\partial \tilde{\phi}_1}{\partial z} \mathbf{e}_z \right\} = \mathbf{n} \cdot \left\{ \frac{\partial \zeta}{\partial t} \mathbf{e}_z \right\} = \mathbf{n} \cdot \left\{ \frac{\partial \tilde{\phi}_2}{\partial x} \mathbf{e}_x + \frac{\partial \tilde{\phi}_2}{\partial z} \mathbf{e}_z \right\} \quad \text{on } z = \zeta, \tag{11.8}$$

where $\mathbf{n} = \nabla f / |\nabla f| = [-(\partial \zeta / \partial x)\mathbf{e}_x + \mathbf{e}_z] / \sqrt{1 + (\partial \zeta / \partial x)^2}$ when $f(x, z, t) = z - \zeta(x, t) = 0$ defines the interface, and $\mathbf{U}_s = (\partial \zeta / \partial t)\mathbf{e}_z$ can be considered purely vertical. When the dot products are performed, the common square-root factor removed, and the derivatives of the potentials evaluated using (11.2), (11.8) reduces to:

$$-\left(U_1 + \frac{\partial \phi_1}{\partial x} \right) \frac{\partial \zeta}{\partial x} + \frac{\partial \phi_1}{\partial z} = \frac{\partial \zeta}{\partial t} = -\left(U_2 + \frac{\partial \phi_2}{\partial x} \right) \frac{\partial \zeta}{\partial x} + \frac{\partial \phi_2}{\partial z} \quad \text{on } z = \zeta.$$

This condition can be linearized by applying it at $z = 0$ instead of at $z = \zeta$ and by neglecting quadratic terms. Thus, the simplified version of (11.6) is:

$$-U_1 \frac{\partial \zeta}{\partial x} + \frac{\partial \phi_1}{\partial z} = \frac{\partial \zeta}{\partial t} = -U_2 \frac{\partial \zeta}{\partial x} + \frac{\partial \phi_2}{\partial z} \quad \text{on } z = 0. \tag{11.9}$$

The dynamic boundary condition at the interface requires the pressure to be continuous across the interface (when surface tension is neglected). The unsteady Bernoulli equations above and below the layer are:

$$\frac{\partial \tilde{\phi}_1}{\partial t} + \frac{1}{2} |\nabla \tilde{\phi}_1|^2 + \frac{p_1}{\rho_1} + gz = C_1, \quad \text{and} \quad \frac{\partial \tilde{\phi}_2}{\partial t} + \frac{1}{2} |\nabla \tilde{\phi}_2|^2 + \frac{p_2}{\rho_2} + gz = C_2. \tag{11.10}$$

So pressure matching requires:

$$p_1 = \rho_1 \left(C_1 - \frac{\partial \tilde{\phi}_1}{\partial t} - \frac{1}{2} |\nabla \tilde{\phi}_1|^2 - gz \right) = \rho_2 \left(C_2 - \frac{\partial \tilde{\phi}_2}{\partial t} - \frac{1}{2} |\nabla \tilde{\phi}_2|^2 - gz \right) = p_2 \text{ on } z = \zeta. \tag{11.11}$$

In the undisturbed state ($\phi_1 = \phi_2 = 0$, and $\zeta = 0$), (11.11) implies:

$$(p_1)_{undisturbed} = \rho_1 \left(C_1 - \frac{1}{2} U_1^2 \right) = \rho_2 \left(C_2 - \frac{1}{2} U_2^2 \right) = (p_2)_{undisturbed}. \tag{11.12}$$

Subtracting (11.11) from (11.12) and inserting (11.2) leads to:

$$\rho_1 \left(\frac{\partial \phi_1}{\partial t} + U_1 \frac{\partial \phi_1}{\partial x} + \frac{1}{2} |\nabla \phi_1|^2 + gz \right) = \rho_2 \left(\frac{\partial \phi_2}{\partial t} + U_2 \frac{\partial \phi_2}{\partial x} + \frac{1}{2} |\nabla \phi_2|^2 + gz \right) \text{ on } z = \zeta,$$

and this condition can be linearized by dropping quadratic terms and evaluating derivatives on $z = 0$ to find:

$$\rho_1 \left(\frac{\partial \phi_1}{\partial t} + U_1 \frac{\partial \phi_1}{\partial x} + g\zeta \right) = \rho_2 \left(\frac{\partial \phi_2}{\partial t} + U_2 \frac{\partial \phi_2}{\partial x} + g\zeta \right) \text{ on } z = 0. \tag{11.13}$$

Thus, field equations (11.3) and the conditions (11.4), (11.5), (11.9), and (11.13) specify the linear stability of a velocity discontinuity between uniform flows of different speeds and densities.

We now apply the method of normal modes and look for oscillatory solutions for ϕ_1' and ϕ_2' in the second exponential form of (11.1) with $\mathbf{K} = (k, 0, 0)$:

$$\phi_1(x, z, t) = A_1(z) \exp\{ik(x - ct)\}, \text{ and } \phi_2(x, z, t) = A_2(z) \exp\{ik(x - ct)\}. \tag{11.14}$$

Insertion of (11.14) into (11.3) produces:

$$-k^2 A_1 + \frac{d^2 A_1}{dz^2} = 0, \quad \text{and} \quad -k^2 A_2 + \frac{d^2 A_2}{dz^2} = 0,$$

after common factors are divided out. These equations have exponential solutions: $A_\pm \exp(\pm kz)$. The boundary conditions (11.4) and (11.5) require the minus sign for $z > 0$, and the positive sign for $z < 0$, so (11.14) reduces to:

$$\phi_1 = A_- \exp\{ik(x - ct) - kz\}, \quad \text{and} \quad \phi_2 = A_+ \exp\{ik(x - ct) + kz\}. \tag{11.15}$$

Inserting these two equations and a matching form for the interface shape, $\zeta = \zeta_o \exp\{ik(x - ct)\}$, into (11.9) and (11.13) leads to:

$$-iU_1 k\zeta_o - kA_- = -ikc\zeta_o = -iU_2 k\zeta_o + kA_+, \text{ and} \tag{11.16}$$

$$\rho_1(-ikcA_- + ikU_1 A_- + g\zeta_o) = \rho_2(-ikcA_+ + ikU_2 A_+ + g\zeta_o). \tag{11.17}$$

The remnant of the kinematic boundary condition (11.16) is sufficient to find $A_\pm$ in terms of ζ_o:

$$kA_- = -(ikU_1 - ikc)\zeta_o, \quad \text{and} \quad kA_+ = (ikU_2 - ikc)\zeta_o.$$

Substituting these into the remnant of the dynamic boundary condition (11.17) leads to a quadratic equation for c:

$$\rho_1 \left(-(-ikc + ikU_1)^2 + gk \right) = \rho_2 \left((-ikc + ikU_2)^2 + gk \right),$$

after the common factor of ζ_o has been divided out. The two solutions for c are:

$$c = \frac{\rho_2 U_2 + \rho_1 U_1}{\rho_2 + \rho_1} \pm \left[\left(\frac{\rho_2 - \rho_1}{\rho_2 + \rho_1} \right) \frac{g}{k} - \frac{\rho_2 \rho_1}{(\rho_2 + \rho_1)^2} (U_2 - U_1)^2 \right]^{1/2}. \tag{11.18}$$

Clearly, both possible values for c imply neutral stability ($c_i = 0$) as long as the second term within the square root is smaller than the first. However, one of these solutions will lead to exponential growth ($c_i > 0$) when

$$\left(\frac{\rho_2 - \rho_1}{\rho_2 + \rho_1}\right)\frac{g}{k} < \frac{\rho_2\rho_1}{(\rho_2 + \rho_1)^2}(U_2 - U_1)^2 \text{ or } g(\rho_2^2 - \rho_1^2) < k\rho_1\rho_2(U_2 - U_1)^2,$$

which occurs when the free-stream velocity difference is high enough, the density difference is small enough, or the wave number k (presumed positive real) is large enough. In addition, for each growing solution there is a corresponding decaying solution. This happens because the coefficients of the differential equation and the boundary conditions are all real (see Section 11.7).

Although it is somewhat complicated, (11.18) includes several limiting cases with simple interpretations. First of all, setting $U_1 = U_2 = 0$ simplifies (11.18) to

$$c = \pm\left[\left(\frac{\rho_2 - \rho_1}{\rho_2 + \rho_1}\right)\frac{g}{k}\right]^{1/2}, \qquad (11.19)$$

which indicates a neutrally stable situation as long as $\rho_2 > \rho_1$. In this case, (11.19) is the dispersion relation for interface waves in an initially static medium; see (7.96). When $U_1 \neq U_2$, one can always find a value of k large enough to satisfy the requirement for instability. Because all wavelengths must be allowed in an instability analysis, we can say that the *flow is always unstable to short waves when* $U_1 \neq U_2$. When $\rho_1 = \rho_2$, the interface becomes a vortex sheet (see Section 5.8) with strength $\gamma = U_2 - U_1$, and (11.18) reduces to

$$c = \left(\frac{U_2 + U_1}{2}\right) \pm i\left(\frac{U_2 - U_1}{2}\right). \qquad (11.20)$$

Here there is always a positive imaginary value of c for every k, so a vortex sheet is unstable to disturbances of any wavelength. It is also seen that the unstable wave moves with a phase velocity, c_r, equal to the average velocity of the basic flow. This must be true from symmetry considerations. In a frame of reference moving with the average velocity, the basic flow is symmetric and the wave therefore should have no preference between the positive and negative x directions (Figure 11.3).

The Kelvin-Helmholtz instability is caused by the destabilizing effect of shear, which overcomes the stabilizing effect of stratification. This kind of instability is easy to generate in the laboratory by filling a horizontal glass tube (of rectangular cross-section) containing two liquids of slightly different densities (one colored) and gently tilting it. This starts a current in the lower layer down the plane and a current in the upper layer up the plane. An example of instability generated in this manner is shown in Figure 11.4.

Shear instability of stratified fluids is ubiquitous in the atmosphere and the ocean and believed to be a major source of internal waves. Figure 11.5 is a striking photograph of a cloud pattern, which is clearly due to the existence of high shear across a sharp density gradient. Similar photographs of injected dye have been recorded in oceanic thermoclines (Woods, 1969).

Figures 11.4 and 11.5 show the advanced nonlinear stage of the instability in which the interface is a rolled-up layer of vorticity. Such an observed evolution of the interface is in agreement with results of numerical calculations in which the nonlinear terms are retained (Figure 11.6).

The source of energy for generating the Kelvin-Helmholtz instability is derived from the kinetic energy of the two streams. The disturbances evolve to smear out the gradients

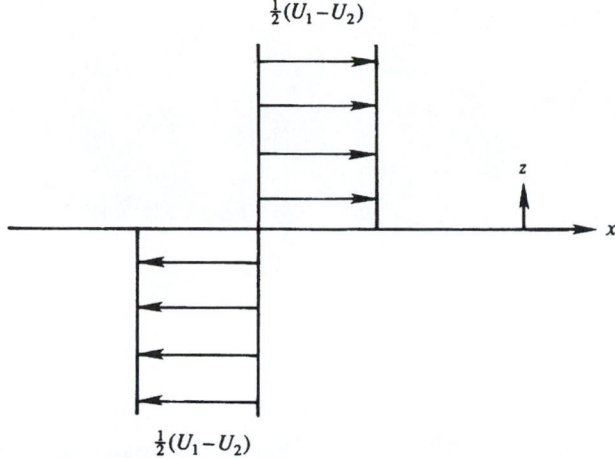

FIGURE 11.3   Background velocity field for the Kelvin-Helmholtz instability as seen by an observer traveling at the average velocity $(U_1 + U_2)/2$ of the two layers. When the densities of the two layers are equal, a disturbance to the interface will be stationary in this frame of reference.

FIGURE 11.4   Kelvin-Helmholtz instability generated by tilting a horizontal channel containing two liquids of different densities. The lower layer is dyed and 18 wavelengths of the developing interfacial disturbance are shown. The mean flow in the lower layer is down the plane (to the left) and that in the upper layer is up the plane (to the right). *S. A. Thorpe, Journal of Fluid Mechanics, 46, 299–319, 1971; reprinted with the permission of Cambridge University Press.*

until they cannot grow any longer. Figure 11.7 shows a typical behavior, in which the unstable waves at the interface have transformed the sharp density profile ACDF to ABEF and the sharp velocity profile MOPR to MNQR. The high-density fluid in the depth range DE has been raised upward (and mixed with the lower density fluid in the depth range BC), which means that the potential energy of the system has increased after the action of the instability. The required energy has been drawn from the kinetic energy of the basic field. It is easy to show that the kinetic energy of the initial profile MOPR is larger than that of the final profile MNQR. To see this, assume that the initial velocity of the lower layer is zero and that of the upper layer is U_1. Then the linear velocity profile after mixing is given by

FIGURE 11.5 Overturning billow cloud near Denver, Colorado. The similarity in shape of the developing instability with that shown in Figure 11.4 is striking. *P. G. Drazin and W. H. Reid,* Hydrodynamic Stability, *1981; reprinted with the permission of Cambridge University Press.*

$$U(z) = U_1 \left(\frac{1}{2} + \frac{z}{2h} \right) \quad \text{for} - h \leq z \leq h.$$

Consider the change in kinetic energy only in the depth range $-h < z < h$, as the energy outside this range does not change. Then the initial and final kinetic energies per unit width are:

$$
\begin{aligned}
E_{\text{initial}} &= \frac{\rho}{2} U_1^2 h, \\
E_{\text{final}} &= \frac{\rho}{2} \int_{-h}^{h} U^2(z) \ dz = \frac{\rho}{3} U_1^2 h.
\end{aligned}
$$

The kinetic energy of the flow has therefore decreased, although the total momentum $(= \int U dz)$ is unchanged. This is a general result: If the integral of $U(z)$ does not change, then the integral of $U^2(z)$ decreases if the gradients decrease.

In this section the case of a discontinuous variation across an infinitely thin interface is considered and the flow is always unstable. The case of continuous variation is considered in Section 11.7, and we shall see that one or more additional conditions must be satisfied in order for the flow to be unstable.

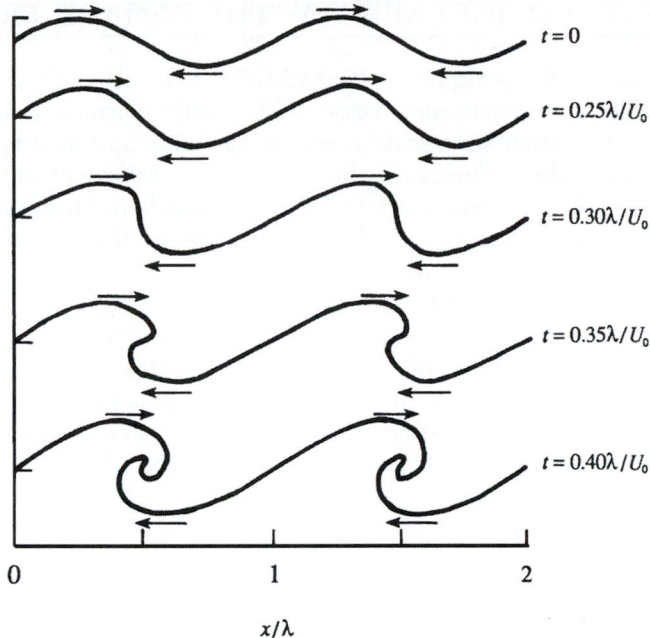

x/λ

FIGURE 11.6 Nonlinear numerical calculation of the evolution of a vortex sheet that has been given a small transverse sinusoidal displacement with wavelength λ. The density difference across the interface is zero, and U_0 is the velocity difference across the vortex sheet. Here again, the similarity of the interface shape at the last time with the results shown in Figures 11.4 and 11.5 is striking. The smaller vertical displacements shown in Figures 11.4 and 11.5 are consistent with the effects of stratification that are absent from the calculations shown in this figure. *J. S. Turner, Buoyancy Effects in Fluids, 1973; reprinted with the permission of Cambridge University Press.*

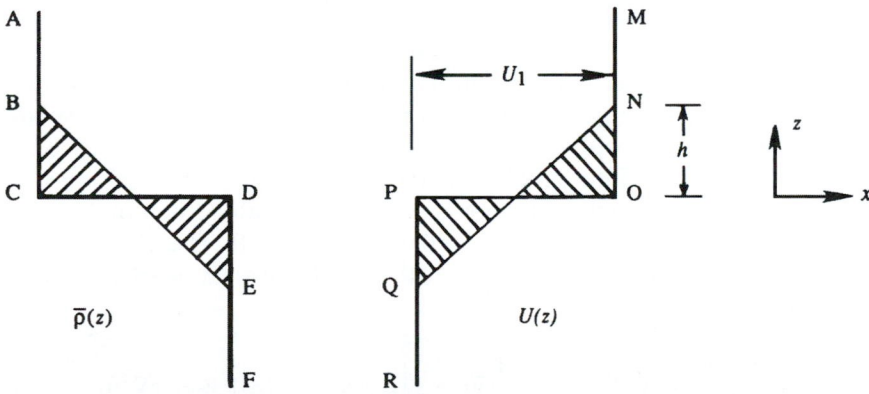

FIGURE 11.7 Smearing out of sharp density and velocity profiles, resulting in an increase of potential energy and a decrease of kinetic energy. When turbulent, the overturning eddies or billows shown in Figures 11.4 and 11.5 lead to cross-stream momentum transport and fluid mixing. The discontinuous profiles ACDF and MOPR evolve toward ABEF and MNQR as the instability develops.

11.4. THERMAL INSTABILITY: THE BÉNARD PROBLEM

In natural flows and engineering flows, heat addition to a nominally quiescent fluid from below can lead to a situation where cool, dense fluid overlies warmer, less dense fluid. Equation (11.19) indicates that such stratification will be unstable and lead to instability-driven motion when the fluid is ideal. However, when viscosity and thermal conduction are active, they may delay the onset of unstable convective motion, and only for large enough temperature gradients is the situation unstable. In this section, the conditions necessary for the onset of thermal instability in a layer of fluid are presented.

The first intensive experiments on instability caused by heating a layer of fluid from below were conducted by Bénard in 1900. Bénard experimented on only very thin layers (a millimeter or less) that had a free surface, and observed beautiful hexagonal cells when the convection developed. Stimulated by these experiments, Rayleigh in 1916 derived the theoretical requirement for the development of convective motion in a layer of fluid with two free surfaces. He showed that the instability would occur when the adverse temperature gradient was large enough to make the ratio,

$$\text{Ra} = g\alpha\Gamma d^4/\kappa\nu, \tag{11.21}$$

exceed a certain critical value. Here, g is the acceleration due to gravity, α is the fluid's coefficient of thermal expansion, $\Gamma = -d\overline{T}/dz$ is the vertical temperature gradient of the background state, d is the depth of the layer, κ is the fluid's thermal diffusivity, and ν is the fluid's kinematic viscosity. The parameter Ra is called the *Rayleigh number*, and it represents a ratio of the destabilizing effect of buoyancy to the stabilizing effect of viscosity. It has been recognized only recently that most of the *motions observed by Bénard were instabilities driven by the variation of surface tension with temperature and not the thermal instability due to a top-heavy density gradient* (Drazin & Reid, 1981, p. 34). The importance of instabilities driven by surface tension decreases as the layer becomes thicker. Later experiments on thermal convection in thicker layers (with or without a free surface) have obtained convective cells of many forms, not just hexagonal. Nevertheless, the phenomenon of thermal convection in a layer of fluid is still commonly called the *Bénard convection*. Rayleigh's solution of the thermal convection problem is considered a major triumph of linear stability theory. The concept of a critical Rayleigh number finds application in such geophysical problems as solar convection, cloud formation in the atmosphere, and the motion of the earth's core.

The formulation of the problem starts with a fluid layer of thickness d confined between two isothermal walls where the lower wall is maintained at a higher temperature, T_0, than the upper wall, $T_0 - \Delta T$, where $\Delta T > 0$ (see Figure 11.8). Use Cartesian coordinates centered in the middle of the fluid layer with the z-axis vertical; start from the Boussinesq set of equations,

$$\nabla\cdot\tilde{\mathbf{u}} = 0, \quad \frac{\partial\tilde{\mathbf{u}}}{\partial t} + (\tilde{\mathbf{u}}\cdot\nabla)\tilde{\mathbf{u}} = -\frac{1}{\rho_0}\nabla\tilde{p} - g\left[1 - \alpha(\tilde{T} - T_0)\right]\mathbf{e}_z + \nu\nabla^2\tilde{\mathbf{u}}, \quad \frac{\partial\tilde{T}}{\partial t} + (\tilde{\mathbf{u}}\cdot\nabla)\tilde{T} = \kappa\nabla^2\tilde{T},$$

$$(4.10, 4.86, 4.89)$$

and the simplified equation for the density in terms of the temperature: $\rho = \rho_0[1 - \alpha(\tilde{T} - T_0)]$, where ρ_0 and T_0 are the reference density and temperature. Here again, the total flow variables

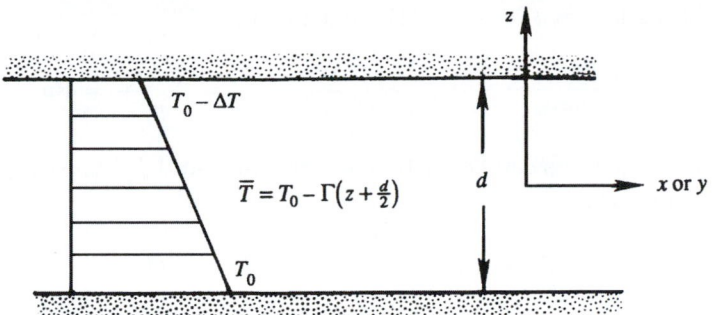

FIGURE 11.8 Flow geometry for the thermal convection between horizontal surfaces separated by a distance d. The lower surface is maintained at a higher temperature than the upper surface, and the coordinates are centered between them. For a given fluid and a fixed geometry, when the temperature difference ΔT is small, the fluid remains motionless and heat is transferred between the plates by thermal conduction. However, a sufficiently high ΔT will cause a cellular flow pattern to appear and thermal convection of heat to occur.

(background plus perturbation) carry a tilde ($\sim$). We decompose the total flow field into a motionless background plus perturbations:

$$\tilde{\mathbf{u}} = 0 + \mathbf{u}(\mathbf{x}, t), \quad \tilde{T} = \overline{T}(z) + T'(\mathbf{x}, t), \text{ and } \tilde{p} = P(z) + p(\mathbf{x}, t). \quad (11.22)$$

The basic state is represented by a quiescent fluid with temperature and pressure distributions $\overline{T}(z)$ and $P(z)$ that satisfy the equations,

$$0 = -\frac{1}{\rho_0}\nabla P - g[1 - \alpha(\overline{T} - T_0)]\,\mathbf{e}_z \quad \text{and} \quad 0 = \kappa\frac{\partial^2 \overline{T}}{\partial z^2}. \quad (11.23)$$

The preceding thermal equation gives the linear vertical temperature distribution:

$$\overline{T}(z) = T_0 - \frac{1}{2}\Delta T - \Gamma z, \quad (11.24)$$

where $\Gamma \equiv \Delta T/d$ is the magnitude of the vertical temperature gradient. Substituting (11.22) into the Boussinesq equation set and subtracting (11.23) produces:

$$\nabla\cdot\mathbf{u} = 0, \quad \frac{\partial\mathbf{u}}{\partial t} + (\mathbf{u}\cdot\nabla)\mathbf{u} = -\frac{1}{\rho_0}\nabla p + g\alpha T'\mathbf{e}_z + \nu\nabla^2\mathbf{u}, \text{ and } \frac{\partial T'}{\partial t} - w\Gamma + (\mathbf{u}\cdot\nabla)T' = \kappa\nabla^2 T',$$

where w is the vertical component of the fluid velocity, and the $-w\Gamma$ term in the final equation comes from evaluating $(\mathbf{u}\cdot\nabla)\overline{T}$ using (11.24). For small perturbations, it is appropriate to linearize the second two equations by dropping quadratic and higher order terms:

$$\nabla\cdot\mathbf{u} = 0, \quad \frac{\partial\mathbf{u}}{\partial t} = -\frac{1}{\rho_0}\nabla p + g\alpha T'\mathbf{e}_z + \nu\nabla^2\mathbf{u}, \text{ and } \frac{\partial T'}{\partial t} - w\Gamma = \kappa\nabla^2 T'. \quad (11.25, 11.26, 11.27)$$

These equations govern the behavior of perturbations to the basic state. A simple scaling analysis based on these equations leads to the Rayleigh number when $T' \sim \Delta T$, and $\nabla \sim 1/d$. From (11.27), the vertical velocity scale is found by equating the advective and diffusion terms:

$$w\Gamma \sim \kappa\nabla^2 T' \sim \kappa\frac{1}{d^2}\Delta T = \kappa\frac{1}{d}\frac{\Delta T}{d} = \kappa\frac{1}{d}\Gamma, \quad \text{so } w \sim \kappa/d.$$

Forming a ratio of the last two terms in (11.26) leads to:

$$\frac{buoyant\ force}{viscous\ force} \sim \frac{g\alpha T'}{\nu(1/d^2)w} \sim \frac{g\alpha(\Delta T/d)d}{\nu(1/d^2)(\kappa/d)} = \frac{g\alpha\Gamma d^4}{\nu\kappa} = \mathrm{Ra}.$$

The perturbation equations can be written in terms of w and T' by taking the Laplacian of the z-component of (11.26):

$$\frac{\partial}{\partial t}\nabla^2 w = -\frac{1}{\rho_0}\nabla^2\frac{\partial p}{\partial z} + g\alpha\nabla^2 T' + \nu\nabla^4 w. \tag{11.28}$$

The pressure term in (11.28) can be eliminated by taking the divergence of (11.26), using (11.25),

$$\frac{\partial}{\partial t}\nabla\cdot\mathbf{u} = -\frac{1}{\rho_0}\nabla^2 p + g\alpha\frac{\partial}{\partial z}T' + \nu\nabla^2\nabla\cdot\mathbf{u}, \text{ or } 0 = -\frac{1}{\rho_0}\nabla^2 p + g\alpha\frac{\partial}{\partial z}T',$$

and then differentiating with respect to z to obtain:

$$0 = -\frac{1}{\rho_0}\nabla^2\frac{\partial p}{\partial z} + g\alpha\frac{\partial^2 T'}{\partial z^2},$$

which can be subtracted from (11.28) to find:

$$\frac{\partial}{\partial t}\nabla^2 w = +g\alpha\nabla_H^2 T' + \nu\nabla^4 w, \tag{11.29}$$

where $\nabla_H^2 = \partial^2/\partial x^2 + \partial^2/\partial y^2$ is the horizontal Laplacian operator.

Equations (11.27) and (11.29) govern the development of perturbations on the system. The boundary conditions on the upper and lower rigid surfaces are that the no-slip condition is satisfied and that the walls are maintained at constant temperatures. These conditions require $u = v = w = T' = 0$ at $z = \pm d/2$. Because the conditions on u and v hold for all x and y, it follows from the continuity equation that $\partial w/\partial z = 0$ at the walls. The boundary conditions therefore can be written as

$$w = \partial w/\partial z = T' = 0 \quad \text{on } z = \pm d/2. \tag{11.30}$$

Dimensionless independent variables are used in the rest of the analysis via the transformation:

$$t \to (d^2/\kappa)t \quad \text{and} \quad (x,y,z) \to (xd,yd,zd),$$

where the dimensional variables are on the left side and the new dimensionless variables are on the right-hand side; note that we are avoiding the introduction of new symbols for the dimensionless variables. Equations (11.27), (11.29), and (11.30) then become:

$$\left(\frac{\partial}{\partial t} - \nabla^2\right)T' = \frac{\Gamma d^2}{\kappa}w, \quad \left(\frac{1}{\mathrm{Pr}}\frac{\partial}{\partial t} - \nabla^2\right)\nabla^2 w = \frac{g\alpha d^2}{\nu}\nabla_H^2 T', \text{ and} \tag{11.31, 11.32, 11.33}$$

$$w = \partial w/\partial z = T' = 0 \quad \text{on } z = \pm\frac{1}{2}.$$

where $\mathrm{Pr} \equiv \nu/\kappa$ is the Prandtl number.

The method of normal modes is now introduced. Because the coefficients in (11.31) and (11.32) are independent of x, y, and t, solutions exponential in these variables are allowed. We therefore assume normal modes given by the first version of (11.1) with $\mathbf{K} = (k, l, 0)$:

$$w = \hat{w}(z)\,\exp\{ikx + ily + \sigma t\}, \quad \text{and} \quad T' = \hat{T}(z)\,\exp\{ikx + ily + \sigma t\}.$$

The requirement that solutions remain bounded as x, $y \to \infty$ implies that the wave numbers k and l must be real. In other words, the normal modes must be oscillatory in the directions of unboundedness. The temporal growth rate $\sigma = \sigma_r + i\sigma_i$ is allowed to be complex. With this dependence, the operators in (11.31) and (11.32) primarily transform to algebraic multipliers via:

$$\partial/\partial t \to \sigma, \quad \nabla_H^2 \to -k^2 - l^2 \equiv -K^2, \quad \text{and} \quad \nabla^2 \to -K^2 + d^2/dz^2,$$

where $K = |\mathbf{K}|$ is the magnitude of the (dimensionless) horizontal wave number. Equations (11.31) and (11.32) then become

$$\left(\sigma + K^2 - \frac{d^2}{dz^2}\right)\hat{T} = \frac{\Gamma d^2}{\kappa}\hat{w} \quad \text{and} \quad \left(\frac{\sigma}{\Pr} + K^2 - \frac{d^2}{dz^2}\right)\left(\frac{d^2}{dz^2} - K^2\right)\hat{w} = -\frac{g\alpha d^2 K^2}{\nu}\hat{T}.$$

$$(11.34, 11.35)$$

Making the substitution $W \equiv (\Gamma d^2/\kappa)\hat{w}$, (11.34) and (11.35) reduce to:

$$\left(\sigma + K^2 - \frac{d^2}{dz^2}\right)\hat{T} = W \quad \text{and} \quad \left(\frac{\sigma}{\Pr} + K^2 - \frac{d^2}{dz^2}\right)\left(\frac{d^2}{dz^2} - K^2\right)W = -\mathrm{Ra}K^2\hat{T}. \quad (11.36, 11.37)$$

The boundary conditions (11.33) become

$$W = \partial W/\partial z = \hat{T} = 0 \quad \text{on } z = \pm 1/2. \qquad (11.38)$$

Here we note that σ is real for $\mathrm{Ra} > 0$ (see Exercise 11.6). The Bénard problem is one of two well-known problems in which σ is real. (The other one is Taylor-Couette flow between rotating cylinders, discussed in the following section.) In most other problems σ is complex, and the marginal state ($\sigma_r = 0$) contains propagating waves (as is true for the Kelvin-Helmholtz instability). In the Bénard and Taylor problems, however, the marginal state corresponds to $\sigma = 0$, and is therefore *stationary* and does not contain propagating waves. In these flows, the onset of instability is marked by a transition from the background state to another *steady* state. In such a case we commonly say that the *principle of exchange of stabilities* is valid, and the instability sets in as a *cellular convection*, which will be explained shortly.

Two solutions for Rayleigh-Bénard flow are presented in the remainder of this section. First, the solution is presented for the case that is easiest to realize in a laboratory experiment, namely, a layer of fluid confined between two rigid plates on which no-slip conditions are satisfied. The solution to this problem was first given by Jeffreys in 1928. The second solution for a layer of fluid with two stress-free surfaces is presented after the first.

For the marginal state $\sigma = 0$, the equation pair (11.36) and (11.37) become

$$\left(\frac{d^2}{dz^2} - K^2\right)\widehat{T} = -W \quad \text{and} \quad \left(\frac{d^2}{dz^2} - K^2\right)^2 W = \text{Ra}K^2\widehat{T}. \tag{11.39}$$

Eliminating $\widehat{T}$, we obtain

$$\left(\frac{d^2}{dz^2} - K^2\right)^3 W = -\text{Ra}K^2 W. \tag{11.40}$$

The boundary condition (11.38) becomes

$$W = \partial W/\partial z = (d^2/dz^2 - K^2)^2 W = 0 \quad \text{on } z = \pm 1/2. \tag{11.41}$$

We have a sixth-order homogeneous differential equation with six homogeneous boundary conditions. Nonzero solutions for such a system can only exist for a particular value of Ra (for a given K). It is therefore an eigenvalue problem. Note that the Prandtl number has dropped out of the marginal state.

The point to observe is that the problem is symmetric with respect to the two boundaries, thus the eigenfunctions fall into two distinct classes—those with the vertical velocity symmetric about the midplane $z = 0$, and those with the vertical velocity antisymmetric about the midplane (Figure 11.9). The gravest even mode therefore has one row of cells, and the gravest odd mode has two rows of cells. It can be shown that the smallest critical Rayleigh number is obtained by assuming disturbances in the form of the gravest even mode, which also agrees with experimental findings of a single row of cells.

Because the coefficients of the governing equation (11.40) are independent of z, the general solution can be expressed as a superposition of solutions of the form: $W \propto \exp(qz)$, where the six roots of q are found from

$$\left(q^2 - K^2\right)^3 = -\text{Ra}K^2.$$

The three roots of this equation for q^2 are:

$$q^2 = -K^2\left[\left(\frac{\text{Ra}}{K^4}\right)^{1/3} - 1\right] \quad \text{and} \quad q^2 = K^2\left[1 + \frac{1}{2}\left(\frac{\text{Ra}}{K^4}\right)^{1/3}(1 \pm i\sqrt{3})\right]. \tag{11.42}$$

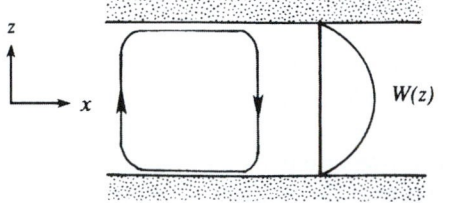
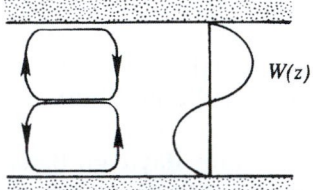

Gravest even mode   Gravest odd mode

FIGURE 11.9 Flow pattern and eigenfunction structure of the gravest even mode and the gravest odd mode in the Bénard problem. The even mode is observed first as the temperature difference between the surfaces is increased.

Taking square roots, the six roots for q are $\pm iq_0$, $\pm q$, and $\pm q^*$, where

$$q_0 = K\left[\left(\frac{Ra}{K^4}\right)^{1/3} - 1\right]^{1/2}$$

and q and its conjugate q^* are given by the two roots of the second part of (11.42).

The even solution of (11.40) is therefore

$$W = A\cos q_0 z + B\cosh qz + C\cosh q^* z,$$

where A, B, and C are constants. To apply the boundary conditions on this solution, we find the following derivatives:

$$dW/dz = -Aq_0\sin q_0 z + Bq\sinh qz + Cq^*\sinh q^* z, \quad \text{and}$$

$$\left(d^2/dz^2 - K^2\right)^2 W = A\left(q_0^2 + K^2\right)^2\cos q_0 z + B\left(q^2 - K^2\right)^2\cosh qz + B\left(q^{*2} - K^2\right)^2\cosh q^* z.$$

The boundary conditions (11.41) then require:

$$\begin{bmatrix} \cos\dfrac{q_0}{2} & \cosh\dfrac{q}{2} & \cosh\dfrac{q^*}{2} \\[2mm] -q_0\sin\dfrac{q_0}{2} & q\sinh\dfrac{q}{2} & q^*\sinh\dfrac{q^*}{2} \\[2mm] \left(q_0^2 + K^2\right)^2\cos\dfrac{q_0}{2} & \left(q^2 - K^2\right)^2\cosh\dfrac{q}{2} & \left(q^{*2} - K^2\right)^2\cosh\dfrac{q^*}{2} \end{bmatrix} \begin{bmatrix} A \\ B \\ C \end{bmatrix} = 0.$$

Here, A, B, and C cannot all be zero if we want to have a nonzero solution, which requires that the determinant of the matrix must vanish. This gives a relation between Ra and the corresponding eigenvalue K (Figure 11.10). Points on the curve $K(Ra)$ represent marginally stable states, which separate regions of stability and instability. The lowest value of Ra for marginal stability is found to be $Ra_{cr} = 1708$, attained at $K_{cr} = 3.12$. As *all* values of K are allowed by the system, the flow first becomes unstable when the Rayleigh number reaches a value of

$$Ra_{cr} = 1708.$$

The wavelength at the onset of instability is: $\lambda_{cr} = 2\pi d/K_{cr} \cong 2d$. Laboratory experiments agree remarkably well with these predictions, and the solution of the Bénard problem is considered one of the major successes of the linear stability theory.

The solution for a fluid layer with stress-free surfaces is somewhat simpler and was first given by Rayleigh. This case can be approximately realized in a laboratory experiment if a layer of liquid is floating on top of a somewhat heavier liquid. Here the boundary conditions are $w = T' = \mu(\partial u/\partial z + \partial w/\partial x) = \mu(\partial v/\partial z + \partial w/\partial y) = 0$ at the surfaces, the latter two conditions resulting from zero stress. Because w vanishes (for all x and y) on the boundaries, it follows that the vanishing stress conditions require $\partial u/\partial z = \partial v/\partial z = 0$ at the boundaries. On differentiating the continuity equation with respect to z, it follows that $\partial^2 w/\partial z^2 = 0$ on the free surfaces. In terms of the complex amplitudes, the eigenvalue problem is therefore defined by (11.39) and with boundary conditions:

$$W = (d^2/dz^2 - K^2)^2 W = d^2 W/dz^2 = 0 \quad \text{on } z = \pm 1/2.$$

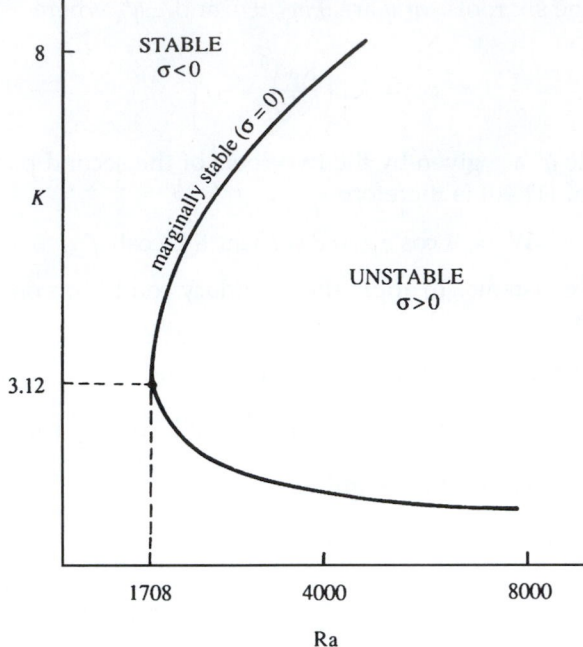

FIGURE 11.10 Stable and unstable regions for Bénard convection in a plot of the dimensionless wave number K vs. Ra, the Rayleigh number (11.21). The lowest possible Ra value for which the flow may be unstable is 1708, and the wave number of the first mode of instability is $3.12/d$, where d is the separation between the horizontal surfaces.

By expanding and simplifying the products of operators, the boundary conditions can be rewritten as

$$W = d^2W/dz^2 = d^4W/dz^4 = 0 \quad \text{on } z = \pm 1/2, \tag{11.43}$$

which should be compared with the conditions (11.41) for rigid boundaries.

Successive differentiation of (11.40) shows that *all* even derivatives of W vanish on the boundaries. The eigenfunctions must therefore be

$$W = A \sin(n\pi z),$$

where A is any constant and n is an integer. Substitution into equation (11.40) leads to the eigenvalue relation

$$Ra = \left(n^2\pi^2 + K^2\right)^3/K^2, \tag{11.44}$$

which gives the Rayleigh number in the marginal state. For a given K^2, the lowest value of Ra occurs when $n = 1$, which is the gravest mode. The critical Rayleigh number is obtained by finding the minimum value of Ra as K^2 is varied, that is, by setting $d\text{Ra}/dK^2 = 0$:

$$\frac{d\text{Ra}}{dK^2} = \frac{3\left(\pi^2 + K^2\right)^2}{K^2} - \frac{3\left(\pi^2 + K^2\right)^3}{K^4} = 0,$$

FIGURE 11.11 Two-dimensional convection rolls in Bénard convection. Fluid alternately ascends and descends between the rolls. The horizontal spacing between roll centers is nearly the same as the spacing between the horizontal surfaces.

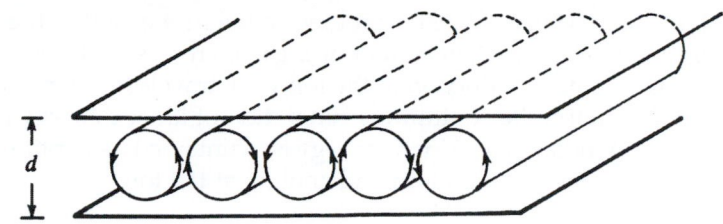

which requires $K_{cr}^2 = \pi^2/2$. The corresponding value of Ra is:

$$\text{Ra}_{cr} = (27/4)\pi^4 = 657.5.$$

For a layer with a free upper surface (where the stress is zero) and a rigid bottom wall, the solution of the eigenvalue problem gives $\text{Ra}_{cr} = 1101$ and $K_{cr} = 2.68$. This case is of interest in laboratory experiments having the most visual effects, as originally conducted by Bénard.

The linear theory specifies the horizontal wavelength at the onset of instability, but not the horizontal pattern of the convective cells. This is because a given wave number vector **K** can be decomposed into two orthogonal components in an infinite number of ways. If we assume that the experimental conditions are horizontally isotropic, with no preferred directions, then regular polygons in the form of equilateral triangles, squares, and regular hexagons are all possible structures. Bénard's original experiments showed only hexagonal patterns, but we now know that he was observing a different phenomenon. The observations summarized in Drazin and Reid (1981) indicate that hexagons frequently predominate initially. As Ra is increased, the cells tend to merge and form rolls, on the walls of which the fluid rises or sinks (Figure 11.11). The cell structure becomes more chaotic as Ra is increased further, and the flow becomes turbulent when $\text{Ra} > 5 \times 10^4$.

The magnitude or direction of flow in the cells cannot be predicted by linear theory. After a short time of exponential growth, the flow becomes fast enough for the nonlinear terms to be important and it reaches a nonlinear equilibrium stage. The flow pattern for a hexagonal cell is sketched in Figure 11.12. Particles in the middle of the cell usually rise in a liquid and

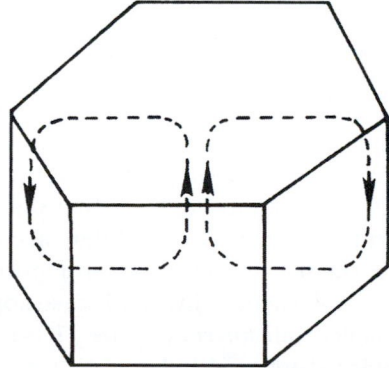

FIGURE 11.12 Above the critical Rayleigh number, complicated flow patterns may exist because a range of wave numbers is unstable for the first mode. A commonly observed Bénard-convection flow pattern involves hexagonal cells. Once such cell is shown here.

fall in a gas. This has been attributed to the property that the viscosity of a liquid decreases with temperature, whereas that of a gas increases with temperature. The rising fluid loses heat by thermal conduction at the top wall, travels horizontally, and then sinks. For a steady cellular pattern, the continuous generation of kinetic energy is balanced by viscous dissipation. The generation of kinetic energy is maintained by continuous release of potential energy due to heating at the bottom and cooling at the top.

11.5. DOUBLE-DIFFUSIVE INSTABILITY

An interesting instability results when the density of the fluid depends on two opposing gradients. The possibility of this phenomenon was first suggested by Stommel et al. (1956), but the dynamics of the process was first explained by Stern (1960). Turner (1973), review articles by Huppert and Turner (1981), and Turner (1985) discuss the dynamics of this phenomenon and its applications to various fields such as astrophysics, engineering, and geology. Historically, the phenomenon was first suggested with oceanic application in mind, and this is how we shall present it. For sea water the density depends on the temperature $\tilde{T}$ and salt content $\tilde{s}$ (kilograms of salt per kilograms of water), so that the density is given by:

$$\tilde{\rho} = \rho_0 \left[1 - \alpha(\tilde{T} - T_0) + \beta(\tilde{s} - s_0) \right],$$

where the value of α determines how fast the density decreases with temperature, and the value of β determines how fast the density increases with salinity. As defined here, both α and β are positive. The key factor in this instability is that the diffusivity κ_s of salt in water is only 1% of the thermal diffusivity κ. *Such a system can be unstable even when the density decreases upwards.* By means of the instability, the flow releases the potential energy of the *component* that is "heavy at the top." Therefore, the effect of diffusion in such a system can be to *destabilize* a stable density gradient. This is in contrast to a medium containing a single diffusing component, for which the analysis of the preceding section shows that the effect of diffusion is to *stabilize* the system even when it is heavy at the top.

Consider the two situations of Figure 11.13, both of which can be unstable although each is stably stratified in density ($d\bar{\rho}/dz < 0$). Consider first the case of hot and salty water lying over cold and fresh water (Figure 11.13a), that is, when the system is top heavy in salt. In this case both $d\bar{T}/dz$ and dS/dz are positive, and we can arrange the composition of water such that the density decreases upward. Because $\kappa_s \ll \kappa$, a displaced particle would be near thermal equilibrium with the surroundings, but would exchange negligible salt. A rising particle therefore would be constantly lighter than the surroundings because of the salinity deficit, and would continue to rise. A parcel displaced downward would similarly continue to plunge downward. The basic state shown in Figure 11.13a is therefore unstable. Laboratory observations show that the instability in this case appears in the form of a forest of long narrow convective cells, called *salt fingers* (Figure 11.14). Shadowgraph images in the deep ocean have confirmed their existence in nature.

We can derive a criterion for instability by generalizing our analysis of the Bénard convection so as to include salt diffusion. Assume a layer of depth d confined between stress-free boundaries maintained at constant temperature and constant salinity. If we repeat the

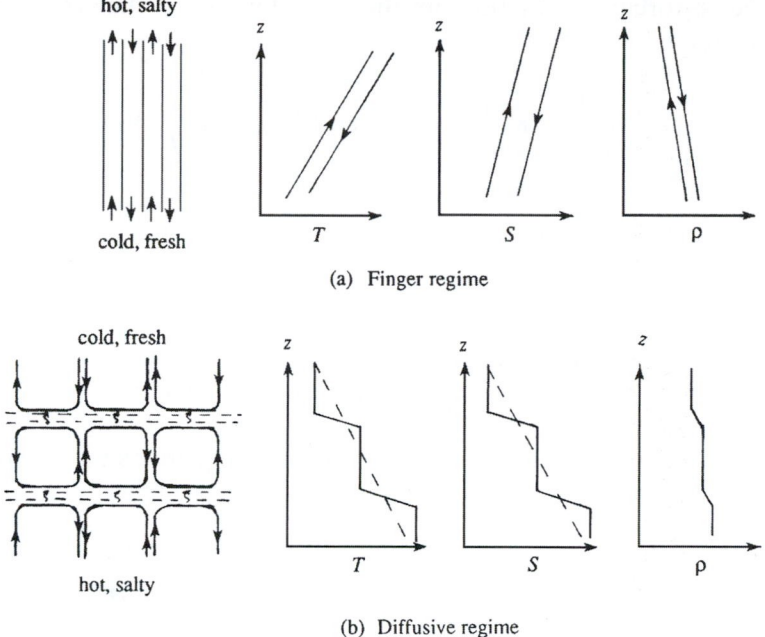

(a) Finger regime

(b) Diffusive regime

FIGURE 11.13 Two kinds of double-diffusive instabilities. (a) Finger instability, showing up- and down-going salt fingers and their temperature, salinity, and density. Arrows indicate the direction of fluid motion. (b) Oscillating instability, finally resulting in a series of convecting layers separated by "diffusive" interfaces. Across these interfaces T and S vary sharply, but heat is transported much faster than salt.

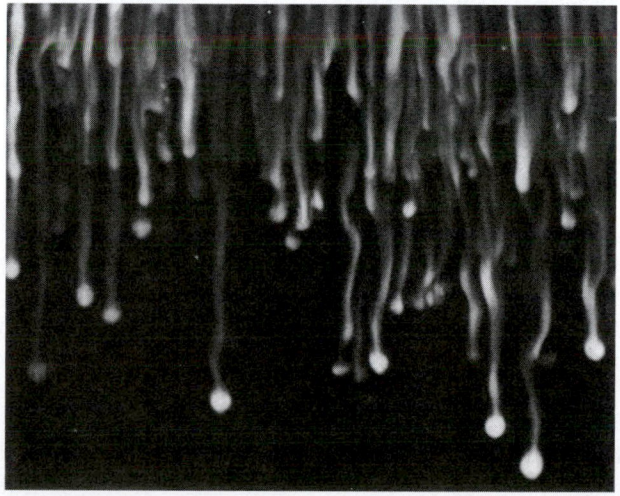

FIGURE 11.14 Salt fingers, produced by pouring a salt solution on top of a stable temperature gradient. Flow visualization by fluorescent dye and a horizontal beam of light. *J. Turner, Naturwissenschaften, 72, 70–75, 1985; reprinted with the permission of Springer-Verlag GmbH & Co.*

derivation of the perturbation equations for the normal modes of the system, the equations that replace (11.39) are found to be:

$$\left(\frac{d^2}{dz^2} - K^2\right)\widehat{T} = -W, \quad \frac{\kappa_s}{\kappa}\left(\frac{d^2}{dz^2} - K^2\right)\widehat{s} = -W, \quad \text{and} \quad \left(\frac{d^2}{dz^2} - K^2\right)^2 W = -\text{Ra}K^2\widehat{T} + \text{Rs}'K^2\widehat{s},$$

$$(11.45)$$

where $\widehat{s}(z)$ is the complex amplitude of the salinity perturbation, and we have defined

$$\text{Ra} \equiv \frac{g\alpha d^4(d\overline{T}/dz)}{\nu\kappa} \quad \text{and} \quad \text{Rs}' \equiv \frac{g\beta d^4(dS/dz)}{\nu\kappa}.$$

Note that κ (and not κ_s) appears in the definition of Rs'. In contrast to (11.45), a positive sign appeared in (11.39) in front of Ra because in the preceding section Ra was defined to be positive for a top-heavy situation.

It is seen from the first two equations of (11.45) that the equations for $\widehat{T}$ and $\widehat{s}\kappa_s/\kappa$ are the same. The boundary conditions are also the same for these variables:

$$\widehat{T} = \widehat{s}\kappa_s/\kappa = 0 \quad \text{at } z = \pm 1/2.$$

It follows that we must have $\widehat{T} = \kappa_s\widehat{s}/\kappa$ everywhere. Equations (11.45) therefore become:

$$(d^2/dz^2 - K^2)\widehat{T} = -W \quad \text{and} \quad (d^2/dz^2 - K^2)^2 W = (\text{Rs} - \text{Ra})K^2\widehat{T},$$

where

$$\text{Rs} \equiv \frac{\kappa}{\kappa_s}\text{Rs}' = \frac{g\beta d^4(dS/dz)}{\nu\kappa_s}.$$

The preceding set is now identical to the set (11.39) for the Bénard convection, with $(\text{Rs} - \text{Ra})$ replacing Ra. For stress-free boundaries, the solution of the preceding section shows that the critical value is

$$\text{Rs} - \text{Ra} = \frac{27}{4}\pi^4 = 657,$$

which can be written as

$$\frac{gd^4}{\nu}\left[\frac{\beta}{\kappa_s}\frac{dS}{dz} - \frac{\alpha}{\kappa}\frac{d\overline{T}}{dz}\right] = 657. \qquad (11.46)$$

Even if $\alpha(d\overline{T}/dz) - \beta(dS/dz) > 0$ (i.e., $\overline{\rho}$ decreases upward), the condition (11.46) can be quite easily satisfied because κ_s is much smaller than κ. The flow can therefore be made unstable simply by ensuring that the factor within [ ] is positive and making d large enough.

The analysis predicts that the lateral width of the cell is of the order of d, but such wide cells are not observed at supercritical stages when $(\text{Rs} - \text{Ra})$ far exceeds 657. Instead, long thin salt fingers are observed, as shown in Figure 11.14. If the salinity gradient is large, then experiments as well as calculations show that a deep layer of salt fingers becomes unstable and breaks down into a series of convective layers, with fingers confined to the interfaces. Oceanographic observations frequently show a series of staircase-shaped vertical distributions of salinity and temperature; with a positive overall dS/dz and $d\overline{T}/dz$ such distributions can indicate salt finger activity.

The double-diffusive instability may also occur when cold fresh water overlays hot salty water (Figure 11.13b). In this case both $d\overline{T}/dz$ and dS/dz are negative, and we can choose their values such that the density decreases upward. Again the system is unstable, but the dynamics are different. A particle displaced upward loses heat but no salt. Thus it becomes heavier than the surroundings and buoyancy forces it back toward its initial position, resulting in an oscillation. However, a stability calculation shows that less than perfect heat conduction results in a growing oscillation, although some energy is dissipated. In this case the growth rate σ is complex, in contrast to the situation of Figure 11.13a where it is real.

Laboratory experiments show that the initial oscillatory instability does not last long, and eventually results in the formation of a number of horizontal *convecting layers*, as sketched in Figure 11.13b. Consider the situation when a stable salinity gradient in an isothermal fluid is heated from below (Figure 11.9). The initial instability starts as a growing oscillation near the bottom. As the heating is continued beyond the initial appearance of the instability, a well-mixed layer develops, capped by a salinity step, a temperature step, and no density step. The heat flux through this step forms a thermal boundary layer, as shown in Figure 11.15. As the well-mixed layer grows, the temperature step across the thermal boundary layer becomes larger. Eventually, the Rayleigh number across the thermal boundary layer becomes critical, and a second convecting layer forms on top of the first. The second layer is maintained by heat flux (and negligible salt flux) across a sharp laminar interface on top of the first layer. This process continues until a stack of horizontal layers forms one upon another. From comparison with the Bénard convection, it is clear that inclusion of a stable salinity gradient has prevented a complete overturning from top to bottom.

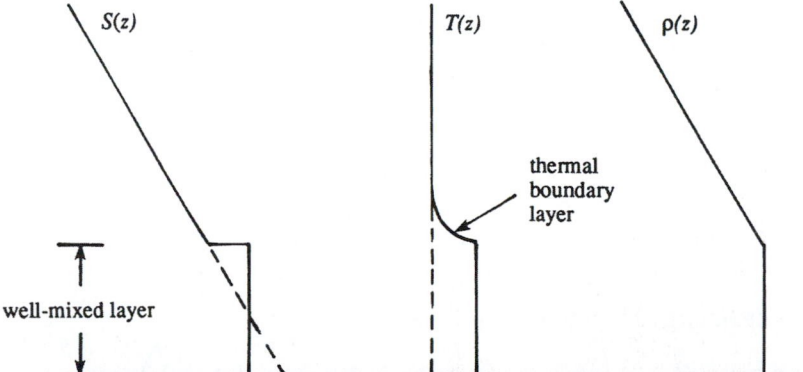

FIGURE 11.15   Distributions of salinity, temperature, and density generated by heating a linear salinity gradient from below. As heating continues the mixed layer depth will increase until a second mixed layer forms. Eventually, the flow pattern sketched and described in Figure 11.13b forms. Top to bottom overturning motion is not possible because of the overall stratification.

The two examples in this section show that in a double-component system in which the diffusivities for the two components are different, the effect of diffusion can be destabilizing, even if the system is judged hydrostatically stable. In contrast, diffusion is stabilizing in a single-component system, such as the Bénard system. The two requirements for the double-diffusive instability are that the diffusivities of the components be different, and that the components make opposite contributions to the vertical density gradient.

11.6. CENTRIFUGAL INSTABILITY: TAYLOR PROBLEM

In this section we shall consider the instability of a Couette flow between concentric rotating cylinders, a problem first solved by G. I. Taylor in 1923. In many ways the problem is similar to the Bénard problem, in which there is a potentially unstable arrangement of temperature. In the Couette-flow problem the source of the instability is the unstable arrangement of angular momentum. Whereas convection in a heated layer is brought about by buoyant forces becoming large enough to overcome the viscous resistance, the convection in a Couette flow is generated by the centrifugal forces being able to overcome the viscous forces. We shall first present Rayleigh's discovery of an inviscid stability criterion for the problem and then outline Taylor's solution of the viscous case. Experiments indicate that the instability initially appears in the form of axisymmetric disturbances, for which $\partial/\partial\theta = 0$. Accordingly, we shall limit ourselves only to the axisymmetric case.

The problem was first considered by Rayleigh in 1888. Neglecting viscous effects, he discovered the source of instability for this problem and demonstrated a necessary and sufficient condition for instability. Let $U_\theta(r)$ be the angular-directed velocity in the r-θ plane at any radial distance from the origin. For inviscid flows $U_\theta(r)$ can be any function, but only certain distributions can be stable. Imagine that two fluid rings with equal mass at radial distances r_1 and r_2 ($>r_1$) are interchanged. As the motion is inviscid, Kelvin's theorem requires that the circulation $\Gamma = 2\pi r U_\theta$ (proportional to the angular momentum rU_θ) should remain constant during the interchange. That is, after the interchange, the fluid at r_2 will have the circulation (namely, Γ_1) that it had at r_1 before the interchange. Similarly, the fluid at r_1 will have the circulation (namely, Γ_2) that it had at r_2 before the interchange. Conservation of circulation requires that the kinetic energy E must change during the interchange. Because $E = U_\theta^2/2 = \Gamma^2/8\pi^2 r^2$ we have

$$E_{\text{final}} = \frac{1}{8\pi^2}\left[\frac{\Gamma_2^2}{r_1^2} + \frac{\Gamma_1^2}{r_2^2}\right],$$

$$E_{\text{initial}} = \frac{1}{8\pi^2}\left[\frac{\Gamma_1^2}{r_1^2} + \frac{\Gamma_2^2}{r_2^2}\right],$$

so that the kinetic energy change per unit mass is:

$$\Delta E = E_{\text{final}} - E_{\text{initial}} = \frac{1}{8\pi^2}(\Gamma_2^2 - \Gamma_1^2)\left(\frac{1}{r_1^2} - \frac{1}{r_2^2}\right).$$

Because $r_2 > r_1$, a velocity distribution for which $\Gamma_2^2 > \Gamma_1^2$ would make ΔE positive, and this implies that an external source of energy would be necessary to perform the interchange of the fluid rings. Under this condition a *spontaneous* interchange of the rings is not possible, and the flow is stable. On the other hand, if Γ^2 decreases with r, then an interchange of rings will result in a release of energy; such a flow is unstable. It can be shown that in this situation the centrifugal force in the new location of an outwardly displaced ring is larger than the prevailing (radially inward) pressure gradient force.

Rayleigh's criterion can therefore be stated as follows: *An inviscid Couette flow is unstable if*

$$d\Gamma^2/dr < 0 \quad \text{(unstable)}.$$

The criterion is analogous to the inviscid requirement for static instability in a density-stratified fluid:

$$d\bar{\rho}/dz > 0 \quad \text{(unstable)}.$$

Therefore, the stratification of angular momentum in a Couette flow is unstable if it decreases radially outwards. When the outer cylinder is held stationary and the inner cylinder is rotated, $d\Gamma^2/dr < 0$ and Rayleigh's criterion implies that the flow is inviscidly unstable. As in the Bénard problem, however, merely having a potentially unstable arrangement does not cause instability in a viscous medium.

This inviscid Rayleigh criterion is modified in a viscous version of the problem. Taylor's solution of the viscous problem is outlined in what follows. Using cylindrical polar coordinates (R, φ, z) and assuming axial symmetry, the equations of motion are:

$$\frac{D\tilde{u}_R}{Dt} - \frac{\tilde{u}_\varphi^2}{R} = -\frac{1}{\rho}\frac{\partial \tilde{p}}{\partial R} + \nu\left(\nabla^2 \tilde{u}_R - \frac{\tilde{u}_R}{R^2}\right), \quad \frac{D\tilde{u}_\varphi}{Dt} + \frac{\tilde{u}_R \tilde{u}_\varphi}{R} = \nu\left(\nabla^2 \tilde{u}_\varphi - \frac{\tilde{u}_\varphi}{R^2}\right),$$

$$\frac{D\tilde{u}_z}{Dt} = -\frac{1}{\rho}\frac{\partial \tilde{p}}{\partial z} + \nu\nabla^2 \tilde{u}_z, \quad \text{and} \quad \frac{\partial}{\partial R}(R\tilde{u}_R) + \frac{\partial \tilde{u}_z}{\partial z} = 0, \qquad (11.47)$$

where

$$\frac{D}{Dt} = \frac{\partial}{\partial t} + \tilde{u}_R\frac{\partial}{\partial R} + \tilde{u}_z\frac{\partial}{\partial z} \quad \text{and} \quad \nabla^2 = \frac{\partial^2}{\partial R^2} + \frac{1}{R}\frac{\partial}{\partial R} + \frac{\partial^2}{\partial z^2}.$$

We decompose the motion into a background state plus perturbation:

$$\tilde{\mathbf{u}} = \mathbf{U} + \mathbf{u} \quad \text{and} \quad \tilde{p} = P + p. \qquad (11.48)$$

The background state is given by (see Section 8.2):

$$U_R = U_z = 0, \; U_\varphi = AR + B/R \quad \text{and} \quad \frac{1}{\rho}\frac{dP}{dR} = \frac{U_\varphi^2}{R}, \qquad (11.49)$$

where

$$A \equiv \frac{\Omega_2 R_2^2 - \Omega_1 R_1^2}{R_2^2 - R_1^2}, \quad B \equiv \frac{(\Omega_1 - \Omega_2)R_1^2 R_2^2}{R_2^2 - R_1^2}.$$

Here, Ω_1 and Ω_2 are the angular speeds of the inner and outer cylinders, respectively, and R_1 and R_2 are their radii (Figure 11.16).

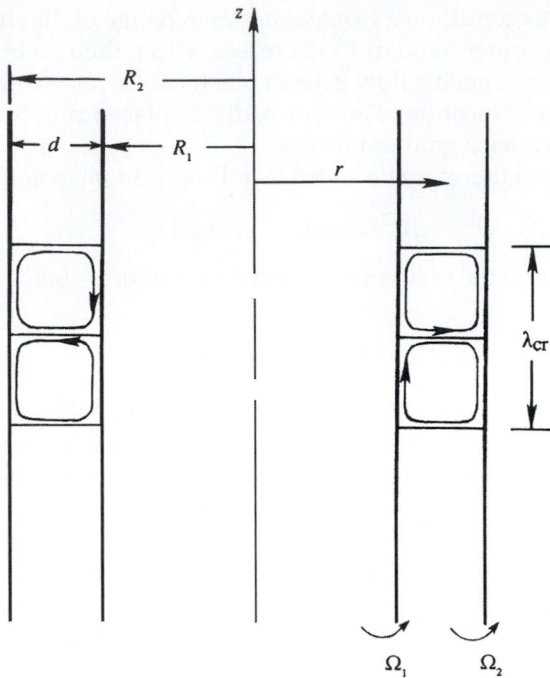

FIGURE 11.16 Geometry of the flow and the instability in rotating Couette flow. The fluid resides between rotating cylinders with radii R_1 and R_2. As for Bénard convection, the resulting instability forms as counter-rotating rolls with a wavelength that is approximately twice the gap between the cylinders.

Substituting (11.48) into (11.47), neglecting nonlinear terms, and subtracting the background state (11.49), we obtain the perturbation equations:

$$\frac{\partial u_R}{\partial t} - \frac{2U_\varphi u_\varphi}{R} = -\frac{1}{\rho}\frac{\partial p}{\partial R} + \nu\left(\nabla^2 u_R - \frac{u_R}{R^2}\right), \quad \frac{\partial u_\varphi}{\partial t} + \left(\frac{dU_\varphi}{dR} + \frac{U_\varphi}{R}\right)u_R = \nu\left(\nabla^2 u_\varphi - \frac{u_\varphi}{R^2}\right),$$

$$\frac{\partial u_z}{\partial t} = -\frac{1}{\rho}\frac{\partial p}{\partial z} + \nu\nabla^2 u_z, \text{ and } \frac{\partial}{\partial R}(R u_R) + \frac{\partial u_z}{\partial z} = 0.$$

(11.50)

As the coefficients in these equations depend only on R, the equations admit solutions that depend on z and t exponentially. We therefore consider normal mode solutions of the form:

$$(u_R, u_\varphi, u_z, p) = \left(\widehat{u}_R(R), \widehat{u}_\varphi(R), \widehat{u}_z(R), \widehat{p}(R)\right)\exp\{ikz + \sigma t\}.$$

The requirement that the solutions remain bounded as $z \to \pm\infty$ implies that the axial wave number k must be real. After substituting the normal modes into (11.50) and eliminating $\widehat{u}_z$ and $\widehat{p}$, we get a coupled system of equations in $\widehat{u}_R$ and $\widehat{u}_\varphi$. Under the *narrow-gap approximation*, for which $d = R_2 - R_1$ is much smaller than $(R_1 + R_2)/2$, these equations finally become (see Chandrasekhar, 1961, for details):

$$(d^2/dR^2 - k^2 - \sigma)(d^2/dR^2 - k^2)\hat{u}_R = (1 + \alpha x)\hat{u}_\varphi, \text{ and}$$
$$(d^2/dR^2 - k^2 - \sigma)\hat{u}_\varphi = -Tak^2\hat{u}_R,$$

(11.51)

where

$$\alpha \equiv (\Omega_2/\Omega_1) - 1, \; x \equiv (R - R_1)/d, \; d \equiv R_2 - R_1,$$

and Ta is the Taylor number

$$Ta \equiv 4\left(\frac{\Omega_1 R_1^2 - \Omega_2 R_2^2}{R_2^2 - R_1^2}\right)\frac{\Omega_1 d^4}{\nu^2}.$$

(11.52)

It is the ratio of the centrifugal force to viscous force, and equals $2(\Omega_1 R_1 d/\nu)^2(d/R_1)$ when only the inner cylinder is rotating and the gap is narrow.

The boundary conditions are

$$\hat{u}_R = d\hat{u}_R/dR = \hat{u}_\varphi = 0 \quad \text{at } x = 0, 1.$$

(11.53)

The eigenvalues k at the marginal state are found by setting the real part of σ to zero. On the basis of experimental evidence, Taylor assumed that the marginal states are given by $\sigma = 0$. This was later proven to be true for cylinders rotating in the same directions, but a general demonstration for all conditions is still lacking.

A solution of the eigenvalue problem (11.51), subject to (11.53), was obtained by Taylor. Figure 11.17 shows the results of his calculations and his own experimental

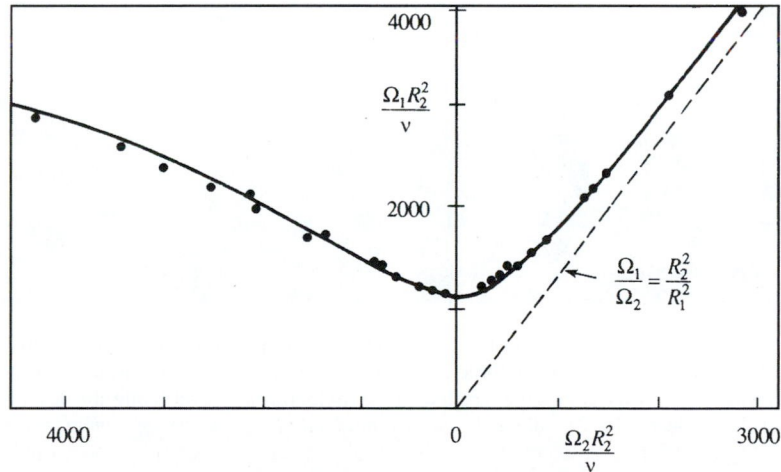

FIGURE 11.17 Taylor's observation and narrow-gap calculation of marginal stability in rotating Couette flow of water. The ratio of radii is $R_2/R_1 = 1.14$. The region above the curve is unstable. The dashed line represents Rayleigh's inviscid criterion, with the region to the left of the line representing instability. The experimental and theoretical results agree well and suggest that viscosity acts to stabilize the flow.

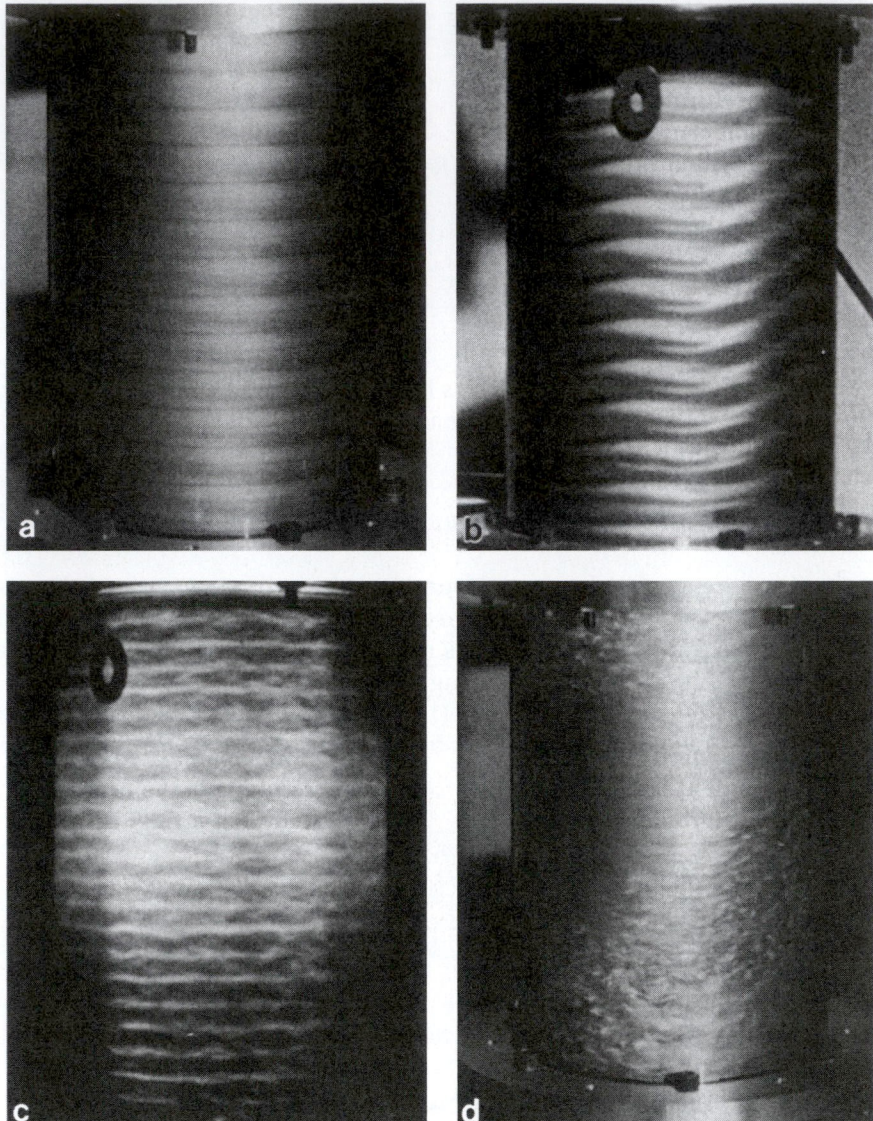

FIGURE 11.18   Instability of rotating Couette flow. Panels (a), (b), (c), and (d) correspond to increasing Taylor number. At first the instability appears as periodic rolls that do not vary with the azimuthal angle. Next, the rolls develop azimuthal waves with wavelengths that depend on the Taylor number. Eventually, the flow becomes turbulent. *D. Coles*, Journal of Fluid Mechanics, 21, *385–425, 1965; reprinted with the permission of Cambridge University Press.*

verification of the analysis. The vertical axis represents the angular velocity of the inner cylinder (taken positive), and the horizontal axis represents the angular velocity of the outer cylinder. Cylinders rotating in opposite directions are represented by a negative Ω_2. Taylor's solution of the marginal state is indicated, with the region above the curve corresponding to instability. Rayleigh's inviscid criterion is also indicated by the straight dashed line. Taylor's viscous solution indicates that the flow remains stable until a critical Taylor number of

$$\text{Ta}_{cr} = \frac{1708}{(1/2)(1 + \Omega_2/\Omega_1)} \tag{11.54}$$

is attained. The nondimensional axial wave number at the onset of instability is found to be $k_{cr} = 3.12$, which implies that the wavelength at onset is $\lambda_{cr} = 2\pi d/k_{cr} \approx 2d$. The height of one cell is therefore nearly equal to d, so that the cross-section of a cell is nearly a square. In the limit $\Omega_2/\Omega_1 \to 1$, the critical Taylor number is identical to the critical Rayleigh number for thermal convection discussed in the preceding section, for which the solution was given by Jeffreys five years later. The agreement is expected, because in this limit $\alpha = 0$, and the eigenvalue problem (11.51) reduces to that of the Bénard problem (11.39). For cylinders rotating in opposite directions the Rayleigh criterion predicts instability, but the viscous solution can be stable.

Taylor's analysis of the problem was enormously satisfying, both experimentally and theoretically. He measured the wavelength at the onset of instability by injecting dye and obtained an almost exact agreement with his calculations. The observed onset of instability in the $\Omega_1\,\Omega_2$ -plane (Figure 11.17) was also in remarkable agreement. This has prompted remarks such as, "the closeness of the agreement between his theoretical and experimental results was without precedent in the history of fluid mechanics" (Drazin & Reid, 1981, p. 105). It even led some people to suggest happily that the agreement can be regarded as a verification of the underlying Navier-Stokes equations, which make a host of assumptions including a linearity between stress and strain rate.

The instability appears in the form of counter-rotating toroidal (or doughnut-shaped) vortices (Figure 11.18a) called *Taylor vortices*. The streamlines are in the form of helixes, with axes wrapping around the annulus, somewhat like the stripes on a barber's pole. These vortices themselves become unstable at higher values of Ta, when they give rise to wavy vortices for which $\partial/\partial\phi \neq 0$ (Figure 11.18b). In effect, the flow has now attained the next higher mode. The number of waves around the annulus depends on the Taylor number, and the wave pattern travels around the annulus. More complicated patterns of vortices result at a higher rate of rotation, finally resulting in the occasional appearance of turbulent patches (Figure 11.18d), and then a fully turbulent flow.

Phenomena analogous to the Taylor vortices are called *secondary flows* because they are superposed on a primary flow (such as the Couette flow in the present case). There are two other situations where a combination of curved streamlines (which give rise to centrifugal forces) and viscosity result in instability and steady secondary flows in the form of vortices. One is the flow through a curved channel, driven by a pressure gradient. The other is the appearance of *Görtler vortices* in a boundary-layer flow along a concave wall (Figure 11.19). The possibility of secondary flows signifies that the *solutions of the Navier-Stokes equations are nonunique* in the sense that more than one steady solution is allowed under the same boundary conditions. We can derive the form of the primary flow only if we exclude the secondary flow by appropriate assumptions. For example, we can derive the expression (11.50) for Couette flow by *assuming $U_r = 0$ and $U_z = 0$* and thereby rule out the secondary flow.

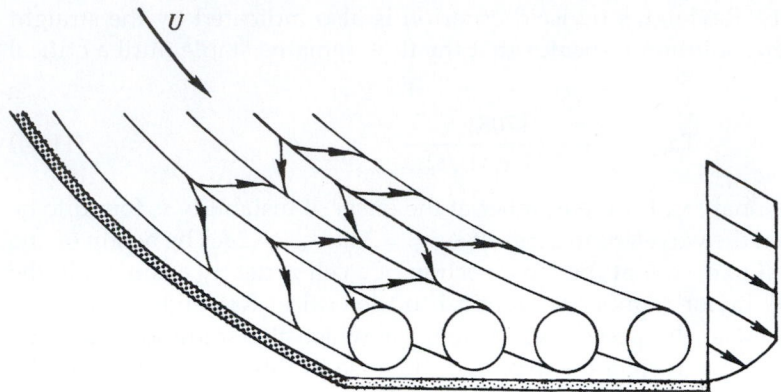

FIGURE 11.19   Görtler vortices in a boundary layer along a concave wall. The instability phenomenon here is essentially the same as that in Taylor-Couette flow, the only difference being the lack of the inner curved surface.

11.7. INSTABILITY OF CONTINUOUSLY STRATIFIED PARALLEL FLOWS

An instability of great geophysical importance is that of an inviscid stratified fluid in horizontal parallel flow. If the density and velocity vary discontinuously across an interface, the analysis in Section 11.3 shows that the flow is unconditionally unstable. Although only the discontinuous case was studied by Kelvin and Helmholtz, the more general case of continuous distribution is also commonly called the *Kelvin-Helmholtz instability*.

The problem has a long history. In 1915, Taylor, on the basis of his calculations with assumed distributions of velocity and density, *conjectured* that a gradient Richardson number (to be defined shortly) must be less than ¼ for instability. Other values of the critical Richardson number (ranging from 2 to ¼) were suggested by Prandtl, Goldstein, Richardson, Synge, and Chandrasekhar. Finally, Miles (1961) was able to prove Taylor's conjecture, and Howard (1961) immediately and elegantly generalized Miles' proof. A short record of the history is given in Miles (1986). In this section we shall prove the Richardson number criterion in the manner given by Howard.

Consider a horizontal parallel flow $U(z)$ directed along the x-axis. The z-axis is taken vertically upward. The basic flow is in equilibrium with the undisturbed density field $\bar{\rho}(z)$ and the basic pressure field $P(z)$. We shall only consider two-dimensional disturbances on this basic state, assuming that they are more unstable than three-dimensional disturbances; this is called *Squires' theorem* and is demonstrated in Section 11.8 in another context. The disturbed state has velocity, pressure, and density fields of:

$$\tilde{\mathbf{u}} = U\mathbf{e}_x + \mathbf{u} = (U + u, 0, w), \quad \tilde{p} = P + p, \text{ and } \tilde{\rho} = \bar{\rho} + \rho,$$

where, as before, the tilde indicates a total flow variable. The continuity equation reduces to $\partial u/\partial x + \partial w/\partial z = 0$, and the disturbed velocity field is assumed to satisfy the inviscid Boussinesq momentum equation:

$$\frac{\partial \tilde{\mathbf{u}}}{\partial t} + (\tilde{\mathbf{u}} \cdot \nabla)\tilde{\mathbf{u}} = -\frac{1}{\rho_0}\nabla\tilde{p} - g\frac{(\bar{\rho} + \rho)}{\rho_0}\mathbf{e}_z,$$

where the density variations are neglected except in the vertical component. Here, ρ_0 is a reference density. The basic flow satisfies

$$0 = -\frac{1}{\rho_0}\frac{\partial P}{\partial z} - g\frac{\bar{\rho}}{\rho_0}.$$

Subtracting the last two equations and dropping nonlinear terms, we obtain the perturbation equation of motion:

$$\frac{\partial \mathbf{u}}{\partial t} + (\mathbf{u}\cdot\nabla)(U\mathbf{e}_x) + U(\mathbf{e}_x\cdot\nabla)\mathbf{u} = -\frac{1}{\rho_0}\nabla p - g\frac{\rho}{\rho_0}\mathbf{e}_z.$$

The horizontal (x) and vertical (z) components of the preceding equation are

$$\frac{\partial u}{\partial t} + w\frac{\partial U}{\partial z} + U\frac{\partial u}{\partial x} = -\frac{1}{\rho_0}\frac{\partial p}{\partial x} \quad\text{and}\quad \frac{\partial w}{\partial t} + U\frac{\partial w}{\partial x} = -\frac{1}{\rho_0}\frac{\partial p}{\partial x} - g\frac{\rho}{\rho_0}. \tag{11.55}$$

In the absence of diffusion the density of fluid particles does not change: $D\tilde{\rho}/Dt = 0$, or

$$\frac{\partial}{\partial t}(\bar{\rho} + \rho) + (U + u)\frac{\partial}{\partial x}(\bar{\rho} + \rho) + w\frac{\partial}{\partial z}(\bar{\rho} + \rho) = 0.$$

Keeping only the linear terms, and using the fact that $\bar{\rho}$ is a function of z only, we obtain

$$\frac{\partial \rho}{\partial t} + U\frac{\partial \rho}{\partial x} + w\frac{\partial \bar{\rho}}{\partial z} = 0,$$

which can be written as

$$\frac{\partial \rho}{\partial t} + U\frac{\partial \rho}{\partial x} - \frac{\rho_0 N^2 w}{g} = 0, \tag{11.56}$$

where N is the buoyancy frequency in an incompressible flow:

$$N^2 \equiv -\frac{g}{\rho_0}\frac{d\bar{\rho}}{dz}. \tag{7.128}$$

The last term in equation (11.56) represents the density change at a point due to the vertical advection of the basic density field across the point.

The continuity equation can be satisfied with a stream function $u = \partial\psi/\partial z$ and $w = -\partial\psi/\partial x$. Equations (11.55) and (11.56) then become

$$\frac{\partial^2\psi}{\partial t\partial z} - \frac{\partial\psi}{\partial x}\frac{dU}{dz} + U\frac{\partial^2\psi}{\partial x\partial z} = -\frac{1}{\rho_0}\frac{\partial p}{\partial x}, \quad -\frac{\partial^2\psi}{\partial t\partial x} - U\frac{\partial^2\psi}{\partial x^2} = -\frac{g\rho}{\rho_0} - \frac{1}{\rho_0}\frac{\partial p}{\partial z},$$

$$\frac{\partial \rho}{\partial t} + U\frac{\partial \rho}{\partial x} + \frac{\rho_0 N^2}{g}\frac{\partial\psi}{\partial x} = 0. \tag{11.57}$$

Since the coefficients of derivatives in (11.57) are independent of x and t, exponential variations in these variables are allowed. Consequently, we assume traveling-wave normal mode solutions of the form:

$$[\rho, p, \psi] = [\hat{\rho}(z), \hat{p}(z), \hat{\psi}(z)]\exp\{ik(x - ct)\},$$

where quantities denoted by (^) are complex amplitudes. Because the flow is unbounded in x, the wave number k must be real. The eigenvalue $c = c_r + ic_i$ can be complex, and the solution is unstable if there exists a $c_i > 0$, similar to the development in Section 11.3. Substituting the normal modes, (11.57) becomes:

$$(U - c)\frac{\partial \widehat{\psi}}{\partial z} - \frac{\partial U}{\partial z}\widehat{\psi} = -\frac{1}{\rho_0}\widehat{p}, \quad k^2(U - c)\widehat{\psi} = -g\frac{\widehat{\rho}}{\rho_0} - \frac{1}{\rho_0}\frac{\partial \widehat{p}}{\partial z}, (U - c)\widehat{\rho} + \frac{\rho_0 N^2}{g}\widehat{\psi} = 0.$$

$$(11.58, \ 11.59, \ 11.60)$$

We seek a single equation for $\widehat{\psi}$. The pressure can be eliminated by taking the z-derivative of (11.58) and subtracting (11.59). The density can be eliminated via substitution from (11.60) to produce:

$$(U - c)\left(\frac{d^2}{dz^2} - k^2\right)\widehat{\psi} - \frac{\partial^2 U}{\partial z^2}\widehat{\psi} + \frac{N^2}{U - c}\widehat{\psi} = 0. \tag{11.61}$$

This is the *Taylor-Goldstein equation*, which governs the behavior of perturbations in a stratified parallel flow. Note that the complex conjugate of (11.61) is also a valid equation because we can take the imaginary part of the equation, change the sign, and add it to the real part of the equation. Now because the Taylor-Goldstein equation does not involve any i, a complex conjugate of the equation shows that if $\widehat{\psi}$ is an eigenfunction with eigenvalue c for some k, then $\widehat{\psi}^*$ is a possible eigenfunction with eigenvalue c^* for the same k. Therefore, to each eigenvalue with a positive c_i there is a corresponding eigenvalue with a negative c_i. In other words, *to each growing mode there is a corresponding decaying mode*. A nonzero c_i therefore ensures instability.

The boundary conditions are that $w = 0$ on rigid boundaries, presuming these are located at $z = 0$ and d. This requires $\partial \psi / \partial x = ik\widehat{\psi}\exp\{ik(x - ct)\} = 0$ at the walls, which is possible only if

$$\widehat{\psi}(0) = \widehat{\psi}(d) = 0. \tag{11.62}$$

A necessary condition involving the Richardson number for linear instability of inviscid stratified parallel flows can be derived by defining a new field variable ϕ (not the velocity potential) by

$$\phi \equiv \frac{\widehat{\psi}}{(U - c)^{1/2}} \quad \text{or} \quad \widehat{\psi} = (U - c)^{1/2}\phi. \tag{11.63}$$

Then we obtain the derivatives:

$$\frac{\partial \widehat{\psi}}{\partial z} = (U - c)^{1/2}\frac{\partial \phi}{\partial z} + \frac{\phi}{2(U - c)^{1/2}}\frac{dU}{dz},$$

$$\frac{\partial^2 \widehat{\psi}}{\partial z^2} = (U - c)^{1/2}\frac{\partial^2 \phi}{\partial z^2} + \frac{1}{(U - c)^{1/2}}\left(\frac{d\phi}{dz}\frac{dU}{dz} + \frac{1}{2}\phi\frac{d^2 U}{dz^2}\right) - \frac{\phi}{4(U - c)^{3/2}}\left(\frac{dU}{dz}\right)^2.$$

The Taylor-Goldstein equation then becomes, after some rearrangement:

$$\frac{d}{dz}\left[(U-c)\frac{d\phi}{dz}\right] - \left\{k^2(U-c)+\frac{1}{2}\frac{d^2U}{dz^2}+\frac{(1/4)(dU/dz)^2-N^2}{U-c}\right\}\phi = 0. \tag{11.64}$$

Now multiply equation (11.64) by ϕ^* (the complex conjugate of ϕ), integrate from $z=0$ to $z=d$, and use the boundary conditions $\phi(0)=\phi(d)=0$. The first term gives:

$$\int_0^d \frac{d}{dz}\left\{(U-c)\frac{d\phi}{dz}\right\}\phi^*dz = \int_0^d \left[\frac{d}{dz}\left\{(U-c)\frac{d\phi}{dz}\phi^*\right\}-(U-c)\frac{d\phi}{dz}\frac{d\phi^*}{dz}\right]dz = -\int_0^d (U-c)\left|\frac{d\phi}{dz}\right|^2dz.$$

Integrals of the other terms in (11.64) are also simple to manipulate. We finally obtain:

$$\int_0^d \left\{\frac{N^2-(1/4)(dU/dz)^2}{U-c}\right\}|\phi|^2dz = \int_0^d (U-c)\left\{\left|\frac{d\phi}{dz}\right|^2+k^2|\phi|^2\right\}dz + \int_0^d \frac{1}{2}\frac{d^2U}{dz^2}|\phi|^2dz. \tag{11.65}$$

The last term in the preceding equation is real. The imaginary part of the first term can be found by noting that:

$$\frac{1}{U-c} = \frac{U-c^*}{|U-c|^2} = \frac{U-c_r+ic_i}{|U-c|^2}.$$

Taking the imaginary part of (11.65) leads to:

$$c_i\int_0^d \left\{\frac{N^2-(1/4)(dU/dz)^2}{|U-c|^2}\right\}|\phi|^2dz = -c_i\int_0^d \left\{\left|\frac{d\phi}{dz}\right|^2+k^2|\phi|^2\right\}dz.$$

The integral on the right side is positive. If the flow is such that $N^2 > (1/4)(dU/dz)^2$ everywhere, then the preceding equation states that c_i times a positive quantity equals c_i times a negative quantity; this is impossible and requires that $c_i=0$ for such a case. Thus, defining the *gradient Richardson number*:

$$Ri(z) \equiv N^2/(dU/dz)^2, \tag{11.66}$$

we can say that linear stability is guaranteed if the inequality

$$Ri > \frac{1}{4}\ (stable) \tag{11.67}$$

is satisfied everywhere in the flow.

Note that the criterion does not state that the flow is necessarily unstable if $Ri < \frac{1}{4}$ somewhere, or even everywhere, in the flow. Thus $Ri < \frac{1}{4}$ is a *necessary* but not sufficient condition for instability. For example, in a jet-like velocity profile $u \propto sech^2z$ and an exponential density profile, the flow does not become unstable until the Richardson number falls below 0.214. A critical Richardson number lower than $\frac{1}{4}$ is also found in the presence of boundaries, which stabilize the flow. In fact, there is no unique critical Richardson number that applies to all distributions of $U(z)$ and $N(z)$. However, several calculations show that in many shear layers (having linear, tanh, or error function profiles for velocity and density), the flow does become unstable to disturbances of certain wavelengths if the minimum value of Ri in the flow (which is generally at the center of the shear layer where $|dU/dz|$ is greatest) is less than ¼. The *most unstable* wave, defined as the first to become unstable as Ri is reduced below ¼, is found to have a wavelength $\lambda \approx 7h$, where h is the thickness of the shear layer. Laboratory (Scotti & Corcos, 1972) as well as geophysical observations (Eriksen, 1978) show that the requirement

$$\text{Ri}_{\min} < \frac{1}{4}$$

is a useful guide for the prediction of instability of a stratified shear layer.

Similar to the previous analysis, another useful result concerning the behavior of the complex phase speed c in an inviscid parallel shear flow can be determined by considering an alternative version of (11.63):

$$F \equiv \frac{\widehat{\psi}}{U - c}, \tag{11.68}$$

which leads to derivatives:

$$\frac{\partial \widehat{\psi}}{\partial z} = (U - c)\frac{\partial F}{\partial z} + \frac{dU}{dz}F,$$

$$\frac{\partial^2 \widehat{\psi}}{\partial z^2} = (U - c)\frac{\partial^2 F}{\partial z^2} + 2\frac{dU}{dz}\frac{dF}{dz} + \frac{d^2 U}{dz^2}F.$$

When (11.68) is substituted into the Taylor-Goldstein equation (11.61), the result is:

$$(U - c)\left[(U - c)\frac{d^2 F}{dz^2} + 2\frac{dU}{dz}\frac{dF}{dz} - k^2(U - c)F\right] + N^2 F = 0,$$

and the terms involving $d^2 U/dz^2$ have canceled out. This can be rearranged in the form

$$\frac{d}{dz}\left[(U - c)^2\frac{dF}{dz}\right] - k^2(U - c)F + N^2 F = 0.$$

Multiplying by F^*, integrating (by parts when necessary) over the depth of flow, and using the boundary conditions, we obtain

$$-\int (U - c)^2\left|\frac{dF}{dz}\right|^2 dz - k^2 \int (U - c)^2|F|^2 dz + \int N^2|F|^2 dz = 0,$$

which can be written as

$$\int (U - c)^2 Q dz = \int N^2|F|^2 dz \quad \text{where } Q \equiv |dF/dz|^2 + k^2|F|^2$$

is positive. Equating real and imaginary parts, we obtain

$$\int \left[(U - c_r)^2 - c_i^2\right]Q dz = \int N^2|F|^2 dz \quad \text{and} \quad c_i \int (U - c_r)Q dz = 0. \tag{11.69, 11.70}$$

For instability $c_i \neq 0$, for which (11.70) shows that $(U - c_r)$ must change sign somewhere in the flow:

$$U_{min} < c_r < U_{max},\tag{11.71}$$

which states that c_r lies in the range of U. Recall that we have assumed solutions of the form

$$\exp\{ik(x - ct)\} = \exp\{ik(x - c_r t\}\exp\{+kc_i t\},$$

which means that c_r is the phase velocity in the positive x direction, and kc_i is the growth rate. Equation (11.71) shows that c_r is positive if U is everywhere positive, and is negative if U is everywhere negative. In these cases we can say that unstable waves propagate in the direction of the background flow.

Limits on the maximum growth rate can also be predicted. Equation (11.69) gives

$$\int \left[U^2 - 2Uc_r + c_r^2 - c_i^2 \right] Q dz > 0,$$

which, on using (11.70), becomes

$$\int \left[U^2 - c_r^2 - c_i^2 \right] Q dz > 0.\tag{11.72}$$

Now because $(U_{min} - U) < 0$ and $U_{max} - U > 0$, it is always true that

$$\int [U_{min} - U][U_{max} - U] dz \leq 0,$$

which can be recast as

$$\int \left[U_{max} U_{min} + U^2 - U(U_{max} + U_{min}) \right] Q dz \leq 0.$$

Using (11.72), this gives

$$\int \left[U_{max} U_{min} + c_r^2 + c_i^2 - U(U_{max} + U_{min}) \right] Q dz \leq 0,$$

and after using (11.70), this becomes

$$\int \left[U_{max} U_{min} + c_r^2 + c_i^2 - c_r(U_{max} + U_{min}) \right] Q dz \leq 0.$$

Because the quantity within [,]-brackets is independent of z, and $\int Q dz > 0$, we must have $[\ ] \leq 0$. With some rearrangement, this condition can be written as

$$\left[c_r - \frac{1}{2}(U_{max} + U_{min}) \right]^2 + c_i^2 \leq \left[\frac{1}{2}(U_{max} - U_{min}) \right]^2.$$

This shows that *the complex wave velocity, c, of any unstable mode of a disturbance in parallel flows of an inviscid fluid must lie inside the semicircle in the upper half of the c-plane, which has the range of U as the diameter* (Figure 11.20). This result was first derived by Howard

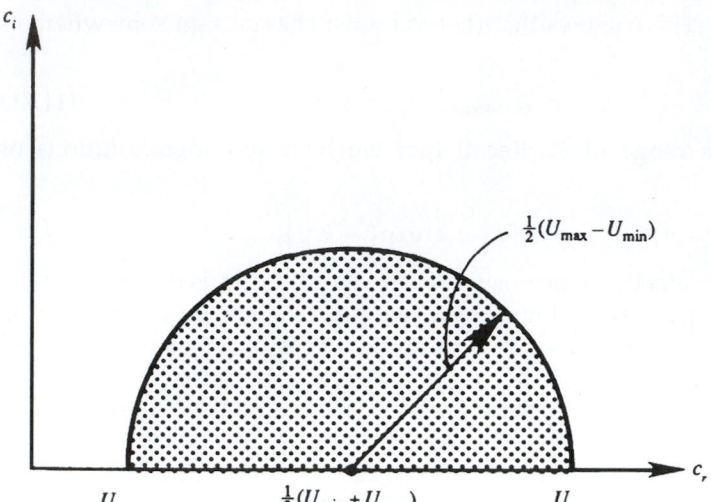

(1961) and is valid for flows with and without stratification. It is called the *Howard semicircle theorem* and states that the maximum growth rate is limited by:

$$kc_i < (k/2)(U_{max} - U_{min}).$$

The theorem is very useful in searching for eigenvalues $c(k)$ in numerical solution of instability problems.

11.8. SQUIRE'S THEOREM AND THE ORR-SOMMERFELD EQUATION

In our studies of the Bénard and Taylor problems, we encountered two flows in which viscosity has a stabilizing effect. Curiously, viscous effects can also be *destabilizing*, as indicated by several calculations of wall-bounded parallel flows. In this section we shall derive the equation governing the stability of parallel flows of a homogeneous viscous fluid. Let the primary flow be directed along the x direction and vary in the y direction so that $\mathbf{U} = (U(y), 0, 0)$. We decompose the total flow as the sum of the basic flow plus the perturbation:

$$\tilde{\mathbf{u}} = (U + u, v, w), \text{ and } \tilde{p} = P + p.$$

Both the background and the perturbed flows satisfy the Navier-Stokes equations. The perturbed flow satisfies the x-momentum equation:

$$\frac{\partial u}{\partial t} + (U + u)\frac{\partial}{\partial x}(U + u) + v\frac{\partial}{\partial y}(U + u) = -\frac{\partial}{\partial x}(P + p) + \frac{1}{Re}\nabla^2(U + u), \qquad (11.73)$$

where the variables have been made dimensionless with a characteristic length scale L (say, the width of flow), and a characteristic velocity U_0 (say, the maximum velocity of the basic flow); time is scaled by L/U_0 and the pressure is scaled by ρU_0^2. The Reynolds number is $Re = U_0 L/\nu$.

The background flow satisfies:

$$0 = -\frac{\partial P}{\partial x} + \frac{1}{Re}\nabla^2 U.$$

Subtracting this from (11.73) and neglecting terms nonlinear in the perturbations, we obtain the x-momentum equation for the perturbations:

$$\frac{\partial u}{\partial t} + U\frac{\partial u}{\partial x} + v\frac{\partial U}{\partial y} = -\frac{\partial p}{\partial x} + \frac{1}{Re}\nabla^2 u. \tag{11.74}$$

Similarly the y-momentum, z-momentum, and continuity equations for the perturbations are:

$$\frac{\partial v}{\partial t} + U\frac{\partial v}{\partial x} = -\frac{\partial p}{\partial y} + \frac{1}{Re}\nabla^2 v, \quad \frac{\partial w}{\partial t} + U\frac{\partial w}{\partial x} = -\frac{\partial p}{\partial z} + \frac{1}{Re}\nabla^2 w, \text{ and } \frac{\partial u}{\partial x} + \frac{\partial v}{\partial y} + \frac{\partial w}{\partial z} = 0. \tag{11.75}$$

The coefficients in (11.74) and (11.75) depend only on y, so that the equations admit solutions exponential in x, z, and t. Accordingly, we assume normal modes of the form:

$$[\mathbf{u}, p] = \left[\hat{\mathbf{u}}(y), \hat{p}(y)\right]\exp\{i(kx + mz - kct)\}. \tag{11.76}$$

As the flow is unbounded in x and z, the wave number components k and m must be real. However, the wave speed $c = c_r + ic_i$ may be complex. Without loss of generality, we can consider only positive values for k and m; the sense of propagation is then left open by keeping the sign of c_r unspecified. The normal modes represent waves that travel obliquely to the basic flow with a wave number magnitude $\sqrt{k^2 + m^2}$ and have an amplitude that varies as $\exp(kc_i t)$. Solutions are therefore stable if $c_i < 0$ and unstable if $c_i > 0$.

Substitution of (11.76) into the perturbation equations (11.74) and (11.75) produces:

$$ik(U - c)\hat{u} + \hat{v}(dU/dy) = -ikp + (1/Re)\left[d^2\hat{u}/dy^2 - (k^2 + m^2)\hat{u}\right],$$

$$ik(U - c)\hat{v} = -d\hat{p}/dy + (1/Re)\left[d^2\hat{v}/dy^2 - (k^2 + m^2)\hat{v}\right],$$

$$ik(U - c)\hat{w} = -im\hat{p} + (1/Re)\left[d^2\hat{w}/dy^2 - (k^2 + m^2)\hat{w}\right], \text{ and}$$

$$ik\hat{u} + d\hat{v}/dy + im\hat{w} = 0. \tag{11.77}$$

These are the normal mode equations for three-dimensional disturbances.

Before proceeding further, we shall first prove Squire's Theorem (1933) which states that *to each unstable three-dimensional disturbance there corresponds a more unstable two-dimensional one*. To prove this theorem, consider the *Squire transformation*:

$$\bar{k} = \sqrt{k^2 + m^2}, \bar{c} = c, \bar{k}\bar{u} = k\hat{u} + m\hat{w}, \bar{v} = \hat{v}, \bar{p}/\bar{k} = \hat{p}/k, \text{ and } \bar{k}\overline{Re} = k Re. \tag{11.78}$$

After substituting (11.78) into (11.77), and adding the first and third equations, the result is

$$\bar{i}k(U-c)\bar{u} + \bar{v}(dU/dy) = -i\bar{k}\bar{p} + (1/\overline{\text{Re}})\left[d^2\bar{u}/dy^2 - \bar{k}^2\bar{u}\right],$$

$$i\bar{k}(U-c)\bar{v} = -d\hat{p}/dy + (1/\overline{\text{Re}})\left[d^2\bar{v}/dy^2 - \bar{k}^2\bar{v}\right], \text{ and}$$

$$i\bar{k}\bar{u} + d\bar{v}/dy = 0.$$

These equations are exactly the same as (11.77), but with $m = \hat{w} = 0$. Thus, to each three-dimensional problem corresponds an equivalent two-dimensional one. Moreover, Squire's transformation (11.78) shows that the equivalent two-dimensional problem is associated with a *lower* Reynolds number since $\bar{k} > k$. It follows that the critical Reynolds number at which the instability starts is lower for two-dimensional disturbances. Therefore, we only need to consider a two-dimensional disturbance if we want to determine the minimum Reynolds number for the onset of instability.

The three-dimensional disturbance (11.76) is a wave propagating obliquely to the basic flow. If we orient the coordinate system with the new x-axis in this direction, the equations of motion are such that only the component of the basic flow in this direction affects the disturbance. Thus, the effective Reynolds number is reduced.

Interestingly, Squire's theorem also holds for several other problems that do not involve the Reynolds number. Equation (11.78) shows that the growth rate for a two-dimensional disturbance is $\exp(\bar{k}\bar{c}_i t)$, whereas (11.76) shows that the growth rate of a three-dimensional disturbance is $\exp(kc_i t)$. The two-dimensional growth rate is therefore larger because Squire's transformation requires $\bar{k} > k$ and $\bar{c} = c$. We can therefore say that the two-dimensional disturbances are more unstable.

Because of Squire's theorem, we only need consider the equation set (11.77) with $m = \hat{w} = 0$. The two-dimensionality allows the use of a stream function $\psi(x,y,t)$ for the perturbation field via the usual relationships: $u = \partial\psi/\partial y$ and $v = -\partial\psi/\partial x$. Again, we assume normal modes of the form:

$$[u, v, \psi] = [\hat{u}(y), \hat{v}(y), \phi(y)]\exp\{ik(x - ct)\}.$$

(To be consistent, we should denote the complex amplitude of ψ by $\hat{\psi}$; we are using ϕ [not the potential] instead to follow the standard notation for this variable in the literature.) Then we must have $\hat{u} = \partial\phi/\partial y$ and $\hat{v} = -ik\phi$, and a single equation in terms of ϕ can now be found by eliminating the pressure from (11.77). This effort yields a fourth-order ordinary differential equation:

$$(U - c)\left(\frac{d^2\phi}{dy^2} - k^2\phi\right) - \frac{d^2U}{dy^2}\phi = \frac{1}{ik\text{Re}}\left(\frac{d^4\phi}{dy^4} - 2k^2\frac{d^2\phi}{dy^2} + k^4\phi\right). \tag{11.79}$$

The boundary conditions are the no-slip conditions which require

$$\phi = d\phi/dy = 0 \quad \text{at } y = y_1 \text{ and } y_2. \tag{11.80}$$

Equation (11.79) is the well-known *Orr-Sommerfeld equation*, which governs the stability of nearly parallel viscous flows such as those in a straight channel or in a boundary layer. It is essentially a vorticity equation because the pressure has been eliminated. Analytical solutions of the Orr-Sommerfeld equations are difficult to obtain, and only the results of some simple flows will be discussed in the later sections. However, we shall first present certain results obtained by ignoring the viscous terms on the right side of this equation.

11.9. INVISCID STABILITY OF PARALLEL FLOWS

Insight into the viscous stability of parallel flows can be obtained by first assuming that the disturbances obey inviscid dynamics. The governing equation can be found by letting Re $\rightarrow \infty$ in the Orr-Sommerfeld equation, giving

$$(U - c)\left(\frac{d^2\phi}{dy^2} - k^2\phi\right) - \frac{d^2U}{dy^2}\phi = 0, \tag{11.81}$$

which is called the *Rayleigh equation*. If the flow is bounded by walls at y_1 and y_2 where $v = 0$, then the boundary conditions are

$$\phi = 0 \quad \text{at } y = y_1 \text{ and } y_2. \tag{11.82}$$

The set (11.81) and (11.82) defines an eigenvalue problem, with $c(k)$ as the eigenvalue and ϕ as the eigenfunction. As these equations do not involve i, taking the complex conjugate shows that if ϕ is an eigenfunction with eigenvalue c for some k, then ϕ^* is also an eigenfunction with eigenvalue c^* for the same k. Therefore, to each eigenvalue with a positive c_i there is a corresponding eigenvalue with a negative c_i. In other words, *to each growing mode there is a corresponding decaying mode*. Stable solutions therefore can have only a real c. Note that this is true of inviscid flows only. The viscous term in the full Orr-Sommerfeld equation (11.79) involves an i, and the foregoing conclusion is no longer valid.

We shall now show that certain velocity distributions $U(y)$ are potentially unstable according to the inviscid Rayleigh equation (11.81). In this discussion it should be noted that we are only assuming that the *disturbances* obey inviscid dynamics; the background flow profile $U(y)$ may be that of a steady laminar viscous flow.

The first deduction that can be made from (11.81) is Rayleigh's inflection point criterion that states that *a necessary (but not sufficient) criterion for instability of an inviscid parallel flow is that the basic velocity profile $U(y)$ has a point of inflection*. To prove the theorem, rewrite the Rayleigh equation (11.81) in the form

$$\frac{d^2\phi}{dy^2} - k^2\phi - \frac{1}{U-c}\frac{d^2U}{dy^2}\phi = 0,$$

and consider the unstable mode for which $c_i > 0$, and therefore $U - c \neq 0$. Multiply this equation by ϕ^*, integrate from y_1 to y_2, by parts where necessary, and apply the boundary condition (11.82). The first term transforms as follows:

$$\int \phi^*(d^2\phi/dy^2)dy = [\phi^*(d\phi/dy)]_{y_1}^{y_2} - \int (d\phi^*/dy)(d\phi/dy)dy = -\int |d\phi/dy|^2 dy,$$

where the limits on the integrals have not been explicitly written. The Rayleigh equation then gives

$$\int (|d\phi/dy|^2 + k^2|\phi|^2)dy + \int \frac{1}{U-c} \frac{d^2U}{dy^2}|\phi|^2 dy = 0.$$ (11.83)

The first term is real. The second term in (11.83) is complex, and its imaginary part can be found by multiplying the numerator and denominator by $(U - c^*)$. Thus, the imaginary part of (11.83) implies:

$$c_i \int \frac{1}{|U-c|^2} \frac{d^2U}{dy^2}|\phi|^2 dy = 0.$$ (11.84)

For the unstable case, for which $c_i \neq 0$, (11.84) can be satisfied only if d^2U/dy^2 changes sign at least once in the *open* interval $y_1 < y < y_2$. In other words, for instability the background velocity distribution must have at least one point of inflection (where $d^2U/dy^2 = 0$) within the flow. Clearly, the existence of a point of inflection does not guarantee a nonzero c_i. The inflection point is therefore a necessary but not sufficient condition for inviscid instability.

Some seventy years after Rayleigh's discovery, the Swedish meteorologist Fjortoft in 1950 discovered a stronger necessary condition for the instability of inviscid parallel flows. He showed that *a necessary condition for instability of inviscid parallel flows is that* $(U - U_1)(d^2U/dy^2) < 0$ *somewhere in the flow*, where U_1 is the value of U at the point of inflection. To prove the theorem, take the real part of (11.83):

$$\int \frac{U-c_r}{|U-c|^2} \frac{d^2U}{dy^2}|\phi|^2 dy = -\int (|d\phi/dy|^2 + k^2|\phi|^2)dy < 0.$$ (11.85)

Suppose that the flow is unstable, so that $c_i \neq 0$, and a point of inflection does exist according to the Rayleigh criterion. Then it follows from (11.84) that

$$(c_r - U_1) \int \frac{1}{|U-c|^2} \frac{d^2U}{dy^2}|\phi|^2 dy = 0.$$ (11.86)

Adding equations (11.85) and (11.86), we obtain

$$\int \frac{U-U_1}{|U-c|^2} \frac{d^2U}{dy^2}|\phi|^2 dy < 0,$$

so that $(U - U_1)(d^2U/dy^2)$ must be negative somewhere in the flow.

Some common velocity profiles are shown in Figure 11.21. Only the two flows shown in the bottom row can possibly be unstable, for only they satisfy Fjortoft's theorem. Flows (a), (b), and (c) do not have an inflection point: flow (d) does satisfy Rayleigh's condition but not Fjortoft's because $(U - U_1)(d^2U/dy^2)$ is positive. Note that an alternate way of stating Fjortoft's theorem is that *the magnitude of vorticity of the basic flow must have a maximum within the region of flow*, not at the boundary. In flow (d), the maximum magnitude of vorticity occurs at the walls.

The criteria of Rayleigh and Fjortoft essentially point to the importance of having a point of inflection in the velocity profile. They show that flows in jets, wakes, shear layers, and boundary layers with adverse pressure gradients, all of which have a point of inflection and satisfy Fjortoft's theorem, are potentially unstable. On the other hand,

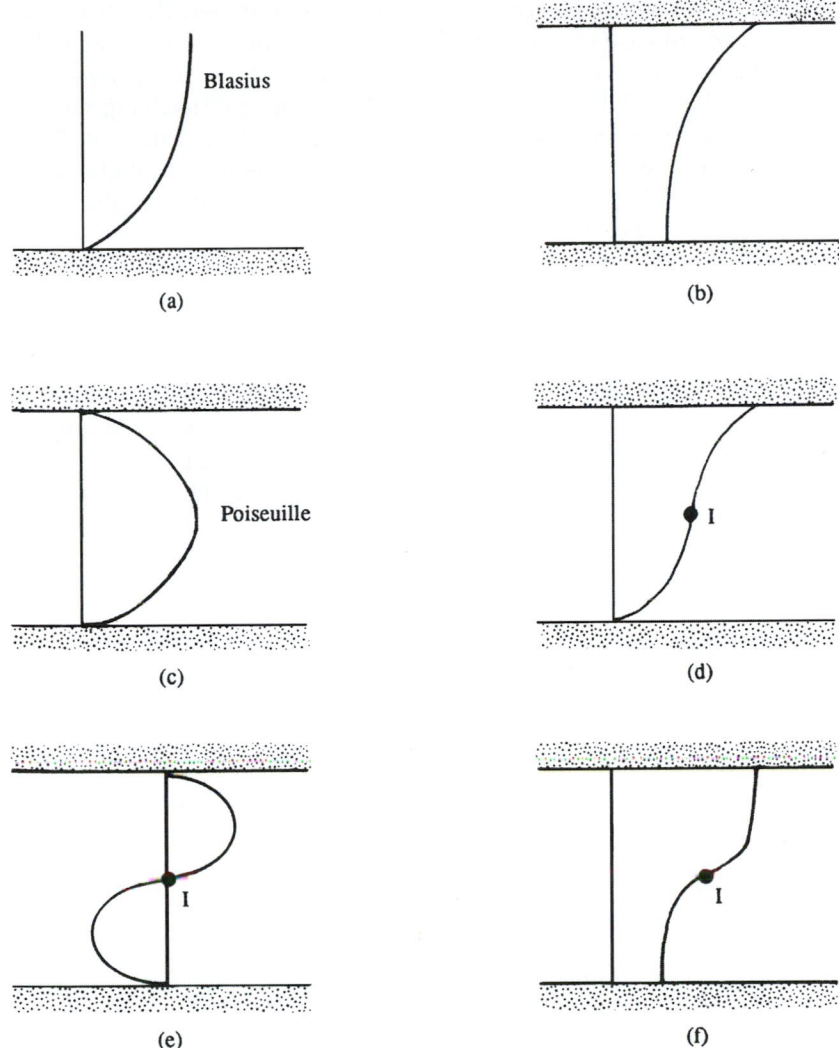

FIGURE 11.21 Examples of parallel flows. Points of inflection are denoted by I. Profiles (a), (b), and (c) are inviscidly stable. Profiles (d), (e), and (f) may be inviscidly unstable by Rayleigh's inflection point criterion. Only profiles (e) and (f) satisfy Fjortoft's criterion of inviscid instability.

plane Couette flow, Poiseuille flow, and a boundary-layer flow with zero or favorable pressure gradient have no point of inflection in the velocity profile and are stable in the inviscid limit.

However, neither of the two conditions is sufficient for instability. An example is the sinusoidal profile $U = \sin(y)$, with boundaries at $y = \pm b$. It has been shown that the flow is stable if the width is restricted to $2b < \pi$, although it has an inflection point at $y = 0$.

Invisicd parallel flows satisfy Howard's semicircle theorem, which was proved in Section 11.7 for the more general case of a stratified shear flow. The theorem states that the phase speed c_r of an unstable mode with wave number k has a value that lies between the minimum and the maximum values of $U(y)$ in the flow field. Growing and decaying modes are characterized by a nonzero c_i, whereas neutral modes can have only a real $c = c_r$. Thus, it follows that neutral modes must have $U = c$ somewhere in the flow field. The neighborhood y around y_c at which $U = c = c_r$ is called a *critical layer*. The location y_c is a critical point of the inviscid governing equation (11.81), because the highest derivative drops out at this value of y, and the eigenfunction for this k and c may be discontinuous across this layer. The full Orr-Sommerfeld equation (11.79) has no such critical layer because the highest-order derivative does not drop out when $U = c$. It is apparent that in a real flow a viscous boundary layer must form at the location where $U = c$, and that the layer becomes thinner as $Re \to \infty$.

The streamline pattern in the neighborhood of the critical layer where $U = c$ was given by Kelvin in 1888, and indicates the nature of the nearby unstable modes having the same k but small positive c_i. The discussion provided here is adapted from Drazin and Reid (1981). Consider a flow viewed by an observer moving with the phase velocity $c = c_r$. Then the basic velocity field seen by this observer is $(U - c)$, so that the stream function due to the basic flow is

$$\Psi = \int (U - c)dy.$$

The total stream function is obtained by adding the perturbation:

$$\widehat{\psi} = \int (U - c)dy + A\phi(y)\exp\{ikx\}. \tag{11.87}$$

where A is an arbitrary constant, and the time factor in the second term is omitted because the disturbance is neutrally stable. Near the critical layer $y = y_c$, a Taylor series expansion of the real part of (11.87) is approximately

$$\widehat{\psi} \cong \frac{(y - y_c)^2}{2}\left[\frac{dU}{dy}\right]_{y=y_c} + A\phi(y_c)\cos(kx),$$

where $\phi(y_c)$ is assumed to be real. The streamline pattern corresponding to this equation is sketched in Figure 11.22, showing the so-called *Kelvin cat's eye pattern* that is

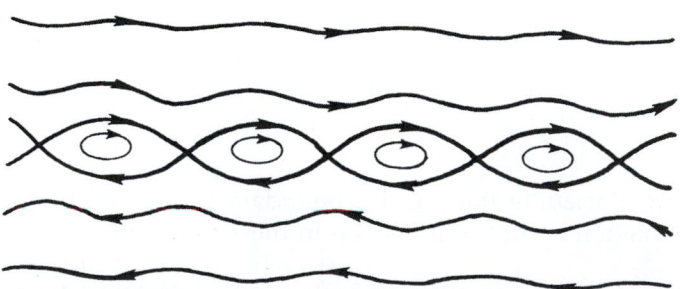

FIGURE 11.22   The Kelvin cat's eye pattern near a critical layer, showing streamlines as seen by an observer moving with a neutrally stable wave having $c = c_r$. This flow pattern is reminiscent of those shown in Figures 11.4–11.6.

visually similar to the illustrations of the Kelvin-Helmholtz instability given in Figures 11.4 through11.6.

11.10. RESULTS FOR PARALLEL AND NEARLY PARALLEL VISCOUS FLOWS

The dominant intuitive expectation is that viscous effects are stabilizing. The stability of thermal and centrifugal convections discussed in Sections 11.4 and 11.6 confirm this expectation. However, the actual situation is more complicated. Consider the Poiseuille-flow and Blasius boundary-layer velocity profiles in Figure 11.21. Neither has an inflection point so both are inviscidly stable. Yet, in experiments, these flows are known to undergo transition to turbulence at some Reynolds number, and this suggests that viscous effects are destabilizing in these flows. Thus, fluid viscosity may be stabilizing as well as destabilizing, a duality confirmed by stability calculations of parallel viscous flows.

Analytical solution of the Orr-Sommerfeld equation is notoriously complicated and will not be presented here. The viscous term in (11.79) contains the highest-order derivative, and therefore the eigenfunction may contain regions of rapid variation in which the viscous effects become important. Sophisticated asymptotic techniques are therefore needed to treat these boundary layers. Alternatively, solutions can be obtained numerically. For our purposes, we shall discuss only certain features of these calculations for the two-stream shear layer, plane Poiseuille flow, plane Couette flow, pipe flow, and boundary layers with pressure gradients. This section concludes with an explanation of how viscosity can act to destabilize a flow. Additional information can be found in Drazin and Reid (1981), and in the review article by Bayly, Orszag, and Herbert (1988).

Two-Stream Shear Layer

Consider a shear layer with the velocity profile $U(y) = U_0\tanh(y/L)$, so that $U(y) \rightarrow \pm U_0$ as $y/L \rightarrow \pm\infty$. This profile has its peak vorticity at its inflection point and is of the type shown in Figure 11.21f. A stability diagram for solution of the Orr-Sommerfeld equation for this velocity distribution is sketched in Figure 11.23. At all Reynolds numbers the flow is unstable to waves having low wave numbers in the range $0 < k < k_u$, where the upper limit k_u depends on the Reynolds number $\text{Re} = U_0L/\nu$. For high values of Re, the range of unstable wave numbers increases to $0 < k < 1/L$, which corresponds to a wavelength range of $\infty > \lambda > 2\pi L$. It is therefore essentially a long-wavelength instability. In the limit $kL \rightarrow 0$, these results simplify to those given in Section 11.3 for a vortex sheet.

Figure 11.23 implies that the critical Reynolds number for the onset of instability in a shear layer is zero. In fact, viscous calculations for all flows with *inflectional profiles* show a small critical Reynolds number; for example, for a jet of the form $u = U\text{sech}^2(y/L)$, it is $\text{Re}_{cr} = 4$. These wall-free shear flows therefore become unstable very quickly, and the inviscid prediction that these flows are always unstable is a fairly good description. The reason the inviscid analysis works well in describing the stability characteristics of free shear flows can be explained as follows. For flows with inflection points the eigenfunction of the inviscid solution is smooth. On this zero-order approximation, the viscous term acts as a *regular*

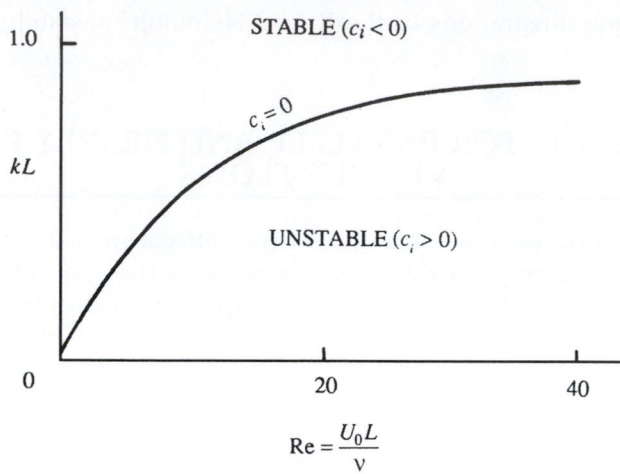

FIGURE 11.23 Marginal stability curve for a shear layer with a velocity profile of $U_0 \tanh(y/L)$ in terms of the Reynolds number $U_0 L/\nu$ and the dimensionless wave number kL of the disturbance. This flow is only unstable to low wave number disturbances.

perturbation, and the resulting correction to the eigenfunction and eigenvalues can be computed as a perturbation expansion in powers of the small parameter $1/\text{Re}$. This is true even though the viscous term in the Orr-Sommerfeld equation contains the highest-order derivative.

The instability in flows with inflection points is observed to form rolled-up regions of vorticity, much like in the calculations of Figure 11.6 or in the photographs in Figures 11.4 and 11.5. This behavior is robust and insensitive to the detailed experimental conditions. They are therefore easily observed. In contrast, the unstable waves in a wall-bounded shear flow are extremely difficult to observe, as discussed in the next section.

Plane Poiseuille Flow

The flow in a channel with parabolic velocity distribution has no point of inflection and is inviscidly stable. However, linear viscous calculations show that the flow becomes unstable at a critical Reynolds number of 5780. Nonlinear calculations, which consider the distortion of the basic profile by the finite amplitude of the perturbations, give a critical Reynolds number of 2510, which agrees better with the observations of transition. In any case, the interesting point is that viscosity is *destabilizing* for this flow. The solution of the Orr-Sommerfeld equation for Poiseuille flow and other parallel flows with rigid boundaries, which do not have an inflection point, is complicated. In contrast to flows with inflection points, the viscosity here acts as a *singular* perturbation, and the eigenfunction has viscous boundary layers on the channel walls and around critical layers where $U = c_r$. The disturbances that cause instability in these flows are called *Tollmien-Schlichting* waves, and their experimental detection is discussed in the next section. In his 1979 text, Yih gives a thorough discussion of the solution of the Orr-Sommerfeld equation using asymptotic expansions in the limit sequence Re $\to \infty$, then $k \to 0$ (but kRe $\gg 1$). He follows closely the analysis of Heisenberg (1924). Yih presents Lin's (1955) improvements on Heisenberg's analysis with Shen's (1954) calculations of the stability curves.

Plane Couette Flow

This is the flow confined between two parallel plates; it is driven by the motion of one of the plates parallel to itself. The basic velocity profile is linear, with $U \propto y$. Contrary to the experimentally observed fact that the flow does become turbulent at high Reynolds numbers, all linear analyses have shown that the flow is stable to small disturbances. It is now believed that the observed instability is caused by disturbances of finite magnitude.

Pipe Flow

The absence of an inflection point in the velocity profile signifies that the flow is inviscidly stable. All linear stability calculations of the *viscous* problem have also shown that the flow is stable to small disturbances. In contrast, most experiments show that the transition to turbulence takes place at a Reynolds number of about Re $= U_{\max}d/v \sim 3000$. However, careful experiments, some of them performed by Reynolds in his classic investigation of the onset of turbulence, have been able to maintain laminar flow up to Re $= 50,000$. Beyond this the observed flow is invariably turbulent. The observed transition has been attributed to one of the following effects: 1) It could be a finite amplitude effect; 2) the turbulence may be initiated at the entrance of the tube by boundary-layer instability (Figure 9.2); and 3) the instability could be caused by a slow rotation of the inlet flow which, when added to the Poiseuille distribution, has been shown to result in instability. This is still under investigation. New insights into the instability and transition of pipe flow were described by Eckhardt et al. (2007) by analysis via dynamical systems theory and comparison with recent very carefully crafted experiments by them and others. They characterized the turbulent state as a *chaotic saddle in state space*. The boundary between laminar and turbulent flow was found to be exquisitely sensitive to initial conditions. Because pipe flow is linearly stable, finite amplitude disturbances are necessary to cause transition, but as Reynolds number increases, the amplitude of the critical disturbance diminishes. The boundary between laminar and turbulent states appears to be characterized by a pair of vortices closer to the walls that give the strongest amplification of the initial disturbance.

Boundary Layers with Pressure Gradients

Recall from Section 9.7 that when pressure decreases in the direction of flow the pressure gradient is said to be *favorable*, and when pressure increases in the direction of flow the pressure gradient is said to be *adverse*. It was shown there that boundary layers with an adverse pressure gradient have a point of inflection in the velocity profile. This has a dramatic effect on stability characteristics. A schematic plot of the marginal stability curve for a boundary layer with favorable and adverse gradients of pressure is shown in Figure 11.24. The ordinate in the plot represents the longitudinal wave number, and the abscissa represents the Reynolds number based on the free-stream velocity and the displacement thickness δ^* of

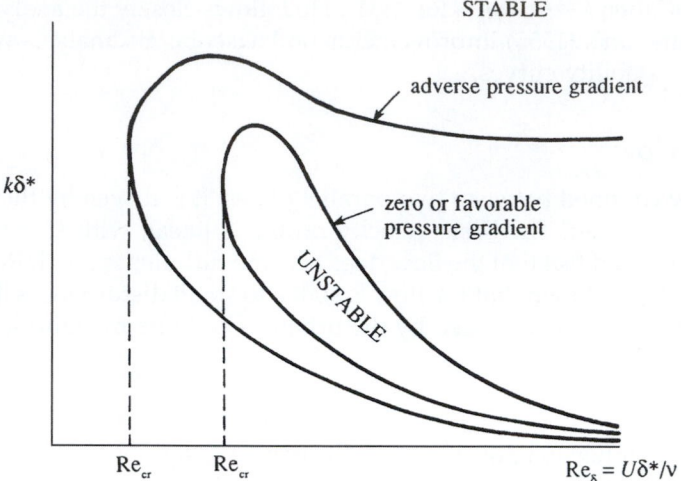

FIGURE 11.24 Sketch of marginal stability curves for laminar boundary layers with favorable and adverse pressure gradients in terms of the displacement-thickness Reynolds number $U_o\delta^*/\nu$ and the dimensionless wave number $k\delta^*$ of the disturbance. The addition of the inflection point in the adverse-pressure gradient case increases the parametric realm of instability.

the boundary layer. The marginal stability curve divides stable and unstable regions, with the region within the loop representing instability. Because the boundary layer thickness grows along the direction of flow, Re_δ increases with x, and points at various downstream distances are represented by larger values of Re_δ.

The following features can be noted in the figure. Boundary-layer flows are stable for low Reynolds numbers, but may become unstable as the Reynolds number increases. The effect of increasing viscosity is therefore stabilizing in this range. For boundary layers with a zero pressure gradient (Blasius flow) or a favorable pressure gradient, the instability loop shrinks to zero as $Re_\delta \to \infty$. This is consistent with the fact that these flows do not have a point of inflection in the velocity profile and are therefore inviscidly stable. In contrast, for boundary layers with an adverse pressure gradient, the instability loop does not shrink to zero; the upper branch of the marginal stability curve now becomes flat with a limiting value of k_∞ as $Re_\delta \to \infty$. The flow is then unstable to disturbances with wave numbers in the range $0 < k < k_\infty$. This is consistent with the existence of a point of inflection in the velocity profile, and the results of the shear layer calculations (Figure 11.23). Note also that the critical Reynolds number is lower for flows with adverse pressure gradients.

Table 11.1 summarizes the results of the linear stability analyses of some common parallel viscous flows. The first two flows in the table have points of inflection in the velocity profile and are inviscidly unstable; the viscous solution shows either a zero or a small critical Reynolds number. The remaining flows are stable in the inviscid limit. Of these, the Blasius boundary layer and the plane Poiseuille flow are unstable in the presence of viscosity, but have high critical Reynolds numbers. Although the idealized *tanh* profile for a shear layer, assuming straight and parallel streamlines, is immediately unstable, more recent work by Bhattacharya et al. (2006), which allowed for the basic

TABLE 11.1   Linear Stability Results of Common Viscous Parallel Flows

Flow	$U\,(y)/U_0$	Re_{cr}	Remarks
Jet	$\text{sech}^2\,(y/L)$	4	
Shear layer	$\tanh\,(y/L)$	0	Always unstable
Blasius		520	Re based on δ^*
Plane Poiseuille	$1 - (y/L)^2$	5780	$L = \text{half} - \text{width}$
Pipe flow	$1 - (r/R)^2$	∞	Always stable
Plane Couette	y/L	∞	Always stable

flow to be two dimensional, has yielded a finite critical Reynolds number, modifying somewhat Table 11.1.

While the results presented in the preceding paragraphs document flows where viscous effects are destabilizing, the mechanism of this destabilization has not been identified. One means of describing the destabilization mechanism relies on use of the equation for integrated kinetic energy of the disturbance:

$$\frac{d}{dt}\int \frac{1}{2}u_i^2 dV = -\int u_i u_j \frac{\partial U_i}{\partial x_j} dV - \Lambda, \tag{11.88}$$

where V is a stationary volume having stream-wise control surfaces chosen to coincide with the walls where no-slip conditions are satisfied or where $u_i \to 0$, and having a length (in the stream-wise direction) that is an integer number of disturbance wavelengths (see Figure 11.25). In (11.88), $\Lambda = \nu\int(\partial u_i/\partial x_j)^2 dV$ is the total viscous dissipation rate of kinetic energy in V. This disturbance kinetic energy equation can be derived from the incompressible Navier-Stokes momentum equation for the flow (see Exercise 11.13).

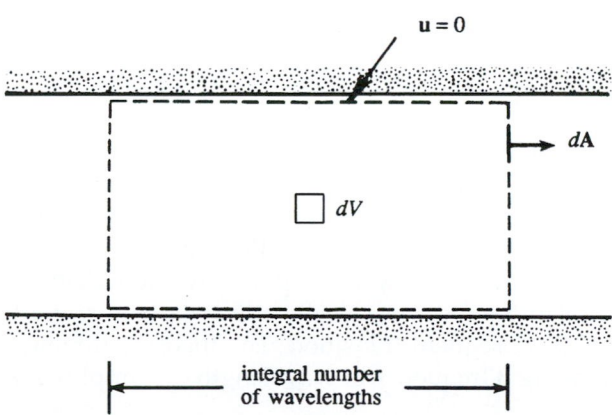

FIGURE 11.25   A control volume for deriving (11.88). Here there is zero net flux across boundaries. This control volume can be extended to boundary-layer flow stability, when the boundary layer forms on the lower wall, by placing the upper control surface far enough from the lower wall so that the disturbance velocity $u_i \to 0$ on this control surface, even if this control surface may not abut the upper wall.

For two-dimensional disturbances in a shear flow defined by $\mathbf{U} = [U(y), 0, 0]$, the disturbance energy equation becomes:

$$\frac{d}{dt}\int \frac{1}{2}(u^2 + v^2)dV = -\int uv\frac{\partial U}{\partial y}dV - \Lambda,$$

and has a simple interpretation. The first term is the rate of change of kinetic energy of the two-dimensional disturbance, and the second term is the rate of production of disturbance energy by the interaction of the product uv (also known as the *Reynolds shear stress*) and the mean shear $\partial U/\partial y$. (The concept of Reynolds stresses is explained in Chapter 12.) The point to note here is that the value of the product uv averaged over a period is zero if the velocity components u and v are out of phase; for example, the mean value of uv is zero if $u = \sin t$ and $v = \cos t$. In inviscid parallel flows without a point of inflection in the velocity profile, the u and v components are such that the disturbance field cannot extract energy from the basic shear flow, thus resulting in stability. The presence of viscosity, however, changes the phase relationship between u and v, which causes the spatial integral of $-uv(\partial U/\partial y)$ to be positive and larger than the viscous dissipation rate. This is how viscous effects can cause instability.

11.11. EXPERIMENTAL VERIFICATION OF BOUNDARY-LAYER INSTABILITY

This section presents the results of stability calculations of the Blasius boundary-layer profile and compares them with experiments. Because of the nearly parallel nature of the Blasius flow, most stability calculations are based on an analysis of the Orr-Sommerfeld equation, which assumes a parallel flow. The first calculations were performed by Tollmien in 1929 and Schlichting in 1933. Instead of assuming exactly the Blasius profile (which can be specified only numerically), they used the profile

$$\frac{U}{U_\infty} = \begin{Bmatrix} 1.7(y/\delta) & 0 \leq y/\delta \leq 0.1724 \\ 1 - 1.03\left[1 - (y/\delta)^2\right] & 0.1724 \leq y/\delta \leq 1 \\ 1 & y/\delta \geq 1 \end{Bmatrix},$$

which, like the Blasius profile, has a zero curvature at the wall. The calculations of Tollmien and Schlichting showed that unstable waves appear when the Reynolds number is high enough; the unstable waves in a viscous boundary layer are called *Tollmien-Schlichting waves*. Until 1947 these waves remained undetected, and the experimentalists of the period believed that the transition in a real boundary layer was probably a finite-amplitude effect. The speculation was that large disturbances cause locally adverse pressure gradients, which resulted in a local separation and consequent transition. The theoretical view, in contrast, was that small disturbances of the right frequency or wavelength can amplify if the Reynolds number is large enough.

Verification of the theory was finally provided by some clever experiments conducted by Schubauer and Skramstad in 1947. The experiments were conducted in a wind tunnel specially designed to suppress fluctuations in the free-stream flow. The experimental technique used was novel. Instead of depending on natural disturbances, they introduced periodic disturbances of known frequency by means of a vibrating metallic ribbon stretched across the flow close to the wall. The ribbon was vibrated by passing an alternating current through it in the field of a magnet. The subsequent development of the disturbance was measured downstream via hot-wire anemometry. Such techniques later become standard.

The experimental data are shown in Figure 11.26, which also shows the calculations of Schlichting and the more accurate calculations of Shen (1954). Instead of the wave number, the ordinate represents the frequency of the disturbance, which is easier to measure. It is apparent that the agreement between Shen's calculations and the experimental data is very good.

The prediction of the Tollmien-Schlichting waves is regarded as a major accomplishment of linear stability theory. The ideal conditions for their existence are two dimensionality and negligible fluctuations in the free stream. These waves have been found to be very sensitive to small deviations from the ideal conditions, and that is why they can be observed only under very carefully controlled experimental conditions with artificial excitation. People who care about historical fairness have suggested that the waves should only be referred to as TS waves, to honor Tollmien, Schlichting, Schubauer, and Skramstad. TS waves have also been observed in natural flow (Bayly et al., 1988).

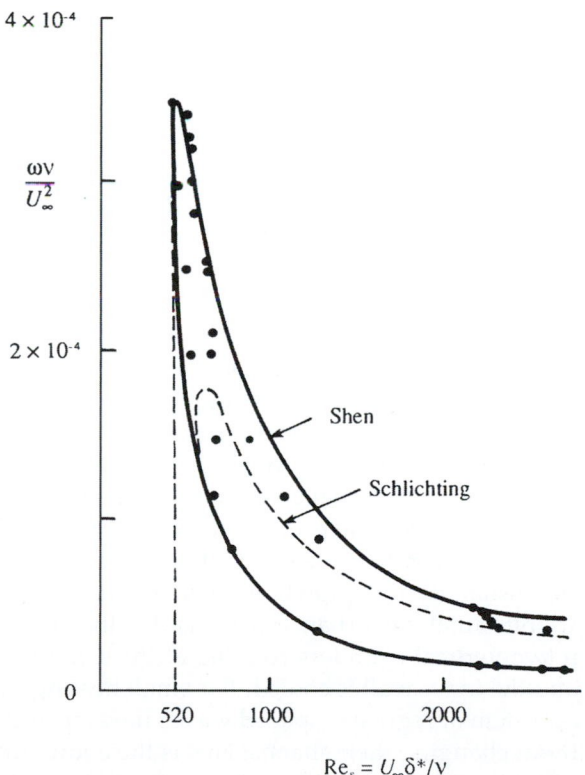

FIGURE 11.26 Marginal stability curve for a Blasius boundary layer. Theoretical solutions of Shen and Schlichting are compared with experimental data of Schubauer and Skramstad.

Nayfeh and Saric (1975) treated Falkner-Skan flows in a study of nonparallel stability and found that generally there is a decrease in the critical Reynolds number. The decrease is least for favorable pressure gradients, about 10% for zero pressure gradients, and grows rapidly as the pressure gradient becomes more adverse. Grabowski (1980) applied linear stability theory to the boundary layer near a stagnation point on a body of revolution. His stability predictions were found to be close to those of parallel-flow stability theory obtained from solutions of the Orr-Sommerfeld equation. Reshotko (2001) provides a review of temporally and spatially transient growth as a path from subcritical (Tollmien-Schlichting) disturbances to transition. Growth or decay is studied from the Orr-Sommerfeld and Squire equations. Growth may occur because eigenfunctions of these equations are not orthogonal as the operators are not self-adjoint. Results for Poiseuille pipe flow and compressible blunt body flows are given.

Fransson and Alfredsson (2003) have shown that the asymptotic suction profile (solved in Exercise 9.26) significantly delays transition stimulated by free-stream turbulence or by Tollmien-Schlichting waves. Specifically, the value of $Re_{cr} = 520$ based on δ^* in Table 11.1 is increased for suction velocity ratio $v_0/U_\infty = -.00288$ to more than 54,000. The very large stabilizing effect is a result of the change in the shape of the stream-wise velocity profile from the Blasius profile to an exponential.

11.12. COMMENTS ON NONLINEAR EFFECTS

To this point we have discussed only linear stability theory, which considers infinitesimal perturbations and predicts exponential growth when the relevant parameter exceeds a critical value. The effect of the perturbations on the basic field is neglected in the linear theory. An examination of (11.88) shows that the perturbation field must be such that the average uv (the average taken over a wavelength) must be nonzero for the perturbations to extract energy from the basic shear; similarly, the heat flux, the average of uT', must be nonzero in a thermal convection problem. These rectified fluxes of momentum and heat change the *basic* velocity and temperature fields. Linear instability theory neglects these changes of the basic state. A consequence of the constancy of the basic state is that the growth rate of the perturbations is also constant, leading to predictions of exponential growth. However, after some time, the perturbations eventually become so large that the rectified fluxes of momentum and heat significantly change the basic state, which in turn alters the growth of the perturbations.

A frequent effect of nonlinearity is to change the basic state in such a way as to arrest the growth of the disturbances after they have reached significant amplitude via their initial exponential growth. (Note, however, that the effect of nonlinearity can sometimes be destabilizing; for example, the instability in a pipe flow may be a finite-amplitude effect because the flow is stable to infinitesimal disturbances.) Consider the thermal convection in the annular space between two vertical cylinders rotating at the same speed. The outer wall of the annulus is heated and the inner wall is cooled. For small heating rates the flow is steady. For large heating rates a system of regularly spaced waves develop and progress azimuthally at a uniform speed without changing their shape. (This is the equilibrated form of baroclinic instability, discussed in Section 13.17.) At still larger heating rates an irregular, aperiodic, or chaotic flow develops. The chaotic response to constant forcing (in this case the heating rate) is an interesting nonlinear effect and is discussed further in Section 11.14. Meanwhile, a brief description of the transition from laminar to turbulent flow is given in the next section.

11.13. TRANSITION

The process by which a laminar flow changes to a turbulent one is called *transition*. Instability of a laminar flow does not immediately lead to turbulence, which is a severely nonlinear and chaotic flow state. After the initial breakdown of laminar flow because of amplification of small disturbances, the flow goes through a complex sequence of changes, finally resulting in the chaotic state we call turbulence. The process of transition is greatly affected by such experimental conditions as intensity of fluctuations of the free stream, roughness of the walls, and shape of the inlet. The sequence of events that leads to turbulence is also greatly dependent on boundary geometry. For example, the scenario of transition in wall-bounded shear flows is different from that in free shear flows such as jets and wakes.

Early stages of transition consist of a succession of instabilities on increasingly complex basic flows, an idea first suggested by Landau in 1944 (see Landau and Lifshitz, 1959). The basic state of wall-bounded parallel shear flows becomes unstable to two-dimensional TS waves, which grow and eventually reach equilibrium at some finite amplitude. This steady state can be considered a new background state, and calculations show that it is generally unstable to *three-dimensional* waves of short wavelength, which vary in the cross-stream or *span-wise* direction. (If x denotes the stream-wise flow direction and y denotes the wall-normal direction, then the z-axis lies in the *span-wise* direction.) We shall call this the *secondary instability*. Interestingly, the secondary instability does not reach equilibrium at finite amplitude but directly evolves to a fully turbulent flow. Recent calculations of the secondary instability have been quite successful in reproducing critical Reynolds numbers for various wall-bounded flows, as well as predicting three-dimensional structures observed in experiments.

A key experiment on the three-dimensional nature of the transition process in a boundary layer was performed by Klebanoff, Tidstrom, and Sargent (1962). They conducted a series of controlled experiments by which they introduced three-dimensional disturbances on a field of TS waves in a boundary layer. The TS waves were as usual artificially generated by an electromagnetically vibrated ribbon, and the three dimensionality of a particular span-wise wavelength was introduced by placing spacers (small pieces of transparent tape) at equal intervals underneath the vibrating ribbon (Figure 11.27). When the amplitude of the TS waves became roughly 1% of the free-stream velocity, the three-dimensional perturbations grew rapidly and resulted in a span-wise irregularity of the stream-wise velocity displaying peaks and valleys in the amplitude of u. The three-dimensional disturbances continued to grow until the boundary layer became fully turbulent. The chaotic flow seems to result from the nonlinear evolution of the secondary instability, and numerical calculations have accurately reproduced several characteristic features of real flows (see Figures 7 and 8 in Bayly et al., 1988).

It is interesting to compare the chaos observed in turbulent shear flows with that in controlled low-order dynamical systems such as the Bérnard convection or Taylor vortex flow. In these low-order flows only a very small number of modes participate in the dynamics because of the strong constraint of the boundary conditions. All but a few low modes are identically zero, and the chaos develops in an orderly way. As the constraints are relaxed (we can think of this as increasing the number of allowed Fourier modes), the evolution toward apparent chaos becomes less orderly.

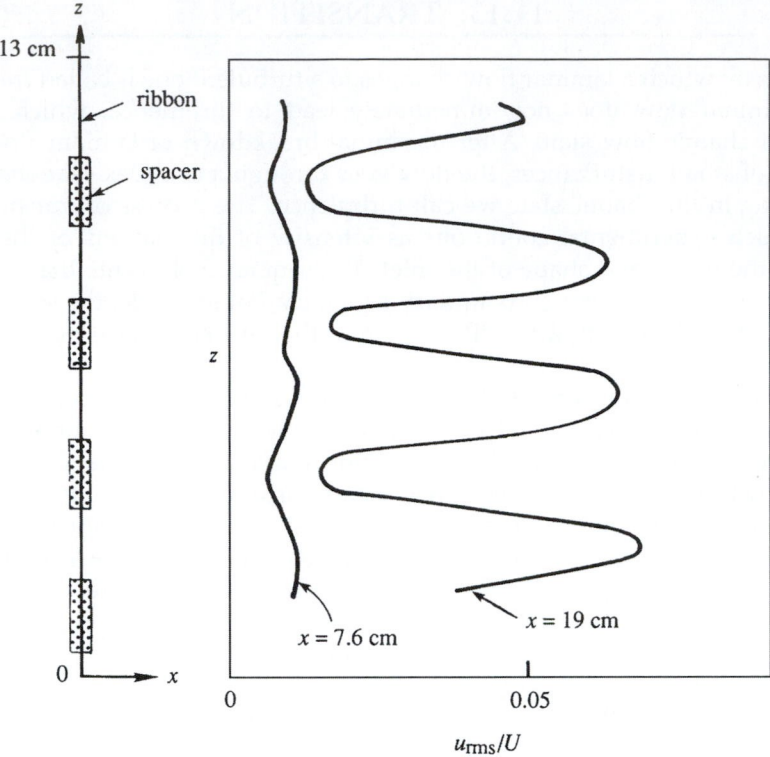

FIGURE 11.27 Three-dimensional unstable waves initiated by a vibrating ribbon. Measured distributions of intensity of the *u*-fluctuation at two distances from the ribbon are shown. Clearly the span-wise variation enhances the signature of the instability. *P. S. Klebanoff et al.,* Journal of Fluid Mechanics, 12, *1–34, 1962; reprinted with the permission of Cambridge University Press.*

Transition in a free shear layer, such as a jet or a wake, occurs in a different manner. Because of the inflectional velocity profiles involved, these flows are unstable at very low Reynolds numbers, that is, of order 10 compared to about 10^3 for wall-bounded flows. The breakdown of the laminar flow therefore occurs quite readily and close to the origin of such a flow. Transition in a free shear layer is characterized by the appearance of a rolled-up row of vortices, whose wavelength corresponds to the one with the largest growth rate. Frequently, these pairs of vortices regroup themselves and result in a dominant wavelength twice that of the original wavelength. Small-scale turbulence develops in the strain fields between and within these larger scale vortices, finally leading to turbulence.

11.14. DETERMINISTIC CHAOS

The discussion in the previous section has shown that dissipative nonlinear systems such as fluid flows reach a random or chaotic state when the parameter measuring nonlinearity (say, the Reynolds number or the Rayleigh number) is large. The evolution from laminar flow to the chaotic state generally takes place through a sequence of transitions, with the exact route depending on the flow geometry and other characteristics. It has been realized that chaotic behavior not only occurs in continuous systems having an infinite number of degrees of freedom, but also in discrete nonlinear systems having only a small number of degrees of freedom, governed by ordinary nonlinear differential equations. In this context, a *chaotic system* is defined as one in which the solution is *extremely sensitive to initial conditions*. That is, solutions with arbitrarily close initial conditions evolve into quite different states. Other symptoms of a chaotic system are that the solutions are *aperiodic*, and that the spectrum of fluctuations is broadband instead of being composed of a few discrete frequencies or wave numbers.

Numerical integrations (to be shown later in this section) have recently demonstrated that nonlinear systems governed by a finite set of deterministic ordinary differential equations allow chaotic solutions in response to a steady forcing. This fact is interesting because in a dissipative *linear* system a constant forcing ultimately (after the decay of the transients) leads to a constant response, a periodic forcing leads to a periodic response, and a random forcing leads to a random response. In the presence of nonlinearity, however, a constant forcing can lead to a variable response, both periodic and aperiodic. Consider again the experiment mentioned in Section 11.12, namely, the thermal convection in the annular space between two vertical cylinders rotating at the same speed. The outer wall of the annulus is heated and the inner wall is cooled. For small heating rates the flow is steady. For large heating rates a system of regularly spaced waves develops and progresses azimuthally at a uniform speed, without the waves changing shape. At still larger heating rates an irregular, aperiodic, or chaotic flow develops. This experiment shows that both periodic and aperiodic flow can result in a nonlinear system even when the forcing (in this case the heating rate) is constant. Another example is the periodic oscillation in the flow behind a blunt body at Re ~40 (associated with the initial appearance of the von Karman vortex street) and the breakdown of the oscillation into turbulent flow at larger values of the Reynolds number.

It has been found that transition to chaos in the solution of ordinary nonlinear differential equations displays a certain *universal* behavior and proceeds in one of a few different ways. Transition to turbulence in fluid flows may be related to the development of chaos in the solutions of these simple systems. In this section we shall discuss some of the elementary ideas involved, starting with the definitions for phase space and attractors, moving on to the Lorenz model of thermal convection and scenarios for transition to chaos, and then concluding with a description of the implications of such phenomena. An introduction to the subject of chaos is given by Bergé, Pomeau, and Vidal (1984); a useful review is given in Lanford (1982). The subject has far-reaching cosmic consequences in physics and evolutionary biology, as discussed by Davies (1988).

Very few nonlinear equations have analytical solutions. For nonlinear systems, a typical procedure is to find a numerical solution and display its properties in a space whose axes are the *dependent* variables. Consider the equation governing the motion of a simple pendulum of length l:

$$\ddot{X} + (g/l)\sin X = 0,$$

where X is the *angular* displacement and $\ddot{X}$ is the angular acceleration. The equation is nonlinear because of the sin$X$ term. The second-order equation can be split into two coupled first-order equations:

$$\dot{X} = Y \quad \text{and} \quad \dot{Y} = -(g/l)\sin X. \tag{11.89}$$

Starting with some initial conditions on X and Y, one can integrate (11.89) forward in time. The behavior of the system can be studied by describing how the variables $Y (= \dot{X})$ and X vary as functions of time. For the pendulum problem, the space whose axes are $\dot{X}$ and X is called a *phase space,* and the evolution of the system is described by a *trajectory* in this space. The dimension of the phase space is called the *degree of freedom* of the system; it equals the number of independent initial conditions necessary to specify the system. For example, the degree of freedom for the set (11.89) is two.

Dissipative systems are characterized by the existence of *attractors,* which are structures in the phase space toward which neighboring trajectories approach as $t \rightarrow \infty$. An attractor can be a *fixed point* representing a stable steady flow or a closed curve (called a *limit cycle*) representing a stable oscillation (Figure 11.28a, b). The nature of the attractor depends on the value of the nonlinearity parameter, which will be denoted by R in this section. As R is increased, the fixed point representing a steady solution may change from being an attractor to a repeller with spirally outgoing trajectories, signifying that the steady flow has become unstable to infinitesimal perturbations. Frequently, the trajectories are then attracted by a limit cycle, which means that the unstable steady solution gives way to a steady oscillation (Figure 11.28b). For example, the steady flow behind a blunt body becomes oscillatory as Re is increased, resulting in the periodic von Karman vortex street (Figure 9.17).

The branching of a solution at a critical value R_{cr} of the nonlinearity parameter is called a *bifurcation.* Thus, we say that the stable steady solution of Figure 11.28a bifurcates to a stable limit cycle as R increases through R_{cr}. This can be represented on the graph of a dependent variable (say, X) versus R (Figure 11.28c). At $R = R_{cr}$, the solution curve branches into two paths; the two values of X on these branches (say, X_1 and X_2) correspond to the maximum and minimum values of X in Figure 11.28b. It is seen that the size of the limit cycle grows larger as $(R - R_{cr})$ becomes larger. Limit cycles, representing oscillatory response with amplitude independent of initial conditions, are characteristic features of nonlinear systems. Linear stability theory predicts an exponential growth of the perturbations if $R > R_{cr}$, but a nonlinear theory frequently shows that the perturbations eventually equilibrate to a steady oscillation whose amplitude increases with $(R - R_{cr})$.

A famous fluid-flow example involving these concepts comes from thermal convection in a layer heated from below (the Bénard problem). Lorenz (1963) demonstrated that the development of chaos is associated with the flow's attractor acquiring certain strange properties. He considered a layer with stress-free boundaries. Assuming nonlinear disturbances in the form of rolls invariant in the y direction, and defining a disturbance stream function in the x-z plane by $u = -\partial\psi/\partial z$ and $w = \partial\psi/\partial x$, he substituted solutions of the form

$$\psi \propto X(t)\cos(\pi z)\sin(kx) \quad \text{and} \quad T' \propto Y(t)\cos(\pi z)\cos(kx) + Z(t)\sin(2\pi z) \tag{11.90}$$

into the equations of motion. Here, T' is the departure of temperature from the state of no convection, k is the wave number of the perturbation, and the boundaries are at $z = \pm\frac{1}{2}$. It is clear that X is proportional to the speed of convective motion, Y is proportional to

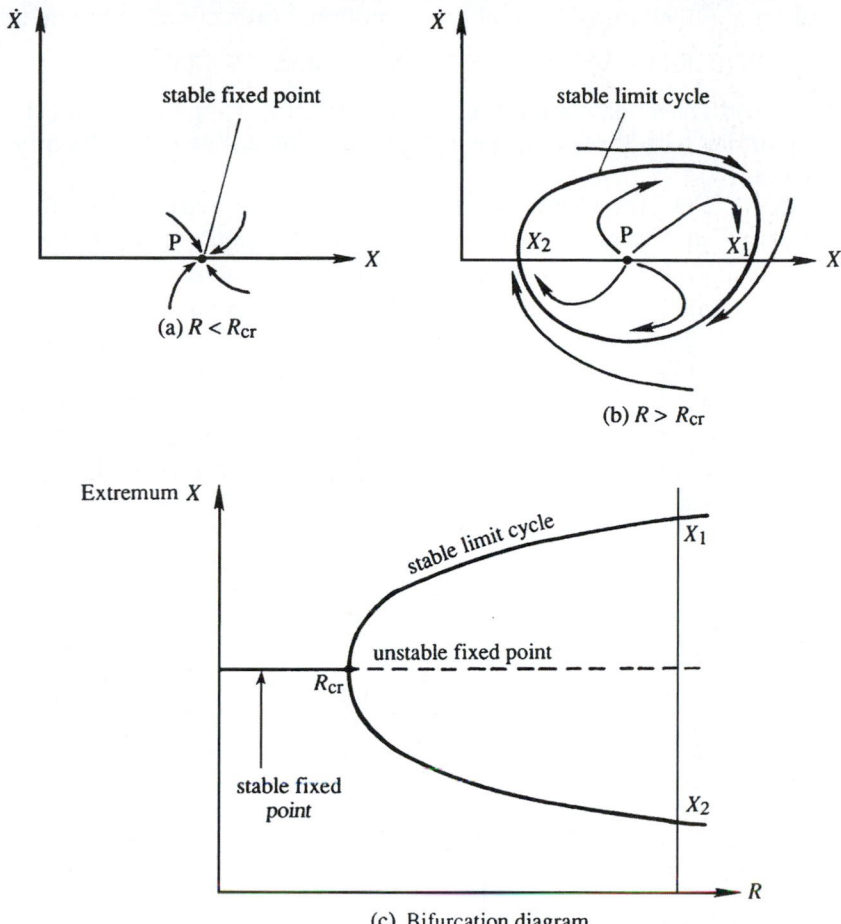

FIGURE 11.28 Attractors in a phase plane of X and $\dot{X}$. In (a), point P is an attractor. For a larger value of R, the nonlinearity parameter, panel (b) shows that P becomes an unstable fixed point (a *repeller*), and the trajectories are attracted to an orbit or limit cycle that encircles P. Panel (c) is the bifurcation diagram corresponding to this situation.

the temperature difference between the ascending and descending currents, and Z is proportional to the distortion of the average vertical profile of temperature from linearity. (Note in (11.90) that the x-average of the term multiplied by $Y(t)$ is zero, so that this term does not cause distortion of the basic temperature profile.) As discussed in Section 11.4, Rayleigh's linear analysis showed that solutions of the form (11.90), with X and Y constants and $Z = 0$, would develop if Ra slightly exceeds the critical value $Ra_{cr} = 27\pi^4/4$. Equations (11.90) are expected to give realistic results when Ra is slightly supercritical but not when strong convection occurs because only the lowest wave number terms are retained.

On substitution of (11.90) into the equations of motion, Lorenz finally obtained the system:

$$\dot{X} = \text{Pr}(Y - X),\ \dot{Y} = -XZ + rX - Y,\ \text{and}\ \dot{Z} = XY - bZ, \tag{11.91}$$

where Pr is the Prandtl number, $r = \text{Ra}/\text{Ra}_{cr}$, and $b = 4\pi^2/(\pi^2 + k^2)$. Equations (11.91) are a set of nonlinear equations with three degrees of freedom, which means that the phase space is three dimensional.

Equations (11.91) allow the steady solution $X = Y = Z = 0$, representing the state of no convection. For $r > 1$ the system possesses two additional steady-state solutions, which we shall denote by $\overline{X} = \overline{Y} = \pm\sqrt{b(r-1)},\ \overline{Z} = r - 1$; the two signs correspond to the two possible senses of rotation of the rolls. (The fact that these steady solutions satisfy (11.91) can easily be checked by substitution and setting $\dot{X} = \dot{Y} = \dot{Z} = 0$.) Lorenz showed that the steady-state convection becomes unstable if r is large. Choosing $\text{Pr} = 10$, $b = 8/3$, and $r = 28$, he numerically integrated the set and found that the solution never repeats itself; it is aperiodic and wanders about in a chaotic manner. Figure 11.29 shows the variation of $X(t)$, starting with some initial conditions. (The variables $Y(t)$ and $Z(t)$ also behave in a similar way.) It is seen that the amplitude of the convecting motion initially oscillates around one of the steady values $\overline{X} = \pm\sqrt{b(r-1)}$, with the oscillations growing in magnitude. When it is large enough, the amplitude suddenly goes through zero to start oscillations of opposite sign about the other value of $\overline{X}$. That is, the motion switches in a chaotic manner between two oscillatory limit cycles, with the number of oscillations between transitions seemingly random. Calculations show that the variables X, Y, and Z have continuous spectra and that the solution is extremely sensitive to initial conditions.

The trajectories in the phase space of the Lorenz model of thermal convection are shown in Figure 11.30. The centers of the two loops represent the two steady convections

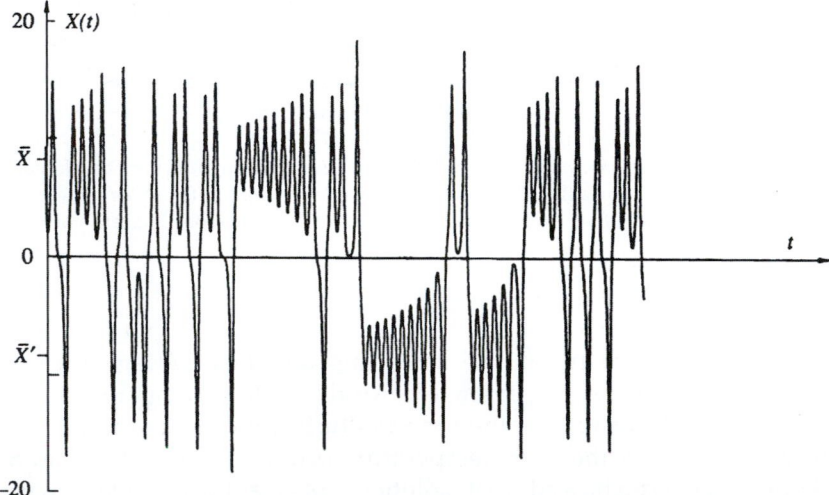

FIGURE 11.29 Variation of $X(t)$ in the Lorenz model. Note that the solution oscillates erratically around the two steady values $\overline{X}$ and $\overline{X}'$ and does not have a reliable period. *P. Bergé, Y. Pomeau, and C. Vidal, Order Within Chaos, 1984; reprinting permitted by Heinemann Educational, a division of Reed Educational & Professional Publishing Ltd.*

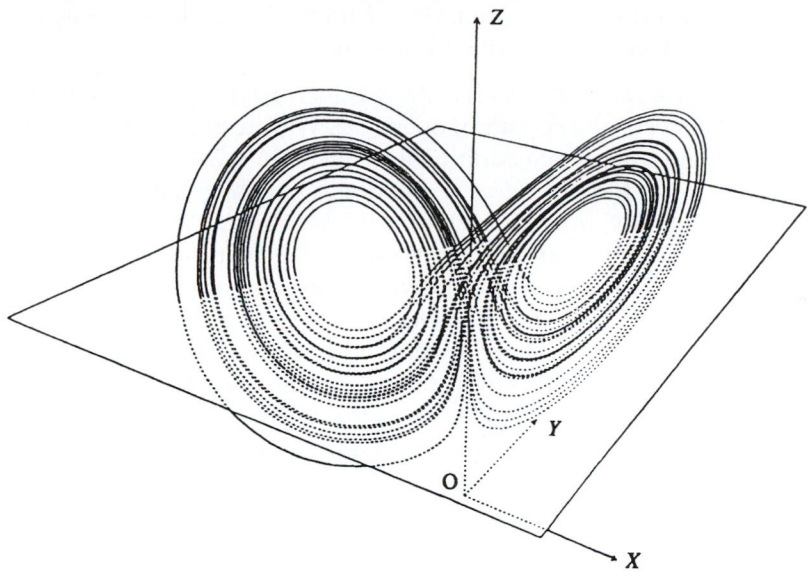

FIGURE 11.30 The Lorenz attractor. All nearby initial conditions are attracted to this double loop structure, but any two such trajectories will eventually diverge, even if they begin very close together. The centers of the two loops represent the two steady solutions $(\overline{X}, \overline{Y}, \overline{Z})$.

$\overline{X} = \overline{Y} = \pm\sqrt{b(r-1)}$ and $\overline{Z} = r - 1$. The structure resembles two rather flat loops of ribbon, one lying slightly in front of the other along a central band, with the two joined together at the bottom of that band. The trajectories go clockwise around the left loop and counterclockwise around the right loop; two trajectories never intersect. The structure shown in Figure 11.30 is an attractor because orbits starting with initial conditions *outside of the attractor* merge onto it and then follow it. The attraction is a result of dissipation in the system. The aperiodic attractor, however, is unlike the normal attractor in the form of a fixed point (representing steady motion) or a closed curve (representing a limit cycle). This is because two trajectories *on the aperiodic attractor*, with infinitesimally different initial conditions, follow each other closely only for a while, eventually diverging to very different final states. This is the basic reason for sensitivity to initial conditions.

For these reasons the aperiodic attractor is called a *strange attractor*. The idea of a strange attractor is not intuitive because it has the dual property of attraction and divergence. Trajectories starting from the neighboring regions in phase space are drawn toward it, but once on the attractor the trajectories eventually diverge and result in chaos. An ordinary attractor in phase space allows the trajectories from slightly different initial conditions to merge, so that the *memory* of initial conditions is lost. However, the strange attractor ultimately accentuates small initial condition differences. The idea of the strange attractor was first conceived by Lorenz, and since then attractors of other chaotic systems have also been studied. They all have the common property of aperiodicity, continuous spectra, and sensitivity to initial conditions.

Thus far we have described a discrete dynamical system having only a small number of degrees of freedom and seen that aperiodic or chaotic solutions result when the nonlinearity parameter is large. Several routes or scenarios of transition to chaos in such systems have been identified. Two of these are described briefly here.

(1) *Transition through subharmonic cascade*: As R is increased, a typical nonlinear system develops a limit cycle of a certain frequency ω. With further increase of R, several systems are found to generate additional frequencies $\omega/2$, $\omega/4$, $\omega/8$, … . The addition of frequencies in the form of *subharmonics* does not change the periodic nature of the solution, but the period doubles each time a lower harmonic is added. The period doubling takes place more and more rapidly as R is increased, until an *accumulation point* (Figure 11.31) is reached, beyond which the solution wanders about in a chaotic manner. At this point the peaks disappear from the temporal-frequency spectrum, which becomes broadband. Many systems approach chaotic behavior through period doubling. Feigenbaum (1978) proved the important result that this kind of transition develops in a *universal* way, independent of the particular nonlinear systems studied. If R_n represents the value for development of a new subharmonic, then R_n converges in a geometric series with

$$\frac{R_n - R_{n-1}}{R_{n+1} - R_n} \rightarrow 4.6692 \quad \text{as } n \rightarrow \infty$$

That is, the horizontal gap between two bifurcation points is about a fifth of the previous gap. The vertical gap between the branches of the bifurcation diagram also decreases, with each gap about two-fifths of the previous gap. In other words, the bifurcation

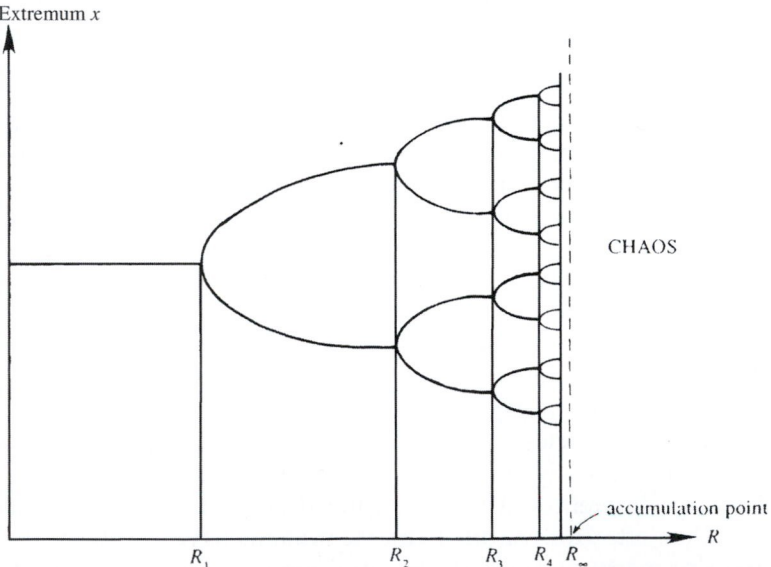

FIGURE 11.31 Bifurcation diagram during period doubling. The period doubles at each value R_n of the nonlinearity parameter. For large n the "bifurcation tree" becomes self-similar. Chaos sets in beyond the accumulation point R_∞. This process may mimic the transition from laminar to turbulent flow under some circumstances.

diagram (Figure 11.31) becomes "self-similar" as the accumulation point is approached. (Note that Figure 11.31 has not been drawn to scale, for illustrative purposes.) Experiments in low Prandtl number fluids (such as liquid metals) indicate that Bénard convection in the form of rolls develops oscillatory motion of a certain frequency ω at $Ra = 2Ra_{cr}$. As Ra is further increased, additional frequencies $\omega/2$, $\omega/4$, $\omega/8$, $\omega/16$, and $\omega/32$ have been observed. The convergence ratio has been measured to be 4.4, close to the value of 4.669 predicted by Feigenbaum's theory. The experimental evidence is discussed further in Bergé, Pomeau, and Vidal (1984).

(2) *Transition through quasi-periodic regime*: Ruelle and Takens (1971) have mathematically proven that certain systems need only a *small number* of bifurcations to produce chaotic solutions. As the nonlinearity parameter is increased, the steady solution loses stability and bifurcates to an oscillatory limit cycle with frequency ω_1. As R is increased, two more frequencies (ω_2 and ω_3) appear through additional bifurcations. In this scenario the ratios of the three frequencies (such as ω_1/ω_2) are *irrational* numbers, so that the motion consisting of the three frequencies is not exactly periodic. (When the ratios are rational numbers, the motion is exactly periodic. To see this, think of the Fourier series of a periodic function in which the various terms represent sinusoids of the fundamental frequency ω and its harmonics 2ω, 3ω, … . Some of the Fourier coefficients could be zero.) The spectrum for these systems suddenly develops broadband characteristics of chaotic motion as soon as the third frequency ω_3 appears. The exact point at which chaos sets in is not easy to detect in a measurement; in fact the third frequency may not be identifiable in the spectrum before it becomes broadband. The Ruelle-Takens theory is fundamentally different from that of Landau, who conjectured that turbulence develops due to an *infinite* number of bifurcations, each generating a new higher frequency, so that the spectrum becomes saturated with peaks and resembles a continuous one. According to Bergé, Pomeau, and Vidal (1984), the Bénard convection experiments in *water* seem to suggest that turbulence in this case probably sets in according to the Ruelle-Takens scenario.

The development of chaos in the Lorenz attractor is more complicated and does not follow either of the two routes mentioned in the preceding discussion.

Closure

Perhaps the most intriguing characteristic of a chaotic system is the extreme *sensitivity to initial conditions*. That is, solutions with arbitrarily close initial conditions evolve into two quite different states. Most nonlinear systems are susceptible to chaotic behavior. The extreme sensitivity to initial conditions implies that nonlinear phenomena (including the weather, in which Lorenz was primarily interested when he studied the convection problem) are essentially unpredictable, no matter how well we know the governing equations or the initial conditions. Although the subject of chaos has become a scientific revolution recently, the central idea was conceived by Henri Poincaré in 1908. He did not, of course, have the computing facilities to demonstrate it through numerical integration.

It is important to realize that the behavior of chaotic systems is not *intrinsically* nondeterministic; as such the implication of deterministic chaos is different from that of the uncertainty principle of quantum mechanics. In any case, the extreme sensitivity to initial conditions implies that the *future is essentially unknowable* because it is never possible to know the initial conditions *exactly*. As discussed by Davies (1988), this fact has interesting philosophical implications regarding the evolution of the universe, including that of living species.

We have examined certain elementary ideas about how chaotic behavior may result in simple nonlinear systems having only a small number of degrees of freedom. Turbulence in a continuous fluid medium is capable of displaying an infinite number of degrees of freedom, and it is unclear whether the study of chaos can throw a great deal of light on more complicated transitions such as those in pipe or boundary-layer flow. However, the fact that nonlinear systems can have chaotic solutions for a large value of the nonlinearity parameter (see Figure 11.29) is an important result by itself.

EXERCISES

11.1. A perturbed vortex sheet nominally located at $y = 0$ separates inviscid flows of differing density in the presence of gravity with downward acceleration g. The upper stream is semi-infinite and has density ρ_1 and horizontal velocity U_1. The lower stream has thickness h density ρ_2, and horizontal velocity U_2. A smooth flat impenetrable surface located at $y = -h$ lies below the second layer. The interfacial tension between the two fluids is σ. Assume a disturbance occurs on the vortex sheet with wave number $k = 2\pi/\lambda$, and complex wave speed c, i.e., $[y]_{sheet} = f(x,t) = f_o \operatorname{Re}\{e^{ik(x-ct)}\}$. The four boundary conditions are:

1) $u_1, v_1 \to 0$ as $y \to +\infty$.
2) $v_2 = 0$ on $y = -h$.
3) $\mathbf{u_1} \cdot \mathbf{n} = \mathbf{u_2} \cdot \mathbf{n}$ = normal velocity of the vortex sheet on both sides of the vortex sheet.
4) $p_1 - p_2 = \sigma \dfrac{\partial^2 f}{\partial x^2}$ on the vortex sheet (σ = interfacial surface tension).

   a) Following the development in Section 11.3, show that:

$$c = \frac{\rho_1 U_1 + \rho_2 U_2 \coth(kh)}{\rho_1 + \rho_2 \coth(kh)} \pm \left[\frac{(g/k)(\rho_2 - \rho_1) + \sigma k}{\rho_1 + \rho_2 \coth(kh)} - \frac{\rho_1 \rho_2 (U_1 - U_2)^2 \coth(kh)}{(\rho_1 + \rho_2 \coth(kh))^2} \right]^{1/2}.$$

   b) Use the result of part a) to show that the vortex sheet is *unstable* when:

$$\left(\tanh(kh) + \frac{\rho_2}{\rho_1} \right) \left(\frac{g}{k} \frac{(\rho_2 - \rho_1)}{\rho_2} + \frac{\sigma k}{\rho_2} \right) < (U_1 - U_2)^2.$$

   c) Will the sheet be stable or unstable to long wavelength disturbances ($k \to 0$) when $\rho_2 > \rho_1$ for a fixed velocity difference?
   d) Will the sheet be stable or unstable to short wavelength disturbances ($k \to \infty$) for a fixed velocity difference?
   e) Will the sheet ever be unstable when $U_1 = U_2$?
   f) Under what conditions will the thickness h matter?

11.2. Consider a fluid layer of depth h and density ρ_2 lying under a lighter, infinitely deep fluid of density $\rho_1 < \rho_2$. By setting $U_1 = U_2 = 0$, in the results of Exercise 11.1, the following formula for the phase speed is found:

$$c = \pm\left[\frac{(g/k)(\rho_2 - \rho_1) + \sigma k}{\rho_1 + \rho_2 \coth(kh)}\right]^{1/2}.$$

Now invert the sign of gravity and consider why drops form when a liquid is splashed on the underside of a flat surface. Are long or short waves more unstable? Does a professional painter want interior ceiling paint with high or low surface tension? For a smooth finish should the painter apply thin or thick coats of paint? Assuming the liquid has the properties of water (surface tension ≈ 0.072 N/m, density $\approx 10^3$ kg/m) and that the lighter fluid is air, what is the longest neutrally stable wavelength on the underside of a horizontal surface? [This is the *Rayleigh-Taylor instability* and it occurs when density and pressure gradients point in opposite directions. It may be readily observed by accelerating rapidly downward an upward-open cup of water.]

11.3. Inviscid horizontal flow in the half space $y > 0$ moves at speed U over a porous surface located at $y = 0$. Here the fluid density ρ is constant and gravity plays no roll. A weak vertical velocity fluctuation occurs at the porous surface: $[v]_{surface} = v_o \operatorname{Re}\{e^{ik(x-ct)}\}$, where $v_o \ll U$.

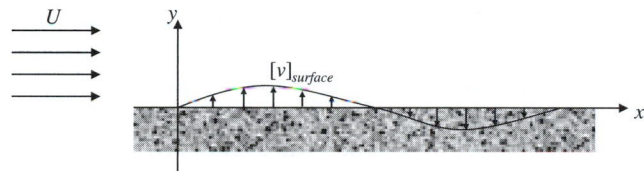

a) The velocity potential for the flow may be written $\tilde{\phi} = Ux + \phi$, where ϕ leads to $[v]_{surface}$ at $y = 0$ and ϕ vanishes as $y \to +\infty$. Determine the perturbation potential ϕ in terms of v_o, U, ρ, k, c, and the independent variables (x,y,t).

b) The porous surface responds to pressure fluctuations in the fluid via: $[p - p_s]_{y=0} = -\gamma[v]_{surface}$, where p is the pressure in the fluid, p_s is the steady static pressure that is felt on the surface when the vertical velocity fluctuations are absent, and γ is a real material parameter that defines the porous surface's flow resistance. Determine a formula for c in terms of U, γ, ρ, and k.

c) What is the propagation velocity, $\operatorname{Re}\{c\}$, of the surface velocity fluctuation?

d) What sign should γ have for the flow to be stable? Interpret your answer.

11.4. Repeat Exercise 11.3 for a compliant surface nominally lying at $y = 0$ that is perturbed from equilibrium by a small surface wave: $[y]_{surface} = \zeta(x,t) = \zeta_o \operatorname{Re}\{e^{ik(x-ct)}\}$.

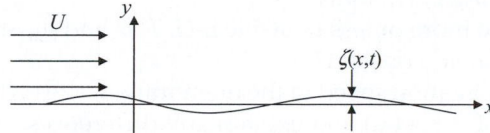

a) Determine the perturbation potential ϕ in terms of U, ρ, k, and c by assuming that ϕ vanishes as $y \to +\infty$, and that there is no flow through the compliant surface. Ignore gravity.

b) The compliant surface responds to pressure fluctuations in the fluid via: $[p - p_s]_{y=0} = -\gamma\zeta(x,t)$, where p is the pressure in the fluid, p_s is the steady pressure that is felt on the surface when the surface wave is absent, and γ is a real material parameter that defines the surface's compliance. Determine a formula for c in terms of U, γ, ρ, and k.

c) What is the propagation velocity, Re$\{c\}$, of the surface waves?

d) If γ is positive, is the flow stable? Interpret your answer.

11.5. As a simplified version of flag waving, consider the stability of a simple membrane in a uniform flow. Here, the undisturbed membrane lies in the x-z plane at $y = 0$, the flow is parallel to the x-axis at speed U, and the fluid has density ρ. The membrane has mass per unit area $= \rho_m$ and uniform tension per unit length $= T$. The membrane satisfies a dynamic equation based on pressure forces and internal tension combined with its local surface curvature:

$$\rho_m \frac{\partial^2 \zeta}{\partial t^2} = p_2 - p_1 + T\left(\frac{\partial^2 \zeta}{\partial x^2} + \frac{\partial^2 \zeta}{\partial z^2}\right).$$

Here, the vertical membrane displacement is given by $y = \zeta(x,z,t)$, and p_1 and p_2 are the pressures acting on the membrane from above and below, respectively. The velocity potentials for the undisturbed flow above (1) and below (2) the membrane are $\phi_1 = \phi_2 = Ux$. For the following items, assume a small amplitude wave is present on the membrane $\zeta(x,t) = \zeta_o \,\text{Re}\{e^{ik(x-ct)}\}$ with k a real parameter, and assume that all deflections and other fluctuations are uniform in the z-direction and small enough for the usual linear simplifications. In addition, assume the static pressures above and below the membrane, in the absence of membrane motion, are matched.

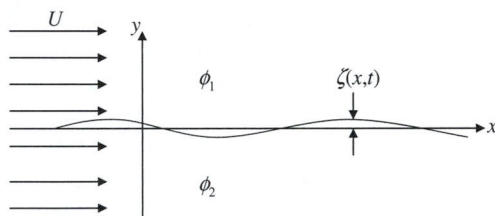

a) Using the membrane equation, determine the propagation speed of the membrane waves, Re$\{c\}$, in the absence of fluid loading (i.e., when $p_2 - p_1 = \rho = 0$).

b) Assuming inviscid flow above and below the membrane, determine a formula for c in terms of T, ρ_m, ρ, U, and k.

c) Is the membrane more or less unstable if U, T, ρ, and ρ_m are individually increased with the others held constant?

d) What is the propagation speed of the membrane waves when $U = 0$? Compare this to your answer for part a) and explain any differences.

11.6. Prove that $\sigma_r > 0$ for the thermal instability discussed in Section 11.4 via the following steps that include integration by parts and use of the boundary conditions (11.38).

   a) Multiply (11.36) by $\widehat{T}^*$ and integrate the result from $z = -1/2$ to $z = +1/2$, where z is the dimensionless vertical coordinate, to find: $\sigma I_1 + I_2 = \int \widehat{T}^* W dz$, where $I_1 \equiv \int |\widehat{T}|^2 dz$, $I_2 \equiv \int [|d\widehat{T}/dz|^2 + K^2 |\widehat{T}|^2] dz$, and the limits of the integrations have been suppressed for clarity.

   b) Multiply (11.37) by W^* and integrate from $z = -\frac{1}{2}$ to $z = +\frac{1}{2}$ to find:

   $$\frac{\sigma}{Pr} J_1 + J_2 = RaK^2 \int W^* \widehat{T} dz \text{ where } J_1 \equiv \int [|dW/dz|^2 + K^2 |W|^2] dz,$$

   $J_2 \equiv \int [|d^2 W/dz^2|^2 + 2K^2 |dW/dz|^2 + K^4 |W|^2] dz$, and again the limits of the integrations have been suppressed.

   c) Combine the results of parts a) and b) to eliminate the mixed integral of W and $\widehat{T}$, and use the result of this combination to show that $\sigma_i = 0$ for Ra > 0. [*Note:* The integrals I_1, I_2, J_1, and J_2 are all positive definite.]

11.7. Consider the thermal instability of a fluid confined between two rigid plates, as discussed in Section 11.4. It was stated there without proof that the minimum critical Rayleigh number of $Ra_{cr} = 1708$ is obtained for the gravest *even* mode. To verify this, consider the gravest *odd* mode for which

$$W = A \sin q_0 z + B \sinh q z + C \sinh q^* z.$$

(Compare this with the gravest even mode structure: $W = A \cos q_0 z + B \cosh q z + C \cosh q^* z$.) Following Chandrasekhar (1961, p. 39), show that the minimum Rayleigh number is now 17,610, reached at the wave number $K_{cr} = 5.365$.

11.8. Consider the centrifugal instability problem of Section 11.6. Making the narrow-gap approximation, work out the algebra of going from (11.50) to (11.51).

11.9. Consider the centrifugal instability problem of Section 11.6. From (11.51) and (11.53), the eigenvalue problem for determining the marginal state ($\sigma = 0$) is

$$(d^2/dR^2 - k^2)^2 \widehat{u}_R = (1 + \alpha x) \widehat{u}_\varphi, \quad \left(d^2/dR^2 - k^2 \right)^2 \widehat{u}_\varphi = -Tak^2 \widehat{u}_R, \qquad \text{(11.92, 11.93)}$$

with $\widehat{u}_R = d\widehat{u}_R/dR = \widehat{u}_\varphi = 0$ at $x = 0$ and 1. Conditions on $\widehat{u}_\varphi$ are satisfied by assuming solutions of the form

$$\widehat{u}_\varphi = \sum_{m=1}^{\infty} C_m \sin(m\pi x). \qquad \text{(11.94)}$$

Inserting this into (11.92), obtain an equation for $\widehat{u}_R$, and arrange so that the solution satisfies the four remaining conditions on $\widehat{u}_R$. With $\widehat{u}_R$ determined in this manner and $\widehat{u}_\varphi$ given by (11.94), (11.93) leads to an eigenvalue problem for Ta(k). Following Chandrasekhar (1961, p. 300), show that the minimum Taylor number is given by (11.54) and is reached at $k_{cr} = 3.12$.

11.10. For a Kelvin-Helmholtz instability in a continuously stratified ocean, obtain a globally integrated energy equation in the form

$$\frac{1}{2}\frac{d}{dt}\int \left(u^2 + w^2 + g^2\rho^2/\rho_0^2 N^2\right)dV = -\int uw\frac{\partial U}{\partial z}dV.$$

(As in Figure 11.25, the integration in x takes place over an integer number of wavelengths.) Discuss the physical meaning of each term and the mechanism of instability.

11.11. [1]Consider the inviscid stability of a constant vorticity layer of thickness h between uniform streams with flow speeds U_1 and U_3. Region 1 lies above the layer, $y > h/2$ with $U(y) = U_1$. Region 2 lies within the layer, $|y| \leq h/2$, $U(y) = 1/2(U_1 + U_3) + (U_1 - U_3)(y/h)$. Region 3 lies below the layer, $y < -h/2$ with $U(y) = U_3$.

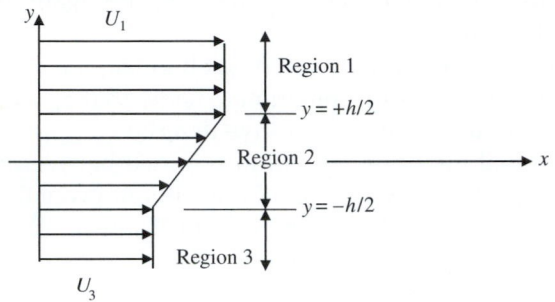

a) Solve the Rayleigh equation, $f'' - k^2 f - \dfrac{fU''}{U - c} = 0$, in each region, then use appropriate boundary and matching conditions to obtain:

$$f_1(y) = (A\cosh(kh/2) + B\sinh(kh/2))e^{-k(y-h/2)} \quad \text{for } y > +h/2,$$
$$f_2(y) = A\cosh(ky) + B\sinh(ky) \quad \text{for } |y| \leq h/2,$$
$$f_3(y) = (A\cosh(kh/2) - B\sinh(kh/2))e^{+k(y+h/2)} \quad \text{for } y < -h/2.$$

where f defines the spatial extent of the disturbance: $v' = f(y)e^{ik(x-ct)}$ and $u' = -(f'/ik)e^{ik(x-ct)}$, and A and B are undetermined constants.

b) The linearized horizontal momentum equation is: $\dfrac{\partial u'}{\partial t} + U\dfrac{\partial u'}{\partial x} + v'\dfrac{\partial U}{\partial y} = -\dfrac{1}{\rho}\dfrac{\partial p'}{\partial x}.$

Integrate this equation with respect to x, require the pressure to be continuous at $y = \pm h/2$, and simplify your results to find two additional constraint equations:

$$(c - U_1)f_1'(+h/2) = (c - U_1)f_2'(+h/2) + \frac{U_1 - U_3}{h}f_2(+h/2), \text{ and}$$
$$(c - U_3)f_3'(-h/2) = (c - U_3)f_2'(-h/2) + \frac{U_1 - U_3}{h}f_2(-h/2).$$

[1]Developed from Sherman, F. S. (1990).*Viscous Fluid Flow*. New York: McGraw-Hill, pp. 466–467.

c) Define $c_o = c - (1/2)(U_1 + U_3)$ (this is the phase speed of the disturbance waves in a frame of reference moving at the average speed), and use the results of parts a) and b) to determine a single equation for c_o:

$$c_o^2 = \left(\frac{U_1 - U_3}{2kh}\right)^2 \left\{(kh - 1)^2 - e^{-2kh}\right\}.$$

[This part of this problem requires patience and algebraic skill.]

d) From the result of part c), c_o will be real for $kh \gg 1$ (short wave disturbances), so the flow is stable or neutrally stable. However, for $kh \ll 1$ (long wave disturbances), use the result of part c) to show that:

$$c_o \cong \pm i\left(\frac{U_1 - U_3}{2}\right)\sqrt{1 - \frac{4}{3}kh + \dots}.$$

e) Determine the largest value of kh at which the flow is unstable.

11.12. Consider the inviscid instability of parallel flows given by the Rayleigh equation:

$$(U - c)\left(\frac{d^2\hat{v}}{dy^2} - k^2\hat{v}\right) - \frac{d^2U}{dy^2}\hat{v} = 0, \tag{11.95}$$

where the y-component of the perturbation velocity is $v = \hat{v}(y) \exp\{ik(x - ct)\}$.

a) Note that this equation is identical to the Rayleigh equation (11.81) for the stream function amplitude ϕ, as it must be because $\hat{v}(y) = -ik\phi$. For a flow bounded by walls at y_1 and y_2, note that the boundary conditions are identical in terms of ϕ and $\hat{v}$.

b) Show that if c is an eigenvalue of (11.95), then so is its conjugate $c^* = c_r - ic_i$. What aspect of (11.95) allows this result to be valid?

c) Let $U(y)$ be *antisymmetric*, so that $U(y) = -U(-y)$. Demonstrate that if $c(k)$ is an eigenvalue, then $-c(k)$ is also an eigenvalue. Explain the result physically in terms of the possible directions of propagation of perturbations in such an antisymmetric flow.

d) Let $U(y)$ be *symmetric* so that $U(y) = U(-y)$. Show that in this case $\hat{v}$ is either symmetric or antisymmetric about $y = 0$.

[*Hint*: Letting $y \rightarrow -y$, show that the solution $\hat{v}(-y)$ satisfies (11.95) with the same eigenvalue c. Form a symmetric solution, $S(y) = \hat{v}(y) + \hat{v}(-y) = S(-y)$, and an antisymmetric solution, $A(y) = \hat{v}(y) - \hat{v}(-y) = -A(-y)$. Then write $A[S\text{-eqn}] - S[A\text{-eqn}] = 0$ where S-eqn indicates the differential equation (11.95) in terms of S. Canceling terms this reduces to $(SA' - AS')' = 0$, where the prime (') indicates a y-derivative. Integration gives $SA' - AS' = 0$, where the constant of integration is zero because of the boundary conditions. Another integration gives $S = bA$, where b is a constant of integration. Because the symmetric and antisymmetric functions cannot be proportional, it follows that one of them must be zero.]

Comments: If v is symmetric, then the cross-stream velocity has the same sign across the entire flow, although the sign alternates every half wavelength along the flow. This mode is consequently called *sinuous*. On the other hand, if v is antisymmetric, then the shape of the jet expands and contracts along the length. This mode is now generally called the *sausage* instability because it resembles a line of linked sausages.

11.13. Derive (11.88) starting from the incompressible Navier-Stokes momentum equation for the disturbed flow:

$$\frac{\partial}{\partial t}(U_i + u_i) + (U_j + u_j)\frac{\partial}{\partial x_j}(U_i + u_i) = -\frac{1}{\rho}\frac{\partial}{\partial x_i}(P + p) + \nu\frac{\partial^2}{\partial x_j \partial x_j}(U_i + u_i), \qquad (11.96)$$

where U_i and u_i represent the basic flow and the disturbance, respectively. Subtract the equation of motion for the basic state from (11.96), multiply by u_i, and integrate the result within a stationary volume having stream-wise control surfaces chosen to coincide with the walls where no-slip conditions are satisfied or where $u_i \to 0$, and having a length (in the stream-wise direction) that is an integer number of disturbance wavelengths.

11.14. [2]The process of transition from laminar to turbulent flow may be driven both by exterior flow fluctuations and nonlinearity. Both of these effects can be simulated with the simple nonlinear logistic map $x_{n+1} = Ax_n(1 - x_n)$ and a computer spreadsheet program. Here, x_n can be considered to be the flow speed at the point of interest with A playing the role of the nonlinearity parameter (Reynolds number), x_0 (the initial condition) playing the role of an external disturbance, and iteration of the equation playing the role of increasing time. The essential feature illustrated by this problem is that increasing the nonlinearity parameter or changing the initial condition in the presence of nonlinearity may fully alter the character of the resulting sequence of x_n values. Plotting x_n vs. n should aid understanding for parts b) through e).

a) Determine the background solution of the logistic map that occurs when $x_{n+1} = x_n$ in terms of A.

Now, using a spreadsheet program, set up a column that computes x_{n+1} for $n = 1$ to 100 for user selectable values of x_0 and A for $0 < x_0 < 1$, and $0 < A < 4$.

b) For $A = 1.0$, 1.5, 2.0, and 2.9, choose a few different values of x_0 and numerically determine if the background solution is reached by $n = 100$. Is the flow *stable* for these values of A, i.e., does it converge toward the background solution?

c) For the slightly larger value, $A = 3.2$, choose $x_0 = 0.6875$, 0.6874, and 0.6876. Is the flow stable or oscillatory in these three cases? If it is oscillatory, how many iterations are needed for it to repeat?

d) For $A = 3.5$, is the flow stable or oscillatory? If it is oscillatory, how many iterations are needed for it to repeat? Does any value of x_0 lead to a stable solution?

e) For $A = 3.9$, is the flow stable, oscillatory, or chaotic? Does any value of x_0 lead to a stable solution?

[2]Provided to the third author by Professor Werner Dahm.

Literature Cited

Bayly, B. J., Orszag, S. A., & Herbert, T. (1988). Instability mechanisms in shear-flow transition. *Annual Review of Fluid Mechanics, 20,* 359–391.

Bergé, P., Pomeau, Y., & Vidal, C. (1984). *Order Within Chaos.* New York: Wiley.

Bhattacharya, P., Manoharan, M. P., Govindarajan, R., & Narasimha, R. (2006). The critical Reynolds number of a laminar incompressible mixing layer from minimal composite theory. *Journal of Fluid Mechanics, 565,* 105–114.

Chandrasekhar, S. (1961). *Hydrodynamic and Hydromagnetic Stability.* London: Oxford University Press. New York: Dover reprint, 1981.

Coles, D. (1965). Transition in circular Couette flow. *Journal of Fluid Mechanics, 21,* 385–425.

Davies, P. (1988). *Cosmic Blueprint.* New York: Simon and Schuster.

Drazin, P. G., & Reid, W. H. (1981). *Hydrodynamic Stability.* London: Cambridge University Press.

Eckhardt, B., Schneider, T. M., Hof, B., & Westerweel, J. (2007). Turbulence transition in pipe flow. *Annual Review of Fluid Mechanics, 39,* 447–468.

Eriksen, C. C. (1978). Measurements and models of fine structure, internal gravity waves, and wave breaking in the deep ocean. *Journal of Geophysical Research, 83,* 2989–3009.

Feigenbaum, M. J. (1978). Quantitative universality for a class of nonlinear transformations. *Journal of Statistical Physics, 19,* 25–52.

Fjørtoft, R. (1950). Application of integral theorems in deriving criteria of instability for laminar flows and for the baroclinic circular vortex. *Geofysiske Publikasjoner Oslo, 17*(6), 1–52.

Fransson, J. H. M., & Alfredsson, P. H. (2003). On the disturbance growth in an asymptotic suction boundary layer. *Journal of Fluid Mechanics, 482,* 51–90.

Grabowski, W. J. (1980). Nonparallel stability analysis of axisymmetric stagnation point flow. *Physics of Fluids, 23,* 1954–1960.

Heisenberg, W. (1924). Über Stabilität und Turbulenz von Flüssigkeitsströmen. *Annalen der Physik (Leipzig), 74*(4), 577–627.

Howard, L. N. (1961). Note on a paper of John W. Miles. *Journal of Fluid Mechanics, 13,* 158–160.

Huerre, P., & Monkewitz, P. A. (1990). Local and global instabilities in spatially developing flows. *Annual Review of Fluid Mechanics, 22,* 473–537.

Huppert, H. E., & Turner, J. S. (1981). Double-diffusive convection. *Journal of Fluid Mechanics, 106,* 299–329.

Jefferies, H. (1928). Some cases of instability in fluid motion. *Proceedings of the Royal Society London A, 118,* 195–208.

Klebanoff, P. S., Tidstrom, K. D., & Sargent, L. H. (1962). The three-dimensional nature of boundary layer instability. *Journal of Fluid Mechanics, 12,* 1–34.

Landau, L. D., & Lifshitz, E. M. (1959). *Fluid Mechanics.* Oxford, England: Pergamon Press.

Lanford, O. E. (1982). The strange attractor theory of turbulence. *Annual Review of Fluid Mechanics, 14,* 347–364.

Lin, C. C. (1955). *The Theory of Hydrodynamic Stability.* London: Cambridge University Press.

Lorenz, E. (1963). Deterministic nonperiodic flows. *Journal of Atmospheric Sciences, 20,* 130–141.

Miles, J. W. (1961). On the stability of heterogeneous shear flows. *Journal of Fluid Mechanics, 10,* 496–508.

Miles, J. W. (1986). Richardson's criterion for the stability of stratified flow. *Physics of Fluids, 29,* 3470–3471.

Nayfeh, A. H., & Saric, W. S. (1975). Nonparallel stability of boundary layer flows. *Physics of Fluids, 18,* 945–950.

Reshotko, E. (2001). Transient growth: A factor in bypass transition. *Physics of Fluids, 13,* 1067–1075.

Ruelle, D., & Takens, F. (1971). On the nature of turbulence. *Communications in Mathematical Physics, 20,* 167–192.

Schubauer, G. B., & Skramstad, H. K. (1947). Laminar boundary-layer oscillations and transition on a flat plate. *Journal of Research of the National Bureau of Standards, 38,* 251–292.

Scotti, R. S., & Corcos, G. M. (1972). An experiment on the stability of small disturbances in a stratified free shear layer. *Journal of Fluid Mechanics, 52,* 499–528.

Shen, S. F. (1954). Calculated amplified oscillations in plane Poiseuille and Blasius Flows. *Journal of the Aeronautical Sciences, 21,* 62–64.

Squire, H. B. (1933). On the stability of three dimensional disturbances of viscous flows between parallel walls. *Proceedings of the Royal Society London A, 142,* 621–628.

Stern, M. E. (1960). The salt fountain and thermohaline convection. *Tellus, 12,* 172–175.

Stommel, H., Arons, A. B., & Blanchard, D. (1956). An oceanographic curiosity: The perpetual salt fountain. *Deep-Sea Research, 3,* 152–153.

Thorpe, S. A. (1971). Experiments on the instability of stratified shear flows: Miscible fluids. *Journal of Fluid Mechanics, 46,* 299–319.

Turner, J. S. (1973). *Buoyancy Effects in Fluids.* London: Cambridge University Press.

Turner, J. S. (1985). Convection in multicomponent systems. *Naturwissenschaften, 72,* 70–75.

Woods, J. D. (1969). On Richardson's number as a criterion for turbulent–laminar transition in the atmosphere and ocean. *Radio Science, 4,* 1289–1298.

Yih, C. S. (1979). *Fluid Mechanics: A Concise Introduction to the Theory.* Ann Arbor, MI: West River Press.

OUTLINE

12.1. Introduction   542

12.2. Historical Notes   544

12.3. Nomenclature and Statistics for Turbulent Flow   545

12.4. Correlations and Spectra   549

12.5. Averaged Equations of Motion   554

12.6. Homogeneous Isotropic Turbulence   560

12.7. Turbulent Energy Cascade and Spectrum   564

12.8. Free Turbulent Shear Flows   571

12.9. Wall-Bounded Turbulent Shear Flows   581

12.10. Turbulence Modeling   591

12.11. Turbulence in a Stratified Medium   596

12.12. Taylor's Theory of Turbulent Dispersion   601

12.13. Concluding Remarks   607

Exercises   608

Literature Cited   618

Supplemental Reading   620

CHAPTER OBJECTIVES

- To introduce and describe turbulent flow

- To define the statistics and functions commonly used to quantify turbulent flow phenomena

- To derive the Reynolds-averaged equations

- To present assumptions and approximations leading to the classical scaling laws for turbulent flow

- To provide useful summaries of mean flow results for free and wall-bounded turbulent shear flows

- To introduce the basic elements of turbulence modeling

- To describe basic turbulence phenomena relevant in atmospheric turbulence

12.1. INTRODUCTION

Nearly all macroscopic flows encountered in the natural world and in engineering practice are turbulent. Winds and currents in the atmosphere and ocean; flows through residential, commercial, and municipal water (and air) delivery systems; flows past transportation devices (cars, trains, aircraft, ships, etc.); and flows through turbines, engines, and reactors used for power generation and conversion are all turbulent. Turbulence is an enigmatic state of fluid flow that may be simultaneously beneficial and problematic. For example, in air-breathing combustion systems, it is exploited for mixing reactants but, within the same device, it also leads to noise and efficiency losses. Within the earth's ocean and atmosphere, turbulence sets the mass, momentum, and heat transfer rates involved in pollutant dispersion and climate regulation.

Turbulence involves fluctuations that are unpredictable in detail, and it has not been conquered by deterministic or statistical analysis. However, useful predictions about it are still possible and these may arise from physical intuition, dimensional arguments, direct numerical simulations, or empirical models and computational schemes. In spite of our everyday experience with it, turbulence is not easy to define precisely and there is a tendency to confuse turbulence with randomness. A turbulent fluid velocity field conserves mass, momentum, and energy while a purely random time-dependent vector field need not. With some humor, Lesieur (1987) said:

> Turbulence is a dangerous topic which is at the origin of serious fights in scientific meetings since it represents extremely different points of view, all of which have in common their complexity, as well as an inability to solve the problem. It is even difficult to agree on what exactly is the problem to be solved. (p. 000)

This chapter presents basic features of turbulence beginning with this listing of generic characteristics.

(1) *Fluctuations*: Turbulent flows contain fluctuations in the dependent-field quantities (velocity, pressure, temperature, etc.) even when the flow's boundary conditions are steady. Turbulent fluctuations appear to be irregular, chaotic, and unpredictable.

(2) *Nonlinearity*: Turbulence is found to occur when the relevant nonlinearity parameter, say the Reynolds number Re, the Rayleigh number Ra, or the inverse Richardson number Ri^{-1}, exceeds a critical value. The nonlinearity of turbulence is evident since it is the final state of a nonlinear transition process. Once the critical parametric value is exceeded small perturbations can grow spontaneously and may equilibrate as finite amplitude disturbances. However, the new equilibrium state can become unstable to more complicated disturbances, and so on, until the flow eventually reaches a nonrepeating unpredictable state (turbulence). The nonlinearity of a turbulent flow is also evident in vortex stretching, a key process by which three-dimensional turbulent flows maintain their fluctuations.

(3) *Vorticity*: Turbulence is characterized by fluctuating vorticity. A cross-section view of a turbulent flow typically appears as a diverse collection of streaks, strain regions, and swirls of various sizes that deform, coalesce, divide, and spin. Identifiable structures in a turbulent flow, particularly those that spin, are called *eddies*. Turbulence always

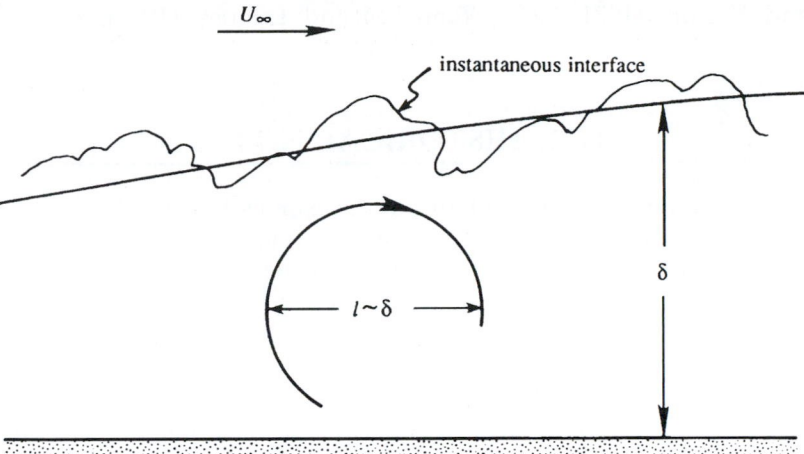

FIGURE 12.1 Turbulent boundary-layer flow showing a typical large eddy of size l, the average layer thickness δ, and the instantaneous interface between turbulent and nonturbulent (typically irrotational) fluid. Here, as in most turbulent flows, the size of the largest eddies is comparable to the overall layer thickness.

involves a range of eddy sizes and the size range increases with increasing Reynolds number. The characteristic size of the largest eddies is the width of the turbulent region; in a turbulent boundary layer this is the thickness of the layer (Figure 12.1). Such layer-spanning eddies commonly contain most of the fluctuation energy in a turbulent flow and may be several orders of magnitude larger than the smallest eddies.

(4) *Dissipation*: On average, the vortex stretching mechanism transfers fluctuation energy and vorticity to smaller and smaller scales via nonlinear interactions, until velocity gradients become so large that the energy is converted into heat (i.e., dissipated) by the action of viscosity and the motion of the smallest eddies. Persistent turbulence therefore requires a continuous supply of energy to make up for this energy loss.

(5) *Diffusivity*: Due to the prevalence of agitation and overturning motions (macroscopic mixing), turbulent flows are characterized by a rapid rate of mixing and diffusion of species, momentum, and heat compared to equivalent laminar flows that lack fluctuations.

These features of turbulence suggest that many flows that seem random, such as wind-driven ocean-surface waves or internal waves in the ocean or the atmosphere, are not turbulent because they are not simultaneously dissipative, vortical, and nonlinear.

Although imperfect, a simple definition of turbulence as *a dissipative flow state characterized by nonlinear fluctuating three-dimensional vorticity* is offered for the reader who may benefit from keeping a concise description in mind while gaining a greater appreciation of this subject. Incompressible turbulent mean flows in systems not large enough to be influenced by the Coriolis force are emphasized in this chapter. The fluctuations in such flows are three dimensional. In large-scale geophysical systems, on the other hand, the existence of stratification and the Coriolis force severely restricts vertical motion and leads to chaotic flow that may be nearly two dimensional or *geostropic*. *Geostrophic turbulence* is briefly mentioned in Chapter 13. More extensive treatments of turbulence are provided in Monin and Yaglom (1971, 1975), Tennekes and Lumley (1972), Hinze (1975), and Pope (2000).

12.2. HISTORICAL NOTES

Turbulence is a leading topic in modern fluid dynamics research, and some of the best-known physicists have worked in this area during the last century. Among them are G. I. Taylor, Kolmogorov, Reynolds, Prandtl, von Karman, Heisenberg, Landau, Millikan, and Onsagar. A brief historical outline is given in what follows; further interesting details can be found in Monin and Yaglom (1971). The reader is expected to fully appreciate these historical remarks only after reading the chapter.

The first systematic work on turbulence was carried out by Osborne Reynolds in 1883. His experiments in pipe flows showed that the flow becomes turbulent or irregular when the dimensionless ratio $Re = UL/\nu$, later named the *Reynolds number* by Sommerfeld, exceeds a certain critical value. (Here ν is the kinematic viscosity, U is the velocity scale, and L is the length scale.) This dimensionless number subsequently proved to be the parameter that determines the dynamic similarity of viscous flows. Reynolds also separated turbulent flow-dependent variables into mean and fluctuating components, and arrived at the concept of turbulent stress. The meaning of the Reynolds number and the existence of turbulent stresses are foundational elements in our present understanding of turbulence.

In 1921 the British physicist G. I. Taylor, in a simple and elegant study of turbulent diffusion, introduced the idea of a correlation function. He showed that the root-mean-square distance of a particle from its source point initially increases with time as t, and subsequently as $t^{1/2}$, as in a random walk. Taylor continued his outstanding work in a series of papers during 1935−1936 in which he laid down the foundation of the statistical theory of turbulence. Among the concepts he introduced were those of homogeneous and isotropic turbulence and of a turbulence spectrum. Although real turbulent flows are not isotropic (turbulent shear stresses, in fact, vanish for isotropic flows), the mathematical techniques involved have proved valuable for describing the *small scales* of turbulence, which are isotropic or nearly so. In 1915 Taylor also introduced the mixing length concept, although it is generally credited to Prandtl for making full use of the idea.

During the 1920s Prandtl and his student von Karman, working in Göttingen, Germany, developed semi-empirical theories of turbulence. The most successful of these was the mixing length theory, which is based on an analogy with the concept of mean free path in the kinetic theory of gases. By guessing at the correct form for the mixing length, Prandtl was able to deduce that the average turbulent velocity profile near a solid wall is logarithmic, one of the most reliable results for turbulent flows. It is for this reason that subsequent textbooks on fluid mechanics have for a long time glorified the mixing length theory. Recently, however, it has become clear that the mixing length theory is not helpful since there is really no rational way of predicting the form of the mixing length. In fact, the logarithmic law can be justified from dimensional considerations alone.

Some very important work was done by the British meteorologist Lewis Richardson. In 1922 he wrote the very first book on numerical weather prediction. In this book he proposed that the turbulent kinetic energy is transferred from large to small eddies, until it is destroyed by viscous dissipation. This idea of a spectral energy cascade is at the heart of our present understanding of turbulence. However, Richardson's work was largely ignored at the time, and it was not until some 20 years later that the idea of a spectral cascade took a quantitative shape in the hands of Kolmogorov and Obukhov in Russia. Richardson also did another important piece of work that displayed his amazing physical intuition. On the basis of experimental data for the movement of balloons in the atmosphere, he proposed that the effective diffusion coefficient of a patch of turbulence is proportional to $l^{4/3}$, where l is the scale of the patch. This is called *Richardson's four-third law*, which has been subsequently found to be in agreement with Kolmogorov's famous five-third law for the energy spectrum.

The Russian mathematician Kolmogorov, generally regarded as the greatest probabilist of the twentieth century, followed up on Richardson's idea of a spectral energy cascade. He hypothesized that the statistics of small scales are isotropic and depend on only two parameters, namely ν, the kinematic viscosity, and $\bar{\varepsilon}$, the average rate of kinetic energy dissipation per unit mass of fluid. On dimensional grounds, he derived that the smallest scales must be of size $\eta = (\nu^3 / \bar{\varepsilon})^{1/4}$. His second hypothesis was that, at scales much smaller than l (see Figure 12.1) and much larger than η, there must exist an inertial subrange of turbulent eddy sizes for which ν plays no role; in this range the statistics depend only on a single parameter $\bar{\varepsilon}$. Using this idea, in 1941 Kolmogorov and Obukhov independently derived that the spectrum in the inertial subrange must be proportional to $\varepsilon^{2/3}k^{-5/3}$, where k is the wave number. The five-thirds law is one of the most important results of turbulence theory and is in agreement with high Reynolds number observations.

Recent decades have seen much progress in theory, calculations, and measurements. Among these may be mentioned the work on the modeling, coherent structures, direct numerical simulations, and multidimensional diagnostics. Observations in the ocean and the atmosphere (which von Karman called "a giant laboratory for turbulence research"), in which the Reynolds numbers are very large, are shedding new light on the structure of stratified turbulence.

12.3. NOMENCLATURE AND STATISTICS FOR TURBULENT FLOW

The dependent-field variables in a turbulent flow (velocity components, pressure, temperature, etc.) are commonly analyzed and described using definitions and nomenclature borrowed from the theory of *stochastic processes* and *random variables* even though fluid-dynamic turbulence is not entirely random. Thus, the characteristics of turbulent-flow field variables are commonly specified in terms of their *statistics* or *moments*. In particular, a turbulent field quantity, $\tilde{\vartheta}$, is commonly separated into is first moment, $\overline{\vartheta}$, and its fluctuations, $\vartheta \equiv \tilde{\vartheta} - \overline{\vartheta}$, which have zero mean. This separation is known as the *Reynolds decomposition* and is further described and utilized in Section 12.5.

To define moments precisely, specific terminology is needed. A *collection* of independent realizations of a random variable, obtained under identical conditions, is called an *ensemble*. The ordinary arithmetic average over the collection is called an *ensemble average* and is denoted herein by an over bar. When the number N of realizations in the ensemble is large, $N \rightarrow \infty$, the ensemble average is called an *expected value* and is denoted with angle brackets.

With this terminology and notation, the *mth-moment*, $\overline{u^m}$, of the random variable u at location $\mathbf{x}$ and time t is defined as the ensemble average of u^m:

$$\langle u^m(\mathbf{x},t) \rangle = \lim_{N\to\infty} \overline{u^m(\mathbf{x},t)} \equiv \lim_{N\to\infty} \frac{1}{N} \sum_{n=1}^{N} (u(\mathbf{x},t{:}n))^m, \tag{12.1}$$

where $u(\mathbf{x},t{:}n)$ is nth the realization in the ensemble. The limit $N\to\infty$ can only be taken formally in theoretical analysis, so when dealing with measurements, $\overline{u^m}$ is commonly used in place of $\langle u^m \rangle$ and good experimental design ensures that N is large enough for reliable determination of the first few moments of u. Thus, the over-bar notation for ensemble average is favored in the remainder of this chapter. Collectively, the moments for integer values of m are known as the *statistics* of $u(\mathbf{x},t)$.

Under certain circumstances, ensemble averaging is not necessary for moment estimation. When u is *stationary in time*, its statistics do not depend on time, and $\overline{u^m}$ at $\mathbf{x}$ can be reliably estimated from time averaging:

$$\overline{u^m(\mathbf{x})} = \frac{1}{\Delta t} \int_{t-\Delta t/2}^{t+\Delta t/2} u^m(\mathbf{x},t)dt, \tag{12.2}$$

when Δt is large enough. Time averages are relevant for turbulent flows that persist with the same boundary conditions for long periods of time, an example being the turbulent boundary-layer flow on the hull of a long-range ship that traverses a calm sea at a constant speed. Example time histories of temporally stationary and nonstationary processes are shown in Figure 12.2. When u is *homogeneous* or *stationary in space*, its statistics do not depend on location, and $\overline{u^m}$ at time t can be reliably estimated from spatial averaging in a volume V,

$$\overline{u^m(t)} = \frac{1}{V} \int_V u^m(\mathbf{x},t)dV, \tag{12.3}$$

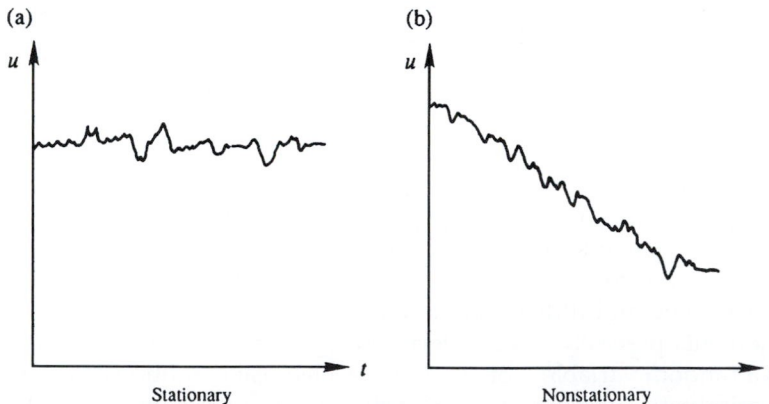

FIGURE 12.2   Sample time series indicating temporally stationary (a) and nonstationary processes (b). The time series in (b) clearly shows that the average value of u decreases with time compared to the time series in (a).

when V is large enough and defined appropriately. This type of average is often relevant in confined turbulent flows subject to externally imposed temporal variations, an example being the in-cylinder swirling and tumbling gas flow driven by piston motion and valve flows in an internal combustion piston engine.

For the discussions in this chapter, all moments denoted by over bars are ensemble averages determined from (12.1), unless otherwise specified. Equations (12.2) and (12.3) are provided here because they are commonly used to convert turbulent flow measurements into moment values. In particular, (12.2) or (12.3) are used in atmospheric and oceanic field measurements because ongoing natural phenomena like weather or the slow meandering of ocean currents make it is practically impossible to precisely repeat field observations under identical circumstances. For such measurements, a judicious selection of Δt or V is necessary; they should be long or large enough for reliable moment estimation but small enough so that the resulting statistics are only weakly influenced by ongoing natural variations.

Before defining and describing specific moments, several important properties of the process of ensemble averaging defined by (12.1) must be mentioned. First, ensemble averaging *commutes* with differentiation, that is, the application order of these two operators can be interchanged:

$$\frac{\overline{\partial u^m}}{\partial t} = \frac{1}{N}\sum_{n=1}^{N}\frac{\partial}{\partial t}(u(\mathbf{x},t{:}n))^m = \frac{\partial}{\partial t}\left(\frac{1}{N}\sum_{n=1}^{N}(u(\mathbf{x},t{:}n))^m\right) = \frac{\partial}{\partial t}\overline{u^m}.$$

Similarly, ensemble averaging commutes with addition, multiplication by a constant, time integration, spatial differentiation, and spatial integration. Thus the following are all true:

$$\overline{u^m + v^m} = \overline{u^m} + \overline{v^m}, \quad \overline{Au^m} = A\overline{u^m}, \quad \frac{\overline{\partial u^m}}{\partial t} = \frac{\partial}{\partial t}\overline{u^m}, \qquad (12.4,\ 12.5,\ 12.6)$$

$$\overline{\int_a^b u^m dt} = \int_a^b \overline{u^m} dt, \quad \frac{\overline{\partial u^m}}{\partial x_j} = \frac{\partial}{\partial x_j}\overline{u^m}, \quad \overline{\int u^m d\mathbf{x}} = \int \overline{u^m}d\mathbf{x}, \qquad (12.7,\ 12.8,\ 12.9)$$

where v is another random variable; a, b, m, and A are all constants; and $d\mathbf{x}$ represents a general spatial increment. In particular, (12.5) with $m = 0$ implies $\overline{A} = A$, so if $A = \overline{u}$ then $\overline{\overline{u}} = \overline{u}$; the ensemble average of an average is just the average. However, the ensemble average of a product of random variables is not necessarily the product of the ensemble averages. In general,

$$\overline{u^m} \neq \overline{u}^m \quad \text{and} \quad \overline{uv} \neq \overline{u}\,\overline{v},$$

when $m \neq 1$, and u and v are different random variables.

The simplest statistic of a random variable u is its *first moment, mean,* or *average,* $\overline{u}$. From (12.1) with $m = 1$, $\overline{u}$ is:

$$\overline{u(\mathbf{x},t)} \equiv \frac{1}{N}\sum_{n=1}^{N} u(\mathbf{x},t{:}n). \qquad (12.10)$$

In general, $\overline{u}$ may depend on both space and time, and is obtained by summing the N separate realizations of the ensemble, $u(\mathbf{x},t{:}n)$ for $1 \leq n \leq N$, at time t and location $\mathbf{x}$, and then dividing the sum by N. A graphical depiction of ensemble averaging, as specified by (12.10), is shown in

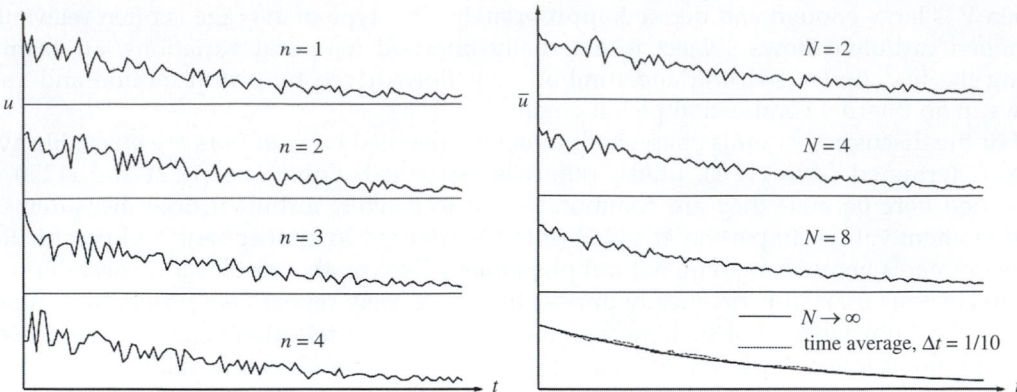

FIGURE 12.3 Illustration of ensemble and temporal averaging. The left panel shows four members of an ensemble of time series for the decaying random variable u. In all four cases, the fluctuations are different but the decreasing trend with increasing N is clearly apparent in each. The right panel shows averages of two, four, and eight members of the ensemble in the upper three plots. As the sample number N increases, fluctuations in the ensemble average decreases. The lowest plot on the right shows the $N \to \infty$ curve—this is the expected value of $u(t)$—and a simple sliding time average of the $n = 4$ curve where the duration of the time average is one-tenth of the time period shown. In this case, time and ensemble averaging produce nearly the same curve.

Figure 12.3 for time-series measurements recorded at the same point $\mathbf{x}$ in space. The left panel of the figure shows four members, $u(\mathbf{x},t{:}n)$ for $1 \leq n \leq 4$, of the ensemble. Here the average value of u decreases with increasing time. Time records such as these might represent atmospheric temperature measurements during the first few hours after sunset on different days, or they might represent a component of the flow velocity from an engine cylinder in the first 10 or 20 milliseconds after an exhaust valve opens. The right panel of Figure 12.3 shows the ensemble average $\overline{u}(\mathbf{x},t)$ obtained from the first two, four, and eight members of the ensemble. The solid smooth curve in the lower right panel of Figure 12.3 is the expected value that would be obtained from ensemble averaging in the limit $N \to \infty$. The dashed curve is a time average computed from only the fourth member of the ensemble using (12.2) with $m = 1$ and Δt equal to one-tenth of the total time displayed for each time history. Figure 12.3 clearly shows the primary effect of averaging is to suppress fluctuations since they become less prominent as N increases and are absent from the expected value. In addition, it shows that differences between an ensemble average of many realizations and a finite-duration temporal average of a single realization may be small, even when the flow is not stationary in time.

Although useful and important in many situations, the average or *first moment* alone does not directly provide information about turbulent fluctuations. Such information is commonly reported in terms of one or more *higher order central moments* defined by:

$$\overline{(u - \langle u \rangle)^m} \equiv \frac{1}{N} \sum_{n=1}^{N} (u(\mathbf{x},t{:}n) - \langle u(\mathbf{x},t) \rangle)^m, \tag{12.11}$$

where in practice $\overline{u}(\mathbf{x},t)$ often replaces $\langle u \rangle$. The central moments primarily carry information about the fluctuations since (12.11) explicitly shows that the mean is removed from each ensemble member. The first central moment is zero by definition. The next three have special names: $\overline{(u - \bar{u})^2}$ is the *variance* of u, $\overline{(u - \bar{u})^3}$ is the *skewness* of u, and $\overline{(u - \bar{u})^4}$ is the *kurtosis* of u. In addition, the square root of the variance, $\sqrt{\overline{(u - \bar{u})^2}}$, is known as the *standard deviation*, the square root of the second moment, $\sqrt{\overline{u^2}} = u_{rms}$, is known as the *root mean square*, and these are equal when $\bar{u} = 0$. In the study of turbulence, a field variable's first moment and variance are most important.

EXAMPLE 12.1

Compute the time average of the function $u(t) = Ae^{-t/\tau} + B\cos(\omega t)$ using (12.2). Presuming this function is meant to represent a turbulent field variable with zero-mean fluctuations, $B\cos(\omega t)$, superimposed on a decaying time-dependent average, $Ae^{-t/\tau}$, what condition on Δt leads to an accurate recovery of the decaying average? And, what condition on Δt leads to suppression of the fluctuations?

Solution

Start by directly substituting the given function into (12.2):

$$\overline{u(t)} = \frac{1}{\Delta t} \int\limits_{t-\Delta t/2}^{t+\Delta t/2} \left(Ae^{-t/\tau} + B\cos(\omega t) \right) dt,$$

and evaluating the integral:

$$\overline{u(t)} = \frac{1}{\Delta t}\left(-A\tau \exp\left(-\frac{t + \Delta t/2}{\tau}\right) + A\tau \exp\left(-\frac{t - \Delta t/2}{\tau}\right) + \frac{B}{\omega}\sin\left[\omega(t + \Delta t/2)\right] \right.$$
$$\left. - \frac{B}{\omega}\sin\left[\omega(t - \Delta t/2)\right] \right).$$

This can be simplified to find:

$$\overline{u(t)} = \left[\frac{\sinh(\Delta t/2\tau)}{\Delta t/2\tau}\right] Ae^{-t/\tau} + \left[\frac{\sin(\omega \Delta t/2)}{\omega \Delta t/2}\right] B\cos(\omega t).$$

In the limit $\Delta t \to 0$, both factors in [,]-braces go to unity and the original function is recovered. Thus, the condition for properly determining the decaying average is $\Delta t \ll \tau$; the averaging interval Δt must be short compared to the time scale for decay, τ. However, to suppress the contribution of the fluctuations represented by the second term, its coefficient must be small. This occurs when $\omega \Delta t \gg 1$ which implies the averaging interval must be many fluctuation time periods long. Therefore, a proper averaging interval should satisfy: $1 \ll \omega \Delta t \ll \omega \tau$, but such a choice for Δt is not possible unless $\omega \tau \gg 1$.

12.4. CORRELATIONS AND SPECTRA

While moments of a random variable are important and interesting, they do not convey information about the temporal duration or spatial extent of fluctuations, nor do they indicate anything about relationships between one or more dependent-field variables at different places and times. In the study of turbulence, correlations and spectra are commonly used to further characterize fluctuations and are described in this section. Furthermore, since we seek to describe fluctuations, all the random variables in this section are assumed to have zero mean, an assumption that is consistent with the Reynolds decomposition. The material presented here starts with general definitions that are simplified for a temporally stationary random variable sampled at the same point in space, or a spatially stationary random variable sampled at different points at the same time. Other approaches to specifying the temporal and spatial character of fluctuations, such as structure functions, fractal dimensions, multi-fractal spectra, and multiplier distributions, etc., are beyond the scope of this text.

In three spatial dimensions, the *correlation function* of the random variable u_i at location $\mathbf{x}_1$ and time t_1 with the random variable u_j at location $\mathbf{x}_2$ and time t_2 is defined as

$$R_{ij}(\mathbf{x}_1, t_1, \mathbf{x}_2, t_2) \equiv \overline{u_i(\mathbf{x}_1, t_1)u_j(\mathbf{x}_2, t_2)}, \tag{12.12}$$

where we will soon interpret u_i and u_j as turbulent-flow velocity-component fluctuations. Note that this R_{ij} is not the rotation tensor defined in Chapter 3. The correlation function R_{ij} can be computed via (12.1) when each realization of the ensemble contains time history pairs: $u_i(\mathbf{x}, t{:}n)$ and $u_j(\mathbf{x}, t{:}n)$. First, the N pairs $u_i(\mathbf{x}_1, t_1{:}n)$ and $u_j(\mathbf{x}_2, t_2{:}n)$ are selected from the realizations and multiplied together. Then the N pair-products are summed and divided by N to complete the calculation of R_{ij}.

The correlation R_{ij} specifies how similar $u_i(\mathbf{x}_1, t_1)$ and $u_j(\mathbf{x}_2, t_2)$ are to each other. The magnitude of R_{ij} is zero when positive values of $u_i(\mathbf{x}_1, t_1{:}n)$ are associated with equal likelihood with both positive and negative values of $u_j(\mathbf{x}_2, t_2{:}n)$. In this case, $u_i(\mathbf{x}_1, t_1)$ and $u_j(\mathbf{x}_2, t_2)$ are said to be *uncorrelated* when $R_{ij} = 0$, or *weakly correlated* if R_{ij} is small and positive. If, a positive value of $u_i(\mathbf{x}_1, t_1{:}n)$ is mostly associated with a positive value of $u_j(\mathbf{x}_2, t_2{:}n)$, and a negative value of $u_i(\mathbf{x}_1, t_1{:}n)$ is mostly associated with a negative value of $u_j(\mathbf{x}_2, t_2{:}n)$, then the magnitude of R_{ij} is large and positive. In this case, $u_i(\mathbf{x}_1, t_1)$ and $u_j(\mathbf{x}_2, t_2)$ are said to be *strongly correlated*. It is also possible for $u_i(\mathbf{x}_1, t_1{:}n)$ to be mostly associated with values of $u_j(\mathbf{x}_2, t_2{:}n)$ having the opposite sign so that R_{ij} is negative. In this case, $u_i(\mathbf{x}_1, t_1)$ and $u_i(\mathbf{x}_2, t_2)$ are said to be *anticorrelated*.

When $i \neq j$ in (12.12) the resulting function is called a *cross-correlation* function. When $i = j$ in (12.12) and $u_j(\mathbf{x}_2, t_2)$ is replaced by $u_i(\mathbf{x}_2, t_2)$, the resulting function is called an *autocorrelation* function; for example, $i = 1$ implies

$$R_{11}(\mathbf{x}_1, t_1, \mathbf{x}_2, t_2) \equiv \overline{u_1(\mathbf{x}_1, t_1)u_1(\mathbf{x}_2, t_2)}. \tag{12.13}$$

The two definitions, (12.12) and (12.13), may be normalized to define the *correlation coefficients*. For example when $i = 1$ and $j = 2$:

$$r_{12}(\mathbf{x}_1, t_1, \mathbf{x}_2, t_2) \equiv \frac{R_{12}(\mathbf{x}_1, t_1, \mathbf{x}_2, t_2)}{\sqrt{R_{11}(\mathbf{x}_1, t_1, \mathbf{x}_1, t_1) R_{22}(\mathbf{x}_2, t_2, \mathbf{x}_2, t_2)}} = \frac{\overline{u_1(\mathbf{x}_1, t_1)u_2(\mathbf{x}_2, t_2)}}{\sqrt{\overline{u_1^2(\mathbf{x}_1, t_1)}}\sqrt{\overline{u_2^2(\mathbf{x}_2, t_2)}}} \quad \text{and} \tag{12.14}$$

$$r_{11}(\mathbf{x}_1, t_1, \mathbf{x}_2, t_2) \equiv \frac{R_{11}(\mathbf{x}_1, t_1, \mathbf{x}_2, t_2)}{\sqrt{R_{11}(\mathbf{x}_1, t_1, \mathbf{x}_1, t_1) R_{11}(\mathbf{x}_2, t_2, \mathbf{x}_2, t_2)}} = \frac{\overline{u_1(\mathbf{x}_1, t_1)u_1(\mathbf{x}_2, t_2)}}{\sqrt{\overline{u_1^2(\mathbf{x}_1, t_1)}}\sqrt{\overline{u_1^2(\mathbf{x}_2, t_2)}}}, \tag{12.15}$$

which are restricted to lie between -1 (perfect anticorrelation) and $+1$ (perfect correlation). For any two functions u and v, it can be proved that

$$\overline{u(\mathbf{x}_1, t_1) v(\mathbf{x}_2, t_2)} \leq \sqrt{\overline{u^2(\mathbf{x}_1, t_1)}} \sqrt{\overline{v^2(\mathbf{x}_2, t_2)}}, \tag{12.16}$$

which is called the *Schwartz inequality*. It is analogous to the rule that the inner product of two vectors cannot be larger than the product of their magnitudes. Obviously, from (12.15), $r_{11}(\mathbf{x}_1, t_1, \mathbf{x}_1, t_1)$ is unity.

For temporally stationary processes that are sampled at the same point in space, $\mathbf{x} = \mathbf{x}_1 = \mathbf{x}_2$, the above formulas simplify, and the listing of $\mathbf{x}$ as an argument may be dropped to streamline the notation. The statistics of temporally stationary random processes are independent of the time origin, so we can shift the time origin to t_1 when computing a correlation so that $\overline{u_i(t_1) u_j(t_2)} = \overline{u_i(0) u_j(t_2 - t_1)} = \overline{u_i(0) u_j(\tau)}$, where $\tau = t_2 - t_1$ is the *time lag*, without changing the correlation. Or, we can change t_1 in (12.13) into t, $\overline{u_i(t) u_j(t_2)} = \overline{u_i(t) u_j(t + \tau)}$, without changing the correlation. Thus, the correlation and autocorrelation functions can be written:

$$R_{ij}(\tau) = \overline{u_i(t) u_j(t + \tau)} \text{ and } R_{11}(\tau) = \overline{u_1(t) u_1(t + \tau)}, \tag{12.17}$$

where the over bar can be regarded as either an ensemble or time average in this case. Furthermore, under these conditions, the autocorrelation is symmetric:

$$R_{11}(\tau) = \overline{u_1(t) u_1(t + \tau)} = \overline{u_1(t - \tau) u_1(t)} = \overline{u_1(t) u_1(t - \tau)} = R_{11}(-\tau).$$

However, this is not the case for cross correlations, $R_{ij}(\tau) \neq R_{ij}(-\tau)$ when $i \neq j$. The value of a cross-correlation function at $\tau = 0$, $\overline{u_i(t) u_j(t)}$, is simply written as $\overline{u_i u_j}$ and called the *correlation* of u_i and u_j.

Figure 12.4 illustrates several of these concepts for two temporally stationary random variables $u(t)$ and $v(t)$. The left panel shows $u(t)$, $u(t + t_o)$, $v(t)$, and the time shift t_o is indicated near the top. The right panel shows the autocorrelation of u, the cross correlation of u and u with an imposed time shift of t_o, and the cross correlation of u and v. The tic-mark spacing represents the same amount of time in both panels. The time shift necessary for $\overline{u(t) u(t + \tau)}$ to reach zero is comparable to the width of peaks or valleys of $u(t)$. As expected, the autocorrelation is maximum when the two time arguments of u under the ensemble average are equal, and this correlation peak is symmetric about this time shift. Correlation is a mathematical shape-comparison indicator that is sensitive to time alignment. Consider $u(t)$, $u(t + t_o)$, and $\overline{u(t + t_o) u(t + \tau)}$. When $\tau = 0$ the peaks and valleys of $u(t)$ and $u(t + t_o)$—which are of course are identical—are not temporally aligned so $\overline{u(t + t_o) u(t + \tau)}$ in Figure 12.4 is nearly zero when $\tau = 0$. However, as τ increases—this corresponds to moving the time history of $u(t)$ to the left—the peaks and valleys of $u(t)$ and $u(t + t_o)$ come closer into temporal alignment. Perfect alignment is reached when $\tau = t_o$ and this produces the correlation maximum in $\overline{u(t + t_o) u(t + \tau)}$ at $\tau = t_o$. The cross-correlation function results in Figure 12.4 can be understood in a similar manner by looking for temporal alignment in u and v as v slides to the left with increasing τ. As shown in the left panel, the largest peak of u is temporally aligned with the largest peak of v when $\tau = t_+$ and this leads to the positive correlation maximum in $\overline{u(t) v(t + \tau)}$ at $\tau = t_+$. However, as τ increases further, the largest peak of u becomes

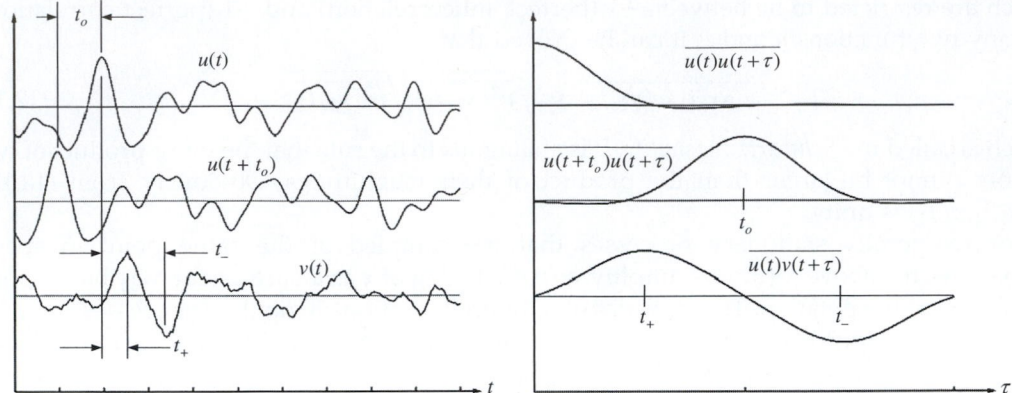

FIGURE 12.4 Sample results for auto- and cross-correlation functions of $u(t)$, $u(t + t_o)$, and $v(t)$. These three time series are shown on the left. The upper curve on the right is the autocorrelation function, $\overline{u(t)u(t + \tau)}$, of the upper time series on the left. The tic marks on the axes represent the same time interval so the width of a peak of $u(t)$ is about equal to the correlation time determined from $\overline{u(t)u(t + \tau)}$. The correlation of $u(t + t_o)$ and $u(t)$ is shown as the middle curve on the right, and it is just a shifted replica of $\overline{u(t)u(t + \tau)}$. The cross correlation of $u(t)$ and $v(t)$ is the lower curve on the right. Here the maximum cross correlation occurs when $\tau = t_+$ and the peaks of u and v coincide. Similarly, u and v are most anti-correlated when peaks in u align with valleys in v at $\tau = t_-$.

temporally aligned with the deepest valley of v, and this leads to the cross-correlation minimum at $\tau = t_-$. Thus, the zeros and extrema of correlation functions indicate the time shifts necessary to temporally misalign, align, or anti-align field-variable fluctuations. Such timing information cannot be obtained from moments.

Several time scales can be determined from the autocorrelation function. For turbulence, the most important of these is the *integral time scale* Λ_t. Under normal conditions R_{11} goes to 0 as $\tau \to \infty$ because the turbulent fluctuation u_1 becomes uncorrelated with itself after a long time. The integral time scale is found by equating the area under the autocorrelation coefficient curve to a rectangle of unity height and duration Λ_t:

$$\Lambda_t \equiv \int_0^\infty r_{11}(\tau)d\tau = (1/R_{11}(0)) \int_0^\infty R_{11}(\tau)d\tau, \tag{12.18}$$

where $r_{11}(\tau) = R_{11}(\tau)/R_{11}(0)$ is the autocorrelation coefficient for the stream-wise velocity fluctuation u_1. Of course, (12.18) can be written in terms of r_{22} or r_{33}, but for the purposes at hand this is not necessary. The calculation in (12.18) is shown graphically in Figure 12.5. The integral time scale is a generic specification of the time over which a turbulent fluctuation is correlated with itself. In other words, Λ_t is a measure of the *memory* of the turbulence. The correlation time t_c is also shown in Figure 12.5 as the time when $r_{11}(\tau)$ first reaches zero. When temporally averaging a single random-variable time history of length Δt to mimic an ensemble average, the equivalent number of ensemble members can be estimated from $N \approx \Delta t/t_c$. A third time scale, the Taylor microscale λ_t, can also be extracted from $r_{11}(\tau)$. It is obtained from the curvature of the autocorrelation peak at $\tau = 0$ and is given by:

$$\lambda_t^2 \equiv -2 \Big/ \left[d^2r_{11}/d\tau^2 \right]_{\tau=0} \tag{12.19}$$

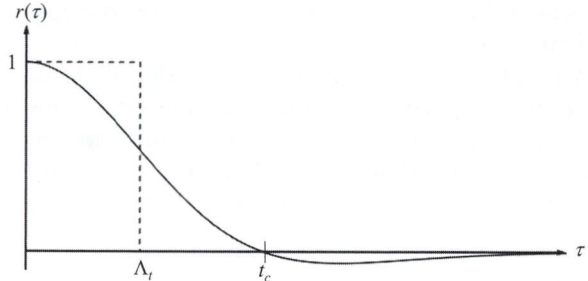

FIGURE 12.5 Sample plot of an autocorrelation coefficient showing the integral time scale Λ_t, and the correlation time t_c. The normalization requires $r(0) = 1$. In the limit $\tau \to \infty$, $r(\tau) \to 0$ and thereby indicates that the random process used to construct r becomes uncorrelated with itself when the time shift τ is large enough.

(see Exercise 12.9). The Taylor microscale λ_t is much less than Λ_t in high Reynolds number turbulence, and it indicates where a turbulent fluctuation, $u_1(t)$ in (12.19), is well correlated with itself.

A second (and equivalent) means of describing the characteristics of turbulent fluctuations, which also complements the information provided by moments, is the *energy spectrum* $S_e(\omega)$ defined as the Fourier transform of the autocorrelation function $R_{11}(\tau)$:

$$S_e(\omega) \equiv \frac{1}{2\pi} \int_{-\infty}^{+\infty} R_{11}(\tau) \exp\{-i\omega\tau\}\, d\tau. \qquad (12.20)$$

Thus, $S_e(\omega)$ and $R_{11}(\tau)$ are a Fourier transform pair:

$$R_{11}(\tau) \equiv \int_{-\infty}^{+\infty} S_e(\omega) \exp\{+i\omega\tau\}\, d\omega. \qquad (12.21)$$

The relationships (12.20) and (12.21) are not special for $S_e(\omega)$ and $R_{11}(\tau)$ alone, but hold for many function pairs for which a Fourier transform exists. Roughly speaking, a Fourier transform can be defined if the function decays to zero fast enough as its argument goes to infinity. Since $R_{11}(\tau)$ is real and symmetric, then $S_e(\omega)$ is real and symmetric (see Exercise 12.6). Substitution of $\tau = 0$ in (12.21) gives

$$\overline{u_1^2} \equiv \int_{-\infty}^{+\infty} S_e(\omega) d\omega. \qquad (12.22)$$

This shows that the integrand increment $S_e(\omega)d\omega$ is the contribution to the variance (or fluctuation energy) of u_1 from the frequency band $d\omega$ centered at ω. Therefore, the function $S_e(\omega)$ represents the way fluctuation energy is distributed across frequency ω. From (12.20) it also follows that

$$S_e(0) = \frac{1}{2\pi} \int_{-\infty}^{\infty} R_{11}(\tau)d\tau = \frac{\overline{u_1^2}}{\pi} \int_0^{\infty} r_{11}(\tau)d\tau = \frac{\overline{u_1^2}}{\pi}\Lambda_t,$$

which shows that the spectral value at zero frequency is proportional to the variance of u_1 and the integral time scale.

From (12.16) to this point, u_i and u_j have been considered stationary functions of time measured at the same point in space. In a similar manner, we now consider u_i and u_j to be stationary functions in space measured at the same instant in time t. For simplicity we drop the listing of t as an independent variable. In this case, the correlation tensor only depends on the vector separation between $\mathbf{x}_1$ and $\mathbf{x}_2$, $\mathbf{r} = \mathbf{x}_2 - \mathbf{x}_1$,

$$R_{ij}(\mathbf{r}) \equiv \overline{u_i(\mathbf{x})u_j(\mathbf{x}+\mathbf{r})}. \tag{12.23}$$

An instantaneous field measurement of $u_i(\mathbf{x})$ is needed to calculate the spatial autocorrelation $R_{ij}(\mathbf{x})$. This is a difficult task in three dimensions, although planar particle imaging velocimetry (PIV) makes it possible in two. However, single-point measurements of a time series $u_1(t)$ in turbulent flows are still quite common and spatial results may be obtained approximately by rapidly moving a probe in a desired direction. If the speed U_0 of the probe is high enough, we can assume that the field of turbulence is *frozen* and does not change while the probe moves through it during the measurement. Although the probe actually records a time series $u_1(t)$, it can be transformed into a spatial series $u_1(x)$ by replacing t by x/U_0. The assumption that the turbulent fluctuations at a point are caused by the advection of a frozen field past the point is called *Taylor's hypothesis*, and the accuracy of this approximation increases as the ratio u_{rms}/U_0 decreases.

12.5. AVERAGED EQUATIONS OF MOTION

In this section, the equations of motion for the mean state in a turbulent flow are derived. The contribution of turbulent fluctuations appears in these equations as a correlation of velocity-component fluctuations. A turbulent flow instantaneously satisfies the Navier-Stokes equations. However, it is virtually impossible to predict the flow in detail at high Reynolds numbers, as there is an enormous range of length and time scales to be resolved. Perhaps more importantly, we seldom want to know all the details. If a commercial aircraft must fly from Los Angeles, California, to Sydney, Australia, and turbulent skin-friction fluctuations occur in a frequency range from a few Hz to more than 10^4 Hz, the economically important parameter is the average skin friction because the time of the flight (many hours) is much longer than even the longest fluctuation time scale. Here, the integrated effect of the fluctuations approaches zero when compared to the integral of the average. This situation where the overall duration of the flow far exceeds turbulent-fluctuation time scales is very common in engineering and geophysical science.

The following development of the mean flow equations is for incompressible turbulent flow with constant viscosity where density fluctuations are caused by temperature fluctuations alone. The first step is to separate the dependent-field quantities into components representing the mean (capital letters and those with over bars) and those representing the deviation from the mean (lowercase letters and those with primes):

$$\tilde{u}_i = U_i + u_i, \; \tilde{p} = P + p, \; \tilde{\rho} = \overline{\rho} + \rho', \text{ and } \tilde{T} = \overline{T} + T' \tag{12.24}$$

where—as in the preceding chapter—the complete field quantities are denoted by a tilde (~). As mentioned in Section 12.3, this separation into mean and fluctuating components is called the *Reynolds decomposition*. Although it doubles the number of dependent-field variables, this decomposition remains useful and relevant more than a century after it was first proposed. However, it leads to a closure problem in the resulting equation set that has still not been resolved. The mean quantities in (12.24) are regarded as expected values,

$$\bar{\tilde{u}}_i = U_i, \ \bar{\tilde{p}} = P, \ \bar{\tilde{\rho}} = \bar{\rho}, \text{ and } \bar{\tilde{T}} = \overline{T}, \tag{12.25}$$

and the fluctuations have zero mean,

$$\overline{u}_i = 0, \ \overline{p} = 0, \ \overline{\rho'} = 0, \text{ and } \overline{T'} = 0. \tag{12.26}$$

The equations satisfied by the mean flow are obtained by substituting (12.24) into the governing equations and averaging. Here, the starting point is the Boussinesq set:

$$\frac{\partial \tilde{u}_i}{\partial t} + \tilde{u}_j \frac{\partial \tilde{u}_i}{\partial x_j} = \frac{\partial \tilde{u}_i}{\partial t} + \frac{\partial}{\partial x_j}(\tilde{u}_j \tilde{u}_i) = -\frac{1}{\rho_0} \frac{\partial \tilde{p}}{\partial x_i} - g\left[1 - \alpha(\tilde{T} - T_0)\right] \delta_{i3} + \nu \frac{\partial^2 \tilde{u}_i}{\partial x_j^2}, \tag{4.86}$$

$$\frac{\partial \tilde{u}_i}{\partial x_i} = 0, \text{ and } \frac{\partial \tilde{T}}{\partial t} + \tilde{u}_j \frac{\partial \tilde{T}}{\partial x_j} = \frac{\partial \tilde{T}}{\partial t} + \frac{\partial}{\partial x_j}(\tilde{u}_j \tilde{T}) = \kappa \frac{\partial^2 \tilde{T}}{\partial x_j^2}, \tag{4.10, 4.89}$$

where the first equality in (4.86) and (4.89) follows from adding $\tilde{u}_i(\partial \tilde{u}_j / \partial x_j) = 0$ and $\tilde{T}(\partial \tilde{u}_j / \partial x_j) = 0$, respectively, to the leftmost sides of these equations. Simplifications for constant-density flow are easily obtained at the end of this effort.

The continuity equation for the mean flow is obtained by putting the velocity decomposition of (12.24) into (4.10) and averaging:

$$\overline{\frac{\partial \tilde{u}_i}{\partial x_i}} = \overline{\frac{\partial}{\partial x_i}(U_i + u_i)} = \frac{\partial}{\partial x_i}\overline{(U_i + u_i)} = \frac{\partial}{\partial x_i}(U_i + \overline{u}_i) = \frac{\partial U_i}{\partial x_i} = 0, \tag{12.27}$$

where (12.8), $\overline{U}_i = U_i$, and $\overline{u}_i = 0$ have been used. Subtracting (12.27) from (4.10) produces:

$$\partial u_i / \partial x_i = 0. \tag{12.28}$$

Thus, the mean and fluctuating velocity fields are each divergence free.

The procedure for the mean momentum equation is similar but requires slightly more effort. Substituting (12.24) into (4.86) produces:

$$\frac{\partial(U_i + u_i)}{\partial t} + \frac{\partial}{\partial x_j}\left((U_j + u_j)(U_i + u_i)\right) = -\frac{1}{\rho_0}\frac{\partial(P + p)}{\partial x_i} - g\left[1 - \alpha(\overline{T} + T' - T_0)\right]\delta_{i3}$$

$$+ \nu \frac{\partial^2(U_i + u_i)}{\partial x_j^2}. \tag{12.29}$$

The averages of each term in this equation can be determined by using (12.26) and the properties of an ensemble average: (12.4) through (12.6) and (12.8). The term-by-term results are:

$$\overline{\frac{\partial(U_i + u_i)}{\partial t}} = \frac{\partial\overline{(U_i + u_i)}}{\partial t} = \frac{\partial(U_i + \overline{u}_i)}{\partial t} = \frac{\partial U_i}{\partial t},$$

$$\overline{\frac{\partial}{\partial x_j}\left((U_j + u_j)(U_i + u_i)\right)} = \frac{\partial}{\partial x_j}\overline{\left(U_iU_j + U_iu_j + u_iU_j + u_iu_j\right)} = \frac{\partial}{\partial x_j}\left(U_iU_j + \overline{u}_iU_j + U_i\overline{u}_j + \overline{u_iu_j}\right)$$

$$= \frac{\partial}{\partial x_j}\left(U_iU_j + \overline{u_iu_j}\right),$$

$$\overline{\frac{1}{\rho_0}\frac{\partial(P + p)}{\partial x_i}} = \frac{1}{\rho_0}\frac{\partial\overline{(P + p)}}{\partial x_i} = \frac{1}{\rho_0}\frac{\partial(P + \overline{p})}{\partial x_i} = \frac{1}{\rho_0}\frac{\partial P}{\partial x_i},$$

$$\overline{g[1 - \alpha(\overline{T} + T' - T_0)]\delta_{i3}} = g\left[1 - \alpha(\overline{T} + \overline{T'} - T_0)\right]\delta_{i3} = g\left[1 - \alpha(\overline{T} - T_0)\right]\delta_{i3}, \text{ and}$$

$$\overline{\nu\frac{\partial^2(U_i + u_i)}{\partial x_j^2}} = \nu\frac{\partial^2\overline{(U_i + u_i)}}{\partial x_j^2} = \nu\frac{\partial^2(U_i + \overline{u}_i)}{\partial x_j^2} = \nu\frac{\partial^2 U}{\partial x_j^2}.$$

Collecting terms, the ensemble average of the momentum equation is:

$$\frac{\partial U_i}{\partial t} + \frac{\partial}{\partial x_j}(U_iU_j) + \frac{\partial}{\partial x_j}(\overline{u_iu_j}) = -\frac{1}{\rho_0}\frac{\partial P}{\partial x_i} - g\left[1 - \alpha(\overline{T} - T_0)\right]\delta_{i3} + \nu\frac{\partial^2 U_i}{\partial x_j^2}.$$

This equation can be mildly rearranged by using the final result of (12.27) and combing the gradient terms together to form the mean stress tensor $\overline{\tau}_{ij}$:

$$\frac{\partial U_i}{\partial t} + U_j\frac{\partial U_i}{\partial x_j} = -g\left[1 - \alpha(\overline{T} - T_0)\right]\delta_{i3} + \frac{1}{\rho_0}\frac{\partial\overline{\tau}_{ij}}{\partial x_j}$$

$$= -g\left[1 - \alpha(\overline{T} - T_0)\right]\delta_{i3} + \frac{1}{\rho_0}\frac{\partial}{\partial x_j}\left(-P\delta_{ij} + 2\mu\overline{S}_{ij} - \rho_0\overline{u_iu_j}\right), \qquad (12.30)$$

where $\overline{S}_{ij} = 1/2(\partial U_i/\partial x_j + \partial U_j/\partial x_i)$ is the mean strain-rate tensor, and (4.40) has been used to put the mean viscous stress in the form shown in (12.30). The correlation tensor $\overline{u_iu_j}$ in (12.30) is generally nonzero even though $\overline{u}_i = 0$. Its presence in (12.30) is important because it has no counterpart in the instantaneous momentum equation (4.86) and it links the character of the fluctuations to the mean flow. Unfortunately, the process of reaching (12.30) does not provide any new equations for this correlation tensor. Thus, the final equality of (12.27), and (12.30) do not comprise a closed system of equations, even when the flow is isothermal.

The new tensor in (12.30), $-\rho_0\overline{u_iu_j}$, plays the role of a stress and is called the *Reynolds stress tensor*. When present, Reynolds stresses are often much larger than viscous stresses, $\mu(\partial U_i/\partial x_j + \partial U_i/\partial x_j)$, except very close to a solid surface where the fluctuations go to zero and mean flow gradients are large. The Reynolds stress tensor is symmetric since $\overline{u_iu_j} = \overline{u_ju_i}$, so it has six independent Cartesian components. Its diagonal components $\overline{u_1^2}$, $\overline{u_2^2}$, and $\overline{u_3^2}$ are normal stresses that augment the mean pressure, while its off-diagonal components $\overline{u_1u_2}$, $\overline{u_1u_3}$, and $\overline{u_2u_3}$ are shear stresses.

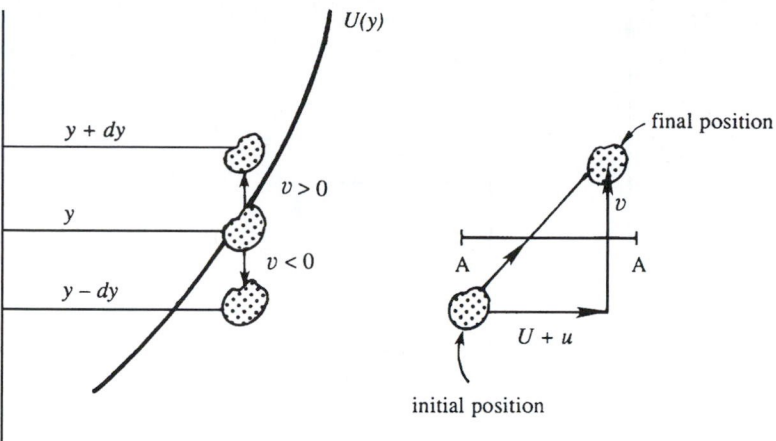

FIGURE 12.6 A schematic illustration of the development of nonzero Reynolds shear stress in a simple shear flow. A fluid particle that starts at y and is displaced upward to $y + dy$ by a positive vertical velocity fluctuation v brings an average horizontal fluid velocity of $U(y)$ that is lower than $U(y + dy)$. Thus, a positive vertical velocity fluctuation v is correlated with negative horizontal velocity fluctuation u, so $\overline{uv} < 0$. Similarly, a negative v displaces the fluid particle to $y - dy$ where it arrives on average with positive u, so again $\overline{uv} < 0$. Thus, turbulent fluctuations in shear flow are likely to produce negative nonzero Reynolds shear stress.

An explanation why the average product of the velocity fluctuations in a turbulent flow is not expected to be zero follows. Consider a shear flow where the mean shear dU/dy is positive (Figure 12.6). Assume that a fluid particle at level y travels upward because of a fluctuation ($v > 0$). On average this particle retains its original horizontal velocity during the migration, so when it arrives at level $y + dy$ it finds itself in a region where a larger horizontal velocity prevails. Thus the particle is on average slower ($u < 0$) than the neighboring fluid particles after it has reached the level $y + dy$. Conversely, fluid particles that travel downward ($v < 0$) tend to cause a positive u at their new level $y - dy$. Taken together, a positive v is associated with a negative u, and a negative v is associated with a positive u. Therefore, the correlation $\overline{uv}$ is negative for the velocity field shown in Figure 12.6, where $dU/dy > 0$. This makes sense, since in this case the x-momentum should tend to flow in the negative y-direction as the turbulence tends to diffuse the gradients and decrease dU/dy.

The Reynolds stresses arise from the nonlinear advection term $u_j(\partial u_i/\partial x_j)$ of the momentum equation, and are the average stress exerted by turbulent fluctuations on the mean flow. Another way to interpret the Reynolds stress is that it is the rate of mean momentum transfer by turbulent fluctuations. Consider again the shear flow $U(y)$ shown in Figure 12.6, where the instantaneous velocity is $(U + u, v, w)$. The fluctuating velocity components constantly transport fluid particles, and associated momentum, across a plane AA normal to the y-direction. The instantaneous rate of mass transfer across a unit area is $\rho_0 v$, and consequently the instantaneous rate of x-momentum transfer is $\rho_0(U + u)v$. Per unit area, the average rate of flow of x-momentum in the y-direction is therefore

$$\rho_0 \overline{(U + u)v} = \rho_0 U\overline{u} + \rho_0 \overline{uv} = \rho_0 \overline{uv}.$$

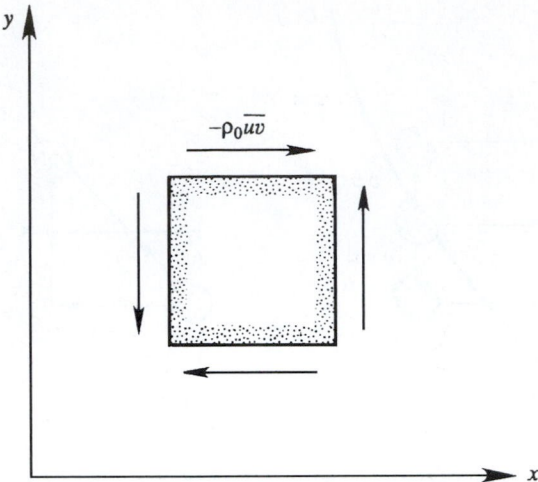

FIGURE 12.7 Positive directions of Reynolds stresses on a square element. These stress components are consistent with those drawn in Figure 2.4.

Generalizing, $\rho_0 \overline{u_i u_j}$ *is the average flux of i-momentum along the j-direction, which also equals the average flux of j-momentum along the i-direction.*

The sign convention for the Reynolds stress is the same as that explained in Section 2.4. On a surface whose outward normal points in the positive i-direction, a positive τ_{ij} points along the j-direction. According to this convention, the Reynolds shear stresses $-\rho_0 \overline{u_i u_j}$ $(i \neq j)$ on a rectangular element are directed as in Figure 12.7, if they are positive. Such a Reynolds stress causes mean transport of x-momentum along the negative y-direction.

The mean flow thermal energy equation comes from substituting the velocity and temperature decompositions of (12.24) into (4.89) and averaging. The substitution step produces:

$$\frac{\partial}{\partial t}(\overline{T} + T') + \frac{\partial}{\partial x_j}\Big((U_j + u_j)(\overline{T} + T')\Big) = \kappa \frac{\partial^2}{\partial x_j^2}(\overline{T} + T').$$

The averages of each term in this equation are:

$$\overline{\frac{\partial}{\partial t}(\overline{T} + T')} = \frac{\partial}{\partial t}(\overline{T} + \overline{T'}) = \frac{\partial \overline{T}}{\partial t},$$

$$\overline{\frac{\partial}{\partial x_j}\Big((U_j + u_j)(\overline{T} + T')\Big)} = \frac{\partial}{\partial x_j}\Big(U_j \overline{T} + \overline{u_j}\overline{T} + U_j \overline{T'} + \overline{u_j T'}\Big) = U_j \frac{\partial \overline{T}}{\partial x_j} + \frac{\partial}{\partial x_j}(\overline{u_j T'}), \text{ and}$$

$$\overline{\kappa \frac{\partial^2}{\partial x_j^2}(\overline{T} + T')} = \kappa \frac{\partial^2}{\partial x_j^2}(\overline{T} + \overline{T'}) = \kappa \frac{\partial^2 \overline{T}}{\partial x_j^2},$$

where the final equality of (12.27), (12.4) through (12.6), and (12.8) have been used. Collecting terms, the mean temperature equation takes the form:

$$\frac{\partial \overline{T}}{\partial t} + U_j \frac{\partial \overline{T}}{\partial x_j} + \frac{\partial}{\partial x_j}(\overline{u_j T'}) = \kappa \frac{\partial^2 \overline{T}}{\partial x_j^2}. \tag{12.31}$$

When multiplied by $\rho_0 C_p$ and rearranged, (12.31) becomes the heat transfer equivalent of (12.30) and can be stated in terms of the mean heat flux Q_j:

$$\rho_0 C_p \left(\frac{\partial \overline{T}}{\partial t} + U_j \frac{\partial \overline{T}}{\partial x_j} \right) = -\frac{\partial Q_j}{\partial x_j} = -\frac{\partial}{\partial x_j} \left(-k \frac{\partial \overline{T}}{\partial x_j} + \rho_0 C_p \overline{u_j T'} \right), \tag{12.32}$$

where $k = \rho_0 C_p \kappa$ is the thermal conductivity. Equation (12.32) shows that the fluctuations cause an additional mean *turbulent heat flux of* $\rho_0 C_p \overline{u_j T'}$ that has no equivalent in (4.89). The turbulent heat flux is the thermal equivalent of the Reynolds stress $-\rho_0 \overline{u_i u_j}$ found in (12.30). Unfortunately, the process of reaching (12.31) and (12.32) has not provided any new equations for the turbulent heat flux. However, some understanding of the turbulent heat flux can be gained by considering diurnal heating of the earth's surface. During daylight hours, the sun may heat the surface of the earth, resulting in a mean temperature that decreases with height and in the potential for turbulent convective air motion. When such motions occur, an upward velocity fluctuation is mostly associated with a positive temperature fluctuation, giving rise to an upward heat flux $\rho_0 C_p \overline{u_3 T'} > 0$.

The final mean-flow equation commonly considered for turbulent flows is that for transport of a dye or a nonreacting molecular species that is merely carried by the turbulent flow without altering the flow. Such passive contaminants are commonly called *passive scalars* or *conserved scalars* and the rate at which they are mixed with nonturbulent fluid is often of significant technological interest for pollutant dispersion and premixed combustion. Consider a simple binary mixture composed of a primary fluid with density ρ and a contaminant fluid (the passive scalar) with density ρ_s. The density ρ_m that results from mixing these two fluids is $\rho_m = \tilde{v}\rho_s + (1 - \tilde{v})\rho$, where $\tilde{v}$ is the volume fraction of the passive scalar. The relevant conservation equation for the passive scalar is:

$$\frac{\partial}{\partial t}(\rho_m \tilde{Y}) + \frac{\partial}{\partial x_j}(\rho_m \tilde{Y}\tilde{u}_j) = \frac{\partial}{\partial x_j} \left(\rho_m \kappa_m \frac{\partial}{\partial x_j} \tilde{Y} \right), \tag{12.33}$$

(see Kuo, 1986) where $\tilde{u}_j$ is the instantaneous mass-averaged velocity of the mixture, $\tilde{Y}$ is the mass fraction of the passive scalar, and κ_m is the mass-based molecular diffusivity of the passive scalar (see (1.1)). If the mean and fluctuating mass fraction of the conserved scalar are $\overline{Y}$ and Y', and the mixture density is constant, then the mean-flow passive scalar conservation equation is (see Exercise 12.12):

$$\frac{\partial \overline{Y}}{\partial t} + U_j \frac{\partial \overline{Y}}{\partial x_j} = \frac{\partial}{\partial x_j} \left(\kappa_m \frac{\partial \overline{Y}}{\partial x_j} - \overline{u_j Y'} \right), \tag{12.34}$$

where $\overline{u_j Y'}$ is the turbulent flux of the passive scalar. This equation is valid when the mixture density is constant, and this occurs when $\rho = \rho_s = $ constant and when the contaminant is dilute so that $\rho_m \approx \rho = $ constant. If the amount of a passive scalar is characterized by a concentration (mass per unit volume), molecular number density, or mole fraction—instead of a mass fraction—the forms of (12.33) and (12.34) are unchanged but the diffusivity may need to be adjusted and molecular number or mass density factors may appear (see Bird et al., 1960; Kuo, 1986). Equation (12.34) is of the same form as (12.32), and temperature may be considered a passive scalar in turbulent flows when it does not induce buoyancy, cause chemical reactions, or lead to significant density changes.

To summarize, (12.27), (12.30), (12.32), and (12.34) are the mean flow equations for incompressible turbulent flow (in the Boussinesq approximation). The process of reaching these equations is known as *Reynolds averaging*, and it may be applied to the full compressible-flow equations of fluid motion as well. The equations that result from **R**eynolds **a**veraging of any form of the **N**avier-**S**tokes equations are commonly known as *RANS equations*. The constant-density mean flow RANS equations commonly used in hydrodynamics are obtained from the results provided in this section by dropping the gravity term and the "0" from ρ_0 in (12.30), and reinterpreting the mean pressure as the deviation from hydrostatic (as explained in Section 4.9, "Neglect of Gravity in Constant Density Flows").

The primary problem with RANS equations is that there are more unknowns than equations. The system of equations for the first moments depends on correlations involving pairs of variables (second moments). And, RANS equations developed for these pair correlations involve triple correlations. For example, the conservation equation for the Reynolds stress correlation, $\overline{u_i u_j}$, is:

$$\frac{\partial \overline{u_i u_j}}{\partial t} + U_k \frac{\partial \overline{u_i u_j}}{\partial x_k} + \frac{\partial \overline{u_i u_j u_k}}{\partial x_k} = -\overline{u_i u_k}\frac{\partial U_j}{\partial x_k} - \overline{u_j u_k}\frac{\partial U_i}{\partial x_k} - \frac{1}{\rho}\left(\overline{u_i \frac{\partial p}{\partial x_j}} + \overline{u_j \frac{\partial p}{\partial x_i}}\right)$$
$$- 2\nu\overline{\frac{\partial u_i}{\partial x_k}\frac{\partial u_j}{\partial x_k}} + \nu\frac{\partial^2}{\partial x_k^2}\overline{u_i u_j} + g\alpha\left(\overline{u_j T'}\delta_{i3} + \overline{u_i T'}\delta_{j3}\right)$$

(12.35)

(see Exercise 12.16), and triple correlations appear in the third term on the left. Similar conservation equations for the triple correlations involve quadruple correlations, and the equations for the quadruple correlations depend on fifth-order correlations, and so on. This problem persists at all correlation levels and is known as the *closure problem* in turbulence. At the present time there are three approaches to the closure problem. The first, known as *RANS closure modeling* (see Section 12.10), involves terminating the equation hierarchy at a given level and closing the resulting system of equations with models developed from dimensional analysis, intuition, symmetry requirements, and experimental results. The second, known as *direct numerical simulations* (DNS) involves numerically solving the time-dependent equations of motion and then Reynolds averaging the computational output to determine mean-flow quantities. The third, known as *large-eddy simulation* (LES), combines elements of the other two and involves some modeling and some numerical simulation of large-scale turbulent fluctuations.

A secondary problem associated with the RANS equations is that the presence of the Reynolds stresses in (12.30) excludes the possibility of converting it into a Bernoulli equation, even when the density is constant and the terms containing $\partial/\partial t$ and ν are zero.

12.6. HOMOGENEOUS ISOTROPIC TURBULENCE

It is clear from (12.27), (12.30), (12.32), and (12.34) that even with suitable boundary conditions the RANS equations for the mean flow are not directly solvable (even numerically) because of the closure problem. However, the idealization of turbulence as being homogeneous (or spatially stationary) and isotropic allows some significant simplifications. Turbulence behind a grid towed through a nominally quiescent fluid bath is approximately

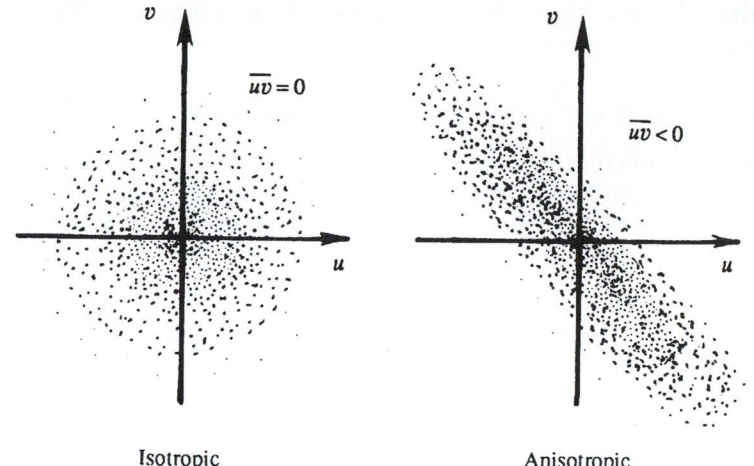

FIGURE 12.8 Scatter plots of velocity fluctuation samples in isotropic and anisotropic turbulent fields. Each dot represents a (u,v)-pair at a sample time and many sample times are represented in each panel. The isotropic case produces a symmetric cloud of points and indicates $\overline{uv} = 0$. The anisotropic case shows the data clustering around the line $v = -u$ and this indicates a negative correlation of u and v; $\overline{uv} < 0$.

homogeneous and isotropic, and turbulence in the interior of a real inhomogeneous turbulent flow is commonly assumed to be homogeneous and isotropic. Homogenous isotropic turbulence is discussed in greater detail in Batchelor (1953) and Hinze (1975).

If the turbulent fluctuations are completely isotropic, that is, if they do not have any directional preference, then the off-diagonal components of $\overline{u_i u_j}$ vanish, and the normal stresses are equal. This is illustrated in Figure 12.8, which shows a cloud of data points (sometimes called a *scatter plot*) on a uv-plane. The dots represent the instantaneous values of the (u,v)-velocity component pair at different times. In the isotropic case there is no directional preference, and the dots form a symmetric pattern. In this case positive u is equally likely to be associated with both positive and negative v. Consequently, *the average value of the product uv is zero when the turbulence is isotropic*. In contrast, the scatter plot in an anisotropic turbulent field has an orientation. The figure shows a case where a positive u is mostly associated with a negative v, giving $\overline{uv} < 0$.

If, in addition, the turbulence is homogeneous, then there are no spatial variations in the flow's statistics and all directions are equivalent:

$$\frac{\partial}{\partial x_i}\overline{u_j^n} = 0, \ \overline{u_1^2} = \overline{u_2^2} = \overline{u_3^2}, \text{ and } \overline{\left(\frac{\partial u_1}{\partial x_1}\right)^n} = \overline{\left(\frac{\partial u_2}{\partial x_2}\right)^n} = \overline{\left(\frac{\partial u_3}{\partial x_3}\right)^n}, \quad (12.36)$$

but relative directions must be respected,

$$\overline{\left(\frac{\partial u_1}{\partial x_2}\right)^n} = \overline{\left(\frac{\partial u_1}{\partial x_3}\right)^n} = \overline{\left(\frac{\partial u_2}{\partial x_1}\right)^n} = \overline{\left(\frac{\partial u_2}{\partial x_3}\right)^n} = \overline{\left(\frac{\partial u_3}{\partial x_1}\right)^n} = \overline{\left(\frac{\partial u_3}{\partial x_2}\right)^n}. \quad (12.37)$$

Note that the continuity equation requires derivatives in the third set of equalities of (12.36) to be zero when $n = 1$.

The spatial structure of the flow may be ascertained by considering the two-point correlation tensor, defined by (12.23), which reduces to the Reynolds stress correlation when $\mathbf{r} = 0$. In homogenous flow, R_{ij} does not depend on $\mathbf{x}$, and can only depend on $\mathbf{r}$. If the turbulence is also isotropic, the direction of $\mathbf{r}$ cannot matter. In this special situation, only two different types of velocity-field correlations survive. These are described by the longitudinal (f) and transverse (g) correlation coefficients defined by:

$$f(r) \equiv \overline{u_{\parallel}(\mathbf{x} + \mathbf{r})u_{\parallel}(\mathbf{x})}/\overline{u_{\parallel}^2} \text{ and } g(r) \equiv \overline{u_{\perp}(\mathbf{x} + \mathbf{r})u_{\perp}(\mathbf{x})}/\overline{u_{\perp}^2}, \tag{12.38}$$

where $u_{\parallel}$ is parallel to $\mathbf{r}$, $u_{\perp}$ is perpendicular to $\mathbf{r}$, $\overline{u_{\parallel}^2} = \overline{u_{\perp}^2} = \overline{u^2}$, and $f(0) = g(0) = 1$. The geometries for these two correlation functions are shown in Figure 12.9, where solid vectors indicate velocities and the dashed vector represents $\mathbf{r}$. Longitudinal and transverse integral scales and Taylor microscales are defined by:

$$\Lambda_f \equiv \int_0^{\infty} f(r)\,dr, \ \Lambda_g \equiv \int_0^{\infty} g(r)\,dr, \ \lambda_f^2 \equiv -2/\left[d^2 f/dr^2\right]_{r=0}, \text{ and } \lambda_g^2 \equiv -2/\left[d^2 g/dr^2\right]_{r=0}, \tag{12.39}$$

similar to the temporal integral scale and temporal Taylor microscale defined in (12.18) and (12.19), respectively. The most general possible form of $R_{ij}(r)$ that satisfies all the necessary symmetries is:

$$R_{ij} = F(r)r_i r_j + G(r)\delta_{ij}, \tag{12.40}$$

where the components of $\mathbf{r}$ are r_i, $|\mathbf{r}| = r$, and the functions $F(r) = \overline{u^2}(f(r) - g(r))r^{-2}$ and $G(r) = \overline{u^2}g(r)$ can be found by equating the diagonal components of R_{ij} (Exercise 12.17). For incompressible flow, $g(r)$ can be eliminated from (12.40) to find:

$$R_{ij} = \overline{u^2}\left\{f(r)\delta_{ij} + \frac{r}{2}\frac{df}{dr}\left(\delta_{ij} - \frac{r_i r_j}{r^2}\right)\right\} \tag{12.41}$$

(Exercise 12.18), and $\Lambda_g = \Lambda_f/2$ and $\lambda_g = \lambda_f/\sqrt{2}$.

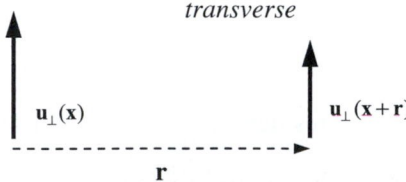

FIGURE 12.9 Longitudinal and transverse correlation geometries. In the longitudinal case, $\mathbf{u}_{\parallel}$ is parallel to the displacement $\mathbf{r}$. In the transverse case, $\mathbf{u}_{\perp}$ is perpendicular to $\mathbf{r}$. Here $\mathbf{r}$ is shown horizontal but it may point in any direction.

Admittedly, the preceding formulae do not readily produce insights; however, the trace of R_{ij} evaluated at $r = 0$ is twice the average kinetic energy $\bar{e}$ (per unit mass) of the turbulent fluctuations

$$R_{ii}(0) = \overline{u_i u_i} = 2 \cdot \frac{1}{2}(\overline{u_1^2} + \overline{u_2^2} + \overline{u_3^2}) = 2\bar{e},$$

and $\bar{e}$ is an important element in understanding and modeling turbulence. We know from Section 4.8 that the kinetic energy of a flowing fluid may be converted into heat (dissipated) by the action of viscosity. Thus, the average kinetic energy dissipation rate $\bar{\varepsilon}$ (per unit mass) in an incompressible turbulent flow comprised entirely of fluctuations is the average of (4.58):

$$\bar{\varepsilon} = \frac{\nu}{2}\overline{\left(\partial u_i/\partial x_j + \partial u_j/\partial x_i\right)^2}. \tag{12.42}$$

When the flow is isotropic, the various directional symmetries, (12.36), and (12.37) imply:

$$\bar{\varepsilon} = 6\nu\left\{\overline{\left(\frac{\partial u_1}{\partial x_1}\right)^2} + \overline{\left(\frac{\partial u_1}{\partial x_2}\right)^2} + \overline{\left(\frac{\partial u_1}{\partial x_2}\right)\left(\frac{\partial u_2}{\partial x_1}\right)}\right\} = -15\nu\overline{u^2}\left[\frac{d^2f}{dr^2}\right]_{r=0} = 30\nu\frac{\overline{u^2}}{\lambda_f^2} = 15\nu\frac{\overline{u^2}}{\lambda_g^2}, \tag{12.43}$$

where everything inside the {,}-braces has been put in terms of the first and second directions, and the second equality follows from the results of Exercise 12.19. Until the development of modern multidimensional measurement techniques, (12.43) was the primary means available for estimating $\bar{\varepsilon}$ from measurements in turbulent flows. Even today, fully resolved three-dimensional turbulent flow measurements are seldom possible, so reduced dimensionality relationships like (12.43), based on some assumed homogeneity and isotropy, commonly appear in the literature. In addition, Taylor-scale Reynolds numbers,

$$R_\lambda \equiv \lambda_{(g \text{ or } f)}\sqrt{\overline{u^2}}/\nu, \tag{12.44}$$

are occasionally quoted with $R_\lambda > 10^2$ being a nominal condition for fully turbulent flow (Dimotakis, 2000).

These concepts from homogeneous isotropic turbulence also allow the energy spectrum $S_{11}(k_1)$ of stream-wise velocity fluctuations along a stream-wise line through the turbulent field to be defined in terms of the autocorrelation function (12.23) when $i = j = 1$:

$$S_{11}(k_1) = \frac{1}{2\pi}\int_{-\infty}^{+\infty} R_{11}(r_1)\exp\{-ik_1 r_1\}dr_1 = \frac{\overline{u_1^2}}{2\pi}\int_{-\infty}^{+\infty} f(r_1)\exp\{-ik_1 r_1\}dr_1, \tag{12.45}$$

where "1" implies the stream-wise flow direction. Measured spectra reported in the turbulence literature are commonly produced using (12.45) or its alternative involving finite-window Fourier transformations (see Exercise 12.8). The basic procedure is to collect time-series measurements of u_1, convert them to spatial measurements using Taylor's frozen turbulence hypothesis, compute R_{11} from the spatial series, and then use (12.45) to determine $S_{11}(k_1)$. As described in the next section, the functional dependence of a portion of S_{11} on k_1 and $\bar{\varepsilon}$ can be anticipated from dimensional analysis and insights derived from the progression or cascade of fluctuation kinetic energy through a turbulent flow. Additional relationships for R_{ij} and its associated spectrum tensor may be found (Hinze, 1975).

12.7. TURBULENT ENERGY CASCADE AND SPECTRUM

As mentioned in the introductory section of this chapter, turbulence rapidly dissipates kinetic energy, and an understanding of how this happens is possible via a term-by-term inspection of the equations that govern the kinetic energy in the mean flow and the average kinetic energy of the fluctuations.

An equation for the mean flow's kinetic energy per unit mass, $\overline{E} = (1/2)U_i^2$, can be obtained by multiplying (12.30) by U_i, and averaging (Exercise 12.15). With $\overline{S}_{ij} = 1/2(\partial U_i/\partial x_j + \partial U_j/\partial x_i)$ defining the mean strain-rate tensor, the resulting *energy-balance* or *energy-budget* equation for $\overline{E}$ is

$$
\underbrace{\frac{\partial \overline{E}}{\partial t} + U_j \frac{\partial \overline{E}}{\partial x_j}}_{\substack{\text{Time rate of}\\\text{change of } \overline{E} \text{ following}\\\text{the mean flow}}} = \underbrace{\frac{\partial}{\partial x_j}\left(-\frac{U_j P}{\rho_0} + 2\nu U_i \overline{S}_{ij} - \overline{u_i u_j} U_i\right)}_{\text{transport}} - \underbrace{2\nu \overline{S}_{ij}\overline{S}_{ij}}_{\substack{\text{viscous}\\\text{dissipation}}} + \underbrace{\overline{u_i u_j}\frac{\partial U_i}{\partial x_j}}_{\substack{\text{loss to}\\\text{turbulence}}} - \underbrace{\frac{g}{\rho_0}\overline{\rho}U_3}_{\substack{\text{loss to}\\\text{potential}\\\text{energy}}}.
$$

$$(12.46)$$

The left side is merely the total time derivative of $\overline{E}$ following a mean-flow fluid particle, while the right side represents the various mechanisms that bring about changes in $\overline{E}$.

The first three divergence terms on the right side of (12.46) represent *transport* of mean kinetic energy by pressure, viscous stresses, and Reynolds stresses. If (12.46) is integrated over the volume occupied by the flow to obtain the rate of change of the total (or global) mean-flow kinetic energy, then these transport terms can be transformed into a surface integral by Gauss' theorem. Thus, these terms do not contribute to the total rate of change of $\overline{E}$ if $U_i = 0$ on the boundaries of the flow. Therefore, these three terms only *transport* or redistribute mean-flow kinetic energy from one region to another; they do not generate it or dissipate it.

The fourth term is the product of the mean flow's viscous stress (per unit mass) $2\nu \overline{S}_{ij}$ and the mean strain rate $\overline{S}_{ij}$. It represents the *direct viscous dissipation* of mean kinetic energy via its conversion into heat.

The fifth term is analogous to the fourth term. It can be written as $\overline{u_i u_j}\,(\partial U_i/\partial x_j) = \overline{u_i u_j}\,\overline{S}_{ij}$ so that it is a product of the turbulent stress (per unit mass) and the mean strain rate. Here, the doubly contracted product of a symmetric tensor $\overline{u_i u_j}$ and the tensor $\partial U_i/\partial x_j$ is equal to the product of $\overline{u_i u_j}$ and the *symmetric* part of $\partial U_i/\partial x_j$, namely $\overline{S}_{ij}$, as proved in Section 2.10. If the mean flow is given by $U(y)$ alone, then $\overline{u_i u_j}\,(\partial U_i/\partial x_j) = \overline{uv}\,(dU/dy)$. From the preceding section, $\overline{uv}$ is likely to be negative if dU/dy is positive. Thus, the fifth term is likely to be negative in shear flows. So, by analogy with the fourth term, it must represent a mean-flow kinetic energy loss to the fluctuating velocity field. Indeed, this term appears on the right-hand side of the equation for the rate of change of the turbulent kinetic energy, but *with the sign reversed*. Therefore, this term generally results in a loss of mean kinetic energy and a gain of turbulent kinetic energy. It is commonly known as the *shear production* term.

The sixth term represents the work done by gravity on the mean vertical motion. For example, an upward mean motion results in a loss of mean kinetic energy, which is accompanied by an increase in the potential energy of the mean field.

The two viscous terms in (12.46), namely, the viscous transport $2\nu\partial(U_i\,\overline{S}_{ij})/\partial x_j$ and the mean-flow viscous dissipation $-2\nu\overline{S}_{ij}\overline{S}_{ij}$, are small compared to the equivalent turbulence terms in a fully turbulent flow at high Reynolds numbers. Compare, for example, the mean-flow viscous dissipation and the shear-production terms:

$$\frac{2\nu\overline{S}_{ij}\overline{S}_{ij}}{\overline{u_iu_j}\,(\partial U_i/\partial x_j)} \sim \frac{\nu(U/L)^2}{u_{rms}^2(U/L)} \sim \frac{\nu}{UL} = \frac{1}{\mathrm{Re}} \ll 1,$$

where U is the velocity scale for the mean flow, L is a length scale for the mean flow (e.g., the overall thickness of a boundary layer), and u_{rms} is presumed to be of the same order of magnitude as U, a presumption commonly supported by experimental evidence. The direct influence of viscous terms is therefore negligible on the *mean* kinetic energy budget. However, this is *not* true for the *turbulent* kinetic energy budget, in which viscous terms play a major role. What happens is as follows: The mean flow loses energy to the turbulent field by means of the shear production term and the *turbulent* kinetic energy so generated is then dissipated by viscosity.

An equation for the mean kinetic energy $\overline{e} = (1/2)\overline{u_i^2}$ of the turbulent velocity fluctuations can be obtained by setting $i = j$ (12.35) and dividing by two. With $S'_{ij} = 1/2(\partial u_i/\partial x_j + \partial u_j/\partial x_i)$ defining the fluctuation strain-rate tensor, the resulting energy-budget equation for $\overline{e}$ is:

$$\underbrace{\frac{\partial\overline{e}}{\partial t} + U_j\frac{\partial\overline{e}}{\partial x_j}}_{\substack{\text{Time rate of}\\\text{change of }\overline{E}\text{ following}\\\text{the mean flow}}} = \underbrace{\frac{\partial}{\partial x_j}\left(-\frac{1}{\rho_0}\overline{pu_j} + 2\nu\overline{u_iS'_{ij}} - \frac{1}{2}\overline{u_i^2u_j}\right)}_{\text{transport}} - \underbrace{2\nu\overline{S'_{ij}S'_{ij}}}_{\substack{\text{viscous}\\\text{dissipation}}} - \underbrace{\overline{u_iu_j}\frac{\partial U_i}{\partial x_j}}_{\substack{\text{gain from}\\\text{mean flow}}} + \underbrace{g\alpha\overline{u_3T'}}_{\substack{\text{buoyant}\\\text{production}}}.$$

$$(12.47)$$

The first three terms on the right side are in divergence form and consequently represent the spatial transport of turbulent kinetic energy via turbulent pressure fluctuations, viscous diffusion, and turbulent stresses.

The fourth term $\overline{\varepsilon} = 2\nu\overline{S'_{ij}S'_{ij}}$ is the *viscous dissipation of turbulent kinetic energy*, and it is *not* negligible in the turbulent kinetic energy budget (12.47), although the analogous term $2\nu\overline{S}_{ij}\overline{S}_{ij}$ is negligible in the mean-flow kinetic energy budget (12.46). In fact, the viscous dissipation $\overline{\varepsilon}$ is always positive and its magnitude is typically similar to that of the turbulence-production terms in most locations.

The fifth term $\overline{u_iu_j}\,(\partial U_i/\partial x_j)$ is the shear-production term and it represents the rate at which kinetic energy is lost by the mean flow and gained by the turbulent fluctuations. It appears in the mean-flow kinetic energy budget with the other sign.

The sixth term $g\alpha\overline{u_3T'}$ can have either sign, depending on the nature of the background temperature distribution $\overline{T}(x_3)$. In a stable situation in which the background temperature increases upward (as found, e.g., in the atmospheric boundary layer at night), rising fluid elements are likely to be associated with a negative temperature fluctuation, resulting in $\overline{u_3 T'} < 0$, which means a downward turbulent heat flux. In such a stable situation $g\alpha\overline{u_3 T'}$ represents the rate of turbulent energy loss via work against the stable background density gradient. In the opposite case, when the background density profile is unstable, the turbulent heat flux correlation $\overline{u_3 T'}$ is upward, and convective motions cause an increase of turbulent kinetic energy (Figure 12.10). Thus, $g\alpha\overline{u_3 T'}$ is the *buoyant production* of turbulent kinetic energy; it can also be a buoyant *destruction* when the turbulent heat flux is downward. In isotropic turbulence, the upward thermal flux correlation $\overline{u_3 T'}$ is zero because there is no preference between the upward and downward directions.

The buoyant generation of turbulent kinetic energy lowers the potential energy of the mean field. This can be understood from Figure 12.10, where it is seen that the heavier fluid has moved downward in the final state as a result of the heat flux. This can also be demonstrated by deriving an equation for the mean potential energy, in which the term $g\alpha\overline{u_3 T'}$ appears with a *negative* sign on the right-hand side. Therefore, the *buoyant generation* of turbulent kinetic energy by the upward heat flux occurs at the expense of the mean *potential* energy. This is in contrast to the *shear production* of turbulent kinetic energy, which occurs at the expense of the mean *kinetic* energy.

The kinetic energy budgets for constant density flow are recovered from (12.46) and (12.47) by dropping the terms with gravity and re-interpreting the mean pressure as the deviation from hydrostatic (see Section 4.9, "Neglect of Gravity in Constant Density Flows").

The shear-production term represents an essential link between the mean and fluctuating fields. For it to be active (or nonzero), the flow must have mean shear and the turbulence must be anisotropic. When the turbulence is isotropic, the off-diagonal components of the

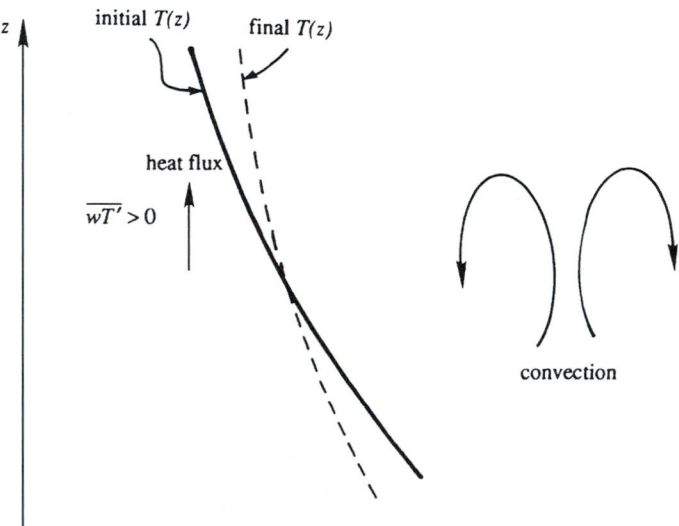

FIGURE 12.10 Heat flux in an unstable environment. Here warm air from below may rise and cool air may sink thereby generating turbulent kinetic energy by lowering the mean potential energy. In the final state, the upper air is warmer and less dense and the lower air is cooler and denser.

Reynolds stress $\overline{u_i u_j}$ are zero (see Section 12.6) and the on-diagonal ones are equal (12.36). Thus the double sum implied by $\overline{u_i u_j}\,(\partial U_i/\partial x_j)$ reduces to:

$$\overline{u_1^2}(\partial U_1/\partial x_1) + \overline{u_2^2}(\partial U_2/\partial x_2) + \overline{u_3^2}(\partial U_3/\partial x_3) = \overline{u_1^2}(\partial U_i/\partial x_i) = 0,$$

where the final equality holds from (12.27). Experimental observations suggest the largest eddies in a turbulent shear flow generally span the cross-stream distance L between those locations in a turbulent flow giving the maximum average velocity difference ΔU (Figure 12.11). In a layer with only one sign for the mean shear, L spans the layer as in Figure 12.11a, but for consistency when the shear has both signs, such as in turbulent pipe flow, L is the pipe radius as in Figure 12.11b. These largest eddies feel the mean shear—which must be of order $\Delta U/L$—and are distorted or made anisotropic by it. Energy is provided to these largest eddies by the mean flow as it forces them to deform and turn over. In this situation, turbulent velocity fluctuations are also of order ΔU, so the energy input rate $\dot{W}$ to a region of turbulence by the mean flow (per unit mass of fluid) is

$$\dot{W} \sim \overline{u_i u_j}\,(\partial U_i/\partial x_j) \sim (\Delta U)^2[\Delta U/L] = (\Delta U)^3/L, \qquad (12.48)$$

where L and ΔU are commonly called the *outer* length scale and velocity difference. Of course the details of $\dot{W}$ will vary with flow geometry but its parametric dependence is set by (12.48). In reaching (12.48), it was implicitly assumed that the outer scale Reynolds number $Re_L = \Delta U L/\nu$ is so large that viscosity plays no role in the interaction between the mean flow and the largest eddies of the turbulent shear flow.

In temporally stationary turbulence, the turbulent kinetic energy $\bar{e}$ cannot build up (or shrink to zero) so the work input at the largest scales from the mean flow must be balanced by the kinetic energy dissipation rate:

$$\dot{W} = \bar{\varepsilon}, \text{ so } \bar{\varepsilon} \sim (\Delta U)^3/L. \qquad (12.49)$$

Thus, $\bar{\varepsilon}$ does not depend on ν, but is determined instead by the *inviscid* properties of the largest eddies, which extract energy from the mean flow. Second-tier eddies that are somewhat smaller than the largest ones are distorted and forced to roll over by the strain field

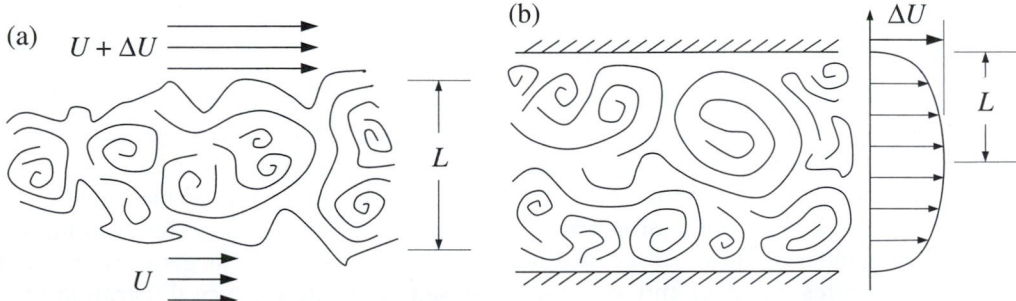

FIGURE 12.11 Schematic drawings of a turbulent flow without boundaries (a), and one with boundaries (b). Here the outer scale of the turbulence L spans the cross-stream distance over which the outer scale velocity difference ΔU develops. Here L may be the half width of the flow when the flow is symmetric as in (b). This choice of L and ΔU ensures that mean-flow velocity gradients will be of order $\Delta U/L$.

of the largest eddies, and these thereby extract energy from the largest eddies by the same mechanism that the largest eddies extract energy from the mean flow. Thus the average turbulent-kinetic-energy cascade pattern is set, and third-tier eddies extract energy from second-tier eddies, fourth-tier eddies extract energy from third-tier eddies, and so on. So, *turbulent kinetic energy is on average cascaded down from large to small eddies by interactions between eddies of neighboring size.* Small eddies are essentially advected in the velocity field of large eddies, since the scales of the strain-rate field of the large eddies are much larger than the size of a small eddy. Therefore, small eddies do not interact directly with the large eddies or the mean field, and are therefore nearly isotropic. The turbulent kinetic-energy cascade process is essentially inviscid with decreasing eddy scale size l' and eddy-velocity u' as long as the eddy Reynolds number $u'l'/v$ is much greater than unity. The cascade terminates when the eddy Reynolds number becomes of order unity and viscous effects are important. This average cascade process was first discussed by Richardson (1922), and is a foundational element in the understanding of turbulence.

In 1941, Kolmogorov suggested that the dissipating eddies are essentially homogeneous and isotropic, and that their size depends on those parameters that are relevant to the smallest eddies. These parameters are the rate $\bar{\varepsilon}$, the rate at which kinetic energy is supplied to the smallest eddies, and v, the kinematic viscosity that smears out the velocity gradients of the smallest eddies. Since the units of $\bar{\varepsilon}$ are m^2/s^3, dimensional analysis only allows one way to construct a length scale η and a velocity scale u_K from $\bar{\varepsilon}$ and v:

$$\eta = \left(v^3/\bar{\varepsilon}\right)^{1/4} \quad \text{and} \quad u_K = \left(v\bar{\varepsilon}\right)^{1/4}. \tag{12.50}$$

These are called the *Kolmogorov microscale* and *velocity scale*, and the Reynolds number determined from them is

$$\eta u_K/v = \left(v\bar{\varepsilon}\right)^{1/4}\left(v^3/\bar{\varepsilon}\right)^{1/4}/v = 1,$$

which appropriately suggests a balance of inertial and viscous effects for Kolmogorov-scale eddies. The relationship (12.50) and the recognition that v does not influence $\bar{\varepsilon}$ suggests that a *decrease of v merely decreases the eddy size at which viscous dissipation takes place.* In particular, the size of η relative to L can be determined by eliminating $\bar{\varepsilon}$ from (12.49) and the first equation of (12.50) to find:

$$\eta/L \sim \mathrm{Re}_L^{-3/4}, \quad \text{where } \mathrm{Re}_L = \Delta U L/v. \tag{12.51}$$

Therefore, the sizes of the largest and smallest eddies in high Reynolds number turbulence potentially differ by many orders of magnitude. For flow in a fixed-size device, the length scale L is fixed, so increasing the input velocity that leads to shear (or decreasing v) leads to an increase in Re_L and a decrease in the size of the Kolmogorov eddies. In the ocean and the atmosphere, the Kolmogorov microscale η is commonly of the order of millimeters. However, in engineering flows η may be much smaller because of the larger power densities and dissipation rates. Landahl and Mollo-Christensen (1986) give a nice illustration of this. Suppose a 100-W household mixer is used to churn 1 kg of water in a cube 0.1 m ($= L$) on a side. Since all the power is used to generate turbulence, the rate of energy dissipation is $\bar{\varepsilon} = 100$ W/kg $= 100$ m^2/s^3. Using $v = 10^{-6}$ m^2/s for water, we obtain $\eta = 10^{-5}$ m from (12.50).

Interestingly, the path that leads to (12.51) can also be used for either of the Taylor microscales (generically labeled λ_T here). Eliminating $\bar{\varepsilon}$ from (12.43) and (12.49) produces:

$$\frac{(\Delta U)^3}{L} \propto \frac{\nu \overline{u^2}}{\lambda_T^2} \rightarrow \frac{\lambda_T^2}{L^2} \propto \frac{\nu \overline{u^2}}{(\Delta U)^3 L} = \frac{\overline{u^2}}{(\Delta U)^2}\left(\frac{\nu}{\Delta U L}\right) \propto \frac{1}{\mathrm{Re}_L}, \text{ or } \frac{\lambda_T}{L} \propto \mathrm{Re}_L^{-1/2}, \tag{12.52}$$

where the final two proportionalities are valid when the fluctuation velocity is proportional to the ΔU. The negative half-power of the outer-scale Reynolds number matches that for laminar boundary-layer thicknesses (see (9.30)). Thus, the Taylor microscale can be interpreted as an internal boundary-layer thickness that develops at the edge of a large eddy during a single rotational movement having a path-length length L. However, it is not a distinguished length scale in the partition of turbulent kinetic energy even though (12.43) associates λ_T with $\bar{\varepsilon}$. The reason for this anonymity is that the velocity fluctuation appearing in (12.43) is not appropriate for eddies that dissipate turbulent kinetic energy. The appropriate dissipation-scale velocity is given by the second equality of (12.50). Thus, in high Reynolds number turbulence, λ_T is larger than η, as is clear from a comparison of (12.51) and (12.52) with $\mathrm{Re} \rightarrow \infty$. In addition, (12.52) implies $\mathrm{Re}_\lambda \sim (\mathrm{Re}_L)^{1/2}$, so a nominal condition for fully turbulent flow is $\mathrm{Re}_L > 10^4$ (Dimotakis, 2000). Above such a Reynolds number, the following ordering of length scales should occur: $\eta < \lambda_T < \Lambda_{(f \text{ or } g)} < L$.

Richardson's cascade, Kolmogorov's insights, the simplicity of homogeneous isotropic turbulence, and dimensional analysis lead to perhaps the most famous and prominent feature of high Reynolds number turbulence: the universal power law form of the energy spectrum in the inertial subrange. Consider the one-dimensional energy spectrum $S_{11}(k_1)$—it is the one most readily determined from experimental measurements—and associate eddy size l with the inverse of the wave number: $l \sim 2\pi/k_1$. For large eddy sizes (small wave numbers), the energy spectrum will not be universal because these eddies are directly influenced by the geometry-dependent mean flow. However, smaller eddies a few tiers down in the cascade may approach isotropy. In this case the mean shear no longer matters, so their spectrum of fluctuations S_{11} can only depend on the kinetic energy cascade rate $\bar{\varepsilon}$, the fluid's kinematic viscosity ν, and the wave number k_1. From (12.45), the units of S_{11} are found to be $\mathrm{m}^3/\mathrm{s}^2$; therefore dimensional analysis using S_{11}, $\bar{\varepsilon}$, ν, and k_1 requires:

$$\frac{S_{11}(k_1)}{\nu^{5/4}\bar{\varepsilon}^{1/4}} = \Phi\left(\frac{k_1 \nu^{3/4}}{\bar{\varepsilon}^{1/4}}\right), \text{ or } \frac{S_{11}(k_1)}{u_K^2 \eta} = \Phi(k_1 \eta) \quad \text{for } k_1 \gg 2\pi/L, \tag{12.53}$$

where Φ is an undetermined function, and both parts of (12.50) have been used to reach the second form of (12.53). Furthermore, for eddy sizes somewhat less than L, but also somewhat greater than η, $2\pi/L \ll k_1 \ll 2\pi/\eta$, the spectrum must be independent of both the mean shear and the kinematic viscosity. This wave number range is known as the *inertial subrange*. Turbulent kinetic energy is transferred through this range of length scales without much loss to viscosity. Thus, the form of the spectrum in the inertial subrange is obtained from dimensional analysis using only S_{11}, $\bar{\varepsilon}$, and k_1:

$$S_{11}(k_1) = const \cdot \bar{\varepsilon}^{2/3} \cdot k_1^{5/3} \quad \text{for } 2\pi/L \ll k_1 \ll 2\pi/\eta. \tag{12.54}$$

The *constant* has been found to be universal for all turbulent flows and is approximately 0.25 for $S_{11}(k_1)$ subject to a *double-sided* normalization:

$$\int_{-\infty}^{+\infty} S_{11}(k_1)dk_1 = \overline{u_1^2} = \int_{0}^{+\infty} 2S_{11}(k_1)dk_1. \qquad (12.55)$$

Equation (12.54) is usually called *Kolmogorov's $k^{-5/3}$ law* and it is one of the most important results of turbulence theory. When the spectral form (12.54) is subject to the normalization,

$$\bar{e} = \int_{0}^{+\infty} S(K)dK,$$

where K is the magnitude of the three-dimensional wave-number vector, the constant in the three-dimensional form of (12.54) is approximately 1.5 (see Pope, 2000). If the Reynolds number of the flow is large, then the dissipating eddies are much smaller than the energy-containing eddies, and the inertial subrange is broad.

Figure 12.12 shows a plot of experimental spectral measurements of $2S_{11}$ from several different types of turbulent flows (Chapman, 1979). The normalizations of the axes follow (12.53), $\bar{\varepsilon}$ is calculated from (12.43), η is calculated from (12.50), and the Taylor-Reynolds numbers (labeled R_λ in the figure) come from the longitudinal autocorrelation $f(r)$. The collapse of the data at high wave numbers to a single curve indicates the universal character of (12.53) at high wave numbers. The spectral form of Pao (1965) adequately fits the data and indicates how the spectral amplitude decreases faster than $k^{-5/3}$ as $k_1\eta$ approaches and then exceeds unity. The scaled wave number at which the data are approximately a factor of two below the $-5/3$ power law is $k_1\eta \approx 0.2$ (dashed vertical line), so the actual eddy size where viscous dissipation is clearly felt is $l_D \approx 30\eta$. The various spectra shown in Figure 12.12 turn horizontal with decreasing $k_1\eta$ where $k_1 L$ is of order unity.

Because very large Reynolds numbers are difficult to generate in an ordinary laboratory, the Kolmogorov spectral law (12.54) was not verified for many years. In fact, doubts were raised about its theoretical validity. The first confirmation of the Kolmogorov law came from the oceanic observations of Grant *et al.* (1962), who obtained a velocity spectrum in a tidal flow through a narrow passage between two islands just off the west coast of Canada. The velocity fluctuations were measured by hanging a hot film anemometer from the bottom of a ship. Based on the water depth and the average flow velocity, the outer-scale Reynolds number was of order 10^8. Such large Reynolds numbers are typical of geophysical flows, since the length scales are very large. Thus, the tidal channel data and results from other geophysical flows prominently display the $k^{-5/3}$ spectral form in Figure 12.12.

For the purpose of formulating predictions, the universality of the high wave number portion of the energy spectrum of turbulent fluctuations suggests that a single-closure model might adequately represent the effects of inertial subrange and smaller eddies on the nonuniversal large-scale eddies. This possibility has inspired the development of a wide variety of RANS equation closure models, and it provides justification for the central idea behind large-eddy simulations (LES) of turbulent flow. Such models are described in Pope (2000).

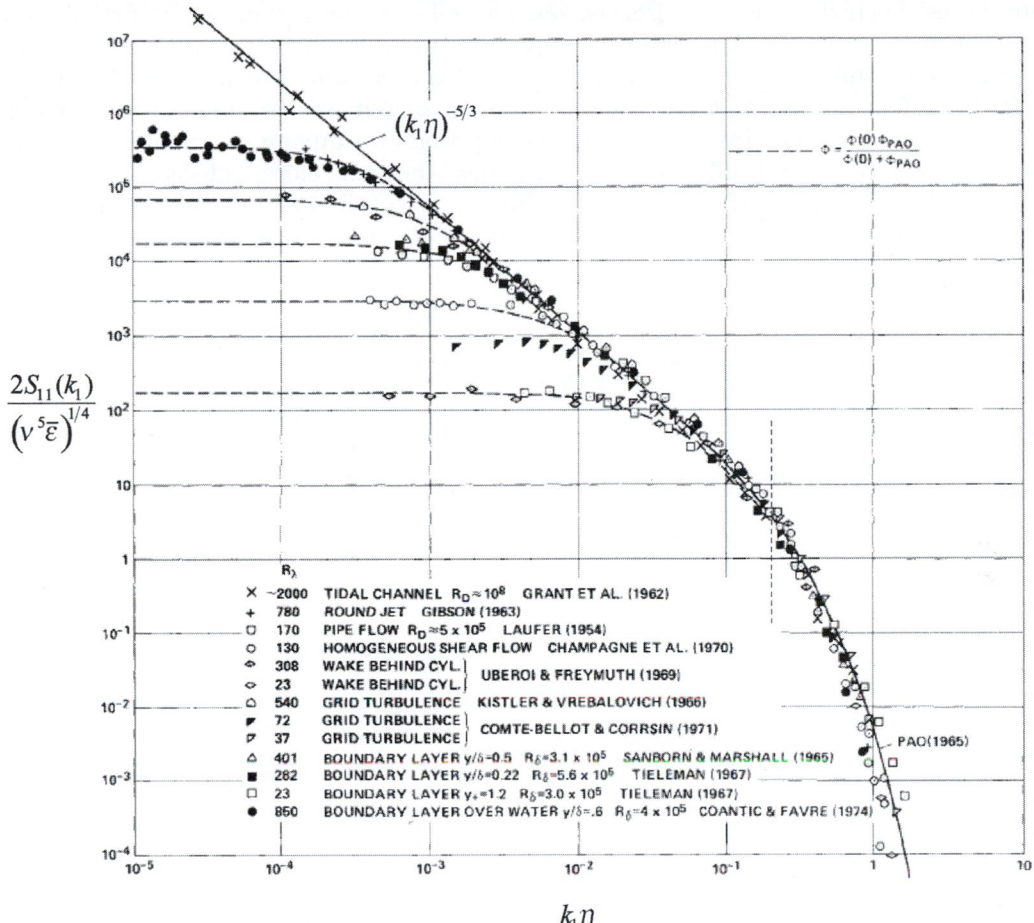

FIGURE 12.12 One-dimensional energy spectra $S_{11}(k_1)$ from a variety of turbulent flows plotted in Kolmogorov normalized form, reproduced from Chapman (1979). Here k_1 is the stream-wise wave number, η is the Kolmogorov scale defined by (12.50), and $\bar{\varepsilon}$ is the average kinetic energy dissipation rate determined from (12.43). Kolmogorov's $-5/3$ power law is indicated by the sloping line. The collapse of the various spectra to this line and to each other as $k_1\eta$ approaches and then exceeds unity strongly suggests that high-wave-number turbulent velocity fluctuations are universal when the Reynolds number is high enough. The dashed vertical line indicates the location where the spectral data are a factor of two below the $-5/3$ line established at lower wave numbers.

12.8. FREE TURBULENT SHEAR FLOWS

Persistent turbulence is maintained by the presence of mean-flow shear. This shear may exist because of a mismatch of fluid momentum within a flow, or because of the presence of one or more solid boundaries near the moving fluid. Turbulent flows in the former category are called *free* turbulent shear flows, and those in the latter are called *wall-bounded* turbulent shear flows. This section covers free turbulent flows that develop away from solid boundaries. Such flows include jets, wakes, shear layers, and plumes; the first three are depicted in Figure 12.13. A plume is a buoyancy-driven jet that develops vertically so its appearance is similar to that shown in Figure 12.13a when the flow direction is rotated to be vertical. Jets, wakes, and plumes may exist in planar and axisymmetric geometries. Although idealized, such free shear flows are important for mixing reactants and in remote sensing, and are scientifically interesting because their development can sometimes be described by a single length scale and one boundary condition or origin parameter. Such a description commonly results from a similarity analysis in which the mean flow is assumed

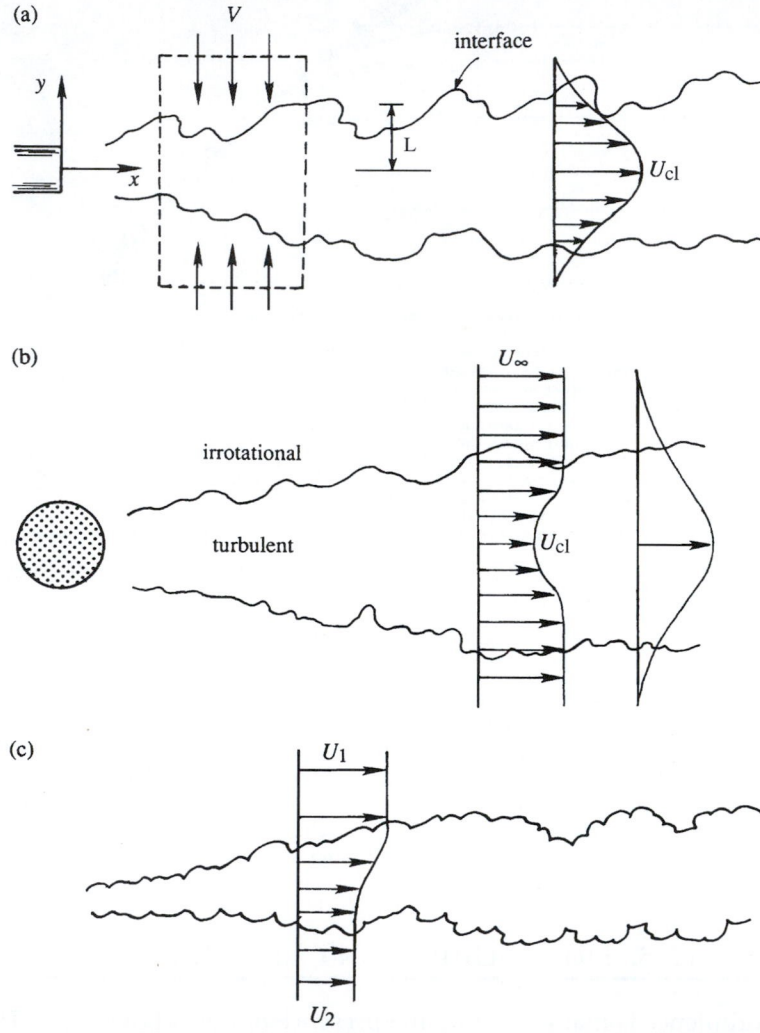

FIGURE 12.13 Three generic free turbulent shear flows: (a) jet, (b) wake, and (c) shear layer. In each case, the region of turbulence coincides with the region of shear in the mean velocity profile, and entrainment causes the cross-stream dimension L or δ of each flow to increase with increasing downstream distance. The fluid outside the region of turbulence is assumed to be irrotational.

to be *self-preserving*. This section presents one such similarity analysis for a single free turbulent shear flow (the planar jet), and then summarizes the similarity characteristics of a variety of planar and axisymmetric free turbulent shear flows. In most circumstances, free turbulent shear flows are simpler than wall-bounded turbulent shear flows. However, the outer portion of a turbulent boundary layer (from $y \sim 0.2\delta$ to its unconstrained edge) is similar to a free shear flow.

In snapshots and laser-pulse images, free shear flows usually appear with an erratic boundary that divides nominally turbulent from irrotational (or nonturbulent) fluid. Locally, the motion of this boundary is determined by the velocity induced by the turbulent vortices inside the region of turbulence. Typically, these vortices induce the surrounding nonturbulent fluid to flow toward the region of turbulence, and this induced flow, commonly called *entrainment*, causes the cross-stream size (L or δ) of the turbulent region to increase with increasing downstream distance. Because of entrainment, a passive scalar in the body of the turbulent flow is diluted with increasing downstream distance. The actual mechanism of entrainment involves both large- and small-eddy motions, and it may be altered within limits in some free shear flows by introducing velocity, pressure, or geometrical perturbations.

When a time-lapse image or an ensemble average of measurements from a free shear flow is examined, the edge of the region of turbulence is diffuse and the average velocity field and average passive scalar fields are found to be smooth functions. Significantly, the shapes of these mean profiles from different downstream locations within the same flow are commonly found to be self-similar when scaled appropriately. When this happens, the flow is in a state of *moving equilibrium*, in which both the mean and the turbulent fields are determined solely by the *local* length and velocity scales, a situation called *self-preservation*.

Some characteristics of the self-preserving state may be determined from a similarity analysis of the mean momentum equation (12.30) for a variety of free turbulent shear flows. The details of such an analysis are provided here for the plane turbulent jet. The scalings for other free turbulent shear flows are listed in Table 12.1, and are covered in this chapter's exercises. A plane turbulent jet is formed by fast-moving fluid that emerges into a quiescent reservoir from a long slot of width d, as shown in Figure 12.13a. Here, the long dimension of the slot is

TABLE 12.1   Self-Similar Far-Field Results for Some Free Turbulent Shear Flows

Flow	Mean Fields	ξ	Profile Widths
Planar Jet	$\dfrac{U(x,y)}{U_0} = \dfrac{U_{CL}(x)}{U_0}F(\xi) = 2.4\left(\dfrac{\rho_s}{\rho}\right)^{1/2}\left(\dfrac{x}{d}\right)^{-1/2}F(\xi)$	y/x	$(\xi_{1/2})_U = 0.11$
	$\dfrac{\overline{Y}(x,y)}{Y_0} = \dfrac{Y_{CL}(x)}{Y_0}H(\xi) = 2.0\left(\dfrac{\rho_s}{\rho}\right)^{1/2}\left(\dfrac{x}{d}\right)^{-1/2}H(\xi)$		$(\xi_{1/2})_Y = 0.14$
Planar Plume	$U(x,y) = U_{CL}(x)F(\xi) = 1.9\left(\dfrac{g(\rho-\rho_s)U_0 d}{\rho}\right)^{1/3}F(\xi)$	y/x	$(\xi_{1/2})_U = 0.12$
	$\dfrac{\overline{Y}(x,y)}{Y_0} = \dfrac{Y_{CL}(x)}{Y_0}H(\xi) = 2.4\left(\dfrac{\rho U_0^2}{g(\rho-\rho_s)d}\right)^{1/3}\left(\dfrac{x}{d}\right)^{-1}H(\xi)$		$(\xi_{1/2})_Y = 0.13$

(Continued)

TABLE 12.1   Self-Similar Far-Field Results for Some Free Turbulent Shear Flows—cont'd

Flow	Mean Fields	ξ	Profile Widths
Round Jet	$\dfrac{U(x,r)}{U_0} = \dfrac{U_{CL}(x)}{U_0}F(\xi) = 6.0\left(\dfrac{\rho_s}{\rho}\right)^{1/2}\left(\dfrac{x}{d}\right)^{-1}F(\xi)$	r/x	$(\xi_{1/2})_U = 0.090$
	$\dfrac{\overline{Y}(x,r)}{Y_0} = \dfrac{Y_{CL}(x)}{Y_0}H(\xi) = 5.0\left(\dfrac{\rho_s}{\rho}\right)^{1/2}\left(\dfrac{x}{d}\right)^{-1}H(\xi)$		$(\xi_{1/2})_Y = 0.11$
Round Plume	$U(x,r) = U_{CL}(x)F(\xi) = 3.5\left(\dfrac{g(\rho-\rho_s)U_0 d}{\rho}\right)^{1/3}\left(\dfrac{x}{d}\right)^{-1/3}F(\xi)$	r/x	$(\xi_{1/2})_U = 0.11$
	$\dfrac{\overline{Y}(x,r)}{Y_0} = \dfrac{Y_{CL}(x)}{Y_0}H(\xi) = 9.4\left(\dfrac{\rho U_0^2}{g(\rho-\rho_s)d}\right)^{1/3}\left(\dfrac{x}{d}\right)^{-5/3}H(\xi)$		$(\xi_{1/2})_Y = 0.10$
Shear Layer	$U(x,y) = U_2 + (U_1 - U_2)\dfrac{\int_{-\infty}^{\xi}F(\xi')d\xi'}{\int_{-\infty}^{+\infty}F(\xi')d\xi'}$	$\dfrac{y - y_{CL}(x)}{x}$	$(\Delta\xi_{80})_U =$ $0.085 \times \dfrac{U_1 - U_2}{\frac{1}{2}(U_1 + U_2)}$
Planar Wake	$U(x,y) = U_\infty - \Delta U_{CL}(x)F(\xi)$	$y/\sqrt{\theta_p x}$	$(\xi_{1/2})_U = 0.31$
	$\dfrac{\Delta U_{CL}(x)}{U_\infty} = 1.8\left(\dfrac{x}{\theta_p}\right)^{-1/2}; \theta_p = \dfrac{\text{drag force}}{\rho U_\infty^2 \cdot \text{span}}$		
Round Wake	$U(x,y) = U_\infty - \Delta U_{CL}(x)F(\xi)$	$r/(\theta_r^2 x)^{1/3}$	$(\xi_{1/2})_U = 0.4 \text{ to } 0.9$
	$\dfrac{\Delta U_{CL}(x)}{U_\infty} = (0.4\text{ to }2.0)\left(\dfrac{x}{\theta_r}\right)^{-2/3}; \quad \theta_r^2 = \dfrac{\text{drag force}}{\rho U_\infty^2}$		

TABLE 12.1 NOMENCLATURE

d = slot- or nozzle-exit width or diameter
U_0 = slot- or nozzle-exit fluid velocity
ρ_s = slot- or nozzle-exit fluid density
ρ = nominally quiescent reservoir fluid density
Y_0 = slot- or nozzle-exit passive scalar mass fraction
x = stream-wise, centerplane, or centerline coordinate
y = distance from the flow's centerplane
r = radial distance from the flow's centerline
$y_{CL}(x)$ = location of the point in the shear layer where $U = (U_1 + U_2)/2$
$(\Delta\xi_{80})_U$ = difference in ξ that spans the central 80% of the velocity difference $U_1 - U_2$
U_∞ = uniform velocity outside the wake
ξ = profile similarity variable
$F(\xi)$, $H(\xi)$ = velocity and mass fraction profiles, approximately = $\exp\{-\ln(2)\xi^2/\xi_{1/2}^2\}$

perpendicular to the page so the mean velocity field has only U and V components. Using the x-y coordinates shown in Figure 12.13a, the self-preserving form for the jet's mean stream-wise velocity and Reynolds shear-stress correlation is:

$$U(x,y) = U_{CL}(x)F(y/\delta(x)), \text{ and } -\overline{uv} = \Psi(x)G(y/\delta(x)), \qquad (12.56, 12.57)$$

where $U_{CL}(x)$ is the mean stream-wise velocity on the centerline of the flow ($y = 0$), $\Psi(x)$ is a function that sets the amplitude of the turbulent shear stress with increasing downstream distance, F and G are undetermined profile functions, and $\delta(x)$ is a characteristic cross-stream length scale. The profile functions must confine the region of turbulence, so $F, G \rightarrow 0$ as $y/\delta \rightarrow \pm\infty$, and they must allow the jet to spread equally upward and downward, so F must be even and G must be odd; thus $F(0) = 1$ and $G(0) = 0$. When the self-preserving forms (12.56) and (12.57) are successful, the turbulence is said to have one characteristic length scale. These two equations are the similarity-solution forms (see (8.32)) for the steady mean-flow RANS equations when x and y are the independent variables.

For two-dimensional, constant-density flow with steady boundary conditions, the mean flow equations are:

$$\frac{\partial U}{\partial x} + \frac{\partial V}{\partial y} = 0, \quad U\frac{\partial U}{\partial x} + V\frac{\partial U}{\partial y} = -\frac{1}{\rho}\frac{\partial P}{\partial x} + \nu\left(\frac{\partial^2 U}{\partial x^2} + \frac{\partial^2 U}{\partial y^2}\right) - \frac{\partial \overline{u^2}}{\partial x} - \frac{\partial \overline{uv}}{\partial y}, \text{ and} \quad (12.58, 12.59)$$

$$U\frac{\partial V}{\partial x} + V\frac{\partial V}{\partial y} = -\frac{1}{\rho}\frac{\partial P}{\partial y} + \nu\left(\frac{\partial^2 V}{\partial x^2} + \frac{\partial^2 V}{\partial y^2}\right) - \frac{\partial \overline{uv}}{\partial x} - \frac{\partial \overline{v^2}}{\partial y}. \quad (12.60)$$

For this analysis, the simplest possible form of these equations is adequate. Thus, the jet flow is assumed to be thin, so the boundary-layer approximations are made: $U \gg V$, and $\partial/\partial y \gg \partial/\partial x$. In addition, pressure gradients are presumed small within the nominally quiescent reservoir fluid so that $\partial P/\partial x \approx 0$, and the jet flow's Reynolds number is assumed to be high enough so that viscous stresses can be ignored compared to Reynolds stresses. With these simplifications, the two momentum equations (12.59) and (12.60) become:

$$U\frac{\partial U}{\partial x} + V\frac{\partial U}{\partial y} \cong -\frac{\partial \overline{uv}}{\partial y}, \text{ and } 0 \cong -\frac{1}{\rho}\frac{\partial}{\partial y}\left(P + \rho\overline{v^2}\right). \quad (12.61)$$

Because the viscous terms have been dropped, these equations are independent of the Reynolds number and should be valid for all Re that are high enough to justify this approximation. Multiplying (12.58) by U and adding it to the first part of (12.61) produces:

$$\frac{\partial}{\partial x}(U^2) + \frac{\partial}{\partial y}(VU + \overline{uv}) \cong 0,$$

which can be integrated in the cross-stream direction between infinite limits to obtain:

$$\frac{\partial}{\partial x}\int_{-\infty}^{+\infty} U^2 dy + [VU + \overline{uv}]_{y=-\infty}^{y=+\infty} \cong 0.$$

When evaluated, the terms in [,]-brackets are zero because U, V, and $\overline{uv}$ are all presumed to go to zero as $y \rightarrow \pm\infty$. This equation can be integrated in the stream-wise direction from 0 to x to find:

$$J_s \equiv \rho_s \int_{-\infty}^{+\infty} [U^2]_{x=0} dy \cong \rho \int_{-\infty}^{+\infty} U^2 dy = const. \quad (12.62)$$

In (12.62), J_s is the momentum injected into the flow per unit span of the slot and the two integrals in (12.62) come from evaluating J_s at $x = 0$ and at a location well downstream in the jet. Here, ρ_s is the density of the fluid that emerges from the nozzle, and any difference between ρ and ρ_s is presumed to be insignificant downstream in the jet because the fluid that comes from the slot is mixed with and diluted by the nominally quiescent fluid entrained into the jet. The basis for this presumption is provided further on in this section. Overall, (12.62) can be regarded as a constraint that requires the turbulent flow to contain the same amount of stream-wise momentum at all locations downstream of the nozzle.

To determine the form of the similarity solution for the plane jet, first eliminate V from (12.61) using an integrated form of (12.58), $V = - \int_0^y (\partial U / \partial x) dy$, to find

$$U\frac{\partial U}{\partial x} - \left[\int_0^y \left(\frac{\partial U}{\partial x} \right) dy \right] \frac{\partial U}{\partial y} \cong - \frac{\partial \overline{uv}}{\partial y},$$

where $V(0) = 0$ by symmetry and y is an integration variable. Then, substitute (12.56) and (12.57) into this equation to reach a single equation involving the two amplitude functions, U_{CL} and Ψ, and the two profile functions, F and G:

$$U_{CL}F\frac{\partial}{\partial x}(U_{CL}F) - \left[\int_0^y \frac{\partial}{\partial x}(U_{CL}F)dy \right] \frac{\partial}{\partial y}(U_{CL}F) \cong - \frac{\partial}{\partial y}(\Psi G).$$

Although somewhat tedious, the terms of this equation can be expanded and simplified to find:

$$\left\{ \frac{\delta U'_{CL}}{U_{CL}} \right\} F^2 - \left\{ \frac{\delta U'_{CL}}{U_{CL}} + \delta' \right\} F' \int_0^\xi F d\xi = \left\{ \frac{\Psi}{U_{CL}^2} \right\} G', \qquad (12.63)$$

where a prime indicates differentiation of a function with respect to its is argument, $\xi = y/\delta$, and ξ is an integration variable. For a simple similarity solution to exist, the coefficients inside {,}-braces in (12.63) should not be functions of x. Setting each equal to a constant produces two ordinary differential equations and an algebraic one:

$$\frac{\delta U'_{CL}}{U_{CL}} = C_1, \quad \frac{\delta U'_{CL}}{U_{CL}} + \delta' = C_2, \text{ and } \frac{\Psi}{U_{CL}^2} = C_3. \qquad (12.64)$$

The first two of these imply $\delta' = C_2 - C_1$, which is readily integrated to find: $\delta(x) = (C_2 - C_1)(x - x_o)$, where x_o is a constant and is known as the virtual origin of the flow. It is traditional to choose $C_2 - C_1 = 1$ and to presume that x_o is small so that $\delta = x$. In experiments, x_o is typically found to be of order d, the width of the slot. With $\delta = x$, the first equation of (12.64) may be integrated to determine: $U_{CL} = C_4 x^\gamma$, where C_4 and γ are constants. Substituting this into (12.62) leads to:

$$J_s = \rho \int_{-\infty}^{+\infty} U^2 dy = \rho U_{CL}^2 \int_{-\infty}^{+\infty} F^2(\xi) dy = \rho U_{CL}^2 \delta \int_{-\infty}^{+\infty} F^2(\xi) d\xi = \rho C_4^2 x^{2\gamma+1} \int_{-\infty}^{+\infty} F^2(\xi) d\xi.$$

$$(12.65)$$

Here the final definite integral is just a dimensionless number, so the final form of (12.65) can only be independent of x when $2\gamma + 1 = 0$, or $\gamma = -1/2$. Thus, the results of (12.64) imply that (12.56) and (12.57) may be rewritten:

$$U(x,y) = C_5(J_s/\rho)^{1/2}x^{-1/2}F(y/x) \quad \text{and}$$

$$-\overline{uv} = C_3 U_{CL}^2 G(y/x) = C_3 C_5^2 (J_s/\rho)x^{-1}G(y/x),$$

(12.66, 12.67)

where the constants C_3 and C_4 and the profile functions F and G must be determined from experimental measurements, direct numerical simulations, or an alternate theory. They cannot be determined from this type of simple similarity analysis because (12.63) is one equation for two unknown profile functions, a situation that is a direct legacy of the closure problem. However, the parametric dependencies shown in (12.66) and (12.67) are those found in experiments, and this is the primary reason for seeking self-preserving forms via a similarity analysis.

The result (12.66) may be used to determine the volume flux (per unit span) $\dot{V}$ in the jet via a simple integration,

$$\dot{V}(x) = \int_{-\infty}^{+\infty} U(x,y)dy = C_5(J_s/\rho)^{1/2}x^{+1/2} \int_{-\infty}^{+\infty} F(\xi)d\xi,$$

(12.68)

where again the definite integral is just a dimensionless number. Therefore, the volume flux in the jet increases with increasing downstream distance like $x^{1/2}$, so the dilution assumption made about J_s in (12.62) should be valid sufficiently far from the jet nozzle. At such distances, commonly known as the *far field* of the jet, the mean mass fraction $\overline{Y}(x,y)$ of slot fluid (or any other suitably defined passive scalar like a dye concentration) will also follow a similarity form:

$$\overline{Y}(x,y) = Y_{CL}(x)H(y/x),$$

(12.69)

where Y_{CL} is the centerline nozzle-fluid mass fraction, and H is another profile function defined so that $H(0) = 1$ and $H \to 0$ as $y/\delta \to \pm\infty$. Conservation of slot fluid requires:

$$\dot{M}_s = \rho_s \int_{-\infty}^{+\infty} [U]_{y=0}dy \cong \rho \int_{-\infty}^{+\infty} \overline{Y}(x,y)U(x,y)dy$$

$$= \rho Y_{CL}C_5(J_s/\rho)^{1/2}x^{+1/2} \int_{-\infty}^{+\infty} H(\xi)F(\xi)d\xi,$$

(12.70)

where $\dot{M}_s$ is the slot-fluid mass injection rate per unit span. In (12.70) the stream-wise turbulent scalar transport term $\overline{uY'}$ has been neglected because it tends to be much smaller than the stream-wise mean scalar transport term $\overline{Y}U$. Reducing (12.70) to a single relationship for Y_{CL}, and substituting this into (12.69) produces:

$$\overline{Y}(x,y) = C_6\left(\dot{M}_s/\sqrt{\rho J_s}\right)x^{-1/2}H(y/x),$$

(12.71)

where C_6 is another constant. The equations (12.66), (12.67), and (12.71) represent the outcomes from this similarity analysis and can be compared with the $u \propto x^{-1/3}$, $\delta \propto x^{2/3}$ behavior of a planar laminar jet derived in Section 9.10.

Over the years some success has been achieved in determining the various profile shapes and the constants. For example, when the slot exit velocity is uniform and equal to U_0, then $J_s = \rho_s U_0^2 d$ and $\dot{M}_s = \rho_s U_0 d$ so (12.66) and (12.71) reduce to:

$$U(x,y) = C_5 U_0 (\rho_s/\rho)^{1/2}(x/d)^{-1/2} F(y/x), \text{ and}$$
$$\overline{Y}(x,y) = C_7 Y_0 (\rho_s/\rho)^{1/2}(x/d)^{-1/2} H(y/x), \tag{12.72, 12.73}$$

where Y_0 is the mass fraction of a passive scalar in the slot fluid. Here $Y_0 = 1$ if the slot fluid is the passive scalar, but Y_0 may be much less than one if it represents a trace contaminant or a dye concentration. In addition, the profile functions F and H are smooth, bell-shaped curves commonly specified by their one-sided half-widths $(\xi_{1/2})_U$ and $(\xi_{1/2})_Y$, the values of y/δ that produce F and $H = 1/2$, respectively. For example when a Gaussian function is fit to mean profiles of U, the function F becomes:

$$F(y/x) = \exp\left\{-\ln(2)(y/x)^2/(\xi_{1/2})_U^2\right\}.$$

Approximate empirical values for C_5, C_6, $(\xi_{1/2})_U$, and $(\xi_{1/2})_Y$ from Chen and Rodi (1980) and Pope (2000) are provided in Table 12.1 for the plane turbulent jet along with results for other free shear flows. The similarity forms shown in this table for the planar and round wakes should also be followed in the far-field of jets in coflowing streams. Unfortunately, variations in the empirical constants between experiments may be $\pm 20\%$ (or even more; see the round-wake results) and these variations are thought to be caused by unintentional experimental artifacts, such as unmeasured vibrations, geometrical imperfections, or fluctuations in one of the input flows.

Interestingly, as pointed out in George (1989), such variation in similarity constants is consistent with the type of similarity analysis presented earlier in this section. The three equations (12.64) determined from the coefficients of the similarity momentum equation specify the simplest possibility leading to self-similarity of the mean flow. A more general version of (12.64) that also leads to self-similarity is:

$$\frac{\delta U_{CL}'}{U_{CL}} = C_8 \left(\frac{\delta U_{CL}'}{U_{CL}} + \delta'\right) = C_9 \frac{\Psi}{U_{CL}^2}, \tag{12.74}$$

which specifies that the x-dependence of the three coefficients must be equal. Here, $\delta \sim x^m$, $U_{CL} \sim x^n$, and $\Psi \sim x^{2n+m-1}$ satisfy (12.74) as do $\delta \sim \exp\{ax\}$, $U_{CL} \sim \exp\{-ax\}$, and $\Psi \sim \exp\{-ax\}$; thus, multiple possibilities are allowed by (12.63) for the plane jet's similarity solution. While the second law of thermodynamics and the constraint (12.62) rule out some of these possibilities, (12.74) or its equivalent for other free shear flows and conditions at the flow's origin ($x = 0$) apparently allow the expected self-similar states for a particular shear flow to vary somewhat from experiment to experiment.

EXAMPLE 12.2

Pure methane gas issues from a round nozzle with diameter $d = 1$ cm at a speed of $U_0 = 20$ m/s into a combustion chamber nominally filled with quiescent air at room temperature and pressure. Assuming the volume fraction of oxygen in air is 0.21, use the round turbulent jet similarity law to estimate the centerline distance from the nozzle exit where the stoichiometric condition is reached, and the centerline speed and nominal width of the jet flow at that point.

Solution

Using subscripts "A" for air and "M" for methane, and molecular weights of 28.96 and 16.04 for air and methane, respectively, the Reynolds number of the flow is:

$$\sqrt{J_s/\rho_A}/\nu \sim (\rho_M/\rho_A)^{1/2} U_0 d/\nu_A = (16.04/28.96)^{1/2}(20\,m/s)(0.01\,m)/(1.5 \times 10^{-5}\,m^2 s^{-1}) \sim 10^4,$$

which is high enough to form a turbulent jet. Since the relevant chemical reaction is $CH_4 + 2O_2 \rightarrow CO_2 + 2H_2O$, the mole fraction of methane is half that of oxygen at the stoichiometric condition. Thus, at the location of interest x, there are two equations relating mean volume fractions, $\bar{v}_A + \bar{v}_M = 1$ and $\bar{v}_M = 0.5(0.21\bar{v}_A)$, that are readily solved to find: $\bar{v}_A = 1/1.105 = 0.905$, and $\bar{v}_M = 1 - (1/1.105) = 0.095$. With this composition, the mixture density is within 4% or so of the density of air. The requisite mass fraction of methane is:

$$Y_M = (0.095)(16.04)/[(0.905)(28.96) + (0.095)(16.04)] = 0.0549.$$

The location x is found from the entry in Table 12.1 for the mass-fraction field of a round turbulent jet. This means setting $Y_M = 5.0\,Y_0(\rho_s/\rho_A)^{1/2}(x/d)^{-1}$ and solving for x to find:

$$x = 5.0\,(Y_0/Y_M)(\rho_s/\rho_A)^{1/2}d = 5.0(1.0/0.0549)(16.04/28.96)^{1/2}(0.01\,m) \cong 0.68\,m.$$

Using the Table 12.1 entry for the velocity field of a round turbulent jet, the jet centerline velocity at this location is:

$$U_{CL}(x) = 6.0U_o\,(\rho_s/\rho_A)^{1/2}(x/d)^{-1} = 6.0(20\,m/s)(16.04/28.96)^{1/2}(68)^{-1} \cong 1.3\,m/s.$$

The nominal width of the flow will be approximately four times larger than the jet's mean concentration profile half radius $r_{1/2}$. From the round jet profile width entry in Table 12.1, we have:

$$(\xi_{1/2})_Y = 0.11 = (r_{1/2}/x)_Y \quad \text{so } 4r_{1/2} = 4(0.11)(0.68\,m) \cong 0.30\,m.$$

Thus, the width of a jet's cone of turbulence is a little less than half the downstream distance.

When the mean velocity and mass fraction fields of a free turbulent shear flow are self-similar, their corresponding fluctuations are commonly self-similar with the same dependence on the downstream coordinate as that found for the mean fields. However, the profile functions for the various Reynolds stress components or the passive scalar variance are typically not bell-shaped curves. Sample free shear flow measurements for the Reynolds stress components for the plane turbulent jet are shown in Figure 12.14. Results such as these indicate how fluctuation energy varies within a turbulent flow and may be used to develop and

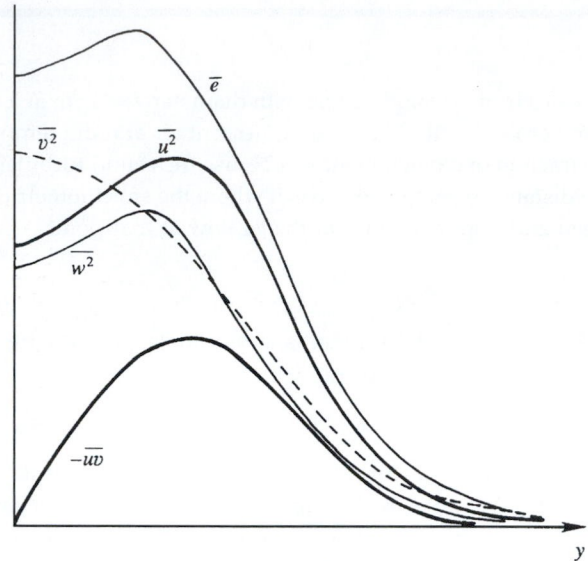

FIGURE 12.14   Sketch of the observed variation of the turbulent kinetic energy $\bar{e}$ and the nonzero Reynolds stress components across a planar jet. Here, $\overline{uv} = 0$ on the jet's centerline ($y = 0$), and $\overline{uw} = \overline{vw} = 0$ throughout the flow because the mean flow is symmetric about $y = 0$, and because the flow is homogeneous in the z-direction.

test closure models for RANS equations. Here $\overline{u^2}$, $\overline{v^2}$, and $\overline{w^2}$ are the velocity component variances (commonly called *turbulent intensities*) in the stream-wise (x), slot-normal (y), and slot-parallel (z) directions, respectively. The Reynolds stresses $\overline{uw}$ and $\overline{vw}$ are zero throughout the planar jet since the flow is homogeneous in the z-direction and there is no reason for w to be mostly of one sign if u or v is either positive or negative. Similarly, the Reynolds stress $\overline{uv}$ is zero on the jet centerline by symmetry. In Figure 12.14, the Reynolds stress reaches a maximum magnitude roughly where $\partial U/\partial y$ is maximum. This is also close to the region where the turbulent kinetic energy $\bar{e}$ reaches a maximum. Such correspondences are commonly exploited in the development of turbulence models.

The terms in the turbulent kinetic energy budget for a two-dimensional jet are shown in Figure 12.15. Under the boundary layer assumption for derivatives, $\partial/\partial y \gg \partial/\partial x$, the budget equation (12.47) becomes

$$0 = -U\frac{\partial \bar{e}}{\partial x} - V\frac{\partial \bar{e}}{\partial y} - \overline{uv}\frac{\partial U}{\partial y} - \frac{\partial}{\partial y}\left(\frac{1}{\rho_0}\overline{pv} + \frac{1}{2}\overline{ev}\right) - \bar{\varepsilon}, \tag{12.75}$$

where the left side represents $\partial \bar{e}/\partial t$. Here, the viscous transport and the term $(\overline{v^2} - \overline{u^2})(\partial U/\partial x)$ arising out of the shear production have been neglected because they are small. The balance of terms shown in this figure is analyzed in Townsend (1976). Here, T denotes turbulent transport represented by the fourth term on the right-hand side of (12.75). The shear production is zero on the jet centerline where both $\partial U/\partial y$ and $\overline{uv}$ are zero, and reaches a maximum close to the position of the maximum Reynolds stress. Near the center of the jet, the dissipation is primarily balanced by the downstream advection

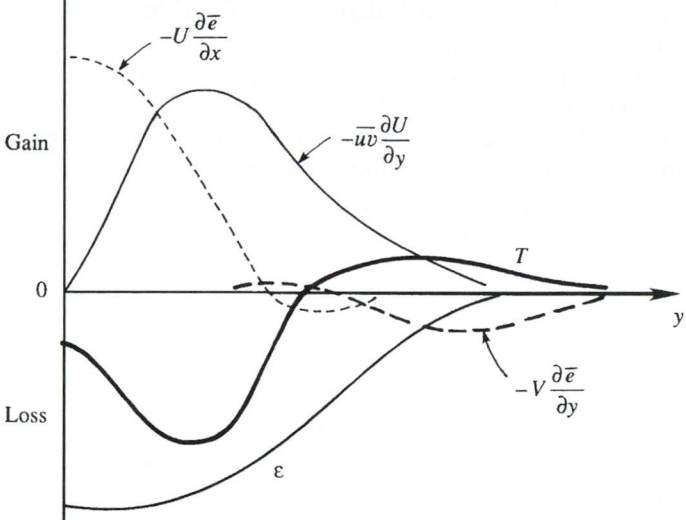

FIGURE 12.15 Sketch of measurements of the terms in the kinetic energy budget of a planar turbulent jet. Here the turbulent transport terms are lumped together and indicated by T. Information of this type is used to build, adjust, and validate closure models for RANS equations.

$-U(\partial \bar{e}/\partial x)$, which is positive because the turbulent kinetic energy $\bar{e}$ decays downstream. Away from the jet's center, but not too close to the jet's outer edge, the production and dissipation terms balance. In the outer parts of the jet, the transport term balances the cross-stream advection. In this region V is negative (i.e., toward the center) due to entrainment of the surrounding fluid, and also $\bar{e}$ decreases with increasing y. Therefore the cross-stream advection $-V(\partial \bar{e}/\partial y)$ is negative, signifying that the entrainment velocity V tends to decrease the turbulent kinetic energy at the outer edge of the jet. A temporally stationary state is therefore maintained by the transport term T carrying $\bar{e}$ away from the jet's center (where $T < 0$) into the outer parts of the jet (where $T > 0$).

12.9. WALL-BOUNDED TURBULENT SHEAR FLOWS

At sufficiently high Reynolds number, the characteristics of free turbulent shear flows discussed in the preceding section are independent of Reynolds number and may be self-similar based on a single length scale. However, neither of these simplifications occurs when the flow is bounded by one or more solid surfaces. The effects of viscosity are always felt near the wall where turbulent fluctuations go to zero, and this gives rise to a second fundamental length scale l_ν that complements the turbulent layer thickness δ. In addition, the persistent effects of viscosity are reflected in the fact that the skin-friction coefficient for a smooth flat plate or smooth round pipe depends on Re, even when Re $\rightarrow \infty$, as seen in Figure 9.11. Therefore, Re independence of the flow as Re $\rightarrow \infty$ does not occur in wall-bounded turbulent shear flows when the wall(s) is(are) smooth.

The importance of wall-bounded turbulence in engineering applications and geophysical situations is hard to overstate since it sets fundamental limits for the efficiency of transportation systems and on the exchange of mass, momentum, and heat at the earth's surface. Thus, the literature on wall-bounded turbulent flows is large and the material provided here merely covers the fundamentals of the mean flow. A more extensive presentation that includes turbulence intensities is provided in Chapter 7 of Pope (2000). Vortical structures in wall-bounded turbulence are discussed in Kline et al. (1967), Cantwell (1981), and Adrian (2007). The review articles by George (2006) and Marusic et al. (2010) are also recommended.

Three generic wall-bounded turbulent shear flows are described in this section: pressure-driven channel flow between smooth stationary parallel plates, pressure-driven flow through a smooth round pipe, and the turbulent boundary-layer flow that develops from nominally uniform flow over a smooth flat plate. The first two are fully confined while the boundary layer has one free edge. The main differences between turbulent and laminar wall-bounded flows are illustrated on Figure 12.16. In general, mean turbulent-flow profiles (solid curves) are blunter, and turbulent-flow wall-shear stresses are higher than those of equivalent steady laminar flows (dashed curves). In addition, a turbulent boundary-layer, mean-velocity profile approaches the free-stream speed very gradually with increasing y so the full thickness of the profile shown in the right panel of Figure 12.16 lies beyond the extent of the figure. Throughout this section, the density of the flow is taken to be constant.

Fully developed channel flow is perhaps the simplest wall-bounded turbulent flow. Here, the modifier *fully developed* implies that the statistics of the flow are independent of the downstream direction. The analysis provided here is readily extended to pipe flow, after a suitable redefinition of coordinates. Further extension of channel flow results to boundary-layer flows is not as direct, but can be made when the boundary-layer approximation replaces the fully

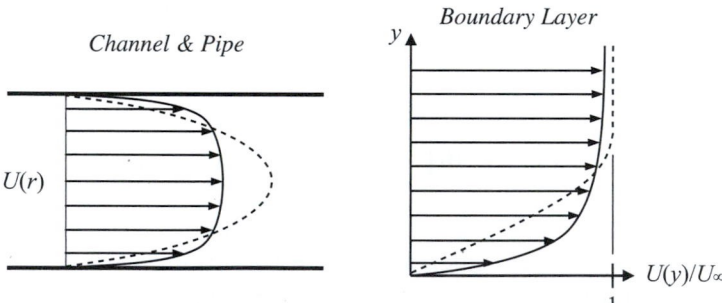

FIGURE 12.16 Sample profiles for wall-bounded turbulent flows (solid curves) compared to equivalent laminar profiles (dashed curves). In general turbulent profiles are blunter with higher skin friction; that is, $\mu(dU/dy)$ evaluated at the wall is greater in turbulent flows than in equivalent laminar ones. In channel and pipe flows, the steady laminar profile is parabolic while a mean turbulent flow profile is more uniform across the central 80% of the channel or pipe. For boundary layers having the same displacement thickness, the steady laminar profile remains linear farther above the wall and converges to the free-stream speed more rapidly than the mean turbulent profile.

developed flow assumption. If we align the x-axis with the flow direction, and chose the y-axis in the cross-stream direction perpendicular to the plates so that $y = 0$ and $y = h$ define the plate surfaces, then fully developed channel flow must have $\partial U/\partial x = 0$. Hence, U can only depend on y, and it is the only mean velocity component because the remainder of (12.27) implies $\partial V/\partial y = 0$, and the boundary conditions $V = 0$ on $y = 0$ and h then require $V = 0$ throughout the channel. Under these circumstances, the mean flow momentum equations are:

$$0 = -\frac{\partial P}{\partial x} + \frac{\partial \bar{\tau}}{\partial y} \quad \text{and} \quad 0 = -\frac{\partial}{\partial y}\left(P + \rho\overline{v^2}\right), \tag{12.76}$$

where $\bar{\tau} = \mu(\partial U/\partial y) - \rho_0\overline{uv}$ is the total average stress and it cannot depend on x. Integrating the second of these equations from the lower wall up to y produces:

$$P(x,y) - P(x,0) = -\rho\overline{v^2} + \rho\left[\overline{v^2}\right]_{y=0} = -\rho\overline{v^2},$$

where the final equality follows because the variance of the vertical velocity fluctuation is zero at the wall ($y = 0$). Differentiating this with respect to x produces:

$$\frac{\partial}{\partial x}P(x,y) - \frac{d}{dx}P(x,0) = -\rho\frac{\partial}{\partial x}\overline{v^2} = 0, \tag{12.77}$$

where $P(x,0)$ is the ensemble-average pressure on $y = 0$ and the final equality follows from the fully developed flow assumption. Thus, the stream-wise pressure gradient is only a function of x; $\partial P(x,y)/\partial x = dP(x,0)/dx$. Therefore, the only way for the first equation of (12.76) to be valid is for $\partial P/\partial x$ and $\partial\bar{\tau}/\partial y$ to each be constant, so the total average stress distribution $\bar{\tau}(y)$ in turbulent channel flow is linear as shown in Figure 12.17a. Away from the wall, $\bar{\tau}$ is due mostly to the Reynolds stress, close to the wall the viscous contribution dominates, and at the wall the stress is entirely viscous.

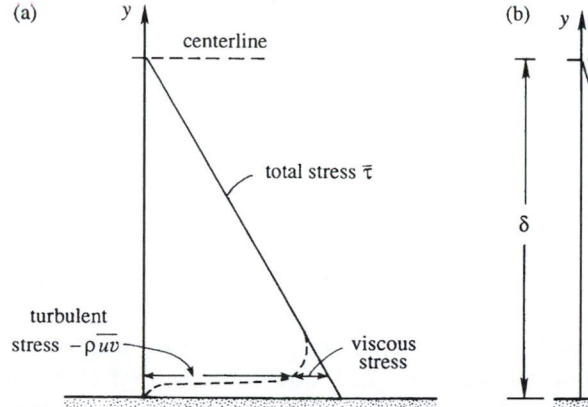
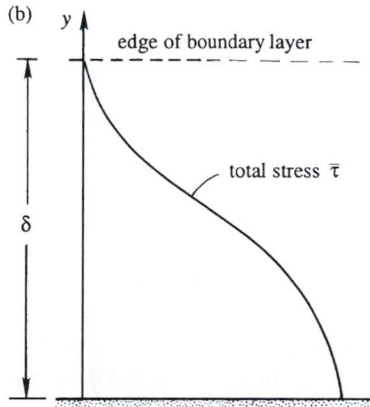

FIGURE 12.17   Variation of total shear stress across a turbulent channel flow (a) and through a zero-pressure-gradient turbulent boundary layer (b). In both cases, the Reynolds shear stress dominates away from the wall but the viscous shear stress takes over close to the wall. The shape of the two stress curves is set by momentum transport between the fast-moving part of the flow and the wall where $U = 0$.

For a boundary layer on a flat plate, the stream-wise mean-flow momentum equation is

$$U\frac{\partial U}{\partial x} + V\frac{\partial U}{\partial y} = -\frac{1}{\rho}\frac{\partial P}{\partial x} + \frac{1}{\rho}\frac{\partial \bar{\tau}}{\partial y}, \tag{12.78}$$

where $\bar{\tau}$ is a function of x and y. The variation of the stress across a boundary layer is sketched in Figure 12.17b for the zero-pressure-gradient (ZPG) condition. Here, a constant stress layer, $\partial\bar{\tau}/\partial y \approx 0$, occurs near the wall since both U and $V \to 0$ as $y \to 0$. When the pressure gradient is not zero, the stress profile approaches the wall with a constant slope. Although it is not shown in the figure, the structure of the near-wall region of the turbulent boundary layer is similar to that depicted for the channel flow in Figure 12.17a with viscous stresses dominating at and near the wall.

The partitioning of the stress based on its viscous and turbulent origins leads to the identification of two different scaling laws for wall-bounded turbulent flows. The first is known as the *law of the wall* and it applies throughout the region of the boundary layer where viscosity matters and the largest relevant length scale is y, the distance from the wall. This region of the flow is typically called the *inner layer*. The second scaling law is known as the *velocity defect law*, and it applies where the flow is largely independent of viscosity and the largest relevant length scale is the overall thickness of the turbulent layer δ. This region of the flow is typically called the *outer layer*. Fortunately, the inner and outer layers of wall-bounded turbulent flow overlap, and in this overlap region the form of the mean stream-wise velocity profile may be deduced from dimensional analysis.

Inner Layer: Law of the Wall

Consider the flow near the wall of a channel, pipe, or boundary layer. Let U_∞ be the centerline velocity in the channel or pipe, or the free-stream velocity outside the boundary layer. Let δ be the thickness of the flow between the wall and the location where $U = U_\infty$. Thus, δ may be the channel half width, the radius of the pipe, or the boundary-layer thickness. Assume that the wall is smooth, so that any surface roughness is too small to affect the flow. Physical considerations suggest that the near-wall velocity profile should depend only on the near-wall parameters and not on U_∞ or the thickness of the flow δ. Thus, very near the smooth surface, we expect

$$U = U(\rho, \tau_0, \nu, y), \tag{12.79}$$

where τ_0 is the shear stress on the smooth surface. This equation may be recast in dimensionless form as:

$$U^+ \equiv \frac{U}{u_*} = f\left(\frac{yu_*}{\nu}\right) = f\left(\frac{y}{l_\nu}\right) = f(y^+) \quad \text{where } u_*^2 \equiv \frac{\tau_0}{\rho}, \tag{12.80, 12.81}$$

f is an undetermined function, u_* is the *friction velocity* or *shear velocity*, and $l_\nu = \nu/u_*$ is the *viscous wall unit*. Equation (12.80) is the *law of the wall* and it states that U/u_* should be a universal function of yu_*/ν near a smooth wall. The superscript plus signs are standard in the literature and indicate a dimensionless law-of-the-wall variable.

The inner part of the wall layer, right next to the wall, is dominated by viscous effects and is called the *viscous sublayer*. In spite of the fact that it contains fluctuations, the Reynolds stresses are small here because the presence of the wall quells wall-normal velocity fluctuations. At high Reynolds numbers, the viscous sublayer is thin enough so that the stress is uniform within the layer and equal to the wall shear stress τ_0. Therefore the mean velocity gradient in the viscous sublayer is given by

$$\mu(dU/dy) = \tau_0 \to U = \tau_0 y/\mu \text{ or } U^+ = y^+, \qquad (12.82)$$

where the second two equalities follow from integrating the first. Equation (12.82) shows that the velocity distribution is linear in the viscous sublayer, and experiments confirm that this linearity holds up to $yu_*/v \sim 5$, which may be taken to be the limit of the viscous sublayer.

Outer Layer: Velocity Defect Law

Now consider the velocity distribution in the outer part of a turbulent boundary layer. The gross characteristics of the turbulence in the outer region are inviscid and resemble those of a free shear flow. The existence of Reynolds stresses in the outer region results in a drag on the flow and generates a *velocity defect* $\Delta U = U_\infty - U$, just like the planar wake. Therefore, in the outer layer we expect,

$$U = U(\rho, \tau_0, \delta, y), \qquad (12.83)$$

and by dimensional analysis can write:

$$\frac{U_\infty - U}{u_*} = F\left(\frac{y}{\delta}\right) = F(\xi) \qquad (12.84)$$

so that the deficit velocity, $U_\infty - U$, is proportional to the friction velocity u_* and a profile function. This is called the *velocity defect law,* and this is its traditional form. In the last two decades, it has been the topic of considerable discussion in the research community, and alternative velocity and length scales have been proposed for use in (12.84), especially for turbulent boundary-layer flows.

Overlap Layer: Logarithmic Law

From the preceding discussion, the mean velocity profiles in the inner and outer layers of a wall-bounded turbulent flow are governed by different laws, (12.80) and (12.84), in which the independent coordinate y is scaled differently. Distances in the outer part are scaled by δ, whereas those in the inner part are scaled by the much smaller viscous wall unit $l_v = v/u_*$. Thus, wall-bounded turbulent flows involve at least two turbulent length scales, and this prevents them from reaching the same type of self-similar form with increasing Reynolds number as that found for simple free turbulent shear flows.

Interestingly, a region of overlap in the two profile forms can be found by taking the limits $y^+ \to \infty$ and $\xi \to 0$ simultaneously. Instead of matching the mean velocity directly, in this case it is more convenient to match mean velocity gradients. (The following short derivation closely follows that in Tennekes and Lumley, 1972.) From (12.80) and (12.84), dU/dy in the inner and outer regions is given by

$$\frac{dU}{dy} = \frac{u_*^2}{v}\frac{df}{dy^+} \quad \text{and} \quad -\frac{dU}{dy} = \frac{u_*}{\delta}\frac{dF}{d\xi}, \qquad (12.85, 12.86)$$

respectively. Equating these and multiplying by y/u_*, produces:

$$-\xi\frac{dF}{d\xi} = y^+\frac{df}{dy^+},\qquad(12.87)$$

an equation that should be valid for large y_+ and small ξ. As the left-hand side can only be a function of ξ and the right-hand side can only be a function of y^+, both sides must be equal to the same universal constant, say $1/\kappa$, where κ is the *von Karman constant* (not the thermal diffusivity). Experiments show that $\kappa \approx 0.4$ with some dependence on flow type and pressure gradient, as is discussed further later on. Setting each side of (12.87) equal to $1/\kappa$, integrating, and using (12.80) gives:

$$U^+ \equiv \frac{U}{u_*} = f(y^+) = \frac{1}{\kappa}\ln(y^+) + B \quad\text{and}\quad F(\xi) = -\frac{1}{\kappa}\ln(\xi) + A,\qquad(12.88, 12.89)$$

where B and A are constants with values around 4 or 5, and 1, respectively, again with some dependence on flow type and pressure gradient. Equation (12.88) or (12.89) is the mean velocity profile in the *overlap layer* or the *logarithmic layer*. In addition, the constants in (12.88), κ and B, are known as the logarithmic-law (or log-law) constants. As the derivation shows, (12.88) and (12.89) are only valid for large y^+ and small y/δ, respectively. The foregoing method of justifying the logarithmic velocity distribution near a wall was first given by Clark B. Millikan in 1938. The logarithmic law, however, was known from experiments conducted by the German researchers, and several derivations based on semi-empirical theories were proposed by Prandtl and von Karman. One such derivation using the so-called mixing length theory is presented in the following section.

The logarithmic law (12.88) may be the best-known and most important result for wall-bounded turbulent flows. Experimental confirmation of this law is shown in Figure 12.18 in law-of-the-wall coordinates for the turbulent boundary-layer data reported in Oweis et al. (2010). Nominal specifications for the extent of the viscous sublayer, the buffer layer, the logarithmic layer, and the wake region are shown there as well. On this log-linear plot, the linear viscous sublayer profile appears as a curve for $y^+ < 5$. However, a logarithmic velocity profile will appear as a straight line on a log-linear plot, and such a linear region is evident for approximately two decades in y^+ starting near $y^+ \sim 10^2$. The extent of this logarithmic region increases in these coordinates with increasing Reynolds number. The region $5 < y^+ < 30$, where the velocity distribution is neither linear nor logarithmic, is called the *buffer layer*. Neither the viscous stress nor the Reynolds stresses are negligible here, and this layer is dynamically important because turbulence production reaches a maximum here. Overall, the measured results collapse well to a single curve below $y^+ \sim 10^4$ (or $y/\delta \sim 0.2$) in conformance with the law of the wall. For larger values of y^+, the collapse ends where the overlap region ends and the boundary layer's wake flow begins. Although the wake region appears to be smaller than the log-region on the plot, this is an artifact of the logarithmic horizontal axis. A turbulent boundary layer's wake region typically occupies the outer 80% of the flow's full thickness. These velocity profiles do not collapse in the wake region when plotted with law-of-the-wall normalizations because the wake-flow similarity variable is y/δ (not y/l_ν) and ratio $\delta^+ = \delta/l_\nu$ is different at the three different Reynolds numbers. The fitted curves shown in Figure 12.18 are mildly adjusted versions of

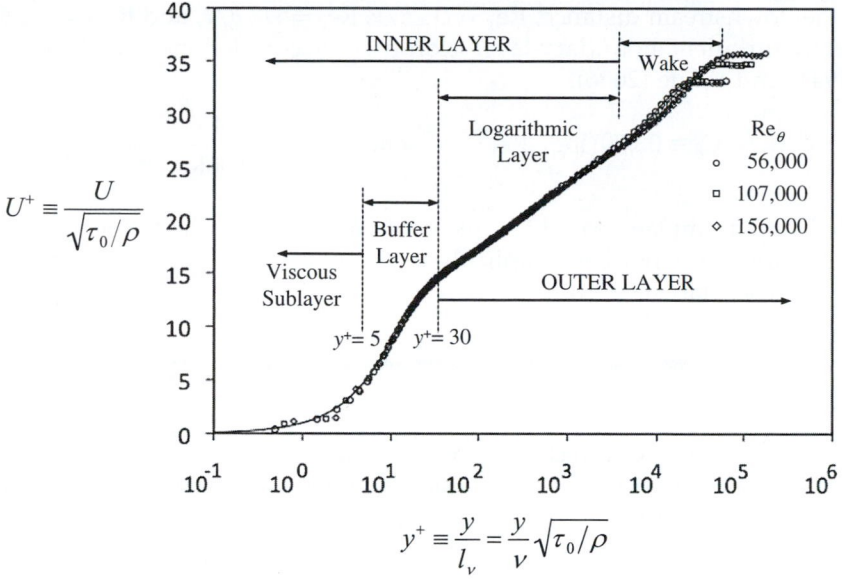

FIGURE 12.18 Mean velocity profile of a smooth-flat-plate turbulent boundary layer plotted in log-linear coordinates with law-of-the-wall normalizations. The data are replotted from Oweis et al. (2010) and represent three Reynolds numbers. The extent of the various layers within a wall-bounded turbulent flow are indicated by vertical dashed lines. The log-layer-to-wake-region boundary is usually assumed to begin at $y/\delta \approx 0.15$ to 0.20 in turbulent boundary layers. Overall the data collapse well for the inner layer region, as expected, and the logarithmic layer extends for approximately two decades. The wake region shows differences between the Reynolds numbers because its similarity variable is y/δ, and δ/l_ν differs between the various Reynolds numbers.

those recommended in Monkewitiz et al. (2007) for smooth-flat-plate ZPG turbulent boundary layers.

For fully developed channel and pipe flows, the mean stream-wise velocity profile does not evolve with increasing downstream distance. However, turbulent boundary layers do thicken. The following parameter results are developed from the systematic fitting and expansion efforts for ZPG turbulent boundary layers described in Monkewitz et al. (2007), and are intended for use when $Re_x > 10^6$:

$$\text{Momentum thickness} = \theta \approx 0.016 \times Re_x^{-0.15},$$

$$\text{Displacement thickness} = \delta^* \approx \theta \exp\left\{\frac{7.11\kappa}{\ln(Re_\theta)}\right\},$$

$$\text{99\% thickness} = \delta_{99} = 0.2\delta^*\left[\kappa^{-1}\ln(Re_{\delta*}) + 3.30\right], \text{ and}$$

$$\text{Skin friction coefficient} = C_f = \frac{\tau_0}{\frac{1}{2}\rho U_\infty^2} \cong \frac{2.0}{\left[\kappa^{-1}\ln(Re_{\delta*}) + 3.30\right]^2},$$

where x is the downstream distance, $\mathrm{Re}_x = U_\infty x/\nu$, $\mathrm{Re}_\theta = U_\infty \theta/\nu$, and $\mathrm{Re}_{\delta*} = U_\infty \delta*/\nu$. Other common ZPG turbulent boundary-layer skin-friction correlations are those by Schultz-Grunow (1941) and White (2006),

$$C_f \cong 0.370\left(\log_{10}\mathrm{Re}_x\right)^{-2.584} \text{ and } C_f \cong \frac{0.455}{[\ln(0.06\mathrm{Re}_x)]^2},$$

respectively. These formulae should be used cautiously because the influence of a boundary layer's virtual origin has not been explicitly included and it may be substantial (Chauhan et al., 2009; see also Marusic, 2010).

EXAMPLE 12.3

Estimate the boundary layer thicknesses on the underside of the wing of a large commercial airliner on its landing approach. Use the flat-plate results provided above, a chord-length distance of $x = 8$ m, a flow speed of 100 m/s, and a nominal value of $\kappa \approx 0.4$.

Solution

First compute the downstream-distance Reynolds number Re_x using the nominal kinematic viscosity air: $\mathrm{Re}_x = (100 \text{ m/s})(8 \text{ m})/(1.5 \times 10^{-5} \text{ m}^2/\text{s}) = 53 \times 10^6$. This Reynolds number is clearly high enough for turbulent flow, so the estimates are:

$$\theta \approx 0.016 \times \mathrm{Re}_x^{-0.15} = 0.016(8 \text{ m})\left(53 \times 10^6\right)^{-0.15} = 0.0089 \text{ m},$$

$$\delta* \approx \theta \exp\left\{\frac{7.11\kappa}{\ln(\mathrm{Re}_\theta)}\right\} = (0.0089 \text{ m})\exp\left\{\frac{7.11(0.4)}{\ln\left((0.0089 \text{ m})(100 \text{ m/s})/1.5 \times 10^{-5} \text{ m}^2/\text{s}\right)}\right\} \cong 0.0115 \text{ m, and}$$

$$\delta_{99} = 0.2\delta*\left[\kappa^{-1}\ln(\mathrm{Re}_{\delta*}) + 3.30\right] = 0.2(0.0115 \text{ m})\left[0.4^{-1}\ln\left(\frac{(0.0115 \text{ m})(100 \text{ m/s})}{1.5 \times 10^{-5} \text{ m}^2/\text{s}}\right) + 3.30\right] \cong 0.072 \text{ m}.$$

Here we note that θ and $\delta*$ are almost an order of magnitude smaller than δ_{99}, and that all three boundary-layer thicknesses are miniscule compared to the wing's chord length of 8 m. The latter finding is a primary reason why boundary-layer thicknesses are commonly ignored in aerodynamic analyses.

For the purpose of completeness, the following approximate mean velocity profile functions are offered for wall-bounded turbulent flows:

$$\text{inner profile } y^+ = U_{inner}^+ + e^{-\kappa B}\left[\exp\left(\kappa U_{inner}^+\right) - 1 - \kappa U_{inner}^+ - \frac{1}{2}\left(\kappa U_{inner}^+\right)^2 - \frac{1}{6}\left(\kappa U_{inner}^+\right)^3\right], \text{ and}$$

$$\text{outer profile } U_{outer}^+ = \frac{1}{\kappa}\ln\left(y^+\right) + B + \frac{2\Pi}{\kappa}W(y/\delta),$$

where the inner profile from Spalding (1961) is specified in implicit form, κ and B are the log-law constants from (12.88), and Π and W are the wake strength parameter and a wake function, respectively, both introduced by Coles (1956). The wake function W and the length scale δ in its argument are empirical and are typically determined by fitting curves to experimental profile data.

Of the three generic wall-bounded turbulent flows, the boundary layer's wake is typically the most prominent. For ZPG boundary layers the wake strength is $\Pi = 0.44$ (Chauhan et al., 2009). When the pressure gradient is favorable, Π is lower, and when the pressure gradient is adverse, Π is higher. The wake function is typically chosen to go smoothly from zero to unity as y goes from zero to δ. Among the simplest possibilities for $W(\xi)$ are $3\xi^2 - 2\xi^3$ and $\sin^2(\pi\xi/2)$, however more sophisticated fits are currently in use (see Monkewitz et al., 2007; Chauhan et al., 2009). In the outer profile form given above, δ is interpreted as the 100% boundary-layer thickness where U first equals the local free-stream velocity as y increases. In practice, this requirement cannot be evaluated with finite-precision experimental data so δ is often approximated as being the 99% or the 99.5% thickness, δ_{99} or $\delta_{99.5}$, respectively. Of course, for channel or pipe flows, δ is half the channel height or the pipe radius, respectively.

As of this writing, new and important concepts and results for wall-bounded turbulence continue to emerge. These include the possibility that the overlap layer might instead be of power law form (Barenblatt, 1993; George & Castillo, 1997) and a reinterpretation of the layer structure in terms of stress gradients (Wei et al., 2005; Fife et al., 2005). The comparisons in Monkewitz et al. (2008) suggest that the logarithmic law should be favored over a power law, while the implications of the stress gradient balance approach are still under consideration. These and other topics in the current wall-bounded turbulent flow literature are discussed in Marusic et al. (2010).

The one additional topic raised here concerns the universality of wall-bounded turbulent flow profiles. Are all wall-bounded turbulent flows universal (statistically the same) when scaled appropriately? To answer this question, consider the inner, outer, and overlap layers separately. First of all, the viscous sublayer profile $U^+ = y^+$ (12.82) is universal using law-of-the-wall normalizations. However, geometrical considerations suggest that the wake flow region is not universal. Consider the zone of maximum average fluid velocity at the outer edge of the wake of a wall-bounded flow. This maximum velocity zone occurs on the centerline of a channel flow (a plane), on the centerline of a pipe flow (a line), and at the edge of a boundary layer (a slightly tilted, nearly planar surface). Thus, the ratio of the maximum-velocity area to the bounding-wall surface area is one-half for channel flow, vanishingly small for pipe flow, and slightly greater than unity for boundary-layer flow. On this basis, the three wake flows are distinguished. Additionally, the boundary layer differs from the other two flows because it is bounded on one side only. The boundary layer's wake-flow region entrains irrotational fluid at its free edge and does not collide or interact with turbulent flow arising from an opposing wall, as is the case for channel and pipe flows. Thus, the wake-flow regions of these wall-bounded turbulent flows should all be different and not universal.

Now consider the overlap layer in which the mean velocity profile takes a logarithmic form. Logarithmic profiles have been observed in all three generic wall-bounded turbulent flows. However, in each circumstance, these layers inherit properties from the universal viscous sublayer and from a nonuniversal wake flow. Thus, the log-law (12.88) may imperfectly approach universality, and this situation is found in experiments. In particular, the current published literature (Nagib & Chauhan, 2008) supports the following values for the logarithmic-law constants at high Reynolds numbers:

Channel flow:	$\kappa = 0.37$	$B = 3.7$
Pipe flow:	$\kappa = 0.41$	$B = 5.0$
ZPG boundary layer:	$\kappa = 0.384$	$B = 4.17$

This observed flow-to-flow variation in log-law constants is not anticipated by the analysis presented earlier in this section because the geometric differences in the wake-flow regions were not accounted for in (12.83). However, the previous analysis remains valid for each outer-layer flow geometry. Thus, the log-law (12.88) does describe the overlap layer of a wall-bounded turbulent flow when the log-law constants are appropriate for that flow's geometry.

Interestingly, there is another issue at play here for turbulent boundary layers. From a flow-parameter perspective, a turbulent boundary layer differs from fully developed channel and pipe flows because the pressure gradient that may exist in a boundary layer flow is not directly linked to the wall shear stress τ_0. In fully developed channel and pipe flow, a stationary control volume calculation (see Exercise 12.31) requires:

$$dP/dx = -2\tau_0/h \text{ or } dP/dx = -4\tau_0/d, \tag{12.90, 12.91}$$

respectively, where h is the channel height and d is the pipe diameter. Thus, the starting points for the dimensional analysis of the inner and outer layers of the mean velocity profile, (12.79) and (12.83), need not include dP/dx for pipe and channel flows because τ_0 is already included. Yet, there is no equivalent to (12.90) or (12.91) for turbulent boundary layers. More general forms of (12.79) and (12.83) that would be applicable to all turbulent boundary layers need to include $\partial P/\partial x$, especially since $\partial P/\partial x$ does not drop from the mean stream-wise momentum equation, (12.78), for any value of y when $\partial P/\partial x$ is nonzero. The apparent outcome of this situation is that the log-law constants in turbulent boundary layers depend on the pressure gradient. Surprisingly, the following empirical correlation, offered by Nagib and Chauhan (2008),

$$\kappa B = 1.6 \left[\exp(0.1663B) - 1\right], \tag{12.92}$$

collapses measured values of κ and B from all three types of wall-bounded shear flows for $0.15 < \kappa < 0.80$, and $-4 < B < 12$. Here, the most extreme values of κ and B arise from turbulent boundary layers in adverse (low values of κ and B) and favorable (high values of κ and B) pressure gradients.

Rough Surfaces

In deriving the logarithmic law (12.88), we assumed that the flow in the inner layer is determined by viscosity. This is true only for *hydrodynamically smooth* surfaces, for which the average height of the surface roughness elements is smaller than the thickness of the

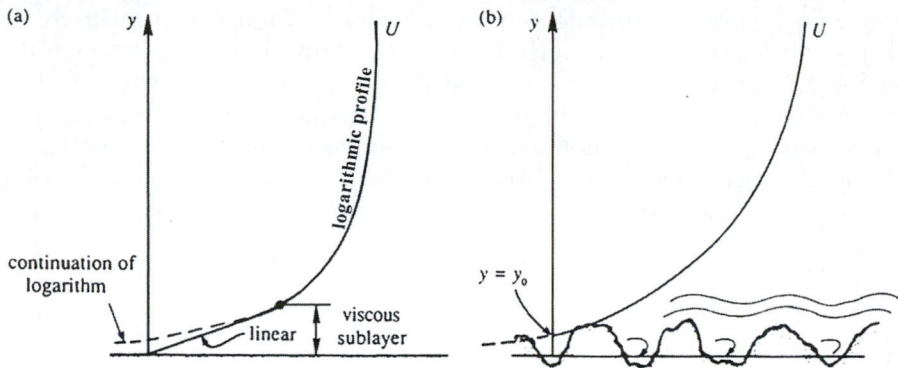

FIGURE 12.19 Logarithmic velocity distributions near smooth (a) and rough (b) surfaces. The presence of roughness may eliminate the viscous sublayer when the roughness elements protrude higher than several l_ν. In this case the log-law may be extended to a virtual wall location y_0 where U appears to go to zero.

viscous sublayer. For a hydrodynamically rough surface, on the other hand, the roughness elements protrude out of the viscous sublayer. An example is the terrestrial boundary layer, where the trees, buildings, etc., act as roughness elements. A wake develops behind each roughness element, and shear stress is transmitted to the wall by the resulting drag on the roughness elements. Viscosity becomes irrelevant for determining either the velocity distribution or the overall drag on the surface. This is why the friction coefficients for a rough pipe and a rough flat surface become constant as Re $\rightarrow \infty$.

The velocity distribution near a rough surface is again logarithmic, but the intercept constant can be set by noting that the mean velocity U is expected to be negligible somewhere within the roughness elements (Figure 12.19b). We can therefore assume that (12.88) applies for $y > y_0$, where y_0 is a measure of the roughness heights and is defined as the value of y at which the logarithmic distribution gives $U = 0$. Appropriately evaluating the constant B in (12.88) then produces:

$$U^+ = \frac{U}{u_*} = \frac{1}{\kappa}\ln\left(\frac{y}{y_0}\right). \tag{12.93}$$

12.10. TURBULENCE MODELING

The closure problem arising from Reynolds-averaging of the equations of fluid motion has lead to the development of approximate models to close systems of RANS equations. Because of the practical importance of such models for weather forecasting and performance prediction for engineered devices, RANS-closure modeling efforts have existed for more than a century and continue to this day. This section presents a truncated overview of the essential elements leading to the so-called k-ε closure model for the RANS mean-flow momentum equation (12.30). Second-order closures, large-eddy simulations, and other RANS closure schemes are described in Part Two of Pope (2000). The review article by Speziale (1991) is also recommended.

The primary purpose of a turbulent-mean-flow closure model is to relate the Reynolds stress correlations $\overline{u_i u_j}$ to the mean velocity field U_i. Prandtl and von Karman developed certain semi-empirical theories that attempted to provide this relationship. These theories are based on drawing an analogy between molecular-motion-based laminar momentum and scalar transport, and eddy-motion-based turbulent momentum and scalar transport. The outcome of such modeling efforts is typically an *eddy viscosity* ν_T (first introduced by Boussinesq in 1877) and *eddy diffusivities* κ_T and κ_{mT} for the closure-model equations:

$$\overline{u_i u_j} = \frac{2}{3}\bar{e}\delta_{ij} - \nu_T\left(\frac{\partial U_i}{\partial x_j} + \frac{\partial U_j}{\partial x_i}\right), \; \overline{u_i T'} = -\kappa_T\frac{\partial \overline{T}}{\partial x_i}, \; \text{and} \; \overline{u_i Y'} = -\kappa_{mT}\frac{\partial \overline{Y}}{\partial x_i}. \quad \text{(12.94, 12.95, 12.96)}$$

Equation (12.94) is mathematically analogous to the stress-rate-of-strain relationship for a Newtonian fluid (4.37) with the term that includes the turbulent kinetic energy $\bar{e}$ playing the role of a turbulent pressure. It represents the *turbulent viscosity hypothesis*. Similarly, (12.95) and (12.96) are mathematically analogous to Fourier's law and Fick's law for molecular diffusion of heat and species, respectively, and these equations represent the *gradient diffusion hypothesis* for turbulent transport of heat and a passive scalar.

To illustrate the implications of such hypotheses, substitute (12.94) into (12.30) to find:

$$\frac{\partial U_i}{\partial t} + U_j\frac{\partial U_i}{\partial x_j} = -\frac{1}{\rho}\frac{\partial P}{\partial x_j} + \frac{\partial}{\partial x_j}\left([\nu + \nu_T]\left(\frac{\partial U_i}{\partial x_j} + \frac{\partial U_j}{\partial x_i}\right) - \frac{2}{3}\bar{e}\delta_{ij}\right) \quad \text{(12.97)}$$

for constant-density flow. The factor in [,]-brackets is commonly known as the *effective viscosity*, and correspondence between this mean-flow equation and its unaveraged counterpart, (4.86) simplified for constant density, is clear and compelling. Mean flow equations for $\overline{T}$ and $\overline{Y}$ similar to (12.97) are readily obtained by substituting (12.95) and (12.96) into (12.32) and (12.34), respectively. Unfortunately, the molecular-dynamics-to-eddy-dynamics analogy is imperfect. Molecular sizes are typically much less than fluid-flow gradient length scales while turbulent eddy sizes are typically comparable to fluid-flow gradient length scales. For ordinary fluid-molecule sizes, averages taken over small volumes include many molecules and these averages converge adequately for macroscopic transport predictions. Equivalent averages over eddies may be unsuccessful because turbulent eddies are so much larger than molecules. Thus, ν_T, κ_T, and κ_{mT} are *not* properties of the fluid or fluid mixture, as ν, κ, and κ_m are. Instead, ν_T, κ_T, and κ_{mT} are properties of the flow, and this transport-flow relationship must be modeled. Hence, (12.97) and its counterparts for $\overline{T}$ and $\overline{Y}$ must be regarded as approximate because (12.94) through (12.96) have inherent limitations. Nevertheless, RANS closure models involving (12.94) through (12.96) are sufficiently accurate for many tasks involving computational fluid dynamics.

From dimensional considerations, ν_T, κ_T, and κ_{mT} should all be proportional to the product of a characteristic turbulent length scale l_T and a characteristic turbulent velocity u_T:

$$\nu_T, \; \kappa_T, \; \text{or} \; \kappa_{mT} \sim l_T u_T. \quad \text{(12.98)}$$

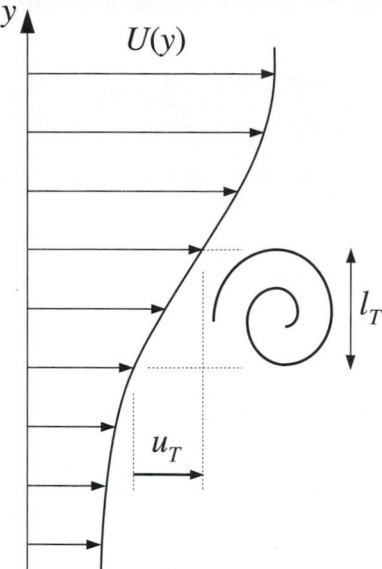

FIGURE 12.20 Schematic drawing of an eddy of size l_T in a shear flow with mean velocity profile $U(y)$. A velocity fluctuation, u or v, that might be produced by this eddy must be of order $l_T(dU/dy)$. Therefore, we expect that the Reynolds shear stress will scale like $\overline{uv} \sim l_T^2(dU/dy)^2$.

For simplicity, consider fully developed, temporally stationary unidirectional shear flow $U(y)$ where y is the cross-stream coordinate (Figure 12.20). The mean-flow momentum equation in this case is:

$$0 = -\frac{1}{\rho}\frac{dP}{dx} + \frac{\partial}{\partial y}\left(\nu\frac{\partial U}{\partial y} - \overline{uv}\right) = -\frac{1}{\rho}\frac{dP}{dx} + \frac{\partial}{\partial y}\left([\nu + \nu_T]\left(\frac{\partial U}{\partial y}\right)\right). \tag{12.99}$$

A Mixing Length Model

An eddy viscosity for this equation can be constructed by interpreting l_T as a *mixing length,* defined as the cross-stream distance traveled by a fluid particle before it gives up its momentum and loses its identity. In this situation, an eddy of size l_T driven by a local shear rate of dU/dy produces a velocity fluctuation of $u_T \sim l_T(dU/dy)$ as it turns over, so

$$-\overline{uv} = \nu_T\frac{dU}{dy} \sim l_T u_T \frac{dU}{dy} \sim l_T\left(l_T\frac{dU}{dy}\right)\frac{dU}{dy} = l_T^2\left(\frac{dU}{dy}\right)^2.$$

The mixing-length concept was first introduced by Taylor (1915), but the approach was fully developed by Prandtl and his coworkers. For a wall-bounded flow it makes sense to assume that l_T is proportional to y when $y = 0$ defines the wall. Thus, setting $l_T = \kappa y$, where κ is presumed to be constant, completes a simple mixing-length turbulence model, and (12.99) becomes:

$$0 = -\frac{1}{\rho}\frac{dP}{dx} + \frac{\partial}{\partial y}\left(\nu\frac{dU}{dy} + \kappa^2 y^2 \left(\frac{dU}{dy}\right)^2\right). \tag{12.100}$$

When the pressure gradient is zero or small enough to be ignored, (12.100) can be integrated once to find:

$$\nu\frac{dU}{dy} + \kappa^2 y^2 \left(\frac{dU}{dy}\right)^2 = const. = \frac{\tau_0}{\rho},$$

where the final equality comes from evaluating the expression on the left at $y = 0$. For points outside the viscous sublayer, where the turbulence term dominates, the last equation reduces to a simple ordinary differential equation that is readily integrated to reach:

$$\frac{dU}{dy} \cong \sqrt{\frac{\tau_0}{\rho}}\frac{1}{\kappa y}, \text{ or } \frac{U}{u_*} \cong \frac{1}{\kappa}\ln y + const., \tag{12.101}$$

which replicates the log-law (12.88). This simplest-level turbulence model is known as an *algebraic* or *zero-equation* model. Such mixing length models can be generalized to a certain extent by using a contracted form of the mean strain-rate tensor or the mean rotation-rate tensor in place of $(dU/dy)^2$. However, there is no rational approach for relating l_T to the mean flow field in general.

Since the development of modern computational techniques for solving partial differential equations, the need for simple intuitive approaches like the mixing length theory has essentially vanished, and Prandtl's derivation of the empirically known logarithmic velocity distribution has only historical value. However, the relationship (12.98) remains useful for estimating the order of magnitude of the eddy diffusivity in turbulent flows, and for development of more sophisticated RANS closure models (see below). Consider the estimation task first via the specific example of thermal convection between two horizontal plates in air when the plates are separated by a distance $L = 3$ m, and the lower plate is warmer by $\Delta T = 1°C$. The equation for the vertical velocity fluctuation gives the vertical acceleration as

$$Dw/Dt \sim g\alpha T' \sim g\Delta T/T, \tag{12.102}$$

since T' is expected to be of order ΔT and $\alpha = 1/T$ for a perfect gas. The time t_r to rise through a height L will be proportional to L/w, so (12.102) gives a characteristic vertical velocity acceleration of

$$w/t_r = w^2/L \sim Dw/Dt \sim g\Delta T/T \rightarrow w \sim \sqrt{gL\Delta T/T} \cong 0.3 \, m/s.$$

The largest eddies will scale with the plate separation L, so the thermal eddy diffusivity, κ_T, is

$$\kappa_T \sim wL \sim 0.9 \, m^2/s,$$

which is significantly larger than the molecular value of 2×10^{-5} m^2/s.

One-Equation Models

Independently, Kolmogorov and Prandtl suggested that the velocity scale in (12.98) should be determined from the turbulent kinetic energy:

$$u_T = c\sqrt{\bar{e}},$$

where c is a model constant. The turbulent viscosity is then obtained from an algebraic specification of the turbulent length scale l_T, and the solution of a transport equation for $\bar{e}$ that is based on its exact transport equation (12.47). In this case, the dissipation rate $\bar{\varepsilon}$ and the transport terms must be modeled. For high Reynolds number turbulence, the scaling relationship (12.48) and the gradient diffusion hypothesis lead to the following model equations for the dissipation and the transport of turbulent kinetic energy:

$$\bar{\varepsilon} = C_\varepsilon (\bar{e})^{3/2}/l_T \ \text{ and } \ -\frac{1}{\rho_0}\overline{pu_j} + 2\nu \overline{u_j S'_{ij}} - \frac{1}{2}\overline{u_i^2 u_j} = \frac{\nu_T}{\sigma_e}\frac{\partial \bar{e}}{\partial x_j},$$

where C_ε and σ_e are model constants. So, for constant density, the turbulent kinetic energy model equation is:

$$\frac{\partial \bar{e}}{\partial t} + U_j \frac{\partial \bar{e}}{\partial x_j} = \frac{\partial}{\partial x_j}\left(\frac{\nu_T}{\sigma_e}\frac{\partial \bar{e}}{\partial x_j}\right) - \bar{\varepsilon} - \overline{u_i u_j}\frac{\partial U_i}{\partial x_j}, \tag{12.103}$$

and this represents *one* additional nonlinear second-order partial-differential equation that must be solved, hence the name *one-equation model*. As mentioned in Pope, one-equation models provide a modest accuracy improvement over the simpler algebraic models.

Two-Equation Models

These models eliminate the need for a specified turbulent length scale by generating l_T from the solutions of transport equations for $\bar{e}$ and $\bar{\varepsilon}$. The popular k-ε closure model of Jones and Launder (1972) is described here. A k-ω closure model also exists. (Throughout much of the turbulence modeling literature "k" is used for the turbulent kinetic energy, so the model name "k-ε " is merely a specification of the dependent-field variables in the two extra partial differential equations.) The k-ε model is based on the turbulent viscosity hypothesis (12.94) with ν_T specified by (12.98), $l_T = (\bar{e})^{3/2}/\bar{\varepsilon}$, and $u_T = (\bar{e})^{1/2}$:

$$\nu_T = C_\mu \left[(\bar{e})^{3/2}/\bar{\varepsilon}\right]\sqrt{\bar{e}} = C_\mu (\bar{e})^2/\bar{\varepsilon}, \tag{12.104}$$

where C_μ is one of five model constants. The first additional partial-differential equation is (12.103) for $\bar{e}$. The second additional partial-differential equation is an empirical construction for the dissipation:

$$\frac{\partial \bar{\varepsilon}}{\partial t} + U_j \frac{\partial \bar{\varepsilon}}{\partial x_j} = \frac{\partial}{\partial x_j}\left(\frac{\nu_T}{\sigma_\varepsilon}\frac{\partial \bar{\varepsilon}}{\partial x_j}\right) - C_{\varepsilon 1}\left(\overline{u_i u_j}\frac{\partial U_i}{\partial x_j}\right)\frac{\bar{\varepsilon}}{\bar{e}} - C_{\varepsilon 2}\frac{(\bar{\varepsilon})^2}{\bar{e}}. \tag{12.105}$$

The standard model constants are from Launder and Sharma (1974):

$$C_\mu = 0.09, \ C_{\varepsilon 1} = 1.44, \ C_{\varepsilon 2} = 1.92, \ C_{\varepsilon 1} = 1.44, \ \sigma_e = 1.0, \ \text{and } \sigma_\varepsilon = 1.3,$$

and these have been set so the model's predictions reasonably conform to experimentally determined mean velocity profiles, fluctuation profiles, and energy budgets of the type shown in Figures 12.14 and 12.15 for a variety of simple turbulent flows. More recently renormalization group theory has been used to justify (12.105) with slightly modified constants (Yakhot & Orszag, 1986; Lam, 1992; see also Smith & Reynolds, 1992).

When the density is constant, (12.27), (12.30), (12.94), and (12.103) through (12.105) represent a closed set of equations. Ideally, the usual viscous boundary conditions would be applied to U_i. However, steep near-wall gradients of the dependent field variables pose a significant computational challenge. Thus, boundary conditions on solid surfaces are commonly applied slightly above the surface using empirical *wall functions* intended to mimic the inner layer of a wall-bounded turbulent flow. Wall functions allow the mean-flow momentum equation (12.30) and the turbulence model equations, (12.103) and (12.105), to be efficiently, but approximately, evaluated near a solid surface. Unfortunately, wall functions that perform well with attached turbulent boundary layers are of questionable validity for separating, impinging, and adverse-pressure-gradient flows. Furthermore, the use of wall functions introduces an additional model parameter, the distance above the wall where boundary conditions are applied.

Overall, the k-ε turbulence model is complete and versatile. It is commonly used to rank the performance of fluid dynamic system designs before experimental tests are undertaken. Limitations on its accuracy arise from the turbulent viscosity hypothesis, the $\bar{\varepsilon}$ equation, and wall functions when they are used. In addition, variations in inlet boundary conditions for $\bar{e}$ and $\bar{\varepsilon}$, which may not be known precisely, can produce changes in predicted results. In recent years, two equation turbulence models based on the eddy viscosity hypothesis have begun to be eclipsed by *Reynolds stress models* or *second-order closures* that directly compute the Reynolds stress tensor from a modeled version of its exact transport equation (12.35).

12.11. TURBULENCE IN A STRATIFIED MEDIUM

Effects of stratification become important in such laboratory flows as heat transfer from a heated plate and in geophysical flows such as those in the atmosphere and in the ocean. Some effects of stratification on turbulent flows will be considered in this section. Further discussion can be found in Tennekes and Lumley (1972), Phillips (1977), and Panofsky and Dutton (1984).

As is customary in the geophysical literature, the z-direction points upward opposing gravity so the mean velocity of a horizontally flowing shear flow will be denoted by $U(z)$. For simplicity, U is assumed to be independent of x and y. Turbulence in a stratified medium depends critically on the stability of the vertical density profile. In the neutrally stable state of a compressible environment the density decreases upward, because of the decrease of pressure, at a rate $d\rho_a/dz$ called the *adiabatic density gradient*, as discussed in Section 1.10. A medium is statically stable if the density decreases faster than the adiabatic decrease. The effective density gradient that determines the stability of the environment is then determined by the sign of $d(\rho - \rho_a)/dz$, where $\rho - \rho_a$ is called the *potential density*. In the following discussion, it is assumed that the adiabatic variations of density have been subtracted out, so that "density" or "temperature" really mean potential density or potential temperature.

The Richardson Numbers

First examine the equation for turbulent kinetic energy (12.47). Omitting the viscous transport and assuming that the flow is independent of x and y, it reduces to

$$\frac{\partial \bar{e}}{\partial t} + U \frac{\partial \bar{e}}{\partial x} = -\frac{\partial}{\partial z}\left(\frac{1}{\rho_0}\overline{pw} + \overline{ew}\right) - \overline{uw}\frac{\partial U}{\partial z} + g\alpha\overline{wT'} - \bar{\varepsilon}, \tag{12.106}$$

where x increases in the downstream direction. The first term on the right side is the transport of turbulent kinetic energy by vertical velocity fluctuations. The second term is the production of turbulent energy by the interaction of Reynolds stress and the mean shear; this term is almost always positive. The third term is the production of turbulent kinetic energy by the vertical heat flux; it is called the *buoyant production*, and was discussed in Section 12.7. In an unstable environment, in which the mean temperature $\bar{T}$ decreases upward, the heat-flux correlation $\overline{wT'}$ is positive (upward), signifying that the turbulence is generated convectively by upward heat fluxes. In the opposite case of a stable environment, the turbulence is suppressed by stratification. The ratio of the buoyant destruction of turbulent kinetic energy to the shear production is called the *flux Richardson number*:

$$\text{Rf} = \frac{-g\alpha\overline{wT'}}{-\overline{uw}(dU/dz)} = \frac{\text{buoyant destruction}}{\text{shear production}}. \tag{12.107}$$

As the shear production is positive with the minus sign displayed, the sign of Rf depends on the sign of $\overline{wT'}$. For an unstable environment in which the heat flux is upward Rf is negative, and for a stable environment it is positive. For Rf > 1, buoyant destruction removes turbulence at a rate larger than the rate at which it is produced by shear production. However, the critical value of Rf at which the turbulence ceases to be self-supporting is less than unity, as dissipation is necessarily a large fraction of the shear production. Observations indicate that the critical value is $\text{Rf}_{cr} \approx 0.25$ (Panofsky & Dutton, 1984, p. 94). If measurements indicate the presence of turbulent fluctuations, but at the same time a value of Rf much larger than 0.25, then a fair conclusion is that the turbulence is decaying. When Rf is negative, a large $-$Rf means strong convection and weak mechanical turbulence.

Instead of Rf, it is easier to measure the *gradient Richardson number*, defined as

$$\text{Ri} \equiv \frac{N^2}{(dU/dz)^2} = \frac{\alpha g(d\bar{T}/dz)}{(dU/dz)^2}, \tag{12.108}$$

where N is the buoyancy frequency and the second equality follows for stratification by thermal variations. If we make the turbulent viscosity and gradient diffusion assumptions (12.94) and (12.95), then the two Richardson numbers are related by

$$\text{Ri} = (\nu_T/\kappa_T)\text{Rf}. \tag{12.109}$$

The ratio ν_T/κ_T is the *turbulent Prandtl number*, which determines the relative efficiency of the vertical turbulent exchanges of momentum and heat. The presence of a stable stratification damps vertical transport of both heat and momentum; however, the momentum flux is reduced less because the internal waves in a stable environment can transfer momentum (by moving vertically from one region to another) but not heat. Therefore, $v_T/\kappa_T > 1$ for a stable environment. Equation (12.109) then shows that turbulence can persist even when Ri > 1, if the critical value of 0.25 applies on the *flux* Richardson number (Turner, 1981; Bradshaw & Woods, 1978). In an unstable environment, on the other hand, v_T/κ_T becomes small. In a neutral environment it is usually found that $v_T \approx \kappa_T$; the idea of equating the eddy coefficients of heat and momentum is called the *Reynolds analogy*.

Monin-Obukhov Length

The Richardson numbers are ratios that compare the relative importance of mechanical and convective turbulence. Another parameter used for the same purpose is not a ratio, but has the unit of length. It is the *Monin-Obukhov length*, defined as

$$L_M \equiv -u_*^3/\kappa\alpha g\overline{wT'},\qquad(12.110)$$

where u_* is the friction velocity, $\overline{wT'}$ is the heat flux correlation, α is the coefficient of thermal expansion, and κ is the von Karman constant introduced for convenience. Although $\overline{wT'}$ is a function of z, the parameter L_M is effectively a constant for the flow, as it is used only in the logarithmic region of the earth's atmospheric boundary layer in which both $\overline{uw}$ and $\overline{wT'}$ are nearly constant. The Monin-Obukhov length then becomes a parameter determined from the boundary conditions of friction and the heat flux at the surface. Like Rf, it is positive for stable conditions and negative for unstable conditions.

The significance of L_M within the atmospheric boundary layer becomes clearer if we write Rf in terms of L_M, using the logarithmic velocity distribution (12.88), from which $dU/dz = u_*/\kappa z$. (Note that z is the distance perpendicular to the surface.) Using $\overline{uw} = u_*^2$ because of the near uniformity of stress in the logarithmic layer, (12.107) becomes

$$\text{Rf} = z/L_M.\qquad(12.111)$$

As Rf is the ratio of buoyant destruction to shear production of turbulence, (12.111) shows that L_M is the height at which these two effects are of the same order. For both stable and unstable conditions, the effects of stratification are slight if $z \ll |L_M|$. At these small heights, then, the velocity profile is logarithmic, as in a neutral environment. This is called a *forced convection* region, because the turbulence is mechanically forced. For $z \gg |L_M|$, the effects of stratification dominate. In an unstable environment, it follows that the turbulence is generated mainly by buoyancy at heights $z \gg -L_M$, and the shear production is negligible. The region beyond the forced convecting layer is therefore called a zone of *free convection* (Figure 12.21), containing thermal plumes (columns of hot rising gases) characteristic of free convection from heated plates in the absence of shear flow.

Observations as well as analysis show that the effect of stratification on the velocity distribution in the surface layer is given by the log-linear profile (Turner, 1973):

$$U = \frac{u_*}{\kappa}\left[\ln\frac{z}{z_o} + 5\frac{z}{L_M}\right].$$

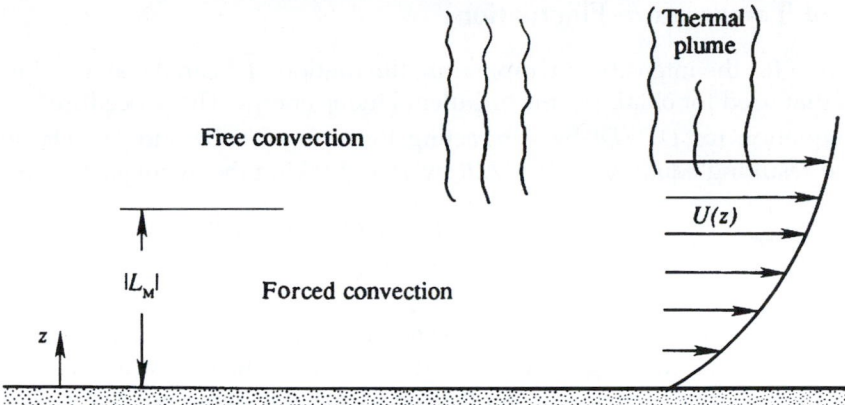

FIGURE 12.21   Forced and free convection zones in an unstable atmosphere. In strongly sheared regions, the turbulence will not include buoyant effects (forced convection). However, where shear is weak, buoyant convection will set the turbulent scales (free convection).

The form of this profile is sketched in Figure 12.22 for stable and unstable conditions. It shows that the velocity is more uniform than $ln(z)$ in the unstable case because of the enhanced vertical mixing due to buoyant convection.

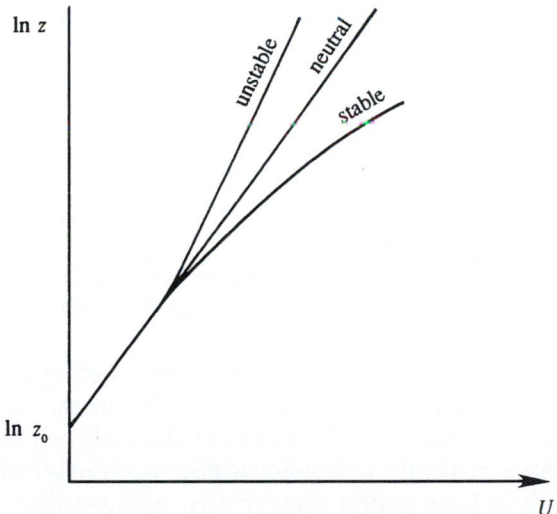

FIGURE 12.22   Effect of stability on velocity profiles in the surface layer. When the atmospheric boundary layer is neutrally stable, the mean velocity profile is logarithmic. When it is stable, vertical turbulent motions are suppressed so higher shear may exist in the mean flow; this is shown as the lower curve labeled *stable*. When the atmospheric boundary layer is unstable, vertical turbulent motions are enhanced, mean flow shear is reduced, and $U(z)$ becomes more nearly uniform; this is shown as the upper curve labeled *unstable*.

Spectrum of Temperature Fluctuations

An equation for the intensity of temperature fluctuations $\overline{T'^2}$ can be obtained in a manner identical to that used for obtaining the turbulent kinetic energy. The procedure is therefore to obtain an equation for DT'/Dt by subtracting those for $D\tilde{T}/Dt$ and $D\overline{T}/Dt$, and then to multiply the resulting equation for DT'/Dt by T' and taking the average. The result is

$$\frac{\partial}{\partial t}\left(\frac{1}{2}\overline{T'^2}\right) + U\frac{\partial}{\partial x}\left(\frac{1}{2}\overline{T'^2}\right) = -\overline{wT'}\frac{d\overline{T}}{dz} - \frac{\partial}{\partial z}\left(\frac{1}{2}\overline{T'^2 w} - \kappa\frac{\partial\overline{T'^2}}{\partial z}\right) - \bar{\varepsilon}_T, \tag{12.112}$$

where $\bar{\varepsilon}_T = \overline{\kappa(\partial T'/\partial x_j)^2}$ is the *dissipation rate of temperature fluctuations*, analogous to the dissipation of turbulent kinetic energy $\bar{\varepsilon}$ defined within (12.47). The first term on the right side is the generation of $\overline{T'^2}$ by the mean temperature gradient, $\overline{wT'}$ being positive if $d\overline{T}/dz$ is negative. The second term on the right side is the turbulent transport of $\overline{T'^2}$.

A wave number spectrum of temperature fluctuations can be defined such that

$$\overline{T'^2} \equiv \int_0^\infty S_T(K)dK,$$

where K is the magnitude of the three-dimensional wave number. As in the case of the kinetic energy spectrum, an inertial range of wave numbers exists in which neither the production by large-scale eddies nor the dissipation by conductive and viscous effects are important. As the temperature fluctuations are intimately associated with velocity fluctuations, $S_T(K)$ in this range must depend not only on ε_T but also on the variables that determine the velocity spectrum, namely ε and K. Therefore

$$S_T = S_T(K, \bar{\varepsilon}, \bar{\varepsilon}_T) \quad \text{for } 2\pi/L \ll K \ll 2\pi/\eta,$$

where L is the size of the largest eddies. The unit of S_T is $°C^2$ m, and the unit of ε_T is $°C^2/s$, so dimensional analysis gives

$$S_T \propto \bar{\varepsilon}_T\bar{\varepsilon}^{-1/3}K^{-5/3} \quad \text{for } 2\pi/L \ll K \ll 2\pi/\eta, \tag{12.113}$$

which was first derived by Obukhov in 1949. Comparing with (12.54), it is apparent that the spectra of both velocity and temperature fluctuations in the inertial subrange have the same $K^{-5/3}$ form.

The spectrum beyond the inertial subrange depends on whether the Prandtl number ν/κ of the fluid is smaller or larger than one. We shall only consider the case of $\nu/\kappa \gg 1$, which applies to water for which the Prandtl number is 7.1. Let η_T be the scale responsible for smearing out the temperature gradients and η be the Kolmogorov microscale at which the velocity gradients are smeared out. For $\nu/\kappa \gg 1$ we expect that $\eta_T \ll \eta$, because then the conductive effects are important at scales smaller than the viscous scales. In fact, Batchelor (1959) showed that $\eta_T = \eta(\kappa/\nu)^{1/2} \ll \eta$. In such a case there exists a range of wave numbers $2\pi/\eta \ll K \ll 2\pi/\eta_T$, in which the scales are not small enough for the thermal diffusivity to smear out the temperature fluctuation. Therefore, $S_T(K)$ continues up to wave numbers of order $2\pi/\eta_T$, although the kinetic energy spectrum has dropped off sharply. This is called the *viscous*

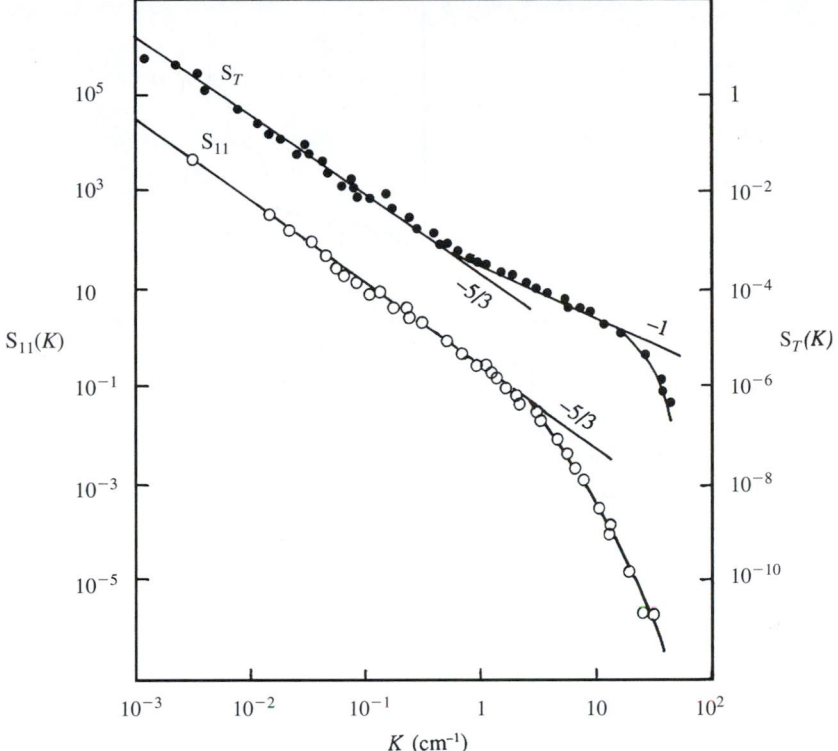

FIGURE 12.23 Temperature and velocity spectra measured by Grant et al. (1968). The measurements were made at a depth of 23 m in a tidal passage through islands near the coast of British Columbia, Canada. The wave number K is in cm^{-1}. Solid points represent S_T in $(°C)^2/cm^{-1}$, and open points represent S_{11} in $(cm/s)^2/cm^{-1}$. Powers of K that fit the observation are indicated by straight lines. *O. M. Phillips, The Dynamics of the Upper Ocean, 1977; reprinted with the permission of Cambridge University Press.*

convective subrange, because the spectrum is dominated by viscosity but is still actively convective. Batchelor (1959) showed that the spectrum in the viscous convective subrange is

$$S_T \propto K^{-1} \text{ for } 2\pi/\eta \ll K \ll 2\pi/\eta_T. \tag{12.114}$$

Figure 12.23 shows a comparison of velocity and temperature spectra, observed in a tidal flow through a narrow channel. The temperature spectrum shows that the spectral slope increases from $-5/3$ in the inertial subrange to -1 in the viscous convective subrange.

12.12. TAYLOR'S THEORY OF TURBULENT DISPERSION

The large mixing rate in a turbulent flow is due to the fact that the fluid particles wander away from their initial location. Taylor (1921) studied this problem and calculated the rate at which a particle disperses (i.e., moves away) from its initial location. The presentation here is

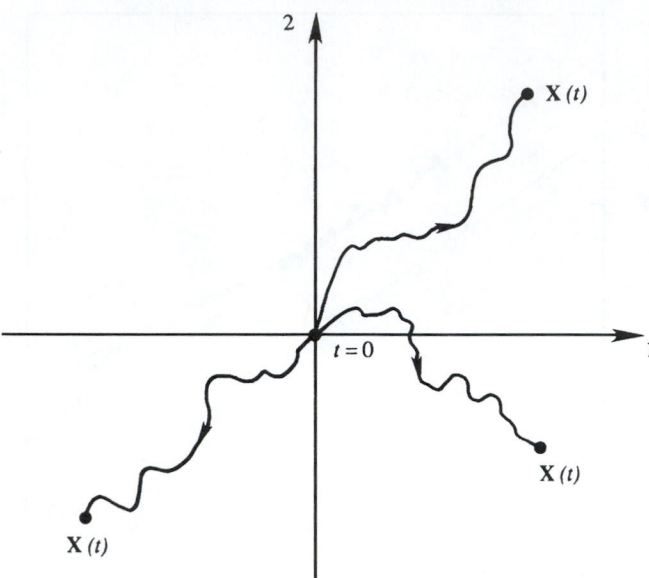

FIGURE 12.24 Three members of an ensemble of particle trajectories, $\mathbf{X}(t)$, at time t for particles released at the origin of coordinates at $t = 0$ in a turbulent flow with zero mean velocity. The distance traveled by the particles indicates how tracer particles disperse in a turbulent flow.

directly adapted from his classic paper. He considered a point source emitting particles, say a chimney emitting smoke. The particles are emitted into a stationary and homogeneous turbulent medium in which the mean velocity is zero. Taylor used Lagrangian coordinates $\mathbf{X}(\mathbf{a}, t)$, which is the present location at time t of a particle that was at location $\mathbf{a}$ at time $t = 0$. We shall take the point source to be the origin of coordinates and consider an ensemble of experiments in which we evaluate the location $\mathbf{X}(\mathbf{0}, t)$ at time t of all the particles that started from the origin (Figure 12.24). For notational simplicity the first argument in $\mathbf{X}(\mathbf{0}, t)$ will be dropped from here on so that $\mathbf{X}(\mathbf{0}, t) = \mathbf{X}(t)$.

Rate of Dispersion of a Single Particle

Consider the behavior of a single component of $\mathbf{X}$, say X_α ($\alpha = 1, 2,$ or 3). (Recall that a Greek subscript means that the summation convention is *not* followed.) The average rate at which the *magnitude* of X_α increases with time can be found by finding $\overline{d(X_\alpha^2)}/dt$, where the over bar denotes an ensemble average and not a time average. We can write

$$\frac{d}{dt}\left(\overline{X_\alpha^2}\right) = \overline{2X_\alpha \frac{dX_\alpha}{dt}}, \tag{12.115}$$

where we have used the commutation rule (12.6). Defining $u_\alpha = dX_\alpha/dt$ as the *Lagrangian* velocity component of a fluid particle at time t, (12.115) becomes

$$\frac{d}{dt}\overline{(X_\alpha^2)} = 2\overline{X_\alpha u_\alpha} = 2\overline{\left[\int_0^t u_\alpha(t')\,dt'\right]u_\alpha}$$

$$= 2\int_0^t \overline{u_\alpha(t')u_\alpha(t)}\,dt',$$

(12.116)

where we have used the commutation rule (12.7) for averaging and integration. We have also written

$$X_\alpha = \int_0^t u_\alpha(t')\,dt',$$

which is valid when X_α and u_α are associated with the same particle. Because the flow is assumed to be stationary, $\overline{u_\alpha^2}$ is independent of time, and the autocorrelation of $u_\alpha(t)$ and $u_\alpha(t')$ is only a function of the time difference $t - t'$. Defining

$$r_\alpha(\tau) \equiv \frac{\overline{u_\alpha(t)u_\alpha(t+\tau)}}{\overline{u_\alpha^2}}$$

to be the autocorrelation coefficient of the Lagrangian velocity components of a particle, (12.116) becomes

$$\frac{d}{dt}\overline{(X_\alpha^2)} = 2\overline{u_\alpha^2}\int_0^t r_\alpha(t'-t)\,dt' = 2\overline{u_\alpha^2}\int_0^t r_\alpha(\tau)\,d\tau,$$

(12.117)

where we have changed the integration variable from t' to $\tau = t - t'$. Integrating, we obtain

$$\overline{X_\alpha^2}(t) = 2\overline{u_\alpha^2}\int_0^t dt'\int_0^{t'} r_\alpha(\tau)\,d\tau,$$

(12.118)

which shows how the variance of the particle position changes with time.

Another useful form of equation (12.118) is obtained by integrating it by parts. We have:

$$\int_0^t dt'\int_0^{t'} r_\alpha(\tau)\,d\tau = \left[t'\int_0^{t'} r_\alpha(\tau)\,d\tau\right]_{t'=0}^t - \int_0^t t'r_\alpha(t')\,dt'$$

$$= t\int_0^t r_\alpha(\tau)\,d\tau - \int_0^t t'r_\alpha(t')\,dt'$$

$$= t\int_0^t \left(1 - \frac{\tau}{t}\right)r_\alpha(\tau)\,d\tau.$$

Equation (12.118) then becomes

$$\overline{X_\alpha^2}(t) = 2\overline{u_\alpha^2}t\int_0^t \left(1 - \frac{\tau}{t}\right)r_\alpha(\tau)\,d\tau.$$

(12.119)

Two limiting cases are examined in what follows.

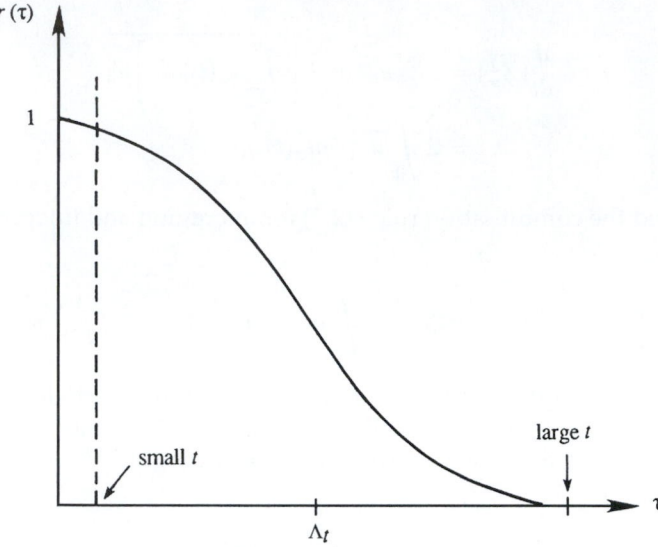

FIGURE 12.25 Small and large values of time on a plot of the correlation function. For small times, $r(\tau)$ is nearly unity, while for large times it is nearly zero.

Behavior for small t

If t is small compared to the correlation scale of $r_\alpha(\tau)$, then $r_\alpha(\tau) \approx 1$ throughout the integral in (12.118) (Figure 12.25). This gives:

$$\overline{X_\alpha^2}(t) \simeq \overline{u_\alpha^2}\, t^2. \tag{12.120}$$

Taking the square root of both sides, we obtain

$$(X_\alpha)_{rms} = (u_\alpha)_{rms}\, t \quad \text{for } t \ll \Lambda_t, \tag{12.121}$$

which shows that the *rms* displacement increases linearly with time and is proportional to the standard deviation of the turbulent fluctuations in the medium.

Behavior for large t

If t is large compared with the correlation scale of $r_\alpha(\tau)$, then τ/t in (11.119) is negligible, giving

$$\overline{X_\alpha^2}(t) = 2\overline{u_\alpha^2}\, \Lambda_t\, t, \tag{12.122}$$

where

$$\Lambda_t = \int_0^\infty r_\alpha(\tau)\, d\tau$$

is the integral time scale determined from the Lagrangian correlation $r_\alpha(\tau)$. Taking the square root of (12.122) gives

$$(X_\alpha)_{rms} = (u_\alpha)_{rms}\sqrt{2\Lambda_t t} \quad \text{for } t \gg \Lambda_t. \tag{12.123}$$

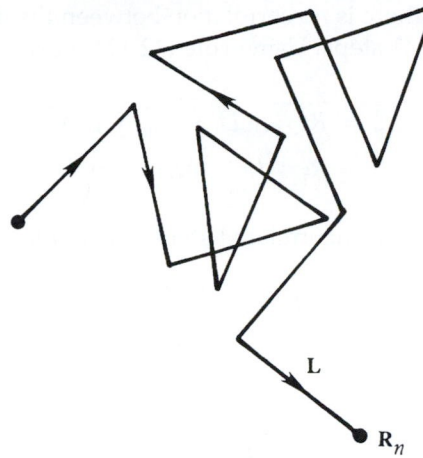

FIGURE 12.26 A sample realization of a random walk where the step length L is a uniform distance, but the step direction is random. After n steps, the vector distance from the starting point is $\mathbf{R}_n$. However, the root-mean-square distance from the starting point is only $L\sqrt{n}$ (not Ln) because many of the n steps lie in nearly opposite directions.

The $t^{1/2}$ behavior of (12.123) at large times is similar to the behavior in a *random walk*, in which the average distance traveled in a series of random (i.e., uncorrelated) steps increases as $t^{1/2}$. This similarity is due to the fact that for large t the fluid particles have *forgotten* their initial behavior at $t = 0$. In contrast, the small time behavior described by (12.121) is due to complete correlation, with *each experiment* giving $X_\alpha \simeq u_\alpha t$. The random walk concept is discussed in what follows.

Random Walk

The description provided here is adapted from Feynman et al. (1963, pp. 5–6, 41–48). Imagine a person walking in a random manner, by which we mean that there is no correlation between the directions of two consecutive steps. Let the vector $\mathbf{R}_n$ represent the distance from the origin after n steps, and the vector $\mathbf{L}$ represent the nth step (Figure 12.26). We assume that each step has the same magnitude L. Then

$$\mathbf{R}_n = \mathbf{R}_{n-1} + \mathbf{L},$$

which gives

$$R_n^2 = \mathbf{R}_n \cdot \mathbf{R}_n = (\mathbf{R}_{n-1} + \mathbf{L}) \cdot (\mathbf{R}_{n-1} + \mathbf{L})$$
$$= R_{n-1}^2 + L^2 + 2\mathbf{R}_{n-1} \cdot \mathbf{L}.$$

Taking the average, we get

$$\overline{R_n^2} = \overline{R_{n-1}^2} + L^2 + 2\overline{\mathbf{R}_{n-1} \cdot \mathbf{L}}. \tag{12.124}$$

The last term is zero because there is no correlation between the direction of the nth step and the location reached after $n - 1$ steps. Using rule (12.124) successively, we get

$$\overline{R_n^2} = \overline{R_{n-1}^2} + L^2 = \overline{R_{n-2}^2} + 2L^2$$
$$= \overline{R_1^2} + (n - 1)L^2 = nL^2.$$

The *rms* distance traveled after n uncorrelated steps, each of length L, is therefore

$$(R_n)_{rms} = L\sqrt{n}, \tag{12.125}$$

which is called a *random walk*.

Behavior of a Smoke Plume in the Wind

Taylor's analysis can be easily adapted to account for the presence of a constant mean velocity. Consider the dispersion of smoke into a wind blowing in the x-direction (Figure 12.27). A photograph of the smoke plume, in which the film is exposed for a long time, would outline the average width Z_{rms}. As the x-direction in this problem is similar to time in Taylor's problem, the limiting behavior in (12.121) and (12.123) shows that the smoke plume is parabolic with a *pointed* vertex.

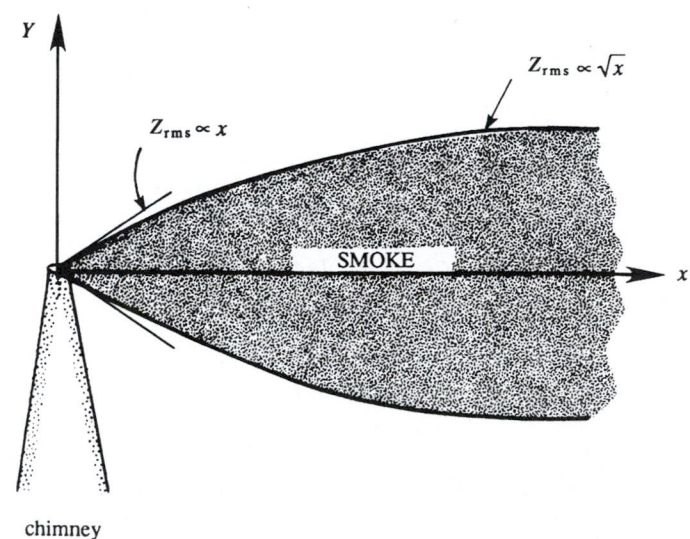

FIGURE 12.27   Average cross-sectional shape of a smoke plume in a turbulent wind blowing uniformly along the x-axis. Close to the chimney outlet, the *rms* width Z_{rms} of the smoke plume is proportional $x^{1/2}$. Far from the chimney, Z_{rms} is proportional to x. G. I. Taylor, Proc. London Mathematical Society, 20, 196–211, 1921.

Turbulent Diffusivity

An equivalent eddy diffusivity can be estimated from Taylor's analysis. The equivalence is based on considering the spreading of a concentrated line source in a fluid of *constant* diffusivity. What should the diffusivity be in order that the spreading rate equals that predicted by (12.117)? The problem of the sudden introduction of a line vortex of strength Γ (Exercise 8.26) is such a problem of diffusion from a concentrated line source. The tangential velocity in this flow is given by

$$u_\theta = (\Gamma/2\pi r) \exp(-r^2/4\nu t).$$

The solution is therefore proportional to $\exp(-r^2/4\nu t)$, which has a Gaussian shape in the radial direction r, with a characteristic width of $\sigma = \sqrt{2\nu t}$. It follows that the momentum diffusivity ν in this problem is related to the variance σ^2 as

$$\nu = \frac{1}{2}\frac{d\sigma^2}{dt}, \tag{12.126}$$

which can be calculated if $\sigma^2(t)$ is known. Generalizing (12.126), we can say that the effective diffusivity D_T in a problem of turbulent dispersion of a patch of particles issuing from a point is given by

$$D_T \equiv \frac{1}{2}\frac{d}{dt}(\overline{X_\alpha^2}) = \overline{u_\alpha^2}\int_0^t r_\alpha(\tau)d\tau, \tag{12.127}$$

where we have used (12.117). From (12.120) and (12.122), the two limiting cases of (12.127) are

$$D_T \cong \overline{u_\alpha^2}\, t \quad \text{for } t \ll \Lambda_t, \quad \text{and} \quad D_T \cong \overline{u_\alpha^2}\, \Lambda_t \quad \text{for } t \ll \Lambda_t. \tag{12.128, 12.129}$$

Equation (12.128) shows the interesting fact that the eddy diffusivity initially increases with time, a behavior different from that in molecular diffusion with constant diffusivity. This can be understood as follows. The dispersion (or separation) of particles in a patch is caused by eddies with scales less than or equal to the scale of the patch, since the larger eddies simply advect the patch and do not cause any separation of the particles. As the patch size becomes larger, an *increasing* range of eddy sizes is able to cause dispersion, giving $D_T \propto t$. This behavior shows that *it is frequently impossible to represent turbulent diffusion by means of a large but constant eddy diffusivity.* Turbulent diffusion does not behave like molecular diffusion. For large times, on the other hand, the patch size becomes larger than the largest eddies present, in which case the diffusive behavior becomes similar to that of molecular diffusion with a constant diffusivity given by (12.129).

12.13. CONCLUDING REMARKS

Turbulence is an area of classical fluid mechanics that is the subject of continuing research. Frequent symposia are held to summarize and communicate new findings and promising approaches and a few are listed in the "Supplementary Reading" section at the end of this chapter.

EXERCISES

12.1. Determine general relationships for the second, third, and fourth central moments (variance $= \sigma^2$, skewness $= S$, and kurtosis $= K$) of the random variable u in terms of its first four ordinary moments: $\overline{u}$, $\overline{u^2}$, $\overline{u^3}$, and $\overline{u^4}$.

12.2. Calculate the mean, mean square, variance, and *rms* value of the periodic time series $u(t) = \overline{U} + U_0 \cos(\omega t)$, where $\overline{U}$, U_0, and ω are positive real constants.

12.3. Show that the autocorrelation function $\overline{u(t)u(t + \tau)}$ of a periodic series $u = U\cos(\omega t)$ is itself periodic.

12.4. Calculate the zero-lag cross-correlation $\overline{u(t)v(t)}$ between two periodic series $u(t) = \cos \omega t$ and $v(t) = \cos(\omega t + \phi)$ by performing at time average over one period $= 2\pi/\omega$. For values of $\phi = 0$, $\pi/4$, and $\pi/2$, plot the scatter diagrams of u vs. v at different times, as in Figure 12.8. Note that the plot is a straight line if $\phi = 0$, an ellipse if $\phi = \pi/4$, and a circle if $\phi = \pi/2$; the straight line, as well as the axes of the ellipse, are inclined at $45°$ to the uv-axes. Argue that the straight line signifies a perfect correlation, the ellipse a partial correlation, and the circle a zero correlation.

12.5. If $u(t)$ is a stationary random signal, show that $u(t)$ and $du(t)/dt$ are uncorrelated.

12.6. Let $R(\tau)$ and $S(\omega)$ be a Fourier transform pair. Show that $S(\omega)$ is real and symmetric if $R(\tau)$ is real and symmetric.

12.7. Compute the power spectrum, integral time scale, and Taylor time scale when $R_{11}(\tau) = \overline{u_1^2}\exp(-\alpha\tau^2)\cos(\omega_o\tau)$, assuming that α and ω_o are real positive constants.

12.8. There are two formulae for the energy spectrum $S_e(\omega)$ of the stationary zero-mean signal $u(t)$:

$$S_e(\omega) = \frac{1}{2\pi} \int_{-\infty}^{+\infty} R_{11}(\tau) \exp\{-i\omega\tau\} \, d\tau \text{ and}$$

$$S_e(\omega) = \lim_{T \to \infty} \frac{1}{2\pi T} \left| \int_{-T/2}^{+T/2} u(t) \exp\{-i\omega t\}dt \right|^2 .$$

Prove that these two are identical *without* requiring the existence of the Fourier transformation of $u(t)$.

12.9. Derive the formula for the temporal Taylor microscale λ_t by expanding the definition of the temporal correlation function (12.17) into a two-term Taylor series and determining the time shift, $\tau = \lambda_t$, where this two-term expansion equals zero.

12.10. When x, r, and k_1 all lie in the stream-wise direction, the wave number spectrum $S_{11}(k_1)$ of the stream-wise velocity fluctuation $u_1(x)$ defined by (12.45) can be interpreted as a distribution function for energy across stream-wise wave number k_1.

Show that the energy-weighted mean-square value of the stream-wise wave number is:

$$\overline{k_1^2} \equiv \frac{1}{\overline{u^2}} \int_{-\infty}^{+\infty} k_1^2 S_{11}(k_1)dk_1 = -\frac{1}{\overline{u^2}}\left[\frac{d^2}{dr^2}R_{11}(r)\right]_{r=0} , \text{ and that } \lambda_f = \sqrt{2/\overline{k_1^2}}.$$

12.11. In many situations, measurements are only possible of one velocity component at one point in a turbulent flow, but consider a flow that has a nonzero mean velocity and moves past the measurement point. Thus, the experimenter obtains a time history of $u_1(t)$ at a fixed point. In order to estimate spatial velocity gradients, Taylor's frozen-turbulence hypothesis can be invoked to estimate a spatial gradient from a time derivative: $\dfrac{\partial u_1}{\partial x_1} \approx -\dfrac{1}{U_1}\dfrac{\partial u_1}{\partial t}$ where the "1"-axis must be aligned with the direction of the average flow, i.e., $U_i = (U_1, 0, 0)$. Show that this approximate relationship is true when $\sqrt{\overline{u_i u_i}}/U_1 \ll 1$, $p \sim \rho u_1^2$, and Re is high enough to neglect the influence of viscosity.

12.12. a) Starting from (12.33), derive (12.34) via an appropriate process of Reynolds decomposition and ensemble averaging.
   b) Determine an equation for the scalar fluctuation energy $= \dfrac{1}{2}\overline{Y'^2}$, one-half the scalar variance.
   c) When the scalar variance goes to zero, the fluid is well mixed. Identify the term in the equation from part b) that dissipates scalar fluctuation energy.

12.13. Measurements in an atmosphere at $20°C$ show an *rms* vertical velocity of $w_{rms} = 1\,\text{m/s}$ and an *rms* temperature fluctuation of $T_{rms} = 0.1°C$. If the correlation coefficient is 0.5, calculate the heat flux $\rho C_p \overline{wT'}$.

12.14. a) Compute the divergence of the constant-density Navier-Stokes momentum equation $\dfrac{\partial u_i}{\partial t} + u_j \dfrac{\partial u_i}{\partial x_j} = -\dfrac{1}{\rho}\dfrac{\partial p}{\partial x_i} + \nu\dfrac{\partial^2 u_i}{\partial x_j \partial x_j}$ to determine a Poisson equation for the pressure.
   b) If the equation $\dfrac{\partial^2 G}{\partial x_j \partial x_j} = \delta(x_j - \tilde{x}_j)$ has solution: $G(x_j, \tilde{x}_j) = \dfrac{-1}{4\pi\sqrt{(x_j - \tilde{x}_j)^2}}$, then use the result from part a) to show that:

$$P(x_j) = \frac{\rho}{4\pi}\int_{\text{all } \tilde{x}} \frac{1}{\sqrt{(x_j - \tilde{x}_j)^2}}\frac{\partial^2}{\partial \tilde{x}_j \partial \tilde{x}_i}(U_i U_j + \overline{u_i u_j})d^3\tilde{x} \text{ in a turbulent flow.}$$

12.15. Starting with the RANS momentum equation (12.30), derive the equation for the kinetic energy of the average flow field (12.46).

12.16. Derive the RANS transport equation for the Reynolds stress correlation (12.35) via the following steps.
   a) By subtracting (12.30) from (4.86), show that the instantaneous momentum equation for the fluctuating turbulent velocity u_i is:

$$\frac{\partial u_i}{\partial t} + u_k\frac{\partial U_i}{\partial x_k} + U_k\frac{\partial u_i}{\partial x_k} + u_k\frac{\partial u_i}{\partial x_k} = -\frac{1}{\rho_0}\frac{\partial p}{\partial x_i} + \nu\frac{\partial^2 u_i}{\partial x_k^2} + g\alpha T'\delta_{i3} + \frac{\partial}{\partial x_k}\overline{u_i u_k}.$$

b) Show that: $\overline{u_i \dfrac{Du_j}{Dt} + u_j \dfrac{Du_i}{Dt}} = \dfrac{\partial}{\partial t}(\overline{u_i u_j}) + U_k \dfrac{\partial}{\partial x_k}(\overline{u_i u_j}) + \dfrac{\partial}{\partial x_k}(\overline{u_i u_j u_k})$

c) Combine and simplify the results of parts a) and b) to reach (12.35).

12.17. Starting from (12.38) and (12.40), set $\mathbf{r} = r\mathbf{e}_1$ and use $R_{11} = \overline{u^2}f(r)$, and $R_{22} = \overline{u^2}g(r)$, to show that $F(r) = \overline{u^2}(f(r) - g(r))r^{-2}$ and $G(r) = \overline{u^2}g(r)$.

12.18. **a)** Starting from R_{ij} from (12.39), compute $\partial R_{ij}/\partial r_j$ for incompressible flow.

b) For homogeneous-isotropic turbulence use the result of part a) to show that the longitudinal, $f(r)$, and transverse, $g(r)$, correlation functions are related by $g(r) = f(r) + (r/2)(df(r)/dr)$.

c) Use part b) and the integral length scale and Taylor microscale definitions to find $2\Lambda_g = \Lambda_f$ and $\sqrt{2}\lambda_g = \lambda_f$.

12.19. In homogeneous turbulence: $R_{ij}(\mathbf{r}_b - \mathbf{r}_a) = \overline{u_i(\mathbf{x} + \mathbf{r}_a)u_j(\mathbf{x} + \mathbf{r}_b)} = R_{ij}(\mathbf{r})$, where $\mathbf{r} = \mathbf{r}_b - \mathbf{r}_a$.

a) Show that $\overline{(\partial u_i(\mathbf{x})/\partial x_k)(\partial u_j(\mathbf{x})/\partial x_l)} = -(\partial^2 R_{ij}/\partial r_k \partial r_l)_{r=0}$.

b) If the flow is incompressible and isotropic, show that

$$-\overline{(\partial u_1(\mathbf{x})/\partial x_1)^2} = -\frac{1}{2}\overline{(\partial u_1(\mathbf{x})/\partial x_2)^2} = +2\overline{(\partial u_1(\mathbf{x})/\partial x_2)(\partial u_2(\mathbf{x})/\partial x_1)}$$

$$= \overline{u^2}\left(d^2 f/dr^2\right)_{r=0}.$$

[*Hint*: Expand $f(r)$ about $r = 0$ *before* taking any derivatives.]

12.20. The turbulent kinetic energy equation contains a pressure-velocity correlation, $K_j = \overline{p(\mathbf{x})u_j(\mathbf{x} + \mathbf{r})}$. In homogeneous isotropic turbulent flow, the most general form of this correlation is: $K_j = K(r)r_j$. If the flow is also incompressible, show that $K(r)$ must be zero.

12.21. The velocity potential for two-dimensional water waves of small amplitude ξ_o on a deep pool can be written:

$$\phi(x_1, x_2, t) = \frac{\omega \xi_o}{k}e^{+kx_2}\cos(\omega t - kx_1),$$

where x_1 and x_2 are the horizontal and vertical coordinates with $x_2 = 0$ defining the average free surface. Here, ω is the temporal radian frequency of the waves and k is the wave number.

a) Compute the two-dimensional velocity field: $\mathbf{u} = (u_1, u_2) = (\partial \phi/\partial x_1, \partial \phi/\partial x_2)$.

b) Show that this velocity field is a solution of the two-dimensional continuity and Navier-Stokes equations for incompressible fluid flow.

c) Compute the strain-rate tensor $S_{ij} = 1/2(\partial u_i/\partial x_j + \partial u_j/\partial x_i)$.

d) Although this flow is not turbulent, it must still satisfy the turbulent kinetic energy equation that contains an energy dissipation term. Denote the kinematic viscosity by ν, and compute the kinetic energy dissipation rate in this flow: $\varepsilon = 2\nu \overline{S_{ij}S_{ij}}$, where the over bar implies a time average over one wave period is $2\pi/\omega$.

Only time averages of even powers of the trig-functions are nonzero, for example: $\overline{\cos^2(\omega t - kx)} = \overline{\sin^2(\omega t - kx)} = 1/2$ while $\overline{\cos(\omega t - kx)} = \overline{\sin(\omega t - kx)} = 0$.

 e) The original potential represents a lossless flow and does not include any viscous effects. Explain how this situation can occur when the kinetic-energy dissipation rate is not zero.

12.22. A mass of 10 kg of water is stirred by a mixer. After one hour of stirring, the temperature of the water rises by 1.0°C. What is the power output of the mixer in watts? What is the size η of the dissipating eddies?

12.23. In locally isotropic turbulence, Kolmogorov determined that the wave number spectrum can be represented by $S_{11}(k)/(\nu^5\overline{\varepsilon})^{1/4} = \Phi(k\nu^{3/4}/\overline{\varepsilon}^{1/4})$ in the inertial subrange and dissipation range of turbulent scales, where Φ is an undetermined function.

 a) Determine the equivalent form for the temporal spectrum $S_e(\omega)$ in terms of the average kinetic energy dissipation rate $\overline{\varepsilon}$, the fluid's kinematic viscosity ν, and the temporal frequency ω.

 b) Simplify the results of part a) for the inertial range of scales where ν is dropped from the dimensional analysis.

 c) To obtain the results for parts a) and b), an implicit assumption has been made that leads to the neglect of an important parameter. Add the missing parameter and redo the dimensional analysis of part a).

 d) Use the missing parameter and ω to develop an equivalent wave number. Insist that your result for S_e only depend on this equivalent wave number and $\overline{\varepsilon}$ to recover the minus-five-thirds law.

12.24.[1] Estimates for the importance of anisotropy in a turbulent flow can be developed by assuming that fluid velocities and spatial derivatives of the average-flow (or RANS) equation are scaled by the average velocity difference ΔU that drives the largest eddies in the flow having a size L, and that the fluctuating velocities and spatial derivatives in the turbulent kinetic energy (TKE) equation are scaled by the kinematic viscosity ν and the Kolmogorov scales η and u_K (see (12.50)). Thus, the scaling for a mean velocity gradient is: $\partial U_i/\partial x_j \sim \Delta U/L$, while the mean-square turbulent velocity gradient scales as: $\overline{(\partial u_i/\partial x_j)^2} \sim (u_K/\eta)^2 = \nu^2/\eta^4$, where the "$\sim$" sign means "scales as." Use these scaling ideas in parts a) and d):

 a) The total energy dissipation rate in a turbulent flow is $2\nu\overline{S}_{ij}\overline{S}_{ij} + 2\nu\overline{S'_{ij}S'_{ij}}$,

  where $\overline{S}_{ij} = \dfrac{1}{2}\left(\dfrac{\partial U_i}{\partial x_j} + \dfrac{\partial U_j}{\partial x_i}\right)$ and $S'_{ij} = \dfrac{1}{2}\left(\dfrac{\partial u_i}{\partial x_j} + \dfrac{\partial u_j}{\partial x_i}\right)$. Determine how the ratio

  $\dfrac{\overline{S'_{ij}S'_{ij}}}{\overline{S}_{ij}\overline{S}_{ij}}$ depends on the outer-scale Reynolds number: $Re_L = \Delta U \cdot L/\nu$.

 b) Is average-flow or fluctuating-flow energy dissipation more important?

---

[1]Obtained from Prof. Werner Dahm.

c) Show that the turbulent kinetic energy dissipation rate, $\bar{\varepsilon} = 2\nu\overline{S'_{ij}S'_{ij}}$ can be written:

$$\bar{\varepsilon} = \nu\left[\overline{\frac{\partial u_i}{\partial x_j}\frac{\partial u_i}{\partial x_j}} + \frac{\partial^2}{\partial x_i\partial x_j}\overline{u_iu_j}\right].$$

d) For homogeneous isotropic turbulence, the second term in the result of part c) is zero but it is nonzero in a turbulent shear flow. Therefore, estimate how

$$\frac{\partial^2}{\partial x_i\partial x_j}\overline{u_iu_j}\left/\overline{\frac{\partial u_i}{\partial x_j}\frac{\partial u_i}{\partial x_j}}\right.$$

depends on Re_L in turbulent shear flow as means of assessing how much impact anisotropy has on the turbulent kinetic energy dissipation rate.

e) Is an isotropic model for the turbulent dissipation appropriate at high Re_L in a turbulent shear flow?

12.25. Determine the self-preserving form of the average stream-wise velocity $U_x(x,r)$ of a round turbulent jet using cylindrical coordinates where x increases along the jet axis and r is the radial coordinate. Ignore gravity in your work. Denote the density of the nominally quiescent reservoir fluid by ρ.

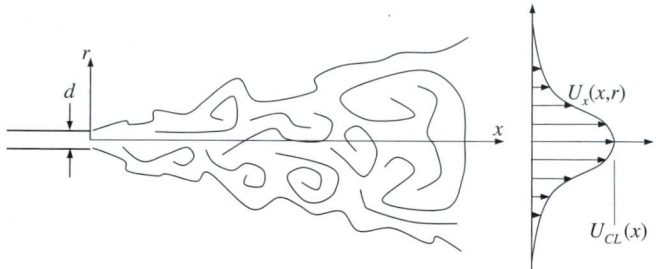

a) Place a stationary cylindrical control volume around the jet's cone of turbulence so that circular control surfaces slice all the way through the jet flow at its origin and at a distance x downstream where the fluid density is ρ. Assuming that the fluid outside the jet is nearly stationary so that pressure does not vary in the axial direction and that the fluid entrained into the volume has negligible x-direction momentum, show

$$J_0 \equiv \int_0^{d/2} \rho_0 U_0^2 2\pi r dr = \int_0^{D/2} \rho U_x^2(x,r) 2\pi r dr,$$

where J_0 is the jet's momentum flux, ρ_0 is the density of the jet fluid, and U_0 is the jet exit velocity.

b) Simplify the exact mean-flow equations

$$\frac{\partial U_x}{\partial x} + \frac{1}{r}\frac{\partial}{\partial r}(rU_r) = 0, \text{ and}$$

$$U_x\frac{\partial U_x}{\partial x} + U_r\frac{\partial U_x}{\partial r} = -\frac{1}{\rho}\frac{\partial P}{\partial x} + \frac{\nu}{r}\frac{\partial}{\partial r}\left(r\frac{\partial U_x}{\partial r}\right) - \frac{1}{r}\frac{\partial}{\partial r}(r\overline{u_xu_r}) - \frac{\partial}{\partial x}(r\overline{u_x^2}),$$

when $\partial P / \partial x \approx 0$, the jet is slender enough for the boundary-layer approximation $\partial / \partial r \gg \partial / \partial x$ to be valid, and the flow is at high Reynolds number so that the viscous terms are negligible.

c) Eliminate the average radial velocity from the simplified equations to find:

$$
U_x \frac{\partial U_x}{\partial x} - \left\{ \frac{1}{r} \int_0^r \not{r} \frac{\partial U_x}{\partial x} d\not{r} \right\} \frac{\partial U_x}{\partial r} = -\frac{1}{r} \frac{\partial}{\partial r} (r \overline{u_x u_r}),
$$

where $\not{r}$ is just an integration variable.

d) Assume a similarity form: $U_x(x,r) = U_{CL}(x) f(\xi)$, $-\overline{u_x u_r} = \Psi(x) g(\xi)$, where $\xi = r / \delta(x)$ and f and g are undetermined functions, use the results of parts a) and c), and choose constant values appropriately to find
$U_x(x,y) = const.(J_0/\rho)^{1/2} x^{-1} F(y/x)$.

e) Determine a formula for the volume flux in the jet. Will the jet fluid be diluted with increasing x?

12.26. Consider the turbulent wake far from a two-dimensional body placed perpendicular to the direction of a uniform flow.

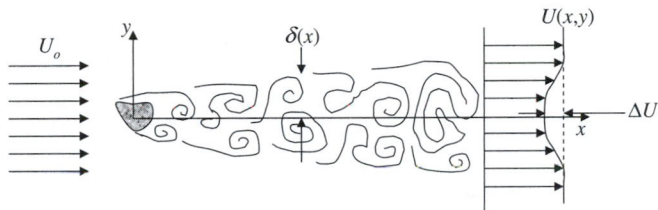

a) Use a large rectangular control volume that encloses the body but only intersects the wake vertically at x somewhere downstream of the body to show that the average fluid-dynamic drag force per unit span, $\overline{F}_D/b$, acting on the body is given by:

$$
\frac{\overline{F}_D/b}{\rho U_o^2} = \theta = \int_{-\infty}^{+\infty} \left[\frac{U(x,y)}{U_o} \left(1 - \frac{U(x,y)}{U_o} \right) - \frac{\overline{u^2}}{U_o^2} \right] dy,
$$

where θ is the momentum thickness of the wake flow (a constant), and $U(x,y)$ is the average horizontal velocity profile a distance x downstream of the body.

b) When $\Delta U \ll U_o$, find the conditions necessary for a self-similar form for the wake's velocity deficit, $U(x,y) = U_o - \Delta U(x) f(\xi)$, to be valid based on the result of part a) and the steady two-dimensional continuity and boundary-layer RANS equations. Here, $\xi = y / \delta(x)$ and δ is the transverse length scale of the wake.

c) Determine how ΔU and δ must depend on x in the self-similar region. State your results in appropriate dimensionless form using θ and U_o as appropriate.

12.27. Consider the two-dimensional shear layer that forms between two steady streams with flow speed U_2 above and U_1 below $y = 0$, that meet at $x = 0$, as shown. Assume a self-similar form for the average horizontal velocity:

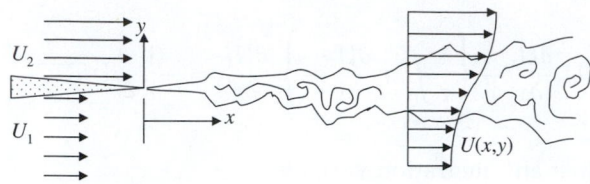

$$U(x,y) = U_1 + (U_2 - U_1)f(\xi) \text{ with } \xi = y/\delta(x).$$

a) What are the boundary conditions on $f(\xi)$ as $y \to \pm\infty$?

b) If the flow is laminar, use $\dfrac{\partial U}{\partial x} + \dfrac{\partial V}{\partial y} = 0$ and $U\dfrac{\partial U}{\partial x} + V\dfrac{\partial U}{\partial y} = \nu\dfrac{\partial^2 U}{\partial y^2}$ with $\delta(x) = \sqrt{\nu x/U_1}$ to obtain a single equation for $f(\xi)$. There is no need to solve this equation.

c) If the flow is turbulent, use: $\dfrac{\partial U}{\partial x} + \dfrac{\partial V}{\partial y} = 0$ and $U\dfrac{\partial U}{\partial x} + V\dfrac{\partial U}{\partial y} = -\dfrac{\partial}{\partial y}(\overline{uv})$ with $-\overline{uv} = (U_2 - U_1)^2 g(\xi)$ to obtain a single equation involving f and g. Determine how δ must depend on x for the flow to be self-similar.

d) Does the laminar or the turbulent mixing layer grow more quickly as x increases?

12.28. Consider an orifice of diameter d that emits an incompressible fluid of density ρ_o at speed U_o into an infinite half space of fluid with density ρ_∞. With gravity acting and $\rho_\infty > \rho_o$, the orifice fluid rises, mixes with the ambient fluid, and forms a buoyant plume with a diameter $D(z)$ that grows with increasing height above the orifice. Assuming that the plume is turbulent and self-similar in the far-field ($z \gg d$), determine how the plume diameter D, the mean centerline velocity U_{cl}, and the mean centerline mass fraction of orifice fluid Y_{cl} depend on the vertical coordinate z via the steps suggested below. Ignore the initial momentum of the orifice fluid. Use both dimensional and control-volume analysis as necessary. Ignore stream-wise turbulent fluxes to simplify your work. Assume uniform flow from the orifice.

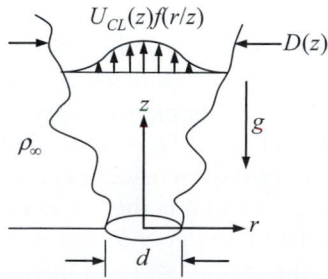

a) Place a stationary cylindrical control volume around the plume with circular control surfaces that slice through the plume at its origin and at height z. Use similarity forms for the average vertical velocity $U_z(z,r) = U_{cl}(z)f(r/z)$ and nozzle fluid mass fraction $\overline{Y}(z,r) = (\rho_\infty - \overline{\rho})/(\rho_\infty - \rho_0) = Y_{cl}(z)h(r/z)$ to conserve the flux of nozzle fluid in the plume, and find:

$\dot{m}_0 = \int_{source} \rho_0 U_0 dA = \int_0^{D/2} \rho_0 \overline{Y}(z,r)U_z(z,r)2\pi r dr$.

b) Conserve vertical momentum using the same control volume assuming that all entrained fluid enters with negligible vertical momentum, to determine:

$$-\int_{source} \rho_0 U_0^2 dA + \int_0^{D/2} \overline{\rho}(z,r)U_z^2(z,r)2\pi r dr = \int_{volume} g[\rho_\infty - \overline{\rho}(z,r)]dV,$$

where $\overline{\rho} = \overline{Y}\rho_0 + (1-\overline{Y})\rho_\infty$.

c) Ignore the source momentum flux, assume z is large enough so that $Y_{CL} \ll 1$, and use the results of parts a) and b) to find: $U_{cl}(z) = C_1 \cdot \sqrt[3]{B/\rho_\infty z}$ and $\frac{\rho_\infty - \rho_0}{\rho_\infty}Y_{cl}(z) = C_2\sqrt[3]{B^2/g^3\rho_\infty^2 z^5}$ where C_1 and C_2 are dimensionless constants, and $B = \int_{source}(\rho_\infty - \rho_0)gU_0 dA$.

12.29. Laminar and turbulent boundary-layer skin friction are very different. Consider skin-friction correlations from zero-pressure-gradient (ZPG) boundary-layer flow over a flat plate placed parallel to the flow.

Laminar boundary layer: $C_f = \dfrac{\tau_0}{(1/2)\rho U^2} = \dfrac{0.664}{\sqrt{Re_x}}$ (Blasius boundary layer).

Turbulent boundary layer: see correlations in Section 12.9.
Create a table of computed results at $Re_x = Ux/\nu = 10^4, 10^5, 10^6, 10^7, 10^8$, and 10^9 for the laminar and turbulent skin-friction coefficients, and the friction force acting on 1.0 m^2 plate surface in sea-level air at 100 m/s and in water at 20 m/s assuming laminar and turbulent flow.

12.30. Derive the following logarithmic velocity profile for a smooth wall: $U^+ = (1/\kappa)\ln y^+ + 5.0$ by starting from $U = (u_*/\kappa)\ln y^+ + const.$ and matching the profile to the edge of the viscous sublayer assuming the viscous sublayer ends at $y = 10.7\,\nu/u_*$.

12.31.[2] Derive the log-law for the mean flow profile in a zero-pressure gradient (ZPG) flat-plate turbulent boundary layer (TBL) through the following mathematical and dimensional arguments.

a) Start with the law of the wall, $U/u^* = f(yu_*/\nu)$ or $U^+ = f(y^+)$, for the near-wall region of the boundary layer, and the defect law for the outer region, $\dfrac{U_e - U}{u_*} = F\left(\dfrac{y}{\delta}\right)$. These formulae must overlap when $y^+ \to +\infty$ and $y/\delta \to 0$. In this matching or overlap region, set U and $\partial U/\partial y$ from both formulas equal to get two equations involving f and F.

b) In the limit as $y^+ \to +\infty$, the kinematic viscosity must drop out of the equation that includes df/dy^+. Use this fact, to show that $U/u_* = A_I \ln(yu_*/\nu) + B_I$ as

[2]Inspired by exercise 7.20 in Pope (2000), p. 311.

$y^+ \to +\infty$ where A_I and B_I are constants for the near-wall or *inner* boundary layer scaling.

c) Use the result of part b) to determine $F(\xi) = -A_I \ln(\xi) - B_O$ where $\xi = y/\delta$, and A_I and B_O are constants for the wake flow or *outer* boundary layer scaling.

d) It is traditional to set $A_I = 1/\kappa$, and to keep B_I but to drop its subscript. Using these new requirements determine the two functions, f_I and F_O, in the matching region. Which function explicitly depends on the Reynolds number of the flow?

12.32. Prove (12.90) and (12.91) by considering a stationary control volume that resides inside the channel or pipe and has stream-normal control surfaces separated by a distance dx and stream-parallel surfaces that coincide with the wall or walls that confine the flow.

12.33. A horizontal smooth pipe 20 cm in diameter carries water at a temperature of 20°C. The drop of pressure is $dp/dx = -8$ N/m² per meter. Assuming turbulent flow, verify that the thickness of the viscous sublayer is ≈ 0.25 mm. [*Hint*: Use dp/dx as given by (12.91) to find $\tau_0 = 0.4$ N/m², and therefore $u_* = 0.02$ m/s.]

12.34. The cross-section averaged flow speed U_{av} in a round pipe of radius a may be written:

$$U_{av} \equiv \frac{volume\ flux}{area} = \frac{1}{\pi a^2} \int_0^a U(y) 2\pi r dr = \frac{2}{a^2} \int_0^a U(y)(a-y) dy,$$

where r is the radial distance from the pipe's centerline, and $y = a - r$ is the distance inward from the pipe's wall. Turbulent pipe flow has very little wake, and the viscous sublayer is very thin at high Reynolds number; therefore assume the log-law profile, $U(y) = (u_*/\kappa)\ln(yu_*/\nu) + u_* B$, holds throughout the pipe to find

$$U_{av} \cong u_*[(1/\kappa)\ln(au_*/\nu) + B - 3/2\kappa].$$

Now use the definitions $C_f = \tau_0/(1/2)\rho U_{av}^2$, $Re_d = 2U_{av}a/\nu$, $\bar{f} = 4C_f =$ Darcy friction factor, $\kappa = 0.41$, and $B = 5.0$, and switch to base-10 logarithms to reach Prandtl's 1935 correlation for turbulent pipe flow friction: $\bar{f}^{-1/2} = 2.0\log_{10}(Re_d\bar{f}^{1/2}) - 1.0$. When the second constant is adjusted from -1.0 to -0.8, this correlation is valid for $Re_d \geq 4000$ (White, 2006) and yields $\bar{f}$-values substantially larger than the laminar pipe flow result $\bar{f} = 64/Re_d$.

12.35. Perhaps the simplest way to model turbulent flow is to develop an eddy viscosity from dimensional analysis and physical reasoning. Consider turbulent Couette flow with wall spacing h. Assume that eddies of size l produce velocity fluctuations of size $l(\partial U/\partial y)$ so that the turbulent shear stress correlation can be modeled as: $-\overline{uv} \propto l^2(\partial U/\partial y)^2$. Unfortunately, l cannot be a constant because it must disappear near the walls. Thus, more educated guessing is needed, so for this problem assume $\partial U/\partial y$ will have some symmetry about the channel centerline (as shown) and try: $l = Cy$ for $0 \le y \le h/2$ where C is a positive dimensionless constant and y is the vertical distance measured from the lower wall. With this turbulence model, the RANS equation for $0 \le y \le h/2$ becomes:

$$U\frac{\partial U}{\partial x} + V\frac{\partial U}{\partial y} = -\frac{1}{\rho}\frac{d\overline{p}}{dx} + \frac{1}{\rho}\frac{\partial \tau_{xy}}{\partial y} \quad \text{where } \tau_{xy} = \mu\frac{\partial U}{\partial y} + \rho C^2 y^2 \left(\frac{\partial U}{\partial y}\right)^2.$$

Determine an analytic form for $U(y)$ after making appropriate simplifications of the RANS equation for fully developed flow assuming the pressure gradient is zero. Check to see that your final answer recovers the appropriate forms as $y \to 0$ and $C \to 0$. Use the fact that $U(y = h/2) = U_o/2$ in your work if necessary.

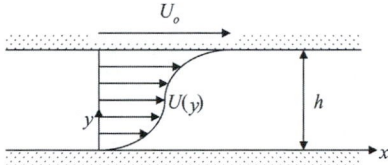

12.36. Turbulence largely governs the mixing and transport of water vapor (and other gases) in the atmosphere. Such processes can sometimes be assessed by considering the conservation law (12.34) for a passive scalar.

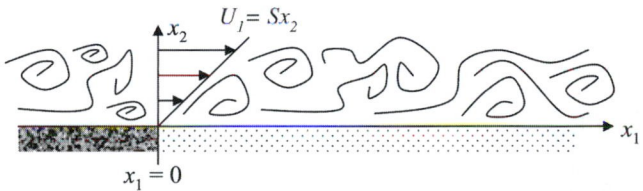

a) Appropriately simplify (12.34) for turbulence at high Reynolds number that is characterized by: an outer length scale of L, a large-eddy turnover time of T, and a mass-fraction magnitude of Y_o. In addition, assume that the molecular diffusivity κ_m is at most as large as $\nu = \mu/\rho =$ the fluid's kinematic viscosity.

b) Now consider a simple model of how a dry turbulent wind collects moisture as it blows over a nominally flat water surface ($x_1 > 0$) from a dry surface ($x_1 < 0$). Assume the mean velocity is steady and has a single component with a linear gradient, $U_j = (Sx_2, 0, 0)$, and use a simple gradient diffusion model: $-\overline{u_j Y'} = \Delta U L(0, \partial \overline{Y}/\partial x_2, 0)$, where ΔU and L are (constant) velocity and length scales that characterize the turbulent diffusion in this case. This turbulence model allows the turbulent mean flow to be treated like a laminar flow with a large diffusivity $= \Delta U L$ (a turbulent diffusivity). For the simple boundary conditions: $\overline{Y}(x_j) = 0$ for $x_1 < 0$, $\overline{Y}(x_j) = 1$ at $x_2 = 0$ for $x_1 > 0$, and $\overline{Y}(x_j) \to 0$ as $x_2 \to \infty$, show that

$$\overline{Y}(x_1, x_2, x_3) = \int\limits_{\xi}^{\infty} \exp\left(-\frac{1}{9}\zeta^3\right) d\zeta \left/ \int\limits_{0}^{\infty} \exp\left(-\frac{1}{9}\zeta^3\right) d\zeta \right.$$

$$\text{where } \xi = x_2 \left(\frac{S}{\Delta U L x_1}\right)^{1/3} \quad \text{for } x_1, x_2 > 0.$$

12.37. Estimate the Monin-Obukhov length in the atmospheric boundary layer if the surface stress is 0.1 N/m^2 and the upward heat flux is 200 W/m^2.

12.38. Consider one-dimensional turbulent diffusion of particles issuing from a point source. Assume a Gaussian-Lagrangian correlation function of particle velocity:

$$r(\tau) = e^{-\tau^2/t_c^2},$$

where t_c is a constant. By integrating the correlation function from $\tau = 0$ to ∞, find the integral time scale Λ_t in terms of t_c. Using the Taylor theory, estimate the eddy diffusivity at large times $t/\Lambda_t \gg 1$, given that the *rms* fluctuating velocity is 1 m/s and $t_c = 1$ s.

Literature Cited

Adrian, R. J. (2007). Hairpin vortex organization in wall turbulence. *Physics of Fluids, 19*, 041301.

Barenblatt, G. I. (1993). Scaling laws for fully developed shear flows. Part I. Basic hypotheses and analysis. *Journal of Fluid Mechanics, 248*, 513—520.

Batchelor, G. K. (1953). *The Theory of Homogeneous Turbulence*. New York: Cambridge University Press.

Batchelor, G. K. (1959). Small scale variation of convected quantities like temperature in turbulent fluid. Part I: General discussion and the case of small conductivity. *Journal of Fluid Mechanics, 5*, 113—133.

Bird, R. B., Stewart, W. E., & Lightfoot, E. N. (1960). *Transport Phenomena*. New York: John Wiley and Sons.

Bradshaw, P., & Woods, J. D. (1978). Geophysical turbulence and buoyant flows. In P. Bradshaw (Ed.), *Turbulence*. New York: Springer-Verlag.

Cantwell, B. J. (1981). Organized motion in turbulent flow. *Annual Review of Fluid Mechanics, 13*, 457—515.

Chapman, D. R. (1979). Computational aerodynamics development and outlook. *AIAA Journal, 17*, 1293—1313.

Chauhan, K. A., Monkewitz, P. A., & Nagib, H. M. (2009). Criteria for assessing experiments in zero pressure gradient boundary layers. *Fluid Dynamics Research, 41*, 021404.

Chen, C. J., & Rodi, W. (1980). *Vertical Turbulent Buoyant Jets—A Review of Experimental Data*. Oxford: Pergamon Press.

Coles, D. E. (1956). The law of the wake in the turbulent boundary layer. *Journal of Fluid Mechanics, 1*, 191—226.

Dimotakis, P. E. (2000). The mixing transition in turbulent flows. *Journal of Fluid Mechanics, 409*, 69—98.

Feynman, R. P., Leighton, R. B., & Sands, M. (1963). *The Feynman Lectures on Physics*. New York: Addison-Wesley.

Fife, P., Wei, T., Klewicki, J., & McMurtry, P. (2005). Stress gradient balance layers and scale hierarchies in wall-bounded turbulent flows. *Journal of Fluid Mechanics, 532*, 165—189.

George, W. K. (1989). The self-preservation of turbulent flows and its relation to initial conditions and coherent structures. In W. K. George, & R. Arndt (Eds.), *Advances in Turbulence* (pp. 39—73). New York: Hemisphere Publishing Corp.

George, W. K. (2006). Recent advancements toward the understanding of turbulent boundary layers. *AIAA Journal, 44*, 2435—2449.

George, W. K., & Castillo, L. (1997). Zero-pressure-gradient turbulent boundary layer. *Applied Mechanics Reviews, 50*, 689—729.

Grant, H. L., Hughes, B. A., Vogel, W. M., & Moilliet, A. (1968). The spectrum of temperature fluctuation in turbulent flow. *Journal of Fluid Mechanics, 34*, 423—442.

Grant, H. L., Stewart, R. W., & Moilliet, A. (1962). The spectrum of a cross-stream component of turbulence in a tidal stream. *Journal of Fluid Mechanics, 13,* 237—240.

Hinze, J. O. (1975). *Turbulence* (2nd ed.). New York: McGraw-Hill.

Jones, W. P., & Launder, B. E. (1972). The prediction of laminarization with a two-equation model of turbulence. *International Journal of Heat and Mass Transfer, 15,* 301—314.

Kline, S. J., Reynolds, W. C., Schraub, F. A., & Runstadler, P. W. (1967). The structure of turbulent boundary layers. *Journal of Fluid Mechanics, 30,* 741—773.

Kolmogorov, A. N. (1941a). Dissipation of energy in locally isotropic turbulence. *Doklady Akademii Nauk SSSR, 32,* 19—21.

Kolmogorov, A. N. (1941b). The local structure of turbulence in incompressible viscous fluid for very large Reynolds numbers. *Doklady Akademii Nauk SSSR, 30,* 299—303.

Kuo, K. K. (1986). *Principles of Combustion*. New York: John Wiley and Sons.

Lam, S. H. (1992). On the RNG theory of turbulence. *The Physics of Fluids A, 4,* 1007—1017.

Landahl, M. T., & Mollo-Christensen, E. (1986). *Turbulence and Random Processes in Fluid Mechanics*. London: Cambridge University Press.

Launder, B. E., & Sharma, B. I. (1974). Application of the energy dissipation model of turbulence to the calculation of flow near a spinning disk. *Letters in Heat and Mass Transfer, 1,* 131—137.

Lesieur, M. (1987). *Turbulence in Fluids*. Dordrecht, Netherlands: Martinus Nijhoff Publishers.

Marusic, I., Mckeon, B. J., Monkewitz, P. A., Nagib, H. M., Smits, A. J., & Sreenivasan, K. R. (2010). Wall bounded turbulent flows at high Reynolds numbers: Recent advances and key issues. *Physics of Fluids, 22,* 065103.

Monin, A. S., & Yaglom, A. M. (1971, 1975). In J. L. Lumley (Ed.), *Statistical Fluid Mechanics, Vol. I and II*. Cambridge, MA: MIT Press.

Monkewitz, P. A., Chauhan, K. A., & Nagib, H. M. (2007). Self-consistent high-Reynolds-number asymptotics for zero-pressure-gradient turbulent boundary layers. *Physics of Fluids, 19,* 115101.

Monkewitz, P. A., Chauhan, K. A., & Nagib, H. M. (2008). Comparison of mean flow similarity laws in zero pressure gradient turbulent boundary layers. *Physics of Fluids, 20,* 105102.

Nagib, H. M., & Chauhan, K. A. (2008). Variations of von Kármán coefficient in canonical flows. *Physics of Fluids, 20,* 101518.

Obukhov, A. M. (1949). Structure of the temperature field in turbulent flow. *Izvestiya Akademii Nauk SSSR, Georgr. and Geophys. Ser., 13*(1), 58—69.

Oweis, G. F., Winkel, E. S., Cutbirth, J. M., Ceccio, S. L., Perlin, M., & Dowling, D. R. (2010). The mean velocity profile of a smooth-flat-plate turbulent boundary layer at high Reynolds number. *Journal of Fluid Mechanics, 665,* 357—381.

Panofsky, H. A., & Dutton, J. A. (1984). *Atmospheric Turbulence*. New York: Wiley.

Pao, Y. H. (1965). Structure of turbulent velocity and scalar fields at large wave numbers. *The Physics of Fluids, 8,* 1063—1075.

Phillips, O. M. (1977). *The Dynamics of the Upper Ocean*. London: Cambridge University Press.

Pope, S. B. (2000). *Turbulent Flows*. London: Cambridge University Press.

Richardson, L. F. (1922). *Weather Prediction by Numerical Process*. Cambridge: Cambridge University Press.

Schultz-Grunow, F. (1941). New frictional resistance law for smooth plates. *NACA Technical Memorandum, 17*(8), 1—24.

Smith, L. M., & Reynolds, W. C. (1992). On the Yakhot-Orszag renormalization group method for deriving turbulence statistics and models. *The Physics of Fluids A, 4,* 364—390.

Spalding, D. B. (1961). A single formula for the law of the wall. *Journal of Applied Mechanics, 28,* 455—457.

Speziale, C. G. (1991). Analytical methods for the development of Reynolds-stress closures in turbulence. *Annual Review of Fluid Mechanics, 23,* 107—157.

Taylor, G. I. (1915). Eddy motion in the atmosphere. *Philosophical Transactions of the Royal Society of London, A215,* 1—26.

Taylor, G. I. (1921). Diffusion by continuous movements. *Proceedings of the London Mathematical Society, 20,* 196—211.

Tennekes, H., & Lumley, J. L. (1972). *A First Course in Turbulence*. Cambridge, MA: MIT Press.

Townsend, A. A. (1976). *The Structure of Turbulent Shear Flow*. London: Cambridge University Press.

Turner, J. S. (1973). *Buoyancy Effects in Fluids*. London: Cambridge University Press.

Turner, J. S. (1981). Small-scale mixing processes. In B. A. Warren, & C. Wunch (Eds.), *Evolution of Physical Ocean-ography*. Cambridge, MA: MIT Press.

Wei, T., Fife, P., Klewicki, J., & McMurtry, P. (2005). Properties of the mean momentum balance in turbulent boundary layer, pipe and channel flows. *Journal of Fluid Mechanics, 522*, 303–327.

White, F. M. (2006). *Viscous Fluid Flow*. Boston: McGraw-Hill.

Yakhot, V., & Orszag, S. A. (1986). Renormalization group analysis of turbulence. I. Basic theory. *Journal of Scientific Computing, 1*, 3–51.

Supplemental Reading

George, W. K., & Arndt, R. (Eds.). (1989). *Advances in Turbulence*. New York: Hemisphere Publishing Corp.

Hunt, J. C. R., Sandham, N. D., Vassilicos, J. C., Launder, B. E., Monkewitz, P. A., & Hewitt, G. F. (2001). Devel-opments in turbulence research: are view based on the 1999 Programme of the Isaac Newton Institute, Cambridge. Published in *Journal of Fluid Mechanics, 436*, 353–391.

Proceedings of the Boeing Symposium on Turbulence. (1970). Published in *Journal of Fluid Mechanics, 41*, Parts 1 (March) and 2 (April).

Symposium on Fluid Mechanics of Stirring and Mixing, IUTAM. (1991). Published in *Physics of Fluids, A*(5), 3, May, Part 2.

The Turbulent Years. (2002). John Lumley at 70, A Symposium in Honor of John L. Lumley on his 70th Birthday. Published in *Physics of Fluids, 14*, 2424–2557.

Geophysical Fluid Dynamics

O U T L I N E

13.1. Introduction   622

13.2. Vertical Variation of Density in
the Atmosphere and Ocean   623

13.3. Equations of Motion   625

13.4. Approximate Equations for a
Thin Layer on a Rotating Sphere   628

13.5. Geostrophic Flow   630

13.6. Ekman Layer at a Free Surface   633

13.7. Ekman Layer on a Rigid Surface   639

13.8. Shallow-Water Equations   642

13.9. Normal Modes in a Continuously
Stratified Layer   644

13.10. High- and Low-Frequency
Regimes in Shallow-Water
Equations   649

13.11. Gravity Waves with Rotation   651

13.12. Kelvin Wave   654

13.13. Potential Vorticity Conservation
in Shallow-Water Theory   658

13.14. Internal Waves   662

13.15. Rossby Wave   671

13.16. Barotropic Instability   676

13.17. Baroclinic Instability   678

13.18. Geostrophic Turbulence   685

Exercises   688

Literature Cited   690

Supplemental Reading   690

CHAPTER OBJECTIVES

- To introduce the approximations and phenomena that are common in geophysical fluid dynamics

- To describe flows near air-water interfaces and solid surfaces in steadily rotating coordinate systems

- To specify the effects of planetary rotation on waves in stratified fluids
- To describe the instabilities of very long waves that span a significant range of latitude
- To provide an introduction to geostrophic turbulence and the reversed energy cascade

13.1. INTRODUCTION

The subject of geophysical fluid dynamics deals with the dynamics of the atmosphere and the ocean. Motions within these fluid masses are intimately connected through continual exchanges of momentum, heat, and moisture, and cannot be considered separately on a global scale. The field has been largely developed by meteorologists and oceanographers, but nonspecialists have also been interested in the subject. Taylor was not a geophysical fluid dynamicist, but he held the position of a meteorologist for some time, and through this involvement he developed a special interest in the problems of turbulence and instability. Although Prandtl was mainly interested in the engineering aspects of fluid mechanics, his well-known textbook (Prandtl, 1952) contains several sections dealing with meteorological aspects of fluid mechanics. Notwithstanding the pressure for technical specialization, it is worthwhile to learn something of this fascinating field even if one's primary interest is in another area of fluid mechanics.

Together the atmosphere and ocean have a large and consequential impact on humanity. The combined dynamics of the atmosphere and ocean are leading contributors to global climate. We all live within the atmosphere and are almost helplessly affected by the weather and its rather chaotic behavior which modulates agricultural success. Ocean currents effect navigation, fisheries, and pollution disposal. Populations that occupy coastlines can do little to prevent hurricanes, typhoons, or tsunamis. Thus, understanding and reliably predicting geophysical fluid dynamic events and trends are scientific, economic, humanitarian, and even political priorities. This chapter provides the basic elements necessary for developing an understanding of geophysical fluid dynamics.

The two features that distinguish geophysical fluid dynamics from other areas of fluid dynamics are the rotation of the earth and the vertical density stratification of the medium. We shall see that these two effects dominate the dynamics to such an extent that entirely new classes of phenomena arise, which have no counterpart in the laboratory-scale flows emphasized in the preceding chapters. (For example, the dominant mode of flow in the atmosphere and the ocean is *along* the lines of constant pressure, not from high to low pressures.) The motion of the atmosphere and the ocean is naturally studied in a coordinate frame rotating with the earth. This gives rise to the Coriolis force (see Section 4.7). The density stratification gives rise to buoyancy forces (Section 4.11 and Chapter 7). In addition, important relevant material includes vorticity, boundary layers, instability, and turbulence (Chapters 5, 9, 11, and 12). The reader should be familiar with these topics before proceeding further with the present chapter.

Because Coriolis forces and stratification effects play dominating roles in both the atmosphere and the ocean, there is a great deal of similarity between the dynamics of these two media; this makes it possible to study them together. There are also significant differences, however. For example the effects of lateral boundaries, due to the presence of continents, are important in the ocean but less so in the atmosphere. The intense currents (like the Gulf Stream and the Kuroshio) along the western ocean boundaries have no atmospheric analog. On the other hand phenomena like cloud formation and latent heat release due to moisture condensation are typically atmospheric phenomena. Plus, processes are generally slower in the ocean, in which a typical horizontal velocity is 0.1 m/s, although velocities of the order of 1−2 m/s are found within the intense western boundary currents. In contrast, typical velocities in the atmosphere are 10−20 m/s. The nomenclature can also be different in the two fields. Meteorologists refer to a flow directed to the west as an "easterly wind" (i.e., *from* the east), while oceanographers refer to such a flow as a "westward current." Atmospheric scientists refer to vertical positions by *heights* measured upward from the earth's surface, while oceanographers refer to *depths* measured downward from the sea surface. However, we shall always take the vertical coordinate z to be upward, so no confusion should arise.

We shall see that rotational effects caused by the presence of the Coriolis force have opposite signs in the two hemispheres. Note that *all figures and descriptions given here are valid for the northern hemisphere.* In some cases the sense of the rotational effect for the southern hemisphere has been explicitly mentioned. When the sense of the rotational effect is left unspecified for the southern hemisphere, it has to be assumed as opposite to that in the northern hemisphere.

13.2. VERTICAL VARIATION OF DENSITY IN THE ATMOSPHERE AND OCEAN

An important variable in the study of geophysical fluid dynamics is the density stratification. In (1.35) we saw that the static stability of a fluid medium is determined by the sign of the potential density gradient:

$$\frac{d\rho_\theta}{dz} = \frac{d\rho}{dz} + \frac{g\rho}{c^2}, \tag{13.1}$$

where c is the speed of sound. A medium is statically stable if the potential density decreases with height. The first term on the right-hand side corresponds to the *in situ* density change due to all sources such as pressure, temperature, and concentration of a constituent such as the salinity in the sea or the water vapor in the atmosphere. The second term on the right side is the density gradient due to the pressure decrease with height in an adiabatic environment and is called the *adiabatic density gradient*. The corresponding temperature gradient is called the *adiabatic temperature gradient*. For incompressible fluids $c = \infty$ and the adiabatic density gradient is zero.

As shown in Section 1.10, the temperature of a dry adiabatic atmosphere decreases upward at the rate of $\approx 10°C/km$; that of a moist atmosphere decreases at the rate of $\approx 5-6°C/km$. In the ocean, the adiabatic density gradient is $g\rho/c^2 \sim 4 \times 10^{-3}$ kg/m^4,

624   13. GEOPHYSICAL FLUID DYNAMICS

for a typical sound speed of $c = 1520$ m/s. The potential density in the ocean increases with depth at a much smaller rate of 0.6×10^{-3} kg/m^4, so it follows that most of the *in situ* density increase with depth in the ocean is due to the compressibility effects and not to changes in temperature or salinity. As potential density is the variable that determines the static stability, oceanographers take into account the compressibility effects by referring all their density measurements to the sea-level pressure. Unless specified otherwise, throughout the present chapter potential density will simply be referred to as *density*, omitting the qualifier *potential*.

The mean vertical distribution of the *in situ* temperature in the lower 50 km of the atmosphere is shown in Figure 13.1. The lowest 10 km is called the *troposphere*, in which the temperature decreases with height at the rate of 6.5°C/km. This is close to the moist adiabatic lapse rate, which means that the troposphere is close to being neutrally stable. The neutral stability is expected because turbulent mixing due to frictional and convective effects in the lower atmosphere keeps it well stirred and therefore close to the neutral stratification. Practically all the clouds, weather changes, and water vapor of the atmosphere are found in the troposphere. The layer is capped by the *tropopause*, at an average height of 10 km, above which the temperature increases. This higher layer is called the *stratosphere*, because it is very stably stratified. The increase of temperature with height in this layer is caused by the absorption of the sun's ultraviolet rays by ozone. The stability of the layer inhibits mixing and

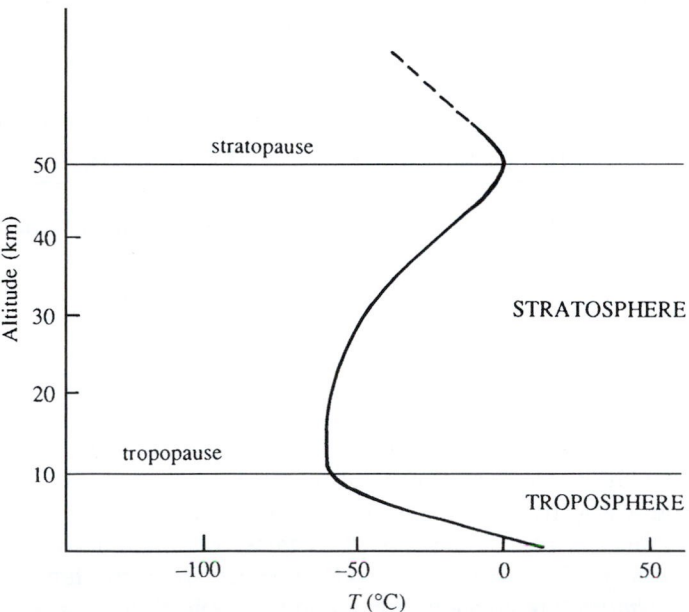

FIGURE 13.1 Sketch of the vertical distribution of temperature in the lower 50 km of the atmosphere. In the lowest layer, the troposphere, the temperature decreases with height and this is where nearly all weather occurs. The next layer is the stratosphere where temperature increases with height. The troposphere is separated from the stratosphere by the tropopause, and the stratosphere ends at the stratopause.

FIGURE 13.2 Typical vertical distributions of: (a) temperature and density, and (b) buoyancy frequency in the ocean. Temperature falls while density increases with increasing depth. The buoyancy frequency peaks in the region of the thermocline where temperature changes most rapidly with depth.

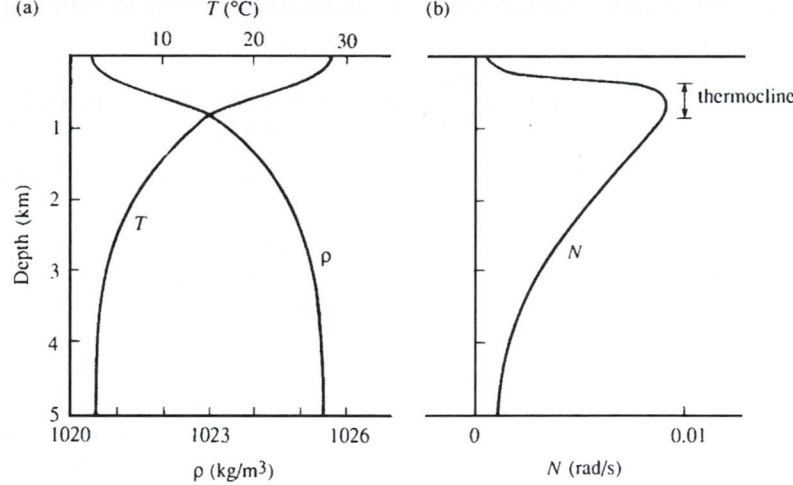

consequently acts as a lid on the turbulence and convective motion of the troposphere. The increase of temperature stops at the *stratopause* at a height of nearly 50 km.

The vertical structure of density in the ocean is sketched in Figure 13.2, showing typical profiles of potential density and temperature. Most of the temperature increase with height is due to the absorption of solar radiation within the upper layer of the ocean. The density distribution in the ocean is also affected by the salinity. However, there is no characteristic variation of salinity with depth, and a decrease with depth is found to be as common as an increase with depth. In most cases, however, the vertical structure of density in the ocean is determined mainly by that of temperature, the salinity effects being secondary. The upper 50–200 m of ocean is well mixed, due to the turbulence generated by the wind, waves, current shear, and the convective overturning caused by surface cooling. Temperature gradients decrease with depth, becoming quite small below a depth of 1500 m. There is usually a large temperature gradient in the depth range of 100–500 m. This layer of high stability is called the *thermocline*. Figure 13.2 also shows the profile of *buoyancy frequency* N, defined by

$$N^2 \equiv -\frac{g}{\rho_0}\frac{d\rho}{dz},$$

where ρ of course stands for the potential density and ρ_0 is a constant reference density (*cf.* (1.29) and (7.128)). The buoyancy frequency reaches a typical maximum value of $N_{max} \sim 0.01\mathrm{s}^{-1}$ (period $\sim$ 10 min) in the thermocline and decreases both upward and downward.

13.3. EQUATIONS OF MOTION

In this section we shall review the relevant equations of motion, which are derived and discussed in Chapter 4. The equations of motion for a stratified medium, observed in a system of coordinates rotating at an angular velocity Ω with respect to the "fixed stars," are:

$$\nabla \cdot \mathbf{u} = 0, \quad \frac{D\mathbf{u}}{Dt} + 2\Omega \times \mathbf{u} = \frac{1}{\rho_0}\nabla p - \frac{g\rho}{\rho_0}\mathbf{e}_z + \mathbf{F}, \quad \text{and} \quad \frac{D\rho}{Dt} = 0, \tag{13.2}$$

where $\mathbf{F}$ is the friction force per unit mass. The diffusive effects in the density equation are omitted in set (13.2) because they are not considered here.

Set (13.2) makes the so-called *Boussinesq approximation*, discussed in Section 4.18, in which the density variations are neglected everywhere except in the gravity term. Along with other restrictions, it assumes that the vertical scale of the motion is less than the "scale height" of the medium c^2/g, where c is the speed of sound. This assumption is very good in the ocean, in which $c^2/g \sim 200$ km. In the atmosphere it is less applicable, because $c^2/g \sim 10$ km. Under the Boussinesq approximation, the principle of mass conservation is expressed by $\nabla \cdot \mathbf{u} = 0$. In contrast, the density equation $D\rho/Dt = 0$ follows from the nondiffusive heat equation $DT/Dt = 0$ and an incompressible equation of state of the form $\delta\rho/\rho_0 = -\alpha\delta T$. (If the density is determined by the concentration S of a constituent, say the water vapor in the atmosphere or the salinity in the ocean, then $D\rho/Dt = 0$ follows from the nondiffusive conservation equation for the constituent in the form $DS/Dt = 0$, plus the incompressible equation of state $\delta\rho/\rho_0 = \beta\delta S$.)

The equations can be written in terms of the pressure and density *perturbations* from a state of rest. In the absence of any motion, suppose the density and pressure have the vertical distributions $\overline{\rho}(z)$ and $\overline{p}(z)$, where the z-axis is taken vertically upward. As this state is hydrostatic, we must have

$$\frac{d\overline{p}}{dz} = -\overline{\rho}g. \tag{13.3}$$

In the presence of a flow field $\mathbf{u}(\mathbf{x},t)$, we can write the density and pressure as

$$\begin{aligned} \rho(\mathbf{x},t) &= \overline{\rho}(z) + \rho'(\mathbf{x},t), \\ p(\mathbf{x},t) &= \overline{p}(z) + p'(\mathbf{x},t), \end{aligned} \tag{13.4}$$

where ρ' and p' are the changes from the state of rest. With this substitution, the first two terms on the right-hand side of the momentum equation in (13.2) give

$$-\frac{1}{\rho_0}\nabla p - \frac{g\rho}{\rho_0}\mathbf{e}_z = -\frac{1}{\rho_0}\nabla(\overline{p} + p') - \frac{g}{\rho_0}(\overline{\rho} + \rho')\mathbf{e}_z = -\frac{1}{\rho_0}\left[\frac{d\overline{p}}{dz}\mathbf{e}_z + \nabla p'\right] - \frac{g}{\rho_0}(\overline{\rho} + \rho')\mathbf{e}_z.$$

Subtracting the hydrostatic state (13.3), this becomes

$$-\frac{1}{\rho_0}\nabla p - \frac{g\rho}{\rho_0}\mathbf{e}_z = -\frac{1}{\rho_0}\nabla p' - \frac{g\rho'}{\rho_0}\mathbf{e}_z,$$

which shows that we can replace p and ρ in (13.2) by the perturbation quantities p' and ρ'.

The friction force per unit mass $\mathbf{F}$ in equation (13.2) needs to be related to the velocity field. From Section 4.4, the friction force is given by $F_i = \partial\tau_{ij}/\partial x_j$, where τ_{ij} is the viscous stress tensor. The stress in a laminar flow is caused by the molecular exchanges of momentum. From (4.35), the viscous stress tensor in an isotropic incompressible medium in laminar flow is given by $\mu(\partial u_i/\partial x_j + \partial u_j/\partial x_i)$. In large-scale geophysical flows, however, the frictional forces are provided by turbulent momentum exchange and viscous effects are negligible. The complexity of turbulent behavior makes it impossible to relate the stress to the velocity field in a simple way. To proceed, then, we adopt the eddy viscosity hypothesis (12.94), which sets the turbulent stress proportional to the velocity gradient field.

Geophysical media are in the form of shallow stratified layers, in which the vertical velocities are much smaller than horizontal velocities. This means that the exchange of momentum across a horizontal surface is much weaker than that across a vertical surface. We expect then that the vertical eddy viscosity ν_v is much smaller than the horizontal eddy viscosity ν_H, and we assume that the turbulent stress components have the form:

$$\tau_{xz} = \tau_{zx} = \rho\nu_v\frac{\partial u}{\partial z} + \rho\nu_H\frac{\partial w}{\partial x},$$

$$\tau_{yz} = \tau_{zy} = \rho\nu_v\frac{\partial v}{\partial z} + \rho\nu_H\frac{\partial w}{\partial y},$$

$$\tau_{xy} = \tau_{yx} = \rho\nu_H\left(\frac{\partial u}{\partial y} + \frac{\partial v}{\partial x}\right),$$

$$\tau_{xx} = 2\rho\nu_H\frac{\partial u}{\partial x}, \quad \tau_{yy} = 2\rho\nu_H\frac{\partial v}{\partial y}, \quad \tau_{zz} = 2\rho\nu_v\frac{\partial w}{\partial z}.$$

(13.5)

The difficulty with set (13.5) is that the expressions for τ_{xz} and τ_{yz} depend on the fluid *rotation* in the vertical plane and not just the deformation. In Section 4.5, we saw that a requirement for a constitutive equation is that the stresses should be independent of fluid rotation and should depend only on the deformation. Therefore, τ_{xz} should depend only on the combination $(\partial u/\partial z + \partial w/\partial x)$, whereas the expression in (13.5) depends on both deformation and rotation. A tensorially correct geophysical treatment of the frictional terms is discussed, for example, in Kamenkovich (1967). However, the assumed form (13.5) leads to a simple formulation for viscous effects, as we shall see shortly. As the eddy viscosity assumption is of questionable validity (which Pedlosky [1971] describes as a "rather disreputable and desperate attempt"), there does not seem to be any purpose in formulating the stress-strain relation in more complicated ways merely to obey the requirement of invariance with respect to rotation.

With the assumed form for the turbulent stress, the components of the frictional force $F_i = \partial\tau_{ij}/\partial x_j$ become:

$$F_x = \frac{\partial\tau_{xx}}{\partial x} + \frac{\partial\tau_{xy}}{\partial y} + \frac{\partial\tau_{xz}}{\partial z} = \nu_H\left(\frac{\partial^2 u}{\partial x^2} + \frac{\partial^2 u}{\partial y^2}\right) + \nu_v\frac{\partial^2 u}{\partial z^2},$$

$$F_y = \frac{\partial\tau_{yx}}{\partial x} + \frac{\partial\tau_{yy}}{\partial y} + \frac{\partial\tau_{yz}}{\partial z} = \nu_H\left(\frac{\partial^2 v}{\partial x^2} + \frac{\partial^2 v}{\partial y^2}\right) + \nu_v\frac{\partial^2 v}{\partial z^2},$$

(13.6)

$$F_z = \frac{\partial\tau_{zx}}{\partial x} + \frac{\partial\tau_{zy}}{\partial y} + \frac{\partial\tau_{zz}}{\partial z} = \nu_H\left(\frac{\partial^2 w}{\partial x^2} + \frac{\partial^2 w}{\partial y^2}\right) + \nu_v\frac{\partial^2 w}{\partial z^2}.$$

Estimates of the eddy coefficients vary greatly. Typical suggested values are $\nu_v \sim 10 \text{ m}^2/\text{s}$ and $\nu_H \sim 10^5 \text{ m}^2/\text{s}$ for the lower atmosphere, and $\nu_v \sim 0.01 \text{ m}^2/\text{s}$ and $\nu_H \sim 100 \text{ m}^2/\text{s}$ for the upper ocean. In comparison, the molecular values are $\nu = 1.5 \times 10^{-5} \text{ m}^2/\text{s}$ for air and $\nu = 10^{-6} \text{ m}^2/\text{s}$ for water.

13.4. APPROXIMATE EQUATIONS FOR A THIN LAYER ON A ROTATING SPHERE

The atmosphere and the ocean are very thin layers in which the depth scale of flow is a few kilometers, whereas the horizontal scale is of the order of hundreds, or even thousands, of kilometers. The trajectories of fluid elements are nearly horizontal while vertical velocities are much smaller than horizontal velocities. In fact, the continuity equation suggests that the scale of the vertical velocity W is related to that of the horizontal velocity U by

$$\frac{W}{U} \sim \frac{H}{L},$$

where H is the depth scale and L is the horizontal length scale. Stratification and Coriolis effects usually constrain the vertical velocity to be even smaller than UH/L.

Large-scale geophysical flow problems should be solved using spherical polar coordinates. If, however, the horizontal length scales are much smaller than the radius of the earth ($= 6371$ km), then the curvature of the earth can be ignored, and the motion can be studied by adopting a *local* Cartesian system on a *tangent plane* (Figure 13.3). On this plane we take an xyz

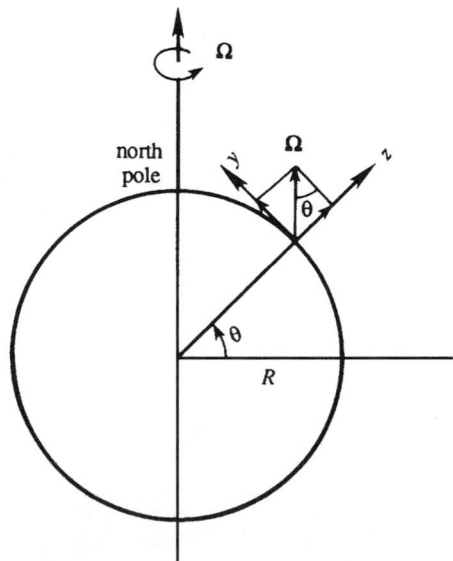

FIGURE 13.3 Local Cartesian coordinates. The x-axis points into the plane of the paper. The y-axis is tangent to the earth's surface and points toward the north pole. The z-axis is vertical, opposing gravity. The earth's angular rotation vector has positive y and z components in the northern hemisphere.

coordinate system, with x increasing eastward, y northward, and z upward. The corresponding velocity components are u (eastward), v (northward), and w (upward).

The earth rotates at a rate,

$$\Omega = 2\pi \, rad/day = 0.73 \times 10^{-4} \, s^{-1},$$

around the polar axis, in a counterclockwise sense looking from above the north pole. From Figure 13.3, the components of angular velocity of the earth in the local Cartesian system are $\Omega = (\Omega_x, \Omega_y, \Omega_z) = (0, \Omega\cos\theta, \Omega\sin\theta)$, where θ is the latitude. The Coriolis acceleration is therefore:

$$2\Omega \times \mathbf{u} = \begin{vmatrix} \mathbf{e}_x & \mathbf{e}_y & \mathbf{e}_z \\ 0 & 2\Omega\cos\theta & 2\Omega\sin\theta \\ u & v & w \end{vmatrix} = 2\Omega\left[\mathbf{e}_x\left(w\cos\theta - v\sin\theta\right) + \mathbf{e}_y u\sin\theta - \mathbf{e}_z u\cos\theta\right].$$

In the term multiplied by $\mathbf{e}_x$ we can use the condition $w\cos\theta \ll v\sin\theta$, because the thin-sheet approximation requires that $w \ll v$. The three components of the Coriolis acceleration are therefore

$$2\Omega \times \mathbf{u} \cong (-2\Omega v \sin\theta, \, 2\Omega u \sin\theta, \, -2\Omega u \cos\theta) = (-fv, fu, -2\Omega u \cos\theta), \tag{13.7}$$

where we have defined

$$f = 2\Omega \sin\theta \tag{13.8}$$

to be twice the *vertical* component of Ω. As vorticity is twice the angular velocity, f is called the *planetary vorticity*. More commonly, f is referred to as the *Coriolis parameter*, or the *Coriolis frequency*. It is positive in the northern hemisphere and negative in the southern hemisphere, varying from $\pm 1.45 \times 10^{-4} \, s^{-1}$ at the poles to zero at the equator. This makes sense, since a person standing at the north pole spins around himself in a counterclockwise sense at a rate Ω, whereas a person standing at the equator does not spin around himself but simply translates. The quantity,

$$T_i = 2\pi/f,$$

is called the *inertial period*, for reasons that will be clear in Section 13.11; it does refer to a component of the vector.

The vertical component of the Coriolis force, namely $-2\Omega u \cos\theta$, is generally negligible compared to the dominant terms in the vertical equation of motion, namely $g\rho'/\rho_0$ and $\rho_0^{-1}(\partial p'/\partial z)$. Using (13.6), (13.7), and (13.8), the equations of motion (13.2) reduce to:

$$\frac{Du}{Dt} - fv = -\frac{1}{\rho_0}\frac{\partial p}{\partial x} + \nu_H\left(\frac{\partial^2 u}{\partial x^2} + \frac{\partial^2 u}{\partial y^2}\right) + \nu_v\frac{\partial^2 u}{\partial z^2},$$

$$\frac{Dv}{Dt} + fu = -\frac{1}{\rho_0}\frac{\partial p}{\partial y} + \nu_H\left(\frac{\partial^2 v}{\partial x^2} + \frac{\partial^2 v}{\partial y^2}\right) + \nu_v\frac{\partial^2 v}{\partial z^2}, \tag{13.9}$$

$$\frac{Dw}{Dt} = -\frac{1}{\rho_0}\frac{\partial p}{\partial z} - \frac{g\rho}{\rho_0} + \nu_H\left(\frac{\partial^2 w}{\partial x^2} + \frac{\partial^2 w}{\partial y^2}\right) + \nu_v\frac{\partial^2 w}{\partial z^2}.$$

These are the equations of motion for a thin shell on a rotating earth. Note that only the *vertical* component of the earth's angular velocity appears as a consequence of the flatness of the fluid trajectories.

f-Plane Model

The Coriolis parameter $f = 2\Omega \sin \theta$ varies with latitude θ. However, this variation is important only for phenomena having very long time scales (several weeks) or very long length scales (thousands of kilometers). For many purposes we can assume f to be a constant, say $f_0 = 2\Omega \sin \theta_0$, where θ_0 is the central latitude of the region under study. A model using a constant Coriolis parameter is called an *f-plane model*.

β-Plane Model

The variation of f with latitude can be approximately represented by expanding f in a Taylor series about the central latitude θ_0:

$$f = f_0 + \beta y, \tag{13.10}$$

where

$$\beta \equiv \left(\frac{df}{dy}\right)_{\theta_0} = \left(\frac{df}{d\theta}\frac{d\theta}{dy}\right)_{\theta_0} = \frac{2\Omega \cos \theta_0}{R}.$$

Here, we have used $f = 2\Omega \sin \theta$ and $d\theta/dy = 1/R$, where the radius of the earth is nearly $R = 6371$ km. A model that takes into account the variation of the Coriolis parameter in the simplified form $f = f_0 + \beta y$, with β as constant, is called a *β-plane model*.

13.5. GEOSTROPHIC FLOW

Consider quasi-steady, large-scale motions in the atmosphere or the ocean, away from boundaries. For these flows an excellent approximation for the horizontal equilibrium is a balance between the Coriolis force and the pressure gradient:

$$-fv = -\frac{1}{\rho_0}\frac{\partial p}{\partial x}, \quad \text{and} \quad fu = -\frac{1}{\rho_0}\frac{\partial p}{\partial y}. \tag{13.11, 13.12}$$

Here we have neglected the nonlinear acceleration terms, which are of order U^2/L, in comparison to the Coriolis force $\sim fU$ (U is the horizontal velocity scale, and L is the horizontal length scale). The ratio of the nonlinear term to the Coriolis term is called the *Rossby number*:

$$\text{Rossby number} = \frac{\text{Nonlinear acceleration}}{\text{Coriolis force}} \sim \frac{U^2/L}{fU} = \frac{U}{fL} = Ro. \tag{13.13}$$

For a typical atmospheric value of $U \sim 10$ m/s, $f \sim 10^{-4}$ s^{-1}, and $L \sim 1000$ km, the Rossby number turns out to be 0.1. The Rossby number is even smaller for many flows in the ocean, so that the neglect of nonlinear terms is justified for many flows.

The balance of forces represented by (13.11), in which the horizontal pressure gradients are balanced by Coriolis forces, is called a *geostrophic balance*. In such a system the velocity distribution can be determined from a measured distribution of the pressure field. The geostrophic equilibrium breaks down near the equator (within a latitude belt of $\pm 3°$), where f becomes small. It also breaks down if the frictional effects or unsteadiness become important.

Velocities in a geostrophic flow are perpendicular to the horizontal pressure gradient. This is because (13.11) implies that $\mathbf{u} \cdot \nabla p = 0$, that is:

$$(u\mathbf{e}_x + v\mathbf{e}_y) \cdot \nabla p = \frac{1}{\rho_0 f}\left(-\mathbf{e}_x \frac{\partial p}{\partial y} + \mathbf{e}_y \frac{\partial p}{\partial x}\right) \cdot \left(\mathbf{e}_x \frac{\partial p}{\partial x} + \mathbf{e}_y \frac{\partial p}{\partial y}\right) = 0.$$

Thus, the horizontal velocity is *along*, and not across, the lines of constant pressure. If f is regarded as constant, then the geostrophic balance (13.11) shows that $p/f\rho_0$ can be regarded as a stream function. The isobars on a weather map are therefore nearly the streamlines of the flow.

Figure 13.4 shows the geostrophic flow around low- and high-pressure centers in the northern hemisphere. Here the Coriolis force acts to the right of the velocity vector. This requires the flow to be counterclockwise (viewed from above) around a low-pressure region and clockwise around a high-pressure region. The sense of circulation is opposite in the southern hemisphere, where the Coriolis force acts to the left of the velocity vector. (Frictional forces become important at lower levels in the atmosphere and result in a flow partially *across* the isobars. This will be discussed in Section 13.7, where it is shown that flow around a low-pressure center spirals *inward* due to frictional effects.)

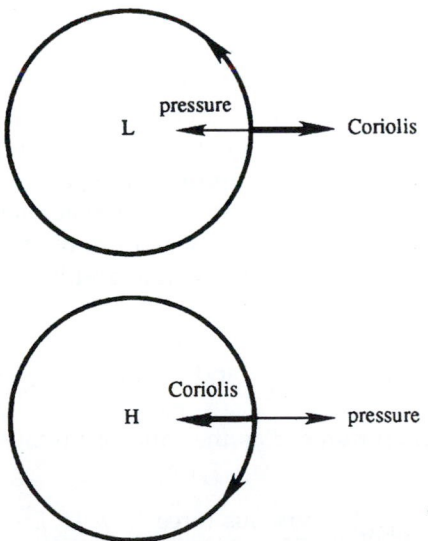

FIGURE 13.4 Circular geostrophic flow around ideal low- and high-pressure centers in the northern hemisphere. The pressure force $(-\nabla p)$ is indicated by a thin arrow, and the Coriolis force is indicated by a thick arrow.

The flow along isobars at first surprises a reader unfamiliar with the effects of the Coriolis force. A question commonly asked is: How is such a motion set up? A typical manner of establishment of such a flow is as follows. Consider a horizontally converging flow in the surface layer of the ocean. The convergent flow sets up the sea surface in the form of a gentle hill, with the sea surface dropping away from the center of the hill. A fluid particle starting to move down the hill is deflected to the right in the northern hemisphere, and a steady state is reached when the particle finally moves *along* the isobars.

Thermal Wind

In the presence of a *horizontal* gradient of density, the geostrophic velocity develops a *vertical* shear. This is easy to demonstrate from an analysis of the geostrophic and hydrostatic balance using (13.11), (13.12), and

$$0 = -\frac{\partial p}{\partial z} - g\rho. \tag{13.14}$$

Eliminating p between (13.11) and (13.14), and also between equations (13.12) and (13.14), we obtain, respectively,

$$\frac{\partial v}{\partial z} = -\frac{g}{\rho_0 f}\frac{\partial \rho}{\partial x}, \quad \text{and} \quad \frac{\partial u}{\partial z} = \frac{g}{\rho_0 f}\frac{\partial \rho}{\partial y}. \tag{13.15}$$

Meteorologists call these the *thermal wind* equations because they give the vertical variation of wind from measurements of horizontal temperature (and pressure) gradients. The thermal wind is a baroclinic phenomenon, because the surfaces of constant p and ρ do not coincide.

Taylor-Proudman Theorem

A striking phenomenon occurs in the geostrophic flow of a *homogeneous* fluid. It can only be observed in a laboratory experiment because stratification effects cannot be avoided in natural flows. Consider then a laboratory experiment in which a tank of fluid is steadily rotated at a high angular speed Ω and a solid body is moved slowly along the bottom of the tank. The purpose of making Ω large and the movement of the solid body slow is to make the Coriolis force much larger than the advective acceleration terms, which must be made negligible for geostrophic equilibrium. Away from the frictional effects of boundaries, the balance is therefore geostrophic in the horizontal and hydrostatic in the vertical. Setting $f = 2\Omega$ in (13.11) and (13.12) produces

$$-2\Omega v = -\frac{1}{\rho}\frac{\partial p}{\partial x} \quad \text{and} \quad -2\Omega u = -\frac{1}{\rho}\frac{\partial p}{\partial y}. \tag{13.16, 13.17}$$

It is useful to define an Ekman number as the ratio of viscous to Coriolis forces (per unit volume):

$$\text{Ekman number} = \frac{\text{viscous force}}{\text{Coriolis force}} = \frac{\rho \nu U/L^2}{\rho f U} = \frac{\nu}{fL^2} = E. \tag{13.18}$$

Under the circumstances already described here, both Ro and E are small.

Elimination of p by cross differentiation of the horizontal momentum equations (13.16) and (13.17) gives

$$2\Omega\left(\frac{\partial v}{\partial y}+\frac{\partial u}{\partial x}\right)=0.$$

Using the continuity equation, this gives

$$\frac{\partial w}{\partial z}=0. \qquad (13.19)$$

Also, differentiating (13.16) and (13.17) with respect to z, and using (13.14), we obtain

$$\frac{\partial v}{\partial z}=\frac{\partial u}{\partial z}=0. \qquad (13.20)$$

Taken together, (13.19) and (13.20) imply

$$\partial\mathbf{u}/\partial z=0. \qquad (13.21)$$

Thus, the fluid velocity cannot vary in the direction of Ω. In other words, steady slow motions in a rotating, homogeneous, inviscid fluid are two dimensional. This is the *Taylor-Proudman theorem*, first derived by Proudman in 1916 and demonstrated experimentally by Taylor soon afterward.

In Taylor's experiment, a tank was made to rotate as a solid body, and a small cylinder was slowly dragged along the bottom of the tank (Figure 13.5). Dye was introduced from point A above the cylinder and directly ahead of it. In a nonrotating fluid the water would pass over the top of the moving cylinder. In the rotating experiment, however, the dye divides at a point S, as if it had been blocked by a vertical extension of the cylinder, and flows around this *imaginary* cylinder, called the *Taylor column*. Dye released from a point B within the Taylor column remained there and moved with the cylinder. The conclusion was that the flow outside the upward extension of the cylinder is the same as if the cylinder extended across the entire water depth and that a column of water directly above the cylinder moves with it. The motion is two dimensional, although the solid body does not extend across the entire water depth. Taylor did a second experiment, in which he dragged a solid body *parallel* to the axis of rotation. In accordance with $\partial w/\partial z=0$, he observed that a column of fluid is pushed ahead. The lateral velocity components u and v were zero. In both of these experiments, there are shear layers at the edge of the Taylor column.

In summary, Taylor's experiment established the following striking fact for steady inviscid motion of homogeneous fluid in a strongly rotating system: Bodies moving either parallel or perpendicular to the axis of rotation carry along with their motion a so-called Taylor column of fluid, oriented parallel to the axis of rotation. The phenomenon is analogous to the horizontal *blocking* caused by a solid body (say a mountain) in a strongly stratified system, shown in Figure 7.30.

13.6. EKMAN LAYER AT A FREE SURFACE

In the preceding section, we discussed a steady linear inviscid motion expected to be valid away from frictional boundary layers. We shall now examine the motion within frictional

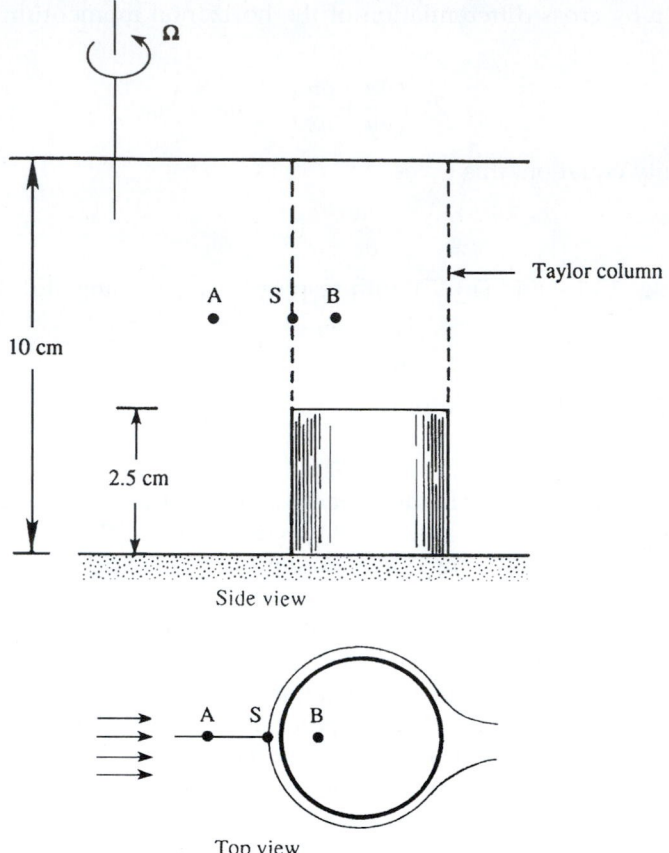

FIGURE 13.5   Taylor's experiment in a strongly rotating flow of a homogeneous fluid. When the short cylinder is moved toward the axis of rotation, an extension of the cylinder forms in the fluid above it. Dye released above the cylinder at point A flows around the extension of cylinder as if it were a solid object. Dye released above the cylinder at point B follows the motion of the short cylinder.

layers over horizontal surfaces. In viscous flows unaffected by Coriolis forces and pressure gradients, the only term which can balance the viscous force is either the time derivative $\partial \mathbf{u}/\partial t$ or the advection $\mathbf{u} \cdot \nabla \mathbf{u}$. The balance of $\partial \mathbf{u}/\partial t$ and the viscous force gives rise to a viscous layer whose thickness increases with time, as in the suddenly accelerated plate discussed in Section 8.7. The balance of $\mathbf{u} \cdot \nabla \mathbf{u}$ and the viscous force give rise to a viscous layer whose thickness increases in the direction of flow, as in the boundary layer over a semi-infinite plate discussed in Sections 9.3 through 9.5. In a rotating flow, however, we can have a balance between the Coriolis and the viscous forces, and the thickness of the viscous layer can be invariant in time and space. Two examples of such layers are given in this and the following sections.

Consider first the case of a frictional layer near the free surface of the ocean, which is acted on by a wind stress τ in the x-direction. We shall not consider how the flow adjusts to the steady state but examine only the steady solution. We shall assume that the horizontal pressure gradients are zero and that the field is horizontally homogeneous. From (13.9), the horizontal equations of motion for flow within the ocean are:

$$-fv = \nu_v \frac{d^2u}{dz^2} \quad \text{and} \quad fu = \nu_v \frac{d^2v}{dz^2}. \qquad (13.22, 13.23)$$

Defining $z = 0$ on the surface of the ocean, the boundary conditions are:

$$\rho\nu_v(du/dz) = \tau \text{ at } z = 0, \, dv/dz = 0 \text{ at } z = 0, \text{ and } u, v \to 0 \text{ as } z \to \infty. \quad (13.24, 13.25, 13.26)$$

Multiplying equation (13.23) by the imaginary root, $i = \sqrt{-1}$, and adding equation (13.22), we obtain

$$\frac{d^2V}{dz^2} = \frac{if}{\nu_v}V, \qquad (13.27)$$

where we have defined the *complex velocity*

$$V \equiv u + iv.$$

The solution of (13.27) is

$$V = A\,e^{(1+i)z/\delta} + B\,e^{-(1+i)z/\delta}, \text{ where } \delta = \sqrt{2\nu_v/f}. \qquad (13.28, 13.29)$$

The constant B is zero because the field must remain finite as $z \to -\infty$, and δ is the thickness of the *Ekman layer*. The surface boundary conditions (13.24) and (13.25) can be combined as $\rho\nu_v(dV/dz) = \tau$ at $z = 0$, from which (13.28) gives

$$A = \frac{\tau\delta(1-i)}{2\rho\nu_v}.$$

Substitution of this into (13.28) gives the velocity components:

$$u = \frac{\tau/\rho}{\sqrt{f\nu_v}}e^{z/\delta}\cos\left(-\frac{z}{\delta} + \frac{\pi}{4}\right),$$

$$v = -\frac{\tau/\rho}{\sqrt{f\nu_v}}e^{z/\delta}\sin\left(-\frac{z}{\delta} + \frac{\pi}{4}\right).$$

The Swedish oceanographer Ekman worked out this solution in 1905. The solution is shown in Figure 13.6 for the case of the northern hemisphere, in which f is positive. The velocities at various depths within the ocean are plotted in Figure 13.6a where each arrow represents the velocity vector at a certain depth. Such a plot of v versus u is sometimes called a *hodograph*. The vertical distributions of u and v are shown in Figure 13.6b. The hodograph shows that the surface velocity is deflected 45° to the right of the applied wind stress. (In the southern hemisphere the deflection is to the left of the surface stress.) The velocity vector rotates clockwise (looking down) with depth, and the magnitude exponentially decays with an e-folding scale of δ, the Ekman layer thickness. The tips of the velocity vector at various depths form a spiral, called the *Ekman spiral*.

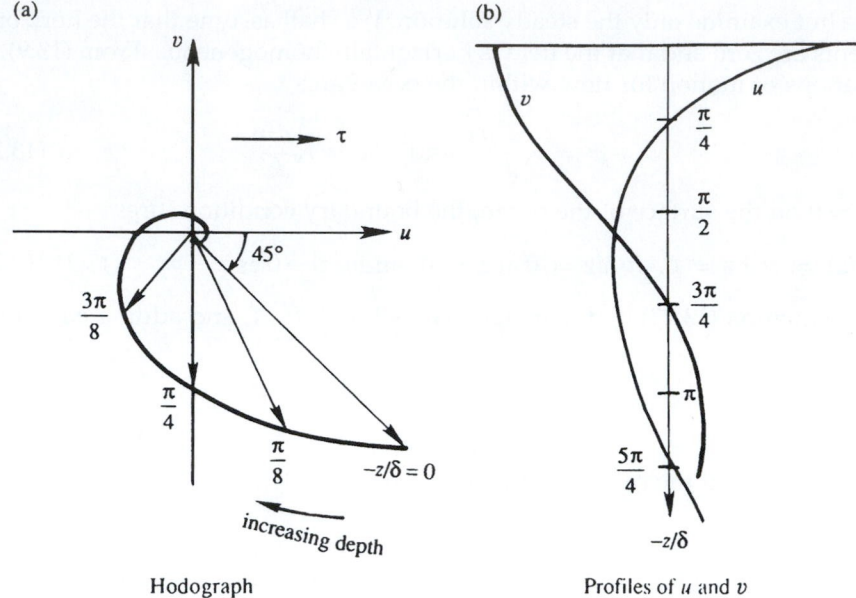

FIGURE 13.6 Ekman layer below a water surface on which a shear stress τ is applied in the x-direction. The left panel (a) shows the horizontal fluid velocity components (u, v) at various depths; values of $-z/\delta$ are indicated along the curve traced out by the tip of the velocity vector. The flow speed is highest near the surface. The right panel (b) shows vertical distributions of u and v. Here, the Coriolis force produces significant depth dependence in the fluid velocity even though τ is constant and unidirectional.

The components of the volume transport in the Ekman layer are:

$$\int_{-\infty}^{0} u \, dz = 0,$$
$$\int_{-\infty}^{0} v \, dz = -\frac{\tau}{\rho f}. \tag{13.30}$$

This shows that the *net transport is to the right of the applied stress and is independent of ν_v*. In fact, the result $\int v \, dz = -\tau/f\rho$ follows directly from a vertical integration of the equation of motion in the form $-\rho f v = d\tau/dz$ so that the result does not depend on the eddy viscosity assumption. The fact that the transport is to the right of the applied stress makes sense because then the net (depth-integrated) Coriolis force, directed to the right of the depth-integrated transport, can balance the wind stress.

The horizontal uniformity assumed in the solution is not a serious limitation. Since Ekman layers near the ocean surface have a thickness (~50 m) much smaller than the scale of horizontal variation ($L > 100$ km), the solution is still locally applicable. The assumed absence of a horizontal pressure gradient can also be reconsidered. Because of the thinness of the layer, any imposed horizontal pressure gradient remains constant across the layer. The presence of a horizontal pressure gradient merely adds a depth-independent geostrophic velocity to the Ekman solution. Suppose the sea surface slopes

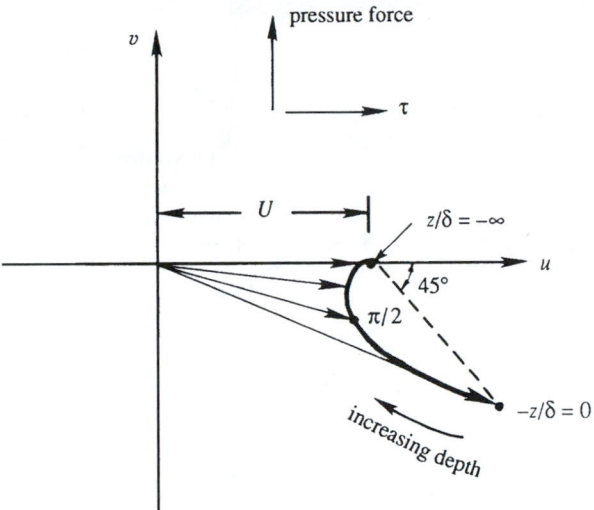

FIGURE 13.7 Ekman layer at a free surface in the presence of a pressure gradient. The geostrophic velocity forced by the pressure gradient is U. The flow profile in this case is the sum of U and the profile shown in Figure 13.6.

down to the north, so that there is a pressure force acting northward throughout the Ekman layer and below (Figure 13.7). This means that at the bottom of the Ekman layer $(z/\delta \rightarrow -\infty)$ there is a geostrophic velocity U to the right of the pressure force. The surface Ekman spiral forced by the wind stress joins smoothly to this geostrophic velocity as $z/\delta \rightarrow -\infty$.

Pure Ekman spirals are not observed in the surface layer of the ocean, mainly because the assumptions of constant eddy viscosity and steadiness are particularly restrictive. When the flow is averaged over a few days, however, several instances have been found in which the current does look like a spiral. One such example is shown in Figure 13.8.

Explanation in Terms of Vortex Tilting

We have seen in previous chapters that the thickness of a viscous layer usually grows in a nonrotating flow, either in time or in the direction of flow. The Ekman solution, in contrast, results in a viscous layer that does not grow either in time or space. This can be explained by examining the vorticity equation (Pedlosky, 1987). The vorticity components in the x- and y-directions are:

$$\omega_x = \frac{\partial w}{\partial y} - \frac{\partial v}{\partial z} = -\frac{dv}{dz},$$

$$\omega_y = \frac{\partial u}{\partial z} - \frac{\partial w}{\partial x} = \frac{du}{dz},$$

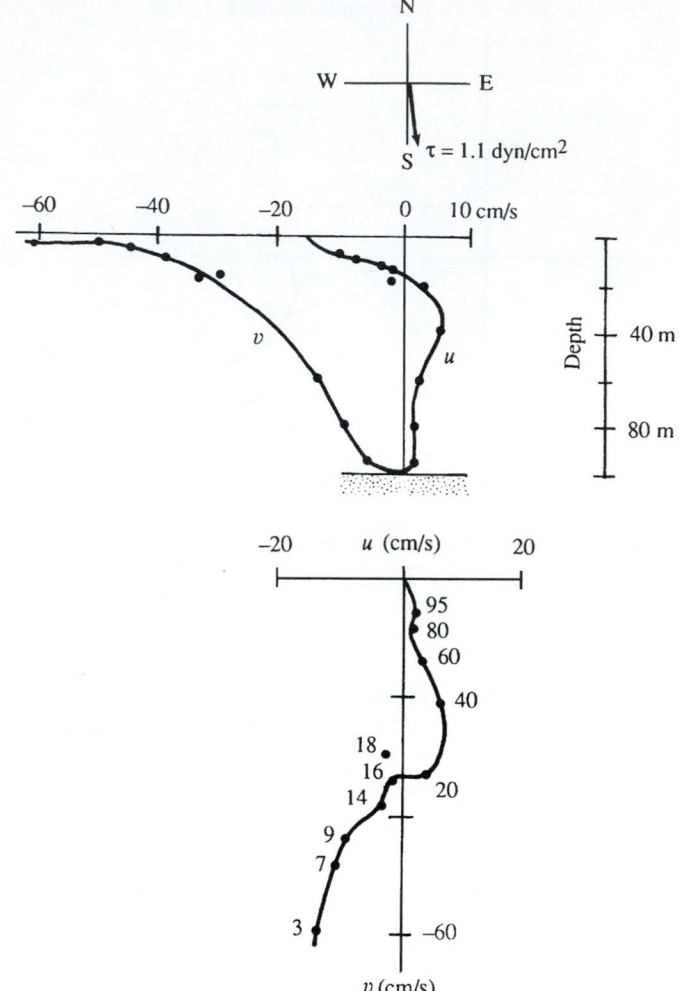

FIGURE 13.8　An observed velocity distribution near the coast of Oregon. Velocity is averaged over 7 days. Wind stress had a magnitude of 1.1 dyn/cm² and was directed nearly southward, as indicated at the top of the figure. The upper panel shows vertical distributions of u and v, and the lower panel shows the hodograph in which depths are indicated in meters. The hodograph is similar to that of a surface Ekman layer (of depth 16 m) lying over the bottom Ekman layer (extending from a depth of 16 m to the ocean bottom). *P. Kundu, in* Bottom Turbulence, *J. C. J. Nihoul, ed., Elsevier, 1977; reprinted with the permission of Jacques C. J. Nihoul.*

where we have used $w = 0$. Using these, the z-derivative of the equations of motion (13.22) and (13.23) gives:

$$-f\frac{dv}{dz} = \nu_v \frac{d^2\omega_y}{dz^2},$$

$$-f\frac{du}{dz} = \nu_v \frac{d^2\omega_x}{dz^2}.$$

(13.31)

The right sides of these equations represent diffusion of vorticity. Without Coriolis forces this diffusion would cause a thickening of the viscous layer. The presence of planetary rotation, however, means that vertical fluid lines coincide with the planetary vortex lines. The tilting of vertical fluid lines, represented by terms on the left-hand sides of equations (13.31), then causes a rate of change of the horizontal component of vorticity that just cancels the diffusion term.

13.7. EKMAN LAYER ON A RIGID SURFACE

Consider a steady viscous layer on a solid surface in a rotating flow that is independent of the horizontal coordinates x and y. This can be the atmospheric boundary layer over the solid earth or the boundary layer over the ocean bottom. We assume that at large distances from the surface the velocity is toward the x-direction and has a magnitude U. Viscous forces are negligible far from the wall, so that the Coriolis force can be balanced only by a pressure gradient:

$$fU = -\frac{1}{\rho}\frac{dp}{dy}. \tag{13.32}$$

This simply states that the flow outside the viscous layer is in geostrophic balance, U being the geostrophic velocity. For our assumed case of positive U and f, we must have $dp/dy < 0$, so that the pressure falls with y—that is, the pressure force is directed along the positive y direction, resulting in a geostrophic flow U to the right of the pressure force in the northern hemisphere. The horizontal pressure gradient remains constant within the thin boundary layer.

Near the solid surface friction forces are important, so that the balance within the boundary layer is

$$-fv = \nu_v\frac{d^2u}{dz^2} \quad \text{and} \quad fu = \nu_v\frac{d^2v}{dz^2} + fU, \tag{13.33, 13.34}$$

where we have replaced $-\rho^{-1}(dp/dy)$ by fU in accordance with (13.32). The boundary conditions are:

$$u = U, \quad v = 0 \quad \text{as } z \to \infty, \tag{13.35}$$

$$u = 0, \quad v = 0 \quad \text{at } z = 0, \tag{13.36}$$

where z is taken vertically upward from the solid surface. Multiplying equation (13.34) by i and adding equation (13.33), the equations of motion become

$$\frac{d^2V}{dz^2} = \frac{if}{\nu_v}(V - U), \tag{13.37}$$

where we have again used the complex velocity $V \equiv u + iv$. The boundary conditions (13.35) and (13.36) in terms of the complex velocity are:

$$V = U \quad \text{as } z \to \infty, \tag{13.38}$$

$$V = 0 \quad \text{at } z = 0. \tag{13.39}$$

The particular solution of (13.37) is $V = U$. The total solution is, therefore,

$$V = A\,e^{-(1+i)z/\delta} + B\,e^{(1+i)z/\delta} + U, \tag{13.40}$$

where $\delta \equiv \sqrt{2\nu_v/f}$, as before. To satisfy (13.38), we must have $B = 0$. Condition (13.39) gives $A = -U$. The velocity components then become:

$$u = U\big[1 - e^{-z/\delta}\cos{(z/\delta)}\big],$$
$$v = U e^{-z/\delta}\sin{(z/\delta)}. \tag{13.41}$$

According to (13.41), the tip of the velocity vector describes a spiral for various values of z (Figure 13.9a). As with the Ekman layer at a free surface, the frictional effects are confined within a layer of thickness $\delta = \sqrt{2\nu_v/f}$, which increases with ν_v and decreases with the rotation rate f. Interestingly, the layer thickness is independent of the magnitude of the free-stream velocity U; this behavior is quite different from that of a steady nonrotating boundary layer on a semi-infinite plate (the Blasius solution of Section 9.3) in which the thickness is proportional to $1/\sqrt{U}$.

Figure 13.9b shows the vertical distribution of the velocity components. Far from the wall the velocity is entirely in the x-direction, and the Coriolis force balances the pressure gradient. As the wall is approached, frictional effects decrease u and the associated Coriolis force, so that the pressure gradient (which is independent of z) forces a component v in the direction of the pressure force. Using (13.41), the net transport in the Ekman layer normal to the uniform stream outside the layer is

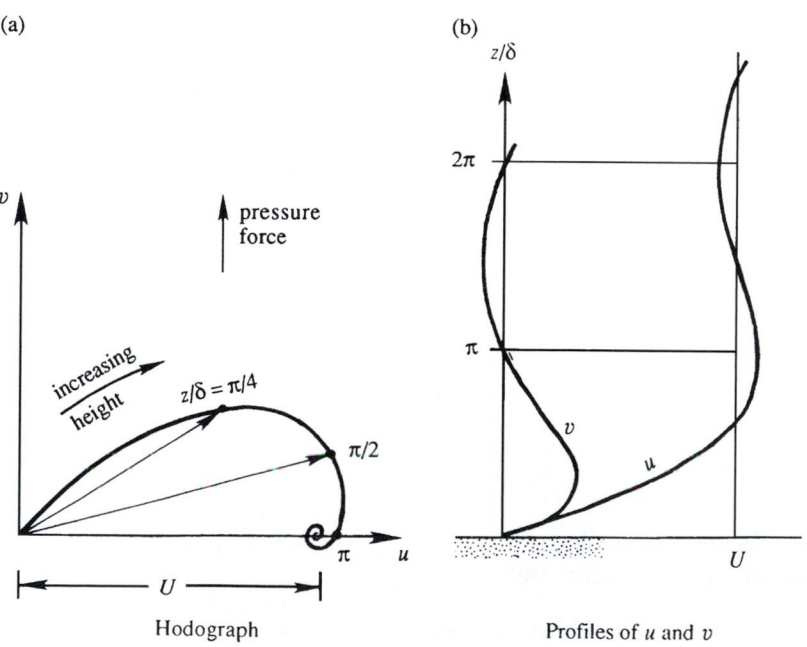

(a)

(b)

FIGURE 13.9 Ekman layer above a rigid surface for a steady outer-flow velocity of U (parallel to the x-axis). The left panel shows velocity vectors at various heights; values of z/δ are indicated along the curve traced out by the tip of the velocity vectors. The right panel shows vertical distributions of u and v.

Hodograph

Profiles of u and v

$$\int_0^\infty v\,dz = U\left[\frac{\nu_v}{2f}\right]^{1/2} = \frac{1}{2}U\delta,$$

which is directed to the *left* of the free-stream velocity, in the direction of the pressure force.

If the atmosphere were in laminar motion, ν_v would be equal to its molecular value for air, and the Ekman layer thickness at a latitude of 45° (where $f \simeq 10^{-4}\,s^{-1}$) would be $\approx \delta \sim 0.4$ m. The observed thickness of the atmospheric boundary layer is of order 1 km, which implies an eddy viscosity of order $\nu_v \sim 50$ m^2/s. In fact, Taylor (1915) tried to estimate the eddy viscosity by matching the predicted velocity distributions (13.41) with the observed wind at various heights.

The Ekman layer solution on a solid surface demonstrates that the three-way balance among the Coriolis force, the pressure force, and the frictional force within the boundary layer results in a component of flow directed toward the lower pressure.

The balance of forces within the boundary layer is illustrated in Figure 13.10. The net frictional force on an element is oriented approximately opposite to the velocity vector **u**. It is clear that a balance of forces is possible only if the velocity vector has a component from high to low pressure, as shown. Frictional forces therefore cause the flow around a low-pressure center to spiral *inward*. Mass conservation requires that the inward converging flow rise within a low-pressure system, resulting in cloud formation and rainfall. This is what happens in a *cyclone*, a low-pressure system. In contrast, within a high-pressure system the air sinks as it spirals outward due to frictional effects. The arrival of high-pressure systems therefore brings in clear skies and fair weather, because the sinking air suppresses cloud formation.

Frictional effects, in particular the Ekman transport by surface winds, play a fundamental role in the theory of wind-driven ocean circulation. Possibly the most important result of

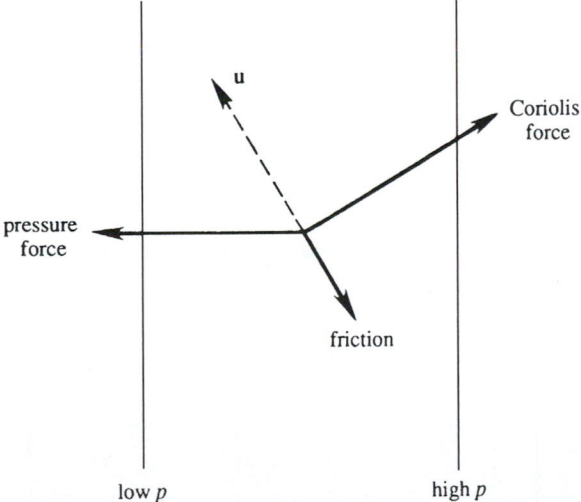

FIGURE 13.10 Balance of forces within an Ekman layer. For steady flow without friction, pressure and Coriolis forces would balance. When friction is added, pressure and Coriolis forces must counteract it. Since friction acts opposite the direction of flow, the velocity **u** must have a component toward lower pressure when friction is present.

such theories was given by Henry Stommel in 1948. He showed that the northward increase of the Coriolis parameter f is responsible for making the currents along western ocean boundaries (e.g., the Gulf Stream in the Atlantic and the Kuroshio in the Pacific) much stronger than the currents on the eastern side. These are discussed in books on physical oceanography and will not be presented here. Instead, we shall now turn our attention to the influence of Coriolis forces on inviscid wave motions.

13.8. SHALLOW-WATER EQUATIONS

Surface and internal gravity waves were discussed in Chapter 7. There the effect of planetary rotation was assumed to be small, which is valid if the frequency ω of the wave is much larger than the Coriolis parameter f. Here, we are considering phenomena slow enough for ω to be comparable to f. Consider surface gravity waves on a shallow layer of homogeneous fluid whose mean depth is H. If we restrict ourselves to wavelengths λ much larger than H, then the vertical velocities are much smaller than the horizontal velocities. In Section 7.2 we determined that the pressure distribution is hydrostatic, and that fluid particles execute a horizontal rectilinear motion that is independent of z. When the effects of planetary rotation are included, the horizontal velocity is still depth independent, although the particle orbits are no longer rectilinear but elliptic on a horizontal plane, as we shall see in the following section.

Consider a layer of fluid of average depth H lying over a flat horizontal bottom (Figure 13.11). Set $z = 0$ on the bottom surface, and let η be the displacement of the free surface. When the pressure on the fluid's surface is set to zero, the pressure at height z from the bottom, which is hydrostatic, is given by

$$p = \rho g(H + \eta - z).$$

The horizontal pressure gradients are therefore

$$\frac{\partial p}{\partial x} = \rho g \frac{\partial \eta}{\partial x}, \quad \frac{\partial p}{\partial y} = \rho g \frac{\partial \eta}{\partial y}. \tag{13.42}$$

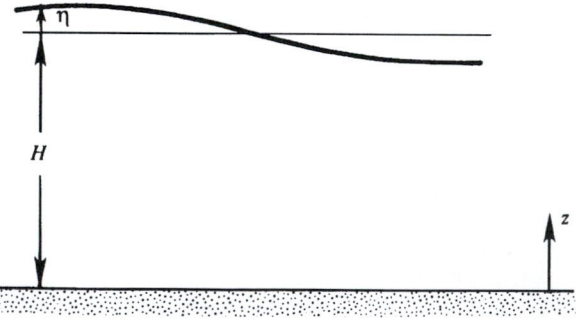

FIGURE 13.11   Geometry for a fluid layer of average thickness H above a flat bottom coincident with $z = 0$. At any horizontal location the liquid's surface height is $H + \eta$.

Since these are independent of z, the resulting horizontal motion is also depth independent so $\partial u/\partial x$ and $\partial v/\partial y$ are independent of z. Therefore, the continuity equation, $\partial u/\partial x + \partial v/\partial y + \partial w/\partial z = 0$, requires that w vary linearly with z, from zero at the bottom to the maximum value at the free surface. Integrating the continuity equation vertically across the water column from $z = 0$ to $z = H + \eta$, and noting that u and v are depth independent, we obtain

$$(H + \eta)\frac{\partial u}{\partial x} + (H + \eta)\frac{\partial v}{\partial y} + w(\eta) - w(0) = 0, \qquad (13.43)$$

where $w(\eta)$ is the vertical velocity at the surface and $w(0) = 0$ is the vertical velocity at the bottom. The surface velocity is given by

$$w(\eta) = \frac{D\eta}{Dt} = \frac{\partial\eta}{\partial t} + u\frac{\partial\eta}{\partial x} + v\frac{\partial\eta}{\partial y},$$

which we recognize as the exact kinematic boundary condition on a free surface with two independent horizontal dimensions (*cf.* (7.17)). The continuity equation (13.43) then becomes

$$(H + \eta)\frac{\partial u}{\partial x} + (H + \eta)\frac{\partial v}{\partial y} + \frac{\partial\eta}{\partial t} + u\frac{\partial\eta}{\partial x} + v\frac{\partial\eta}{\partial y} = 0,$$

which can be written as

$$\frac{\partial\eta}{\partial t} + \frac{\partial}{\partial x}\left[u(H + \eta)\right] + \frac{\partial}{\partial y}\left[v(H + \eta)\right] = 0. \qquad (13.44)$$

This says simply that the divergence of the horizontal fluid transport depresses the free surface. For small amplitude waves, the quadratic nonlinear terms can be neglected in comparison to the linear terms, so that the divergence term in (13.44) simplifies to $H\nabla\cdot\mathbf{u}$.

The linearized continuity and momentum equations are then:

$$\frac{\partial\eta}{\partial t} + H\left(\frac{\partial u}{\partial x} + \frac{\partial v}{\partial y}\right) = 0, \quad \frac{\partial u}{\partial t} - fv = -g\frac{\partial\eta}{\partial x}, \text{ and } \frac{\partial v}{\partial t} + fu = -g\frac{\partial\eta}{\partial y}. \qquad (13.45)$$

In the momentum equations of (13.45), the pressure gradient terms are written in the form (13.42) and the nonlinear advective terms have been neglected under the small amplitude assumption. Equations (13.45), called the *shallow water equations*, govern the motion of a layer of fluid in which the horizontal scale is much larger than the depth of the layer. These equations will be used in the following sections for studying various types of gravity waves.

Although the preceding analysis has been formulated for a layer of *homogeneous* fluid, (13.45) are applicable to internal waves in a stratified medium, if we replace H by the *equivalent depth* H_e, defined by

$$c^2 = gH_e, \qquad (13.46)$$

where c is the speed of long nonrotating *internal* gravity waves. This will be demonstrated in the following section.

13.9. NORMAL MODES IN A CONTINUOUSLY STRATIFIED LAYER

In the preceding section we considered a homogeneous medium and derived the governing equations for waves of wavelength larger than the depth of the fluid layer. Now consider a continuously stratified medium and assume that the horizontal scale of motion is much larger than the vertical scale. The pressure distribution is therefore hydrostatic, and the linearized equations of motion are:

$$\frac{\partial u}{\partial x} + \frac{\partial v}{\partial y} + \frac{\partial w}{\partial z} = 0, \tag{13.47}$$

$$\frac{\partial u}{\partial t} - fv = -\frac{1}{\rho_0}\frac{\partial p}{\partial x}, \quad \frac{\partial v}{\partial t} + fu = -\frac{1}{\rho_0}\frac{\partial p}{\partial y}, \tag{13.48, 13.49}$$

$$0 = -\frac{\partial p}{\partial z} - g\rho, \quad \frac{\partial \rho}{\partial t} - \frac{\rho_0 N^2}{g}w = 0, \tag{13.50, 13.51}$$

where p and ρ represent *perturbations* of pressure and density from the state of rest. The advective term in the density equation (13.51) is written in the linearized form $w(d\bar{\rho}/dz) = -\rho_0 N^2 w/g$, where $N(z)$ is the buoyancy frequency. In this form the rate of change of density at a point is assumed to be due only to the vertical advection of the background density distribution $\bar{\rho}(z)$, as discussed in Section 7.8.

In a continuously stratified medium, it is convenient to use the method of separation of variables and write $q = \sum q_n(x, y, t)\psi_n(z)$ for a dependent-field variable q. The solution is thus written as the sum of various vertical modes $\psi_n(z)$, which are called *normal modes* because they turn out to be orthogonal to each other. The vertical structure of a mode is described by ψ_n while q_n describes the horizontal propagation of the mode. Although each mode propagates only horizontally, the *sum* of a number of modes can also propagate vertically if the various q_n are out of phase.

We assume separable solutions of the form:

$$[u, v, p/\rho_0] = \sum_{n=0}^{\infty} [u_n, v_n, p_n]\psi_n(z), \tag{13.52}$$

$$w = \sum_{n=0}^{\infty} w_n \int_{-H}^{z} \psi_n(z)\, dz, \tag{13.53}$$

$$\rho = \sum_{n=0}^{\infty} \rho_n \frac{d\psi_n}{dz}, \tag{13.54}$$

where the amplitudes u_n, v_n, p_n, w_n, and ρ_n are functions of (x,y,t). The z-axis is measured from the upper free surface of the fluid layer, and $z = -H$ represents the bottom wall. The reasons for assuming the various forms of z-dependence in (13.52) through (13.54) are the following: Variables u, v, and p have the same vertical structure in order to be consistent with (13.48) and (13.49). The continuity equation (13.47) requires that the vertical structure of w should be the integral of $\psi_n(z)$. Equation (13.50) requires that the vertical structure of ρ must be the z-derivative of the vertical structure of p.

Substitution of (13.53) and (13.54) into (13.51) gives

$$\sum_{n=0}^{\infty} \left[\frac{\partial \rho_n}{\partial t} \frac{d\psi_n}{dz} - \frac{\rho_0 N^2}{g} w_n \int_{-H}^{z} \psi_n \, dz \right] = 0.$$

This is valid for all values of z, and the modes are linearly independent, so the quantity within brackets must vanish for each mode, which implies

$$\frac{d\psi_n/dz}{N^2 \int_{-H}^{z} \psi_n \, dz} = \frac{\rho_0}{g} \frac{w_n}{\partial \rho_n/\partial t} \equiv -\frac{1}{c_n^2}. \tag{13.55}$$

As the first term is a function of z alone and the second term is a function of (x,y,t) alone, for consistency both terms must be equal to a constant that we take to be $-1/c_n^2$. The vertical structure is then given by

$$\frac{1}{N^2} \frac{d\psi_n}{dz} = -\frac{1}{c_n^2} \int_{-H}^{z} \psi_n \, dz.$$

Taking the z-derivative,

$$\frac{d}{dz} \left(\frac{1}{N^2} \frac{d\psi_n}{dz} \right) + \frac{1}{c_n^2} \psi_n = 0, \tag{13.56}$$

which is the differential equation governing the vertical structure of the normal modes. Equation (13.56) has the so-called Sturm-Liouville form, for which the various solutions are orthogonal.

Equation (13.55) also gives

$$w_n = -\frac{g}{\rho_0 c_n^2} \frac{\partial \rho_n}{\partial t}.$$

Substitution of (13.52) through (13.54) into (13.47) through (13.51) finally gives the normal mode equations:

$$\frac{\partial u_n}{\partial x} + \frac{\partial v_n}{\partial y} + \frac{1}{c_n^2} \frac{\partial p_n}{\partial t} = 0, \tag{13.57}$$

$$\frac{\partial u_n}{\partial t} - f v_n = -\frac{\partial p_n}{\partial x}, \quad \frac{\partial v_n}{\partial t} + f u_n = -\frac{\partial p_n}{\partial y}, \tag{13.58, 13.59}$$

$$p_n = -\frac{g}{\rho_0} \rho_n, \quad w_n = \frac{1}{c_n^2} \frac{\partial p_n}{\partial t}. \tag{13.60, 13.61}$$

Once (13.57) through (13.59) have been solved for u_n, v_n, and p_n, the amplitudes ρ_n and w_n can be obtained from (13.60) and (13.61). The set (13.57) through (13.59) is identical to the set (13.45) governing the motion of a *homogeneous* layer, provided p_n is identified with $g\eta$ and c_n^2 is identified with gH. In a stratified flow each mode (having a fixed vertical structure) behaves, in the horizontal dimensions and in time, just like a homogeneous layer, with an *equivalent depth* H_e defined by

$$c_n^2 \equiv gH_e. \tag{13.62}$$

Boundary Conditions on ψ_n

At the bottom of the fluid layer, the boundary condition is

$$w = 0 \quad \text{at } z = -H.$$

To write this condition in terms of ψ_n, we first combine the hydrostatic equation (13.50) and the density equation (13.51) to give w in terms of p:

$$w = \frac{g(\partial \rho / \partial t)}{\rho_0 N^2} = -\frac{1}{\rho_0 N^2} \frac{\partial^2 p}{\partial z \, \partial t} = -\frac{1}{N^2} \sum_{n=0}^{\infty} \frac{\partial p_n}{\partial t} \frac{d\psi_n}{dz}. \tag{13.63}$$

The requirement $w = 0$ then yields the bottom boundary condition:

$$\frac{d\psi_n}{dz} = 0 \quad \text{at } z = -H. \tag{13.64}$$

We now formulate the surface boundary condition. The linearized surface boundary conditions are:

$$w = \frac{\partial \eta}{\partial t}, \quad p = \rho_0 g \eta \quad \text{at } z = 0, \tag{13.65'}$$

where η is the free surface displacement. These conditions can be combined into:

$$\frac{\partial p}{\partial t} = \rho_0 g w \quad \text{at } z = 0.$$

Using (13.63) this becomes:

$$\frac{g}{N^2} \frac{\partial^2 p}{\partial z \, \partial t} + \frac{\partial p}{\partial t} = 0 \quad \text{at } z = 0.$$

Substitution of the normal mode decomposition (13.52) gives:

$$\frac{d\psi_n}{dz} + \frac{N^2}{g} \psi_n = 0 \quad \text{at } z = 0. \tag{13.65}$$

The boundary conditions on ψ_n are therefore (13.64) and (13.65).

Vertical Mode Solution for Uniform N

For a medium of uniform N, a simple solution can be found for ψ_n. From (13.56), (13.64), and (13.65), the vertical structure of the normal modes is given by

$$\frac{d^2 \psi_n}{dz^2} + \frac{N^2}{c_n^2} \psi_n = 0, \tag{13.66}$$

with the boundary conditions (13.64) and (13.65). The set (13.64) through (13.66) defines an eigenvalue problem, with ψ_n as the eigenfunction and c_n as the eigenvalue. The solution of (13.66) is

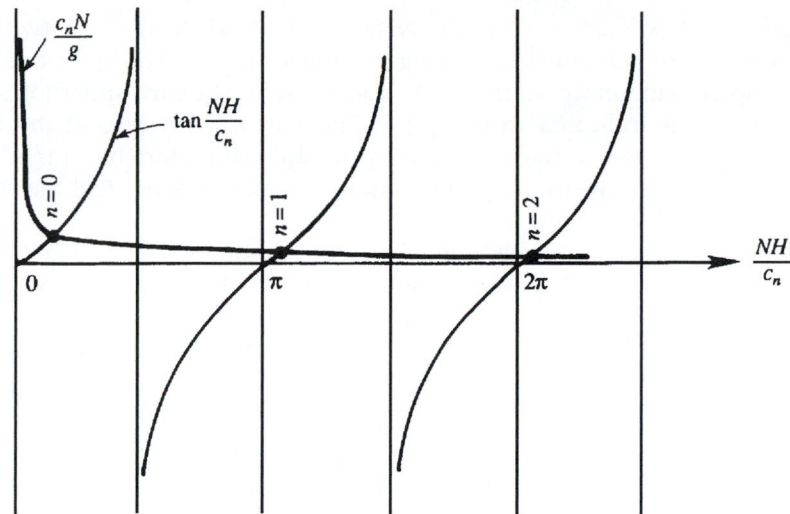

FIGURE 13.12 Calculation of eigenvalues c_n of vertical normal modes in a fluid layer of depth H and uniform stratification N. The eigenvalues occur where the curves defined by $c_n N/g$ and $\tan(NH/c_n)$ cross. As drawn, these crossing points lie slightly above $n\pi$ for $n = 0, 1,$ and 2.

$$\psi_n = A_n \cos\frac{Nz}{c_n} + B_n \sin\frac{Nz}{c_n}. \tag{13.67}$$

Application of the surface boundary condition (13.65) gives

$$B_n = -\frac{c_n N}{g}A_n. \tag{13.68}$$

The bottom boundary condition (13.64) then gives

$$\tan\frac{NH}{c_n} = \frac{c_n N}{g}, \tag{13.69}$$

whose roots define the eigenvalues of the problem.

The solution of (13.69) is indicated graphically in Figure 13.12. The first root occurs for $NH/c_n = 1$, for which we can write $\tan(NH/c_n) \approx NH/c_n$, so that (13.69) gives (indicating this root by $n = 0$):

$$c_0 = \sqrt{gH}. \tag{13.70}$$

The vertical modal structure is found from (13.67). Because the magnitude of an eigenfunction is arbitrary, we can set $A_0 = 1$, obtaining

$$\psi_0 = \cos\frac{Nz}{c_0} - \frac{c_0 N}{g}\sin\frac{Nz}{c_0} \simeq 1 - \frac{N^2 z}{g} \simeq 1,$$

where we have used $N|z|/c_0 \ll 1$ (with $NH/c_0 \ll 1$), and $N^2 z/g \ll 1$ (with $N^2 H/g = (NH/c_0)(c_0 N/g) \ll 1$, both sides of (13.69) being much less than 1). For this mode the vertical structure of u, v, and p is therefore nearly depth independent. The corresponding structure for w (given by $\int \psi_0 \, dz$, as indicated in (13.53)) is linear in z, with zero at the bottom and a maximum at the upper free surface. A stratified medium therefore has a mode of motion that behaves like that in an unstratified medium; this mode does not feel the stratification. The $n = 0$ mode is called the *barotropic mode*.

The remaining modes $n \geq 1$ are *baroclinic*. For these modes $c_n N/g \ll 1$ but NH/c_n is not small, as can be seen in Figure 13.12, so that the baroclinic roots of (13.69) are nearly given by

$$\tan \frac{NH}{c_n} = 0,$$

which gives

$$c_n = \frac{NH}{n\pi}, \qquad n = 1, 2, 3, \ldots. \tag{13.71}$$

Taking a typical depth-average oceanic value of $N \sim 10^{-3}\ \mathrm{s}^{-1}$ and $H \sim 5$ km, the eigenvalue for the first baroclinic mode is $c_1 \sim 2$ m/s. The corresponding equivalent depth is $H_e = c_1^2/g \sim 0.4$ m.

An examination of the algebraic steps leading to (13.69) shows that neglecting the right-hand side is equivalent to replacing the upper boundary condition (13.65) by $w = 0$ at $z = 0$. This is called the *rigid lid approximation*. The *baroclinic modes are negligibly distorted by the rigid lid approximation*. In contrast, the rigid lid approximation applied to the *barotropic* mode would yield $c_0 = \infty$, as (13.71) shows for $n = 0$. Note that the rigid lid approximation does *not* imply that the free surface displacement corresponding to the baroclinic modes is negligible in the ocean. In fact, excluding the wind waves and tides, much of the free surface displacements in the ocean are due to baroclinic motions. The rigid lid approximation merely implies that, for baroclinic motions, the vertical displacements at the surface are much smaller than those within the fluid column. A valid baroclinic solution can therefore be obtained by setting $w = 0$ at $z = 0$. Further, the rigid lid approximation does not imply that the pressure is constant at the level surface $z = 0$; if a rigid lid were actually imposed at $z = 0$, then the pressure on the lid would vary due to the baroclinic motions.

The vertical mode shape under the rigid lid approximation is given by the cosine distribution

$$\psi_n = \cos \frac{n\pi z}{H}, \quad n = 0, 1, 2, \ldots,$$

because it satisfies $d\psi_n/dz = 0$ at $z = 0, -H$. The nth mode ψ_n has n zero crossings within the layer (Figure 13.13).

A decomposition into normal modes is only possible in the absence of topographic variations and mean currents with shear. It is valid with or without Coriolis forces and with or without the β-effect. However, the hydrostatic approximation here means that the frequencies are much smaller than N. Under this condition the eigenfunctions are independent of the frequency, as (13.56) shows. Without the hydrostatic approximation the eigenfunctions ψ_n become dependent on the frequency ω. This is discussed, for example, in LeBlond and Mysak (1978).

FIGURE 13.13 Vertical distributions of the first three normal modes in a stratified medium of uniform buoyancy frequency for a fluid layer of depth H. The first mode ($n = 0$) is nearly uniform through the depth. The second mode ($n = 1$) shows one-half wavelength in $-H < z < 0$. The third mode ($n = 2$) shows one full wavelength in $-H < z < 0$. Note that all modes must have $d\psi_n/dz = 0$ on $z = -H$, while $d\psi_n/dz$ is only approximately zero a $z = 0$.

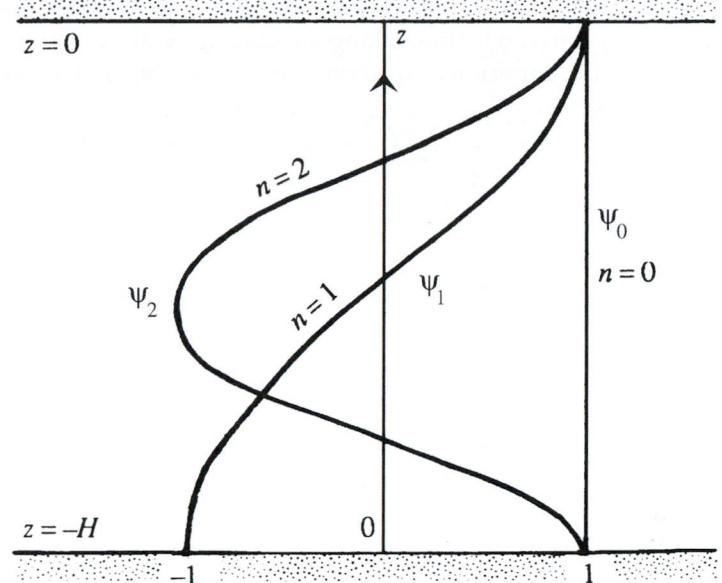

Summary

Small amplitude motion in a frictionless continuously stratified ocean can be decomposed in terms of noninteracting vertical normal modes. The vertical structure of each mode is defined by an eigenfunction $\psi_n(z)$. If the horizontal scale of the waves is much larger than the vertical scale, then the equations governing the horizontal propagation of each mode are identical to those of a shallow *homogeneous* layer, with the layer depth H replaced by an equivalent depth H_e defined by $c_n^2 = gH_e$. For a medium of constant N, the baroclinic ($n \geq 1$) eigenvalues are given by $c_n = NH/\pi n$, while the barotropic eigenvalue is $c_0 = \sqrt{gH}$. The rigid lid approximation is quite good for the baroclinic modes.

13.10. HIGH- AND LOW-FREQUENCY REGIMES IN SHALLOW-WATER EQUATIONS

We shall now examine what terms are negligible in the shallow-water equations for the various frequency ranges. Our analysis is valid for a single homogeneous layer or for a stratified medium. In the latter case H has to be interpreted as the *equivalent* depth, and c has to be interpreted as the speed of long nonrotating *internal* gravity waves. The β-effect is considered in this section. As f varies only northward, horizontal isotropy is lost whenever the β-effect is included, and it becomes necessary to distinguish between the different horizontal directions. We shall follow the usual geophysical convention that the x-axis is directed eastward and the y-axis is directed northward, with u and v the corresponding velocity components.

The simplest way to perform the analysis is to examine the v-equation. A single equation for v can be derived by first taking the time derivatives of the momentum equations in (13.45) and using the continuity equation to eliminate $\partial \eta / \partial t$. This gives

$$\frac{\partial^2 u}{\partial t^2} - f\frac{\partial v}{\partial t} = gH\frac{\partial}{\partial x}\left(\frac{\partial u}{\partial x} + \frac{\partial v}{\partial y}\right), \quad \frac{\partial^2 v}{\partial t^2} + f\frac{\partial u}{\partial t} = gH\frac{\partial}{\partial y}\left(\frac{\partial u}{\partial x} + \frac{\partial v}{\partial y}\right). \qquad (13.72, 13.73)$$

Now take $\partial/\partial t$ of (13.73) and use (13.72) to obtain

$$\frac{\partial^3 v}{\partial t^3} + f\left[f\frac{\partial v}{\partial t} + gH\frac{\partial}{\partial x}\left(\frac{\partial u}{\partial x} + \frac{\partial v}{\partial y}\right)\right] = gH\frac{\partial^2}{\partial y\,\partial t}\left(\frac{\partial u}{\partial x} + \frac{\partial v}{\partial y}\right). \qquad (13.74)$$

To eliminate u, we first obtain a vorticity equation by cross differentiating and subtracting the momentum equations in (13.45):

$$\frac{\partial}{\partial t}\left(\frac{\partial u}{\partial y} - \frac{\partial v}{\partial x}\right) - f_0\left(\frac{\partial u}{\partial x} + \frac{\partial v}{\partial y}\right) - \beta v = 0.$$

Here, we have made the customary β-plane approximation valid if the y-scale is small enough so that $\Delta f/f \ll 1$. Accordingly, we have treated f as constant (and replaced it by an average value f_0) *except* when df/dy appears; this is why we have written f_0 in the second term of the preceding equation. Taking the x-derivative, multiplying by gH, and adding to (13.74), we finally obtain a vorticity equation in terms of v only:

$$\frac{\partial^3 v}{\partial t^3} - gH\frac{\partial}{\partial t}\nabla_H^2 v + f_0^2\frac{\partial v}{\partial t} - gH\beta\frac{\partial v}{\partial x} = 0, \qquad (13.75)$$

where $\nabla_H^2 = \partial^2/\partial x^2 + \partial^2/\partial y^2$ is the horizontal Laplacian operator.

Equation (13.75) is Boussinesq, linear, and hydrostatic, but otherwise quite general in the sense that it is applicable to both high and low frequencies. Consider wave solutions of the form

$$v = \hat{v}\, e^{i(kx + ly - \omega t)},$$

where k is the eastward wave number and l is the northward wave number. Then (13.75) gives

$$\omega^3 - c^2\omega K^2 - f_0^2\omega - c^2\beta k = 0, \qquad (13.76)$$

where $K^2 = k^2 + l^2$ and $c = \sqrt{gH}$. It can be shown that all roots of (13.76) are real, two of the roots being superinertial ($\omega \gg f$) and the third being subinertial ($\omega \ll f$). Equation (13.76) is the complete dispersion relation for linear shallow-water equations. In various parametric ranges it takes simpler forms, representing simpler waves.

First, consider high-frequency waves $\omega \gg f$. The third term of (13.76) is negligible compared to the first term. Moreover, the fourth term is also negligible in this range. Compare, for example, the fourth and second terms:

$$\frac{c^2\beta k}{c^2\omega K^2} \sim \frac{\beta}{\omega K} \sim 10^{-3},$$

where we have assumed typical values of $\beta = 2 \times 10^{-11}$ m^{-1} s^{-1}, $\omega = 3f \sim 3 \times 10^{-4}$ s^{-1}, and $2\pi/K \sim 100$ km. For $\omega \gg f$, therefore, the balance is between the first and second terms in (13.76), and the roots are $\omega = \pm K\sqrt{gH}$, which correspond to a propagation speed of $\omega/K = \sqrt{gH}$. The effects of both f and β are therefore negligible for high-frequency waves, as is expected as they are too fast to be affected by the Coriolis effects.

Next consider $\omega > f$, but $\omega \sim f$. Then the third term in (13.76) is not negligible, but the β-effect is. These are gravity waves influenced by Coriolis forces; gravity waves are discussed in the next section. However, the time scales are still too short for the motion to be affected by the β-effect.

Last, consider very slow waves for which $\omega \ll f$. Then the β-effect becomes important, and the first term in (13.76) becomes negligible. Compare, for example, the first and the last terms:

$$\frac{\omega^3}{c^2 \beta k} \ll 1.$$

Typical values for the ocean are $c \sim 200$ m/s for the barotropic mode, $c \sim 2$ m/s for the baroclinic mode, $\beta = 2 \times 10^{-11}$ m^{-1} s^{-1}, $2\pi/k \sim 100$ km, and $\omega \sim 10^{-5}$ s^{-1}. This makes the aforementioned ratio about 0.2×10^{-4} for the barotropic mode and 0.2 for the baroclinic mode. The first term in (13.76) is therefore negligible for $\omega \gg f$.

Equation (13.75) governs the dynamics of a variety of wave motions in the ocean and the atmosphere, and the discussion in this section shows what terms can be dropped under various limiting conditions. An understanding of these limiting conditions will be useful in the following sections.

13.11. GRAVITY WAVES WITH ROTATION

In this chapter we examine several free-wave solutions of the shallow-water equations. In this section we focus on gravity waves with frequencies in the range $\omega > f$, for which the β-effect is negligible, as demonstrated in the preceding section. Consequently, the Coriolis frequency f is regarded as constant here. Consider progressive waves of the form

$$(u, v, \eta) = (\widehat{u}, \widehat{v}, \widehat{\eta})e^{i(kx+ly-\omega t)},$$

where $\widehat{u}$, $\widehat{v}$, and $\widehat{\eta}$ are the complex amplitudes, and the real part of the right-hand side is meant. Then (13.45) gives

$$-i\omega\widehat{u} - f\widehat{v} = -ikg\widehat{\eta}, \quad -i\omega\widehat{v} + f\widehat{u} = -ilg\widehat{\eta}, \quad -i\omega\widehat{\eta} + iH(k\widehat{u} + l\widehat{v}) = 0. \quad (13.77, 13.78, 13.79)$$

Solving for $\widehat{u}$ and $\widehat{v}$ between (13.77) and (13.78), we obtain:

$$\widehat{u} = \frac{g\widehat{\eta}}{\omega^2 - f^2}(\omega k + ifl),$$

$$\widehat{v} = \frac{g\widehat{\eta}}{\omega^2 - f^2}(-ifk + \omega l).$$

(13.80)

Substituting these in (13.79), we obtain

$$\omega^2 - f^2 = gH(k^2 + l^2). \tag{13.81}$$

This is the dispersion relation of gravity waves in the presence of Coriolis forces. (The relation can be most simply derived by setting the determinant of the set of linear homogeneous equations (13.77) through (13.79) to zero.) It can be written as

$$\omega^2 = f^2 + gHK^2, \tag{13.82}$$

where $K = \sqrt{k^2 + l^2}$ is the magnitude of the horizontal wave number. The dispersion relation shows that the waves can propagate in any horizontal direction and have $\omega > f$. Gravity waves affected by Coriolis forces are called *Poincaré waves*, *Sverdrup waves*, or simply *rotational gravity waves*. (Sometimes the name "Poincaré wave" is used to describe those rotational gravity waves that satisfy the boundary conditions in a channel.) In spite of its names, the solution was first worked out by Kelvin (Gill, 1982, p. 197). A plot of (13.82) is shown in Figure 13.14. It is seen that the waves are dispersive except for $\omega \gg f$ when equation (13.82) gives $\omega^2 \approx gHK^2$, so that the propagation speed is $\omega/K = \sqrt{gH}$. The high-frequency limit agrees with our previous discussion of surface gravity waves unaffected by Coriolis forces.

Particle Orbit

The symmetry of the dispersion relation (13.81) with respect to k and l means that the x- and y-directions are not felt differently by the wave field. The horizontal isotropy is a result of treating f as constant. (We shall see later that Rossby waves, which depend on the β-effect, are not horizontally isotropic.) We can therefore orient the x-axis along the wave number

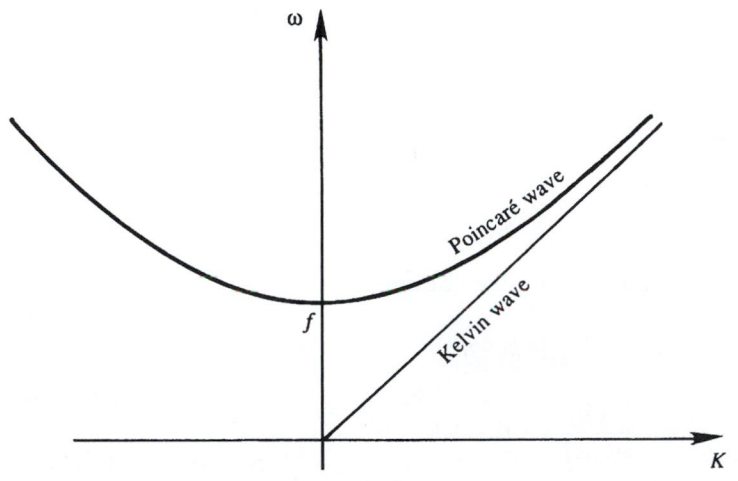

FIGURE 13.14 Dispersion relations for Poincaré and Kelvin waves. Here, ω is the wave frequency. K is the magnitude of the wave number, and f is the local inertial frequency. At frequencies $\omega \gg f$, the ordinary shallow-water wave dispersion relationship $\omega^2 = gHK^2$ is recovered.

FIGURE 13.15 Particle orbit in a gravity wave traveling in the positive x-direction. Looking down on the surface, the orbit is an ellipse having major and minor axes proportional to the wave frequency ω and the inertial frequency f. Velocity components corresponding to $\omega t = 0$, $\pi/2$, and π are indicated.

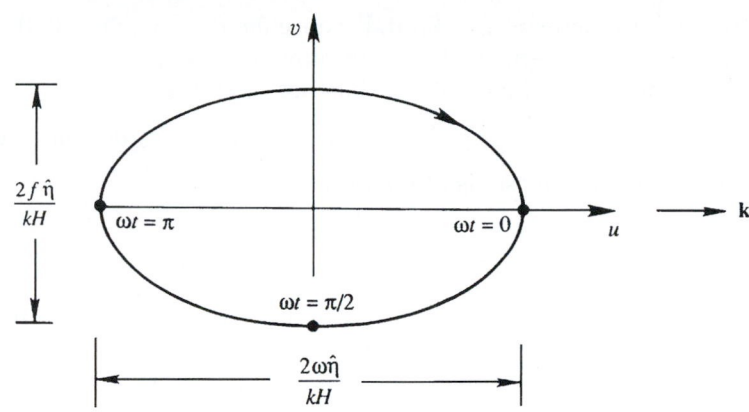

vector and set $l = 0$, so that the wave field is invariant along the y-axis. To find the particle orbits, it is convenient to work with real quantities. Let the displacement be

$$\eta = \widehat{\eta} \cos(kx - \omega t),$$

where $\widehat{\eta}$ is real. The corresponding velocity components can be found by multiplying (13.80) by $\exp(ikx - i\omega t)$ and taking the real part of both sides. This gives

$$u = \frac{\omega \widehat{\eta}}{kH} \cos(kx - \omega t),$$

$$v = \frac{f \widehat{\eta}}{kH} \sin(kx - \omega t).$$

(13.83)

To find the particle paths, take $x = 0$ and consider three values of time corresponding to $\omega t = 0$, $\pi/2$, and π. The corresponding values of u and v from (13.83) show that the velocity vector rotates clockwise (in the northern hemisphere) in elliptic paths (Figure 13.15). The ellipticity is expected, since the presence of Coriolis forces means that fu must generate $\partial v/\partial t$ according to the equation of motion (13.45). (In equation (13.45), $\partial \eta/\partial y = 0$ due to our orienting the x-axis along the direction of propagation of the wave.) Particles are therefore constantly deflected to the right by the Coriolis force, resulting in elliptic orbits. *The ellipses have an axis ratio of ω/f and the major axis is oriented in the direction of wave propagation.* The ellipses become narrower as ω/f increases, approaching the rectilinear orbit of gravity waves unaffected by planetary rotation. However, the sea surface in a rotational gravity wave is no different from that for ordinary gravity waves, namely oscillatory in the direction of propagation and invariant in the perpendicular direction.

Inertial Motion

Consider the limit $\omega \to f$, that is, when the particle paths are circular. The dispersion relation (13.82) then shows that $K \to 0$, implying a horizontal uniformity of the flow field. Equation (13.79) shows that $\widehat{\eta}$ must tend to zero in this limit, so that there are no horizontal pressure gradients in this limit. Because $\partial u/\partial x = \partial v/\partial y = 0$, the continuity equation shows that $w = 0$. The particles therefore move on horizontal sheets, each layer decoupled from the one above and below it. The balance of forces is

$$\partial u/\partial t - fv = 0 \quad \text{and} \quad \partial v/\partial t + fu = 0.$$

The solution of this set is of the form

$$u = q\cos(ft) \quad \text{and} \quad v = -q\sin(ft),$$

where the speed $q = \sqrt{u^2 + v^2}$ is constant along the path. The radius r of the orbit can be found by adopting a Lagrangian point of view, and noting that the equilibrium of forces is between the Coriolis acceleration fq and the centrifugal acceleration $r\omega^2 = rf^2$, giving $r = q/f$. The limiting case of motion in circular orbits at a frequency f is called *inertial motion*, because in the absence of pressure gradients a particle moves by virtue of its inertia alone. The corresponding period $2\pi/f$ is called the *inertial period*. In the absence of planetary rotation such motion would be along straight lines; in the presence of Coriolis forces the motion is along circular paths, called *inertial circles*. Near-inertial motion is frequently generated in the surface layer of the ocean by sudden changes of the wind field, essentially because the equations of motion (13.45) have a natural frequency f. Taking a typical current magnitude of $q \sim 0.1$ m/s, the radius of the orbit is $r \sim 1$ km.

13.12. KELVIN WAVE

In the preceding section we considered a shallow-water gravity wave propagating in a horizontally *unbounded* ocean. We saw that the crests are horizontal and oriented in a direction perpendicular to the direction of propagation. The *absence* of a transverse pressure gradient $\partial \eta/\partial y$ resulted in a transverse flow and elliptic orbits. This is clear from the third equation in (13.45), which shows that the presence of fu must result in $\partial v/\partial t$ if $\partial \eta/\partial y = 0$. In this section we consider a gravity wave propagating parallel to a wall, whose presence allows a pressure gradient $\partial \eta/\partial y$ that can decay away from the wall. We shall see that this allows a gravity wave in which fu is geostrophically balanced by $-g(\partial \eta/\partial y)$, and $v = 0$. Consequently the particle orbits are not elliptic but rectilinear.

Consider first a gravity wave propagating in a channel. From Figure 7.5 we know that the fluid velocity under a crest is in the direction of wave propagation, and that under a trough it is the opposite. Figure 13.16 shows two transverse sections of the wave, one through a crest (left panel) and the other through a trough (right panel). The wave is propagating into the plane of the paper, along the x-direction. Then the fluid velocity under the crest is into the plane of the paper and that under the trough is out of the plane of the paper. The constraints of the sidewalls require that $v = 0$ at the walls, and we are exploring the possibility of a wave motion in which v is zero everywhere. Then the equation of motion along the y-direction requires that fu can only be geostrophically balanced by a transverse slope of the sea surface across the channel:

$$fu = -g\frac{\partial \eta}{\partial y}.$$

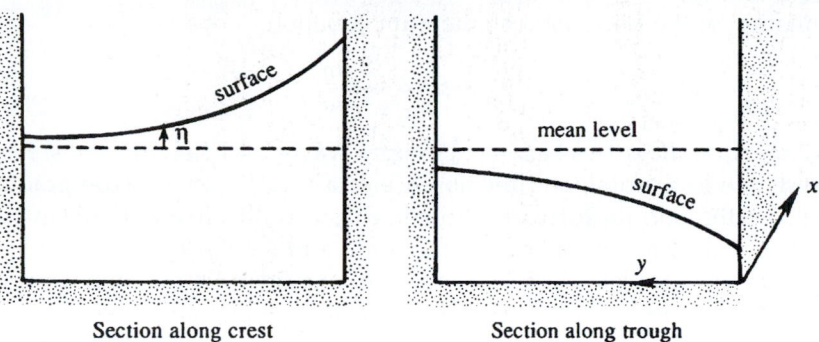

Section along crest   **Section along trough**

FIGURE 13.16 Free surface distribution in a Kelvin gravity wave propagating into the plane of the paper (the x-direction) within a channel. The wave crests and troughs are enhanced on the right side of the channel.

In the northern hemisphere, the surface must slope as indicated in the figure, that is, downward to the left under the crest and upward to the left under the trough, so that the pressure force has the current directed to its right. The result is that the amplitude of the wave is larger on the right-hand side of the channel, looking into the direction of propagation, as indicated in Figure 13.16. The current amplitude, like the surface displacement, also decays to the left.

If the left wall in Figure 13.16 is moved away to infinity, we get a gravity wave trapped to the coast (Figure 13.17). A coastally trapped long gravity wave, in which the transverse velocity $v = 0$ everywhere, is called a *Kelvin wave*. It is clear that it can propagate only in a direction such that the coast is to the right (looking in the direction of propagation) in the northern hemisphere and to the left in the southern hemisphere. The opposite direction of propagation would result in a sea surface displacement increasing exponentially away from the coast, which is not possible.

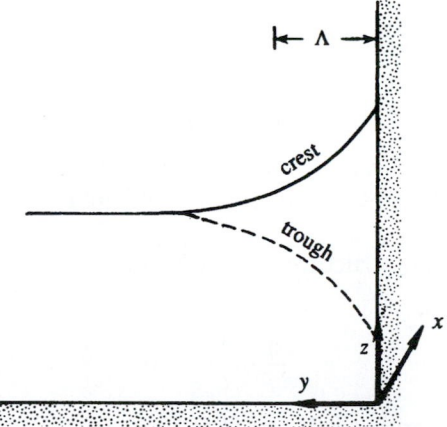

FIGURE 13.17 Coastal Kelvin wave propagating along the x-axis. The sea surface shape across a section through a crest is indicated by the continuous line, and that along a trough is indicated by the dashed line.

An examination of the transverse momentum equation,

$$\frac{\partial v}{\partial t} + fu = -g\frac{\partial \eta}{\partial y},$$

reveals fundamental differences between Poincaré waves and Kelvin waves. For a Poincaré wave the crests are horizontal, and the absence of a transverse pressure gradient requires a $\partial v/\partial t$ to balance the Coriolis force, resulting in elliptic orbits. In a Kelvin wave a transverse velocity is prevented by a geostrophic balance of fu and $-g(\partial \eta/\partial y)$.

From the shallow-water set (13.45), the equations of motion for a Kelvin wave propagating along a coast aligned with the x-axis (Figure 13.17) are:

$$\frac{\partial \eta}{\partial t} + H\frac{\partial u}{\partial x} = 0, \quad \frac{\partial u}{\partial t} = -g\frac{\partial \eta}{\partial x}, \text{ and } fu = -g\frac{\partial \eta}{\partial y}. \tag{13.84}$$

Assume a solution of the form:

$$[u, \eta] = \left[\widehat{u}(y), \widehat{\eta}(y)\right]e^{i(kx-\omega t)}.$$

Then (13.84) gives

$$-i\omega\widehat{\eta} + iHk\widehat{u} = 0,$$
$$-i\omega\widehat{u} = -igk\widehat{\eta},$$
$$f\widehat{u} = -g\frac{d\widehat{\eta}}{dy}. \tag{13.85}$$

The dispersion relation can be found solely from the first two of these equations; the third equation then determines the transverse structure. Eliminating $\widehat{u}$ between the first two, we obtain

$$\widehat{\eta}\left[\omega^2 - gHk^2\right] = 0.$$

A nontrivial solution is therefore possible only if $\omega = \pm k\sqrt{gH}$, so that the wave propagates with a nondispersive speed:

$$c = \sqrt{gH}. \tag{13.86}$$

The propagation speed of a Kelvin wave is therefore identical to that of nonrotating gravity waves. Its dispersion equation is a straight line and is shown in Figure 13.14. All frequencies are possible.

To determine the transverse structure, eliminate $\widehat{u}$ between the first and third equation of (13.85), giving

$$\frac{d\widehat{\eta}}{dy} \pm \frac{f}{c}\widehat{\eta} = 0.$$

The solution that decays away from the coast is

$$\widehat{\eta} = \eta_0\,e^{-fy/c},$$

where η_0 is the amplitude at the coast. Therefore, the sea surface slope and the velocity field for a Kelvin wave have the form:

$$\eta = \eta_0 \, e^{-fy/c} \cos k(x - ct),$$

$$u = \eta_0 \sqrt{\frac{g}{H}} e^{-fy/c} \cos k(x - ct),$$

(13.87)

where we have taken the real parts, and have used equation (13.85) in obtaining the u field.

Equations (13.87) show that the transverse decay scale of the Kelvin wave is

$$\Lambda \equiv \frac{c}{f'}$$

which is called the *Rossby radius of deformation*. For a deep sea of depth $H = 5$ km, and a mid-latitude value of $f = 10^{-4}\,\text{s}^{-1}$, we obtain $c = \sqrt{gH} = 220$ m/s and $\Lambda = c/f = 2200$ km. Tides are frequently in the form of coastal Kelvin waves of semidiurnal frequency. The tides are forced by the periodic changes in the gravitational attraction of the moon and the sun. These waves propagate along the boundaries of an ocean basin and cause sea level fluctuations at coastal stations.

Analogous to the surface or "external" Kelvin waves discussed in the preceding paragraphs, we can have *internal Kelvin waves* at the interface between two fluids of different densities (Figure 13.18). If the lower layer is very deep, then the speed of propagation is given by (see (7.117)):

$$c = \sqrt{g'H},$$

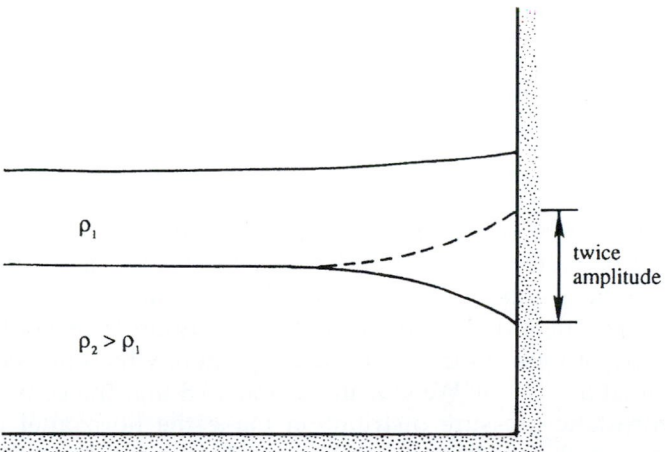

FIGURE 13.18   Internal Kelvin wave at an interface. Dashed line indicates position of the interface when it is at its maximum height. Displacement of the free surface is much smaller than that of the interface and is oppositely directed.

where H is the thickness of the upper layer and $g' = g(\rho_2 - \rho_1)/\rho_2$ is the reduced gravity. For a continuously stratified medium of depth H and buoyancy frequency N internal Kelvin waves can propagate at any of the normal mode speeds

$$c = NH/n\pi, \qquad n = 1, 2, \dots .$$

The decay scale for *internal* Kelvin waves, $\Lambda = c/f$, is called the *internal Rossby radius of deformation*, whose value is much smaller than that for the external Rossby radius of deformation. For $n = 1$, a typical value in the ocean is $\Lambda = NH/\pi f \sim 50$ km; a typical atmospheric value is much larger, being of order $\Lambda \sim 1000$ km.

Internal Kelvin waves in the ocean are frequently forced by wind changes near coastal areas. For example, a southward wind along the west coast of a continent in the northern hemisphere (say, California) generates an Ekman layer at the ocean surface, in which the mass flow is *away* from the coast (to the right of the applied wind stress). The mass flux in the near-surface layer is compensated by the movement of deeper water toward the coast, which raises the thermocline. An upward movement of the thermocline, as indicated by the dashed line in Figure 13.18, is called *upwelling*. The vertical movement of the thermocline in the wind-forced region then propagates poleward along the coast as an internal Kelvin wave.

13.13. POTENTIAL VORTICITY CONSERVATION IN SHALLOW-WATER THEORY

In this section we shall derive a useful conservation law for the vorticity of a shallow layer of fluid. From Section 13.8, the equations of motion for a shallow layer of homogeneous fluid are:

$$\frac{\partial u}{\partial t} + u\frac{\partial u}{\partial x} + v\frac{\partial u}{\partial y} - fv = -g\frac{\partial \eta}{\partial x}, \tag{13.88}$$

$$\frac{\partial v}{\partial t} + u\frac{\partial v}{\partial x} + v\frac{\partial v}{\partial y} + fu = -g\frac{\partial \eta}{\partial y}, \tag{13.89}$$

$$\frac{\partial h}{\partial t} + \frac{\partial}{\partial x}(uh) + \frac{\partial}{\partial y}(vh) = 0, \tag{13.90}$$

where $h(x,y,t)$ is the depth of flow and η is the height of the sea surface measured from an arbitrary horizontal plane (Figure 13.19). The x-axis is taken eastward and the y-axis is taken northward, with u and v the corresponding velocity components. The Coriolis frequency $f = f_0 + \beta y$ is regarded as dependent on latitude. The nonlinear terms have been retained, including those in the continuity equation, which has been written in the form (13.44); note that $h = H + \eta$. We saw in Section 13.8 that the constant density of the layer and the hydrostatic pressure distribution make the horizontal pressure gradient depth independent, so that only a depth-independent current can be generated. The vertical velocity is linear in z.

A vorticity equation can be derived by differentiating (13.88) with respect to y, (13.89) with respect to x, and subtracting. As expected, these steps eliminate the pressure, and we obtain:

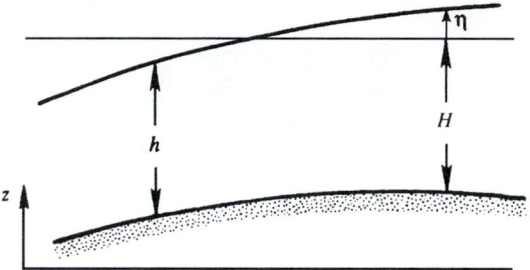

FIGURE 13.19 Shallow layer of instantaneous depth $h(x,y,t)$ when the ocean bottom is not flat. Here η is the sea surface deflection measured from a conveniently chosen horizontal plane.

$$\frac{\partial}{\partial t}\left(\frac{\partial v}{\partial x} - \frac{\partial u}{\partial y}\right) + \frac{\partial}{\partial x}\left[u\frac{\partial v}{\partial x} + v\frac{\partial v}{\partial y}\right] - \frac{\partial}{\partial y}\left[u\frac{\partial u}{\partial x} + v\frac{\partial u}{\partial y}\right] + f_0\left(\frac{\partial u}{\partial x} + \frac{\partial v}{\partial y}\right) + \beta v = 0. \qquad (13.91)$$

Following the customary β-plane approximation, we have treated f as constant (and replaced it by an average value f_0) *except* when df/dy appears. We now introduce

$$\zeta \equiv \frac{\partial v}{\partial x} - \frac{\partial u}{\partial y}$$

as the vertical component of *relative vorticity*, that is, the vorticity measured relative to the rotating earth. Then the nonlinear terms in (13.91) can easily be rearranged in the form

$$u\frac{\partial \zeta}{\partial x} + v\frac{\partial \zeta}{\partial y} + \left(\frac{\partial u}{\partial x} + \frac{\partial v}{\partial y}\right)\zeta.$$

Equation (13.91) then becomes

$$\frac{\partial \zeta}{\partial t} + u\frac{\partial \zeta}{\partial x} + v\frac{\partial \zeta}{\partial y} + \left(\frac{\partial u}{\partial x} + \frac{\partial v}{\partial y}\right)(\zeta + f_0) + \beta v = 0,$$

which can be written as

$$\frac{D\zeta}{Dt} + (\zeta + f_0)\left(\frac{\partial u}{\partial x} + \frac{\partial v}{\partial y}\right) + \beta v = 0, \qquad (13.92)$$

where D/Dt is the derivative following the horizontal motion of the layer:

$$\frac{D}{Dt} \equiv \frac{\partial}{\partial t} + u\frac{\partial}{\partial x} + v\frac{\partial}{\partial y}.$$

The horizontal divergence $(\partial u/\partial x + \partial v/\partial y)$ in (13.92) can be eliminated by using the continuity equation (13.90), which can be written as

$$\frac{Dh}{Dt} + h\left(\frac{\partial u}{\partial x} + \frac{\partial v}{\partial y}\right) = 0.$$

Equation (13.92) then becomes

$$\frac{D\zeta}{Dt} = \frac{\zeta + f_0}{h}\frac{Dh}{Dt} - \beta v.$$

This can be written as

$$\frac{D(\zeta + f)}{Dt} = \frac{\zeta + f_0}{h}\frac{Dh}{Dt}, \tag{13.93}$$

where we have used

$$\frac{Df}{Dt} = \frac{\partial f}{\partial t} + u\frac{\partial f}{\partial x} + v\frac{\partial f}{\partial y} = v\beta.$$

Because of the absence of vertical shear, the vorticity in a shallow-water model is purely vertical and independent of depth. The relative vorticity measured with respect to the rotating earth is ζ, while f is the planetary vorticity, so that the *absolute vorticity* is $(\zeta + f)$. Equation (13.93) shows that the rate of change of absolute vorticity is proportional to the absolute vorticity times the vertical stretching Dh/Dt of the water column. It is apparent that $D\zeta/Dt$ can be nonzero even if $\zeta = 0$ initially. This is different from a nonrotating flow in which stretching a fluid line changes its vorticity only if the line has an *initial* vorticity. (This is why the process was called the *vortex* stretching; see Section 5.7.) The difference arises because vertical lines in a rotating earth contain the planetary vorticity even when $\zeta = 0$. Note that the vortex *tilting* term, discussed in Section 5.6, is absent in the shallow-water theory because the water moves in the form of vertical columns without ever tilting.

Equation (13.93) can be written in the compact form

$$\frac{D}{Dt}\left(\frac{\zeta + f}{h}\right) = 0, \tag{13.94}$$

where $f = f_0 + \beta y$, and we have assumed $\beta y \ll f_0$. The ratio $(\zeta + f)/h$ is called the *potential vorticity* in shallow-water theory. Equation (13.94) shows that the *potential vorticity is conserved along the motion*, an important principle in geophysical fluid dynamics. In the ocean, outside regions of strong current vorticity such as coastal boundaries, the magnitude of ζ is much smaller than that of f. In such a case $(\zeta + f)$ has the sign of f. The principle of conservation of potential vorticity means that an increase in h must make $(\zeta + f)$ more positive in the northern hemisphere and more negative in the southern hemisphere.

As an example of application of the potential vorticity equation, consider an eastward flow over a step (at $x = 0$) running north–south, across which the layer thickness changes discontinuously from h_0 to h_1 (Figure 13.20). The flow upstream of the step has a uniform speed U, so that the oncoming stream has no relative vorticity. To conserve the ratio $(\zeta + f)/h$, the flow must suddenly acquire negative (clockwise) relative vorticity due to the sudden decrease in layer thickness. The relative vorticity of a fluid element just after passing the step can be found from

$$\frac{f}{h_0} = \frac{\zeta + f}{h_1},$$

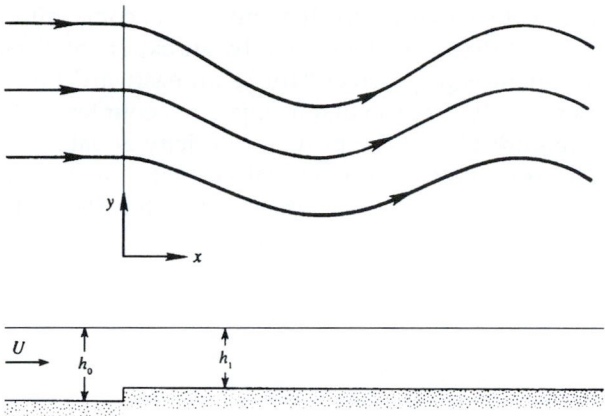

FIGURE 13.20 Eastward flow over a step change in depth. Looking down from above, the step causes southward deflection of the streamlines that is eventually countered by latitude change and results in stationary spatial oscillations of wavelength $2\pi\sqrt{U/\beta}$.

giving $\zeta = f(h_1 - h_0)/h_0 < 0$, where f is evaluated at the upstream latitude of the streamline. Because of the clockwise vorticity, the fluid starts to move south at $x = 0$. The southward movement decreases f, so that ζ must correspondingly increase to keep $(f + \zeta)$ constant. This means that the clockwise curvature of the stream reduces, and eventually becomes a counterclockwise curvature. In this manner an eastward flow over a step generates stationary undulatory flow on the downstream side. In Section 13.15 we shall see that the stationary oscillation is due to a Rossby wave generated at the step whose westward phase velocity is canceled by the eastward current. We shall see that the wavelength is $2\pi\sqrt{U/\beta}$.

Suppose we try the same argument for a *westward* flow over a step. Then a particle should suddenly acquire clockwise vorticity as the depth of flow decreases at $x = 0$, which

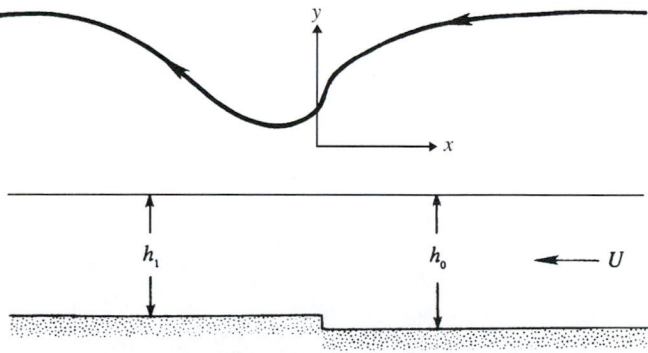

FIGURE 13.21 Westward flow over a step change in depth. Unlike the eastward flow depicted in Figure 13.20, the westward flow is not oscillatory and feels the upstream influence of the step. Looking down from above, the step causes one southward deflection that starts before the step and recovers after it.

would require the particle to move north. It would then come into a region of larger f, which would require ζ to decrease further. Clearly, an exponential behavior is predicted, suggesting that the argument is not correct. Unlike an eastward flow, a westward current feels the *upstream* influence of the step so that it acquires a counterclockwise curvature *before* it encounters the step (Figure 13.21). The positive vorticity is balanced by a reduction in f, which is consistent with conservation of potential vorticity. At the location of the step the vorticity decreases suddenly. Finally, far downstream of the step a fluid particle is again moving westward at its original latitude. The westward flow over a topography is *not* oscillatory.

13.14. INTERNAL WAVES

In Chapter 7.8 we studied internal gravity waves unaffected by Coriolis forces. We saw that they are not isotropic; in fact the direction of propagation with respect to the vertical determines their frequency. We also saw that their frequency satisfies the inequality $\omega \leq N$, where N is the buoyancy frequency. Their phase-velocity vector $\mathbf{c}$ and the group-velocity vector $\mathbf{c}_g$ are perpendicular and have oppositely directed vertical components (Figure 7.29 and Figure 7.31). That is, phases propagate upward if the groups propagate downward, and vice versa. In this section we shall study the effect of Coriolis forces on internal waves, assuming that f is independent of latitude.

Internal waves are ubiquitous in the atmosphere and the ocean. In the lower atmosphere turbulent motions dominate, so that internal wave activity represents a minor component of the motion. In contrast, the stratosphere contains a great deal of internal wave activity, and very little convective motion, because of its stable density distribution. They generally propagate upward from the lower atmosphere, where they are generated. In the ocean they may be as common as the waves on the surface, and measurements show that they can cause the isotherms to go up and down by as much as $50-100$ m. Sometimes the internal waves break and generate smaller-scale turbulence in a somewhat similar manner to bubble and foam generation by breaking surface waves.

We shall now examine the nature of the fluid motion in internal waves. The equations of motion are:

$$\frac{\partial u}{\partial x} + \frac{\partial v}{\partial y} + \frac{\partial w}{\partial z} = 0,$$

$$\frac{\partial u}{\partial t} - fv = -\frac{1}{\rho_0}\frac{\partial p}{\partial x},$$

$$\frac{\partial v}{\partial t} + fu = -\frac{1}{\rho_0}\frac{\partial p}{\partial y}, \tag{13.95}$$

$$\frac{\partial w}{\partial t} = -\frac{1}{\rho_0}\frac{\partial p}{\partial z} - \frac{\rho g}{\rho_0},$$

$$\frac{\partial \rho}{\partial t} - \frac{\rho_0 N^2}{g}w = 0.$$

We have not made the hydrostatic assumption because we are *not* assuming that the horizontal wavelength is long compared to the vertical wavelength. The advective term in the density equation is written in a linearized form $w(d\bar{\rho}/dz) = -\rho_0 N^2 w/g$. Thus the rate of change of density at a point is assumed to be due only to the vertical advection of the background density distribution $\bar{\rho}(z)$. Because internal wave activity is more intense in the thermocline where N varies appreciably (Figure 13.2), we shall be somewhat more general than in Chapter 7 and let N be depth independent.

An equation for w can be formed from the set (13.95) by eliminating all other variables. The algebraic steps of such a procedure are shown in Section 7.8 without the Coriolis forces. This gives

$$\frac{\partial^2}{\partial t^2}\nabla^2 w + N^2 \nabla_{H}^2 w + f^2\frac{\partial^2 w}{\partial z^2} = 0, \tag{13.96}$$

where $\nabla^2 \equiv \partial^2/\partial x^2 + \partial^2/\partial y^2 + \partial^2/\partial z^2$ and $\nabla_{H}^2 \equiv \partial^2/\partial x^2 + \partial^2/\partial y^2$. Because the coefficients in (13.96) are independent of the horizontal directions, equation (13.96) can have solutions that are trigonometric in x and y. We therefore assume a solution of the form

$$[u, v, w] = [\hat{u}(z), \hat{v}(z), \hat{w}(z)]\, e^{i(kx+ly-\omega t)}. \tag{13.97}$$

Substitution into (13.96) gives

$$(-i\omega)^2\left[(ik)^2 + (il)^2 + \frac{d^2}{dz^2}\right]\hat{w} + N^2\left[(ik)^2 + (il)^2\right]\hat{w} + f^2\frac{d^2\hat{w}}{dz^2} = 0,$$

from which we obtain

$$\frac{d^2\hat{w}}{dz^2} + \frac{(N^2 - \omega^2)(k^2 + l^2)}{\omega^2 - f^2}\hat{w} = 0. \tag{13.98}$$

Defining

$$m^2(z) \equiv \frac{(k^2 + l^2)[N^2(z) - \omega^2]}{\omega^2 - f^2}, \tag{13.99}$$

equation (13.98) becomes

$$\frac{d^2\hat{w}}{dz^2} + m^2\hat{w} = 0. \tag{13.100}$$

For $m^2 < 0$, the solutions of (13.100) are exponential in z signifying that the resulting motion is surface-trapped. It represents a surface wave propagating horizontally. For a positive m^2, on the other hand, solutions are trigonometric in z, giving internal waves propagating vertically as well as horizontally. From (13.99), therefore, internal waves are possible only in the frequency range:

$$f < \omega < N,$$

where we have assumed $N > f$, as is true for much of the atmosphere and the ocean.

WKB Solution

To proceed further, we assume that $N(z)$ is a slowly varying function in that its fractional change over a vertical wavelength is much less than unity. We are therefore considering only those internal waves whose vertical wavelength is short compared to the scale of variation of N. If H is a characteristic vertical distance over which N varies appreciably, then we are assuming that

$$Hm \gg 1.$$

For such slowly varying $N(z)$, we expect that $m(z)$ given by (13.99) is also a slowly varying function, that is, $m(z)$ changes by a small fraction in a distance $1/m$. Under this assumption the waves *locally* behave like plane waves, as if m is constant. This is the so-called *WKB approximation* (after Wentzel-Kramers-Brillouin), which applies when the properties of the medium (in this case N) are slowly varying.

To derive the approximate WKB solution of equation (13.100), we look for a solution in the form

$$\widehat{w} = A(z)e^{i\phi(z)},$$

where the phase ϕ and the (slowly varying) amplitude A are real. (No generality is lost by assuming A to be real. Suppose it is complex and of the form $A = \overline{A}exp(i\alpha)$, where $\overline{A}$ and α are real. Then $\widehat{w} = \overline{A}exp\,[i(\phi + \alpha)]$, a form in which $(\phi + \alpha)$ is the phase.) Substitution into (13.100) gives

$$\frac{d^2A}{dz^2} + A\left[m^2 - \left(\frac{d\phi}{dz}\right)^2\right] + i2\frac{dA}{dz}\frac{d\phi}{dz} + iA\frac{d^2\phi}{dz^2} = 0.$$

Equating the real and imaginary parts, we obtain

$$\frac{d^2A}{dz^2} + A\left[m^2 - \left(\frac{d\phi}{dz}\right)^2\right] = 0, \quad 2\frac{dA}{dz}\frac{d\phi}{dz} + A\frac{d^2\phi}{dz^2} = 0. \qquad (13.101, 13.102)$$

In (13.101) the term d^2A/dz^2 is negligible because its ratio with the second term is

$$\frac{d^2A/dz^2}{Am^2} \sim \frac{1}{H^2m^2} \ll 1.$$

Equation (13.101) then becomes approximately

$$\frac{d\phi}{dz} = \pm m, \qquad (13.103)$$

whose solution is

$$\phi = \pm \int^z m\,dz,$$

the lower limit of the integral being arbitrary.

The amplitude is determined by writing (13.102) in the form

$$\frac{dA}{A} = -\frac{(d^2\phi/dz^2)dz}{2(d\phi/dz)} = -\frac{(dm/dz)dz}{2m} = -\frac{1}{2}\frac{dm}{m},$$

where (13.103) has been used. Integrating, we obtain $\ln A = -\frac{1}{2}\ln m + \text{const.}$, that is,

$$A = \frac{A_0}{\sqrt{m}},$$

where A_0 is a constant. The WKB solution of (13.100) is therefore

$$\widehat{w} = \frac{A_0}{\sqrt{m}}e^{\pm i\int^z m\,dz}. \tag{13.104}$$

Because of neglect of the β-effect, the waves must behave similarly in x and y, as indicated by the symmetry of the dispersion relation (13.99) in k and l. Therefore, we lose no generality by orienting the x-axis in the direction of propagation and taking:

$$k > 0 \quad l = 0 \quad \omega > 0.$$

To find u and v in terms of w, use the continuity equation $\partial u/\partial x + \partial w/\partial z = 0$, noting that the y-derivatives are zero because of our setting $l = 0$. Substituting the wave solution (13.97) into the continuity equation gives

$$ik\widehat{u} + \frac{d\widehat{w}}{dz} = 0. \tag{13.105}$$

The z-derivative of $\widehat{w}$ in (13.104) can be obtained by treating the denominator $\sqrt{m}$ as approximately constant because the variation of $\widehat{w}$ is dominated by the wiggly behavior of the local plane wave solution. This gives

$$\frac{d\widehat{w}}{dz} = \frac{A_0}{\sqrt{m}}(\pm im)e^{\pm i\int^z m\,dz} = \pm iA_0\sqrt{m}e^{\pm i\int^z m\,dz},$$

so that equation (13.105) becomes

$$\widehat{u} = \mp\frac{A_0\sqrt{m}}{k}e^{\pm i\int^z m\,dz}. \tag{13.106}$$

An expression for $\widehat{v}$ can now be obtained from the horizontal equations of motion in (13.95). Cross differentiating, we obtain the vorticity equation

$$\frac{\partial}{\partial t}\left(\frac{\partial u}{\partial y} - \frac{\partial v}{\partial x}\right) = f\left(\frac{\partial u}{\partial x} + \frac{\partial v}{\partial y}\right).$$

Using the wave solution (13.97), this gives

$$\frac{\widehat{u}}{\widehat{v}} = \frac{i\omega}{f}.$$

Equation (13.106) then gives

$$\hat{v} = \pm \frac{if}{\omega} \frac{A_0\sqrt{m}}{k} e^{\pm i \int^z m\, dz}. \tag{13.107}$$

Taking real parts of equations (13.104), (13.106), and (13.107), we obtain the velocity field:

$$u = \mp \frac{A_0\sqrt{m}}{k} \cos\left(kx \pm \int^z m\, dz - \omega t \right),$$

$$v = \mp \frac{A_0 f \sqrt{m}}{\omega k} \sin\left(kx \pm \int^z m\, dz - \omega t \right), \tag{13.108}$$

$$w = \frac{A_0}{\sqrt{m}} \cos\left(kx \pm \int^z m\, dz - \omega t \right),$$

where the dispersion relation is

$$m^2 = \frac{k^2(N^2 - \omega^2)}{\omega^2 - f^2}. \tag{13.109}$$

The meaning of $m(z)$ is clear from (13.108). If we call the argument of the trigonometric terms the *phase*, then it is apparent that $\partial(phase)/\partial z = m(z)$, so that $m(z)$ is the *local* vertical wave number. Because we are treating $k, m, \omega > 0$, it is also apparent that the *upper signs represent waves with upward phase propagation, and the lower signs represent downward phase propagation*.

Particle Orbit

To find the shape of the hodograph in the horizontal plane, consider the point $x = z = 0$. Then (13.108) gives:

$$u = \mp \cos \omega t,$$

$$v = \pm \frac{f}{\omega} \sin \omega t, \tag{13.110}$$

where the amplitude of u has been arbitrarily set to one. Taking the upper signs in (13.110), the values of u and v are indicated in Figure 13.22a for three values of time corresponding to $\omega t = 0, \pi/2$, and π. It is clear that the horizontal hodographs are clockwise ellipses, with the major axis in the direction of propagation x, and the axis ratio is f/ω. The same conclusion applies for the lower signs in (13.110). The particle orbits in the horizontal plane are therefore identical to those of Poincaré waves (Figure 13.15).

However, the plane of the motion is no longer horizontal. From the velocity components equation (13.108), we note that

$$\frac{u}{w} = \mp \frac{m}{k} = \mp \tan \theta, \tag{13.111}$$

where $\theta = \tan^{-1}(m/k)$ is the angle made by the wave number vector $\mathbf{K}$ with the horizontal (Figure 13.23). For upward phase propagation, equation (13.111) gives $u/w = -\tan \theta$, so that w is negative if u is positive, as indicated in Figure 13.23. A three-dimensional sketch of the particle orbit is shown in Figure 13.22b. It is easy to show (Exercise 13.6) that the phase

(a)

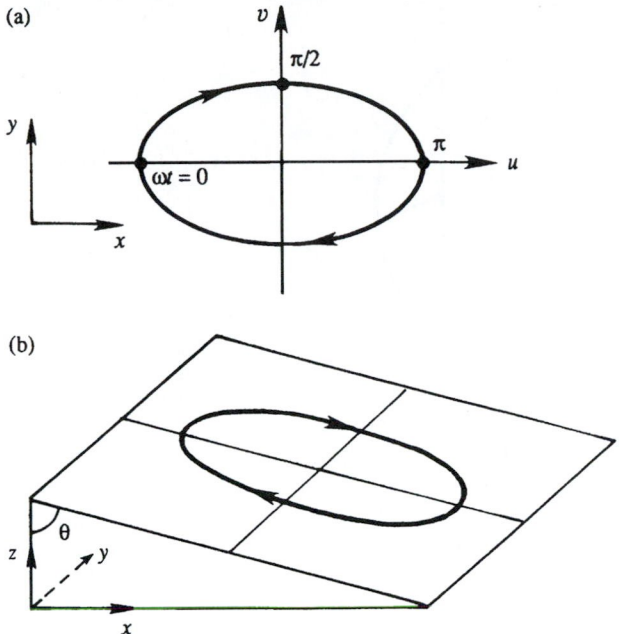

(b)

FIGURE 13.22 Particle orbit in an internal wave having x-direction wave number $k \neq 0$, and y-direction wave number $l = 0$. The upper panel (a) shows a projection on a horizontal plane; points corresponding to $\omega t = 0, \pi/2$, and π are indicated. The sense of rotation is the same as that of the surface-gravity-wave particle orbit shown in Figure 13.15 and is valid for the northern hemisphere. The lower panel (b) shows a three-dimensional view of the orbit.

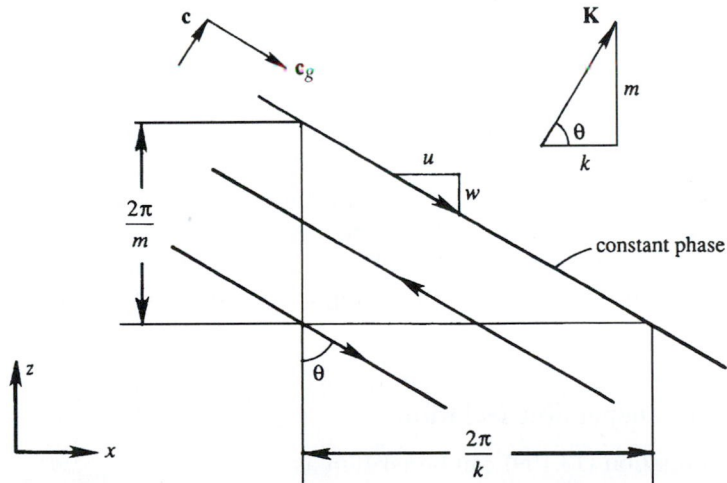

FIGURE 13.23 Vertical section of an internal wave. The three parallel lines are constant phase lines corresponding to one full wavelength, with the arrows indicating fluid motion along the lines. The phase velocity is perpendicular to the crests. The group velocity is parallel to the crests. The angle θ of the wave number with respect to the horizontal depends on the wave frequency ω, the buoyancy frequency N, and the local inertial frequency f.

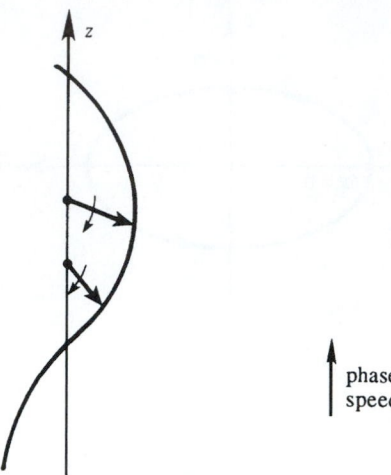

FIGURE 13.24 Helical-spiral traced out by the tips of instantaneous velocity vectors in an internal wave with upward phase speed. Heavy arrows show the velocity vectors at two depths, and light arrows indicate that they are rotating clockwise with increasing time. Note that the instantaneous vectors turn clockwise with increasing depth.

velocity vector c is in the direction of $\mathbf{K}$, that $\mathbf{c}$ and $\mathbf{c}_g$ are perpendicular, and that the fluid motion $\mathbf{u}$ is parallel to $\mathbf{c}_g$; these facts are discussed in Chapter 7 for internal waves unaffected by Coriolis forces.

The velocity vector at any location rotates clockwise with time. Because of the vertical propagation of phase, the tips of the *instantaneous* vectors also turn with *depth*. Consider the turning of the velocity vectors with depth when the phase velocity is upward, so that the deeper currents have a phase lead over the shallower currents (Figure 13.24). Because the currents at all depths rotate clockwise in *time* (whether the vertical component of $\mathbf{c}$ is upward or downward), it follows that the tips of the instantaneous velocity vectors should fall on a helical spiral that turns clockwise with *depth*. Only such a turning in depth, coupled with a clockwise rotation of the velocity vectors with time, can result in a phase lead of the deeper currents. In the opposite case of a *downward* phase propagation, the helix turns *counterclockwise* with depth. The direction of turning of the velocity vectors can also be found from (13.108), by considering $x = t = 0$ and finding u and v at various values of z.

Discussion of the Dispersion Relation

The dispersion relation (13.109) can be written as

$$\omega^2 - f^2 = \frac{k^2}{m^2}(N^2 - \omega^2). \tag{13.112}$$

Introducing $\tan \theta = m/k$, (13.112) becomes

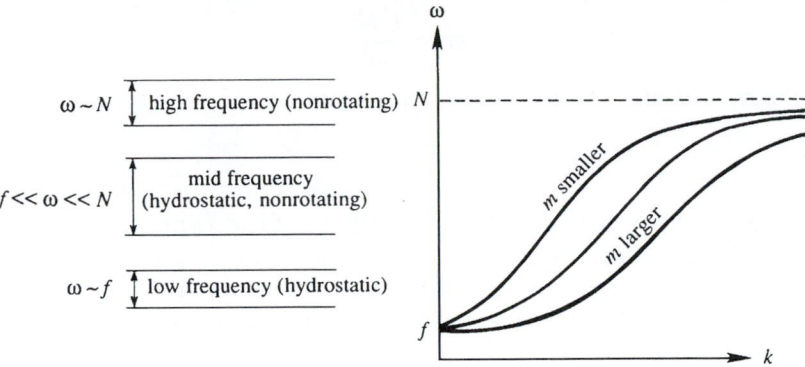

FIGURE 13.25 Dispersion relation for internal waves. The different regimes are indicated on the left-hand side of the figure. The wave frequency ω increases monotonically with increasing horizontal wave number k. The buoyancy frequency N and the local inertial frequency f set the upper and lower limits for ω.

$$\omega^2 = f^2\sin^2\theta + N^2\cos^2\theta,$$

which shows that ω is a function of the angle made by the wave number with the horizontal and is not a function of the magnitude of $\mathbf{K}$. For $f = 0$ the aforementioned expression reduces to $\omega = N \cos \theta$, derived in Section 7.8 without Coriolis forces.

A plot of the dispersion relation (13.112) is presented in Figure 13.25, showing ω as a function of k for various values of m. All curves pass through the point $\omega = f$, which represents inertial oscillations. Typically, $N \gg f$ in most of the atmosphere and the ocean. Because of the wide separation of the upper and lower limits of the internal wave range $f \le \omega \le N$, various limiting cases are possible, as indicated in Figure 13.25. They are

(1) *High-frequency regime ($\omega \sim N$, but $\omega \le N$):* In this range f^2 is negligible in comparison with ω^2 in the denominator of the dispersion relation (13.109), which reduces to:

$$m^2 \simeq \frac{k^2(N^2 - \omega^2)}{\omega^2}, \quad \text{that is,} \quad \omega^2 \simeq \frac{N^2k^2}{m^2 + k^2}.$$

Using $\tan \theta = m/k$, this gives $\omega = N \cos \theta$. Thus, the high-frequency internal waves are the same as the nonrotating internal waves discussed in Chapter 7.

(2) *Low-frequency regime ($\omega \sim f$, but $\omega \ge f$):* In this range ω^2 can be neglected in comparison to N^2 in the dispersion relation (13.109), which becomes:

$$m^2 \simeq \frac{k^2N^2}{\omega^2 - f^2}, \quad \text{that is,} \quad \omega^2 \simeq f^2 + \frac{k^2N^2}{m^2}.$$

The low-frequency limit is obtained by making the hydrostatic assumption, that is, neglecting $\partial w/\partial t$ in the vertical equation of motion.

(3) *Mid-frequency regime ($f \ll \omega \ll N$):* In this range the dispersion relation (13.109) simplifies to

$$m^2 \simeq \frac{k^2 N^2}{\omega^2},$$

so that *both* the hydrostatic and the nonrotating assumptions are applicable.

Lee Wave

Internal waves are frequently found in the *lee* (that is, the downstream side) of mountains. In stably stratified conditions, the flow of air over a mountain causes a vertical displacement of fluid particles, which sets up internal waves as it moves downstream of the mountain. If the amplitude is large and the air is moist, the upward motion causes condensation and cloud formation.

Due to the effect of a mean flow, the lee waves are stationary with respect to the ground. This is shown in Figure 13.26, where the westward phase speed is canceled by the eastward mean flow. We shall determine what wave parameters make this cancellation possible. The frequency of lee waves is much larger than f, so that rotational effects are negligible. The dispersion relation is therefore

$$\omega^2 = \frac{N^2 k^2}{m^2 + k^2}. \tag{13.113}$$

However, we now have to introduce the effects of the mean flow. The dispersion relation (13.113) is still valid if ω is interpreted as the *intrinsic frequency*, that is, the frequency measured in a frame of reference moving with the mean flow. In a medium moving with a velocity $\mathbf{U}$, the *observed frequency* of waves at a fixed point is Doppler shifted to

$$\omega_0 = \omega + \mathbf{K} \cdot \mathbf{U},$$

where ω is the intrinsic frequency; this is discussed further in Section 7.1. For a stationary wave $\omega_0 = 0$, which requires that the intrinsic frequency is $\omega = -\mathbf{K} \cdot \mathbf{U} = kU$. (Here $-\mathbf{K} \cdot \mathbf{U}$ is

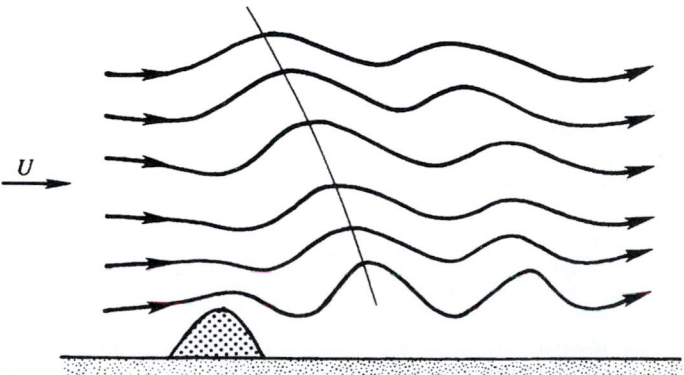

FIGURE 13.26 Schematic streamlines in a lee wave downstream of a mountain. The thin line drawn through crests shows that the phase propagates downward and westward when the eastward velocity U is accounted for.

positive because **K** is westward and **U** is eastward.) The dispersion relation (13.113) then gives

$$U = \frac{N}{\sqrt{k^2 + m^2}}.$$

If the flow speed U is given, and the mountain introduces a typical horizontal wave number k, then the preceding equation determines the vertical wave number m that generates stationary waves. Waves that do not satisfy this condition would radiate away.

The energy source of lee waves is at the surface. The energy therefore must propagate upward, and consequently the phases propagate downward. The intrinsic phase speed is therefore westward and downward in Figure 13.26. With this information, we can determine which way the constant phase lines should tilt in a stationary lee wave. Note that the wave pattern in Figure 13.26 would propagate to the left in the absence of a mean velocity, and only with the constant phase lines tilting backward with height would the flow at larger height lead the flow at a lower height.

Further discussion of internal waves can be found in Phillips (1977) and Munk (1981); lee waves are discussed in Holton (1979).

13.15. ROSSBY WAVE

To this point we have discussed wave motions that are possible with a constant Coriolis frequency f and found that these waves have frequencies larger than f. We shall now consider wave motions that owe their existence to the variation of f with latitude. With such a variable f, the equations of motion allow a very important type of wave motion called the *Rossby wave*. Their spatial scales are so large in the atmosphere that they usually have only a few wavelengths around the entire globe (Figure 13.27). This is why Rossby waves are also called *planetary waves*. In the ocean, however, their wavelengths are only about 100 km. Rossby-wave frequencies obey the inequality $\omega \ll f$. Because of this slowness the time derivative terms are an order of magnitude smaller than the Coriolis acceleration and the pressure gradients in the horizontal equations of motion. Such *nearly* geostrophic flows are called *quasi-geostrophic motions*.

Quasi-Geostrophic Vorticity Equation

We shall first derive the governing equation for quasi-geostrophic motions. For simplicity, we shall make the customary β-plane approximation valid for $\beta y \ll f_0$, keeping in mind that the approximation is not a good one for atmospheric Rossby waves, which have planetary scales. Although Rossby waves are frequently superposed on a mean flow, we shall derive the equations without a mean flow, and superpose a uniform mean flow at the end, assuming that the perturbations are small and that a linear superposition is valid. The first step is to simplify the vorticity equation for quasi-geostrophic motions, assuming that the *velocity is geostrophic to the lowest order*. The small departures from geostrophy, however, are important because they determine the *evolution* of the flow with time.

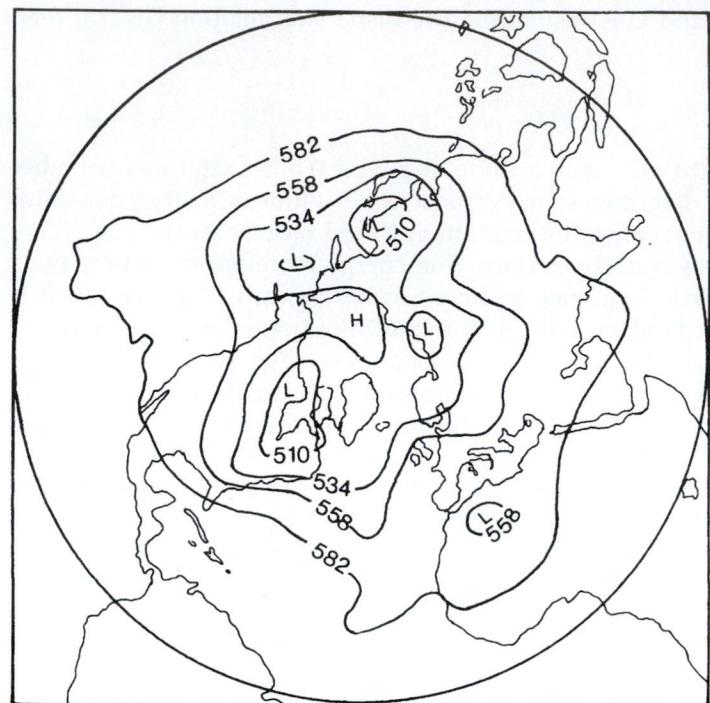

FIGURE 13.27 Observed height (in decameters = km/100) of the 50 kPa pressure surface in the northern hemisphere. The center of the picture represents the north pole. The undulations are due to Rossby waves. *J. T. Houghton, The Physics of the Atmosphere, 1986; reprinted with the permission of Cambridge University Press.*

We start with the shallow-water potential vorticity equation:

$$\frac{D}{Dt}\left(\frac{\zeta + f}{h}\right) = 0,$$

which can be written as

$$h\frac{D}{Dt}(\zeta + f) - (\zeta + f)\frac{Dh}{Dt} = 0.$$

We now expand the material derivative and substitute $h = H + \eta$, where H is the uniform undisturbed depth of the layer, and η is the surface displacement. This gives

$$(H + \eta)\left(\frac{\partial \zeta}{\partial t} + u\frac{\partial \zeta}{\partial x} + v\frac{\partial \zeta}{\partial y} + \beta v\right) - (\zeta + f_0)\left(\frac{\partial \eta}{\partial t} + u\frac{\partial \eta}{\partial x} + v\frac{\partial \eta}{\partial y}\right) = 0. \qquad (13.114)$$

Here, we have used $Df/Dt = v(df/dy) = \beta v$. We have also replaced f by f_0 in the second term because the β-plane approximation neglects the variation of f *except* when it involves df/dy. For small perturbations we can neglect the quadratic nonlinear terms in (13.114), obtaining

$$H\frac{\partial \zeta}{\partial t} + H\beta v - f_0\frac{\partial \eta}{\partial t} = 0. \qquad (13.115)$$

This is the linearized form of the potential vorticity equation. Its quasi-geostrophic version is obtained if we substitute the approximate geostrophic expressions for velocity:

$$u \simeq -\frac{g}{f_0}\frac{\partial \eta}{\partial y},$$

$$v \simeq \frac{g}{f_0}\frac{\partial \eta}{\partial x}. \tag{13.116}$$

From this the vorticity is found as

$$\zeta = \frac{g}{f_0}\left(\frac{\partial^2 \eta}{\partial x^2} + \frac{\partial^2 \eta}{\partial y^2}\right),$$

so that the vorticity equation (13.115) becomes

$$\frac{gH}{f_0}\frac{\partial}{\partial t}\left(\frac{\partial^2 \eta}{\partial x^2} + \frac{\partial^2 \eta}{\partial y^2}\right) + \frac{gH\beta}{f_0}\frac{\partial \eta}{\partial x} - f_0\frac{\partial \eta}{\partial t} = 0.$$

Denoting $c = \sqrt{gH}$, this becomes

$$\frac{\partial}{\partial t}\left(\frac{\partial^2 \eta}{\partial x^2} + \frac{\partial^2 \eta}{\partial y^2} - \frac{f_0^2}{c^2}\eta\right) + \beta\frac{\partial \eta}{\partial x} = 0. \tag{13.117}$$

This is the quasi-geostrophic form of the linearized vorticity equation, which governs the flow of large-scale motions. The ratio c/f_0 is recognized as the Rossby radius. Note that we have not set $\partial \eta/\partial t = 0$ in (13.115) during the derivation of (13.117), although a strict validity of the geostrophic relations (13.116) would require that the horizontal divergence, and hence $\partial \eta/\partial t$, be zero. This is because the *departure* from strict geostrophy determines the evolution of the flow described by (13.117). We can therefore use the geostrophic relations for velocity everywhere except in the horizontal divergence term in the vorticity equation.

Dispersion Relation

Assume solutions of the form:

$$\eta = \hat{\eta}\, e^{i(kx+ly-\omega t)}.$$

We shall regard ω as positive; the signs of k and l then determine the direction of phase propagation. A substitution into the vorticity equation (13.117) gives

$$\omega = -\frac{\beta k}{k^2 + l^2 + f_0^2/c^2}. \tag{13.118}$$

This is the dispersion relation for *Rossby waves*. The asymmetry of the dispersion relation with respect to k and l signifies that the wave motion is not isotropic in the horizontal, which is expected because of the β-effect. Although we have derived it for a single homogeneous layer, it is equally applicable to stratified flows if c is replaced by the corresponding *internal*

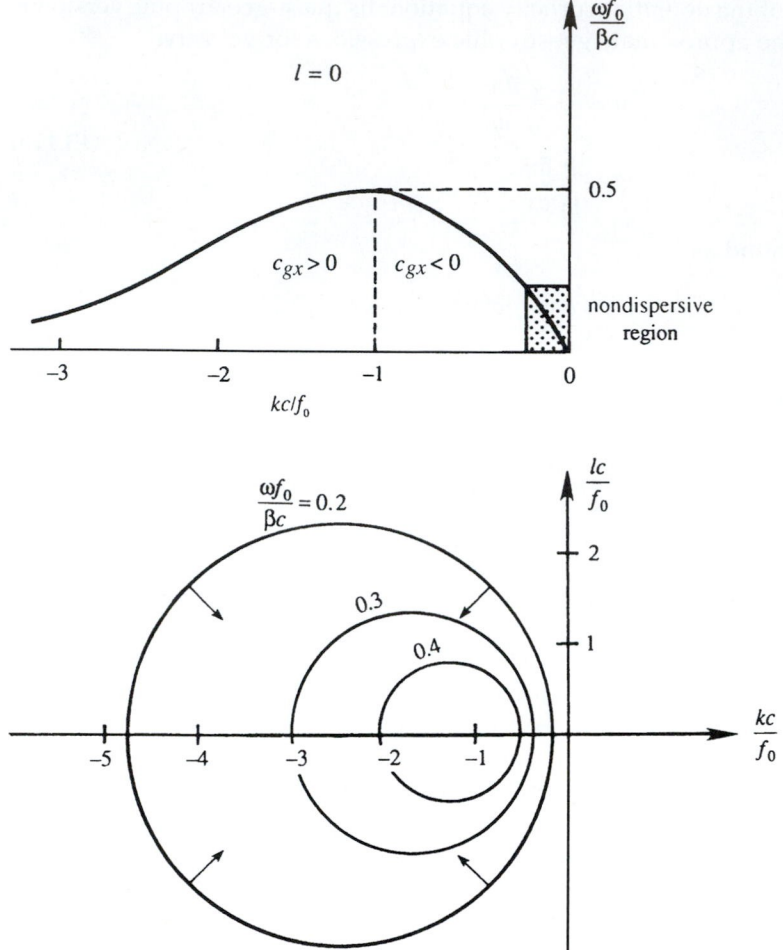

FIGURE 13.28 Dispersion relation $\omega(k,l)$ for a Rossby wave. The upper panel shows ω versus k for $l=0$. Regions of positive and negative group velocity c_{gx} are indicated. The lower panel shows a plane view of the surface $\omega(k,l)$, showing contours of constant ω on a kl-plane. The values of $\omega f_0/\beta c$ for the three circles are 0.2, 0.3, and 0.4. Arrows perpendicular to ω contours indicate directions of group velocity vector $\mathbf{c}_g$. A. E. Gill, Atmosphere–Ocean Dynamics, 1982; reprinted with the permission of Academic Press and Mrs. Helen Saunders-Gill.

value, which is $c = \sqrt{g'H}$ for the reduced-gravity model (see Section 7.7) and $c = NH/n\pi$ for the nth mode of a continuously stratified model. For the barotropic mode c is very large, and f_0^2/c^2 is usually negligible in the denominator of (13.118).

The dispersion relation $\omega(k,l)$ in (13.118) can be displayed as a surface, taking k and l along the horizontal axes and ω along the vertical axis. The section of this surface along $l=0$ is indicated in the upper panel of Figure 13.28, and sections of the surface for three values of ω are indicated in the bottom panel. The contours of constant ω are circles because the dispersion relation (13.118) can be written as

$$\left(k + \frac{\beta}{2\omega}\right)^2 + l^2 = \left(\frac{\beta}{2\omega}\right)^2 - \frac{f_0^2}{c^2}.$$

The definition of group velocity,

$$c_g = \mathbf{e}_x \frac{\partial \omega}{\partial k} + \mathbf{e}_y \frac{\partial \omega}{\partial l},$$

shows that the group velocity vector is the gradient of ω in the wave number space. The direction of $\mathbf{c}_g$ is therefore perpendicular to the ω contours, as indicated in the lower panel of Figure 13.28. For $l = 0$, the maximum frequency and zero group speed are attained at $kc/f_0 = -1$, corresponding to $\omega_{max} f_0/\beta c = 0.5$. The maximum frequency is much smaller than the Coriolis frequency. For example, in the ocean the ratio $\omega_{max}/f_0 = 0.5\beta c/f_0^2$ is of order 0.1 for the barotropic mode, and of order 0.001 for a baroclinic mode, taking a typical mid-latitude value of $f_0 \sim 10^{-4}$ s^{-1}, a barotropic gravity wave speed of $c \sim 200$ m/s, and a baroclinic gravity wave speed of $c \sim 2$ m/s. The shortest period of mid-latitude baroclinic Rossby waves in the ocean can therefore be more than a year.

The eastward phase speed is

$$c_x = \frac{\omega}{k} = -\frac{\beta}{k^2 + l^2 + f_0^2/c^2}. \tag{13.119}$$

The negative sign shows that the *phase propagation is always westward*. The phase speed reaches a maximum when $k^2 + l^2 \to 0$, corresponding to very large wavelengths represented by the region near the origin of Figure 13.28. In this region the waves are nearly nondispersive and have an eastward phase speed:

$$c_x \simeq -\frac{\beta c^2}{f_0^2}.$$

With $\beta = 2 \times 10^{-11}$ m^{-1} s^{-1}, a typical baroclinic value of $c \sim 2$ m/s, and a mid-latitude value of $f_0 \sim 10^{-4}$ s^{-1}, this gives $c_x \sim 10^{-2}$ m/s. At these slow speeds the Rossby waves would take years to cross the width of the ocean at mid-latitudes. The Rossby waves in the ocean are therefore more important at lower latitudes, where they propagate faster. (The dispersion relation (13.118), however, is not valid within a latitude band of 3° from the equator, for then the assumption of a near geostrophic balance breaks down. A different analysis is needed in the tropics. A discussion of the wave dynamics of the tropics is given in Gill (1982) and in the review paper by McCreary (1985).) In the atmosphere c is much larger, and consequently the Rossby waves propagate faster. A typical large atmospheric disturbance can propagate as a Rossby wave at a speed of several meters per second.

Frequently, the Rossby waves are superposed on a strong eastward mean current, such as the atmospheric jet stream. If U is the speed of this eastward current, then the observed eastward phase speed is

$$c_x = U - \frac{\beta}{k^2 + l^2 + f_0^2/c^2}. \tag{13.120}$$

Stationary Rossby waves can therefore form when the eastward current cancels the westward phase speed, giving $c_x = 0$. This is how stationary waves are formed downstream of the topographic step in Figure 13.20. A simple expression for the wavelength results if we assume

$l = 0$ and the flow is barotropic, so that f_0^2/c^2 is negligible in (13.120). This gives $U = \beta/k^2$ for stationary solutions, so that the wavelength is $2\pi\sqrt{U/\beta}$.

Finally, note that we have been rather cavalier in deriving the quasi-geostrophic vorticity equation in this section, in the sense that we have substituted the approximate geostrophic expressions for velocity without a formal ordering of the scales. Gill (1982) has given a more precise derivation, expanding in terms of a small parameter. Another way to justify the dispersion relation (13.118) is to obtain it from the general dispersion relation (13.76) derived in Section 13.10:

$$\omega^3 - c^2\omega(k^2 + l^2) - f_0^2\omega - c^2\beta k = 0. \tag{13.121}$$

For $\omega \ll f$, the first term is negligible compared to the third, reducing (13.121) to (13.118).

13.16. BAROTROPIC INSTABILITY

In Section 11.9 we discussed the inviscid stability of a shear flow $U(y)$ in a nonrotating system, and demonstrated that a necessary condition for its instability is that d^2U/dy^2 must change sign somewhere in the flow. This was called *Rayleigh's inflection point criterion*. In terms of vorticity $\overline{\zeta} = -dU/dy$, the criterion states that $d\overline{\zeta}/dy$ must change sign somewhere in the flow. We shall now show that, on a rotating earth, the criterion requires that $d(\overline{\zeta}+f)/dy$ must change sign somewhere within the flow.

Consider a horizontal current $U(y)$ in a medium of uniform density. In the absence of horizontal density gradients only the barotropic mode is allowed, and $U(y)$ does not vary with depth. The vorticity equation is

$$\left(\frac{\partial}{\partial t} + \mathbf{u}\cdot\nabla\right)(\zeta + f) = 0. \tag{13.122}$$

This is identical to the potential vorticity equation $D/Dt[(\zeta + f)/h] = 0$, with the added simplification that the layer depth is constant because $w = 0$. Let the total flow be decomposed into background flow plus a disturbance:

$$u = U(y) + u',$$
$$v = v'.$$

The total vorticity is then

$$\zeta = \overline{\zeta} + \zeta' = -\frac{dU}{dy} + \left(\frac{\partial v'}{\partial x} - \frac{\partial u'}{\partial y}\right) = -\frac{dU}{dy} + \nabla^2\psi,$$

where we have defined the perturbation stream function:

$$u' = -\frac{\partial\psi}{\partial y}, \qquad v' = \frac{\partial\psi}{\partial x}.$$

Substituting into (13.122) and linearizing, we obtain the perturbation vorticity equation:

$$\frac{\partial}{\partial t}(\nabla^2 \psi) + U\frac{\partial}{\partial x}(\nabla^2 \psi) + \left(\beta - \frac{d^2 U}{dy^2}\right)\frac{\partial \psi}{\partial x} = 0. \tag{13.123}$$

Because the coefficients of (13.123) are independent of x and t, there can be solutions of the form:

$$\psi = \widehat{\psi}(y)\, e^{ik(x-ct)}.$$

The phase speed c is complex and solutions are unstable if its imaginary part $c_i > 0$. The perturbation vorticity equation (13.123) then becomes

$$(U - c)\left[\frac{d^2}{dy^2} - k^2\right]\widehat{\psi} + \left[\beta - \frac{d^2 U}{dy^2}\right]\widehat{\psi} = 0.$$

Comparing this with (11.81) derived without Coriolis forces, it is seen that the effect of planetary rotation is the replacement of $-d^2 U/dy^2$ by $(\beta - d^2 U/dy^2)$. The analysis of the section therefore carries over to the present case, resulting in the following criterion: *A necessary condition for the inviscid instability of a barotropic current $U(y)$ is that the gradient of the absolute vorticity,*

$$\frac{d}{dy}(\bar{\zeta} + f) = \beta - \frac{d^2 U}{dy^2}, \tag{13.124}$$

must change sign somewhere in the flow. This result was first derived by Kuo (1949).

Barotropic instability quite possibly plays an important role in the instability of currents in the atmosphere and in the ocean. The instability has no preference for any latitude, because

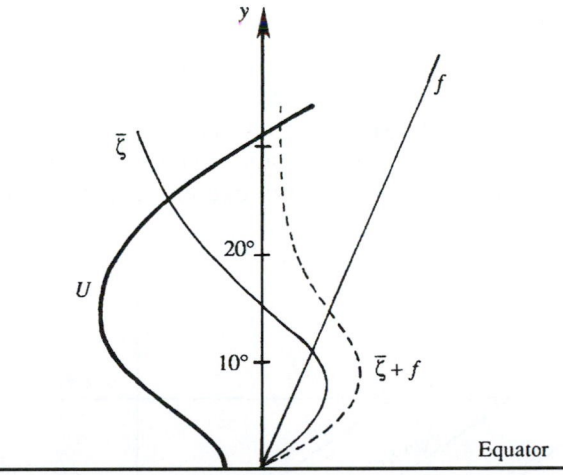

FIGURE 13.29 Profiles of velocity and vorticity in a westward tropical wind. The velocity distribution is barotropically unstable as $d(\bar{\zeta} + f)/dy$ (shown as the dashed curve) changes sign within the flow. *J. T. Houghton, The Physics of the Atmosphere, 1986; reprinted with the permission of Cambridge University Press.*

the criterion involves β and not f. However, the mechanism presumably dominates in the tropics because mid-latitude disturbances prefer the *baroclinic instability* mechanism discussed in the following section. An unstable distribution of westward tropical wind is shown in Figure 13.29.

13.17. BAROCLINIC INSTABILITY

The weather maps at mid-latitudes invariably show the presence of wavelike horizontal excursions of temperature and pressure contours, superposed on eastward mean flows such as the jet stream. Similar undulations are also found in the ocean on eastward currents such as the Gulf Stream in the north Atlantic. A typical wavelength of these disturbances is observed to be of the order of the internal Rossby radius, that is, about 4000 km in the atmosphere and 100 km in the ocean. They seem to be propagating as Rossby waves, but their erratic and unexpected appearance suggests that they are not forced by any external agency, but are due to an inherent *instability* of mid-latitude eastward flows. In other words, the eastward flows have a spontaneous tendency to develop wavelike disturbances. In this section we shall investigate the instability mechanism that is responsible for the spontaneous relaxation of eastward jets into a meandering state.

The poleward decrease of the solar irradiation results in a poleward decrease of air temperature and a consequent increase of air density. An idealized distribution of the atmospheric density in the northern hemisphere is shown in Figure 13.30. The density increases northward due to the lower temperatures near the poles and decreases upward because of static stability. According to the thermal wind relation (13.15), an eastward flow (such as the jet stream in the atmosphere or the Gulf Stream in the Atlantic) in equilibrium with such a density structure must have a velocity that increases with height. A system with inclined density surfaces, such as the one in Figure 13.30, has more potential energy than a system with horizontal density surfaces, just as a system with an inclined free surface has more potential energy than a system with a horizontal free surface. It is therefore potentially unstable because it can release the stored potential energy by means of an instability that would cause the density surfaces to flatten out. In the process, vertical shear of the mean flow $U(z)$ would decrease, and perturbations would gain kinetic energy.

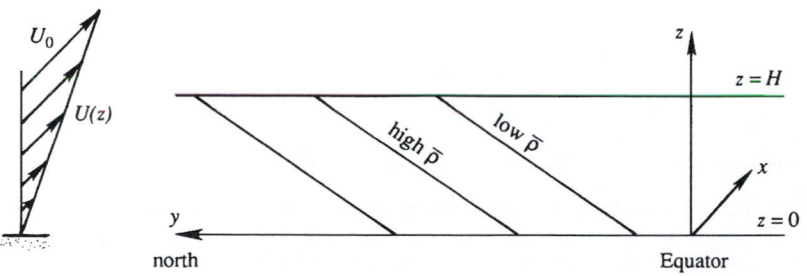

FIGURE 13.30 Lines of constant density in the northern hemispheric atmosphere. The lines are nearly horizontal and the slopes are greatly exaggerated in the figure. The velocity $U(z)$ shown at the left is into the plane of paper.

Instability of baroclinic jets that release potential energy by flattening out the density surfaces is called the *baroclinic instability*. Our analysis would show that the preferred scale of the unstable waves is indeed of the order of the Rossby radius, as observed for the mid-latitude weather disturbances. The theory of baroclinic instability was developed in the 1940s by Vilhem Bjerknes and others, and is considered one of the major triumphs of geophysical fluid mechanics. Our presentation is essentially based on the review article by Pedlosky (1971).

Consider a basic state in which the density is stably stratified in the vertical with a *uniform* buoyancy frequency N, and increases northward at a *constant* rate $\partial \bar{\rho}/\partial y$. According to the thermal wind relation, the constancy of $\partial \bar{\rho}/\partial y$ requires that the vertical shear of the basic eastward flow $U(z)$ also be constant. The β-effect is neglected as it is not an essential requirement of the instability. (The β-effect does modify the instability, however.) This is borne out by the spontaneous appearance of undulations in laboratory experiments in a rotating annulus, in which the inner wall is maintained at a higher temperature than the outer wall. The β-effect is absent in such an experiment.

Perturbation Vorticity Equation

The equations for total flow are:

$$\frac{\partial u}{\partial t} + u\frac{\partial u}{\partial x} + v\frac{\partial u}{\partial y} - fv = -\frac{1}{\rho_0}\frac{\partial p}{\partial x},$$

$$\frac{\partial v}{\partial t} + u\frac{\partial v}{\partial x} + v\frac{\partial v}{\partial y} + fu = -\frac{1}{\rho_0}\frac{\partial p}{\partial y},$$

$$0 = -\frac{\partial p}{\partial z} - \rho g, \qquad (13.125)$$

$$\frac{\partial u}{\partial x} + \frac{\partial v}{\partial y} + \frac{\partial w}{\partial z} = 0,$$

$$\frac{\partial \rho}{\partial t} + u\frac{\partial \rho}{\partial x} + v\frac{\partial \rho}{\partial y} + w\frac{\partial \rho}{\partial z} = 0,$$

where ρ_0 is a constant reference density. We assume that the total flow is composed of a basic eastward jet $U(z)$ in geostrophic equilibrium with the basic density structure $\bar{\rho}(y, z)$ shown in Figure 13.30, plus perturbations, that is:

$$u = U(z) + u'(x, y, z),$$

$$v = v'(x, y, z),$$

$$w = w'(x, y, z), \qquad (13.126)$$

$$\rho = \bar{\rho}(y, z) + \rho'(x, y, z),$$

$$p = \bar{p}(y, z) + p'(x, y, z).$$

The basic flow is in geostrophic and hydrostatic balance:

$$fU = -\frac{1}{\rho_0}\frac{\partial \bar{p}}{\partial y},$$

$$0 = -\frac{\partial \bar{p}}{\partial z} - \bar{\rho}g. \tag{13.127}$$

Eliminating the pressure, we obtain the thermal wind relation:

$$\frac{dU}{dz} = \frac{g}{f\rho_0}\frac{\partial \bar{\rho}}{\partial y}, \tag{13.128}$$

which states that the eastward flow must increase with height because $\partial\bar{\rho}/\partial y > 0$. For simplicity, we assume that $\partial\bar{\rho}/\partial y$ is constant, and that $U = 0$ at the surface $z = 0$. Thus the background flow is

$$U = \frac{U_0 z}{H},$$

where U_0 is the velocity at the top of the layer at $z = H$.

We first form a vorticity equation by cross differentiating the horizontal equations of motion in (13.125), obtaining

$$\frac{\partial \zeta}{\partial t} + u\frac{\partial \zeta}{\partial x} + v\frac{\partial \zeta}{\partial y} - (\zeta + f)\frac{\partial w}{\partial z} = 0. \tag{13.129}$$

This is identical to (13.92), except for the exclusion of the β-effect here; the algebraic steps are therefore not repeated. Substituting the decomposition (13.126), and noting that $\zeta = \zeta'$ because the basic flow $U = U_0 z/H$ has no vertical component of vorticity, (13.129) becomes

$$\frac{\partial \zeta'}{\partial t} + U\frac{\partial \zeta'}{\partial x} - f\frac{\partial w'}{\partial z} = 0, \tag{13.130}$$

where the nonlinear terms have been neglected. This is the perturbation vorticity equation, which we shall now write in terms of p'.

Assume that the perturbations are large-scale and slow, so that the velocity is nearly geostrophic:

$$u' \simeq -\frac{1}{\rho_0 f}\frac{\partial p'}{\partial y}, \quad v' \simeq \frac{1}{\rho_0 f}\frac{\partial p'}{\partial x}, \tag{13.131}$$

from which the perturbation vorticity is found as

$$\zeta' = \frac{1}{\rho_0 f}\nabla_H^2 p'. \tag{13.132}$$

We now express w' in (13.130) in terms of p'. The density equation gives

$$\frac{\partial}{\partial t}(\bar{\rho} + \rho') + (U + u')\frac{\partial}{\partial x}(\bar{\rho} + \rho') + v'\frac{\partial}{\partial y}(\bar{\rho} + \rho') + w'\frac{\partial}{\partial z}(\bar{\rho} + \rho') = 0.$$

Linearizing, we obtain

$$\frac{\partial \rho'}{\partial t} + U \frac{\partial \rho'}{\partial x} + v' \frac{\partial \bar{\rho}}{\partial y} - \frac{\rho_0 N^2 w'}{g} = 0, \tag{13.133}$$

where $N^2 = -g\rho_0^{-1}(\partial \bar{\rho}/\partial z)$. The perturbation density ρ' can be written in terms of p' by using the hydrostatic balance in (13.125), and subtracting the basic state (13.127). This gives

$$0 = -\frac{\partial p'}{\partial z} - \rho' g, \tag{13.134}$$

which states that the perturbations are hydrostatic. Equation (13.133) then gives

$$w' = -\frac{1}{\rho_0 N^2} \left[\left(\frac{\partial}{\partial t} + U \frac{\partial}{\partial x} \right) \frac{\partial p'}{\partial z} - \frac{dU}{dz} \frac{\partial p'}{\partial x} \right], \tag{13.135}$$

where we have written $\partial \bar{\rho}/\partial y$ in terms of the thermal wind dU/dz. Using (13.132) and (13.135), the perturbation vorticity equation (13.130) becomes:

$$\left(\frac{\partial}{\partial t} + U \frac{\partial}{\partial x} \right) \left[\nabla_H^2 p' + \frac{f^2}{N^2} \frac{\partial^2 p'}{\partial z^2} \right] = 0. \tag{13.136}$$

This is the equation that governs the quasi-geostrophic perturbations on an eastward current $U(z)$.

Wave Solution

We assume that the flow is confined between two horizontal planes at $z = 0$ and $z = H$ and that it is unbounded in x and y. Real flows are likely to be bounded in the y direction, especially in a laboratory situation of flow in an annular region, where the walls set boundary conditions parallel to the flow. The boundedness in y, however, simply sets up normal modes in the form $\sin(n\pi y/L)$, where L is the width of the channel. Each of these modes can be replaced by a periodicity in y. Accordingly, we assume wavelike solutions:

$$p' = \hat{p}(z) e^{i(kx+ly-\omega t)}. \tag{13.137}$$

The perturbation vorticity equation (13.136) then gives

$$\frac{d^2 \hat{p}}{dz^2} - \alpha^2 \hat{p} = 0, \tag{13.138}$$

where

$$\alpha^2 \equiv \frac{N^2}{f^2}(k^2 + l^2). \tag{13.139}$$

The solution of (13.138) can be written as

$$\hat{p} = A \cosh \alpha \left(z - \frac{H}{2} \right) + B \sinh \alpha \left(z - \frac{H}{2} \right). \tag{13.140}$$

Boundary conditions have to be imposed on (13.140) in order to derive an instability criterion. These are:

$$w' = 0 \quad \text{at } z = 0, H.$$

The corresponding conditions on p' can be found from (13.135) and $U = U_0 z/H$. We obtain:

$$-\frac{\partial^2 p'}{\partial t \, \partial z} - \frac{U_0 z}{H} \frac{\partial^2 p'}{\partial x \, \partial z} + \frac{U_0}{H} \frac{\partial p'}{\partial x} = 0 \quad \text{at } z = 0, H,$$

where we have also used $U = U_0 z/H$. The two boundary conditions are therefore:

$$\frac{\partial^2 p'}{\partial t \, \partial z} - \frac{U_0}{H} \frac{\partial p'}{\partial x} = 0 \quad \text{at } z = 0,$$

$$\frac{\partial^2 p'}{\partial t \, \partial z} - \frac{U_0}{H} \frac{\partial p'}{\partial x} + U_0 \frac{\partial^2 p'}{\partial x \, \partial z} = 0 \quad \text{at } z = H.$$

Instability Criterion

Using (13.137) and (13.140), the foregoing boundary conditions require:

$$A\left[\alpha c \sinh \frac{\alpha H}{2} - \frac{U_0}{H} \cosh \frac{\alpha H}{2}\right] + B\left[-\alpha c \cosh \frac{\alpha H}{2} + \frac{U_0}{H} \sinh \frac{\alpha H}{2}\right] = 0,$$

$$A\left[\alpha(U_0 - c) \sinh \frac{\alpha H}{2} - \frac{U_0}{H} \cosh \frac{\alpha H}{2}\right] + B\left[\alpha(U_0 - c) \cosh \frac{\alpha H}{2} - \frac{U_0}{H} \sinh \frac{\alpha H}{2}\right] = 0,$$

where $c = \omega/k$ is the eastward phase velocity.

This is a pair of homogeneous equations for the constants A and B. For nontrivial solutions to exist, the determinant of the coefficients must vanish. This gives, after some straightforward algebra, the phase velocity:

$$c = \frac{U_0}{2} \pm \frac{U_0}{\alpha H} \sqrt{\left(\frac{\alpha H}{2} - \tanh \frac{\alpha H}{2}\right)\left(\frac{\alpha H}{2} - \coth \frac{\alpha H}{2}\right)}. \tag{13.141}$$

Whether the solution grows with time depends on the sign of the radicand. The behavior of the functions under the radical sign is sketched in Figure 13.31. It is apparent that the first factor in the radicand is positive because $\alpha H/2 > \tanh(\alpha H/2)$ for all values of αH. However, the second factor is negative for small values of αH for which $\alpha H/2 < \coth(\alpha H/2)$. In this range the roots of c are complex conjugates, with $c = U_0/2 \pm i c_i$. Because we have assumed that the perturbations are of the form $\exp(-ikct)$, the existence of a nonzero c_i implies the possibility of a perturbation that grows as $\exp(kc_i t)$, and the solution is unstable. The marginal stability is given by the critical value of α satisfying:

$$\frac{\alpha_c H}{2} = \coth\left(\frac{\alpha_c H}{2}\right),$$

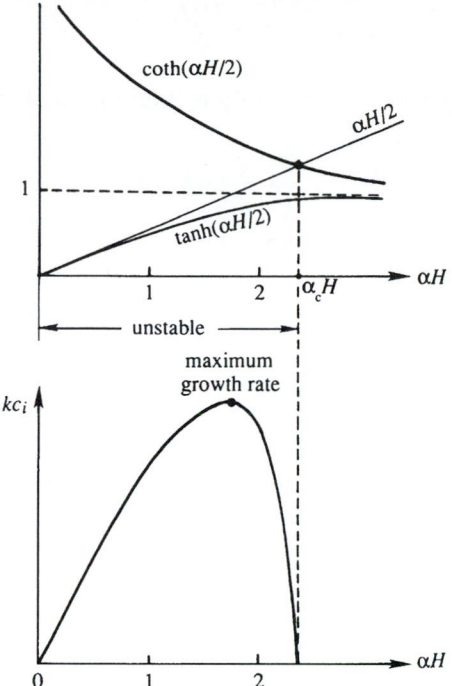

FIGURE 13.31   Baroclinic instability. The upper panel shows behavior of the functions in (13.141), and the lower panel shows growth rates of unstable waves.

whose solution is

$$\alpha_c H = 2.4,$$

and the flow is unstable if $\alpha H < 2.4$. Using the definition of α in (13.139), it follows that the flow is unstable if

$$\frac{HN}{f} < \frac{2.4}{\sqrt{k^2 + l^2}}.$$

As all values of k and l are allowed, we can always find a value of $k^2 + l^2$ low enough to satisfy the aforementioned inequality. *The flow is therefore always unstable* (*to low wave numbers*). For a north-south wave number $l = 0$, instability is ensured if the east-west wave number k is small enough such that

$$\frac{HN}{f} < \frac{2.4}{k}. \tag{13.142}$$

In a continuously stratified ocean, the speed of a long internal wave for the $n = 1$ baroclinic mode is $c = NH/\pi$, so that the corresponding internal Rossby radius is $c/f = NH/\pi f$. It is usual to omit the factor π and define the Rossby radius in a continuously stratified fluid as

$$\Lambda \equiv \frac{HN}{f}.$$

The condition (13.142) for baroclinic instability is therefore that the east-west wavelength be large enough so that

$$\lambda > 2.6\Lambda.$$

However, the wavelength $\lambda = 2.6\Lambda$ does not grow at the fastest rate. It can be shown from (13.141) that the wavelength with the largest growth rate is

$$\lambda_{max} = 3.9\Lambda.$$

This is therefore the wavelength that is observed when the instability develops. Typical values for f, N, and H suggest that $\lambda_{max} \sim 4000$ km in the atmosphere and 200 km in the ocean, which agree with observations. Waves much smaller than the Rossby radius do not grow, and the ones much larger than the Rossby radius grow very slowly.

Energetics

The foregoing analysis suggests that the existence of planet-encircling weather waves is due to the fact that small perturbations can grow spontaneously when superposed on an eastward current maintained by the sloping density surfaces (Figure 13.30). Although the basic current does have a vertical shear, the perturbations do not grow by extracting energy from the vertical shear field. Instead, they extract their energy from the *potential energy* stored in the system of sloping density surfaces. The energetics of the baroclinic instability is therefore quite different than that of the Kelvin-Helmholtz instability (which also has a vertical shear of the mean flow), where the perturbation Reynolds stress $\overline{u'w'}$ interacts with the vertical shear and extracts energy from the mean shear flow. The baroclinic instability is *not* a shear flow instability; the Reynolds stresses are too small because of the small w in quasi-geostrophic large-scale flows.

The energetics of the baroclinic instability can be understood by examining the equation for the perturbation kinetic energy. Such an equation can be derived by multiplying the equations for $\partial u'/\partial t$ and $\partial v'/\partial t$ by u' and v', respectively, adding the two, and integrating over the region of flow. Because of the assumed periodicity in x and y, the extent of the region of integration is chosen to be one wavelength in either direction. During this integration, the boundary conditions of zero normal flow on the walls and periodicity in x and y are used repeatedly. The procedure is similar to that for the derivation of (11.88) and is not repeated here. The result is

$$\frac{dK}{dt} = -g \int w'\rho' \, dx \, dy \, dz,$$

where K is the global perturbation kinetic energy:

$$K \equiv \frac{\rho_0}{2} \int (u'^2 + v'^2) \, dx \, dy \, dz.$$

FIGURE 13.32 Wedge of instability (shaded) in a baroclinic instability. The wedge is bounded by constant density lines and the horizontal. Unstable waves have a particle trajectory that falls within the wedge and causes lighter fluid particles to move upward and northward, and heaver fluid particles to move downward and southward.

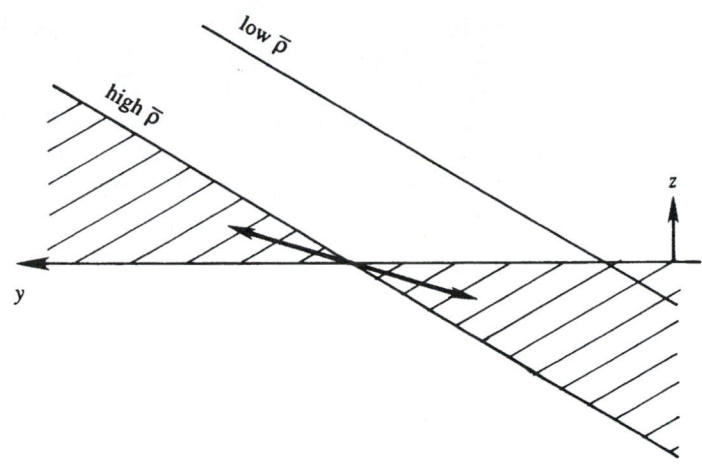

In unstable flows we must have $dK/dt > 0$, which requires that the volume integral of $w'\rho'$ must be negative. Let us denote the volume average of $w'\rho'$ by $\overline{w'\rho'}$. A negative $\overline{w'\rho'}$ means that on average the lighter fluid rises and the heavier fluid sinks. By such an interchange the center of gravity of the system, and therefore its potential energy, is lowered. The interesting point is that this cannot happen in a stably stratified system with *horizontal* density surfaces; in that case an exchange of fluid particles *raises* the potential energy. Moreover, a basic state with inclined density surfaces (Figure 13.30) cannot have $\overline{w'\rho'} < 0$ if the particle excursions are vertical. If, however, the particle excursions fall within the wedge formed by the constant density lines and the horizontal (Figure 13.32), then an exchange of fluid particles takes lighter particles upward (and northward) and denser particles downward (and southward). Such an interchange would tend to make the density surfaces more horizontal, releasing potential energy from the mean density field with a consequent growth of the perturbation energy. This type of convection is called *sloping convection*. According to Figure 13.32 the exchange of fluid particles within the *wedge of instability* results in a net poleward transport of heat from the tropics, which serves to redistribute the larger solar heat received by the tropics.

In summary, baroclinic instability draws energy from the potential energy of the mean density field. The resulting eddy motion has particle trajectories that are oriented at a small angle with the horizontal, so that the resulting heat transfer has a poleward component. The preferred scale of the disturbance is the Rossby radius.

13.18. GEOSTROPHIC TURBULENCE

Two common modes of instability of a large-scale current system were presented in the preceding sections. When the flow is strong enough, such instabilities can make a flow chaotic or turbulent. A peculiarity of large-scale turbulence in the atmosphere or the ocean is that it is essentially two dimensional in nature. The existence of the

Coriolis force, stratification, and small thickness of geophysical media severely restricts the vertical velocity in large-scale flows, which tend to be quasi-geostrophic, with the Coriolis force balancing the horizontal pressure gradient to the lowest order. Because vortex stretching, a key mechanism by which ordinary three-dimensional turbulent flows transfer energy from large to small scales, is absent in two-dimensional flow, one expects that the dynamics of geostrophic turbulence are likely to be fundamentally different from that of three-dimensional, laboratory-scale turbulence discussed in Chapter 12. However, we can still call the motion *turbulent* because it is unpredictable and diffusive.

A key result on the subject was discovered by the meteorologist Fjortoft (1953), and since then Kraichnan, Leith, Batchelor, and others have contributed to various aspects of the problem. A good discussion is given in Pedlosky (1987), to which the reader is referred for a fuller treatment. Here, we shall only point out a few important results.

An important variable in the discussion of two-dimensional turbulence is *enstrophy*, which is the mean square vorticity $\overline{\zeta^2}$. In an isotropic turbulent field we can define an energy spectrum $S(K)$, a function of the magnitude of the wave number K, as

$$\overline{u^2} = \int_0^\infty S(K)\, dK.$$

It can be shown that the enstrophy spectrum is $K^2 S(K)$, that is,

$$\overline{\zeta^2} = \int_0^\infty K^2 S(K)\, dK,$$

which makes sense because vorticity involves the spatial derivatives of velocity.

We consider a freely evolving turbulent field in which the shape of the velocity spectrum changes with time. The large scales are essentially inviscid, so that both energy and enstrophy are nearly conserved:

$$\frac{d}{dt}\int_0^\infty S(K)\, dK = 0, \quad \frac{d}{dt}\int_0^\infty K^2 S(K)\, dK = 0, \qquad (13.143, 13.144)$$

where terms proportional to the molecular viscosity v have been neglected on the right-hand sides of the equations. The enstrophy conservation is unique to two-dimensional turbulence because of the absence of vortex stretching.

Suppose that the energy spectrum initially contains all its energy at wave number K_0. Nonlinear interactions transfer this energy to other wave numbers, so that the sharp spectral peak smears out. For the sake of argument, suppose that all of the initial energy goes to two neighboring wave numbers K_1 and K_2, with $K_1 < K_0 < K_2$. Conservation of energy and enstrophy requires that:

$$S_0 = S_1 + S_2,$$
$$K_0^2 S_0 = K_1^2 S_1 + K_2^2 S_2,$$

where S_n is the spectral energy at K_n. From this we can find the ratios of energy and enstrophy spectra before and after the transfer:

$$\frac{S_1}{S_2} = \frac{K_2 - K_0}{K_0 - K_1} \frac{K_2 + K_0}{K_1 + K_0},$$

$$\frac{K_1^2 S_1}{K_2^2 S_2} = \frac{K_1^2}{K_2^2} \frac{K_2^2 - K_0^2}{K_0^2 - K_1^2}.$$

(13.145)

As an example, suppose that nonlinear smearing transfers energy to wave numbers $K_1 = K_0/2$ and $K_2 = 2K_0$. Then (13.145) shows that $S_1/S_2 = 4$ and $K_1^2 S_1/K_2^2 S_2 = \frac{1}{4}$, so that more energy goes to lower wave numbers (large scales), whereas more enstrophy goes to higher wave numbers (smaller scales). This important result on two-dimensional turbulence was derived by Fjortoft (1953). Clearly, the constraint of enstrophy conservation in two-dimensional turbulence has prevented a symmetric spreading of the initial energy peak at K_0.

The unique character of two-dimensional turbulence is evident here. In three-dimensional turbulence studied in Chapter 12, the energy goes to smaller and smaller scales until it is dissipated by viscosity. In geostrophic turbulence, on the other hand, the energy goes to larger scales, where it is less susceptible to viscous dissipation. Numerical calculations are indeed in agreement with this behavior and show that energy-containing eddies grow in size by coalescing. On the other hand, the vorticity becomes increasingly confined to thin shear layers on the eddy boundaries; these shear layers contain very little energy. The backward (or inverse) energy cascade and forward enstrophy cascade are represented schematically in Figure 13.33. It is clear that there are two *inertial* regions in the spectrum of a two-dimensional turbulent flow, namely, the energy cascade region and the enstrophy cascade region. If energy is injected into the system at a rate ε, then the energy spectrum in the energy cascade region has the form $S(K) \propto \varepsilon^{2/3} K^{-5/3}$; the argument is essentially the same as in the case of the Kolmogorov spectrum in three-dimensional turbulence (Section 12.7), except that the transfer is backward. A dimensional argument also shows that the

FIGURE 13.33 Energy and enstrophy cascade in two-dimensional turbulence. Here the two-dimensional character of the turbulence causes turbulent kinetic energy to cascade upward to larger scales, while enstrophy cascades to smaller scales.

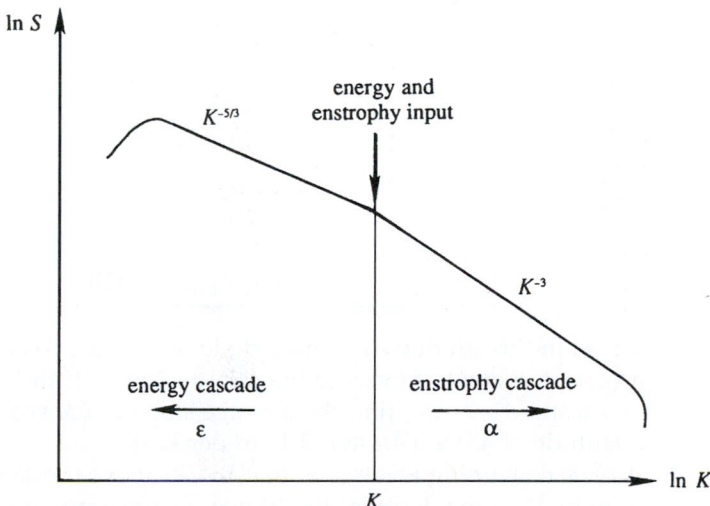

energy spectrum in the enstrophy cascade region is of the form $S(K) \propto \alpha^{2/3}K^{-3}$, where α is the forward enstrophy flux to higher wave numbers. There is negligible energy flux in the enstrophy cascade region.

As the eddies grow in size, they become increasingly immune to viscous dissipation, and the inviscid assumption implied in (13.143) becomes increasingly applicable. (This would not be the case in three-dimensional turbulence in which the eddies continue to decrease in size until viscous effects drain energy out of the flow.) In contrast, the corresponding assumption in the enstrophy conservation equation (13.144) becomes less and less valid as enstrophy goes to smaller scales, where viscous dissipation drains enstrophy out of the system. At later stages in the evolution, then, (13.144) may not be a good assumption. However, it can be shown (see Pedlosky, 1987) that the dissipation of enstrophy actually *intensifies* the process of energy transfer to larger scales, so that the *red* cascade (that is, transfer to larger scales) of energy is a general result of two-dimensional turbulence.

The eddies, however, do not grow in size indefinitely. They become increasingly slower as their length scale l increases, while their velocity scale u remains constant. The slower dynamics makes them increasingly wavelike, and the eddies transform into Rossby-wave packets as their length scale becomes of order (Rhines, 1975):

$$l \sim \sqrt{\frac{u}{\beta}} \quad \text{(Rhines length),}$$

where $\beta = df/dy$ and u is the rms fluctuating speed. The Rossby-wave propagation results in an anisotropic elongation of the eddies in the east–west ("zonal") direction, while the eddy size in the north–south direction stops growing at $\sqrt{u/\beta}$. Finally, the velocity field consists of zonally directed jets whose north–south extent is of order $\sqrt{u/\beta}$. This has been suggested as an explanation for the existence of zonal jets in the atmosphere of the planet Jupiter (Williams, 1979). The inverse energy cascade regime may not occur in the earth's atmosphere and the ocean at mid-latitudes because the Rhines length (about 1000 km in the atmosphere and 100 km in the ocean) is of the order of the internal Rossby radius, where the energy is injected by baroclinic instability. (For the inverse cascade to occur, $\sqrt{u/\beta}$ needs to be larger than the scale at which energy is injected.)

Eventually, however, the kinetic energy has to be dissipated by molecular effects at the Kolmogorov microscale η, which is of the order of a few millimeters in the ocean and the atmosphere. A fair hypothesis is that processes such as internal waves drain energy out of the mesoscale eddies, and breaking internal waves generate three-dimensional turbulence that finally cascades energy to molecular scales.

EXERCISES

13.1. The Gulf Stream flows northward along the east coast of the United States with a surface current of average magnitude 2 m/s. If the flow is assumed to be in geostrophic balance, find the average slope of the sea surface across the current at a latitude of 45°N. [*Answer:* 2.1 cm per km]

13.2. A plate containing water ($v = 10^{-6}$ m^2/s) above it rotates at a rate of 10 revolutions per minute. Find the depth of the Ekman layer, assuming that the flow is laminar.

13.3. Assume that the atmospheric Ekman layer over the earth's surface at a latitude of 45°N can be approximated by an eddy viscosity of $v_v = 10 \ m^2/s$. If the geostrophic velocity above the Ekman layer is 10 m/s, what is the Ekman transport across isobars? [*Answer*: 2203 m^2/s]

13.4. Find the axis ratio of a hodograph plot for a semidiurnal tide in the middle of the ocean at a latitude of 45°N. Assume that the mid-ocean tides are rotational surface gravity waves of long wavelength and are unaffected by the proximity of coastal boundaries. If the depth of the ocean is 4 km, find the wavelength, the phase velocity, and the group velocity. Note, however, that the wavelength is comparable to the width of the ocean, so that the neglect of coastal boundaries is not very realistic.

13.5. An internal Kelvin wave on the thermocline of the ocean propagates along the west coast of Australia. The thermocline has a depth of 50 m and has a nearly discontinuous density change of 2 kg/m^3 across it. The layer below the thermocline is deep. At a latitude of 30°S, find the direction and magnitude of the propagation speed and the decay scale perpendicular to the coast.

13.6. Using the dispersion relation $m^2 = k^2(N^2 - \omega^2)/(\omega^2 - f^2)$ for internal waves, show that the group velocity vector is given by $[c_{gx}, c_{gz}] = \dfrac{(N^2 - f^2) \, km}{(m^2 + k^2)^{3/2}(m^2 f^2 + k^2 N^2)^{1/2}}[m, -k]$.

[*Hint*: Differentiate the dispersion relation partially with respect to k and m.] Show that c_g and c are perpendicular and have oppositely directed vertical components. Verify that c_g is parallel to u.

13.7. Suppose the atmosphere at a latitude of 45°N is idealized by a uniformly stratified layer of height 10 km, across which the potential temperature increases by 50°C.
a) What is the value of the buoyancy frequency N?
b) Find the speed of a long gravity wave corresponding to the $n = 1$ baroclinic mode.
c) For the $n = 1$ mode, find the westward speed of nondispersive (i.e., very large wavelength) Rossby waves. [*Answer*: $N = 0.01279 \ s^{-1}; c_1 = 40.71 \ m/s; c_x = -3.12 \ m/s$]

13.8. Consider a steady flow rotating between plane parallel boundaries a distance L apart. The angular velocity is Ω and a small rectilinear velocity U is superposed. There is a protuberance of height $h \ll L$ in the flow. The Ekman and Rossby numbers are both small: $Ro \ll 1, E \ll 1$. Obtain an integral of the relevant equations of motion that relates the modified pressure and the streamfunction for the motion, and show that the modified pressure is constant on streamlines.

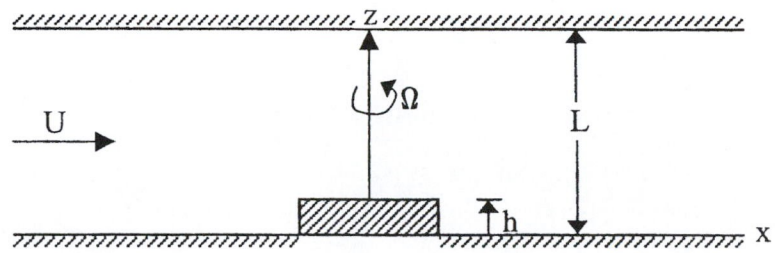

Literature Cited

Fjortoft, R. (1953). On the changes in the spectral distributions of kinetic energy for two-dimensional non-divergent flow. *Tellus, 5*, 225–230.

Gill, A. E. (1982). *Atmosphere–Ocean Dynamics*. New York: Academic Press.

Holton, J. R. (1979). *An Introduction to Dynamic Meteorology*. New York: Academic Press.

Houghton, J. T. (1986). *The Physics of the Atmosphere*. London: Cambridge University Press.

Kamenkovich, V. M. (1967). On the coefficients of eddy diffusion and eddy viscosity in large-scale oceanic and atmospheric motions. *Izvestiya, Atmospheric and Oceanic Physics, 3*, 1326–1333.

Kundu, P. K. (1977). On the importance of friction in two typical continental waters: Off Oregon and Spanish Sahara. In J. C. J. Nihoul (Ed.), *Bottom Turbulence*. Amsterdam: Elsevier.

Kuo, H. L. (1949). Dynamic instability of two-dimensional nondivergent flow in a barotropic atmosphere. *Journal of Meteorology, 6*, 105–122.

LeBlond, P. H., & Mysak, L. A. (1978). *Waves in the Ocean*. Amsterdam: Elsevier.

McCreary, J. P. (1985). Modeling equatorial ocean circulation. *Annual Review of Fluid Mechanics, 17*, 359–409.

Munk, W. (1981). Internal waves and small-scale processes. In B. A. Warren, & C. Wunch (Eds.), *Evolution of Physical Oceanography*. Cambridge, MA: MIT Press.

Pedlosky, J. (1971). Geophysical fluid dynamics. In W. H. Reid (Ed.), *Mathematical Problems in the Geophysical Sciences*. Providence, Rhode Island: American Mathematical Society.

Pedlosky, J. (1987). *Geophysical Fluid Dynamics*. New York: Springer-Verlag.

Phillips, O. M. (1977). *The Dynamics of the Upper Ocean*. London: Cambridge University Press.

Prandtl, L. (1952). *Essentials of Fluid Dynamics*. New York: Hafner Publ. Co.

Rhines, P. B. (1975). Waves and turbulence on a β-plane. *Journal of Fluid Mechanics, 69*, 417–443.

Stommel, H. (1948). The westward intensification of wind-driven ocean currents. *Transactions, American Geophysical Union, 29*(2), 202–206.

Taylor, G. I. (1915). Eddy motion in the atmosphere. *Philosophical Transactions of the Royal Society of London, A215*, 1–26.

Williams, G. P. (1979). Planetary circulations: 2. The Jovian quasi-geostrophic regime. *Journal of Atmospheric Sciences, 36*, 932–968.

Supplemental Reading

Chan, J. C. L. (2005). The physics of tropical cyclone motion. *Annual Review of Fluid Mechanics, 37*, 99–128.

Haynes, P. (2005). Stratospheric dynamics. *Annual Review of Fluid Mechanics, 37*, 263–293.

Wiggins, S. (2005). The dynamical systems approach to Lagrangian transport in oceanic flows. *Annual Review of Fluid Mechanics, 37*, 295–328.

O U T L I N E

14.1. Introduction   692

14.2. Aircraft Terminology   692

14.3. Characteristics of Airfoil Sections   696

14.4. Conformal Transformation for Generating Airfoil Shapes   702

14.5. Lift of a Zhukovsky Airfoil   706

14.6. Elementary Lifting Line Theory for Wings of Finite Span   708

14.7. Lift and Drag Characteristics of Airfoils   717

14.8. Propulsive Mechanisms of Fish and Birds   719

14.9. Sailing against the Wind   721

Exercises   722

Literature Cited   728

Supplemental Reading   728

CHAPTER OBJECTIVES

- To introduce the fundamental concepts and vocabulary associated with aircraft and aerodynamics

- To quantify the ideal-flow performance of simple two-dimensional airfoil sections

- To present the lifting line theory of Prandtl and Lanchester for a finite-span wing

- To describe the means by which fish, birds, insects, and sails exploit lift forces for flight and/or propulsion

14.1. INTRODUCTION

Aerodynamics is the branch of fluid mechanics that deals with the determination of the fluid mechanical forces and moments on bodies of interest. The subject is called *incompressible aerodynamics* if the flow speeds are low enough (Mach number < 0.3) for compressibility effects to be negligible. At larger Mach numbers where fluid-compressibility effects are important the subject is normally called *gas dynamics*. Aerodynamic parametric ranges of interest are usually consistent with: 1) neglecting buoyancy forces and fluid stratification, 2) assuming uniform constant-density flow upstream of the body, and 3) presuming viscous effects are confined to thin boundary layers adjacent to the body surface (Figure 9.1). Airfoil stall is a notable exception to this last presumption.

This chapter presents the elementary aspects of incompressible aerodynamics of aircraft wing shapes. Thus, with the simplifications just stated, the flows considered here are primarily ideal flows, and a significant portion of the material in Chapter 6 is relevant here. In addition, much of the material in this chapter also applies to ship propellers and to turbomachines (e.g., fans, turbines, compressors, and pumps) since the blades of these devices may all have similar cross sections.

14.2. AIRCRAFT TERMINOLOGY

Modern commercial aircraft embody nearly all the principles of aerodynamics presented in this chapter. Thus, a review of aircraft terminology and control strategies is provided in this section. Figure 14.1 shows three views of a commercial airliner. The body of the aircraft, which houses the passengers and other payload, is called the *fuselage*. The engines (jet or propeller) are often attached to the wings but they may be mounted on the fuselage too. Figure 14.2 shows an overhead (or planform) view of an airliner wing. The location where a wing attaches to the fuselage is called the *wing root*. The outer end of a wing furthest from the fuselage is called the *wing tip*, and the distance between the wing tips is called the *wingspan, s*. The distance between the leading and trailing edges of the wing is called the *chord length, c*, and it varies in the span-wise direction. The area of the wing when viewed from above is called the *planform area*, A. The slenderness of the wing planform is measured by its *aspect ratio*:

$$\Lambda \equiv s^2/A = s/\bar{c}, \quad \text{where } \bar{c} = A/s \qquad (14.1, 14.2)$$

is the average chord length.

The various possible rotational motions of an aircraft can be referred to three aircraft-fixed axes, called the *pitch axis*, the *roll axis*, and the *yaw axis* (Figure 14.3). A positive aerodynamic drag force points opposite to the direction of flight. Negative drag is called *thrust* and it must be produced by the aircraft's engines for full execution of the aircraft's flight envelope (take-off, cruise, landing, etc.). Lift is the aerodynamic force that points perpendicular to the direction of flight. It must be generated by the wings to counter the weight of the aircraft in flight. Movable surfaces on the wings and tail fin, known as *control surfaces*,

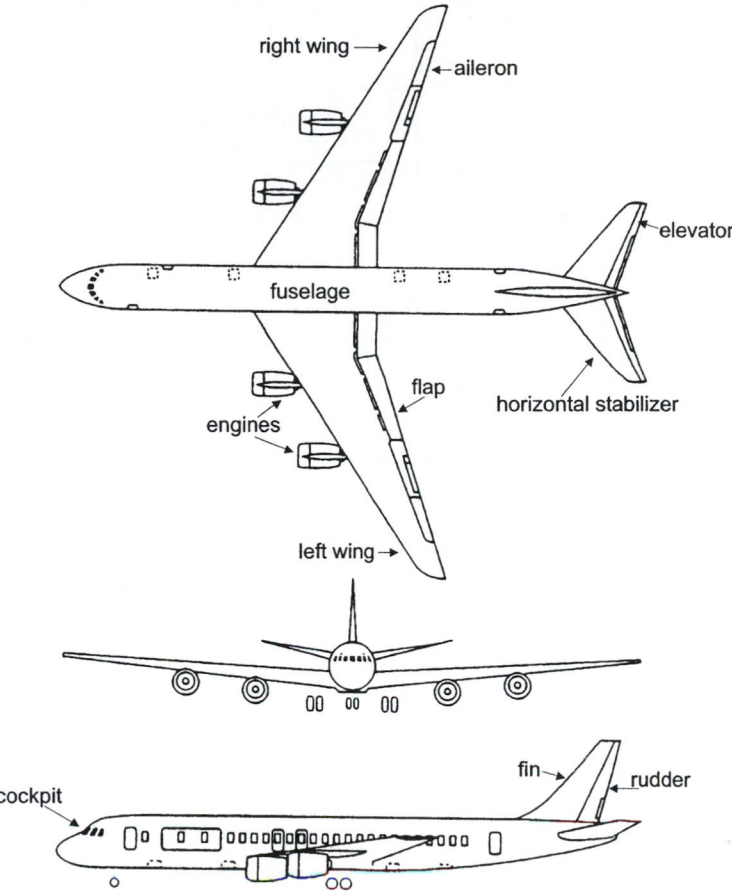

FIGURE 14.1   Three views of a commercial airliner and its control surfaces (NASA). The top view shows the wing planform. The wing is both backward swept and tapered. The various control surfaces shown modify the trailing edge geometry of the wing and tail fins. Landing gear details have been omitted.

can alter the distribution of lift and drag forces on the aircraft and provide the primary means for controlling the direction of flight. However, variation of engine thrust can also be used to steer the aircraft.

Control Surfaces

The aircraft is controlled by the pilot seated in the *cockpit*, who—with hydraulic assistance—sets the engine thrust and moves the control surfaces described in the following paragraphs. For the most part, these control surfaces act to change the local *camber* or curvature of the wings or fins to alter the lift force generated in the vicinity of the control surface.

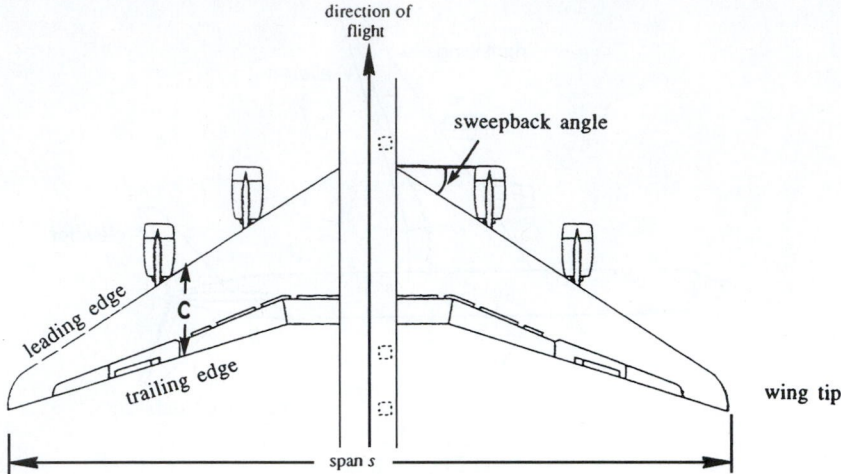

FIGURE 14.2 Wing planform geometry. The span, s, is the straight-line distance between wing tips and is shown at the bottom of the figure. The sweepback angle is shown near the starboard wing root. The chord c depends on location along the span.

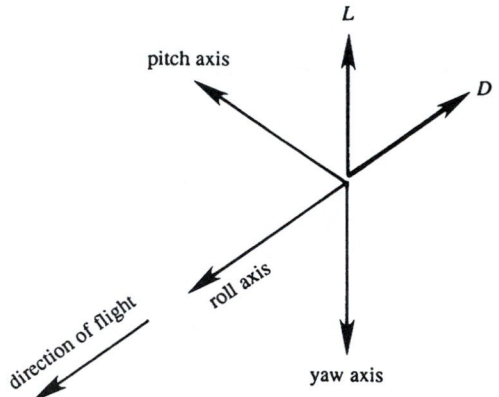

FIGURE 14.3 Aircraft axes. These are defined by the names of aircraft rotations about these axes. Positive pitch raises the aircraft's nose. Positive roll banks the aircraft for a right turn. Positive yaw moves the aircraft's nose to the right from the point of view of the pilot.

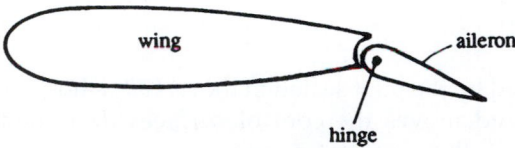

FIGURE 14.4 The aileron. As shown, this aileron deflection would increase lift by increasing the camber of the effective foil shape.

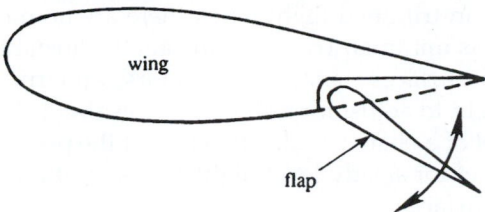

FIGURE 14.5 The flap. As shown, this flat deflection would increase lift by increasing the camber of the effective foil shape. The design of flaps often exploits the flow in the slot formed between the main wing and the rotating flap element to increase lift and delay stall.

Aileron: These are flaps near each wing tip (Figure 14.1), joined to the main wing by a hinged connection, as shown in Figure 14.4. They move differentially in the sense that one moves up while the other moves down. A depressed aileron increases the lift, and a raised aileron decreases the lift, so that a rolling moment results when they are differentially actuated. Ailerons are located near each wing tip to generate a large rolling moment with minimal angular deflection. The pilot generally controls the ailerons by moving a control stick, whose movement to the left or right causes a roll to the left or right. In larger aircraft the aileron motion is controlled by rotating a small wheel that resembles half of an automobile steering wheel.

Elevator: The elevators are hinged to the trailing edge of the horizontal stabilizers (tail fins). Unlike ailerons they move together, and their movement generates a pitching motion of the aircraft. The elevator movements are imparted by the forward and backward movement of a control stick, so that a backward pull lifts the nose of the aircraft.

Rudder: The yawing motion of the aircraft is governed by the hinged rear portion of the vertical tail fin, called the *rudder*. The pilot controls the rudder by pressing his feet against two rudder pedals so arranged that moving the left pedal forward moves the aircraft's nose to the left.

Flap: During takeoff, the speed of the aircraft is too small to generate enough lift to support the weight of the aircraft. To overcome this, a section of the rear of the wing is split, so that it can be rotated downward and moved aft to increase the lift (Figure 14.5). A further function of the flap is to increase both lift and drag during landing.

Modern airliners also have *spoilers* on the top surface of each wing. When raised slightly, they cause early boundary-layer separation on part of the top of the wing and this decreases or *spoils* the wing's lift. They can be deployed together or individually. Reducing the lift on one wing will bank the aircraft so that it will turn in the direction of the lowered wing. When deployed together, overall lift is decreased and the aircraft descends. Spoilers have another function as well. During landing immediately after touchdown, they are deployed fully to eliminate a significant fraction of the aircraft's wing lift and thereby ensure that the aircraft stays on the ground and does not become unintentionally airborne again, even in gusty winds. In addition, the spoilers increase drag and slow the aircraft to shorten the length of its landing roll.

An aircraft is said to be in trimmed flight when there are no moments about its center of gravity and the drag force is minimal. Trim tabs are small adjustable surfaces within or adjacent to the major control surfaces—ailerons, elevators, and rudder. Deflections of these surfaces may be set and held to adjust for a change in the aircraft's center of gravity in flight due to consumption of fuel or a change in the direction of the prevailing wind with respect to the flight path. These are set for steady-level flight on a straight path with minimum deflection of the major control surfaces.

14.3. CHARACTERISTICS OF AIRFOIL SECTIONS

Figure 14.6 shows the shape of the cross section of a wing, called an *airfoil* section (spelled *aerofoil* in the British literature). The leading edge of the profile is generally rounded, whereas the trailing edge is sharp. The straight line joining the centers of curvature of the leading and trailing edges is called the *chord*. The meridian line of the section passing midway between the upper and lower surfaces is called the *camber line*. The maximum height of the camber line above the chord line is called the *camber* of the section. Normally the camber varies from nearly zero for high-speed supersonic wings, to ≈ 5% of chord length for low-speed wings. The angle α between the chord line and the direction of flight (i.e., the direction of the undisturbed stream) is called the *angle of attack* or *angle of incidence*.

The forces on airfoils are usually studied in a foil-fixed frame of reference with a uniform flow approaching the foil along the *x*-axis with the *y*-axis pointing vertically upward. The aerodynamic force *F* on an airfoil can be resolved into a *drag force D* parallel to the oncoming stream, and a *lift force L* perpendicular to the oncoming stream (see Figure 14.7). In steady-level flight the lift equals the weight of the aircraft while its drag is balanced by engine thrust. These forces may be expressed in dimensionless form via the coefficients of lift and drag:

$$C_D \equiv \frac{D}{(1/2)\rho U^2 A}, \quad \text{and} \quad C_L \equiv \frac{L}{(1/2)\rho U^2 A}. \tag{4.107, 4.108}$$

Measurements or specifications of C_D and C_L are the primary means for stating airfoil performance. The drag results from the stress and pressure distributions on the foil's surface. These are called the *friction drag* and the *pressure drag* (or *form drag*), respectively. The lift is almost entirely due to the pressure distribution. Figure 14.8 shows the distribution of the pressure

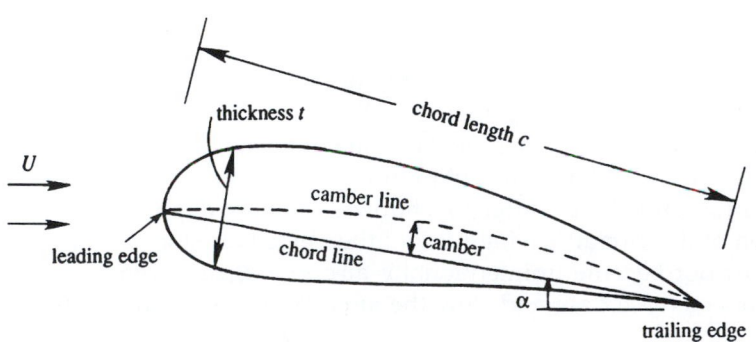

FIGURE 14.6 Airfoil geometry. A rounded leading edge and a sharp trailing edge are essential geometrical features of airfoils. For the material discussed in this chapter, the most important parameters are: the angle of attack α, the chord length c, and the maximum camber. An airfoil's thickness distribution is often modified to minimize drag and/or prevent stall.

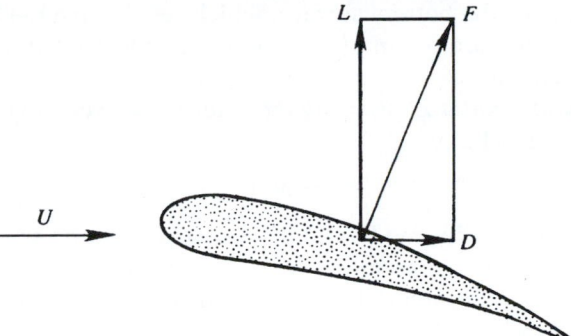

FIGURE 14.7   Forces on an airfoil. Lift L acts perpendicular to the oncoming stream and may be positive or negative. Drag D acts parallel to the oncoming stream and is positive for passive objects.

coefficient $C_p = (p - p_\infty)/(1/2)\rho U^2$ on an airfoil at a moderate angle of attack. The outward arrows correspond to a negative C_p, while a positive C_p is represented by inward arrows. It is seen that the pressure coefficient is negative over most of the surface, except over small regions near the nose and the tail. However, the pressures over most of the upper surface

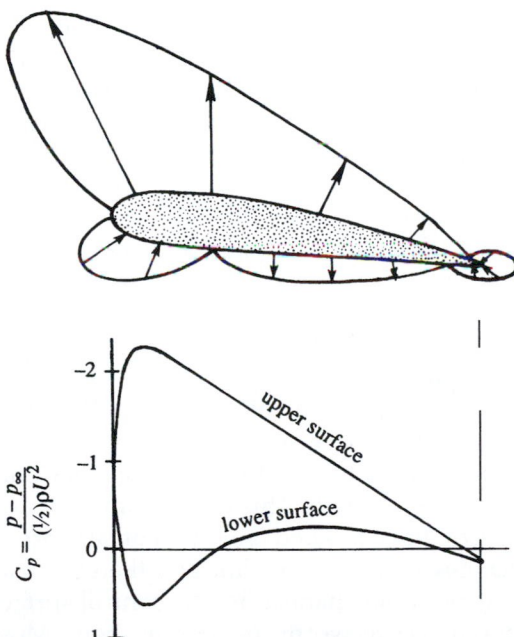

FIGURE 14.8   Distribution of the pressure coefficient C_p over an airfoil. The upper panel shows C_p plotted normal to the surface and the lower panel shows C_p plotted normal to the chord line. Note that negative values appear on the upper half of the vertical axis in the lower panel. And, on the upper or suction foil surface, a pressure minimum occurs near the foil's leading edge. Thus, the suction-side boundary layer faces an adverse pressure gradient over most of the upper surface of the foil.

are smaller than those over the bottom surface, which results in a net lift force. The top and bottom surfaces of an airfoil are popularly referred to as the *suction side* and the *compression* (or *pressure*) *side*, respectively.

In steady ideal flow, the *Kutta-Zhukhovsky lift theorem* (see Section 6.5) requires the lift on a two-dimensional airfoil to be

$$L = \rho U \Gamma, \tag{6.62}$$

where U is the free-stream velocity and Γ is the clockwise circulation around the body. Thus, lift development on an airfoil is synonymous with circulation development. As was seen in Section 6.3 for $0 < \Gamma < 4\pi a U$, the amount of circulation held by a cylinder determines the location of stagnation points where the oncoming stream attaches and separates from the cylinder's surface. This is also true for an airfoil with circulation; foil-surface flow attachment and separation locations are set by the foil's circulation strength. In subsonic aerodynamics, airfoil circulation is determined by the net amount of vorticity trapped in the foil's viscous boundary layers. Thus, asymmetrical foil shapes intended for positive lift generation are designed to place more vorticity in the suction-side boundary layer than in the pressure-side boundary layer. For fixed chord length and free-stream flow speed, three common strategies are followed for robust lift generation and control. The first allows the other two to be effective.

For reliable subsonic lift generation, a foil should have a sharp tailing edge. At low to moderate angles of attack, $|\alpha|$ up to approximately 15° to 20°, a sharp trailing edge causes the suction and pressure side boundary layers to separate together at the foil's trailing edge. Thus, a sharp trailing edge becomes the downstream flow separation point, so its location thereby determines the foil's circulation for a given foil shape and free-stream speed. The actual fluid-dynamic interaction leading to this situation involves the foil's viscous boundary layers and is described later. However, this possibility for controlling circulation was experimentally observed before the development of boundary-layer theory. In 1902, the German aerodynamicist Wilhelm Kutta proposed the following rule: *in flow over a two-dimensional body with a sharp trailing edge, there develops a circulation of magnitude just sufficient to move the rear stagnation point to the trailing edge*. This statement is called the *Kutta condition*, sometimes also called the *Zhukhovsky hypothesis*. It is applied in ideal-flow aerodynamics as a simple means of capturing the viscous flow effects of a foil's attached boundary layers.

A second strategy for controlling a foil's lift is to change its angle of attack α. For $|\alpha| < 15°$ to 20°, increasing α increases the lift, and nominal extreme C_L values from -2 to $+2$ can be obtained from well-designed, single-piece airfoils.

The final strategy for controlling a foil's lift is to change its camber. For a fixed angle of attack, increasing camber increases the lift. This is the primary reason for moveable control surfaces at the trailing edges of the wings and tail fins of an aircraft. Angular rotation of such control surfaces locally changes a foil's camber line and thereby changes the lift force generated by the portion of the wing or fin spanned by the control surface.

Two additional considerations are worth mentioning here. Most foils have a rounded leading edge to keep the foil's suction-side boundary layer attached, and this increases lift and decreases drag. A properly designed leading edge recovers nearly all of the ideal-flow leading edge suction that occurs on foils of negligible thickness (see Exercise 14.7). And, when a foil is pitched upward to a sufficiently high angle of attack, the Kutta condition will fail and the foil's suction-side boundary layer will separate upstream of the foil's trailing edge. This situation is called *stall* and its onset depends on: the foil's shape, the Reynolds number of the flow, the foil's surface roughness, and other three-dimensional effects. Stall occurs when the suction-side boundary layer cannot overcome the adverse pressure gradient aft of the pressure minimum on the foil's suction side. For small violations of the Kutta condition where the suction-side boundary layer separates at ~80% or 90% of the chord length, a typical foil's lift is not strongly affected but its drag increases. For more severe violations of the Kutta condition, where the suction-side boundary layer separates upstream of the mid-chord location, the foil's lift is noticeably reduced and its drag is greatly increased. In nearly all cases, stall leads to such undesirable foil performance that its onset places important limitations on an aircraft's operating envelope.

The physical reason for the Kutta condition is illustrated in Figure 14.9 where the same simple airfoil and nearby streamlines are shown at three different times. Here, the foil is held fixed and flow is impulsively accelerated to speed U at $t = 0$. Figure 14.9a shows streamlines near the foil immediately after the fluid has started moving but before boundary layers have developed on either its suction or pressure sides. The fluid velocity at this stage has a near discontinuity adjacent to the foil's surface. And, the fluid goes around the foil's trailing edge with a very high velocity and overcomes a steep deceleration and pressure rise from the

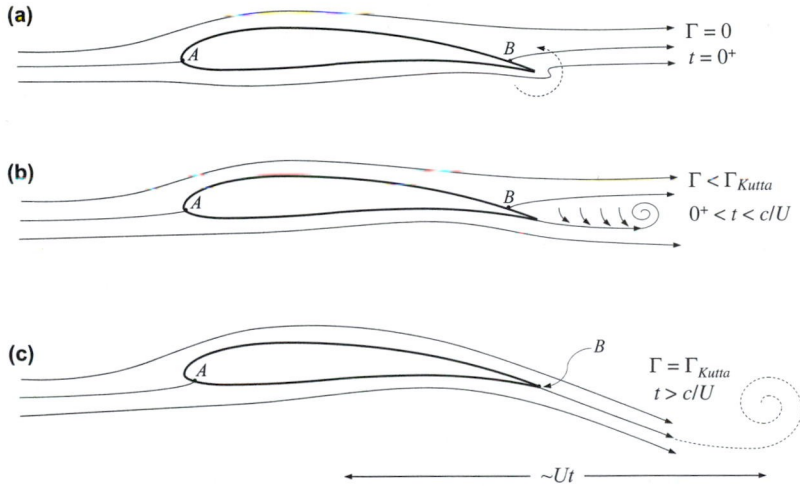

FIGURE 14.9   Flow patterns over a stationary airfoil at a low angle of attack in an impulsively started horizontal flow. (a) Streamlines immediately after the velocity jumps to a positive value. Here the boundary layers on the foil have not had a chance to develop and the rear stagnation (separation) point B occurs on the suction surface of the foil. The foil-surface vorticity at the trailing edge is nearly singular and induces a counterclockwise fluid velocity that draws fluid around the sharp trailing edge. (b) If the pressure-side boundary layer develops first, it will separate from the trailing edge as shown. However, the pressure distribution near the trailing edge and the induced velocities from the foil's near-wake vorticity both act to bring B to the trailing edge. (c) Steady-flow pattern established after the flow has moved a chord length or two. Here the leading edge stagnation point A has moved onto the pressure side of the foil and the net circulation trapped in the foil's boundary layers satisfies the Kutta condition. In this case the rear stagnation (separation) point lies at the foil's trailing edge. The net circulation of the whole flow field remains zero because the unsteady flow process leading to this flow pattern produces a counter-rotating starting vortex, shown in (c) as a dashed spiral.

trailing edge to the rear flow-separation (and -stagnation) point at B. The flow is able to turn the sharp trailing-edge corner because the vorticity on the foil's surface near the trailing edge at this instant is nearly singular at the trailing edge and it induces counterclockwise fluid motion (shown in Figure 14.9a by a dashed arrow). Overall at this time, the flow is irrotational away from the foil's surface, the foil's net circulation is zero, it generates no lift, the forward flow-attachment stagnation point at A is very close to the nose of the foil, and the rear stagnation point at B resides on the foil's suction surface.

Figure 14.9b shows the flow a short time later in a hypothetical situation where the foil's pressure-side boundary layer has developed first. In this case, the points A and B have not moved much. However, the pressure-side boundary layer now separates at the sharp trailing edge because the slowly moving boundary-layer fluid near the foil's surface does not have sufficient kinetic energy to negotiate the steep pressure rise near the stagnation point B nor can it turn the sharp trailing-edge corner. Furthermore, the separated pressure-side boundary-layer flow has carried the near singularity of vorticity, which initially resided on the foil's surface at its trailing edge, into the foil's near wake as a concentrated vortex. Two phenomena near the trailing edge now act to eliminate the zone of separated flow caused by pressure-side boundary-layer separation at the trailing edge. First, the Bernoulli equation ensures that the stagnation pressure at B is higher than the pressure in the moving fluid that is leaving the trailing edge from the pressure side of the foil. The resulting pressure gradient between B and the trailing edge pushes the stationary fluid near B toward the foil's trailing edge. Second, the induced velocities from the vorticity in the separated pressure-side boundary layer and from the near-wake concentrated vortex both induce the stationary fluid near B to move toward the foil's trailing edge. Together these two phenomena cause the rear stagnation point at B to move to the foil's trailing edge. Although an actual impulsively started flow involves simultaneous suction- and pressure-side boundary layer development, the outcome is the same; the rear stagnation point winds up at the trailing edge.

Figure 14.9c shows the final condition after the flow has traveled a chord length or two past the foil. The leading-edge stagnation point has traveled under the nose of the foil and onto the foil's pressure side, and the suction-surface separation point B has been drawn to the foil's trailing edge. (The ideal airfoil trailing edge is a perfect cusp with zero included angle that allows the pressure and suction side flows to meet and separate from the foil without changing direction and without a stagnation point. However, structural requirements cause real foils to have finite included-angle trailing edges, thus point B is a stagnation point even when the trailing edge's included angle is very small; see Section 6.4 and Exercise 14.2.) Once the flow shown in Figure 14.9c is established, the foil now carries more vorticity in its suction-side boundary layer than it does in its pressure-side boundary layer. This difference causes the flow to sweep upward ahead of the foil and downward behind it. The foil's net circulation is that necessary to satisfy the Kutta condition, Γ_{Kutta}. If the foil's circulation is further increased beyond Γ_{Kutta}, the rear stagnation point moves under the foil and onto the *pressure* surface. Although it is an ideal-flow possibility, $\Gamma > \Gamma_{Kutta}$ is not observed in real airfoil flows.

The net circulation in the impulsively started flow described in this section and illustrated in Figure 14.9 is maintained at zero by the presence of an opposite sign vortex, known as a *starting vortex*, in the fluid that was near the foil when the flow began moving. In the scenario described earlier, this vortex is the remnant of the vorticity shed by the pressure-side boundary layer

FIGURE 14.10 A material circuit in a stationary fluid that contains an impulsively started airfoil moving to the left. The entire outer part of the circuit was initially in stationary fluid. Thus, the circulation on ABCD must be zero. Therefore, if the sub-circuit ABD contains the airfoil with circulation Γ, then the other sub-circuit BCD must contain a starting vortex with circulation −Γ.

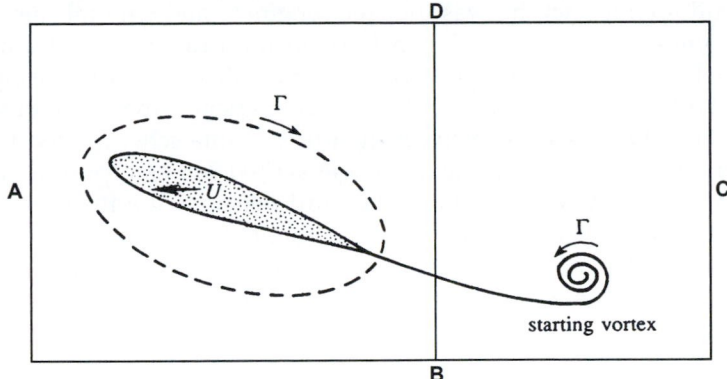

before point B moved to the foil's trailing edge and the cast-off concentrated vorticity that initially caused the flow to fully turn the foil's sharp trailing-edge corner.

The equivalence of the final circulation magnitude bound to the foil and that in the starting vortex is illustrated in Figure 14.10 where the sense of the foil's circulation is clockwise and that in the starting vortex is counterclockwise. For the flow shown in this figure, imagine that the fluid is stationary and the airfoil is moving to the left. Consider a material circuit ABCD large enough to enclose both the initial and final locations of the airfoil. Initially the trailing edge was within the region BCD, which now contains the starting vortex only. According to Kelvin's circulation theorem, the circulation around any material circuit remains constant, if the circuit remains in a region of inviscid flow (although viscous processes may go on *inside* the region enclosed by the circuit). The circulation around the large circuit ABCD therefore remains zero, since it was zero initially. Consequently the counterclockwise circulation of the starting vortex around DBC is balanced by an equal clockwise circulation around ADB. The wing is therefore left with a circulation Γ equal and opposite to the circulation of the starting vortex.

It is clear from the discussion and illustrations in Figure 14.9 that a value of circulation other than Γ_{Kutta} would result a readjustment of the flow. Thus, with every change in flow speed, angle of attack, or airfoil camber (via flap deflection) a new starting vortex is cast off and left behind the foil. A new value of circulation around the airfoil is established to once again place the rear stagnation point at the foil's trailing edge.

Interestingly, *fluid viscosity is not only responsible for the drag, but also for the development of circulation and lift.* In developing the circulation, the flow leads to a steady state where further boundary-layer separation is prevented. The establishment of circulation around an airfoil-shaped body in a real fluid is truly remarkable.

Historical Notes

According to von Karman (1954), the connection between the lift of airplane wings and the circulation around them was recognized and developed by three persons. One of them was the Englishman Frederick Lanchester (1887–1946). He was a multisided and imaginative person, a practical engineer as well as an amateur mathematician. His trade was automobile building; in fact, he was the chief engineer and general manager of the Lanchester Motor Company. He once took von Karman for a ride around Cambridge in an automobile that he built himself, but von Karman "felt a little uneasy discussing aerodynamics at such rather frightening speed" (p. 34). The second person is the German mathematician Wilhelm Kutta (1867–1944), well known for the Runge-Kutta scheme used in the numerical integration of ordinary differential equations. He started out as a pure mathematician, but later became interested in aerodynamics. The third person is the Russian physicist Nikolai Zhukhovsky, who developed the mathematical foundations of the theory of lift for wings of infinite span, independently of Lanchester and Kutta. An excellent history of flight and the science of aerodynamics is provided by Anderson (1998).

14.4. CONFORMAL TRANSFORMATION FOR GENERATING AIRFOIL SHAPES

In the study of airfoils, one is interested in finding the flow pattern and the surface-pressure distribution. The *direct* solution of the Laplace equation for the prescribed boundary shape of the airfoil is straightforward using a computer, but analytically it is more difficult. In general, analytical solutions are possible only when the airfoil is assumed thin. This is called *thin airfoil theory*, in which the airfoil is replaced by a vortex sheet coinciding with the camber line. An integral equation is developed for the local vorticity distribution from the condition that the camber line be a streamline (velocity tangent to the camber line). The velocity at each point on the camber line is the superposition (i.e., integral) of velocities induced at that point due to the vorticity distribution at all other points on the camber line plus that from the oncoming stream (at infinity). Since the maximum camber is small, evaluations are made on the x-axis of the x–y-plane. The Kutta condition is enforced by requiring the strength of the vortex sheet at the trailing edge to be zero. Thin airfoil theory is treated in detail in Kuethe and Chow (1998, Chapter 5) and Anderson (2007, Chapter 4). An *indirect* way to solve the problem involves the method of conformal transformation, in which a mapping function is determined such that the airfoil shape is transformed into a circle. Then a study of the flow around the circle determines the flow pattern around the airfoil. This is called *Theodorsen's method*, which is complicated and will not be discussed here.

Instead, we shall deal with the case in which a *given* transformation maps a circle into an airfoil-like shape and determines the properties of the airfoil generated thereby. This is the *Zhukhovsky transformation*:

$$z = \zeta + \frac{b^2}{\zeta},$$
(14.3)

where b is a constant. It maps regions of the ζ-plane into the z-plane, some examples of which are discussed in Section 6.6. Here, we shall consider circles in different configurations in the ζ-plane and examine their transformed shapes in the z-plane. It will be seen that one of them will result in an airfoil shape.

First consider the transformation of a circle into a straight line. Start from a circle, centered at the origin in the ζ-plane, whose radius b is the same as the constant in the Zhukhovsky transformation (Figure 14.11). For a point $\zeta = be^{i\theta}$ on the circle, the corresponding point in the z-plane is

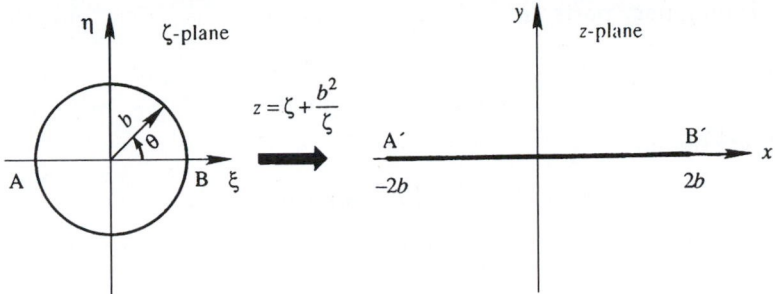

FIGURE 14.11 Transformation of a circle into a straight line. Here the ζ-plane contains the circle of radius b and the transformation $z = \zeta + b^2/\zeta$ converts it into a line segment of length $4b$ in the z-plane.

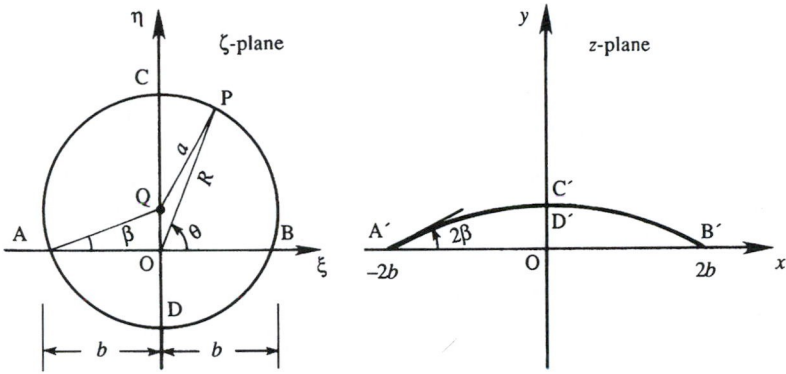

FIGURE 14.12 Transformation of a circle into a circular arc. This situation is similar to that shown in Figure 14.11 except that here the circle is displaced upward and its radius is larger. The object created in the z-plane is a circular arc.

$$z = b\,e^{i\theta} + b\,e^{-i\theta} = 2b\cos\theta.$$

As θ varies from 0 to π, z goes along the x-axis from $2b$ to $-2b$. As θ varies from π to 2π, z goes from $-2b$ to $2b$. The circle of radius b in the ζ-plane is thus transformed into a straight line of length $4b$ in the z-plane. It is clear that the region *outside* the circle in the ζ-plane is mapped into the *entire* z-plane. (It can be shown that the region inside the circle is also transformed into the entire z-plane. This, however, is of no concern to us, since we shall not consider the interior of the circle in the ζ-plane.)

Next consider the transformation of a circle into a circular arc. Again start with a circle in the ζ-plane, but this time let its radius be a ($>b$), let it be centered at point Q along the vertical the η-axis, and let it cut the horizontal ξ-axis at $(\pm b, 0)$, as shown in Figure 14.12. If a point on the circle in the ζ-plane is represented by $\zeta = Re^{i\theta}$, then the corresponding point in the z-plane is

$$z = Re^{i\theta} + \frac{b^2}{R}e^{-i\theta},$$

whose real and imaginary parts are:

$$x = (R + b^2/R)\cos\theta,$$
$$y = (R - b^2/R)\sin\theta. \tag{14.4}$$

Eliminating R, we obtain

$$x^2 \sin^2\theta - y^2 \cos^2\theta = 4b^2 \sin^2\theta \cos^2\theta. \tag{14.5}$$

To understand the shape of the curve represented by (14.5) we must express θ in terms of x, y, and the known constants. From triangle OQP, we obtain

$$QP^2 = OP^2 + OQ^2 - 2(OQ)(OP)\cos(Q\hat{O}P).$$

Using $QP = a = b/\cos\beta$ and $OQ = b\tan\beta$, this becomes

$$\frac{b^2}{\cos^2\beta} = R^2 + b^2 \tan^2\beta - 2Rb\tan\beta\cos(90° - \theta),$$

which simplifies to

$$2b\tan\beta\sin\theta = R - b^2/R = y/\sin\theta, \tag{14.6}$$

where (14.4) has been used. We now eliminate θ between (14.5) and (14.6). First note from (14.6) that $\cos^2\theta = (2b\tan\beta - y)/2b\tan\beta$, and $\cot^2\theta = (2b\tan\beta - y)/y$. Then divide (14.5) by $\sin^2\theta$, and substitute these expressions for $\cos^2\theta$ and $\cot^2\theta$. This gives

$$x^2 + (y + 2b\cot 2\beta)^2 = (2b\csc 2\beta)^2,$$

where β is known from $\cos\beta = b/a$. This is the equation of a circle in the z-plane, having the center at $(0, -2b\cot 2\beta)$ and a radius of $2b\csc 2\beta$. The Zhukhovsky transformation has thus mapped a complete circle into a circular arc.

Now consider what happens when the center of the circle in the ζ-plane is displaced to a point Q on the real axis (Figure 14.13). The radius of the circle is again a ($>b$), and we assume that a is slightly larger than b:

$$a \equiv b(1 + e) \qquad e \ll 1. \tag{14.7}$$

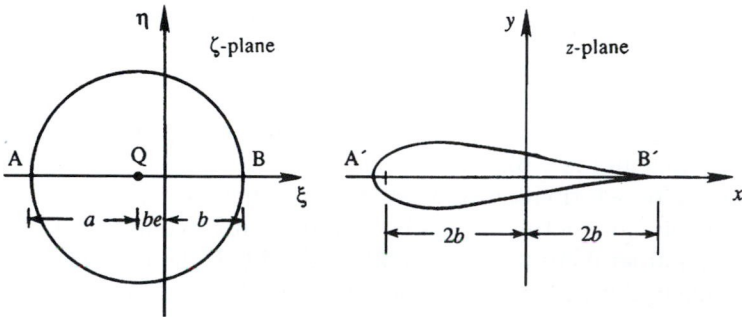

FIGURE 14.13 Transformation of a circle into a symmetric airfoil. This situation is similar to that shown in Figure 14.11 except that here the circle is displaced to the left and its radius is larger. The object created in the z-plane has a symmetric (zero camber) airfoil shape.

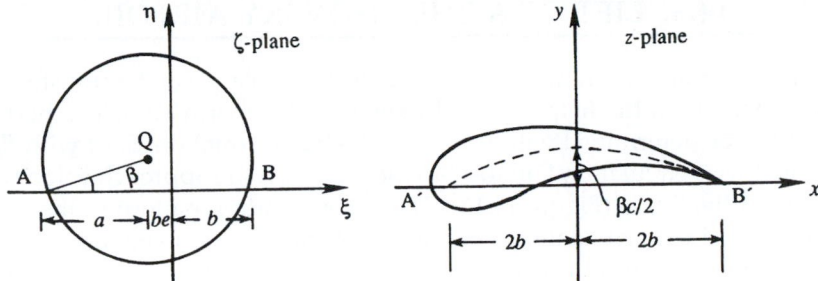

FIGURE 14.14 Transformation of a circle into a cambered airfoil. This situation combines the effects illustrated in Figures 14.11–14.13. The circle is displaced upward and leftward, and its radius is larger. The resulting shape in the z-plane is that of an airfoil.

A numerical evaluation of the Zhukhovsky transformation (14.3), with assumed values for a and b, shows that the corresponding shape in the z-plane is a symmetrical airfoil shape, a streamlined body that is symmetrical about the x-axis. Note that the airfoil in Figure 14.13 has a rounded nose and thickness, while the one in Figure 14.12 has camber but no thickness.

Therefore, a potentially realistic airfoil shape with both thickness and camber can be generated by starting from a circle in the ζ-plane that is displaced in both η and ξ directions (Figure 14.14). The following relations can be proved for $e \ll 1$:

$$c \cong 4b, \; camber = \; \cong \frac{1}{2}\beta c, \quad \text{and} \quad t_{max}/c \cong 1.3e. \qquad (14.8)$$

Here t_{max} is the maximum thickness, which is reached nearly at the quarter chord position $x = -b$, and *camber* as defined in Figure 14.6 is indicated in Figure 14.14.

Such airfoils generated from the Zhukhovsky transformation are called *Zhukhovsky airfoils*. They have the property that the trailing edge is a *cusp*, which means that the upper and lower surfaces are tangent to each other at the trailing edge. Without the Kutta condition, the trailing edge is a point of infinite velocity. If the trailing edge angle is nonzero (Figure 14.15a), then a stagnation point occurs at the trailing edge because the suction and pressure side flows must change direction when they meet (Exercise 14.2). However, the cusped trailing edge of a Zhukhovsky airfoil (Figure 14.15b) does not require any flow deflection so it is not a stagnation point. In that case the tangents to the upper and lower surfaces coincide at the trailing edge, and the fluid leaves the trailing edge smoothly. The trailing edge for the Zhukhovsky airfoil is simply an ordinary point where the velocity is neither zero nor infinite.

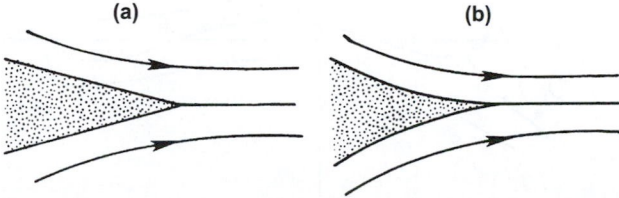

FIGURE 14.15 Shapes of the trailing edge: (a) trailing edge with finite angle; and (b) cusped trailing edge. Application of the Kutta condition to a trailing edge with a finite included angle results in a stagnation point at the trailing edge. A cusped trailing edge avoids the stagnation point.

14.5. LIFT OF A ZHUKHOVSKY AIRFOIL

The preceding section has shown how a circle in the ζ-plane can be transformed into an airfoil in the z-plane with the help of the Zhukhovsky transformation. The performance of such an airfoil can be determined with the aid of the transformation. Start with flow around a circle with clockwise circulation Γ in the ζ-plane, in which the approach velocity is inclined at an angle α with the ξ-axis (Figure 14.16). The corresponding pattern in the z-plane is the flow around an airfoil with circulation Γ and angle of attack α. It can be shown that the circulation does not change during a conformal transformation. If $w = \phi + i\psi$ is the complex potential, then the velocities in the two planes are related by

$$\frac{dw}{dz} = \frac{dw}{d\zeta}\frac{d\zeta}{dz}.$$

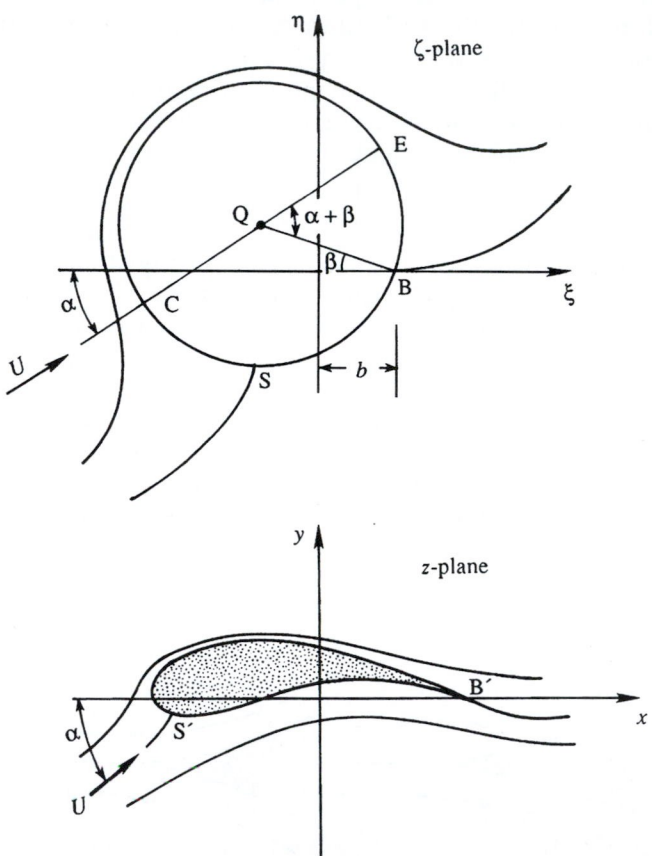

FIGURE 14.16 Transformation of flow around a circle with circulation in the ζ-plane into flow around a Zhukhovsky airfoil in the z-plane. The stagnation points S and B in the upper panel are mapped into the stagnation points S' and B' in the lower panel. The angle of attack α is the same in both complex planes.

Using the Zhukhovsky transformation (14.3), this becomes

$$\frac{dw}{dz} = \frac{dw}{d\zeta} \frac{\zeta^2}{\zeta^2 - b^2}. \tag{14.9}$$

Here $dw/dz = u - iv$ is the complex velocity in the z-plane, and $dw/d\zeta$ is the complex velocity in the ζ-plane. Equation (14.9) shows that the velocities in the two planes become equal as $\zeta \to \infty$, which means that the free-stream velocities are inclined at the same angle α in the two planes.

Point B with coordinates $(b, 0)$ in the ζ-plane is transformed into the trailing edge B' of the airfoil. Because $\zeta^2 - b^2$ vanishes there, it follows from (14.9) that the velocity at the trailing edge will in general be infinite. If, however, we arrange that B is a stagnation point in the ζ-plane at which $dw/d\zeta = 0$, then dw/dz at the trailing edge will have a zero-over-zero form. Our discussion of Figure 14.15b has shown that this will in fact result in a finite velocity at B'.

From (6.37), the tangential velocity at the surface of the circle in the ζ-plane is given by

$$u_\theta = -2U \sin\theta - \frac{\Gamma}{2\pi a}, \tag{14.10}$$

where θ is measured from the free-stream-aligned diameter CQE. At point B, we have $u_\theta = 0$ and $\theta = -(\alpha + \beta)$. Therefore (14.10) gives

$$\Gamma = 4\pi U a \sin(\alpha + \beta), \tag{14.11}$$

which is the clockwise circulation required by the Kutta condition. It shows that the circulation around an airfoil depends on the speed U, the chord length c $(\approx 4a)$, the angle of attack α, and the camber/chord ratio $\beta/2$. The coefficient of lift is

$$C_L = \frac{L}{(1/2)\,\rho U^2 c} \approx 2\pi(\alpha + \beta), \tag{14.12}$$

where we have used $4a \approx c$, $L = \rho U\Gamma$, and $\sin(\alpha + \beta) \approx (\alpha + \beta)$ for small angles of attack. Equation (14.12) shows that the lift can be increased by adding a certain amount of camber. The lift is zero at a negative angle of attack $\alpha = -\beta$, so that the angle $(\alpha + \beta)$ can be called the *absolute angle of attack*. The fact that the lift of an airfoil is proportional to the angle of attack allows the pilot to control the lift simply by adjusting the attitude (orientation) of the airfoil with respect to its flight direction.

A comparison of the theoretical lift equation (14.12) with typical experimental results for a Zhukhovsky airfoil is shown in Figure 14.17. The small disagreement can be attributed to the finite thickness of the foil-surface boundary layers whose displacement thicknesses change the effective shape of the airfoil. The sudden drop of the lift at $\alpha + \beta \approx 20°$ is the signature of stall, and it is caused by early suction-side boundary-layer separation that worsens with increasing angle of attack. Stall is further discussed in Section 14.7.

Zhukhovsky airfoils are not practical for two basic reasons. First, they demand a cusped trailing edge, which cannot be practically constructed or maintained. Second, the camber line in a Zhukhovsky airfoil is nearly a circular arc, and therefore the maximum camber lies close to the center of the chord. However, a maximum camber within the forward portion of the chord is usually preferred so as to obtain a desirable pressure distribution. To get

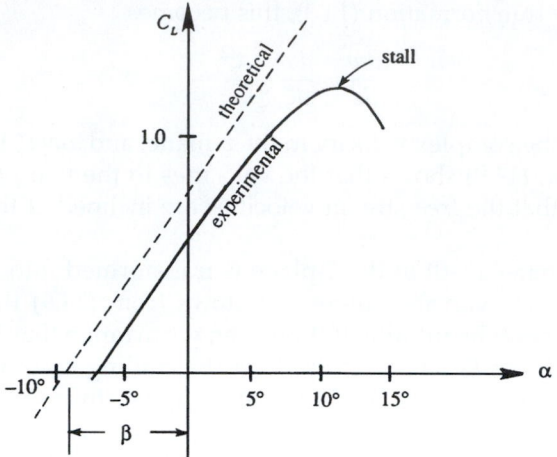

FIGURE 14.17 Comparison of theoretical and experimental lift coefficients for a cambered Zhukhovsky airfoil. The lift curve slopes match well and boundary-layer thicknesses may account for the offset between theoretical and measured curves. The most important difference is that the real airfoil stalls while the ideal one does not.

around these difficulties, other families of airfoils have been generated from circles by means of more complicated transformations. Nevertheless, the results for a Zhukhovsky airfoil given here have considerable application as reference values, and the conformal mapping technique remains an efficient means for assessing airfoil designs.

14.6. ELEMENTARY LIFTING LINE THEORY FOR WINGS OF FINITE SPAN

The foregoing two-dimensional results apply only to wings of infinite span. However, many of the concepts of two-dimensional aerodynamics can be extrapolated to three-dimensional flow and wings of finite span when the vorticity shed from a three-dimensional wing is accounted for. The lifting line theory of Prandtl and Lanchester is the simplest means for accomplishing this task and it provides useful insights into how lift and drag develop on finite span wings. Lifting line theory is based on several approximations to the three-dimensional flow field of a finite wing, so our starting point is a description of such a flow.

Figure 14.18 shows a schematic view of a finite-span wing, looking downstream from the aircraft. As the pressure on the lower surface of the wing is greater than that on the upper surface, air flows around the wing tips from the lower into the upper side. Therefore, there is a span-wise component of velocity toward the wing tip on the underside of the wing and toward the wing root on the upper side, as shown by the streamlines in Figure 14.19a. The span-wise momentum acquired as the fluid passes the wing continues into the wake downstream of the trailing edge. On the stream surface extending downstream from the wing, therefore, the lateral component of the flow is outward (toward the wing tips) on the underside and inward on the upper side. On this surface, then, there is vorticity oriented in the stream-wise direction. This stream-wise vorticity has opposite signs on the two sides of

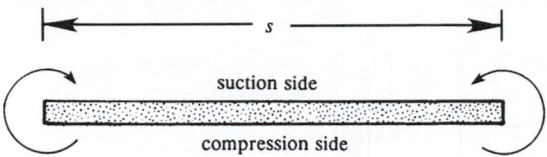

FIGURE 14.18 Flow around wing tips. Low suction-side pressures and high pressure-side pressures cause fluid to move toward the wing tips on the underside of a finite wing, and to move away from the wing tips on the topside of a finite wing. This three-dimensional flow eventually produces the tip vortices.

the wing-center axis OQ. The stream-wise vortex filaments downstream of the wing are called *trailing vortices*, which form a *vortex sheet* (Figure 14.19b) in the near wake of the wing. As discussed in Section 5.8, a vortex sheet is composed of closely spaced vortex filaments that generate a discontinuity in tangential velocity.

Downstream of the wing, each half of the vortex sheet rolls up on itself and forms two distinct counter-rotating vortices called *tip vortices* (Figure 14.20). The circulation of each tip vortex is equal to Γ_0, the circulation at the center of the wing. Tip vortices may become visually evident when an aircraft flies in humid air. The decreased pressure (due to the high velocity) and temperature in the core of the tip vortices may cause atmospheric moisture to condense into droplets or ice crystals, which may be seen in the form of *vapor trails* extending for many kilometers behind an aircraft traversing a clear sky. This textbook's cover shows the trailing-vortex-induced distortion of a cloud layer behind a commercial airliner. Here the aircraft's mass is more than 100,000 kilograms. Thus, the strength of its trailing vortices is sufficient to cause substantial cloud motion on a scale comparable to the aircraft's wingspan. As an aircraft proceeds after takeoff, the tip vortices get longer, which means that kinetic energy is being constantly supplied to generate them. Thus, an additional drag force must be experienced by a wing of finite span. This is called the *induced drag*, and it can be predicted with lifting line theory.

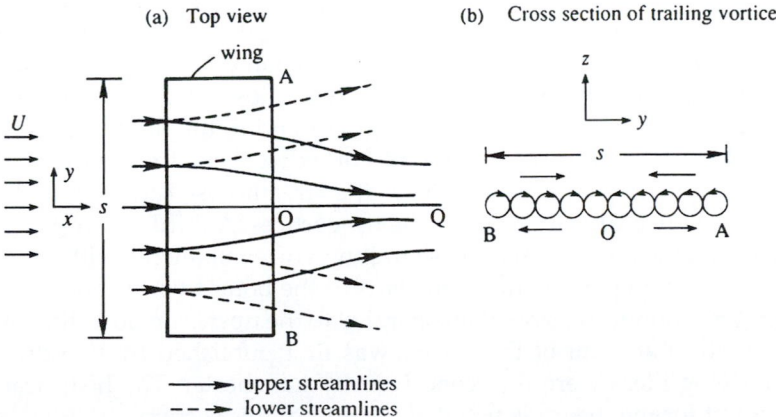

FIGURE 14.19 Flow over a wing of finite span: (a) top view of streamline patterns on the upper and lower surfaces of the wing; and (b) cross section of trailing vortices behind the wing. The trailing vortices change sign at O, the center of the wing.

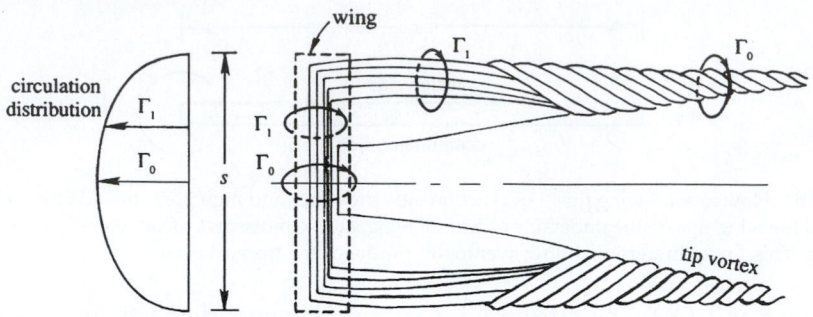

FIGURE 14.20　Rolling up of trailing vortices to form tip vortices. The mutual interaction of the trailing vortices eventually produces two counter-rotating wing-tip vortices having the same circulation as that bound to the center of the main wing. The effect of such vortices on a cloud layer is shown on the cover of this textbook.

One of Helmholtz's vortex theorems states that a vortex filament cannot end in the fluid, but must either end at a solid surface or form a closed vortex loop or ring. In the case of the finite wing, the tip vortices are the extension of the vorticity trapped in the wing's boundary layers. The tip vortices start at the wing and are joined together downstream of the aircraft by the various starting vortices of the wing. Starting vortices are left behind at the point where the aircraft took off and where the wing's lift was changed for aircraft maneuvers (ascent, descent, turns, etc.). In any case, the starting vortices are usually so far behind the wing that their effect on the wing's performance may be neglected and the tip vortices may be regarded as extending an infinite distance aft of the wing.

Three assumptions are needed for the simple version of lifting line theory presented here. The first is that the wing's aspect ratio, span/(average chord), is so large that the flow at any span-wise location may be treated as two dimensional. A second assumption is that the actual physical structure of the aircraft does not matter and that the aircraft's main wing may be replaced by a single (straight) vortex segment of variable strength. This vortex segment is called the *bound vortex*. It moves with the aircraft and lies along the aircraft's wings, nominally located at the center of lift at any span-wise location along the wing. The bound vortex forms the *lifting line* segment from which the theory draws its name. In general, the bound vortex is strongest near the midspan and weakest near the wing tips. According to one of the Helmholtz theorems (Section 5.3), a vortex cannot begin or end in the fluid; it must end at a wall or form a closed loop. Therefore, as the bound vortex weakens from wing root to wing tip it releases vortex filaments that turn parallel to the stream-wise direction and are advected downstream, eventually coalescing to form the tip vortices. A third assumption made in lifting line theory is that the interaction of these trailing vortex filaments with each other can be ignored. Thus, each trailing vortex filament starts at the bound vortex and is assumed to lie along a straight semi-infinite horizontal line parallel to the upstream flow direction. Although a formal mathematical account of the theory was first published by Prandtl, many of the important underlying ideas were first conceived by Lanchester. The historical controversy regarding the credit for the theory is noted at the end of this section.

With these assumptions and the geometry shown in Figure 14.21, a relation can be derived between the distribution of circulation along the wingspan and the strength of the trailing vortex filaments. Suppose that the clockwise circulation of the bound vortex changes from

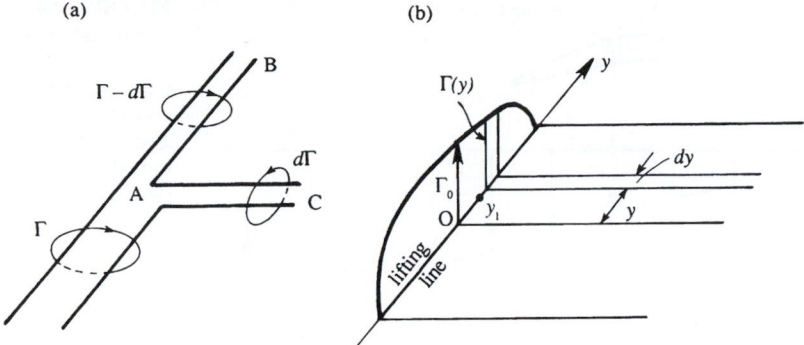

FIGURE 14.21 The mechanism leading to trailing vortices. (a) When the bound vortex having strength Γ weakens, it sheds a vortex filament AC of strength $d\Gamma$ into the wing's wake and continues along the wing as the vortex AB with strength $\Gamma - d\Gamma$. (b) The shed vortex filament that leaves the bound vortex at location y induces a downward velocity at location y_1 of the bound vortex when $y > y_1$. The induced velocity from all trailing vortex filaments is known as *downwash*.

Γ to $\Gamma - d\Gamma$ at a certain point (Figure 14.21a). Then another vortex AC of strength $d\Gamma$ must emerge from the location of the change. In fact, the strength and sign of the circulation around AC is such that, when AC is folded back onto AB, the circulation is uniform along the composite vortex tube. (Recall the vortex theorem of Helmholtz, which says that the strength of a vortex tube is constant along its length.) Now consider the vortex strength or circulation distribution $\Gamma(y)$ that represents the main wing (Figure 14.21b). The change in circulation in length dy is $d\Gamma$, which is a decrease if $dy > 0$. It follows that the magnitude of the trailing vortex filament of width dy is $-(d\Gamma/dy)dy$. For simple wings, the trailing vortices will be stronger near the wing tips where $d\Gamma/dy$ is the largest.

The critical contribution of lifting line theory is that it allows an approximate means of assessing the impact of the trailing vortex filaments on the performance of the bound vortex representing the aircraft's wing. The simplest means of assessing this impact is to determine the velocity induced at a point y_1 on the lifting line by the trailing vortex filament that leaves the wing at location y, and then integrating over the trailing filament contributions from all possible y values. Based on the Biot-Savart law (5.17), a straight semi-infinite trailing vortex filament that leaves the wing at y with strength $-(d\Gamma/dy)dy$ and remains horizontal induces a downward velocity of magnitude:

$$dw(y_1) = \frac{-(d\Gamma/dy)dy}{4\pi(y - y_1)}$$

at location y ($< y_1$) along the lifting line (Exercise 14.10 with $\theta_1 = 0$ and $\theta_2 = 90°$). This velocity increment is *half* the velocity induced by an infinitely long vortex element. The bound vortex does not induce a velocity on itself, so for a wing of span s, the total downward velocity at y_1 due to the entire trailing vortex sheet is therefore

$$w(y_1) = \frac{1}{4\pi}\int_{-s/2}^{s/2} \frac{d\Gamma}{dy}\frac{dy}{(y_1 - y)}, \tag{14.13}$$

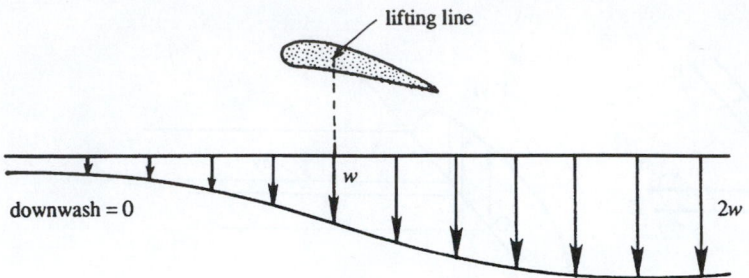

FIGURE 14.22 Variation of downwash ahead of and behind an airfoil. The downwash is weaker upstream of the wing and stronger downstream of it. The actual profile can be determined from the Biot-Savart law (see (5.17) and Exercise 14.10).

which is called the *downwash* at y_1 on the lifting line. The vortex sheet also induces a smaller downward velocity in front of the airfoil and a larger one behind the airfoil (Figure 14.22).

This downwash velocity adds to the free-stream velocity so that the incident flow at any location along the wing is the vector resultant of U and w (Figure 14.23). The downwash therefore changes the local angle of attack of the airfoil, decreasing it by the angle

$$\varepsilon = \tan\frac{w}{U} \simeq \frac{w}{U},$$

where the approximate equality follows when $w \ll U$, the most common situation in applications. Thus, the *effective angle of attack* at any span-wise location is

$$\alpha_e = \alpha - \varepsilon = \alpha - \frac{w}{U}. \tag{14.14}$$

Because the aspect ratio is assumed large, ε is assumed to be small. Each element dy of the finite wing may then be assumed to act as though it is an isolated two-dimensional section set in a stream of uniform velocity U_e, at an angle of attack α_e. According to the Kutta-Zhukhovsky lift theorem, a circulation Γ superimposed on the actual resultant velocity U_e generates an elemental aerodynamic force $dL_e = \rho U_e \Gamma dy$, which acts normal to U_e. This force

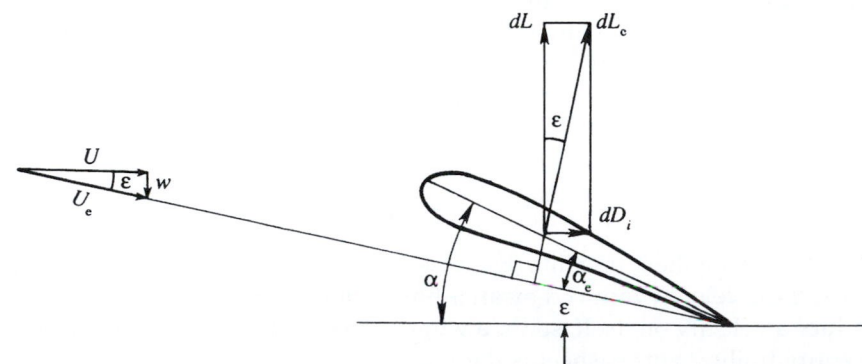

FIGURE 14.23 Lift and lift-induced drag on a wing element dy in the presence of a downwash velocity w. The downwash velocity locally lowers the angle of attack of the free stream and rotates the lift vector backward to produce the lift-induced drag.

may be resolved into two components, the conventional lift force dL normal to the direction of flight and a component dD_i parallel to the direction of flight (Figure 14.23). Therefore

$$dL = dL_e \cos \varepsilon = \rho U_e \Gamma \, dy \cos \varepsilon \simeq \rho U \Gamma \, dy,$$

$$dD_i = dL_e \sin \varepsilon = \rho U_e \Gamma \, dy \sin \varepsilon \simeq \rho w \Gamma \, dy.$$

In general w, Γ, U_e, ε, and α_e are all functions of y, so that for the entire wing:

$$L = \int_{-s/2}^{s/2} \rho U \Gamma \, dy,$$

$$D_i = \int_{-s/2}^{s/2} \rho w \Gamma \, dy. \qquad (14.15)$$

These expressions have a simple interpretation: whereas the interaction of U and Γ generates L, which acts normal to U, the interaction of w and Γ generates D_i, which acts normal to w.

The drag force D_i induced by the trailing vortices is called the *induced drag* and is zero for a wing of infinite span. It arises on a wing of finite span because it continuously creates trailing vortices and the rate of generation of trailing-vortex kinetic energy must equal the rate of work done against the induced drag, namely $D_i U$. For this reason, the induced drag is also known as the *vortex drag*. It is analogous to the *wave drag* experienced by a ship, which continuously radiates gravity waves during its motion. As we shall see, the induced drag is the largest part of the total drag experienced by an airfoil (away from stall).

A basic reason why there must be a downward velocity behind the wing is the following: The fluid exerts an upward lift force on the wing, and therefore the wing exerts a downward force on the fluid. The fluid must therefore constantly gain downward momentum as it goes past the wing.

For a given $\Gamma(y)$, $w(y)$ can be determined from (14.13) and D_i can then be determined from (14.15). However, $\Gamma(y)$ itself depends on the distribution of $w(y)$ because the effective angle of attack is changed due to $w(y)$. To see how $\Gamma(y)$ may be estimated, first note that the lift coefficient for a two-dimensional Zhukhovsky airfoil is nearly $C_L = 2\pi (\alpha + \beta)$. For a finite wing we may assume

$$C_L = K\left[\alpha - \frac{w(y)}{U} + \beta(y)\right], \qquad (14.16)$$

where $(\alpha - w/U)$ is the effective angle of attack, $-\beta(y)$ is the angle of attack for zero lift (found from experimental data such as Figure 14.17), and K is the lift-curve slope, a constant whose value is nearly six for most airfoils ($K = 2\pi$ for Zhukhovsky and thin airfoils). An expression for the circulation can be obtained by noting that the lift coefficient is related to the circulation as $C_L = L/((1/2)\rho U^2 c) = \Gamma/((1/2)Uc)$, so that $\Gamma = (1/2)Uc C_L$. Equation (14.16) is then equivalent to the assumption that the circulation for a wing of finite span is

$$\Gamma(y) = \frac{K}{2}Uc(y)\left[\alpha - \frac{w(y)}{U} + \beta(y)\right]. \qquad (14.17)$$

For a given $U, \alpha, c(y),$ and $\beta(y)$, (14.13) and (14.17) define an integral equation for determining $\Gamma(y)$.

An approximate solution to these two equations can be obtained by changing y and y_1 to angular variables γ and γ_1:

$$y = -(s/2)\cos\gamma \quad \text{and} \quad y_1 = -(s/2)\cos\gamma_1,$$

so that $\gamma = 0$ and $\gamma = \pi$ correspond to the left (port) and right (starboard) wing tips, respectively, and then assuming a Fourier series form for the circulation strength of the lifting line:

$$\Gamma = \sum_{n=1}^{\infty} \Gamma_n \sin(n\gamma), \tag{14.18}$$

where the Γ_n are undetermined coefficients. When (14.18) is substituted into (14.13), the resulting equation is:

$$w(y_1) = \frac{1}{2\pi s} \int_0^\pi \sum_{n=1}^{\infty} n\Gamma_n \frac{\cos(n\gamma)d\gamma}{\cos\gamma_1 - \cos\gamma} = \frac{1}{2\pi s} \sum_{n=1}^{\infty} n\Gamma_n \int_0^\pi \frac{\cos(n\gamma)d\gamma}{\cos\gamma_1 - \cos\gamma} = \frac{1}{2s} \sum_{n=1}^{\infty} n\Gamma_n \frac{\sin(n\gamma_1)}{\sin\gamma_1},$$

$$\tag{14.19}$$

where the final equality comes from evaluating the integral. Combing (14.17) through (14.19) and dropping the subscript "1" from γ, produces a single equation for the coefficients Γ_n:

$$\frac{K}{2}Uc(\alpha + \beta) = \sum_{n=1}^{\infty} \left(1 + \frac{nKc}{4s\sin\gamma}\right)\Gamma_n \sin(n\gamma), \tag{14.20}$$

where K, c, α, and β may all be functions of the transformed span coordinate γ. Thus, (14.20) is not a typical Fourier series solution because the coefficients of $\sin(n\gamma)$ inside the sum depend on γ. In practice, (14.20) can be solved approximately by truncating the sum after N terms, and then requiring its validity at N points along the wing to convert it into N algebraic equations for $\Gamma_1, \Gamma_2, \dots \Gamma_N$. Fortunately in many circumstances, just a few terms in the sum are needed to adequately represent $\Gamma(y)$.

With an approximate solution for $\Gamma(y)$ provided by several Γ_n computed algebraically from (14.20), the wing's lift and induced drag computed from (14.15) are:

$$L = \frac{\pi s}{4}\rho U\Gamma_1, \quad \text{and} \quad D_i = \frac{\pi}{8}\rho \sum_{n=1}^{N} n\Gamma_n^2. \tag{14.21, 14.22}$$

Thus, the wing's performance is maximized when $\Gamma_1 \neq 0$ and $\Gamma_n = 0$ for all $n > 1$, because this produces the maximum lift-to-drag ratio. In this case (14.18) reduces to:

$$\Gamma = \Gamma_1\sin(\gamma) = \Gamma_1\sqrt{1 - (2y/s)^2}, \tag{14.23}$$

which is known as an *elliptical lift distribution*. The downwash for an elliptical lift distribution is constant across the wingspan,

$$w(y) = \Gamma_1/2s, \tag{14.24}$$

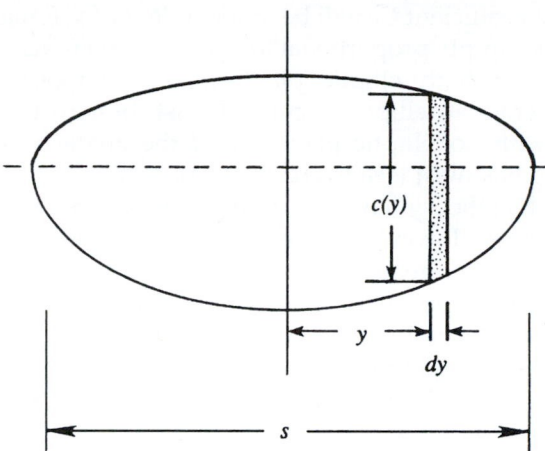

FIGURE 14.24 Wing with an elliptic planform. Here the variation in the chord over the span can produce an elliptical lift distribution. This planform is similar to that of the *British Spitfire*, a WWII combat aircraft.

as can be found from substituting (14.23) into (14.19). The induced drag for an elliptical lift distribution is

$$D_i = \frac{\pi}{8}\rho\Gamma_1^2 = \frac{2L^2}{\pi\rho U^2 s^2},$$

(14.25)

where (14.21) has been used to introduce L in the second equality. Thus, the induced drag coefficient for an elliptical lift distribution is:

$$C_{D_i} = \frac{D_i}{(1/2)\rho U^2 A} = \frac{C_L^2}{\pi(s^2/A)} = \frac{C_L^2}{\pi\Lambda},$$

(14.26)

where C_L and C_D are given by (4.107) and (4.108) in Section 4.3, A is the wing's planform area, and Λ is the wing's aspect ratio. Equation (14.26) shows that $C_{D_i} \to 0$ when the flow is two dimensional, that is, in the limit $\Lambda \to \infty$. More importantly, it shows that the *induced drag coefficient increases as the square of the lift coefficient*. We shall see in the following section that the induced drag generally makes the largest contribution to the total drag of an airfoil.

Since an elliptic circulation distribution minimizes the induced drag, it is of interest to determine the circumstances under which such a circulation can be established. Consider an element dy of the wing (Figure 14.24). The lift on the element is:

$$dL = \rho U \Gamma \, dy = C_L \frac{1}{2}\rho U^2 c \, dy,$$

(14.27)

where cdy is a wing area element. If the circulation distribution is elliptic, then the downwash is independent of y. In addition, if the wing profile is geometrically similar at every point along the span and has the same geometrical angle of attack α, then the effective angle of attack and hence the lift coefficient C_L will be independent of y. Equation (14.27) shows that the chord length c is then simply proportional to Γ, and so $c(y)$ is also elliptically distributed. Thus, an untwisted wing with elliptic planform, or composed of two semi-ellipses (Figure 14.24), will generate an elliptic circulation distribution. However, the same effect can also be achieved with nonelliptic planforms if the angle of attack varies along the span, that is, if the wing has twist (see Exercise 14.14).

The results of lifting line theory have had an enormous impact on the design and development of subsonic aircraft. However, the results presented here are approximate because of the geometrical assumptions made about the aircraft's wings, its trailing vortices, and the tip vortices. Thus, an elliptical lift distribution is only approximately optimal, and a more general theory would produce refinements. Yet, with suitable geometric modifications lifting line theory can be applied to multiple-wing aircraft and rotating propellers. Furthermore, its implications help explain near-ground effects for landing aircraft, and the Λ-pattern commonly formed by flocks of migrating birds.

Lanchester Versus Prandtl

There is some controversy in the literature about who should get more credit for developing lifting line theory. Since Prandtl in 1918 first published the theory in a mathematical form, textbooks for a long time have called it the *Prandtl Lifting Line Theory*. Lanchester was bitter about this, because he felt that his contributions were not adequately recognized. The controversy has been discussed by von Karman (1954, p. 50), who witnessed the development of the theory. He gives a lot of credit to Lanchester, but falls short of accusing his teacher Prandtl of being deliberately unfair. Here we shall note a few facts that von Karman brings up.

Lanchester was the first person to study a wing of finite span. He was also the first person to conceive that a wing can be replaced by a bound vortex, which bends backward to form the tip vortices. Last, Lanchester was the first to recognize that the minimum power necessary to fly is that required to generate the kinetic energy field of the downwash field. It seems, then, that Lanchester had conceived all of the basic ideas of the wing theory, which he published in 1907 in the form of a book called *Aerodynamics*. In fact, a figure from his book looks very similar to the current Figure 14.20.

Many of these ideas were explained by Lanchester in his talk at Göttingen, long before Prandtl published his theory. Prandtl, his graduate student von Karman, and Carl Runge were all present. Runge, well known for his numerical integration scheme of ordinary differential equations, served as an interpreter, because neither Lanchester nor Prandtl could speak the other's language. As von Karman said, "both Prandtl and Runge learned very much from these discussions."

However, Prandtl did not want to recognize Lanchester for priority of ideas, saying that he conceived of them before he saw Lanchester's book. Such controversies cannot be settled, and great intellects have been involved in controversies before.

In view of the fact that Lanchester's book was already in print when Prandtl published his theory, and the fact that Lanchester had all the ideas but not a formal mathematical theory, we have called it the *Lifting Line Theory of Prandtl and Lanchester* at the outset of this section.

14.7. LIFT AND DRAG CHARACTERISTICS OF AIRFOILS

Before an aircraft is built its wing design is tested in a wind tunnel, and the results are generally given as plots of C_L and C_D versus the angle of attack α. A typical plot is shown in Figure 14.25 where it is seen that, for $-4° < \alpha < 12°$, the variation of C_L with α is approximately linear, a typical value of $dC_L/d\alpha$ ($= K$) being ≈ 0.1 per degree. The lift reaches a maximum value at $\alpha \approx 15°$. If the angle of attack is increased further, the steep adverse pressure gradient on the upper surface of the airfoil causes the flow to separate before reaching the wing's trailing edge, and a large wake is formed (Figure 14.26). The drag coefficient increases and the lift coefficient drops. The wing is said to *stall* as the suction-side boundary-layer separation point moves toward the leading edge. Beyond the stalling incidence angle the lift coefficient levels off again and remains at $\approx 0.7-0.8$ up to α values of 10s of degrees.

For a fixed-shape wing, the maximum possible lift coefficient depends largely on the Reynolds number Re. For chord-based Reynolds numbers of Re $\sim 10^5-10^6$, the suction-side

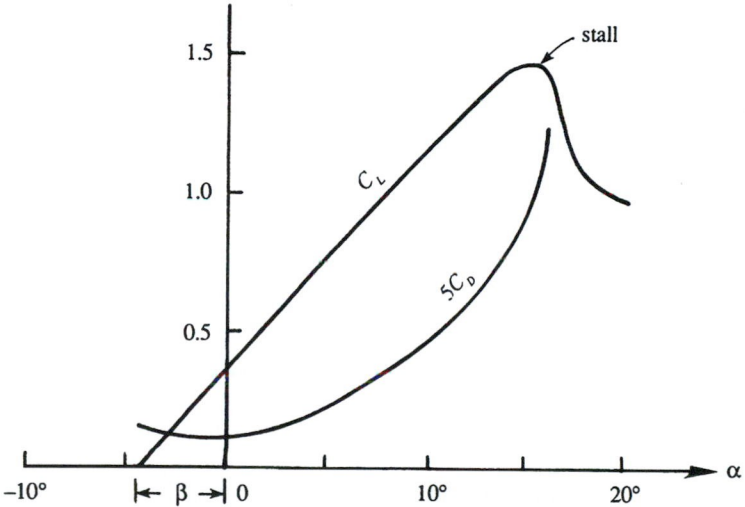

FIGURE 14.25 Generic lift and drag coefficients vs. angle of attack. There is lift at $\alpha = 0$ so the foil shape has nonzero camber. The drag increase is almost quadratic with increasing angle of attack in accordance with (14.26).

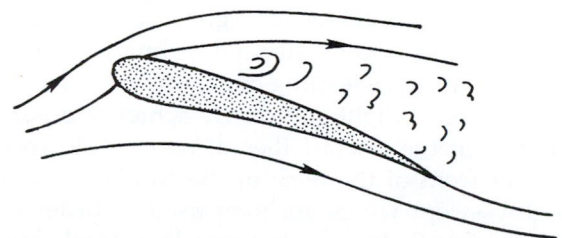

FIGURE 14.26 Stalling of an airfoil. Here the Kutta condition is no longer satisfied, and the flow separates near the leading edge on the foil's suction side. In this situation, the foil's lift and drag are comparable.

boundary layer may separate before it undergoes transition, and stall may begin before α reaches $10°$ leading to maximum lift coefficients < 0.9. At larger Reynolds numbers, say $Re > 10^7$, the suction-side boundary layer transitions to turbulence before it separates and is therefore able to stay attached up to α-values approaching or exceeding $20°$. Maximum lift coefficients near or even slightly above two may be obtained at the highest Reynolds numbers.

The angle of attack at zero lift, denoted by $-\beta$ here, is a function of the airfoilsection's camber. (For a Zhukhovsky airfoil, $\beta = 2(camber)/chord$.) The effect of increasing the airfoil camber is to raise the entire graph of C_L versus α, thus increasing the maximum values of C_L without stalling. A cambered profile delays stall because its leading edge points into the airstream while the rest of the airfoil is inclined to the stream. Rounding the airfoil nose is also essential, since an airfoil of zero thickness would undergo separation at the leading edge. Trailing edge flaps act to increase the camber and thereby the lift coefficient when they are deployed, and this allows lower aircraft landing speeds.

Various terms are in common usage to describe the different components of the drag. The total drag of a body can be divided into a *friction drag* due to the tangential stresses on the surface and *pressure drag* due to the normal stresses. The pressure drag can be further subdivided into an *induced drag* and a *form drag*. The induced drag is the drag that results from the work done by the body to supply the kinetic energy of the downwash field as the trailing vortices increase in length. The form drag is defined as the part of the total pressure drag that remains after the induced drag is subtracted out. (Sometimes the skin friction and form drags are grouped together and called the *profile drag*, which represents the drag due to the wing's geometrical profile alone and not due to the finiteness of the wing.) The form drag depends strongly on the shape and orientation of the airfoil and can be minimized by good design. In contrast, relatively little can be done about the induced drag if the wing's aspect ratio is fixed.

Normally the induced drag constitutes the major part of the total drag of a wing. As C_{D_i} is nearly proportional to C_L^2, and C_L is nearly proportional to α, it follows that $C_{D_i} \propto \alpha^2$. This is why the drag coefficient in Figure 14.25 seems to increase quadratically with angle of attack.

For high-speed aircraft, the appearance of shock waves can adversely affect the behavior of the lift and drag characteristics. In such cases the maximum *flow* speeds can be close to or higher than the speed of sound even when the aircraft is flying at subsonic speeds. Shock waves can form when the local flow speed exceeds the local speed of sound. To reduce their effect, the wings are given a *sweepback angle*, as shown in Figure 14.2. The maximum flow speeds depend primarily on the component of the oncoming stream perpendicular to the leading edge; this component is reduced as a result of the sweepback. Thus, increased flight speeds are achievable with highly swept wings. This is particularly true when the aircraft flies at supersonic speeds in which there is invariably a shock wave in front of the nose of the fuselage, extending downstream in the form of a cone. Highly swept wings are then used in order that the wing does not penetrate this shock wave. For flight speeds exceeding Mach numbers of order 2, the wings have such large sweepback angles that they resemble the Greek letter Δ; these wings are sometimes called *delta wings*.

14.8. PROPULSIVE MECHANISMS OF FISH AND BIRDS

The propulsive mechanisms of many animals are based on lift generation by wing-like surfaces. Just the basic ideas of this interesting subject are presented here. More detail is provided by Lighthill (1986).

First consider swimming fish. They develop *forward* thrust by horizontally oscillating their tails from *side to side*. Fish tails like that shown in Figure 14.27a have a cross section resembling that of a symmetric airfoil. One-half of the oscillation is represented in Figure 14.27b, which shows the top view of the tail. The sequence 1 to 5 represents the positions of the tail during the tail's motion to the left. A quick change of *orientation* occurs at one extreme position of the oscillation during 1 to 2; the tail then moves to the left during 2 to 4, and another quick change of orientation occurs at the other extreme during 4 to 5.

Suppose the tail is moving to the left at speed V, and the fish is moving forward at speed U. The fish controls these magnitudes so that the resultant fluid velocity U_r (relative to the tail) is inclined to the tail surface at a positive angle of attack. The resulting lift L is perpendicular to U_r and has a forward component $L \sin \theta$. (It is easy to verify that there is a similar forward propulsive force when the tail moves from left to right.) This thrust, working at the rate $UL \sin \theta$, propels the fish. To achieve this propulsion, the tail of the fish pushes sideways on the water against a force of $L \cos \theta$, which requires work at the rate $VL \cos \theta$. Since $V/U = \tan \theta$, the conversion of energy is ideally perfect—all of the oscillatory work done by the fish tail goes into the translation. In practice, however, this is not the case because of the presence of induced drag and other effects that generate a wake.

Most fish stay afloat by controlling the buoyancy of an internal swim bladder. In contrast, some large marine mammals such as whales and dolphins develop *both* a forward thrust and a vertical lift by moving their tails *vertically*. They are able to do this because their tail surface is *horizontal*, in contrast to the vertical tail shown in Figure 14.27. A review by Fish and Lauder (2006) provided evidence that leading-edge tubercles as seen on humpback whale flippers increase lift and reduce drag at high angles of attack. This is because separation is

FIGURE 14.27   Propulsion of fish. (a) The cross section of the tail along AA is that of a symmetric airfoil. Five positions of the tail during its motion to the left are shown in (b). The lift force L is normal to the resultant speed U_r of water with respect to the tail.

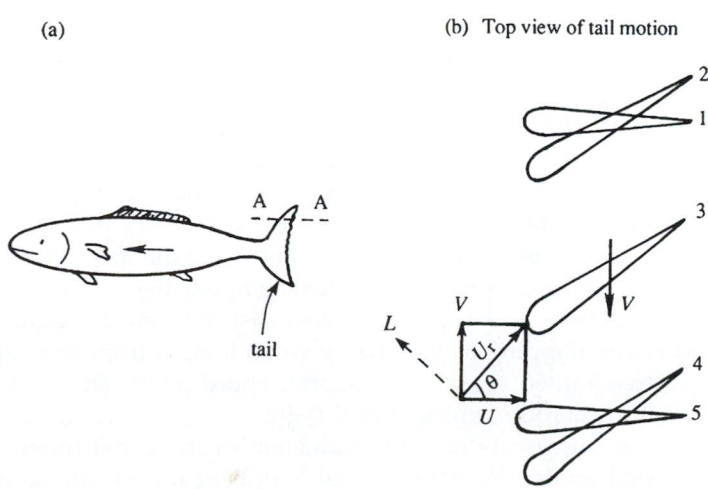

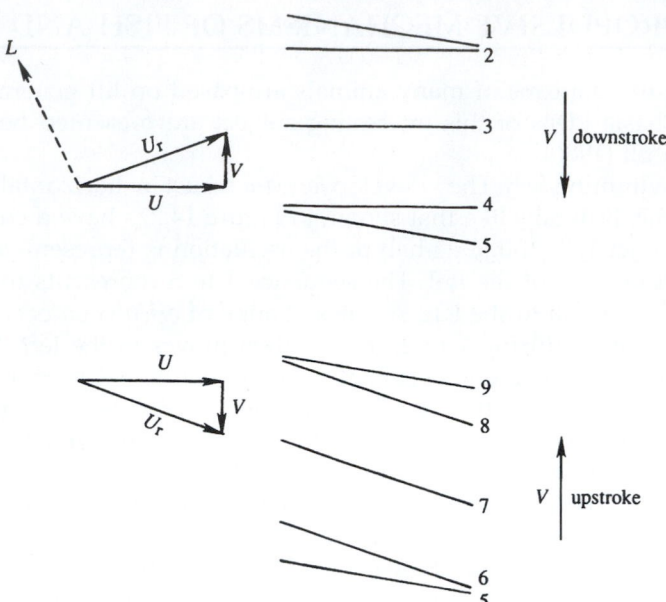

FIGURE 14.28 Propulsion of a bird. A cross section of the wing is shown during upstroke and downstroke. During the downstroke, a lift force L acts normal to the resultant speed U_r of air with respect to the wing. During the upstroke, U_r is nearly parallel to the wing and very little aerodynamic force is generated.

delayed due to the creation of stream-wise vortices on the suction side. Cetacean flukes or flippers and fish tail fins as well as dorsal and pectoral fins are flexible and can vary their camber during a stroke. As a result they are very efficient propulsive devices.

Now consider flying birds, who flap their wings to generate *both* the lift to support their body weight and the forward thrust to overcome drag. Figure 14.28 shows a vertical section of the wing positions during the upstroke and downstroke of the wing. (Birds have cambered wings, but this is not shown in the figure.) The angle of inclination of the wing with the airstream changes suddenly at the end of each stroke, as shown. The important point is that the upstroke is inclined at a greater angle to the airstream than the downstroke. As the figure shows, the downstroke develops a lift force L perpendicular to the resultant velocity of the air relative to the wing. Both a forward thrust and an upward force result from the downstroke. In contrast, very little aerodynamic force is developed during the upstroke, as the resultant velocity is then nearly parallel to the wing. Birds therefore do most of the work necessary for flight during the downstroke.

Liu et al. (2006) provide the most complete description to date of wing planform, camber, airfoil section, and span-wise twist distribution of seagulls, mergansers, teals, and owls. Moreover, flapping as viewed by video images from free flight was digitized and modeled by a two-jointed wing at the quarter chord point. The data from this paper can be used to model the aerodynamics of bird flight.

Using previously measured kinematics and experiments on an approximately 100-times upscaled model, Ramamurti and Sandberg (2001) calculated the flow about a Drosophila (fruit fly) in flight. They matched Reynolds number (based on wing-tip speed and average chord) and found that viscosity had negligible effect on thrust and drag at a flight Reynolds number of 120. The wings were near elliptical plates with axis ratio 3:1.2 and thickness about 1/80 of the span. Averaged over a cycle, the mean thrust coefficient (thrust/[dynamic pressure × wing surface]) was 1.3 and the mean drag coefficient close to 1.5.

14.9. SAILING AGAINST THE WIND

People have sailed without the aid of an engine for thousands of years and have known how to reach an upwind destination. Actually, it is not possible to sail exactly against the wind, but it is possible to sail at ≈ 40–45° to the wind. Figure 14.29 shows how this is made possible by the aerodynamic lift on the sail, which is a piece of stretched and stiffened cloth. The wind speed is U, and the sailing speed is V, so that the apparent wind speed relative to the boat is U_r. If the sail is properly oriented, this gives rise to a lift force perpendicular to U_r and a drag force parallel to U_r. The resultant force F can be resolved into a driving component (thrust) along the motion of the boat and a lateral component. The driving component performs work in moving the boat; most of this work goes into overcoming the frictional drag and in generating the gravity waves that radiate outward from the hull. The lateral component does not cause much sideways drift because of the shape of the hull. It is clear that the thrust decreases as the angle θ decreases and normally vanishes when θ is ≈ 40–45°. The energy for sailing comes from the wind field, which loses kinetic energy after passing the sail.

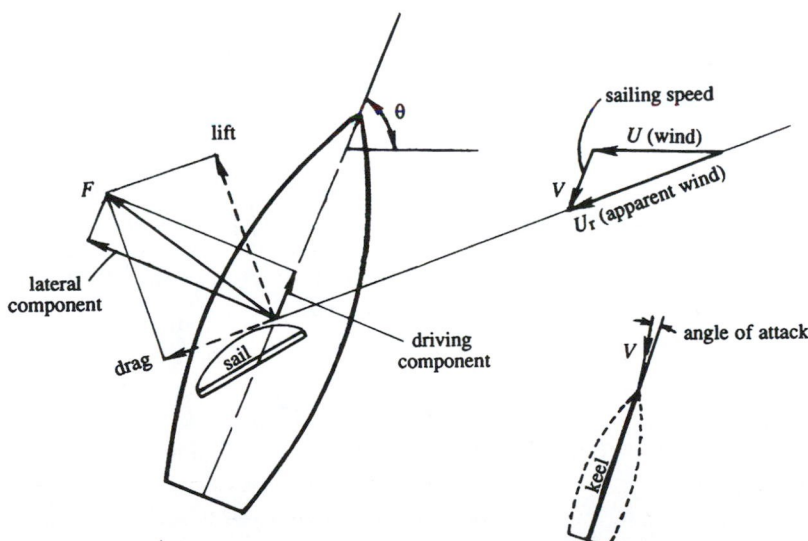

FIGURE 14.29 Principle of sailing against the wind. A small component of the sail's lift pushes the boat forward at an angle $\theta < 90°$ to the wind. Thus by traversing a zig-zag course at angles $\pm\theta$, a sailboat can reach an upwind destination. A sailboat's keel may make a contribution to its upwind progress too.

In the foregoing discussion we have not considered the hydrodynamic forces exerted by the water on the hull. At constant sailing speed the net hydrodynamic force must be equal and opposite to the net aerodynamic force on the sail. The hydrodynamic force can be decomposed into a drag (parallel to the direction of motion) and a lift. The lift is provided by the sailboat's *keel*, which is a thin vertical surface extending downward from the bottom of the hull. For the keel to act as a lifting surface, the longitudinal axis of the boat points at a small angle to the direction of motion of the boat, as indicated near the bottom right part of Figure 14.29. This keel-angle of attack is generally $< 3°$ and is not noticeable. The hydrodynamic lift developed by the keel opposes the aerodynamic lateral force on the sail. It is clear that without the keel the lateral aerodynamic force on the sail would topple the boat around its longitudinal axis.

To arrive at a destination directly against the wind, one has to sail in a zig-zag path, always maintaining an angle of $\approx 45°$ to the wind. For example, if the wind is coming from the east, we can first proceed northeastward as shown, then change the orientation of the sail to proceed southeastward, and so on. In practice, a combination of a number of sails is used for effective maneuvering. The mechanics of sailing yachts is discussed in Herreshoff and Newman (1966).

EXERCISES

14.1. Consider the elementary aerodynamics of a projectile of mass m with $C_L = 0$ and $C_D =$ constant. In Cartesian coordinates with gravity g acting downward along the y-axis, a set of equations for such a projectile's motion are:

$$m\frac{dV_x}{dt} = -D\cos\theta, m\frac{dV_y}{dt} = -mg - D\sin\theta, \tan\theta = V_y/V_x, \text{ and } D = \frac{1}{2}\rho\left(V_x^2 + V_y^2\right)AC_D,$$

where V_x and V_y are the horizontal and vertical components of the projectile's velocity, θ is the angle of the projectile's trajectory with respect to the horizontal, D is the drag force on the projectile, ρ is the air density, and A is projectile's frontal area. Assuming a shallow trajectory, where $V_x^2 \gg V_y^2$ and $mg \gg D\sin\theta$, show that the distance traveled by the projectile over level ground is: $x \cong \dfrac{2m}{\rho AC_D}\ln\left(1 + \dfrac{\rho AC_D V_o^2 \cos\theta_o \sin\theta_o}{mg}\right)$ if it is launched from ground level with speed of V_o at an angle of θ_o with respect to the horizontal. Does this answer make sense as $C_D \to 0$?

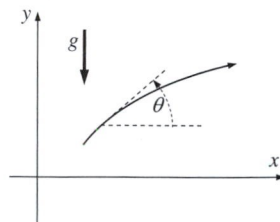

14.2. As a model of a two-dimensional airfoil's trailing edge flow, consider the potential $\phi(r,\theta) = (Ud/n)(r/d)^n\cos(n\theta)$ in the usual r-θ coordinates (Figure 3.3a). Here U, d, and

n are positive constants, the fluid has density ρ, and the foil's trailing edge lies at the origin of coordinates.

a) Sketch the flow for $n = 3/2$, $5/4$, and $9/8$ in the angle range $|\theta| < \pi/n$, and determine the full included angle of the foil's trailing edge in terms of n.

b) Determine the fluid velocity at $r = d$ and $\theta = 0$.

c) If p_0 is the pressure at the origin of coordinates and p_d is the pressure at $r = d$ and $\theta = 0$, determine the pressure coefficient: $C_p = (p_0 - p_d)/(1/2)\rho U^2$ as a function of n. In particular, what is C_p when $n = 1$ and when $n > 1$?

14.3. Consider an airfoil section in the xy-plane, the x-axis being aligned with the chord line. Examine the pressure forces on an element $ds = (dx, dy)$ on the surface, and show that the net force (per unit span) in the y-direction is

$$F_y = -\int_0^c p_u \, dx + \int_0^c p_1 \, dx,$$

where p_u and p_1 are the pressure on the upper and the lower surfaces and c is the chord length. Show that this relation can be rearranged in the form

$$C_y \equiv \frac{F_y}{(1/2)\rho U^2 c} = \oint C_p d\left(\frac{x}{c}\right),$$

where $C_p = (p_0 - p_\infty)/(1/2)\rho U^2$, and the integral represents the area enclosed in a C_p versus x/c diagram, such as Figure 14.8. Neglect shear stresses. [Note that C_y is not exactly the lift coefficient, since the airstream is inclined at a small angle α with respect to the x-axis.]

14.4. The measured pressure distribution over a section of a two-dimensional airfoil at $4°$ incidence has the following form:

Upper Surface: C_p is constant at -0.8 from the leading edge to a distance equal to 60% of chord and then increases linearly to 0.1 at the trailing edge.

Lower Surface: C_p is constant at -0.4 from the leading edge to a distance equal to 60% of chord and then increases linearly to 0.1 at the trailing edge.

Using the results of Exercise 14.3, show that the lift coefficient is nearly 0.32.

14.5. The Zhukhovsky transformation $z = \zeta + b^2/\zeta$ transforms a circle of radius b, centered at the origin of the ζ-plane, into a flat plate of length $4b$ in the z-plane. The circulation around the cylinder is such that the Kutta condition is satisfied at the trailing edge of the flat plate. If the plate is inclined at an angle α to a uniform stream U, show that:

(i) The complex potential in the ζ-plane is $w = U(\zeta e^{-i\alpha} + 1/\zeta b^2 e^{i\alpha}) + \dfrac{i\Gamma}{2\pi}\ln(\zeta e^{-i\alpha})$,

where $\Gamma = 4\pi Ub \sin \alpha$. Note that this represents flow over a circular cylinder with circulation in which the oncoming velocity is oriented at an angle α.

(ii) The velocity components at point $P(-2b, 0)$ in the ζ-plane are $\left[\dfrac{3}{4}U \cos \alpha, \dfrac{9}{4}U \sin \alpha\right]$.

(iii) The coordinates of the transformed point P' in the xy-plane are $[-5b/2, 0]$.

(iv) The velocity components at $[-5b/2, 0]$ in the xy-plane are $[U\cos\alpha, 3U\sin\alpha]$.

14.6. In Figure 14.12, the angle at A' has been marked 2β. Prove this. [*Hint*: Locate the center of the circular arc in the z-plane.]

14.7. Ideal flow past a flat plate inclined at angle α with respect to a horizontal free stream produces lift but no drag when the Kutta condition is applied at the plate's trailing edge. However, pressure forces can only act in the plate-normal direction and this direction is *not* perpendicular to the flow. Therefore, to achieve zero drag, another force must act on the plate. This extra force is known as *leading-edge suction* and its existence can be assessed from the potential for flow around the tip of a flat plate that is coincident with the x-axis for $x > 0$. In two-dimensional polar coordinates, this velocity potential is $\phi = 2U_o\sqrt{ar}\cos(\theta/2)$ where U_o and a are velocity and length scales, respectively, that characterize the flow.

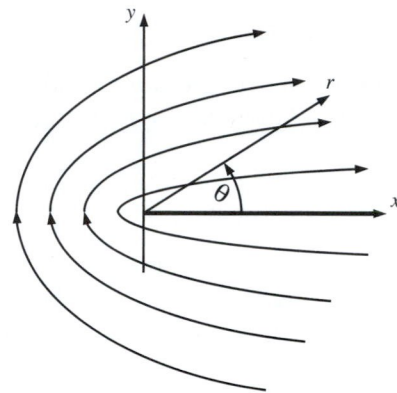

a) Determine u_r and u_θ, the radial and angular-directed velocity components, respectively.

b) If the pressure far from the origin is p_∞, determine the pressure p at any location (r, θ).

c) Use the given potential, a circular control volume of radius ε centered at the origin of coordinates, and the control volume version of the ideal flow momentum equation, $\int_C \rho\mathbf{u}(\mathbf{u}\cdot\mathbf{n})d\xi = -\int_C p\mathbf{n}d\xi + \mathbf{F}$, to determine the force $\mathbf{F}$ (per unit depth into the page) that holds the plate stationary when $\varepsilon \to 0$. Here, $\mathbf{n}$ is the outward unit normal vector to the control volume surface, and $d\xi$ is the length increment of the circular control surface.

d) If the plate is released from rest, in what direction will it initially accelerate?

14.8. Consider a cambered Zhukhovsky airfoil determined by the following parameters: $a = 1.1$, $b = 1.0$, and $\beta = 0.1$. Using a computer, plot its contour by evaluating the Zhukhovsky transformation. Also plot a few streamlines, assuming an angle of attack of $5°$.

14.9. A thin Zhukhovsky airfoil has a lift coefficient of 0.3 at zero incidence. What is the lift coefficient at $5°$ incidence?

14.10. Lifting line theory involves calculating vortex-induced velocities from the Biot-Savart induction law, (5.17). Consider an idealized vortex segment of uniform strength Γ that lies along the z-axis between z_1 and z_2, and has a sense of rotation that points along the z-axis. In this case the induced velocity $\mathbf{u}$ at the location (R, φ, z) will be given by the integral:

$$\mathbf{u}(\mathbf{x}, t) = \int_{z_1}^{z_2} d\mathbf{u} = \frac{\Gamma}{4\pi} \int_{z_1}^{z_2} \mathbf{e}_z \times \frac{(\mathbf{x} - \mathbf{x}')}{|\mathbf{x} - \mathbf{x}'|^3} dz.$$

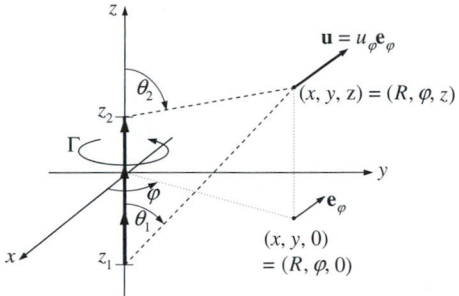

a) Evaluate this integral to show that $\mathbf{u}(\mathbf{x}, t) = (\Gamma/4\pi R)(\cos\theta_1 - \cos\theta_2)\mathbf{e}_\varphi$ where the angles θ_1 and θ_2 are defined in the figure.

b) Show that the velocity induced by an infinite ideal-line vortex is recovered from the part a) result for an appropriate choice of angles.

c) What is the induced velocity on the z-axis ($R = 0$) when $z < z_1$ or $z > z_2$?

14.11.[1] The simplest representation of a three-dimensional aircraft wing in flight is the rectangular horseshoe vortex.

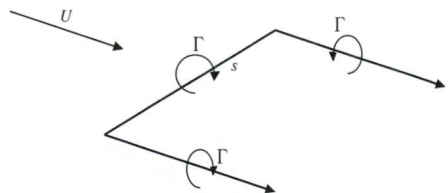

a) Calculate the induced downwash at the center of the wing.

b) Assuming the result of part a) applies along the entire wingspan, estimate C_{D_i}, the lift-induced coefficient of drag, in terms of the wing's aspect ratio: $AR = s^2/A$, and the wing's coefficient of lift $C_L = L/(1/2)\rho U^2 A$, where A is the planform area of the wing.

c) Explain why the result of part b) appears to surpass the performance of the optimal elliptic lift distribution.

14.12. The circulation across the span of a wing follows the parabolic law $\Gamma = \Gamma_0(1 - (2y/s)^2)$. Calculate the induced velocity w at midspan, and compare the value with that obtained when the distribution is elliptic.

14.13. An untwisted elliptic wing of 20-m span supports a weight of 80,000 N in a level flight at 300 km/hr. Assuming sea level conditions, find a) the induced drag and b) the circulation around sections halfway along each wing.

14.14.[1] A wing with a rectangular planform (span $= s$, chord $= c$) and uniform airfoil section without camber is twisted so that its geometrical angle, α_w, decreases from α_r at the root ($y = 0$) to zero at the wing tips ($y = \pm s/2$) according to the distribution:

$$\alpha_w(y) = \alpha_r\sqrt{1 - (2y/s)^2}.$$

a) At what global angle of attack, α_t, should this wing be flown so that it has an elliptical lift distribution? The local angle of attack at any location along the span will be $\alpha_t + \alpha_w$. Assume the two-dimensional lift curve slope of the foil section is K.

b) Evaluate the lift and the lift-induced drag forces on the wing at the angle of attack determined in part a) when: $\alpha_r = 2°$, $K = 5.8$ rad.$^{-1}$, $c = 1.5$ m, $b = 9$ m, the air density is 1 kg/m^3, and the airspeed is 150 m/s.

14.15. Consider the wing shown in Figure 14.24. If the foil section is uniform along the span and the wing is not twisted, show that the three-dimensional lift coefficient, $C_{L,3D}$ is related to the two-dimensional lift coefficient of the foil section, $C_{L,2D}$, by: $C_{L,3D} = C_{L,2D}/(1 + 2/\Lambda)$, where $\Lambda = s^2/A$ is the aspect ratio of the wing.

14.16. The wing-tip vortices from large, heavy aircraft can cause a disruptive rolling torque on smaller, lighter ones. Lifting line theory allows the roll torque to be estimated when the small airplane's wing is modeled as a single linear vortex with strength $\Gamma(y)$ that resides at $x = 0$ between $y = -s/2$ and $y = +s/2$. Here, the small airplane's wing will be presumed rectangular (span s, chord c) with constant foil-shape, and the trailing vortex from the heavy airplane's wing will be assumed to lie along the x-axis and produce a vertical velocity distribution at $x = 0$ given by: $w(y) = \dfrac{\Gamma'}{2\pi y}[1 - \exp(-|y|/\ell)]$. To simplify your work for the following items, ignore the trailing vortices (shown as dashed lines) from the small airplane's wing and assume $U \gg w$. [Note this w differs by a sign from that specified in (14.17).]

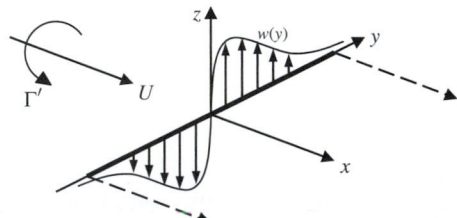

a) Determine a formula for the rolling moment, $M = \int_{-s/2}^{+s/2} \rho U y \Gamma(y) dy$, on the small aircraft's wing in terms of Γ', s, c, ℓ, the air density ρ, the flight speed of the small aircraft U, and the lift-curve slope of the small aircraft's wing section $K = dC_{L,2D}/d\alpha$, where α is the small-aircraft-wing angle of attack.

[1]Obtained by the third author while a student in a course taught by Professor Fred Culick.

b) Calculate M when $\rho = 1.2 \text{ kg/m}^3$, $U = 150 \text{ m/s}$, $K = 6.0/\text{rad}$, $s = 9 \text{ m}$, $c = 1.5 \text{ m}$, $\Gamma' = 50 \text{ m}^2/\text{s}$, and $s/(2\ell) = 1$. Comment on the magnitude of this torque.

14.17. Consider the ideal rectilinear horseshoe vortex of a simple wing having span s. Use the (x, y, z) coordinates shown for the following items.

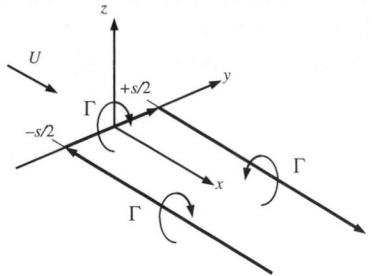

a) Determine a formula for the induced vertical velocity w at $(x, y, 0)$ for $x > 0$ and $y > 0$.

b) Using the results of part a), evaluate the induced vertical velocity at the following three locations: $(s, 0, 0)$, $(0, s, 0)$, and $(s, s, 0)$.

c) Imagine that you are an efficiency-minded migrating bird and that the rectilinear horseshoe vortex shown is produced by another member of your flock. Describe where you would choose to center your own wings. List the coordinates of the part b) location that is closest to your chosen location.

14.18. As an airplane lands, the presence of the ground changes the plane's aerodynamic performance. To address the essential features of this situation, consider uniform flow past a horseshoe vortex (heavy solid lines below) with wingspan b located a distance h above a large, flat boundary defined by $z = 0$. From the method of images, the presence of the boundary can be accounted for by an image horseshoe vortex (heavy dashed lines below) of opposite strength located a distance h below the boundary.

a) Determine the direction and the magnitude of the induced velocity at $\mathbf{x} = (0, 0, h)$, the center of the wing.

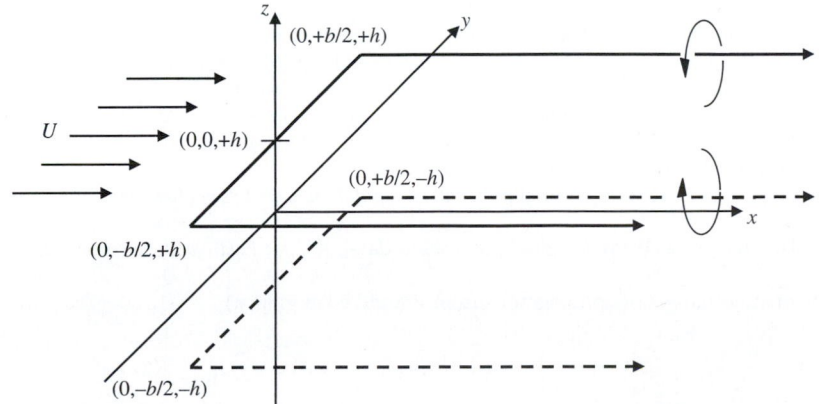

    b) Assuming the result of part a) applies along the entire wingspan, estimate L and D_i, the lift and lift-induced drag, respectively, in terms of b, h, Γ, and $\rho =$ fluid density.

    c) Compare the result of part b) to that obtained for the horseshoe vortex without a large, flat surface: $L = \rho U \Gamma b$ and $D_i = \rho \Gamma^2/\pi$. Which configuration has more lift? Which one has less drag? Why?

14.19. Before modifications, an ordinary commercial airliner with wingspan $s = 30$ m generates two tip vortices of equal and opposite circulation having Rankine velocity profiles (see (3.28)) and a core size $\sigma_o = 0.5$ m for test-flight conditions. The addition of wing-tip treatments (sometimes known as *winglets*) to both of the aircraft's wing tips doubles the tip vortex core size at the test condition. If the aircraft's weight is negligibly affected by the change, has the lift-induced drag of the aircraft been increased or decreased? Justify your answer. Estimate the percentage change in the induced drag.

Literature Cited

Anderson, John D., Jr. (1998). *A History of Aerodynamics*. London: Cambridge University Press.

Anderson, John D., Jr. (2007). *Fundamentals of Aerodynamics*. New York: McGraw-Hill.

Fish, F. E., & Lauder, G. V. (2006). Passive and Active Control by Swimming Fishes and Mammals. *Annual Rev. Fluid Mech, 38,* 193—224.

Herreshoff, H. C., & Newman, J. N. (1966). The study of sailing yachts. *Scientific American, 215,* August issue, 61—68.

von Karman, T. (1954). *Aerodynamics*. New York: McGraw-Hill.

(A delightful little book, written for the nonspecialist, full of historical anecdotes and at the same time explaining aerodynamics in the easiest way.)

Kuethe, A. M., & Chow, C. Y. (1998). *Foundations of Aerodynamics: Basis of Aerodynamic Design*. New York: Wiley.

Lighthill, M. J. (1986). *An Informal Introduction to Theoretical Fluid Mechanics*. Oxford, England: Clarendon Press.

Liu, T., Kuykendoll, K., Rhew, R., & Jones, S. (2006). Avian Wing Geometry and Kinematics. *AIAA J, 44,* 954—963.

Ramamurti, R., & Sandberg, W. C. (2001). Computational Study of 3-D Flapping Foil Flows. AIAA Paper. 2001—0605.

Supplemental Reading

Ashley, H., & Landahl, M. (1965). *Aerodynamics of Wings and Bodies*. Reading, MA: Addison-Wesley.

Batchelor, G. K. (1967). *An Introduction to Fluid Dynamics*. London: Cambridge University Press.

Karamcheti, K. (1980). *Principles of Ideal-Fluid Aerodynamics*. Melbourne, FL: Krieger Publishing Co.

Millikan, C. B. (1941). *Aerodynamics of the Airplane*. New York: John Wiley & Sons.

Prandtl, L. (1952). *Essentials of Fluid Dynamics*. London: Blackie & Sons Ltd.

(This is the English edition of the original German edition. It is very easy to understand, and much of it is still relevant today. Printed in New York by Hafner Publishing Co. If this is unavailable, see the following reprints in paperback that contain much if not all of this material.)

Prandtl, L., & Tietjens, O. G. (1934). [original publication date]. *Fundamentals of Hydro and Aero-mechanics*. New York: Dover Publ. Co.

Prandtl, L., & Tietjens, O. G. (1934). [original publication date]. *Applied Hydro and Aeromechanics*. New York: Dover Publ. Co.

(This contains many original flow photographs from Prandtl's laboratory.)

15.1.	Introduction	730	
15.2.	Acoustics	732	
15.3.	Basic Equations for One-Dimensional Flow	736	
15.4.	Reference Properties in Compressible Flow	738	
15.5.	Area-Velocity Relationship in One-Dimensional Isentropic Flow	740	
15.6.	Normal Shock Waves	748	
15.7.	Operation of Nozzles at Different Back Pressures	755	
15.8.	Effects of Friction and Heating in Constant-Area Ducts	761	
15.9.	Pressure Waves in Planar Compressible Flow	765	
15.10.	Thin Airfoil Theory in Supersonic Flow	773	
	Exercises	775	
	Literature Cited	778	
	Supplemental Reading	778	

CHAPTER OBJECTIVES

- To introduce the fundamental compressible flow interactions between velocity, density, pressure, and temperature
- To describe the features of isentropic flows in ducts with smoothly varying cross-sectional area
- To derive the jump conditions across normal shock waves from the conservation equations

- To describe the effects of friction and heat transfer in compressible flows through constant-area ducts
- To indicate how wall geometry produces pressure waves that cause fluid compression, expansion, and turning in steady supersonic flows near walls

15.1. INTRODUCTION

Up to this point, this text has primarily covered incompressible flows. This chapter presents some of the elementary aspects of flows in which pressure-induced changes in density are important. The subject of compressible flow is also called *gas dynamics*, and it has wide applications in high-speed flows around objects of engineering interest. These include *external flows* such as those around projectiles, rockets, re-entry vehicles, and airplanes; and *internal flows* in ducts and passages such as nozzles and diffusers used in jet engines and rocket motors. Compressibility effects are also important in astrophysics. Recommended gas dynamics texts that further discuss the material presented here are Shapiro (1953), Liepmann and Roshko (1957), and Thompson (1972).

Several startling and fascinating phenomena arise in compressible flows (especially in the supersonic range) that defy expectations developed from incompressible flows. Discontinuities (shock waves) appear within the flow, and a rather strange circumstance arises in which an increase of flow area *accelerates* a supersonic stream. And, in subsonic compressible duct flow, friction can increase the flow's speed and heat addition can lower the flow's temperature. These phenomena, which have no counterparts in low-speed flows, are therefore worthy of our attention. Except for the treatment of friction in constant area ducts in Section 15.8, the material presented here is limited to that of frictionless flows outside boundary layers. In spite of this simplification, the results presented here have a great deal of practical value because boundary layers are especially thin in high-speed flows. Gravitational effects, which are of minor importance in high-speed flows, are also neglected.

As discussed in Section 4.11, the importance of compressibility in the equations of motion can be assessed by considering the Mach number M, defined as

$$M \equiv U/c, \tag{4.111}$$

where U is a representative flow speed, and c is the speed of sound, a thermodynamic quantity defined by:

$$c^2 \equiv (\partial p/\partial \rho)_s. \tag{1.19}$$

Here the subscript s signifies that the partial derivative is taken at constant entropy. In particular, the dimensionless scaling (4.109) of the compressible-flow continuity equation for isentropic conditions leads to:

$$\nabla \cdot u = -M^2 \left(\frac{\rho_0}{\rho}\right) \frac{D}{Dt}\left(\frac{p - p_0}{\rho_0 U^2}\right), \tag{4.110}$$

where ρ_0 and p_0 are appropriately chosen reference values for density and pressure. In (4.110), the pressure is scaled by fluid inertia parameters as is appropriate for primarily frictionless high-speed flow. In engineering practice, the incompressible flow assumption is presumed valid *if $M < 0.3$, but not at higher Mach numbers*. Equation (4.110) suggests that $M = 0.3$ corresponds to ~10% departure from perfectly incompressible flow behavior when the remainder of the right side of (4.110) is of order unity.

Although the significance of the ratio U/c was known for a long time, the Swiss aerodynamist Jacob Ackeret introduced the term *Mach number*, just as the term *Reynolds number* was introduced by Sommerfeld many years after Reynolds' experiments. The name of the Austrian physicist Ernst Mach (1836–1916) was chosen because of his pioneering studies on supersonic motion and his invention of the so-called *Schlieren method* for optical visualization of flows involving density changes; see von Karman (1954, p. 106). (Mach distinguished himself equally well in philosophy. Einstein acknowledged that his own thoughts on relativity were influenced by "Mach's principle," which states that properties of space have no independent existence but are determined by the mass distribution within it. Strangely, Mach never accepted either the theory of relativity or the atomic structure of matter.)

Using the Mach number, compressible flows can be nominally classified as follows:

(i) *Incompressible flow*: $M = 0$. Fluid density does not vary with pressure in the flow field. The flowing fluid may be a compressible gas but its density may be regarded as constant.

(ii) *Subsonic flow*: $0 < M < 1$. The Mach number does not exceed unity anywhere in the flow field. Shock waves do not appear in the flow. In engineering practice, subsonic flows for which $M < 0.3$ are often treated as being incompressible.

(iii) *Transonic flow*: The Mach number in the flow lies in the range 0.8–1.2. Shock waves may appear. Analysis of transonic flows is difficult because the governing equations are inherently nonlinear, and also because a separation of the inviscid and viscous aspects of the flow is often impossible. (The word *transonic* was invented by von Karman and Hugh Dryden, although the latter argued in favor of spelling it *transsonic*. Von Karman [1954] stated, "I first introduced the term in a report to the U.S. Air Force. I am not sure whether the general who read the word knew what it meant, but his answer contained the word, so it seemed to be officially accepted" [p. 116].)

(iv) *Supersonic flow*: $M > 1$. Shock waves are generally present. In many ways analysis of a flow that is supersonic everywhere is easier than an analysis of a subsonic or incompressible flow as we shall see. This is because information propagates along certain directions, called *characteristics*, and a determination of these directions greatly facilitates the computation of the flow field.

(v) *Hypersonic flow*: $M > 3$. Very high flow speeds combined with friction or shock waves may lead to sufficiently large increases in a fluid's temperature that molecular dissociation and other chemical effects occur.

Perfect Gas Thermodynamic Relations

As density changes are accompanied by temperature changes, thermodynamic principles are constantly used throughout this chapter. Most of the necessary concepts and relations have been summarized in Sections 1.8 and 1.9, which may be reviewed before proceeding further. The most frequently used relations, valid for a perfect gas with constant specific heats, are listed here for quick reference:

$$\text{\textit{Internal energy. }} e = C_v T,$$

$$\text{\textit{Enthalpy. }} h = C_p T,$$

$$\text{Thermal equation of state } p = \rho RT,$$

$$\text{Specific heats } C_v = \frac{R}{\gamma - 1}, \quad C_p = \frac{\gamma R}{\gamma - 1}, \quad C_p - C_v = R, \quad \text{and} \quad \gamma = C_p/C_v, \tag{15.1}$$

$$\text{Speed of sound } c = \sqrt{\gamma RT} = \sqrt{\gamma p/\rho},$$

$$\text{Entropy change } S_2 - S_1 = C_p \ln\left(\frac{T_2}{T_1}\right) - R \ln\left(\frac{p_2}{p_1}\right) = C_v \ln\left(\frac{T_2}{T_1}\right) - R \ln\left(\frac{\rho_2}{\rho_1}\right).$$

An isentropic process of a perfect gas between states 1 and 2 obeys the following relations:

$$\frac{p_2}{p_1} = \left(\frac{\rho_2}{\rho_1}\right)^\gamma, \quad \text{and} \quad \frac{T_2}{T_1} = \left(\frac{\rho_2}{\rho_1}\right)^{\gamma-1} = \left(\frac{p_2}{p_1}\right)^{(\gamma-1)/\gamma}. \tag{15.2}$$

Some important properties of air at ordinary temperatures and pressures are:

$$R = 287 \text{ m}^2/(\text{s}^2 \text{ K}), \quad C_v = 717 \text{ m}^2/(\text{s}^2 \text{ K}), \quad C_p = 1004 \text{ m}^2/(\text{s}^2 \text{ K}), \quad \text{and} \quad \gamma = 1.40; \tag{15.3}$$

these values are useful for solution of the exercises at the end of this chapter.

15.2. ACOUSTICS

Perhaps the simplest and most common form of compressible flow is found when the pressure and velocity variations are small compared to steady reference values and the variations in pressure are isentropic. This branch of compressible flow is known as *acoustics* and is concerned with the study of sound waves. Acoustics is the linearized theory of compressible fluid dynamics and is a broad field with its own rich history (see Pierce, 1989). Our primary concern here is to deduce how the speed of sound enters the inviscid equations of fluid motion and to develop some insight into the behavior of pressure disturbances in compressible flow.

To determine the field equation governing acoustic phenomena, the dependent field variables may be separated into nominally steady and fluctuating values:

$$u_i = U_i + u_i', \ p = p_0 + p', \ \rho = \rho_0 + \rho', \quad \text{and} \quad T = T_0 + T', \tag{15.4}$$

where U_i, p_0, ρ_0, and T_0 are constants applicable to the region of interest, and all the fluctuating quantities—denoted by primes in (15.4)—are considered to be small compared to these. In addition, the isentropic condition allows the pressure to be Taylor expanded about the reference thermodynamic state specified by p_0 and ρ_0:

$$p = p_0 + p' = p_0 + \left(\frac{\partial p}{\partial \rho}\right)_s (\rho - \rho_0) + \frac{1}{2}\left(\frac{\partial^2 p}{\partial \rho^2}\right)_s (\rho - \rho_0)^2 + \ldots = p_0 + c^2 \rho' + \frac{1}{2}\left(\frac{\partial^2 p}{\partial \rho^2}\right)_s \rho'^2 + \ldots$$

For small isentropic variations, the second-order and higher terms can be neglected, and this leads to a simple relationship between acoustic pressure and density fluctuations:

$$p' \cong c^2 \rho', \tag{15.5}$$

where c^2 is given by (1.19). Using (15.2) and evaluating the derivative at the local reference values yields

$$c^2 = \gamma p_0/\rho_0 = \gamma R T_0,$$ (15.6)

where T_0 is the local reference temperature. Thus, we see that c is larger in monotonic and low-molecular weight gases (where γ and R are larger), and that it increases with increasing local temperature. Equation (15.5) is valid when the fractional density change or *condensation* $\equiv \rho'/\rho_0 = p'/\rho_0 c^2$ is much less than unity. Thus, the primary parametric requirement for the validity of acoustic theory is:

$$p'/\rho_0 c^2 \ll 1.$$ (15.7)

For ordinary sound levels in air, acoustic-pressure magnitudes are of order 1 Pa or less, so the ratio specified in (15.7) is typically less than 10^{-5} since $\rho_0 c^2 = \gamma p_0 \approx 1.4 \times 10^5$ Pa. Additionally, positive p' is called *compression* while negative p' is called *expansion* (or *rarefaction*). Acoustic pressure disturbances are commonly composed of equal amounts of compression and expansion.

The first approximate expression for c was found by Newton, who assumed that p'/p_0 was equal to ρ'/ρ_0 (Boyle's law) as would be true if the process undergone by a fluid particle was isothermal. In this manner Newton arrived at the expression $c = \sqrt{R T_0}$. He attributed the discrepancy of this formula with experimental measurements as due to "unclean air." However, the science of thermodynamics was virtually nonexistent at the time, so that the idea of an isentropic process was unknown to Newton. The correct expression for the sound speed was first given by Laplace.

The field equation for acoustic pressure disturbances is obtained from the continuity equation (4.7) and the Euler equation (4.41) with $\mathbf{g} = 0$, by linearizing these equations and then combining them to reach a single equation for p'. The linearization is accomplished by inserting (15.4) into (4.7) and (4.41) with $\mathbf{g} = 0$, and dropping the terms that include products of primed field variables:

$$\frac{\partial \rho}{\partial t} + \frac{\partial \rho u_j}{\partial x_j} = \frac{\partial(\rho_0 + \rho')}{\partial t} + \frac{\partial\left(\rho_0 U_j + \rho' U_j + \rho_0 u_j' + \rho' u_j'\right)}{\partial x_j} \cong \frac{\partial \rho'}{\partial t} + U_j \frac{\partial \rho'}{\partial x_j} + \rho_0 \frac{\partial u_j'}{\partial x_j} = 0,$$ (15.8)

$$\frac{\partial u_i}{\partial t} + u_j \frac{\partial u_i}{\partial x_j} + \frac{1}{\rho} \frac{\partial p}{\partial x_i} = \frac{\partial(U_i + u_i')}{\partial t} + (U_j + u_j') \frac{\partial(U_i + u_i')}{\partial x_j} + \frac{1}{\rho_0 + \rho'} \frac{\partial(P + p')}{\partial x_i}$$
$$\cong \frac{\partial u_i'}{\partial t} + U_j \frac{\partial u_i'}{\partial x_j} + \frac{1}{\rho_0} \frac{\partial p'}{\partial x_i} = 0.$$ (15.9)

The next steps involve using (15.6) to eliminate ρ' from (15.8), and mildly rewriting (15.8) and (15.9),

$$\frac{1}{c^2}\left(\frac{\partial}{\partial t} + U_j \frac{\partial}{\partial x_j}\right)p' + \rho_0\left(\frac{\partial}{\partial x_j}\right)u_j' = 0 \quad \text{and} \quad \left(\frac{\partial}{\partial t} + U_j \frac{\partial}{\partial x_j}\right)u_i' + \frac{1}{\rho_0}\left(\frac{\partial}{\partial x_i}\right)p' = 0,$$ (15.10, 15.11)

to see that the differentiation operations in the two equations are identical but act on different field variables. In this case cross-differentiation can be used to eliminate u_i'. Applying

$(\partial/\partial t + U_j \partial/\partial x_j)$ to (15.10) and $-\rho_0 \partial/\partial x_j$ to (15.11) and adding the resulting equations leads to:

$$\frac{1}{c^2}\left(\frac{\partial}{\partial t} + U_j \frac{\partial}{\partial x_j}\right)^2 p' - \frac{\partial^2 p'}{\partial x_j \partial x_j} = 0, \tag{15.12}$$

which is the field equation for acoustic pressure disturbances in a uniform flow.

To highlight the importance of the sound speed, consider one-dimensional pressure disturbances $p'(x,t)$ that only vary along the x-axis in a stationary fluid ($U_j = 0$). For this case, the simplified forms of (15.12) and (15.11) are:

$$\frac{1}{c^2}\frac{\partial^2 p'}{\partial t^2} - \frac{\partial^2 p'}{\partial x^2} = 0, \quad \text{and} \quad u_1'(x,t) = -\frac{1}{\rho_0}\int \frac{\partial p'}{\partial x} dt, \tag{15.13, 15.14}$$

where (15.11) has been integrated to show how u_1' and p' are related. Equation (15.13) is the one-dimensional wave equation, and its solutions are of the form:

$$p'(x,t) = f(x - ct) + g(x + ct), \tag{15.15}$$

where f and g are functions determined by initial conditions (see Exercise 15.1). Equation (15.15) is known as *d'Alembert's solution*, and $f(x - ct)$ and $g(x + ct)$ represent traveling pressure disturbances that propagate to the right and left, respectively, with increasing time. Consider a pressure pulse $p'(x,t)$ that propagates to the right and is centered at $x = 0$ with shape $f(x)$ at $t = 0$ as shown in Figure 15.1. An arbitrary time t later, the wave is centered at $x = ct$ and its shape is described by $f(x - ct)$. Similarly, when $p'(x,t) = g(x + ct)$, the pressure disturbance propagates to the left and is located at $x = -ct$ at time t. Thus, the speed at which acoustic pressure disturbances travel is c, and this is independent of the shape of the pressure disturbance waveform.

However, the disturbance waveform does influence the fluid velocity u_1'. It can be determined from (15.14) and (15.15), and is given by

$$u_1'(x,t) = \frac{1}{\rho_0 c}(f(x - ct) - g(x + ct)) \tag{15.16}$$

(see Exercise 15.2). Thus, the fluid velocity includes rightward- and leftward-propagating components that are matched to the pressure disturbance. Moreover, (15.16) shows that

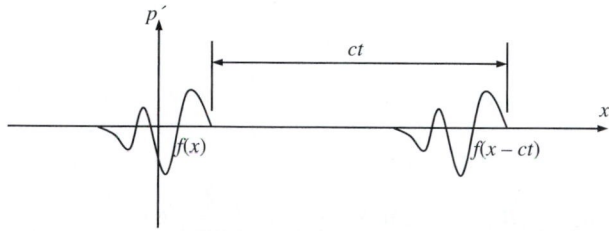

FIGURE 15.1 Propagation of an acoustic pressure disturbance p' that travels to the right with increasing time. At $t = 0$ the disturbance is centered at $x = 0$ and has waveform $f(x)$. At time t later, the disturbance has moved a distance ct but its waveform shape has not changed.

the compression portions of f and g lead to fluid velocity in the same direction as wave propagation; the fluid velocity from $f(x - ct)$ is to the right when $f > 0$, and the fluid velocity from $g(x + ct)$ is to the left when $g > 0$. Similarly, the expansion portions of f and g lead to fluid velocity in the direction opposite of wave propagation; the fluid velocity from $f(x - ct)$ is to the left when $f < 0$, and the fluid velocity from $g(x + ct)$ is to the right when $g < 0$. These fluid velocity directions are worth noting because they persist with the same signs when the wave amplitudes exceed those allowed by the approximation (15.7).

Now consider one-dimensional pressure waves $p'(x, t)$ when $U_j = (U, 0, 0)$ so that (15.12) becomes

$$\frac{1}{c^2}\left(\frac{\partial}{\partial t} + U\frac{\partial}{\partial x}\right)^2 p' - \frac{\partial^2 p'}{\partial x^2} = 0.$$

The general solution of this equation is:

$$p'(x, t) = f(x - (c + U)t) + g(x + (c - U)t). \tag{15.17}$$

When $U > 0$, the travel speed of the downstream-propagating waves is enhanced and that of the upstream-propagating waves is reduced. However, when the flow is supersonic, $U > c$, both portions of (15.17) travel downstream, and this represents a major change in the character of the flow. In subsonic flow, both upstream and downstream pressure disturbances may influence the flow at the location of interest, while in supersonic flow only upstream disturbances may influence the flow. For aircraft moving through a nominally quiescent atmosphere, this means that a ground-based observer below the aircraft's flight path may hear a subsonic aircraft before it is overhead. However, a supersonic aircraft does not radiate sound forward in the direction of flight so the same ground-based observer will only hear a supersonic aircraft after it has passed overhead (see Section 15.9).

Linear acoustic theory is valuable and effective for weak pressure disturbances, but it also indicates how nonlinear phenomena arise as pressure-disturbance amplitudes increase. The speed of sound in gases depends on the local temperature, $c = \sqrt{\gamma R T}$. For air at 15°C, this gives $c = 340$ m/s. The nonlinear terms that were dropped in the linearization (15.8) and (15.9) may change the waveform of a propagating nonlinear pressure disturbance depending on whether it is a compression or expansion. Because $\gamma > 1$, the isentropic relations show that if $p' > 0$ (compression), then $T' > 0$ so the sound speed c increases within a compression disturbance. Therefore, pressure variations within a region of nonlinear compression travel faster than a zero-crossing of p' where $c = \sqrt{\gamma R T_0}$ and therefore may catch up with the leading edge of the compressed region. Such compression-induced changes in c cause nonlinear compression waves to spontaneously steepen as they travel. The opposite is true for nonlinear expansion waves where $p' < 0$ and $T' < 0$, so c decreases. Here, any pressure variations within the region of expansion fall farther behind the leading edge of the expansion. This causes nonlinear expansion waves to spontaneously flatten as they travel. When combined these effects cause a nonlinear sinusoidal pressure disturbance involving equal amounts of compression and expansion to evolve into a saw-tooth shape (see Chapter 11 in Pierce, 1989). Pressure disturbances that do not satisfy the approximation (15.7) are called *finite* amplitude waves.

The limiting form of a finite-amplitude compression wave is a discontinuous change of pressure, commonly known as a *shock wave*. In Section 15.6 it will be shown that finite-amplitude compression waves are not isentropic and that they propagate through a still fluid *faster* than acoustic waves.

15.3. BASIC EQUATIONS FOR ONE-DIMENSIONAL FLOW

This section presents fundamental results for steady compressible flows that can be analyzed using one spatial dimension. The specific emphasis here is on flow through a duct whose centerline may be treated as being straight and whose cross section varies slowly enough so that all dependent flow-field variables (u, p, ρ, T) are well approximated at any location as being equal to their cross-section-averaged values. If the duct area $A(x)$ varies with the distance x along the duct, as shown in Figure 15.2, the dependent flow-field variables are taken as $u(x)$, $p(x)$, $\rho(x)$, and $T(x)$. Unsteadiness (and much complexity) can be introduced by including t as an additional independent variable.

In this situation a control volume development of the basic equations is appropriate. Start with scalar equations representing conservation of mass and energy using the stationary control volume shown in Figure 15.2. For steady flow within this control volume, the integral form of the continuity equation (4.5) requires:

$$\rho_1 u_1 A_1 = \rho_2 u_2 A_2, \quad \text{or} \quad \rho u A = \dot{m} = const., \tag{15.18}$$

where $\dot{m}$ is the mass flow rate in the duct, and the second form follows from the first because the locations 1 and 2 are arbitrary. Forming a general differential of the second form and dividing the result by $\dot{m}$ produces:

$$\frac{d\rho}{\rho} + \frac{du}{u} + \frac{dA}{A} = 0. \tag{15.19}$$

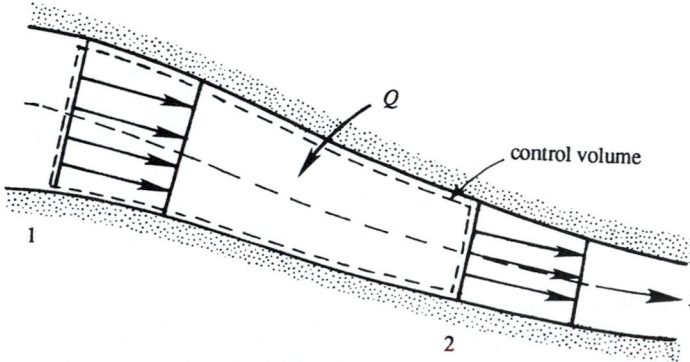

FIGURE 15.2 One-dimensional compressible flow in a duct with smoothly varying centerline direction and cross-sectional area. A stationary control volume in this duct is indicated by dotted lines. Conditions at the upstream and downstream control surfaces are denoted by "1" and "2," respectively. In some circumstances, heat Q may be added to the fluid in the volume. When the control surfaces normal to the flow are only a differential distance apart, then $x_2 = x_1 + dx$, $A_2 = A_1 + dA$, $u_2 = u_1 + du$, $p_2 = p_1 + dp$, $\rho_2 = \rho_1 + d\rho$, etc., where x is the duct's centerline coordinate, A is the duct's cross-sectional area, and u, p, and ρ are the cross-section averaged flow speed, pressure, and density.

For steady flow in a stationary control volume, the integral form of the energy equation (4.48) simplifies to:

$$\int_{A*} \rho\left(e + \frac{1}{2}u_i^2\right)u_j n_j dA = \int_{A*} u_i \tau_{ij} n_j dA - \int_{A*} q_j n_j dA, \tag{15.20}$$

where e is the internal energy per unit mass, $A*$ is the control surface, n_j is the outward normal on the control surface, the body force has been neglected, τ_{ij} is the stress tensor, and q_j is heat flux vector. The term on the left side represents the net flux of internal and kinetic energy out of the control volume. The first term on the right side represents the rate of work done on the control surface, and the second term on the right-hand side represents the heat *input* through the control surface. Here the minus sign in front of the final term occurs because $q_j n_j$ is positive when heat leaves the control volume. A term-by-term evaluation of (15.20) with the chosen control volume produces:

$$-\left(e + \frac{1}{2}u^2\right)_1 \dot{m} + \left(e + \frac{1}{2}u^2\right)_2 \dot{m} = (upA)_1 - (upA)_2 + \dot{m}Q, \tag{15.21}$$

where $\dot{m} = \rho_1 u_1 A_1 = \rho_2 u_2 A_2$ has been used, and Q is the heat added per unit mass of flowing fluid so that:

$$-\int_{A*} q_j n_j dA = \dot{m}Q.$$

Here the wall shear stress does no work, because $u_i = 0$ in (15.20) at the wall. Thus the surface work done on the control volume comes from the pressure on the control surfaces lying perpendicular to the flow direction. Dividing (15.21) by $\dot{m}$ and noting that $upA/\dot{m} = p/\rho$ allows it to be simplified to:

$$\left(e + \frac{p}{\rho} + \frac{1}{2}u^2\right)_2 - \left(e + \frac{p}{\rho} + \frac{1}{2}u^2\right)_1 = Q, \quad \text{or} \quad h_2 + \frac{1}{2}u_2^2 - h_1 - \frac{1}{2}u_1^2 = Q, \tag{15.22}$$

where $h = e + p/\rho$, is the enthalpy per unit mass. This energy equation is valid even if there are frictional or nonequilibrium conditions (e.g., shock waves) between sections 1 and 2. It implies that the *sum of enthalpy and kinetic energy remains constant in an adiabatic flow*. Therefore, enthalpy plays the same role in a flowing system that internal energy plays in a nonflowing system. The difference between the two types of systems is the *flow work* required to push matter along the duct.

Now consider momentum conservation without the body force using the same control volume. The simplified version of (4.17) is:

$$\int_{A*} \rho u_i u_j n_j \, dA = \int_{A*} \tau_{ij} n_j \, dA. \tag{15.23}$$

The term on the left side represents the net flux of momentum out of the control volume and the term on the right side represents forces on the control surface. When applied to the control volume in Figure 15.2 for the x-direction, (15.23) becomes

$$-\dot{m}u_1 + \dot{m}u_2 = (pA)_1 - (pA)_2 + F, \tag{15.24}$$

where F is the x-component of the force exerted on the fluid in the control volume by the walls of the duct between locations 1 and 2. When the control volume has differential length, $x_2 = x_1 + dx$, then (15.24) can be written:

$$\dot{m}\frac{du}{dx} = -\frac{d}{dx}(pA) + p\frac{dA}{dx} - F_f = -A\frac{dp}{dx} - F_f, \tag{15.25}$$

where F_f is the perimeter friction force per unit length along the duct, and the second term in the middle portion of (15.25) is the pressure force on the control volume that occurs when the duct walls expand or contract. This term also appears in the derivation of (4.19), the inviscid steady-flow constant-density Bernoulli equation. For inviscid flow, F_f is zero and (15.25) simplifies to

$$\rho u A\frac{du}{dx} = -A\frac{dp}{dx}, \quad \text{or} \quad u\,du + \frac{dp}{\rho} = 0, \tag{15.26}$$

where $\dot{m}$ in (15.25) has been replaced by $\rho u A$. The second equation of (15.26) is the Euler equation without a body force. A frictionless and adiabatic flow is isentropic, so the property relation (1.18) implies:

$$TdS = dh - dp/\rho = 0, \quad \text{so } dh = dp/\rho.$$

Inserting the last relationship into the second equation of (15.26) and integrating produces:

$$h + \frac{1}{2}u^2 = const.$$

This is the steady Bernoulli equation for isentropic compressible flow (4.78) without the body force term. It is identical to (15.22) when $Q = 0$.

To summarize, the equations for steady one-dimensional compressible flow in a duct with slowly varying area are (15.19), (15.22), and (15.25).

15.4. REFERENCE PROPERTIES IN COMPRESSIBLE FLOW

In incompressible flows, boundary conditions or known properties or profiles typically provide reference values for h, c, T, p, and ρ. In compressible flows, these thermodynamic variables depend on the flow's speed. Thus, reference values for thermodynamic variables must include a specification of the flow speed. The most common reference conditions are the stagnation state ($u = 0$) and the sonic condition ($u = c$), and these are discussed in turn in this section.

If the properties of a compressible flow (h, ρ, u, etc.) are known at a certain point, the reference stagnation properties at that point are defined as those that would be obtained if the local flow were *imagined* to slow down to zero velocity *isentropically*. Stagnation properties are denoted by a subscript zero in gas dynamics. Thus the *stagnation enthalpy* is defined as

$$h_0 \equiv h + \frac{1}{2}u^2.$$

For a perfect gas, this implies

$$C_pT_0 = C_pT + \frac{1}{2}u^2, \tag{15.27}$$

which defines the *stagnation temperature*. Ratios of local and stagnation variables are often sought, and these can be expressed in terms of the Mach number, M. For example (15.27) can be rearranged to find:

$$\frac{T_0}{T} = 1 + \frac{u^2}{2C_pT} = 1 + \frac{\gamma - 1}{2}\frac{u^2}{\gamma RT} = 1 + \frac{\gamma - 1}{2}M^2, \tag{15.28}$$

where $C_p = \gamma R/(\gamma - 1)$ from (15.1) has been used. Thus the stagnation temperature T_0 can be found for a given T and M. The isentropic relations (15.2) can then be used to obtain the *stagnation pressure* and *stagnation density*:

$$\frac{p_0}{p} = \left(\frac{T_0}{T}\right)^{\gamma/(\gamma-1)} = \left[1 + \frac{\gamma - 1}{2}M^2\right]^{\gamma/(\gamma-1)}, \quad \text{and}$$

$$\frac{\rho_0}{\rho} = \left(\frac{T_0}{T}\right)^{1/(\gamma-1)} = \left[1 + \frac{\gamma - 1}{2}M^2\right]^{1/(\gamma-1)}. \tag{15.29, 15.30}$$

In a general flow the stagnation properties can vary throughout the flow field. If, however, the flow is adiabatic (but not necessarily isentropic), then $h + u^2/2$ is constant throughout the flow as shown by (15.22). It follows that h_0, T_0, and $c_0 \ (= \sqrt{\gamma RT_0})$ *are constant throughout an adiabatic flow, even in the presence of friction. In contrast, the stagnation pressure p_0 and density ρ_0 decrease if there is friction.* To see this, consider the entropy change in an adiabatic flow between sections 1 and 2 in a smoothly varying duct, with 2 being the downstream section. Let the flow at both sections hypothetically be brought to rest by isentropic processes, giving the local stagnation conditions p_{01}, p_{02}, T_{01}, and T_{02}. Then the entropy change between the two sections can be expressed as

$$S_2 - S_1 = S_{02} - S_{01} = -R \ln\frac{p_{02}}{p_{01}} + C_p \ln\frac{T_{02}}{T_{01}},$$

from the final equation of (15.1). The last term is zero for an adiabatic flow in which $T_{02} = T_{01}$. As the second law of thermodynamics requires that $S_2 > S_1$, it follows that

$$p_{02} < p_{01},$$

which shows that the stagnation pressure falls due to friction. And, from $p_0 = \rho_0 RT_0$, ρ_0 must fall too for constant T_0.

It is apparent that all stagnation properties are constant along an isentropic flow. If such a flow happens to start from a large reservoir where the fluid is practically at rest, then the properties in the reservoir equal the stagnation properties everywhere in the flow (Figure 15.3).

In addition to the stagnation properties, there is another useful set of reference quantities. These are called *sonic* or *critical* conditions and are commonly denoted by an asterisk. Thus, p^*, ρ^*, c^*, and T^* are properties attained if the local fluid is imagined to expand or compress

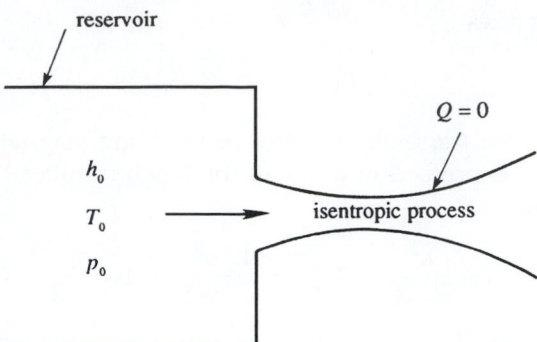

FIGURE 15.3 Schematic of an isentropic compressible-flow process starting from a reservoir. An isentropic process is both adiabatic (no heat exchange) and frictionless. Stagnation properties, indicated with a subscript 0, are uniform everywhere and are equal to the properties in the reservoir.

isentropically until it reaches $M = 1$. The sonic area A^* is often the most useful or important because the stagnation area is infinite for any compressible duct flow. If M is known where the duct area is A, the passage area, A^*, at which the sonic conditions are attained can be determined to be

$$\frac{A}{A^*} = \frac{1}{M}\left[\frac{2}{\gamma+1}\left(1 + \frac{\gamma-1}{2}M^2\right)\right]^{(1/2)(\gamma+1)/(\gamma-1)} \tag{15.31}$$

(see Exercise 15.4).

We shall see in the following section that sonic conditions can only be reached at the *throat* of a duct, where the area is minimum. However, a throat need not actually exist in the flow; the sonic variables are simply reference values that are reached *if* the flow were brought to the sonic state isentropically. From its definition it is clear that the value of A^* in a flow remains constant in isentropic flow. The presence of shock waves, friction, or heat transfer changes the value of A^* along the flow.

The values of T_0/T, P_0/P, ρ_0/ρ, and A/A^* at a point can be determined from (15.28) through (15.31) if the local Mach number is known. For $\gamma = 1.4$, these ratios are tabulated in Table 15.1. The reader should examine this table at this point. Examples 15.1 and 15.2 illustrate the use of this table.

15.5. AREA-VELOCITY RELATIONSHIP IN ONE-DIMENSIONAL ISENTROPIC FLOW

Some surprising consequences of compressibility are found in isentropic flow through a duct of varying area. The natural application area for this topic is in the design of *nozzles* and *diffusers*. A nozzle is a device through which the flow expands from high to low pressure to generate a high-speed jet. Examples of nozzles are the exit ducts of a fireman's hose or a rocket motor. A diffuser's function is opposite that of a nozzle (and it has little or nothing to do with the diffusive transport of heat or species by molecular motion). In a diffuser

TABLE 15.1 Isentropic Flow of a Perfect Gas ($\gamma = 1.4$)

M	p/p_0	ρ/ρ_0	T/T_0	$A/A*$	M	p/p_0	ρ/ρ_0	T/T_0	$A/A*$
0	1	1	1	∞	0.62	0.7716	0.8310	0.9286	1.1656
0.02	0.9997	0.9998	0.9999	28.9421	0.64	0.7591	0.8213	0.9243	1.1451
0.04	0.9989	0.9992	0.9997	14.4815	0.66	0.7465	0.8115	0.9199	1.1265
0.06	0.9975	0.9982	0.9993	9.6659	0.68	0.7338	0.8016	0.9153	1.1097
0.08	0.9955	0.9968	0.9987	7.2616	0.7	0.7209	0.7916	0.9107	1.0944
0.1	0.9930	0.9950	0.9980	5.8218	0.72	0.7080	0.7814	0.9061	1.0806
0.12	0.9900	0.9928	0.9971	4.8643	0.74	0.6951	0.7712	0.9013	1.0681
0.14	0.9864	0.9903	0.9961	4.1824	0.76	0.6821	0.7609	0.8964	1.0570
0.16	0.9823	0.9873	0.9949	3.6727	0.78	0.6690	0.7505	0.8915	1.0471
0.18	0.9776	0.9840	0.9936	3.2779	0.8	0.6560	0.7400	0.8865	1.0382
0.2	0.9725	0.9803	0.9921	2.9635	0.82	0.6430	0.7295	0.8815	1.0305
0.22	0.9668	0.9762	0.9904	2.7076	0.84	0.6300	0.7189	0.8763	1.0237
0.24	0.9607	0.9718	0.9886	2.4956	0.86	0.6170	0.7083	0.8711	1.0179
0.26	0.9541	0.9670	0.9867	2.3173	0.88	0.6041	0.6977	0.8659	1.0129
0.28	0.9470	0.9619	0.9846	2.1656	0.9	0.5913	0.6870	0.8606	1.0089
0.3	0.9395	0.9564	0.9823	2.0351	0.92	0.5785	0.6764	0.8552	1.0056
0.32	0.9315	0.9506	0.9799	1.9219	0.94	0.5658	0.6658	0.8498	1.0031
0.34	0.9231	0.9445	0.9774	1.8229	0.96	0.5532	0.6551	0.8444	1.0014
0.36	0.9143	0.9380	0.9747	1.7358	0.98	0.5407	0.6445	0.8389	1.0003
0.38	0.9052	0.9313	0.9719	1.6587	1.0	0.5283	0.6339	0.8333	1
0.4	0.8956	0.9243	0.9690	1.5901	1.02	0.5160	0.6234	0.8278	1.0003
0.42	0.8857	0.9170	0.9659	1.5289	1.04	0.5039	0.6129	0.8222	1.0013
0.44	0.8755	0.9094	0.9627	1.4740	1.06	0.4919	0.6024	0.8165	1.0029
0.46	0.8650	0.9016	0.9594	1.4246	1.08	0.4800	0.5920	0.8108	1.0051
0.48	0.8541	0.8935	0.9559	1.3801	1.1	0.4684	0.5817	0.8052	1.0079
0.5	0.8430	0.8852	0.9524	1.3398	1.12	0.4568	0.5714	0.7994	1.0113
0.52	0.8317	0.8766	0.9487	1.3034	1.14	0.4455	0.5612	0.7937	1.0153
0.54	0.8201	0.8679	0.9449	1.2703	1.16	0.4343	0.5511	0.7879	1.0198
0.56	0.8082	0.8589	0.9410	1.2403	1.18	0.4232	0.5411	0.7822	1.0248
0.58	0.7962	0.8498	0.9370	1.2130	1.2	0.4124	0.5311	0.7764	1.0304
0.6	0.7840	0.8405	0.9328	1.1882	1.22	0.4017	0.5213	0.7706	1.0366

(Continued)

TABLE 15.1   Isentropic Flow of a Perfect Gas ($\gamma = 1.4$)—cont'd

M	p/p_0	ρ/ρ_0	T/T_0	$A/A*$	M	p/p_0	ρ/ρ_0	T/T_0	$A/A*$
1.24	0.3912	0.5115	0.7648	1.0432	1.88	0.1539	0.2627	0.5859	1.5308
1.26	0.3809	0.5019	0.7590	1.0504	1.9	0.1492	0.2570	0.5807	1.5553
1.28	0.3708	0.4923	0.7532	1.0581	1.92	0.1447	0.2514	0.5756	1.5804
1.3	0.3609	0.4829	0.7474	1.0663	1.94	0.1403	0.2459	0.5705	1.6062
1.32	0.3512	0.4736	0.7416	1.0750	1.96	0.1360	0.2405	0.5655	1.6326
1.34	0.3417	0.4644	0.7358	1.0842	1.98	0.1318	0.2352	0.5605	1.6597
1.36	0.3323	0.4553	0.7300	1.0940	2.0	0.1278	0.2300	0.5556	1.6875
1.38	0.3232	0.4463	0.7242	1.1042	2.02	0.1239	0.2250	0.5506	1.7160
1.4	0.3142	0.4374	0.7184	1.1149	2.04	0.1201	0.2200	0.5458	1.7451
1.42	0.3055	0.4287	0.7126	1.1262	2.06	0.1164	0.2152	0.5409	1.7750
1.44	0.2969	0.4201	0.7069	1.1379	2.08	0.1128	0.2104	0.5361	1.8056
1.46	0.2886	0.4116	0.7011	1.1501	2.1	0.1094	0.2058	0.5313	1.8369
1.48	0.2804	0.4032	0.6954	1.1629	2.12	0.1060	0.2013	0.5266	1.8690
1.5	0.2724	0.3950	0.6897	1.1762	2.14	0.1027	0.1968	0.5219	1.9018
1.52	0.2646	0.3869	0.6840	1.1899	2.16	0.0996	0.1925	0.5173	1.9354
1.54	0.2570	0.3789	0.6783	1.2042	2.18	0.0965	0.1882	0.5127	1.9698
1.56	0.2496	0.3710	0.6726	1.2190	2.2	0.0935	0.1841	0.5081	2.0050
1.58	0.2423	0.3633	0.6670	1.2344	2.22	0.0906	0.1800	0.5036	2.0409
1.6	0.2353	0.3557	0.6614	1.2502	2.24	0.0878	0.1760	0.4991	2.0777
1.62	0.2284	0.3483	0.6558	1.2666	2.26	0.0851	0.1721	0.4947	2.1153
1.64	0.2217	0.3409	0.6502	1.2836	2.28	0.0825	0.1683	0.4903	2.1538
1.66	0.2151	0.3337	0.6447	1.3010	2.3	0.0800	0.1646	0.4859	2.1931
1.68	0.2088	0.3266	0.6392	1.3190	2.32	0.0775	1.1609	0.4816	2.2333
1.7	0.2026	0.3197	0.6337	1.3376	2.34	0.0751	0.1574	0.4773	2.2744
1.72	0.1966	0.3129	0.6283	1.3567	2.36	0.0728	0.1539	0.4731	2.3164
1.74	0.1907	0.3062	0.6229	1.3764	2.38	0.0706	0.1505	0.4688	2.3593
1.76	0.1850	0.2996	0.6175	1.3967	2.4	0.0684	0.1472	0.4647	2.4031
1.78	0.1794	0.2931	0.6121	1.4175	2.42	0.0663	0.1439	0.4606	2.4479
1.8	0.1740	0.2868	0.6068	1.4390	2.44	0.0643	0.1408	0.4565	2.4936
1.82	0.1688	0.2806	0.6015	1.4610	2.46	0.0623	0.1377	0.4524	2.5403
1.84	0.1637	0.2745	0.5963	1.4836	2.48	0.0604	0.1346	0.4484	2.5880
1.86	0.1587	0.2686	0.5910	1.5069	2.5	0.0585	0.1317	0.4444	2.6367

TABLE 15.1 Isentropic Flow of a Perfect Gas ($\gamma = 1.4$)—cont'd

M	p/p_0	ρ/ρ_0	T/T_0	$A/A*$	M	p/p_0	ρ/ρ_0	T/T_0	$A/A*$
2.52	0.0567	0.1288	0.4405	2.6865	3.16	0.0215	0.0643	0.3337	4.9304
2.54	0.0550	0.1260	0.4366	2.7372	3.18	0.0208	0.0630	0.3309	5.0248
2.56	0.0533	0.1232	0.4328	2.7891	3.2	0.0202	0.0617	0.3281	5.1210
2.58	0.0517	0.1205	0.4289	2.8420	3.22	0.0196	0.0604	0.3253	5.2189
2.6	0.0501	0.1179	0.4252	2.8960	3.24	0.0191	0.0591	0.3226	5.3186
2.62	0.0486	0.1153	0.4214	2.9511	3.26	0.0185	0.0579	0.3199	5.4201
2.64	0.0471	0.1128	0.4177	3.0073	3.28	0.0180	0.0567	0.3173	5.5234
2.66	0.0457	0.1103	0.4141	3.0647	3.3	0.0175	0.0555	0.3147	5.6286
2.68	0.0443	0.1079	0.4104	3.1233	3.32	0.0170	0.0544	0.3121	5.7358
2.7	0.0430	0.1056	0.4068	3.1830	3.34	0.0165	0.0533	0.3095	5.8448
2.72	0.0417	0.1033	0.4033	3.2440	3.36	0.0160	0.0522	0.3069	5.9558
2.74	0.0404	0.1010	0.3998	3.3061	3.38	0.0156	0.0511	0.3044	6.0687
2.76	0.0392	0.0989	0.3963	3.3695	3.4	0.0151	0.0501	0.3019	6.1837
2.78	0.0380	0.0967	0.3928	3.4342	3.42	0.0147	0.0491	0.2995	6.3007
2.8	0.0368	0.0946	0.3894	3.5001	3.44	0.0143	0.0481	0.2970	6.4198
2.82	0.0357	0.0926	0.3860	3.5674	3.46	0.0139	0.0471	0.2946	6.5409
2.84	0.0347	0.0906	0.3827	3.6359	3.48	0.0135	0.0462	0.2922	6.6642
2.86	0.0336	0.0886	0.3794	3.7058	3.5	0.0131	0.0452	0.2899	6.7896
2.88	0.0326	0.0867	0.3761	3.7771	3.52	0.0127	0.0443	0.2875	6.9172
2.9	0.0317	0.0849	0.3729	3.8498	3.54	0.0124	0.0434	0.2852	7.0471
2.92	0.0307	0.0831	0.3696	3.9238	3.56	0.0120	0.0426	0.2829	7.1791
2.94	0.0298	0.0813	0.3665	3.9993	3.58	0.0117	0.0417	0.2806	7.3135
2.96	0.0289	0.0796	0.3633	4.0763	3.6	0.0114	0.0409	0.2784	7.4501
2.98	0.0281	0.0779	0.3602	4.1547	3.62	0.0111	0.0401	0.2762	7.5891
3.0	0.0272	0.0762	0.3571	4.2346	3.64	0.0108	0.0393	0.2740	7.7305
3.02	0.0264	0.0746	0.3541	4.3160	3.66	0.0105	0.0385	0.2718	7.8742
3.04	0.0256	0.0730	0.3511	4.3990	3.68	0.0102	0.0378	0.2697	8.0204
3.06	0.0249	0.0715	0.3481	4.4835	3.7	0.0099	0.0370	0.2675	8.1691
3.08	0.0242	0.0700	0.3452	4.5696	3.72	0.0096	0.0363	0.2654	8.3202
3.1	0.0234	0.0685	0.3422	4.6573	3.74	0.0094	0.0356	0.2633	8.4739
3.12	0.0228	0.0671	0.3393	4.7467	3.76	0.0091	0.0349	0.2613	8.6302
3.14	0.0221	0.0657	0.3365	4.8377	3.78	0.0089	0.0342	0.2592	8.7891

(Continued)

TABLE 15.1 Isentropic Flow of a Perfect Gas ($\gamma = 1.4$)—cont'd

M	p/p_0	ρ/ρ_0	T/T_0	$A/A*$	M	p/p_0	ρ/ρ_0	T/T_0	$A/A*$
3.8	0.0086	0.0335	0.2572	8.9506	4.42	0.0038	0.0187	0.2038	15.4724
3.82	0.0084	0.0329	0.2552	9.1148	4.44	0.0037	0.0184	0.2023	15.7388
3.84	0.0082	0.0323	0.2532	0.2817	4.46	0.0036	0.0181	0.2009	16.0092
3.86	0.0080	0.0316	0.2513	9.4513	4.48	0.0035	0.0178	0.1994	16.2837
3.88	0.0077	0.0310	0.2493	9.6237	4.5	0.0035	0.0174	0.1980	16.5622
3.9	0.0075	0.0304	0.2474	9.7990	4.52	0.0034	0.0171	0.1966	16.8449
3.92	0.0073	0.0299	0.2455	9.9771	4.54	0.0033	0.0168	0.1952	17.1317
3.94	0.0071	0.0293	0.2436	10.1581	4.56	0.0032	0.0165	0.1938	17.4228
3.96	0.0069	0.0287	0.2418	10.3420	4.58	0.0031	0.0163	0.1925	17.7181
3.98	0.0068	0.0282	0.2399	10.5289	4.6	0.0031	0.0160	0.1911	18.0178
4.0	0.0066	0.0277	0.2381	10.7188	4.62	0.0030	0.0157	0.1898	18.3218
4.02	0.0064	0.0271	0.2363	10.9117	4.64	0.0029	0.0154	0.1885	18.6303
4.04	0.0062	0.0266	0.2345	11.1077	4.66	0.0028	0.0152	0.1872	18.9433
4.06	0.0061	0.0261	0.2327	11.3068	4.68	0.0028	0.0149	0.1859	19.2608
4.08	0.0059	0.0256	0.2310	11.5091	4.7	0.0027	0.0146	0.1846	19.5828
4.1	0.0058	0.0252	0.2293	11.7147	4.72	0.0026	0.0144	0.1833	19.9095
4.12	0.0056	0.0247	0.2275	11.9234	4.74	0.0026	0.0141	0.1820	20.2409
4.14	0.0055	0.0242	0.2258	12.1354	4.76	0.0025	0.0139	0.1808	20.5770
4.16	0.0053	0.0238	0.2242	12.3508	4.78	0.0025	0.0137	0.1795	20.9179
4.18	0.0052	0.0234	0.2225	12.5695	4.8	0.0024	0.0134	0.1783	21.2637
4.2	0.0051	0.0229	0.2208	12.7916	4.82	0.0023	0.0132	0.1771	21.6144
4.22	0.0049	0.0225	0.2192	13.0172	4.84	0.0023	0.0130	0.1759	21.9700
4.24	0.0048	0.0221	0.2176	13.2463	4.86	0.0022	0.0128	0.1747	22.3306
4.26	0.0047	0.0217	0.2160	13.4789	4.88	0.0022	0.0125	0.1735	22.6963
4.28	0.0046	0.0213	0.2144	13.7151	4.9	0.0021	0.0123	0.1724	23.0671
4.3	0.0044	0.0209	0.2129	13.9549	4.92	0.0021	0.0121	0.1712	23.4431
4.32	0.0043	0.0205	0.2113	14.1984	4.94	0.0020	0.0119	0.1700	23.8243
4.34	0.0042	0.0202	0.2098	14.4456	4.96	0.0020	0.0117	0.1689	24.2109
4.36	0.0041	0.0198	0.2083	14.6965	4.98	0.0019	0.0115	0.1678	24.6027
4.38	0.0040	0.0194	0.2067	14.9513	5.0	0.0019	0.0113	0.1667	25.0000
4.4	0.0039	0.0191	0.2053	15.2099					

a high-speed stream is decelerated and compressed. For example, air may enter the jet engine of an aircraft after passing through a diffuser, which raises the pressure and temperature of the air. In incompressible flow, a nozzle profile converges in the direction of flow to increase the flow velocity, while a diffuser profile diverges. We shall see that such convergence and divergence must be reversed for supersonic flows in nozzles and diffusers.

Conservation of mass for compressible flow in a duct with smoothly varying area is specified by (15.19). For constant density flow $d\rho/dx = 0$ and (15.19) implies $dA/dx + du/dx = 0$, so a decreasing area leads to an increase of velocity. When the flow is compressible, frictionless, and adiabatic then (15.26) implies

$$u\,du = -dp/\rho = c^2 d\rho/\rho, \tag{15.32}$$

because the flow is isentropic under these circumstances. Thus, the Euler equation requires that an increasing speed ($du > 0$) in the direction of flow must be accompanied by a fall of pressure ($dp < 0$). In terms of the Mach number, (15.32) becomes

$$d\rho/\rho = -M^2 du/u. \tag{15.33}$$

This shows that for $M \ll 1$, the percentage change of density is much smaller than the percentage change of velocity. The density changes in the continuity equation (15.19) can therefore be neglected in low Mach number flows, a fact also mentioned in Section 15.1. Substituting (15.33) into (15.19), we obtain a velocity-area differential relationship that is valid in compressible flow:

$$\frac{du}{u} = -\frac{1}{1-M^2}\frac{dA}{A}. \tag{15.34}$$

This relation leads to the following important conclusions about compressible flows:

(i) At subsonic speeds ($M < 1$) a decrease of area increases the speed of flow. A subsonic nozzle therefore must have a convergent profile, and a subsonic diffuser must have a divergent profile (upper row of Figure 15.4). The behavior is qualitatively the same as in incompressible ($M = 0$) flows.

(ii) At supersonic speeds ($M > 1$) the denominator in (15.34) is negative, and we arrive at the conclusion that an *increase* in area leads to an increase of speed. The reason for such a behavior can be understood from (15.33), which shows that for $M > 1$ the density decreases faster than the velocity increases, thus the area must increase in an accelerating flow in order for $\rho u A$ to remain constant.

Therefore, the supersonic portion of a nozzle must have a divergent profile, and the supersonic part of a diffuser must have a convergent profile (bottom row of Figure 15.4).

Suppose a nozzle is used to generate a supersonic stream, starting from a low-speed, high-pressure air stream at its inlet (Figure 15.5). Then the Mach number must increase continuously from $M = 0$ near the inlet to $M > 1$ at the exit. The foregoing discussion shows that the nozzle must converge in the subsonic portion and diverge in the supersonic portion. Such a nozzle is called a *convergent–divergent nozzle*. From Figure 15.5 it is clear that the Mach number must be unity at the *throat*, where the area is neither increasing nor decreasing ($dA \to 0$). This is consistent with (15.34), which shows that du can be nonzero at the throat

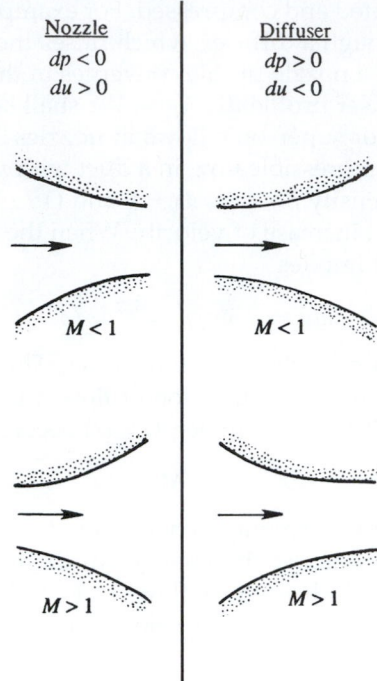

FIGURE 15.4 Shapes of nozzles and diffusers in subsonic and supersonic regimes. Nozzles are devices that accelerate the flow and are shown in the left column. Diffusers are devices that decelerate the flow and are shown in the right column. The area change with increasing downstream distance, dA/dx, switches sign for nozzles and diffusers and when the flow switches from subsonic to supersonic.

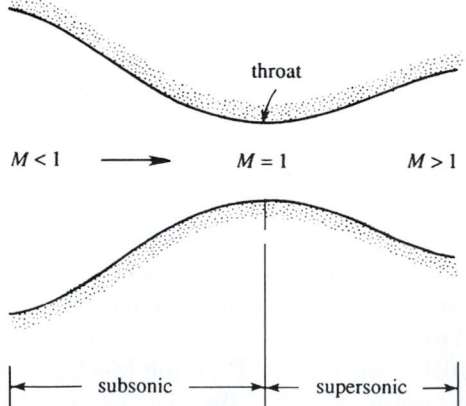

FIGURE 15.5 A convergent–divergent nozzle. When the pressure difference between the nozzle inlet and outlet is large enough, a compressible flow may be continuously accelerated from low speed to a supersonic Mach number through such a nozzle. When this happens the Mach number is unity at the minimum area, known as the nozzle's *throat*.

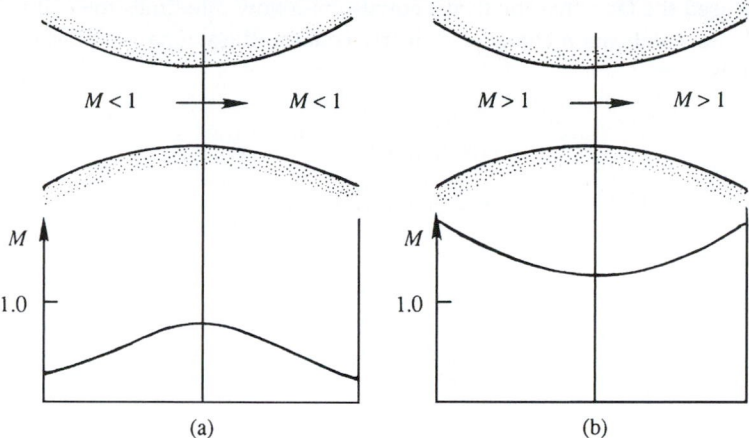

FIGURE 15.6 Convergent–divergent passages in which the condition at the throat is not sonic. This occurs when the flow is entirely subsonic as in (a), and when it is entirely supersonic as in (b).

only if $M = 1$. It follows that the *sonic velocity can be achieved only at the throat of a nozzle or a diffuser and nowhere else.*

It does not, however, follow that M must necessarily be unity at the throat. According to (15.34), we may have a case where $M \neq 1$ at the throat if $du = 0$ there. As an example, the flow in a convergent–divergent tube may be subsonic everywhere, with M increasing in the convergent portion and decreasing in the divergent portion, with $M \neq 1$ at the throat (Figure 15.6a). In this case the nozzle may also be known as a *venturi tube*. For entirely subsonic flow, the first half of the tube here is acting as a nozzle, whereas the second half is acting as a diffuser. Alternatively, we may have a convergent–divergent tube in which the flow is supersonic everywhere, with M decreasing in the convergent portion and increasing in the divergent portion, and again $M \neq 1$ at the throat (Figure 15.6b).

EXAMPLE 15.1

The nozzle of a rocket motor is designed to generate a thrust of 30,000 N when operating at an altitude of 20 km. The pressure and temperature inside the combustion chamber are 1000 kPa and 2500 K. The gas constant of the fluid in the jet is $R = 280$ m^2/(s^{2}K), and $\gamma = 1.4$. Assuming that the flow in the nozzle is isentropic, calculate the throat and exit areas. Use the isentropic table (Table 15.1).

Solution

At an altitude of 20 km, the pressure of the standard atmosphere (Section A.4 in Appendix A) is 5467 Pa. If subscripts 0 and e refer to the stagnation and exit conditions, then a summary of the information given is as follows:

$$p_e = 5467 \text{ Pa}, p_0 = 1000 \text{ kPa}, T_0 = 2500 \text{ K}, \quad \text{and} \quad \text{Thrust} = \rho_e u_e^2 A_e = 30 \text{ kN}.$$

Here, we have used the facts that the thrust equals mass flow rate times the exit velocity, and the pressure inside the combustion chamber is nearly equal to the stagnation pressure. The pressure ratio at the exit is

$$\frac{p_e}{p_0} = \frac{5467}{(1000)(1000)} = 5.467 \times 10^{-3}.$$

For this ratio of p_e/p_0, the isentropic table (Table 15.1) gives:

$$m_e = 4.15, \quad A_e/A^* = 12.2, \quad \text{and} \quad T_e/T_0 = 0.225.$$

The exit temperature and density are therefore:

$$\begin{aligned} T_e &= (0.225)(2500) = 562 \text{ K}, \\ \rho_e &= p_e/RT_e = 5467/(280)(562) = 0.0347 \text{ kg}/\text{m}^3. \end{aligned}$$

The exit velocity is

$$u_e = M_e\sqrt{\gamma RT_e} = 4.15\sqrt{(1.4)(280)(562)} = 1948 \text{ m/s}.$$

The exit area is found from the expression for thrust:

$$A_e = \frac{\text{Thrust}}{\rho_e u_e^2} = \frac{30,000}{(0.0347)(1948)^2} = 0.228 \text{ m}^2.$$

Because $A_e/A^* = 12.2$, the throat area is

$$A^* = \frac{0.228}{12.2} = 0.0187 \text{ m}^2.$$

15.6. NORMAL SHOCK WAVES

A shock wave is similar to a step-change compression sound wave except that it has finite strength. The thickness of such waves is typically of the order of micrometers, so that fluid properties vary almost discontinuously across a shock wave. The high gradients of velocity and temperature result in entropy production within the wave so isentropic relations cannot be used across a shock. This section presents the relationships between properties of the flow upstream and downstream of a *normal shock*, where the shock is perpendicular to the direction of flow. The shock wave is treated as a discontinuity and the actual process by which entropy is generated is not addressed. However, the entropy rise across the shock predicted by this analysis is correct. The internal structure of a shock as predicted by the Navier-Stokes equations under certain simplifying assumptions is given at the end of this section.

Stationary Normal Shock Wave in a Moving Medium

To get started, consider a thin control volume shown in Figure 15.7 that encloses a stationary shock wave. The control surface locations 1 and 2, shown as dashed lines in

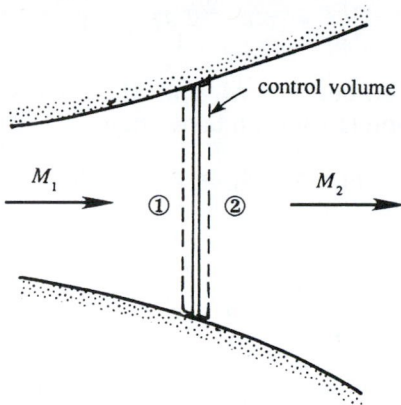

FIGURE 15.7 A normal shock wave trapped in a steady nozzle flow. Here a control volume is shown that has control surfaces immediately upstream (1) and downstream (2) of the shock wave. Shock waves are very thin in most gases, so the area change and wall friction of the duct need not be considered as the flow traverses the shock wave.

the figure, can be taken close to each other because of the discontinuous nature of the wave. In this case, the area change and the wall-surface friction between the upstream and the downstream control volume surfaces can be neglected. Furthermore, external heat addition is not of interest here so the basic equations are (15.18) and (15.24) with $F = 0$, both simplified for constant area, and (15.22) with $Q = 0$:

$$\rho_1 u_1 = \rho_2 u_2, \quad p_1 - p_2 = -\rho_1 u_1^2 + \rho_2 u_2^2, \quad \text{and} \quad h_1 + \frac{1}{2}u_1^2 = h_2 + \frac{1}{2}u_2^2. \quad \text{(15.35, 15.36, 15.37)}$$

The Bernoulli equation cannot be used here because the process inside the shock wave is dissipative. The equations (15.35) through (15.37) contain four unknowns (h_2, u_2, p_2, ρ_2). The additional relationship comes from the thermodynamics of a perfect gas (15.1):

$$h = C_p T = \frac{\gamma R}{\gamma - 1} \frac{p}{\rho R} = \frac{\gamma p}{(\gamma - 1)\rho},$$

so that (15.37) becomes

$$\frac{\gamma}{\gamma - 1} \frac{p_1}{\rho_1} + \frac{1}{2}u_1^2 = \frac{\gamma}{\gamma - 1} \frac{p_2}{\rho_2} + \frac{1}{2}u_2^2. \quad \text{(15.38)}$$

There are now three unknowns (u_2, p_2, ρ_2) and three equations: (15.35), (15.36), and (15.38), so the remainder of the effort to link the conditions upstream and downstream of a shock is primarily algebraic. Elimination of ρ_2 and u_2 from these gives, after some algebra:

$$\frac{p_2}{p_1} = 1 + \frac{2\gamma}{\gamma + 1}\left[\frac{\rho_1 u_1^2}{\gamma p_1} - 1\right].$$

This can be expressed in terms of the upstream Mach number M_1 by noting that $\rho u^2/\gamma p = u^2/\gamma RT = M^2$. The pressure ratio then becomes

$$\frac{p_2}{p_1} = 1 + \frac{2\gamma}{\gamma+1}(M_1^2 - 1).\qquad(15.39)$$

Let us now derive a relation between M_1 and M_2. Because $\rho u^2 = \rho c^2 M^2 = \rho(\gamma p/\rho)M^2 = \gamma p M^2$, the momentum equation (15.36) can be written

$$p_1 + \gamma p_1 M_1^2 = p_2 + \gamma p_2 M_2^2.$$

Using (15.39), this gives

$$M_2^2 = \frac{(\gamma-1)M_1^2 + 2}{2\gamma M_1^2 + 1 - \gamma},\qquad(15.40)$$

which is plotted in Figure 15.8. Because $M_2 = M_1$ (state 2 = state 1) is a solution of (15.35), (15.36), and (15.38), that is shown as well, indicating two possible solutions for M_2 for all $M_1 > [(\gamma-1)/2\gamma]^{1/2}$. As is shown below, M_1 must be greater than unity to avoid violation of the second law of thermodynamics, so the two possibilities for the downstream state are: 1) no change from upstream, and 2) a sudden transition from supersonic to subsonic flow with consequent increases in pressure, density, and temperature. The density, velocity, and temperature ratios can be similarly obtained from the equations provided so far. They are:

$$\frac{\rho_2}{\rho_1} = \frac{u_1}{u_2} = \frac{(\gamma+1)M_1^2}{(\gamma-1)M_1^2 + 2},\qquad(15.41)$$

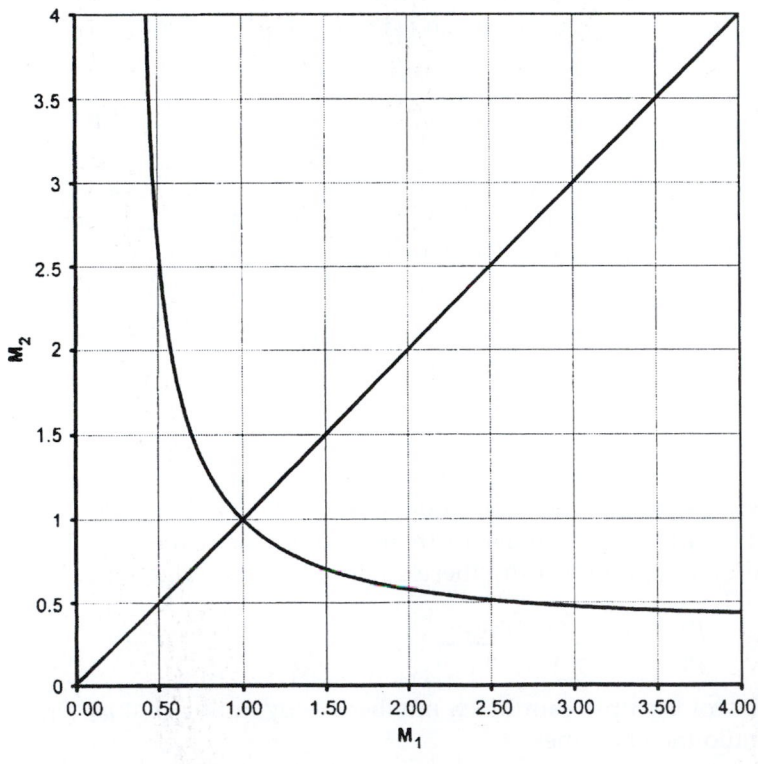

FIGURE 15.8 Normal shock-wave solution for M_2 as function of M_1 for $\gamma = 1.4$. The trivial (no change) solution is also shown as the straight line with unity slope. Asymptotes are $[(\gamma-1)/2\gamma]^{1/2} = 0.378$ for M_1 or $M_2 \to \infty$. The second law of thermodynamics limits valid shock-wave solutions to those having $M_1 > 1$.

$$\frac{T_2}{T_1} = 1 + \frac{2(\gamma - 1)}{(\gamma + 1)^2} \frac{\gamma M_1^2 + 1}{M_1^2}(M_1^2 - 1). \tag{15.42}$$

The normal shock relations (15.39) through (15.43) were worked out independently by the British engineer W. J. M. Rankine (1820–1872) and the French ballistician Pierre Henry Hugoniot (1851–1887). These equations are sometimes known as the *Rankine-Hugoniot relations*.

An important quantity is the change of entropy across the shock. Using (15.1), the entropy change is

$$\frac{S_2 - S_1}{C_v} = \ln\left[\frac{p_2}{p_1}\left(\frac{\rho_1}{\rho_2}\right)^{\gamma}\right] = \ln\left\{\left[1 + \frac{2\gamma}{\gamma + 1}(M_1^2 - 1)\right]\left[\frac{(\gamma - 1)M_1^2 + 2}{(\gamma + 1)M_1^2}\right]^{\gamma}\right\}, \tag{15.43}$$

which is plotted in Figure 15.9. This figure shows that the entropy change across an expansion shock in a perfect gas would decrease, which is impermissible. However, expansion shocks may be possible when the gas follows a different equation of state (Fergason et al., 2001). When the upstream Mach number is close to unity, Figure 15.9 shows that the entropy change may be very small. The dependence of $S_2 - S_1$ on M_1 in the neighborhood of $M_1 = 1$ can be ascertained by treating $M_1^2 - 1$ as a small quantity and expanding (15.43) in terms of it (see Exercise 15.5) to find:

$$\frac{S_2 - S_1}{C_v} \cong \frac{2\gamma(\gamma - 1)}{3(\gamma + 1)^2}(M_1^2 - 1)^3. \tag{15.44a}$$

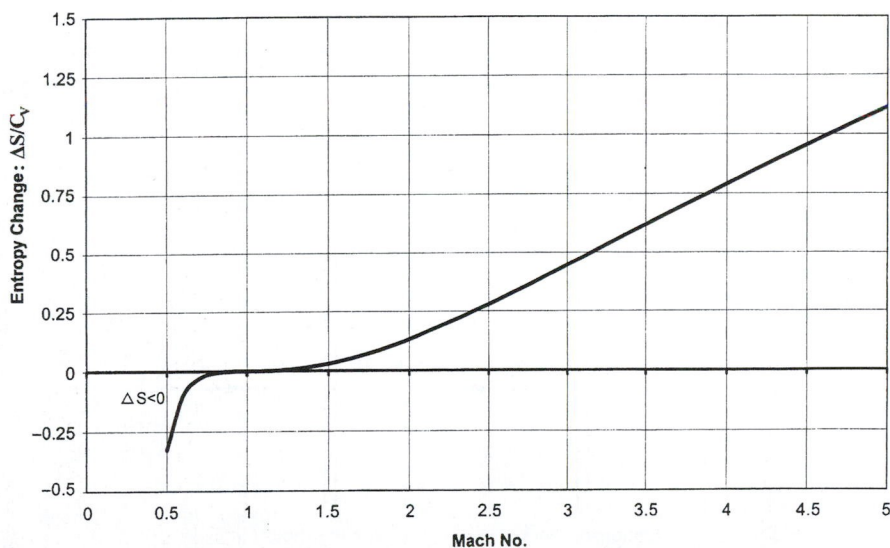

FIGURE 15.9 Entropy change $(S_2 - S_1)/C_v$ as a function of M_1 for $\gamma = 1.4$. Note higher-order contact at $M = 1$ to the horizontal line corresponding to zero entropy change as $M_1 \to 1$ from above. Negative entropy changes are predicted for $M_1 < 1$, so shock waves do not occur unless the upstream speed is supersonic, $M_1 > 1$.

This equation explicitly shows that $S_2 - S_1$ will only be positive for a perfect gas when $M_1 > 1$. Thus, stationary shock waves do not occur when $M_1 < 1$ because of the second law of thermodynamics. However, when $M_1 > 1$, then (15.40) requires that $M_2 < 1$. Thus, the *Mach number changes from supersonic to subsonic values across a normal shock*, and this is the only possibility. A shock wave is therefore analogous to a hydraulic jump (see Section 7.6) in a gravity current, in which the Froude number jumps from supercritical to subcritical values; see Figure 7.20. Equations (15.39), (15.41), and (15.42) then show that the jumps in p, ρ, and T are also from lower to higher values, so that a shock wave leads to compression and heating of a fluid at the expense of streamwise velocity.

Interestingly, terms involving the first two powers of $(M_1^2 - 1)$ do not appear in (15.44a). Using the pressure ratio from (15.39), (15.44a) can be rewritten:

$$\frac{S_2 - S_1}{C_v} \cong \frac{\gamma^2 - 1}{12\gamma^2} \left(\frac{p_2 - p_1}{p_1} \right)^3 . \tag{15.44b}$$

This shows that as the wave amplitude $\Delta p = p_2 - p_1$ decreases the entropy jump goes to zero like $(\Delta p)^3$. Thus weak shock waves are nearly isentropic and this is the primary reason that loud acoustic disturbances are successfully treated as isentropic. Because of the adiabatic nature of the process, the stagnation properties T_0 and h_0 are constant across the shock. In contrast, the stagnation properties p_0 and ρ_0 decrease across the shock due to the dissipative processes inside the wave front.

Moving Normal Shock Wave in a Stationary Medium

Frequently, one needs to calculate the properties of flow due to the propagation of a shock wave through a still medium, for example, that caused by an explosion. The Galilean transformation necessary to analyze this problem is indicated in Figure 15.10. The left panel shows a stationary shock, with incoming and outgoing velocities u_1 and

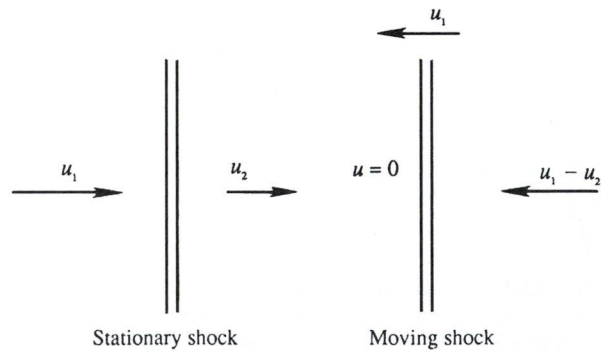

Stationary shock   Moving shock

FIGURE 15.10   Stationary and moving shocks. The stationary shock shown in the left panel corresponds to a situation like that depicted in Figure 15.7 where the incoming flow moves toward the shock. The moving shock situation shown on the right corresponds to a blast wave that propagates away from an explosion into still air.

u_2, respectively. To this flow we add a velocity u_1 directed to the left, so that the fluid entering the shock is stationary, and the fluid downstream of the shock is moving to the *left* at a speed $u_1 - u_2$, as shown in the right panel of the figure. This is consistent with acoustic results in Section 15.2 where it was found that the fluid within a compression wave moves in the direction of wave propagation. The shock speed is therefore u_1, with a supersonic Mach number $M_1 = u_1/c_1 > 1$. It follows that a *finite pressure disturbance propagates through a still fluid at supersonic speed*, in contrast to infinitesimal waves that propagate at the sonic speed. The expressions for all the thermodynamic properties of the flow, such as (15.39) through (15.44), are still applicable.

Normal Shock Structure

We conclude this section on normal shock waves with a look into the structure of a shock wave. The viscous and heat conductive processes within the shock wave result in an entropy increase across the wave. However, the magnitudes of the viscosity μ and thermal conductivity k only determine the thickness of the shock wave and not the magnitude of the entropy increase. The entropy increase is determined solely by the upstream Mach number as shown by (15.43). We shall also see later that the *wave drag* experienced by a body due to the appearance of a shock wave is independent of viscosity or thermal conductivity. (The situation here is analogous to the viscous dissipation in fully turbulent flows, Section 12.7, in which the average kinetic-energy dissipation rate $\bar{\varepsilon}$ is determined by the velocity and length scales of a large-scale turbulence field (12.49) and not by the magnitude of the viscosity; a change in viscosity merely changes the length scale at which the dissipation takes place, namely, the Kolmogorov microscale.)

A shock wave can be considered a very thin boundary layer involving a large stream-wise velocity gradient du/dx, in contrast to the cross-stream (or wall-normal) velocity gradient involved in a viscous boundary layer near a solid surface. Analysis shows that the thickness δ of a shock wave is given by

$$(u_1 - u_2)\delta/\nu \sim 1,$$

where the left side is a Reynolds number based on the velocity change across the shock, its thickness, and the average kinematic viscosity. Taking a typical value for air of $\nu \sim 10^{-5}$ m^2/s, and a velocity jump of $\Delta u \sim 100$ m/s, we obtain a shock thickness of 10^{-7} m. This is not much larger than the mean free path (average distance traveled by a molecule between collisions), which suggests that the continuum hypothesis and the assumption of local thermodynamic equilibrium are both of questionable validity in analyzing shock structure.

With these limitations noted, some insight into the structure of shock waves may be gained by considering the one-dimensional steady Navier-Stokes equations, including heat conduction and Newtonian viscous stresses, in a shock-fixed coordinate system. The solution we obtain provides a smooth transition between upstream and downstream states, looks reasonable, and agrees with experiments and kinetic theory models for upstream Mach numbers less than about 2. The equations for conservation of mass, momentum, and energy, respectively, are the steady one-dimensional versions of (4.7), (4.38) without a body force, and (4.60) written in terms of enthalpy h:

$$\frac{d(\rho u)}{dx} = 0, \quad \rho u \frac{du}{dx} + \frac{dp}{dx} = \frac{d}{dx}\left(\left(\frac{4}{3}\mu + \mu_v\right)\frac{du}{dx}\right), \quad \text{and}$$

$$\rho u \frac{dh}{dx} - u\frac{dp}{dx} = \left(\frac{4}{3}\mu + \mu_v\right)\left(\frac{du}{dx}\right)^2 + \frac{d}{dx}\left(k\frac{dT}{dx}\right).$$

By adding the product of u and the momentum equation to the energy equation, these can be integrated once to find:

$$\rho u = m, \quad mu + p = \mu''\frac{du}{dx} + mV, \quad \text{and} \quad m\left(h + \frac{1}{2}u^2\right) = \mu''u\frac{du}{dx} + k\frac{dT}{dx} + mI,$$

where m, V, and I are the constants of integration and $\mu'' = (4/3)\mu + \mu_v$. When these are evaluated upstream (state 1) and downstream (state 2) of the shock where gradients vanish, they yield the Rankine-Hugoniot relations derived earlier. We also need the equations of state for a perfect gas with constant specific heats to solve for the structure: $h = C_pT$, and $p = \rho RT$. Multiplying the energy equation by C_p/k we obtain the form:

$$m\frac{C_p}{k}\left(C_pT + \frac{1}{2}u^2\right) = \frac{\mu''C_p}{2k}\frac{du^2}{dx} + C_p\frac{dT}{dx} + m\frac{C_p}{k}I.$$

This equation has an exact integral in the special case $Pr'' \equiv \mu''C_p/k = 1$ that was found by Becker in 1922. For most simple gases, Pr'' is likely to be near unity so it is reasonable to proceed assuming $Pr'' = 1$. The Becker integral is $C_pT + u^2/2 = I$. Eliminating all variables but u from the momentum equation, using the equations of state, mass conservation, and the energy integral, we reach:

$$mu + (m/u)\left(R/C_p\right)\left(I - u^2/2\right) - \mu''\, du/dx = mV.$$

With $C_p/R = \gamma/(\gamma - 1)$, multiplying by u/m leads to

$$-\left[2\gamma/(\gamma + 1)\right]\left(\mu''/m\right)u\, du/dx = -u^2 + \left[2\gamma/(\gamma + 1)\right]uV - 2I\left(\gamma - 1\right)/(\gamma + 1)$$

$$\equiv \left(U_1 - U\right)\left(U - U_2\right).$$

Divide by V^2 and let $u/V = U$. The equation for the structure becomes

$$-U(U_1 - U)^{-1}(U - U_2)^{-1}dU = \left[(\gamma + 1)/2\gamma\right](m/\mu'')dx,$$

where the roots of the quadratic are

$$U_{1,2} = \left[\gamma/(\gamma + 1)\right]\left\{1 \pm \left[1 - 2(\gamma^2 - 1)I/(\gamma^2 V^2)\right]^{1/2}\right\},$$

the dimensionless speeds far up- and downstream of the shock. The left-hand side of the equation for the structure is rewritten in terms of partial fractions and then integrated to obtain

$$\left[U_1 \ln(U_1 - U) - U_2 \ln(U - U_2)\right]/(U_1 - U_2) = \left[(\gamma + 1)/(2\gamma)\right]m\int dx/\mu'' \equiv \left[(\gamma + 1)/(2\gamma)\right]\eta.$$

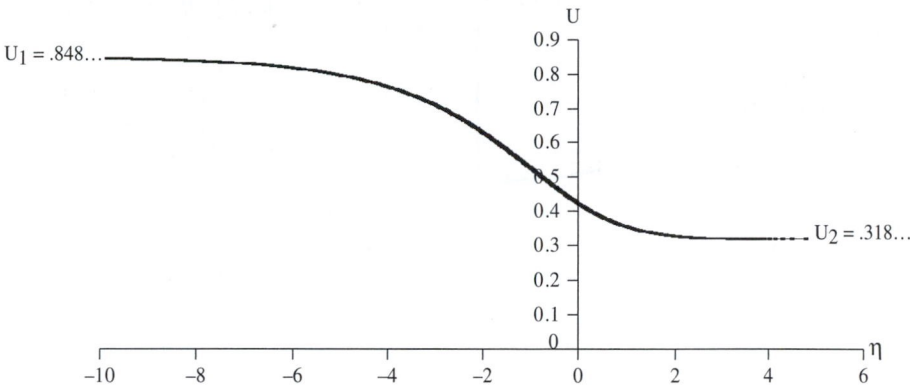

FIGURE 15.11   Shock structure velocity profile for the case $U_1 = 0.848485$, $U_2 = 0.31818$, corresponding to $M_1 = 2.187$. The units of the horizontal coordinate may be approximately interpreted as mean-free paths. Thus, a shock wave is typically a small countable number of mean-free-paths thick.

The resulting shock structure is shown in Figure 15.11 in terms of the stretched coordinate $\eta = \int (m/\mu'') dx$ where μ'' is often a strong function of temperature and thus of x. A similar structure is obtained for all except quite small values of Pr''. In the limit $\mathrm{Pr}'' \to 0$, Hayes (1958) points out that there must be a "shock within a shock" because heat conduction alone cannot provide the entire structure. In fact, Becker (1922, footnote, p. 341) credits Prandtl for originating this idea. Cohen and Moraff (1971) provided the structure of both the outer (heat conducting) and inner (isothermal viscous) shocks. Here, the variable η is a dimensionless length scale measured very roughly in units of mean free paths. We see that a measure of shock thickness is of the order of 5 mean free paths from this analysis.

15.7. OPERATION OF NOZZLES AT DIFFERENT BACK PRESSURES

Nozzles are used to accelerate a fluid stream and are employed in such systems as wind tunnels, rocket motors, ejector pumps, and steam turbines. A pressure drop is maintained across the nozzle to accelerate fluid through it. This section presents the behavior of the flow through a nozzle as the back pressure p_B on the nozzle is varied when the nozzle-supply pressure is maintained at a constant value p_0 (the stagnation pressure). Here the p_B is the pressure in the nominally quiescent environment into which the nozzle flow is directed. In the following discussion, the pressure p_{exit} at the exit plane of the nozzle equals the back pressure p_B if the flow at the exit plane is subsonic, but *not* if it is supersonic. This must be true because subsonic flow allows the downstream pressure p_B to be communicated up into the nozzle exit, and sharp pressure changes are only allowed in a supersonic flow.

Convergent Nozzle

Consider first the case of a convergent nozzle shown in Figure 15.12, which examines a sequence of states *a* through *c* during which the back pressure is gradually lowered.

15. COMPRESSIBLE FLOW

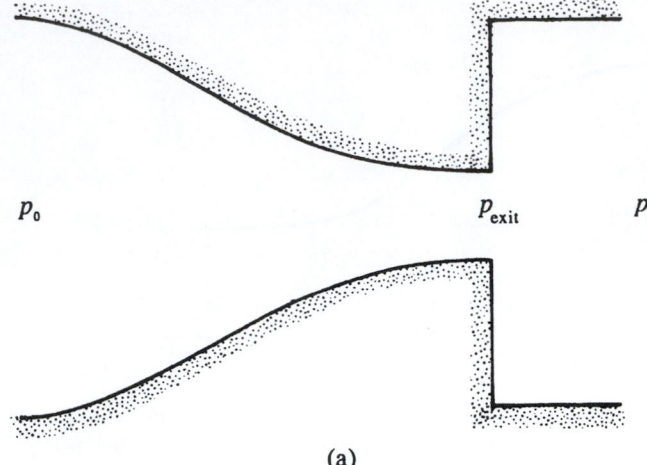

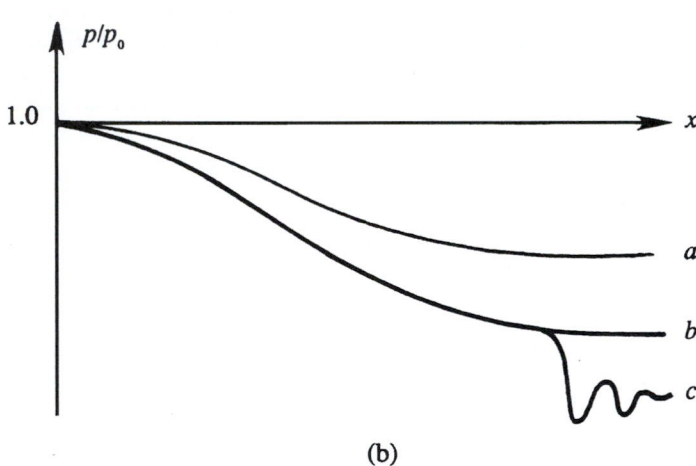

(b)

FIGURE 15.12 Pressure distribution along a convergent nozzle for different values of back pressure p_B: (a) diagram of the nozzle, and (b) pressure distributions as p_B is lowered. Here the highest possible flow speed at the nozzle exit is sonic. When p_B is lowered beyond the point of sonic flow at the nozzle exit, the flow continues to accelerate outside the nozzle via expansion waves that lead to nonuniform pressures (curve c).

For curve a, the flow throughout the nozzle is subsonic. As p_B is lowered, the Mach number increases everywhere and the mass flux through the nozzle also increases. This continues until sonic conditions are reached at the exit, as represented by curve b. Further lowering of the back pressure has no effect on the flow inside the nozzle. This is because the fluid at the exit is now moving downstream at the velocity at which no pressure changes can propagate upstream. Changes in p_B therefore cannot propagate upstream after sonic conditions are reached at the nozzle exit. We say that the nozzle at this stage is *choked* because the mass flux cannot be increased by further lowering of back pressure. If p_B is lowered further (curve c in Figure 15.12), supersonic flow is generated downstream of the nozzle, and the jet pressure adjusts to p_B by means of a series of oblique compression and expansion waves, as schematically indicated by the oscillating pressure distribution for curve *c*. Oblique compression and expansion waves are explained in Sections 15.9 and 15.10. It is only necessary to note here that they are oriented at an angle to the direction of flow, and that the pressure increases through an oblique compression wave and decreases through an oblique expansion wave.

Convergent–Divergent Nozzle

Now consider the case of a convergent–divergent passage, also known as a Laval nozzle (Figure 15.13). Completely subsonic flow applies to curve *a*. As p_B is lowered to p_b, the sonic

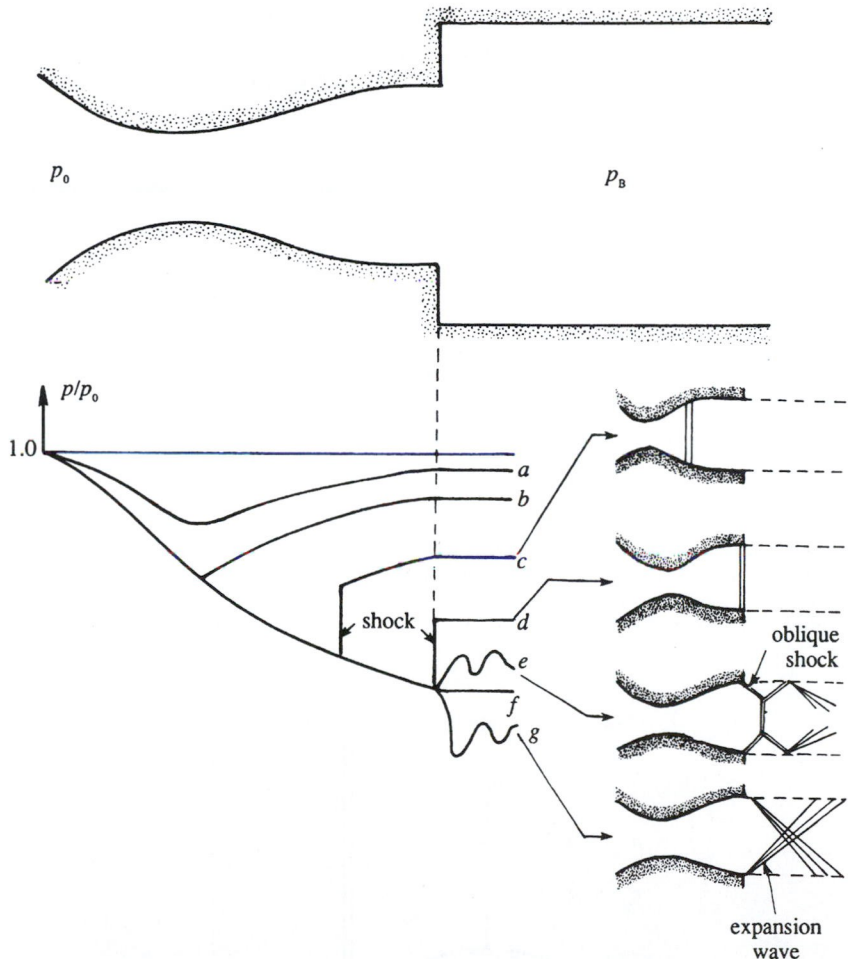

FIGURE 15.13 Pressure distribution along a convergent–divergent (aka Laval) nozzle for different values of the back pressure p_B. Flow patterns for cases *c*, *d*, *e*, and *g* are indicated schematically on the right. The condition *f* is the *pressure matched* case and usually corresponds to the nozzle's design condition. For this case, the flow looks like that of *c* or *d* without the shock wave. *H. W. Liepmann and A. Roshko,* Elements of Gas Dynamics, *Wiley, New York, 1957; reprinted with the permission of Dr. Anatol Roshko.*

condition is reached at the throat. On further reduction of the back pressure, the flow upstream of the throat does not respond, and the nozzle flow is choked in the sense that it has reached the maximum mass flow rate for the given values of p_0 and throat area. There is a range of back pressures, shown by curves c and d, in which the flow initially becomes supersonic in the divergent portion, but then adjusts to the back pressure by means of a normal shock standing inside the nozzle. The flow downstream of the shock is, of course, subsonic. In this range the position of the shock moves downstream as p_B is decreased, and for curve d the normal shock stands right at the exit plane. The flow in the entire divergent portion up to the exit plane is now supersonic and remains so on further reduction of p_B. When the back pressure is further reduced to p_e, there is no normal shock anywhere within the nozzle, and the jet pressure adjusts to p_B by means of oblique compression waves downstream of the nozzle's exit plane. These oblique waves vanish when $p_B = p_f$. On further reduction of the back pressure, the adjustment to p_B takes place outside the exit plane by means of oblique expansion waves.

EXAMPLE 15.2

A convergent–divergent nozzle is operating under off-design conditions, resulting in the presence of a shock wave in the diverging portion. A reservoir containing air at 400 kPa and 800 K supplies the nozzle, whose throat area is 0.2 m². The Mach number upstream of the shock is $M_1 = 2.44$. The area at the nozzle exit is 0.7 m². Find the area at the location of the shock and the exit temperature.

Solution

Figure 15.14 shows the profile of the nozzle, where sections 1 and 2 represent conditions across the shock. As a shock wave can exist only in a supersonic stream, we know that sonic conditions are reached at the throat, and the throat area equals the critical area A^*. The values given are therefore:

$$p_0 = 400 \text{ kPa}, \ T_0 = 800 \text{ K}, \ A_{throat} = A_1^* = 0.2 \text{ m}^2, \ M_1 = 2.44, \quad \text{and} \quad A_3 = 0.7 \text{ m}^2.$$

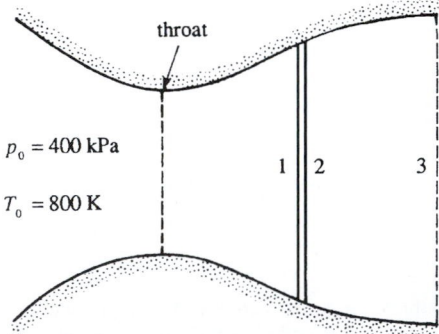

$p_0 = 400$ kPa

$T_0 = 800$ K

FIGURE 15.14 Drawing for Example 15.2. This is case c from Figure 15.13 where a normal shock occurs in the nozzle.

Note that A^* is constant upstream of the shock because the flow is isentropic there; this is why $A_{\text{throat}} = A_1^*$.

The technique of solving this problem is to proceed downstream from the given stagnation conditions. For $M_1 = 2.44$, the isentropic table Table 15.1 gives:

$$A_1/A_1^* = 2.5, \quad \text{so that } A_1 = A_2 = (2.5)(0.2) = 0.5\,\text{m}^2.$$

This is the nozzle's cross-section area at the location of the shock. For $M_1 = 2.44$, the normal shock Table 15.2 gives:

$$M_2 = 0.519, \quad \text{and} \quad p_{02}/p_{01} = 0.523.$$

There is no loss of stagnation pressure up to section 1, so $p_{01} = p_0$, which implies

$$p_{02} = 0.523p_0 = 0.523(400) = 209.2\,\text{kPa}.$$

TABLE 15.2 One-Dimensional Normal-Shock Relations ($\gamma = 1.4$)

M_1	M_2	p_2/p_1	T_2/T_1	$(p_0)_2/(p_0)_1$	M_1	M_2	p_2/p_1	T_2/T_1	$(p_0)_2/(p_0)_1$
1	1	1	1	1	1.4	0.74	2.12	1.255	0.958
1.02	0.98	1.047	1.013	1	1.42	0.731	2.186	1.268	0.953
1.04	0.962	1.095	1.026	1	1.44	0.723	2.253	1.281	0.948
1.06	0.944	1.144	1.039	1	1.46	0.716	2.32	1.294	0.942
1.08	0.928	1.194	1.052	0.999	1.48	0.708	2.389	1.307	0.936
1.1	0.912	1.245	1.065	0.999	1.5	0.701	2.458	1.32	0.93
1.12	0.896	1.297	1.078	0.998	1.52	0.694	2.529	1.334	0.923
1.14	0.882	1.35	1.09	0.997	1.54	0.687	2.6	1.347	0.917
1.16	0.868	1.403	1.103	0.996	1.56	0.681	2.673	1.361	0.91
1.18	0.855	1.458	1.115	0.995	1.58	0.675	2.746	1.374	0.903
1.2	0.842	1.513	1.128	0.993	1.6	0.668	2.82	1.388	0.895
1.22	0.83	1.57	1.14	0.991	1.62	0.663	2.895	1.402	0.888
1.24	0.818	1.627	1.153	0.988	1.64	0.657	2.971	1.416	0.88
1.26	0.807	1.686	1.166	0.986	1.66	0.651	3.048	1.43	0.872
1.28	0.796	1.745	1.178	0.983	1.68	0.646	3.126	1.444	0.864
1.3	0.786	1.805	1.191	0.979	1.7	0.641	3.205	1.458	0.856
1.32	0.776	1.866	1.204	0.976	1.72	0.635	3.285	1.473	0.847
1.34	0.766	1.928	1.216	0.972	1.74	0.631	3.366	1.487	0.839
1.36	0.757	1.991	1.229	0.968	1.76	0.626	3.447	1.502	0.83
1.38	0.748	2.055	1.242	0.963	1.78	0.621	3.53	1.517	0.821

(Continued)

TABLE 15.2 One-Dimensional Normal-Shock Relations ($\gamma = 1.4$)—cont'd

M_1	M_2	p_2/p_1	T_2/T_1	$(p_0)_2/(p_0)_1$	M_1	M_2	p_2/p_1	T_2/T_1	$(p_0)_2/(p_0)_1$
1.8	0.617	3.613	1.532	0.813	2.42	0.521	6.666	2.06	0.532
1.82	0.612	3.698	1.547	0.804	2.44	0.519	6.779	2.079	0.523
1.84	0.608	3.783	1.562	0.795	2.46	0.517	6.894	2.098	0.515
1.86	0.604	3.869	1.577	0.786	2.48	0.515	7.009	2.118	0.507
1.88	0.6	3.957	1.592	0.777	2.5	0.513	7.125	2.138	0.499
1.9	0.596	4.045	1.608	0.767	2.52	0.511	7.242	2.157	0.491
1.92	0.592	4.134	1.624	0.758	2.54	0.509	7.36	2.177	0.483
1.94	0.588	4.224	1.639	0.749	2.56	0.507	7.479	2.198	0.475
1.96	0.584	4.315	1.655	0.74	2.58	0.506	7.599	2.218	0.468
1.98	0.581	4.407	1.671	0.73	2.6	0.504	7.72	2.238	0.46
2	0.577	4.5	1.688	0.721	2.62	0.502	7.842	2.26	0.453
2.02	0.574	4.594	1.704	0.711	2.64	0.5	7.965	2.28	0.445
2.04	0.571	4.689	1.72	0.702	2.66	0.499	8.088	2.301	0.438
2.06	0.567	4.784	1.737	0.693	2.68	0.497	8.213	2.322	0.431
2.08	0.564	4.881	1.754	0.683	2.7	0.496	8.338	2.343	0.424
2.1	0.561	4.978	1.77	0.674	2.72	0.494	8.465	2.364	0.417
2.12	0.558	5.077	1.787	0.665	2.74	0.493	8.592	2.386	0.41
2.14	0.555	5.176	1.805	0.656	2.76	0.491	8.721	2.407	0.403
2.16	0.553	5.277	1.822	0.646	2.78	0.49	8.85	2.429	0.396
2.18	0.55	5.378	1.837	0.637	2.8	0.488	8.98	2.451	0.389
2.2	0.547	5.48	1.857	0.628	2.82	0.487	9.111	2.473	0.383
2.22	0.544	5.583	1.875	0.619	2.84	0.485	9.243	2.496	0.376
2.24	0.542	5.687	1.892	0.61	2.86	0.484	9.376	2.518	0.37
2.26	0.539	5.792	1.91	0.601	2.88	0.483	9.51	2.541	0.364
2.28	0.537	5.898	1.929	0.592	2.9	0.481	9.645	2.563	0.358
2.3	0.534	6.005	1.947	0.583	2.92	0.48	9.781	2.586	0.352
2.32	0.532	6.113	1.965	0.575	2.94	0.479	9.918	2.609	0.346
2.34	0.53	6.222	1.984	0.566	2.96	0.478	10.055	2.632	0.34
2.36	0.527	6.331	2.003	0.557	2.98	0.476	10.194	2.656	0.334
2.38	0.525	6.442	2.021	0.549	3	0.475	10.333	2.679	0.328
2.4	0.523	6.553	2.04	0.54					

The value of A^* changes across a shock wave. The ratio A_2/A_2^* can be found from the *isentropic* table (Table 15.1) corresponding to a Mach number of $M_2 = 0.519$. (Note that A_2^* simply denotes the area that would be reached if the flow from state 2 were accelerated isentropically to sonic conditions.) For $M_2 = 0.519$, Table 15.1 gives:

$$A_2/A_2^* = 1.3, \quad \text{which leads to} \quad A_2^* = A_2/1.3 = 0.5/1.3 = 0.3846 \text{ m}^2.$$

The flow from section 2 to section 3 is isentropic, during which A^* remains constant, so

$$A_3/A_3^* = A_3/A_2^* = 0.7/0.3846 = 1.82.$$

Now find the conditions at the nozzle exit from the isentropic table (Table 15.1). However, the value of $A/A^* = 1.82$ may be found either in the supersonic or the subsonic branch of the table. Since the flow downstream of a normal shock can only be subsonic, use the subsonic branch. For $A/A^* = 1.82$, Table 15.1 gives

$$T_3 = T_{03} = 0.977.$$

The stagnation temperature remains constant in an adiabatic process, so that $T_{03} = T_0$. Thus

$$T_3 = 0.977\,(800) = 782 \text{ K}.$$

15.8. EFFECTS OF FRICTION AND HEATING IN CONSTANT-AREA DUCTS

For steady one-dimensional compressible flow in a duct of constant cross-sectional area, the equations of mass, momentum, and energy conservation are:

$$\rho_1 u_1 = \rho_2 u_2, \quad p_1 + \rho_1 u_1^2 = p_2 + \rho_2 u_2^2 + p_1 f, \quad \text{and} \quad h_1 + \frac{1}{2}u_1^2 + h_1 q = h_2 + \frac{1}{2}u_2^2, \quad (15.45)$$

where $f = F/(p_1 A)$ is a dimensionless friction parameter and $q = Q/h_1$ is a dimensionless heat transfer parameter. In terms of Mach number, for a perfect gas with constant specific heats, the momentum and energy equations become, respectively:

$$p_1(1 + \gamma M_1^2 - f) = p_2(1 + \gamma M_2^2), \quad \text{and} \quad h_1\left(1 + \frac{\gamma - 1}{2}M_1^2 + q\right) = h_2\left(1 + \frac{\gamma - 1}{2}M_2^2\right).$$

Using mass conservation, the thermal equation of state $p = \rho RT$, and the definition of the Mach number, all thermodynamic variables can be eliminated resulting in

$$\frac{M_2}{M_1} = \frac{1 + \gamma M_2^2}{1 + \gamma M_1^2 - f}\left[\frac{1 + ((\gamma - 1)/2)M_1^2 + q}{1 + ((\gamma - 1)/2)M_2^2}\right]^{1/2}.$$

Bringing the unknown M_2 to the left-hand side and assuming q and f are specified along with M_1,

$$\frac{M_2^2\left(1+\frac{1}{2}(\gamma-1)M_2^2\right)}{\left(1+\gamma M_2^2\right)^2}=\frac{M_1^2\left(1+\frac{1}{2}(\gamma-1)M_1^2+q\right)}{\left(1+\gamma M_1^2-f\right)^2}\equiv A.$$

This is a biquadratic equation for M_2 with the solution:

$$M_2^2=\frac{-(1-2A\gamma)\pm[1-2A(\gamma+1)]^{1/2}}{(\gamma-1)-2A\gamma^2}. \tag{15.46}$$

Figures 15.15 and 15.16 are plots of M_2 versus M_1 from (15.46), first with f as a parameter and $q=0$ (Figure 15.15), and then with q as a parameter and $f=0$ (Figure 15.16). Generally, flow properties are known at the inlet station (1) and the flow properties at the outlet station (2) are

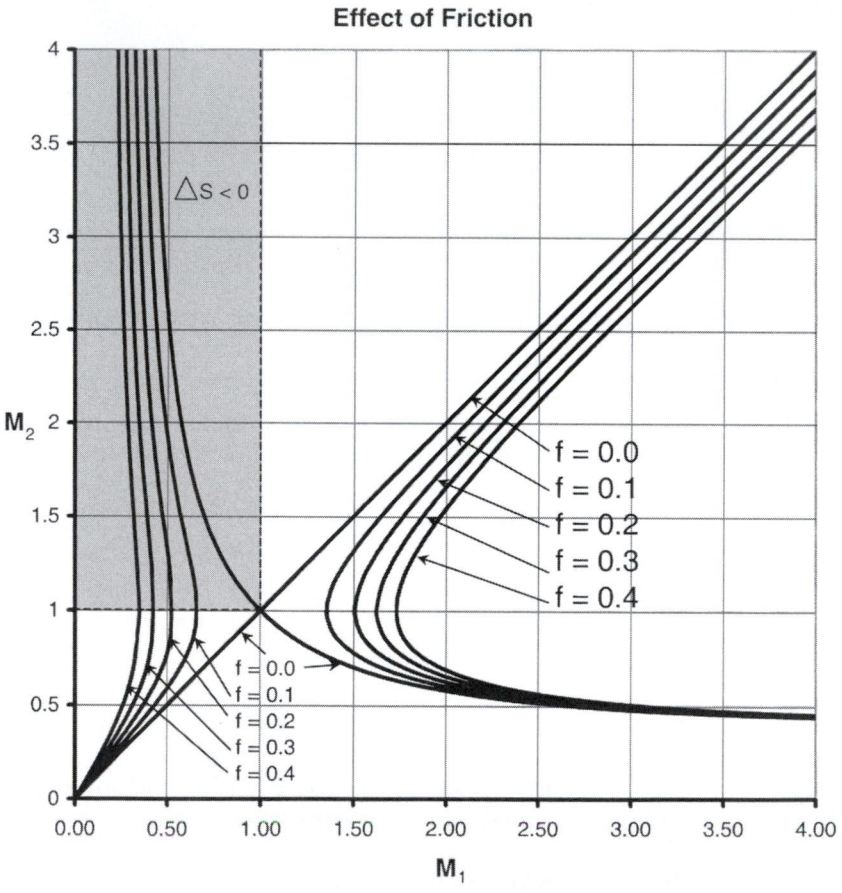

Effect of Friction

FIGURE 15.15 Flow in a constant-area duct with the dimensionless friction f as a parameter without heat exchange, $q=0$, at $\gamma=1.4$. The shaded region in the upper left is inaccessible because $\Delta S<0$. For any duct inlet value of M_1 the curves indicate possible outlet states. Interestingly, for $M_1<1$, all possible M_2 values are at a higher Mach number. For $M_1>1$, the two possible final states are both at lower Mach numbers.

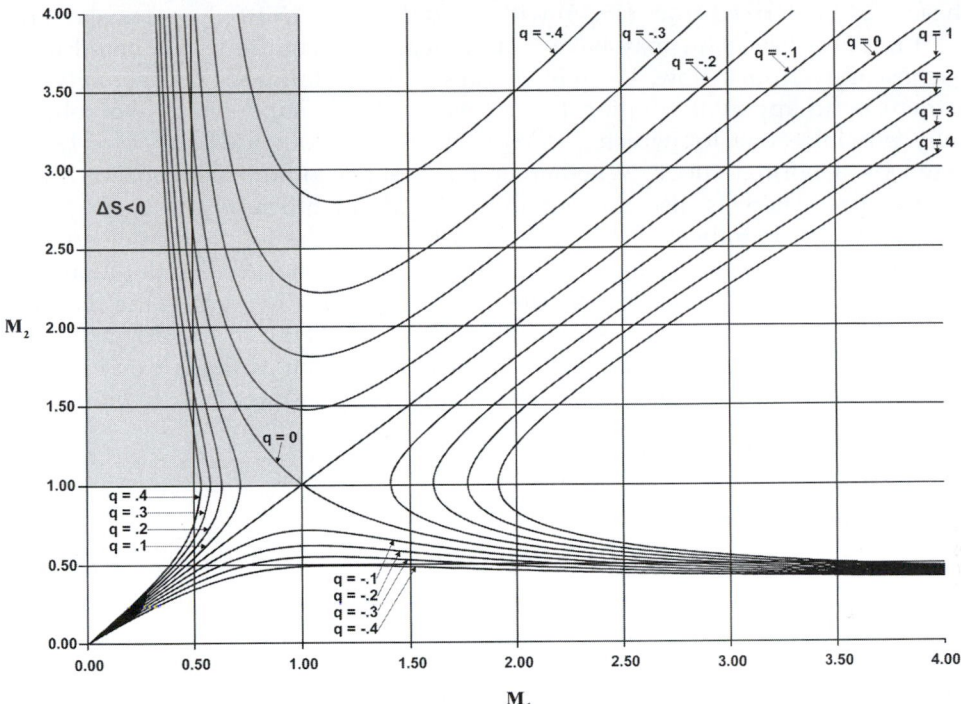

FIGURE 15.16 Flow in a constant-area duct with the dimensionless heat exchange q as a parameter without friction, $f = 0$, at $\gamma = 1.4$. The shaded region in the upper left is inaccessible because $\Delta S < 0$. Here, heat addition is seen to have much the same effect as friction.

sought. Here, the dimensionless friction f and heat transfer q are presumed to be specified. Thus, once M_2 is calculated from (15.45), all of the other properties may be obtained from the dimensionless formulation of the conservation laws above. When q and $f = 0$, two solutions are possible: the trivial solution $M_1 = M_2$ and the normal shock solution given in Section 15.6. We also showed that the upper left branch of the solution $M_2 > 1$ when $M_1 < 1$ is inaccessible because it violates the second law of thermodynamics, that is, it results in a spontaneous decrease of entropy.

Effect of Friction

Referring to the left branch of Figure 15.15, the solution indicates the surprising result that friction accelerates a subsonic flow leading to $M_2 > M_1$. This happens because friction causes the pressure, and therefore the density, to drop rapidly enough so that the fluid velocity must increase to maintain a constant mass flow. For this case of adiabatic flow with friction, the relevant equations for differential changes in pressure, velocity, and density in terms of the local Mach number $M = u/c$ are:

$$-\frac{dp}{p_1} = \frac{1 - (\gamma - 1)M^2}{1 - M^2}df, \quad \text{and} \quad \frac{du}{u} = -\frac{d\rho}{\rho} = \frac{p_1}{p}\frac{df}{1 - M^2}, \tag{15.47}$$

and these may be derived from (15.45) with $q = 0$ (Exercise 15.10). In particular since $df > 0$, (15.47) implies that dp/p_1 may have a large negative magnitude compared to df as M approaches unity from below. We will discuss in what follows what actually happens when there is no apparent solution for M_2. When M_1 is supersonic, two solutions are generally possible—one for which $1 < M_2 < M_1$ and the other where $M_2 < 1$. They are connected by a normal shock. Whether or not a shock occurs depends on the downstream pressure. There is also the possibility of M_1 insufficiently large or f too large so that no solution is indicated. We will discuss that in the following but note that the two solutions coalesce when $M_2 = 1$ and the flow is choked. At this condition the maximum mass flow is passed by the duct. In the case $1 < M_2 < M_1$, the flow is decelerated and the pressure, density, and temperature all increase in the downstream direction. The stagnation pressure is always decreased by friction as the entropy is increased. In summary, friction's net effect is to drive a compressible duct flow toward $M_2 = 1$ for any value of M_1.

Effect of Heat Transfer

The range of solutions is twice as rich in this case as q may take both signs while f must be positive. Figure 15.16 shows that for $q > 0$ solutions are similar in most respects to those with friction ($f > 0$). Heating accelerates a subsonic flow and lowers the pressure and density. However, heating generally increases the fluid temperature except in the limited range $1/\gamma < M_1^2 < 1$ in which the fluid temperature decreases with heat addition. The relevant equations for differential changes in temperature and flow speed in terms of the local Mach number $M = u/c$ are:

$$\frac{dT}{T_1} = \left(\frac{1 - \gamma M^2}{1 - M^2}\right) dq, \quad \text{and} \quad \frac{u\,du}{h_1} = \frac{(1 - \gamma)M^2}{1 - M^2} dq, \tag{15.48}$$

and these can be derived from (15.45) with $f = 0$ (Exercise 15.11). When $1/\gamma < M_1^2 < 1$, the energy from heat addition goes preferentially into increasing the velocity of the fluid. The supersonic branch $M_2 > 1$ when $M_1 < 1$ is inaccessible because those solutions violate the second law of thermodynamics. Again, as with f too large or M_1 too close to 1, there is a possibility no indicated solution when q is too large; this is discussed in what follows. When $M_1 > 1$, two solutions for M_2 are generally possible and they are connected by a normal shock. The shock is absent if the downstream pressure is low and present if the downstream pressure is high. Although $q > 0$ (and $f > 0$) does not always indicate a solution (if the flow has been choked), there will always be a solution for $q < 0$. Cooling a supersonic flow accelerates it, thus decreasing its pressure, temperature, and density. If no shock occurs, $M_2 > M_1$. Conversely, cooling a subsonic flow decelerates it so that the pressure and density increase. The temperature decreases when heat is removed from the flow except in the limited range $1/\gamma < M_1^2 < 1$ in which the heat removal decelerates the flow so rapidly that the temperature increases.

For high molecular-weight gases, near critical conditions (high pressure, low temperature), the gas dynamic relationships may be completely different from those developed here for perfect gases. Cramer and Fry (1993) found that nonperfect gases may support expansion shocks, accelerated flow through "antithroats," and generally behave in unfamiliar ways.

Figures 15.15 and 15.16 show that friction or heat input in a constant-area duct both drive a compressible flow in the duct toward the sonic condition. For any given M_1, the maximum f or $q > 0$ that is permissible is the one for which $M = 1$ at the exit station. The flow is then said to be choked, and the mass flow rate through that duct cannot be increased without increasing p_1 or decreasing p_2. This is analogous to flow in a convergent duct. Imagine pouring liquid through a funnel from one container into another. There is a maximum volumetric flow rate that can be passed by the funnel, and beyond that flow rate, the funnel overflows. The same thing happens here. If f or q is too large, such that no (steady-state) solution is possible, there is an external adjustment that reduces the mass flow rate to that for which the exit speed is just sonic. For $M_1 < 1$ and $M_1 > 1$ the limiting curves for f and q indicating choked flow intersect $M_2 = 1$ at right angles. Qualitatively, the effect is the same as choking by area contraction.

15.9. PRESSURE WAVES IN PLANAR COMPRESSIBLE FLOW

To this point, the emphasis in this chapter has been on steady one-dimensional flows in which flow properties vary only in the direction of flow. This section presents compressible flow results for more than one spatial dimension. To get started, consider a point source emitting infinitesimal pressure (acoustic) disturbances in a still compressible fluid in which the speed of sound is c. If the point source is stationary, then the pressure-disturbance wave fronts are concentric spheres. Figure 15.17a shows the intersection of these wave fronts with a plane containing the source at times corresponding to integer multiples of Δt.

When the source propagates to the left at speed $U < c$, the wave-front diagram changes to look like Figure 15.17b, which shows four locations of the source separated by equal time intervals Δt, with point 4 being the present location of the source. At the first point, the source emitted a wave that has spherically expanded to a radius of $3c\Delta t$ in the time interval $3\Delta t$. During this time the source has moved to the fourth location, a distance of $3U\Delta t$ from the first point of wave-front emission. The figure also shows the locations of the wave fronts emitted while the source was at the second and third points. Here, the wave fronts do not intersect because $U < c$. As in the case of the stationary source, the wave fronts propagate vertically upward and downward, and horizontally upstream and downstream from the source. Thus, *a body moving at a subsonic speed influences the entire flow field.*

Now consider the case depicted in Figure 15.17c where the source moves supersonically, $U > c$. Here, the centers of the spherically expanding wave fronts are separated by more than $c\Delta t$, and no pressure disturbance propagates ahead of the source. Instead, the edges of the wave fronts form a conical tangent surface called the *Mach cone.* In planar two-dimensional flow, the tangent surface is in the form of a wedge, and the tangent lines are called *Mach lines.* An examination of the figure shows that the half-angle of the Mach cone (or wedge), called the *Mach angle* μ, is given by $\sin \mu = (c\Delta t)/(U\Delta t)$, so that

$$\sin \mu = 1/M. \tag{15.49}$$

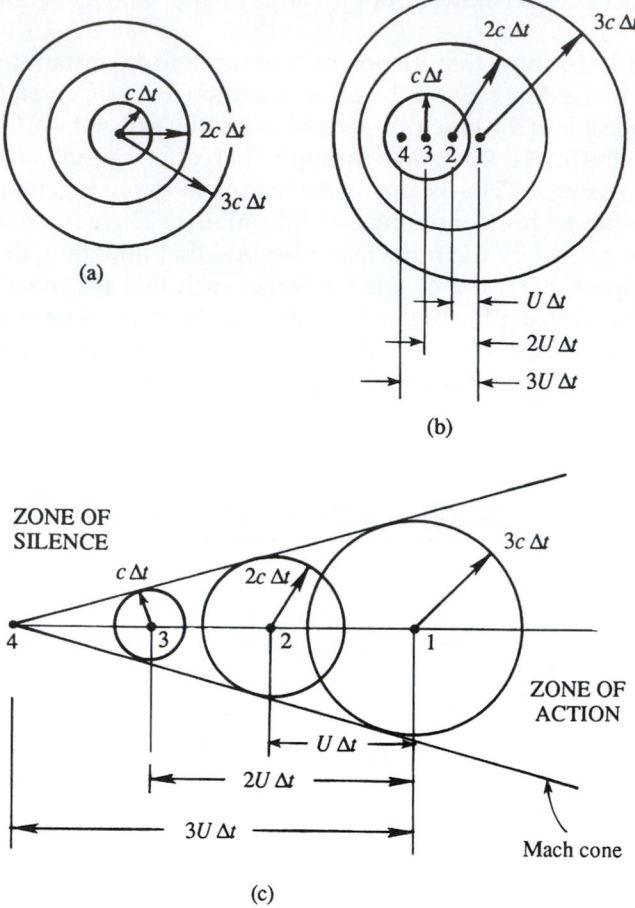

FIGURE 15.17 Wave fronts emitted by a point source in a still fluid when the source speed U is: (a) $U = 0$; (b) $U < c$; and (c) $U > c$. In each case the wave fronts are emitted at integer multiples of Δt. At subsonic source speeds, the wave fronts do not overlap and they spread ahead of the source. At supersonic source speeds, all the wave fronts lie behind the source within the Mach cone having a half-angle $\sin^{-1}(1/M)$.

The Mach cone becomes wider as M decreases and becomes a plane front (that is, $\mu = 90°$) when $M = 1$.

The point source considered here could be part of a solid body, which sends out pressure waves as it moves through the fluid. Moreover, after a simple Galilean transformation, Figure 15.17b and c apply equally well to a stationary point source with a compressible fluid moving past it at speed U. From Figure 15.17c it is clear that in a supersonic flow an observer outside the source's Mach cone would not detect or hear a pressure signal emitted by the source, hence this region is called the *zone of silence*. In contrast, the region inside the Mach cone is called the *zone of action*, within which the effects of the disturbance are felt. Thus, the sound of a supersonic aircraft passing overhead does not reach an observer on the ground

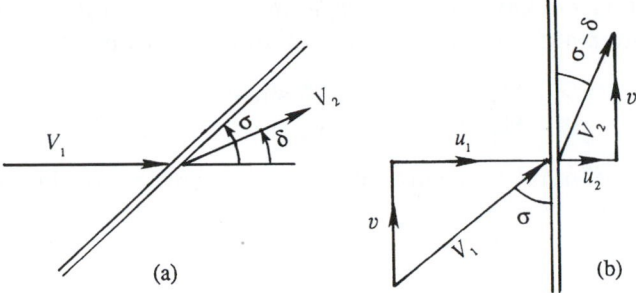

FIGURE 15.18 Two coordinate systems for an oblique shock wave. (a) Stream-aligned coordinates where the oblique shock wave lies at shock angle $= \sigma$ and produces a flow-deflection of angle $= \delta$. (b) Shock-normal coordinates which are preferred for analysis because an oblique shock wave is merely a normal shock with a super-imposed shock-parallel velocity v.

until its Mach cone reaches the observer, and this arrival occurs *after* the aircraft has passed overhead.

At every point in a planar supersonic flow there are two Mach lines, oriented at $\pm \mu$ to the local direction of flow. Pressure disturbance information propagates along these lines, which are the *characteristics* of the governing differential equation. It can be shown that the nature of the governing differential equation is hyperbolic in a supersonic flow and elliptic in a subsonic flow.

When the pressure disturbances from the source are of finite amplitude, they may evolve into a shock wave that is not normal to the flow direction. Such *oblique* shock waves are commonly encountered in ballistics and supersonic flight, and differ from normal shock waves because they change the upstream flow velocity's magnitude *and* direction. A generic depiction of an oblique shock wave is provided in Figure 15.18 in two coordinate systems. Part a of this figure shows the stream-aligned coordinate system where the shock wave resides at an angle σ from the horizontal. Here the velocity upstream of the shock is horizontal with magnitude V_1, while the velocity downstream of the shock is deflected from the horizontal by an angle δ and has magnitude V_2. Part b of Figure 15.18 shows the same shock wave in a shock-aligned coordinate system where the shock wave is vertical, and the fluid velocities upstream and downstream of the shock are (u_1, v) and (u_2, v), respectively. Here v is parallel to the shock wave and is not influenced by it (see Exercise 15.18). Thus an oblique shock may be analyzed as a normal shock involving u_1 and u_2 to which a constant shock-parallel velocity v is added. Using the Cartesian coordinates in Figure 15.18b where the shock coincides with the vertical axis, the relationships between the various components and angles are:

$$(u_1, v) = \sqrt{u_1^2 + v^2}\,(\sin \sigma, \cos \sigma) = V_1(\sin \sigma, \cos \sigma),$$

and

$$(u_2, v) = \sqrt{u_2^2 + v^2}(\sin(\sigma - \delta), \cos(\sigma - \delta)) = V_2(\sin(\sigma - \delta), \cos(\sigma - \delta)).$$

The angle σ is called the *shock angle* or *wave angle* and δ is called the *deflection angle*. The normal Mach numbers upstream (1) and downstream (2) of the shock are:

$$M_{n1} = u_1/c_1 = M_1 \sin \sigma > 1,$$
$$M_{n2} = u_2/c_2 = M_2 \sin(\sigma - \delta) < 1.$$

Because $u_2 < u_1$, there is a sudden change of direction of flow across the shock and the flow is turned *toward* the shock by angle δ.

Superposition of the tangential velocity v does not affect the *static* properties, which are therefore the same as those for a normal shock. The expressions for the ratios p_2/p_1, ρ_2/ρ_1, T_2/T_1, and $(S_2 - S_1)/C_v$ are therefore those given by (15.39) and (15.41) through (15.43), if M_1 is replaced by $M_{n1} = M_1 \sin \sigma$. For example:

$$\frac{p_2}{p_1} = 1 + \frac{2\gamma}{\gamma + 1}(M_1^2 \sin^2\sigma - 1), \tag{15.50}$$

$$\frac{\rho_2}{\rho_1} = \frac{(\gamma + 1)M_1^2 \sin^2\sigma}{(\gamma - 1)M_1^2 \sin^2\sigma + 2} = \frac{u_1}{u_2} = \frac{\tan \sigma}{\tan (\sigma - \delta)}. \tag{15.51}$$

Thus, the normal shock table, Table 15.2, is applicable to oblique shock waves if we use $M_1 \sin\sigma$ in place of M_1.

The relation between the upstream and downstream Mach numbers can be found from (15.40) by replacing M_1 by $M_1 \sin \sigma$ and M_2 by $M_2 \sin (\sigma - \delta)$. This gives

$$M_2^2 \sin^2(\sigma - \delta) = \frac{(\gamma - 1)M_1^2 \sin^2 \sigma + 2}{2\gamma M_1^2 \sin^2 \sigma + 1 - \gamma}. \tag{15.52}$$

An important relation is that between the deflection angle δ and the shock angle σ for a given M_1, given in (15.51). Using the trigonometric identity for $\tan (\sigma - \delta)$, this becomes

$$\tan \delta = 2 \cot \sigma \frac{M_1^2 \sin^2 \sigma - 1}{M_1^2(\gamma + \cos 2\sigma) + 2}. \tag{15.53}$$

A plot of this relation is given in Figure 15.19. The curves represent δ versus σ for constant M_1. The value of M_2 varies along the curves, and the locus of points corresponding to $M_2 = 1$ is indicated. It is apparent that there is a maximum deflection angle δ_{max} for oblique shock solutions to be possible; for example, $\delta_{max} = 23°$ for $M_1 = 2$. For a given M_1, δ becomes zero at $\sigma = \pi/2$ corresponding to a normal shock, and at $\sigma = \mu = \sin^{-1}(1/M_1)$ corresponding to the Mach angle. For a fixed M_1 and $\delta < \delta_{max}$, there are two possible solutions: a *weak shock* corresponding to a smaller σ and a *strong shock* corresponding to a larger σ. It is clear that the flow downstream of a strong shock is always subsonic; in contrast, the flow downstream of a weak shock is generally supersonic, except in a small range in which δ is slightly smaller than δ_{max}.

Oblique shock waves are commonly generated when a supersonic flow is forced to change direction to go around a structure where the flow area cross section is reduced.

FIGURE 15.19 Plot of oblique shock solutions. The strong shock branch is indicated by dashed lines on the right, and the heavy dotted line indicates the maximum deflection angle δ_{max}. (*From Ames Research Staff, 1953, NACA Report 1135.*)

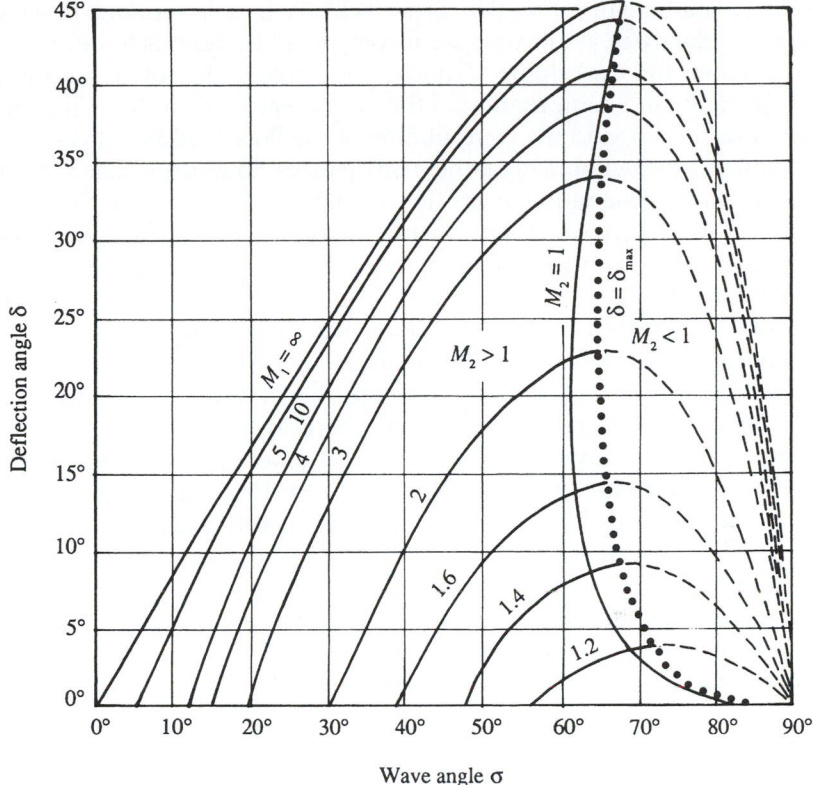

Wave angle σ

Two examples are shown in Figure 15.20 that show supersonic flow past a wedge of half-angle δ, or the flow past a compression bend where the wall turns into the flow by an angle δ. If M_1 and δ are known, then σ can be obtained from Figure 15.19, and M_{n2} (and therefore $M_2 = M_{n2}/\sin(\sigma - \delta)$) can be obtained from the shock table (Table 15.2). An attached shock wave, corresponding to the weak solution, forms at the nose of the wedge, such that the flow is parallel to the wedge after turning through an angle δ. The shock angle σ decreases to the Mach angle $\mu_1 = \sin^{-1}(1/M_1)$ as the deflection δ tends to

FIGURE 15.20 Two possible means for producing oblique shocks in a supersonic flow. In both cases a solid surface causes the flow to turn, and the flow area is reduced. The geometry shown in the right panel is sometimes called a *compression corner*.

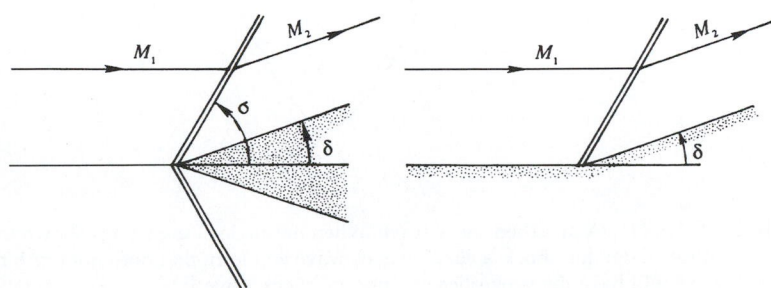

zero. It is interesting that the corner velocity in a supersonic flow is finite. In contrast, the corner velocity in a subsonic (or incompressible) flow is either zero or infinite, depending on whether the wall shape is concave or convex. Moreover, the streamlines in Figure 15.20 are straight, and computation of the field is easy. By contrast, the streamlines in a subsonic flow are curved, and the computation of the flow field is not easy. The basic reason for this is that, in a supersonic flow, the disturbances do not propagate upstream of Mach lines or shock waves emanating from the disturbances, hence the flow field can be constructed step by step, *proceeding downstream*. In contrast, disturbances propagate both upstream and downstream in a subsonic flow so that all features in the entire flow field are related to each other.

As δ is increased beyond δ_{max}, attached oblique shocks are not possible, and a detached curved shock stands in front of the body (Figure 15.21). The central streamline goes through a normal shock and generates a subsonic flow in front of the wedge. The *strong* shock solution of Figure 15.19 therefore holds near the nose of the body. Farther out, the shock angle decreases, and the weak shock solution applies. If the wedge angle is not too large, then the curved detached shock in Figure 15.21 becomes an oblique attached shock as the Mach number is increased. In the case of a blunt-nosed body, however, the shock at the leading edge is always detached, although it moves closer to the body as the Mach number is increased.

We see that shock waves may exist in supersonic flows and their location and orientation adjust to satisfy boundary conditions. In external flows, such as those just described, the boundary condition is that streamlines at a solid surface must be tangent to that

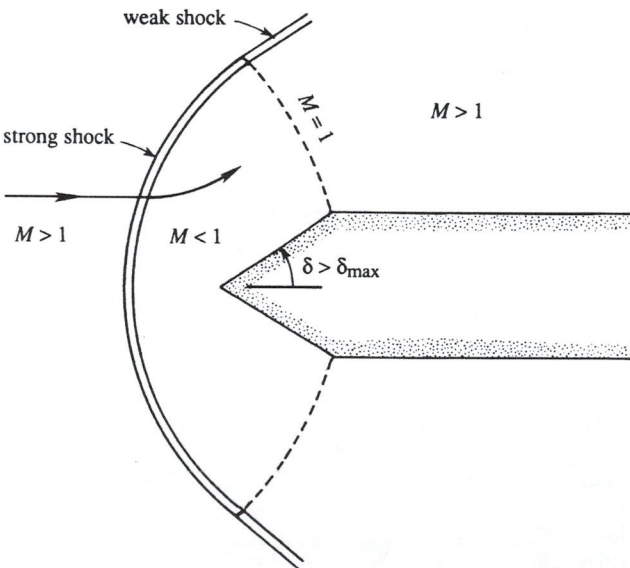

FIGURE 15.21 A detached shock wave. When the angle of the wedge shown in the left panel of Figure 15.20 is too great for an oblique shock, a curved shock wave will form that does not touch body. A portion of this detached shock wave will have the properties of a normal shock wave.

surface. In duct flows the boundary condition locating the shock is usually the downstream pressure.

From the foregoing analysis, it is clear that large-angle supersonic flow deflections should be avoided when designing efficient devices that produce minimal total pressure losses. Efficient devices tend to be slender and thin, and their performance may be analyzed using a weak oblique shock approximation that can be obtained from the results above in the limit of small flow deflection angle, $\delta \ll 1$. To obtain this expression, simplify (15.53) by noting that as $\delta \to 0$, the shock angle σ tends to the Mach angle $\mu_1 = \sin^{-1}(1/M_1)$. And, from (15.50) we note that $(p_2 - p_1)/p_1 \to 0$ as $M_1^2 \sin^2 \sigma - 1 \to 0$ (as $\sigma \to \mu$ and $\delta \to 0$). Then from (15.50) and (15.53):

$$\tan \delta = 2 \cot \sigma \frac{\gamma + 1}{2\gamma} \left(\frac{p_2 - p_1}{p_1} \right) \frac{1}{M_1^2(\gamma + 1 - 2 \sin^2 \sigma) + 2}. \tag{15.54}$$

As $\delta \to 0$, $\tan \delta \approx \delta$, $\cot \mu = \sqrt{M_1^2 - 1}$, $\sin \sigma \approx 1/M_1$, and

$$\frac{p_2 - p_1}{p_1} = \frac{\gamma M_1^2}{\sqrt{M_1^2 - 1}} \delta. \tag{15.55}$$

The interesting point is that the relation (15.55) is also applicable to weak *expansion* waves and not just weak compression waves. By this we mean that the pressure increase due to a small deflection of the wall toward the flow is the same as the pressure *decrease* due to a small deflection of the wall *away* from the flow. This extended range of validity of (15.55) occurs because the entropy change across a weak shock may be negligible even when the pressure change is appreciable (see (15.44b) and the related discussion). Thus, weak shock waves can be treated as isentropic or reversible. Relationships for a weak shock wave can therefore be applied to a weak expansion wave, except for some sign changes. In the final section of this chapter, (15.55) is used to estimate the lift and drag of a thin airfoil in supersonic flow.

When an initially horizontal supersonic flow follows a curving wall, the wall radiates compression and expansion waves into the flow that modulate the flow's direction and Mach number. When the wall is smoothly curved these compression and expansion waves follow Mach lines, inclined at an angle of $\mu = \sin^{-1}(1/M)$ to the *local* direction of flow (Figure 15.22). In this simple circumstance where there is no upper wall that radiates compression or expansion waves downward into the region of interest, the flow's orientation and Mach number are constant on each Mach line. In the case of compression, the Mach number decreases along the flow, so that the Mach angle increases. The Mach lines may therefore coalesce and form an oblique shock as in Figure 15.22a. In the case of a gradual expansion, the Mach number increases along the flow and the Mach lines diverge as in Figure 15.22b.

If the wall has a sharp deflection (a corner) away from the approaching stream, then the pattern of Figure 15.22b takes the form of Figure 15.23 where all the Mach lines originate from the corner. In this case, this portion of the flow where it expands and turns, and is not parallel to the wall upstream or downstream of the corner, is known as a *Prandtl-Meyer expansion fan*. The Mach number increases through the fan, with $M_2 > M_1$. The first Mach line

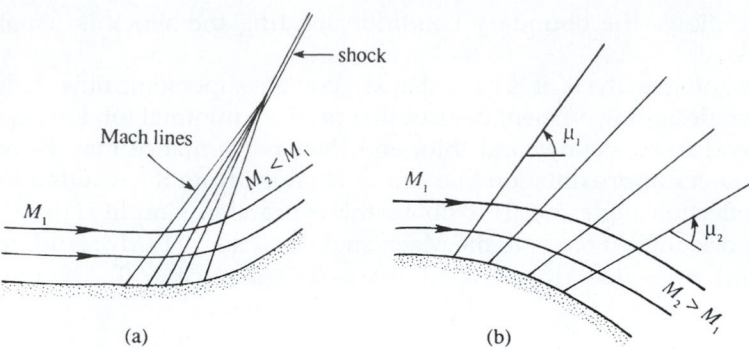

FIGURE 15.22 Gradual compression and expansion in supersonic flow. (a) A gradual compression corner like the one shown will eventually result in an oblique shock wave as the various Mach lines merge, each carrying a fraction of the overall compression. (b) A gradual expansion corner like the one shown produces Mach lines that diverge so the expansion becomes even more gradual farther from the wall.

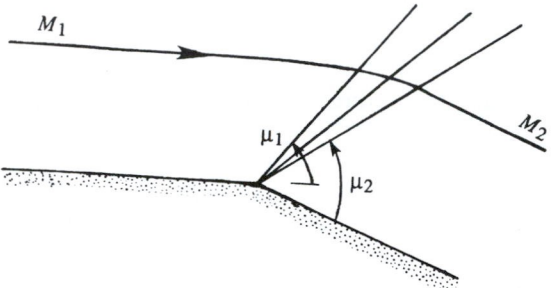

FIGURE 15.23 The Prandtl-Meyer expansion fan. This is the flow field developed by a sharp expansion corner. Here the flow area increases downstream of the corner so it accelerates a supersonic flow.

is inclined at an angle of μ_1 to the upstream wall direction, while the last Mach line is inclined at an angle of μ_2 to the downstream wall direction. The pressure falls gradually along any streamline through the fan. Along the wall, however, the pressure remains constant along the upstream wall, falls discontinuously at the corner, and then remains constant along the downstream wall. Figure 15.23 should be compared with Figure 15.20, in which the wall turns *inward* and generates an oblique shock wave. By contrast, the expansion in Figure 15.23 is gradual and isentropic away from the wall.

The flow through a Prandtl-Meyer expansion fan is calculated as follows. From Figure 15.18b, conservation of momentum tangential to the shock shows that the tangential velocity is unchanged, or:

$$V_1 \cos \sigma = V_2 \cos(\sigma - \delta) = V_2 \left(\cos \sigma \cos \delta + \sin \sigma \sin \delta \right).$$

We are concerned here with very small deflections, $\delta \to 0$ so $\sigma \to \mu$. Here, $\cos \delta \approx 1$, $\sin \delta \approx \delta$, $V_1 \approx V_2(1 + \delta \tan \sigma)$, so $(V_2 - V_1)/V_1 \approx -\delta \tan \sigma \approx -\delta/\sqrt{M_1^2 - 1}$, where $\tan \sigma \approx 1/\sqrt{M_1^2 - 1}$. Thus, the velocity change dV for an infinitesimal wall deflection $d\delta$ can be written as

$d\delta = -(dV/V)\sqrt{M^2-1}$ (first quadrant deflection). Because $V = Mc$, $dV/V = dM/M + dc/c$. With $c = \sqrt{\gamma RT}$ for a perfect gas, $dc/c = dT/2T$. Using (15.28) for adiabatic flow of a perfect gas, $dT/T = -(\gamma-1)M\, dM/[\,1 + ((\gamma-1)/2)M^2]$, then

$$d\delta = -\frac{\sqrt{M^2-1}}{M}\frac{dM}{1+\frac{1}{2}(\gamma-1)M^2}.$$

Integrating δ from 0 (radians) and M from 1 gives $\delta + \nu(M) = const.$, where

$$\nu(M) = \int_1^M \frac{\sqrt{M^2-1}}{1+\frac{1}{2}(\gamma-1)M^2}\frac{dM}{M} = \sqrt{\frac{\gamma+1}{\gamma-1}}\tan^{-1}\sqrt{\frac{\gamma-1}{\gamma+1}(M^2-1)} - \tan^{-1}\sqrt{M^2-1} \quad (15.56)$$

is called the *Prandtl-Meyer function*. The sign of $\sqrt{M^2-1}$ originates from the identification of $\tan\sigma = \tan\mu = 1/\sqrt{M^2-1}$ for a first quadrant deflection (upper half-plane). For a fourth quadrant deflection (lower half-plane), $\tan\mu = -1/\sqrt{M^2-1}$. For example, for Figure 15.22a or b with δ_1, δ_2, and M_1 given, we would write

$$\delta_1 + \nu(M_1) = \delta_2 + \nu(M_2), \text{ and then } \nu(M_2) = \delta_1 - \delta_2 + \nu(M_1)$$

would determine M_2. In Figure 15.22a, $\delta_1 - \delta_2 < 0$, so $v_2 < v_1$ and $M_2 < M_1$. In Figure 15.22b, $\delta_1 - \delta_2 > 0$, so $v_2 > v_1$ and $M_2 > M_1$.

15.10. THIN AIRFOIL THEORY IN SUPERSONIC FLOW

Simple expressions can be derived for the lift and drag coefficients of an airfoil in supersonic flow if the thickness and angle of attack are small. Under these circumstances the pressure disturbances caused by the airfoil are small, and the total flow can be built up by superposition of small disturbances emanating from points on the body. Such a linearized theory of lift and drag was developed by Ackeret. Because all flow inclinations are small, we can use the relation (15.55) to calculate the pressure changes due to a change in flow direction. We can write this relation as

$$\frac{p - p_\infty}{p_\infty} = \frac{\gamma M_\infty^2 \delta}{\sqrt{M_\infty^2 - 1}}, \quad (15.57)$$

where p_∞ and M_∞ refer to the properties of the free stream, and p is the pressure at a point where the flow is inclined at an angle δ to the free-stream direction. The sign of δ in (15.57) determines the sign of $(p - p_\infty)$.

To see how the lift and drag of a thin body in a supersonic stream can be estimated, consider a flat plate inclined at a small angle α to a horizontal stream (Figure 15.24). At the leading edge there is a weak expansion fan above the top surface and a weak oblique shock below the bottom surface. The streamlines ahead of these waves are straight. The streamlines above the plate turn through an angle α by expanding through an expansion fan, downstream of which they become parallel to the plate with a pressure $p_2 < p_\infty$. The upper streamlines

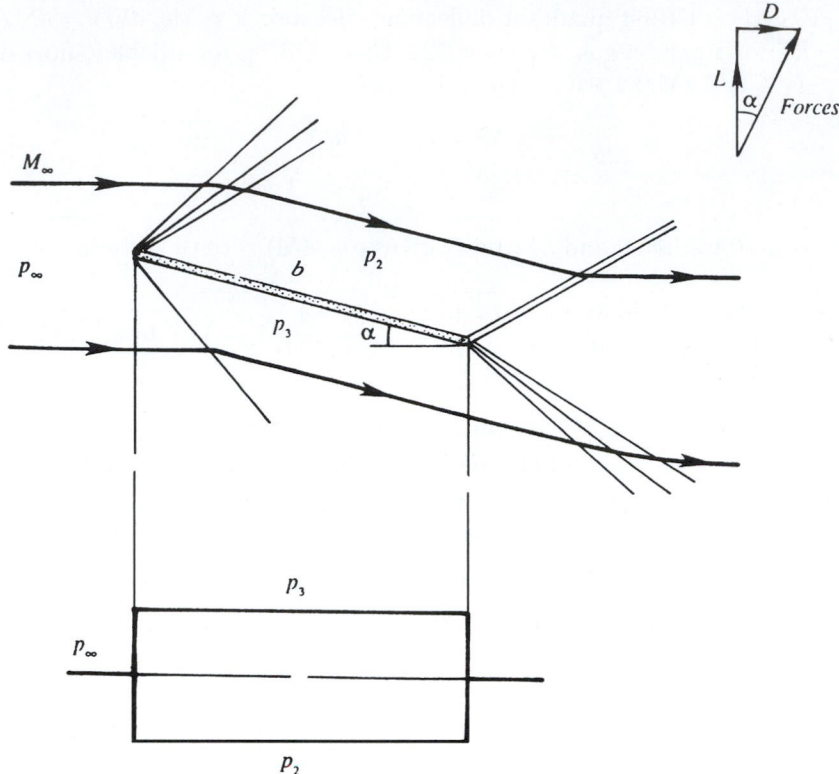

FIGURE 15.24 Inclined flat plate in a supersonic stream as a simple illustration of supersonic aerodynamics. The upper panel shows the flow pattern and the lower panel shows the pressure distribution on the suction and pressure sides of the simple foil. Here, an ideal compressible flow analysis does predict a drag component, unlike an equivalent ideal incompressible flow.

then turn sharply across an oblique shock emanating from the trailing edge, becoming parallel to the free stream once again. Opposite features occur for the streamlines below the plate where the flow first undergoes compression across an oblique shock coming from the leading edge, which results in a pressure $p_3 > p_\infty$. It is, however, not important to distinguish between shock and expansion waves in Figure 15.24, because the linearized theory treats them the same way, except for the sign of the pressure changes they produce.

The pressures above and below the plate can be found from (15.47), giving:

$$\frac{p_2 - p_\infty}{p_\infty} = -\frac{\gamma M_\infty^2 \alpha}{\sqrt{M_\infty^2 - 1}}, \quad \text{and} \quad \frac{p_3 - p_\infty}{p_\infty} = \frac{\gamma M_\infty^2 \alpha}{\sqrt{M_\infty^2 - 1}}.$$

The pressure difference across the plate is therefore

$$\frac{p_3 - p_2}{p_\infty} = \frac{2\alpha\gamma M_\infty^2}{\sqrt{M_\infty^2 - 1}}.$$

If b is the chord length, then the lift L and drag D forces per unit span are:

$$L = (p_3 - p_2)\, b \cos \alpha \cong \frac{2\alpha\gamma M_\infty^2 p_\infty b}{\sqrt{M_\infty^2 - 1}}, \quad \text{and} \quad D = (p_3 - p_2)\, b \sin \alpha \cong \frac{2\alpha^2 \gamma M_\infty^2 p_\infty b}{\sqrt{M_\infty^2 - 1}}.$$

$$(15.58)$$

Using the relationship $\rho U^2 = \gamma p M^2$, the lift and drag coefficients are:

$$C_L = \frac{L}{(1/2)\rho_\infty U_\infty^2 b} \cong \frac{4\alpha}{\sqrt{M_\infty^2 - 1}}, \quad \text{and} \quad C_D = \frac{D}{(1/2)\rho_\infty U_\infty^2 b} \cong \frac{4\alpha^2}{\sqrt{M_\infty^2 - 1}}. \qquad (15.59)$$

These expressions do not hold at transonic speeds $M_\infty \to 1$, when the process of linearization used here breaks down. The expression for the lift coefficient should be compared to the incompressible expression $C_L = 2\pi\alpha$ derived in the preceding chapter. Note that the flow in Figure 15.24 does have a circulation because the velocities at the upper and lower surfaces are parallel but have different magnitudes. However, in a supersonic flow it is not necessary to invoke the Kutta condition (discussed in the preceding chapter) to determine the magnitude of the circulation. The flow in Figure 15.24 does leave the trailing edge smoothly.

The drag in (15.59) is the *wave drag* experienced by a body in a supersonic stream, and exists even in an inviscid flow. The d'Alembert paradox therefore does not apply in a supersonic flow. The supersonic wave drag is analogous to the gravity wave drag experienced by a ship moving at a speed greater than the velocity of surface gravity waves, in which a system of bow waves is carried with the ship. The magnitude of the supersonic wave drag is independent of the value of the viscosity, although the energy spent in overcoming this drag is finally dissipated through viscous effects within the shock waves. In addition to the wave drag, additional drags due to viscous and finite-span effects, considered in the preceding chapter, act on a real wing.

In this connection, it is worth noting the difference between the airfoil shapes used in subsonic and supersonic airplanes. Low-speed airfoils have a streamlined shape, with a rounded nose and a sharp trailing edge. These features are not helpful in supersonic airfoils. The most effective way to reduce the drag of a supersonic airfoil is to reduce its thickness. Supersonic wings are characteristically thin and have sharp leading edges.

EXERCISES

15.1. The field equation for one-dimensional acoustic pressure fluctuations in an ideal compressible fluid is (15.13).

a) Change the independent variables x and t to $\xi = x - ct$ and $\zeta = x + ct$ to simplify (15.13) to $\dfrac{\partial^2 p'}{\partial \xi \partial \zeta} = 0$.

b) Use the simplified equation in part b) to find the general solution to the original field equation: $p'(x, t) = f(x - ct) + g(x + ct)$ where f and g are undetermined functions.

c) When the initial conditions are: $p' = F(x)$ and $\partial p'/\partial t = G(x)$ at $t = 0$, show that:

$$f(x) = \frac{1}{2}\left[F(x) - \frac{1}{c}\int_0^x G(\maltese)d\maltese\right], \quad \text{and} \quad g(x) = \frac{1}{2}\left[F(x) + \frac{1}{c}\int_0^x G(\maltese)d\maltese\right],$$

where $\maltese$ is just an integration variable.

15.2. Starting from (15.15) use (15.14) to prove (15.16).

15.3. Consider two approaches to determining the upper Mach number limit for incompressible flow.

　　a) First consider pressure errors in the simplest possible steady-flow Bernoulli equation. Expand (15.29) for small Mach number to determine the next term in the expansion: $p_0 = p + (1/2)\rho u^2 + \ldots$ and determine the Mach number at which this next term is 5% of p when $\gamma = 1.4$.

　　b) Second consider changes to the density. Expand (15.30) for small Mach number and determine the Mach number at which the density ratio ρ_0 / ρ differs from unity by 5% when $\gamma = 1.4$.

　　c) Which criterion is correct? Explain why the criteria for incompressibility determined in parts a) and b) differ, and reconcile them if you can.

15.4. The critical area A^* of a duct flow was defined in Section 15.4. Show that the relation between A^* and the actual area A at a section, where the Mach number equals M, is that given by (15.31). This relation was not proved in the text. [*Hint*: Write

$$\frac{A}{A^*} = \frac{\rho^* c^*}{\rho u} = \frac{\rho^*}{\rho_0} \frac{\rho_0}{\rho} \frac{c^*}{c} \frac{c}{u} = \frac{\rho^*}{\rho_0} \frac{\rho_0}{\rho} \sqrt{\frac{T^*}{T_0} \frac{T_0}{T}} \frac{1}{M}.$$

Then use the other relations given in Section 15.4.]

15.5. The entropy change across a normal shock is given by (15.43). Show that this reduces to expressions (15.44a and b) for weak shocks. [*Hint*: Let $M_1^2 - 1 \ll 1$. Write the terms within the two sets of brackets in equation (15.43) in the form $[1 + \varepsilon_1][1 + \varepsilon_2]^\gamma$, where ε_1 and ε_2 are small quantities. Then use series expansion $\ln(1 + \varepsilon) = \varepsilon - \varepsilon^2/2 + \varepsilon^3/3 + \ldots$. This gives equation (15.44) times a function of M_1 in which we can set $M_1 = 1$.]

15.6. Show that the maximum velocity generated from a reservoir in which the stagnation temperature equals T_0 is $u_{max} = \sqrt{2C_p T_0}$. What are the corresponding values of T and M?

15.7. In an adiabatic flow of air through a duct, the conditions at two points are $u_1 = 250$ m/s, $T_1 = 300$ K, $p_1 = 200$ kPa, $u_2 = 300$ m/s, and $p_2 = 150$ kPa. Show that the loss of stagnation pressure is nearly 34.2 kPa. What is the entropy increase?

15.8. A shock wave generated by an explosion propagates through a still atmosphere. If the pressure downstream of the shock wave is 700 kPa, estimate the shock speed and the flow velocity downstream of the shock.

15.9. Using dimensional analysis, G. I. Taylor deduced that the radius $R(t)$ of the blast wave from a large explosion would be proportional to $(E/\rho_1)^{1/5} t^{2/5}$ where E is the explosive energy, ρ_1 is the quiescent air density ahead of the blast wave, and t is the time since the blast (see Example 1.4). The goal of this problem is to (approximately) determine the constant of proportionality assuming perfect-gas thermodynamics.

　　a) For the strong shock limit where $M_1^2 \gg 1$, show:

$$\frac{\rho_2}{\rho_1} \cong \frac{\gamma + 1}{\gamma - 1}, \quad \frac{T_2}{T_1} \cong \frac{\gamma - 1}{\gamma + 1} \frac{p_2}{p_1}, \quad \text{and} \quad u_1 = M_1 c_1 \cong \sqrt{\frac{\gamma + 1}{2} \frac{p_2}{\rho_1}}.$$

b) For a perfect gas with internal energy per unit mass e, the internal energy per unit volume is ρe. For a hemispherical blast wave, the volume inside the blast wave will be $(2/3)\pi R^3$. Thus, set $\rho_2 e_2 = E/(2/3)\pi R^3$, determine p_2, set $u_1 = dR/dt$, and integrate the resulting first-order differential equation to show that $R(t) = K(E/\rho_1)^{1/5} t^{2/5}$ when $R(0) = 0$ and K is a constant that depends on "gamma".

c) Evaluate K for $\gamma = 1.4$. A similarity solution of the non-linear gas-dynamic equations in spherical coordinates produces $K = 1.033$ for $\gamma = 1.4$ (see Thompson 1972, p. 501). What is the percentage error in this exercise's approximate analysis?

15.10. Starting from the set (15.45) with $q = 0$, derive (15.47) by letting station (2) be a differential distance downstream of station (1).

15.11. Starting from the set (15.45) with $f = 0$, derive (15.48) by letting station (2) be a differential distance downstream of station (1).

15.12. A wedge has a half-angle of 50°. Moving through air, can it ever have an attached shock? What if the half-angle were 40°? [*Hint*: The argument is based entirely on Figure 15.19.]

15.13. Air at standard atmospheric conditions is flowing over a surface at a Mach number of $M_1 = 2$. At a downstream location, the surface takes a sharp inward turn by an angle of 20°. Find the wave angle σ and the downstream Mach number. Repeat the calculation by using the weak shock assumption and determine its accuracy by comparison with the first method.

15.14. A flat plate is inclined at 10° to an airstream moving at $M_\infty = 2$. If the chord length is $b = 3$ m, find the lift and wave drag per unit span.

15.15. A perfect gas is stored in a large tank at the conditions specified by p_0, T_0. Calculate the maximum mass flow rate that can exhaust through a duct of cross-sectional area A. Assume that A is small enough that during the time of interest p_0 and T_0 do not change significantly and that the flow is adiabatic.

15.16. For flow of a perfect gas entering a constant area duct at Mach number M_1, calculate the maximum admissible values of f, q for the same mass flow rate. Case (a) $f = 0$; case (b) $q = 0$.

15.17. Using thin airfoil theory calculate the lift and drag on the airfoil shape given by $y_u = t \sin(\pi x/c)$ for the upper surface and $y_1 = 0$ for the lower surface. Assume a supersonic stream parallel to the x-axis. The thickness ratio $t/c \ll 1$.

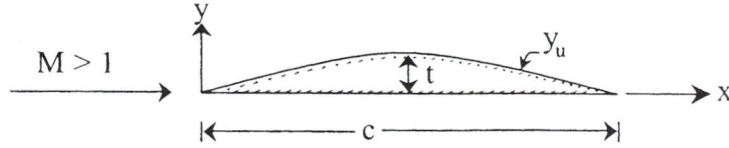

15.18. Write momentum conservation for the volume of the small cylindrical control volume shown in Figure 4.17 where the interface is a shock with flow from side 1 to side 2. Let the two end faces approach each other as the shock thickness $\to 0$ and assume viscous stresses may be neglected on these end faces (outside the structure). Show that the n component of momentum conservation yields (15.36) and the t component gives $\boldsymbol{u} \cdot \boldsymbol{t}$ is conserved or v is continuous across the shock.

Literature Cited

Ames Research Staff (1953). NACA Report 1135: Equations, Tables, and Charts for Compressible Flow.

Becker, R. (1922). Stosswelle und Detonation. *Z. Physik, 8*, 321–362.

Cohen, I. M., & Moraff, C. A. (1971). Viscous inner structure of zero Prandtl number shocks. *Phys. Fluids, 14*, 1279–1280.

Cramer, M. S., & Fry, R. N. (1993). Nozzle flows of dense gases. *The Physics of Fluids A, 5*, 1246–1259.

Fergason, S. H., Ho, T. L., Argrow, B. M., & Emanuel, G. (2001). Theory for producing a single-phase rarefaction shock wave in a shock tube. *Journal of Fluid Mechanics, 445*, 37–54.

Hayes, W. D. (1958). The basic theory of gasdynamic discontinuities, Sect. D of *Fundamentals of Gasdynamics*. In H. W. Emmons (Ed.), Vol. III of *High Speed Aerodynamics and Jet Propulsion*. Princeton, NJ: Princeton University Press.

Liepmann, H. W., & Roshko, A. (1957). *Elements of Gas Dynamics*. New York: Wiley.

Pierce, A. D. (1989). *Acoustics*. New York: Acoustical Society of America.

Shapiro, A. H. (1953). *The Dynamics and Thermodynamics of Compressible Fluid Flow*, 2 volumes. New York: Ronald.

Thompson, P. A. (1972). *Compressible Fluid Dynamics*. New York: McGraw-Hill.

von Karman, T. (1954). *Aerodynamics*. New York: McGraw-Hill.

Supplemental Reading

Courant, R., & Friedrichs, K. O. (1977). *Supersonic Flow and Shock Waves*. New York: Springer-Verlag.

Yahya, S. M. (1982). *Fundamentals of Compressible Flow*. New Delhi: Wiley Eastern.

CHAPTER

16

Introduction to Biofluid Mechanics

Portonovo S. Ayyaswamy

OUTLINE

16.1. Introduction   779

16.2. The Circulatory System in the
      Human Body   780

16.3. Modeling of Flow in Blood Vessels 796

16.4. Introduction to the Fluid
      Mechanics of Plants   844

Exercises   849

Acknowledgment   850

Literature Cited   851

Supplemental Reading   852

CHAPTER OBJECTIVES

- To properly introduce the subject of biofluid mechanics including the necessary language

- To describe the components of the human circulation system and document their nominal characteristics

- To present analytical results of relevant models of steady and pulsatile blood flow

- To review the parametric impact of the properties of rigid, flexible, branched, and curved tubes on blood flow

- To provide an overview of fluid transport in plants.

16.1. INTRODUCTION

This chapter is intended to be of an introductory nature to the vast field of biofluid mechanics. Here, we shall consider the ideas and principles of the preceding chapters in the context of fluid motion in biological systems. Topical emphasis is placed on fluid motion in the human body, and some aspects of the fluid mechanics of plants.

The human body is a complex system that requires materials such as air, water, minerals, and nutrients for survival and function. Upon intake, these materials have to be transported and distributed around the body as required. The associated biotransport and distribution processes involve interactions with membranes, cells, tissues, and organs comprising the body. Subsequent to cellular metabolism in the tissues, waste byproducts have to be transported to the excretory organs for synthesis and removal. In addition to these functions, biotransport systems and processes are required for homeostasis (physiological regulation—for example, maintenance of pH and of body temperature), and for enabling the movement of immune substances to aid in the body's defense and recovery from infection and injury. Furthermore, in certain other specialized systems such as the cochlea in the ear, fluid transport enables hearing and motion sensing. Evidently, in the human body, there are multiple types of fluid dynamic systems that operate at macro-, micro-, nano-, and pico-scales. Systems at the micro and macro levels, for example, include cells (micro), tissue (micro–macro), and organs (macro). Transport at the micro, nano, and pico levels include ion channeling, binding, signaling, endocytosis, and so on. Tissues constitute organs, and organs as systems perform various functions. For example, the cardiovascular system consists of the heart, blood vessels (arteries, arterioles, venules, veins, capillaries), lymphatic vessels, and the lungs. Its function is to provide adequate blood flow and to regulate that flow as required by the various organs of the body. In this chapter, as related to the human body, we shall restrict attention to some aspects of the cardiovascular system for blood circulation.

16.2. THE CIRCULATORY SYSTEM IN THE HUMAN BODY

The primary functions of the cardiovascular system are: 1) to pick up oxygen and nutrients from the lungs and the intestine, respectively, and deliver them to tissues (cells) of the body, 2) to remove waste and carbon dioxide from the body for excretion through the kidneys and the lungs, respectively, and 3) to regulate body temperature by advecting the heat generated and transferring it to the environment outside the skin. The circulatory system in a normal human body (as in all vertebrates and some other select groups of species) can be considered as a closed system, meaning that the blood never leaves the system of blood vessels. The motive mechanism for blood flow is the prevailing pressure gradient.

The circulations associated with the cardiovascular system may be considered under three subsystems. These are the 1) systemic circulation, 2) pulmonary circulation, and 3) coronory circulation (see Figure 16.1). In the systemic circulation, blood flows to all of the tissues in the body except the lungs. Contraction of the left ventricle of the heart pumps oxygen-rich blood to a relatively high pressure and ejects it through the aortic valve into the aorta. Branches from the aorta supply blood to the various organs via systemic arteries and arterioles. These, in turn, carry blood to the capillaries in the tissues of various organs. Oxygen and nutrients are transported by diffusion across the walls of the capillaries to the tissues. Cellular metabolism in the tissues generates carbon dioxide and byproducts (waste). Carbon dioxide dissolves in the blood and waste is carried by the bloodstream. Blood drains into venules and veins. These vessels ultimately empty into two large veins called the *superior vena cava (SVC)* and

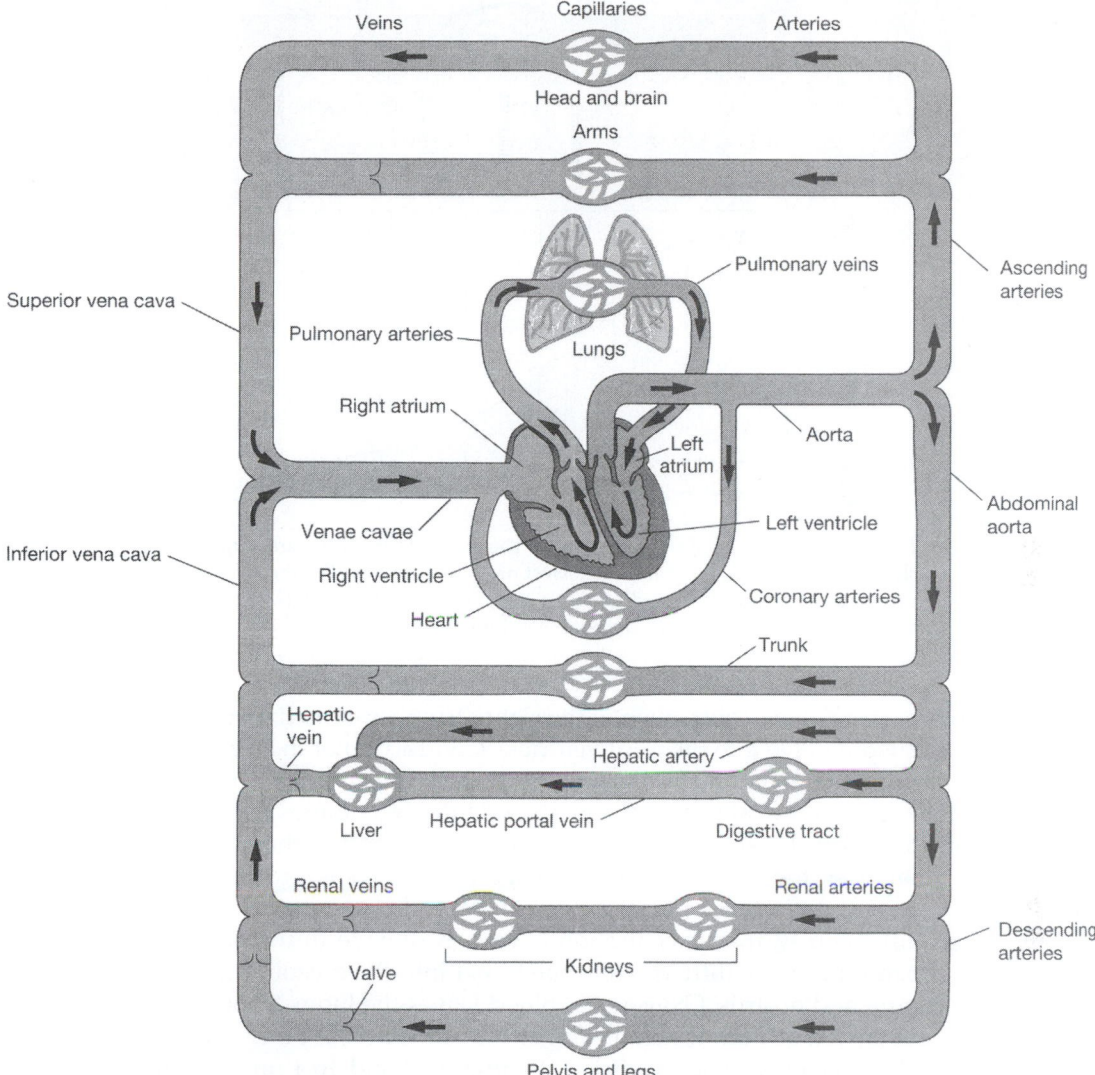

FIGURE 16.1   Schematic of blood flow in systemic and pulmonary circulation showing the major branches. *Reproduced with permission from Silverthorn, D. U. (2001), Human Physiology: An Integrated Approach, 2nd ed., Prentice Hall, Upper Saddle River, NJ.*

inferior vena cava (IVC) that return carbon dioxide–rich blood to the right atrium. The mean blood pressure of the systemic circulation ranges from a high of 93 mm Hg in the arteries to a low of a few mm Hg in the venae cavae. Figure 16.2 shows that pressure falls continuously as blood moves farther from the heart. The highest pressure in the vessels of the circulatory system is in the aorta and in the systemic arteries while the lowest pressure is in the venae cavae.

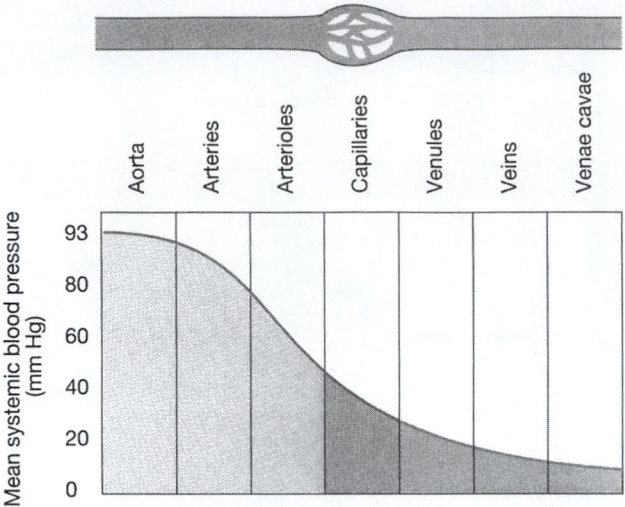

FIGURE 16.2 Pressure gradient in the blood vessels. The highest pressures are found in the aorta which conveys oxygen-rich blood away from the heart. The lowest pressures are found in the largest veins which convey oxygen-poor blood toward the heart. *Reproduced with permission from Silverthorn, D. U. (2001),* Human Physiology: An Integrated Approach, *2nd ed., Prentice Hall, Upper Saddle River, NJ.*

In pulmonary circulation, contraction of the right atrium ejects carbon dioxide—rich blood through the tricuspid valve into the right ventricle. Contraction of the right ventricle pumps the blood through the pulmonic valve (also called *semilunar valve*) into the pulmonary arteries. These arteries bifurcate and transport blood into the complex network of pulmonary capillaries in the lungs. These capillaries lie between and around the alveoli walls. During respiratory inhalation, the concentration of oxygen in the air is greater in the air sacs of the alveolar region than in the capillary blood. Oxygen diffuses across capillary walls into the blood. Simultaneously, the concentration of carbon dioxide in the blood is higher than in the air and carbon dioxide diffuses from the blood into the alveoli. Carbon dioxide exits through the mouth and nostrils. Oxygenated blood leaves the lungs through the pulmonary veins and enters the left atrium. When the left atrium contracts, it pumps blood through the bicuspid (mitral) valve into the left ventricle. Figures 16.3 and 16.4 provide an overview of external and cellular respiration and the branching of the airways, respectively.

Blood is pumped through the systemic and pulmonary circulations at a rate of about 5.2 liters per minute under normal conditions. The systemic and pulmonary circulations described above constitute one cardiac cycle. The cardiac cycle denotes any one or all of such events related to the flow of blood that occur from the beginning of one heartbeat to the beginning of the next. Throughout the cardiac cycle, the blood pressure increases and decreases. The frequency of the cardiac cycle is the heart rate. The cardiac cycle is controlled by a portion of the autonomic nervous system (that part of the nervous system which does not require the brain's involvement in order to function).

In coronary circulation, blood is supplied to and from the heart muscle itself. The muscle tissue of the heart, or myocardium, is thick and it requires coronary blood vessels

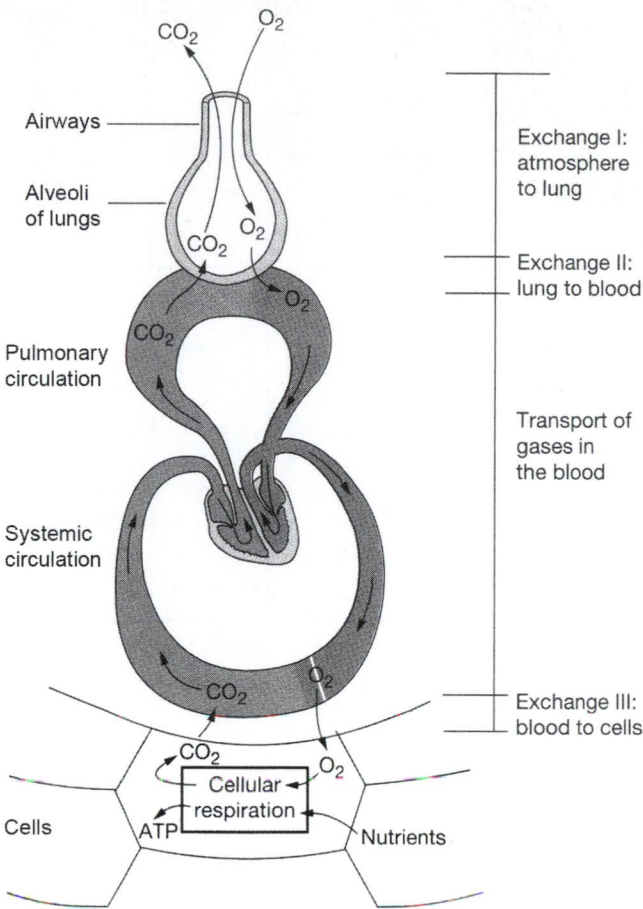

Airways

Alveoli
of lungs

Pulmonary
circulation

Systemic
circulation

Cells

Exchange I:
atmosphere
to lung

Exchange II:
lung to blood

Transport of
gases in
the blood

Exchange III:
blood to cells

FIGURE 16.3 Overview of external and cellular respiration. Cells collect oxygen and nutrients from the stream blood and discard carbon dioxide and wastes into the bloodstream. *Reproduced with permission from Silverthorn, D. U. (2001),* Human Physiology: An Integrated Approach, *2nd ed., Prentice Hall, Upper Saddle River, NJ.*

to deliver blood deep into the myocardium. The vessels that supply blood with a high concentration of oxygen to the myocardium are known as *coronary arteries*. The main coronary artery arises from the root of the aorta and branches into the left and right coronary arteries. Up to about seventy-five percent of the coronary blood supply goes to the left coronary artery, with the remainder going to the right coronary artery. Blood flows through the capillaries of the heart and returns through the cardiac veins which remove the deoxygenated blood from the heart muscle. The coronary arteries that run on the surface of the heart are relatively narrow vessels and are commonly affected by atherosclerosis and can become blocked, causing angina or a heart attack. The coronary arteries are classified as *end circulation*, since they represent the only source of blood supply to the myocardium.

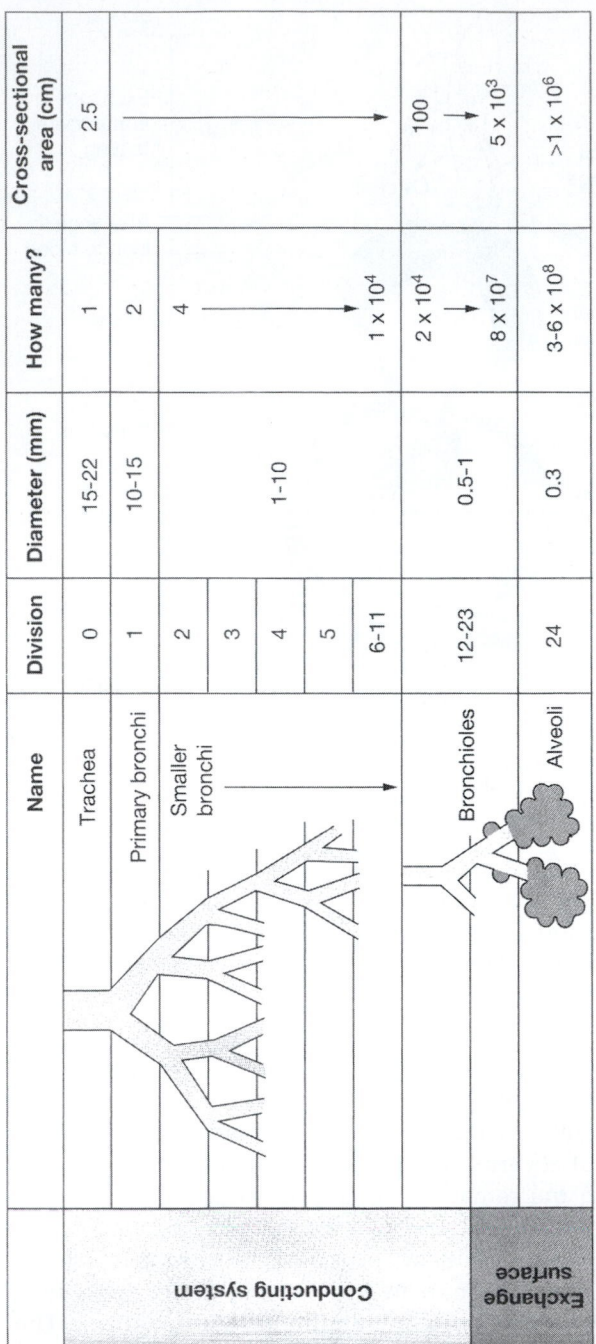

FIGURE 16.4 Branching of the airways in the human lungs. Areas have units of cm². *Reproduced with permission from Silverthorn, D. U. (2001), Human Physiology: An Integrated Approach, 2nd ed., Prentice Hall, Upper Saddle River, NJ.*

The Heart as a Pump

The heart has four pumping chambers—two atria (upper) and two ventricles (lower). The left and right parts of the heart are separated by a muscle called the *septum* which keeps the blood volumes in each part separate. The upper chambers interact with the lower chambers via the heart valves. The heart has four valves which ensure that blood flows only in the desired direction. The atrio-ventricular valves (AV) consist of the tricuspid (three flaps) valve between the right atrium and the right ventricle, and the bicuspid (two flaps, also called the *mitral*) valve between the left atrium and the left ventricle. The pulmonary valve is between the right ventricle and the pulmonary artery, and the aortic valve is between the left ventricle and the aorta. Both the pulmonary and aortic valves have three symmetrical half-moon-shaped valve flaps (cusps), and are called the *semilunar valves*. The function of the four chambers in the heart is to pump blood through pulmonary and systemic circulations. The atria receive blood from the veins—the right atrium receives carbon dioxide–rich blood from the SVC and IVC, and the left atrium receives oxygen-rich blood from the pulmonary veins. The heart is controlled by a single electrical impulse and both sides of the heart act synchronously. Electrical activity stimulates the heart muscle (myocardium) of the chambers of the heart to make them contract. This is immediately followed by mechanical contraction of the heart. Both atria contract at the same time. The contraction of the atria moves the blood from the upper chambers through the valves into the ventricles. The atrial muscles are electrically separated from the ventricular muscles except for one pathway through which an electrical impulse is conducted from the atria to the ventricles. The impulse reaching the ventricles is delayed by about 110 ms while the conduction occurs through the pathway. This delay allows the ventricles to be filled before they contract. The left ventricle is a high-pressure pump and its contraction supplies systemic circulation while the right ventricle is a low-pressure pump supplying pulmonary circulation (lungs offer much less resistance to flow than systemic organs).

From the above discussions, we see that the pumping action of the heart can be regarded as a two-step process—a contraction step (systole) and a filling (relaxation) step (diastole). Systole describes that portion of the heartbeat during which contraction of the heart muscle and hence ejection of blood takes place. A single *beat* of the heart involves three operations: atrial systole, ventricular systole, and complete cardiac diastole. Atrial systole is the contraction of the heart muscle of the left and right atria, and occurs over a period of 0.1 s. As the atria contract, the blood pressure in each atrium increases, which forces the mitral and tricuspid valves to open, forcing blood into the ventricles. The AV valves remain open during atrial systole. Following atrial systole, ventricular systole, which is the contraction of the muscles of the left and right ventricles, occurs over a period of 0.3 s. The ventricular systole generates enough pressure to force the AV valves to close, and the aortic and pulmonic valves open. (The aortic and pulmonic valves are always closed except for the short period of ventricular systole when the pressure in the ventricle rises above the pressure in the aorta for the left ventricle and above the pressure in the pulmonary artery for the right ventricle.) During systole, the typical pressures in the aorta and the pulmonary artery rise to 120 mm Hg and 24 mm Hg, respectively (1 mm Hg = 133 Pa). In normal adults, blood flow through the aortic valve begins at the start of ventricular systole and rapidly accelerates to a peak value of approximately 1.35 m/s during the first one-third of systole. Thereafter, the blood flow begins to decelerate. Pulmonic valve peak velocities are lower, and in normal adults they are about 0.75 m/s. Contraction of the ventricles in systole ejects about two-thirds of the blood from these chambers. As the left ventricle empties, its pressure falls below the pressure in the aorta, and the aortic valve closes. Similarly, as the pressure in the right ventricle falls below the pressure in the pulmonary artery, the pulmonic valve closes. Thus, at the end of the ventricular systole, the aortic and pulmonic valves close, with the aortic valve closing a little earlier than the pulmonic valve. Diastole describes that portion of the heartbeat during which the chamber refilling takes place. The cardiac diastole is the period of time when the heart relaxes after contraction in preparation for refilling with circulating blood. The ventricles refill or ventricular diastole occurs during atrial systole. When the ventricle is filled and ventricular systole begins, then the AV valves are closed and the atria begin refilling with blood, or atrial diastole occurs. About a period of 0.4 s following ventricular systole, both the atria and the ventricles begin refilling and both chambers are in diastole. During this period, both AV valves are open and aortic and pulmonic valves are closed. The typical diastolic pressure in the aorta is 80 mm Hg, and in the pulmonary artery it is 8 mm Hg. Thus, the typical systolic and diastolic pressure ratios are 120/80 mm Hg for the aorta and 24/8 mm Hg for the pulmonary artery. The systolic pressure minus the diastolic pressure is called the *pressure pulse*, and for the aorta (left ventricle) it is 40 mm Hg. The pulse pressure is a measure of the strength of the pressure wave. It increases with increased stroke volume (say, due to activity or exercise). Pressure waves created by the ventricular contraction diminish in amplitude with the distance from the heart and are not perceptible in the capillaries. Figure 16.5 shows the pressure throughout the systemic circulation.

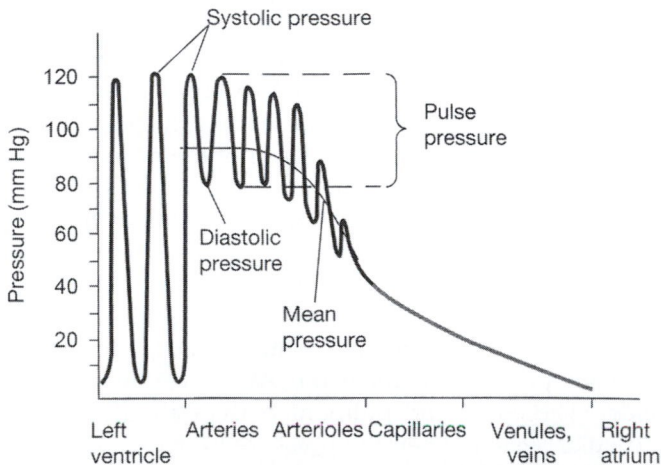

FIGURE 16.5 Pressure variations throughout the systemic circulation. The largest pressure fluctuations occur in the left ventricle. These are gradually damped out by the flexibility of the arteries, blood viscosity, and the branched nature of the system. *Reproduced with permission from Silverthorn, D. U. (2001),* Human Physiology: An Integrated Approach, *2nd ed., Prentice Hall, Upper Saddle River, NJ.*

Net Work Done by the Ventricle on the Blood During One Cardiac Cycle

The work done by the ventricle on blood may be calculated from the area enclosed by the pressure–volume curve for the ventricle. Consider, for example, the left ventricle (LV). Figure 16.6 shows the pressure–volume curve for the LV. Blood pressure is measured in mm of Hg, and the volume in mL. At A, the ventricular pressure and volume are at their lowest values. With the increase of atrial pressure, the bicuspid valve will open and let blood flow into the ventricle. AB represents diastolic ventricular filling. During AB work is being done by the blood in the LV to increase the volume. At B, the ventricular volume is filled to its maximum and this volume is called the *end diastolic volume (EDV)*. The ventricular muscles begin to contract, pressure increases, and the bicuspid valve closes. BC is the constant-volume contraction of the ventricle. No work is done during BC but energy is stored as elastic energy in the muscles. At C, ventricular pressure is greater than that in the aorta, the aortic valve opens, and blood is ejected into the aorta. Ventricular volume decreases, but the ventricle continues to contract and the pressure increases. However, at D, pressure in the aorta exceeds that of the ventricular pressure and the aortic valve closes. During CD, work is done by the heart muscles on blood. The volume in the LV at D is at its lowest value,

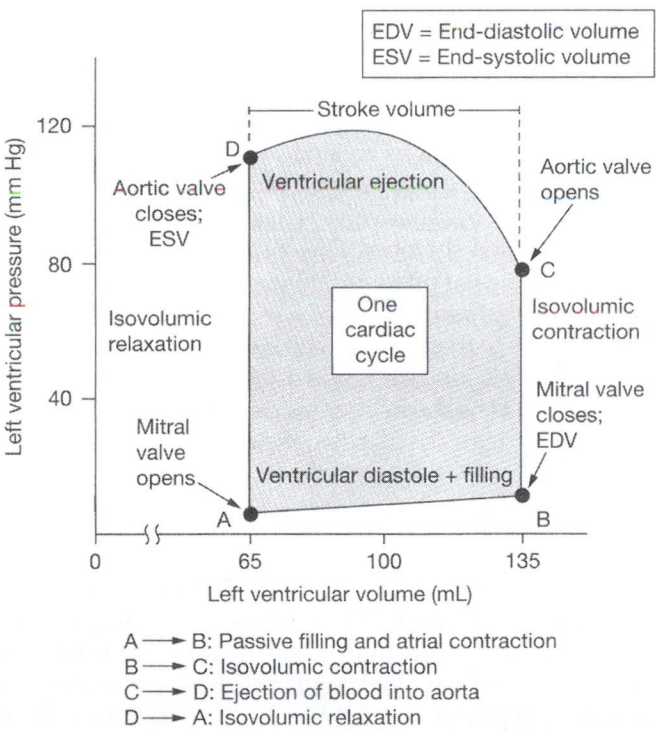

FIGURE 16.6   Left ventricular pressure–volume curve for one cardiac cycle. The work done by the left ventricle is the shaded area. The cardiac cycle follows the edge of the shaded area in the counterclockwise direction. *Reproduced with permission from Silverthorn, D. U. (2001),* Human Physiology: An Integrated Approach, *2nd ed., Prentice Hall, Upper Saddle River, NJ.*

and this is called the *end systolic volume* (*ESV*). DA is the constant-volume pressure decrease in the ventricle due to muscle relaxation and no work is done during this process. Ventricular pressure falls below that in the aorta causing the aortic valve to close. ABCD constitutes one cardiac cycle, and the area within the pressure-volume diagram represents the net work done by the LV on blood. The energy required to perform this work is derived from the oxygen in the blood. A similar development applies for the right ventricle.

Typically, the work done by the heart is only about 10–15% of the total input energy. The remainder is dissipated as heat.

The volume of blood pumped by the LV into the systemic circulation in a cardiac cycle is called the *stroke volume* (*SV*), and it is expressed in mL/beat. The normal stroke volume is 70 mL/beat.

$$SV = EDV - ESV \tag{16.1}$$

A parameter that is related to stroke volume is ejection fraction (*EF*). *EF* is the fraction of blood ejected by the LV during systole. At the start of systole, the LV is filled with blood to the *EDV*. During systole, the LV contracts and ejects blood until it reaches *ESV*. *EF* is given by

$$EF = (SV/EDV) \times 100\%. \tag{16.2}$$

Cardiac output (*CO*) is the volume of blood being pumped by the heart (in particular, by a ventricle) in a minute. It is the time-averaged flow rate. It is equal to the heart rate multiplied by the stroke volume. Thus,

$$CO = SV \times HR, \tag{16.3}$$

where *HR* is the heart rate in beats/min. For a normal adult, the typical *HR* is between 70 and 75 beats per minute. With 70 beats per minute, and 70 mL blood ejection with each beat of the heart, the *CO* is 4900 mL/m. This value is typical for a normal adult at rest, although *CO* may reach up to 30 L/m during extreme activity (say, exercise). Heart rate can vary by a factor of approximately 3, between 60 and 180 beats per minute, while the stroke volume can vary between 70 and 120 mL, a factor of only 1.7. The cardiac index (*CI*) relates *CO* with the body surface area, *BSA*, as given by:

$$CI = CO/BSA = SV \times HR/BSA, \tag{16.4}$$

where *BSA* is in square meters.

Nature of Blood

Blood is about 7% of the human body weight. Its density is approximately 1054 kg/m^3. The pH of normal blood is in the range 7.35 < pH < 7.45. The normal adult has a blood volume of about 5 liters. At any given time, about 13% of the total blood volume resides in the arteries and about 7% resides in the capillaries. Blood is a complex circulating liquid tissue consisting of several types of formed elements (corpuscles or cells; about 45% by volume) suspended in a fluid medium known as *plasma* (about 55% by volume; 2.7–3.0 liters in a normal human). The plasma is a dilute electrolyte solution (almost 92% water) containing, about 8% by weight, three major types of blood proteins—fibrinogen (5%), globulin (45%), and albumin (50%) in water. Beta lipoprotein and lipalbumin are also present in trace amounts. Plasma proteins are large molecules with high molecular weight and do not pass through the capillary wall. The formed elements (cells) consist of red blood cells (erythrocytes; about 45% of blood volume), white blood cells (leukocytes; about 1% of blood volume), and platelets (thrombocytes; <1% of blood volume). Thus, the formed elements in blood consist of 95% red blood cells, 0.13% white blood cells, and about 4.9% platelets. The specific gravity of red blood cells is about 1.06. The white blood cells further consist of monocytes, lymphocytes, neutrophils, eosinophils, and basophils.

In humans, mature red blood cells lack a nucleus and organelles. They are produced in the bone marrow, and the cell life span is about 125 days. The red blood cell is biconcave in shape. It consists of a concentrated solution of hemoglobin, an oxygen-carrying protein, surrounded by a flexible membrane. The hemoglobin transports oxygen (and some carbon dioxide) from the lungs to capillaries in various tissues. The cell is about 8.5 μm in diameter with transverse dimensions of 2.5 μm at the thickest portion and about 1 μm at the thinnest portion. However, its flexibility is such that it can bend and pass through capillaries as small as 5 μm in diameter. The surface area of the cell is about 163 $(\mu m)^2$, and the intracellular fluid volume is about 87 $(\mu m)^3$. There are approximately 5×10^6 red blood cells in each mm^3 of blood. The biconcave shape of the cell provides it with a very large ratio of surface area to volume. This enables efficient gas exchange in the capillaries. The percentage of blood volume made up by red blood cells is referred to as the *hematocrit*. Hematocrit ranges from 42 to 45 in normal blood, and plays a major role in determining the rheological properties of blood. White blood cells, or leukocytes, are cells of the immune system which defend the body against infectious disease and foreign materials. Several different and diverse types of leukocytes exist and they are all produced in the bone marrow. There are normally about 10^4 white blood cells in each mm^3 of blood. Platelets or thrombocytes are cell fragments circulating in blood that are involved in the cellular mechanisms of hemostasis leading to the formation of blood clots. They are smaller in size than red or white blood cells. Low levels of platelets predisposes to a person bleeding, while high levels increase the risk of thrombosis (coagulation of blood in the heart or a blood vessel).

Blood is a non-Newtonian fluid. Its viscosity depends on the viscosity of the plasma, its protein content, the hematocrit, the temperature, the shear rate (also called the *rate of shearing strain*), and the narrowness of the vessel in which it is flowing (for example, a narrow diameter capillary). The dependence on the narrowness of the vessel diameter is called the *Fahraeus-Lindqvist effect*. The presence of white cells and platelets does not significantly affect the viscosity since they are such a small fraction of the formed elements. We will briefly discuss the various dependencies of blood viscosity.

The viscosity of plasma and blood are often given in terms of relative viscosity as compared to that of water (viscosity of water is about 0.8 centipoise at 30°C; 1 centipoise (1 cP) = 0.01 Poise = 1 dyne s/cm^2 = 0.1 N s/m^2). The viscosity of plasma depends on its protein content and ranges between 1.1 and 1.6 centipoise. The viscosity of whole blood at a physiological hematocrit of 45% is about 3.2 cP. Higher hematocrit results in higher viscosity. At a hematocrit of 60%, the relative viscosity of blood is about 8. Blood viscosity increases with decreasing temperature, and the increase is approximately 2% for each °C. The dependence of viscosity on flow rate in vessels is complicated. As noted in earlier chapters, flow rates through tubes are significantly influenced by the shear stress, τ, and the associated rate of shearing strain (or shear rate), $\dot{\gamma}$. For Newtonian fluids, τ is linearly related to $\dot{\gamma}$.

For example, $\tau = \mu\dot{\gamma}$ and the slope of this characteristic is the viscosity, μ. For whole blood, this relationship between τ and $\dot{\gamma}$ is complicated for the following reasons. In a blood volume at rest, above a minimum hematocrit of about 5–8%, blood cells form a continuous structure. A finite stress (called the *yield stress*), τ_y, is required to break this continuous structure into a suspension of aggregates in the plasma. This yield stress also depends on the concentration of plasma proteins, in particular, fibrinogen. An empirical correlation for the yield stress is given by the expression:

$$\sqrt{\tau_y} = (H - 0.1)(C_F + 0.5), \tag{16.5}$$

where H is the hematocrit expressed as a fraction and it is > 0.1, and C_F is the fibrinogen content in grams per 100 mL and $0.21 < C_F < 0.46$. For 45% hematocrit blood, the yield stress is in the range $0.01 < \tau_y < 0.06$ dyne/cm^2 (1 dyne/cm^2 = 0.1 N/m^2). Beyond the yield stress, when sheared in the bulk, up to about $\dot{\gamma} < 50$ sec^{-1}, the aggregates in blood break into smaller units called *rouleaux formations*. For shear rates up to about 200 sec^{-1}, the rouleaux progressively break into individual cells. Beyond this, no further reduction in structure is noted to occur with an increase in the shearing rate.

For whole blood, at low shear rates, $\dot{\gamma} < 200$ sec^{-1}, the variation of τ with $\dot{\gamma}$ is noted to be nonlinear. This behavior at low $\dot{\gamma}$ is non-Newtonian. Low $\dot{\gamma}$ values are associated with flows in small arteries and capillaries (microcirculation). At higher shear rates, $\dot{\gamma} > 200$ sec^{-1}, the relationship between τ and $\dot{\gamma}$ is linear, and the viscosity approaches an asymptotic value of about 3.5 cP. Blood flows in large arteries have such high shear rates, and the viscosity in such cases may be assumed as constant and equal to 3.5 cP. Since whole blood behaves like a non-Newtonian yield stress fluid, the slope of the shear stress—rate of strain characteristic at any given point on the curve—is defined as the apparent viscosity of blood at that point, μ_{app}. Clearly, μ_{app} is not a constant but depends on the prevailing $\dot{\gamma}$ at that point (see Figure 16.7). There are a number of constitutive equations available in the literature that attempt to model the relationship between shear stress and shear rate of flowing blood. A commonly used one is called the *Casson model* and it is expressed as follows:

$$\sqrt{\frac{\tau}{\mu_p}} = k_c\sqrt{\dot{\gamma}} + \sqrt{\frac{\tau_y}{\mu_p}}, \tag{16.6}$$

where μ_p is plasma viscosity and k_c is the Casson viscosity coefficient (a dimensionless number). An expression based on a least squares fit of the experimental data and expressed in Casson form is that of Whitmore (1968):

$$\sqrt{\frac{\tau}{\mu_p}} = 1.53\sqrt{\dot{\gamma}} + 2.0. \tag{16.7}$$

This expression is plotted in Figure 16.8. Apparent viscosity significantly increases at low rates of shear. It must be noted that although the Casson model is suitable at low shear rates, it still assumes that blood can be modelled as a homogeneous fluid.

In blood vessels of less than about 500 μm in diameter, the inhomogeneous nature of blood starts to have an effect on the apparent viscosity. This feature will be discussed next.

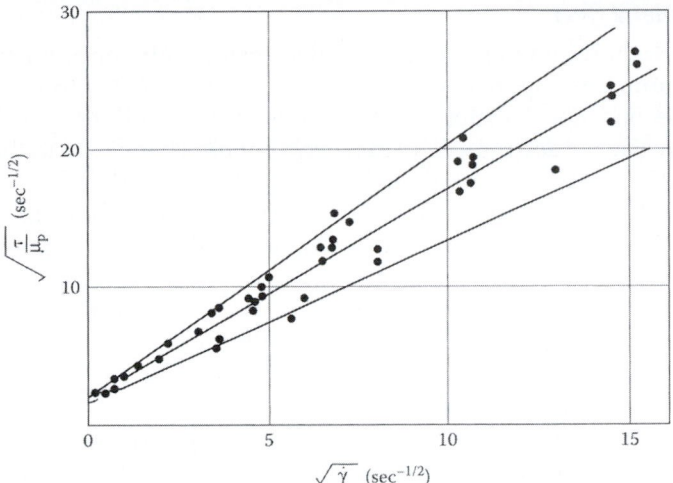

FIGURE 16.7 Shear stress versus shear rate for blood flow. Note that the shear stress is finite at small (approximately zero) shear rates. Blood is a non-Newtonian fluid. *Reproduced with permission from Whitmore, R. L. (1968),* Rheology of Circulation, *Pergamon Press, New York.*

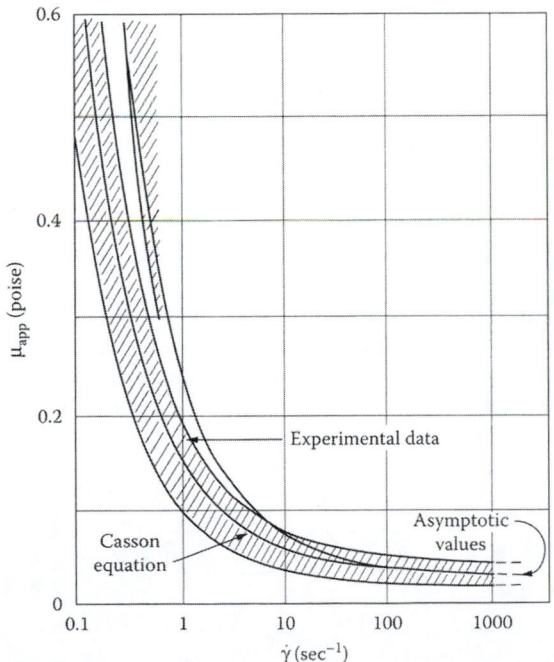

FIGURE 16.8 A least square fit of apparent viscosity as a function of shear rate in Casson form. The apparent viscosity of blood falls with increasing shear rate making it a shear-thinning fluid. *Reproduced with permission from Whitmore, R. L. (1968),* Rheology of Circulation, *Pergamon Press, New York.*

Fahraeus-Lindqvist Effect

When blood flows through narrow tubes of decreasing radii, approximately in the range 15 μm $< d <$ 500 μm, the apparent viscosity, μ_{app}, decreases with decreasing radius of the vessel. This is a second non-Newtonian characteristic of blood and is called the *Fahraeus-Lindqvist (FL) effect*. The reduced viscosity in narrow tubes is beneficial to the pumping action of the heart.

The basis for the FL effect is the Fahraeus effect. When blood of constant hematocrit (feed hematocrit or bulk hematocrit, H_F) flows from a large vessel into a small vessel (vessel sizes in the ranges cited earlier), the hematocrit in the small vessel (dynamic or tube hematocrit, H_T) decreases as the tube diameter decreases (see Figure 16.9). This phenomenon is called the *Fahraeus effect* and must not be confused with a diminution of particle concentration in the smaller vessel because of an entrance effect whereby particle entry into the smaller vessel is hindered (see Goldsmith et al., 1989, for detailed discussions). To separate such an entry-screening effect and confirm the Fahraeus effect, H_T may be compared with the hematocrit in the blood flowing out (discharge hematocrit, H_D) from the smaller tube into a discharge vessel of comparable size to the feed vessel. In the steady state, $H_F = H_D$. In vivo and in vitro experiments show that $H_T < H_D$ in tubes up to about 15 μm in diameter. The H_T/H_D ratio decreases from about 1 to about 0.46 as the capillary diameter decreases from about 600 μm to about 15 μm. While the discharge hematocrit value may be 45%, the corresponding dynamic hematocrit in a narrow-sized vessel such as an arteriole may be just 20%. As a consequence, the apparent viscosity decreases in the diameter range 15 μm $< d <$ 500 μm. However, for tubes less than about 15 μm in diameter, the ratio H_T/H_D starts to increase.

Why does the hematocrit decrease in small blood vessels? The reason for this effect is not fully understood at this time. In blood vessel flow, there seems to be a tendency for the red cells to move toward the axis of the tube, leaving a layer of plasma, whose width, usually designated by δ, increases with increasing shear rate. This tendency to move away from the wall is not observed with rigid particles; thus, the deformability of the red cell appears to be the reason for lateral migration. Deformable particles are noted to experience a net radial hydrodynamic force even at low Reynolds numbers and tend to migrate toward the tube axis (see Fung, 1993, for detailed discussions). Chandran et al. (2007) state that as the blood flows through a tube, the blood cells (with their deformable biconcave shape) rotate (spin) in the shear field. Due to this spinning, they tend to move away from the wall and toward the center of the tube. The cell-free plasma layer reduces the tube hematocrit. As the size of the vessel gets smaller, the fraction of the volume occupied by the cell-free layer increases, and the tube

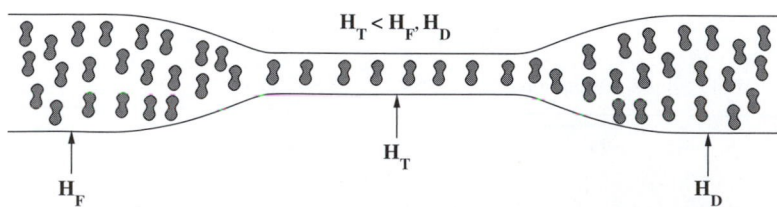

FIGURE 16.9 The Fahraeus effect. Here the hematocrit falls as blood moves from larger to smaller vessels because of non-Newtonian effects, the particulate nature of blood cells, and other factors.

hematocrit is further lowered. A numerical validation of this reasoning is available in a paper by Liu and Liu (2006). There is yet another reason. Blood vessels have many smaller-sized branches. If a branching daughter vessel is so located that it draws blood from the larger parent vessel, mainly from the cell-free layer, then the hematocrit in the branch will end up being lower. This is called *plasma skimming*. In all these circumstances, the tube hematocrit is lowered. The viscosity of blood at the core may be higher due to a higher core hematocrit, H_c, there, but the overall apparent viscosity in the tube flow is lower.

As the tube diameter becomes less than about 6 μm, the apparent viscosity increases dramatically. The erythrocyte is about 8 μm in diameter and can enter tubes somewhat smaller in size, and a tube of about 2.7 microns is about the smallest size that an RBC can enter (Fournier, 2007; Fung, 1993). When the tube diameter becomes very small, the pressure drop associated with the flow increases greatly and there is increase in apparent viscosity.

If we consider laminar blood flow in straight, horizontal, circular, feed, and capillary tubes, a number of straightforward relationships among Q_F, Q_c, Q_p, H_F, H_T, H_c, δ, and a may be established based on the law of conservation of blood cells. Here, Q denotes flow rate, subscripts c and p denote core and plasma regions, respectively, and a is the radius of the capillary tube. Thus,

$$Q_F H_F = Q_c H_c, \quad Q_c + Q_p = Q_F, \quad \text{and} \quad H_T a^2 = H_c (a - \delta)^2, \tag{16.8}$$

where a is the radius of the capillary tube. Equation (16.8) will be useful in modeling the FL phenomenon. A simple mathematical model for the FL effect is included in a subsequent section.

Nature of Blood Vessels

All blood vessels other than capillaries are usually composed of three layers: the tunica intima, tunica media, and tunica adventitia. The tunica intima consists of a layer of endothelial cells lining the lumen of the vessel (the hollow internal cavity in which the blood flows), as well as a subendothelial layer made up of mostly loose connective tissue. The endothelial cells are in direct contact with the blood flow. An internal elastic lamina often separates the tunica intima from the tunica media. The tunica media is composed chiefly of circumferentially arranged smooth muscle cells. Again, an external elastic lamina often separates the tunica media from the tunica adventitia. The tunica adventitia is primarily composed of loose connective tissue made up of fibroblasts and associated collagen fibers. In the largest arteries, such as the aorta, the amount of elastic tissue is considerable. Veins have the same three layers as arteries, but boundaries are indistinct, walls are thinner, and elastic components are not as well developed.

Blood flows under high pressure in the aorta (about 120 mm Hg systolic, 80 mm Hg diastolic, pressure pulse of 40 mm Hg at the root) and the major arteries. These vessels have strong walls. The aorta is an elastic artery, about 25 mm in diameter with a wall thickness of about 2 mm, and is quite distensible. During left ventricular systole (about one-third of the cardiac cycle), the aorta expands. This stretching provides the potential energy that will help maintain blood pressure during diastole. During the diastole (about two-thirds of the cardiac cycle), the pressure pulse decays exponentially and the aorta contracts passively. Medium arteries are about 4 mm in diameter with a wall thickness of about 1 mm. Arterioles are about 50 μm in diameter and have thin muscular walls (usually only one to two layers of smooth muscle) of about 20 μm thickness. Their vascular tone is controlled by regulatory mechanisms, and they constrict or relax as needed to maintain blood pressure. Arterioles are the primary site of vascular resistance and blood-flow distribution to various regions is controlled by changes in resistance offered by various arterioles. True capillaries average from 9 to 12 μm in diameter, just large enough to permit passage of cellular components of blood. The thin wall consists of extremely attenuated endothelial cells. In cross section, the lumen of small capillaries may be encircled by a single endothelial cell, while larger capillaries may be made up of portions of 2 or 3 cells. No smooth muscle is present. Venules are about 20 μm in diameter and allow deoxygenated blood to return from the capillary beds to the larger veins. They have three layers: an inner endothelium layer which acts as a membrane, a middle layer of muscle and elastic tissue, and an outer layer of fibrous connective tissue. The middle layer is poorly developed. The walls of venules are about 2 μm in thickness, and thus are very much thinner than those of arterioles. Veins are thin-walled, distensible, and collapsible tubes. Some of them may be collapsed in normal function. They transport blood at a lower pressure than the arteries. They are about 5 mm in diameter and the wall thickness is about 500 μm. They are surrounded by helical bands of smooth muscles which help maintain blood flow to the right atrium. Most veins have one-way flaps called *venous valves*. These valves prevent gravity from causing blood to flow back and collect in the lower extremities. Veins more distal to the heart have more valves. Pulmonary veins and the smallest venules have no valves. Veins also have a thick collagen outer layer, which helps maintain blood pressure. In the venous system, a large increase in the blood volume results in a relatively small increase in pressure compared to the arterial system (see Chandran et al., 2007). The veins act as the main reservoir for blood in the circulatory system and the total capacity of the veins is more than sufficient to hold the entire blood volume of the body. This capacity is reduced through the constriction of smooth muscles, minimizing the cross-sectional area (and hence volume) of the individual veins and therefore the total venous system. The superior vena cava is a large, yet short vein that carries deoxygenated blood from the upper half of the body to the heart's right atrium. The inferior vena cava is the large vein that carries deoxygenated blood from the lower half of the body into the heart. The vena cava is about 30 mm in diameter with a wall thickness of about 1.5 mm. The venae cavae have no valves. Figure 16.10 shows the cross-sectional areas of different parts of the systemic circulation with velocity of blood flow in each part. The fastest flow is in the arterial system. The slowest flow is in the capillaries and venules.

As stated earlier, arterioles are the primary site of vascular flow resistance, and blood-flow distribution to various regions is controlled by changes in resistance offered by various arterioles. To quantify the resistance of the arterioles in an averaged sense, the concept of *total peripheral resistance* is introduced. Total peripheral resistance essentially refers to the cumulative resistance of the thousands of arterioles involved in the systemic or pulmonary circulation, respectively. For systemic circulation, with time averaging of quantities over a cardiac cycle:

$$\text{Total Peripheral Resistance} = R = \frac{\Delta \bar{p}}{Q},$$

(16.9)

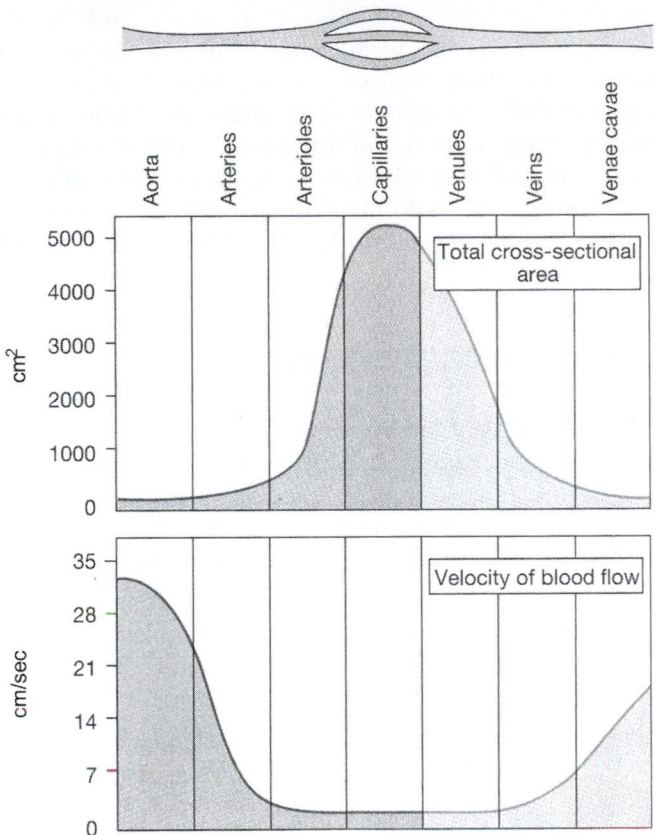

FIGURE 16.10   Vessel diameter, total cross-sectional area, and velocity of flow. The total cross-sectional area available for flow is largest at the capillary size because there are so many. The highest blood-flow speeds are found in the largest arteries and veins. *Reproduced with permission from Silverthorn, D. U. (2001),* Human Physiology: An Integrated Approach, *2nd ed., Prentice Hall, Upper Saddle River, NJ.*

where R denotes resistance, $\Delta \bar{p}$ is the difference between the time-averaged pressure at the aortic valve and the time-averaged venous pressure at the right atrium, and Q is the time-averaged flow rate (cardiac output). The units of peripheral resistance would therefore be in mm Hg per cm^3/s. This unit of measuring resistance is called the *peripheral resistance unit (PRU)*. Letting $\bar{p}_A$ and $\bar{p}_V$ denote the time-averaged pressures at the aortic valve and at the right atrium, respectively:

$$\Delta p = \bar{p}_A - \bar{p}_V, \tag{16.10}$$

and, with $\bar{p}_V = 0$, $\Delta \bar{p} = \bar{p}_A$, the time-averaged arterial pressure. Then, $\bar{p}_A = QR$. The average pressure, $\bar{p}_A$, may be estimated as:

$$\bar{p}_A = \frac{1}{3}p_S + \frac{2}{3}p_D = p_D + \frac{1}{3}(p_S - p_D), \tag{16.11}$$

where p_s is the systolic pressure, p_D is the diastolic pressure, and $(p_s - p_D)$ is the pressure pulse (see Kleinstreuer, 2006). For a normal person at rest, with $\bar{p}_A = 100$ mmHg and $Q = 86.6$ cm^3/s, then $R = 1.2$ PRU. An expression similar to that in (16.9) would apply for pulmonary circulation and would involve the difference between time-averaged pressures at the pulmonary artery and at the left atrium, and the flow rate in pulmonary circulation (same as that in systemic circulation). Since the difference between time-averaged pressures in pulmonary circulation is about an order of magnitude smaller than in the systemic circulation, the corresponding PRU would be an order of magnitude smaller.

16.3. MODELING OF FLOW IN BLOOD VESSELS

There are approximately 100,000 km of blood vessels in the adult human body (Brown et al., 1999). In this section, we examine several models for describing blood flow in some important vessels.

Blood flow in the circulatory system is in general unsteady. In most regions it is pulsatile due to the systolic and diastolic pumping. In pulsatile flow, the flow has a periodic behavior and a net directional motion over the cycle. Pressure and velocity profiles vary periodically with time, over the duration of a cardiac cycle. A dimensionless parameter called the *Womersley number*, α, is used to characterize the pulsatile nature of blood flow, and it is defined by:

$$\alpha = a\sqrt{\frac{\omega}{\nu}},$$ (16.12)

where a is the radius of the tube, ω is the frequency of the pulse wave (heart rate expressed in radians/sec), and ν is the kinematic viscosity. This definition shows that the Womersley number is a composite parameter of the Reynolds number, $Re = 2au/\nu$, and the Strouhal number, $St = 2a\omega/u$. The square of the Womersley number is called the *Stokes number*. The Womersley number denotes the ratio of unsteady inertial forces to viscous forces in the flow. It ranges from as large as about 20 in the aorta, to significantly greater than 1 in all large arteries, to as small as about 10^{-3} in the capillaries.

Let us estimate the Womersley number for an illustration. With a normal heart rate of 72 beats per minute, $\omega = (2\pi\, 72/60) \approx 8$ rad/s. Take $\rho = 1.05$ g cm^{-3}, $\mu = 0.04$ g cm^{-1} s^{-1}, and an artery of radius $a = 0.5$ cm. Then $\alpha \approx 7$. Decreasing α values correspond to increasing role of viscous forces and, for $\alpha < 1$, viscous effects are dominant. In that highly viscous regime, the flow may be regarded as quasi-steady. With increasing α, inertial forces become important. In pulsatile flows, flow separation may occur both by a geometric adverse pressure gradient, and/or by time-varying changes in the driving pressure. Geometric adverse-pressure gradients may arise due to varying cross-sectional areas through which the flow passes. On the other hand, time-varying changes in a cardiac cycle result in acceleration and deceleration during the cycle. An adverse-pressure gradient during the deceleration phase may result in flow separation.

Blood vessel walls are viscoelastic in their behavior. The ability of a blood vessel wall to expand and contract passively with changes in pressure is an important function of large arteries and veins. This ability of a vessel to distend and increase volume with increasing transmural pressure difference (inside minus outside pressure) is quantified as vessel compliance. During systole, pressure from the left ventricle is transmitted as a wave due to the elasticity of the arteries. Due to the compliant nature of the arteries and their finite thickness, the pressure travels like a wave at a speed much faster than the flow velocity. Since blood vessels may have many branches, the reflection and transmission of waves in such branching vessels significantly complicate the understanding of such flows. In this chapter, a reasonably simplified picture of these various complex features will be presented. Further reading in advanced treatments such as the book by Fung (1997) will be necessary to obtain a comprehensive understanding.

Steady Blood Flow Theory

First we start with the study of laminar, steady flow of blood in circular tubes, and in subsequent sections, we shall consider more realistic models. In the simplest model, blood flow in a vessel is modeled as a laminar, steady, incompressible, fully developed flow of a Newtonian fluid through a straight, rigid, cylindrical, horizontal tube of constant circular cross section (see Figure 16.11). Such a flow is called *circular Poiseuille flow* or more commonly *Hagen-Poiseuille flow*, and is covered in Section 8.2. The primary question here is: How valid is a Hagen-Poiseuille model for blood flow? Issues related to the assumptions inherent in Hagen-Poiseuille flow are summarized in the following paragraphs.

In the normal body, blood flow in vessels is generally laminar. However, at high flow rates, particularly in the ascending aorta, the flow may become turbulent at or near peak systole. Disturbed flow may occur during the deceleration phase of the cardiac cycle (Chandran et al., 2007). Turbulent flow may also occur in large arteries at branch points. However, under normal conditions, the critical Reynolds number, Re_c, for transition of blood flow in long, straight, smooth blood vessels is relatively high, and the blood flow remains laminar. Let us consider some estimates. The aorta is about 40 cm long and the average velocity u of flow in it is about 40 cm/s. The lumen diameter at the root of the aorta is $d = 25$ mm, and the corresponding $Re = \rho u d / \mu$ is ~3000. The maximum Reynolds number may be as high as 9000. The average value for Re in the vena cava is also about 3000. Arteries have varying sizes and the maximum Re is about 1000. For Newtonian fluid flow in a straight cylindrical rigid tube, Re_c is about ~3000. However, aorta and arteries are distensible tubes, and this Re_c criterion does not apply. In the case of blood flow, laminar flow conditions generally prevail even at Reynolds numbers as high as 10,000 (Mazumdar, 2004). In summary, the laminar flow assumption is reasonable in many cases.

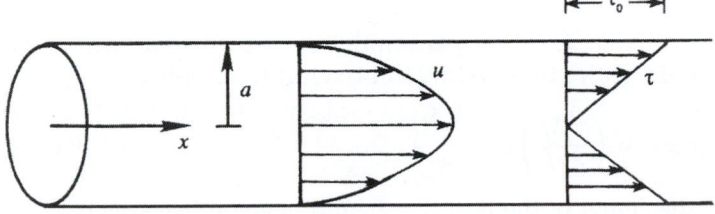

FIGURE 16.11 Poiseuille flow. Here the x-axis is coincident with the tube axis, and the radial coordinate is r. The fluid velocity has a parabolic profile and is entirely in the x-direction. The shear within the flow is zero at $r = 0$ and increases linearly with r, reaching a maximum at $r = a$, the tube radius.

Blood flow in the circulatory system is generally unsteady and pulsatile. The large arteries have elastic walls and are subject to substantially pulsatile flow. The steady-flow assumption is inapplicable until the flow has reached smaller muscular arteries and arterioles in the circulatory system. Blood flow in arteries has been described by several authors (see McDonald, 1974; Pedley, 1980; Ku, 1997).

In the heart chambers and blood vessels, blood may be considered incompressible. In the walls of the heart and in the blood vessel walls, it may not be considered as incompressible (Fung, 1997).

The fully developed flow assumption is very restrictive in describing blood flow in vessels. Since blood flow remains laminar at very high Reynolds numbers, the entry length is very large in many cases, and branches and curved vessels hinder flow development.

Flow in large blood vessels may be generally regarded as Newtonian. The Newtonian fluid assumption is inapplicable at low shear rates such as those that would occur in arterioles and capillaries.

Many blood vessels are not straight but are curved and have branches. However, flow may be regarded to occur in straight sections in many cases of interest.

Arterial walls are not rigid but are viscoelastic and distensible. The pressure pulse generated during left ventricular contraction travels through the arterial wall. The speed of wave propagation depends upon the elastic properties of the wall and the fluid-structure interaction. Arterial branches and curves may cause reflections of the wave.

Gravitational and hydrostatic effects become very important for orientations of the body other than the supine position.

Systemic arteries are generally circular tubes but may have tapering cross sections, while the veins and pulmonary arteries tend to be elliptical.

However, there remain many situations where the Hagen-Poiseuille model is reasonably applicable. Thus, a recapitulation of the results from Chapter 8 is provided here using cylindrical coordinates (r, θ, x) where x is the axial coordinate, r is the radial distance from the x-axis, and θ is the circumferential (azimuthal) angle. The axial flow velocity, $u = u(r)$, in a pipe of radius, a (see (8.6)) is:

$$u(r) = \frac{r^2 - a^2}{4\mu}\left(\frac{dp}{dx}\right). \tag{16.13}$$

In a fully developed flow, the pressure gradient, (dp/dx), is a constant, and it may be expressed in terms of the overall pressure difference:

$$\left(\frac{dp}{dx}\right) = -\frac{\Delta p}{L} = -\frac{(p_1 - p_2)}{L}, \tag{16.14}$$

where Δp is the imposed pressure difference, subscripts 1 and 2 denote inlet and exit ends, respectively, and L is the length of the entire tube. With (16.14), (16.13) becomes

$$u(r) = \left(\frac{a^2 \Delta p}{4\mu L}\right)\left(1 - \frac{r^2}{a^2}\right). \tag{16.15}$$

The maximum velocity occurs at the center of the tube, $r = 0$, and is given by

$$u_{\max} = \left[\frac{\Delta p \; a^2}{4\mu L} \right].$$

(16.16)

The volume flow rate is:

$$Q = \int_0^a u 2\pi r dr = -\frac{\pi a^4}{8\mu}\left(\frac{dp}{dx}\right) = \frac{\pi a^4}{8\mu}\frac{(p_1 - p_2)}{L} = \frac{\pi \; a^4}{8\mu}\frac{\Delta p}{L} = \frac{u_{\max}}{2}\pi a^2.$$

(16.17)

Equation (16.17) is called the *Poiseuille formula*. The average velocity over the cross section is:

$$V = \frac{Q}{A} = \frac{Q}{\pi a^2} = \frac{u_{\max}}{2},$$

(16.18)

where A is the cross section of the tube. The shear stress at the tube wall is:

$$\tau_{xr}|_{r=a} = \tau = -\mu\left(\frac{du}{dr}\right)\bigg|_{r=a} = -\frac{a}{2}\left(\frac{dp}{dx}\right) = -\frac{a}{2}\frac{\Delta p}{L},$$

(16.19)

where the negative sign has been included to give $\tau > 0$ with $(du/dr) < 0$ (the velocity decreases from the tube centerline to the tube wall). The maximum shear stress occurs at the walls, and the stress decreases toward the center of the vessel.

The Hagen-Poiseuille equation and its derivatives are most applicable to flow in the muscular arteries. Modifications are likely to be required outside this range (see Brown et al., 1999). For an application of Poiseuille flow relationships in the context of perfused tissue heat transfer and thermally significant blood vessels, see Baish et al. (1986a, 1986b).

With the results for the Hagen-Poiseuille flow, we have from (16.9):

$$\text{Total Peripheral Resistance} = R = \frac{\Delta \bar{p}}{Q} = \frac{8\mu L}{\pi a^4}$$

(16.20)

Equation (16.20) shows that peripheral resistance to the flow of blood is inversely proportional to the fourth power of vessel diameter.

Hagen-Poiseuille Flow and the Fahraeus-Lindqvist Effect

Consider laminar, steady flow of blood through a straight, rigid, cylindrical, horizontal tube of constant circular cross section and radius a, as shown in Figure 16.12, and let the flow be divided into two regions: a central core containing RBCs with axial velocity u_c and a cell-free plasma layer of thickness δ surrounding the core with axial velocity u_p. In addition, let the viscosities of the core and the plasma layer be μ_c and μ_p, respectively. Let the shear rates be such that each region can be considered Newtonian, and that we could employ Hagen-Poiseuille theory.

The shear stress distribution in the core region is governed by:

$$\tau_{xr} = -\mu_c\frac{du^c}{dr} = -\frac{r}{2}\frac{\Delta p}{L},$$

(16.21)

FIGURE 16.12 Fahraeus-Lindqvist effect. When the core and the plasma flows have different viscosities and occupy different regions of the tube, the relationship between the volume flow rate and the pressure drop in a round tube can be found in terms of the geometry and the viscosities.

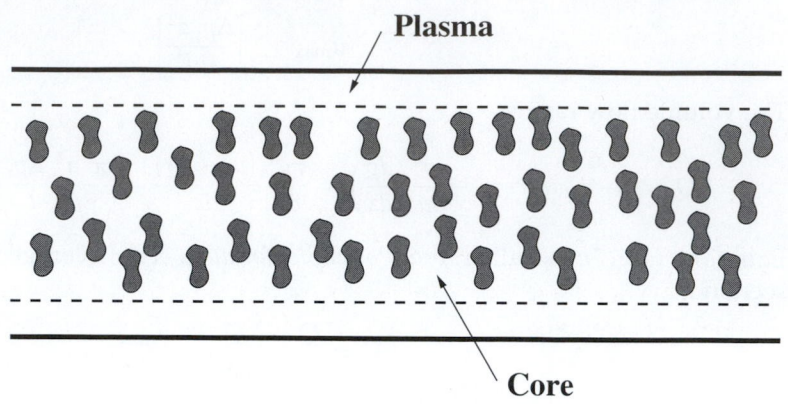

subject to conditions:

$$\frac{du^c}{dr} = 0, \quad \text{at } r = 0, \tag{16.22}$$

$$\tau_{xr}|_c = \tau_{xr}|_p, \quad \text{at } r = (a - \delta). \tag{16.23}$$

The shear stress distribution in the plasma region is governed by:

$$\tau_{xr} = -\mu_p \frac{du^p}{dr} = -\frac{r}{2}\frac{\Delta p}{L}, \tag{16.24}$$

subject to conditions:

$$u^c = u^p, \quad \text{at } r = (a - \delta), \tag{16.25}$$

$$u^p = 0, \quad \text{at } r = a. \tag{16.26}$$

Integration of (16.21) and (16.24) subject to the indicated conditions yields the following expressions for the axial velocities in the plasma and core regions:

$$u^p = \frac{a^2}{4\mu_p}\frac{\Delta p}{L}\left[1 - \left(\frac{r}{a}\right)^2\right], \quad \text{for } a - \delta \le r \le a, \tag{16.27}$$

and

$$u^c = \frac{a^2}{4\mu_p}\frac{\Delta p}{L}\left[1 - \left(\frac{a-\delta}{a}\right)^2 - \frac{\mu_p}{\mu_c}\left(\frac{r}{a}\right)^2 + \frac{\mu_p}{\mu_c}\left(\frac{a-\delta}{a}\right)^2\right], \quad \text{for } 0 \le r \le a - \delta. \tag{16.28}$$

The volume flow rates in the plasma, Q_p, and core region, Q_c, are:

$$Q_p = 2\pi \int_{a-\delta}^{a} u^p r\,dr = \frac{\pi\,\Delta p}{8\mu_p L}\left[a^2 - (a-\delta)^2\right]^2, \tag{16.29}$$

and

$$Q_c = 2\pi \int_0^{a-\delta} u^c r dr = \frac{\pi a^2 \Delta p}{4\mu_p L}\left[a^2 - \left(1 - \frac{\mu_p}{2\mu_c}\right)\frac{(a-\delta)^4}{a^2}\right]. \tag{16.30}$$

The total flow rate of blood within the tube, Q, is the sum of the flow rates in the plasma and core regions. Therefore:

$$Q = Q_p + Q_c = \frac{\pi a^4 \Delta p}{8\mu_p L}\left[1 - \left(1 - \frac{\delta}{a}\right)^4\left(1 - \frac{\mu_p}{\mu_c}\right)\right]. \tag{16.31}$$

From (16.31), we could calculate the apparent viscosity of the two-region fluid by measuring Q and $\Delta P/L$. Define μ_{app}, by analogy with Hagen-Poiseuille flow, as given by:

$$Q = \frac{\pi a^4 \Delta p}{8\mu_{app}L}. \tag{16.32}$$

From (16.31) and (16.32), the apparent viscosity, μ_{app}, may be expressed in terms of μ_p as:

$$\mu_{app} = \mu_p\left[1 - \left(1 - \frac{\delta}{a}\right)^4\left(1 - \frac{\mu_p}{\mu_c}\right)\right]^{-1}. \tag{16.33}$$

In the limit $(\delta/a) \ll 1$, $\left(1 - \frac{\delta}{a}\right)^4 \approx (1 - 4\delta/a)$. Then, (16.33) reduces to:

$$\mu_{app} = \mu_c\left[1 + 4\frac{\delta}{a}\left(\frac{\mu_c}{\mu_p} - 1\right)\right]^{-1} \rightarrow \mu_c \rightarrow \mu. \tag{16.34}$$

In (16.31) and (16.33), δ and μ_c are unknown. From (16.8), we have $H_c/H = 1 + (Q_p/Q_C)$. We still need input from experimental data to set up a modeling procedure for the Fahraeus-Lindqvist effect. Fournier (2007) recommends the use of Charm and Kurland's (1974) equation for this purpose (see reference for details):

$$\mu_c = \mu_p\frac{1}{1 - \alpha_c H_c}, \tag{16.35}$$

where,

$$\alpha_c = 0.070 \exp\left[2.49H_c + \frac{1107}{T}\exp(-1.69H_c)\right], \tag{16.36}$$

and T is temperature in K. Equation (16.36) may be used to a hematocrit of 0.60. With this input, a modeling procedure can be developed for various flow and tube parameters.

Effect of Developing Flow

When we discussed Poiseuille flow, we noted that the fully developed flow assumption that is often invoked in the study of blood flow in vessels is very restrictive. We will now learn about some of the limitations of this assumption.

When a fluid under the action of a pressure gradient enters a cylindrical tube, it takes a certain distance called the *inlet* or *entrance length, l,* before the flow in the tube becomes

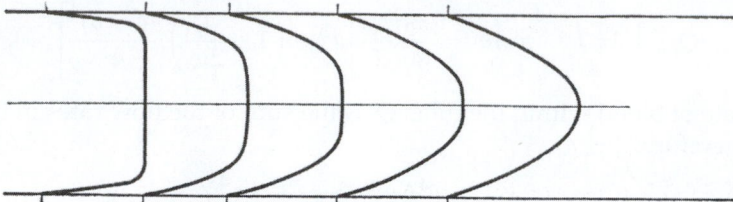

FIGURE 16.13 Developing velocity profile in a tube flow. The first profile on the left corresponds to the beginning of flow development: The wall shear stress is high and a large fraction of the flow is still at a uniform speed. As the fluid moves down the tube, the influence of the wall shear stress spreads toward the tube centerline and eventually the flow reaches a smooth, unchanging profile that is parabolic for Newtonian fluids.

steady and fully developed. When the flow is fully developed and laminar, the velocity profile is parabolic. Within the inlet length, the velocity profile changes in the direction of the flow and the fluid accelerates or decelerates as it flows. There is a balance among pressure, viscous, and inertia (acceleration) forces. Compared to fully developed flow, the entrance region is subject to large velocity gradients near the wall and these result in high wall-shear stresses. The entry of blood from the ventricular reservoir into the aortic tube or from a large artery into a smaller branch will involve an entrance length. It must be understood, however, that the inlet length with pulsating flow (say, in the proximal aorta) is different from that for a steady flow.

If we assume that the fluid enters the tube from a reservoir, the profile at the inlet is virtually flat. The transition from a flat velocity distribution, at the entrance of a tube, to the fully developed parabolic velocity profile is illustrated in Figure 16.13. Once inside the tube, the layer of fluid immediately in contact with the wall will become stationary (no-slip condition) and the laminae adjacent to it slide on it subject to viscous forces and a boundary layer is formed. The presence of the endothelial lining on the inside of a blood vessel wall does not negate the no-slip condition. The motion of the bulk of the fluid in the central region of the tube will not be affected by the viscous forces and will have a flat velocity profile. As flow progresses down the tube, the boundary layer will grow in thickness as the viscous shear stress slows more and more of the fluid.

Eventually, the boundary layer fills the whole of the tube and steady viscous flow is established or the flow is fully developed. In the literature (see, for example, Mohanty & Asthana, 1979), there are discussions which divide the entrance region into two parts, the inlet region and the filled region. At the end of the inlet region, the boundary layers meet at the tube axis but the velocity profiles are not yet similar. In the filled region, adjustment of the completely viscous profile takes place until the Poiseuille similar profile is attained at the end of it. In our discussion here, we will treat the entrance region as a region with a potential core and a developing boundary layer at the wall. The shape of the velocity profile in the tube depends on whether the flow is laminar or turbulent, as does the length of the entrance region, l. This is a direct consequence of the differences in the nature of the shear stress in laminar and turbulent flows. The magnitude of the pressure gradient, $\partial p/\partial x$, is larger in the entrance region than in the fully developed region. There is also an expenditure of kinetic energy involved in transition from a flat to a parabolic profile. For steady flow of a Newtonian fluid in a rigid-walled horizontal circular tube, the entrance length may be estimated from:

$$\frac{\ell}{d} = 0.06\,Re \quad \text{for}\quad \text{laminar flow}\quad \text{and}\quad Re > 50,$$

$$\frac{\ell}{d} = 0.693\,Re^{1/4} \quad \text{for}\quad \text{turbulent flow}.$$

(16.37)

For steady flow at low Reynolds number, the entrance region is approximately one tube radius long (for $Re \leq 0.001$, say in capillaries, $\ell/d = 0.65$). In large arteries, the entrance length is relatively long and over a significant length of the artery the velocity gradients are high near the wall. This affects the mass exchange of gas and nutrient molecules between the blood and artery wall.

Unsteady flow through the entrance region with a pulsating flow depends on the Womersley and Reynolds numbers. For a medium-sized artery, the Reynolds number is typically on the order of 100 to 1000, and the Womersley number ranges from 1 to 10. Pedley (1980) has estimated the wall shear stress in the entrance region for pulsatile flow using asymptotic boundary-layer theory while He and Ku (1994) have employed a spectral element simulation to investigate unsteady entrance flow in a straight tube. For a mean Re of 200 and α varying from 1.8–12.5 and an inlet waveform $1 + \sin\,\omega t$, He and Ku have computed variations in entrance length during the pulsatile cycle. The amplitude of the entrance-length variation decreases with an increase in α. The phase lag between the entrance length and the inlet flow waveform increases for α up to 5.0 and decreases for larger values of α. For low α, the maximum entrance length during pulsatile flow is approximately the same as the steady entrance length for the peak flow and is primarily dependent on the Reynolds number. For high α, the Stokes boundary-layer growth is faster and the entrance length is more uniform during the cycle. For $\alpha \geq 12.5$, the pulsatile entrance length is approximately the same length as the entrance length of the mean flow. At all α, the wall-shear rate converges to its fully developed value at about half the length at which the centerline velocity converges to its fully developed value. This leads to the conclusion that the upstream flow conditions leading to a specific artery may or may not be fully developed and can be predicted only by the magnitudes of the Reynolds number and Womersley number.

Effect of Tube Wall Elasticity on Poiseuille Flow

Here, we will include the elastic behavior of the vessel wall and examine the effect on the Hagen-Poiseuille model. Consider a pressure-gradient-driven, laminar, steady flow of a Newtonian fluid in a long, circular, cylindrical, thin-walled, elastic tube. Let the initial radius of the tube be a_0, and h be the wall thickness, and it is small compared to a_0. Because the tube is elastic, it will distend more at the high pressure end (inlet) than at the outlet end. The tube radius, a, will now be a function of x.

The variation in tube radius due to wall elasticity has to be ascertained. The difference between the exterior pressure on the outside of the tube, p_e, and the pressure inside the tube, $p(x)$, at any cross section of the tube (the negative of transmural pressure difference), is $(p_e - p(x))$. This pressure difference acts across h at every cross section, and will induce a circumferential stress. There will be a corresponding circumferential strain. This strain is the ratio of the change in radius to the original radius of the tube. In this way, we can ascertain the cross section at x.

Consider the static force equilibrium on a cylindrical segment of the blood vessel consisting of the top half cross section and of unit length. Let $\sigma_{\theta\theta}$ denote the average circumferential (hoop) stress in the tube wall. The net downward force due to the pressure difference will be balanced by the net upward force; this balance is:

$$2\sigma_{\theta\theta}h = \int_0^\pi (p(x) - p_e)\, a(x)\, \sin\theta\, d\theta,$$

(16.38)

which results in:

$$\sigma_{\theta\theta} = \frac{(p(x) - p_e)a(x)}{h}.$$

(16.39)

From Hooke's law, the circumferential strain $\sigma_{\theta\theta}$ is given by:

$$e_{\theta\theta} = \frac{\sigma_{\theta\theta}}{E} = \frac{(a(x) - a_0)}{a_0} = \left(\frac{a(x)}{a_0}\right) - 1,$$

(16.40)

where E is the Young's modulus of the tube wall material, and we have neglected the radial stress σ_{rr} as compared to $\sigma_{\theta\theta}$ in the thin-walled tube. The wall is considered thin if $(h/a) \ll 1$. From (16.39) and (16.40), we get the pressure-radius relationship:

$$a(x) = a_0 \left[1 - \frac{a_0}{Eh}(p(x) - p_e)\right]^{-1}.$$

(16.41)

Since the flow is laminar and steady, we can still apply the Hagen-Poiseuille formula, (16.17), to the flow. Thus,

$$Q = -\frac{\pi}{8\mu}\left(\frac{dp}{dx}\right)(a(x))^4.$$

(16.42)

Therefore,

$$\frac{dp}{dx} = -\frac{8\mu Q}{\pi(a(x))^4}.$$

(16.43)

With (16.41):

$$\left[1 - \frac{a_0}{Eh}(p(x) - p_e)\right]^{-4} dp = -\frac{8\mu}{\pi a_0}Q\, dx.$$

(16.44)

This is subject to the conditions, $P = P_1$ at $x = 0$, and $P = P_2$ at $x = L$. By integration of (16.44) and from the boundary conditions:

$$\frac{Eh}{3a_0}\left\{\left[1 - \frac{a_0}{Eh}(p_2 - p_e)\right]^{-3} - \left[1 - \frac{a_0}{Eh}(p_1 - p_e)\right]^{-3}\right\} = -\frac{8\mu}{\pi a_0}\, L\, Q.$$

(16.45)

Solving for Q,

$$Q = \frac{\pi a_0^3 Eh}{24\mu L}\left\{\left[1 - \frac{a_0}{Eh}(p_1 - p_e)\right]^{-3} - \left[1 - \frac{a_0}{Eh}(p_2 - p_e)\right]^{-3}\right\}.$$

(16.46)

From (16.46), we see that the flow is a nonlinear function of pressure drop if wall elasticity is taken into account. In the above development, we have assumed Hookean behavior for the stress-strain relationship. However, blood vessels do not necessarily obey Hooke's law, their zero-stress states are open sectors, and their constitutive equations may be nonlinear (see Zhou & Fung, 1997).

Pulsatile Blood Flow Theory

As stated earlier, blood flow in the arteries is pulsatile in nature. One of the earliest attempts to model pulsatile flow was carried out by Otto Frank in 1899 (see Fung, 1997).

Elasticity of the Aorta and the Windkessel Theory

Recall that when the left ventricle contracts during systole, pressure within the chamber increases until it is greater than the pressure in the aorta, leading to the opening of the aortic valve. The ventricular muscles continue to contract, increasing the chamber pressure while ejecting blood into the aorta. As a result, the ventricular volume decreases. The pressure in the aorta starts to build up and the aorta begins to distend due to wall elasticity. At the end of the systole, ventricular muscles start to relax, the ventricular pressure rapidly falls below that of the aorta, and the aortic valve closes. Not all of the blood pumped into the aorta, however, immediately goes into systemic circulation. A part of the blood is used to distend the aorta and a part of the blood is sent to peripheral vessels. The distended aorta acts as an elastic reservoir or a Windkessel (the name in German for an elastic reservoir), the rate of outflow from which is determined by the total peripheral resistance of the system. As the distended aorta contracts, the pressure diminishes in the aorta. The rate of pressure decrease in the aorta is much slower compared to that in the heart chamber. In other words, during the systole part of the heart pumping cycle, the large fluctuation of blood pressure in the left ventricle is converted to a pressure wave with a high mean value and a smaller fluctuation in the distended aorta (Fung, 1997). This behavior of the distended aorta was thought to be analogous to the high-pressure air chamber (Windkessel) of nineteenth-century fire engines in Germany, and hence the name Windkessel theory was used by Otto Frank to describe this phenomenon.

In the Windkessel theory, blood flow at a rate $Q(t)$ from the left ventricle enters an elastic chamber (the aorta) and a part of this flows out into a single rigid tube representative of all of the peripheral vessels. The rigid tube offers constant resistance, R, equal to the total peripheral resistance that was evaluated in the Hagen-Poiseuille model, (16.9). From the law of conservation of mass, assuming blood is incompressible:

Rate of Inflow into Aorta = Rate of change of volume of elastic chamber

$$+ \text{Rate of outflow into rigid tube.} \qquad (16.47)$$

Let the instantaneous blood pressure in the elastic chamber be $p(t)$, and its volume be $v(t)$. The pressure on the outside of the aorta is taken to be zero. The rate of change of volume of an elastic chamber is given by:

$$\frac{dv}{dt} = \left(\frac{dv}{dp}\right)\left(\frac{dp}{dt}\right). \qquad (16.48)$$

In (16.48), the quantity (dv/dp) is the compliance, K, of the vessel and is a measure of the distensibility. Compliance at a given pressure is the change in volume for a change in pressure. Here pressures are always understood to be transmural pressure differences. Compliance essentially represents the distensibility of the vascular walls in response to a certain pressure. Also, from (16.9), the rate of flow into peripherals is given by $(p(t)/R)$, where we have assumed $\bar{p}_V = 0$. Therefore, (16.47) becomes

$$Q(t) = K\left(\frac{dp}{dt}\right) + \left(\frac{p(t)}{R}\right). \tag{16.49}$$

Equation (16.49) is a linear equation of the form:

$$Q = \frac{dy}{dx} + Py, \tag{16.50}$$

whose solution is

$$ye^{\int Pdx} = A + \int Qe^{\int Pdx}\,dx. \tag{16.51}$$

From (16.49) and (16.51), with p_0 denoting p at $t = 0$, the instantaneous pressure p in the aorta as a function of the left ventricular ejection rate $Q(t)$ is given by

$$p(t) = \frac{1}{K}e^{-t/RK}\int_0^t Q(\tau)e^{\tau/RK}d\tau + p_0e^{-t/RK}. \tag{16.52}$$

In (16.52), p_0 would be the aortic pressure at the end of diastolic phase.

A fundamental assumption in the Windkessel theory is that the pressure pulse wave generated by the heart is transmitted instantaneously throughout the arterial system and disappears before the next cardiac cycle. In reality, pressure waves require finite but small transmission times, and are modified by reflection at bifurcations, bends, tapers, and at the end of short tubes of finite length, and so on. We will now account for some of these features.

Pulse Wave Propagation in an Elastic Tube: Inviscid Theory

Consider a homogeneous, incompressible, and inviscid fluid in an infinitely long, horizontal, cylindrical, thin-walled, elastic tube. Let the fluid be initially at rest. The propagation of a disturbance wave of small amplitude and long wavelength compared to the tube radius is of interest to us. In particular, we wish to calculate the wave speed. Since the disturbance wavelength is much greater than the tube diameter, the time-dependent internal pressure can be taken to be a function only of (x, t).

Before we embark on developing the solution, we need to understand the inviscid approximation. For flow in large arteries, the Reynolds and Womersley numbers are large; the wall boundary layers are very thin compared to the radius of the vessel. The inviscid approximation may be useful in giving us insights in understanding such flows. Clearly, this will not be the case with arterioles, venules, and capillaries. However, the inviscid analysis is strictly of limited use since it is the viscous stress that is dominant in determining flow stability in large arteries.

Under the various conditions prescribed, the resulting flow may be treated as one dimensional.

Let $A(x,t)$ and $u(x,t)$ denote the cross-sectional area of the tube and the longitudinal velocity component, respectively. The continuity equation is:

$$\frac{\partial A}{\partial t} + \frac{\partial (Au)}{\partial x} = 0, \tag{16.53}$$

and the equation for the conservation of momentum is:

$$\rho A\left(\frac{\partial u}{\partial t} + u\frac{\partial u}{\partial x}\right) = -\frac{\partial\left((p - p_e)A\right)}{\partial x}, \tag{16.54}$$

where $(p - p_e)$ is the transmural pressure difference. Since the tube wall is assumed to be elastic (not viscoelastic), under the further assumption that A depends on the transmural pressure difference $(p - p_e)$ alone, and the material obeys Hooke's law, we have from (16.41) the pressure-radius relationship (referred to as the *tube law*):

$$p - p_e = \frac{Eh}{a_0}\left(1 - \frac{a_0}{a}\right) = \frac{Eh}{a_0}\left[1 - \left(\frac{A_0}{A}\right)^{\frac{1}{2}}\right], \tag{16.55}$$

where $A = \pi a^2$, and $A_0 = \pi a_0^2$. The equations (16.53), (16.54), and (16.55) govern the wave propagation. We may simplify this equation system further by linearizing it. This is possible if the pressure amplitude $(p - p_e)$ compared to p_0, the induced fluid speed u, and $(A - A_0)$ compared to A_0, and their derivatives are all small. If the pulse is moving slowly relative to the speed of sound in the fluid, the wave amplitude is much smaller than the wavelength, and the distension at one cross section has no effect on the distension elsewhere, the assumptions are reasonable. As discussed by Pedley (2000), in normal human beings, the mean blood pressure, relative to atmospheric, at the level of the heart is about 100 mm Hg, and there is a cyclical variation between 80 and 120 mm Hg, so the amplitude-to-mean ratio is 0.2, which is reasonably small. Also, in the ascending aorta, the pulse wave speed, C, is about 5 m/s, and the maximum value of u is about 1 m/s, and (u/c) is also around 0.2. In that case, the system of equations reduce to

$$\frac{\partial A}{\partial t} + A_0\frac{\partial u}{\partial x} = 0, \tag{16.56}$$

and

$$\rho\frac{\partial u}{\partial t} = -\frac{\partial p}{\partial x}, \tag{16.57}$$

and

$$p - p_e = \frac{Eh}{2a_0 A_0}(A - A_0), \quad \text{and} \quad \frac{\partial p}{\partial A} = \frac{Eh}{2a_0 A_0}. \tag{16.58}$$

Differentiating (16.56) with respect to t and (16.57) with respect to x, and subtracting the resulting equations, we get

$$\frac{\partial^2 A}{\partial t^2} = \frac{A_0}{\rho} \frac{\partial^2 p}{\partial x^2},$$

(16.59)

and with (16.58), we obtain:

$$\frac{\partial^2 p}{\partial t^2} = \frac{Eh}{2a_0 A_0} \frac{\partial^2 A}{\partial t^2} = \frac{\partial p}{\partial A} \frac{A_0}{\rho} \frac{\partial^2 p}{\partial x^2}.$$

(16.60)

Combining (16.59) and (16.60), we produce:

$$\frac{\partial^2 p}{\partial x^2} = \frac{1}{c^2} \frac{\partial^2 p}{\partial t^2}, \quad \text{or,} \quad \frac{\partial^2 p}{\partial t^2} = c^2(A_0)\frac{\partial^2 p}{\partial x^2},$$

(16.61)

where $c^2 = \dfrac{Eh}{2\rho a_0} = \dfrac{A}{\rho}\dfrac{dp}{dA}$. Equation (16.61) is the wave equation, and the quantity,

$$c = \sqrt{\frac{Eh}{2\rho a_0}} = \sqrt{\frac{A}{\rho}\frac{dp}{dA}},$$

(16.62)

is the speed of propagation of the pressure pulse. This is known as the *Moens-Korteweg wave speed*. If the thin wall assumption is not made, following Fung (1997), by evaluating the strain on the midwall of the tube:

$$c = \sqrt{\frac{Eh}{2\rho(a_0 + h/2)}}.$$

(16.63)

Next, similar to (16.61), we can develop:

$$\frac{\partial^2 u}{\partial x^2} = \frac{1}{c^2} \frac{\partial^2 u}{\partial t^2},$$

(16.64)

for the velocity component u. The wave equation (16.61) has the general solution:

$$p = f_1\left(t - \frac{x}{c}\right) + f_2\left(t + \frac{x}{c}\right),$$

(16.65)

where f_1 and f_2 are arbitrary functions; f_2 is zero if the wave propagates only in the $+x$ direction. This result states that the small-amplitude disturbance can propagate along the tube, in either direction, without change of shape of the waveform, at speed $c(a_0)$. Also, the velocity waveform is predicted to be of the same shape as the pressure waveform.

In principle, the Moens-Korteweg wave speed given in (16.63) must enable the determination of the arterial modulus E as a function of a by noninvasive measurement of the values of arterial dimensions (a, h), the waveforms of the arterial inner radius at two sites, the transit time (as the time interval between the waveform peaks), and hence the pulse-wave velocity. More details in this regard are available in the book by Mazumdar (1999).

Next, consider the solutions of wave equations (16.61) and (16.64):

$$p = \hat{p}_1 f(x - ct) + \hat{p}_2 g(x + ct),$$

(16.66)

and

$$u = \widehat{u}_1 f(x - ct) + \widehat{u}_2 g(x + ct), \tag{16.67}$$

where $\widehat{p}_1, \widehat{u}_1, \widehat{p}_2$, and $\widehat{u}_2$ are the pressure and velocity amplitudes for waves traveling in the positive x-direction and negative x-direction, respectively. From (16.57):

$$\widehat{p}_1 = \rho c \widehat{u}_1, \quad \text{and} \quad \widehat{p}_2 = -\rho c \widehat{u}_2. \tag{16.68}$$

This equation (16.68) relates the amplitudes of the pressure and velocity waves.

The above analysis would equally apply if the inviscid fluid in the tube was initially in steady motion, say from left to right. In that case, u would have to be regarded as a small perturbation superposed on the steady flow, and c would be the speed of the perturbation wave relative to the undisturbed flow.

Let us now examine the limitations of this model. For typical flow in the aorta, the speed of propagation of the pulse is about 4 m/s (Brown et al., 1999), about 5 m/s in the ascending aorta, rising to about 8 m/s in more peripheral arteries. These predictions are very close to measured values in normal subjects, either dogs or humans (Pedley, 2000). The peak flow speed is about 1 m/s. The speed of propagation in a collapsible vein might be as low as 1 m/s, and this may lead to phenomena analogous to sonic flow (Brown et al., 1999). From (16.62), for given E, h, ρ, and size of vessel, the wave speed is a constant. Experimental studies indicate, however, that the wave speed is a function of frequency. The shape of the waveform does not remain the same. The theory must be modified to account for peaking of the pressure pulse due to wave reflection from arterial junctions, wave-front steepening due to nonlinear dispersion effects (Lighthill, 1978), and observed velocity waveform by including dissipative effects due to viscosity (Lighthill, 1978; Pedley, 2000). The neglect of the inertial terms and the effects of viscosity have therefore to be examined to address these concerns and to develop a systematic understanding. These issues will be addressed in later sections in the following order. First, we will learn about pulsatile viscous flow in a single rigid-walled, straight tube. This implies the assumption of an infinite wave speed. Subsequent to that, we will examine the effects of wall elasticity on pulsatile viscous flow in a single tube to gain a more realistic understanding. This allows us to understand wave transmission at finite speed. Following this, we will study blood vessel bifurcation. This will be extended to understand the effects of wave reflection from arterial junctions under the inviscid flow approximation.

Pulsatile Flow in a Rigid Cylindrical Tube: Viscous Effects Included, Infinite Wave Speed Assumption

Consider the axisymmetric flow of a Newtonian incompressible fluid in a long, thin, circular, cylindrical, horizontal, rigid-walled tube. Clearly, the assumption of a rigid wall implies that the speed of wave propagation is infinite and unrealistic. However, the development presented here will provide us with useful insights and these will be helpful in formulating a much improved theory in the next section.

We again employ the cylindrical coordinates (r, θ, x) with velocity components $(u_r, u_\theta,$ and $u_x)$, respectively. Let λ be the wavelength of the pulse. This is long, and $a \ll \lambda$. Since the wave speed is infinite, all the velocity components are very much smaller than the wave speed. These assumptions enable us to drop the inertial terms in the momentum equations. With the additional assumptions of axisymmetry ($u_\theta = 0$, and $\partial/\partial\theta = 0$), and rigid tube wall ($u_r = 0$), and omitting the subscript x in u_x for convenience, the continuity equation may be written:

$$\frac{\partial u}{\partial x} = 0, \tag{16.69}$$

and the r-momentum equation is:

$$0 = -\frac{\partial p}{\partial r}; \tag{16.70}$$

the x-momentum equation is:

$$\rho\frac{\partial u}{\partial t} = -\frac{\partial p}{\partial x} + \mu\left[\frac{\partial^2 u}{\partial r^2} + \frac{1}{r}\frac{\partial u}{\partial r}\right]. \tag{16.71}$$

We see that $u = u(r, t)$ and $p = p(x, t)$. Therefore, we are left with just one equation:

$$\mu\left[\frac{\partial^2 u}{\partial r^2} + \frac{1}{r}\frac{\partial u}{\partial r}\right] - \rho\frac{\partial u}{\partial t} = \frac{\partial p}{\partial x}. \tag{16.72}$$

In (16.72), since $p = p(x, t)$, $\partial p/\partial x$ will be a function only of t. Since the pressure waveform is periodic, it is convenient to express the partial derivative of pressure using a Fourier series. Such a periodic function depends on the fundamental frequency of the signal, ω, heart rate (unit, rad/s), and the time t. Recall that ω is also called the *circular frequency*, $\omega/2\pi$ is the frequency (unit, Hz), and λ is the wavelength (unit, m). Also, $\lambda = c/(\omega/2\pi)$, where c is wave speed. The wavelength is the wave speed divided by frequency, or the distance traveled per cycle.

We set

$$\frac{\partial p}{\partial x} = -Ge^{i\omega t}, \tag{16.73}$$

where G is a constant denoting the amplitude of the pressure gradient pulse and $e^{i\omega t} = \cos \omega t + i \sin \omega t$. With this representation for $P(t)$, (16.72) becomes:

$$\mu\left[\frac{\partial^2 u}{\partial r^2} + \frac{1}{r}\frac{\partial u}{\partial r}\right] - \rho\left[\frac{\partial u}{\partial t}\right] = \frac{\partial p}{\partial x} = -Ge^{i\omega t}. \tag{16.74}$$

This is a linear, second-order, partial differential equation with a forcing function. For $\omega = 0$, the flow is described by the Hagen-Poiseuille model. Womersley (1955a, 1955b), has solved this problem, and we will provide essential details.

For $\omega \neq 0$, we may try solutions of the form:

$$u(r, t) = U(r)e^{i\omega t}, \tag{16.75}$$

where $U(r)$ is the velocity profile in any cross section of the tube. The real part in (16.75) gives the velocity for the pressure gradient $G \cos \omega t$ and the imaginary part gives the velocity for the pressure gradient G sin ωt. Assume that the flow is identical at each cross section along the tube. From (16.74) and (16.75), we get:

$$\frac{d^2U}{dr^2} + \frac{1}{r}\frac{dU}{dr} - \frac{i\omega\rho}{\mu}U = \frac{G}{\mu}. \tag{16.76}$$

This is a Bessel's differential equation, and the solution involves Bessel functions of zeroth order and complex arguments. Thus,

$$U(r) = C_1 J_0\left(i\sqrt{(i\omega\rho/\mu)}\ r\right) + C_2 Y_0\left(i\sqrt{(i\omega\rho/\mu)}\ r\right) + \frac{G}{\omega\rho i}, \tag{16.77}$$

where C_1 and C_2 are constants. In (16.77), from the requirement that U is finite at $r = 0$, $C_2 = 0$. For a rigid-walled tube, $U = 0$ at $r = a$. Therefore,

$$C_1 J_0\left(i^{3/2}\sqrt{(\omega\rho/\mu)}\ a\right) + \frac{G}{\omega\rho i} = 0. \tag{16.78}$$

From (16.12), the Womersley number is defined by $\alpha = a\sqrt{\omega/\nu}$. Therefore, from (16.78), we may write,

$$C_1 = \frac{iG}{\omega\rho}\frac{1}{J_0(i^{3/2}\alpha)}. \tag{16.79}$$

Therefore, from (16.77),

$$U(r) = -\frac{iG}{\omega\rho}\left(1 - \frac{J_0(i^{3/2}\alpha\ r/a)}{J_0(i^{3/2}\alpha)}\right). \tag{16.80}$$

Introduce, for convenience:

$$F_1(\alpha) = \frac{J_0(i^{3/2}\alpha\ r/a)}{J_0(i^{3/2}\alpha)}. \tag{16.81}$$

Now, from (16.75):

$$u(r, t) = U(r)e^{i\omega t} = -\frac{iG}{\omega\rho}(1 - F_1(\alpha))e^{i\omega t} = \frac{Ga^2}{i\mu\alpha^2}(1 - F_1(\alpha))e^{i\omega t}. \tag{16.82}$$

In the above development, we have found the velocity as a function of radius r and time t for the entire driving-pressure gradient. Since we have represented both $\partial p/\partial x$ and $u(r, t)$ in terms of Fourier modes, we could also express the solution for both these quantities in terms of individual Fourier modes or harmonics explicitly as:

$$\frac{\partial p}{\partial x} = -\sum_{n=0}^{N} G_n e^{in\omega t}, \tag{16.83}$$

where N is the number of modes (harmonics), and the $n = 0$ term represents the mean pressure gradient. Similarly, for velocity,

$$u(r, t) = u_0(r) + \sum_{1}^{N} u_n(r)e^{in\omega t}. \tag{16.84}$$

In (16.84),

$$u_0(r) = \frac{G_0 a^2}{4\mu}\left(1 - \frac{r^2}{a^2}\right) \tag{16.85}$$

is the mean flow and is recognized as the steady Hagen-Poiseuille flow with G_0 as the mean pressure gradient, and for each harmonic:

$$u_n(r) = \frac{G_n a^2}{i\mu\alpha_n^2}(1 - F_1(\alpha_n)). \tag{16.86}$$

We can now write down the expressions for $u_n(r)$ in the limits of α_n small and large. These are, for α_n small:

$$u_n(r) \approx \frac{G_n a^2}{4\mu}\left(1 - \frac{r^2}{a^2}\right), \tag{16.87}$$

which represents a quasi-steady flow, and for α_n large:

$$u_n(r) \approx \frac{G_n a^2}{i\mu\alpha_n^2}\left\{1 - \exp\left[-\sqrt{\frac{\omega}{2\nu}}(1+i)(a-r)\right]\right\}, \tag{16.88}$$

which is the velocity boundary layer on a plane wall in an oscillating flow. This flow was discussed in Chapter 8 (Stokes' second problem).

The volume flow rate, $Q(t)$, may be obtained by integrating the velocity profile across the cross section. Thus, from (16.85) and (16.86):

$$Q(t) = \int_0^a u 2\pi r dr = \pi a^2 \left\{\frac{G_0 a^2}{8\mu} + \frac{a^2}{i\mu}\sum_{1}^{\infty}\frac{G_n}{\alpha_n^2}[1 - F_2(\alpha_n)]e^{in\omega t}\right\}, \tag{16.89}$$

or equivalently, with (16.82),

$$Q(t) = \int_0^a 2\pi e^{i\omega t}\frac{Ga^2}{i\mu\alpha^2}(1 - F_1(\alpha))\, rdr = \frac{\pi a^4}{i\mu\alpha^2}G(1 - F_2(\alpha))\,e^{i\omega t}, \tag{16.90}$$

where

$$F_2(\alpha) = \frac{2J_1(i^{3/2}\alpha)}{i^{3/2}\alpha J_0(i^{3/2}\alpha)}. \tag{16.91}$$

The real part of (16.90) gives the volume flow rate when the pressure gradient is $G\cos\omega t$ and the imaginary part gives the rate when the pressure gradient is $G\sin\omega t$.

Next, the wall shear rate $\tau(t)|_{r=a}$ is given by:

$$\tau(t)|_{r=a} = \left.\frac{\partial u}{\partial r}\right|_{r=a} = \frac{G_0 a}{2} + \frac{a}{2}\sum_{1}^{N} G_n F(\alpha_n)e^{in\omega t}. \tag{16.92}$$

We may now examine the flow rates in the limit cases of $\alpha \to 0$ and $\alpha \to \infty$. As $\alpha \to 0$, by Taylor's expansion,

$$F_2(\alpha) \approx 1 - \frac{i\alpha^2}{8} - O(\alpha^4), \tag{16.93}$$

and from (16.90) in the limit as $\alpha \to 0$,

$$Q = \frac{\pi G a^4}{8\mu} e^{in\omega t}, \tag{16.94}$$

and the magnitude of the volumetric flow rate Q_0 in the limit as $\alpha \to 0$ is

$$|Q_0| = \frac{\pi G a^4}{8\mu}, \tag{16.95}$$

as would be expected (the Hagen-Poiseuille result). As $\alpha \to \infty$:

$$F_2(\alpha) \approx \frac{2}{i^{1/2}\alpha}\left(1 + \frac{1}{2\alpha}\right). \tag{16.96}$$

Next, in Hagen-Poiseuille flow, the steady flow rate is the maximum attainable and there is no phase lag between the applied pressure gradient and the flow. To understand the phase difference between the applied pressure gradient pulse and the flow rate in the present flow model, we set

$$(1 - F_2(\alpha)) = Z(\alpha), \quad Z(\alpha) = X(\alpha) + iY(\alpha). \tag{16.97}$$

Then from (16.90):

$$Q = \frac{\pi G a^4}{\mu\alpha^2}\left\{ [Y\cos(\omega t) + X\sin(\omega t)] - i[X\cos(\omega t) - Y\sin(\omega t)] \right\}. \tag{16.98}$$

The magnitude of Q is

$$|Q| = \frac{\pi G a^4}{\mu\alpha^2}\sqrt{X^2 + Y^2}. \tag{16.99}$$

The phase angle between the applied pressure gradient $Ge^{i\omega t}$ and the flow rate (16.90) is now noted to be

$$\tan\phi = \frac{X}{Y}. \tag{16.100}$$

With increasing ω, the phase lag between the pressure gradient and the flow rate increases, and the flow rate decreases. Thus, the magnitude of the volumetric flow rate, $|Q|$, given by (16.99) will be considerably less than the magnitude $|Q_0|$ given by (16.95) as would be expected. For an arterial flow, with $\alpha = 8$, $X \approx 0.85$, $Y \approx 0.16$, the pulsed volumetric flow rate, $|Q|$, would be about one-tenth of the steady value, $|Q_0|$. For more detailed discussions and comparisons with measured values of pressure gradients and flow rates in blood vessels, see Nichols and O'Rourke (1998).

The preceding analysis assumed an infinite wave speed of propagation. In order to accommodate the requirement of wave transmission at a finite wave speed, we need to account for vessel wall elasticity. This is discussed in the next section.

Wave Propagation in a Viscous Liquid Contained in an Elastic Cylindrical Tube

Blood vessel walls are viscoelastic. But in large arteries the effect of nonlinear viscoelasticity on wave propagation is not so severe (Fung, 1997). Even where viscoelastic effects are important, an understanding based on elastic walls will be useful. In this section, we will first study the effects of elastic walls. Then, we will briefly discuss the effects of wall viscoelasticity.

Consider a long, thin, circular, cylindrical, horizontal elastic tube containing a Newtonian, incompressible fluid. Let this system be set in motion solely due to a pressure wave, and the amplitude of the disturbance be small enough so that quadratic terms in the amplitude are negligible compared with linear ones.

In the formulation, we have to consider the fluid flow equations together with the equations of motion governing tube wall displacements. Assume that the tube wall material obeys Hooke's law. Since the tube is thin, membrane theory for modeling the wall displacements is adequately accurate, and we will neglect bending stresses.

The primary question is, how does viscosity attenuate velocity and pressure in this flow?

We shall employ the cylindrical coordinates (r, θ, x) with velocity components $(u_r, u_\theta,$ and $u_x)$, respectively. With the assumption of axisymmetry, $u_\theta = 0$ and $\partial/\partial\theta = 0$. For convenience, we write the u_r component as v, and we omit the subscript x in u_x.

Restricting the analysis to small disturbances, the governing equations for the fluid are:

$$\frac{\partial u}{\partial x} + \frac{1}{r}\frac{\partial (rv)}{\partial r} = 0, \tag{16.101}$$

$$\rho\frac{\partial u}{\partial t} = -\frac{\partial p}{\partial x} + \mu\left(\frac{\partial^2 u}{\partial r^2} + \frac{1}{r}\frac{\partial u}{\partial r} + \frac{\partial^2 u}{\partial x^2}\right), \tag{16.102}$$

$$\rho\frac{\partial v}{\partial t} = -\frac{\partial p}{\partial r} + \mu\left(\frac{\partial^2 v}{\partial r^2} + \frac{1}{r}\frac{\partial v}{\partial r} + \frac{\partial^2 v}{\partial x^2} - \frac{v}{r^2}\right), \tag{16.103}$$

where u and v are the velocity components in the axial and radial directions, respectively.

These have to be supplemented with the tube wall displacement equations. Let the tube wall displacements in the (r, θ, x) directions be $(\eta, \zeta,$ and $\xi)$, respectively, and the tube material density be ρ_w. The initial radius of the tube is a_0, and the wall thickness is h.

For this thin elastic tube, the circumferential (hoop) tension and the tension in the axial direction are related by Hooke's law as follows:

$$T_\theta = \frac{Eh}{1-\hat{\nu}^2}\left(\frac{\eta}{a_0} + \hat{\nu}\frac{\partial\xi}{\partial x}\right), \tag{16.104}$$

and

$$T_x = \frac{Eh}{1 - \hat{\nu}^2}\left(\frac{\partial \xi}{\partial x} + \hat{\nu}\frac{\eta}{a_0}\right),$$ (16.105)

where $\hat{\nu}$ is Poisson's ratio.

By a force balance on a wall element of volume ($h\, rd\theta\, dx$), the equations governing wall displacements may be written as:

- r-direction

$$\rho_w h\, \frac{\partial^2 \eta}{\partial t^2} = \sigma_{rr}|_{r=a} - \frac{T_\theta}{a_0},$$ (16.106)

and

- x-direction

$$\rho_w h\, \frac{\partial^2 \xi}{\partial t^2} = +\frac{\partial T_x}{\partial x} - \sigma_{rx}|_{r=a}.$$ (16.107)

There is no displacement equation for the θ direction. In (16.106) and (16.107), $\sigma_{rr}|_{r=a}$ and $\sigma_{rx}|_{r=a}$ refer to radial and shear stresses, respectively, which the fluid exerts on the tube wall. These equations are based on the assumptions that shear and bending stresses in the tube wall material are negligible and the slope of the disturbed tube wall ($\partial a/\partial x$) is small. These also imply that the ratios (a/λ) and (h/λ), where λ is the wavelength of disturbance, are small.

From (16.104) through (16.107), we obtain

$$\rho_w h\frac{\partial^2 \eta}{\partial t^2} = \sigma_{rr}|_{r=a} - \frac{Eh}{1-\hat{\nu}}\left(\frac{\eta}{a_0^2} + \frac{\hat{\nu}}{a_0}\frac{\partial \xi}{\partial x}\right),$$ (16.108)

and

$$\rho_w h\frac{\partial^2 \xi}{\partial t^2} = -\mu\left(\frac{\partial u}{\partial r} + \frac{\partial v}{\partial x}\right)\bigg|_{r=a} + \frac{Eh}{1-\hat{\nu}^2}\left(\frac{\partial^2 \xi}{\partial x^2} + \frac{\hat{\nu}}{a_0}\frac{\partial \eta}{\partial x}\right).$$ (16.109)

In the above equations, from the theory of fluid flow, the normal compressive stress due to fluid flow on an area element perpendicular to the tube's radius is given by:

$$\sigma_{rr} = +p - 2\mu\frac{\partial v}{\partial r},$$ (16.110)

and the shear stress due to fluid flow acting in a direction parallel to the axis of the tube on an element of area perpendicular to a radius is

$$\sigma_{rx} = \mu\left(\frac{\partial u}{\partial r} + \frac{\partial v}{\partial x}\right).$$ (16.111)

These are the radial and shear stresses exerted by the fluid on the wall of the vessel. With (16.110) and (16.111), (16.108) and (16.109) become

$$\rho_w h \frac{\partial^2 \eta}{\partial t^2} = +p|_{r=a} - 2\mu \frac{\partial v}{\partial r}\Big|_{r=a} - \frac{Eh}{1-\hat{\nu}^2}\left(\frac{\eta}{a_0^2} + \frac{\hat{\nu}}{a_0}\frac{\partial \xi}{\partial x}\right), \tag{16.112}$$

and

$$\rho_w h \frac{\partial^2 \xi}{\partial t^2} = -\mu \left(\frac{\partial u}{\partial r} + \frac{\partial v}{\partial x}\right)\Big|_{r=a} + \frac{Eh}{1-\hat{\nu}^2}\left(\frac{\partial^2 \xi}{\partial x^2} + \frac{\hat{\nu}}{a_0}\frac{\partial \eta}{\partial x}\right). \tag{16.113}$$

We have to solve (16.101) through (16.103), together with (16.112), and (16.113) subject to prescribed conditions. The boundary conditions at the wall are that the velocity components of the fluid be equal to those of the wall. Thus,

$$u|_{r=a_0} = \frac{\partial \xi}{\partial t}\Big|_{r=a_0}, \tag{16.114}$$

and

$$v|_{r=a_0} = \frac{\partial \eta}{\partial t}\Big|_{r=a_0}. \tag{16.115}$$

We note that the boundary conditions given in (16.114) and (16.115) are linearized conditions, since we are evaluating u and v at the undisturbed radius a_0.

We now represent the various quantities in terms of Fourier modes. Thus,

$$u(x, r, t) = \hat{u}(r)e^{i(kx-\omega t)}, \quad v(x, r, t) = \hat{v}(r)e^{i(kx-\omega t)},$$
$$p(x, t) = \hat{p}e^{i(kx-\omega t)}, \quad \xi(x, t) = \hat{\xi}e^{i(kx-\omega t)}, \tag{16.116}$$
$$\eta(x, t) = \hat{\eta}e^{i(kx-\omega t)},$$

where $\hat{u}(r)$, $\hat{v}(r)$, $\hat{p}$, $\hat{\xi}$, and $\hat{\eta}$ are the amplitudes, $\omega = 2\pi/T$ is a real constant, the frequency of the forced disturbance, T is the period of the heart cycle, $k = k_1 + ik_2$ is a complex constant, with k_1 being the wave number and k_2 a measure of the decay of the disturbance as it travels along the vessel (damping constant), $|k| = \sqrt{k_1 + k_2} = 2\pi/\lambda$, where λ is the wavelength of the disturbance, and $c = \omega/k_1$ is the wave speed.

The above formulation has been solved by Morgan and Kiely (1954) and by Womersley (1957a, 1957b), and we will provide the essential details here. The analysis will be restricted to disturbances of long wavelength, that is, $a/\lambda \ll 1$, and large Womersley number, $\alpha \gg 1$.

From (16.101):

$$\left|\frac{v}{u}\right| = \left|\frac{\hat{v}(r)}{\hat{u}(r)}\right| = O(|ak|). \tag{16.117}$$

For small damping, we note that $|k| \approx k_1 = 2\pi/\lambda$, and $c = \omega/k_1$ is the wave speed.

From (16.102) and (16.103), we may make the following observations. In (16.102), $\partial^2 u/\partial x^2$ may be neglected in comparison with the other terms since $a/\lambda \ll 1$ and $\lambda\alpha \gg 1$. In (16.103), $\partial p/\partial r$ is of a higher order of magnitude in a/λ than is $\partial p/\partial x$. In fact, we may neglect all terms that are of order a/λ. In effect, we are neglecting radial acceleration and damping terms and taking the pressure to be uniform over each cross section. The fluid equations become:

$$\frac{\partial u}{\partial x} + \frac{1}{r}\frac{\partial (rv)}{\partial r} = 0, \tag{16.118}$$

$$\rho\frac{\partial u}{\partial t} = -\frac{\partial p}{\partial x} + \mu\left(\frac{\partial^2 u}{\partial r^2} + \frac{1}{r}\frac{\partial u}{\partial r}\right), \tag{16.119}$$

$$\frac{\partial p}{\partial r} = 0, \tag{16.120}$$

$$p = \widehat{p}e^{i(kx-\omega t)}. \tag{16.121}$$

Now substitute the assumed forms given in (16.116) into (16.118) and (16.119) to produce:

$$\frac{d(r\widehat{v})}{dr} = -ikr\widehat{u}, \tag{16.122}$$

$$\frac{d^2\widehat{u}}{dr^2} + \frac{1}{r}\frac{d\widehat{u}}{dr} + \frac{i\omega\rho}{\mu}\widehat{u} = \frac{ik\widehat{p}}{\mu}. \tag{16.123}$$

The boundary conditions given by (16.114) and (16.115) become:

$$\widehat{u}(a_0)e^{i(kx-\omega t)} = -i\omega\widehat{\xi}e^{i(kx-\omega t)}, \tag{16.124}$$

$$\widehat{v}(a_0)e^{i(kx-\omega t)} = -i\omega\widehat{\eta}e^{i(kx-\omega t)}. \tag{16.125}$$

We may now note that the linearization of the boundary conditions will involve an error of the same order as that caused by neglecting the nonlinear terms in the equations. The error would be small if $\widehat{\xi}$ and $\widehat{\eta}$ are very small compared to a.

Next, introduce the assumed form given in (16.116), and use (16.120) in the displacement equations (16.112) and (16.113) to develop:

$$-\rho_w h\omega^2\widehat{\eta} = \widehat{p} - 2\mu\left(\frac{d\widehat{v}}{dr}\right)\bigg|_{r=a_0} - \frac{Eh}{1-\widehat{\nu}^2}\left(\frac{\widehat{\eta}}{a_0^2} + \frac{i\nu k}{a_0}\widehat{\xi}\right), \tag{16.126}$$

$$-\rho_w h\omega^2\widehat{\xi} = -\mu\left(\frac{d\widehat{u}}{dr} + ik\widehat{v}\right)\bigg|_{r=a_0} + \frac{Eh}{1-\widehat{\nu}^2}\left(-k^2\widehat{\xi} + \frac{i\nu k}{a_0}\widehat{\eta}\right). \tag{16.127}$$

Now invoke the assumptions that $h/a \ll 1$, ρ is of the same order of magnitude as ρ_w, and $a^2/\lambda^2 \ll 1$ in (16.126) and (16.127). This amounts to neglecting the terms which represent tube inertia, and approximating σ_{rx} in (16.111) by $\mu(\partial v/\partial x)$ and σ_{rr} in (16.110) by p. After considerable algebra, (16.126) and (16.127) reduce to:

$$\widehat{p} = \frac{Eh}{a_0^2}\widehat{\eta} - \frac{i\widehat{\nu}}{a_0 k}\mu\frac{d\widehat{u}}{dr}\bigg|_{r=a_0}, \tag{16.128}$$

$$\widehat{\xi} = \frac{i\widehat{v}}{ka_0}\widehat{\eta} - \frac{1 - \widehat{v}^2}{Ehk^2}\mu\frac{d\widehat{u}}{dr}\Big|_{r=a_0}. \tag{16.129}$$

We are now left with (16.122), (16.123), (16.128), and (16.129), subject to boundary conditions given by (16.124) and (16.125) and the pseudo boundary condition that $u(r)$ be nonsingular at $r = 0$.

Equations (16.123) and (16.128) can be combined to give:

$$\frac{d^2\widehat{u}}{dr^2} + \frac{1}{r}\frac{d\widehat{u}}{dr} + \frac{i\omega\rho}{\mu}\widehat{u} = \frac{ik}{\mu}\frac{Eh}{a_0^2}\widehat{\eta} + \frac{\widehat{v}}{a_0}\frac{d\widehat{u}}{dr}\Big|_{r=a_0}. \tag{16.130}$$

Satisfying the pseudo boundary condition, the solution to this Bessel's differential equation is given by:

$$\widehat{u}(r) = AJ_0(\beta r) + \frac{k}{\omega}\frac{Eh}{\rho a_0^2}\widehat{\eta} - \frac{\widehat{v}}{\beta a_0}AJ_1(\beta a_0), \tag{16.131}$$

where $\beta = \sqrt{i\omega/\nu}$, and A is an arbitrary constant. Next, from (16.122):

$$\widehat{v} = -\frac{ik}{r}\int_0^r r\widehat{u}(r)dr. \tag{16.132}$$

From (16.131) and (16.132):

$$\widehat{v}(r) = -\frac{ikA}{\beta}J_1(\beta r) - \frac{ik^2}{\omega}\frac{Eh\widehat{\eta}}{\rho a_0^2}\frac{r}{2} + \frac{ik\widehat{v}}{\beta a_0}A\frac{r}{2}J_1(\beta a_0). \tag{16.133}$$

Equations (16.131) and (16.133) give the expressions for $\widehat{u}(r)$ and $\widehat{v}(r)$, respectively. Subjecting them to the boundary conditions given in (16.124) and (16.125), introducing $\widehat{\beta} = \beta a_0$, and eliminating $\widehat{\xi}$ by the use of (16.129), the following two linear homogeneous equations for $\widehat{\eta}$ are developed:

$$\widehat{\eta}\left[\frac{\omega}{k}\frac{\widehat{v}}{a_0} - \frac{kEh}{\omega\rho a_0^2}\right] = A\left[J_0(\widehat{\beta}) + J_1(\widehat{\beta})\left\{\frac{i\beta\omega\mu(1 - \widehat{v}^2)}{Ehk^2} - \frac{\widehat{v}}{\widehat{\beta}}\right\}\right], \tag{16.134}$$

$$\widehat{\eta}\left[1 - \frac{k^2Eh}{\omega^2 2\rho a_0}\right] = AJ_1(\widehat{\beta})\left[\frac{k}{\omega\beta} - \frac{k\widehat{v}}{2\omega\beta}\right]. \tag{16.135}$$

For nonzero solutions, the determinant of the above set of linear algebraic equations in $\widehat{\eta}$ and A must be zero. As a result, the following characteristic equation is developed:

$$\left(\frac{k^2}{\omega^2}\frac{Eh}{2\rho a_0}\right)^2\left[2\widehat{\beta}\frac{J_0(\widehat{\beta})}{J_1(\widehat{\beta})} - 4\right] + \left(\frac{k^2}{\omega^2}\frac{Eh}{2\rho a_0}\right)\left[4\widehat{v} - 1 - 2\widehat{\beta}\frac{J_0(\widehat{\beta})}{J_1(\widehat{\beta})}\right] + (1 - \widehat{v}^2) = 0. \tag{16.136}$$

The solution to this quadratic equation will give k^2/ω^2 in terms of known quantities. Then we can find $k/\omega = (k_1 + ik_2)/\omega$. The wave speed, ω/k_1, and the damping factor may be evaluated by determining the real and imaginary parts of k/ω.

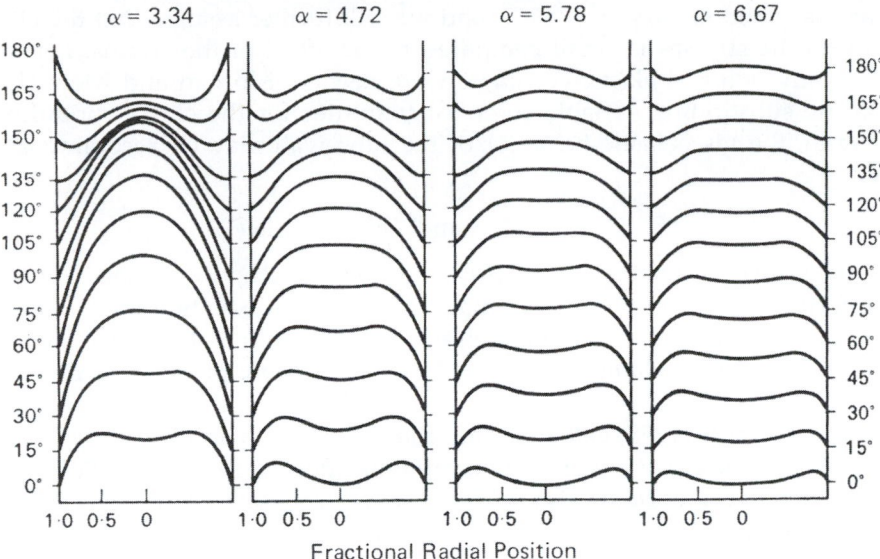

FIGURE 16.14 Velocity profiles of a sinusoidally oscillating flow in a pipe. A the lowest value of α, the Womersley number, the flow oscillations are slow enough so that the flow becomes fully developed, at least momentarily, during an oscillation. At higher values of α, the flow is slower in the center of the tube but it moves like a solid object. *Reproduced from McDonald, D. A. (1974),* Blood Flow in Arteries, *The Williams & Wilkins Company, Baltimore.*

Morgan and Kiely (1954) have provided explicit results for the wave speed, c, and the damping constant, k_2, in the limits of small and large α. Mazumdar (1999) has indicated that by an *in vivo* study, the wave speed, ω/k_1, can be evaluated noninvasively by monitoring the transit time as the time interval between the peaks of ultrasonically measured waveforms of the arterial diameter at two arterial sites at a known distance apart. Then from (16.136), E can be calculated. From either (16.134) or (16.135), A can be expressed in terms of $\hat{\eta}$, and with that, $\hat{u}(r)$ can be related to $\hat{p}$. Mazumdar gives details as to how the cardiac output may be calculated with the information developed in conjunction with pulsed Doppler flowmetry.

Figure 16.14 shows velocity profiles at intervals of $\Delta \omega t = 15°$ of the flow resulting from a pressure gradient varying as $\cos(\omega t)$ in a tube. As this is harmonic motion, only half a cycle is illustrated, and for $\omega t > 180°$, the velocity profiles are of the same form but opposite in sign. The Womersley number is α. The reversal of flow starts in the laminae near the wall. As the Womersley number increases, the profiles become flatter in the central region, there is a reduction in the amplitudes of the flow, and the rate of reversal of flow increases close to the wall. At $\alpha = 6.67$, the central mass of the fluid is seen to reciprocate like a solid core.

Effect of Viscoelasticity of Tube Material

In general, the wall of a blood vessel must be treated as viscoelastic. This means that the relations given in (16.104) and (16.105) must be replaced by corresponding relations for a tube of viscoelastic material. In this problem, all the stresses and strains in the problem are assumed to vary as $e^{i(kx-\omega t)}$, and we will further assume that the effect of the strain rates on the stresses is small compared to the effect of the strains. For the purely elastic case, only two real elastic constants were needed. Morgan and Kiely (1954) have shown that by substituting suitable complex quantities for the elastic modulus and the Poisson's ratio, the viscoelastic behavior of the tube wall may be accommodated. They introduce:

$$E^* = E - i\omega E', \quad \text{and} \quad \widehat{\nu}^* = \widehat{\nu} - i\omega\widehat{\nu}', \tag{16.137}$$

where, E' and $\widehat{\nu}'$ are new constants. In (16.104) and (16.105), E^* and $\widehat{\nu}^*$ replace E and $\widehat{\nu}$, respectively. The formulation will otherwise remain the same. An equation for k/ω will arise as before. The fact that E^* and $\widehat{\nu}^*$ are complex has to be taken into account while evaluating the wave velocity and the damping factor. Morgan and Kiely provide results appropriate for small and large α.

Morgan and Ferrante (1955) extended the study by Morgan and Kiely (1954) to the situation for small α values where there is Poiseuille-like flow in the thin, elastic-walled tube. The flow oscillations are small and they are superimposed on a large steady-stream velocity. The steady flow modifies the wave velocity. The wave velocity in the presence of a steady flow is the algebraic sum of the normal wave velocity and the steady-flow velocity. Morgan and Ferrante predict a decrease in the damping of a wave propagated in the direction of the stream and an increase in the damping when propagated upstream. However, the steady-flow component in arteries is so small in comparison with the pulse wave velocity that its role in damping is of little importance (see McDonald, 1974). Womersley (1957a) considered the situation where the flow oscillations are large in amplitude compared to the mean stream velocity (this is similar to the situation in an artery), predicting that the presence of a steady-stream velocity would produce a small increase in the damping.

Blood Vessel Bifurcation: An Application of Poiseuille's Formula and Murray's Law

Blood vessels bifurcate into smaller daughter vessels which in turn bifurcate to even smaller ones. On the basis that the flow satisfies Poiseuille's formula in the parent and all the daughter vessels, and by invoking the principle of minimization of energy dissipation in the flow, we can determine the optimal size of the vessels and the geometry of bifurcation. We recall that Hagen-Poiseuille flow involves established (fully developed) flow in a long tube. Here, for simplicity, we will assume that established Poiseuille flow exists in all the vessels. This is obviously a drastic assumption but the analysis will provide us with some useful insights.

Let the parent and daughter vessels be straight, circular in cross section, and lie in a plane.

Consider a parent vessel AB of length L_0 and radius a_0 in which the flow rate is Q, which bifurcates into two daughter vessels BC and BD with lengths L_1 and L_2, radii a_1 and a_2, and flow rates Q_1 and Q_2, respectively. The axes of vessels BC and BD are inclined at angles θ and ϕ with respect to the axis of AB, as shown in Figure 16.15. Points A, C, and D are fixed. The optimal sizes of the vessels and the optimal location of B have to determined from the principle of minimization of energy dissipation.

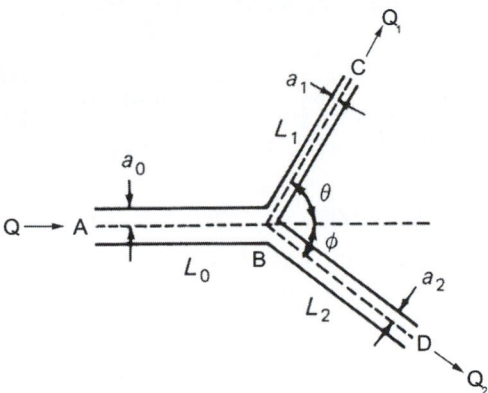

FIGURE 16.15 Schematic of an arterial bifurcation from one large vessel into two smaller ones. Here a_0, a_1, and a_2 are the vessel radii and the branching angles with respect to the incoming flow direction are θ and ϕ.

The total rate of energy dissipation by flow rate Q in a blood vessel of length L and radius a is equal to sum of the rate at which work is done on the blood, $Q\Delta p$, and the rate at which energy is used up by the blood vessel by metabolism, $K\pi a^2 L$, where K is a constant. For Hagen-Poiseuille flow, from (16.17): $Q = \dfrac{\pi}{8\mu}\dfrac{a^4}{L}\dfrac{\Delta p}{L}$. Therefore,

$$\text{Total energy dissipation} = \frac{8\mu L}{\pi a^4}Q^2 + K\pi a^2 L = \widehat{E}_1 (\text{say}). \tag{16.138}$$

To obtain the optimal size of a vessel for transport, for a given length of vessel, we need to minimize this quantity with respect to radius of the vessel. Thus,

$$\frac{\partial \widehat{E}_1}{\partial a} = -\frac{32\mu L}{\pi}Q^2 a^{-5} + 2K\pi L a = 0. \tag{16.139}$$

Solving for a:

$$a = \left[\frac{16\mu}{\pi^2}K\right]^{1/6} Q^{1/3}. \tag{16.140}$$

Equation (16.140) gives the optimal radius for the blood vessel indicating that minimum energy dissipation occurs under this condition. The optimal relationship, $Q \sim a^3$, is called *Murray's Law*.

With (16.140), the minimum value for energy dissipation is

$$\widehat{E}_{1,\,min} = \frac{3\pi}{2}KLa^2. \tag{16.141}$$

Next, consider the flow with the branches. The minimum value for energy dissipation with branches is

$$\widehat{E}_{2,\,min} = \frac{3\pi}{2}K\left(L_0 a_0^2 + L_1 a_1^2 + L_2 a_2^2\right). \tag{16.142}$$

Also,

$$a_0 = \left[\frac{16\mu}{\pi^2}K\right]^{1/6} Q_0^{1/3}, \quad a_1 = \left[\frac{16\mu}{\pi^2}K\right]^{1/6} Q_1^{1/3}, \quad \text{and} \quad a_2 = \left[\frac{16\mu}{\pi^2}K\right]^{1/6} Q_2^{1/3}, \quad (16.143)$$

and from mass conservation:

$$Q = Q_1 + Q_2 \rightarrow a_0^3 = a_1^3 + a_2^3. \quad (16.144)$$

The lengths L_0, L_1, and L_2 depend on the location of point B. The optimum location of point B is determined by examining associated variational problems (see Fung, 1997).

Any small movement of B changes $\widehat{E}_{2,\,min}$ by $\delta\widehat{E}_{2,\,min}$ and

$$\delta\widehat{E}_{2,\,min} = \frac{3\pi}{2}K\left(\delta L_0 \; a_0^2 + \delta L_1 \; a_1^2 + \delta L_2 \; a_2^2\right). \quad (16.145)$$

The optimal location of B would make $\delta\widehat{E}_{2,\,min} = 0$ for arbitrary small movement δL of point B. By making such displacements of B, one at a time, in the direction of AB, in the direction of BC, and finally in the direction of DB, and setting the value of the corresponding $\delta\widehat{E}_{2,\,min}$ to zero, we develop a set of three conditions governing optimization. These are:

$$\cos\theta = \frac{a_0^4 + a_1^4 - a_2^4}{2a_0^2a_1^2}, \quad \cos\phi = \frac{a_0^4 - a_1^4 + a_2^4}{2a_0^2a_2^2}, \quad \cos(\theta+\phi) = \frac{a_0^4 - a_1^4 - a_2^4}{2a_1^2a_2^2}. \quad (16.146)$$

Together with (16.144), the set (16.146) may be solved for the optimum angle θ as

$$\cos\theta = \frac{a_0^4 + a_1^4 - \left(a_0^3 - a_1^3\right)^{4/3}}{2a_0^2a_1^2}, \quad (16.147)$$

and a similar equation for ϕ. Comparison of these optimization results with experimental data are noted to be excellent (see Fung, 1997).

Reflection of Waves at Arterial Junctions: Inviscid Flow and Long Wavelength Approximation

Arteries have branches. When a pressure or a velocity wave reaches a junction where the parent artery 1 bifurcates into daughter tubes 2 and 3 as shown in Figure 16.16, the incident wave is partially reflected at the junction into the parent tube and partially transmitted down the daughters. In the long wavelength approximation, we may neglect the flow at the junction. Let the longitudinal coordinate in each tube be x, with $x = 0$ at the bifurcation. The incident wave in the parent tube comes from $x = -\infty$.

Let p_I be the oscillatory pressure associated with the incident wave, let p_R be associated with the reflected wave, and let p_{T1} and p_{T2} be associated with the transmitted waves. Let the pressure be a single valued and continuous function at the junction for all time t. The continuity requirement ensures that there are no local accelerations. Under these conditions, at the junction:

$$p_I + p_R = p_{T1} = p_{T2}. \quad (16.148)$$

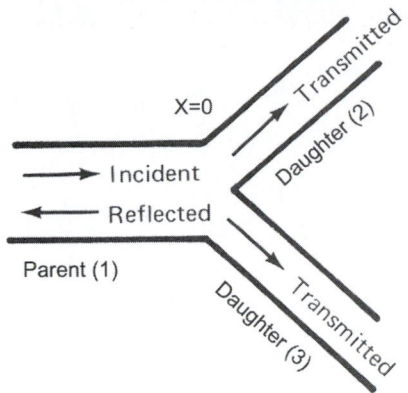

FIGURE 16.16 Schematic of an arterial bifurcation: reflection. Here the change in impedance at the junction can cause a reflected wave to travel backward along the parent artery.

Next, let Q_I be the flow rate associated with the incident wave, let Q_R be associated with the reflected wave, and let Q_{T1} and Q_{T2} be associated with the transmitted waves. The flow rate is also taken to be single valued and continuous at the junction for all time t. The continuity requirement ensures conservation of mass. At the junction,

$$Q_I - Q_R = Q_{T1} + Q_{T2}. \tag{16.149}$$

Let the undisturbed cross-sectional areas of the tubes be A_1, A_2, and A_3, and the intrinsic wave speeds be c_1, c_2, c_3, respectively. In general, for a fluid of density ρ flowing under the influence of a wave with intrinsic wave speed c, through a tube of cross-sectional area A, the flow rate Q is related to the mean velocity u by

$$Q = Au = \pm \frac{A}{\rho c} p, \tag{16.150}$$

where we have employed the relationship given in (16.68). The plus or the minus sign applies depending on whether the wave is going in the positive x-direction or in the negative x-direction. The quantity $A/\rho c$ is called the *characteristic admittance* of the tube and is denoted by Y, while $\rho c/A$ is called the *characteristic impedance* of the tube and is denoted by Z. Admittance is seen to be the ratio of the oscillatory flow to the oscillatory pressure when the wave goes in the direction of the $+x$ axis. With these definitions,

$$Q = Au = \pm Yp = \pm \frac{p}{Z}. \tag{16.151}$$

Equation (16.149) may be written in terms of admittances or impedances as:

$$Y_1(p_I - p_R) = \sum_{j=2}^{3} Y_j p_{Tj}, \quad \text{or} \quad \frac{(p_I - p_R)}{Z_1} = \sum_{j=2}^{3} \frac{p_{Tj}}{Z_j}. \tag{16.152}$$

We can simultaneously solve (16.148) and (16.152) to produce:

$$\frac{p_R}{p_I} = \frac{Y_1 - \sum Y_j}{Y_1 + \sum Y_j} = R, \quad \text{and} \quad \frac{p_{Tj}}{p_I} = \frac{2Y_1}{Y_1 + \sum Y_j} = T, \tag{16.153}$$

or,

$$\frac{p_R}{p_I} = \frac{Z_1^{-1} - \sum Z_j^{-1}}{Z_1^{-1} + \sum Z_j^{-1}}, \quad \text{and} \quad \frac{p_{Tj}}{p_I} = \frac{2Z_1^{-1}}{Z_1^{-1} + \sum Z_j^{-1}}. \tag{16.154}$$

In (16.153), R and T are called the *reflection* and *transmission coefficients*, respectively. From (16.153), the amplitudes of the reflected and transmitted pressure waves are R and T times the amplitude of the incident pressure wave. These relations can be written in more explicit manner as follows (see Lighthill, 1978).

The contribution of the incident wave to the pressure in the parent tube is given by

$$p_I = P_I\, f\!\left(t - \frac{x}{c_1}\right), \tag{16.155}$$

where P_I is an amplitude parameter and f is a continuous, periodic function whose maximum value is 1. The corresponding contribution to the flow rate is

$$Q_I = A_1 u = Y_1 P_I f\!\left(t - \frac{x}{c_1}\right). \tag{16.156}$$

The contributions to pressure from the reflected and transmitted waves to the parent and daughter tubes, respectively, are:

$$p_R = P_R\, g\!\left(t + \frac{x}{c_1}\right), \quad \text{and} \quad p_{Tj} = P_{Tj} h_j\!\left(t - \frac{x}{c_j}\right), \quad (j = 2, 3), \tag{16.157}$$

where P_R and P_T are amplitude parameters, and g and h are continuous, periodic functions. The corresponding contributions to the flow rates are:

$$Q_R = -Y_1 P_R\, g\!\left(t + \frac{x}{c_1}\right), \quad \text{and} \quad Q_{Tj} = Y_j P_{Tj} h_j\!\left(t - \frac{x}{c_j}\right), \quad (j = 2, 3). \tag{16.158}$$

Therefore, the pressure perturbation in the parent tube is given by (16.155) and (16.157) to be:

$$\frac{p}{P_I} = f\!\left(t - \frac{x}{c_1}\right) + \frac{P_R}{P_I} f\!\left(t + \frac{x}{c_1}\right), \tag{16.159}$$

and the flow rate, from (16.156) and (16.158), is:

$$Q = Y_1 P_I \left[f\!\left(t - \frac{x}{c_1}\right) - \frac{P_R}{P_I} f\!\left(t + \frac{x}{c_1}\right) \right]. \tag{16.160}$$

The transmission of energy by the pressure waves is of interest. The rate of work done by the wave motion through the cross section of the tube or, equivalently, the rate of transmission of energy by the wave is clearly pAu or pQ, which is the same as P^2/Z from (16.151). Now we can calculate the incident, reflected, and transmitted quantities at the junction. Thus,

$$\text{Rate of energy transmission by incident wave} = \frac{p_I^2}{Z_1}, \qquad (16.161)$$

$$\text{Rate of energy transmission by reflected wave} = \frac{(Rp_I)^2}{Z_1} = R^2 \frac{p_I^2}{Z_1} \qquad (16.162)$$

The quantity R^2 is called the *energy reflection coefficient*. Similarly, the energy transmission coefficient, which is the rate of energy transfer in the two transmitted waves compared with that in the incident wave, may be defined by

$$\frac{\dfrac{p_{T2}^2}{Z_2} + \dfrac{p_{T3}^2}{Z_3}}{\dfrac{p_I^2}{Z_1}} = \frac{Z_2^{-1} + Z_3^{-1}}{Z_1^{-1}} \left(\frac{p_{T2}}{p_I}\right)^2 = \frac{Z_2^{-1} + Z_3^{-1}}{Z_1^{-1}} T^2, \qquad (16.163)$$

where we have noted that in our case $P_{T2} = P_{T3}$.

A comparison of (16.159) and (16.160) shows that, if we include reflection at bifurcations, the pressure and flow waveforms are no longer of the same shape. Pedley (1980) has offered interesting discussions about the behavior of the waves at the junction. From (16.153), for real values of c_j and Y_j, if $\sum Y_j < Y_1$, then the reflected wave has the same sign as the incident wave, and the pressures in the two waves are in phase at $x = 0$. They combine additively to form a large-amplitude fluctuation at the junction, and the effect of the junction is similar to that of a closed-end ($P_R = P_I$). If $\sum Y_j > Y_1$, there is a phase change at $x = 0$, the smallest-amplitude pressure fluctuation occurs there, and the junction resembles an open end ($P_R = -P_I$). If $\sum Y_j = Y_1$, there is no reflected wave, and the junction is said to be perfectly matched. Pedley (2000) has noted that the increase in the pressure wave amplitude in the aorta with distance down the vessel may indicate that there is a closed-end type of reflection at (or beyond) the iliac bifurcation. Peaking of the pressure pulse is a consequence of the closed-end type of reflection in a blood vessel.

Waves in more complex systems consisting of many branches may be analyzed by repeated application of the results presented in this section.

Next, we will study blood flow in curved tubes. Almost all blood vessels have curvature and the curvature affects both the nature (stability) and volume flow rate.

Flow in a Rigid-Walled Curved Tube

Blood vessels are typically curved and the curvature effects have to be accounted for in modeling in order to get a realistic understanding. The aortic arch is a 3D bend twisting through more than 180° (Ku, 1997). In a curved tube, fluid motion is not everywhere parallel to the curved axis of the tube (see Figure 16.17), secondary motions are generated, the velocity profile is distorted, and there is increased energy dissipation. However, curving of a tube increases the stability of flow, and the critical Reynolds number increases significantly,

FIGURE 16.17 Schematic of flow in a curved tube. Here the radius of curvature is R and the curve of the tube causes a secondary flow within the tube.

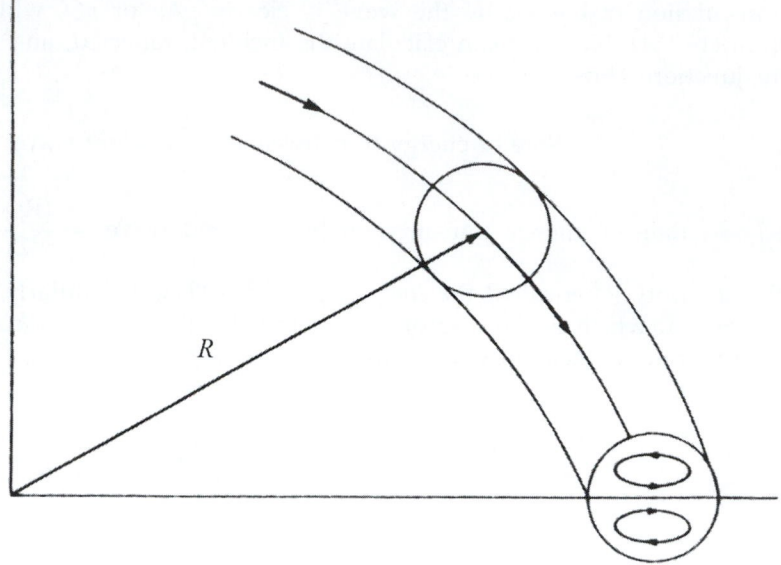

and a critical Reynolds number of 5000 is easily obtained (see McDonald, 1974). Flows in curved tubes are discussed in detail by McConalogue and Srivastava (1968), Singh (1974), Pedley (1980), and Berger et al. (1983). In this section, we concentrate on some of the most important aspects of flow in a uniformly curved vessel of small curvature. The wall is considered to be rigid. Pulsatile flow through a curved tube can induce complicated secondary flows with flow reversals and is very difficult to analyze. It may be noted that steady viscous flow in a symmetrical bifurcation resembles that in two curved tubes stuck together. Thus, an understanding gained in studying curved flows will be beneficial in that regard as well.

Consider fully developed, steady, laminar, viscous flow in a curved tube of radius a and a uniform radius of curvature R. Let us employ the toroidal coordinate system (r', α, θ), where r' denotes the distance from the center of the circular cross section of the pipe, α is the angle between the radius vector and the plane of symmetry, and θ is the angular distance of the cross section from the entry of the pipe (see Figure 16.18). Let the corresponding dimensional velocity components be (u', v', w'). As a fluid particle traverses a curved path of radius R (radius of curvature) with a (longitudinal) speed w' along the θ direction, it will experience a lateral (centrifugal) acceleration of w'^2/R, and a lateral force equal to $m_p w'^2/R$, where m_p is the mass of the particle. The radii of curvature of the particle paths near the inner bend, the central axis, and the outer bend will be of increasing magnitude as we move away from the inner bend. Also, due to the no-slip condition, the velocities, w', of particles near the inner and outer bends will be lower, while that of the particle at the central axis will be the highest. The particle at the central axis will experience the highest centrifugal force while that near the outer bend will experience the least. A lateral pressure gradient will cause the faster flowing fluid near the center to be swept toward the outside of the bend and to be replaced at the inside by the slower moving fluid near the wall. In effect, a secondary circulation will be set up resulting in two vortices,

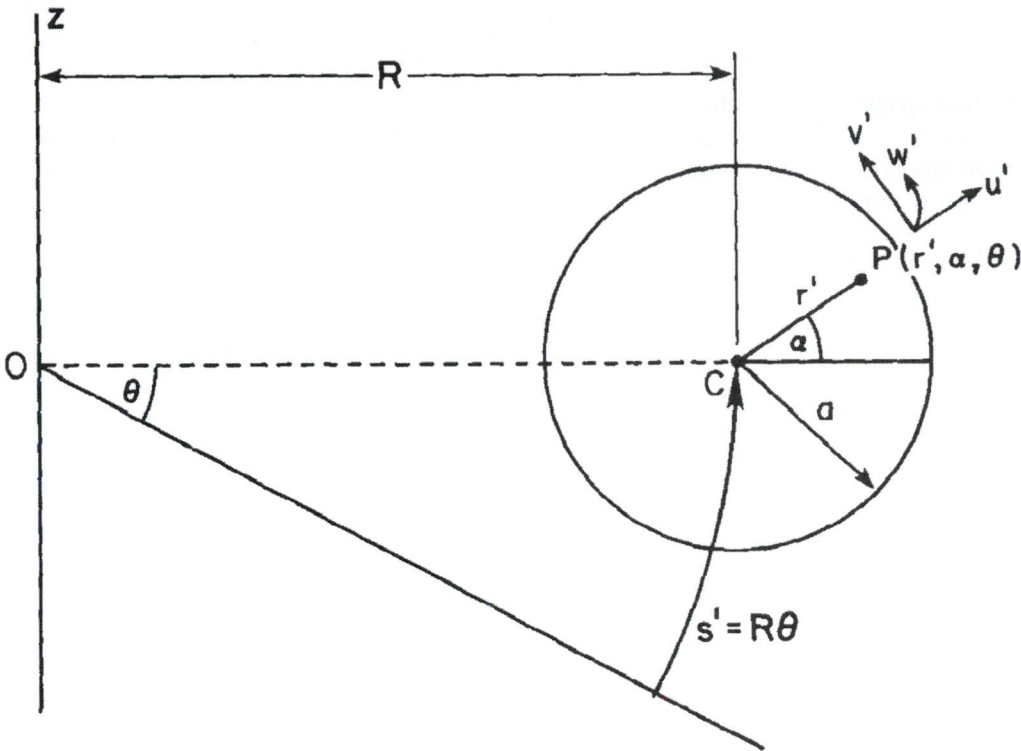

FIGURE 16.18 Toroidal coordinate system. This coordinate system is needed to analyze the flow in a round tube with radius a that has a constant radius of curvature R.

called *Dean vortices* because Dean (1928) was the first to systematically study these secondary motions in curved tubes (see Figure 16.17). Dean vortices significantly influence the axial flow. The wall shear near the outside of the bend is relatively higher than the (much reduced) wall shear on the inside of the bend. Fully developed flow upstream of or through curved tubes exhibits velocity that skews toward the outer wall of the bend. For most arterial flows, skewing will be toward the outer wall. If the flow into the entrance region of a curved tube is not developed, then the inviscid core of the fluid in the curve can act like a potential vortex with velocity skewing toward the inner wall.

Secondary flow in curved tubes is utilized in heart-lung machines to promote oxygenation of blood (Fung, 1997). In the machine, blood flows inside the curved tube and oxygen flows on the outside. The tube is permeable to oxygen. The secondary flow in the tube stirs up the blood and results in faster oxygenation.

Let us now analyze the flow in a curved tube to better understand the salient features. Introduce nondimensional variables, $r = r'/a, s = R\theta/a, \mathbf{u} = \mathbf{u}'/W_0$, and $p = p'/\rho\overline{W_0}^2$, where $\mathbf{u} = (u, v, w)$ is the velocity vector, P is the pressure, ρ is the density, and $\overline{W_0}$ is the mean axial velocity in the pipe. Restrict consideration to the case where the flow is fully developed ($\partial\mathbf{u}/\partial s = 0$). Introduce the dimensionless ratio,

$$\delta = \frac{\text{radius of tube cross section}}{\text{radius of curvature of the centerline}} = \frac{a}{R}. \tag{16.164}$$

We restrict consideration to a uniformly curved tube, $\delta =$ constant, and with a slight curvature (weakly curved) $\delta \ll 1$. Since δ is a constant, the velocity field is independent of s, the components are functions only of r and α, and the pressure gradient $\partial p / \partial s$ is independent of s. With δ constant, the only way that the transverse velocities are affected by the axial velocity is through the centrifugal force, and it is the centrifugal force that drives the secondary motion. This means that the centrifugal force terms must be of the same order of magnitude as the viscous and inertial terms in the momentum equation, and this requires rescaling the velocities. The transformation that accomplishes this is $(u, v, w) \rightarrow (\sqrt{\delta}\hat{u}, \sqrt{\delta}\hat{v}, \hat{w})$. We will also let $s = R\theta / a = \sqrt{1/\delta}\, \tilde{s}$ for convenience.

In the following, we shall omit writing the "^" on u, v, w, and the "~" on s for convenience. When $\delta \ll 1$, the major contribution to the axial pressure gradient may be separated from the transverse component, and we may write

$$p = p_0(s) + \delta p_1(r, \alpha, s) + \dots. \tag{16.165}$$

Under all these restrictions, the governing equations become:

$$\frac{\partial u}{\partial r} + \frac{u}{r} + \frac{1}{r}\frac{\partial v}{\partial \alpha} = 0, \tag{16.166}$$

$$u\frac{\partial u}{\partial r} + \frac{v}{r}\frac{\partial u}{\partial \alpha} - \frac{v^2}{r} - w^2\cos\alpha = -\frac{\partial p_1}{\partial r} - \frac{2}{\kappa}\frac{1}{r}\frac{\partial}{\partial \alpha}\left(\frac{\partial v}{\partial r} + \frac{v}{r} - \frac{1}{r}\frac{\partial u}{\partial \alpha}\right), \tag{16.167}$$

$$u\frac{\partial v}{\partial r} + \frac{v}{r}\frac{\partial v}{\partial \alpha} + \frac{uv}{r} + w^2\sin\alpha = -\frac{1}{r}\frac{\partial p_1}{\partial \alpha} + \frac{2}{\kappa}\frac{\partial}{\partial r}\left(\frac{\partial v}{\partial r} + \frac{v}{r} - \frac{1}{r}\frac{\partial u}{\partial \alpha}\right), \tag{16.168}$$

$$u\frac{\partial w}{\partial r} + \frac{v}{r}\frac{\partial w}{\partial \alpha} = -\frac{\partial p_0}{\partial s} + \frac{2}{\kappa}\left(\frac{\partial^2 w}{\partial r^2} + \frac{1}{r}\frac{\partial w}{\partial r} + \frac{1}{r^2}\frac{\partial^2 w}{\partial \alpha^2}\right). \tag{16.169}$$

The boundary conditions are:

$$u = v = w = 0 \quad \text{at } r = 1, \quad \text{no singularity} \quad \text{at } r = 0. \tag{16.170}$$

The flow is governed by just one parameter κ in the equations, and it is called the *Dean number*. It is given by

$$\kappa = \sqrt{\delta}\,\frac{2a\overline{W}_0}{\nu} = \sqrt{\delta}\, 2Re, \tag{16.171}$$

where $\overline{W}_0$ is the mean axial velocity in the pipe. The Dean number is the Reynolds number modified by the pipe curvature. The appearance of the 2 in the definition of the Dean number is by convention. At higher Dean numbers, the flow can separate along the inner boundary curve.

There are many different definitions of Dean number in the literature and the reader must be careful to see which particular form is being used in any given discussion.

From (16.169), $\partial p_0 / \partial s$ is independent of s, and P_0 can be written as $P_0(s) = -G\,s$, where G is a constant. Equation (16.166) admits the existence of a stream function for the secondary flow, ψ, defined by

$$u = \frac{1}{r}\frac{\partial \psi}{\partial \alpha}, \qquad v = -\frac{\partial \psi}{\partial r}. \tag{16.172}$$

Substitution of (16.172) into (16.169) yields

$$\nabla_1^2 w - \frac{\kappa}{2}\frac{\partial p_0}{\partial s} = \frac{\kappa}{2r}\left(\frac{\partial \psi}{\partial \alpha}\frac{\partial w}{\partial r} - \frac{\partial \psi}{\partial r}\frac{\partial w}{\partial \alpha}\right), \tag{16.173}$$

while elimination of pressure from (16.167) and (16.168) yields

$$\frac{2}{\kappa}\nabla_1^4 \psi - \frac{1}{r}\left(\frac{\partial \psi}{\partial r}\frac{\partial}{\partial \alpha} - \frac{\partial \psi}{\partial \alpha}\frac{\partial}{\partial r}\right)\nabla_1^2 \psi = -2w\left(\sin\alpha\frac{\partial w}{\partial r} + \frac{\cos\alpha}{r}\frac{\partial w}{\partial \alpha}\right), \tag{16.174}$$

where

$$\nabla_1^2 \psi = \frac{\partial^2}{\partial r^2} + \frac{1}{r}\frac{\partial}{\partial r} + \frac{1}{r^2}\frac{\partial^2}{\partial \alpha^2}. \tag{16.175}$$

The boundary conditions are:

$$\psi = \frac{\partial \psi}{\partial r} = w = 0, \quad \text{at } r = 1. \tag{16.176}$$

Equations (16.173) and (16.174) subject to conditions (16.176) have to be solved.

For small Dean number, following Dean (1928), we expand w and ψ in terms of a series in powers of the Dean number as follows:

$$w = \sum_{n=0}^{\infty}\kappa^{2n}w_n(r,\alpha), \quad \text{and} \quad \psi = \kappa\sum_{n=0}^{\infty}\kappa^{2n}\psi_n(r,\alpha). \tag{16.177}$$

The w_0 term corresponds to Poiseuille flow in a straight tube with rigid walls. The ψ_0 term is $O(\kappa)$. The series expansion in κ is equivalent to the successive approximation of inertia terms in lubrication theory. The leading term in the secondary flow takes the form of a pair of counter-rotating helical vortices, placed symmetrically with respect to the plane of symmetry. This flow pattern arises because of a centrifugally induced pressure gradient, approximately uniform over the cross section. The dimensionless volume flux is

$$\frac{Q}{\pi a^2 \overline{W}} = 1 - 0.0306\left(\frac{K}{576}\right)^2 + 0.0120\left(\frac{K}{576}\right)^4 + O(K^6), \tag{16.178}$$

where $K = (2a/R)(W_{max}a/\nu)^2 = 2(\kappa)^2$ is another frequently used definition of Dean's number. Here, $W_{max} = 2\overline{W}$; W_{max} and $\overline{W}$ are the maximum and mean velocities, respectively, in a straight pipe of the same radius under the same axial pressure gradient and under fully developed flow conditions. The first term corresponds to the Poiseuille straight pipe solution. The effect of curvature is seen to reduce the flux.

Many other authors define Dean's number by

$$D = \sqrt{2\delta} \frac{\widehat{G}a^2}{\mu} \frac{a}{\nu},$$

(16.179)

where $-\widehat{G}$ is the dimensional pressure gradient,

$$\widehat{G} = -\frac{8\mu\overline{W}}{a^2}.$$

(16.180)

In terms of D, (16.178) becomes

$$\frac{Q}{\pi a^2 \overline{W}} = 1 - 0.0306 \left(\frac{D}{96}\right)^4 + 0.0120 \left(\frac{D}{96}\right)^8 + O(D^{12}).$$

(16.181)

Next, consider the friction factor for flow in a curved tube. Let λ_c and λ_s denote the flow resistance in a curved and a straight pipe, respectively, while the flows are subject to pressure gradients equal in magnitude. The ratio λ is

$$\lambda = \frac{\lambda_c}{\lambda_s} = \left(\frac{Q_c}{Q_s}\right)^{-1} = 1 + 0.0306 \left(\frac{K}{576}\right)^2 - 0.0110 \left(\frac{K}{576}\right)^4 + \cdots,$$

(16.182)

where Q_c and Q_s are the fluxes in straight and curved pipes, respectively. The flow resistance in a curved tube is not affected by the first-order terms and is increased only by higher order terms. With regard to shear stress, the curvature increases axial wall shear on the outside wall and decreases it on the inside, and it also generates a positive secondary shear in the α direction.

The size of the coefficients suggests that the small D expansion is valid for values of D up to about 100 or $K \approx 600$, and the results here are useful only for smaller blood vessels. Pedley points out that in the canine aorta, where $\delta \approx 0.2$, the mean D is greater than 2000. As mentioned earlier, flow in a curved tube is much more stable than that in a straight tube and the critical Reynolds number could be as high as 5000 which corresponds to $K \approx 1.6 \times 10^6$.

For intermediate values of D, only numerical solutions are possible due to the importance of nonlinear terms. Numerical results of Collins and Dennis (1975) for developed flow up to a Dean number of 5000 are stated to compare very well with experimental results. At intermediate values of D, a boundary layer develops on the outside wall of the bend where the axial shear is high. The secondary flow in the core is approximately uniform and continues to manifest a two-vortex structure. At higher values of D, there is greater distortion of the secondary streamlines. The wall shear at $r = 1$, $\alpha = 0$, is proportional to D ($\approx 0.85D$); see Pedley (2000).

At large Dean numbers, the centers of the two vortices move toward the outer bend, $\alpha = 0$, and the flow is very much reduced compared with a straight pipe for equal magnitude pressure gradients. Detailed studies using advanced computational methods are required to resolve the flow structure at large D. They are as yet unavailable in the published literature.

Pedley (2000) discusses nonuniqueness of curved-tube flow results. When D is sufficiently small, the steady-flow equations have just one solution and there is a single secondary flow vortex in each half of the tube. However, there is a critical value of D, above which more than one steady solution exists and these may correspond to four vortices, two in each half. Again, detailed computational studies are necessary to resolve these features.

We will next study the flow of blood in collapsible tubes. The role of pressure difference, $(P_e - P(x))$, on the vessel wall will be significant in such flows.

Flow in Collapsible Tubes

At large negative values of the transmural pressure difference (the difference between the pressure inside and the pressure outside), the cross-sectional area of a blood vessel is either very small—the lumen being reduced to two narrow channels separated by a flat region of contact between the opposite walls—or it may even fall to zero. There is an intermediate range of values of transmural pressure difference in which the cross section is very compliant and even the small viscous or inertial pressure drop of the flow may be enough to cause a large reduction in area, that is, collapse. Collapse occurs in a number of situations; a listing is given by Kamm and Pedley (1989). Collapse occurs, for example, in systemic veins above the heart (and outside the skull) as a result of the gravitational decrease in internal pressure with height; intramyocardial coronary blood vessels during systole; systemic arteries compressed by a sphygmomanometer cuff, or within the chest during cardiopulmonary resuscitation; pulmonary blood vessels in the upper levels of the lung; large intrathoracic airways during forced expiration or coughing; and the urethra during micturition and in the ureter during peristaltic pumping. Collapse, therefore occurs both in small and large blood vessels, and as a result both at low and high Reynolds numbers. In certain cases, at high Reynolds number, collapse is accompanied by self-excited, flow-induced oscillations. There is audible sound. For example, Korotkoff sounds heard during sphygmomanometry are associated with this.

A Note on Korotkoff Sounds

Korotkoff sounds, named after Dr. Nikolai Korotkoff, a physician who described them in 1905, are sounds that physicians listen for when they are taking blood pressure. When the cuff of a sphygmomanometer is placed around the upper arm and inflated to a pressure above the systolic pressure, there will be no sound audible because the pressure in the cuff would be high enough to completely occlude the blood flow. If the pressure is now dropped, the first Korotkoff sound will be heard. As the pressure in the cuff is the same as the pressure produced by the heart, some blood will be able to pass through the upper arm when the pressure in the artery rises during systole. This blood flows in spurts as the pressure in the artery rises above the pressure in the cuff and then drops back down, resulting in turbulence that results in audible sound. As the pressure in the cuff is allowed to fall further, thumping sounds continue to be heard as long as the pressure in the cuff is between the systolic and diastolic pressures, as the arterial pressure keeps on rising above and dropping back below the pressure in the cuff. Eventually, as the pressure in the cuff drops further, the sounds change in quality, then become muted, then disappear altogether when the pressure in the cuff drops below the diastolic pressure. Korotkoff described five types of Korotkoff sounds. The first Korotkoff sound is the snapping sound first heard at the systolic pressure. The second sounds are the murmurs heard for most of the area between the systolic and diastolic pressures. The third and the fourth sounds appear at pressures within 10 mm Hg above the diastolic blood pressure, and are described as "thumping" and "muting." The fifth Korotkoff sound is silence as the cuff pressure drops below the diastolic pressure. Traditionally, the systolic blood pressure is taken to be the pressure at which the first Korotkoff sound is first heard and the diastolic blood pressure is the pressure at which the fourth Korotkoff sound is just barely audible. There has recently been a move toward the use of the fifth Korotkoff sound (i.e., silence) as the diastolic pressure, as this has been felt to be more reproducible.

Starling Resistor: A Motivating Experiment for Flow in Collapsible Tubes

The study of flows in collapsible tubes is facilitated by a well-known experiment carried out under varying conditions by different researchers. In the experiment, a length of uniform collapsible tube is mounted at each end to a shorter length of rigid tube and is enclosed in a chamber whose pressure p_e can be adjusted. Fluid, say water, flows through the tube. The inlet and outlet pressures at the ends of the collapsible tube are p_1 and p_2. The volume rate of flow is Q. The pressures and the flow rate are next varied in a systematic way and the results are noted. The setup described is called a *Starling resistor* after physiologist Starling (see Fung, 1997). This experiment will enable us understand some aspects of actual flows in physiological systems. There are many different versions of the description of the Starling resistor experiment in the literature. The experiments have been carried out under both steady flow and unsteady flow conditions. We will describe the experiments as reported by Kamm and Pedley (1989).

CASE (1): $(P_1 - P_2)$ IS INCREASED WHILE $(P_1 - P_E)$ IS HELD CONSTANT

This is accomplished either by reducing p_2 with p_1 and p_e fixed, or by simultaneously increasing p_1 and p_e while p_2 is held constant. With either procedure, Q at first increases, but above a critical value it levels off and the condition of *flow limitation* is reached. In this condition, however much the driving pressure is increased the flow rate remains constant, or it may even fall as a result of increasingly severe tube collapse. This experiment is relevant to forced expiration from the lung, to venous return, and to micturition.

CASE (2): $(P_1 - P_2)$ OR Q IS INCREASED WHILE $(P_2 - P_E)$ IS HELD CONSTANT AT SOME NEGATIVE VALUE

In this case, the tube is collapsed at low flow rates, but starts to open up from the upstream end as Q increases above a critical value, so that the resistance falls and $(p_1 - p_2)$ ceases to rise. This is termed *pressure-drop limitation*. This experiment does not seem to apply to any particular physiological condition.

CASE (3): $(P_1 - P_2)$ IS HELD CONSTANT WHILE $(P_2 - P_E)$ IS DECREASED FROM A LARGE POSITIVE VALUE

In this case, the tube first behaves as though it were rigid and the flow rate is nearly constant. Then as $(p_2 - p_e)$ becomes sufficiently negative to produce partial collapse, the resistance rises and Q begins to fall. This experiment is relevant to pulmonary capillary flows.

CASE (4): P_E FIXED

The outlet end is connected to a flow resistor. The pressure downstream of the flow resistor is fixed (flow is exposed to atmosphere). Thus p_2 is equal to atmospheric pressure plus Q times the fixed resistance; p_1 is varied.

In this case, p_2 varies with Q due to the presence of a fixed downstream resistance. The degree of tube collapse (progressive collapse) also varies with Q for the same reason. At high flow rates, the tube is distended and its resistance is low. As the flow rate is reduced below a critical value the tube starts to collapse. Its resistance and $(p_1 - p_2)$ both increase as Q is decreased. Only when the tube is severely collapsed along most of its length does $(p_1 - p_2)$ start to decrease again as Q approaches zero. When p_1 is approximately equal to p_e, virtually the entire tube is collapsed (Fung, 1997). The tube often flutters in Case 4 (see discussions in Fung).

CASE (5): UNSTEADY FLOW EXPERIMENTS

Excepting at small Reynolds numbers, there is always some parameter range where flow oscillations occur. The oscillations have a wide variety of modes.

The experiments reveal the importance of a tube law relating transmural pressure difference with the area of cross section of the collapsible tube and the flow and pressure drop limitations when analyzing collapsible tubes. Shapiro (1977a, 1977b) has developed a comprehensive one-dimensional theory for steady flow based on a suitable tube law. Kamm and Shapiro (1979) have extended it to unsteady flow in a collapsible tube. In the following, we shall discuss the steady-flow theory.

One-Dimensional Flow Treatment

The equations describing one-dimensional flow in a collapsible tube are similar to those in gas dynamics or channel flow of a liquid with a free surface (see Shapiro, 1977a). Here, we will study the one-dimensional, steady-flow formulation for the collapsible tube. However, first let us recapitulate the traditional basic equations for one-dimensional flow in a smoothly varying elastic tube (see "Pulse Wave Propagation in an Elastic Tube: Inviscid Theory" in Section 16.3).

We studied flow in an elastic tube with cross section $A(x, t)$ and longitudinal velocity $u(x, t)$. The constant external pressure on the tube was set at p_e. The primary mechanism of unsteady flow in the tube was wave propagation. The transmural pressure difference $(p - p_e)$ was related to the local cross-sectional area by a "tube law" which involved hoop tension, which may be expressed as

$$(p - p_e) = \widehat{P}(A), \tag{16.183}$$

where the functional form $\widehat{P}$ depends on data. For disturbances of small amplitude and long wavelength compared to the tube diameter:

$$A = A_0 + A', \quad p - p_e = \widehat{P}(A_0) + p', \quad |A'| \ll A_0, \quad |p'| \ll \widehat{P}(A_0), \tag{16.184}$$

and the wave speed is given by

$$c^2 = \frac{A}{\rho}\frac{d\widehat{P}}{dA} = \frac{A}{\rho}\frac{d(p - p_e)}{dA}. \tag{16.185}$$

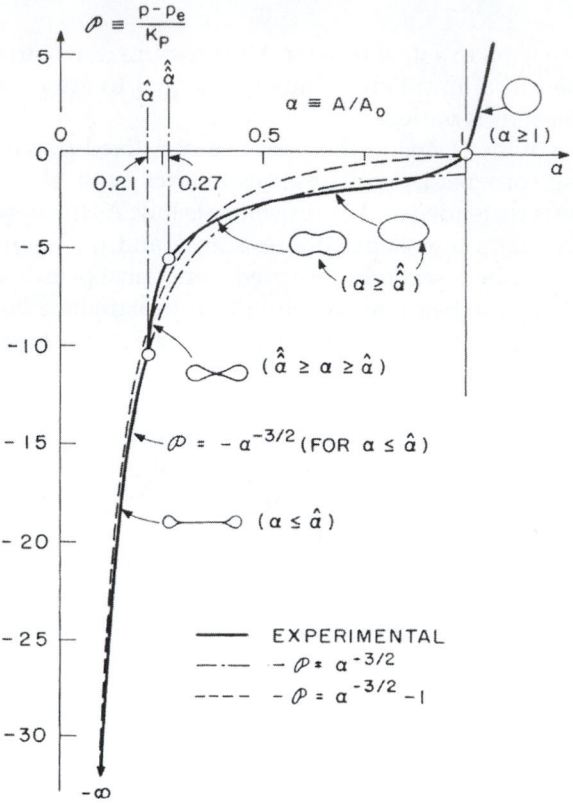

FIGURE 16.19 Behavior of a collapsible tube. Here α is the tube area ratio and is 1 when the pressure inside the tube is greater than the pressure outside the tube. The vertical axis is proportional to the interior minus exterior pressure difference. As the pressure in the tube decreases, the available cross-sectional area is reduced, and this reduction takes place rapidly when the tube collapses. *Reproduced with permission from the American Society of Mechanical Engineers, NY.*

Tube collapse is associated with negative transmural pressure difference, and the pressure difference is supported by bending stiffness of the tube wall (see Figure 16.19). Contrast this with positive transmural pressure difference discussed earlier, which was supported by hoop tension. Following Shapiro (1977a), introduce

$$P = \frac{(p - p_e)}{K_p}, \quad \text{and} \quad \alpha = \frac{A}{A_0}, \tag{16.186}$$

where K_p is a parameter proportional to the bending stiffness of the wall material, and A_0 is the reference area of the tube for zero transmural pressure difference. The pressure difference is supported primarily by the bending stiffness of the tube wall. For a linear elastic tube wall material, K_p is proportional to the modulus of elasticity E, and the bending moment of inertia I, as in

$$K_p \propto EI, \qquad I = (h/a_0)^3/(1 - \hat{\nu}^2), \tag{16.187}$$

where h is wall thickness and $\hat{\nu}$ is Poisson's ratio.

From a fit of experimental data (see Shapiro, 1977a), the tube law for flow in a collapsible tube is taken to be

$$-P \approx \alpha^{-n} - 1, \quad \text{and} \quad n = \frac{3}{2}. \tag{16.188}$$

For P $<$ 0, the tube is partially collapsed. If the tube is in longitudinal tension, say, T_L, then there will be a local curvature R_L in the longitudinal plane. The effect of T_L is to change P_e by T_L/R_L, and the tube law (16.188) will not hold (see Cancelli & Pedley, 1985). We will here assume that $T_L/R \ll (p - p_e)$. Now, if the tube law (16.188) and transmural pressure difference are assumed to be uniform along the length of the tube, then with (16.185), at any location x, the phase velocity of long area waves is given by

$$c^2 = \frac{A}{\rho} \frac{\partial(p - p_e)}{\partial A} = \left[\frac{nK_p\alpha^{-n}}{\rho} \right], \tag{16.189}$$

for the square of the wave speed.

The assumptions of uniformity of tube law and transmural pressure difference are not valid under most physiological circumstances and these have to be relaxed. The physical causes that negate uniformity include: friction, gravity, variations of external pressure or of muscular tone, longitudinal variations in A_0, and longitudinal changes in the mechanical properties of the tube. To address some of these issues, we consider a more general formulation given by Shapiro.

The flow will still be considered steady, one dimensional, and incompressible.

The governing equations now are:

$$\frac{dA}{A} + \frac{du}{u} = 0, \tag{16.190}$$

and

$$-Adp - \tau_w s dx - \rho g A dz = \rho A u du = \rho A u^2 \frac{du}{u}, \tag{16.191}$$

where τ_w is the wall shear stress, s is the perimeter of the tube, and z is the elevation in the gravity field g. For the shear stress, Shapiro (1977a) considers the cases of fully developed turbulent flow and fully developed laminar Poiseuille flow in the tube. For turbulent flow,

$$\frac{\tau_w s dx}{A} = \frac{1}{2}\rho u^2 \frac{4f_T dx}{d_e}, \tag{16.192}$$

where $d_e = 4A/s$ is the equivalent hydraulic diameter and f_T is skin friction coefficient for turbulent flow, while for laminar flow:

$$\frac{\tau_w s dx}{A} = \frac{\mu u}{d_0} \frac{1}{\alpha} \frac{4f_L' dx}{d_0}, \quad \text{where} \quad f_L'(\alpha) = \left(\frac{A}{A_e}\right)f_L, \tag{16.193}$$

and d_0 is the diameter for A_0, and f_L is laminar skin friction coefficient.

With (16.190), (16.191) may be written

$$d\,(p + \rho g z) + \frac{\tau_w s dx}{A} - \rho u^2 \frac{dA}{A} = 0, \tag{16.194}$$

where the appropriate expression for the shear stress must be introduced depending on the nature of the flow.

Shapiro (1977a) introduces a dimensionless speed index, S:

$$S = \frac{u}{c}, \quad \text{so that} \quad \left(\frac{dS^2}{S^2}\right) = 2\,\frac{du}{u} - 2\,\frac{dc}{c}. \tag{16.195}$$

This index facilitates in the development of the theory and in the interpretation of results. Its role is comparable in significance to that of Mach number and Froude number in gas dynamics and in free-surface channel flow, respectively (Shapiro, 1977a). By analogy with gas dynamics, in steady flow, when $S < 1$ (subcritical), friction causes the area and pressure to decrease in the downstream direction, and the velocity to increase. When $S > 1$ (supercritical), the area and pressure increase along the tube, while the velocity decreases. In general, whatever the effect of changes of A_0, P_e, z, etc., in a subcritical flow, the effect is of opposite sign in supercritical flow. For example, let P_e be increased while all other independent variables such as A_0, elasticity, etc., are held constant. Then A and p will decrease for $S < 1$, but they will increase for $S > 1$. When $S = 1$, choking of flow and flow limitation as at the throat of a Laval nozzle will occur. Again, as in gas dynamics, there is the possibility of continuous transitions from supercritical to subcritical flow, and also rapid transitions from supercritical to subcritical as in shock waves.

In the steady-flow problem, the known quantities are dA_0, dP_e, gdz, fdx, dK_p, $\partial P/\partial x$, and $\partial P/\partial a$, while the unknowns are du, dA, dp, da, dS, and so on.

In order to develop the final set of equations relating the dependent and independent variables, a number of useful relationships may be established between the differential quantities.

The external pressure is $p_e(x)$, $dp_e = (dp_e/dx)\,dx$, the area $A_0 = A_0(x)$, and $dA_e = (dA_0/dx)\,dx$. Since $\alpha = A/A_0$,

$$\frac{d\alpha}{\alpha} = \left(\frac{dA}{A} - \frac{dA_0}{A_0}\right). \tag{16.196}$$

The bending stiffness parameter is $K_p = K_p(x)$, $dK_p = (dK_p/dx)\,dx$, and the tube law is

$$\mathrm{P} = \frac{p - p_e}{K_p(x)} = \mathrm{P}\,(\alpha, x), \quad \to dp = dp_e + K_p d\mathrm{P} + \mathrm{P}\,dK_p. \tag{16.197}$$

The appropriate form of (16.185) is

$$c^2(A, x) = \frac{A}{\rho}\left[\frac{\partial (p - p_e)}{\partial A}\right]_x \to c^2(\alpha, x) = \frac{\alpha}{\rho}\,K_p\,\left.\frac{\partial \mathrm{P}}{\partial \alpha}\right|_{x=\text{constant}}. \tag{16.198}$$

In (16.197),

$$d\mathrm{P} = \frac{\partial \mathrm{P}}{\partial \alpha}d\alpha + \frac{\partial \mathrm{P}}{\partial x}dx. \tag{16.199}$$

With (16.198) and (16.199), (16.197) becomes

$$dp = dp_e + \rho c^2 \frac{d\alpha}{\alpha} + K_p \frac{\partial P}{\partial x} dx + P d\, K_p. \tag{16.200}$$

With (16.198) and (16.197), we obtain

$$2 \frac{dc}{c} = \left(1 + \frac{\alpha \partial^2 P / \partial \alpha^2}{\partial P / \partial \alpha}\right) \frac{d\alpha}{\alpha} + \frac{dK_p}{K_p} + \frac{\alpha K_p}{\rho c^2} \frac{\partial}{\partial x}\left(\frac{\partial P}{\partial x}\right) dx, \tag{16.201}$$

and, with (16.195), (16.196) becomes

$$\left(\frac{dS^2}{S^2}\right) = -2 \frac{d\alpha}{\alpha} - 2 \frac{dA_0}{A_0} - 2 \frac{dc}{c}. \tag{16.202}$$

We now have (16.194), (16.196), (16.200), (16.201), and (16.202). With these, Shapiro (1977a) developed a series of equations that relate each dependent variable as a linear sum of terms, each containing an independent variable multiplied by appropriate coefficients (influence coefficients by analogy with one-dimensional gas dynamics). A comprehensive listing of equations is provided in the paper by Shapiro. From the listing, the most important dependent variables turn out to be $d\alpha/dx$ and dS^2/dx. Once these are known, other dependent quantities such as P, u, and c may be calculated easily. We now list these equations.

Let us consider cases where P is just a function of α alone, that is, $P(\alpha)$. For the tube law,

$$p - p_e(x) = K_p(x)P(\alpha), \tag{16.203}$$

the equation governing the variation in α is

$$(1 - S^2)\frac{1}{\alpha}\frac{d\alpha}{dx} = \frac{S^2}{A_0}\frac{dA_0}{dx} - \frac{1}{\rho c^2}\left[\frac{dp_e}{dx} + \rho g\frac{dz}{dx} + RQ + P\frac{dK_p}{dx}\right], \tag{16.204}$$

where R is viscous resistance per unit length (laminar or turbulent) and Q is flow rate, and the equation governing the speed index is

$$\begin{aligned}
(1 - S^2)\frac{1}{S^2}\frac{dS^2}{dx} &= \frac{1}{A_0}\frac{dA_0}{dx}[-2 + (2 - M)S^2] \\
&\quad + \frac{M}{\rho c^2}\left[\frac{dp_e}{dx} + \rho g\frac{dz}{dx} + RQ\right], \\
&\quad + \frac{1}{\rho c^2}\frac{dK_p}{dx}\left[MP - (1 - S^2)\alpha\frac{dP}{d\alpha}\right],
\end{aligned} \tag{16.205}$$

where

$$M = 3 + \frac{\alpha \partial^2 P / \partial \alpha^2}{\partial P / \partial \alpha}. \tag{16.206}$$

The equations for da/dx and dS^2/dx are coupled and must be solved simultaneously by using numerical procedures. Shapiro (1977a) has included results for several limit cases. These include several examples in which a smooth transition through the critical condition $S = 1$ is possible, that is, continuous passage of flow from regime $S < 1$ through $S = 1$ into $S > 1$ might occur. Figure 16.20 shows the transition from subcritical to supercritical flow by means of a minimum in the neutral area A_0. The pressure decreases continuously in the axial direction, and the area A of the deformed cross section would also decrease continuously in the axial direction. Figure 16.20 shows the transition through $S = 1$ caused by a weight or clamp, a sphincter or pressurized cuff, due to changing p_e. The fluid pressure and the area both decrease continuously in the axial direction. $S = 1$ occurs in the region where a sharp constriction exists.

Pedley (2000) points out that when $S = 1$, the right-hand side of (16.205) is $-M$ times that of (16.204). Therefore, at $S = 1$, it is possible for da/dx or dS^2/dx to be nonzero as long as the right-hand sides are zero. Of the terms on the right-hand side, RQ is associated with friction and is always positive. This means that at least one of $d\,(p_e + \rho g z)/dx$, dK_p/dx, or $-dA_0/dx$

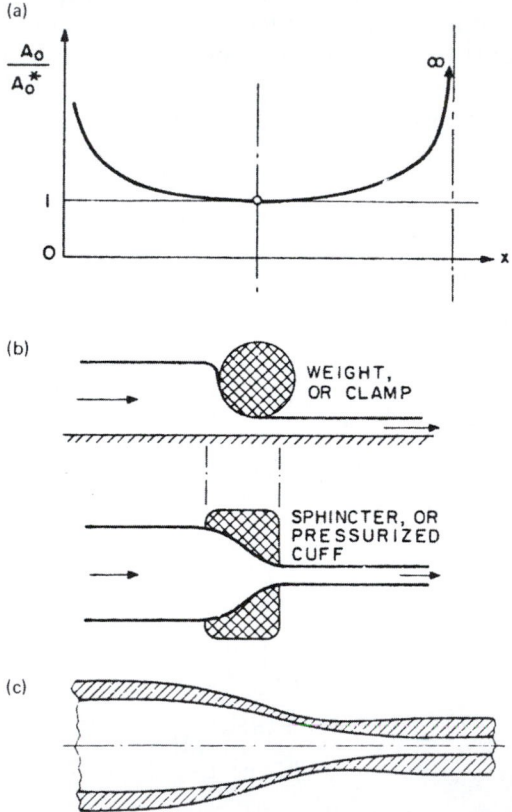

FIGURE: 16.20   Smooth transition through the critical condition. In each case the fluid speed increases and the pressure drops continuously as the area decreases. *Reproduced with permission from the American Society of Mechanical Engineers, NY.*

should be negative, that is, the external pressure, the height, or the stiffness should decrease with x or the undisturbed cross-sectional area should increase. An example where dz/dx in a vertical collapsible tube is negative ($= -1$) is the jugular vein of an upright giraffe and this problem has been discussed in detail by Pedley. Apparently, the giraffe jugular vein is normally partially collapsed!

In the next section, we learn about the modeling of a Casson fluid flow in a tube. We recall that blood behaves as a non-Newtonian fluid at low shear rates below about 200/s, and the apparent viscosity increases to relatively large magnitudes at low rates of shear. The modeling of such a fluid flow is important and will enable us to understand blood flow at various shear rates.

Laminar Flow of a Casson Fluid in a Rigid-Walled Tube

As shear rates decrease below about 200/s, the apparent viscosity of blood rapidly increases (see Figure 16.7). As mentioned earlier, the variation of shear stress in blood flow with shear rate is accurately expressed by (16.6):

$$\tau^{1/2} = \tau_y^{1/2} + K_c \dot{\gamma}^{1/2}, \quad \text{for } \tau \geq \tau_y, \quad \text{and} \quad \dot{\gamma} = 0, \quad \text{for } \tau < \tau_y, \tag{16.207}$$

where τ_y and K_c are determined from viscometer data. The yield stress τ_y for normal blood at 37°C is about 0.04 $dynes/cm^2$. In modeling the flow, this behavior must be included.

Consider the steady laminar axisymmetric flow of a Casson fluid in a rigid-walled, horizontal, cylindrical tube under the action of an imposed pressure gradient, $(p_1 - p_2)/L$. We shall employ cylindrical coordinates (r, θ, x) with velocity components $(u_y, u_\theta, \text{and } u_x)$, respectively. With the assumption of axisymmetry, $(u_\theta = 0, \text{ and } \partial/\partial\theta = 0)$ For convenience, we write the u_y component as v, and we omit the subscript x in u_x.

The maximum shear stress in the flow, τ_w, would be at the vessel wall. If the magnitude of τ_w is equal to or greater than the yield stress, τ_y, then there will be flow. We may estimate the minimum pressure gradient required to cause flow of a yield stress fluid in a cylindrical tube by a straightforward force balance on a cylindrical volume of fluid of radius r and length Δx. For steady flow, the viscous force opposing motion must be balanced by the force due to the applied pressure gradient. Thus,

$$\tau_{rx} \, 2\pi r \, \Delta x = -\pi r^2 \left(p|_{x+\Delta x} - p|_x \right), \tag{16.208}$$

and, as $\Delta x \to 0$,

$$\tau_{rx}(r) = \frac{r}{2} \frac{dp}{dx} = \frac{(p_1 - p_2)r}{2L}. \tag{16.209}$$

The shear stress at the wall, $\tau_w = -(a/2)(dp/dx) = (p_1 - p_2)a/2L$. When τ_y is equal to or less than τ_w, there will be fluid motion. The minimum pressure differential to cause flow is given by $(p_1 - p_2)|_{min} = 2L\tau_y/a$. With $\tau_y = 0.04$ $dynes/cm^2$, for a blood vessel of $L/a = 500$, the minimum pressure drop required for flow is 0.04 $dynes/cm^2$ or 0.03 mm Hg. Recall that during systole, the typical pressures in the aorta and the pulmonary artery rise to 120 mm Hg and 24 mm Hg, respectively.

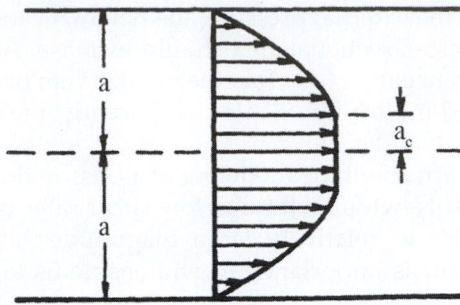

FIGURE 16.21   Velocity profile for axisymmetric blood flow in a circular tube. Here the profile is flattened in the center of the tube because of the non-Newtonian character of blood.

For axisymmetric blood flow in a cylindrical tube, at low shear rates, the fully developed flow is noted to consist of a central core region where the shear rate is zero and the velocity profile is flat, surrounded by a region where the flow has a varying velocity profile (see Figure 16.21). In the core, the fluid moves as if it were a solid body (also called *plug flow*).

Let the radius of this core region be a_c. Then,

$$\tau = \tau_y \quad \text{at } r = a_c, \text{ and } \quad \dot{\gamma} = 0 \quad \text{for } 0 \le r < a_c,$$

$$a_c = 2L\tau_y/(p_1 - p_2) = a\left(\frac{\tau_y}{\tau_w}\right),$$

$$\tau^{1/2} = \tau_y^{1/2} + K_c \dot{\gamma}^{1/2} \quad \text{for } a_c < r \le a. \tag{16.210}$$

In the core region, $\dot{\gamma} = 0 \Rightarrow (du/dr) = 0 \Rightarrow u = \text{constant} = u_c$ (say).

Outside the core region, the velocity is a function of r only, and

$$\dot{\gamma} = -\frac{du}{dr} = \frac{\left[\tau + \tau_y - 2\sqrt{\tau\tau_y}\right]}{K_c^2}. \tag{16.211}$$

Let $(p_1 - p_2) = \Delta p$, $\tau = \Delta p \, r/2L$, and $\tau_y = \Delta p \, a_c/2L$. From (16.211),

$$-\frac{du}{dr} = \frac{1}{2K_c^2}\frac{\Delta p}{L}(r + a_c - 2\sqrt{ra_c}). \tag{16.212}$$

By integration,

$$u = \frac{1}{2K_c^2}\frac{\Delta p}{L}\left(\frac{4}{3}\sqrt{a_c r^3} - \frac{r^2}{2} - a_c r + C\right), \tag{16.213}$$

where C is the integration constant. With the no-slip boundary condition at the wall of the vessel, $u = 0$ at $r = a$:

$$C = -\left(\frac{4}{3}\sqrt{a_c a^3} - \frac{a^2}{2} - a_c a\right). \tag{16.214}$$

Therefore,

$$u = \frac{1}{4K_c^2}\frac{\Delta p}{L}\left[(a^2 - r^2) - \frac{8}{3}\sqrt{a_c}\left(\sqrt{a^3} - \sqrt{r^3}\right) + 2a_c(a - r)\right], \tag{16.215}$$

in $(a_c \le r \le a)$. With $u = u_c$ at $r = r_c$, in terms of τ_w and τ_y, (16.215) becomes

$$u = \frac{a\tau_w}{2K_c^2}\left\{\left[1 - \left(\frac{r}{a}\right)^2\right] - \frac{8}{3}\sqrt{\frac{\tau_y}{\tau_w}}\left[1 - \left(\frac{r}{a}\right)^{3/2}\right] + 2\left(\frac{\tau_y}{\tau_w}\right)\left(1 - \frac{r}{a}\right)\right\}, \tag{16.216}$$

in $(a_c \le r \le a)$. We get the velocity in the core, u_c, by setting:

$$\left(\frac{r}{a}\right) = \left(\frac{a_c}{a}\right) = \left(\frac{\tau_y}{\tau_w}\right), \tag{16.217}$$

in (16.216). In terms of pressure gradient, a and a_c, u_c becomes

$$u_c = \frac{1}{4K_c^2}\frac{\Delta p}{L}(\sqrt{a} - \sqrt{a_c})^3\left(\sqrt{a} + \frac{1}{3}\sqrt{a_c}\right). \tag{16.218}$$

The volume rate of flow is given by

$$Q = \pi a_c^2 u_c + \int_{a_c}^{a} 2\pi r u\, dr. \tag{16.219}$$

After considerable algebra,

$$Q = \frac{\pi}{8}\frac{1}{K_c^2}\frac{\Delta p}{L}a^4\left[1 - \frac{16}{7}\left(\frac{a_c}{a}\right)^{1/2} + \frac{4}{3}\left(\frac{a_c}{a}\right) - \frac{1}{21}\left(\frac{a_c}{a}\right)^4\right]. \tag{16.220}$$

The Casson model predicts results that are in very good agreement with experimental results for blood flow over a large range of shear rates (see Charm & Kurland, 1974).

Pulmonary Circulation

Pulmonary circulation is the movement of blood from the heart, to the lungs, and back to the heart again. The veins bring oxygen-depleted blood back to the right atrium. The contraction of the right ventricle ejects blood into the pulmonary artery. In the human heart, the main pulmonary artery begins at the base of the right ventricle. It is short and wide—approximately 5 cm in length and 3 cm in diameter, and extends about 4 cm before it branches into the right and left pulmonary arteries that feed the two lungs. The pulmonary arteries are larger in size and more distensible than the systemic arteries and the resistance in pulmonary circulation is lower. In the lungs, red blood cells release carbon dioxide and pick up oxygen during respiration. The oxygenated blood then leaves the lungs through the pulmonary veins, which return it to the left heart, completing the pulmonary cycle. The pulmonary veins, like the pulmonary arteries, are also short, but their distensibility characteristics are similar to those of the systemic circulation (Guyton, 1968). The blood is then distributed to the body through the systemic circulation before returning again to the pulmonary circulation. The pulmonary circulation loop is virtually bypassed in fetal circulation. The fetal lungs are collapsed, and blood passes from the right atrium directly into the left atrium through the foramen ovale, an open passage between the two atria. When the lungs expand at birth, the pulmonary pressure drops and blood is drawn from the right atrium into the right ventricle and through the pulmonary circuit.

The rate of blood flow through the lungs is equal to the cardiac output except for the one to two percent that goes through the bronchial circulation (Guyton, 1968). Since almost the entire cardiac output flows through the lungs, the flow rate is very high. However, the low pulmonic pressures generated by the right ventricle are still sufficient to maintain this flow rate because pulmonary circulation involves a much shorter flow path than systemic circulation, and the pulmonary arteries are, as noted earlier, larger and more distensible.

The nutrition to lungs themselves are supplied by bronchial arteries which are a part of systemic circulation. The bronchial circulation empties into pulmonary veins and returns to the left atrium by passing alveoli.

The Pressure Pulse Curve in the Right Ventricle

The pressure pulse curves of the right ventricle and pulmonary artery are illustrated in Figure 16.22. As described by Guyton (1968), approximately 0.16 second prior to ventricular systole, the atrium contracts, pumping a small quantity of blood into the right ventricle, and thereby causing about 4 mm Hg initial rise in the right ventricular diastolic pressure even before the ventricle contracts. Following this, the right ventricle contracts, and the right ventricular pressure rises rapidly until it equals the pressure in the pulmonary artery. The pulmonary valve opens, and for approximately 0.3 second blood flows from the right ventricle into the pulmonary artery. When the right ventricle relaxes, the pulmonary valve closes, and the right ventricular pressure falls to its diastolic level of about zero. The systolic

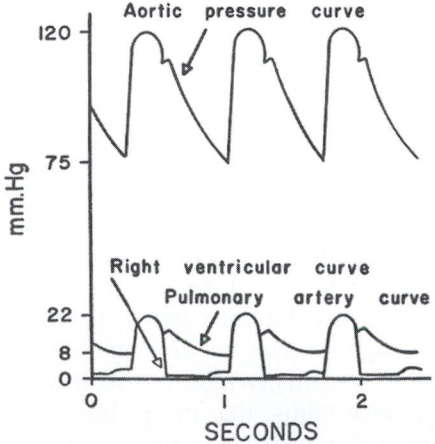

FIGURE 16.22   Pressure pulse contours in the right ventricle, and pulmonary artery. *Reproduced with permission from Guyton, A. C. and Hall, J. E. (2000), Textbook of Medical Physiology, W. B. Saunders Company, Philadelphia.*

pressure in the right ventricle of the normal human being averages approximately 22 mm Hg, and the diastolic pressure averages about 0 to 1 mm Hg.

Effect of Pulmonary Arterial Pressure on Pulmonary Resistance

At the end of systole, the ventricular pressure falls while the pulmonary arterial pressure remains elevated, then falls gradually as blood flows through the capillaries of the lungs. The pulse pressure in the pulmonary arteries averages 14 mm Hg which is almost two-thirds as much as the systolic pressure. Figure 16.23 shows the variation in pulmonary resistance with pulmonary arterial pressure. At low arterial pressures, pulmonary resistance is very high and at high pressures the resistance falls to low values. The rapid fall is due to the high distensibility of the pulmonary vessels.

The ability of lungs to accommodate greatly increased blood flow with little increase in pulmonary arterial pressure helps to conserve the energy of the heart. As described by Guyton, the only reason for flow of blood through the lungs is to pick up oxygen and to release carbon dioxide. The ability of pulmonary vessels to accommodate greatly increased blood flow without an increase in pulmonary arterial pressure accomplishes the required gaseous exchange without overworking the right ventricle.

In the earlier sections, we discussed several modeling procedures in relation to systemic blood circulation. The modeling of the blood flow in pulmonary vessels is similar to what we studied in those sections.

A discussion of gas and material exchange in the capillary beds is beyond the scope of this introductory chapter. Additional information on this topic can be found in Grotberg (1994).

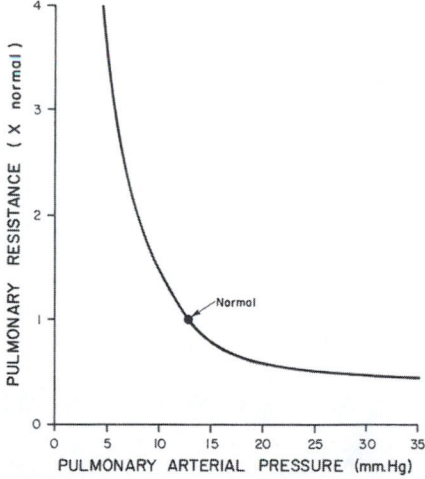

FIGURE 16.23 Effect of pulmonary arterial pressure on pulmonary resistance. At low pressures, the lungs' resistance drops dramatically, and this allows increased blood flow rates for moderate increases in pulmonary arterial pressure. *Reproduced with permission from Guyton, A. C. and Hall, J. E. (2000), Textbook of Medical Physiology, W. B. Saunders Company, Philadelphia.*

16.4. INTRODUCTION TO THE FLUID MECHANICS OF PLANTS

Plant life comprises 99% of the earth's biomass (Bidwell, 1974; Rand, 1983).

The basic unit of a plant is a plant cell. Plant cells are formed at meristems, and then develop into cell types which are grouped into tissues. Plants have three tissue types: 1) dermal; 2) ground; and 3) vascular. Dermal tissue covers the outer surface and is composed of closely packed epidermal cells that secrete a waxy material that aids in the prevention of water loss. The ground tissue comprises the bulk of the primary plant body. Parenchyma, collenchyma, and sclerenchyma cells are common in the ground tissue. Vascular tissue transports food, water, hormones, and minerals within the plant.

Basically, a plant has two organ systems: 1) the shoot system, and 2) the root system. The shoot system is above ground and includes the organs such as leaves, buds, stems, flowers, and fruits. The root system includes those parts of the plant below ground, such as the roots, tubers, and rhizomes. There is transport between the roots and the shoots (see Figure 16.24).

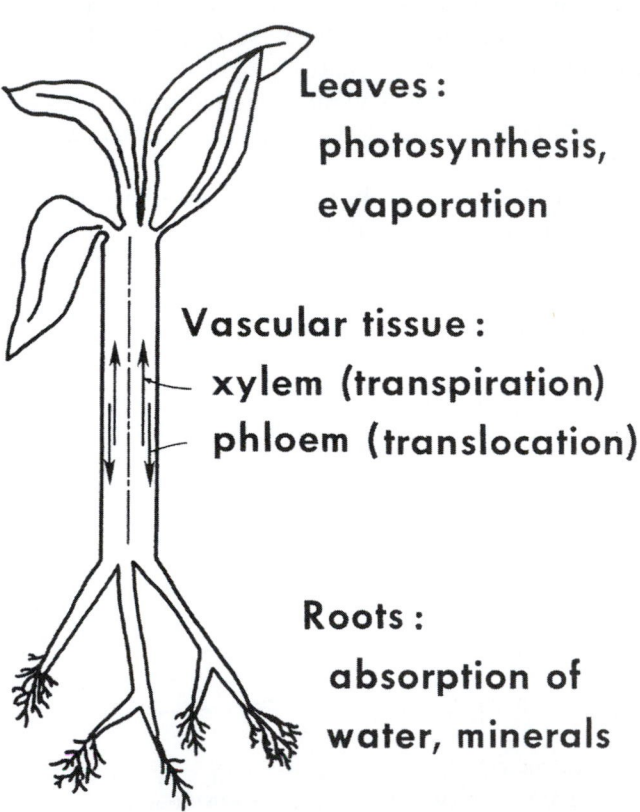

FIGURE 16.24 Overview of plant fluid mechanics. Transport of water and solutes between the leaves and the roots through the vascular tissues is essential. Transpiration of water at the leaves actually helps to lift sap from the roots. *Reproduced with permission from Annual Review of Fluid Mechanics, Vol. 15 © 1983. Annual Reviews: www.AnnualReviews.org.*

Transport in plants occurs on three levels: 1) the uptake and loss of water and solutes by individual cells, 2) short-distance transport of substances from cell to cell at the level of tissues or organs, and 3) long-distance transport of sap within xylem and phloem at the level of the whole plant.

The transport occurs as a result of gradients in chemical concentration (Fickian diffusion), hydrostatic pressure, and gravitational potential. These three driving potentials are grouped under one single quantity, the water potential. The water potential is designated ψ, and

$$\psi = p - RTc + \rho g z, \tag{16.221}$$

where p is hydrostatic pressure (bar), R is gas constant ($= 83.141 \mathrm{cm}^3 \, \mathrm{bar/mole} \, K$), T is temperature (K), c is the concentration of all solutes in assumed dilute solution ($\mathrm{mole/cm}^3$), ρ is density of water ($\mathrm{g/cm}^3$), g is acceleration due to gravity ($= 980 \, \mathrm{cm/sec}^2$), and z is height (cm); ψ is in bars (Conversion: 1 bar $= 10^6 \, \mathrm{dyne/cm}^2$).

Transport at the cellular level in a plant depends on the selective permeability of plasma membranes which controls the movement of solutes between the cell and the extracellular solution. Molecules move down their concentration gradient across a membrane without the direct expenditure of metabolic energy (Fickian diffusion). Transport proteins embedded in the membrane speed up the movement across the membrane. Differences in water potential, ψ, drive water transport in plant cells. Uptake or loss of water by a cell occurs by osmosis across a membrane. Water moves across a membrane from a higher water potential to a lower water potential. If a plant cell is introduced into a solution with a higher water potential than that of the cell, osmotic uptake of water will cause the cell to swell. As the cell swells, it will push against the elastic wall, creating a "turgor" pressure inside the cell. Loss of water causes loss of turgor pressure and may result in wilting.

In contrast to the human circulatory system, the vascular system of plants is open. Unlike the blood vessels of human physiology, the vessels (conduits) of plants are formed of individual plant cells placed adjacent to one another. During cell differentiation the common walls of two adjacent cells develop pores which permit fluid to pass between them. Vascular tissue includes xylem, phloem, parenchyma, and cambium cells. Xylem and phloem make up the big transportation system of vascular plants. Long-distance transport of materials (such as nutrients) in plants is driven by the prevailing pressure gradient.

In this section we restrict attention to the vascular system that includes xylem and phloem cells.

Xylem

The term *xylem* applies to woody walls of certain cells of plants. Xylem cells tend to conduct water and minerals from roots to leaves. Generally speaking, the xylem of a plant is the system of tubes and transport cells that circulates water and dissolved minerals. Xylem is made of vessels that are connected end to end to enable efficient transport. The xylem contains tracheids and vessel elements (see Figure 16.25, from Rand, 1983). Xylem tissue dies after one year and then develops anew (e.g., rings in the tree trunk).

Water and mineral salts from soil enter the plant through the epidermis of roots, cross the root cortex, pass into the stele, and then flow up xylem vessels to the shoot system. The xylem

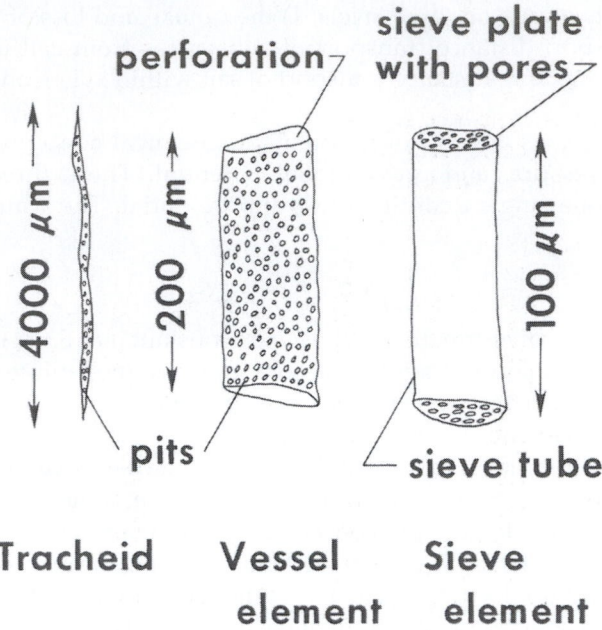

FIGURE 16.25   Fluid-conducting cells in the vascular tissue of plants. *Reproduced with permission from Annual Review of Fluid Mechanics, Vol. 15 © 1983.* Annual Reviews: *www.AnnualReviews.org.*

flow is also called *transpirational flow*. Perforated end walls of xylem vessel elements enhance the bulk flow.

The movement of water and solutes through xylem vessels occurs due to a pressure gradient. In xylem, it is actually tension (negative pressure) that drives long-distance transport. Transpiration (evaporation of water from a leaf) reduces pressure in the leaf xylem and creates a tension that pulls xylem sap upward from the roots. While transpiration enables the pull, the cohesion of water due to hydrogen bonding transmits the upward pull along the entire length of the xylem from the leaves to the root tips. The pull extends down only through an unbroken chain of water molecules. Cavitation, formation of water vapor pockets in the xylem vessel, may break the chain. Cavitation will occur when xylem sap freezes in water and as a result the vessel function will be compromised. Absorption of solar energy drives transpiration by causing water to evaporate from the moist walls of mesophyll cells of a leaf and by maintaining a high humidity in the air spaces within the leaf. To facilitate gas exchange between the inner parts of leaves, stems, and fruits, plants have a series of openings known as *stomata*. These enable exchange of water vapor, oxygen, and carbon dioxide.

The pressure gradient for transpiration flow is essentially created by solar power, and in principle, a plant expends no energy in transporting xylem sap up to the leaves by bulk flow. The detailed mechanism of transpiration from a leaf is very complicated and depends on the interplay of adhesive and cohesive forces of water molecules at mesophyll cell–air space interfaces, resulting in surface tension gradients and capillary forces. This will not be discussed in this section.

Xylem sap flows upward to veins that branch throughout each leaf, providing each with water. Plants lose a huge amount of water by transpiration—an average-sized maple tree loses more than about 200 liters of water per hour during the summer. Flow of water up the xylem replaces water lost by transpiration and carries minerals to the shoots. At night, when transpiration is very low, root cells are still expending energy to pump mineral ions into the xylem, accumulation of minerals in the stele lowers water potential, generating a positive pressure, called *root pressure*, that forces fluid up the xylem. It is the root pressure that is responsible for guttation, the exudation of water droplets that can be seen in the morning on tips of grass blades or leaf margins of some plants. Root pressure is not the main mechanism driving the ascent of xylem sap. It can force water upward by only a few meters, and many plants generate no root pressure at all. Small plants may use root pressure to refill xylem vessels in spring. Thus, for the most part, xylem sap is not pushed from below but pulled upward by the leaves.

Xylem Flow

Water and minerals absorbed in the roots are brought up to the leaves through the xylem. The upward flow in the xylem (also called the *transpiration flow*) is driven by evaporation at the leaves. In the xylem, the flow may be treated as quasi-steady. The rigid tube model for flow description is appropriate because plant cells have stiff walls. The xylem is about 0.02 mm in radius and the typical values for flow are velocity 0.1 cm/s, the kinematic viscosity of the fluid 0.1 cm^2/s, and the Reynolds number, $Re = ud/v$ is 0.04. In view of the low Reynolds number, the Stokes flow in a rigid tube approximation is appropriate.

Phloem

Phloem cells are usually located outside the xylem and conduct food from the leaves to the rest of the plant. The two most common cells in the phloem are the companion cells and sieve cells. Phloem cells are laid out end-to-end throughout the plant to form long tubes with porous cross walls between cells. These tubes enable translocation of the sugars and other molecules created by the plant during photosynthesis. Phloem flow is also called *translocation flow*. Phloem sap is an aqueous solution with sucrose as the most prevalent solute. It also contains minerals, amino acids, and hormones. Dissolved food, such as sucrose, flows through the sieve cells. In general, sieve tubes carry food from a sugar source (for example, mature leaves) to a sugar sink (roots, shoots, or fruits). A tuber or a bulb may be either a source or a sink, depending on the season. Sugar must be loaded into sieve-tube members before it can be exported to sugar sinks. Companion cells pass sugar they accumulate into the sieve-tube members via plasmodesmata. Translocation through the phloem is dependent on metabolic activity of the phloem cells (in contrast to transport in the xylem).

Unlike the xylem, phloem is always alive. In contrast to xylem sap, the direction that phloem sap travels is variable depending on locations of source and sink.

The pressure-flow hypothesis is employed to explain the movement of nutrients through the phloem. It proposes that water-containing nutrient molecules flow under pressure through the phloem. The pressure is created by the difference in water concentration of the solution in the phloem and the relatively pure water in the nearby xylem ducts.

At their "source"—the leaves—sugars are pumped by active transport into the companion cells and sieve elements of the phloem. The exact mechanism of sugar transport in the phloem is not known, but it cannot be simple diffusion. As sugars and other products of photosynthesis accumulate in the phloem, the water potential in the leaf phloem is decreased and water diffuses from the neighboring xylem vessels by osmosis. This increases the hydrostatic pressure in the phloem. Turgor pressure builds up in the sieve tubes (similar to the creation of root pressure). Water and dissolved solutes are forced downward to relieve the pressure. As the fluid is pushed down (and up) the phloem, sugars are removed by the cortex cells of both stem and root (the "sinks") and consumed or converted into starch. Starch is insoluble and exerts no osmotic effect. Therefore, the osmotic pressure of the contents of the phloem decreases. Finally, relatively pure water is left in the phloem. At the same time, ions are being pumped into the xylem from the soil by active transport, reducing the water potential in the xylem. The xylem now has a lower water potential than the phloem, so water diffuses by osmosis from the phloem to the xylem. Water and its dissolved ions are pulled up the xylem by tension from the leaves. Thus it is the pressure gradient between "source" (leaves) and "sink" (shoot and roots) that drives the contents of the phloem up and down through the sieve tubes.

Phloem Flow

Phloem flow occurs mainly through cells called *sieve tubes* which are arranged end to end and are joined by perforated cell walls called *sieve plates* (see Figure 16.26, from Rand & Cooke, 1978). As a model of Phloem flow, Rand et al. (1980) have derived an approximate formula for the pressure drop for flow through a series of sieve tubes with periodically placed sieve plates with pores (see Figure 16.27, from Rand et al., 1980). The approximation arises from treating the transport through the pore as creeping conical flow (see Happel & Brenner, 1983).

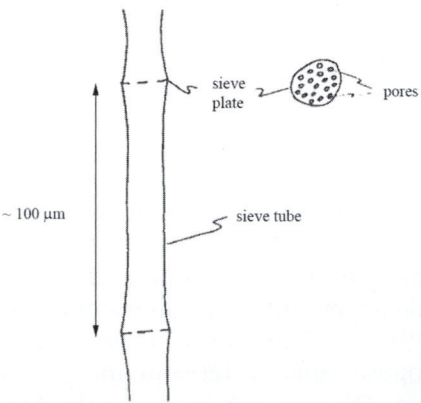

FIGURE 16.26   Sieve tube with sieve plate. These cells and cell structure convey phloem through the plant. *Reproduced with permission from the American Society of Agricultural Engineers, MI.*

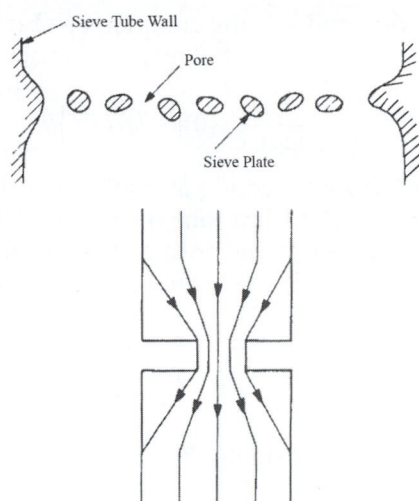

FIGURE 16.27   Sieve tube with pores and stream lines for conical flow through one pore. This geometry was used to derive the pressure drop formula (16.222) which is based on creeping conical flow. *Reproduced with permission from the American Society of Agricultural Engineers, MI.*

The approximate formula given by Rand et al. (1980) is:

$$\Delta p = \frac{8\mu Q}{\pi a^4}\left[L + \frac{\ell}{N}\left(\frac{a}{r}\right)^4\right] + 2\,\Delta p', \quad \text{where}$$

$$\Delta p' = \frac{8\mu Q}{\pi a^3}\left(\frac{a_e}{r}\right)\left\{0.57N\left[\left(\frac{a_e}{r}\right)^3 - 1\right] - 1.5\left(1 - \frac{r}{a_e}\right)\right\}.$$

(16.222)

In (16.222), Δp is the pressure drop due to one sieve tube and one sieve plate, μ is the viscosity of fluid in (g/cms), Q is the flow rate in (cm^3/s), N is the number of pores in the sieve plate, a is sieve tube radius in cm, r is average radius of sieve pore in cm, L is the sieve tube length in cm, ℓ is sieve plate thickness in *cm*, and the effective tube radius $a_e = a/\sqrt{N}$.

Rand et al. (1980) note that the approximate formula has not been tested for $N \neq 1$.

EXERCISES

16.1. Consider steady laminar flow of a Newtonian fluid in a long, cylindrical, elastic tube of length L. The radius of the tube at any cross section is $a = a(x)$. Poiseuille's formula for the flow rate is a good approximation in this case.

  a) Develop an expression for the outlet pressure $p(L)$ in terms of the higher inlet pressure, the flow rate $\dot{Q}$, fluid viscosity μ, and $a(x)$.

  b) For a pulmonary blood vessel, we may assume that the pressure-radius relationship is linear: $a(x) = a_0 + \alpha p/2$, where a_0 is the tube radius when the transmural pressure is zero and α is the compliance of the tube. For a tube of length L, show that

$$\dot{Q} = \frac{\pi}{20\mu\alpha L}\left\{[a(0)]^5 - [a(L)]^5\right\},$$

where $a(0)$ and $a(L)$ are the values of $a(x)$ at $x = 0$ and $x = L$, respectively.

16.2. For pulsatile flow in a rigid cylindrical tube of length L, the pressure drop Δp may be expressed as: $\Delta p = f(L, a, \rho, \mu, \omega, U)$, where a is tube radius, ρ is density, μ is viscosity, ω is frequency, and U is the average velocity of flow. Using dimensional analysis, show that

$$\frac{\Delta p}{\rho U^2} = C_1\left(\frac{L}{a}\right)^{C_2}(Re)^{C_3}(St)^{C_4},$$

where C_1, C_2, C_3, and C_4 are constants, Re is Reynolds number, and St is Strouhal number defined as $a\omega/U$.

16.3. Localized narrowing of an artery may be caused by the formation of artherosclerotic plaque in that region. Such localized narrowing is called *stenosis*. It is important to understand the flow characteristics in the vicinity of a stenosis. Flow in a tube with mild stenosis may be approximated by axisymmetric flow through a converging-diverging tube. In this context, follow the details described in Morgan & Young (1974) and obtain expressions for the velocity profile and wall shear stress.

16.4. Shapiro (1997a) in his analysis of the steady flow in collapsible tubes has developed a series of equations that relate the dependent variables $du, dA, dp, d\alpha, dS$, etc., with the independent variables such as $dA_0, dp_e, g\,dz, f_T\,dx$, etc. In Section IV of that study, explicit calculations of certain simple flows are presented. In particular, consider pure pressure-gravity flows. Discuss the flow behavior patterns in this case.

16.5. Consider the Power-law model to describe the non-Newtonian behavior of blood. In this model, $\tau = \mu\gamma^n$, where τ is the shear stress and the $\dot{\gamma}$ is the rate of shearing strain. Determine the flux for the flow of such a fluid in a rigid cylindrical tube of radius R. Show that when $n = 1$, the results correspond to the Poiseuille flow.

16.6. Consider the Herschel-Bulkley model to describe the non-Newtonian behavior of blood. In this model,

$$\tau = \mu\dot{\gamma}^n + \tau_0, \quad \tau \geq \tau_0$$

$$\dot{\gamma} = 0, \qquad\qquad \tau < \tau_0$$

Determine the flux for the flow of such a fluid in a rigid cylindrical tube of radius R. Show that in the limit $\tau_0 = 0$, the results for the Herschel-Bulkley model coincide with those for the Power-law model.

Acknowledgment

The help received from Dr. K. Mukundakrishnan and Mrs. Olivia Brubaker during the development of this chapter is gratefully acknowledged.

Literature Cited

Baish, J. W., Ayyaswamy, P. S., & Foster, K. R. (1986a). Small scale temperature fluctuations in perfused tissue during local hyperthermia. *J. BioMech. Eng., 108*, 246–250.

Baish, J. W., Ayyaswamy, P. S., & Foster, K. R. (1986b). Heat transport mechanisms in vascular tissues: A model comparison. *J. BioMech. Eng., 108*, 324–331.

Berger, S. A., Talbot, L., & Yao, L.-S. (1983). Flow in curved pipes. *Annual Review of Fluid Mechanics, 15*, 461–512.

Bidwell, R. G. S. (1974). *Plant Physiology.* New York: MacMillan.

Brown, B. H., Smallwood, R. H., Barber, D. C., Lawford, P. V., & Hose, D. R. (1999). *Medical Physics and Biomedical Engineering.* London: Institute of Physics Publishing.

Cancelli, C., & Pedley, T. J. (1985). A separated flow model for collapsible tube oscillations. *Journal of Fluid Mechanics, 157*, 375–404.

Chandran, K. B., Yoganathan, A. P., & Rittgers, S. E. (2007). *Biofluid Mechanics—The Human Circulation.* Boca Raton, FL: Taylor & Francis.

Charm, S. E., & Kurland, G. S. (1974). *Blood Flow and Microcirculation.* New York: John Wiley & Sons.

Collins, W. M., & Dennis, S. C. R. (1975). The steady motion of a viscous fluid in a curved tube. *Q.J. Mech. Appl. Math., 28*, 133–156.

Dean, W. R. (1928). The streamline motion of fluid in a curved pipe. *Philosophical Magazine, Series, 7*(30), 673–693.

Fournier, R. L. (2007). *Basic Transport Phenomena in Biomedical Engineering.* New York: Taylor & Francis.

Fung, Y. C. (1993). *Biomechanics: Mechanical Properties of Living Tissues, Second Edition.* Boca Raton, FL: Springer.

Fung, Y. C. (1997). *Biomechanics: Circulation, Second Edition.* Boca Raton, FL: Springer.

Goldsmith, H. L., Cokelet, G. R., & Gaehtgens, P. (1989). Robin Fahraeus: Evolution of his concepts in cardiovascular physiology. *Am. J of Physiology—Heart and Circulatory Physiology, 257*(3), H1005–H1015.

Grotberg, J. B. (1994). Pulmonary flow and transport phenomena. *Annual Review of Fluid Mechanics, 26*, 529–571.

Guyton, A. C. (1968). *Textbook of Medical Physiology.* Philadelphia: W.B. Saunders Company.

Happel, J., & Brenner, H. (1983). *Low Reynolds Number Hydrodynamics.* New York: McGraw-Hill.

He, X., & Ku, D. N. (1994). Unsteady entrance flow development in a straight tube. *Journal of Biomechanical Engineering, 116*, 355–360.

Kamm, R. D., & Pedley, T. J. (1989). Flow in collapsible tubes: A brief review. *Journal of Biomechanical Engineering, 111*, 177–179.

Kamm, R. D., & Shapiro, A. H. (1979). Unsteady flow in a collapsible tube subjected to external pressure or body forces. *Journal of Fluid Mechanics, 95*, 1–78.

Kleinstreuer, C. (2006). *Biofluid Dynamics: Principles and Selected Applications.* Boca Raton, FL: Taylor & Francis.

Ku, D. N. (1997). Blood flow in arteries. *Annual Review of Fluid Mechanics, 29*, 399–434.

Lighthill, M. J. (1978). *Waves in Fluids.* Cambridge: Cambridge University Press.

Liu, Y., & Liu, W. K. (2006). Rheology of red blood cell aggregation by computer simulation. *J. Comput. Phys., 220*(1), 139–154.

Mazumdar, J. N. (1999). *An Introduction to Mathematical Physiology and Biology* (second ed.). Cambridge: Cambridge University Press.

Mazumdar, J. N. (2004). *Biofluid Mechanics* (third ed.). Singapore: World Scientific.

McConalogue, D. J., & Srivastava, R. S. (1968). Motion of a fluid in a curved tube. *Proc. Roy. Soc. A., 307*, 37–53.

McDonald, D. A. (1974). *Blood Flow in Arteries, Second Edition.* Baltimore: The Williams & Wilkins Company.

Mohanty, A. K., & Asthana, S. B. L. (1979). Laminar flow in the entrance region of a smooth pipe. *Journal of Fluid Mechanics, 90*, 433–447.

Morgan, B. E., & Young, D. F. (1974). *Bull. Math. Biol., 36*, 39–53.

Morgan, G. W., & Ferrante, W. R. (1955). Wave propagation in elastic tubes filled with streaming liquid. *The Journal of the Acoustical Society of America, 27*(4), 715–725.

Morgan, G. W., & Kiely, J. P. (1954). Wave propagation in a viscous liquid contained in a flexible tube. *The Journal of the Acoustical Society of America, 26*(3), 323–328.

Nichols, W. W., & O'Rourke, M. F. (1998). *McDonald's Blood Flow in Arteries: Theoretical, Experimental and Clinical Principles.* London: Arnold.

Pedley, T. J. (1980). *The Fluid Mechanics of Large Blood Vessels.* Cambridge: Cambridge University Press.

Pedley, T. J. (2000). Blood flow in arteries and veins. In G. K. Batchelor, H. K. Moffat, & M. G. Worster (Eds.), *Perspectives in Fluid Dynamics*. Cambridge: Cambridge University Press.

Rand, R. H. (1983). Fluid mechanics of green plants. *Annual Review of Fluid Mechanics, 15*, 29–45.

Rand, R. H., & Cooke, J. R. (1978). Fluid dynamics of phloem flow: An axi-symmetric model. *Trans. ASAE., 21*, 898–900, 906.

Rand, R. H., Upadhyaya, S. K., & Cooke, J. R. (1980). Fluid dynamics of phloem flow. II. An approximate formula. *Trans. ASAE., 23*, 581–584.

Shapiro, A. H. (1977a). Steady flow in collapsible tubes. *Journal of Biomechanical Engineering, 99*, 126–147.

Shapiro, A. H. (1977b). Physiologic and medical aspects of flow in collapsible tubes. *Proc. 6th Can. Congr. Appl. Mech.* 883–906.

Singh, M. P. (1974). Entry flow in a curved pipe. *Journal of Fluid Mechanics, 65*, 517–539.

Whitmore, R. L. (1968). *Rheology of the Circulation*. Oxford: Pergamon Press.

Womersley, J. R. (1955a). Method for the calculation of velocity, rate of flow and viscous drag in arteries when the pressure gradient is known. *Journal of Physiology, 127*, 553–563.

Womersley, J. R. (1955b). Oscillatory motion of a viscous liquid in a thin-walled elastic tube. I. The linear approximation for long waves. *Philosophical Magazine, 46*, 199–221.

Womersley, J. R. (1957a). The mathematical analysis of arterial circulation in a state of oscillatory motion. Technical Report WADC-TR-56-614. Dayton, OH: Wright Air Development Center.

Womersley, J. R. (1957b). Oscillatory flow in arteries: The constrained elastic tube as a model of arterial flow and pulse transmission. *Physics in Medicine and Biology, 2*, 178–187.

Zhou, J., & Fung, Y. C. (1997). The degree of nonlinearity and anisotropy of blood vessel elasticity. *Proc. Natl. Acad. Sci. USA, 94*, 14255–14260.

Supplemental Reading

Lighthill, M. J. (1975). *Mathematical Biofluiddynamics*. Philadelphia: Soc. Ind. Appl. Math.

APPENDIX A

Conversion Factors, Constants, and Fluid Properties

OUTLINE

A.1. Conversion Factors   853

A.2. Physical Constants   854

A.3. Properties of Pure Water at Atmospheric Pressure   854

A.4. Properties of Dry Air at Atmospheric Pressure   855

A.5. The Standard Atmosphere   855

A.1. CONVERSION FACTORS

Length:	1 m = 3.2808 ft
	1 in. = 2.540 cm
	1 mile = 1.609 km
	1 nautical mile = 1.852 km
Mass[1]:	1 kg = 0.06854 slug = 1000 g $\leftrightarrow$ 2.205 lbs
	1 metric ton = 1000 kg
Time:	1 day = 86,400 s
Density[1]:	1 kg m^{-3} = 1.941 $\times$ 10^{-3} slugs ft^{-3} $\leftrightarrow$ 0.06244 lbs/ft^3
Velocity:	1 knot = 0.5144 m/s
Force:	1 N = 10^5 dyn = 0.2248 lbs
Pressure:	1 dyn cm^{-2} = 0.1 N/m^2 = 0.1 Pa
	1 bar = 10^5 Pa
Energy:	1 J = 10^7 erg = 0.2389 cal
	1 cal = 4.186 J
Energy flux:	1 W m^{-2} = 2.39 $\times$ 10^{-5} cal cm^{-2} s^{-1}

[1]At the earth's surface, the weight of a 1 kg mass is 2.205 lbs.

A.2. PHYSICAL CONSTANTS

Avogadro's Number:	6.023×10^{23} gmole^{-1}
Boltzmann's Constant:	1.381×10^{-23} J K^{-1}
Gravitational Acceleration:	9.807 m s$^{-2} = 32.17$ ft s^{-2} (at the surface of the earth)
Graviational Constant:	6.67×10^{-11} m^3 kg^{-1} s^{-2}
Planck's Constant:	6.626×10^{-34} J s
Speed of Light in Vacuum:	2.998×10^8 m s^{-1}
Universal Gas Constant:	8.314 J gmole^{-1} K^{-1}

A.3. PROPERTIES OF PURE WATER AT ATMOSPHERIC PRESSURE

Here, ρ = density, α = coefficient of thermal expansion, μ = shear viscosity, ν = kinematic viscosity = μ/ρ, κ = thermal diffusivity = $k/(\rho C_p)$, (k is first defined in Section 1.5) Pr = Prandtl number, and 1.0×10^{-n} is written as $1.0E - n$.

T °C	ρ kg/m^3	α K^{-1}	μ kg m^{-1} s^{-1}	ν m^2/s	κ m^2/s	C_p J kg^{-1} K^{-1}	Pr ν/κ
0	1000	$-0.6E - 4$	$1.787E - 3$	$1.787E - 6$	$1.33E - 7$	4217	13.4
10	1000	$+0.9E - 4$	$1.307E - 3$	$1.307E - 6$	$1.38E - 7$	4192	9.5
20	998	$2.1E - 4$	$1.002E - 3$	$1.004E - 6$	$1.42E - 7$	4182	7.1
30	996	$3.0E - 4$	$0.799E - 3$	$0.802E - 6$	$1.46E - 7$	4178	5.5
40	992	$3.8E - 4$	$0.653E - 3$	$0.658E - 6$	$1.52E - 7$	4178	4.3
50	988	$4.5E - 4$	$0.548E - 3$	$0.555E - 6$	$1.58E - 7$	4180	3.5

Latent heat of vaporization at 100 °C = 2.257×10^6 J/kg.
Latent heat of melting of ice at 0 °C = 0.334×10^6 J/kg.
Density of ice = 920 kg/m^3.
Surface tension between water and air at 20 °C = 0.0728 N/m.
Sound speed at 20 °C = 1481 m/s.

A.4. PROPERTIES OF DRY AIR AT ATMOSPHERIC PRESSURE

$T\,°C$	$\rho\ \text{kg/m}^3$	$\mu\ \text{kg m}^{-1}\,\text{s}^{-1}$	$\nu\ \text{m}^2\text{/s}$	$\kappa\ \text{m}^2\text{/s}$	$\text{Pr}\ \nu/\kappa$
0	1.293	1.71E − 5	1.33E − 5	1.84E − 5	0.72
10	1.247	1.76E − 5	1.41E − 5	1.96E − 5	0.72
20	1.200	1.81E − 5	1.50E − 5	2.08E − 5	0.72
30	1.165	1.86E − 5	1.60E − 5	2.25E − 5	0.71
40	1.127	1.87E − 5	1.66E − 5	2.38E − 5	0.71
60	1.060	1.97E − 5	1.86E − 5	2.65E − 5	0.71
80	1.000	2.07E − 5	2.07E − 5	2.99E − 5	0.70
100	0.946	2.17E − 5	2.29E − 5	3.28E − 5	0.70

At 20°C and 1 atm:
Specific heat capacity at constant pressure: $C_p = 1004\ \text{J kg}^{-1}\,\text{K}^{-1}$
Specific heat capacity at constant volume: $C_v = 717\ \text{J kg}^{-1}\,\text{K}^{-1}$
Ratio of specific heat capacities: $\gamma = 1.40$
Coefficient of thermal expansion: $\alpha = 3.41 \times 10^{-3}\ \text{K}^{-1}$
Speed of sound: $c = 343\ \text{m s}^{-1}$
Constants for dry air: Gas constant: $R = 287\ \text{J kg}^{-1}\,\text{K}^{-1}$
Molecular mass: $28.966\ \text{g gmole}^{-1}$
or kg kmole^{-1}

A.5. THE STANDARD ATMOSPHERE

The following average values are accepted by international agreement. Here, z is the height above sea level.

$z\ \text{km}$	$T\,°C$	$p\ \text{kPa}$	$\rho\ \text{kg/m}^3$
0	15.0	101.3	1.225
0.5	11.5	95.5	1.168
1	8.5	89.9	1.112
2	2.0	79.5	1.007
3	−4.5	70.1	0.909

(Continued)

z km	T °C	p kPa	ρ kg/m³
4	−11.0	61.6	0.819
5	−17.5	54.0	0.736
6	−24.0	47.2	0.660
8	−37.0	35.6	0.525
10	−50.0	26.4	0.413
12	−56.5	19.3	0.311
14	−56.5	14.1	0.226
16	−56.5	10.3	0.165
18	−56.5	7.5	0.120
20	−56.5	5.5	0.088

Mathematical Tools and Resources

OUTLINE

B.1. Partial and Total Differentiation   857

B.2. Changing Independent Variables   860

B.3. Basic Vector Calculus   861

B.4. The Dirac Delta function   863

B.5. Common Three-Dimensional Coordinate Systems   863

B.6. Equations in Curvilinear Coordinate Systems   866

B.1. PARTIAL AND TOTAL DIFFERENTIATION

In fluid mechanics, the field quantities like fluid velocity, fluid density, pressure, etc. may vary in time, t, and across three-dimensional space, herein specified by three coordinates as a vector $\mathbf{x} = (x, y, z)$ or (x_1, x_2, x_3). For multivariable functions, such as $f(x_1, x_2, x_3, t)$, there are important differences between partial and total derivatives, for example between $\partial f / \partial t$ and df / dt.

Partial Differentiation

$(\partial / \partial t) f(x_1, x_2, x_3, t)$ means differentiate the function $f(x_1, x_2, x_3, t)$ with respect to time, t, treating all other independent variables as constants. Additional information and specifications are not needed. And, multiple partial derivatives that operate on different variables can be applied in either order, that is, $(\partial / \partial x_i)(\partial f / \partial t) = (\partial / \partial t)(\partial f / \partial x_i)$ and $(\partial / \partial x_i)(\partial f / \partial x_j) = (\partial / \partial x_j)(\partial f / \partial x_i)$.

Total Differentiation

$(d/dt) f(x_1, x_2, x_3, t)$ means differentiate the function $f(x_1, x_2, x_3, t)$ with respect to time, t, including the time variation of the spatial coordinates. This total time derivative has meaning along a time-space path specified through the three-dimensional domain. Such a path specification may be given as a vector function of time, for example $\mathbf{x} = (X_1(t), X_2(t), X_3(t))$.

Without such a path specification, the total time derivative of f is not fully defined; however, when the path is specified, then:

$$\frac{d}{dt} f(x_1, x_2, x_3, t) = \frac{\partial f}{\partial x_1} \frac{dX_1}{dt} + \frac{\partial f}{\partial x_2} \frac{dX_2}{dt} + \frac{\partial f}{\partial x_3} \frac{dX_3}{dt} + \frac{\partial f}{\partial t}.$$

When studying fluid mechanics, the time-space path, $\mathbf{x}(t)$, most commonly chosen is that of a fluid particle. This path specification is commonly denoted by use of capital Ds:

$$\frac{D}{Dt} f(\mathbf{x}, t) \equiv \left[\frac{d}{dt} f(\mathbf{x}, t) \right]_{\text{following a fluid particle}}$$

$$= \left[\frac{\partial f}{\partial x_1} \frac{dX_1}{dt} + \frac{\partial f}{\partial x_2} \frac{dX_2}{dt} + \frac{\partial f}{\partial x_3} \frac{dX_3}{dt} + \frac{\partial f}{\partial t} \right]_{\text{following a fluid particle}}. \tag{B.1.1}$$

Here, the evaluation of the total derivative *following a fluid particle* can be formally completed by using the fluid-particle velocity matching condition specified above:

$$\textit{fluid particle velocity} \equiv \frac{d}{dt} \mathbf{x}(t) = \left(\frac{dX_1(t)}{dt}, \frac{dX_2(t)}{dt}, \frac{dX_3(t)}{dt} \right) = (u_1, u_2, u_3)|_{\mathbf{x}(t)} = \mathbf{u}(\mathbf{x}, t), \tag{B.1.2}$$

where $\mathbf{u}(\mathbf{x},t)$ is the fluid velocity at the particle location, and u_1, u_2, and u_3 are the Cartesian components of the fluid velocity. The third equality in (B.1.2) provides three velocity-component matching conditions:

$$dX_1/dt = u_1, \ dX_2/dt = u_2, \ \text{and} \ dX_3/dt = u_3. \tag{B.1.3}$$

When the various parts of (B.1.3) are substituted into (B.1.1), a final form for Df/Dt emerges:

$$\frac{D}{Dt} f(\mathbf{x}, t) = \frac{\partial f}{\partial t} + u_1 \frac{\partial f}{\partial x_1} + u_2 \frac{\partial f}{\partial x_2} + u_3 \frac{\partial f}{\partial x_3} = \frac{\partial f}{\partial t} + \mathbf{u} \cdot \nabla f = \frac{\partial f}{\partial t} + u_i \frac{\partial f}{\partial x_i}, \tag{B.1.4}$$

which is the same as (3.4). Here the final two equalities involve vector and index notation, respectively. These notations are described in Chapter 2. All three forms of Df/Dt are used in this text. *Total* and *partial* differentiation are the same when they operate on the same independent variable and this independent variable is the only independent variable.

Uses of Partial and Total Derivatives

There are situations in the study of fluid mechanics where a first-order partial differential equation, involving both time and space derivatives, like:

$$A(x, t) \frac{\partial f(x, t)}{\partial t} + B(x, t) \frac{\partial f(x, t)}{\partial x} = g(x, t, f) \tag{B.1.5}$$

needs to be solved to find $f(x, t)$. To accomplish this task, let's assume there exists a curve C in x-t space described by equations $x = X(s)$ and $t = T(s)$ that allows (B.1.5) to be recast as a total derivative with respect to s. Here s is the *arc length* in x-t space along the curve C. The total derivative of f along s is:

$$\frac{df}{ds} = \frac{\partial f(x,t)}{\partial t}\frac{dT(s)}{ds} + \frac{\partial f(x,t)}{\partial x}\frac{dX(s)}{ds}. \tag{B.1.6}$$

Thus, (B.1.5) can be simplified to

$$\frac{df}{ds} = g \quad \text{when } \frac{dT}{ds} = A \text{ and } \frac{dX}{ds} = B. \tag{B.1.7}$$

Taking a ratio of the last two equations produces:

$$\frac{dX}{dT} = \frac{B(X,T)}{A(X,T)}, \tag{B.1.8}$$

which parametrically specifies a set of curves C. Along any such curve, $df/ds = g$ and this equation can be integrated starting from an initial condition or boundary condition to determine f.

EXAMPLE B.1

Consider one-dimensional unidirectional wave propagation as specified by:

$$\frac{\partial f(x,t)}{\partial t} + U(t)\frac{\partial f(x,t)}{\partial x} = 0 \quad \text{where } f(x,0) = \phi(x), \tag{B.1.9, B.1.10}$$

f represents a traveling disturbance of some type, and U is the propagation velocity. In this case $A = 1$ and $B = U$; thus, (B.1.8) specifies the C curves via

$$\frac{dX}{dT} = U(T), \text{ or } X(T) = X_o + \int_0^T U(\tau)d\tau. \tag{B.1.11}$$

With $A = 1$, the middle equation of (B.1.7) implies $T = T_o + s$, so (B.1.11) leads to:

$$x = X(s) = X_o + \int_o^{T_o+s} U(\tau)d\tau, \text{ and } t = T(s) = T_o + s. \tag{B.1.12, B.1.13}$$

These two equations define the set of C curves in x-t space along which the behavior of f is easily determined from the first equation of (B.1.7) with $g = 0$:

$$\frac{df}{ds} = 0, \text{ or } f_o = f(x,t) = f(X(s), T(s)) = f\left(X_o + \int_0^{T_o+s} U(\tau)d\tau, T_o + s\right). \tag{B.1.14}$$

Here f_o is the constant value of $f(x,t)$ that is found when s varies along a particular C curve, and X_o and T_o are constants of integration that specify the x-t location of $s = 0$ on this C curve. These constants can be evaluated using the initial condition specified in (B.1.10) in terms of ϕ at $t = T_o + s = 0$, and the last form for f in (B.1.14):

$$f_o = f(X_o, 0) = \phi(X_o). \tag{B.1.15}$$

Here it is important to note that the constant f_o may be different for the various C curves that start from different x-t locations. To reach the final solution of (B.1.9), eliminate f_o and X_o from (B.1.15) using (B.1.12) through (B.1.14) in favor of x, t, and $f(x, t)$:

$$f(x, t) = \phi\left(x - \int_o^t U(\tau)d\tau\right). \tag{B.1.16}$$

This approach to differential equation solving where special paths are found that simplify the governing equation (or equations) can be formalized and generalized; it is called the *method of characteristics*. But, independent of this and perhaps more important, the two fundamental and enduring features of partial differential equation solving are displayed here.

i) Partial differential equations are solved by rearrangement and *integration*. Extra differentiation is typically not useful; always look for ways to *integrate* to find a solution.

ii) Difficulty is not entirely eliminated by changing from partial to total derivatives or vice versa. In the above example, there is initially one unknown function, f, and two independent coordinates, x and t, but this is transformed (via the method of characteristics) into a problem with two unknown functions, f and X, and one independent variable, s or t.

Integration of Partial Derivatives

There is really nothing special here except to note that constants of integration turn into functions that may depend on all the not-integrated-over independent variables. For example, consider $f(x, y, z, t)$ that solves the partial differential equation: $\partial f/\partial x = Ax + By$. Direct integration with y, z, and t treated as constants produces:

$$f = \int (Ax + By)dx = Ax^2/2 + Byx + C(y, z, t),$$

where $C(y, z, t)$ is an unknown function that does not depend on x; it replaces the usual constant of integration in one-variable indefinite integration.

B.2. CHANGING INDEPENDENT VARIABLES

Two situations commonly arise in the study of fluid mechanics where changing the independent variable(s) is advantageous. The first situation is changing coordinate systems. Here the number of new and old independent variables will usually be the same. Consider the situation where a partial differential equation is known in Cartesian-time coordinates (x, y, z, t), but it will be easier to solve in another coordinate system (ξ, ψ, ζ, τ). Assume the transformation between the two coordinate systems is given by: $\xi = X(x, y, z, t)$, $\psi = Y(x, y, z, t)$, $\zeta = Z(x, y, z, t)$, and $\tau = T(x, y, z, t)$. Cartesian and temporal partial derivatives can be transformed as follows:

$$\frac{\partial}{\partial x} = \frac{\partial X}{\partial x}\frac{\partial}{\partial \xi} + \frac{\partial Y}{\partial x}\frac{\partial}{\partial \psi} + \frac{\partial Z}{\partial x}\frac{\partial}{\partial \zeta} + \frac{\partial T}{\partial x}\frac{\partial}{\partial \tau}, \quad \frac{\partial}{\partial y} = \frac{\partial X}{\partial y}\frac{\partial}{\partial \xi} + \frac{\partial Y}{\partial y}\frac{\partial}{\partial \psi} + \frac{\partial Z}{\partial y}\frac{\partial}{\partial \zeta} + \frac{\partial T}{\partial y}\frac{\partial}{\partial \tau},$$
$$\frac{\partial}{\partial z} = \frac{\partial X}{\partial z}\frac{\partial}{\partial \xi} + \frac{\partial Y}{\partial z}\frac{\partial}{\partial \psi} + \frac{\partial Z}{\partial z}\frac{\partial}{\partial \zeta} + \frac{\partial T}{\partial z}\frac{\partial}{\partial \tau}, \quad \text{and} \quad \frac{\partial}{\partial t} = \frac{\partial X}{\partial t}\frac{\partial}{\partial \xi} + \frac{\partial Y}{\partial t}\frac{\partial}{\partial \psi} + \frac{\partial Z}{\partial t}\frac{\partial}{\partial \zeta} + \frac{\partial T}{\partial t}\frac{\partial}{\partial \tau}. \tag{B.2.1}$$

EXAMPLE B.2

Consider the case where (x, y, z, t) and (ξ, ψ, ζ, τ) represent Cartesian systems with parallel axes that are moving with respect to each other at a constant velocity (U, V, W) when observed in (x, y, z, t), so that $\xi = x - Ut$, $\psi = y - Vt$, $\zeta = z - Wt$, and $\tau = t$. Application of the above derivative transformations (B.2.1) produces:

$$\frac{\partial}{\partial x} = \frac{\partial}{\partial \xi}, \quad \frac{\partial}{\partial y} = \frac{\partial}{\partial \psi}, \quad \frac{\partial}{\partial z} = \frac{\partial}{\partial \zeta}, \quad \text{and} \quad \frac{\partial}{\partial t} = -U\frac{\partial}{\partial \xi} - V\frac{\partial}{\partial \psi} - W\frac{\partial}{\partial \zeta} + \frac{\partial}{\partial \tau}. \qquad \text{(B.2.2)}$$

Perhaps unexpectedly, extra differentiations only appear in the transformed time derivative, even though the time variable transformation equation was simplest.

The second situation that requires changing independent variables occurs when a combination of independent variables (and parameters) is found that might simplify a partial differential equation. Here the usual goal is to convert a partial differential equation having multiple independent variables into a total differential equation with one independent variable. If $\eta = H(x, y, z, t)$ is the combination variable, then a straightforward application of the chain rule for partial differentiation produces:

$$\frac{\partial}{\partial x} = \frac{\partial H}{\partial x}\frac{d}{d\eta}, \quad \frac{\partial}{\partial y} = \frac{\partial H}{\partial y}\frac{d}{d\eta}, \quad \frac{\partial}{\partial z} = \frac{\partial H}{\partial z}\frac{d}{d\eta}, \quad \text{and} \quad \frac{\partial}{\partial t} = \frac{\partial H}{\partial t}\frac{d}{d\eta}. \qquad \text{(B.2.3)}$$

EXAMPLE B.3

Consider a function with two independent variables, $f(x, t)$, for which we hypothesize the existence of a special combination (or similarity) variable $\eta = xt^{\alpha}$, where α is a real number, that facilitates the solution of the partial differential equation for $f(x, t)$. Mathematically, this hypothesis can be stated as: $f(x, t) = f(\eta) = f(xt^{\alpha})$, and partial derivatives of f can be obtained from the first and last equations of (B.2.3) with $H = xt^{\alpha}$:

$$\frac{\partial}{\partial x}f(x, t) = \frac{\partial(xt^{\alpha})}{\partial x}\frac{d}{d\eta}f(\eta) = t^{\alpha}\frac{df}{d\eta}, \quad \text{and} \quad \frac{\partial}{\partial t}f(x, t) = \frac{\partial(xt^{\alpha})}{\partial t}\frac{d}{d\eta}f(\eta) = \alpha xt^{\alpha-1}\frac{df}{d\eta} = \frac{\alpha}{t}\eta\frac{df}{d\eta}.$$

Second-order derivatives are generated by appropriately differentiating these first-order results.

B.3. BASIC VECTOR CALCULUS

The gradient operator, ∇, is the general-purpose directional derivative for multiple spatial coordinates. It is a vector operator, and it exists in all suitably defined coordinate systems. Its properties are a combination of those of ordinary partial derivatives and ordinary vectors. It has components and its position and operation character (multiply, dot product, cross product, etc.) matter within a set or grouping of functions or variables. For example, $(\mathbf{u} \cdot \nabla)\mathbf{v} \neq \mathbf{v}(\nabla \cdot \mathbf{u})$ in general, even though these two expressions would be an equal if ∇ were replaced by a constant factor. Some properties of ∇ are listed here:

- In Cartesian coordinates, $\mathbf{x} = (x, y, z)$: $\nabla = \mathbf{e}_x \frac{\partial}{\partial x} + \mathbf{e}_y \frac{\partial}{\partial y} + \mathbf{e}_z \frac{\partial}{\partial z}$ where the $\mathbf{e}$s are unit vectors.

- The *gradient* of the scalar field ρ is: $\nabla \rho = \mathbf{e}_x \frac{\partial \rho}{\partial x} + \mathbf{e}_y \frac{\partial \rho}{\partial y} + \mathbf{e}_z \frac{\partial \rho}{\partial z}$.

- The *divergence* of a vector field $\mathbf{u} = (u, v, w)$ is: $\nabla \cdot \mathbf{u} = \frac{\partial u}{\partial x} + \frac{\partial v}{\partial y} + \frac{\partial w}{\partial z}$.

- The *curl* of a vector field $\mathbf{u} = (u, v, w)$ is: $\nabla \times \mathbf{u} = \det \begin{vmatrix} \mathbf{e}_x & \mathbf{e}_y & \mathbf{e}_z \\ \partial/\partial x & \partial/\partial y & \partial/\partial z \\ u & v & w \end{vmatrix}$.

Vector Identities Involving ∇

Here ρ and ϕ are scalar functions, $\mathbf{u}$ and $\mathbf{F}$ are vector functions, and $\mathbf{x}$ is the position vector.

$$\nabla \cdot \mathbf{x} = 3 \tag{B.3.1}$$

$$\nabla \times \mathbf{x} = 0 \tag{B.3.2}$$

$$\nabla \cdot (\mathbf{x}/|\mathbf{x}|^3) = 0 \tag{B.3.3}$$

$$(\mathbf{u} \cdot \nabla)\mathbf{x} = \mathbf{u} \tag{B.3.4}$$

$$\nabla(\rho \phi) = \rho \nabla \phi + \phi \nabla \rho \tag{B.3.5}$$

$$\nabla \cdot (\rho \mathbf{u}) = \rho \nabla \cdot \mathbf{u} + (\mathbf{u} \cdot \nabla)\rho \tag{B.3.6}$$

$$\nabla \times (\rho \mathbf{u}) = \rho \nabla \times \mathbf{u} + (\nabla \rho) \times \mathbf{u} \tag{B.3.7}$$

$$\nabla \cdot (\mathbf{u} \times \mathbf{F}) = (\nabla \times \mathbf{u}) \cdot \mathbf{F} - \mathbf{u} \cdot (\nabla \times \mathbf{F}) \tag{B.3.8}$$

$$\nabla(\mathbf{u} \cdot \mathbf{F}) = \mathbf{u} \times (\nabla \times \mathbf{F}) + \mathbf{F} \times (\nabla \times \mathbf{u}) + (\mathbf{u} \cdot \nabla)\mathbf{F} + (\mathbf{F} \cdot \nabla)\mathbf{u} \tag{B.3.9}$$

$$\nabla \times (\mathbf{u} \times \mathbf{F}) = (\mathbf{F} \cdot \nabla)\mathbf{u} - \mathbf{F}(\nabla \cdot \mathbf{u}) + \mathbf{u}(\nabla \cdot \mathbf{F}) - (\mathbf{u} \cdot \nabla)\mathbf{F} \tag{B.3.10}$$

$$\nabla \times (\nabla \rho) = 0 \tag{B.3.11}$$

$$\nabla \cdot (\nabla \times \mathbf{u}) = 0 \tag{B.3.12}$$

$$\nabla \times (\nabla \times \mathbf{u}) = \nabla(\nabla \cdot \mathbf{u}) - \nabla^2 \mathbf{u} \tag{B.3.13}$$

Integral Theorems Involving ∇

These are discussed in Sections 2.12 and 2.13.

- For a closed surface A that contains volume V with $\mathbf{n}$ = the outward normal on A, Gauss' Theorem is:

$$\int_A \rho\, \mathbf{n} dA = \int_V \nabla \rho\, dV \text{ for scalars, and } \int_A \mathbf{u} \cdot \mathbf{n} dA = \int_V \nabla \cdot \mathbf{u} dV \text{ for vectors.}$$

- For a closed curve C that bounds surface A with $\mathbf{n}$ = the normal to A and t the tangent to C, Stokes' Theorem is: $\oint_C \mathbf{u} \cdot \mathbf{t}\, ds = \int_A (\nabla \times \mathbf{u}) \cdot \mathbf{n} dA$, where s is the arc length along C.

B.4. THE DIRAC DELTA FUNCTION

The Dirac delta function is commonly denoted $\delta(x)$, where x is a real variable. It is a unit-area impulse that exists at only one point in space; it is zero everywhere except where its argument is zero. The Dirac delta-function can be defined as a limit of a smooth function, such as:

$$\delta(x) = \lim_{\sigma \to 0} (\sqrt{2\pi}\sigma)^{-1} \exp\{-x^2/2\sigma^2\}. \tag{B.4.1}$$

The value of $\delta(x)$ is infinite at $x = 0$ but its integral is unity. Here are a few properties of $\delta(x)$ for a, b, and x_o real constants and $b > a$:

$$x\delta(x - a) = a\delta(x - a), \tag{B.4.2}$$

$$\int_a^b \delta(x - x_o)dx = \begin{cases} 1 & \text{for } a \le x_o \le b \\ 0 & \text{for } x_o < a \quad \text{or} \quad b < x_o \end{cases}, \tag{B.4.3}$$

$$\int_{-\infty}^{+\infty} f(x)\delta(x - x_o)dx = f(x_o). \tag{B.4.4}$$

These properties ease the evaluation of complicated integrals when a Dirac delta function appears in the integrand. In more dimensions where $\mathbf{x} = (x, y, z)$, the following notation is common:

$$\delta(\mathbf{x} - \mathbf{x}_o) = \delta(x - x_o)\delta(y - y_o)\delta(z - z_o).$$

In the study of fluid mechanics, the usual notation for the Dirac delta-function is potentially confusing because δ is also commonly used to denote a length scale of interest in the flow field, such as a boundary-layer thickness or the length scale of a similarity variable. Thus, specific mention of the Dirac delta function is made where it is used in the text.

EXAMPLE B.4

Evaluate the integral: $I = \int_{-\infty}^{+\infty} F(x)[(x_o - x)^2 + r_o^2]^{-1/2} e^{ikx} \delta(x - ct)dx$. Here the limits of integration ensure that x will equal ct somewhere in the integration. Equation (B.4.4) implies that the value of this integral is determined by replacing x with ct in the integrand; therefore:
$I = F(ct)[(x_o - ct)^2 + r_o^2]^{-1/2} e^{ikct}$.

B.5. COMMON THREE-DIMENSIONAL COORDINATE SYSTEMS

In all cases that follow, ξ, ψ, and ζ are constants.

Cartesian Coordinates (Figure B.1)

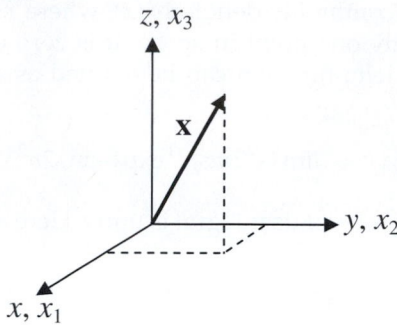

Position: $\mathbf{x} = (x, y, z) = (x_1, x_2, x_3) = x_1\mathbf{e}_1 + x_2\mathbf{e}_2 + x_3\mathbf{e}_3$

Unit vectors: $\mathbf{e}_x$, $\mathbf{e}_y$, and $\mathbf{e}_z$, or $\mathbf{e}_1$, $\mathbf{e}_2$, and $\mathbf{e}_3$

Unit vector dependencies: $\partial\mathbf{e}_i/\partial x_j = 0$ for i and $j = 1, 2$, or 3; that is, Cartesian unit vectors are independent of the coordinate values

Gradient operator: $\nabla = \mathbf{e}_x\frac{\partial}{\partial x} + \mathbf{e}_y\frac{\partial}{\partial y} + \mathbf{e}_z\frac{\partial}{\partial z} = \mathbf{e}_1\frac{\partial}{\partial x_1} + \mathbf{e}_2\frac{\partial}{\partial x_2} + \mathbf{e}_3\frac{\partial}{\partial x_3}$

Surface integral, S, of $f(x,y,z)$ over the plane defined by $x = \xi$: $S = \int\limits_{y=-\infty}^{+\infty}\int\limits_{z=-\infty}^{+\infty} f(\xi, y, z)dzdy$

Surface integral, S, of $f(x,y,z)$ over the plane defined by $y = \psi$: $S = \int\limits_{x=-\infty}^{+\infty}\int\limits_{z=-\infty}^{+\infty} f(x, \psi, z)dzdx$

Surface integral, S, of $f(x,y,z)$ over the plane defined by $z = \zeta$: $S = \int\limits_{x=-\infty}^{+\infty}\int\limits_{y=-\infty}^{+\infty} f(x, y, \zeta)dydx$

Volume integral, V, of $f(x,y,z)$ over all space: $V = \int\limits_{x=-\infty}^{+\infty}\int\limits_{y=-\infty}^{+\infty}\int\limits_{z=-\infty}^{+\infty} f(x, y, z)dzdydx$

Cylindrical Coordinates (Figure B.2)

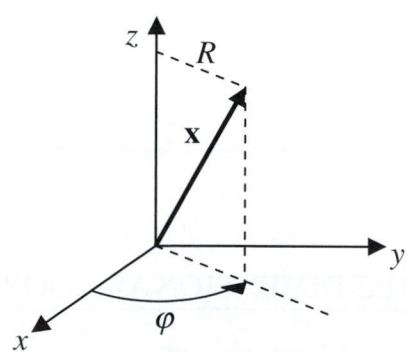

Position: $\mathbf{x} = (R, \varphi, z) = R\mathbf{e}_R + z\mathbf{e}_z$; $x = R\cos\varphi$, $y = R\sin\varphi$; $z = z$, or $R = \sqrt{x^2 + y^2}$, $\varphi = \tan^{-1}(y/x)$

Unit vectors: $\mathbf{e}_R = \mathbf{e}_x\cos\varphi + \mathbf{e}_y\sin\varphi$, $\mathbf{e}_\varphi = -\mathbf{e}_x\sin\varphi + \mathbf{e}_y\cos\varphi$, $\mathbf{e}_z = $ same as Cartesian

Unit vector dependencies: $\partial\mathbf{e}_R/\partial R = 0$, $\partial\mathbf{e}_R/\partial\varphi = \mathbf{e}_\varphi$, $\partial\mathbf{e}_R/\partial z = 0$

$$\partial\mathbf{e}_\varphi/\partial R = 0, \quad \partial\mathbf{e}_\varphi/\partial\varphi = -\mathbf{e}_R, \quad \partial\mathbf{e}_\varphi/\partial z = 0$$

$$\partial\mathbf{e}_z/\partial R = 0, \quad \partial\mathbf{e}_z/\partial\varphi = 0, \quad \partial\mathbf{e}_z/\partial z = 0$$

Gradient operator: $\nabla = \mathbf{e}_R\frac{\partial}{\partial R} + \mathbf{e}_\varphi\frac{1}{R}\frac{\partial}{\partial\varphi} + \mathbf{e}_z\frac{\partial}{\partial z}$

Surface integral, S, of $f(R,\theta,z)$ over the cylinder defined by $R = \xi$: $S = \int_{\varphi=0}^{2\pi}\int_{z=-\infty}^{+\infty} f(\xi, \varphi, z)\xi\, dz\, d\varphi$

Surface integral, S, of $f(R,\theta,z)$ over the half plane defined by $\varphi = \psi$: $S = \int_{R=0}^{+\infty}\int_{z=-\infty}^{+\infty} f(R, \psi, z)\, dz\, dR$

Surface integral, S, of $f(R,\theta,z)$ over the plane defined by $z = \zeta$: $S = \int_{R=0}^{+\infty}\int_{\varphi=0}^{2\pi} f(R, \varphi, \zeta)R\, d\varphi\, dR$

Volume integral, V, of $f(R,\theta,z)$ over all space: $V = \int_{z=-\infty}^{+\infty}\int_{R=0}^{+\infty}\int_{\varphi=0}^{2\pi} f(R, \varphi, z)R\, d\varphi\, dR\, dz$

Spherical Coordinates (Figure B.3)

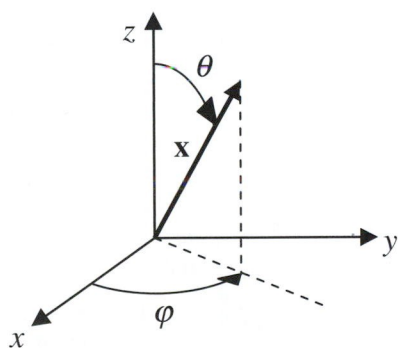

Position: $\mathbf{x} = (r, \theta, \varphi) = r\mathbf{e}_r$; $x = r\cos\varphi\sin\theta$, $y = r\sin\varphi\sin\theta$, $z = r\cos\theta$; or $r = \sqrt{x^2 + y^2 + z^2}$, $\theta = \tan^{-1}(\sqrt{x^2 + y^2}/z)$, and $\varphi = \tan^{-1}(y/x)$

Unit vectors: $\mathbf{e}_r = \mathbf{e}_x\sin\theta\cos\varphi + \mathbf{e}_y\sin\theta\sin\varphi + \mathbf{e}_z\cos\theta$,

$\mathbf{e}_\theta = \mathbf{e}_x\cos\theta\cos\varphi + \mathbf{e}_y\cos\theta\sin\varphi - \mathbf{e}_z\sin\theta$, $\mathbf{e}_\varphi = -\mathbf{e}_x\sin\varphi + \mathbf{e}_y\cos\varphi$

Unit vector dependencies: $\partial\mathbf{e}_r/\partial r = 0$, $\partial\mathbf{e}_r/\partial\theta = \mathbf{e}_\theta$, $\partial\mathbf{e}_r/\partial\varphi = \mathbf{e}_\varphi\sin\theta$

$$\partial\mathbf{e}_\theta/\partial r = 0, \quad \partial\mathbf{e}_\theta/\partial\theta = -\mathbf{e}_r, \quad \partial\mathbf{e}_\theta/\partial\varphi = \mathbf{e}_\varphi\cos\theta$$

$$\partial\mathbf{e}_\varphi/\partial r = 0, \quad \partial\mathbf{e}_\varphi/\partial\theta = 0, \quad \partial\mathbf{e}_\varphi/\partial\varphi = -\mathbf{e}_r\sin\theta - \mathbf{e}_\theta\cos\theta$$

Gradient operator: $\nabla = \mathbf{e}_r\frac{\partial}{\partial r} + \mathbf{e}_\theta\frac{1}{r}\frac{\partial}{\partial\theta} + \mathbf{e}_\varphi\frac{1}{r\sin\theta}\frac{\partial}{\partial\varphi}$

Surface integral, S, of $f(r,\theta,\varphi)$ over the sphere defined by $r = \xi$: $S = \int_{\theta=0}^{\pi}\int_{\varphi=0}^{2\pi} f(\xi, \theta, \varphi)\xi^2\sin\theta\, d\varphi\, d\theta$

Surface integral, S, of $f(r,\theta,\varphi)$ over the cone defined by $\theta = \psi$: $S = \int\limits_{r=0}^{+\infty} \int\limits_{\varphi=0}^{2\pi} f(r, \psi, \varphi) r \sin \psi \, d\varphi \, dr$

Surface integral, S, of $f(r,\theta,\varphi)$ over the half plane defined by $\varphi = \zeta$: $S = \int\limits_{r=0}^{+\infty} \int\limits_{\theta=0}^{\pi} f(r, \theta, \zeta) r \, d\theta \, dr$

Volume integral, V, of $f(r,\theta,\varphi)$ over all space: $V = \int\limits_{r=0}^{+\infty} \int\limits_{\theta=0}^{\pi} \int\limits_{\varphi=0}^{2\pi} f(r, \theta, \varphi) r^2 \sin \theta \, d\varphi \, d\theta \, dr$

B.6. EQUATIONS IN CURVILINEAR COORDINATE SYSTEMS

Plane Polar Coordinates (Figure 3.3a)

Position and velocity vectors $\mathbf{x} = (r, \theta) = r\mathbf{e}_r$; $\mathbf{u} = (u_r, u_\theta) = u_r \mathbf{e}_r + u_\theta \mathbf{e}_\theta$

Gradient of a scalar ψ: $\nabla \psi = \mathbf{e}_r \dfrac{\partial \psi}{\partial r} + \mathbf{e}_\theta \dfrac{1}{r} \dfrac{\partial \psi}{\partial \theta}$

Laplacian of a scalar ψ: $\nabla^2 \psi = \dfrac{1}{r} \dfrac{\partial}{\partial r}\left(r \dfrac{\partial \psi}{\partial r}\right) + \dfrac{1}{r^2} \dfrac{\partial^2 \psi}{\partial \theta^2}$

Divergence of a vector: $\nabla \cdot \mathbf{u} = \dfrac{1}{r} \dfrac{\partial}{\partial r}(r u_r) + \dfrac{1}{r} \dfrac{\partial u_\theta}{\partial \theta}$

Curl of a vector, vorticity: $\boldsymbol{\omega} = \nabla \times \mathbf{u} = \mathbf{e}_z \left(\dfrac{1}{r} \dfrac{\partial (r u_\theta)}{\partial r} - \dfrac{1}{r} \dfrac{\partial u_r}{\partial \theta}\right)$

Laplacian of a vector: $\nabla^2 \mathbf{u} = \mathbf{e}_r \left(\nabla^2 u_r - \dfrac{u_r}{r^2} - \dfrac{2}{r^2} \dfrac{\partial u_\theta}{\partial \theta}\right) + \mathbf{e}_\theta \left(\nabla^2 u_\theta + \dfrac{2}{r^2} \dfrac{\partial u_r}{\partial \theta} - \dfrac{u_\theta}{r^2}\right)$

Strain rate S_{ij} and viscous stress σ_{ij} for an incompressible fluid where $\sigma_{ij} = 2\mu S_{ij}$:

$S_{rr} = \dfrac{\partial u_r}{\partial r} = \dfrac{1}{2\mu}\sigma_{rr}, S_{\theta\theta} = \dfrac{1}{r} \dfrac{\partial u_\theta}{\partial \theta} + \dfrac{u_r}{r} = \dfrac{1}{2\mu}\sigma_{\theta\theta}, S_{r\theta} = \dfrac{r}{2} \dfrac{\partial}{\partial r}\left(\dfrac{u_\theta}{r}\right) + \dfrac{1}{2r} \dfrac{\partial u_r}{\partial \theta} = \dfrac{1}{2\mu}\sigma_{r\theta}$

Equation of continuity: $\dfrac{\partial \rho}{\partial t} + \dfrac{1}{r \cdot \partial r}(r \rho u_r) + \dfrac{1}{r} \dfrac{\partial}{\partial \theta}(\rho u_\theta) = 0$

Navier-Stokes equations with constant ρ, constant ν, and no body force:

$\dfrac{\partial u_r}{\partial t} + u_r \dfrac{\partial u_r}{\partial r} + \dfrac{u_\theta}{r} \dfrac{\partial u_r}{\partial \theta} - \dfrac{u_\theta^2}{r} = -\dfrac{1}{\rho} \dfrac{\partial p}{\partial r} + \nu\left(\nabla^2 u_r - \dfrac{u_r}{r^2} - \dfrac{2}{r^2} \dfrac{\partial u_\theta}{\partial \theta}\right),$

$\dfrac{\partial u_\theta}{\partial t} + u_r \dfrac{\partial u_\theta}{\partial r} + \dfrac{u_\theta}{r} \dfrac{\partial u_\theta}{\partial \theta} + \dfrac{u_r u_\theta}{r} = -\dfrac{1}{\rho r} \dfrac{\partial p}{\partial \theta} + \nu\left(\nabla^2 u_\theta + \dfrac{2}{r^2} \dfrac{\partial u_r}{\partial \theta} - \dfrac{u_\theta}{r^2}\right),$

where $\nabla^2 = \dfrac{1}{r} \dfrac{\partial}{\partial r}\left(r \dfrac{\partial}{\partial r}\right) + \dfrac{1}{r^2} \dfrac{\partial^2}{\partial \theta^2}$.

Cylindrical Coordinates (Figure B.2)

Position and velocity vectors: $\mathbf{x} = (R, \varphi, z) = R\mathbf{e}_R + z\mathbf{e}_z$; $\mathbf{u} = (u_R, u_\varphi, u_z) = u_R \mathbf{e}_R + u_\varphi \mathbf{e}_\varphi + u_z \mathbf{e}_z$

Gradient of a scalar ψ: $\nabla \psi = \mathbf{e}_R \dfrac{\partial \psi}{\partial R} + \mathbf{e}_\varphi \dfrac{1}{R} \dfrac{\partial \psi}{\partial \varphi} + \mathbf{e}_z \dfrac{\partial \psi}{\partial z}$

Laplacian of a scalar ψ: $\nabla^2 \psi = \dfrac{1}{R} \dfrac{\partial}{\partial R}\left(R \dfrac{\partial \psi}{\partial R}\right) + \dfrac{1}{R^2} \dfrac{\partial^2 \psi}{\partial \varphi^2} + \dfrac{\partial^2 \psi}{\partial z^2}$

Divergence of a vector: $\nabla \cdot \mathbf{u} = \dfrac{1}{R}\dfrac{\partial}{\partial R}(Ru_R) + \dfrac{1}{R}\dfrac{\partial u_\varphi}{\partial \varphi} + \dfrac{\partial u_z}{\partial z}$

Curl of a vector, vorticity: $\boldsymbol{\omega} = \nabla \times \mathbf{u} = \mathbf{e}_R\left(\dfrac{1}{R}\dfrac{\partial u_z}{\partial \varphi} - \dfrac{\partial u_\varphi}{\partial z}\right) + \mathbf{e}_\varphi\left(\dfrac{\partial u_R}{\partial z} - \dfrac{\partial u_z}{\partial R}\right)$

$$+ \mathbf{e}_z\left(\dfrac{1}{R}\dfrac{\partial(Ru_\varphi)}{\partial R} - \dfrac{1}{R}\dfrac{\partial u_R}{\partial \varphi}\right)$$

Laplacian of a vector: $\nabla^2 \mathbf{u} = \mathbf{e}_R\left(\nabla^2 u_R - \dfrac{u_R}{R^2} - \dfrac{2}{R^2}\dfrac{\partial u_\varphi}{\partial \varphi}\right) + \mathbf{e}_\varphi\left(\nabla^2 u_\varphi + \dfrac{2}{R^2}\dfrac{\partial u_R}{\partial \varphi} - \dfrac{u_\varphi}{R^2}\right)$

$$+ \mathbf{e}_z\nabla^2 u_z$$

Strain rate S_{ij} and viscous stress σ_{ij} for an incompressible fluid where $\sigma_{ij} = 2\mu S_{ij}$:

$$S_{RR} = \dfrac{\partial u_R}{\partial R} = \dfrac{1}{2\mu}\sigma_{RR}, \; S_{\varphi\varphi} = \dfrac{1}{R}\dfrac{\partial u_\varphi}{\partial \varphi} + \dfrac{u_R}{R} = \dfrac{1}{2\mu}\sigma_{\varphi\varphi}, \; S_{zz} = \dfrac{\partial u_z}{\partial z} = \dfrac{1}{2\mu}\sigma_{zz}$$

$$S_{R\varphi} = \dfrac{R}{2}\dfrac{\partial}{\partial R}\left(\dfrac{u_\varphi}{R}\right) + \dfrac{1}{2R}\dfrac{\partial u_R}{\partial \varphi} = \dfrac{1}{2\mu}\sigma_{R\varphi}, \; S_{\varphi z} = \dfrac{1}{2R}\dfrac{\partial u_z}{\partial \varphi} + \dfrac{1}{2}\dfrac{\partial u_\varphi}{\partial z} = \dfrac{1}{2\mu}\sigma_{\varphi z},$$

$$S_{zR} = \dfrac{1}{2}\left(\dfrac{\partial u_R}{\partial z} + \dfrac{\partial u_z}{\partial R}\right) = \dfrac{1}{2\mu}\sigma_{zR}$$

Equation of continuity: $\dfrac{\partial \rho}{\partial t} + \dfrac{1}{R}\dfrac{\partial}{\partial R}(R\rho u_R) + \dfrac{1}{R}\dfrac{\partial}{\partial \varphi}(\rho u_\varphi) + \dfrac{\partial}{\partial z}(\rho u_z) = 0$

Navier-Stokes equations with constant ρ, constant ν, and no body force:

$$\dfrac{\partial u_R}{\partial t} + (\mathbf{u}\cdot\nabla)u_R - \dfrac{u_\varphi^2}{R} = -\dfrac{1}{\rho}\dfrac{\partial p}{\partial R} + \nu\left(\nabla^2 u_R - \dfrac{u_R}{R^2} - \dfrac{2}{R^2}\dfrac{\partial u_\varphi}{\partial \varphi}\right),$$

$$\dfrac{\partial u_\varphi}{\partial t} + (\mathbf{u}\cdot\nabla)u_\varphi + \dfrac{u_R u_\varphi}{R} = -\dfrac{1}{\rho R}\dfrac{\partial p}{\partial \varphi} + \nu\left(\nabla^2 u_\varphi + \dfrac{2}{R^2}\dfrac{\partial u_R}{\partial \varphi} - \dfrac{u_\varphi}{R^2}\right),$$

$$\dfrac{\partial u_z}{\partial t} + (\mathbf{u}\cdot\nabla)u_z = -\dfrac{1}{\rho}\dfrac{\partial p}{\partial z} + \nu\nabla^2 u_z,$$

where: $\mathbf{u}\cdot\nabla = u_R\dfrac{\partial}{\partial R} + \dfrac{u_\varphi}{R}\dfrac{\partial}{\partial \varphi} + u_z\dfrac{\partial}{\partial z}$ and $\nabla^2 = \dfrac{1}{R}\dfrac{\partial}{\partial R}\left(R\dfrac{\partial}{\partial R}\right) + \dfrac{1}{R^2}\dfrac{\partial^2}{\partial \varphi^2} + \dfrac{\partial^2}{\partial z^2}$.

Spherical Coordinates (Figure B.3)

Position and velocity vectors: $\mathbf{x} = (r, \theta, \varphi) = r\mathbf{e}_r$; $\mathbf{u} = (u_r, u_\theta, u_\varphi) = u_r\mathbf{e}_r + u_\theta\mathbf{e}_\theta + u_\varphi\mathbf{e}_\varphi$

Gradient of a scalar ψ: $\nabla\psi = \mathbf{e}_r\dfrac{\partial \psi}{\partial r} + \mathbf{e}_\varphi\dfrac{1}{r}\dfrac{\partial \psi}{\partial \theta} + \mathbf{e}_\varphi\dfrac{1}{r\sin\theta}\dfrac{\partial \psi}{\partial \varphi}$

Laplacian of a scalar ψ: $\nabla^2\psi = \dfrac{1}{r^2}\dfrac{\partial}{\partial r}\left(r^2\dfrac{\partial \psi}{\partial r}\right) + \dfrac{1}{r^2\sin\theta}\dfrac{\partial}{\partial \theta}\left(\sin\theta\dfrac{\partial \psi}{\partial \theta}\right) + \dfrac{1}{r^2\sin^2\theta}\dfrac{\partial^2\psi}{\partial \varphi^2}$

Divergence of a vector: $\nabla\cdot\mathbf{u} = \dfrac{1}{r^2}\dfrac{\partial}{\partial r}(r^2 u_r) + \dfrac{1}{r\sin\theta}\dfrac{\partial(u_\theta\sin\theta)}{\partial \theta} + \dfrac{1}{r\sin\theta}\dfrac{\partial u_\varphi}{\partial \varphi}$

Curl of a vector, vorticity: $\boldsymbol{\omega} = \nabla \times \mathbf{u} = \dfrac{\mathbf{e}_r}{r \sin \theta}\left(\dfrac{\partial(u_\varphi \sin \theta)}{\partial \theta} - \dfrac{\partial u_\theta}{\partial \varphi}\right) + \dfrac{\mathbf{e}_\theta}{r}\left(\dfrac{1}{\sin \theta}\dfrac{\partial u_r}{\partial \varphi} - \dfrac{\partial(r u_\varphi)}{\partial r}\right)$

$$+ \dfrac{\mathbf{e}_\varphi}{r}\left(\dfrac{\partial(r u_\theta)}{\partial r} - \dfrac{\partial u_r}{\partial \theta}\right)$$

Laplacian of a vector: $\nabla^2 \mathbf{u} = \mathbf{e}_r\left(\nabla^2 u_r - \dfrac{2u_r}{r^2} - \dfrac{2}{r^2 \sin \theta}\dfrac{\partial(u_\theta \sin \theta)}{\partial \theta} - \dfrac{2}{r^2 \sin \theta}\dfrac{\partial u_\varphi}{\partial \varphi}\right) +$

$\mathbf{e}_\theta\left(\nabla^2 u_\theta + \dfrac{2}{r^2}\dfrac{\partial u_r}{\partial \theta} - \dfrac{u_\theta}{r^2 \sin^2 \theta} - \dfrac{2 \cos \theta}{r^2 \sin^2 \theta}\dfrac{\partial u_\varphi}{\partial \varphi}\right) + \mathbf{e}_\varphi\left(\nabla^2 u_\varphi + \dfrac{2}{r^2 \sin \theta}\dfrac{\partial u_r}{\partial \varphi} + \dfrac{2 \cos \theta}{r^2 \sin^2 \theta}\dfrac{\partial u_\theta}{\partial \varphi} - \dfrac{u_\varphi}{r^2 \sin^2 \theta}\right)$

Strain rate S_{ij} and viscous stress σ_{ij} for an incompressible fluid where $\sigma_{ij} = 2\mu S_{ij}$:

$$S_{rr} = \dfrac{\partial u_r}{\partial r} = \dfrac{1}{2\mu}\sigma rr, \; S_{\theta\theta} = \dfrac{1}{r}\dfrac{\partial u_\theta}{\partial \theta} + \dfrac{u_r}{r} = \dfrac{1}{2\mu}\sigma_{\theta\theta}, \; S_{\varphi\varphi} = \dfrac{1}{r \sin \theta}\dfrac{\partial u_\varphi}{\partial \varphi} + \dfrac{u_r}{r} + \dfrac{u_\theta \cot \theta}{r} = \dfrac{1}{2\mu}\sigma\varphi\varphi,$$

$$S_{\theta\varphi} = \dfrac{\sin \theta}{2r}\dfrac{\partial}{\partial \theta}\left(\dfrac{u_\varphi}{\sin \theta}\right) + \dfrac{1}{2r \sin \theta}\dfrac{\partial u_\theta}{\partial \varphi} = \dfrac{1}{2\mu}\sigma_{\theta\varphi}, \; S_{\varphi r} = \dfrac{1}{2r \sin \theta}\dfrac{\partial u_r}{\partial \varphi} + \dfrac{r}{2}\dfrac{\partial}{\partial r}\left(\dfrac{u_\varphi}{r}\right) = \dfrac{1}{2\mu}\sigma\varphi r,$$

$$S_{r\theta} = \dfrac{r}{2}\dfrac{\partial}{\partial r}\left(\dfrac{u_\theta}{r}\right) + \dfrac{1}{2r}\dfrac{\partial u_r}{\partial \theta} = \dfrac{1}{2\mu}\sigma r\theta$$

Equation of continuity: $\dfrac{\partial \rho}{\partial t} + \dfrac{1}{r^2}\dfrac{\partial}{\partial r}(\rho r^2 u_r) + \dfrac{1}{r \sin \theta}\dfrac{\partial}{\partial \theta}(\rho u_\theta \sin \theta) + \dfrac{1}{r \sin \theta}\dfrac{\partial}{\partial \varphi}(\rho u_\varphi) = 0.$

Navier-Stokes equations with constant ρ, constant ν, and no body force:

$$\dfrac{\partial u_r}{\partial t} + (\mathbf{u} \cdot \nabla)u_r - \dfrac{u_\theta^2 + u_\varphi^2}{r}$$
$$= -\dfrac{1}{\rho}\dfrac{\partial p}{\partial r} + \nu\left[\nabla^2 u_r - \dfrac{2u_r}{r^2} - \dfrac{2}{r^2 \sin \theta}\dfrac{\partial(u_\theta \sin \theta)}{\partial \theta} - \dfrac{2}{r^2 \sin \theta}\dfrac{\partial u_\varphi}{\partial \varphi}\right],$$

$$\dfrac{\partial u_\theta}{\partial t} + (\mathbf{u} \cdot \nabla)u_\theta + \dfrac{u_r u_\theta}{r} - \dfrac{u_\varphi^2 \cot \theta}{r}$$
$$= -\dfrac{1}{\rho r}\dfrac{\partial p}{\partial \theta} + \nu\left[\nabla^2 u_\theta + \dfrac{2}{r^2}\dfrac{\partial u_r}{\partial \theta} - \dfrac{u_\theta}{r^2 \sin^2 \theta} - \dfrac{2 \cos \theta}{r^2 \sin^2 \theta}\dfrac{\partial u_\varphi}{\partial \varphi}\right],$$

$$\dfrac{\partial u_\varphi}{\partial t} + (\mathbf{u} \cdot \nabla)u_\varphi + \dfrac{u_\varphi u_r}{r} + \dfrac{u_\theta u_\varphi \cot \theta}{r}$$
$$= -\dfrac{1}{\rho r \sin \theta}\dfrac{\partial p}{\partial \varphi} + \nu\left[\nabla^2 u_\varphi + \dfrac{2}{r^2 \sin \theta}\dfrac{\partial u_r}{\partial \varphi} + \dfrac{2 \cos \theta}{r^2 \sin^2 \theta}\dfrac{\partial u_\theta}{\partial \varphi} - \dfrac{u_\varphi}{r^2 \sin^2 \theta}\right],$$

where

$$\mathbf{u} \cdot \nabla = u_r\dfrac{\partial}{\partial r} + \dfrac{u_\theta}{r}\dfrac{\partial}{\partial \theta} + \dfrac{u_\varphi}{r \sin \theta}\dfrac{\partial}{\partial \varphi},$$

$$\nabla^2 = \dfrac{1}{r^2}\dfrac{\partial}{\partial r}\left(r^2\dfrac{\partial}{\partial r}\right) + \dfrac{1}{r^2 \sin \theta}\dfrac{\partial}{\partial \theta}\left(\sin \theta\dfrac{\partial}{\partial \theta}\right) + \dfrac{1}{r^2 \sin^2 \theta}\dfrac{\partial^2}{\partial \varphi^2}.$$

Founders of Modern Fluid Dynamics

O U T L I N E

Ludwig Prandtl (1875–1953)	869	Supplemental Reading	871
Geoffrey Ingram Taylor (1886–1975)	870		

LUDWIG PRANDTL (1875–1953)

Ludwig Prandtl was born in Freising, Germany, in 1875. He studied mechanical engineering in Munich. For his doctoral thesis he worked on a problem on elasticity under August Föppl, who himself did pioneering work in bringing together applied and theoretical mechanics. Later, Prandtl became Föppl's son-in-law, following the good German academic tradition in those days. In 1901, he became professor of mechanics at the University of Hanover, where he continued his earlier efforts to provide a sound theoretical basis for fluid mechanics. The famous mathematician Felix Klein, who stressed the use of mathematics in engineering education, became interested in Prandtl and enticed him to come to the University of Göttingen. Prandtl was a great admirer of Klein and kept a large portrait of him in his office. He served as professor of applied mechanics at Göttingen from 1904 to 1953; the quiet university town of Göttingen became an international center of aerodynamic research.

In 1904, Prandtl conceived the idea of a boundary layer, which adjoins the surface of a body moving through a fluid, and is perhaps the greatest single discovery in the history of fluid mechanics. He showed that frictional effects in a slightly viscous fluid are confined to a thin layer near the surface of the body; the rest of the flow can be considered inviscid. The idea led to a rational way of simplifying the equations of motion in the different regions of the flow field. Since then the boundary-layer technique has been generalized and has become a most useful tool in many branches of science.

Prandtl's work on wings of finite span (the Prandtl-Lanchester wing theory) elucidated the generation of induced drag. In compressible fluid motions he contributed the Prandtl-Glauert rule of subsonic flow and the Prandtl-Meyer expansion fan in supersonic flow around a corner, and published the first estimate of the thickness of a shock wave. He made notable innovations in the design of wind tunnels and other aerodynamic equipment. His advocacy of monoplanes greatly advanced heavier-than-air aviation. In experimental fluid mechanics he designed the Pitot-static tube for measuring velocity. In turbulence theory he contributed the mixing length theory.

Prandtl liked to describe himself as a plain mechanical engineer. So naturally he was also interested in solid mechanics; for example, he devised a soap-film analogy for analyzing the torsion stresses of structures with noncircular cross sections. In this respect he was like G. I. Taylor, and his famous student von Karman; all three of them did a considerable amount of work on solid mechanics. Toward the end of his career Prandtl became interested in dynamic meteorology and published a paper generalizing the Ekman spiral for turbulent flows.

Prandtl was endowed with rare vision for understanding physical phenomena. His mastery of mathematical tricks was limited; indeed many of his collaborators were better mathematicians. However, Prandtl had an unusual ability of putting ideas in simple mathematical forms. In 1948, Prandtl published a simple and popular textbook on fluid mechanics, which has been referred to in several places here. His varied interest and simplicity of analysis is evident throughout this book. Prandtl died in Göttingen in 1953.

GEOFFREY INGRAM TAYLOR (1886—1975)

Geoffrey Ingram Taylor's name almost always includes his initials G. I. in references, and his associates and friends simply refer to him as "G. I." He was born in 1886 in London. He apparently inherited a bent toward mathematics from his mother, who was the daughter of George Boole, the originator of "Boolean algebra." After graduating from the University of Cambridge, Taylor started to work with J. J. Thomson in pure physics.

He soon gave up pure physics and changed his interest to mechanics of fluids and solids. At this time a research position in dynamic meteorology was created at Cambridge and it was awarded to Taylor, although he had no knowledge of meteorology! At the age of 27 he was invited to serve as meteorologist on a British ship that sailed to Newfoundland to investigate the sinking of the *Titanic*. He took the opportunity to make measurements of velocity, temperature, and humidity profiles up to 2000 m by flying kites and releasing balloons from the ship. These were the very first measurements on the turbulent transfers of momentum and heat in the frictional layer of the atmosphere. This activity started his lifelong interest in turbulent flows.

During World War I he was commissioned as a meteorologist by the British Air Force. He learned to fly and became interested in aeronautics. He made the first measurements of the pressure distribution over a wing in full-scale flight. Involvement in aeronautics led him to an analysis of the stress distribution in propeller shafts. This work finally resulted in a fundamental advance in solid mechanics, the "Taylor dislocation theory."

Taylor had an extraordinarily long and productive research career (1909—1972). The amount and versatility of his work can be illustrated by the size and range of his *Collected Works* published in 1954: Volume I contains "Mechanics of Solids" (41 papers, 593 pages); Volume II contains "Meteorology, Oceanography, and Turbulent Flow" (45 papers, 515 pages); Volume III contains "Aerodynamics and the Mechanics of Projectiles and Explosions" (58 papers, 559 pages); and Volume IV contains "Miscellaneous Papers on Mechanics of Fluids" (49 papers, 579 pages). Perhaps G. I. Taylor is best known for his work on turbulence.

When asked, however, what gave him maximum *satisfaction*, Taylor singled out his work on the stability of Couette flow.

Professor George Batchelor, who has encountered many great physicists at Cambridge, described G. I. Taylor as one of the greatest physicists of the century. He combined a remarkable capacity for analytical thought with physical insight by which he knew "how things worked." He loved to conduct simple experiments, not to gather data to understand a phenomenon, but to demonstrate his theoretical calculations; in most cases he already knew what the experiment would show. Professor Batchelor has stated that Taylor was a thoroughly lovable man who did not suffer from the maladjustment and self-concern that many of today's institutional scientists seem to suffer (because of pressure!), and this allowed his creative energy to be used to the fullest extent.

He thought of himself as an amateur, and worked for pleasure alone. He did not take up a regular faculty position at Cambridge, had no teaching responsibilities, and did not visit another institution to pursue his research. He never had a secretary or applied for a research grant; the only facility he needed was a one-room laboratory and one technical assistant. He did not "keep up with the literature," tended to take up problems that were entirely new, and chose to work alone. Instead of mastering tensor notation, electronics, or numerical computations, G. I. Taylor chose to do things his own way, and did them better than anybody else.

Supplemental Reading

Batchelor, G. K. (1976). Geoffrey Ingram Taylor, 1886–1975. *Biographical Memoirs of Fellows of the Royal Society, 22*, 565–633

Batchelor, G. K. (1986). Geoffrey Ingram Taylor, 7 March 1886–27 June 1975. *Journal of Fluid Mechanics, 173*, 1–14

Oswatitsch, K., & Wieghardt, K. (1987). Ludwig Prandtl and his Kaiser-Wilhelm-Institute. *Annual Review of Fluid Mechanics, 19*, 1–25

Von Karman, T. (1954). *Aerodynamics*. New York: McGraw-Hill.

Visual Resources

Following is a list of films, all but the first by the National Committee for Fluid Mechanics Films (NCFMF), founded in 1961 by the late Ascher H. Shapiro, then professor of Mechanical Engineering at the Massachusetts Institute of Technology. Descriptive text for the films was published separately as described below.

The Fluid Dynamics of Drag, Parts I, II, III, IV (1960)
Text: Ascher H. Shapiro, *Shape and Flow: The Fluid Dynamics of Drag*, Doubleday and Co., New York (1961).
Vorticity, Parts I, II (1961)
The text for this and all following films is: NCFMF, *Illustrated Experiments in Fluid Mechanics*, MIT Press, Cambridge, MA (1972).
Deformation of Continuous Media (1963)
Flow Visualization (1963)
Pressure Fields and Fluid Acceleration (1963)
Surface Tension in Fluid Mechanics (1964)
Waves in Fluids (1964)
*Boundary Layer Control (1965)
Rheological Behavior of Fluids (1965)
Secondary Flow (1965)
Channel Flow of a Compressible Fluid (1967)
Low-Reynolds-Number Flows (1967)
Magnetohydrodynamics (1967)
Cavitation (1968)
Eulerian and Lagrangian Descriptions in Fluid Mechanics (1968)
Flow Instabilities (1968)
Fundamentals of Boundary Layers (1968)
Rarefied Gas Dynamics (1968)
Stratified Flow (1968)
Aerodynamic Generation of Sound (1969)
Rotating Flows (1969)
Turbulence (1969)

Although these films are decades old, they remain excellent visualizations of the principles of fluid mechanics. All but the one marked with an asterisk are available for viewing on the MIT website: *http://web.mit.edu/fluids/www/Shapiro/ncfmf.html*. It would be very beneficial to view the film appropriate to the corresponding section of the text.

Index

Note: Page numbers followed by f indicate figures and t indicate tables.

A

Ackeret, Jacob, 730–731, 773
Acceleration
  advective, 70–71
  fluid particle, 70–71
  unsteady, 70–71
Acoustics, 732–736
Added or Apparent mass, 236–237
Adiabatic density gradient, 596–597, 623
Adiabatic process, 17, 761, 763–764
Adiabatic temperature gradient, 19, 623
Advection, 176
Advective derivative, 176
Aerodynamics
  aircraft parts and controls, 692–693
  airfoil forces, 696–698, 697f
  airfoil geometry, 697f
  conformal transformation, 702–705
  defined, 692
  finite wing span, 708–716
  gas, 692
  generation of circulation, 698, 700, 701
  incompressible, 692
  Kutta condition, 698–700
  lift and drag characteristics, 717–718
  Prandtl and Lanchester lifting line theory, 716
  propulsive mechanisms of fish and birds, 719–721
  sailing, 721–722
  Zhukhovsky airfoil lift, 706–708
Air, physical properties of, 731–732
Aircraft, parts and controls, 692–696
Airfoil(s)
  angle of attack/incidence, 696
  camber line, 696
  chord, 696
  compression side, 696–698
  conformal transformation, 702–705
  drag, induced/vortex, 709, 712f, 713, 714–715
  finite span, 708–716
  forces, 696–698, 697f
  geometry, 696f
  lift and drag characteristics, 717–718
  stall, 707, 717
  suction side, 696–698
  supersonic flow, 773
  thin airfoil theory, 702
  Zhukhovsky airfoil lift, 706–708
Alternating tensor, 50–51
Analytic function, 216–217
Anderson, John D., Jr., 701–702
Angle of attack/incidence, 696, 712–713
Angular momentum principle/theorem, for fixed volume, 125–127
Antisymmetric tensors, 55–58
Aorta, elasticity, 805–806
Apparent or Added mass, 233
Arterioles, 793–794
  resistance, 794–796
Aris, R., 112–113
Aspect ratio of wing, 692
Asymptotic expansion, 516–517
Atmosphere
  properties of standard, 855–856
  scale height of, 21
Attractors, 526
  aperiodic, 528–529
  dissipative systems and, 526
  fixed point, 526
  limit cycle, 526
  strange, 529
Autocorrelation function, 550–551
  normalized, 550–551
  of a stationary process, 551
Averages, 545–549
Axisymmetric irrotational flow, 231–236

B

Babuska-Brezzi stability condition, 447–448
Baroclinic flow, 178–179
Baroclinic instability, 678–679
Baroclinic/internal mode, 291–292
Barotropic flow, 128–129, 176, 178–179
Barotropic instability, 676–678
Barotropic/surface mode, 291–292, 647–648
Baseball dynamics, 399
Batchelor, G. K., 142, 374–375, 560–561, 600–601, 686, 871

Bayly, B. J., 475, 515, 521, 523
Bearing number, 319—320
Becker, R., 753—755
Bergé, P., 525, 528f, 531
Bénard, H., 491
    convection, 484
    thermal instability, 484—492
Bender, C. M., 362
Bernoulli equations, 128—134
    applications of, 131—134
    energy, 130
    one-dimensional, 737—738
    steady flow, 128—129
    unsteady irrotational flow, 143—169
β-plane model, 630
Bifurcation, 526
Biofluid mechanics
    flow in blood vessels, 796—843
    human circulatory system, 780—796
    plants, 844—849
Biot and Savart, law of, 181—183
Bird, R. B., 114, 149, 559
Birds, flight of, 719
Blasius solution, boundary layer, 369—373
Blasius theorem, 219—221
Blast wave, 28f, 776—777
Blocking, in stratified flow, 298—299
Blood
    composition, 788—793
    coronary circulation, 782—784
    Fahraeus-Lindqvist effect, 789, 792,
        799—801
    flow, 796
    flow in vessels, modelling of, 796—797
    plasma, 788—792
    pulmonary circulation, 782, 841—842
    systemic circulation, 794—796
    total peripheral resistance, 794—796
    viscosity, 789—790
Blood vessels
    bifurcation, 820—822
    Casson fluid flow in rigid tube, 839—841
    composition of, 793
    flow in, 796
    flow in collapsible tube, 831
    flow in rigid walled curved tube, 825
    Hagen-Poiseuille flow, 797
    nature, 793—796
    pulsatile flow, 805
Body forces, 102
Body of revolution
    flow around arbitrary, 236
    flow around streamlined, 235

Bohlen, T., 377
Bond number, 150
Boundary conditions, 137—139, 681—682
    geophysical fluids, 646
    at infinity, 202
    kinematic, 257
    on solid surface, 202
Boundary layer
    approximation, 368—369, 380, 401
    Blasius solution, 369—373
    breakdown of laminar solution, 404
    closed form solution, 367—369
    concept, 362
    displacement thickness, 367—369
    drag coefficient, 373
    dynamics of sports balls, 395—396
    effect of pressure gradient, 384—387, 517—520
    Falkner-Skan solution, 373—375
    flat plate and, 369—373
    flow past a circular cylinder, 388—389
    flow past a sphere, 395—396
    instability, 520—522
    Karman momentum integral, 375—377
    momentum thickness, 369
    perturbation techniques, 475
    secondary flows, 407—408
    separation, 384—388
    simplification of equations, 314
    skin friction coefficient, 372—373
    technique, 2
    Thwaites method, 377—380
    transition to turbulence, 382—383
    two-dimensional jets, 399—407
    u = 0.99U thickness, 367
Bound vortices, 710
Boussinesq approximation, 125,
    135—137
    continuity equation and, 135
    geophysical fluid and, 626
    heat equation and, 137
    momentum equation and, 136
Bradshaw, P., 597—598
Brauer, H., 472
Breach, D. R., 347
Bridgeman, P. W., 37
Brooks, A. N., 439
Brunt-Väisälä frequency, 294—295
Buckingham's pi theorem, 22
Buffer layer, 586—587
Bulk strain rate, 78
Bulk viscosity, coefficient of, 113—114
Buoyancy frequency, 294—295, 625
Buoyant production, 565—566, 597

C

Camber line, airfoil, 696

Cantwell, B. J., 582

Capillarity, 9

Capillary number, 150

Capillary waves, 269, 270—271

Cardiac cycle, 782

  net work done by ventricle on blood in one, 787

Cardiac output, 788

Cardiovascular system (human), functions, 780

Carey, G. F., 447—448

Cascade, enstrophy, 687—688

Casson fluid, laminar flow in a rigid walled tube, 839—841

Casson model, 790, 791f

Casten, R. G., 419

Castillo, L., 589

Cauchy-Riemann conditions, 216—217

Cauchy's equation of motion, 111

Cavitation, 846

Central moments, 548—549

Centrifugal force, effect of, 119—121

Centrifugal instability (Taylor), 496—501

Chandrasekhar, S., 475, 498—499, 502

Chang, G. Z., 472

Chaos, deterministic, 524—540

Characteristics, method of, 279

Chester, W., 347

Chord, airfoil, 696

Chorin, A. J., 439—440, 445

Chow, C. Y., 702

Circular Couette flow, 316—317

Circular cylinder

  flow at various Re, 388—395

  flow past, boundary layer, 388—395

  flow past, with circulation, 210—211

  flow past, without circulation, 208—209

Circular Poiseuille flow, 315—316

Circulation, 79—80

  Kelvin's theorem, 96—99

Closure problem in turbulence, 560

Cnoidal waves, 286

Coefficient of bulk viscosity, 113—114

Cohen, I. M., 753—755

Coles, D., 500f, 588

Collapsible tubes

  flow in, 831

  one-dimensional steady flow in, 833

  Starling resistor experiment, 832

Comma notation, 55, 183

Complex potential, 216—219

Complex variables, 216—219

Complex velocity, 217

Compressible flow

  classification of, 731

  friction and heating effects, 761—765

  internal versus external, 730

  Mach cone, 765—766

  Mach number, 730, 731

  one-dimensional, 736—738, 740—748

  shock waves, normal, 748—753

  shock waves, oblique, 767—768, 767f

  speed of sound, 732—736

  stagnation and sonic properties, 738—740

  supersonic, 773—775

Compressible medium, static equilibrium of, 18, 18f

  potential temperature and density, 19—21

  scale height of atmosphere, 21

Compression waves, 254, 280—282

Computational fluid dynamics (CFD)

  advantages of, 422—423

  conclusions, 470

  defined, 421—422

  examples of, 449

  finite difference method, 423—428

  finite element method, 429—436

  incompressible viscous fluid flow, 436—448

  sources of error, 422

Concentric cylinders, laminar flow between, 316—318

Conformal mapping, 222—225

  application to airfoil, 702—705

Conservation laws

  Bernoulli equation, 128—134

  boundary conditions, 137—143

  Boussinesq approximation, 135—137

  differential form, 96

  integral form, 96

  of mass, 96—99

  mechanical energy equation, 123—124

  of momentum, 101—111

  Navier-Stokes equation, 114—115

  rotating frame, 116—121

  thermal energy equation, 123—124

  time derivatives of volume integrals, 86—88

Conservative body forces, 102, 178—179

Consistency, 426—428

Constitutive equation, for Newtonian fluid, 111—114

Continuity equation, 98—99

  Boussinesq approximation and, 135

  one-dimensional, 736

Continuum hypothesis, 5

Control surfaces, 96—97

Control volume, 97—98

Convection, 70—71

  -dominated problems, 437—439

  forced, 598

Convection (*Continued*)
  free, 598
  sloping, 684—685
Convergence, 426—428
Conversion factors, 853
Corcos, G. M., 505—506
Coriolis force, effect of, 118—119
Coriolis frequency, 629
Coriolis parameter, 629
Coronary arteries, 782—784
Coronary circulation, 782—784
Correlation, auto- and cross-, 550—551
Correlation coefficient, 550—551
Couette flow
  circular, 316—317
  plane, 314, 517
Courant, R., 778
Cramer, M. S., 764—765
Creeping flow, around a sphere, 340, 347
Creeping motions, 340
Cricket ball dynamics, 396—398
Critical layers, 514
Critical Re
  blood flow, 797
Critical Re for transition
  over circular cylinder, 393—394
  over flat plate, 395—396
  over sphere, 395—396
Cross-correlation function, 550—551
Cross product, vector, 51—52
Curl, vector, 54
Curtiss, C. F., 149
Curvilinear coordinates, 866

D

D'Alembert's paradox, 208—209,
    221—222
D'Alembert's solution, 734
Davies, P., 525, 531—532
Dead water phenomenon, 289, 290
Dean number, 828—829
Defect law, velocity, 584, 585
Deflection angle, 767—768
Deformation
  of fluid elements, 123—124
  Rossby radius of, 657
Degree of freedom, 525—526
Delta wings, 718
Dennis, S. C. R., 830
Density
  adiabatic density gradient, 596, 623
  potential, 19—21
  stagnation, 738—739

Derivatives
  advective, 176
  material, 176
  particle, 176
  substantial, 176
  time derivatives of volume integrals,
      86—88
Deviatoric stress tensor, 112
Diastole, 785—786
Differential equations, nondimensional parameters
    determined from, 143—151
Diffuser flow, 740—747
Diffusion of vorticity
  from impulsively started plate,
      326—330
  from line vortex, 335
  from vortex sheet, 333
Diffusivity eddy, 594
  effective, 607
  heat, 311
  momentum, 311
  thermal, 137
  vorticity, 178—179, 333
Dimensional analysis, 21—22
Dimensional homogeneity, 21—22
Dimensional matrix, 23—24
Dipole. *See* Doublet
Dirichlet problem, 429
Discretization error, 422
  of transport equation, 425
Dispersion
  of particles, 602—603
  relation, 259, 275, 668—670, 673—676
  Taylor's theory, 601—602
Dispersive wave, 273—278, 283
Displacement thickness, 367—369
Dissipation
  of mean kinetic energy, 519
  of temperature fluctuation, 600
  of turbulent kinetic energy, 564—569,
      595—600
  viscous, 137
Divergence
  flux, 98
  tensor, 53
  theorem, 58, 98
  vector, 52—53
Doppler shift of frequency, 255—256
Dot product, vector, 41—52
Double-diffusive instability, 492—496
Doublet
  in axisymmetric flow, 205—206
  in plane flow, 205—206

Downwash, 711—712
Drag
    characteristics for airfoils, 717
    on circular cylinder, 394—395
    coefficient, 343, 373
    on flat plate, 373
    force, 696—698
    form, 387—388, 718
    induced/vortex, 712f, 713—715
    pressure, 696—698, 718
    profile, 718
    skin friction, 373, 696—698, 718
    on sphere, 395—396
    wave, 713, 775
Drazin, P. G., 475, 482f, 484, 491, 501, 514, 515
Dussan, V., E. B., 168
Dutton, J. A., 596, 597
Dynamic pressure, 133, 145—146
Dynamic similarity
    nondimensional parameters and, 143—151
Dynamic viscosity, 7—8

E

Eddy diffusivity, 594
Eddy viscosity, 593—594
Effective gravity force, 119—121
Eigenvalues and eigenvectors of symmetric tensors,
    56—58
Einstein summation convention, 41
Ekman layer
    at free surface, 633—637
    on rigid surface, 639—642
    thickness, 639—640, 641
Ekman number, 632
Ekman spiral, 635
Ekman transport at a free surface, 636
Element point of view, 434—436
Elliptic circulation, 715—716
Elliptic cylinder, ideal flow, 224—225
Elliptic equation, 224—225
End diastolic volume (EDV), 787—788
End systolic volume (ESV), 787—788
Energy
    baroclinic instability, 684
    Bernoulli equation, 125—133, 153—167
    spectrum, 553—554
Energy equation
    integral form, 96
    mechanical, 123—124
    one-dimensional, 736—738
    thermal, 114—115
Energy flux
    group velocity and, 273—278

    in internal gravity wave, 302—304
    in surface gravity wave, 264—265
Ensemble average, 545—546
Enstrophy, 686
Enstrophy cascade, 687—688
Enthalpy
    defined, 14
    stagnation, 738—739
Entrainment
    in laminar jet, 399—400, 403
    turbulent, 573
Entropy
    defined, 15
    production, 215
Epsilon delta relation, 51
Equations of motion
    averaged, 554—560
    Boussinesq, 135, 626
    Cauchy's, 111
    for Newtonian fluid, 111—115
    in rotating frame, 116—120
    for stratified medium, 625—626
    for thin layer on rotating sphere, 628—630
Equations of state, 14—15
    for perfect gas, 16—17
Equipartition of energy, 264
Equivalent depth, 643
Eriksen, C. C., 505—506
Euler equation, 115, 128
    one-dimensional, 737—738
Euler momentum integral, 153
Eulerian description, 70
Eulerian specifications, 70
Exchange of stabilities, principle of, 487
Expansion coefficient, thermal, 16

F

Fahraeus effect, 792
Fahraeus-Lindqvist (FL) effect, 789, 792—793
    mathematical model, 796—797
Falkner, V. W., 373—374
Falkner-Skan solution, 373—375
Far-field of a turbulent flow, 573t—574t
Feigenbaum, M. J., 530—531
Fermi, E., 140
Feynman, R. P., 605
Fick's law of mass diffusion, 6—7
Finite difference method, 423, 425
Finite element method
    element point of view, 434—436
    Galerkin's approximation, 430—431
    matrix equations, 431—434
    weak or variational form, 429—430

First law of thermodynamics, 13—14
  thermal energy equation and, 123—124
Fish, locomotion of, 719—721
Fixed point, 526
Fixed volume
  angular momentum principle for, 125—127
Fjortoft, R., 512, 686—687
Fjortoft's theorem, 512—513
Flat plate, boundary layer and
  Blasius solution, 369—373
  closed form solution, 367—369
  drag coefficient, 373
Fletcher, C. A. J., 436, 439
Flow limitation, 832
Fluid mechanics, applications, 2—3
Fluid mechanics, visual resources, 873
Fluid, definition, 3—4
Fluid particle, 13
Fluid statics, 9—12
Flux divergence, 98
Flux of vorticity, 79—80
Force field, 102
Force potential, 102
Forces
  conservative body, 102, 178—179
  Coriolis, 118—119
  on a surface, 48—50
Forces in fluid
  body, 102
  line, 102
  surface, 102
Form drag, 387—388, 718
Fourier's law of heat conduction, 7
f-plane model, 630
Franca, L. P., 439, 448
Frank, Otto, 805
Frequency, wave
  circular or radian, 254
  Doppler shifted, 255—256
  intrinsic, 255—256
  observed, 255—256
Free turbulent shear flow, 571—581
Frey, S. L., 448
Friction drag, 373, 696—698, 718
Friction, effects in constant-area ducts, 761—763
Friedrichs, K. O., 778
Froude number, 146, 282, 836
  internal, 147
Fry, R. N., 764
Fully developed flow, 312
Fuselage, 692

G

Galerkin least squares (GLS), 448
Galerkin's approximation, 430—431
Galilean Transformation, 75
Gallo, W. F., 370
Gas constant
  defined, 16—17
  universal, 16—17
Gas dynamics, 692
  See also Compressible flow
Gases, 3—5
Gauge pressure, defined, 9—10
Gauss' theorem, 58—60, 98
Gaussian vortex, 17
Geophysical fluid dynamics
  approximate equations for thin layer on rotating sphere, 628—630
  background information, 622—623
  baroclinic instability, 678—679
  barotropic instability, 676—678
  Ekman layer at free surface, 633—637
  Ekman layer on rigid surface, 639—642
  equations of motion, 625—628
  geostrophic flow, 630—632
  gravity waves with rotation, 651—652
  Kelvin waves, 654—658
  normal modes in continuous stratified layer, 644—645
  Rossby waves, 671
  shallow-water equations, 642—643, 649—651
  vertical variations of density, 623—625
  vorticity conservation in shallow-water theory, 658—662
George, W. K., 578, 582
Geostrophic balance, 631
Geostrophic flow, 630—632
Geostrophic turbulence, 685—688
Ghia, U., 449
Ghia, K. N., 449
Gill, A. E., 289, 298, 652, 674f, 675, 676
Glauert, M. B., 404
Glowinski scheme, 446
Glowinski, R., 446, 448
Gnos, A. V., 418
Goldstein, S., 363, 502
Görtler vortices, 501
Gower, J. F. R., 392f
Grabowski, W. J., 522
Gradient operator, 52—54
Gravity force, effective, 119—121
Gravity waves
  deep water, 265—269
  at density interface, 286—293

dispersion, 267, 273—278, 299—300
energy issues, 302—304
equation, 257—262
finite amplitude, 279—280
group velocity and energy flux, 273—278
hydraulic jump, 280—283
infinite layer, 289, 290f
internal, 296—299
motion equations, 293—296
nonlinear steepening, 280—282
parameters, 254—255
refraction, 267—268
with rotation, 651—652
shallow water, 265—269, 279—286, 292—293
standing, 271—273
Stokes' drift, 284f, 285—286
in stratified fluid, 296—299
surface, 254, 256—265
surface tension, 269—271
Gresho, P. M., 437
Group velocity
concept, 273—278
of deep water wave, 275
energy flux and, 273—278
Rossby waves, 671
wave dispersion and, 273—278

H

Hagen-Poiseuille flow, 797
application, 820
effect of developing flow, 801—803
effect of vessel wall elasticity, 801—803
Fahraeus-Lindqvist effect and, 796—797
Half-body, flow past a, 207—208
Harlow, F. H., 443
Harmonic function, 202
Hatsopoulos, G. N., 37
Hayes, W. D., 755
Heart, pumping action, 785—786
Heat diffusion, 311
Heat equation, 137
Boussinesq equation and, 137
Heat flux, turbulent, 565—566
Heating, effects in constant-area ducts, 761—763
Heisenberg, W., 516—517
Hele-Shaw, H. S., 322, 324, 357, 358, 359
Hele-Shaw flow, 324, 357—358
Helmholtz vortex theorems, 179—180
Hematocrit, 789—790
plasma skimming, 792—793
Herbert, T., 475, 515
Herreshoff, H. C., 722
Hinze, J. O., 544, 561, 563

Hodograph plot, 635
Holstein, H., 377
Holton, J. R., 671
Homogeneous isotropic turbulence, 560—563
Hooke's law, blood vessels and, 804
Hou, S., 452, 453f
Houghton, J. T., 672f, 677f
Howard, L. N., 502, 507—508, 514
Howard's semicircle theorem, 507—508
Hughes, T. J. R., 436, 439, 466
Hugoniot, Pierre Henry, 750—751
Human body, biotransport and distribution processes, 780
Huppert, H. E., 492
Hydraulic jump, 280—282
Hydrostatics, 11—12
Hydrostatic waves, 267
Hypersonic flow, 731

I

Images, method of, 188—189, 213—214
Incompressible aerodynamics. *See* Aerodynamics
Incompressible fluids, 113—114, 115
Incompressible viscous fluid flow, 436—448
convection-dominated problems, 437—439
Glowinski scheme, 446
incompressibility condition, 439—440
MAC scheme, 442—446
mixed finite element, 447—448
Induced/vortex drag, 712f, 713—715
coefficient, 717
Inertia forces, 338—339
Inertial circles, 653—654
Inertial motion, 653—654
Inertial period, 629, 653—654
Inertial sublayer, 584—585
Inertial subrange, 569—570
Inflection point criterion, Rayleigh, 511—512, 676
inf-sup condition, 447—448
Initial and boundary condition error, 422
Inlet (entrance) length, 801—802
Inner layer, law of the wall, 584—585
Input data error, 422
Instability
background information, 475
baroclinic, 678—685
barotropic, 676—678
boundary layer, 517—522
centrifugal (Taylor), 496—501
of continuously stratified parallel flows, 502—508
destabilizing effect of viscosity, 516—517
double-diffusive, 492—496
inviscid stability of parallel flows, 511—515

Instability (*Continued*)
    Kelvin-Helmholtz instability, 477—483
    marginal versus neutral state, 476
    method of normal modes, 475—476
    mixing layer, 515—516
    nonlinear effects, 522
    Orr-Sommerfeld equation, 508—511
    oscillatory mode, 476
    pipe flow, 517
    plane Couette flow, 517
    plane Poiseuille flow, 516—517
    principle of exchange of stabilities, 487
    results of parallel viscous flows, 515—520
    salt finger, 492, 494—496
    sausage instability, 538
    secondary, 523
    sinuous mode, 537
    Squire's theorem, 508—511
    thermal (Bénard), 484—492
Integral time scale, 552—553
Interface, conditions at, 137—138
Internal energy, 13, 124
Internal Froude number, 147
Internal gravity waves, 254
    See also Gravity waves
    energy flux, 302—304
    at interface, 287—288, 287f
    in stratified fluid, 296—299
    in stratified fluid with rotation, 662—671
    WKB solution, 664—666
Internal Rossby radius of deformation, 657—658
Intrinsic frequency, 255—256, 588
Inversion, atmospheric, 17
Inviscid stability of parallel flows, 511—515
Inviscid theory
    application of complex variables, 216—219
    around body of revolution, 233
    axisymmetric, 231—236
    blood flow, 806—809
    conformal mapping, 222—225
    doublet/dipole, 205—206
    forces on two-dimensional body, 219—222
    images, method of, 188—189, 213—214
    irrotational flow, 79
    long wave length approximation, 822—825
    numerical solution of plane, 225—230
    over elliptic cylinder, 224—225
    past circular cylinder with circulation, 210—211
    past circular cylinder without circulation,
        208—209
    past half-body, 207—208
    relevance of, 198—200
    sources and sinks, 201

    uniqueness of, 213
    unsteady, 130
    velocity potential and Laplace equation, 191
    at wall angle, 217—218
Irrotational vector, 54
Irrotational vortex, 84, 171—172, 218
Isentropic flow, one-dimensional, 740—748
Isentropic process, 17
Isotropic tensors, 50—51, 111—113
Isotropic turbulence, 560—563
Iteration method, 225—230

J

Jets, two-dimensional laminar, 399—407

K

Kaplun, S., 314
Karamcheti, K., 728
Karman. *See under* von Karman
Keenan, J. H., 37
Keller, H. B., 472
Kelvin-Helmholtz instability, 477—483
Kelvin's circulation theorem, 96—99
Kelvin waves
    external, 654—658
    internal, 657—658
Kinematics
    defined, 65—66
    Lagrangian and Eulerian specifications, 69—70
    linear strain rate, 76—77
    material derivative, 176
    one-, two-, and three-dimensional flows, 66—67
    parallel shear flows and, 82
    path lines, 72
    polar coordinates, 66—67
    reference frames and streamline pattern, 75
    relative motion near a point, 76
    shear strain rate, 82—83
    streak lines, 72—73
    stream function, 99—100
    streamlines, 71—72
    viscosity, 8
    vortex flows and, 83—84
    vorticity and circulation, 78—79
Kinetic energy
    of mean flow, 564—565
    of turbulent flow, 565—566
Kinsman, B., 268f
Klebanoff, P. S., 484, 523
Kline, S. J., 582
Kolmogorov, A. N., 544—545, 595
    microscale, 568
    spectral law, 569—570

Korotkoff sounds, 831—832
Korteweg-deVries equation, 286
Knudsen number, 5
Kronecker delta, 50—51
Krylov, V. S. , 169
Kuethe, A. M., 702
Kundu, P. K., 638f
Kuo, H. L., 559, 676—677
Kurtosis, 548—549
Kutta condition, 698—699
Kutta, Wilhelm, 211—212
Kutta-Zhukovsky lift theorem, 211—212, 221—222, 698

L

Lagrangian description, 69—70
Lagrangian specifications, 69—70
Lam, S. H., 595—596
Lamb, H., 129, 139—140
Lamb surfaces, 129
Laminar boundary layer equations, Falkner-Skan solution, 373—375
Laminar flow
    creeping flow, around a sphere, 314—315
    defined, 310—311
    diffusion of vortex sheet, 333f
    Hele-Shaw, 314
    high and low Reynolds number flows, 338—347
    oscillating plate, 337—338
    pressure change, 311
    similarity solutions, 326—337
    steady flow between concentric cylinders, 316—318
    steady flow between parallel plates, 312—314
    steady flow in a pipe, 315—316
Laminar flow, of a Casson fluid in a rigid walled tube, 839—841
Laminar jet, 399—407
Lanchester, Frederick, 701—702, 707, 716
    lifting line theory, 708—716
Landahl, M., 568
Lanford, O. E., 525
Laplace equation, 191
    numerical solution, 225—230
Law of the wall, 584—585
LeBlond, P. H., 283, 648
Lee wave, 670—671
Leibniz theorem, 85, 86
Lesieur, M., 542
Levich, V. G., 139
Liepmann, H. W., 279, 730

Lift force, airfoil, 696—698
    characteristics for airfoils, 717—718
    Zhukhovsky, 706—708
Lifting line theory
    Prandtl and Lanchester, 716
    results for elliptic circulation, 715—716
Lift theorem, Kutta-Zhukhovsky, 211—212, 221—222, 698, 712—713
Light scattering , 30
Lighthill, M. J., 188—189, 278f—279f, 279, 283, 719
Limit cycle, 526
Lin, C.-Y., 391, 468—470
Linear strain rate, 76—77
Line forces, 102
Line vortex, 171—172, 335f
Liquids, 3—5
Logarithmic law, 585—590
Long-wave approximation. *See* Shallow-water approximation
Lorenz, E., 475—476, 526—529, 531
    model of thermal convection, 526—527
    strange attractor, 529
Lubrication theory, elementary, 318—326
Lumley, J. L., 543, 585—586, 596

M

MacCormack, R. W., 440—442, 449, 452—453, 455, 456—459
McCreary, J. P., 675
Mach, Ernst, 730—731
    angle, 765—766
    cone, 765—766
    line, 765—766
    number, 282, 692
MAC (marker-and-cell) scheme, 442—446
Magnus effect, 212—213
Marchuk, G. I., 442—443
Marginal state, 476
Mass, conservation of, 96—99
Mass transport velocity, 285
Material derivative, 70
Material volume, 96—97
Matrices
    dimensional, 23
    multiplication of, 44—45
    rank of, 23—24
    transpose of, 40—41
Matrix equations, 431—434
Mean continuity equation, 555
Mean heat equation, 558—559
Mean momentum equation, 555—556
Measurement, units of SI, 3
    conversion factors, 853

Mechanical energy equation, 123—124
Mehta, R., 396, 397f
Miles, J. W., 502
Millikan, R. A., 314, 544
Milne-Thompson, L. M., 251
Mixed finite element, 447—448
Mixing layer, 515—516
Mixing length, 544, 593—594
Modeling error, 422
Moens-Korteweg wave speed, 807—808
Mollo-Christensen, E., 568
Moments, 545—549, 560
Momentum
    conservation of, 101—111
    diffusivity, 311
    thickness, 369
Momentum equation, Boussinesq equation and, 136
Momentum integral, von Karman, 375—377
Momentum principle, for control volume, 737—738
    angular, 125—127
Monin, A. S., 543, 544
Monin-Obukhov length, 598—599
Moore, D. W., 118
Moraff, C. A., 753—755
Morton, K. W., 428
Munk, W., 671
Murray's Law, 821
    application, 820—822
Mysak, L. A., 283, 648

N

Narrow-gap approximation, 498—499
National Committee for Fluid Mechanics Films
    (NCFMF), 873
Navier-Stokes equation, 114—115, 175—176, 310, 347,
    383, 436—437, 439, 440, 441—442, 443, 447, 470, 501,
    508—509, 748, 753—755
    convection-dominated problems, 437—439
    incompressibility condition, 439—440
Nayfeh, A. H., 362, 522
Neutral state, 476
Newman, J. N., 722
Newtonian fluid, 111—114
    non-, 114
Newton's law of friction, 7—8
Nondimensional parameters
    determined from differential equations, 144
    dynamic similarity and, 151
Non-Newtonian fluid, 114
Nonrotating frame, vorticity equation in, 180—181
Nonuniform expansion
    at low Reynolds number, 339, 345
Nonuniformity
    See also Boundary layers

high and low Reynolds number flows, 338—347
    Oseen's equation, 345—347
    region of, 339
    of Stokes' solution, 345
Normal modes
    in continuous stratified layer, 644—649
    instability, 475—476
    for uniform N, 646—649
Normal shock waves, 748—755
Normal strain rate, 76—77
Normalized autocorrelation function, 550—551
No-slip condition, 177—178, 315, 362, 437, 486, 487, 510,
    519, 802, 826—827
Noye, J., 427—428
Nozzle flow, compressible, 748—750
Numerical solution
    Laplace equation, 225—230
    of plane flow, 225—230

O

Oblique shock waves, 767—770
Observed frequency, 670—671
Oden, J. T., 447—448
One-dimensional approximation, 66
One-dimensional flow
    area/velocity relations, 740—748
    equations for, 736—738
One-dimensional flow, in a collapsible tube, 833—839
Ordinary differential equations (ODEs), 431—432
Orifice flow, 133—134
Orr-Sommerfeld equation, 510—511
Orszag, S. A., 362, 475, 515, 595—596
Oscillating plate, flow due to, 337—338
Oscillatory mode, 476, 495
Oseen, C. W., 345—347, 389
Oseen's approximation, 345
Oseen's equation, 345
Outer layer, velocity defect law, 585
Overlap layer, logarithmic law, 585—590

P

Panofsky, H. A., 596, 597
Parallel flows
    instability of continuously stratified, 502—508
    inviscid stability of, 511—515
    results of viscous, 515—520
Parallel plates, steady flow between, 312—315
Parallel shear flows, 82
Particle derivative, 176
Particle orbit, 652—653, 666—668
Pascal's law, 11
Path functions, 14
Path lines, 72

Pearson, J. R. A., 347
Pedlosky, J., 627, 637—639, 679, 686, 688
Peletier, L. A., 370
Perfect differential, 128
Perfect gas, 16—17
Peripheral resistance unit (PRU), 794—796
Permutation symbol, 50—51
Perturbation pressure, 260, 266—267
Perturbation techniques, 475
   asymptotic expansion, 516—517
   nonuniform expansion, 345
   regular, 515—516
   singular, 516—517
Perturbation vorticity equation, 679—681
Petrov-Galerkin methods, 431
Peyret, R., 445, 446
Phase propagation, 675
Phase space, 525—526
Phenomenological laws, 6—7
Phillips, O. M., 274—275, 596, 671
Phloem, 847—848
   flow, 848—850
Pipe flow, dimensional analysis
   instability and, 516
Pipe, steady laminar flow in a, 315—316
Pitch axis of aircraft, 692—693
Pi theorem, Buckingham's, 22
Pitot tube, 131—132
Plane Couette flow, 314, 517
Plane irrotational flow, 200–203
Plane jet
   self-preservation, 571—573
   turbulent kinetic energy, 573
Plane Poiseuille flow, 314
   instability of, 516—517
Planetary vorticity, 185—187, 629
Planetary waves. *See* Rossby waves
Plants
   fluid mechanics, 844—849
   physiology, 844—845
Plasma, blood, 788—789
   skimming, 792—793
   viscosity, 789—790
Plastic state, 4
Platelets (thrombocytes), 788—789
Pohlhausen, K., 375—377
Poincaré, Pitot, Henri, 531
Poincaré waves, 652, 656
Point of inflection criterion, 384—385
Poiseuille flow
   circular, 315
   instability of, 516—517
   plane laminar, 314

Polar coordinates, 66—67
   cylindrical, 839
Pomeau, Y., 525, 531
Potential, body force, 119—121
Potential, complex, 216—219
Potential density gradient, 17, 596
Potential energy
   baroclinic instability, 678—685
   mechanical energy equation and, 123—124
   of surface gravity wave, 264
Potential flow. *See* Irrotational flow
Potential temperature and density, 19—21
Potential vorticity, 660
Prager, W., 64
Prandtl, L., 40—41, 212, 362, 502, 528, 529—531,
    544—545, 595, 600—601, 622, 753—755,
    869—870
   mixing length, 544, 593—594
Prandtl and Lanchester lifting line theory, 716
Prandtl-Meyer expansion fan, 771—773
Prandtl number, 149, 486
   turbulent, 597—598
Pressure
   absolute, 9—10
   coefficient, 147, 207—208
   defined, 5, 9—10
   drag, 696—698, 718
   dynamic, 133, 145—146
   gauge, 9—10
   Laplace, 9
   stagnation, 133
   waves, 258, 765—773
Pressure-drop limitation, 832
Pressure gradient
   boundary layer and effect of, 384—387, 517
   constant, 312—314
Pressure pulse, 785—786
Principal axes, 56, 80—81
Principle of exchange of stabilities, 487
Probstein, R. F., 139
Profile drag, 718
Proudman, I., 347, 633
Pulmonary circulation, 782, 841—842
Pulsatile flow, 796, 805—806
   aorta elasticity and Windkessel theory, 805—806
   inviscid theory, 806—809
   in rigid cylindrical tube, 809—814
   tube material viscoelasticity, 819—820
   wall viscoelasticity, 814—819

Q

Quasi-geostrophic motion, 671—673
Quasi-periodic regime, 529—531

R

Random walk, 604—606
Rankine, W. J. M., 750—751
  vortex, 84—85
Rankine-Hugoniot relations, 753—755
RANS equations, 560
Rayleigh
  equation, 511
  inflection point criterion, 511—512, 676
  inviscid criterion, 497, 499f
  number, 490—491
Rayleigh, Lord (J. W. Strutt), 142
Red blood cells, 789
Reduced gravity, 292
Refraction, shallow-water wave, 267—268
Regular perturbation, 515—516
Reid, W. H., 475, 482f, 484, 491, 501, 514, 515
Relative vorticity, 659—661
Relaxation time, molecular, 13
Renormalization group theories, 595—596
Reshotko, E., 522
Reversible processes, 14
Reynolds
  analogy, 597—598
  averaging, 560
  decomposition, 554—555
  experiment on flows, 310—311
  similarity, 579—580
  stress, 556—557
  transport theorem, 97
Reynolds, W. C., 542—543
Reynolds, O., 517, 544
Reynolds number, 144—146, 198, 389
  high and low flows, 338—347, 389—395
Rhines, P. B., 688
Rhines length, 688
Richardson, L. F., 502, 544—545, 567—568
Richardson number, 147, 542—543, 597
  criterion, 502
  flux, 597—598
  gradient, 147, 504—505, 597—598
Richtmyer, R. D., 428
Rigid lid approximation, 648, 649
Ripples, 270—271
Roll axis of aircraft, 692—693
Root-mean-square (rms), 544, 548—549
Roshko, A., 279, 730
Rossby number, 630
Rossby radius of deformation, 657,
    682—684
  internal, 657—658
Rossby waves, 671—676

Rotating cylinder
  flow inside, 318
  flow outside, 317—318
Rotating frame
  vorticity equation in, 183—187
Rotation, gravity waves with, 651—654
Rotation tensor, 76
Rough surface turbulence, 590—591
Ruelle, D., 530—531
Runge-Kutta technique, 432—433

S

Saad, Y., 448
Sailing, 721—728
Salinity, 20
Salt finger instability, 492, 494—496
Sargent, L. H., 523
Saric, W. S., 522
Scalars, defined, 40
Scale height, atmosphere, 21
Schlichting, H., 363, 520
Schlieren method, 730—731
Schwartz inequality, 550—551
Scotti, R. S., 505—506
Secondary flows, 407—419, 522
Secondary instability, 523
Second law of thermodynamics, 15
  entropy production and, 125
Second-order tensors, 45—47
Seiche, 271—273
Self-preservation, turbulence and, 571—573
Separation, 384—385, 388
Separated flow, 198—199
Serrin, J., 370
Shallow-water approximation, 292—293
Shallow-water equations, 642—643
  high and low frequencies, 649—651
Shallow-water theory, vorticity conservation in, 658—662
Shames, I. H., 251
Shapiro, A. H., 730, 833—838
Shear flow
  wall-bounded turbulent, 581—591
  free turbulent, 571—581
Shear production of turbulence, 564—567, 597
Shear strain rate, 82—83
Shear stress, 7—8
Shen, S. F., 516—517, 521
Sherman, F. S., 318
Shin, C. T., 472
Shock angle, 767—768
Shock structure, 753—755
Shock waves
  normal, 748—755

oblique, 767—770
structure of, 753—755
SI (système international d'unités), units of measurement, 3
conversion factors, 853
Similarity
See also Dynamic similarity
geometric, 222
Similarity solution, 326
for boundary layer, 369—373
decay of line vortex, 335f
diffusion of vortex sheet, 333f
for impulsively started plate, 326—337
for laminar jet, 399—407
Singly connected region, 213
Singularities, 216—217
Singular perturbation, 516—517
Sink, boundary layer, 412
Skan, S. W., 373—374, 377
Skewness, 548—549
Skin friction coefficient, 372—373
Sloping convection, 684—685
Smith, L. M., 595—596
Smits, A. J., 582
Solenoidal vector, 54
Solid-body rotation, 83—84, 83f, 171—172
Solids, 3—5
Solitons, 286—287
Sommerfeld, A., 45—46, 179—180, 191, 544, 730—731
Sonic conditions, 739—740
Sonic properties, compressible flow, 738—740
Sound
speed of, 16, 17, 731—732
waves, 732—736
Source-sink
axisymmetric, 233
near a wall, 213
plane, 201
Spalding, D. B., 589
Spatial distribution, 11
Specific heats, 14
Specific impulse, 109
Spectrum
energy, 553—554
as function of frequency, 553—554
as function of wave number, 545
in inertial subrange, 545, 569—570
temperature fluctuations, 600—601
Speziale, C. G., 591
Sphere
creeping flow around, 340, 347
flow around, 233
flow at various Re, 395—399

Oseen's approximation, 345
Stokes' creeping flow around, 340, 347
Spiegel, E. A., 135
Sports balls, dynamics of, 395—399
Squire's theorem, 502—503, 508—511
Stability, 426—428
See also Instability
Stagnation density, 738—739
Stagnation flow, 217—218
Stagnation points, 698
Stagnation pressure, 133, 738—739
Stagnation properties, compressible flow, 738—740
Stagnation temperature, 738—739
Standard deviation, 548—549
Standing waves, 271—273
Starling resistor, flow in a collapsible tube, 832
State functions, 14, 16
surface tension, 8—9
Stationary turbulent flow, 546—547
Statistics of a variable, 545—546
Steady flow
Bernoulli equation and, 128—129
between concentric cylinders, 316—318
between parallel plates, 312—314
in a pipe, 315—316
Steady flow, in a collapsible tube, 833—839
Stern, M. E., 492
Stokes' assumption, 113—114
Stokes' creeping flow around spheres, 340, 347
Stokes' drift, 284f, 285—286
Stokes' first problem, 326—327
Stokes' law of resistance, 342
Stokes' second problem, 337—338
Stokes' stream function, 231
Stokes' theorem, 60—61, 79—80
Stokes' waves, 283
Stommel, H. M., 118, 492, 641—642
Strain rate
linear/normal, 76—77
shear, 82—83
tensor, 76
Strange attractors, 529
Stratified layer, normal modes in continuous, 644—649
Stratified turbulence, 545
Stratopause, 624
Stratosphere, 622, 624
Streak lines, 72—73
Stream function
in axisymmetric flow, 231—236
generalized, 99—100
in plane flow, 83—84
Stokes, 231
Streamlines, 71—72

Stress, at a point, 111
Stress tensor
 deviatoric, 112
 normal or shear, 102
 Reynolds, 556
 symmetric, 111
Strouhal number, 391
Sturm-Liouville form, 645
Subcritical gravity flow, 282
Subharmonic cascade, 529–531
Sublayer
 inertial, 569–570
 streaks, 542–543
 viscous, 584–585
Subrange
 inertial, 545, 569–570
 viscous convective, 600–601
Subsonic flow, 148–149, 731
Substantial derivative, 176
Sucker, D., 472
Supercritical gravity flow, 282
Supersonic flow, 148–149, 731
 airfoil theory, 773–775
 expansion and compression, 771
Surface forces, 102
Surface gravity waves, 254
 See also Gravity waves
 in deep water, 265–269
 features of, 256–269
 in shallow water, 265–269
Surface tension, 8–9
Surface tension, generalized, 139
Sverdrup waves, 652
Sweepback angle, 718
Symmetric tensors, 55–56
 eigenvalues and eigenvectors of, 56
Systemic circulation, 780–781
 pressure throughout, 785–786
Systole, 785–786
 systolic blood pressure, measurement, 831–832

T

Takami, H., 472
Takens, F., 529–531
Taneda, S., 390f
Tannehill, J. C., 442
Taylor, T. D., 445, 499–501, 544, 622
Taylor, G. I., 544, 593–594, 641
 centrifugal instability, 496–501
 column, 633
 hypothesis, 554, 563
 number, 498–501

theory of turbulent dispersion, 601–607
 vortices, 501
Taylor-Goldstein equation, 503–504
Taylor microscale, 552–553, 561–562, 569
Taylor-Proudman theorem, 632–633
TdS relations, 16–17
Temam, R., 445
Temperature
 adiabatic temperature gradient, 17, 623
 fluctuations, spectrum, 600–601
 potential, 19–21
 stagnation, 738–739
Tennekes, H., 543, 585–586, 596
Tennis ball dynamics, 398–399
Tensors, Cartesian
 boldface versus indicial notation, 41–42
 comma notation, 62
 contraction and multiplication, 47–48
 cross product, 51–52
 dot product, 51–52
 eigenvalues and eigenvectors of symmetric, 56
 force on a surface, 48–49
 Gauss' theorem, 58–59
 invariants of, 47
 isotropic, 50–51, 112–113
 Kronecker delta and alternating, 50–51
 multiplication of matrices, 44–45
 operator del, 52–53
 rotation of axes, 42–43
 scalars and vectors, 39–41
 second-order, 45–47
 Stokes' theorem, 60–61
 strain rate, 56
 symmetric and antisymmetric, 55–56
 vector or dyadic notation, 41
Tezduyar, T. E., 448
Theodorsen's method, 702
Thermal conductivity, 7
Thermal convection, Lorenz model of, 526–527
Thermal diffusivity, 137
Thermal energy equation, 123–124
 Boussinesq equation and, 135–137
Thermal energy, 13
Thermal expansion coefficient, 16
Thermal instability (Bénard), 484–492
Thermal wind, 632
Thermocline, 625
Thermodynamic pressure, 111–112
Thermodynamics
 entropy relations, 16–17
 equations of state, 14, 16
 first law of, 13–14, 123–124
 second law of, 15, 125

specific heats, 14—15
speed of sound, 16
thermal expansion coefficient, 16
Thin airfoil theory, 702, 773—775
Thomson, R. E., 392f
Thorpe, S. A., 481f
Three-dimensional flows, 66—67
Thwaites, B., 376—380
Thwaites method, 377—380
Tidstrom, K. D., 523
Tietjens, O. C., 37, 728
Time derivatives of volume integrals
    general case, 86—87
    material volume, 96—97
Time lag, 551
Tip vortices, 709
Tollmien-Schlichting wave, 422, 516—517
Total peripheral resistance, 794—796, 799
Townsend, A. A., 580—581
Trace velocity, 255
Trailing vortices, 708—709, 713
Transition to turbulence, 382—383, 523—524
Translocation, 847
Transonic flow, 731
Transpiration, 846—847
Transport phenomena, 5—8
Transport terms, 98
Transpose, 40—41
Tropopause, 624
Troposphere, 624
Truesdell, C. A., 113—114
Tube collapse, 833, 834
Turbulent flow/turbulence
    averaged equations of motion, 554—560
    averages, 545
    buoyant production, 565—566, 597
    cascade of energy, 564—570
    characteristics of, 542—543
    commutation rules, 602—603
    correlations and spectra, 549—554
    defined, 310—311
    dispersion of particles, 601—607
    dissipating scales, 567—568
    dissipation of mean kinetic energy, 563
    dissipation of turbulent kinetic energy, 563
    eddy diffusivity, 594, 607
    eddy viscosity, 592—594
    entrainment, 573
    free shear , 571—581
    geostrophic, 685—688
    heat flux, 558—559
    homogeneous, 546—547
    inertial subrange, 545, 569—570
    integral time scale, 552—553
    intensity variations, 578—579
    isotropic, 560—563
    in a jet, 571—575
    kinetic energy of, 563
    kinetic energy of mean flow, 563
    law of the wall, 584—585
    logarithmic law, 585—590
    mean continuity equation, 555
    mean heat equation, 558—559
    mean momentum equation, 555—556
    mixing length, 544, 593—594
    Monin-Obukhov length, 598—599
    research on, 545
    Reynolds analogy, 597—598
    Reynolds stress, 556—557
    rough surface, 590—591
    self-preservation, 571—573
    shear production, 564—567
    stationary, 546—547
    stratified, 596—601
    Taylor theory of, 601—607
    temperature fluctuations, 600—601
    transition to, 382—383, 523—524
    velocity defect law, 585
    viscous convective subrange,
        600—601
    viscous sublayer, 584—585
    wall-bounded flow, 581—591
Turner, J. S., 283, 286, 298—299, 483f, 492, 597—599
Two-dimensional flows, 66, 219—222
Two-dimensional jets. *See* Jets, two-dimensional
    laminar

U

Unbounded ocean, 654
Uniform flow, axisymmetric flow, 218, 221, 223, 232
Unsteady irrotational flow, 130
Upwelling, 658

V

Vallentine, H. R., 251
Van Dyke, M., 71, 362
Vapor trails, 709
Variables, random, 549—550
Variance, 548—549
Vascular system, plant, 845
    phloem, 847—849
    xylem, 845—847
Vector(s)
    cross product, 51—52
    curl of, 54
    defined, 40—43, 45

Vector(s) (*Continued*)
  divergence of, 52—53
  dot product, 51—52
  operator del, 52—53
Velocity defect law, 585
Velocity gradient tensor, 76
Velocity potential, 130, 200—203
Ventricles, work done on blood, 785
Veronis, G., 135
Vertical shear, 632
Vidal, C., 525, 531
Viscoelastic, 4
Viscosity
  coefficient of bulk, 113—114
  destabilizing, 508—509
  dynamic, 7—8
  eddy, 592—594
  irrotational vortices and, 171—172
  kinematic, 8
  net force, 174—176
  rotational vortices and, 173
Viscosity, blood, 789—790
Viscous convective subrange, 600—601
Viscous dissipation, 137
Viscous fluid flow, incompressible, 436—448
Viscous sublayer, 584—585
Vogel, W. M., 618
Volumetric strain rate, 78
  von Karman, 701—702, 716
  constant, 585—586
  momentum integral, 375—377
  vortex streets, 298, 389—392
von Karman, T., 2, 375, 544, 585—586, 592, 716,
      730—731, 870
Vortex
  bound, 710
  decay, 335f
  drag, 713
  Görtler, 501
  Helmholtz theorems, 179—180
  interactions, 187—191
  irrotational, 218
  lines, 172, 314
  sheet, 191, 289, 480, 708—709
  starting, 699—701
  stretching, 186, 660, 685—686
  Taylor, 499—501
  tilting, 186, 637—639, 660
  tip, 709
  trailing, 708—709, 711—712
  tubes, 172—173
  von Karman vortex streets, 298,
      389—392

Vortex flows
  irrotational, 84
  Rankine, 84—85
  solid-body rotation, 83—84, 83f
Vorticity, 78—79
  absolute, 185, 660
  baroclinic flow and, 178—179
  diffusion, 178—179, 311, 333f
  equation in nonrotating frame, 180—181
  equation in rotating frame, 183—187
  flux of, 79—80
  Helmholtz vortex theorems, 179—180
  Kelvin's circulation theorem, 96—99
  perturbation vorticity equation, 679—681
  planetary, 185—187, 629
  potential, 660
  quasi-geostrophic, 671—673
  relative, 659—661
  shallow-water theory, 658—662

W

Wake, 198—199
Wall angle, flow at, 217—218
Wall-bounded turbulent shear flow, 581—591
Wall jet, 404—405
Wall, law of the, 584—585
Water, physical properties of, 854
Wavelength, 255
Wave number, 255
Waves
  See also Internal gravity waves;
      Surface gravity waves
  acoustic, 732—736
  amplitude of, 254
  angle, 767—768
  capillary, 269
  cnoidal, 286
  compression, 254
  deep-water, 265—269
  at density interface, 286—293
  dispersive, 273—278, 299—300
  drag, 713, 753
  energy flux, 264—265, 273—278
  equation, 257—258
  group speed, 264—265, 273
  hydrostatic, 267
  Kelvin, 654—658
  lee, 670—671
  packet, 274, 275f
  parameters, 254—255
  particle path and streamline, 262, 263f
  phase of, 255
  phase speed of, 255

Poincaré, 652, 656
potential energy, 264
pressure, 258, 765—773
pressure change, 267
refraction, 267—268
Rossby, 671—676
shallow-water, 265—269
shock, 748—755
solitons, 286—287
solution, 681—682
sound, 732—736
standing, 271—273
Stokes', 283
surface tension effects, 269—271
Sverdrup, 652
Wedge of instability, 684—685
Welch, J. E., 443
Wen, C. Y., 391, 468—470
White blood cells (leukocytes), 789
Whitham, G. B., 283
Williams, G. P., 688
Windkessel theory, 805—806
Wing(s)
    aspect ratio, 692
    bound vortices, 710
    drag, induced/vortex, 709, 713
    delta, 718
    finite span, 708—716
    lift and drag characteristics, 717—718

Prandtl and Lanchester lifting line theory, 716
    span, 708
    tip, 692
    tip vortices, 709
    trailing vortices, 708—709, 713
WKB approximation, 664
Womersley number, 796
Woods, J. D., 480, 597—598
Work, 14—15

X

Xylem, 845—847
    flow, 847

Y

Yaglom, A. M., 543, 544
Yakhot, V., 595—596
Yanenko, N. N., 442—443
Yaw axis of aircraft, 692—693
Yih, C. S., 516—517

Z

Zhukhovsky, N.
    airfoil lift, 706—708
    hypothesis, 698
    lift theorem, 211—212, 221—222, 698, 701—702
    transformation, 702—703, 706—707
Zone of action, 766—767
Zone of silence, 766—767